Organic & Biochemistry, FIFTH EDITION

Selected material from
General, Organic, and Biochemistry, FIFTH EDITION

Katherine J. Denniston
Towson University

Joseph J. Topping
Towson University

Robert L. Caret
Towson University

Boston   Burr Ridge, IL   Dubuque, IA   New York   San Francisco   St. Louis
Bangkok   Bogotá   Caracas   Lisbon   London   Madrid
Mexico City   Milan   New Delhi   Seoul   Singapore   Sydney   Taipei   Toronto

The McGraw·Hill Companies

ORGANIC & BIOCHEMISTRY

Copyright © 2007 by The McGraw-Hill Companies, Inc. All rights reserved. Printed in the United States of America. Except as permitted under the United States Copyright Act of 1976, no part of this publication may be reproduced or distributed in any form or by any means, or stored in a data base retrieval system, without prior written permission of the publisher.

This book is a McGraw-Hill Custom Publishing textbook and contains select material from *General, Organic, and Biochemistry*, Fifth Edition by Katherine J. Denniston, Joseph J. Topping, and Robert L. Caret. Copyright © 2007 by The McGraw-Hill Companies, Inc. Reprinted with permission of the publisher. Many custom published texts are modified versions or adaptations of our best-selling textbooks. Some adaptations are printed in black and white to keep prices at a minimum, while others are in color.

4 5 6 7 8 9 0 PLK PLK 0 9 8 7 6

ISBN 13: 978-0-07-327429-4
ISBN 10: 0-07-327429-1

Editor: Shirley Grall
Production Editor: Jessica Boatman
Cover Photo: Molecule: PhotoLink/Getty Images
Cover Design: Fairfax Hutter
Printer/Binder: Plastikoil of Pennsylvania

ORGANIC CHEMISTRY

10 An Introduction to Organic Chemistry: The Saturated Hydrocarbons 303

Chemistry Connection: *The Origin of Organic Compounds* 304

10.1 The Chemistry of Carbon 305
　Important Differences Between Organic and Inorganic Compounds 306

An Environmental Perspective: *Frozen Methane: Treasure or Threat?* 307
　Families of Organic Compounds 308

10.2 Alkanes 310
　Structure and Physical Properties 310
　Alkyl Groups 313
　Nomenclature 315

An Environmental Perspective: *Oil-Eating Microbes* 317
　Constitutional or Structural Isomers 319

10.3 Cycloalkanes 321
　cis-trans Isomerism in Cycloalkanes 322

10.4 Conformations of Alkanes and Cycloalkanes 325
　Alkanes 325
　Cycloalkanes 325

An Environmental Perspective: *The Petroleum Industry and Gasoline Production* 326

10.5 Reactions of Alkanes and Cycloalkanes 327
　Combustion 327

A Medical Perspective: *Polyhalogenated Hydrocarbons Used as Anesthetics* 328
　Halogenation 329

A Medical Perspective: *Chloroform in Your Swimming Pool?* 331

Summary of Reactions 332
Summary 332
Key Terms 332
Questions and Problems 333
Critical Thinking Problems 337

11 The Unsaturated Hydrocarbons: Alkenes, Alkynes, and Aromatics 339

Chemistry Connection: *A Cautionary Tale: DDT and Biological Magnification* 340

11.1 Alkenes and Alkynes: Structure and Physical Properties 342

11.2 Alkenes and Alkynes: Nomenclature 343

11.3 Geometric Isomers: A Consequence of Unsaturation 346

A Medical Perspective: *Killer Alkynes in Nature* 346

11.4 Alkenes in Nature 352

11.5 Reactions Involving Alkenes and Alkynes 354
　Hydrogenation: Addition of H_2 354
　Halogenation: Addition of X_2 356
　Hydration: Addition of H_2O 358
　Hydrohalogenation: Addition of HX 361

A Human Perspective: *Folklore, Science, and Technology* 362
　Addition Polymers of Alkenes 364

A Human Perspective: *Life Without Polymers?* 364

11.6 Aromatic Hydrocarbons 366

An Environmental Perspective: *Plastic Recycling* 366
　Structure and Properties 368
　Nomenclature 368
　Polynuclear Aromatic Hydrocarbons 371
　Reactions Involving Benzene 372

11.7 Heterocyclic Aromatic Compounds 373

Summary of Reactions 374
Summary 375
Key Terms 375
Questions and Problems 375
Critical Thinking Problems 379

12 Alcohols, Phenols, Thiols, and Ethers 381

Chemistry Connection: *Polyols for the Sweet Tooth* 382

12.1 Alcohols: Structure and Physical Properties 384

12.2 Alcohols: Nomenclature 385
　I.U.P.A.C. Names 385
　Common Names 386

12.3 Medically Important Alcohols 387

A Medical Perspective: *Fetal Alcohol Syndrome* 388

12.4 Classification of Alcohols 389

12.5 Reactions Involving Alcohols 391
　Preparation of Alcohols 391
　Dehydration of Alcohols 394
　Oxidation Reactions 397

12.6 Oxidation and Reduction in Living Systems 400

A Human Perspective: *Alcohol Consumption and the Breathalyzer Test* 401

12.7 Phenols 402

12.8 Ethers 403

12.9 Thiols 405

Summary of Reactions 409
Summary 409
Key Terms 410
Questions and Problems 410
Critical Thinking Problems 413

13 Aldehydes and Ketones 415

Chemistry Connection: *Genetic Complexity from Simple Molecules* 416

13.1 Structure and Physical Properties 417
13.2 I.U.P.A.C. Nomenclature and Common Names 419
Naming Aldehydes 419
Naming Ketones 422
13.3 Important Aldehydes and Ketones 424
13.4 Reactions Involving Aldehydes and Ketones 424
Preparation of Aldehydes and Ketones 424

A Medical Perspective: *Formaldehyde and Methanol Poisoning* 426

Oxidation Reactions 427

A Human Perspective: *Alcohol Abuse and Antabuse* 430

Reduction Reactions 431

A Medical Perspective: *That Golden Tan Without the Fear of Skin Cancer* 432

Addition Reactions 434
Keto-Enol Tautomers 436
Aldol Condensation 438

A Human Perspective: *The Chemistry of Vision* 440

Summary of Reactions 440
Summary 442
Key Terms 443
Questions and Problems 443
Critical Thinking Problems 445

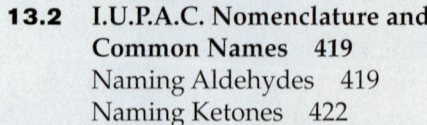

14 Carboxylic Acids and Carboxylic Acid Derivatives 447

Chemistry Connection: *Wake Up, Sleeping Gene* 448

14.1 Carboxylic Acids 449
Structure and Physical Properties 449
Nomenclature 451
Some Important Carboxylic Acids 455

An Environmental Perspective: *Garbage Bags from Potato Peels* 456

Reactions Involving Carboxylic Acids 457

14.2 Esters 461
Structure and Physical Properties 461
Nomenclature 461
Reactions Involving Esters 462

A Human Perspective: *The Chemistry of Flavor and Fragrance* 464

14.3 Acid Chlorides and Acid Anhydrides 470
Acid Chlorides 470
Acid Anhydrides 473
14.4 Nature's High-Energy Compounds: Phosphoesters and Thioesters 476

A Human Perspective: *Carboxylic Acid Derivatives of Special Interest* 478

Summary of Reactions 480
Summary 481
Key Terms 481
Questions and Problems 481
Critical Thinking Problems 485

15 Amines and Amides 487

Chemistry Connection: *The Nicotine Patch* 488

15.1 Amines 489
Structure and Physical Properties 489
Nomenclature 493
Medically Important Amines 495
Reactions Involving Amines 497

A Human Perspective: *Methamphetamine* 500

Quaternary Ammonium Salts 501

15.2 Heterocyclic Amines 502
15.3 Amides 504
Structure and Physical Properties 504
Nomenclature 505
Medically Important Amides 505
Reactions Involving Amides 506

A Medical Perspective: *Semisynthetic Penicillins* 507

15.4 A Preview of Amino Acids, Proteins, and Protein Synthesis 510
15.5 Neurotransmitters 511
Catecholamines 511
Serotonin 511

A Medical Perspective: *Opiate Biosynthesis and the Mutant Poppy* 512

Histamine 514
γ-Aminobutyric Acid and Glycine 515
Acetylcholine 515
Nitric Oxide and Glutamate 516

Summary of Reactions 517
Summary 518
Key Terms 518
Questions and Problems 518
Critical Thinking Problems 521

BIOCHEMISTRY

16 Carbohydrates 523

Chemistry Connection: *Chemistry Through the Looking Glass* 524

- 16.1 Types of Carbohydrates 526
- 16.2 Monosaccharides 527

A Human Perspective: *Tooth Decay and Simple Sugars* 528

- 16.3 Stereoisomers and Stereochemistry 529
  - Stereoisomers 529
  - Rotation of Plane-Polarized Light 530
  - The Relationship Between Molecular Structure and Optical Activity 532
  - Fischer Projection Formulas 532
  - The D- and L- System of Nomenclature 534
- 16.4 Biologically Important Monosaccharides 535
  - Glucose 535
  - Fructose 539
  - Galactose 540
  - Ribose and Deoxyribose, Five-Carbon Sugars 541
  - Reducing Sugars 541
- 16.5 Biologically Important Disaccharides 543
  - Maltose 543
  - Lactose 544
  - Sucrose 545

A Human Perspective: *Blood Transfusions and the Blood Group Antigens* 546

- 16.6 Polysaccharides 548
  - Starch 548
  - Glycogen 548
  - Cellulose 549

A Medical Perspective: *Monosaccharide Derivatives and Heteropolysaccharides of Medical Interest* 550

Summary 552
Key Terms 552
Questions and Problems 553
Critical Thinking Problems 554

17 Lipids and Their Functions in Biochemical Systems 555

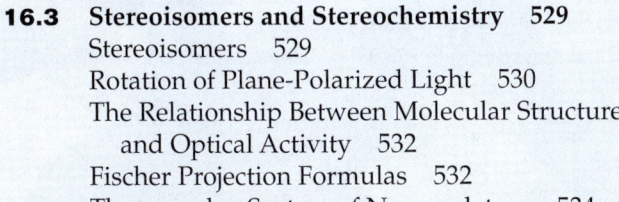

Chemistry Connection: *Lifesaving Lipids* 556

- 17.1 Biological Functions of Lipids 557
- 17.2 Fatty Acids 558
  - Structure and Properties 558
  - Chemical Reactions of Fatty Acids 561

A Human Perspective: *Mummies Made of Soap* 563
- Eicosanoids: Prostaglandins, Leukotrienes, and Thromboxanes 564

- 17.3 Glycerides 567
  - Neutral Glycerides 567
  - Phosphoglycerides 568
- 17.4 Nonglyceride Lipids 570
  - Sphingolipids 570
  - Steroids 572

A Medical Perspective: *Disorders of Sphingolipid Metabolism* 573

A Medical Perspective: *Steroids and the Treatment of Heart Disease* 574
- Waxes 577

- 17.5 Complex Lipids 577
- 17.6 The Structure of Biological Membranes 581
  - Fluid Mosaic Structure of Biological Membranes 582
  - Membrane Transport 583

A Medical Perspective: *Liposome Delivery Systems* 586

A Medical Perspective: *Antibiotics That Destroy Membrane Integrity* 588
- Energy Requirements for Transport 590

Summary 591
Key Terms 591
Questions and Problems 592
Critical Thinking Problems 593

18 Protein Structure and Function 595

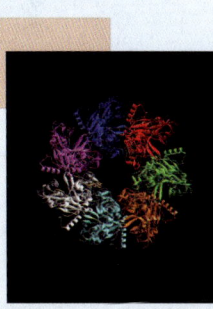

Chemistry Connection: *Angiogenesis Inhibitors: Proteins That Inhibit Tumor Growth* 596

- 18.1 Cellular Functions of Proteins 597
- 18.2 The α-Amino Acids 597
  - Structure of Amino Acids 597
  - Stereoisomers of Amino Acids 598

A Medical Perspective: *Proteins in the Blood* 599
- Classes of Amino Acids 600

- 18.3 The Peptide Bond 602

A Human Perspective: *The Opium Poppy and Peptides in the Brain* 604

- 18.4 The Primary Structure of Proteins 606
- 18.5 The Secondary Structure of Proteins 606
  - α-Helix 607
  - β-Pleated Sheet 609
- 18.6 The Tertiary Structure of Proteins 609

18.7 The Quaternary Structure of Proteins 611
A Human Perspective: *Collagen: A Protein That Holds Us Together* 612

18.8 An Overview of Protein Structure and Function 613

18.9 Myoglobin and Hemoglobin 615
Myoglobin and Oxygen Storage 615
Hemoglobin and Oxygen Transport 616
Oxygen Transport from Mother to Fetus 616
Sickle Cell Anemia 617

18.10 Denaturation of Proteins 617
A Medical Perspective: *Immunoglobulins: Proteins That Defend the Body* 618
Temperature 618
pH 620
Organic Solvents 621
Detergents 621
Heavy Metals 622
Mechanical Stress 622

18.11 Dietary Protein and Protein Digestion 622

Summary 624
Key Terms 625
Questions and Problems 625
Critical Thinking Problems 627

19 Enzymes 629

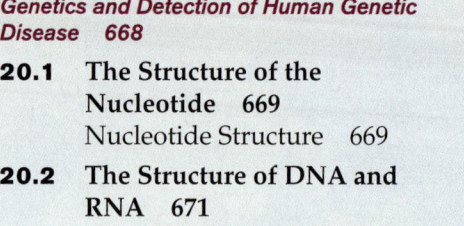

Chemistry Connection: *Super Hot Enzymes and the Origin of Life* 630

19.1 Nomenclature and Classification 631
Classification of Enzymes 631
Nomenclature of Enzymes 635

19.2 The Effect of Enzymes on the Activation Energy of a Reaction 636

19.3 The Effect of Substrate Concentration on Enzyme-Catalyzed Reactions 637

19.4 The Enzyme-Substrate Complex 637

19.5 Specificity of the Enzyme-Substrate Complex 639

19.6 The Transition State and Product Formation 639
A Medical Perspective: *HIV Protease Inhibitors and Pharmaceutical Drug Design* 641

19.7 Cofactors and Coenzymes 643

19.8 Environmental Effects 646
Effect of pH 646
Effect of Temperature 647
A Medical Perspective: α_1-*Antitrypsin and Familial Emphysema* 648

19.9 Regulation of Enzyme Activity 649
Allosteric Enzymes 649
Feedback Inhibition 650
Proenzymes 651
Protein Modification 651

19.10 Inhibition of Enzyme Activity 652
Irreversible Inhibitors 652
Reversible, Competitive Inhibitors 652
Reversible, Noncompetitive Inhibitors 653
A Medical Perspective: *Enzymes, Nerve Transmission, and Nerve Agents* 654

19.11 Proteolytic Enzymes 656
A Medical Perspective: *Enzymes, Isoenzymes, and Myocardial Infarction* 658

19.12 Uses of Enzymes in Medicine 660

Summary 661
Key Terms 662
Questions and Problems 663
Critical Thinking Problems 665

20 Introduction to Molecular Genetics 667

Chemistry Connection: *Molecular Genetics and Detection of Human Genetic Disease* 668

20.1 The Structure of the Nucleotide 669
Nucleotide Structure 669

20.2 The Structure of DNA and RNA 671
DNA Structure: The Double Helix 671
Chromosomes 673
RNA Structure 674

20.3 DNA Replication 674
A Medical Perspective: *Fooling the AIDS Virus with "Look-Alike" Nucleotides* 676
Bacterial DNA Replication 677
Eukaryotic DNA Replication 681

20.4 Information Flow in Biological Systems 681
Classes of RNA Molecules 681
Transcription 682
Post-transcriptional Processing of RNA 684

20.5 The Genetic Code 685

20.6 Protein Synthesis 687
The Role of Transfer RNA 688
The Process of Translation 690

20.7 Mutation, Ultraviolet Light, and DNA Repair 692
The Nature of Mutations 692
The Results of Mutations 692
Mutagens and Carcinogens 693
Ultraviolet Light Damage and DNA Repair 694

A Medical Perspective: *The Ames Test for Carcinogens* **694**
    Consequences of Defects in DNA Repair 695
20.8 Recombinant DNA **696**
    Tools Used in the Study of DNA 696
    Genetic Engineering 699
20.9 Polymerase Chain Reaction **701**
A Human Perspective: *DNA Fingerprinting* **702**
20.10 The Human Genome Project **703**
    Genetic Strategies for Genome Analysis 704
    DNA Sequencing 704
A Medical Perspective: *A Genetic Approach to Familial Emphysema* **705**

Summary 706
Key Terms 707
Questions and Problems 707
Critical Thinking Problems 709

21 Carbohydrate Metabolism 711

Chemistry Connection: *The Man Who Got Tipsy from Eating Pasta* **712**
21.1 ATP: The Cellular Energy Currency **713**
21.2 Overview of Catabolic Processes **715**
    Stage I: Hydrolysis of Dietary Macromolecules into Small Subunits 715
    Stage II: Conversion of Monomers into a Form That Can Be Completely Oxidized 717
    Stage III: The Complete Oxidation of Nutrients and the Production of ATP 718
21.3 Glycolysis **719**
    An Overview 719
    Reactions of Glycolysis 721
A Medical Perspective: *Genetic Disorders of Glycolysis* **722**
    Regulation of Glycolysis 726
21.4 Fermentations **726**
    Lactate Fermentation 727
    Alcohol Fermentation 727
A Human Perspective: *Fermentations: The Good, the Bad, and the Ugly* **728**
21.5 The Pentose Phosphate Pathway **730**
21.6 Gluconeogenesis: The Synthesis of Glucose **730**
21.7 Glycogen Synthesis and Degradation **733**
    The Structure of Glycogen 733
    Glycogenolysis: Glycogen Degradation 733
    Glycogenesis: Glycogen Synthesis 735
A Medical Perspective: *Diagnosing Diabetes* **738**
    Compatibility of Glycogenesis and Glycogenolysis 739
A Human Perspective: *Glycogen Storage Diseases* **740**

Summary 741
Key Terms 742
Questions and Problems 742
Critical Thinking Problems 744

22 Aerobic Respiration and Energy Production 745

Chemistry Connection: *Mitochondria from Mom* **746**
22.1 The Mitochondria **747**
    Structure and Function 747
    Origin of the Mitochondria 747
A Human Perspective: *Exercise and Energy Metabolism* **748**
22.2 Conversion of Pyruvate to Acetyl CoA **750**
22.3 An Overview of Aerobic Respiration **752**
22.4 The Citric Acid Cycle (The Krebs Cycle) **753**
    Reactions of the Citric Acid Cycle 753
22.5 Control of the Citric Acid Cycle **756**
22.6 Oxidative Phosphorylation **757**
A Human Perspective: *Brown Fat: The Fat That Makes You Thin?* **758**
    Electron Transport Systems and the Hydrogen Ion Gradient 760
    ATP Synthase and the Production of ATP 760
    Summary of the Energy Yield 761
22.7 The Degradation of Amino Acids **762**
    Removal of α-Amino Groups: Transamination 762
    Removal of α-Amino Groups: Oxidative Deamination 764
    The Fate of Amino Acid Carbon Skeletons 766
22.8 The Urea Cycle **766**
    Reactions of the Urea Cycle 766
A Medical Perspective: *Pyruvate Carboxylase Deficiency* **769**
22.9 Overview of Anabolism: The Citric Acid Cycle as a Source of Biosynthetic Intermediates **770**

Summary 772
Key Terms 773
Questions and Problems 773
Critical Thinking Problems 775

23 Fatty Acid Metabolism 777

Chemistry Connection: *Obesity: A Genetic Disorder?* **778**
23.1 Lipid Metabolism in Animals **779**
    Digestion and Absorption of Dietary Triglycerides 779
    Lipid Storage 781

23.2 Fatty Acid Degradation 782
 An Overview of Fatty Acid Degradation 782

A Human Perspective: *Losing Those Unwanted Pounds of Adipose Tissue* 784
 The Reactions of β-Oxidation 786

23.3 Ketone Bodies 789
 Ketosis 789
 Ketogenesis 789

23.4 Fatty Acid Synthesis 791
 A Comparison of Fatty Acid Synthesis and Degradation 791

23.5 The Regulation of Lipid and Carbohydrate Metabolism 793
 The Liver 793

A Medical Perspective: *Diabetes Mellitus and Ketone Bodies* 794
 Adipose Tissue 796
 Muscle Tissue 796
 The Brain 796

23.6 The Effects of Insulin and Glucagon on Cellular Metabolism 797

Summary 799
Key Terms 799
Questions and Problems 799
Critical Thinking Problems 801

Glossary G-1
Answers to Odd-Numbered Problems AP-1
Credits C-1
Index I-1

Chemistry Connections and Perspectives

Chemistry Connection

Chance Favors the Prepared Mind	2
Managing Mountains of Information	38
Magnets and Migration	78
The Chemistry of Automobile Air Bags	118
The Demise of the Hindenburg	152
Seeing a Thought	178
The Cost of Energy? More Than You Imagine	208
Drug Delivery	240
An Extraordinary Woman in Science	276
The Origin of Organic Compounds	304
A Cautionary Tale: DDT and Biological Magnification	340
Polyols for the Sweet Tooth	382
Genetic Complexity from Simple Molecules	416
Wake Up, Sleeping Gene	448
The Nicotine Patch	488
Chemistry Through the Looking Glass	524
Lifesaving Lipids	556
Angiogenesis Inhibitors: Proteins That Inhibit Tumor Growth	596
Super Hot Enzymes and the Origin of Life	630
Molecular Genetics and Detection of Human Genetic Disease	668
The Man Who Got Tipsy from Eating Pasta	712
Mitochondria from Mom	746
Obesity: A Genetic Disorder?	778

A Human Perspective

The Scientific Method	4
Food Calories	28
Atomic Spectra and the Fourth of July	52
Origin of the Elements	92
Scuba Diving: Nitrogen and the Bends	183
An Extraordinary Molecule	199
Triboluminescence: Sparks in the Dark with Candy	213
Folklore, Science, and Technology	362
Life Without Polymers?	364
Alcohol Consumption and the Breathalyzer Test	401
Alcohol Abuse and Antabuse	430
The Chemistry of Vision	440
The Chemistry of Flavor and Fragrance	464
Carboxylic Acid Derivatives of Special Interest	478
Methamphetamine	500
Tooth Decay and Simple Sugars	528
Blood Transfusions and the Blood Group Antigens	546
Mummies Made of Soap	563
The Opium Poppy and Peptides in the Brain	604
Collagen: A Protein That Holds Us Together	612
DNA Fingerprinting	702
Fermentations: The Good, the Bad, and the Ugly	728
Glycogen Storage Diseases	740
Exercise and Energy Metabolism	748
Brown Fat: The Fat That Makes You Thin?	758
Losing Those Unwanted Pounds of Adipose Tissue	784

A Medical Perspective

Curiosity, Science, and Medicine	5
Diagnosis Based on Waste	31
Copper Deficiency and Wilson's Disease	57
Dietary Calcium	67
Blood Pressure and the Sodium Ion/Potassium Ion Ratio	94
Carbon Monoxide Poisoning: A Case of Combining Ratios	136
Pharmaceutical Chemistry: The Practical Significance of Percent Yield	144
Blood Gases and Respiration	168
Oral Rehydration Therapy	197
Hemodialysis	201
Hot and Cold Packs	221
Control of Blood pH	262
Oxidizing Agents for Chemical Control of Microbes	264
Electrochemical Reactions in the Statue of Liberty and in Dental Fillings	266
Turning the Human Body into a Battery	270
Magnetic Resonance Imaging	292
Polyhalogenated Hydrocarbons Used as Anesthetics	328
Chloroform in Your Swimming Pool?	331
Killer Alkynes in Nature	346
Fetal Alcohol Syndrome	388
Formaldehyde and Methanol Poisoning	426
That Golden Tan Without the Fear of Skin Cancer	432
Semisynthetic Penicillins	507
Opiate Biosynthesis and the Mutant Poppy	512
Monosaccharide Derivatives and Heteropolysaccharides of Medical Interest	550
Disorders of Sphingolipid Metabolism	573
Steroids and the Treatment of Heart Disease	574
Liposome Delivery Systems	586
Antibiotics That Destroy Membrane Integrity	588
Proteins in the Blood	599
Immunoglobulins: Proteins That Defend the Body	618
HIV Protease Inhibitors and Pharmaceutical Drug Design	641
α_1-Antitrypsin and Familial Emphysema	648
Enzymes, Nerve Transmission, and Nerve Agents	654
Enzymes, Isoenzymes, and Myocardial Infarction	658
Fooling the AIDS Virus with "Look-Alike" Nucleotides	676
The Ames Test for Carcinogens	694
A Genetic Approach to Familial Emphysema	705
Genetic Disorders of Glycolysis	722
Diagnosing Diabetes	738
Pyruvate Carboxylase Deficiency	769
Diabetes Mellitus and Ketone Bodies	794

An Environmental Perspective

Electromagnetic Radiation and Its Effects on Our Everyday Lives	48
The Greenhouse Effect and Global Warming	166
Acid Rain	256
Nuclear Waste Disposal	285
Radon and Indoor Air Pollution	295
Frozen Methane: Treasure or Threat?	307
Oil-Eating Microbes	317
The Petroleum Industry and Gasoline Production	326
Plastic Recycling	366
Garbage Bags from Potato Peels	456

Preface

The fifth edition of *General, Organic, and Biochemistry*, like our earlier editions, has been designed to help undergraduate majors in health-related fields understand key concepts and appreciate the significant connections between chemistry, health, and the treatment of disease. We have tried to strike a balance between theoretical and practical chemistry, while emphasizing material that is unique to health-related studies. We have written at a level intended for students whose professional goals do not include a mastery of chemistry, but for whom an understanding of the principles and practice of chemistry is a necessity.

While we have stressed the importance of chemistry to the health-related professions, this book was written for all students who need a one- or two-semester introduction to chemistry. Our focus on the relationship between chemistry, the environment, medicine, and the function of the human body is an approach that can engage students in a variety of majors. We have integrated the individual disciplines of inorganic, organic, and biochemistry to emphasize their interrelatedness rather than their differences. This approach provides a sound foundation in chemistry and teaches students that life is not a magical property, but rather the result of a set of chemical reactions that obey the scientific laws.

Key Features of the Fifth Edition

In preparing the fifth edition, we have been guided by the collective wisdom of over fifty reviewers who are experts in one of the three sub-disciplines covered in the book and who represent a diversity of experience, in community colleges and in four-year colleges and universities. We have retained the core approach of our successful earlier editions, updated material where necessary, and expanded or removed material consistent with retention of the original focus and mission of the book. Throughout the project, we have been careful to ensure that the final product is as student-oriented and readable as its predecessors.

New Features

- Chapters 2 and 3 of the fourth edition have been combined in this edition to provide more integrated and concise coverage of atomic structure and periodicity.
- Each boxed topic has been enhanced by questions intended to motivate the student to go beyond what is written and/or solidify the relationship between the boxed topic and the chapter material.
- Twenty to thirty new in-chapter and end-of-chapter questions have been added to each chapter to allow instructors greater flexibility in assigning problems and to give students more opportunity to test themselves. Most chapters now include at least 100 questions and problems.
- End-of-chapter problems are now organized according to the level of understanding required of the student. *Foundations* questions provide students the opportunity to review factual information, emphasizing basic concepts, definitions, and drill. *Applications* questions are more complex or relate more directly to real-world situations. They require students to understand the information and to apply that knowledge to higher-order problems.
- The art program has undergone significant revision. New figures have been added and many others revised to create a common style and pedagogical strategy. The efforts of our Art Consultant, Dr. Ann Eakes of Northwest Vista College, have been invaluable in providing a new perspective on the art program.
- The ARIS website and other media supplements, as described later in this Preface, have been enhanced. Appendices, formerly at the end of the textbook, can now be readily accessed on the website. The instructors' Digital Content Manager CD-ROM contains electronic files of text figures and tables as well as PowerPoint lecture slides.

We designed the fifth edition to promote student learning and facilitate teaching. It is important to engage students, to appeal to visual learners, and to provide a variety of pedagogical tools to help them organize and summarize information. We have utilized a variety of strategies to accomplish our goals.

Engaging Students

Students learn better when they can see a clear relationship between the subject material they are studying and real life. We wrote the text to help students make connections between the principles of chemistry and their previous life experiences or their future professional experiences. Our strategy to accomplish this integration includes the following:

- **Boxed Readings—"Chemistry Connection":** Introductory vignettes allow students to see the significance of chemistry in their daily lives and in their future professions.
- **Boxed Perspectives:** These short stories present real-world situations that involve one or more topics that students will encounter in the chapter. The "Medical Perspectives" relate chemistry to a health concern or a diagnostic application. The "Environmental Perspectives" deal with issues, including the impact of chemistry on the ecosystem and how these environmental changes affect human health. "Human Perspectives" delve into chemistry and society and include such topics as gender issues in science and historical viewpoints. In the fifth edition, we have added a number of new boxed topics and, where needed, updated all of those perspectives retained from the fourth edition. We have included topics, such as self-tanning lotions and sugar substitutes, which are of interest to students today, as well as the most recent strategies for the treatment of HIV/AIDS. New perspectives on opioid drugs, methamphetamines, alcohol abuse, and drugs to treat chemical addiction have been added to this edition.

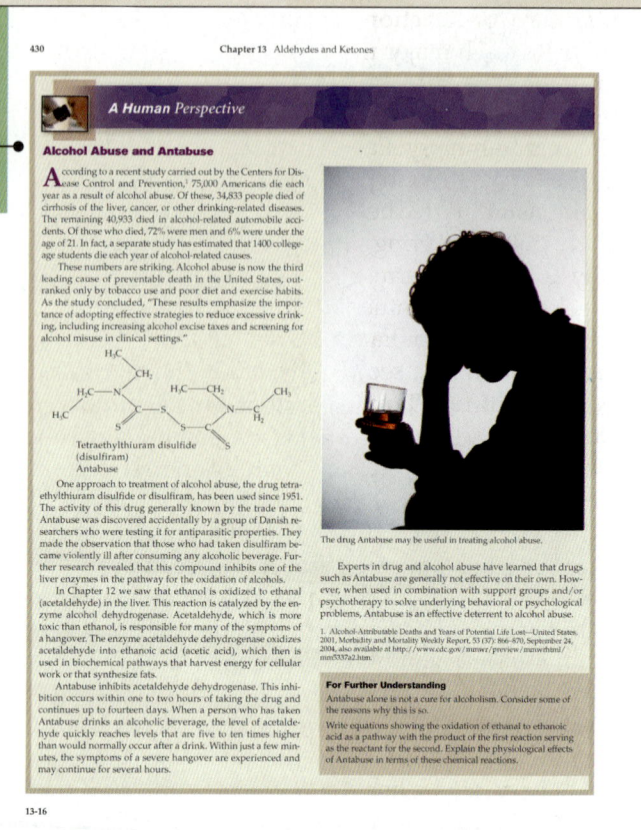

Learning Tools

In designing the original learning system we asked ourselves the question: "If we were students, what would help us organize and understand the material covered in this chapter?" With valuable suggestions from our reviewers, we have made some modifications to improve the learning system. However, with the blessings of those reviewers, we have retained all the elements of the system that have been shown to support student learning:

- **Learning Goals:** A set of chapter objectives at the beginning of each chapter previews concepts that will be covered in the chapter. Icons  locate text material that supports the learning goals.
- **Detailed Chapter Outline:** A listing of topic headings is provided for each chapter. Topics are arranged in outline form to help students organize the material in their own minds.
- **Chapter Cross-References:** To help students locate the pertinent background material, references to previous chapters, sections, and perspectives are noted in the margins of the text. These marginal cross-references also alert students to upcoming topics related to the information currently being studied.

Entropy

The first law of thermodynamics considers the enthalpy of chemical reactions. The second law states that the universe spontaneously tends toward increasing disorder or randomness.

A measure of the randomness of a chemical system is its **entropy**. The entropy of a substance is represented by the symbol S. A random, or disordered, system is characterized by *high entropy*; a well-organized system has *low entropy*.

What do we mean by disorder in chemical systems? Disorder is simply the absence of a regular repeating pattern. Disorder or randomness increases as we convert from the solid to the liquid to the gaseous state. As we have seen, solids often have an ordered crystalline structure, liquids have, at best, a loose arrangement, and gas particles are virtually random in their distribution. Therefore gases have high entropy, and crystalline solids have very low entropy. Figures 7.3 and 7.4 illustrate properties of entropy in systems.

2 LEARNING GOAL

A *system* is a part of the universe upon which we wish to focus our attention. For example, it may be a beaker containing reactants and products.

Chapter 5 compares the physical properties of solids, liquids, and gases.

Figure 7.3

(a) Gas particles, trapped in the left chamber, spontaneously diffuse into the right chamber, initially under vacuum,

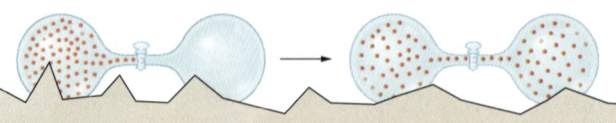

- **Summary of Reactions:** In the organic chemistry chapters, each major reaction type is highlighted on a green background. Major equations are summarized at the end of the chapter, facilitating review.
- **Chapter Summary:** Each major topic of the chapter is briefly reviewed in paragraph form in the end-of-chapter summary. These summaries serve as a mini-study guide, covering the major concepts in the chapter.
- **Key Terms:** Key terms are printed in boldface in the text, defined immediately, and listed at the end of the chapter. Each end-of-chapter key term is accompanied by a section number for rapid reference.
- **Glossary of Key Terms:** In addition to being listed at the end of the chapter, each key term from the text is also defined in the alphabetical glossary at the end of the book.

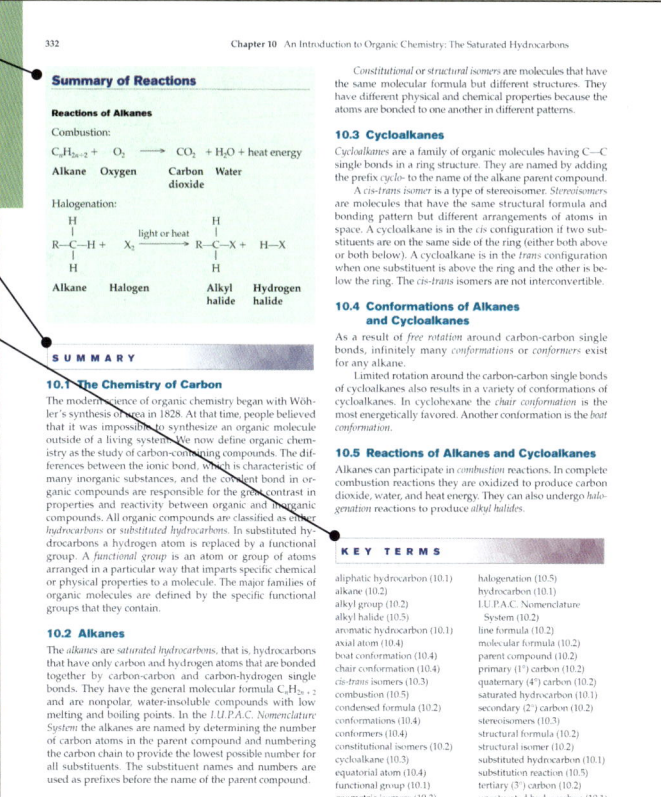

Detailed List of Changes

Changes and updates are evident in every chapter of this fifth edition. Major changes to individual chapters include:

- Chapters 2 and 3 have been combined to give seamless coverage of atomic structure and periodicity, all in one unit.
- The balancing of chemical equations, previously touched on in two sections of the book, has been combined and placed in one section of the book.
- Chapter 10, "Alkanes," now includes an updated perspective on oil-eating microbes and an additional example problem.
- Chapter 11, "Alkenes and Alkynes," now has reactions of alkynes added, including example problems of each.
- Chapter 12, "Alcohols," now includes new perspectives on methanol poisoning, alcohol abuse, and the use of Antabuse.
- Chapter 14, "Carboxylic Acids," offers students an updated perspective on biodegradable plastics.
- Chapter 15, "Amines and Amides," provides new perspectives on Methamphetamine, as well as one on opiate biosynthesis and mutant poppies.
- Chapter 16, "Carbohydrates," gives an updated perspective on the sucrose/tooth decay connection. We have also removed the old food pyramid from the chapter.
- Chapter 18, "Proteins," gives students improved coverage of amino acid structure, properties, and stereoisomers, as well as properties of the peptide bond.
- Chapter 19, "Enzymes," includes a completely reworked section on enzyme nomenclature.
- Chapter 20, "Molecular Genetics," has been highly reorganized. This had been the last chapter in the fourth edition, but is now placed with the other chapters devoted to macromolecules. This chapter also includes a new, more detailed section on bacterial DNA replication.

The Art Program

Today's students are much more visually oriented than any previous generation. Television and the computer represent alternate modes of learning. We have built upon this observation through expanded use of color, figures, and three-dimensional computer-generated models. This art program enhances the readability of the text and provides alternative pathways to learning.

Dynamic Illustrations

Each chapter is amply illustrated using figures, tables, and chemical formulas. All of these illustrations are carefully annotated for clarity. Approximately 220 full-color illustrations have been revised for this edition, in addition to over 30 new illustrations and 50 new photos, to help students better understand difficult concepts. In many cases, illustrations have been redrawn to be more realistic, and have been color-enhanced.

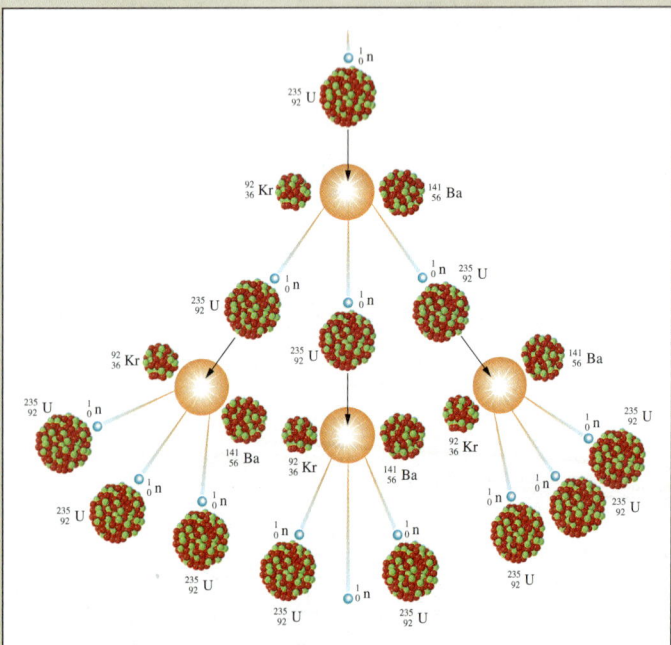

Color-Coding Scheme

We have color-coded reactions so that chemical groups being added or removed in a reaction can be quickly recognized.

- Red print is used in chemical equations or formulas to draw the reader's eye to key elements or properties in a reaction or structure.
- Blue print is used when additional features must be highlighted.

> Aldehydes and ketones can be distinguished on the basis of differences in their reactivity. The most common laboratory test for aldehydes is the **Tollens' test**. When exposed to the Tollens' reagent, a basic solution of $Ag(NH_3)_2^+$, an aldehyde undergoes oxidation. The silver ion (Ag^+) is reduced to silver metal (Ag^0) as the aldehyde is oxidized to a carboxylic acid anion.
>
> $$R-\underset{\underset{O}{\|}}{C}-H + Ag(NH_3)_2^+ \longrightarrow R-\underset{\underset{O}{\|}}{C}-O^- + Ag^0$$
>
> Aldehyde — Silver ammonia complex—Tollens' reagent — Carboxylate anion — Silver metal mirror
>
> Silver metal precipitates from solution and coats the flask, producing a smooth silver mirror, as seen in Figure 13.4. The test is therefore often called the Tollens' silver mirror test. The commercial manufacture of silver mirrors uses a similar process. Ketones cannot be oxidized to carboxylic acids and do not react with the Tollens' reagent.

- Green background screens denote generalized chemical and mathematical equations. In the organic chemistry chapters, the Summary of Reactions at the end of these chapters is also highlighted with a green background screen for ease of recognition.

> **Beta Particles**
>
> The **beta particle** (β), in contrast, is a fast-moving electron traveling at approximately 90% of the speed of light as it leaves the nucleus. It is formed in the nucleus by the conversion of a neutron into a proton. The beta particle is represented as
>
>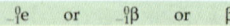

- Yellow background in the general and biochemistry sections of the text illustrates energy, either as energy stored in electrons or in groups of atoms. In the organic chemistry section of the text, yellow background screens also reveal the parent chain of an organic compound.

> **Ionization Energy**
>
> The energy required to remove an electron from an isolated atom is the **ionization energy**. The process for sodium is represented as follows:
>
>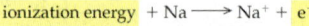

- There are certain situations in which it is necessary to adopt a unique color convention tailored to the material in a particular chapter. For example, in Chapter 18, the structures of amino acids require four colors to draw attention to key features of these molecules. For consistency, red is used to denote the acid portion of an amino acid, and blue is used to denote the basic portion of an amino acid. Green print is used to denote the R groups, and a yellow background screen directs the eye to the α-carbon.

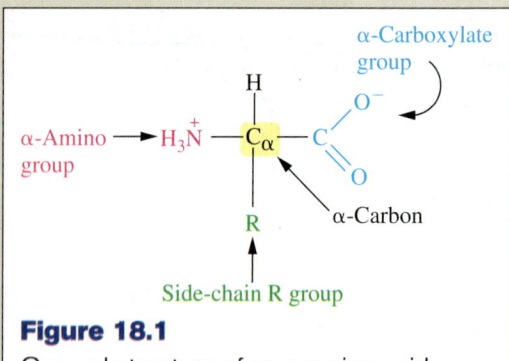

Figure 18.1
General structure of an α-amino acid. All amino acids isolated from proteins, with the exception of proline, have this general structure.

Computer-Generated Models

The students' ability to understand the geometry and three-dimensional structure of molecules is essential to the understanding of organic and biochemical reactions. Computer generated models are used throughout the text because they are both accurate and easily visualized.

Problem Solving and Critical Thinking

Perhaps the best preparation for a successful and productive career is the development of problem-solving and critical thinking skills. To this end, we created a variety of problems that require recall, fundamental calculations, and complex reasoning. In this edition, we have used suggestions from our reviewers, as well as from our own experience, to enhance the problem sets to include more practice problems for difficult concepts and further integration of the subject areas.

In-Chapter Examples, Solutions, and Questions

Each chapter includes a number of examples that show the student, step-by-step, how to properly reach the correct solution to model problems. Whenever possible, the examples are followed by in-text questions that allow students to test their mastery of information and to build self-confidence.

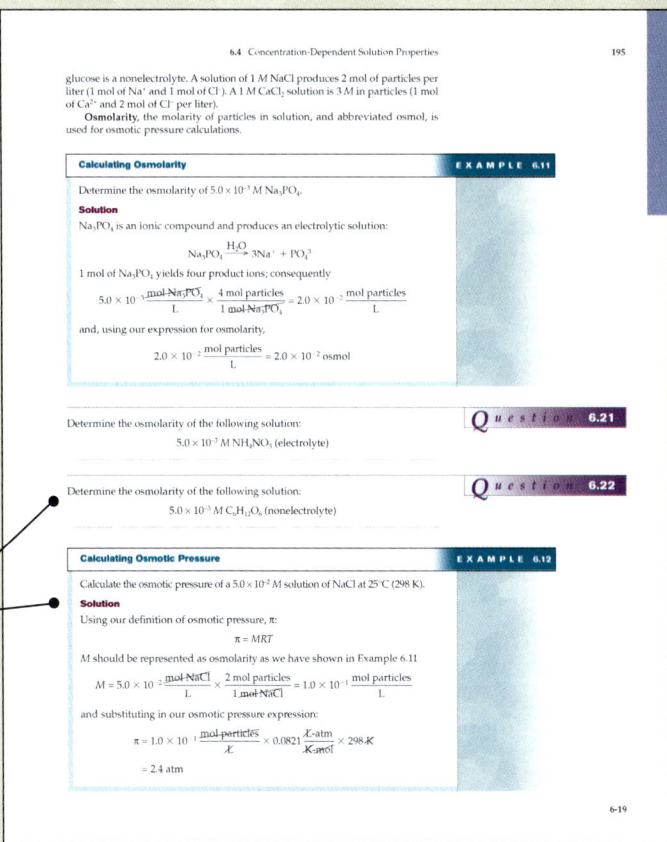

In-Chapter and End-of-Chapter Problems

We have created a wide variety of paired concept problems. The answers to the odd-numbered questions are found at the back of the book as reinforcement for students as they develop problem-solving skills. However, students must then be able to apply the same principles to the related even-numbered problems.

Critical Thinking Problems

Each chapter includes a set of critical thinking problems. These problems are intended to challenge students to integrate concepts to solve more complex problems. They make a perfect complement to the classroom lecture because they provide an opportunity for in-class discussion of complex problems dealing with daily life and the health care sciences.

Over the course of the last four editions, hundreds of reviewers have shared their knowledge and wisdom with us, as well as the reaction of their students to elements of this book. Their contributions, as well as our own continuing experience in the area of teaching and learning science, have resulted in a text that we are confident will provide a strong foundation in chemistry, while enhancing the learning experience of students.

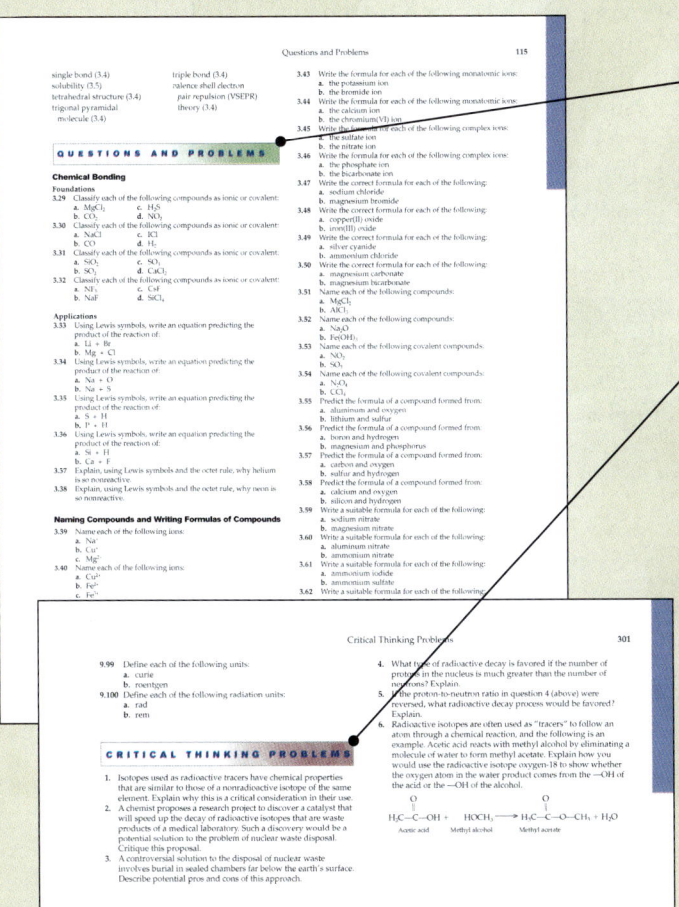

Supplementary Materials

This text is supported by a complete package for instructors and students. Several print and media supplements have been prepared to accompany the text and make learning as meaningful and up-to-date as possible.

For the Instructor

- **Digital Content Manager CD/DVD:** This primary instructor supplement offers over 800 visual images, including illustrations, photos, examples, boxed readings, and tables from the text. These images are in full color and can be readily incorporated into lecture presentations, exams, or classroom materials. Also on the Digital Content Manager are PowerPoint Lecture Outline slides, prepared by Dr. Ann Eakes of Northwest Vista College, that cover all 23 chapters and can be modified according to instructor preference.

- **Instructor's Manual:** Written by the authors and also Dr. Timothy Dwyer of Towson University, this ancillary contains suggestions for organizing lectures, instructional objectives, perspectives on boxed readings from the text, a list of each chapter's key problems and concepts, and more. The Instructor's Manual is a part of the Instructor's Testing and Resource CD, and is also available through the ARIS website for this text.
- **Transparencies:** A set of 100 transparencies is available to help the instructor coordinate the lecture with key illustrations from the text.
- **Computerized Classroom Management System:** This Instructor's Testing and Resource CD includes a database of test questions, reproducible student self-quizzes, and a grade-recording program. Also found on this CD is the Instructor's Manual to accompany this text.

- *A Laboratory Manual for General, Organic, and Biochemistry*, Fifth Edition, by Charles H. Henrickson, Larry C. Byrd, and Norman W. Hunter of Western Kentucky University, offers clear and concise laboratory experiments that reinforce students' understanding of concepts. Prelaboratory exercises, questions, and report sheets are coordinated with each experiment to ensure active student involvement and comprehension. A new student tutorial on graphing with Excel has been added to this edition.
- **Laboratory Resource Guide:** Written by Charles H. Henrickson, Larry C. Byrd, and Norman W. Hunter of Western Kentucky University, this helpful prep guide contains the hints that the authors have learned over the years to ensure students' success in the laboratory. This Resource Guide is available through the ARIS course website for this text.
- **ARIS** (McGraw-Hill's Assessment, Review and Instruction System) for *General, Organic, and Biochemistry*—a complete electronic homework and course management system—is designed for greater ease of use than any other system available. ARIS enables instructors to create and share course materials and assignments with colleagues with a few clicks of the mouse. Instructors can edit questions, import their own content, and create announcements and due dates for assignments. ARIS has automatic grading and reporting of easy-to-assign homework, quizzing, and testing. Once a student is registered in the course, all student activity within McGraw-Hill's ARIS is automatically recorded and available to the instructor through a fully integrated grade book that can be downloaded to Excel. This book-specific website is found at www.mhhe.com/denniston5e.

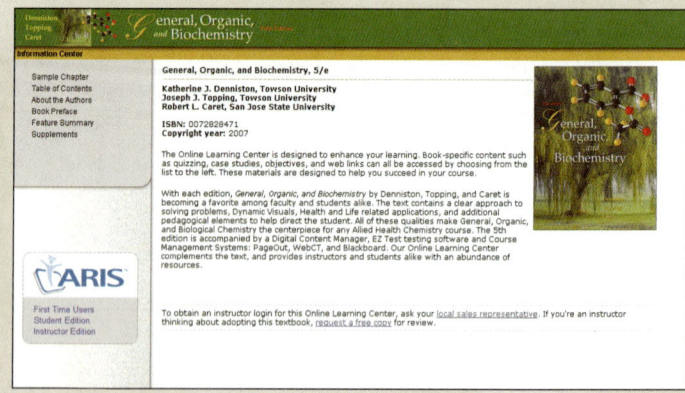

- **Course Management Systems—PageOut, WebCT, and Blackboard:** The course cartridge that accompanies *General, Organic, and Biochemistry*, Fifth Edition, includes all ARIS website content, and the entire test bank that accompanies this new edition.

For the Student

- **Student Study Guide/Solutions Manual:** A separate Student Study Guide/Solutions Manual, prepared by Dr. Timothy Dwyer and the authors of this text, is available. It contains the answers and complete solutions for the odd-numbered problems. It also offers students a variety of exercises and keys for testing their comprehension of basic, as well as difficult, concepts.
- **Schaum's Outline of General, Organic, and Biological Chemistry:** Written by George Odian and Ira Blei, this supplement provides students with over 1400 solved problems with complete solutions. It also teaches effective problem-solving techniques.
- **ARIS:** McGraw-Hill's Assessment, Review, and Instruction System for *General, Organic, and Biochemistry* is available to students and instructors using this text. The website offers quizzes, key definitions, a review of mathematics applied to problem solving, important tables, definitions, and more. This book-specific website can be found at www.mhhe.com/denniston5e.

Acknowledgments

We are thankful to our families, whose patience and support made it possible for us to undertake this project. We are also grateful to our many colleagues at McGraw-Hill for their support, guidance, and assistance.

A revision cannot move forward without the feedback of professors teaching the course. The reviewers have our gratitude and assurance that their comments received serious consideration. The following professors provided reviews, participated in a focus group, or gave valuable advice for the preparation of the fifth edition:

Ayoni F. Akinyele   *Howard University, Washington, D.C.*
James Armstrong   *City College of San Francisco*
Maher Atteya   *Georgia Perimeter College*
Teresa L. Brown   *Rochester Community and Technical College*
Alan J. Bruha   *Lewis and Clark Community College*
Kathleen Brunke   *Christopher Newport University*
Derald Chriss   *Southern University*
Sharon W. Chriss   *Southern University*
Ana Ciereszko   *Miami–Dade College*
Rajeev B. Dabke   *Columbus State University*
Philip Denton   *Fresno City College*
Brahmadeo Dewprashad   *Borough of Manhattan Community College*
Ronald P. Drucker   *City College of San Francisco*
Fredesvinda B. Cand. Dura   *LaGuardia Community College/CUNY*
Ann Eakes   *Northwest Vista College*
Barbara L. Edgar   *University of Minnesota, Twin Cities*
Amber Flynn-Charlebois   *William Paterson University*
Wes Fritz   *College of DuPage*
Elizabeth Gardner   *University of Texas at El Paso*
Galen G. George   *Santa Rosa Junior College*
Walter E. Godwin   *University of Arkansas at Monticello*
Cliff Gottlieb   *Shasta College*
James K. Hardy   *The University of Akron*
Jonathan Heath   *Horry-Georgetown Technical College*
T.G. Jackson   *University of South Alabama*
Richard H. Jarman   *College of DuPage*
James T. Johnson   *Sinclair Community College*
David A. Katz   *Pima Community College*
Colleen Kelley   *Pima Community College*
Laura Kibler-Herzog   *Georgia State University*
Edith Preciosa Klingberg   *University of Toledo*
Terry L. Lampe   *Georgia Perimeter College*
Richard H. Langley   *Stephen F. Austin State University*
David Lippmann   *Texas State University, San Marcos*
Jeanette C. Madea   *Broward Community College*
Tammy J. Melton   *Middle Tennessee State University*
Melvin Merken   *Worcester State College*
Li-June Ming   *University of South Florida*
John T. Moore   *Stephen F. Austin State University*
Joseph C. Muscarella   *Henry Ford Community College*
Elva Mae Nicholson   *Eastern Michigan University*
Thomas J. Nycz   *Broward Community College, North Campus*
Beng Guat Ooi   *Middle Tennessee State University*
Alicia Paterno Parsi   *Duquesne University*
Jerry Poteat   *Georgia Perimeter College*
Parris F. Powers   *Volunteer State Community College*
Betsy Ratcliff   *West Virginia University*
Douglas E. Raynie   *South Dakota State University*
Susan S. Reid   *North Hennepin Community College*
Lynette M. Rushton   *South Puget Sound Community College*
Howard Theodore Silverstein   *Georgia Perimeter College*
Melinda M. Sorensson   *University of Louisana*
Luise Strange de Soria   *Georgia Perimeter College*
Melissa R. Synder   *Rochester Community and Technical College*
Sheryl K. Wallace   *South Plains College*
Marcy Whitney   *University of Alabama*
Catherine Woytowicz   *The George Washington University*
Clara Wu   *LaGuardia Community College, City University of New York*
Vaneica Young   *University of Florida*

ORGANIC CHEMISTRY

10

An Introduction to Organic Chemistry

The Saturated Hydrocarbons

Learning Goals

1. Compare and contrast organic and inorganic compounds.
2. Draw structures that represent each of the families of organic compounds.
3. Write the names and draw the structures of the common functional groups.
4. Write condensed and structural formulas for saturated hydrocarbons.
5. Describe the relationship between the structure and physical properties of saturated hydrocarbons.
6. Use the basic rules of the I.U.P.A.C. Nomenclature System to name alkanes and substituted alkanes.
7. Draw constitutional (structural) isomers of simple organic compounds.
8. Write the names and draw the structures of simple cycloalkanes.
9. Draw *cis* and *trans* isomers of cycloalkanes.
10. Describe conformations of alkanes.
11. Draw the chair and boat conformations of cyclohexane.
12. Write equations for combustion reactions of alkanes.
13. Write equations for halogenation reactions of alkanes.

The origins of fossil fuels.

Outline

Chemistry Connection:
The Origin of Organic Compounds

10.1 The Chemistry of Carbon

An Environmental Perspective:
Frozen Methane: Treasure or Threat?

10.2 Alkanes

An Environmental Perspective:
Oil-Eating Microbes

10.3 Cycloalkanes

10.4 Conformations of Alkanes and Cycloalkanes

An Environmental Perspective:
The Petroleum Industry and Gasoline Production

10.5 Reactions of Alkanes and Cycloalkanes

A Medical Perspective:
Polyhalogenated Hydrocarbons Used as Anesthetics

A Medical Perspective:
Chloroform in Your Swimming Pool?

303

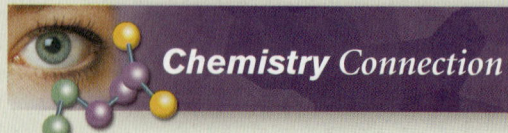

Chemistry Connection

The Origin of Organic Compounds

About 425 million years ago, mountain ranges rose, and enormous inland seas emptied, producing new and fertile lands. In the next 70 million years the simple aquatic plants evolved into land plants, and huge forests of ferns, trees, and shrubs flourished. Reptiles roamed the forests. During the period between 360 and 280 million years ago the seas rose and fell at least fifty times. During periods of flood, the forests were buried under sediments. When the seas fell again, the forests were reestablished. The cycle was repeated over and over. Each flood period deposited a new layer of peat—partially decayed, sodden, compressed plant matter. These layers of peat were compacted by the pressure of the new sediments forming above them. Much of the sulfur and hydrogen was literally squeezed out of the peat, increasing the percentage of carbon. Slowly, the peat was compacted into seams of coal, which is 55–95% carbon. Oil, consisting of a variety of hydrocarbons, formed on the bottoms of ancient oceans from the remains of marine plants and animals.

Together, coal and oil are the "fossil fuels" that we use to generate energy for transportation, industry, and our homes. In the last two centuries we have extracted many of the known coal reserves from the earth and have become ever more dependent on the world's oil reserves. Coal and oil are products of the chemical reactions of photosynthesis that occurred over millions of years of the earth's history. Our society must recognize that they are nonrenewable resources. We must actively work to conserve the supply that remains and to develop alternative energy sources for the future.

In this chapter we take a closer look at the structure and properties of the hydrocarbons, such as those that make up oil. In this and later chapters we will study the amazing array of organic molecules (molecules made up of carbon, hydrogen, and a few other elements), many of which are essential to life. As we will see, all the structural and functional molecules of the cell, including the phospholipids that make up the cell membrane and the enzymes that speed up biological reactions, are organic molecules. Smaller organic molecules, such as the sugars glucose and fructose, are used as fuel by our cells, whereas others, such as penicillin and aspirin, are useful in the treatment of disease. All these organic compounds, and many more, are the subject of the remaining chapters of this text.

Introduction

Organic chemistry is the study of carbon-containing compounds. The term organic was coined in 1807 by the Swedish chemist Jöns Jakob Berzelius. At that time it was thought that all organic compounds, such as fats, sugars, coal, and petroleum, were formed by living or once living organisms. All early attempts to synthesize these compounds in the laboratory failed, and it was thought that a vital force, available only in living cells, was needed for their formation.

This idea began to change in 1828 when a twenty-seven-year-old German physician, whose first love was chemistry, synthesized the organic molecule urea from inorganic starting materials. This man was Friedrich Wöhler, the "father of organic chemistry."

As a child, Wöhler didn't do particularly well in school because he spent so much time doing chemistry experiments at home. Eventually, he did earn his medical degree, but he decided to study chemistry in the laboratory of Berzelius rather than practice medicine.

After a year he returned to Germany to teach and, as it turned out, to do the experiment that made him famous. The goal of the experiment was to prepare ammonium cyanate from a mixture of potassium cyanate and ammonium sulfate. He heated a solution of the two salts and crystallized the product. But the product didn't look like ammonium cyanate. It was a white crystalline material that looked exactly like urea! Urea is a waste product of protein breakdown in the body and is excreted in the urine. Wöhler recognized urea crystals because he had previously purified them from the urine of dogs and humans. Excited about his accidental discovery, he wrote to his teacher and friend Berzelius, "I can make urea without the necessity of a kidney, or even an animal, whether man or dog."

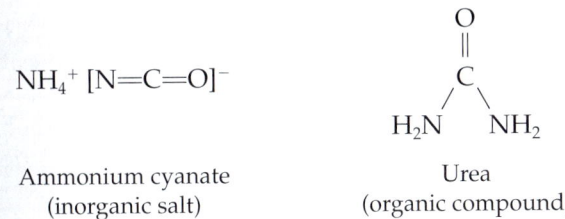

Ammonium cyanate (inorganic salt)

Urea (organic compound)

Ironically, Wöhler, the first person to synthesize an organic compound from inorganic substances, devoted the rest of his career to inorganic chemistry. However, other chemists continued this work, and as a result, the "vital force theory" was laid to rest, and modern organic chemistry was born.

10.1 The Chemistry of Carbon

The number of possible carbon-containing compounds is almost limitless. The importance of these organic compounds is reflected in the fact that over half of this book is devoted to the study of molecules made with this single element.

Why are there so many organic compounds? There are several reasons. First, carbon can form *stable, covalent* bonds with other carbon atoms. Consider three of the *allotropic forms* of elemental carbon: graphite, diamond, and buckminsterfullerene. Models of these allotropes are shown in Figure 10.1.

Graphite consists of planar layers in which all carbon-to-carbon bonds extend in two dimensions. Because the planar units can slide over one another, graphite is an excellent lubricant. In contrast, diamond consists of a large, three-dimensional network of carbon-to-carbon bonds. As a result, it is an extremely hard substance used in jewelry and cutting tools.

The third allotropic form of carbon is buckminsterfullerene, affectionately called the *buckey ball*. The buckey ball consists of sixty carbon atoms in the shape of a soccer ball. Discovered in the 1980s, buckminsterfullerene was named for Buckminster Fuller, who used such shapes in the design of geodesic domes.

A second reason for the vast number of organic compounds is that carbon atoms can form stable bonds with other elements. Several families of organic compounds (alcohols, aldehydes, ketones, esters, and ethers) contain oxygen atoms

> Allotropes are forms of an element that have the same physical state but different properties.

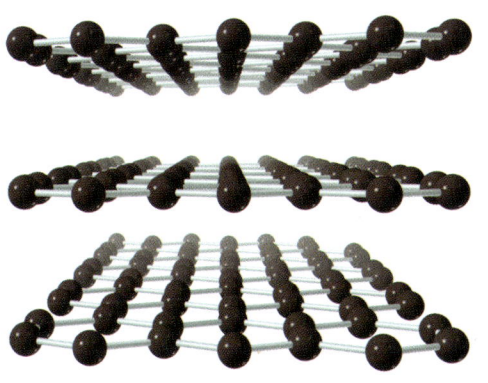
(a) Graphite

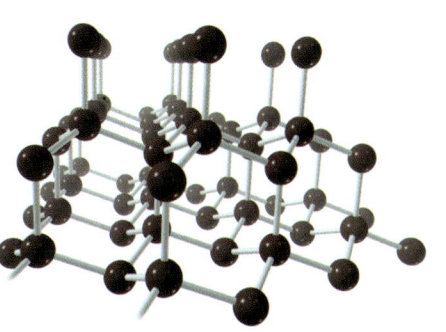
(b) Diamond

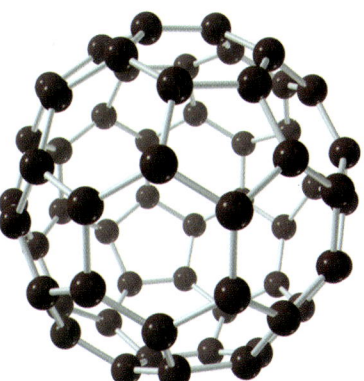
(c) Buckminsterfullerene

Figure 10.1
Three allotropic forms of elemental carbon.

In future chapters we will discuss families of organic molecules containing oxygen atoms (alcohols, aldehydes, ketones, carboxylic acids, ethers, and esters), nitrogen atoms (amides and amines), sulfur atoms, and halogen atoms.

bonded to carbon. Others contain nitrogen, sulfur, or halogens. The presence of these elements confers a wide variety of new chemical and physical properties on an organic compound.

Third, carbon can form double or triple bonds with other carbon atoms to produce a variety of organic molecules with very different properties. Finally, the number of ways in which carbon and other atoms can be arranged is nearly limitless. In addition to linear chains of carbon atoms, ring structures and branched chains are common. Two organic compounds may even have the same number and kinds of atoms but completely different structures and thus different properties. Such organic molecules are called *isomers*.

Important Differences Between Organic and Inorganic Compounds

LEARNING GOAL

The bonds between carbon and another atom are almost always *covalent bonds*, whereas the bonds in many inorganic compounds are *ionic bonds*. Differences between these two types of bonding are responsible for most of the differences between inorganic and organic substances. Ionic bonds result from the *transfer* of one or more electrons from one atom to another. Thus, ionic bonds are electrostatic, resulting from the attraction between the positive and negative ions formed by the electron transfer. Covalent bonds are formed by sharing one or more pairs of electrons.

Ionic compounds often form three-dimensional crystals made up of many positive and negative ions. Covalent compounds exist as individual units called molecules. Water-soluble ionic compounds often dissociate in water to form ions and are called electrolytes. Most covalent compounds are nonelectrolytes, keeping their identity in solution.

Polar covalent compounds, such as HCl, dissociate in water and, thus, are electrolytes. Carboxylic acids, the family of organic compounds we will study in Chapter 14, are weak electrolytes when dissolved in water.

As a result of these differences, ionic substances usually have much higher melting and boiling points than covalent compounds. They are more likely to dissolve in water than in a less-polar solvent, whereas organic compounds, which are typically nonpolar or only moderately polar, are less soluble, or insoluble in water. In Table 10.1, the physical properties of the organic compound butane are compared with those of sodium chloride, an inorganic compound of similar formula weight.

TABLE 10.1 Comparison of the Major Properties of a Typical Organic and an Inorganic Compound: Butane Versus Sodium Chloride

Property	Organic Compounds (e.g., Butane)	Inorganic Compounds (e.g., Sodium Chloride)
Formula weight	58	58.5
Bonding	Covalent (C_4H_{10})	Ionic (Na^+ and Cl^- ions)
Physical state at room temperature and atmospheric pressure	Gas	Solid
Boiling point	Low (−0.5°C)	High (1413°C)
Melting point	Low (−139°C)	High (801°C)
Solubility in water	Insoluble	High (36 g/100 mL)
Solubility in organic solvents (e.g., hexane)	High	Insoluble
Flammability	Flammable	Nonflammable
Electrical conductivity	Nonconductor	Conducts electricity in solution and in molten liquid

An Environmental *Perspective*

Frozen Methane: Treasure or Threat?

Methane is the simplest hydrocarbon, but it has some unusual behaviors. One of these is the ability to form a clathrate, which is an unusual type of matter in which molecules of one substance form a cage around molecules of another substance. For instance, water molecules can form a latticework around methane molecules to form frozen methane hydrate, possibly one of the biggest reservoirs of fossil fuel on earth.

Typically we wouldn't expect a nonpolar molecule, such as methane, to interact with a polar molecule, such as water. So, then, how is this structure formed? As we have studied earlier, water molecules interact with one another by strong hydrogen bonding. In the frozen state, these hydrogen-bonded water molecules form an open latticework. The nonpolar methane molecule is simply trapped inside one of the spaces within the lattice.

Frozen methane is found on the ocean floor. Formed by animals and decaying plant life, there are large pockets of oil and natural gas all over the earth. Methane hydrate forms when methane from one of these pockets under the ocean seeps up through the sea sediments. When the gas reaches the ocean floor, it expands and freezes. Vast regions of the ocean floor are covered by such ice fields.

Fascinating communities of methanogens, organisms that can use methane for a food source, have developed on these ice fields. These creatures live under great pressures, at extremely low temperatures, and with no light. But these unusual communities are not the major focus of interest in the methane ice fields. Scientists would like to "mine" this ice to use the methane as a fuel. In fact, the U.S. Geological Survey estimates that the amount of methane hydrate in the United States is worth over two hundred times the conventional natural gas resources in this country!

But is it safe to harvest the methane from this ice? Caution will certainly be required. Methane is flammable, and, like carbon dioxide, it is a greenhouse gas. In fact, it is about twenty times more efficient at trapping heat than carbon dioxide. (Greenhouse gases are discussed in greater detail in An Environmental Perspective: The Greenhouse Effect and Global Warming in Chapter 5.) So the U.S. Department of Energy, which is working with industry to develop ways to harvest the methane, must figure out how to do that without releasing much into the atmosphere where it could intensify global warming.

It may be that a huge release of methane from these frozen reserves was responsible for a major global warming that occurred fifty-five million years ago and lasted for one hundred thousand years. NASA scientists using computer simulations hypothesize that a shift of the continental plates may have released vast amounts of methane gas from the ocean floor. This methane raised the temperature of earth by about 13°F. In fact the persistence of the methane in the atmosphere warmed earth enough to melt the ice in the oceans and at polar caps and completely change the global climate. This theory, if it turns out to be true, highlights the importance of controlling the amount of methane, as well as carbon dioxide, that we release into the air. Certainly, harvesting the frozen methane of the oceans, if we chose to do it, must be done with great care.

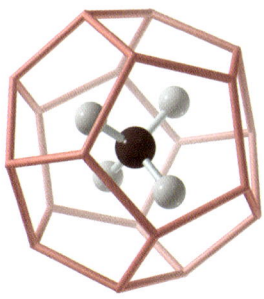

Three-dimensional structure of methane hydrate.

For Further Understanding

News stories alternatively describe frozen methane as a "New Frontier" and "Armageddon." Explain these opposing views.

What are the ethical considerations involved in mining for frozen methane?

Question 10.1

A student is presented with a sample of an unknown substance and asked to classify the substance as organic or inorganic. What tests should the student carry out?

Question 10.2

What results would the student expect if the sample in Question 10.1 were an inorganic compound? What results would the student expect if it were an organic compound?

LEARNING GOAL

Alkanes contain only carbon-to-carbon single bonds (C—C); alkenes have at least one carbon-to-carbon double bond (C=C); and alkynes have at least one carbon-to-carbon triple bond (C≡C).

Recall that each of the lines in these structures represents a shared pair of electrons. See Chapter 3.

We will learn a more accurate way to represent benzene in Section 11.6.

LEARNING GOAL

Families of Organic Compounds

The most general classification of organic compounds divides them into hydrocarbons and substituted hydrocarbons. A **hydrocarbon** molecule contains only carbon and hydrogen. A **substituted hydrocarbon** is one in which one or more hydrogen atoms is replaced by another atom or group of atoms.

The hydrocarbons can be further subdivided into aliphatic and aromatic hydrocarbons (Figure 10.2). The four families of **aliphatic hydrocarbons** are the alkanes, cycloalkanes, alkenes, and alkynes.

Alkanes are **saturated hydrocarbons** because they contain only carbon and hydrogen and have only carbon-to-hydrogen and carbon-to-carbon single bonds. The alkenes and alkynes are **unsaturated hydrocarbons** because they contain at least one carbon-to-carbon double or triple bond, respectively.

Saturated Hydrocarbon

Unsaturated Hydrocarbon

Some hydrocarbons are cyclic. Cycloalkanes consist of carbon atoms bonded to one another to produce a ring. **Aromatic hydrocarbons** contain a benzene ring or a derivative of the benzene ring.

A Cycloalkane (Cyclohexane)

Benzene

A substituted hydrocarbon is produced by replacing one or more hydrogen atoms with a functional group. A **functional group** is an atom or group of atoms arranged in a particular way that is primarily responsible for the chemical and physical properties of the molecule in which it is found. The importance of functional groups becomes more apparent when we consider that hydrocarbons have little biological activity. However, the addition of a functional group confers unique and interesting properties that give the molecule important biological or medical properties.

Figure 10.2
The family of hydrocarbons is divided into two major classes: aliphatic and aromatic. The aliphatic hydrocarbons are further subdivided into four major subclasses: alkanes, cycloalkanes, alkenes, and alkynes.

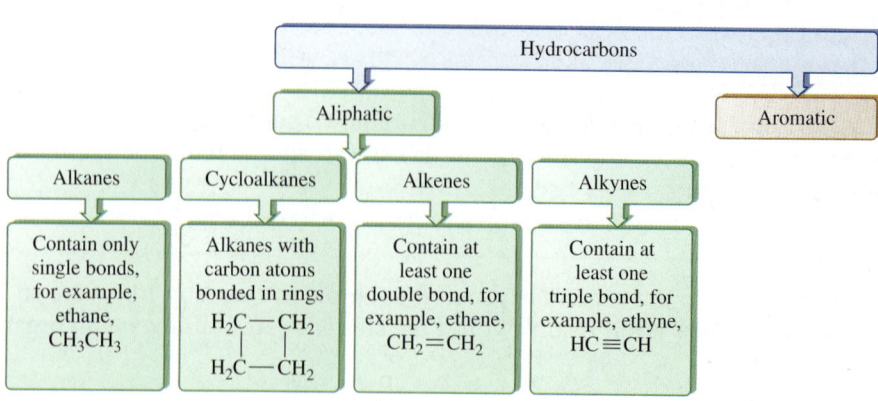

TABLE 10.2 Common Functional Groups

Type of compound	Structural formula	Condensed formula	Structural formula	Example I.U.P.A.C. name	Common name
Alcohol	R—O—H	ROH	CH_3CH_2—O—H	Ethanol	Ethyl alcohol
Aldehyde	R—C(=O)—H	RCHO	CH_3C(=O)—H	Ethanal	Acetaldehyde
Amide	R—C(=O)—N(H)—H	$RCONH_2$	CH_3C(=O)—N(H)—H	Ethanamide	Acetamide
Amine	R—N(H)—H	RNH_2	CH_3CH_2N(H)—H	Aminoethane	Ethyl amine
Carboxylic acid	R—C(=O)—O—H	RCOOH	CH_3C(=O)—O—H	Ethanoic acid	Acetic acid
Ester	R—C(=O)—O—R'	RCOOR'	CH_3C(=O)—OCH_3	Methyl ethanoate	Methyl acetate
Ether	R—O—R'	ROR'	CH_3OCH_3	Methoxymethane	Dimethyl ether
Halide	R—Cl (or —Br, —F, —I)	RCl	CH_3CH_2Cl	Chloroethane	Ethyl chloride
Ketone	R—C(=O)—R'	RCOR'	CH_3CCH_3	Propanone	Acetone
Alkene	R$_2$C=CR$_2$ (H)	RCHCHR	$CH_3CH=CH_2$	Propene	Propylene
Alkyne	R—C≡C—R	RCCR	$CH_3C≡CH$	Propyne	Methyl acetylene

All compounds that have a particular functional group are members of the same family. We have just seen that the alkenes are characterized by the presence of carbon-to-carbon double bonds. Similarly, all alcohols contain a hydroxyl group (-OH). Since this group is polar, an alcohol such as ethanol (CH_3CH_2OH) is liquid at room temperature and highly soluble in water, while the alkane of similar molecular weight, propane ($CH_3CH_2CH_3$), is a gas at room temperature and is completely insoluble in water. Other common functional groups are shown in Table 10.2, along with an example of a molecule from each family.

The chemistry of organic and biological molecules is usually controlled by the functional group found in the molecule. Just as members of the same family of the periodic table exhibit similar chemistry, organic molecules with the same functional group exhibit similar chemistry. Although it would be impossible to learn the chemistry of each organic molecule, it is relatively easy to learn the chemistry of each functional group. In this way you can learn the chemistry of all members of a family of organic compounds, or biological molecules, just by learning the chemistry of its characteristic functional group or groups.

3 LEARNING GOAL

This is analogous to the classification of the elements within the periodic table. See Chapter 2.

10.2 Alkanes

Alkanes are saturated hydrocarbons; that is, alkanes contain only carbon and hydrogen bonded together through carbon-hydrogen and carbon-carbon single bonds. C_nH_{2n+2} is the general formula for alkanes. In this formula, n is the number of carbon atoms in the molecule.

Structure and Physical Properties

LEARNING GOAL 4

LEARNING GOAL 5

Four types of formulas, each providing different information, are used in organic chemistry: the molecular formula, the structural formula, the condensed formula, and the line formula.

The **molecular formula** tells the kind and number of each type of atom in a molecule but does not show the bonding pattern. Consider the molecular formulas for simple alkanes:

CH_4	C_2H_6	C_3H_8	C_4H_{10}
Methane	Ethane	Propane	Butane

For the first three compounds, there is only one possible arrangement of the atoms. However, for C_4H_{10} there are two possible arrangements. How do we know which is correct? The problem is solved by using the **structural formula,** which shows each atom and bond in a molecule. The following are the structural formulas for methane, ethane, propane, and the two isomers of butane:

Recall that a covalent bond, representing a pair of shared electrons, can be drawn as a line between two atoms. For the structure to be correct, each carbon atom must show four pairs of shared electrons.

Methane Ethane Propane Butane 2-Methylpropane (isobutane)

The advantage of a structural formula is that it shows the complete structure, but for large molecules it is time-consuming to draw and requires too much space. The compromise is the **condensed formula.** It shows all the atoms in a molecule and places them in a sequential order that indicates which atoms are bonded to which. The following are the condensed formulas for the preceding five compounds.

CH_4	CH_3CH_3	$CH_3CH_2CH_3$	$CH_3(CH_2)_2CH_3$	$(CH_3)_3CH$
Methane	Ethane	Propane	Butane	2-Methylpropane (isobutane)

The names and formulas of the first ten straight-chain alkanes are shown in Table 10.3.

The simplest representation of a molecule is the **line formula.** In the line formula we assume that there is a carbon atom at any location where two or more lines intersect. We also assume that there is a carbon at the end of any line and that each carbon in the structure is bonded to the correct number of hydrogen atoms. Compare the structural and line formulas for butane and 2-methylpropane, shown here:

Butane 2-Methylpropane

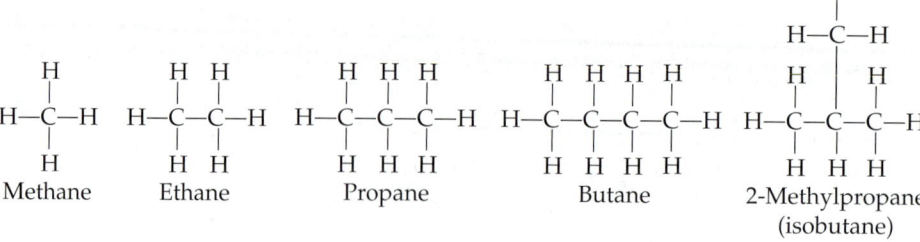

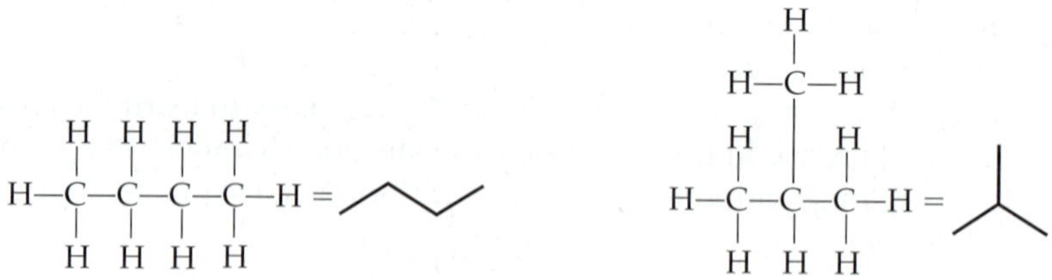

TABLE 10.3 Names and Formulas of the First Ten Straight-Chain Alkanes

Name	Molecular Formula	Condensed Formula	Melting Point, °C*	Boiling Point, °C*
Alkanes	C_nH_{2n+2}			
Methane	CH_4	CH_4	−182.5	−162.2
Ethane	C_2H_6	CH_3CH_3	−183.9	−88.6
Propane	C_3H_8	$CH_3CH_2CH_3$	−187.6	−42.1
Butane	C_4H_{10}	$CH_3CH_2CH_2CH_3$ or $CH_3(CH_2)_2CH_3$	−137.2	−0.5
Pentane	C_5H_{12}	$CH_3CH_2CH_2CH_2CH_3$ or $CH_3(CH_2)_3CH_3$	−129.8	36.1
Hexane	C_6H_{14}	$CH_3CH_2CH_2CH_2CH_2CH_3$ or $CH_3(CH_2)_4CH_3$	−95.2	68.8
Heptane	C_7H_{16}	$CH_3CH_2CH_2CH_2CH_2CH_2CH_3$ or $CH_3(CH_2)_5CH_3$	−90.6	98.4
Octane	C_8H_{18}	$CH_3CH_2CH_2CH_2CH_2CH_2CH_2CH_3$ or $CH_3(CH_2)_6CH_3$	−56.9	125.6
Nonane	C_9H_{20}	$CH_3CH_2CH_2CH_2CH_2CH_2CH_2CH_2CH_3$ or $CH_3(CH_2)_7CH_3$	−53.6	150.7
Decane	$C_{10}H_{22}$	$CH_3CH_2CH_2CH_2CH_2CH_2CH_2CH_2CH_2CH_3$ or $CH_3(CH_2)_8CH_3$	−29.8	174.0

*Melting and boiling points as reported in the National Institute of Standards and Technology Chemistry Webbook, which can be found at http://webbook.nist.gov/.

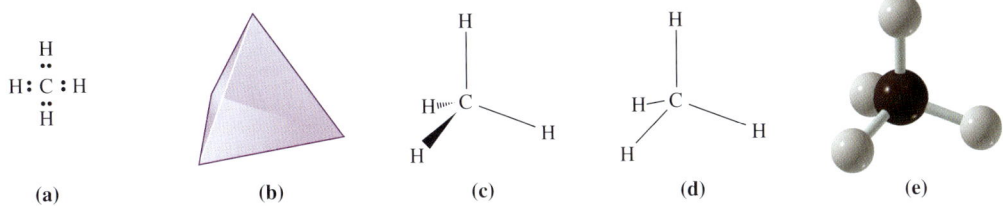

Figure 10.3
The tetrahedral carbon atom: (a) Lewis dot structure; (b) a tetrahedron; (c) the tetrahedral carbon drawn with dashes and wedges; (d) the stick drawing of the tetrahedral carbon atom; (e) ball-and-stick model of methane.

Each carbon atom forms four single covalent bonds, but each hydrogen atom has only a single covalent bond. Although a carbon atom may be involved in single, double, or triple bonds, it always shares four pairs of electrons. The Lewis Dot structure of the simplest alkane, methane, shows the four shared pairs of electrons (Figure 10.3a). When carbon is involved in four single bonds, the *bond angle,* the angle between two atoms or substituents attached to carbon, is 109.5°, as predicted by the valence shell electron pair repulsion (VSEPR) theory. Thus, alkanes contain carbon atoms that have tetrahedral geometry.

A tetrahedron is a geometric solid having the structure shown in Figure 10.3b. There are many different ways to draw the tetrahedral carbon (Figures 10.3c–10.3e). In Figure 10.3c, solid lines, dashes, and wedges are used to represent the structure of methane. Dashes go back into the page away from you; wedges come out of the page toward you; and solid lines are in the plane of the page. The structure in Figure 10.3d is the same as that in Figure 10.3c; it just leaves a lot more to the imagination. Figure 10.3e is a ball-and-stick model of the methane molecule. Three-dimensional drawings of two other simple alkanes are shown in Figure 10.4.

All hydrocarbons are nonpolar molecules. As a result they are not water soluble but are soluble in nonpolar organic solvents. Furthermore, they have relatively low melting points and boiling points and are generally less dense than water. In general, the longer the hydrocarbon chain (greater the molecular weight), the higher the melting and boiling points and the greater the density (see Table 10.3).

Molecular geometry is described in Section 3.4

See Section 5.2 for a discussion of the forces responsible for the physical properties of a substance.

Figure 10.4
(a) Drawing and (b) ball-and-stick model of ethane. All the carbon atoms have a tetrahedral arrangement, and all bond angles are approximately 109.5°. (c) Drawing and (d) ball-and-stick model of a more complex alkane, butane.

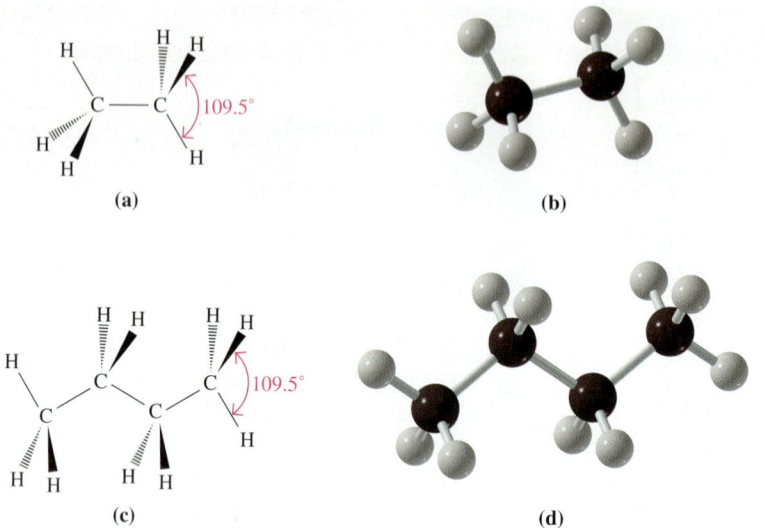

EXAMPLE 10.1 Using Different Types of Formulas to Represent Organic Compounds

Draw the structural and condensed formulas of the following line structure:

Solution

Remember that each intersection of lines represents a carbon atom and that each line ends in a carbon atom. This gives us the following carbon skeleton:

$$\begin{array}{ccccccc} & & & C & & C & \\ & & & | & & | & \\ C-C-C-C & & C-C-C \end{array}$$

By adding the correct number of hydrogen atoms to the carbon skeleton, we are able to complete the structural formula of this compound.

From the structural formula we can write the condensed formula as follows:

$$CH_3CH_2CH_2CH(CH_3)CH_2CH(CH_3)CH_3$$

Structural or constitutional isomers are explored in detail in Section 10.2.

Molecules having the same molecular formula but different arrangements of atoms are called structural or constitutional isomers. The carbon chains of these isomers may be straight (Table 10.3) or branched. Compare the following structural isomers of C_6H_{14}. Hexane has a straight chain or carbon backbone:

$$CH_3CH_2CH_2CH_2CH_2CH_3$$
Hexane

TABLE 10.4 Melting and Boiling Points of Five Alkanes of Molecular Formula C_6H_{14}

Name	Condensed Formula	Boiling Point* °C	Melting Point* °C
Hexane	$CH_3CH_2CH_2CH_2CH_2CH_3$	68.8	−95.2
2-Methylpentane	$CH_3CH(CH_3)CH_2CH_2CH_3$	60.9	−153.2
3-Methylpentane	$CH_3CH_2CH(CH_3)CH_2CH_3$	63.3	−118
2,3-Dimethylbutane	$CH_3CH(CH_3)CH(CH_3)CH_3$	58.1	−130.2
2,2-Dimethylbutane	$CH_3C(CH_3)_2CH_2CH_3$	49.8	−100.2

*Melting and boiling points as reported in the National Institute of Standards and Technology Chemistry Webbook, which can be found at http://webbook.nist.gov/.

All the other isomers have one or more of the carbon atoms branching from the main carbon chain:

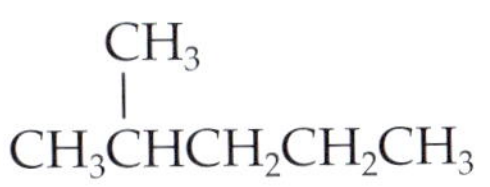

$$\underset{\text{2-Methylpentane}}{CH_3CH(CH_3)CH_2CH_2CH_3} \qquad \underset{\text{3-Methylpentane}}{CH_3CH_2CH(CH_3)CH_2CH_3} \qquad \underset{\text{2,3-Dimethylbutane}}{CH_3CH(CH_3)CH(CH_3)CH_3} \qquad \underset{\text{2,2-Dimethylbutane}}{CH_3C(CH_3)_2CH_2CH_3}$$

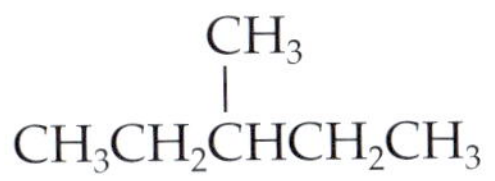

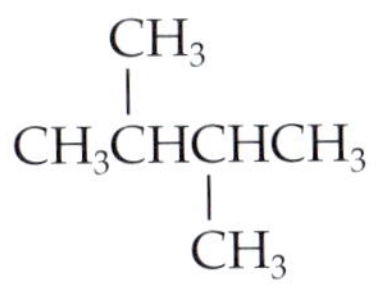

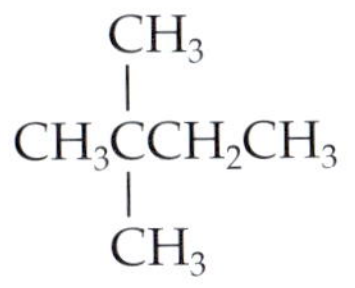

These branched-chain forms of the molecule have a much smaller surface area than the straight-chain. As a result, the van der Waals forces attracting the molecules to one another are less strong and these molecules have lower melting and boiling points than the straight-chain isomers. The melting and boiling points of the five structural isomers of C_6H_{14} are found in Table 10.4.

Van der Waals forces are discussed in Section 5.2.

Question 10.3 Draw the structural formula for each of the molecules in Table 10.4.

Question 10.4 Draw the line formula for each of the molecules in Table 10.4.

Alkyl Groups

Alkyl groups are alkanes with one fewer hydrogen atom. The name of the alkyl group is derived from the name of the alkane containing the same number of carbon atoms. The *-ane* ending of the alkane name is replaced by the *-yl* ending. Thus, CH_3- is a methyl group and CH_3CH_2- is an ethyl group. The dash at the end of these two structures represents the point at which the alkyl group can bond to another atom. The first five continuous-chain alkyl groups are presented in Table 10.5.

Carbon atoms are classified according to the number of other carbon atoms to which they are attached. A **primary carbon (1°)** is directly bonded to one other carbon. A **secondary carbon (2°)** is bonded to two other carbon atoms; a **tertiary carbon (3°)** is bonded to three other carbon atoms, and a **quaternary carbon (4°)** to four.

Alkyl groups are classified according to the number of carbons attached to the carbon atom that joins the alkyl group to a molecule.

Chapter 10 An Introduction to Organic Chemistry: The Saturated Hydrocarbons

TABLE 10.5 Names and Formulas of the First Five Continuous-Chain Alkyl Groups

Alkyl Group Structure	Name
CH_3-	Methyl
CH_3CH_2-	Ethyl
$CH_3CH_2CH_2-$	Propyl
$CH_3CH_2CH_2CH_2-$	Butyl
$CH_3CH_2CH_2CH_2CH_2-$	Pentyl

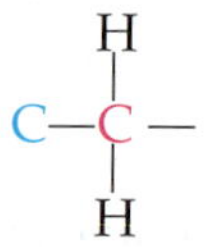

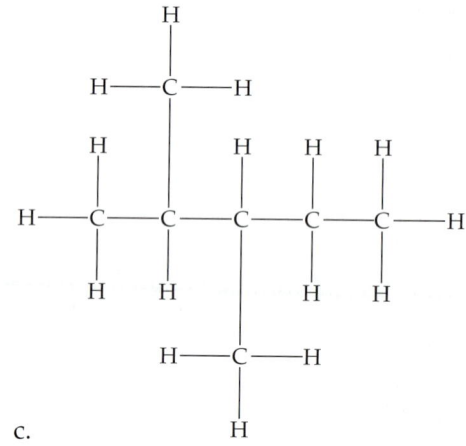

Primary alkyl group 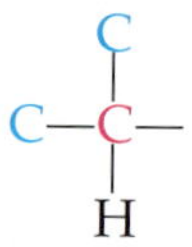   Secondary alkyl group 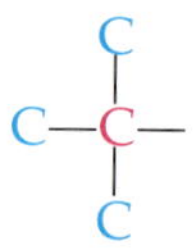   Tertiary alkyl group

All of the continuous-chain alkyl groups are primary alkyl groups (see Table 10.5). Several branched-chain alkyl groups are shown in Table 10.6. Notice that the isopropyl and *sec*-butyl groups are secondary alkyl groups; the isobutyl group is a primary alkyl group; and the *t*-butyl (*tert*-butyl) is a tertiary alkyl group.

Question 10.5

Classify each of the carbon atoms in the following structures as either primary, secondary, or tertiary.

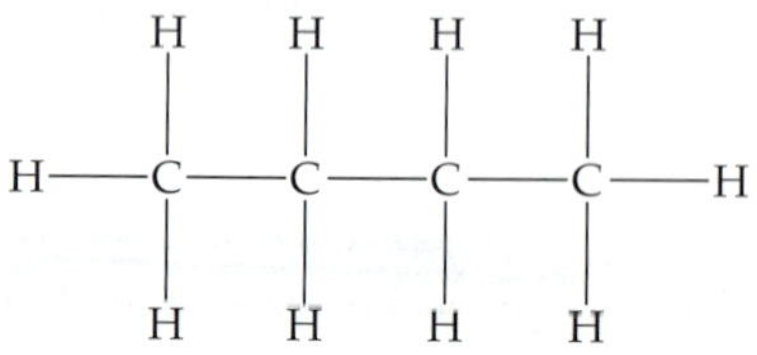

a. [structure: straight chain of 4 carbons with H's]

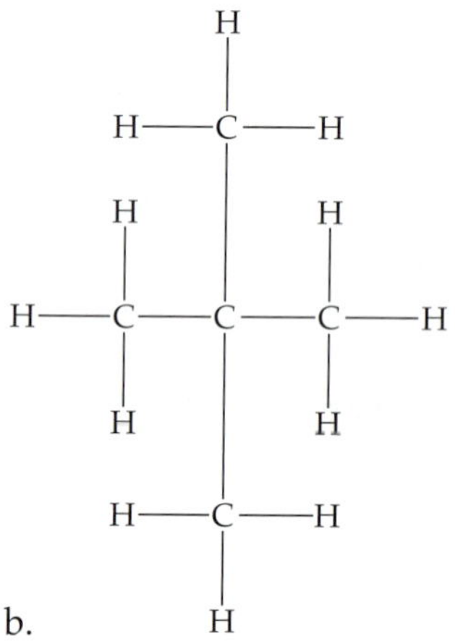

b. [structure: central chain of 3 carbons with CH₃ groups above and below the middle carbon]

c. [structure: chain of 5 carbons with a CH₃ branch]

Question 10.6

Classify each of the carbon atoms in the following structures as either primary, secondary, or tertiary.

a. $CH_3CH_2C(CH_3)_2CH_2CH_3$
b. $CH_3CH_2CH_2CH_2CH(CH_3)CH(CH_3)CH_3$

TABLE 10.6 Structures and Names of Some Branched-Chain Alkyl Groups

Structure	Classification	Common Name	I.U.P.A.C. Name
CH$_3$CH— | CH$_3$	2°	Isopropyl*	1-Methylethyl
CH$_3$ | CH$_3$CHCH$_2$—	1°	Isobutyl*	2-Methylpropyl
CH$_3$CH$_2$CH— | CH$_3$	2°	sec-Butyl†	1-Methylpropyl
CH$_3$ | CH$_3$C— | CH$_3$	3°	t-Butyl or tert-Butyl‡	1,1-Dimethylethyl

*The prefix iso- (isomeric) is used when there are two methyl groups at the end of the alkyl group.
†The prefix sec- (secondary) indicates that there are two carbons bonded to the carbon that attaches the alkyl group to the parent molecule.
‡The prefix t- or tert- (tertiary) means that three carbons are attached to the carbon that attaches the alkyl group to the parent molecule.

Nomenclature

Historically, organic compounds were named by the chemist who discovered them. Often the names reflected the source of the compound. For instance, the antibiotic penicillin is named for the mold *Penicillium notatum*, which produces it. The pain reliever aspirin was made by adding an acetate group to a compound first purified from the bark of a willow tree, shown on the cover of this book, and later from the meadowsweet plant (*Spirea ulmaria*). Thus, the name aspirin comes from *a* (acetate) and *spir* (genus of meadowsweet).

These names are easy for us to remember because we come into contact with these compounds often. However, as the number of compounds increased, organic chemists realized that historical names were not adequate because they revealed nothing about the structure of a compound. Thousands of such compounds and their common names had to be memorized! What was needed was a set of nomenclature (naming) rules that would produce a unique name for every organic compound. Furthermore, the name should be so descriptive that, by knowing the name, a student or scientist could write the structure.

The International Union of Pure and Applied Chemistry (I.U.P.A.C.) is the organization responsible for establishing and maintaining a standard, universal system for naming organic compounds. The system of nomenclature developed by this group is called the **I.U.P.A.C. Nomenclature System.** The following rules are used for naming alkanes by the I.U.P.A.C. system.

1. Determine the name of the **parent compound,** the longest continuous carbon chain in the compound. Refer to Tables 10.3 and 10.7 to determine the parent name. Notice that these names are made up of a prefix related to the number of carbons in the chain and the suffix *-ane,* indicating that the molecule is an alkane (Table 10.7). Write down the name of the parent compound, leaving space before the name to identify the substituents. Parent chains are highlighted in yellow in the following examples:

6 LEARNING GOAL

It is important to learn the prefixes for the carbon chain lengths. We will use them in the nomenclature for all organic molecules.

TABLE 10.7 Carbon Chain Length and Prefixes Used in the I.U.P.A.C. Nomenclature System

Carbon Chain Length	Prefix	Alkane Name
1	Meth-	Meth*ane*
2	Eth-	Eth*ane*
3	Prop-	Prop*ane*
4	But-	But*ane*
5	Pent-	Pent*ane*
6	Hex-	Hex*ane*
7	Hept-	Hept*ane*
8	Oct-	Oct*ane*
9	Non-	Non*ane*
10	Dec-	Dec*ane*

```
     1  2  3              5   4   3  2   1              9   4   7   6   5   4  3
    CH₃CHCH₃           CH₃CH₂CHCH₂CH₃           CH₃CH₂CH₂CH₂CH₂CH₂CHCH₃
        |                       |                                       |
       CH₃                     CH₃                                    CH₂CH₃
                                                                       2   1

Parent name: Propane        Pentane                              Nonane
```

2. Number the parent chain to give the lowest number to the carbon bonded to the first group encountered on the parent chain, regardless of the numbers that result for the other substituents.

3. Name and number each atom or group attached to the parent compound. The number tells you the position of the group on the main chain, and the name tells you what type of substituent is present at that position. For example, it may be one of the halogens [F-(fluoro), Cl-(chloro), Br-(bromo), and I-(iodo)] or an alkyl group (Tables 10.5 and 10.6). In the following examples the parent chain is highlighted in yellow:

```
      1  2  3              3   2   1              1    2    3   4
     CH₃CHCH₃            CH₃CHCH₂CH₃          CH₃CH₂CH₂CHCH₂CH₃
         |                     |                            |
        Br                   CH₂CH₃                      CH₂CH₂CH₂CH₃
                              4   5                       5   6   7  8

Substituent:      2-Bromo             3-Methyl                  4-Ethyl
I.U.P.A.C. name:  2-Bromopropane      3-Methylpentane           4-Ethyloctane
```

4. If the same substituent occurs more than once in the compound, a separate position number is given for each, and the prefixes *di-*, *tri-*, *tetra-*, *penta-*, and so forth are used, as shown in the following examples:

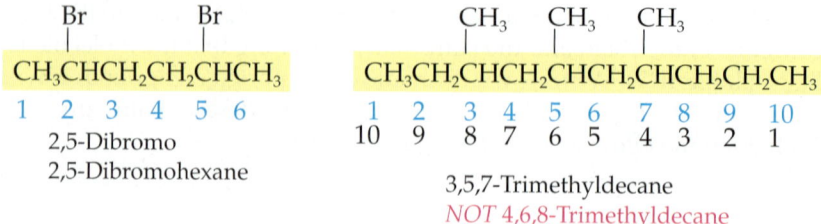

2,5-Dibromo
2,5-Dibromohexane

3,5,7-Trimethyldecane
NOT 4,6,8-Trimethyldecane

Throughout this book we will primarily use the I.U.P.A.C. Nomenclature System. When common names are used, they will be shown in parentheses beneath the I.U.P.A.C. name.

An Environmental Perspective

Oil-Eating Microbes

Our highly industrialized society has come to rely more and more on petroleum as a source of energy and a raw material for the manufacture of plastics, drugs, and a host of other consumables. In fact, the United States uses approximately 710 million gallons of oil each day and over half of that is imported from overseas in supertankers.

The EPA reports as many as 14,000 oil spills each year. By some estimates, that amounts to as much as 100 million gallons of this U.S. oil spilled into the environment. The largest spill occurred in 1991 in the Persian Gulf. Two days after the Coalition forces began their air attack, Iraqi forces invading Kuwait released 240 million gallons of oil from onshore terminals and from tankers moored offshore. By comparison, the well-publicized oil spill from the *Exxon Valdez* was only 11 million gallons!

Oil spills can have disastrous effects on the environment. We see film footage of oil spills on the evening news that show us birds and animals such as sea otters whose feathers or fur become saturated with oil and lose their ability to insulate them. These animals may die of exposure if they are not rescued and decontaminated. Other animals are accidentally poisoned because of the compounds they ingest while trying to clean themselves. Underwater vegetation, shellfish, and fish may also be destroyed by an oil spill.

To try to reduce the environmental damage, floating barriers may be used to prevent the spread of the oil. These may be used along with skimmers that separate the oil from the water. In some cases, wood chips, straw, or other absorbent materials may be introduced to "sop up" the oil. Dispersants may be introduced to break the oil slick up into small droplets. The benefit of dispersing the oil into droplets is that they can then be more effectively attacked by microbes in the environment that have enzymes that allow them to use the molecules in oil as a food source.

These oil-eating microbes, or OEMs, require oxygen, water, inorganic nutrients, and a source of organic compounds, such as the hydrocarbons found in oil, in order to survive. Their value in cleaning up oil spills is apparent when you consider that the products of their metabolic breakdown of oil are harmless carbon compounds, carbon dioxide, and water.

Oil spills cause extensive environmental damage each year.

There are at least seventy genera of OEMs in nature. Among these are bacteria and a variety of fungi. They have the ability to survive and break down oil in temperatures ranging from -2 to $60°C$, and they are active in environments ranging from polar regions to the equator. Although scientists have created genetically engineered microorganisms that have an enhanced ability to metabolize petroleum compounds, it turns out that those that are present in nature can do the job perfectly well, as long as they have enough oxygen and inorganic nutrients to support their metabolism.

For Further Understanding

List some of the many uses for petroleum products on which the U.S. economy is dependent.

What advantage might there be of relying on the OEMs in the environment rather than introducing genetically modified laboratory strains?

---

5. Place the names of the substituents in alphabetical order before the name of the parent compound, which you wrote down in Step 1. Numbers are separated by commas, and numbers are separated from names by hyphens. By convention, halogen substituents are placed before alkyl substituents.

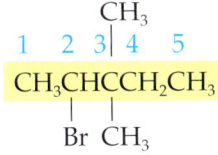

2-Bromo-3,3-dimethylpentane
NOT 3,3-Dimethyl-2-bromopentane

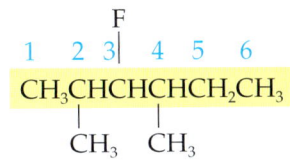

3-Fluoro-2,4-dimethylhexane
NOT 2,4-Dimethyl-3-fluorohexane

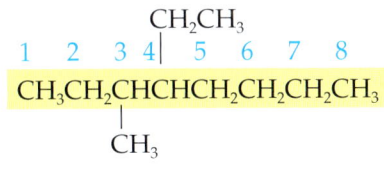

4-Ethyl-3-methyloctane
NOT 3-Methyl-4-ethyloctane

EXAMPLE 10.2 Naming Substituted Alkanes Using the I.U.P.A.C. System

Name the following alkanes using I.U.P.A.C. nomenclature.

Solution

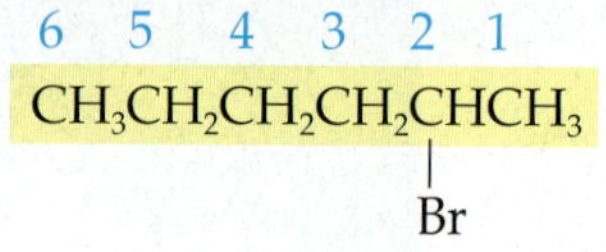

$$\overset{6}{C}H_3\overset{5}{C}H_2\overset{4}{C}H_2\overset{3}{C}H_2\underset{|}{\overset{2}{C}H}\overset{1}{C}H_3$$
$$\quad\quad\quad\quad\quad\quad Br$$

Parent chain: hexane
Substituent: 2-bromo (*not* 5-bromo)
Name: 2-Bromohexane

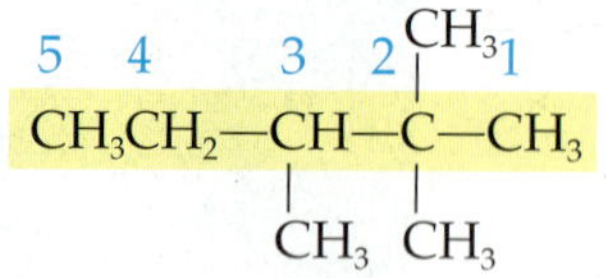

$$\overset{5}{C}H_3\overset{4}{C}H_2-\underset{\underset{CH_3}{|}}{\overset{3}{C}H}-\underset{\underset{CH_3}{|}}{\overset{2}{\underset{|}{C}}}\overset{CH_3}{-}\overset{1}{C}H_3$$

Parent chain: pentane
Substituent: 2,2,3-trimethyl (*not* 3,4,4-trimethyl)
Name: 2,2,3-Trimethylpentane

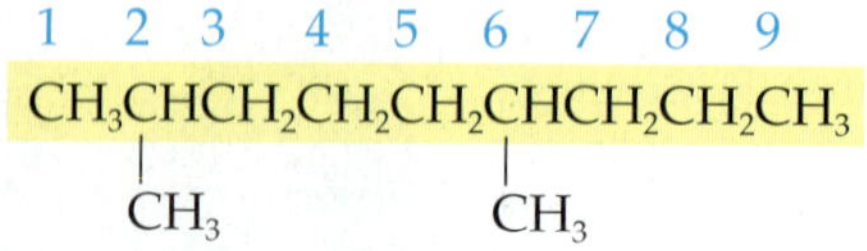

$$\overset{1}{C}H_3\underset{\underset{CH_3}{|}}{\overset{2}{C}H}\overset{3}{C}H_2\overset{4}{C}H_2\overset{5}{C}H_2\underset{\underset{CH_3}{|}}{\overset{6}{C}H}\overset{7}{C}H_2\overset{8}{C}H_2\overset{9}{C}H_3$$

Parent chain: nonane
Substituent: 2,6-dimethyl
(*not* 4,8-dimethyl)
Name: 2,6-Dimethylnonane

Question 10.7

Name the following compounds, using the I.U.P.A.C. Nomenclature System:

a. 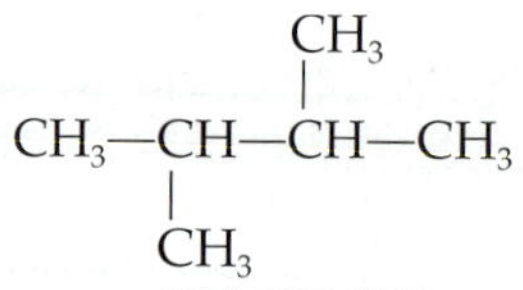 $CH_3-CH-CH-CH_3$ with CH_3 up on C2 and CH_3 down on C3

b. 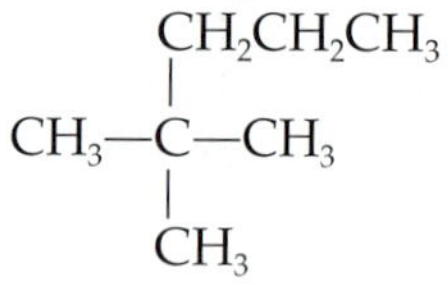 CH_3-C-CH_3 with CH_3 down and $CH_2CH_2CH_3$ up

c. CH_3-C-CH_3 with CH_3 up and CH_3 down

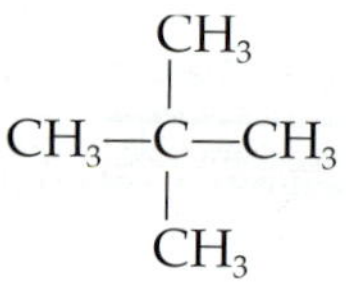

d. 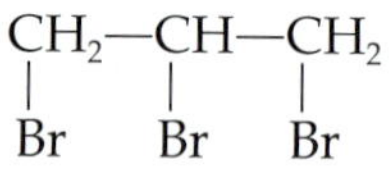 $CH_2-CH-CH_2$ with Br, Br, Br

---

Question 10.8

Name the following compounds, using the I.U.P.A.C. Nomenclature System:

a. 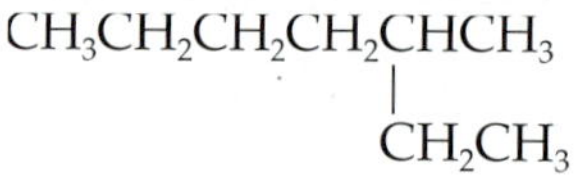 $CH_3CH_2CH_2CH_2CHCH_3$ with CH_2CH_3 substituent

b. 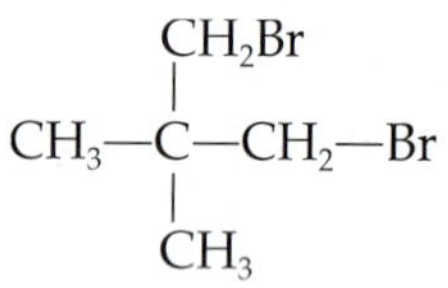 CH_3-C-CH_2-Br with CH_2Br up and CH_3 down

c. $CH_3CHCH_2CH_2CHCH_2-Br$ with I and CH_3 substituents

d. 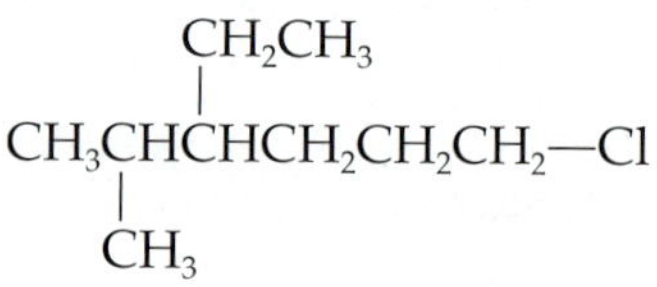 $CH_3CHCHCH_2CH_2-Cl$ with CH_2CH_3 and CH_3 substituents

---

Having learned to name a compound using the I.U.P.A.C. system, we can easily write the structural formula of a compound, given its name. First, draw and number the parent carbon chain. Add the substituent groups to the correct carbon and finish the structure by adding the correct number of hydrogen atoms.

Drawing the Structure of a Compound Using the I.U.P.A.C. Name

EXAMPLE 10.3

Draw the structural formula for 1-bromo-4-methylhexane.

Solution

Begin by drawing the six-carbon parent chain and indicating the four bonds for each carbon atom.

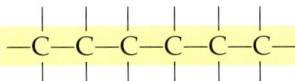

Next, number each carbon atom:

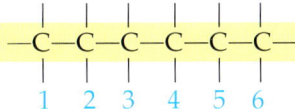

Now add the substituents. In this example a bromine atom is bonded to carbon-1, and a methyl group is bonded to carbon-4:

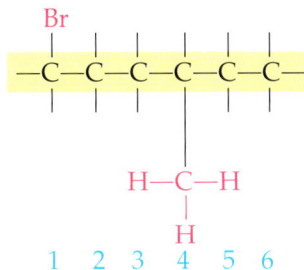

Finally, add the correct number of hydrogen atoms so that each carbon has four covalent bonds:

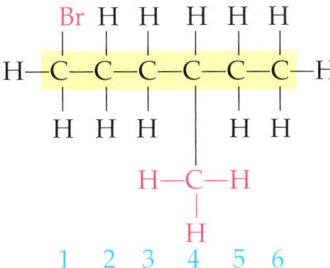

As a final check of your accuracy, use the I.U.P.A.C. system to name the compound that you have just drawn, and compare the name with that in the original problem.

The molecular formula and condensed formula can be written from the structural formula shown. The molecular formula is $C_7H_{15}Br$, and the condensed formula is $BrCH_2CH_2CH_2CH(CH_3)CH_2CH_3$.

Constitutional or Structural Isomers

As we saw earlier, there are two arrangements of the atoms represented by the molecular formula C_4H_{10}: butane and 2-methylpropane. Molecules having the same molecular formula but a different arrangement of atoms are called **constitutional**, or **structural, isomers**. These isomers are unique compounds because of their

LEARNING GOAL 7

structural differences, and they have different physical and chemical properties. For instance, the data in Table 10.4 show that the branched-chain isomers of C_6H_{14} have lower boiling and melting points than the straight-chain isomer, hexane. These differences reflect the different shapes of the molecules.

EXAMPLE 10.4 Drawing Constitutional or Structural Isomers of Alkanes

Write all the constitutional isomers having the molecular formula C_6H_{14}.

Solution

1. Begin with the continuous six-carbon chain structure:

$$\overset{1}{C}H_3-\overset{2}{C}H_2-\overset{3}{C}H_2-\overset{4}{C}H_2-\overset{5}{C}H_2-\overset{6}{C}H_3$$

Isomer A

2. Now try five-carbon chain structures with a methyl group attached to one of the internal carbon atoms of the chain:

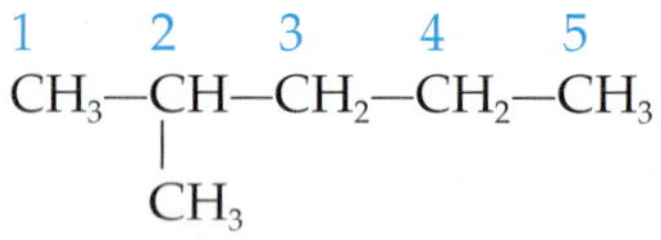

$$\overset{1}{C}H_3-\underset{\underset{CH_3}{|}}{\overset{2}{C}H}-\overset{3}{C}H_2-\overset{4}{C}H_2-\overset{5}{C}H_3 \quad \text{and} \quad \overset{1}{C}H_3-\overset{2}{C}H_2-\underset{\underset{CH_3}{|}}{\overset{3}{C}H}-\overset{4}{C}H_2-\overset{5}{C}H_3$$

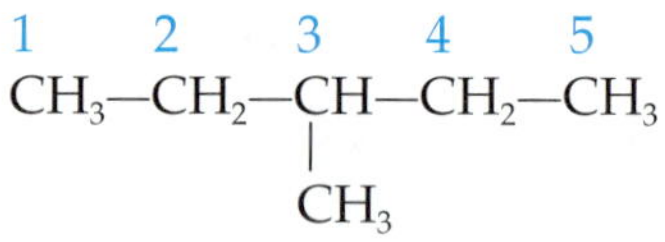

Isomer B Isomer C

3. Next consider the possibilities for a four-carbon structure to which two methyl groups (—CH_3) may be attached:

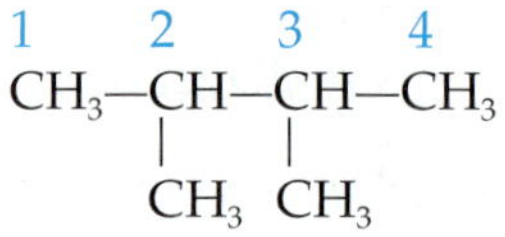

$$\overset{1}{C}H_3-\underset{\underset{CH_3}{|}}{\overset{2}{C}H}-\underset{\underset{CH_3}{|}}{\overset{3}{C}H}-\overset{4}{C}H_3 \quad \text{and} \quad \overset{1}{C}H_3-\underset{\underset{CH_3}{|}}{\overset{2}{\overset{\overset{CH_3}{|}}{C}}}-\overset{3}{C}H_2-\overset{4}{C}H_3$$

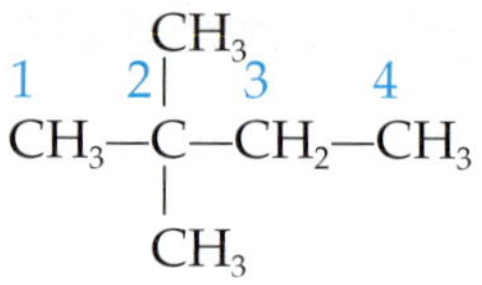

Isomer D Isomer E

These are the five possible constitutional isomers of C_6H_{14}. At first it may seem that other isomers are also possible. But careful comparison will show that they are duplicates of those already constructed. For example, rather than add two methyl groups, a single ethyl group (—CH_2CH_3) could be added to the four-carbon chain:

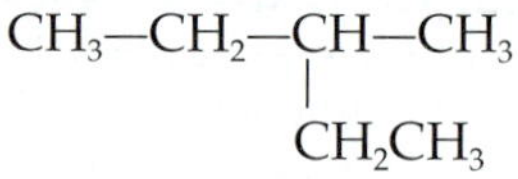

$$CH_3-CH_2-\underset{\underset{CH_2CH_3}{|}}{CH}-CH_3$$

But close examination will show that this is identical to isomer C. Perhaps we could add one ethyl group and one methyl group to a three-carbon parent chain, with the following result:

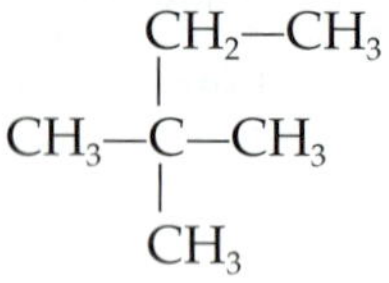

$$CH_3-\underset{\underset{CH_3}{|}}{\overset{\overset{CH_2-CH_3}{|}}{C}}-CH_3$$

Again we find that this structure is the same as one of the isomers we have already identified, isomer E.

Continued—

EXAMPLE 10.4 —Continued

To check whether you have accidentally made duplicate isomers, name them using the I.U.P.A.C. system. All isomers must have different I.U.P.A.C. names. So if two names are identical, the structures are also identical. Use the I.U.P.A.C. system to name the isomers in this example, and prove to yourself that the last two structures are simply duplicates of two of the original five isomers.

Question 10.9

Draw a complete structural formula for each of the straight-chain isomers of the following alkanes:

a. C_4H_9Br
b. $C_4H_8Br_2$

Question 10.10

Name all of the isomers that you obtained in Question 10.9.

10.3 Cycloalkanes

The **cycloalkanes** are a family having C—C single bonds in a ring structure. They have the general molecular formula C_nH_{2n} and thus have two fewer hydrogen atoms than the corresponding alkane (C_nH_{2n+2}). The structures and names of some simple cycloalkanes are shown in Figure 10.5.

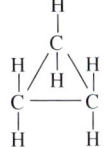
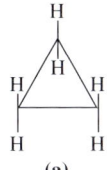

(a)

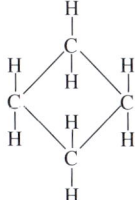
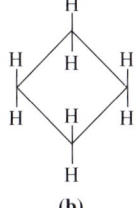
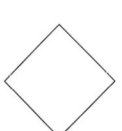
(b)

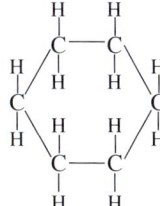
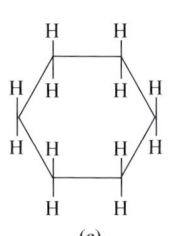
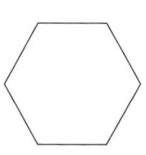
(c)

Figure 10.5
Cycloalkanes: (a) cyclopropane; (b) cyclobutane; (c) cyclohexane. All of the cycloalkanes are shown using structural formulas (left column), condensed structural formulas (center column), and line formulas (right column).

In the I.U.P.A.C. system, the cycloalkanes are named by applying the following simple rules.

- Determine the name of the alkane with the same number of carbon atoms as there are within the ring and add the prefix *cyclo-*. For example, cyclopentane is the cycloalkane that has five carbon atoms.
- If the cycloalkane is substituted, place the names of the groups in alphabetical order before the name of the cycloalkane. No number is needed if there is only one substituent.
- If more than one group is present, use numbers that result in the *lowest possible position numbers.*

EXAMPLE 10.5 Naming a Substituted Cycloalkane Using the I.U.P.A.C. Nomenclature System

Name the following cycloalkanes using I.U.P.A.C. nomenclature.

Solution

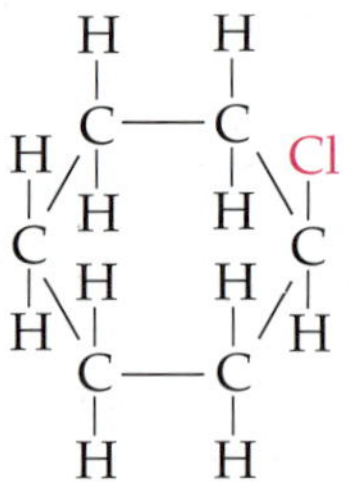

Parent chain: cyclohexane
Substituent: chloro (no number is required because there is only one substituent)
Name: Chlorocyclohexane

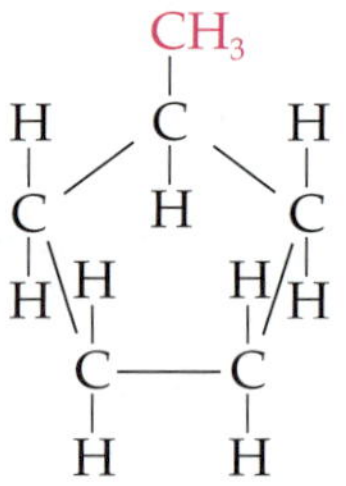

Parent chain: cyclopentane
Substituent: methyl (no number is required because there is only one substituent)
Name: Methylcyclopentane

These cycloalkanes could also be shown as line formulas, as shown below. Each line represents a carbon-carbon bond. A carbon atom and the correct number of hydrogen atoms are assumed to be at the point where the lines meet and at the end of a line.

Chlorocyclohexane

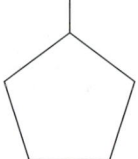
Methylcyclopentane

cis-trans Isomerism in Cycloalkanes

Atoms of an alkane can rotate freely around the carbon-carbon single bond, resulting in an unlimited number of arrangements. However, rotation around the bonds in a cyclic structure is limited by the fact that the carbons of the ring are all bonded to another carbon within the ring. The formation of **cis-trans isomers,** or **geometric isomers,** is a consequence of the absence of free rotation.

Geometric isomers are a type of *stereoisomer.* **Stereoisomers** are molecules that have the same structural formulas and bonding patterns but different arrangements of atoms in space. The *cis-trans* isomers of cycloalkanes are stereoisomers that differ from one another in the arrangement of substituents in space. Consider the following two views of the *cis-* and *trans-*isomers of 1,2-dichlorocyclohexane:

Stereoisomers are discussed in detail in Section 16.3 and online at www.mhhe.com/denniston5e in "Stereochemistry and Stereoisomers Revisited."

cis-1,2-Dichlorocyclohexane   *trans*-1,2-Dichlorocyclohexane

cis-1,2-Dichlorocyclohexane   *trans*-1,2-Dichlorocyclohexane

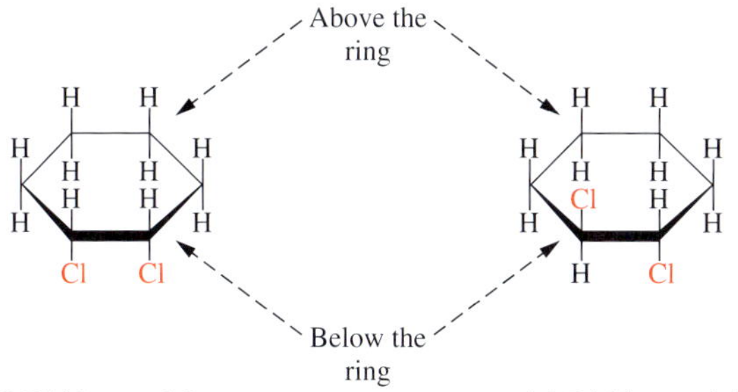

In the wedge and line diagram at the top, it is easy to imagine that you are viewing the ring structures as if an edge were projecting toward you. This will help you understand the more common structural formulas, shown beneath them. In the structure on the left, both Cl atoms are beneath the ring. They are termed *cis* (*L.*, "on the same side"). The complete name for this compound is *cis*-1,2-dichlorocyclohexane. In the structure on the right, one Cl is above the ring and the other is below it. They are said to be *trans* (*L.*, "across from") to one another and the complete name of this compound is *trans*-1,2-dichlorocyclohexane.

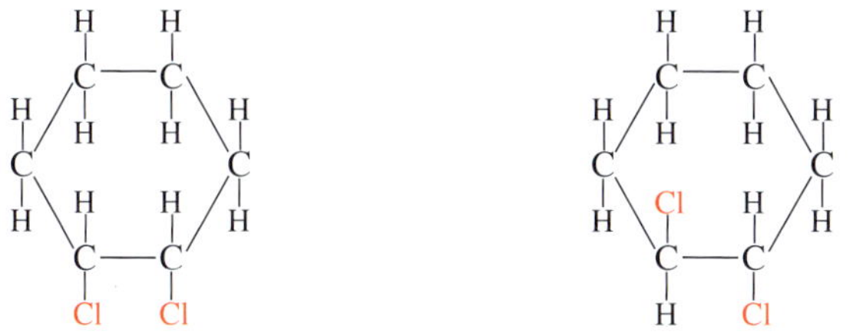

Geometric isomers do not readily interconvert. The cyclic structure prevents unrestricted free rotation and, thus, prevents interconversion. Only by breaking carbon-carbon bonds of the ring could interconversion occur. As a result, geometric isomers may be separated from one another in the laboratory.

Naming *cis-trans* Isomers of Substituted Cycloalkanes — EXAMPLE 10.6

Determine whether the following substituted cycloalkanes are *cis* or *trans* isomers and write the complete name for each.

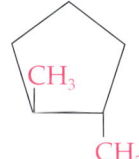

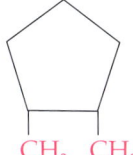

Solution

Both molecules are cyclopentanes having two methyl group substituents. Thus both would be named 1,2-dimethylcyclopentane. In the structure on the left, one methyl group is above the ring and the other is below the ring; they are in the *trans* configuration, and the structure is named *trans*-1,2-dimethylcyclopentane. In the structure on the right, both methyl groups are on the same side of the ring (below it, in this case); they are *cis* to one another, and the complete name of this compound is *cis*-1,2-dimethylcyclopentane.

Chapter 10 An Introduction to Organic Chemistry: The Saturated Hydrocarbons

---

EXAMPLE 10.7 **Naming a Cycloalkane Having Two Substituents Using the I.U.P.A.C. Nomenclature System**

Name the following cycloalkanes using I.U.P.A.C. nomenclature.

Solution

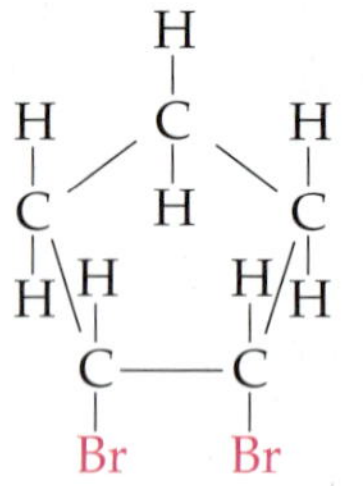

Parent chain: cyclopentane
Substituent: 1,2-dibromo
Isomer: *cis*
Name: *cis*-1,2-Dibromocyclopentane

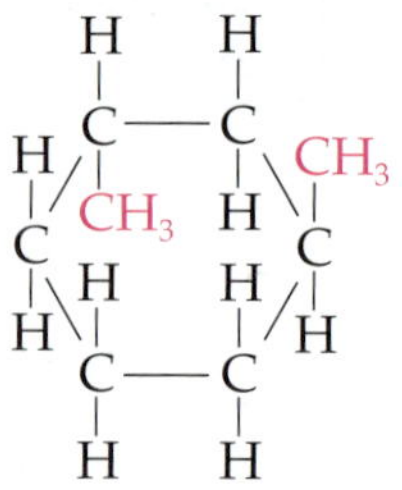

Parent chain: cyclohexane
Substituent: 1,3-dimethyl
Isomer: *trans*
Name: *trans*-1,3-Dimethylcyclohexane

Question 10.11 Name each of the following substituted cycloalkanes using the I.U.P.A.C. Nomenclature System:

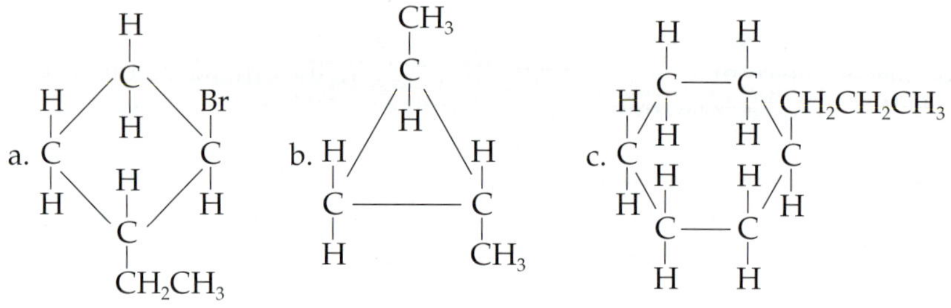

Question 10.12 Name each of the following substituted cycloalkanes using the I.U.P.A.C. Nomenclature System:

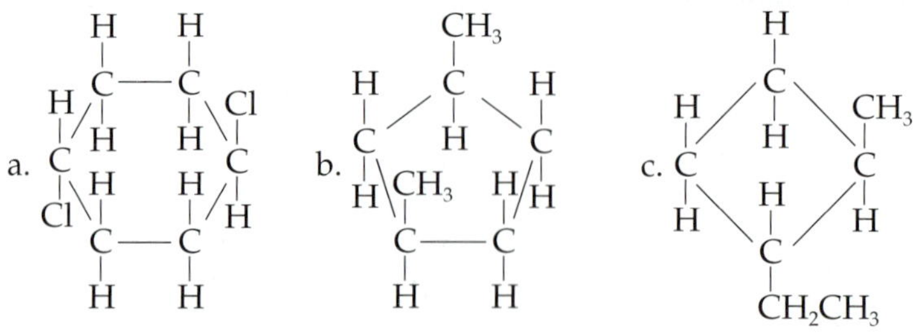

Question 10.13 How many geometric and structural isomers of dichlorocyclopropane can you construct? Use a set of molecular models to construct the isomers and to contrast their differences. Draw all these isomers.

Question 10.14

How many isomers of dibromocyclobutane can you construct? As in Question 10.13, use a set of molecular models to construct the isomers and then draw them.

10.4 Conformations of Alkanes and Cycloalkanes

Because there is *free rotation* around a carbon-carbon single bond, even a very simple alkane, like ethane, can exist in an unlimited number of forms (Figure 10.6a and 10.6b). These different arrangements are called **conformations**, or **conformers**.

10 LEARNING GOAL

11 LEARNING GOAL

Alkanes

Figure 10.6 shows two conformations of a more complex alkane, butane. In addition to these two conformations, an infinite number of intermediate conformers exist. Keep in mind that all these conformations are simply different forms of the same molecule produced by rotation around the carbon-carbon single bonds. Even at room temperature these conformers interconvert rapidly. As a result, they cannot be separated from one another.

Although all conformations can be found in a population of molecules, the *staggered* conformation (see Figure 10.6a and 10.6c) is the most common. One reason for this is that the bonding electrons are farthest from one another in this conformation. This minimizes the repulsion between these bonding electrons.

Make a ball-and-stick model of butane and demonstrate these rotational changes for yourself.

Cycloalkanes

Cycloalkanes also exist in different conformations. The only exception to this is cyclopropane. Because it has only three carbon atoms, it is always planar.

The conformations of six-member rings have been the most thoroughly studied. One reason is that many important and abundant biological molecules have six-member ring structures. Among these is the simple sugar glucose, also called *blood sugar*. Glucose is the most important sugar in the human body. It is absorbed by the cells of the body and broken down to provide energy for the cells.

The most energetically favorable conformation for a six-member ring is the **chair conformation**. In this conformation the hydrogen atoms are perfectly staggered; that is, they are as far from one another as possible. In addition, the bond angle between carbons is 109.5°, exactly the angle expected for tetrahedral carbon atoms.

The structure of glucose is found in Section 16.4. The physiological roles of glucose are discussed in Chapters 21 and 23.

This conformation gets its name because it resembles a lawn chair.

Figure 10.6
Conformational isomers of ethane. The hydrogen atoms are much more crowded in the eclipsed conformation, depicted in (b) compared with the staggered conformation shown in (a). The staggered form is energetically favored. The staggered and eclipsed conformations of butane are shown in (c) and (d).

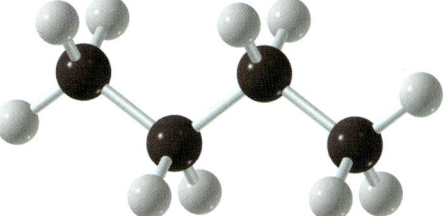

(a) Staggered conformation of ethane

(b) Eclipsed conformation of ethane

(c) Staggered conformation of butane

(d) Eclipsed conformation of butane

An Environmental Perspective

The Petroleum Industry and Gasoline Production

Petroleum consists primarily of alkanes and small amounts of alkenes and aromatic hydrocarbons. Substituted hydrocarbons, such as phenol, are also present in very small quantities. Although the composition of petroleum varies with the source of the petroleum (United States, Persian Gulf, *etc.*), the mixture of hydrocarbons can be separated into its component parts on the basis of differences in the boiling points of various hydrocarbons (distillation). Often several successive distillations of various fractions of the original mixture are required to completely purify the desired component. In the first distillation, the petroleum is separated into several fractions, each of which consists of a mix of hydrocarbons. Each fraction can be further purified by successive distillations. On an industrial scale, these distillations are carried out in columns that may be hundreds of feet in height.

The gasoline fraction of petroleum, called straight-run gasoline, consists primarily of alkanes and cycloalkanes with six to twelve carbon atoms in the skeleton. This fraction has very poor fuel performance. In fact, branched-chain alkanes are superior to straight-chain alkanes as fuels because they are more volatile, burn less rapidly in the cylinder, and thus reduce "knocking." Alkenes and aromatic hydrocarbons are also good fuels. Methods have been developed to convert hydrocarbons of higher and lower molecular weights than gasoline to the appropriate molecular weight range and to convert straight-chain hydrocarbons into branched ones. *Catalytic cracking* fragments a large hydrocarbon into smaller ones. *Catalytic reforming* results in the rearrangement of a hydrocarbon into a more useful form.

The antiknock quality of a fuel is measured as its octane rating. Heptane is a very poor fuel and is given an octane rating of zero. 2,2,4-Trimethylpentane (commonly called isooctane) is an excellent fuel and is given an octane rating of one hundred. Gasoline octane ratings are experimentally determined by comparison with these two compounds in test engines.

Mining the sea for hydrocarbons.

For Further Understanding

Explain why the mixture of hydrocarbons in crude oil can be separated by distillation.

Draw the structures of heptane and 2,2,4-trimethylpentane (isooctane).

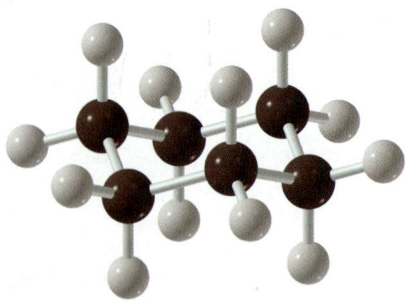

Chair conformation

Six-member rings can also exist in a **boat conformation,** so-called because it resembles a rowboat. This form is much less stable than the chair conformation because the hydrogen atoms are not perfectly staggered.

10.5 Reactions of Alkanes and Cycloalkanes

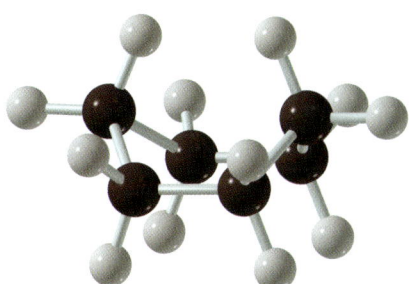

Boat conformation

The hydrogen atoms of cyclohexane are described according to their position relative to the ring. Those that lie above or below the ring are said to be **axial atoms.** Those that lie roughly in the plane of the ring are called **equatorial atoms.**

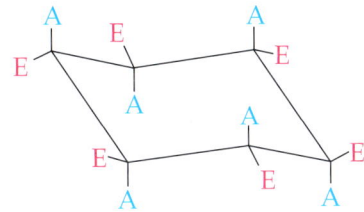

Question 10.15
Describe the positions of the six axial hydrogens of the chair conformation of cyclohexane.

Question 10.16
Describe the positions of the six equatorial hydrogens of the chair conformation of cyclohexane.

10.5 Reactions of Alkanes and Cycloalkanes

Combustion

Alkanes, cycloalkanes, and other hydrocarbons can be oxidized (by burning) in the presence of excess molecular oxygen. In this reaction, called **combustion,** they burn at high temperatures, producing carbon dioxide and water and releasing large amounts of energy as heat.

12 LEARNING GOAL

$$C_nH_{2n+2} + O_2 \longrightarrow CO_2 + H_2O + \text{heat energy}$$

Alkane   Oxygen ⟶ Carbon dioxide   Water

The following examples show a combustion reaction for a simple alkane and a simple cycloalkane:

Combustion reactions are discussed in Section 4.3.

$$CH_4 + 2O_2 \longrightarrow CO_2 + 2H_2O + \text{heat energy}$$
Methane

 (or C_6H_{12}) + $9O_2 \longrightarrow 6CO_2 + 6H_2O$ + heat energy

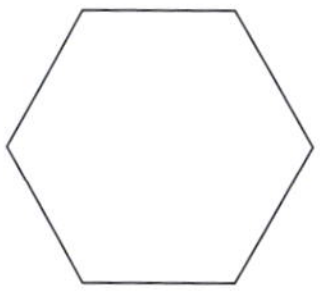

Cyclohexane

A Medical Perspective

Polyhalogenated Hydrocarbons Used as Anesthetics

Polyhalogenated hydrocarbons are hydrocarbons containing two or more halogen atoms. Some polyhalogenated compounds are notorious for the problems they have caused humankind. For instance, some insecticides such as DDT, chlordane, kepone, and lindane do not break down rapidly in the environment. As a result, these toxic compounds accumulate in biological tissue of a variety of animals, including humans, and may cause neurological damage, birth defects, or even death.

Other halogenated hydrocarbons are very useful in medicine. They were among the first anesthetics (pain relievers) used routinely in medical practice. These chemicals played a central role as the studies of medicine and dentistry advanced into modern times.

$$CH_3CH_2{-}Cl \qquad CH_3{-}Cl$$

Chloroethane (ethyl chloride)   Chloromethane (methyl chloride)

Chloroethane and chloromethane are local anesthetics. A local anesthetic deadens the feeling in a portion of the body. Applied topically (on the skin), chloroethane and chloromethane numb the area. Rapid evaporation of these anesthetics lowers the skin temperature, deadening the local nerve endings. They act rapidly, but the effect is brief, and feeling is restored quickly.

$$CHCl_3$$

Trichloromethane (chloroform)

In the past, chloroform was used as both a general and a local anesthetic. When administered through inhalation, it rapidly causes loss of consciousness. However, the effects of this powerful anesthetic are of short duration. Chloroform is no longer used because it was shown to be carcinogenic.

$$\begin{array}{c} \text{F} \quad \text{H} \\ || \\ \text{F}{-}\text{C}{-}\text{C}{-}\text{Br} \\ || \\ \text{F} \quad \text{Cl} \end{array}$$

2-Bromo-2-chloro-1,1,1-trifluoroethane (Halothane)

Halothane is a general anesthetic that is administered by inhalation. It is considered to be a very safe anesthetic and is widely used.

For Further Understanding

In the first 24 hours following administration, 70% of the halothane is eliminated from the body in exhaled gases. Explain why halothane is so readily eliminated in exhaled gases.

Rapid evaporation of chloroethane from the skin surface causes cooling that causes local deadening of nerve endings. Explain why the skin surface cools dramatically as a result of evaporation of chloroethane.

EXAMPLE 10.8 Balancing Equations for the Combustion of Alkanes

Balance the following equation for the combustion of hexane:

$$C_6H_{14} + O_2 \longrightarrow CO_2 + H_2O$$

Solution

First, balance the carbon atoms; there are 6 mol of carbon atoms on the left and only 1 mol of carbon atoms on the right:

$$C_6H_{14} + O_2 \longrightarrow 6CO_2 + H_2O$$

Next, balance hydrogen atoms; there are 14 mol of hydrogen atoms on the left and only 2 mol of hydrogen atoms on the right:

$$C_6H_{14} + O_2 \longrightarrow 6CO_2 + 7H_2O$$

Continued—

EXAMPLE 10.8 —Continued

Now there are 19 mol of oxygen atoms on the right and only 2 mol of oxygen atoms on the left. Therefore, a coefficient of 9.5 is needed for O_2.

$$C_6H_{14} + 9.5O_2 \longrightarrow 6CO_2 + 7H_2O$$

Although decimal coefficients are sometimes used, it is preferable to have all integer coefficients. Multiplying each term in the equation by 2 will satisfy this requirement, giving us the following balanced equation:

$$2C_6H_{14} + 19O_2 \longrightarrow 12CO_2 + 14H_2O$$

The equation is now balanced with 12 mol of carbon atoms, 28 mol of hydrogen atoms and 38 mol of oxygen atoms on each side of the equation.

The energy released, along with their availability and relatively low cost, makes hydrocarbons very useful as fuels. In fact, combustion is essential to our very existence. It is the process by which we heat our homes, run our cars, and generate electricity. Although combustion of fossil fuels is vital to industry and society, it also represents a threat to the environment. The buildup of CO_2 may contribute to global warming and change the face of the earth in future generations.

See An Environmental Perspective: The Greenhouse Effect and Global Warming in Chapter 5.

Halogenation

Alkanes and cycloalkanes can also react with a halogen (usually chlorine or bromine) in a reaction called **halogenation**. Halogenation is a **substitution reaction,** that is, a reaction that results in the replacement of one group for another. In this reaction a halogen atom is substituted for one of the hydrogen atoms in the alkane. The products of this reaction are an **alkyl halide** or *haloalkane* and a hydrogen halide. Alkanes are not very reactive molecules. However, alkyl halides are very useful reactants for the synthesis of other organic compounds. Thus, the halogenation reaction is of great value because it converts unreactive alkanes into versatile starting materials for the synthesis of desired compounds. This is important in the pharmaceutical industry for the synthesis of some drugs. In addition, alkyl halides having two or more halogen atoms are useful solvents, refrigerants, insecticides, and herbicides.

Halogenation can occur only in the presence of heat and/or light, as indicated by the reaction conditions noted over the reaction arrows. The general equation for the halogenation of an alkane follows. The R in the general structure for the alkane may be either a hydrogen atom or an alkyl group.

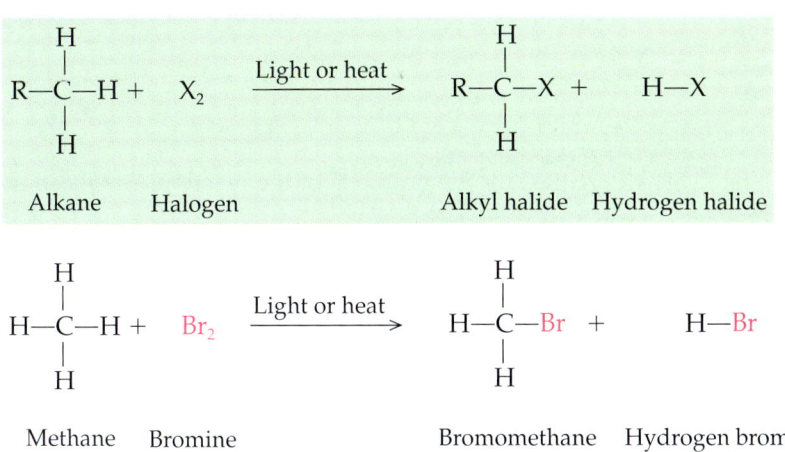

Chapter 10 An Introduction to Organic Chemistry: The Saturated Hydrocarbons

$$CH_3CH_3 + Cl_2 \xrightarrow{Light} CH_3CH_2-Cl + H-Cl$$
Ethane   Chlorine   Chloroethane   Hydrogen chloride

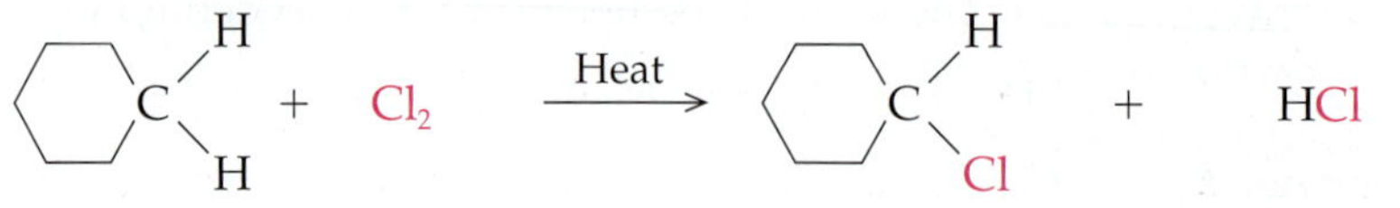

Cyclohexane + Cl_2 $\xrightarrow{Heat}$ Chlorocyclohexane + HCl

The alkyl halide may continue to react forming a mixture of products substituted at multiple sites or substituted multiple times at the same site.

If the halogenation reaction is allowed to continue, the alkyl halide formed may react with other halogen atoms. When this happens, a mixture of products may be formed. For instance, bromination of methane will produce bromomethane (CH_3Br), dibromomethane (CH_2Br_2), tribromomethane ($CHBr_3$), and tetrabromomethane (CBr_4).

In more complex alkanes, halogenation can occur to some extent at all positions to give a mixture of monosubstituted products. For example, bromination of propane produces a mixture of 1-bromopropane and 2-bromopropane.

Question 10.17

Write a balanced equation for each of the following reactions. Show all possible products.

a. the complete combustion of cyclobutane
b. the monobromination of propane
c. the complete combustion of ethane
d. the monochlorination of butane

Question 10.18

Write a balanced equation for each of the following reactions. Show all possible products.

a. the complete combustion of decane
b. the monochlorination of cyclobutane
c. the monobromination of pentane
d. the complete combustion of hexane

Question 10.19

Provide the I.U.P.A.C. names for the products of the reactions in Question 10.17b and 10.17d.

Question 10.20

Provide the I.U.P.A.C. names for the products of the reactions in Question 10.18b and 10.18c.

A Medical Perspective

Chloroform in Your Swimming Pool?

Has anyone ever asked you to take a shower before swimming in an indoor pool? Perhaps that sounds a bit silly, but there may be a very good reason for doing just that. A team of British researchers has identified the suspected carcinogen chloroform and some other potentially hazardous compounds called trihalomethanes (THMs) in indoor swimming pools in London.

Several questions arise. How did these trihalomethanes get into the pool and why are they a problem in indoor pools? THMs are products of chemical reactions between the chlorine used to disinfect the pool and organic substances from the swimmers themselves. Skin cells are shed into the pool, along with lotions and other body care products. Organic molecules from these substances are chlorinated to produce the trihalomethanes. THMs are volatile compounds. In an outdoor pool, they would evaporate and be blown away by the breeze. In an indoor pool, where there is less air circulation, THMs tend to build to higher concentrations in the air above the pool. These fumes are inhaled by the swimmers and the THMs diffuse into the blood.

It seems logical that some of the factors favoring the production of these compounds include warmer water temperatures and larger numbers of swimmers, which means more organic material in the water. In some of the public pools the level of THMs was as high as 132 µg/L. Compare this to the levels the researchers found in drinking water, only about 3.5 µg/L.

In the past, chloroform was used as an anesthetic. This is no longer the case, since many safer alternatives are available. One reason for discontinuing the use of chloroform as an anesthetic is the determination by the U.S. Department of Health and Human Services that chloroform is a potential carcinogen. Rats and mice exposed to chloroform in their food or water developed liver and kidney cancers.

How concerned should we be about these levels of THMs? Studies have shown that a one-hour swim can increase the blood concentrations of chloroform as much as tenfold. Animal studies have shown that miscarriages occurred in rats and mice that breathed air containing 30–300 parts per million of chloroform during pregnancy. Some studies have suggested that miscarriages, birth defects, and low birthrate might be associated with human exposure to chloroform, as well. But at the current time, researchers feel that these results are inconsistent and that further research needs to be done before any conclusions can be reached. They do recommend, however, that the amount of THMs be reduced as much as possible, while maintaining a high enough level of chlorine to control waterborne infectious diseases. Keeping the water at cooler temperatures and asking swimmers to shower before entering the pool are effective measures to help reduce the production of toxic THMs.

Swimmers enjoying an indoor pool.

For Further Understanding

Explain in chemical terms why keeping swimming pool water at lower temperatures and asking swimmers to shower before entering the pool help reduce the levels of THMs.

Four trihalomethanes have been identified by the Environmental Protection Agency in water disinfected with chlorine and other disinfectants. These are chloroform, bromodichloromethane, dibromochloromethane, and bromoform. Draw the structures of each of these compounds.

Summary of Reactions

Reactions of Alkanes

Combustion:

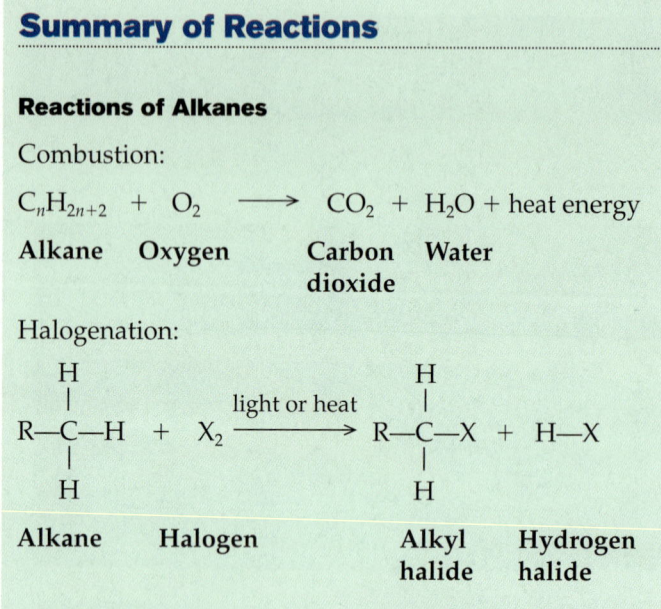

SUMMARY

10.1 The Chemistry of Carbon

The modern science of organic chemistry began with Wöhler's synthesis of urea in 1828. At that time, people believed that it was impossible to synthesize an organic molecule outside of a living system. We now define organic chemistry as the study of carbon-containing compounds. The differences between the ionic bond, which is characteristic of many inorganic substances, and the covalent bond in organic compounds are responsible for the great contrast in properties and reactivity between organic and inorganic compounds. All organic compounds are classified as either *hydrocarbons* or *substituted hydrocarbons*. In substituted hydrocarbons a hydrogen atom is replaced by a functional group. A *functional group* is an atom or group of atoms arranged in a particular way that imparts specific chemical or physical properties to a molecule. The major families of organic molecules are defined by the specific functional groups that they contain.

10.2 Alkanes

The *alkanes* are *saturated hydrocarbons*, that is, hydrocarbons that have only carbon and hydrogen atoms that are bonded together by carbon-carbon and carbon-hydrogen single bonds. They have the general molecular formula C_nH_{2n+2} and are nonpolar, water-insoluble compounds with low melting and boiling points. In the *I.U.P.A.C. Nomenclature System* the alkanes are named by determining the number of carbon atoms in the parent compound and numbering the carbon chain to provide the lowest possible number for all substituents. The substituent names and numbers are used as prefixes before the name of the parent compound.

Constitutional or *structural isomers* are molecules that have the same molecular formula but different structures. They have different physical and chemical properties because the atoms are bonded to one another in different patterns.

10.3 Cycloalkanes

Cycloalkanes are a family of organic molecules having C—C single bonds in a ring structure. They are named by adding the prefix *cyclo-* to the name of the alkane parent compound.

A *cis-trans isomer* is a type of stereoisomer. *Stereoisomers* are molecules that have the same structural formula and bonding pattern but different arrangements of atoms in space. A cycloalkane is in the *cis* configuration if two substituents are on the same side of the ring (either both above or both below). A cycloalkane is in the *trans* configuration when one substituent is above the ring and the other is below the ring. The *cis-trans* isomers are not interconvertible.

10.4 Conformations of Alkanes and Cycloalkanes

As a result of *free rotation* around carbon-carbon single bonds, infinitely many *conformations* or *conformers* exist for any alkane.

Limited rotation around the carbon-carbon single bonds of cycloalkanes also results in a variety of conformations of cycloalkanes. In cyclohexane the *chair conformation* is the most energetically favored. Another conformation is the *boat conformation*.

10.5 Reactions of Alkanes and Cycloalkanes

Alkanes can participate in *combustion* reactions. In complete combustion reactions they are oxidized to produce carbon dioxide, water, and heat energy. They can also undergo *halogenation* reactions to produce *alkyl halides*.

KEY TERMS

aliphatic hydrocarbon (10.1)
alkane (10.2)
alkyl group (10.2)
alkyl halide (10.5)
aromatic hydrocarbon (10.1)
axial atom (10.4)
boat conformation (10.4)
chair conformation (10.4)
cis-trans isomers (10.3)
combustion (10.5)
condensed formula (10.2)
conformations (10.4)
conformers (10.4)
constitutional isomers (10.2)
cycloalkane (10.3)
equatorial atom (10.4)
functional group (10.1)
geometric isomers (10.3)
halogenation (10.5)
hydrocarbon (10.1)
I.U.P.A.C. Nomenclature System (10.2)
line formula (10.2)
molecular formula (10.2)
parent compound (10.2)
primary (1°) carbon (10.2)
quaternary (4°) carbon (10.2)
saturated hydrocarbon (10.1)
secondary (2°) carbon (10.2)
stereoisomers (10.3)
structural formula (10.2)
structural isomer (10.2)
substituted hydrocarbon (10.1)
substitution reaction (10.5)
tertiary (3°) carbon (10.2)
unsaturated hydrocarbon (10.1)

QUESTIONS AND PROBLEMS

The Chemistry of Carbon

Foundations

10.21 Why is the number of organic compounds nearly limitless?
10.22 What are allotropes?
10.23 What are the three allotropic forms of carbon?
10.24 Describe the three allotropes of carbon.
10.25 Why do ionic substances generally have higher melting and boiling points than covalent substances?
10.26 Why are ionic substances more likely to be water-soluble?

Applications

10.27 Rank the following compounds from highest to lowest boiling points:
  a. H_2O CH_4 LiCl
  b. C_2H_6 C_3H_8 NaCl

10.28 Rank the following compounds from highest to lowest melting points:
  a. H_2O CH_4 KCl
  b. C_6H_{14} $C_{16}H_{38}$ NaCl

10.29 What would the physical state of each of the compounds in Question 10.27 be at room temperature?

10.30 Which of the compounds in Question 10.28 would be soluble in water?

10.31 Consider the differences between organic and inorganic compounds as you answer each of the following questions.
  a. Which compounds make good electrolytes?
  b. Which compounds exhibit ionic bonding?
  c. Which compounds have lower melting points?
  d. Which compounds are more likely to be soluble in water?
  e. Which compounds are flammable?

10.32 Describe the major differences between ionic and covalent bonds.

10.33 Give the structural formula for each of the following:
  a. 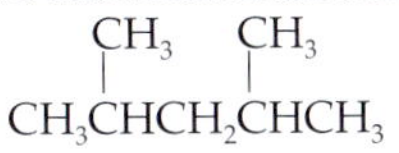 $CH_3CH(CH_3)CH_2CH(CH_3)CH_3$
  b. 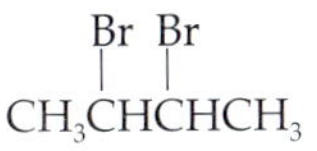 $CH_3CH(Br)CH(Br)CH_3$

10.34 Give the structural formula for each of the following:
  a. 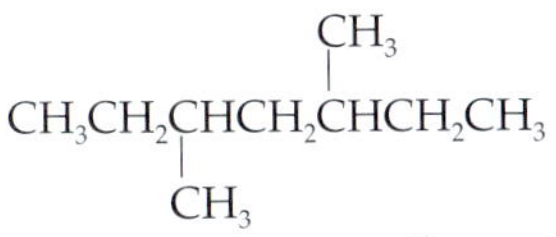 $CH_3CH_2CH(CH_3)CH_2CH(CH_3)CH_2CH_3$
  b. 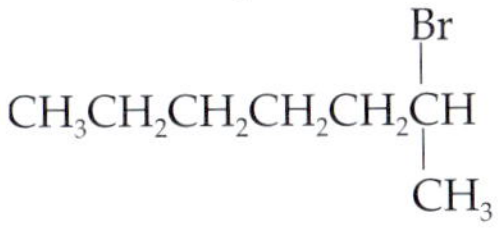 $CH_3CH_2CH_2CH_2CH_2CH(Br)CH_3$

10.35 Condense each of the following structural formulas:

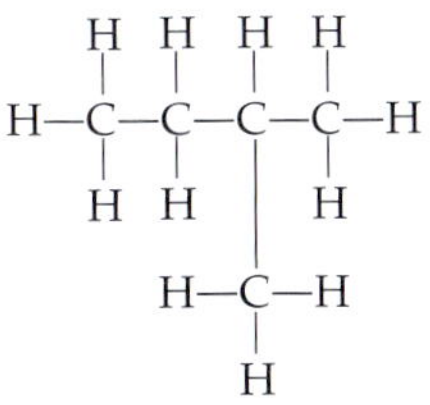

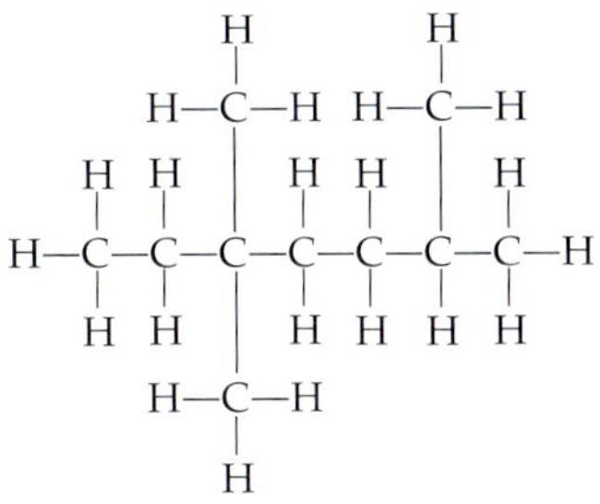

10.36 Condense each of the following structural formulas:

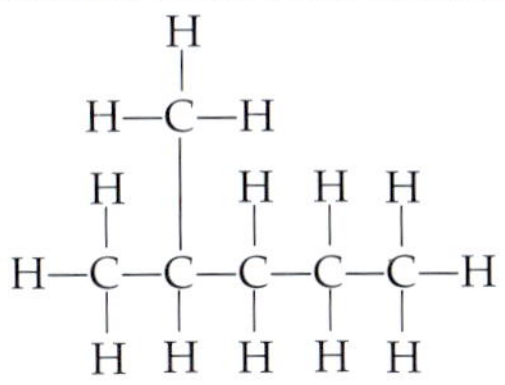

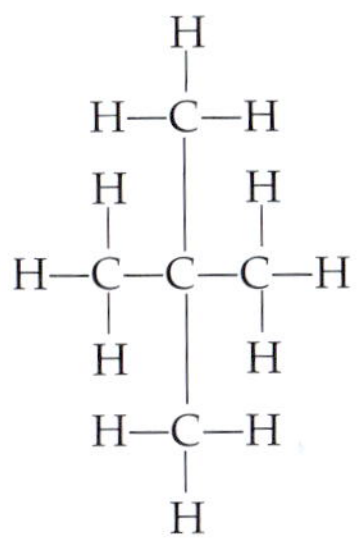

10.37 Convert the following structural formulas into line formulas:

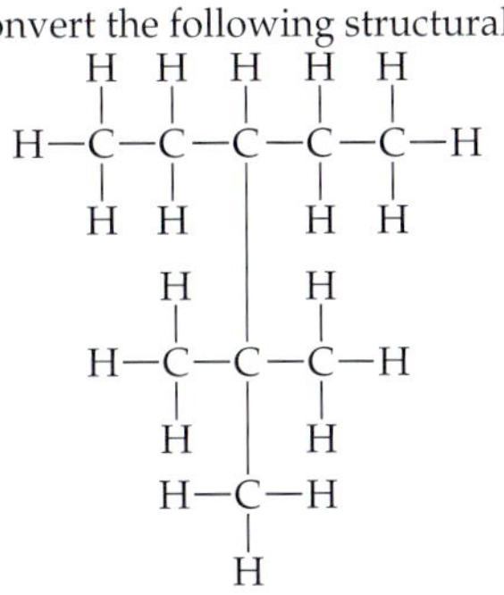

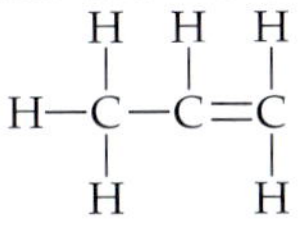

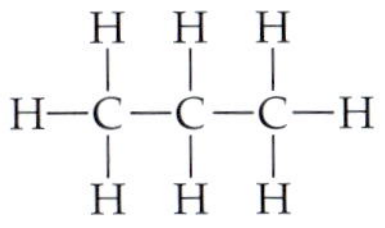

10.38 Convert the structural formulas in question 10.37 into condensed structural formulas.

10.39 Convert the following structural formulas into line formulas:

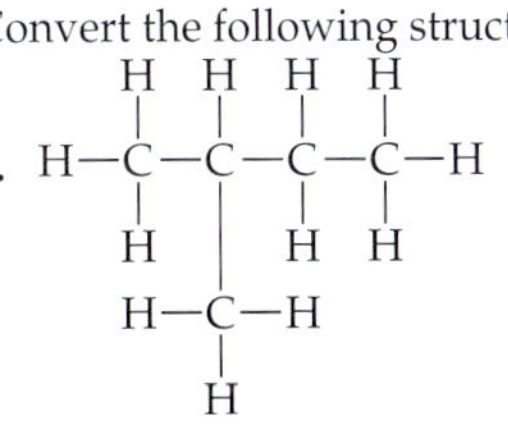

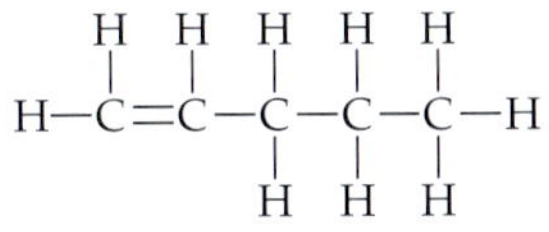

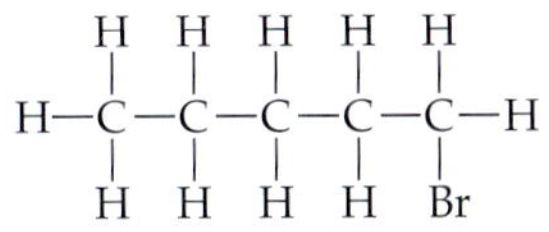

10.40 Convert the following line formulas into condensed structural formulas:

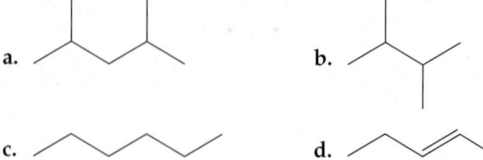

10.41 Which of the following structures are not possible? State your reasons.

a. CH₃CHCH₂CH₃
   |
   CH₃

b. CH₃CHCH₂CH₃
   |
   CH₃
   |
   CH₃

c. CH₃CHCH₂CHCH₃
   | |
   CH₃   CH₃

d. CH₃CH₂CH₂CH₂CH₃
   |
   CH₃

e. CH₂CH₃CH₂CH₃
   |
   CH₃

f. CH₃CH₂CH₂CH₃
   |
   CH₂CH₃

10.42 Using the octet rule, explain why carbon forms four bonds in a stable compound.

10.43 Convert the following condensed structural formulas into structural formulas:
a. CH₃CH(CH₃)CH(CH₃)CH₂CH₃
b. CH₃CH₂CH₂CH₂CH₃

10.44 Convert the following condensed structural formulas into structural formulas:
a. CH₃CH₂CH(CH₂CH₃)CH₂CH₂CH₃
b. CH₃CH(CH₃)CH(CH₃)CH₂CH₂CH₂CH₃

10.45 Convert the following condensed structural formulas into structural formulas:
a. CH₃CH₂CH(CH₃)CH₂CH₂CH₂CH(CH₂CH₃)CH₃
b. CH₃C(CH₃)₂CH₂CH₃

10.46 Convert the following condensed structural formulas into structural formulas:
a. CH₃CH(CH₃)CH(CH₃)CH(CH₃)CH₂CH₃
b. CH₃C(CH₃)₂CH(CH₂CH₃)CH₂CH₂CH₃

10.47 Using structural formulas, draw a typical alcohol, aldehyde, ketone, carboxylic acid, and amine. (*Hint:* Refer to Table 10.2).

10.48 Name the functional group in each of the following molecules:
a. CH₃CH₂CH₂—OH
b. CH₃CH₂CH₂—NH₂
c. CH₃CH₂CH₂—C=O
            |
            H
d. CH₃CH₂CH₂—C=O
            |
            OH
e. CH₃CH₂CH₂—C=O
            |
            OCH₂CH₃
f. CH₃CH₂—O—CH₂CH₃
g. CH₃CH₂CH₂—I

10.49 Give the general formula for each of the following:
a. An alkane
b. An alkyne
c. An alkene
d. A cycloalkane
e. A cycloalkene

10.50 Of the classes of compounds listed in Problem 10.49, which are saturated? Which are unsaturated?

10.51 What major structural feature distinguishes the alkanes, alkenes, and alkynes? Give examples.

10.52 What is the major structural feature that distinguishes between saturated and unsaturated hydrocarbons?

10.53 Give an example, using structural formulas, of each of the following families of organic compounds. Each of your examples should contain a minimum of three carbons. (*Hint:* Refer to Table 10.2.)
a. A carboxylic acid
b. An amine
c. An alcohol
d. An ether

10.54 Folic acid is a vitamin required by the body for nucleic acid synthesis. The structure of folic acid is given below. Circle and identify as many functional groups as possible.

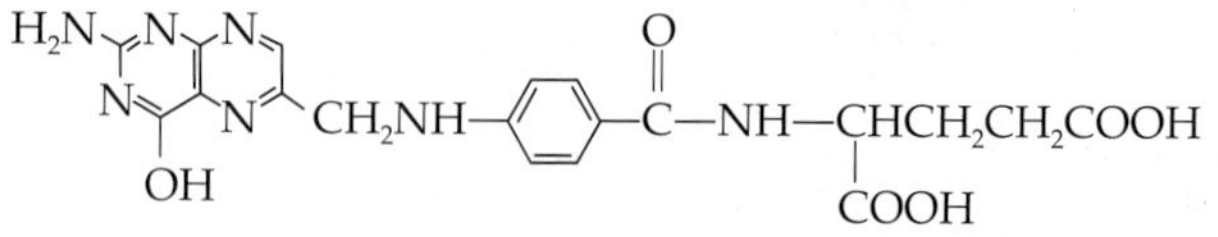

Folic acid

10.55 Aspirin is a pain reliever that has been used for over a hundred years. Today, small doses of aspirin are recommended to reduce the risk of heart attack. The structure of aspirin is given below. Circle and identify the functional groups found in the molecule:

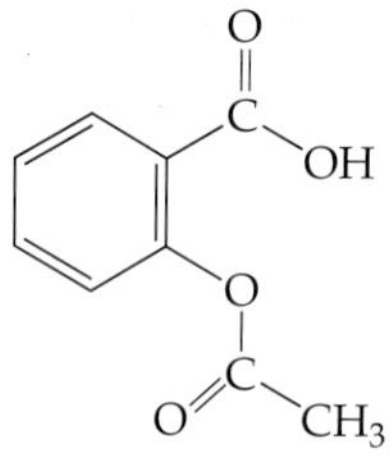

Aspirin
Acetylsalicylic acid

10.56 Vitamin C is one of the water-soluble vitamins. Although most animals can produce vitamin C, humans are not able to do so. We must get our supply of the vitamin from our diet. The structure of vitamin C is shown below. Circle and identify the functional groups found in the molecule:

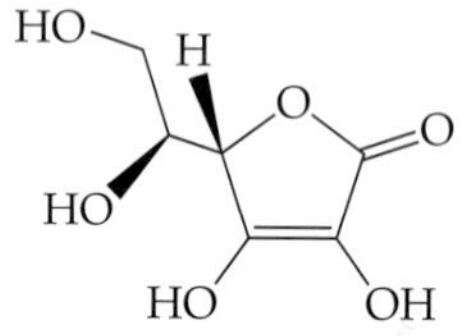

Vitamin C
Ascorbic acid

Alkanes

Foundations

10.57 Why are hydrocarbons not water soluble?

10.58 Describe the relationship between the length of hydrocarbon chains and the melting points of the compounds.

10.59 Rank the following compounds from highest to lowest boiling points:
a. heptane   butane   hexane   ethane
b. CH₃CH₂CH₂CH₂CH₃   CH₃CH₂CH₃
   CH₃CH₂CH₂CH₂CH₂CH₂CH₂CH₃

10.60 Rank the following compounds from highest to lowest melting points:
a. decane   propane   methane   ethane
b. CH₃CH₂CH₂CH₂CH₃   CH₃(CH₂)₈CH₃
   CH₃(CH₂)₆CH₃

10.61 What would the physical state of each of the compounds in Problem 10.59 be at room temperature?

10.62 What would the physical state of each of the compounds in Problem 10.60 be at room temperature?

10.63 Name each of the compounds in Problem 10.59b.

10.64 Name each of the compounds in Problem 10.60b.

Applications

10.65 Draw each of the following:
   a. 2-Bromobutane
   b. 2-Chloro-2-methylpropane
   c. 2,2-Dimethylhexane

10.66 Draw each of the following:
   a. Dichlorodiiodomethane
   b. 1,4-Diethylcyclohexane
   c. 2-Iodo-2,4,4-trimethylpentane

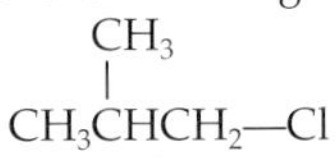

10.67 Draw each of the following compounds using structural formulas:
   a. 2,2-Dibromobutane
   b. 2-Iododecane
   c. 1,2-Dichloropentane
   d. 1-Bromo-2-methylpentane

10.68 Draw each of the following compounds using structural formulas:
   a. 1,1,1-Trichlorodecane
   b. 1,2-Dibromo-1,1,2-trifluoroethane
   c. 3,3,5-Trimethylheptane
   d. 1,3,5-Trifluoropentane

10.69 Name each of the following using the I.U.P.A.C. Nomenclature System:
   a. 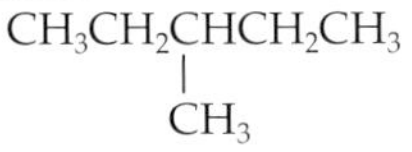 $CH_3CH_2CHCH_2CH_3$ with CH_3 substituent
   b.  $CH_3CHCH_2CH_2CHCH_3$ with two CH_3 substituents
   c. 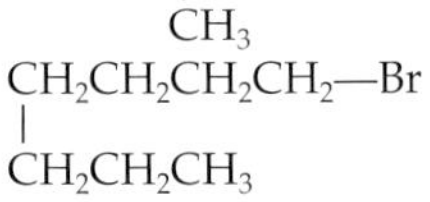 $CH_2CH_2CH_2CH_2$—Br with $CH_2CH_2CH_3$ substituent
   d. 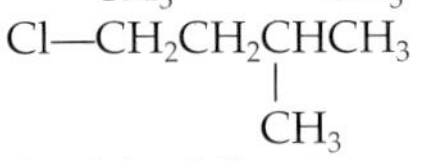 Cl—$CH_2CH_2CHCH_3$ with CH_3 substituent

10.70 Provide the I.U.P.A.C. name for each of the following compounds:
   a. 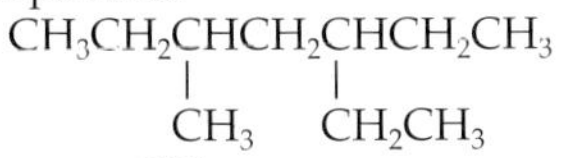 $CH_3CH_2CHCH_2CHCH_2CH_3$ with CH_3 and CH_2CH_3 substituents
   b.  $CH_3CHCH_2CH_2CH_2$—Cl with Cl substituent
   c. 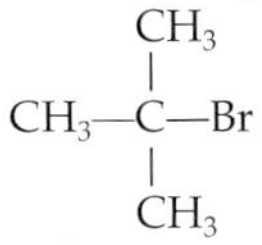 CH_3—C(CH_3)—Br with CH_3 substituents

10.71 Give the I.U.P.A.C. name for each of the following:
   a. 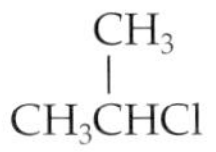 CH_3CHCl with CH_3 substituent
   b.  $CH_3CHCH_2CH_3$ with I substituent
   c. 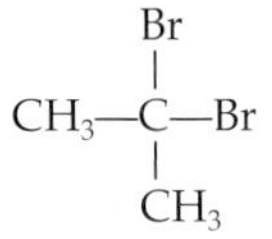 CH_3—C(Br)(Br)—CH_3
   d. CH_3CHCH_2—Cl with CH_3 substituent
   e. CH_3—C(CH_3)(I)—CH_3

10.72 Name the following using the I.U.P.A.C. Nomenclature System:
   a. 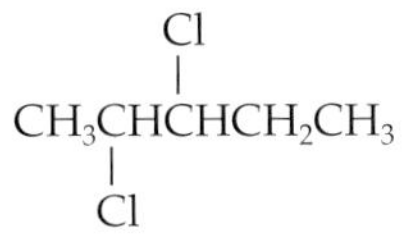 $CH_3CHCHCH_2CH_3$ with Cl and Cl substituents
   b. 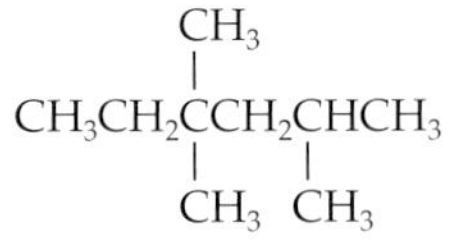 $CH_3CH_2CCH_2CHCH_3$ with CH_3, CH_3 substituents
   c. $CH_3CH_2CHCHCHCH_3$ with CH_3, CH_3, CH_3 substituents
   d. 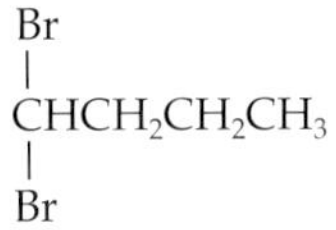 CHCH_2CH_2CH_3 with Br substituent

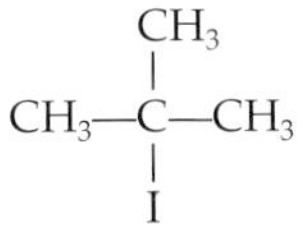

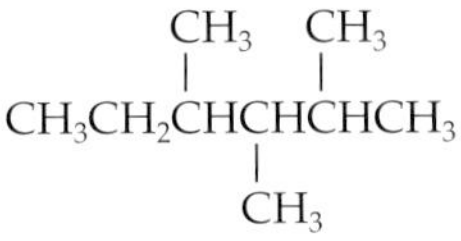

10.73 Name the following using the I.U.P.A.C. Nomenclature System:
   a. $CH_3(CH_2)_3CH(Cl)CH_3$
   b. $CH_2(Br)(CH_2)_2CH_2Br$
   c. $CH_3CH_2CH(Cl)CH_2CH_3$
   d. $CH_3CH(CH_3)(CH_2)_4CH_3$

10.74 Which of the following pairs of compounds are identical? Which are constitutional isomers? Which are completely unrelated?
   a. 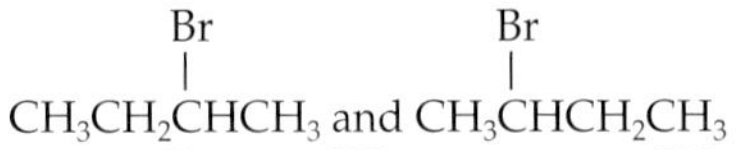 $CH_3CH_2CHCH_3$ and $CH_3CHCH_2CH_3$ (both with Br)
   b. 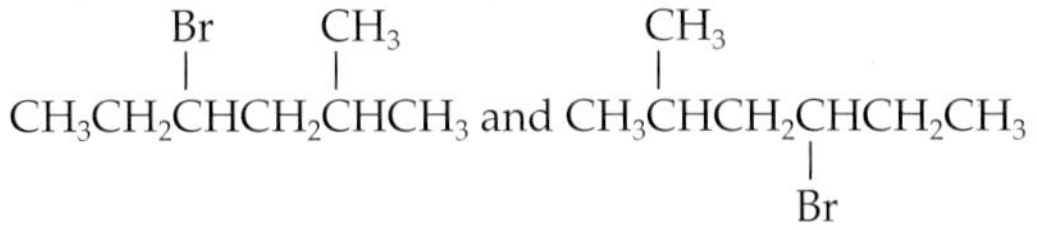 $CH_3CH_2CHCH_2CHCH_3$ (Br, CH_3) and $CH_3CHCH_2CHCH_2CH_3$ (CH_3, Br)
   c. 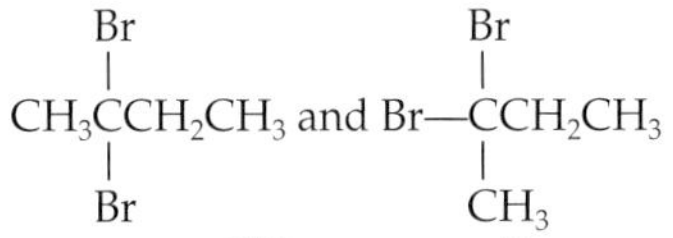 $CH_3CCH_2CH_3$ (Br, Br) and Br—CCH_2CH_3 (Br, CH_3)
   d. 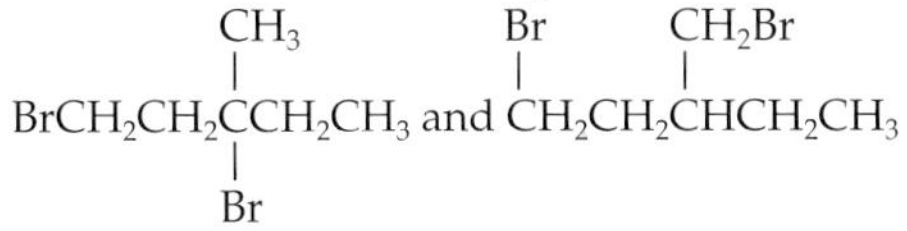 $BrCH_2CH_2CCH_2CH_3$ (CH_3, Br) and $CH_2CH_2CHCH_2CH_3$ (CH_2Br, Br)

10.75 Which of the following pairs of molecules are identical compounds? Which are constitutional isomers?
   a. 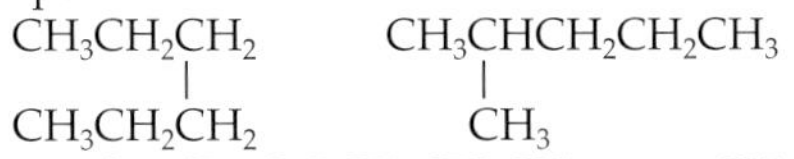 $CH_3CH_2CH_2$ / $CH_3CH_2CH_2$ and $CH_3CHCH_2CH_2CH_3$ / CH_3
   b. $CH_3CH_2CH_2CH_2CH_2CH_2CH_3$ and $CH_3CH_2CH_2CH_2CH_2$ / CH_3CH_2

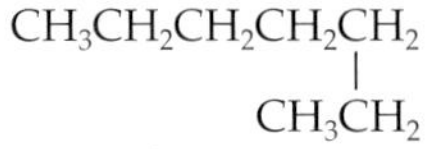

10.76 Which of the following structures are incorrect?
   a. 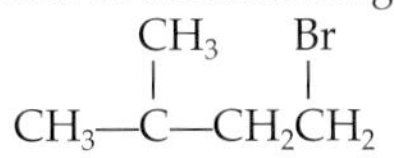 CH_3—C(CH_3)(Br)—CH_2CH_2
   b. 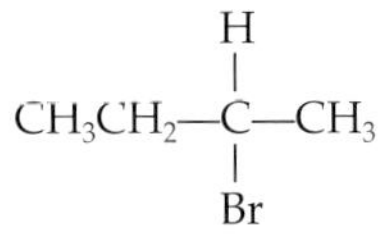 CH_3CH_2—C(H)(Br)—CH_3
   c. $CH_3CH_2CH_2CH_3$ with CH_3 and Br substituents
   d. 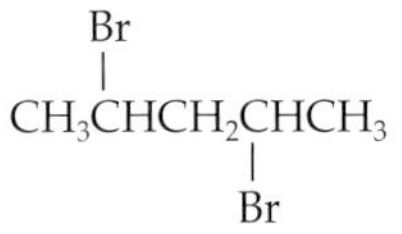 $CH_3CHCH_2CHCH_3$ with Br, Br

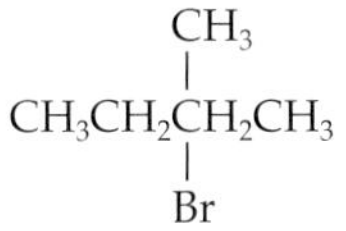

10.77 Are the following names correct or incorrect? If they are incorrect, give the correct name.
   a. 1,3-Dimethylpentane
   b. 2-Ethylpropane
   c. 3-Butylbutane
   d. 3-Ethyl-4-methyloctane

10.78 In your own words, describe the steps used to name a compound, using I.U.P.A.C. nomenclature.

10.79 Draw the structures of the following compounds. Are the names provided correct or incorrect? If they are incorrect, give the correct name.
   a. 2,4-Dimethylpentane
   b. 1,3-Dimethylpentane
   c. 1,5-Diiodopentane
   d. 1,4-Diethylheptane
   e. 1,6-Dibromo-6-methyloctane

10.80 Draw the structures of the following compounds. Are the names provided correct or incorrect? If they are incorrect, give the correct name.
   a. 1,4-Dimethylbutane
   b. 1,2-Dichlorohexane
   c. 2,3-Dimethylbutane
   d. 1,2-Diethylethane

Cycloalkanes

Foundations
10.81 Describe the structure of a cycloalkane.
10.82 Describe the I.U.P.A.C. rules for naming cycloalkanes.
10.83 What is the general formula for a cycloalkane?
10.84 How does the general formula of a cycloalkane compare with that of an alkane?

Applications
10.85 Name each of the following cycloalkanes, using the I.U.P.A.C. system:

a.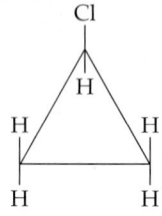

b.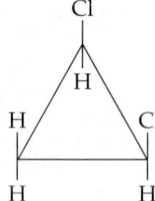

c.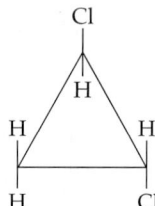

d.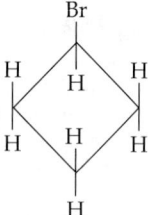

10.86 Name each of the following cycloalkanes using the I.U.P.A.C. system:

a.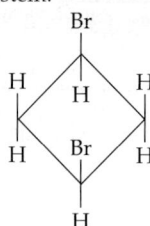

b.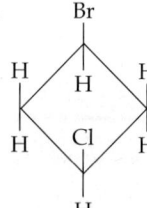

c.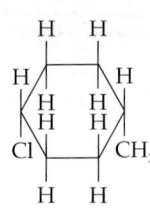

d.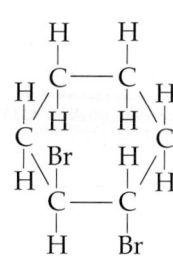

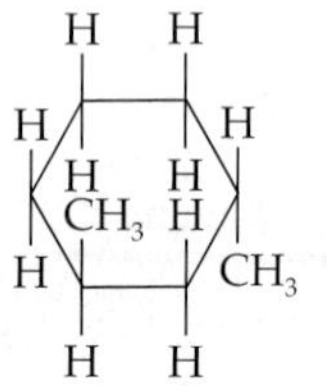

10.87 Draw the structure of each of the following cycloalkanes:
  a. 1-Bromo-2-methylcyclobutane
  b. Iodocyclopropane
10.88 Draw the structure of each of the following cycloalkanes:
  a. 1-Bromo-3-chlorocyclopentane
  b. 1,2-Dibromo-3-methycyclohexane
10.89 Which of the following names are correct and which are incorrect? If incorrect, write the correct name.
  a. 2,3-Dibromocyclobutane
  b. 1,4-Diethylcyclobutane
  c. 1,2-Dimethylcyclopropane
  d. 4,5,6-Trichlorocyclohexane

10.90 Which of the following names are correct and which are incorrect? If incorrect, write the correct name.
  a. 1,4,5-Tetrabromocyclohexane
  b. 1,3-Dimethylcyclobutane
  c. 1,2-Dichlorocyclopentane
  d. 3-Bromocyclopentane
10.91 Draw the structures of each of the following compounds:
  a. cis-1,3-Dibromocyclopentane
  b. trans-1,2-Dimethylcyclobutane
  c. cis-1,2-Dichlorocyclopropane
  d. trans-1,4-Diethylcyclohexane
10.92 Draw the structures of each of the following compounds:
  a. trans-1,4-Dimethylcyclooctane
  b. cis-1,3-Dichlorocyclohexane
  c. cis-1,3-Dibromocyclobutane
10.93 Name each of the following compounds:

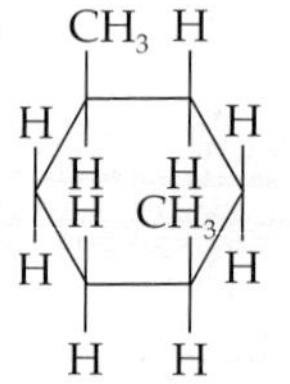

a.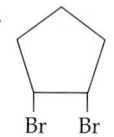

b.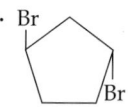

c.

d.

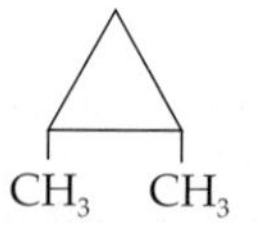

10.94 Name each of the following compounds:

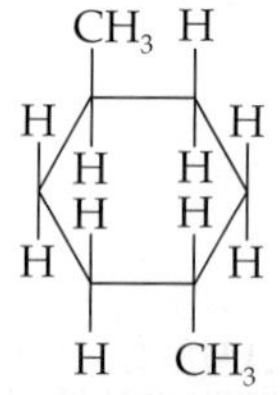

Conformations of Alkanes and Cycloalkanes

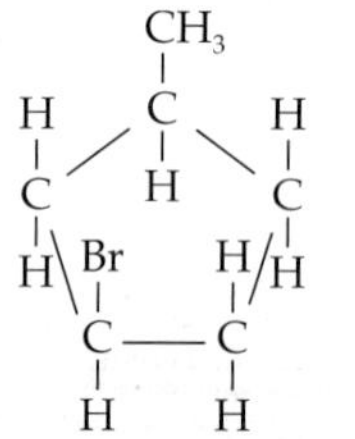

Foundations
10.95 What are conformational isomers?
10.96 Why is the staggered conformation of ethane more stable than the eclipsed conformation?

Applications
10.97 Make a model of cyclohexane and compare the boat and chair conformations. Use your model to explain why the chair conformation is more energetically favored.
10.98 Why would the ethyl group of ethylcyclohexane generally be found in the equatorial position?
10.99 Why can't conformations be separated from one another?
10.100 What is meant by free rotation around a carbon-carbon single bond?
10.101 Explain why one conformation is more stable than another. (*Hint:* refer to Figure 10.6).
10.102 Explain why a substituent on a cyclohexane ring would tend to be located in the equatorial position.

Reactions of Alkanes and Cycloalkanes

Foundations

10.103 Write a balanced equation for the complete combustion of each of the following:
a. propane
b. heptane
c. nonane
d. decane

10.104 Write a balanced equation for the complete combustion of each of the following:
a. pentane
b. hexane
c. octane
d. ethane

Applications

10.105 Complete each of the following reactions by supplying the missing reactant or product as indicated by a question mark:

a. $2CH_3CH_2CH_2CH_3 + 13O_2 \xrightarrow{\text{Heat}}$? (Complete combustion)

b. $CH_3-\underset{\underset{CH_3}{|}}{\overset{\overset{CH_3}{|}}{C}}-H + Br_2 \xrightarrow{\text{Light}}$? (Give all possible monobrominated products)

c. 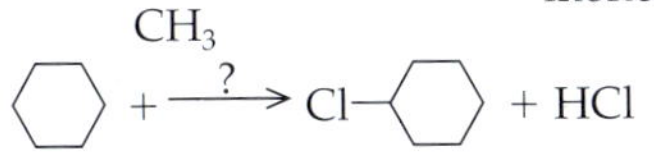 ⬡ + ? $\longrightarrow$ Cl–⬡ + HCl

10.106 Give all the possible monochlorinated products for the following reaction:

$$CH_3CH(CH_3)CH_2CH_3 + Cl_2 \xrightarrow{\text{Light}} ?$$

Name the products, using I.U.P.A.C. nomenclature.

10.107 Draw the constitutional isomers of molecular formula C_6H_{14} and name each using the I.U.P.A.C. system:
a. Which one gives two and only two monobromo derivatives when it reacts with Br_2 and light? Name the products, using the I.U.P.A.C. system.
b. Which give three and only three monobromo products? Name the products, using the I.U.P.A.C. system.
c. Which give four and only four monobromo products? Name the products, using the I.U.P.A.C. system.

10.108 a. Draw and name all of the isomeric products obtained from the monobromination of propane with Br_2/light. If halogenation were a completely random reaction and had an equal probability of occurring at any of the C—H bonds in a molecule, what percentage of each of these monobromo products would be expected?
b. Answer part (a) using 2-methylpropane as the starting material.

10.109 A mole of hydrocarbon formed eight moles of CO_2 and eight moles of H_2O upon combustion. Determine the molecular formula of the hydrocarbon and give the balanced combustion reaction.

10.110 Highly substituted alkyl fluorides, called perfluoroalkanes, are often used as artificial blood substitutes. These perfluoroalkanes have the ability to transport O_2 through the bloodstream as blood does. Some even have twice the O_2 transport capability and are used to treat gangrenous tissue. The structure of perfluorodecalin is shown below. How many moles of fluorine must be reacted with one mole of decalin to produce perfluorodecalin?

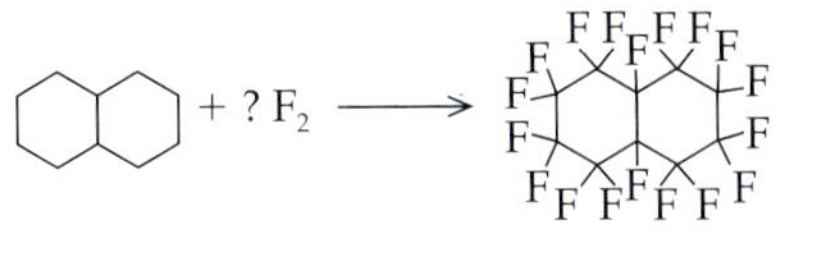

Decalin + ? $F_2 \longrightarrow$ Perfluorodecalin

CRITICAL THINKING PROBLEMS

1. You are given two unlabeled bottles, each of which contains a colorless liquid. One contains hexane and the other contains water. What physical properties could you use to identify the two liquids? What chemical property could you use to identify them?

2. You are given two beakers, each of which contains a white crystalline solid. Both are soluble in water. How would you determine which of the two solids is an ionic compound and which is a covalent compound?

3. Chlorofluorocarbons (CFCs) are man-made compounds made up of carbon and the halogens fluorine and chlorine. One of the most widely used is Freon-12 (CCl_2F_2). It was introduced as a refrigerant in the 1930s. This was an important advance because Freon-12 replaced ammonia and sulfur dioxide, two toxic chemicals that were previously used in refrigeration systems. Freon-12 was hailed as a perfect replacement because it has a boiling point of $-30°C$ and is almost completely inert. To what family of organic molecules do CFCs belong? Design a strategy for the synthesis of Freon-12.

4. Over time, CFC production increased dramatically as their uses increased. They were used as propellants in spray cans, as gases to expand plastic foam, and in many other applications. By 1985 production of CFCs reached 850,000 tons. Much of this leaked into the atmosphere and in that year the concentration of CFCs reached 0.6 parts per billion. Another observation was made by groups of concerned scientists: as the level of CFCs rose, the ozone level in the upper atmosphere declined. Does this correlation between CFC levels and ozone levels prove a relationship between these two phenomena? Explain your reasoning.

5. Although manufacture of CFCs was banned on December 31, 1995, the C—F and C—Cl bonds of CFCs are so strong that the molecules may remain in the atmosphere for 120 years. Within 5 years they diffuse into the upper stratosphere where ultraviolet photons can break the C—Cl bonds. This process releases chlorine atoms, as shown here for Freon-12:

$$CCl_2F_2 + \text{photon} \longrightarrow CClF_2 + Cl$$

The chlorine atoms are extremely reactive because of their strong tendency to acquire a stable octet of electrons. The following reactions occur when a chlorine atom reacts with an ozone molecule (O_3). First, chlorine pulls an oxygen atom away from ozone:

$$Cl + O_3 \longrightarrow ClO + O_2$$

Then ClO, a highly reactive molecule, reacts with an oxygen atom:

$$ClO + O \longrightarrow Cl + O_2$$

Write an equation representing the overall reaction (sum of the two reactions). How would you describe the role of Cl in these reactions?

ORGANIC CHEMISTRY

11

The Unsaturated Hydrocarbons

Alkenes, Alkynes, and Aromatics

Learning Goals

1. Describe the physical properties of alkenes and alkynes.

2. Draw the structures and write the I.U.P.A.C. names for simple alkenes and alkynes.

3. Write the names and draw the structures of simple geometric isomers of alkenes.

4. Write equations predicting the products of addition reactions of alkenes and alkynes: hydrogenation, halogenation, hydration, and hydrohalogenation.

5. Apply Markovnikov's rule to predict the major and minor products of the hydration and hydrohalogenation reactions of unsymmetrical alkenes.

6. Write equations representing the formation of addition polymers of alkenes.

7. Draw the structures and write the names of common aromatic hydrocarbons.

8. Write equations for substitution reactions involving benzene.

9. Describe heterocyclic aromatic compounds and list several biological molecules in which they are found.

Insect pests destroy millions of dollars in crops each year.

Outline

Chemistry Connection:
A Cautionary Tale: DDT and Biological Magnification

11.1 Alkenes and Alkynes: Structure and Physical Properties

11.2 Alkenes and Alkynes: Nomenclature

A Medical Perspective:
Killer Alkynes in Nature

11.3 Geometric Isomers: A Consequence of Unsaturation

11.4 Alkenes in Nature

11.5 Reactions Involving Alkenes and Alkynes

A Human Perspective:
Folklore, Science, and Technology

A Human Perspective:
Life Without Polymers?

An Environmental Perspective:
Plastic Recycling

11.6 Aromatic Hydrocarbons

11.7 Heterocyclic Aromatic Compounds

339

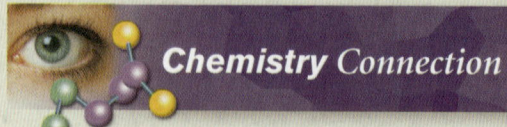

Chemistry Connection

A Cautionary Tale: DDT and Biological Magnification

We have heard the warnings for years: Stop using non-biodegradable insecticides because they are killing many animals other than their intended victims! Are these chemicals not specifically targeted to poison insects? How then can they be considered a threat to humans and other animals?

DDT, a polyhalogenated hydrocarbon, was discovered in the early 1940s by Paul Müller, a Swiss chemist. Müller showed that DDT is a nerve poison that causes convulsions, paralysis, and eventually death in insects. From the 1940s until 1972, when it was banned in the United States, DDT was sprayed on crops to kill insect pests, sprayed on people as a delousing agent, and sprayed in and on homes to destroy mosquitoes carrying malaria. At first, DDT appeared to be a miraculous chemical, saving literally millions of lives and billions of dollars in crops. However, as time went by, more and more evidence of a dark side of DDT use accumulated. Over time, the chemical had to be sprayed in greater doses as the insect populations evolved to become more and more resistant to it. In 1962, Rachel Carson published her classic work, *Silent Spring*, which revealed that DDT was accumulating in the environment. In particular, high levels of DDT in birds interfered with their calcium metabolism. As a result, the eggshells produced by the birds were too thin to support development of the chick within. It was feared that in spring, when the air should have been filled with bird song, there would be silence. This is the "silent spring" referred to in the title of Carson's book.

DDT is not biodegradable; furthermore, it is not water-soluble, but it is soluble in nonpolar solvents. Thus if DDT is ingested by an animal, it will dissolve in fat tissue and accumulate there, rather than being excreted in the urine. When DDT is introduced into the food chain, which is inevitable when it is sprayed over vast areas of the country, the result is *biological magnification*. This stepwise process begins when DDT applied to crops is ingested by insects. The insects, in turn, are eaten by birds, and the birds are eaten by a hawk. We can imagine another food chain: Perhaps the insects are eaten by mice, which are in turn eaten by a fox, which is then eaten by an owl. Or to make it more personal, perhaps the grass is eaten by a steer, which then becomes your dinner. With each step up one of these food chains, the concentration of DDT in the tissues becomes higher and higher because it is not degraded, it is simply stored. Eventually, the concentration may reach toxic levels in some of the animals in the food chain.

DDT: Dichlorodiphenyltrichloroethane

Consider for a moment the series of events that occurred in Borneo in 1955. The World Health Organization elected to spray DDT in Borneo because 90% of the inhabitants were infected with malaria. As a result of massive spraying, the mosquitoes bearing the malaria parasite were killed. If this sounds like the proverbial happy ending, read on. This is just the beginning of the story. In addition to the mosquitoes, millions of other household insects were killed. In tropical areas it is common for small lizards to live in homes, eating insects found there. The lizards ate the dead and dying DDT-contaminated insects and were killed by the neurotoxic effects of DDT. The house cats ate the lizards, and they, too, died. The number of rats increased dramatically because there were no cats to control the population. The rats and their fleas carried sylvatic plague, a form of bubonic plague. With more rats in contact with humans came the threat of a bubonic plague epidemic. Happily, cats were parachuted into the affected areas of Borneo, and the epidemic was avoided.

The story has one further twist. Many of the islanders lived in homes with thatched roofs. The vegetation used to make these roofs was the preferred food source for a caterpillar that was not affected by DDT. Normally, the wasp population preyed on these caterpillars and kept the population under control. Unfortunately, the wasps were killed by the DDT. The caterpillars prospered, devouring the thatched roofs, which collapsed on the inhabitants.

Every good story has a moral, and this one is not difficult to decipher. The introduction of large amounts of any chemical into the environment, even to eradicate disease, has the potential for long-term and far-reaching effects that may be very difficult to predict. We must be cautious with our fragile environment. Our well-intentioned intervention all too often upsets the critical balance of nature, and in the end we inadvertently do more harm than good.

Introduction

Unsaturated hydrocarbons *are those that contain at least one carbon-carbon double or triple bond. They include the alkenes, alkynes, and aromatic compounds. All alkenes have at least one carbon-carbon double bond; all alkynes have at least one carbon-carbon triple bond. Aromatic compounds are particularly stable cyclic compounds and sometimes are depicted as having alternating single and double carbon-carbon bonds. This arrangement of alternating single and double bonds is called a conjugated system of double bonds.*

Many important biological molecules are characterized by the presence of double bonds or a linear or cyclic conjugated system of double bonds (Figure 11.1). For instance, we classify fatty acids as either monounsaturated (having one double bond), polyunsaturated (having two or more double bonds), or saturated (having single bonds only). Vitamin A (retinol), a vitamin required for vision, contains a nine-carbon conjugated hydrocarbon chain. Vitamin K, a vitamin required for blood clotting, contains an aromatic ring.

Fatty acids are long hydrocarbon chains having a carboxyl group at the end. Thus by definition they are carboxylic acids. See Chapters 14 and 17.

You can find further information on lipid-soluble vitamins online at www.mhhe.com/denniston5e in "Lipid-Soluble Vitamins."

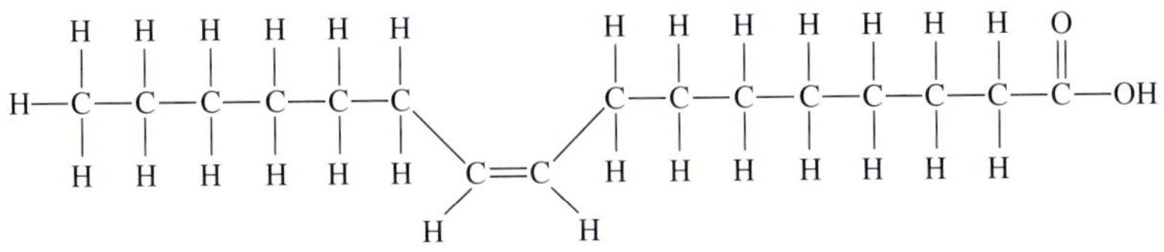

Palmitoleic acid

(a)

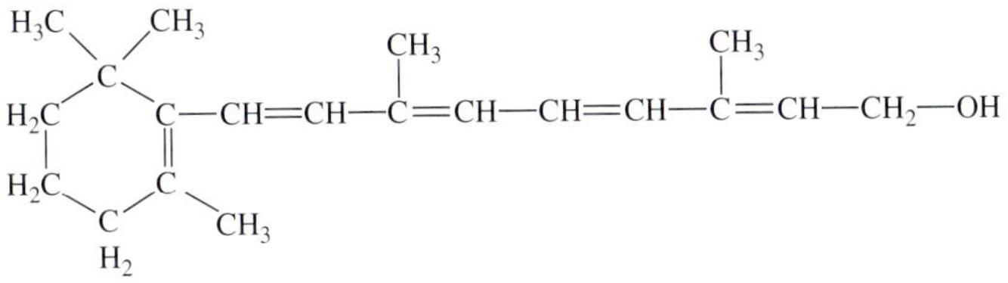

Vitamin A

(b)

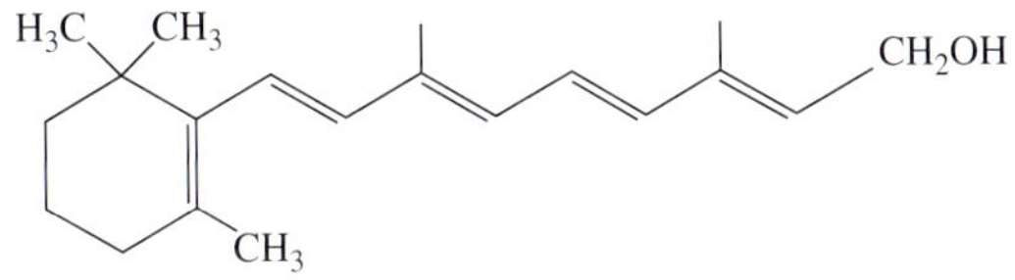

Vitamin A

(c)

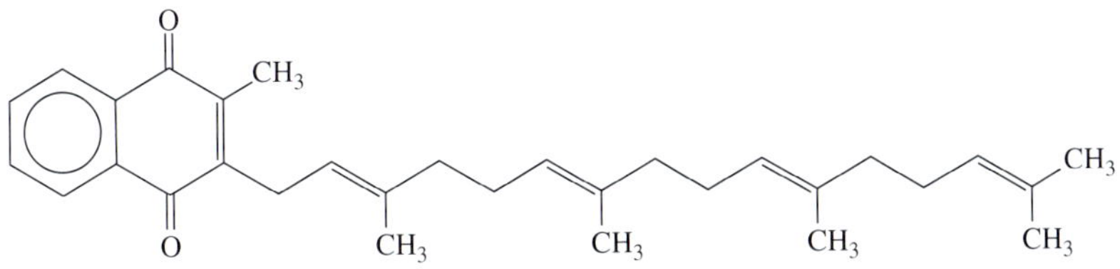

Vitamin K

(d)

Figure 11.1
(a) Structural formula of the sixteen-carbon monounsaturated fatty acid palmitoleic acid. (b) Condensed formula of vitamin A, which is required for vision. Notice that the carbon chain of vitamin A is a conjugated system of double bonds. (c) Line formula of vitamin A. (d) Line formula of vitamin K, a lipid-soluble vitamin required for blood clotting. The six-member ring with the circle represents a benzene ring. See Figure 11.6 for other representations of the benzene ring.

11.1 Alkenes and Alkynes: Structure and Physical Properties

LEARNING GOAL

Alkenes and **alkynes** are unsaturated hydrocarbons. The characteristic functional group of an alkene is the carbon-carbon double bond. The functional group that characterizes the alkynes is the carbon-carbon triple bond. The following general formulas compare the structures of alkanes, alkenes, and alkynes.

	Alkane	Alkene	Alkyne
General formulas:	C_nH_{2n+2}	C_nH_{2n}	C_nH_{2n-2}
Structural formulas:	H—C(H)(H)—C(H)(H)—H	H₂C=CH₂	H—C≡C—H
	Ethane (ethane)	Ethene (ethylene)	Ethyne (acetylene)
Molecular formulas:	C_2H_6	C_2H_4	C_2H_2
Condensed formulas:	CH_3CH_3	$H_2C=CH_2$	$HC≡CH$

These compounds have the same number of carbon atoms but differ in the number of hydrogen atoms, a feature of all alkanes, alkenes, and alkynes that contain the same number of carbon atoms. Alkenes contain two fewer hydrogens than the corresponding alkanes, and alkynes contain two fewer hydrogens than the corresponding alkenes.

In alkanes, the four bonds to the central carbon have tetrahedral geometry. When carbon is bonded by one double bond and two single bonds, as in ethene (an alkene), the molecule is *planar*, because all atoms lie in a single plane. Each bond angle is approximately 120°. When two carbon atoms are bonded by a triple bond, as in ethyne (an alkyne), each bond angle is 180°. Thus, the molecule is linear, and all atoms are positioned in a straight line. For comparison, examples of a five-carbon alkane, alkene, and alkyne are shown in Figure 11.2.

The VSEPR theory and molecular geometry are covered in more detail in Section 3.4.

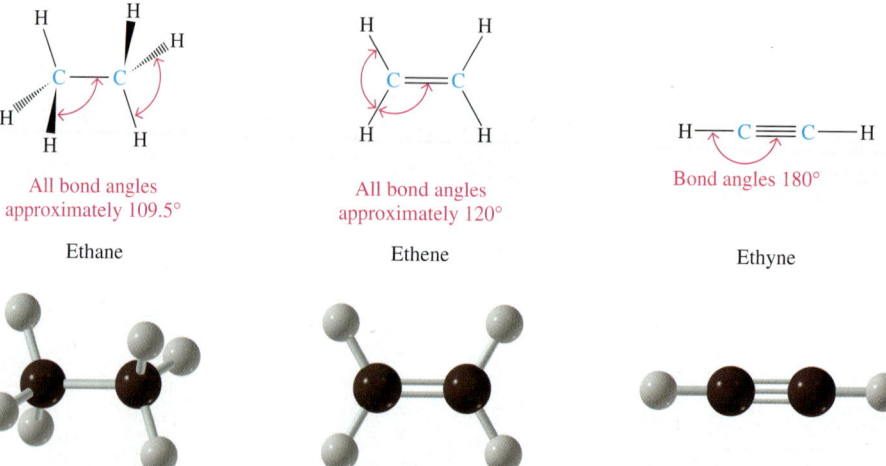

Ethane — All bond angles approximately 109.5°
Ethene — All bond angles approximately 120°
Ethyne — Bond angles 180°

Like alkanes, alkenes, alkynes, and aromatic compounds are nonpolar and have properties similar to alkanes of the same carbon chain length. For comparison, the melting points and boiling points of several alkenes and alkynes are presented in Table 11.1. Since they are nonpolar, the "like dissolves like" rule tells us that they are not soluble in water, but they are very soluble in nonpolar solvents such as other hydrocarbons.

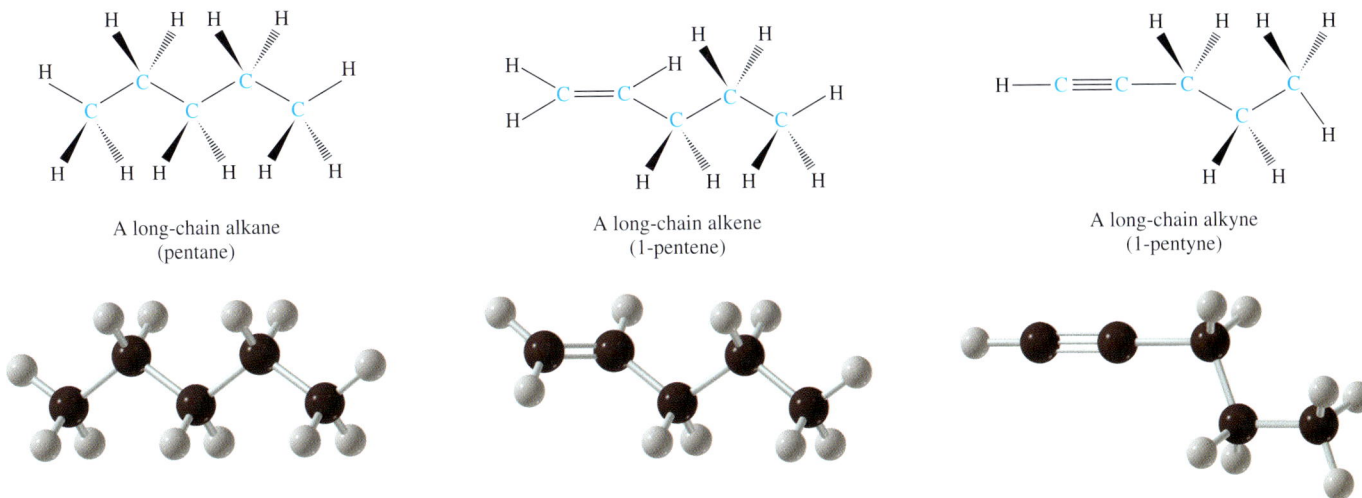

Figure 11.2
Three-dimensional drawings and ball-and-stick models of typical alkanes, alkenes, and alkynes.

TABLE 11.1 Physical Properties of Selected Alkenes and Alkynes

Name	Molecular Formula	Structural Formula	Melting Point (°C)	Boiling Point (°C)
Ethene	C_2H_4	$CH_2=CH_2$	−169.1	−103.7
Propene	C_3H_6	$CH_2=CHCH_3$	−185.0	−47.6
1-Butene	C_4H_8	$CH_2=CHCH_2CH_3$	−185.0	−6.1
2-Methylpropene	C_4H_8	$CH_2=C(CH_3)_2$	−140.0	−6.6
Ethyne	C_2H_2	$HC\equiv CH$	−81.8	−84.0
Propyne	C_3H_4	$HC\equiv CCH_3$	−101.5	−23.2
1-Butyne	C_4H_6	$HC\equiv CCH_2CH_3$	−125.9	8.1
2-Butyne	C_4H_6	$CH_3C\equiv CCH_3$	−32.3	27.0

11.2 Alkenes and Alkynes: Nomenclature

To determine the name of an alkene or alkyne using the I.U.P.A.C. Nomenclature System, use the following simple rules:

- Name the parent compound using the longest continuous carbon chain containing the double bond (alkenes) or triple bond (alkynes).
- Replace the *-ane* ending of the alkane with the *-ene* ending for an alkene or the *-yne* ending for an alkyne. For example:

$$CH_3-CH_3 \qquad CH_2=CH_2 \qquad CH\equiv CH$$
$$\text{Ethane} \qquad \text{Ethene} \qquad \text{Ethyne}$$

$$CH_3-CH_2-CH_3 \qquad CH_2=CH-CH_3 \qquad CH\equiv C-CH_3$$
$$\text{Propane} \qquad \text{Propene} \qquad \text{Propyne}$$

- Number the chain to give the lowest number for the first of the two carbons containing the double bond or triple bond. For example:

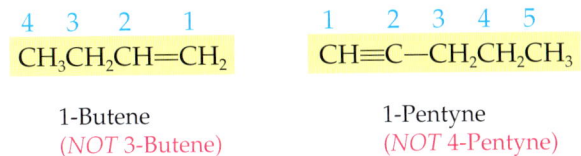

1-Butene
(*NOT* 3-Butene)

1-Pentyne
(*NOT* 4-Pentyne)

2 LEARNING GOAL

Remember, it is the position of the double bond, not the substituent, that determines the numbering of the carbon chain.

- Determine the name and carbon number of each group bonded to the parent alkene or alkyne, and place the name and number in front of the name of the parent compound. Remember that with alkenes and alkynes the double or triple bond takes precedence over a halogen or alkyl group, as shown in the following examples:

2-Chloro-2-butene

2-Bromo-3-hexyne

2,6-Dimethyl-3-octene

3-Chloro-4-methyl-3-hexene

Alkenes with many double bonds are often referred to as polyenes (*poly*—many *enes*—double bonds).

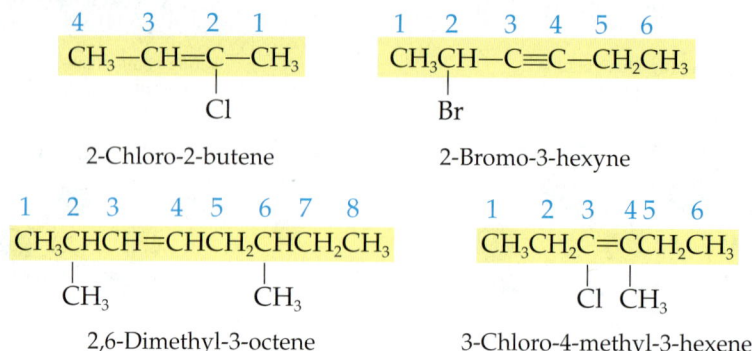

- Alkenes having more than one double bond are called alkadienes (two double bonds) or alkatrienes (three double bonds), as seen in these examples:

1 2 3 4 5 6
$CH_3CH{=}CH{-}CH{=}CHCH_3$

2,4-Hexadiene

1 2 3 4 5
$CH_2{=}CHCH_2CH{=}CH_2$

1,4-Pentadiene

3-Methyl-1, 4-cyclohexadiene

1 2 3 4 5 6 7 8 9
$CH_3CH{=}CHCH_2CH{=}CHCH_2CH_2CH_3$

2,5-Nonadiene

1 2 3 4 5 6 7
$CH_3CH{=}CHCH{=}CHCH_2CH_3$

2,4-Heptadiene

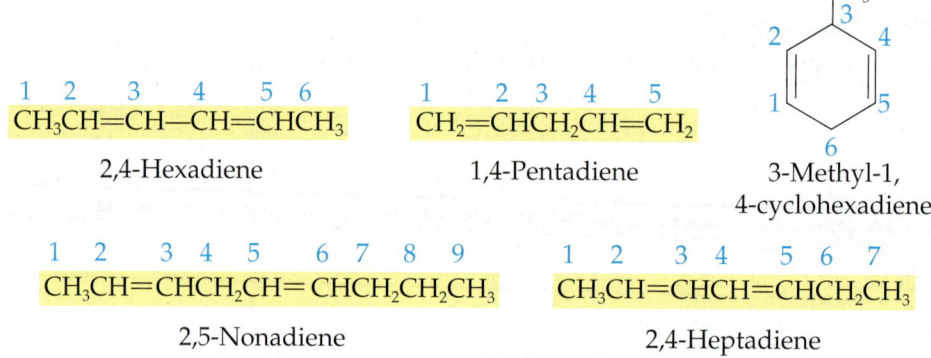

EXAMPLE 11.1 Naming Alkenes and Alkynes Using I.U.P.A.C. Nomenclature

LEARNING GOAL 2

Name the following alkene and alkyne using I.U.P.A.C. nomenclature.

Solution

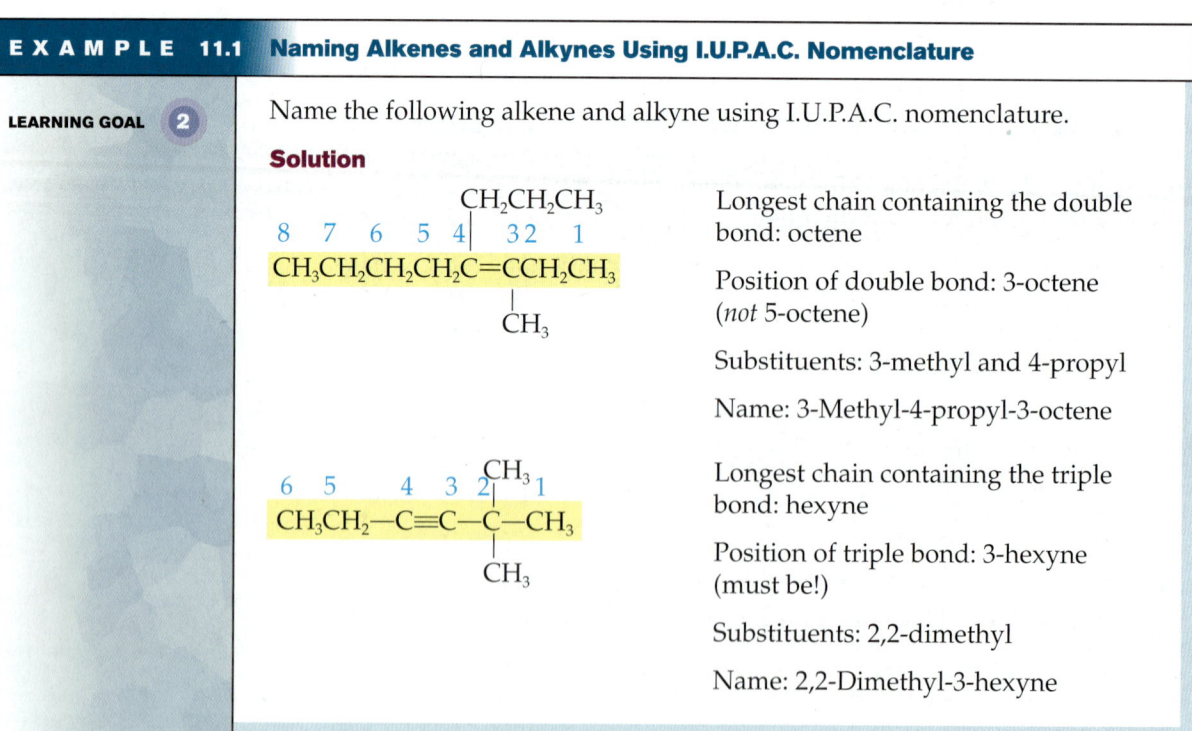

Longest chain containing the double bond: octene

Position of double bond: 3-octene (*not* 5-octene)

Substituents: 3-methyl and 4-propyl

Name: 3-Methyl-4-propyl-3-octene

Longest chain containing the triple bond: hexyne

Position of triple bond: 3-hexyne (must be!)

Substituents: 2,2-dimethyl

Name: 2,2-Dimethyl-3-hexyne

11.2 Alkenes and Alkynes: Nomenclature

EXAMPLE 11.2

Naming Cycloalkenes Using I.U.P.A.C. Nomenclature

Name the following cycloalkenes using I.U.P.A.C. nomenclature.

Solution

Parent chain: cyclohexene
Position of double bond: carbon-1 (carbons of the double bond are numbered 1 and 2)
Substituents: 4-chloro
Name: 4-Chlorocyclohexene

Parent chain: cyclopentene
Position of double bond: carbon-1
Substituent: 3-methyl
Name: 3-Methylcyclopentene

Question 11.1

Draw a complete structural formula for each of the following compounds:

a. 1-Bromo-3-hexyne
b. 2-Butyne

Question 11.2

Name the following compounds using the I.U.P.A.C. Nomenclature System:

a. $CH_3-C\equiv C-CH_2CH_3$
b. $CH_3CH_2CHCHCH_2C\equiv CH$
       $\;\;\;\;\;\;\;\;\;\;\;\;\;\;|\;\;\;|$
       $\;\;\;\;\;\;\;\;\;\;\;\;\;Br\;\;Br$

Question 11.3

Draw a complete structural formula for each of the following compounds:

a. Dichloroethyne
b. 9-Iodo-1-nonyne

Question 11.4

Name the following compounds using the I.U.P.A.C. Nomenclature System:

a.
```
              Cl   CH3
              |    |
   CH3CH — C = C — CHCH3
       |         |
       CH3      CH3
       |
       CH2CH3
```

b.
```
   CH3CH — C≡C — CHCH3
                  |
                  Cl
```

A Medical Perspective

Killer Alkynes in Nature

There are many examples of alkynes that are beneficial to humans. Among these are *parasalamide*, a pain reliever, *paragyline*, an antihypertensive, and *17-ethynylestradiol*, a synthetic estrogen that is used as an oral contraceptive.

But in addition to these medically useful alkynes, there are in nature a number that are toxic. Some are extremely toxic to mammals, including humans; others are toxic to fungi, fish, or insects. All of these compounds are plant products that may help protect the plant from destruction by predators.

Capillin is produced by the oriental wormwood plant. Research has shown that a dilute solution of capillin inhibits the growth of certain fungi. Since fungal growth can damage or destroy a plant, the ability to make capillin may provide a survival advantage to the plants. Perhaps it may one day be developed to combat fungal infections in humans.

Ichthyothereol is a fast-acting poison commonly found in plants referred to as fish-poison plants. Ichthyothereol is a very toxic polyacetylenic alcohol that inhibits energy production in the mitochondria. Latin American native tribes use these plants to coat the tips of the arrows used to catch fish. Although ichthyothereol is poisonous to the fish, fish caught by this method pose no risk to the people who eat them!

An extract of the leaves of English ivy has been reported to have antibacterial, analgesic, and sedative effects. The compound thought to be responsible for these characteristics, as well as antifungal activity, is *falcarinol*. Falcarinol, isolated from a tree in Panama, also has been reported by the Molecular Targets Drug Discovery Program, to have antitumor activity. Perhaps one day this compound, or a derivative of it, will be useful in treating cancer in humans.

Cicutoxin has been described as the most lethal toxin native to North America. It is a neurotoxin that is produced by the water hemlock (*Cicuta maculata*), which is in the same family of plants as parsley, celery, and carrots. Cicutoxin is present in all parts of the plants, but is most concentrated in the root. Eating a portion as small as 2–3 cm² can be fatal to adults. Cicutoxin acts directly on the nervous system. Signs and symptoms of cicutoxin poisoning include dilation of pupils, muscle twitching, rapid pulse and breathing, violent convulsions, coma, and death. Onset of symptoms is rapid and death may occur within two to

Parasalamide

Paragyline

17-Ethynylestradiol

Alkynes used for medicinal purposes.

11.3 Geometric Isomers: A Consequence of Unsaturation

LEARNING GOAL 3

The carbon-carbon double bond is rigid because of the shapes of the orbitals involved in its formation. As a result, rotation around the carbon-carbon double bond is restricted. In Section 10.3, we saw that the rotation around the carbon-carbon

11.3 Geometric Isomers: A Consequence of Unsaturation

Alkynes that exhibit toxic activity.

Cicuta maculata, or water hemlock, produces the most deadly toxin indigenous to North America.

three hours. No antidote exists for cicutoxin poisoning. The only treatment involves controlling convulsions and seizures in order to preserve normal heart and lung function. Fortunately, cicutoxin poisoning is a very rare occurrence. Occasionally animals may graze on the plants in the spring, resulting in death within fifteen minutes. Humans seldom come into contact with the water hemlock. The most recent cases have involved individuals foraging for wild ginseng, or other wild roots, and mistaking the water hemlock root for an edible plant.

For Further Understanding

Circle and name the functional groups in parasalamide and paragyline.

The fungus *Tinea pedis* causes athlete's foot. Describe an experiment you might carry out to determine whether capillin might be effective against athlete's foot.

---

bonds of cycloalkanes was also restricted. As a consequence, these molecules form geometric or *cis-trans* isomers. The *cis* isomers of cycloalkanes have substituent groups on the same side of the ring (L., *cis*, "on the same side"). The *trans* isomers of cycloalkanes have substituent groups located on opposite sides of the ring (L., *trans*, "across from").

In alkenes, **geometric isomers** occur when there are two different groups on each of the carbon atoms attached by the double bond. If both

Restricted rotation around double bonds is partially responsible for the conformation and hence the activity of many biological molecules that we will study later.

groups are on the same side of the double bond, the molecule is a *cis* isomer. If the groups are on opposite sides of the double bond, the molecule is a *trans* isomer.

Consider the two isomers of 1,2-dichloroethene:

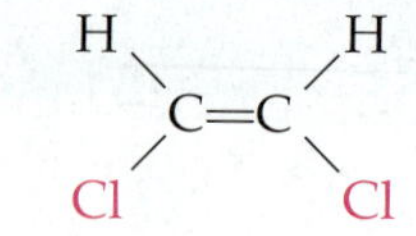

$$\underset{\text{cis-1, 2-Dichloroethene}}{\overset{H}{\underset{Cl}{>}}C=C\overset{H}{\underset{Cl}{<}}} \qquad \underset{\text{trans-1, 2-Dichloroethene}}{\overset{H}{\underset{Cl}{>}}C=C\overset{Cl}{\underset{H}{<}}}$$

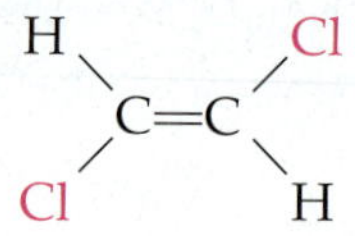

In these molecules, each carbon atom of the double bond is also bonded to two different atoms: a hydrogen atom and a chlorine atom. In the molecule on the left, both chlorine atoms are on the same side of the double bond; this is the *cis* isomer and the complete name for this molecule is *cis*-1,2-dichloroethene. In the molecule on the right, the chlorine atoms are on opposite sides of the double bond; this is the *trans* isomer and the complete name of this molecule is *trans*-1,2-dichloroethene.

If one of the two carbon atoms of the double bond has two identical substituents, there are no *cis-trans* isomers for that molecule. Consider the example of 1,1-dichloroethene:

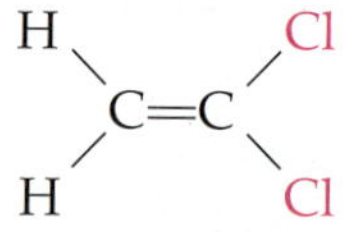

$$\overset{H}{\underset{H}{>}}C=C\overset{Cl}{\underset{Cl}{<}}$$

1, 1-Dichloroethene

EXAMPLE 11.3 Identifying *cis* and *trans* Isomers of Alkenes

LEARNING GOAL 3

Two isomers of 2-butene are shown below. Which is the *cis* isomer and which is the *trans* isomer?

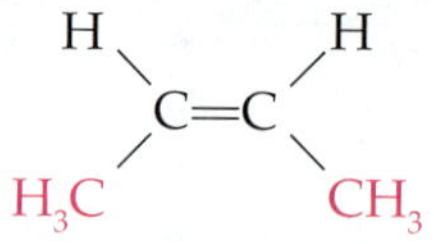

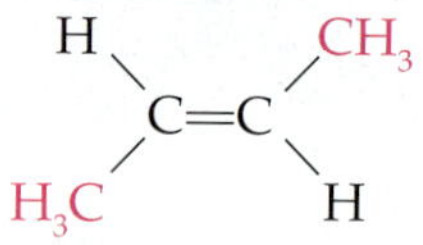

$$\overset{H}{\underset{H_3C}{>}}C=C\overset{H}{\underset{CH_3}{<}} \qquad \overset{H}{\underset{H_3C}{>}}C=C\overset{CH_3}{\underset{H}{<}}$$

Solution

As we saw with cycloalkanes, the prefixes *cis* and *trans* refer to the placement of the substituents attached to a bond that cannot undergo free rotation. In the case of alkenes, it is the groups attached to the carbon-carbon double bond (in this example, the H and CH_3 groups). When the groups are on the same side of the double bond, as in the structure on the left, the prefix *cis* is used. When the groups are on the opposite sides of the double bond, as in the structure on the right, *trans* is the appropriate prefix.

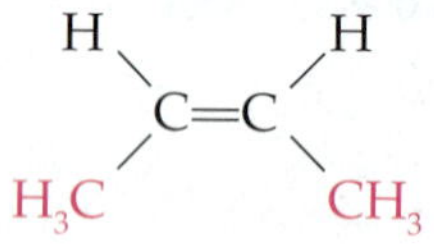

$$\underset{\text{cis-2-Butene}}{\overset{H}{\underset{H_3C}{>}}C=C\overset{H}{\underset{CH_3}{<}}} \qquad \underset{\text{trans-2-Butene}}{\overset{H}{\underset{H_3C}{>}}C=C\overset{CH_3}{\underset{H}{<}}}$$

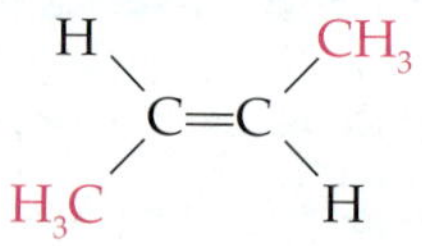

11.3 Geometric Isomers: A Consequence of Unsaturation

EXAMPLE 11.4

Naming cis and trans Compounds

Name the following geometric isomers.

Solution

The longest chain of carbon atoms in each of the following molecules is highlighted in yellow. *The chain must also contain the carbon-carbon double bond.* The location of functional groups relative to the double bond is used in determining the appropriate prefix, *cis* or *trans*, to be used in naming each of the molecules.

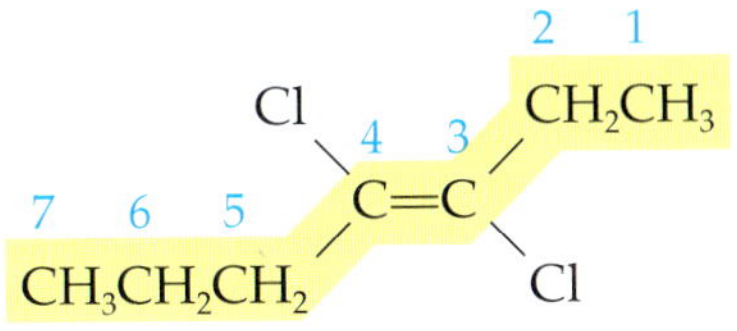

Parent chain: heptene
Position of double bond: 3-
Substituents: 3,4-dichloro
Configuration: *trans*
Name: *trans*-3,4-Dichloro-3-heptene

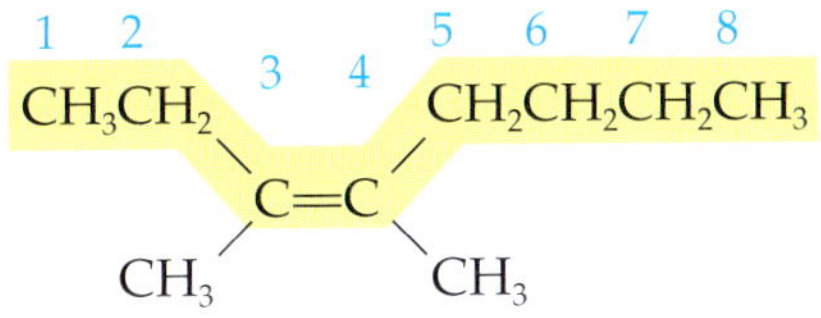

Parent chain: octene
Position of double bond: 3-
Substituents: 3,4-dimethyl
Configuration: *cis*
Name: *cis*-3,4-Dimethyl-3-octene

Question 11.5

In each of the following pairs of molecules, identify the *cis* isomer and the *trans* isomer.

a.
b.

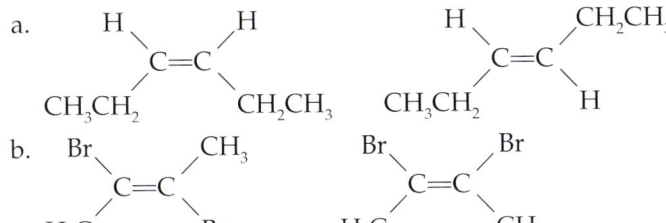

Question 11.6

Provide the complete I.U.P.A.C. name for each of the compounds in Question 11.5.

EXAMPLE 11.5

Identifying Geometric Isomers

Determine whether each of the following molecules can exist as *cis-trans* isomers: (1) 1-pentene, (b) 3-ethyl-3-hexene, and (c) 3-methyl-2-pentene.

Continued—

EXAMPLE 11.5 —Continued

Solution

a. Examine the structure of 1-pentene,

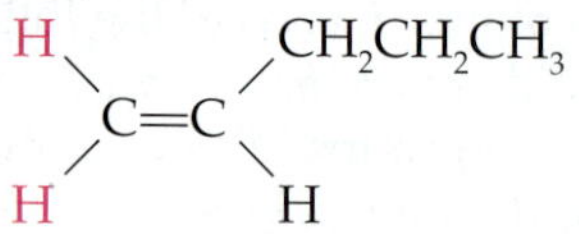

$$\begin{array}{c} H \\ \backslash \\ C=C \\ / \backslash \\ H H \end{array} \begin{array}{c} CH_2CH_2CH_3 \end{array}$$

We see that carbon-1 is bonded to two hydrogen atoms, rather than to two different substituents. In this case there can be no *cis-trans* isomers.

b. Examine the structure of 3-ethyl-3-hexene:

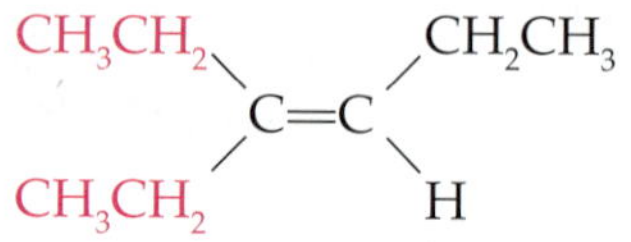

$$\begin{array}{c} CH_3CH_2 CH_2CH_3 \\ C=C \\ CH_3CH_2 H \end{array}$$

We see that one of the carbons of the carbon-carbon double bond is bonded to two ethyl groups. As in example (a), because this carbon is bonded to two identical groups, there can be no *cis* or *trans* isomers of this compound.

c. Finally, examination of the structure of 3-methyl-2-pentene reveals that both a *cis* and *trans* isomer can be drawn.

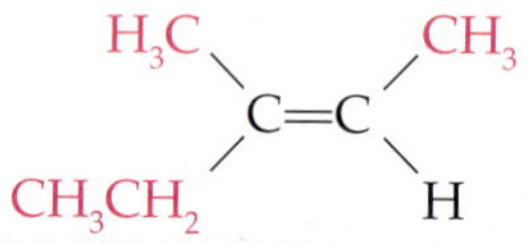

$$\begin{array}{cc} H_3C CH_3 & H_3C H \\ C=C & C=C \\ CH_3CH_2 H & CH_3CH_2 CH_3 \end{array}$$

cis-3-Methyl-2-pentene 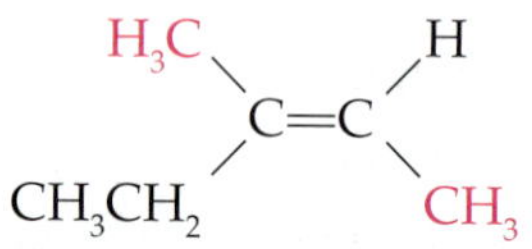 *trans*-3-Methyl-2-pentene

Each of the carbon atoms involved in the double bond is attached to two different groups. As a result, we can determine which is the *cis* isomer and which is the *trans* isomer based on the positions of the methyl groups relative to the double bond.

Question 11.7

Which of the following molecules can exist as both *cis* and *trans* isomers? Explain your reasoning.

a. 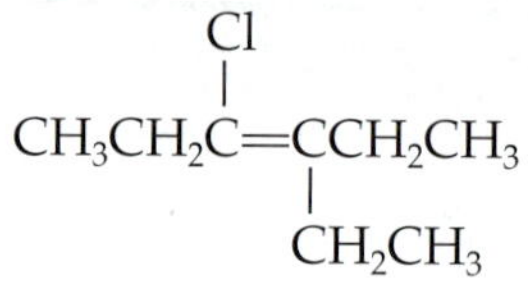 $CH_3CH_2C(Cl)=CCH_2CH_3$ with CH_2CH_3 below

b. 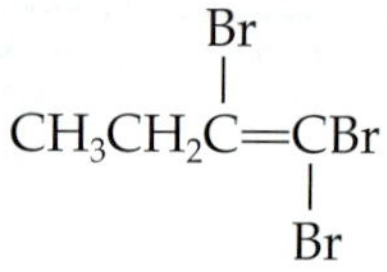 $CH_3CH_2C(Br)=CBr$ with Br below

c. 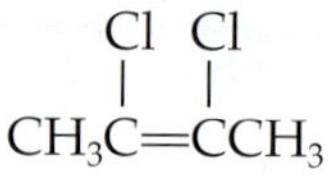 $CH_3C(Cl)=CCH_3$ with Cl on both

---

Question 11.8

Draw each of the *cis-trans* isomers in Question 11.7 and provide the complete names using the I.U.P.A.C. Nomenclature System.

Question 11.9

Draw condensed formulas for each of the following compounds:

a. *cis*-3-Octene
b. *trans*-5-Chloro-2-hexene
c. *trans*-2,3-Dichloro-2-butene

11.3 Geometric Isomers: A Consequence of Unsaturation

Question 11.10

Name each of the following compounds, using the I.U.P.A.C. system. Be sure to indicate *cis* or *trans* where applicable.

a. 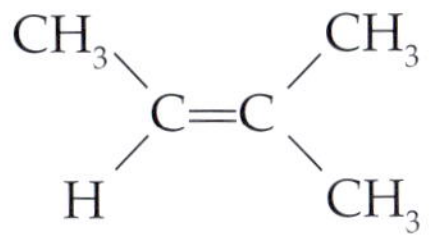
$$\begin{array}{c} CH_3 \\ \diagdown \diagup CH_3 \\ C=C \\ \diagup \diagdown \\ H CH_3 \end{array}$$

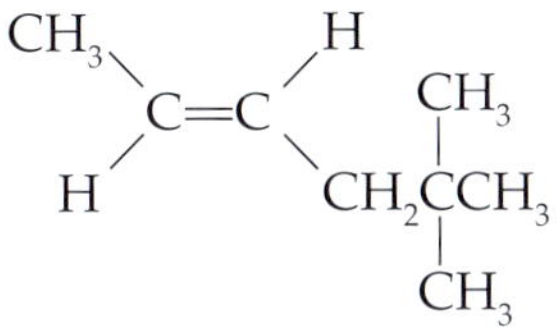

b. 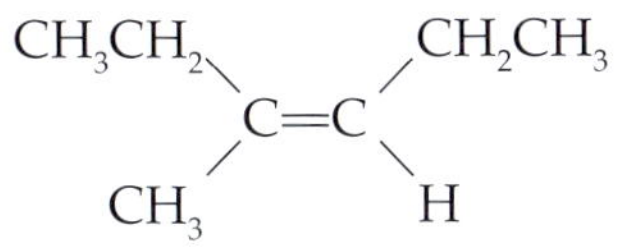
$$\begin{array}{c} CH_3CH_2 CH_2CH_3 \\ \diagdown \diagup \\ C=C \\ \diagup \diagdown \\ CH_3 H \end{array}$$

c.
$$\begin{array}{c} CH_3 H \\ \diagdown \diagup \\ C=C CH_3 \\ \diagup \diagdown | \\ H CH_2CCH_3 \\ | \\ CH_3 \end{array}$$

The recent debate over the presence of *cis* and *trans* isomers of fatty acids in our diet points out the relevance of geometric isomers to our lives. Fatty acids are long-chain carboxylic acids found in vegetable oils (unsaturated fats) and animal fats (saturated fats). Oleic acid,

$$CH_3(CH_2)_7CH=CH(CH_2)_7\overset{\overset{\displaystyle O}{\|}}{C}-OH$$

is the naturally occurring fatty acid in olive oil. Its I.U.P.A.C. name, *cis*-9-octadecenoic acid, reveals that this is *cis*-fatty acid.

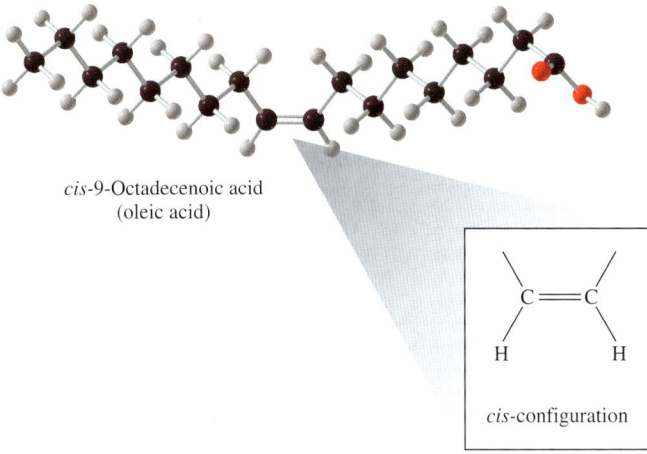

cis-9-Octadecenoic acid (oleic acid)

cis-configuration

While oleic acid is V-shaped as a result of the configuration of the double bond, its geometric isomer, *trans*-9-octadecenoic acid, is a rigid linear molecule.

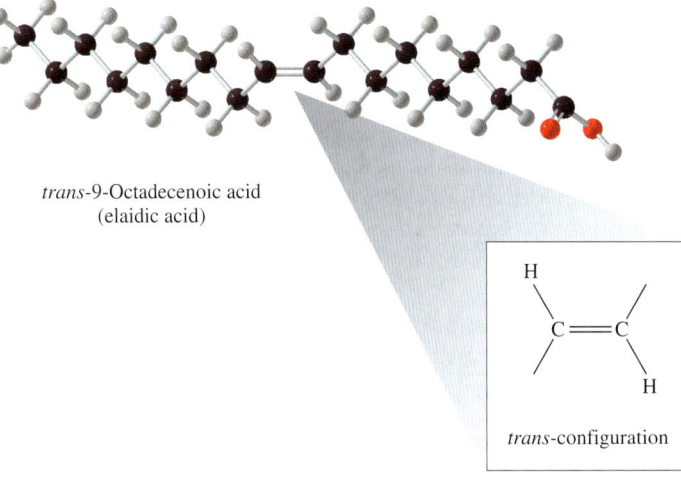

trans-9-Octadecenoic acid (elaidic acid)

trans-configuration

cis- and trans-*Fatty acids will be discussed in greater detail in Chapter 17. Hydrogenation is described in Section 11.5.*

The majority of *trans*-fatty acids found in the diet result from hydrogenation, which is a reaction used to convert oils into solid fats, such as margarine. It has recently been reported that *trans*-fatty acids in the diet elevate levels of "bad" or LDL cholesterol and lower the levels of "good" or HDL cholesterol, thereby increasing the risk of heart disease. Other studies suggest that *trans*-fatty acids may also increase the risk of type 2 diabetes.

11.4 Alkenes in Nature

Folklore tells us that placing a ripe banana among green tomatoes will speed up the ripening process. In fact, this phenomenon has been demonstrated experimentally. The key to the reaction is *ethene,* the simplest alkene. Ethene, produced by ripening fruit, is a plant growth substance. It is produced in the greatest abundance in areas of the plant where cell division is occurring. It is produced during fruit ripening, during leaf fall and flower senescence, as well as under conditions of stress, including wounding, heat, cold, or water stress, and disease.

There are surprising numbers of polyenes, alkenes with several double bonds, found in nature. These molecules, which have wildly different properties and functions, are built from one or more five-carbon units called *isoprene.*

$$CH_2=C(CH_3)-CH=CH_2$$
Isoprene

The molecules that are produced are called *isoprenoids,* or *terpenes* (Figure 11.3). Terpenes include the steroids; chlorophyll and carotenoid pigments that function in photosynthesis; and the lipid-soluble vitamins A, D, E, and K (Figure 11.1).

Many other terpenes are plant products familiar to us because of their distinctive aromas. *Geraniol,* the familiar scent of geraniums, is a molecule made up of two isoprene units. Purified from plant sources, geraniol is the active ingredient in several natural insect repellants. These can be applied directly to the skin to provide four hours of protection against a variety of insects, including mosquitoes, ticks, and fire ants.

D-*Limonene* is the most abundant component of the oil extracted from the rind of citrus fruits. Because of its pleasing orange aroma, D-limonene is used as a flavor and fragrance additive in foods. However, the most rapidly expanding use of the compound is as a solvent. In this role, D-limonene can be used in place of more toxic solvents, such as mineral spirits, methyl ethyl ketone, acetone, toluene, and fluorinated and chlorinated organic solvents. It can also be formulated as a water-based cleaning product, such as Orange Glo, that can be used in place of more caustic cleaning solutions. There is a form of limonene that is a molecular mirror image of D-limonene. It is called L-limonene and has a pine or turpentine aroma.

The terpene *myrcene* is found in bayberry. It is used in perfumes and scented candles because it adds a refreshing, spicy aroma to them. Trace amounts of myrcene may be used as a flavor component in root beer.

Farnesol is a terpene found in roses, orange blossom, wild cyclamen, and lily of the valley. Cosmetics companies began to use farnesol in skin care products in the early 1990s. It is claimed that farnesol smoothes wrinkles and increases skin elasticity. It is also thought to reduce skin aging by promoting regeneration of cells and activation of the synthesis of molecules, such as collagen, that are required for healthy skin.

Another terpene, *retinol,* is a form of vitamin A (Figure 11.1). It is able to penetrate the outer layers of skin and stimulate the formation of collagen and elastin. This reduces wrinkles by creating skin that is firmer and smoother.

11.4 Alkenes in Nature

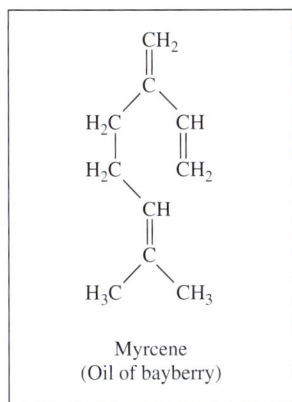

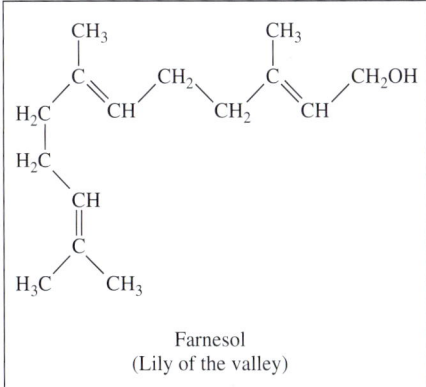

Figure 11.3
Many plant products, familiar to us because of their distinctive aromas, are isoprenoids, which are alkenes having several double bonds.

11.5 Reactions Involving Alkenes and Alkynes

LEARNING GOAL

Reactions of alkenes involve the carbon-carbon double bond. The key reaction of the double bond is the **addition reaction.** This involves the addition of two atoms or groups of atoms to a double bond. The major alkene addition reactions include addition of hydrogen (H_2), halogens (Cl_2 or Br_2), water (HOH), or hydrogen halides (HBr or HCl). A generalized addition reaction is shown here. The R in these structures represents any alkyl group.

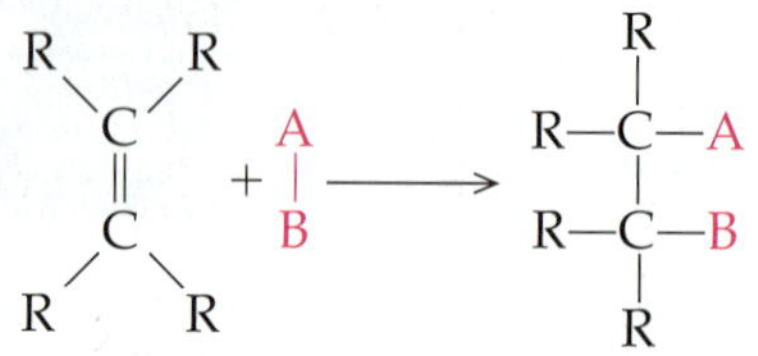

$$\begin{array}{c} R \\ \diagdown \\ C \\ \parallel \\ C \\ \diagup \\ R \end{array} \begin{array}{c} R \\ \diagup \\ \\ \\ \diagdown \\ R \end{array} + \begin{array}{c} A \\ | \\ B \end{array} \longrightarrow \begin{array}{c} R \\ | \\ R\!-\!C\!-\!A \\ | \\ R\!-\!C\!-\!B \\ | \\ R \end{array}$$

Note that the double bond is replaced by a single bond. The former double bond carbons receive a new single bond to a new atom, producing either an alkane or a substituted alkane.

Hydrogenation: Addition of H_2

Recall that a catalyst itself undergoes no net change in the course of a chemical reaction (see Section 7.3).

Hydrogenation is the addition of a molecule of hydrogen (H_2) to a carbon-carbon double bond to give an alkane. In this reaction the double bond is broken, and two new C—H single bonds result. Platinum, palladium, or nickel is needed as a catalyst to speed up the reaction. Heat and/or pressure may also be required.

Note that the alkene is gaining two hydrogens. Thus, hydrogenation is a reduction reaction (see Sections 8.5 and 12.6).

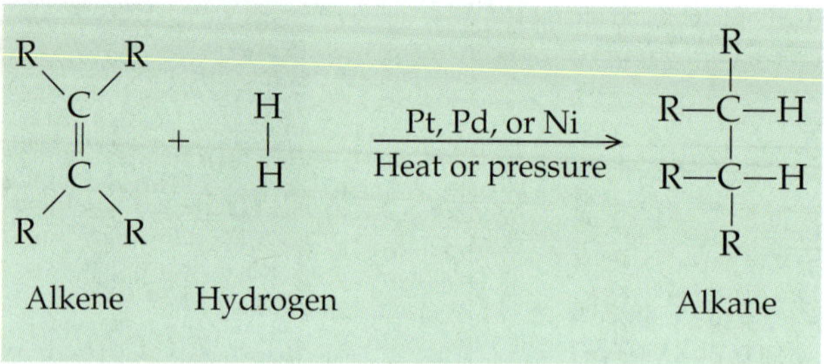

$$\begin{array}{c} R \\ \diagdown \\ C \\ \parallel \\ C \\ \diagup \\ R \end{array} \begin{array}{c} R \\ \diagup \\ \\ \\ \diagdown \\ R \end{array} + \begin{array}{c} H \\ | \\ H \end{array} \xrightarrow[\text{Heat or pressure}]{\text{Pt, Pd, or Ni}} \begin{array}{c} R \\ | \\ R\!-\!C\!-\!H \\ | \\ R\!-\!C\!-\!H \\ | \\ R \end{array}$$

Alkene   Hydrogen   Alkane

EXAMPLE 11.6   **Writing Equations for the Hydrogenation of Alkenes**

LEARNING GOAL

Write a balanced equation showing the hydrogenation of (a) 1-pentene and (b) *trans*-2-pentene.

Solution

(a) Begin by drawing the structure of 1-pentene and of diatomic hydrogen (H_2) and indicating the catalyst.

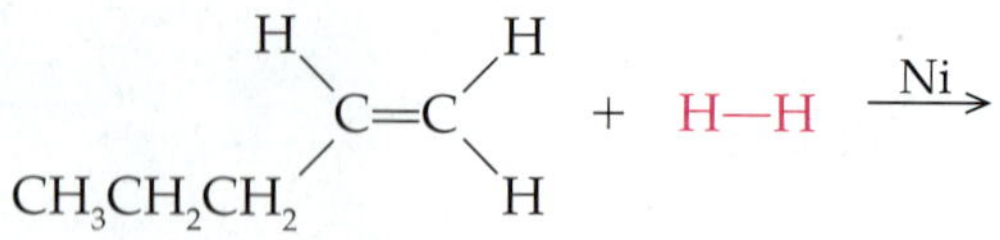

$$\begin{array}{c} \\ CH_3CH_2CH_2 \end{array} \!\!\diagdown\!\! \begin{array}{c} H \\ \\ C\!=\!C \\ \\ \end{array} \!\!\diagup\!\! \begin{array}{c} H \\ \\ \\ H \end{array} + H\!-\!H \xrightarrow{\text{Ni}}$$

1-Pentene   Hydrogen

Knowing that one hydrogen atom will form a covalent bond with each of the carbon atoms of the carbon-carbon double bond, we can write the product and complete the equation.

Continued—

EXAMPLE 11.6 —Continued

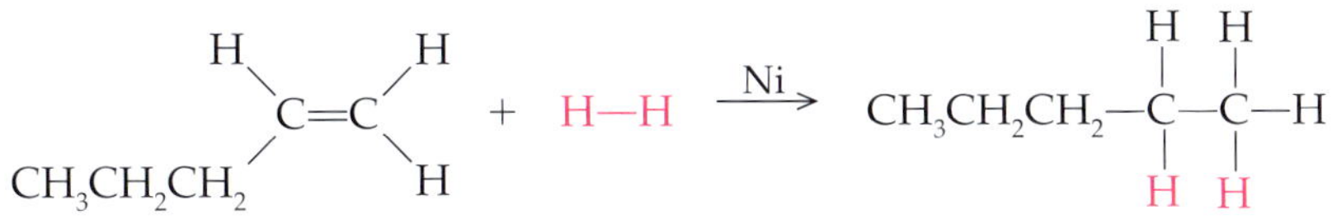

$$CH_3CH_2CH_2-CH=CH_2 + H-H \xrightarrow{Ni} CH_3CH_2CH_2-CH_2-CH_3$$

1-Pentene   Hydrogen   Pentane

(b) Begin by drawing the structure of *trans*-2-pentene and of diatomic hydrogen (H_2) and indicating the catalyst.

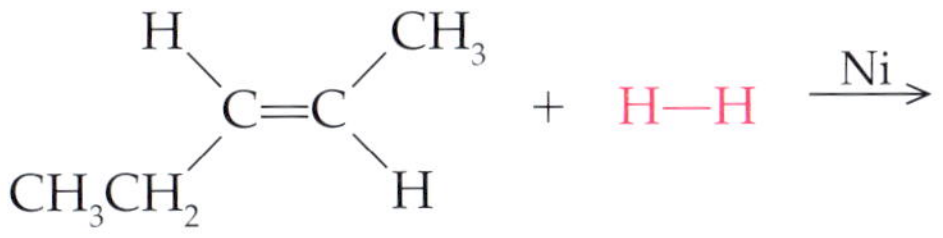

trans-2-Pentene + H—H $\xrightarrow{Ni}$

Knowing that one hydrogen atom will form a covalent bond with each of the carbon atoms of the carbon-carbon double bond, we can write the product and complete the equation.

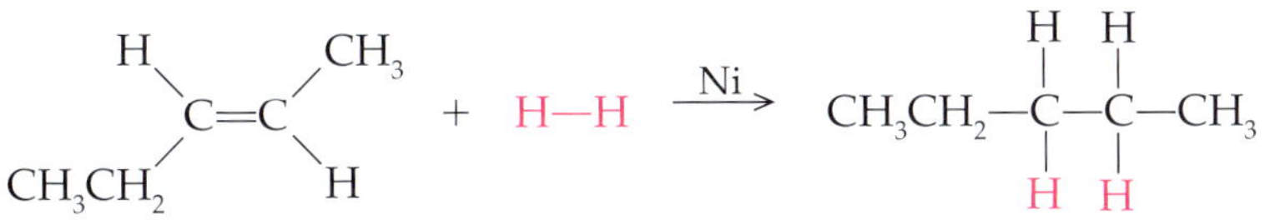

trans-2-Pentene + H—H $\xrightarrow{Ni}$ $CH_3CH_2-CH_2-CH_2-CH_3$

trans-2-Pentene   Hydrogen   Pentane

Question 11.11

The *trans* isomer of 2-pentene was used in Example 11.6. Would the result be any different if the *cis* isomer had been used?

Question 11.12

Write balanced equations for the hydrogenation of 1-butene and *cis*-2-butene.

The conditions for the hydrogenation of alkynes are very similar to those for the hydrogenation of alkenes. Two moles of hydrogen add to the triple bond of the alkyne to produce an alkane, as seen in the following general reaction:

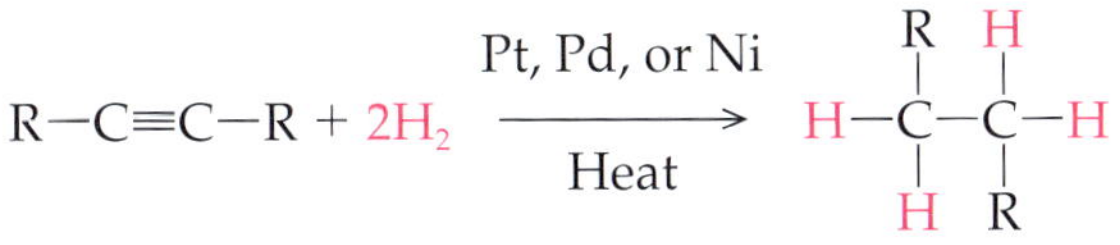

$$R-C\equiv C-R + 2H_2 \xrightarrow[\text{Heat}]{\text{Pt, Pd, or Ni}} H-\underset{H}{\underset{|}{\overset{R}{\overset{|}{C}}}}-\underset{R}{\underset{|}{\overset{H}{\overset{|}{C}}}}-H$$

Alkyne   Hydrogen (2 mol)   Alkane

Question 11.13

Write balanced equations for the complete hydrogenation of each of the following alkynes:

a. $H_3C-C\equiv C-CH_3$
b. $H_3C-C\equiv C-CH_2CH_3$

Figure 11.4
Conversion of a typical oil to a fat involves hydrogenation. In this example, triolein (an oil) is converted to tristearin (a fat).

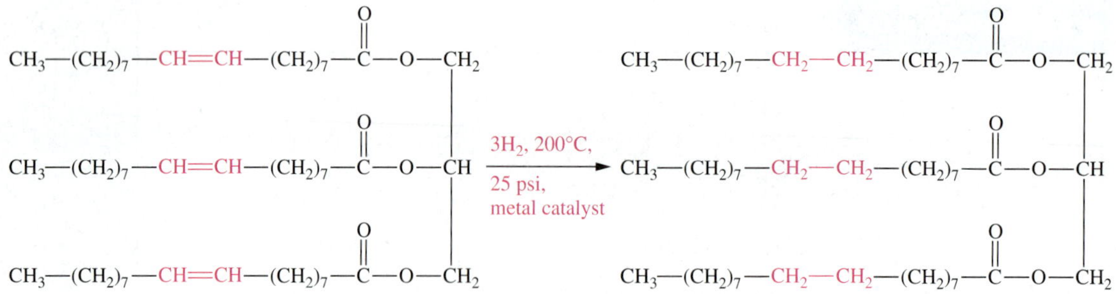

$$\begin{array}{c}
CH_3-(CH_2)_7-CH=CH-(CH_2)_7-\overset{O}{\underset{\parallel}{C}}-O-CH_2 \\
CH_3-(CH_2)_7-CH=CH-(CH_2)_7-\overset{O}{\underset{\parallel}{C}}-O-CH \\
CH_3-(CH_2)_7-CH=CH-(CH_2)_7-\overset{O}{\underset{\parallel}{C}}-O-CH_2
\end{array} \xrightarrow[\text{metal catalyst}]{3H_2,\ 200°C,\ 25\ \text{psi},} \begin{array}{c}
CH_3-(CH_2)_7-CH_2-CH_2-(CH_2)_7-\overset{O}{\underset{\parallel}{C}}-O-CH_2 \\
CH_3-(CH_2)_7-CH_2-CH_2-(CH_2)_7-\overset{O}{\underset{\parallel}{C}}-O-CH \\
CH_3-(CH_2)_7-CH_2-CH_2-(CH_2)_7-\overset{O}{\underset{\parallel}{C}}-O-CH_2
\end{array}$$

An *oil* → A *fat*

Question 11.14

Using the I.U.P.A.C. Nomenclature System, name each of the products and reactants in the reactions described in Question 11.13.

Saturated and unsaturated dietary fats are discussed in Section 17.2

Hydrogenation is used in the food industry to produce margarine, which is a mixture of hydrogenated vegetable oils (Figure 11.4). Vegetable oils are unsaturated, that is, they contain many double bonds and as a result have low melting points and are liquid at room temperature. The hydrogenation of these double bonds to single bonds increases the melting point of these oils and results in a fat, such as Crisco, that remains solid at room temperature. A similar process with highly refined corn oil and added milk solids produces corn oil margarine. As we saw in the previous section, such margarine may contain *trans*-fatty acids as a result of hydrogenation.

Halogenation: Addition of X_2

Chlorine (Cl_2) or bromine (Br_2) can be added to a double bond. This reaction, called **halogenation**, proceeds readily and does not require a catalyst:

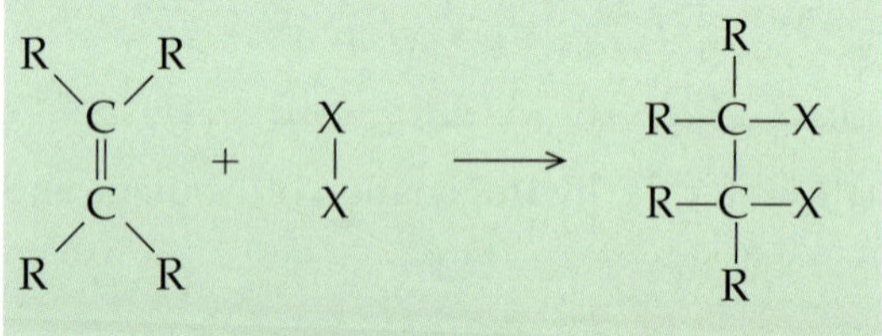

$$\underset{\text{Alkene}}{\begin{array}{c} R \\ \diagdown \\ C \\ \parallel \\ C \\ \diagup \\ R \end{array} \begin{array}{c} R \\ \diagup \\ \\ \\ \\ \diagdown \\ R \end{array}} + \underset{\text{Halogen}}{\begin{array}{c} X \\ | \\ X \end{array}} \longrightarrow \underset{\text{Alkyl dihalide}}{\begin{array}{c} R \\ | \\ R-C-X \\ | \\ R-C-X \\ | \\ R \end{array}}$$

EXAMPLE 11.7   Writing Equations for the Halogenation of Alkenes

LEARNING GOAL 4

Write a balanced equation showing (a) the chlorination of 1-pentene and (b) the bromination of *trans*-2-butene.

Solution

(a) Begin by drawing the structure of 1-pentene and of diatomic chlorine (Cl_2).

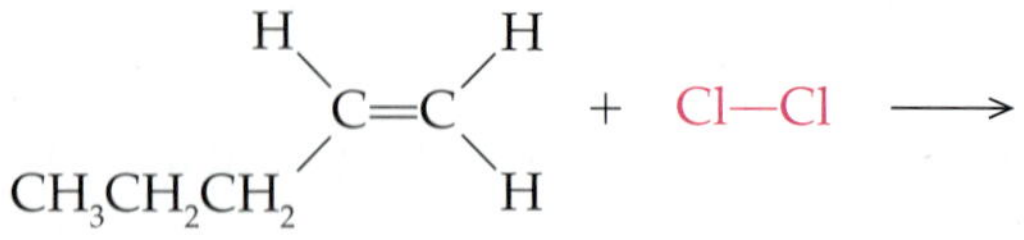

$$\underset{\text{1-Pentene}}{\begin{array}{c} H \\ \diagdown \\ CH_3CH_2CH_2 \end{array} C=C \begin{array}{c} H \\ \diagup \\ \diagdown \\ H \end{array}} + \underset{\text{Chlorine}}{Cl-Cl} \longrightarrow$$

Continued—

EXAMPLE 11.7 —Continued

Knowing that one chlorine atom will form a covalent bond with each of the carbon atoms of the carbon-carbon double bond, we can write the product and complete the equation.

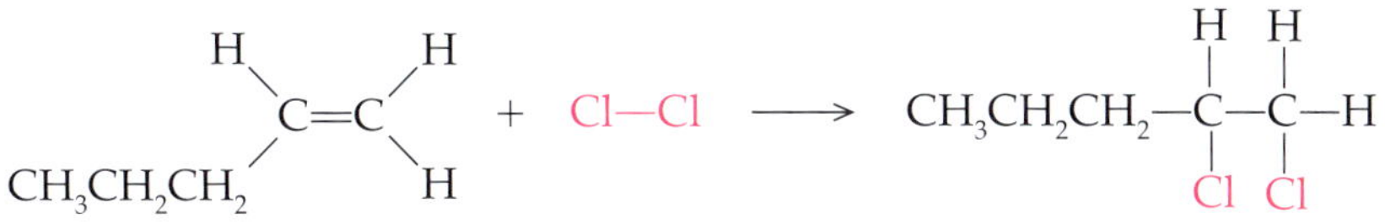

$$\underset{\text{1-Pentene}}{\underset{CH_3CH_2CH_2}{H}\!\!>\!\!C\!=\!C\!\!<\!\!\underset{H}{H}} + \underset{\text{Chlorine}}{Cl\!-\!Cl} \longrightarrow \underset{\text{1,2-Dichloropentane}}{CH_3CH_2CH_2\!-\!\underset{\underset{Cl}{|}}{\overset{\overset{H}{|}}{C}}\!-\!\underset{\underset{Cl}{|}}{\overset{\overset{H}{|}}{C}}\!-\!H}$$

(b) Begin by drawing the structure of *trans*-2-butene and of diatomic bromine (Br$_2$).

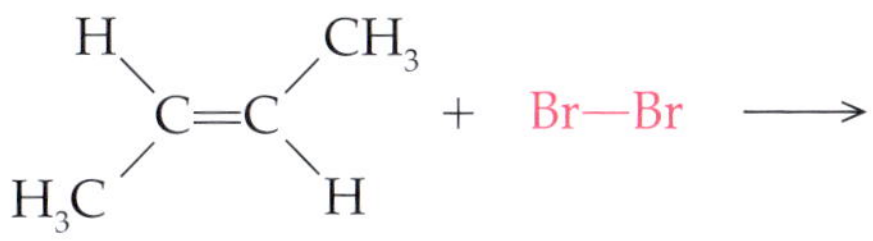

$$\underset{\text{\textit{trans}-2-Butene}}{\underset{H_3C}{H}\!\!>\!\!C\!=\!C\!\!<\!\!\underset{H}{CH_3}} + \underset{\text{Bromine}}{Br\!-\!Br} \longrightarrow$$

Knowing that one bromine atom will form a covalent bond with each of the carbon atoms of the carbon-carbon double bond, we can write the product and complete the equation.

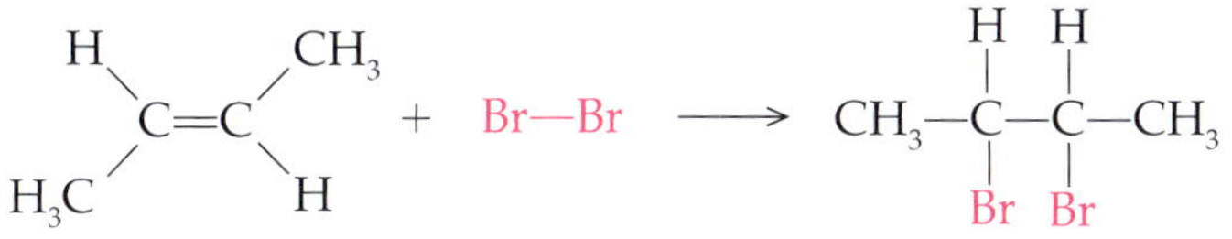

$$\underset{\text{\textit{trans}-2-Butene}}{\underset{H_3C}{H}\!\!>\!\!C\!=\!C\!\!<\!\!\underset{H}{CH_3}} + \underset{\text{Bromine}}{Br\!-\!Br} \longrightarrow \underset{\text{2,3-Dibromobutane}}{CH_3\!-\!\underset{\underset{Br}{|}}{\overset{\overset{H}{|}}{C}}\!-\!\underset{\underset{Br}{|}}{\overset{\overset{H}{|}}{C}}\!-\!CH_3}$$

Question 11.15

Write a balanced equation for the addition of bromine to each of the following alkenes. Draw and name the products and reactants for each reaction.

a. $CH_3CH=CH_2$
b. $CH_3CH=CHCH_3$

Question 11.16

Using the I.U.P.A.C. Nomenclature System, name each of the products and reactants in the reactions described in Question 11.15.

Alkynes also react with the halogens bromine or chlorine. Two moles of halogen add to the triple bond to produce a tetrahaloalkane:

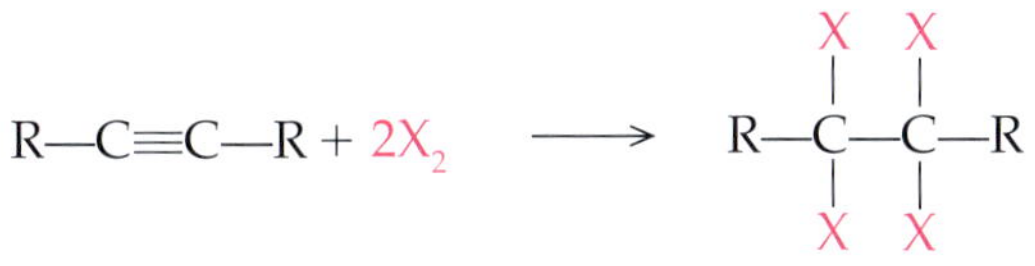

$$\underset{\text{Alkyne}}{R\!-\!C\!\equiv\!C\!-\!R} + \underset{\substack{\text{Halogen}\\\text{(2 mol)}}}{2X_2} \longrightarrow \underset{\text{Tetrahaloalkane}}{R\!-\!\underset{\underset{X}{|}}{\overset{\overset{X}{|}}{C}}\!-\!\underset{\underset{X}{|}}{\overset{\overset{X}{|}}{C}}\!-\!R}$$

Question 11.17

Write balanced equations for the complete chlorination of each of the following alkynes:

a. $H_3C-C\equiv C-CH_3$
b. $H_3C-C\equiv C-CH_2CH_3$

Question 11.18

Using the I.U.P.A.C. Nomenclature System, name each of the products and reactants in the reactions described in Question 11.17.

Below we see an equation representing the bromination of 1-pentene. Notice that the solution of reactants is red because of the presence of bromine. However, the product is colorless (Figure 11.5).

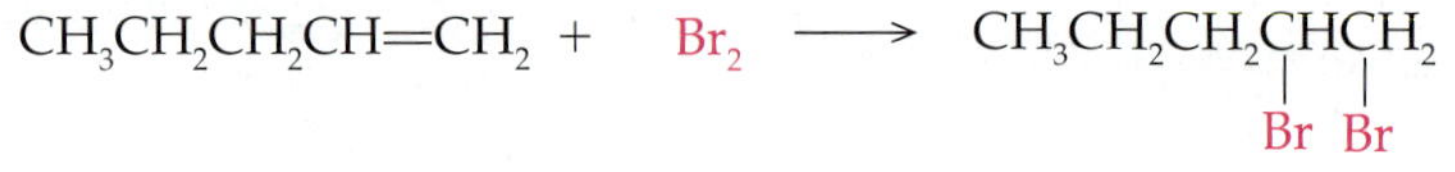

$$CH_3CH_2CH_2CH=CH_2 + Br_2 \longrightarrow CH_3CH_2CH_2CHCH_2$$
$$\underset{Br\ \ Br}{|\ \ \ |}$$

1-Pentene   Bromine   1,2-Dibromopentane
(colorless)   (red)   (colorless)

This bromination reaction can be used to show the presence of double or triple bonds in an organic compound. The reaction mixture is red because of the presence of dissolved bromine. If the red color is lost, the bromine has been consumed. Thus bromination has occurred, and the compound must have had a carbon-carbon double or triple bond. The greater the amount of bromine that must be added to the reaction, the more unsaturated the compound is. For instance, a diene or an alkyne would consume twice as much bromine as an alkene with a single double bond.

Hydration: Addition of H₂O

A water molecule can be added to an alkene. This reaction, termed **hydration**, requires a trace of strong acid (H⁺) as a catalyst. The product is an alcohol, as shown in the following reaction:

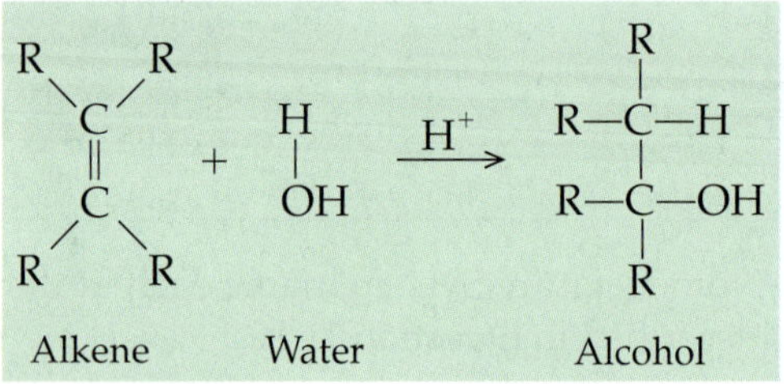

$$\underset{\text{Alkene}}{\overset{R}{\underset{R}{>}}C=C\overset{R}{\underset{R}{<}}} + \underset{\text{Water}}{\overset{H}{\underset{OH}{|}}} \xrightarrow{H^+} \underset{\text{Alcohol}}{\overset{R}{\underset{R}{\overset{|}{\underset{|}{R-C-H}}}}\overset{}{\underset{}{R-C-OH}}}$$

The following equation shows the hydration of ethene to produce ethanol.

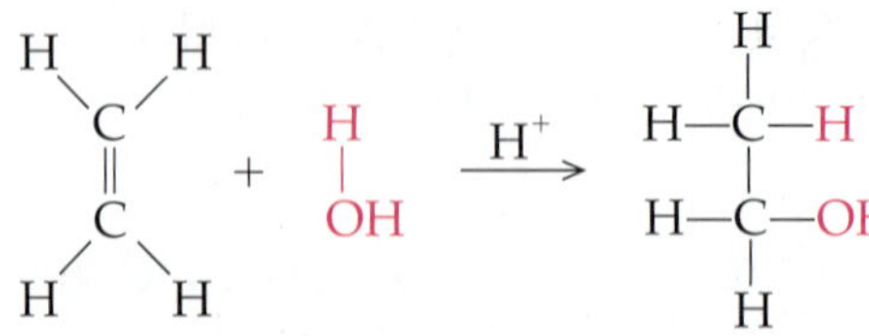

$$\underset{\text{Ethene}}{\overset{H}{\underset{H}{>}}C=C\overset{H}{\underset{H}{<}}} + \underset{\text{Water}}{\overset{H}{\underset{OH}{|}}} \xrightarrow{H^+} \underset{\substack{\text{Ethanol}\\\text{(ethyl alcohol)}}}{\overset{H}{\underset{H}{\overset{|}{\underset{|}{H-C-H}}}}\overset{}{\underset{}{H-C-OH}}}$$

11.5 Reactions Involving Alkenes and Alkynes

Figure 11.5
Bromination of an alkene. The solution on the left is red because of the presence of bromine and absence of an alkene or alkyne. In the presence of an unsaturated hydrocarbon, the bromine is used in the reaction and the solution becomes colorless.

With alkenes in which the groups attached to the two carbons of the double bond are different (unsymmetrical alkenes), two products are possible. For example:

```
   3   2   1                        3   2   1              3   2   1
   H   H                            H   H   H              H   H   H
   |   |                            |   |   |              |   |   |
H—C — C = C—H  +  H—OH   →H+→  H—C — C — C—H  +  H—C — C — C—H
   |   |                            |   |   |              |   |   |
   H   H                            H   OH  H              H   H   OH

Propene                        Major product            Minor product
(propylene)                    2-Propanol               1-Propanol
                               (isopropyl alcohol)      (propyl alcohol)
```

5 LEARNING GOAL

When hydration of an unsymmetrical alkene, such as propene, is carried out in the laboratory, one product (2-propanol) is favored over the other. The Russian chemist Vladimir Markovnikov studied many such reactions and came up with a rule that can be used to predict the major product of such a reaction. **Markovnikov's rule** tells us that the carbon of the carbon-carbon double bond that originally has more hydrogen atoms receives the hydrogen atom being added to the double bond. The remaining carbon forms a bond with the —OH. Simply stated, "the rich get richer"—the carbon with the greater number of hydrogens gets the new one as well. In the preceding example, carbon-1 has two C—H bonds originally, and carbon-2 has only one. The major product, 2-propanol, results from the new C—H bond forming on carbon-1 and the new C—OH bond on carbon-2.

Addition of water to a double bond is a reaction that we find in several biochemical pathways. For instance, the citric acid cycle is a key metabolic pathway in the complete oxidation of the sugar glucose and the release of the majority of the energy used by the body. The citric acid cycle is also the source of starting materials for the synthesis of the biological molecules needed for life. The next-to-last reaction in the citric acid cycle is the hydration of a molecule of fumarate to produce a molecule called malate.

```
        COO⁻                              COO⁻
         |                                 |
         C—H                          HO—C—H
         ||          Fumarase              |
   H—C         + H₂O    →             H—C—H
         |                                 |
        COO⁻                              COO⁻

       Fumarate                          Malate
```

We have seen that hydration of a double bond requires a trace of acid as a catalyst. In the cell, this reaction is catalyzed by an enzyme, or biological catalyst, called fumarase.

Chapter 11 The Unsaturated Hydrocarbons: Alkenes, Alkynes, and Aromatics

Hydration of a double bond also occurs in the β-oxidation pathway (see Section 23.2). This pathway carries out the oxidation of dietary fatty acids. Like the citric acid cycle, β-oxidation harvests the energy of the food molecules to use for body functions.

EXAMPLE 11.8 Writing Equations for the Hydration of Alkenes

LEARNING GOAL 4

LEARNING GOAL 5

Write an equation showing all the products of the hydration of 1-pentene.

Solution

Begin by drawing the structure of 1-pentene and of water and indicating the catalyst.

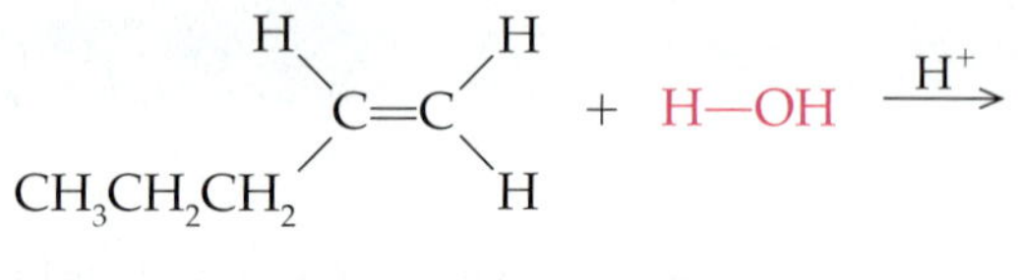

$$\underset{\text{1-Pentene}}{\underset{CH_3CH_2CH_2}{H}}C=C\underset{H}{\overset{H}{\diagup}} + \underset{\text{Water}}{H-OH} \xrightarrow{H^+}$$

Markovnikov's rule tells us that the carbon atom that is already bonded to the greater number of hydrogen atoms is more likely to receive the hydrogen atom from the water molecule. The other carbon atom is more likely to become bonded to the hydroxyl group. Thus we can predict that the major product of this reaction will be 2-pentanol and that the minor product will be 1-pentanol. Now we can complete the equation by showing the products:

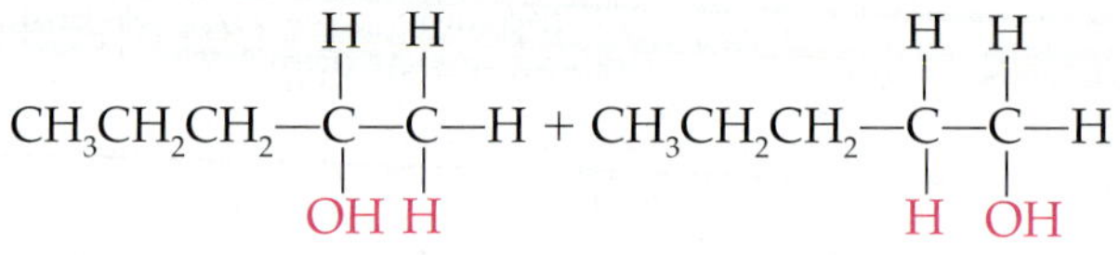

$$\underset{\substack{\text{2-Pentanol}\\\text{(major product)}}}{CH_3CH_2CH_2-\underset{\underset{OH}{|}}{\overset{\overset{H}{|}}{C}}-\underset{\underset{H}{|}}{\overset{\overset{H}{|}}{C}}-H} + \underset{\substack{\text{1-Pentanol}\\\text{(minor product)}}}{CH_3CH_2CH_2-\underset{\underset{H}{|}}{\overset{\overset{H}{|}}{C}}-\underset{\underset{OH}{|}}{\overset{\overset{H}{|}}{C}}-H}$$

Question 11.19

Write a balanced equation for the hydration of each of the following alkenes. Predict the major product of each of the reactions.

a. $CH_3CH=CHCH_3$
b. $CH_2=CHCH_2CH_2CHCH_3$
                            $|$
                            CH_3
c. $CH_3CH_2CH_2CH=CHCH_2CH_3$
d. $CH_3CHClCH=CHCHClCH_3$

Question 11.20

Write a balanced equation for the hydration of each of the following alkenes. Predict the major product of each of the reactions.

a. $CH_2=CHCH_2CH_2CH_3$
b. $CH_3CH_2CH_2CH=CHCH_3$
c. $CH_3CHBrCH_2CH=CHCH_2Cl$
d. $CH_3CH_2CH_2CH_2CH_2CH=CHCH_3$

Hydration of an alkyne is a more complex process because the initial product is not stable and is rapidly isomerized. As you would expect, the product is an alcohol but, in this case, one in which the hydroxyl group is bonded to one of the carbons of a carbon-carbon double bond. This type of molecule is called an *enol* because it is both an alkene *(ene)* and an alcohol *(ol)*. The enol cannot be isolated from the reaction mixture because it is so quickly isomerized into either an aldehyde or ketone, as shown in the following general reaction:

The chemistry of enols, aldehydes, and ketones is found in Chapter 13. In Chapter 21, we will study metabolic reactions involving the enol, phosphoenolpyruvate, which is an essential intermediate in the energy-harvesting pathway called glycolysis.

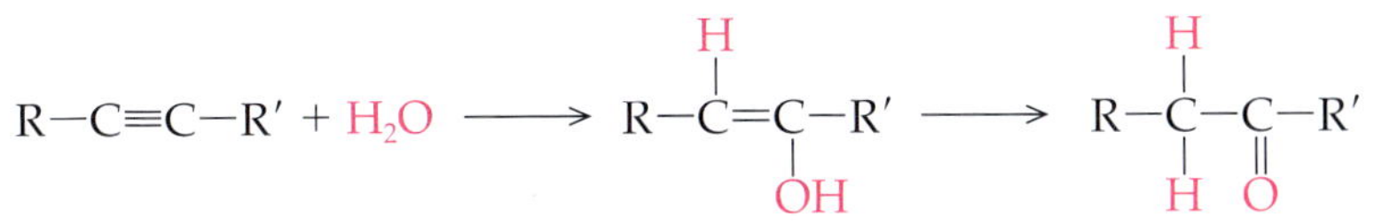

$$R-C \equiv C-R' + H_2O \longrightarrow R-\underset{OH}{\overset{H}{C}}=C-R' \longrightarrow R-\underset{H}{\overset{H}{C}}-\underset{O}{\overset{\|}{C}}-R'$$

Alkyne   Water   Enol   Aldehyde if R' = H
                                               Ketone if R' = alkyl group

Question 11.21

Write balanced equations for the complete hydration of each of the following alkynes:

a. $H_3C-C \equiv CH$
b. $H_3C-C \equiv CCH_2CH_3$

Question 11.22

Is the final product in each of the reactions in Question 11.21 an aldehyde or a ketone?

Hydrohalogenation: Addition of HX

A hydrogen halide (HBr, HCl, or HI) also can be added to an alkene. The product of this reaction, called **hydrohalogenation,** is an alkyl halide:

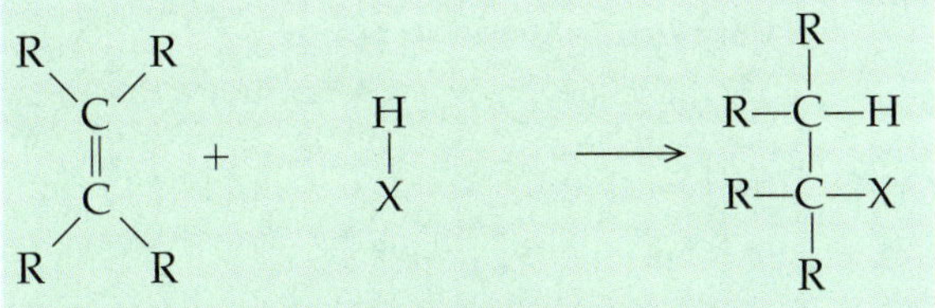

$$\underset{\underset{R}{|}}{\overset{\overset{R}{|}}{C}}=\underset{\underset{R}{|}}{\overset{\overset{R}{|}}{C}} + \underset{X}{\overset{H}{|}} \longrightarrow \underset{\underset{R}{|}}{\overset{\overset{R}{|}}{C}}-H \atop \underset{\underset{R}{|}}{\overset{|}{C}}-X$$

Alkene   Hydrogen halide   Alkyl halide

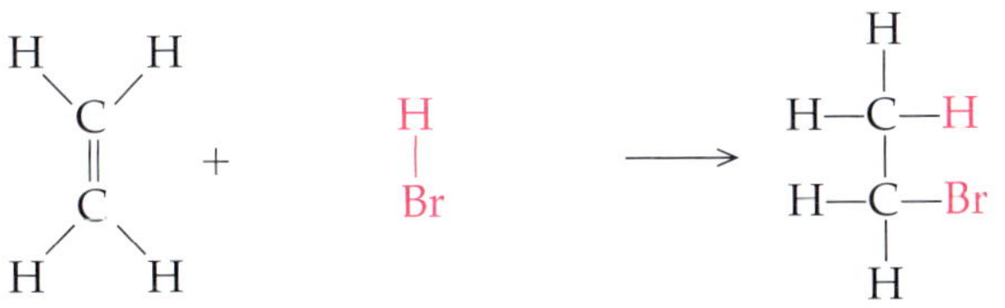

$$\underset{\underset{H}{|}}{\overset{\overset{H}{|}}{C}}=\underset{\underset{H}{|}}{\overset{\overset{H}{|}}{C}} + \underset{Br}{\overset{H}{|}} \longrightarrow H-\underset{|}{\overset{|}{C}}-H \atop H-\underset{H}{\overset{|}{C}}-Br$$

Ethene   Hydrogen bromide   Bromoethane

This reaction also follows Markovnikov's rule. That is, if HX is added to an unsymmetrical alkene, the hydrogen atom will be added preferentially to the carbon atom that originally had the most hydrogen atoms. Consider the following example:

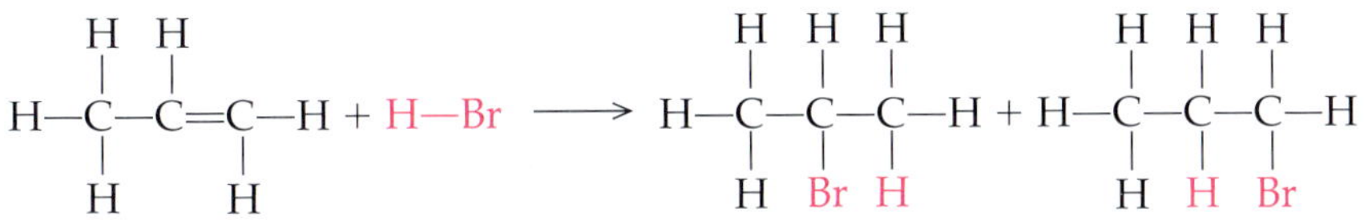

$$H-\underset{H}{\overset{H}{\underset{|}{C}}}-\underset{H}{\overset{H}{C}}=\underset{}{\overset{}{C}}-H + H-Br \longrightarrow H-\underset{H}{\overset{H}{\underset{|}{C}}}-\underset{Br}{\overset{H}{\underset{|}{C}}}-\underset{H}{\overset{H}{\underset{|}{C}}}-H + H-\underset{H}{\overset{H}{\underset{|}{C}}}-\underset{H}{\overset{H}{\underset{|}{C}}}-\underset{Br}{\overset{H}{\underset{|}{C}}}-H$$

Propene   Major product   Minor product
                               2-Bromopropane   1-Bromopropane

A Human Perspective

Folklore, Science, and Technology

For many years it was suspected that there existed a gas that stimulated fruit ripening and had other effects on plants. The ancient Chinese observed that their fruit ripened more quickly if incense was burned in the room. Early in this century, shippers realized that they could not store oranges and bananas on the same ships because some "emanation" given off by the oranges caused the bananas to ripen too early.

Puerto Rican pineapple growers and Philippine mango growers independently developed a traditional practice of building bonfires near their crops. They believed that the smoke caused the plants to bloom synchronously.

In the mid–nineteenth century, streetlights were fueled with natural gas. Occasionally the pipes leaked, releasing gas into the atmosphere. On some of these occasions, the leaves fell from all the shade trees in the region surrounding the gas leak.

What is the gas responsible for these diverse effects on plants? In 1934, R. Gane demonstrated that the simple alkene ethylene was the "emanation" responsible for fruit ripening. More recently, it has been shown that ethylene induces and synchronizes flowering in pineapples and mangos, induces senescence (aging) and loss of leaves in trees, and effects a wide variety of other responses in various plants.

We can be grateful to ethylene for the fresh, unbruised fruits that we can purchase at the grocery store. These fruits are picked when they are not yet ripe, while they are still firm. They then can be shipped great distances and gassed with ethylene when they reach their destination. Under the influence of the ethylene, the fruit ripens and is displayed in the store.

The history of the use of ethylene to bring fresh ripe fruits and vegetables to markets thousands of miles from the farms is an interesting example of the scientific process and its application for the benefit of society. Scientists began with the curious observations of Chinese, Puerto Rican, and Filipino farmers. Through experimentation they came to understand the phenomenon that caused the observations. Finally, through technology, scientists have made it possible to harness the power of ethylene so that grocers can "artificially" ripen the fruits and vegetables they sell to us.

Ethylene, the simplest alkene, is used to ripen fruits after shipment to market. Many vegetables must be protected from ethylene because it causes them to yellow or wilt.

For Further Understanding

Try the following experiment: place a ripe banana in a paper bag with an unripe tomato. Place a second unripe tomato on the countertop. Check daily to see which ripens more quickly.

Investigate the development of the painkiller aspirin to learn about another molecule whose biological activity was discovered accidentally.

EXAMPLE 11.9 Writing Equations for the Hydrohalogenation of Alkenes

LEARNING GOAL 4

LEARNING GOAL 5

Write an equation showing all the products of the hydrohalogenation of 1-pentene with HCl.

Solution

Begin by drawing the structure of 1-pentene and of hydrochloric acid.

Continued—

EXAMPLE 11.9 —Continued

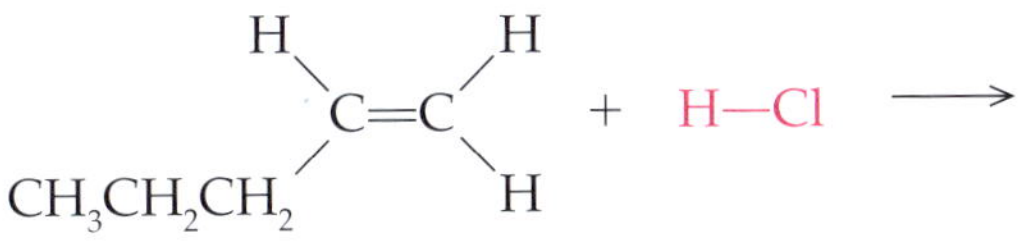

$$CH_3CH_2CH_2\text{-}CH=CH_2 + H\text{-}Cl \longrightarrow$$

1-Pentene   Hydrochloric acid

Markovnikov's rule tells us that the carbon atom that is already bonded to the greater number of hydrogen atoms is more likely to receive the hydrogen atom of the hydrochloric acid molecule. The other carbon atom is more likely to become bonded to the chlorine atom. Thus, we can complete the equation by writing the major and minor products.

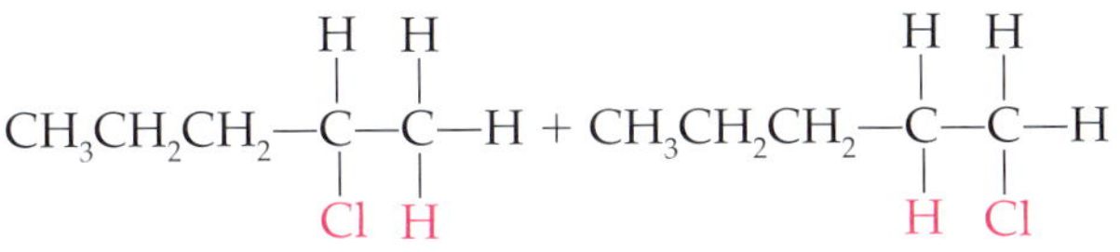

$$CH_3CH_2CH_2\text{-}CHCl\text{-}CH_3 \quad + \quad CH_3CH_2CH_2\text{-}CH_2\text{-}CH_2Cl$$

2-Chloropentane   1-Chloropentane
(major product)   (minor product)

As predicted by Markovnikov's rule, the major product is 2-chloropentane and the minor product is 1-chloropentane.

Question 11.23

Predict the major product in each of the following reactions. Name the alkene reactant and the product, using I.U.P.A.C. nomenclature.

a. 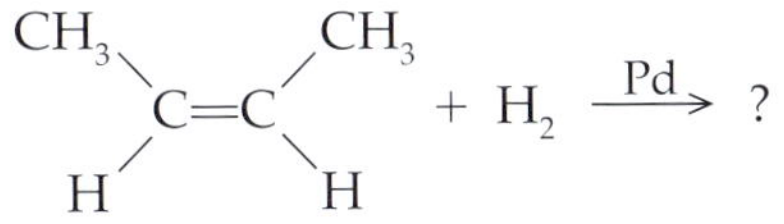 $(CH_3)(H)C=C(CH_3)(H) + H_2 \xrightarrow{Pd} ?$

b. $CH_3CH_2CH=CH_2 + H_2O \xrightarrow{H^+} ?$

c. $CH_3CH=CHCH_3 + Cl_2 \longrightarrow ?$

d. $CH_3CH_2CH_2CH=CH_2 + HBr \longrightarrow ?$

Question 11.24

Predict the major product in each of the following reactions. Name the alkene reactant and the product, using I.U.P.A.C. nomenclature.

a. 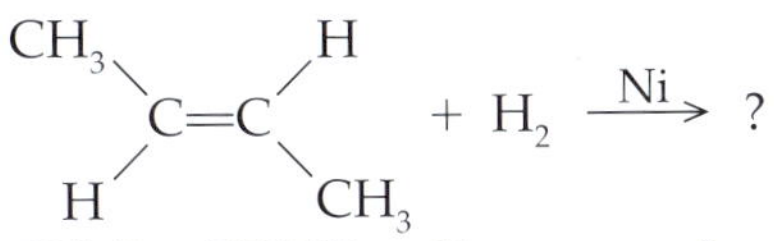 $(CH_3)(H)C=C(H)(CH_3) + H_2 \xrightarrow{Ni} ?$

b. $CH_3\text{-}C(CH_3)=CHCH_2CH_2CH_3 + H_2O \xrightarrow{H^+} ?$

c. $CH_3C(CH_3)=CHCH_3 + Br_2 \longrightarrow ?$

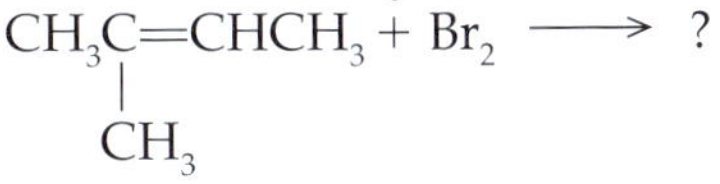

d. 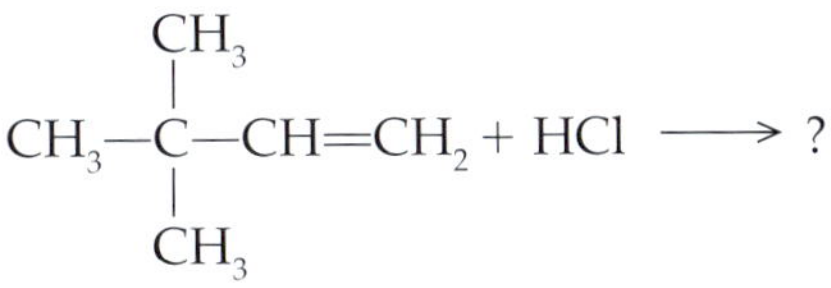 $(CH_3)_3C\text{-}CH=CH_2 + HCl \longrightarrow ?$

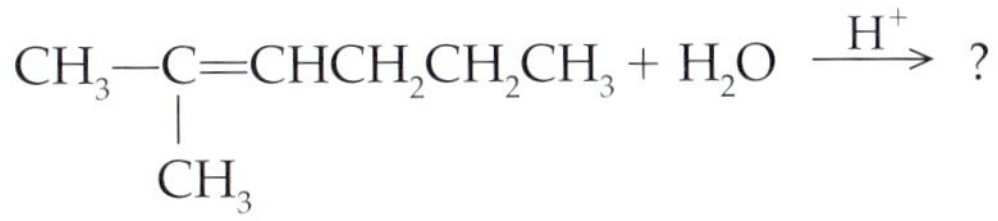

A Human Perspective

Life Without Polymers?

What do Nike Air-Sole shoes, Saturn automobiles, disposable diapers, tires, shampoo, and artificial joints and skin share in common? These products and a great many other items we use every day are composed of synthetic or natural polymers. Indeed, the field of polymer chemistry has come a long way since the 1920s and 1930s when DuPont chemists invented nylon and Teflon.

Consider the disposable diaper. The outer, waterproof layer is composed of polyethylene. The polymerization reaction that produces polyethylene is shown in Section 11.5. The diapers have elastic to prevent leaking. The elastic is made of a natural polymer, rubber. The monomer from which natural rubber is formed is 2-methyl-1,3-butadiene. The common name of this monomer is *isoprene*. As we will see in coming chapters, isoprene is an important monomer in the synthesis of many natural polymers.

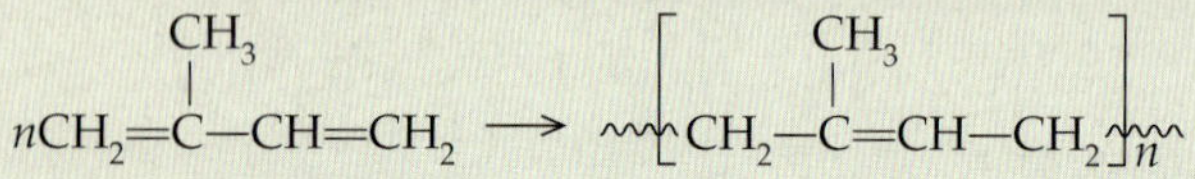

$$n\text{CH}_2=\overset{\overset{\displaystyle\text{CH}_3}{|}}{\text{C}}-\text{CH}=\text{CH}_2 \longrightarrow \left[\text{CH}_2-\overset{\overset{\displaystyle\text{CH}_3}{|}}{\text{C}}=\text{CH}-\text{CH}_2\right]_n$$

2-Methyl-1, 3-butadiene   Rubber polymer
(isoprene)

The diaper is filled with a synthetic polymer called poly(acrylic acid). This polymer has the remarkable ability to absorb many times its own weight in liquid. Polymers that have this ability are called superabsorbers, but polymer chemists have no idea why they have this property! The acrylate monomer and resulting poly(acrylic acid) polymer are shown here:

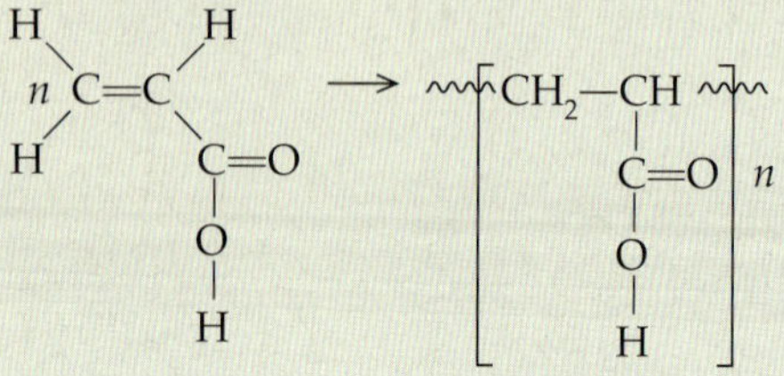

$$n\ \overset{\text{H}}{\underset{\text{H}}{\text{C}}}=\overset{\text{H}}{\underset{\underset{\text{O}-\text{H}}{\text{C}=\text{O}}}{\text{C}}} \longrightarrow \left[\text{CH}_2-\underset{\underset{\text{O}-\text{H}}{\text{C}=\text{O}}}{\text{CH}}\right]_n$$

Acrylate monomer   Poly(acrylic acid)

Another example of a useful polymer is Gore-Tex. This amazing polymer is made by stretching Teflon. Teflon is produced from the monomer tetrafluoroethene, as seen in the following reaction:

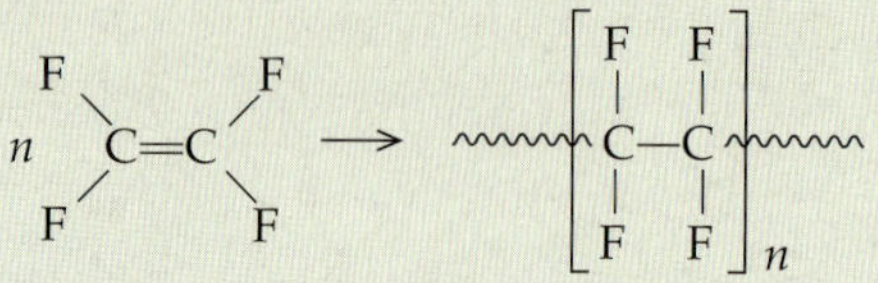

$$n\ \overset{\text{F}}{\underset{\text{F}}{\text{C}}}=\overset{\text{F}}{\underset{\text{F}}{\text{C}}} \longrightarrow \left[\overset{\overset{\text{F}}{|}}{\underset{\underset{\text{F}}{|}}{\text{C}}}-\overset{\overset{\text{F}}{|}}{\underset{\underset{\text{F}}{|}}{\text{C}}}\right]_n$$

Tetrafluoroethene   Teflon

Clothing made from this fabric is used to protect firefighters because of its fire resistance. Because it also insulates, Gore-Tex clothing is used by military forces and by many amateur athletes, for protection during strenuous activity in the cold. In addition to its use in protective clothing, Gore-Tex has been used in millions of medical procedures for sutures, synthetic blood vessels, and tissue reconstruction.

For Further Understanding

Visit The Macrogalleria, www.psrc.usm.edu/macrog/index.htm, an Internet site maintained by the Department of Polymer Science of the University of Southern Mississippi, to help you answer these questions:

Why does shrink wrap shrink?

What are optical fibers made of and how do they transmit light?

LEARNING GOAL

Addition Polymers of Alkenes

Polymers are macromolecules composed of repeating structural units called **monomers**. A polymer may be made up of several thousand monomers. Many commercially important plastics and fibers are addition polymers made from alkenes or substituted alkenes. They are called **addition polymers** because they are made by the sequential addition of the alkene monomer. The general formula for this addition reaction follows:

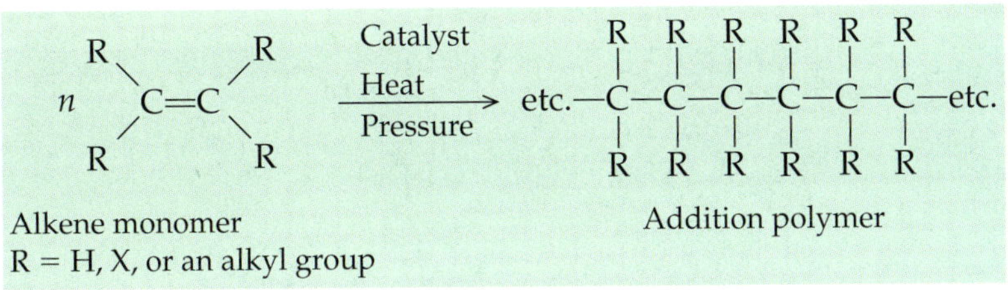

11.5 Reactions Involving Alkenes and Alkynes

$$n \underset{R}{\overset{R}{\diagdown}}C=C\underset{R}{\overset{R}{\diagup}} \xrightarrow[\text{Pressure}]{\text{Catalyst, Heat}} \text{etc.}-\underset{R}{\overset{R}{\underset{|}{\overset{|}{C}}}}-\underset{R}{\overset{R}{\underset{|}{\overset{|}{C}}}}-\underset{R}{\overset{R}{\underset{|}{\overset{|}{C}}}}-\underset{R}{\overset{R}{\underset{|}{\overset{|}{C}}}}-\underset{R}{\overset{R}{\underset{|}{\overset{|}{C}}}}-\underset{R}{\overset{R}{\underset{|}{\overset{|}{C}}}}-\text{etc.}$$

Alkene monomer
R = H, X, or an alkyl group

Addition polymer

The product of the reaction is generally represented in a simplified manner:

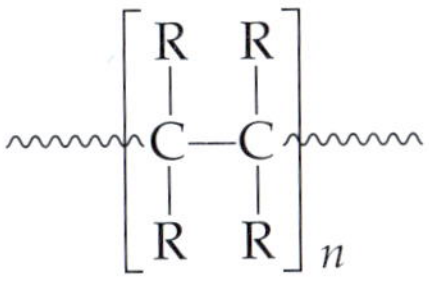

$$\left[\begin{array}{c} R \;\; R \\ | \;\; | \\ \sim\sim C-C \sim\sim \\ | \;\; | \\ R \;\; R \end{array} \right]_n$$

Polyethylene is a polymer made from the monomer ethylene (ethene):

$$n\text{CH}_2=\text{CH}_2 \longrightarrow \sim\sim[\text{CH}_2-\text{CH}_2]_n\sim\sim$$

Ethene
(ethylene)

Polyethylene

It is used to make bottles, injection-molded toys and housewares, and wire coverings.

Polypropylene is a plastic made from propylene (propene). It is used to make indoor-outdoor carpeting, packaging materials, toys, and housewares. When propylene polymerizes, a methyl group is located on every other carbon of the main chain:

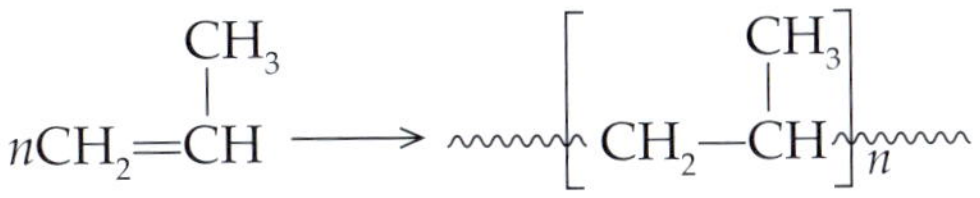

$$n\text{CH}_2=\overset{\overset{\displaystyle \text{CH}_3}{|}}{\text{CH}} \longrightarrow \sim\sim\left[\text{CH}_2-\overset{\overset{\displaystyle \text{CH}_3}{|}}{\text{CH}}\right]_n\sim\sim$$

or

$$\sim\sim\text{CH}_2-\overset{\overset{\displaystyle \text{CH}_3}{|}}{\text{CH}}-\text{CH}_2-\overset{\overset{\displaystyle \text{CH}_3}{|}}{\text{CH}}-\text{CH}_2-\overset{\overset{\displaystyle \text{CH}_3}{|}}{\text{CH}}\sim\sim$$

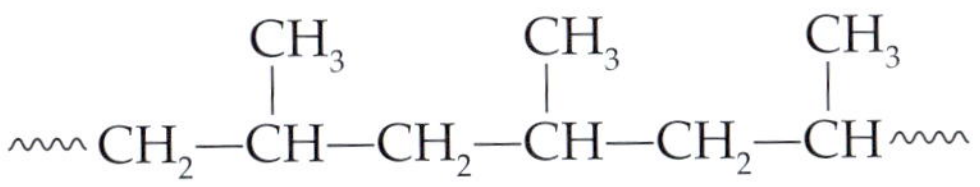

Polymers made from alkenes or substituted alkenes are simply very large alkanes or substituted alkanes. Like the alkanes, they are typically inert. This chemical inertness makes these polymers ideal for making containers to hold juices, chemicals, and fluids used medically. They are also used to make sutures, catheters, and other indwelling devices. A variety of polymers made from substituted alkenes are listed in Table 11.2.

TABLE 11.2 Some Important Addition Polymers of Alkenes

Monomer Name	Formula	Polymer	Uses
Styrene	$CH_2=CH-C_6H_5$	Polystyrene	Styrofoam containers
Acrylonitrile	$CH_2=CHCN$	Polyacrylonitrile (Orlon)	Clothing
Methyl methacrylate	$CH_2=C(CH_3)-COOCH_3$	Polymethyl methacrylate (Plexiglas, Lucite)	Basketball backboards
Vinyl chloride	$CH_2=CHCl$	Polyvinyl chloride (PVC)	Plastic pipe, credit cards
Tetrafluoroethene	$CF_2=CF_2$	Polytetrafluoroethylene (Teflon)	Nonstick surfaces

An Environmental Perspective

Plastic Recycling

Plastics, first developed by British inventor Alexander Parkes in 1862, are amazing substances. Some serve as containers for many of our foods and drinks, keeping them fresh for long periods of time. Other plastics serve as containers for detergents and cleansers or are formed into pipes for our plumbing systems. We have learned to make strong, clear sheets of plastic that can be used as windows, and feather-light plastics that can be used as packaging materials. In the United States alone, seventy-five billion pounds of plastics are produced each year.

But plastics, amazing in their versatility, are a mixed blessing. One characteristic that makes them so useful, their stability, has created an environmental problem. It may take forty to fifty years for plastics discarded into landfill sites to degrade. Concern that we could soon be knee-deep in plastic worldwide has resulted in a creative new industry: plastic recycling.

Since there are so many types of plastics, it is necessary to identify, sort, and recycle them separately. To help with this sorting process, manufacturers place recycling symbols on their plastic wares. As you can see in the accompanying table, each symbol corresponds to a different type of plastic.

Polyethylene terephthalate, also known as PETE or simply #1, is a form of polyester often used to make bottles and jars to contain food. When collected, it is ground up into flakes and formed into pellets. The most common use for recycled PETE is the manufacture of polyester carpets. But it may also be spun into a cotton-candy-like form that can be used as a fiber filling for pillows or sleeping bags. It may also be rolled into thin sheets or ribbons and used as tapes for VCRs or tape decks. Reuse to produce bottles and jars is also common.

HDPE, or #2, is high-density polyethylene. Originally used for milk and detergent bottles, recycled HDPE is used to produce pipes, plastic lumber, trash cans, or bottles for storage of materials other than food. LDPE, #4, is very similar to HDPE chemically. Because it is a more highly branched polymer, it is less-dense and more flexible. Originally used to produce plastic bags, recycled LDPE is also used to make trash bags, grocery bags, and plastic tubing and lumber.

PVC, or #3, is one of the less commonly recycled plastics in the United States, although it is actively recycled in Europe. The recycled material is used to make non-food-bearing containers, shoe soles, flooring, sweaters, and pipes. Polypropy-

11.6 Aromatic Hydrocarbons

LEARNING GOAL 7

In the early part of the nineteenth century, chemists began to discover organic compounds with chemical properties quite distinct from the alkanes, alkenes, and alkynes. They called these substances *aromatic compounds* because many of the first examples were isolated from the pleasant-smelling resins of tropical trees. The carbon:hydrogen ratio of these compounds suggested a very high degree of unsaturation, similar to the alkenes and alkynes. Imagine, then, how puzzled these early organic chemists must have been when they discovered that these compounds do not undergo the kinds of addition reactions common for the alkenes and alkynes.

11.6 Aromatic Hydrocarbons

Code	Type	Name	Formula	Description	Examples
PETE	△1△	Polyethylene terephthalate	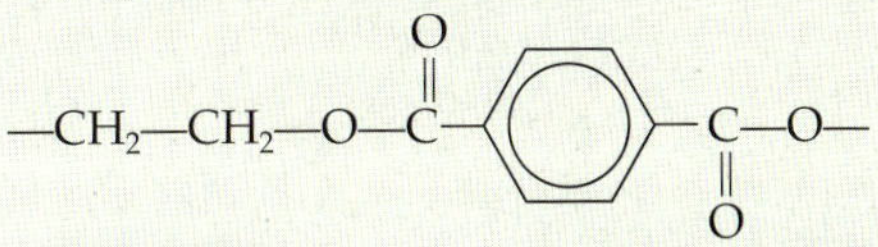 —CH$_2$—CH$_2$—O—C(=O)—⬡—C(=O)—O—	Usually clear or green, rigid	Soda bottles, peanut butter jars, vegetable oil bottles
HDPE	△2△	High-density polyethylene	—CH$_2$—CH$_2$—	Semirigid	Milk and water jugs, juice and bleach bottles
PVC	△3△	Polyvinyl chloride	—CH(Cl)—CH$_2$—	Semirigid	Detergent and cleanser bottles, pipes
LDPE	△4△	Low-density polyethylene	—CH$_2$—CH$_2$—	Flexible, not crinkly	Six-pack rings, bread bags, sandwich bags
PP	△5△	Polypropylene	—CH(CH$_3$)—CH$_2$—	Semirigid	Margarine tubs, straws, screw-on lids
PS	△6△	Polystyrene	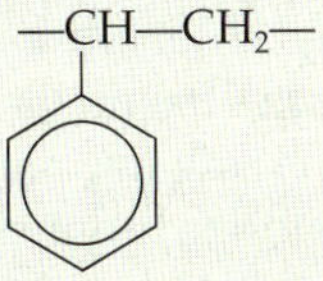 —CH(C$_6$H$_5$)—CH$_2$—	Often brittle	Styrofoam, packing peanuts, egg cartons, foam cups
Other	△7△	Multilayer plastics	N/A	Squeezable	Ketchup and syrup bottles

lene, PP or #5, is found in margarine tubs, fabrics, and carpets. Recycled polypropylene has many uses, including fabrication of gardening implements.

You probably come into contact with polystyrene, PS or #6, almost every day. It is used to make foam egg cartons and meat trays, serving containers for fast food chains, CD "jewel boxes," and "peanuts" used as packing material. At the current time, polystyrene food containers are not recycled. PS from nonfood products can be melted down and converted into pellets that are used to manufacture office desktop accessories, hangers, video and audio cassette housings, and plastic trays used to hold plants.

For Further Understanding

Use the Internet or other resources to investigate recycling efforts in your area.

In some areas, efforts to recycle plastics and paper have been abandoned. What factors contributed to this?

$$CH_2=CH_2 + Br_2 \longrightarrow CH_2(Br)-CH_2(Br)$$

benzene + Br$_2$ ⟶ No reaction

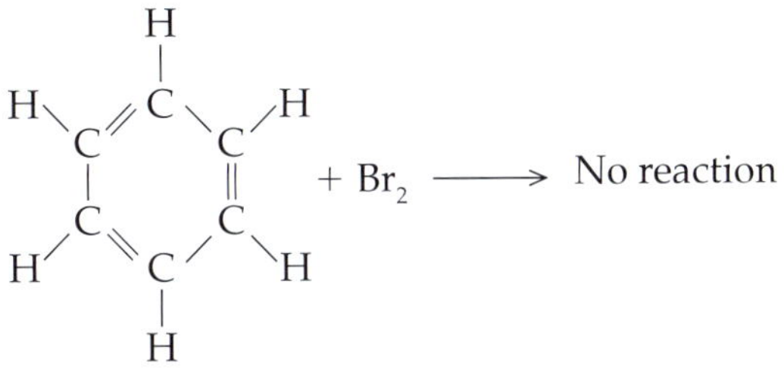

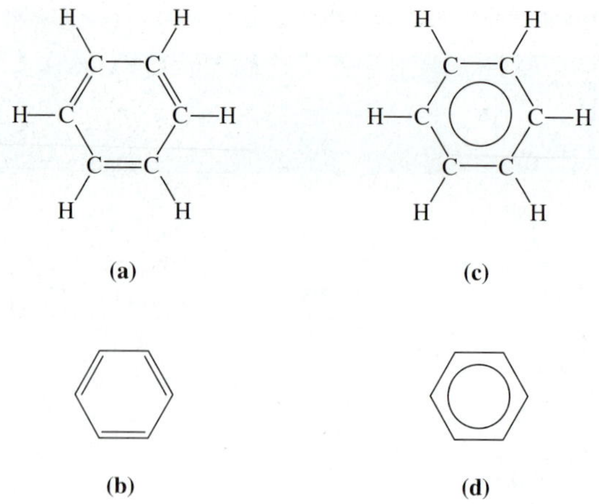

Figure 11.6

Four ways to represent the benzene molecule. Structure (b) is a simplified diagram of structure (a). Structure (d), a simplified diagram of structure (c), is the most commonly used representation.

We no longer define aromatic compounds as those having a pleasant aroma; in fact, many do not. We now recognize **aromatic hydrocarbons** as those that exhibit a much higher degree of chemical stability than their chemical composition would predict. The most common group of aromatic compounds is based on the six-member aromatic ring, the benzene ring. The structure of the benzene ring is represented in various ways in Figure 11.6.

Structure and Properties

The benzene ring consists of six carbon atoms joined in a planar hexagonal arrangement. Each carbon atom is bonded to one hydrogen atom. Friedrich Kekulé proposed a model for the structure of benzene in 1865. He proposed that single and double bonds alternated around the ring (a conjugated system of double bonds). To explain why benzene did not decolorize bromine—in other words, didn't react like an unsaturated compound—he suggested that the double and single bonds shift positions rapidly.

Resonance models are described in Section 3.4.

Actually, the most accurate way to represent the benzene molecules is as a resonance hybrid of two Keluké structures:

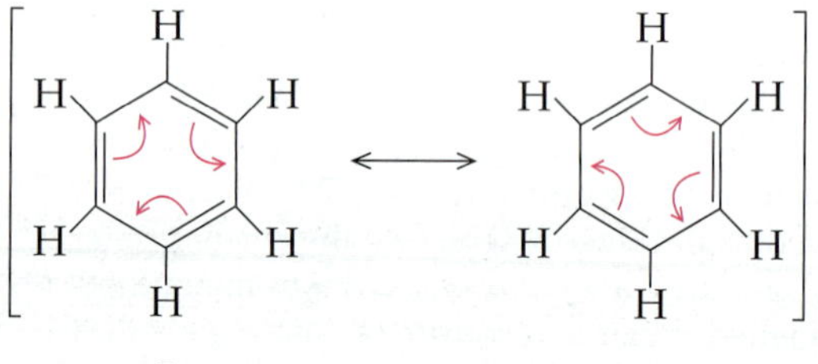

Benzene as a resonance hybrid

The current model of the structure of benzene is based on the idea of overlapping orbitals. Each carbon is bonded to two others by sharing a pair of electrons. Each carbon atom also shares a pair of electrons with a hydrogen atom. The remaining six electrons are located in p orbitals that are perpendicular to the plane of the ring. These p orbitals overlap laterally to form a cloud of electrons above and below the ring (Figure 11.7).

Two symbols are commonly used to represent the benzene ring. The representation in Figure 11.6b is the structure proposed by Kekulé. The structure in Figure 11.6d uses a circle to represent the electron cloud.

Nomenclature

LEARNING GOAL 7

Most simple aromatic compounds are named as derivatives of benzene. Thus benzene is the parent compound, and the name of any atom or group bonded to benzene is used as a prefix, as in these examples:

11.6 Aromatic Hydrocarbons

Figure 11.7
The current model of the bonding in benzene.

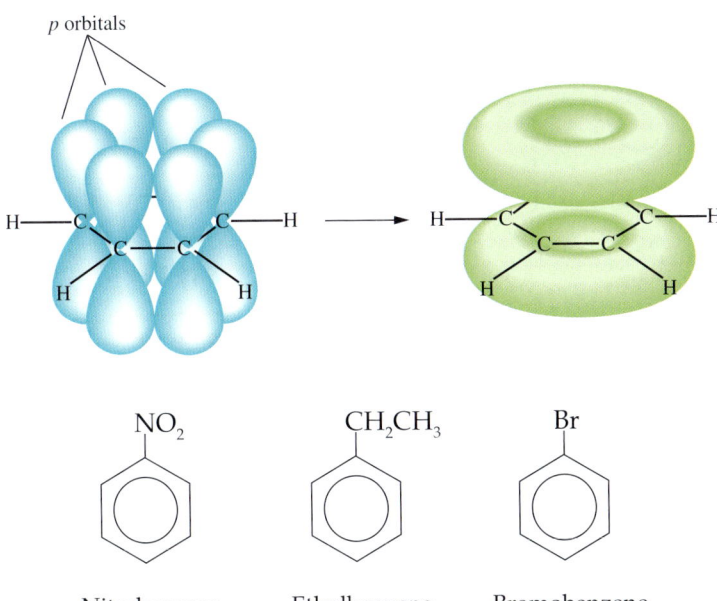

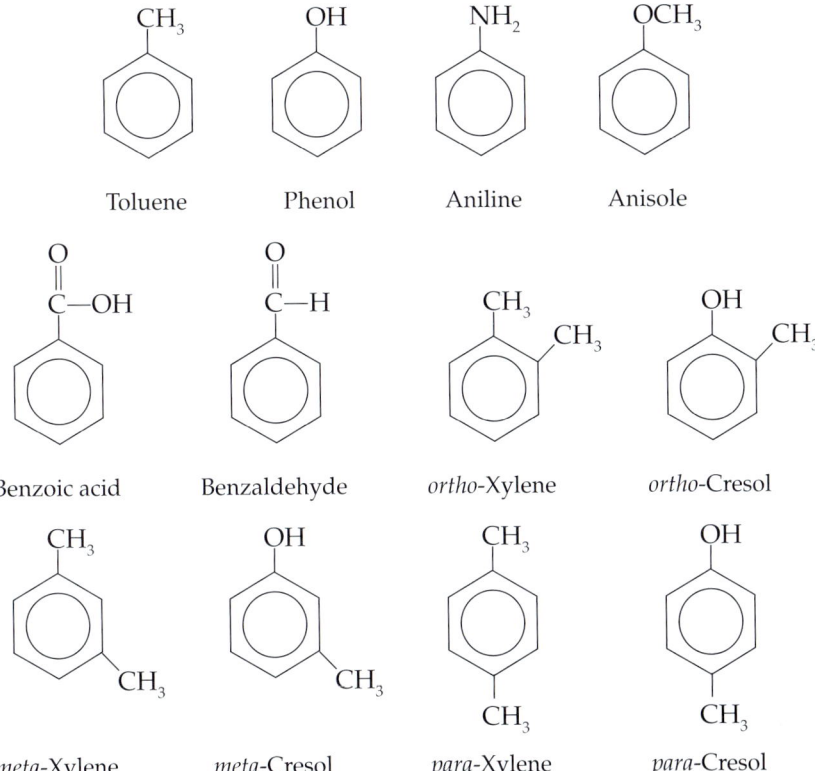

Other members of this family have unique names based on history rather than logic:

When two groups are present on the ring, three possible orientations exist, and they may be named by either the I.U.P.A.C. Nomenclature System or the common system of nomenclature. If the groups or atoms are located on two adjacent carbons, they are referred to as *ortho* (*o*) in the common system or with the prefix 1,2- in the I.U.P.A.C. system. If they are on carbons separated by one carbon atom, they are termed *meta* (*m*) in the common system or 1,3- in the I.U.P.A.C. system. Finally, if the substituents are on carbons separated by two carbon atoms, they are said to be *para* (*p*) in the common system or 1,4- in the I.U.P.A.C. system. The following examples demonstrate both of these systems:

Chapter 11 The Unsaturated Hydrocarbons: Alkenes, Alkynes, and Aromatics

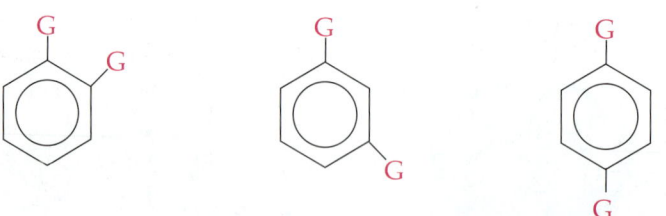

Two groups 1,2 or *ortho*   Two groups 1,3 or *meta*   Two groups 1,4 or *para*
G = Any group

If three or more groups are attached to the benzene ring, numbers must be used to describe their location. The names of the substituents are given in alphabetical order.

EXAMPLE 11.10 Naming Derivatives of Benzene

LEARNING GOAL 7

Name the following compounds using the I.U.P.A.C. Nomenclature System.

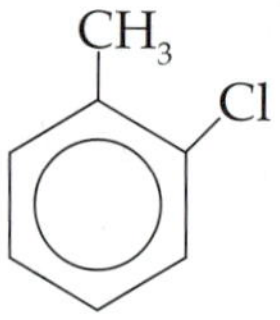

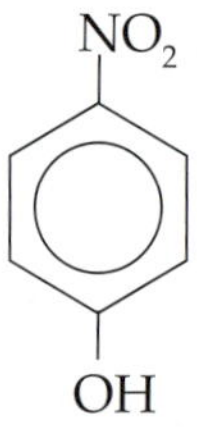

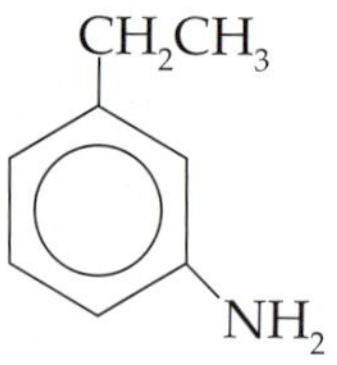

Solution

	a.	b.	c.
Parent compound:	toluene	phenol	aniline
Substituents:	2-chloro	4-nitro	3-ethyl
Name:	2-Chlorotoluene	4-Nitrophenol	3-Ethylaniline

EXAMPLE 11.11 Naming Derivatives of Benzene

LEARNING GOAL 7

Name the following compounds using the common system of nomenclature.

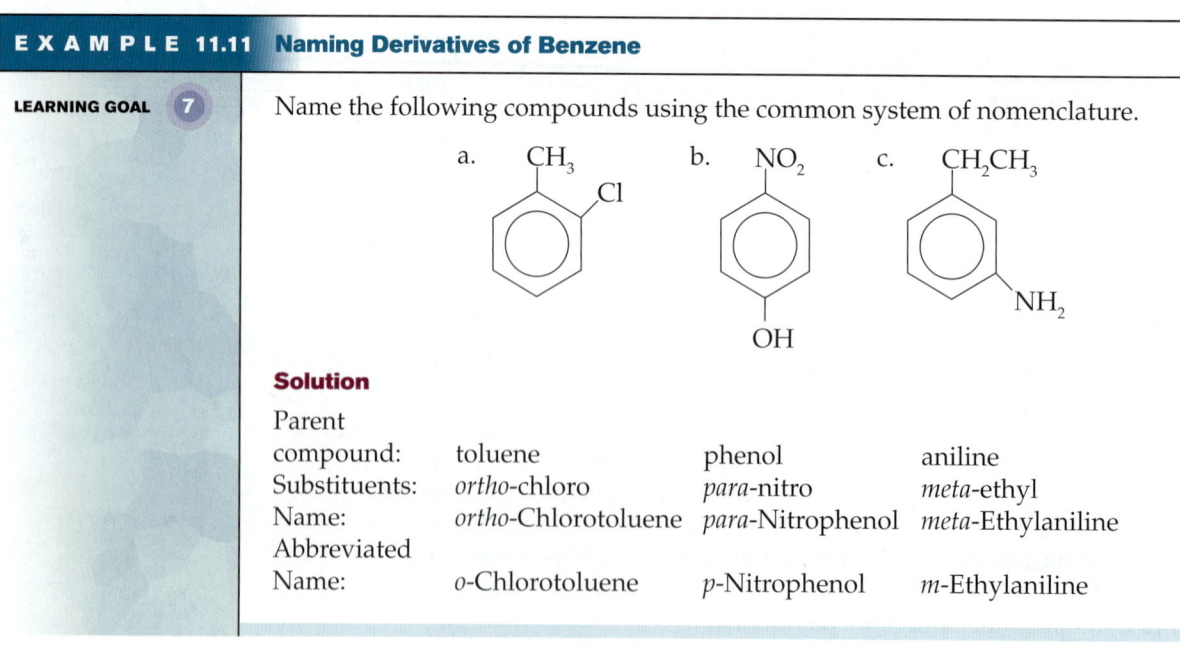

Solution

	a.	b.	c.
Parent compound:	toluene	phenol	aniline
Substituents:	ortho-chloro	para-nitro	meta-ethyl
Name:	ortho-Chlorotoluene	para-Nitrophenol	meta-Ethylaniline
Abbreviated Name:	o-Chlorotoluene	p-Nitrophenol	m-Ethylaniline

11.6 Aromatic Hydrocarbons

In I.U.P.A.C. nomenclature, the group derived by removing one hydrogen from benzene (—C$_6$H$_5$), is called the **phenyl group**. An aromatic hydrocarbon with an aliphatic side chain is named as a phenyl substituted hydrocarbon. For example:

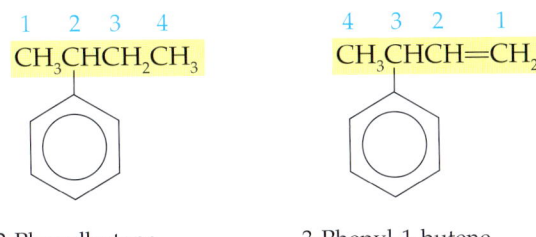

One final special name that occurs frequently in aromatic compounds is the benzyl group:

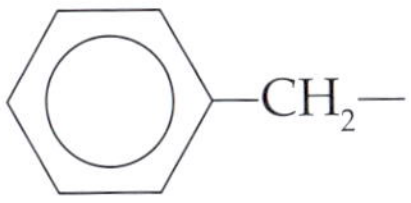

The use of this group name is illustrated by:

Benzyl chloride   Benzyl alcohol

Question 11.25

Draw each of the following compounds:

a. 1,3,5-Trichlorobenzene
b. *ortho*-Cresol
c. 2,5-Dibromophenol
d. *para*-Dinitrobenzene
e. 2-Nitroaniline
f. *meta*-Nitrotoluene

Question 11.26

Draw each of the following compounds:

a. 2,3-Dichlorotoluene
b. 3-Bromoaniline
c. 1-Bromo-3-ethylbenzene
d. *o*-Nitrotoluene
e. *p*-Xylene
f. *o*-Dibromobenzene

Polynuclear Aromatic Hydrocarbons

The polynuclear aromatic hydrocarbons (PAH) are composed of two or more aromatic rings joined together. Many of them have been shown to be carcinogenic, that is, they cause cancer.

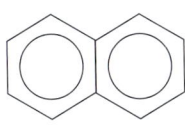  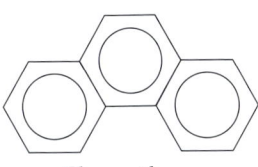 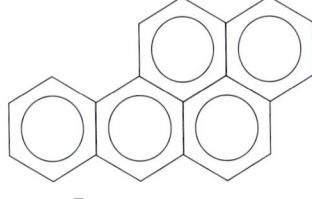

Naphthalene   Anthracene   Phenanthrene   Benzopyrene

Naphthalene has a distinctive aroma. It has been frequently used as mothballs and may cause hemolytic anemia (a condition causing breakdown of red blood cells) in humans, but has not been associated with human or animal cancers. Anthracene, derived from coal tar, is the parent compound of many dyes and pigments. Phenanthrene is common in the environment, although there is no industrial use of the compound. It is a product of incomplete combustion of fossil fuels and wood. Although anthracene is a suspected carcinogen, phenanthrene has not been shown to be one. Benzopyrene is found in tobacco smoke, smokestack effluents, charcoal-grilled meat, and automobile exhaust. It is one of the most potent carcinogens known.

Reactions Involving Benzene

LEARNING GOAL 8

As we have noted, benzene does not readily undergo addition reactions. The typical reactions of benzene are **substitution reactions,** in which a hydrogen atom is replaced by another atom or group of atoms.

Benzene can react (by substitution) with Cl_2 or Br_2. These reactions require either iron or an iron halide as a catalyst. For example:

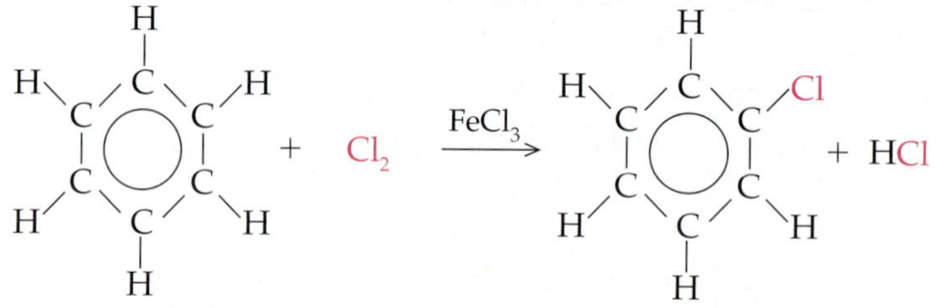

$$\text{Benzene} + Cl_2 \xrightarrow{FeCl_3} \text{Chlorobenzene} + HCl$$

$$\text{Benzene} + Br_2 \xrightarrow{FeBr_3} \text{Bromobenzene} + HBr$$

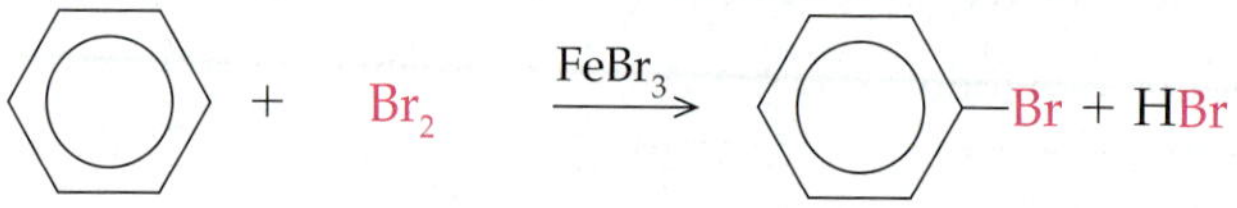

When a second equivalent of the halogen is added, three isomers—*para, ortho,* and *meta*—are formed.

Benzene also reacts with sulfur trioxide by substitution. Concentrated sulfuric acid is required as the catalyst. Benzenesulfonic acid, a strong acid, is the product:

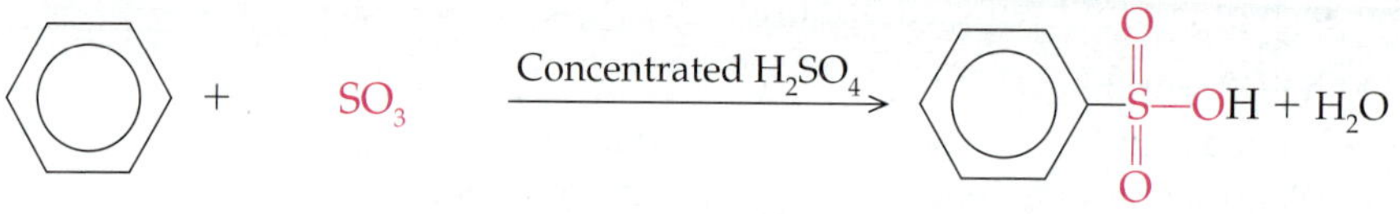

$$\text{Benzene} + SO_3 \xrightarrow{\text{Concentrated } H_2SO_4} \text{Benzenesulfonic acid} + H_2O$$

Benzene can also undergo nitration with concentrated nitric acid dissolved in concentrated sulfuric acid. This reaction requires temperatures in the range of 50–55°C.

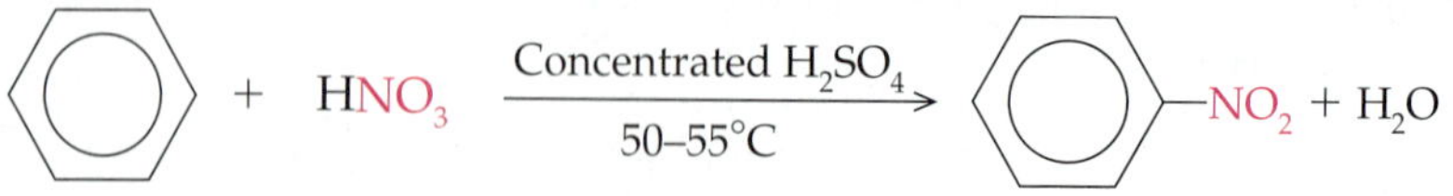

$$\text{Benzene} + HNO_3 \xrightarrow[50-55°C]{\text{Concentrated } H_2SO_4} \text{Nitrobenzene} + H_2O$$

11.7 Heterocyclic Aromatic Compounds

9 LEARNING GOAL

Heterocyclic aromatic compounds are those having at least one atom other than carbon as part of the structure of the aromatic ring. The structures and common names of several heterocyclic aromatic compounds are shown:

Pyridine

Pyrimidine

Purine

Imidazole

Furan

Pyrrole

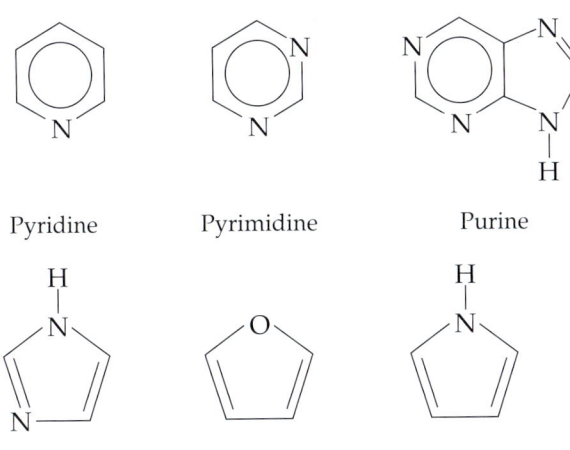

All these compounds are more similar to benzene in stability and chemical behavior than they are to the alkenes. Many of these compounds are components of molecules that have significant effects on biological systems. For instance, the purines and pyrimidines are components of DNA (deoxyribonucleic acid) and RNA (ribonucleic acid). DNA and RNA are the molecules responsible for storing and expressing the genetic information of an organism. The pyridine ring is found in nicotine, the addictive compound in tobacco. The pyrrole ring is a component of the porphyrin ring found in hemoglobin and chlorophyll.

See the Chemistry Connection: The Nicotine Patch in Chapter 15.

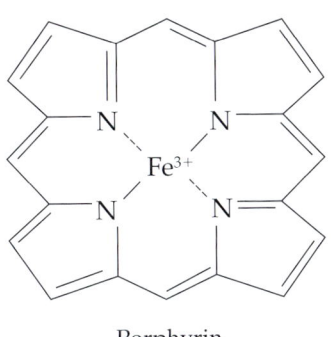

Porphyrin

The imidazole ring is a component of cimetidine, a drug used in the treatment of stomach ulcers. The structure of cimetidine is shown below:

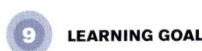

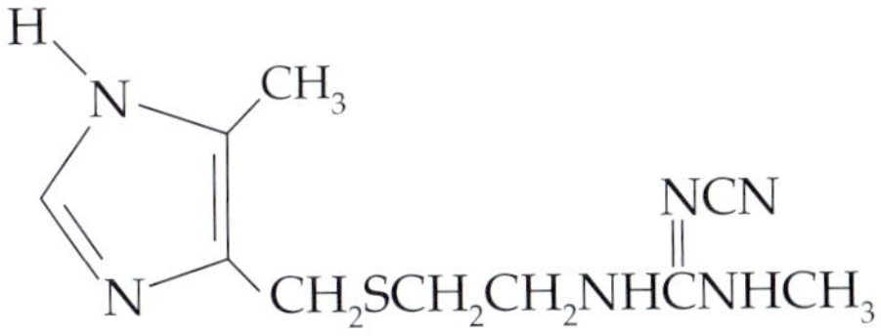

Cimetidine

We will discuss a subset of the heterocyclic aromatic compounds, the heterocyclic amines, in Chapter 15.

Summary of Reactions

Addition Reactions of Alkenes

Hydrogenation:

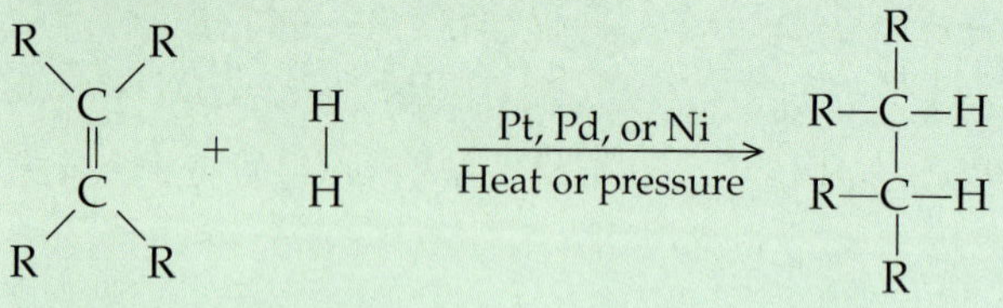

$$\underset{\text{Alkene}}{\underset{R}{\overset{R}{\diagdown}}C=C\underset{R}{\overset{R}{\diagup}}} + \underset{\text{Hydrogen}}{\overset{H}{\underset{H}{|}}} \xrightarrow[\text{Heat or pressure}]{\text{Pt, Pd, or Ni}} \underset{\text{Alkane}}{\overset{R-\overset{R}{\underset{|}{C}}-H}{\underset{R-\overset{|}{\underset{R}{C}}-H}{}}}$$

Hydration:

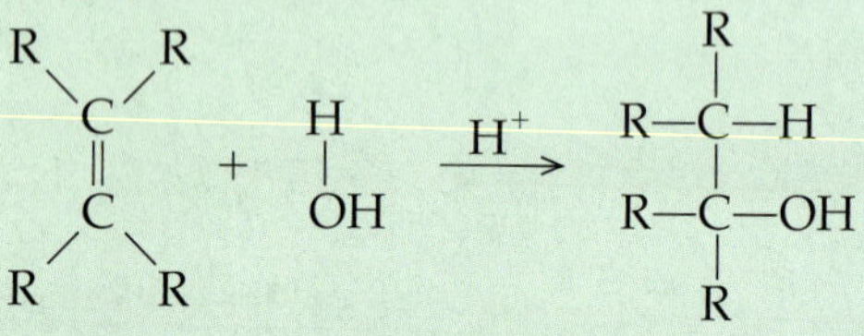

$$\underset{\text{Alkene}}{\underset{R}{\overset{R}{\diagdown}}C=C\underset{R}{\overset{R}{\diagup}}} + \underset{\text{Water}}{\overset{H}{\underset{OH}{|}}} \xrightarrow{H^+} \underset{\text{Alcohol}}{\overset{R-\overset{R}{\underset{|}{C}}-H}{\underset{R-\overset{|}{\underset{R}{C}}-OH}{}}}$$

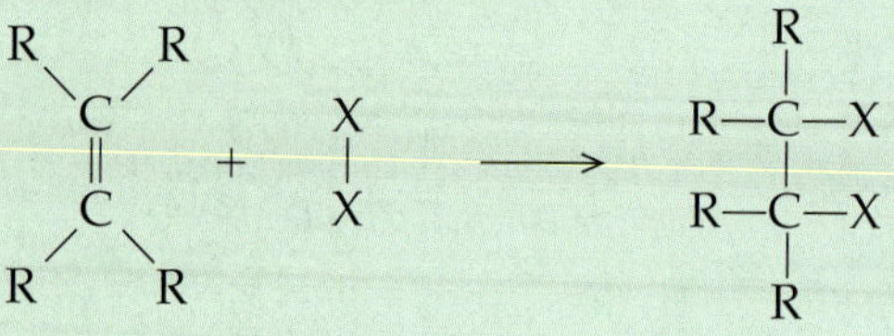

Halogenation:

$$\underset{\text{Alkene}}{\underset{R}{\overset{R}{\diagdown}}C=C\underset{R}{\overset{R}{\diagup}}} + \underset{\text{Halogen}}{\overset{X}{\underset{X}{|}}} \longrightarrow \underset{\text{Alkyl dihalide}}{\overset{R-\overset{R}{\underset{|}{C}}-X}{\underset{R-\overset{|}{\underset{R}{C}}-X}{}}}$$

Hydrohalogenation:

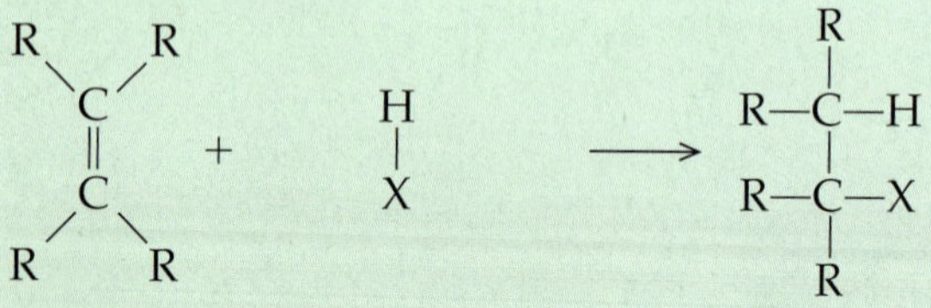

$$\underset{\text{Alkene}}{\underset{R}{\overset{R}{\diagdown}}C=C\underset{R}{\overset{R}{\diagup}}} + \underset{\text{Hydrogen halide}}{\overset{H}{\underset{X}{|}}} \longrightarrow \underset{\text{Alkyl halide}}{\overset{R-\overset{R}{\underset{|}{C}}-H}{\underset{R-\overset{|}{\underset{R}{C}}-X}{}}}$$

Addition Polymers of Alkenes

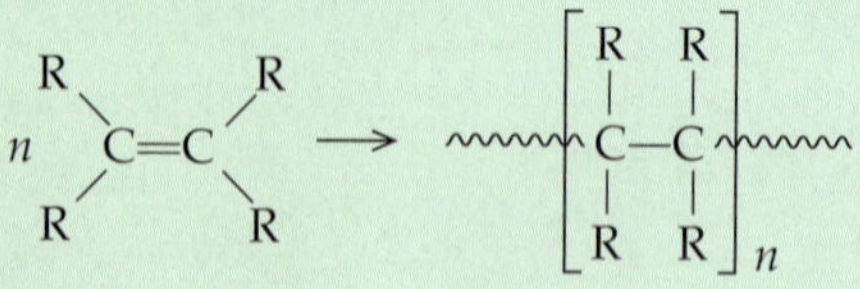

$$n\underset{R}{\overset{R}{\diagdown}}C=C\underset{R}{\overset{R}{\diagup}} \longrightarrow \sim\sim\sim\left[\overset{R}{\underset{R}{\overset{|}{C}}}-\overset{R}{\underset{R}{\overset{|}{C}}}\right]_n\sim\sim\sim$$

Alkene monomer → Addition polymer

Reactions of Benzene

Halogenation

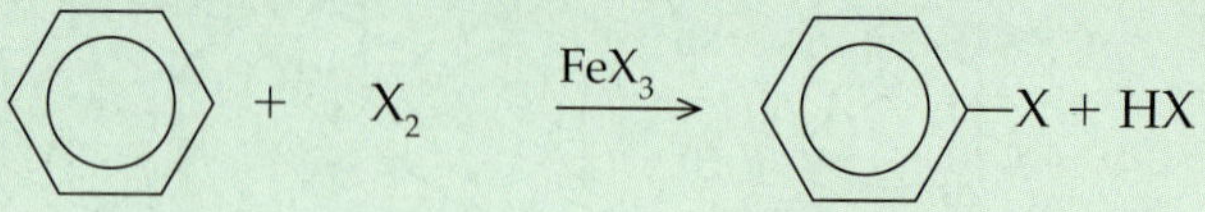

Benzene + X_2 $\xrightarrow{FeX_3}$ Halobenzene (Ph—X) + HX

Sulfonation:

Benzene + SO_3 $\xrightarrow{\text{Concentrated } H_2SO_4}$ Benzenesulfonic acid (Ph—SO_2—OH) + H_2O

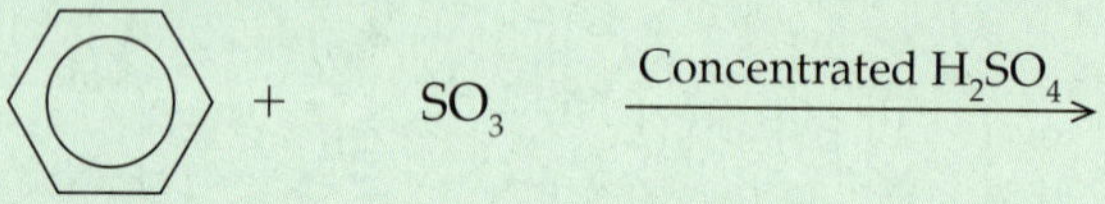

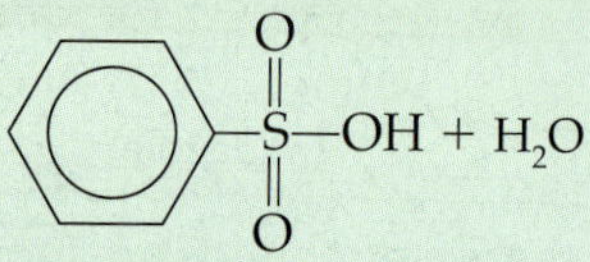

Nitration:

Benzene + HNO_3 $\xrightarrow[50-55°C]{\text{Concentrated } H_2SO_4}$ Nitrobenzene (Ph—NO_2) + H_2O

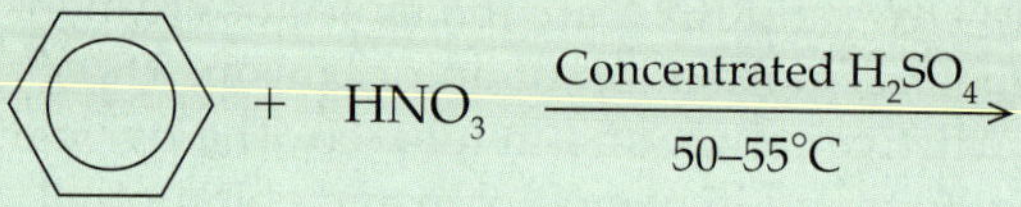

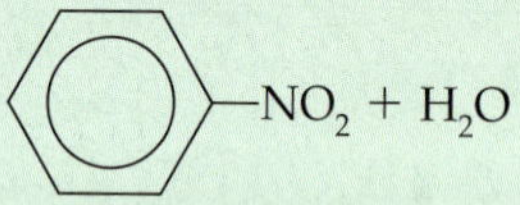

SUMMARY

11.1 Alkenes and Alkynes: Structure and Physical Properties

Alkenes and *alkynes* are *unsaturated hydrocarbons*. Alkenes are characterized by the presence of at least one carbon-carbon double bond and have the general molecular formula C_nH_{2n}. Alkynes are characterized by the presence of at least one carbon-carbon triple bond and have the general molecular formula C_nH_{2n-2}. The physical properties of the alkenes and alkynes are similar to those of alkanes, but their chemical properties are quite different.

11.2 Alkenes and Alkynes: Nomenclature

Alkenes and alkynes are named by identifying the parent compound and replacing the *-ane* ending of the alkane with *-ene* (for an alkene) or *-yne* (for an alkyne). The parent chain is numbered to give the lowest number to the first of the two carbons involved in the double bond (or triple bond). Finally, all groups are named and numbered.

11.3 Geometric Isomers: A Consequence of Unsaturation

The carbon-carbon double bond is rigid. This allows the formation of *geometric isomers,* or isomers that occur when two different groups are bonded to each carbon of the double bond. When groups are on the same side of a double bond, the prefix *cis* is used to describe the compound. When groups are on opposite sides of a double bond, the prefix *trans* is used.

11.4 Alkenes in Nature

Alkenes and polyenes (alkenes with several carbon-carbon double bonds) are common in nature. Ethene, the simplest alkene, is a plant growth substance involved in fruit ripening, senescence and leaf fall, and responses to environmental stresses. Isoprenoids, or terpenes, are polyenes built from one or more isoprene units. Isoprenoids include steroids, chlorophyll and other photosynthetic pigments, and vitamins A, D, E, and K.

11.5 Reactions Involving Alkenes

Whereas alkanes undergo *substitution reactions*, alkenes and alkynes undergo *addition reactions*. The principal addition reactions of the unsaturated hydrocarbons are *halogenation, hydration, hydrohalogenation,* and *hydrogenation. Polymers* can be made from alkenes or substituted alkenes.

11.6 Aromatic Hydrocarbons

Aromatic hydrocarbons contain benzene rings. The rings can be represented as having alternating double and single bonds. However, the most accurate representation is as a resonance hybrid between two Kekulé structures. Simple aromatic compounds are named as derivatives of benzene. Several members of this family have historical common names, such as aniline, phenol, and toluene. Polynuclear aromatic hydrocarbons consist of two or more aromatic molecules bonded together. Aromatic compounds do not undergo addition reactions. The typical reactions of benzene are *substitution* reactions: halogenation, nitration, and sulfonation.

11.7 Heterocyclic Aromatic Compounds

Heterocyclic aromatic compounds are those having at least one atom other than carbon as part of the structure of the aromatic ring. They are more similar to benzene in stability and chemical behavior than they are to the alkenes. Many of these compounds are components of molecules that have significant effects on biological systems, including DNA, RNA, hemoglobin, and nicotine.

KEY TERMS

addition polymer (11.5)
addition reaction (11.5)
alkene (11.1)
alkyne (11.1)
aromatic hydrocarbon (11.6)
geometric isomers (11.3)
halogenation (11.5)
heterocyclic aromatic compound (11.7)
hydration (11.5)
hydrogenation (11.5)
hydrohalogenation (11.5)
Markovnikov's rule (11.5)
monomer (11.5)
phenyl group (11.6)
polymer (11.5)
substitution reaction (11.6)
unsaturated hydrocarbon (Intro)

QUESTIONS AND PROBLEMS

Alkenes and Alkynes: Structure and Physical Properties

Foundations

11.27 What is the general relationship between the carbon chain length of an alkene and its boiling point?
11.28 Describe and explain the reason for the water solubility of an alkyne.
11.29 Write the general formulas for alkanes, alkenes, and alkynes.
11.30 What are the characteristic functional groups of alkenes and alkynes?

Applications

11.31 Describe the geometry of ethene.
11.32 What are the bond angles in ethene?
11.33 Compare the bond angles in ethane with those in ethene.
11.34 Explain the bond angles of ethene in terms of the valence shell electron pair repulsion (VSEPR) theory.
11.35 Describe the geometry of ethyne.
11.36 What are the bond angles in ethyne?
11.37 Compare the bond angles in ethane, ethene, and ethyne.
11.38 Explain the bond angles of ethyne in terms of the valence shell electron pair repulsion (VSEPR) theory.
11.39 Arrange the following groups of molecules from the highest to lowest boiling points:
  a. ethyne   propyne   2-pentyne
  b. 2-butene   3-decene   ethene

11.40 Arrange the following groups of molecules from the highest to the lowest melting points.

a.

b.

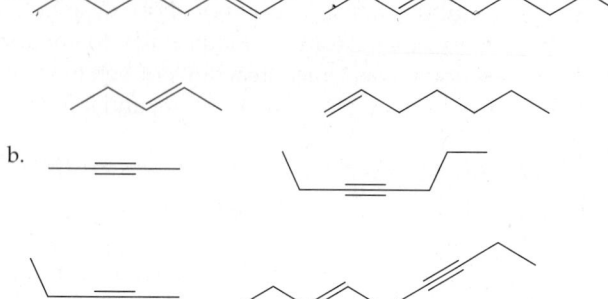

Alkenes and Alkynes: Nomenclature

Foundations
11.41 Briefly describe the rules for naming alkenes and alkynes.
11.42 What is meant by a geometric isomer?
11.43 Describe what is meant by a *cis* isomer of an alkene.
11.44 Describe what is meant by a *trans* isomer of an alkene.

Applications
11.45 Draw a condensed formula for each of the following compounds:
  a. 2-Methyl-2-hexene
  b. *trans*-3-Heptene
  c. *cis*-1-Chloro-2-pentene
  d. *cis*-2-Chloro-2-methyl-3-heptene
  e. *trans*-5-Bromo-2,6-dimethyl-3-octene

11.46 Draw a condensed formula for each of the following compounds:
  a. 2-Hexyne
  b. 4-Methyl-1-pentyne
  c. 1-Chloro-4,4,5-trimethyl-2-heptyne
  d. 2-Bromo-3-chloro-7,8-dimethyl-4-decyne

11.47 Name each of the following using the I.U.P.A.C. Nomenclature System:

a. 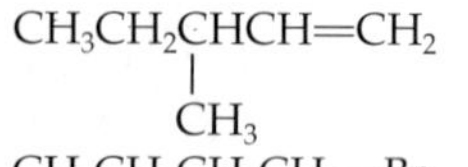 $CH_3CH_2CHCH=CH_2$
         $|$
         CH_3

b. 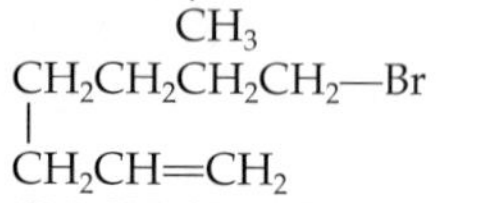 $CH_2CH_2CH_2CH_2-Br$
    $|$
    $CH_2CH=CH_2$

c. 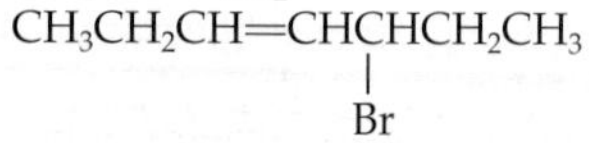 $CH_3CH_2CH=CHCHCH_2CH_3$
                  $|$
                  Br

d.

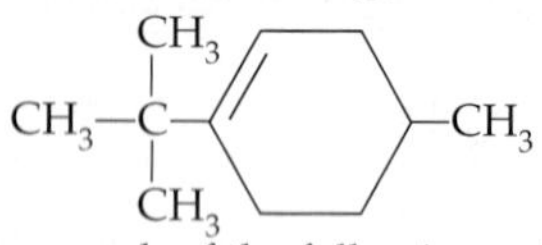

11.48 Name each of the following using the I.U.P.A.C. Nomenclature System:

a. 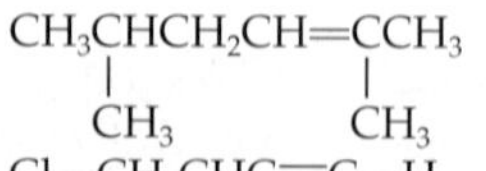 $CH_3CHCH_2CH=CCH_3$
      $|$ $|$
      CH_3 CH_3

b. 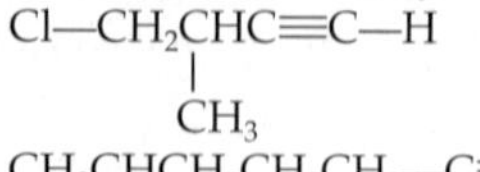 $Cl-CH_2CHC\equiv C-H$
         $|$
         CH_3

c. $CH_3CHCH_2CH_2CH_2-C\equiv C-H$
      $|$
      Cl

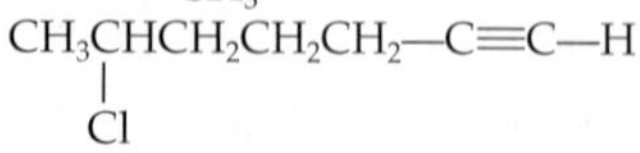

d.

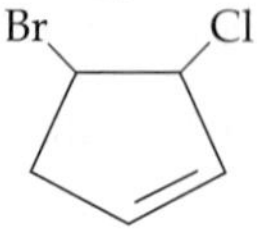

11.49 Draw each of the following compounds using condensed formulas:
  a. 1,3,5-Trifluoropentane
  b. *cis*-2-Octene
  c. Dipropylacetylene

11.50 Draw each of the following compounds using condensed formulas:
  a. 3,3,5-Trimethyl-1-hexene
  b. 1-Bromo-3-chloro-1-heptyne
  c. 3-Heptyne

11.51 Of the following compounds, which can exist as *cis-trans* geometric isomers? Draw the two geometric isomers.
  a. 2,3-Dibromobutane
  b. 2-Heptene
  c. 2,3-Dibromo-2-butene
  d. Propene

11.52 Of the following compounds, which can exist as geometric isomers?
  a. 1-Bromo-1-chloro-2-methylpropene
  b. 1,1-Dichloroethene
  c. 1,2-Dibromoethene
  d. 3-Ethyl-2-methyl-2-hexene

11.53 Which of the following alkenes would not exhibit *cis-trans* geometric isomerism?

a. 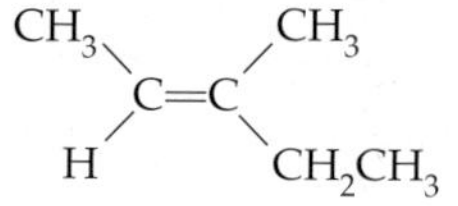 CH_3 CH_3
        $C=C$
    H CH_2CH_3

b. 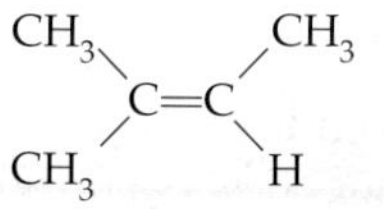 CH_3 CH_3
        $C=C$
    CH_3 H

c. CH_3CH_2 CH_2CH_3
                       $|$
                       $CHCH_3$
        $C=C$
    CH_3 $CHCH_2CH_3$
                  $|$
                  CH_3

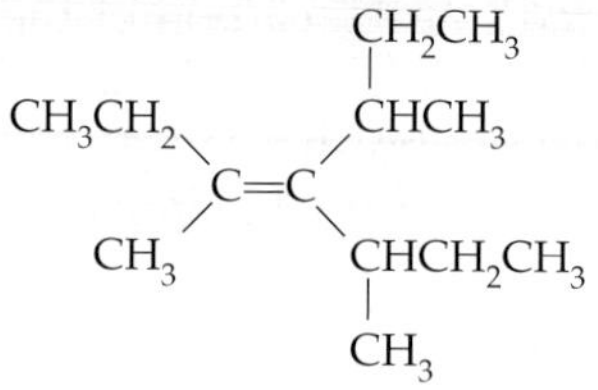

d. 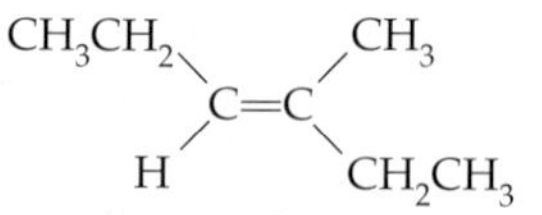 CH_3CH_2 CH_3
        $C=C$
    H CH_2CH_3

11.54 Which of the following structures have incorrect I.U.P.A.C. names? If incorrect, give the correct I.U.P.A.C. name.

a. 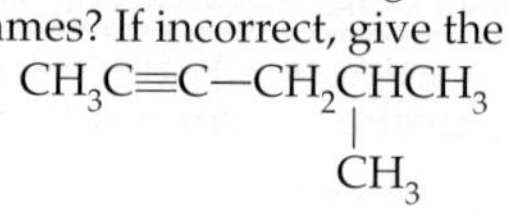 $CH_3C\equiv C-CH_2CHCH_3$
                       $|$
                       CH_3

   2-Methyl-4-hexyne

b. 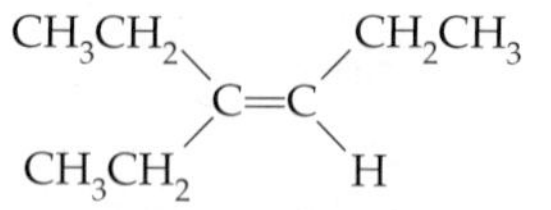 CH_3CH_2 CH_2CH_3
        $C=C$
    CH_3CH_2 H

   3-Ethyl-3-hexyne

c. 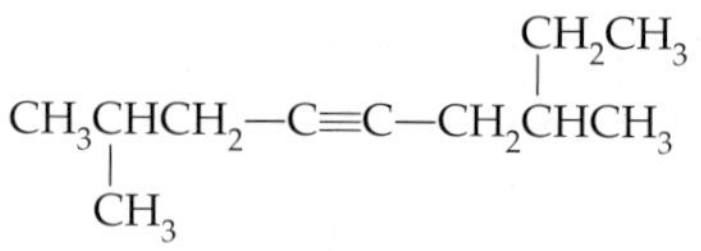 $CH_3CHCH_2-C\equiv C-CH_2CHCH_3$
      $|$ $|$
      CH_3 CH_2CH_3

   2-Ethyl-7-methyl-4-octyne

d. 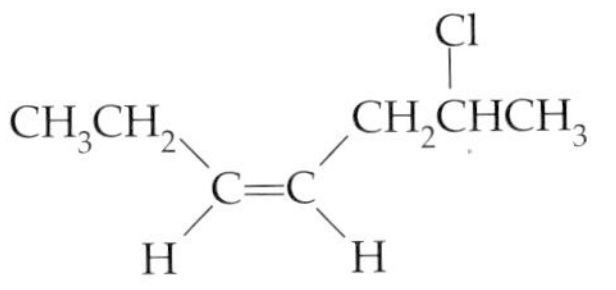 CH_3CH_2\\C=C//CH_2CHCH_3 with Cl on CHCH3, H, H

trans-6-Chloro-3-heptene

e. 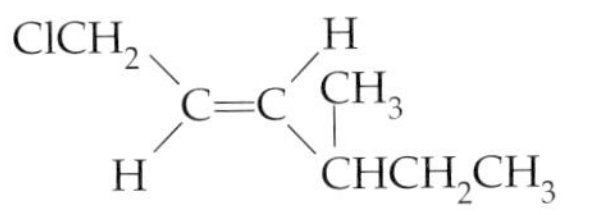 $ClCH_2$\\C=C//H, H, CH_3, $CHCH_2CH_3$

1-Chloro-5-methyl-2-hexene

11.55 Which of the following can exist as *cis* and *trans* isomers?
  a. $H_2C=CH_2$
  b. $CH_3CH=CHCH_3$
  c. $Cl_2C=CBr_2$
  d. $ClBrC=CClBr$
  e. $(CH_3)_2C=C(CH_3)_2$

11.56 Draw and name all the *cis* and *trans* isomers in Question 11.55.

11.57 Provide the I.U.P.A.C. name for each of the following molecules:
  a. $CH_2=CHCH_2CH_2CH=CHCH_2CH_2CH_3$
  b. $CH_2=CHCH_2CH=CHCH_2CH=CHCH_3$
  c. $CH_3CH=CHCH_2CH=CHCH_2CH_3$
  d. 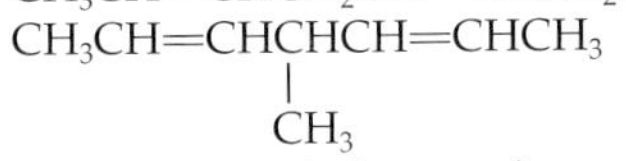 $CH_3CH=CHCHCH=CHCH_3$ with CH_3 substituent

11.58 Provide the I.U.P.A.C. name for each of the following molecules:
  a. 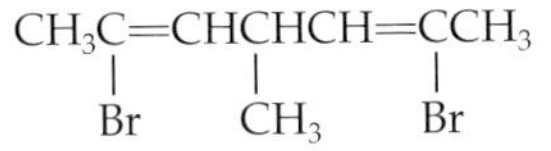 $CH_3C=CHCHCH=CCH_3$ with Br, CH_3, Br substituents
  b. 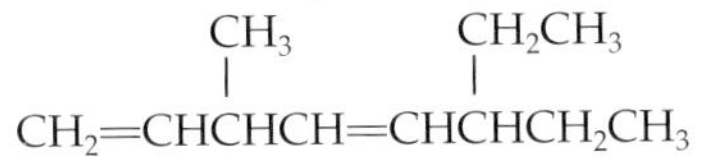 $CH_2=CHCHCH=CHCHCH_2CH_3$ with CH_3, CH_2CH_3 substituents
  c. 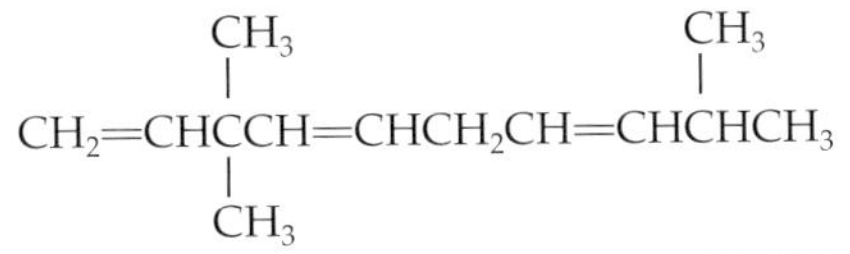 $CH_2=CHCCH=CHCH_2CH=CHCHCH_3$ with CH_3, CH_3 substituents
  d. 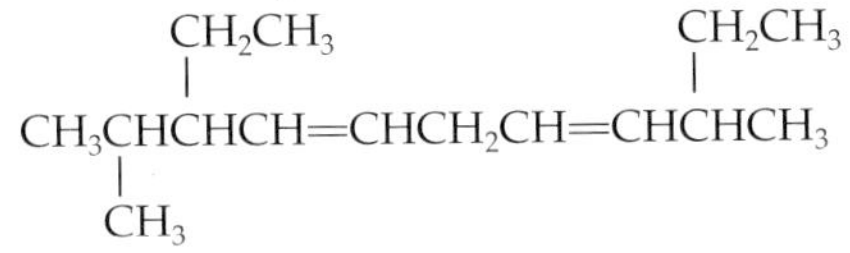 $CH_3CHCHCH=CHCH_2CH=CHCHCH_3$ with CH_2CH_3, CH_3, CH_2CH_3 substituents

Reactions Involving Alkenes and Alkynes

Foundations

11.59 Write a general equation representing the hydrogenation of an alkene.

11.60 Write a general equation representing the hydrogenation of an alkyne.

11.61 Write a general equation representing the halogenation of an alkene.

11.62 Write a general equation representing the halogenation of an alkyne.

11.63 Write a general equation representing the hydration of an alkene.

11.64 Write a general equation representing the hydration of an alkyne.

11.65 What is the principal difference between the hydrogenation of an alkene and hydrogenation of an alkyne?

11.66 What is the major difference between the hydration of an alkene and hydration of an alkyne?

Applications

11.67 How could you distinguish between a sample of cyclohexane and a sample of hexene (both C_6H_{12}) using a simple chemical test? (*Hint:* Refer to the subsection entitled "Halogenation: Addition of X_2.")

11.68 Quantitatively, 1 mol of Br_2 is consumed per mole of alkene, and 2 mol of Br_2 are consumed per mole of alkyne. How many moles of Br_2 would be consumed for 1 mol of each of the following:
  a. 2-Hexyne
  b. Cyclohexene
  c. 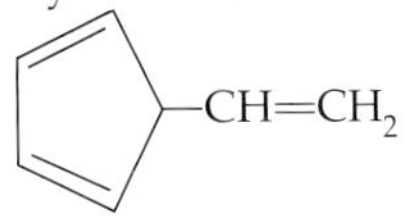 cyclopentadiene–$CH=CH_2$
  d. 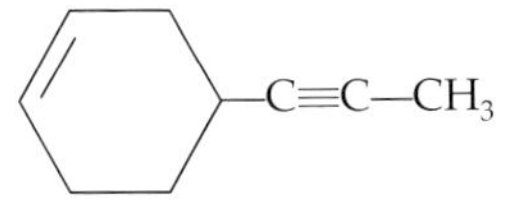 cyclohexene–$C\equiv C-CH_3$

11.69 Complete each of the following reactions by supplying the missing reactant or product(s) as indicated by question marks:
  a. $CH_3CH_2CH=CHCH_2CH_3 + ? \longrightarrow CH_3CH_2CH_2CH_2CH_2CH_3$
  b. 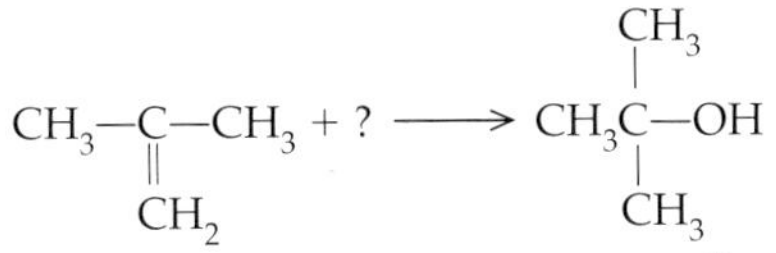 $CH_3-\underset{CH_2}{\overset{\|}{C}}-CH_3 + ? \longrightarrow CH_3\underset{CH_3}{\overset{CH_3}{\underset{|}{C}}}-OH$
  c. ? + cyclohexene $\longrightarrow$ cyclohexane with Br and H
  d. $2CH_3CH_2CH_2CH_2CH_2CH_3 + ?O_2 \xrightarrow{\text{Heat}}$? + ? (complete combustion)
  e. 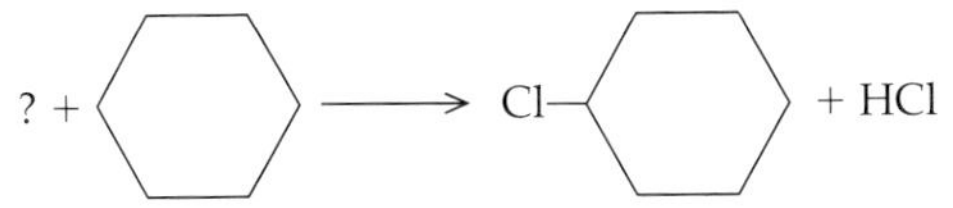 ? + benzene $\longrightarrow$ Cl–benzene + HCl
  f. 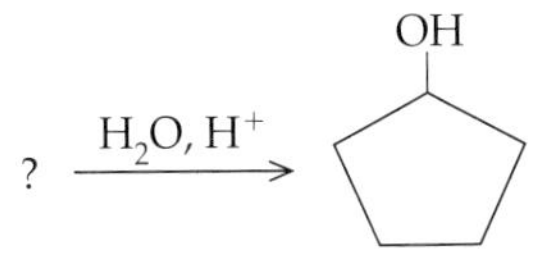 ? $\xrightarrow{H_2O, H^+}$ cyclopentanol

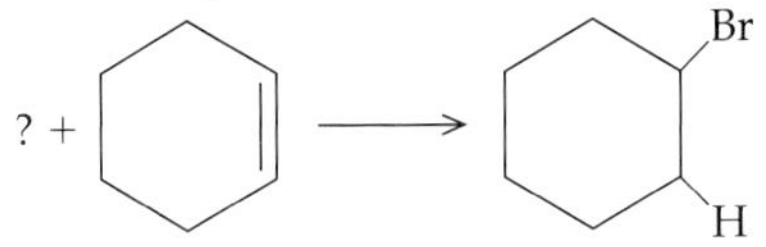

11.70 Draw and name the product in each of the following reactions:
  a. Cyclopentene + H_2O (H^+)
  b. Cyclopentene + HCl
  c. Cyclopentene + H_2
  d. Cyclopentene + HI

11.71 Write a balanced equation for each of the following reactions:
  a. Hydrogenation of 2-butyne
  b. Halogenation of 3-pentyne

11.72 Write a balanced equation for each of the following reactions:
  a. Hydration of 1-butyne
  b. Hydration of 2-butyne

11.73 A hydrocarbon with a formula C_5H_{10} decolorized Br_2 and consumed 1 mol of hydrogen upon hydrogenation. Draw all the isomers of C_5H_{10} that are possible based on the above information.

11.74 Triple bonds react in a manner analogous to that of double bonds. The extra pair of electrons in the triple bond, however, generally allows 2 mol of a given reactant to *add* to the triple bond in contrast to 1 mol with the double bond. The "rich get richer" rule holds. Predict the major product in each of the following reactions:
  a. Acetylene with 2 mol HCl
  b. Propyne with 2 mol HBr
  c. 2-Butyne with 2 mol HI

11.75 Complete each of the following by supplying the missing product indicated by the question mark:
  a. 2-Butene $\xrightarrow{HBr}$?
  b. 3-Methyl-2-hexene $\xrightarrow{HI}$?
  c. (cyclopentene) $\xrightarrow{HCl}$?

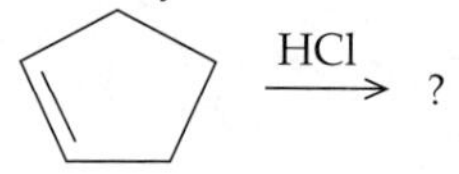

11.76 Bromine is often used as a laboratory spot test for unsaturation in an aliphatic hydrocarbon. Bromine in CCl_4 is red. When bromine reacts with an alkene or alkyne, the alkyl halide formed is colorless; hence, a disappearance of the red color is a positive test for unsaturation. A student tested the contents of two vials, A and B, both containing compounds with a molecular formula, C_6H_{12}. Vial A decolorized bromine, but vial B did not. How may the results for vial B be explained? What class of compound would account for this?

11.77 What is meant by the term *polymer*?

11.78 What is meant by the term *monomer*?

11.79 Write an equation representing the synthesis of Teflon from tetrafluoroethene. (*Hint:* Refer to Table 11.2.)

11.80 Write an equation representing the synthesis of polystyrene. (*Hint:* Refer to Table 11.2.)

11.81 Provide the I.U.P.A.C. name for each of the following molecules. Write a balanced equation for the hydration of each.
  a. $CH_3CH=CHCH_2CH_3$
  b. $CH_2CH=CH_2$
       $\ \ |$
       $\ Br$
  c. (cyclohexene with two CH₃ groups)

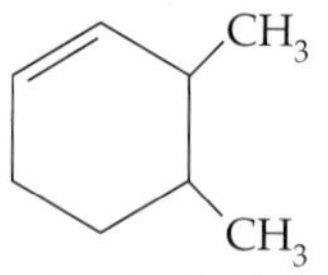

11.82 Provide the I.U.P.A.C. name for each of the following molecules. Write a balanced equation for the hydration of each.
  a. (cyclopentadiene with $-CH_2CH_3$)
  b. $CH_3CH=CHCH_2CH=CHCH_2CH=CHCH_3$
  c. $CH_3CH=CHCCH_3$
          $\ \ \ \ \ \ \ \ \ |$
          $\ \ \ \ \ \ \ CH_2CH_3$
          $\ \ \ \ \ \ \ \ \ |$
          $\ \ \ \ \ \ \ \ CH_3$

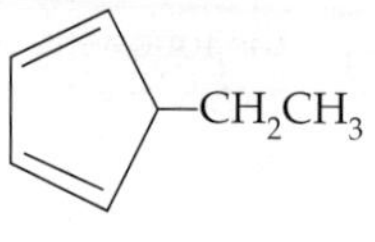

11.83 Write an equation for the addition reaction that produced each of the following molecules:
  a. $CH_3CH_2CH_2CHCH_3$
                  $\ \ \ \ \ \ \ \ \ \ |$
                  $\ \ \ \ \ \ \ \ \ CH_3$
        $\ \ |$
        OH
  b. $CH_3CH_2CHCH_2CH_2CH_3$
              $\ \ \ \ \ \ \ \ |$
              $\ \ \ \ \ \ \ Br$
  c. (bromocyclohexane with CH_3)
  d. (cyclopentane with $-CH_2CH_3$ and OH)

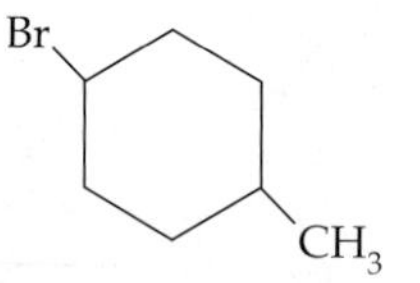

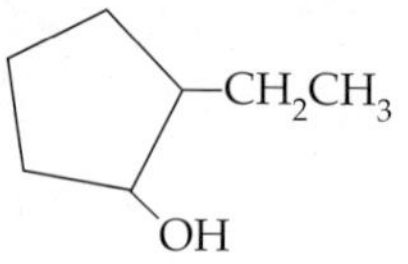

11.84 Write an equation for the addition reaction that produced each of the following molecules:
  a. $CH_3CH_2CHCHCH_3$
            $\ \ \ \ \ |\ \ \ \ \ \ \ |$
            $\ \ OH\ \ CH_2CH_3$
  b. $CH_3CHCH_2CH_3$
           $\ \ \ \ |$
           $\ \ OH$
  c. (cyclohexane with HO, CH_3, CH_3)

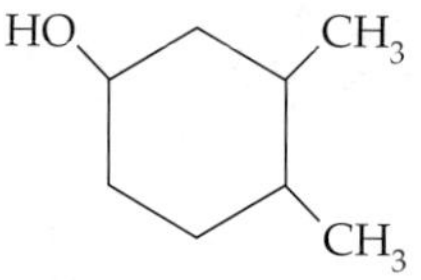

11.85 Draw the structure of each of the following compounds and write a balanced equation for the complete hydrogenation of each:
  a. 1,4-Hexadiene
  b. 2,4,6-Octatriene
  c. 1,3 Cyclohexadiene
  d. 1,3,5-Cyclooctatriene

11.86 Draw the structure of each of the following compounds and write a balanced equation for the bromination of each:
  a. 3-Methyl-1,4-hexadiene
  b. 4-Bromo-1,3-pentadiene
  c. 3-Chloro-2,4-hexadiene
  d. 3-Bromo-1,3-Cyclohexadiene

Aromatic Hydrocarbons

Foundations

11.87 Where did the term *aromatic hydrocarbon* first originate?

11.88 What chemical characteristic of the aromatic hydrocarbons is most distinctive?

11.89 What is meant by the term *resonance hybrid*?

11.90 Draw a pair of structures to represent the benzene resonance hybrid.

Applications

11.91 Draw the structure for each of the following compounds:
  a. 2,4-Dibromotoluene
  b. 1,2,4-Triethylbenzene
  c. Isopropylbenzene
  d. 2-Bromo-5-chlorotoluene

11.92 Name each of the following compounds, using the I.U.P.A.C. system.

a.

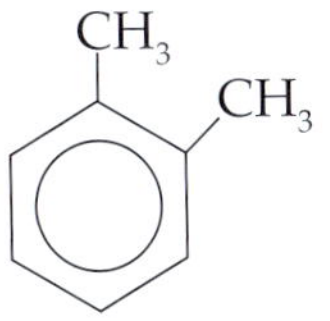
CH3
CH3

b.

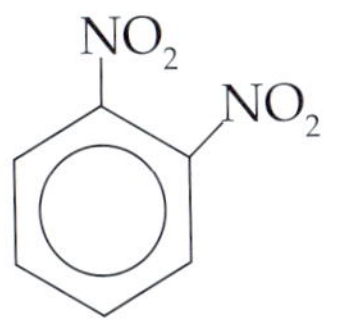
NO2
NO2

c.

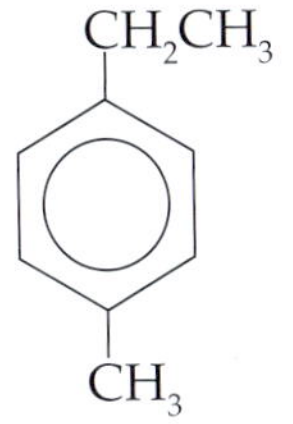
CH2CH3
CH3

d.

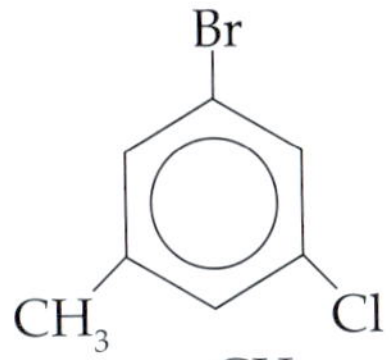
Br
CH3
Cl

e.

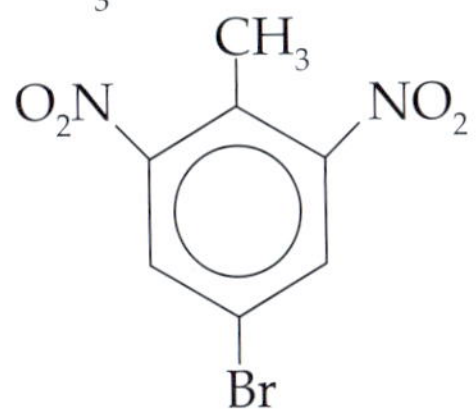
CH3
O2N
NO2
Br

11.93 Draw each of the following compounds, using condensed formulas:

a. *meta*-Cresol
b. Propylbenzene
c. 1,3,5-Trinitrobenzene
d. *m*-Chlorotoluene

11.94 Draw each of the following compounds, using condensed formulas:

a. *p*-Xylene
b. Isopropylbenzene
c. *m*-Nitroanisole
d. *p*-Methylbenzaldehyde

11.95 Describe the Kekulé model for the structure of benzene.

11.96 Describe the current model for the structure of benzene.

11.97 How does a substitution reaction differ from an addition reaction?

11.98 Give an example of a substitution reaction and of an addition reaction.

11.99 Write an equation showing the reaction of benzene with Cl_2 and $FeCl_3$.

11.100 Write an equation showing the reaction of benzene with SO_3. Be sure to note the catalyst required.

Heterocyclic Aromatic Compounds

11.101 Draw the general structure of a pyrimidine.

11.102 What biological molecules contain pyrimidine rings?

11.103 Draw the general structure of a purine.

11.104 What biological molecules contain purine rings?

CRITICAL THINKING PROBLEMS

1. There is a plastic polymer called polyvinylidene difluoride (PVDF) that can be used to sense a baby's breath and thus be used to prevent sudden infant death syndrome (SIDS). The secret is that this polymer can be specially processed so that it becomes piezoelectric (produces an electrical current when it is physically deformed) and pyroelectric (develops an electrical potential when its temperature changes). When a PVDF film is placed beside a sleeping baby, it will set off an alarm if the baby stops breathing. The structure of this polymer is shown here:

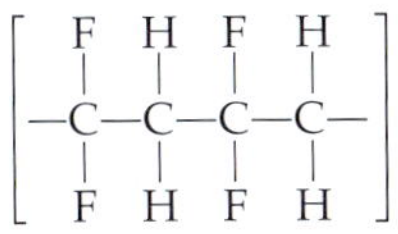
F H F H
C C C C
F H F H

Go to the library and investigate some of the other amazing uses of PVDF. Draw the structure of the alkene from which this compound is produced.

2. Isoprene is the repeating unit of the natural polymer rubber. It is also the starting material for the synthesis of cholesterol and several of the lipid-soluble vitamins, including vitamin A and vitamin K. The structure of isoprene is seen below.

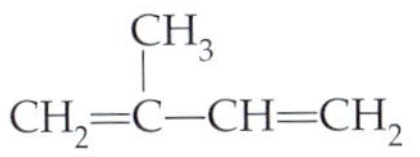
CH3
CH2=C—CH=CH2

What is the I.U.P.A.C. name for isoprene?

3. When polyacrylonitrile is burned, toxic gases are released. In fact, in airplane fires, more passengers die from inhalation of toxic fumes than from burns. Refer to Table 11.2 for the structure of acrylonitrile. What toxic gas would you predict to be the product of the combustion of these polymers?
4. If a molecule of polystyrene consists of 25,000 monomers, what is the molar mass of the molecule?
5. A factory produces one million tons of polypropylene. How many moles of propene would be required to produce this amount? What is the volume of this amount of propene at 25°C and 1 atm?

ORGANIC CHEMISTRY

12

Alcohols, Phenols, Thiols, and Ethers

Sugar-free jelly beans

Learning Goals

1. Rank selected alcohols by relative water solubility, boiling points, or melting points.
2. Write the names and draw the structures for common alcohols.
3. Discuss the biological, medical, or environmental significance of several alcohols.
4. Classify alcohols as primary, secondary, or tertiary.
5. Write equations representing the preparation of alcohols by the hydration of an alkene.
6. Write equations representing the preparation of alcohols by hydrogenation (reduction) of aldehydes or ketones.
7. Write equations showing the dehydration of an alcohol.
8. Write equations representing the oxidation of alcohols.
9. Discuss the role of oxidation and reduction reactions in the chemistry of living systems.
10. Discuss the use of phenols as germicides.
11. Write names and draw structures for common ethers and discuss their use in medicine.
12. Write equations representing the dehydration reaction between two alcohol molecules.
13. Write names and draw structures for simple thiols and discuss their biological significance.

Outline

Chemistry Connection:
Polyols for the Sweet Tooth

12.1 Alcohols: Structure and Physical Properties
12.2 Alcohols: Nomenclature
12.3 Medically Important Alcohols

A Medical Perspective:
Fetal Alcohol Syndrome

12.4 Classification of Alcohols

12.5 Reactions Involving Alcohols
12.6 Oxidation and Reduction in Living Systems

A Human Perspective:
Alcohol Consumption and the Breathalyzer Test

12.7 Phenols
12.8 Ethers
12.9 Thiols

381

Chemistry Connection

Polyols for the Sweet Tooth

Do you crave sweets, but worry about the empty calories in sugary treats? If so, you are not alone. Research tells us that, even as babies, we demonstrate preference for sweet tastes over all others. But there are many reasons to reduce our intake of refined sugars, in particular sucrose or table sugar. Too many people eat high-calorie, low-nutrition snacks rather than more nutritious foods. This can lead to obesity, a problem that is very common in our society. In addition, sucrose is responsible for tooth decay. Lactic acid, one of the products of the metabolism of sucrose by bacteria on our teeth, dissolves the tooth enamel, which results in a cavity. For those with diabetes, glucose intolerance, or hypoglycemia, sucrose in the diet makes it difficult to maintain a constant blood sugar level.

The food chemistry industry has invested billions of dollars in the synthesis of sugar substitutes. We recognize names such as aspartame (Equal or Nutrasweet), saccharin (Sweet & Low), and SPLENDA®, because they are the most common non-nutritive sweeteners worldwide. We also buy products, including candies, soft drinks, and gums, which are advertised to be "sugar-free." But you might be surprised to find that many of these products are not free of calories. A check of the nutritional label may reveal that these products contain sorbitol, mannitol, or one or more other members of a class of compounds called sugar alcohols, or *polyols* (*poly*—many; *ols*—alcohol or hydroxyl groups).

Sugar alcohols are found in many foods, including fruits, vegetables, and mushrooms. Others are made by hydrogenation or fermentation of carbohydrates from wheat or corn. But all are natural products. Compared to sucrose, they range in sweetness from about half to nearly the same; they also have fewer calories per gram than sucrose (about one-third to one-half the calories). Polyols also cause a cooling sensation in the mouth. This cooling is caused by a negative heat of solution (they must absorb heat from the surroundings in order to dissolve) and is used to advantage in breath freshening mints and gums.

Sorbitol, the most commonly used sugar alcohol, is about 0.6 times the sweetness of sucrose. While sucrose contains four calories per gram, sorbitol is only about 2.6 C/g. Discovered in 1872 in the berries of Mountain Ash trees, sorbitol has a smooth mouthfeel, which makes it ideal as a texturizing agent in foods. It also has a pleasant, cool, sweet flavor and acts as a humectant, keeping foods from losing moisture. No acceptable daily intake (ADI) has been specified for sorbitol, which is an indication that it is considered to be a very safe food additive. However, it has been observed that ingestion of more than 50–80 g/day may have a laxative effect.

Mannitol, a structural isomer of sorbitol, is found naturally in asparagus, olives, pineapple, and carrots. For use in the food industry, it is extracted from seaweed. Mannitol has about 0.7 times the sweetness of sucrose and only about 1.6 C/g. As for sorbitol, no ADI has been specified; but ingestion of more than 20 g/day may cause diarrhea and bloating.

Xylitol was discovered in 1891 and has been used as a sweetener since the 1960s. Found in many fruits and vegetables, it has about the same sweetness as sucrose, but only one-third of the caloric value. Its high cooling effect, as well as sweetness, contribute to its popularity as a sweetening agent in hard candies and gums, as well as in oral health products. In fact, extensive studies suggest that use of xylitol-sweetened gum (7–10 g of xylitol per day) between meals results in a 30–60% decrease in dental cavities.

While polyols give us the sweetness that we enjoy without all of the calories, cavities, and blood sugar peaks of sucrose, they are not without a negative side. Some studies have reported weight gain by individuals who overeat these "sugar-free" foods. The American Diabetes Association has reported that these foods are "acceptable in moderate amounts but should not be eaten in excess." In fact, some diabetics have suffered elevated blood sugar after overeating foods containing polyols. Finally, as we noted above, when ingested in excess, sugar alcohols may cause bloating and diarrhea.

The use of these natural products continues to be investigated by the food and pharmaceutical industries. As we learn more about them, sugar alcohols continue to be versatile food additives. Using them in moderation, we can enjoy the benefits that they confer, without suffering uncomfortable side effects.

$$\begin{array}{ccc}
\text{CH}_2\text{OH} & \text{CH}_2\text{OH} & \text{CH}_2\text{OH} \\
| & | & | \\
\text{H}-\text{C}-\text{OH} & \text{HO}-\text{C}-\text{H} & \text{H}-\text{C}-\text{OH} \\
| & | & | \\
\text{HO}-\text{C}-\text{H} & \text{HO}-\text{C}-\text{H} & \text{HO}-\text{C}-\text{H} \\
| & | & | \\
\text{H}-\text{C}-\text{OH} & \text{H}-\text{C}-\text{OH} & \text{H}-\text{C}-\text{OH} \\
| & | & | \\
\text{H}-\text{C}-\text{OH} & \text{H}-\text{C}-\text{OH} & \text{CH}_2\text{OH} \\
| & | & \\
\text{CH}_2\text{OH} & \text{CH}_2\text{OH} & \\
\text{Sorbitol} & \text{Mannitol} & \text{Xylitol}
\end{array}$$

Structures of three of the sugar alcohols used in the food industry.

Introduction

The characteristic functional group of the alcohols and phenols is the hydroxyl group (—OH). Alcohols have the general structure R—OH, in which R is any alkyl group. Phenols are similar in structure but contain an aryl group in place of the alkyl group. Both can be viewed as substituted water molecules in which one of the hydrogen atoms has been replaced by an alkyl or aryl group.

> An aryl group is an aromatic ring with one hydrogen atom removed.

General formulas: Alcohol (R—O—H), Phenol (C$_6$H$_5$—O—H)

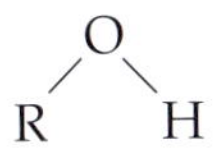

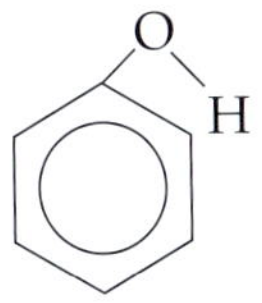

Example: Methanol (methyl alcohol) (CH$_3$—O—H)

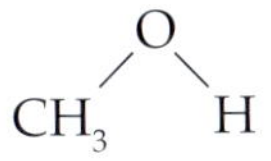

Ethers have two alkyl or aryl groups attached to the oxygen atom and may be thought of as substituted alcohols. The functional group characteristic of an ether is R—O—R. Thiols are a family of compounds that contain the sulfhydryl group (—SH). They, too, have a structure similar to that of alcohols.

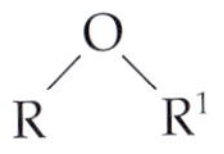

Ethers (R—O—R^1), Methoxymethane (dimethyl ether) (CH$_3$—O—CH$_3$), Thiol (R—SH), Methanethiol (CH$_3$—SH)

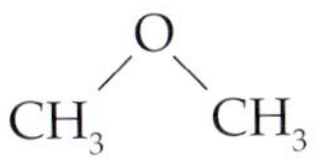

> R and R^1 = alkyl or aryl group

Many important biological molecules, including sugars (carbohydrates), fats (lipids), and proteins, contain hydroxyl and/or thiol groups.

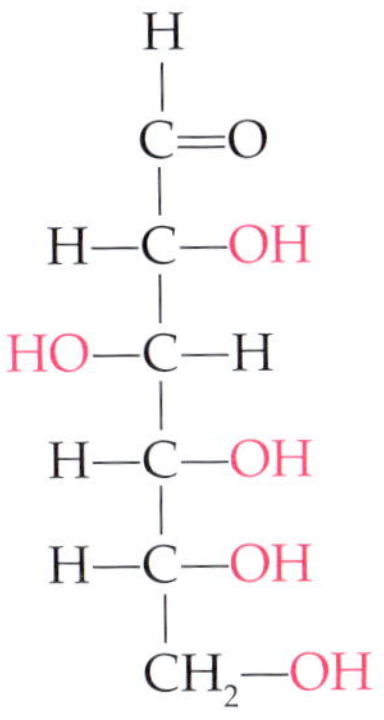

D-Glucose, *a sugar*

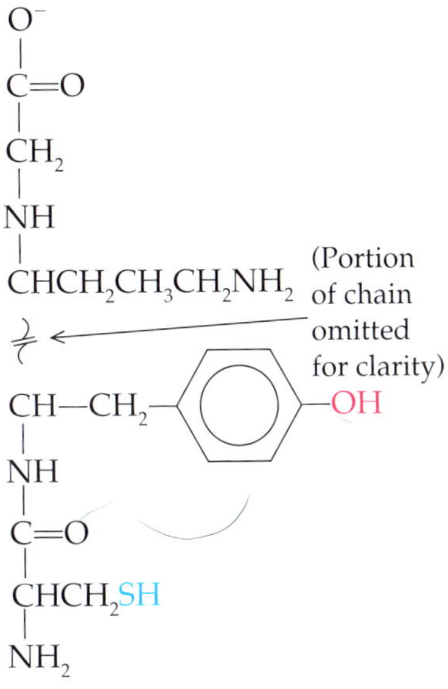

Lysine vasopressin (partial structure), *a protein*

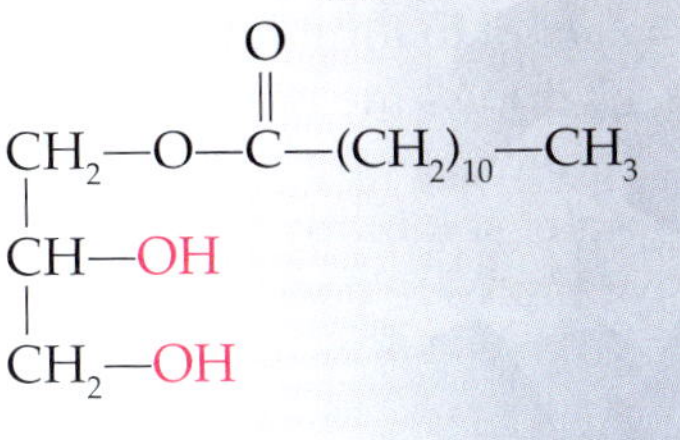

Monolaurin, *a lipid*

In biological systems, the hydroxyl group is often involved in a variety of reactions such as oxidation, reduction, hydration, and dehydration. In glycolysis (a metabolic pathway by which glucose is degraded and energy is harvested in the form of ATP), several steps center on the reactivity of the hydroxyl group. The majority of the consumable alcohol in the world (ethanol) is produced by fermentation reactions carried out by yeasts.

> Glycolysis and fermentation are discussed in Chapter 21.

The thiol group is found in the structure of some amino acids and is essential for keeping proteins in the proper three-dimensional shape required for their biological function. Thus, these functional groups play a central role in the structure and chemical properties of biological molecules. The thiol group of the amino acid cysteine is highlighted in blue in the structure of lysine vasopressin presented above.

12.1 Alcohols: Structure and Physical Properties

LEARNING GOAL 1

An **alcohol** is an organic compound that contains a **hydroxyl group** (—OH) attached to an alkyl group (Figure 12.1). The R—O—H portion of an alcohol is similar to the structure of water. The oxygen and the two atoms bonded to it lie in the same plane, and the R—O—H bond angle is approximately 104°, which is very similar to the H—O—H bond angle of water.

The hydroxyl groups of alcohols are very polar because the oxygen and hydrogen atoms have significantly different electronegativities. Because the two atoms involved in this polar bond are oxygen and hydrogen, hydrogen bonds can form between alcohol molecules (Figure 12.2).

As a result of this intermolecular hydrogen bonding, alcohols boil at much higher temperatures than hydrocarbons of similar molecular weight. These higher boiling points are caused by the large amount of heat needed to break the hydrogen bonds that attract the alcohol molecules to one another. Compare the boiling points of butane and propanol, which have similar molecular weights:

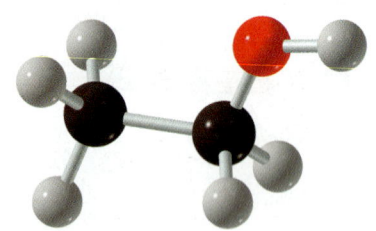

Figure 12.1
Ball-and-stick model of the simple alcohol ethanol.

Electronegativity is discussed in Section 3.1. Hydrogen bonding is described in detail in Section 5.2.

$CH_3CH_2CH_2CH_3$	$CH_3CH_2CH_2OH$
Butane	1-Propanol
M.W. = 58	M.W. = 60
b.p. = −0.5°C	b.p. = 97.2°C

Alcohols of one to four carbon atoms are very soluble in water, and those with five or six carbons are moderately soluble in water. This is due to the ability of the alcohol to form intermolecular hydrogen bonds with water molecules (see Figure 12.2b). As the nonpolar, or hydrophobic, portion of an alcohol (the carbon chain) becomes larger relative to the polar, hydrophilic, region (the hydroxyl group), the water solubility of an alcohol decreases. As a result alcohols of seven carbon atoms or more are nearly insoluble in water. The term *hydrophobic*, which literally means

Figure 12.2
(a) Hydrogen bonding between alcohol molecules. (b) Hydrogen bonding between alcohol molecules and water molecules.

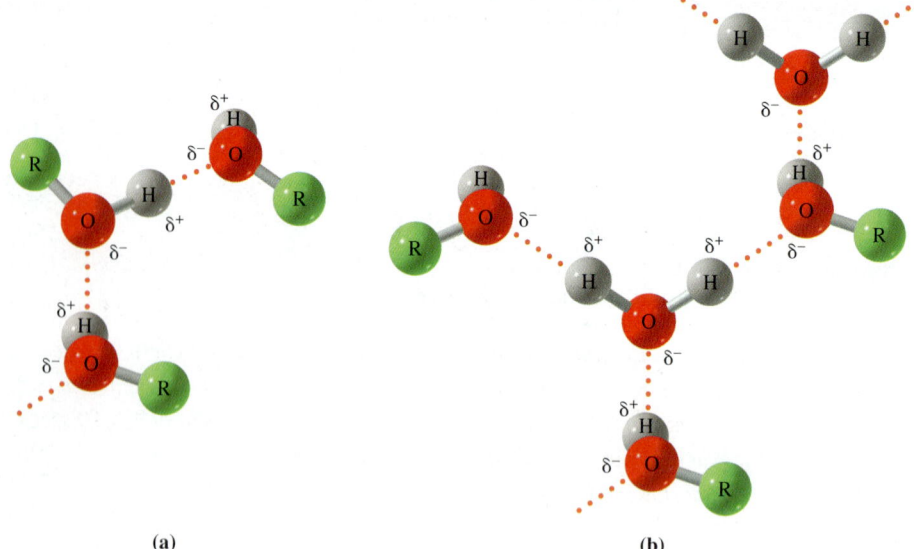

(a)   (b)

"water fearing," is used to describe a molecule or a region of a molecule that is nonpolar and, thus, more soluble in nonpolar solvents than in water. Similarly, the term *hydrophilic,* meaning water loving, is used to describe a polar molecule or region of a molecule that is more soluble in the polar solvent water than in a nonpolar solvent.

An increase in the number of hydroxyl groups along a carbon chain will increase the influence of the polar hydroxyl group. It follows, then that diols and triols are more water soluble than alcohols with only a single hydroxyl group.

The presence of polar hydroxyl groups in large biological molecules—for instance, proteins and nucleic acids—allows intramolecular hydrogen bonding that keeps these molecules in the shapes needed for biological function.

Intermolecular hydrogen bonds are attractive forces between two molecules. *Intramolecular* hydrogen bonds are attractive forces between polar groups within the same molecule.

12.2 Alcohols: Nomenclature

I.U.P.A.C. Names

In the I.U.P.A.C. Nomenclature System, alcohols are named according to the following steps:

- Determine the name of the *parent compound,* the longest continuous carbon chain containing the —OH group.
- Replace the *-e* ending of the alkane chain with the *-ol* ending of the alcohol. Following this pattern, an alkane becomes an alkanol. For instance, etha*ne* becomes etha*nol*, and propa*ne* becomes propa*nol*.
- Number the parent chain to give the carbon bearing the hydroxyl group the lowest possible number.
- Name and number all substituents, and add them as prefixes to the "alkanol" name.
- Alcohols containing two hydroxyl groups are named *-diols*. Those bearing three hydroxyl groups are called *-triols*. A number giving the position of each of the hydroxyl groups is needed in these cases.

LEARNING GOAL 2

The way to determine the parent compound was described in Section 10.2.

Using I.U.P.A.C. Nomenclature to Name an Alcohol

EXAMPLE 12.1

Name the following alcohol using I.U.P.A.C. nomenclature.

LEARNING GOAL 2

Solution

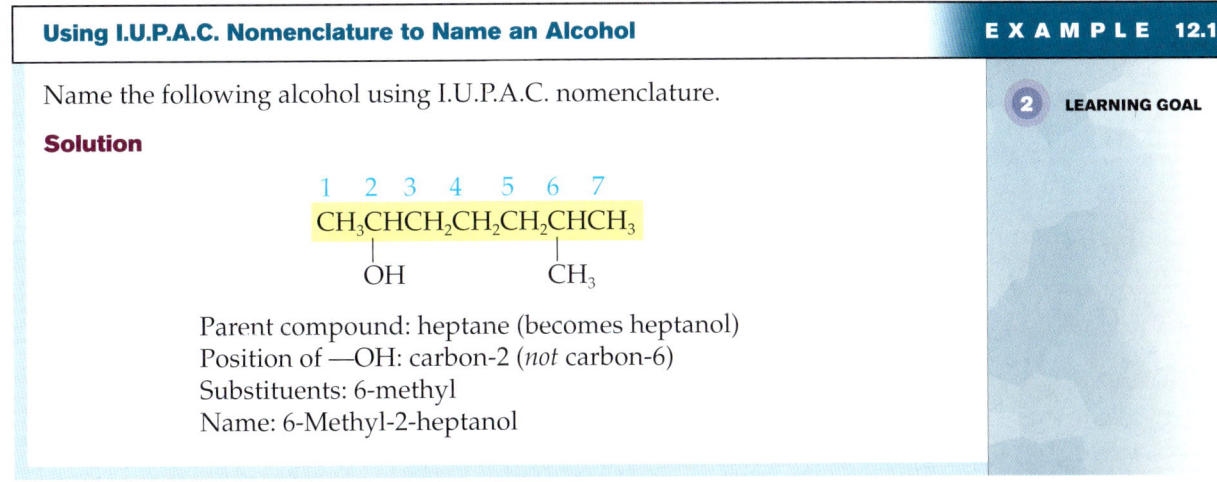

Parent compound: heptane (becomes heptanol)
Position of —OH: carbon-2 (*not* carbon-6)
Substituents: 6-methyl
Name: 6-Methyl-2-heptanol

Using I.U.P.A.C. Nomenclature to Name Alcohols

EXAMPLE 12.2

Name the following cyclic alcohol using I.U.P.A.C. nomenclature.

LEARNING GOAL 2

Continued—

EXAMPLE 12.2 —Continued

Solution

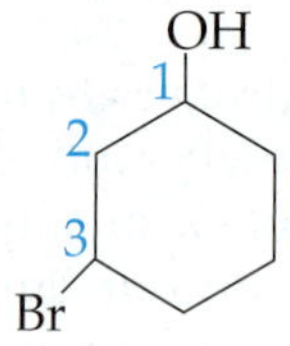

[Structure: cyclohexane ring with OH on carbon-1 and Br on carbon-3, numbered 1, 2, 3]

Remember that this line structure represents a cyclic molecule composed of six carbon atoms and associated hydrogen atoms, as follows:

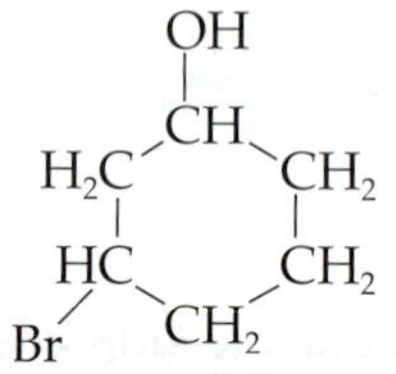

[Expanded cyclohexane structure showing CH(OH), H₂C, CH₂, HC(Br), CH₂, CH₂]

Parent compound: cyclohexane (becomes cyclohexanol)
Position of —OH: carbon-1 (*not* carbon-3)
Substituents: 3-bromo (*not* 5-bromo)
Name: 3-Bromocyclohexanol (it is assumed that the
—OH is on carbon-1 in cyclic structures)

Common Names

See Section 10.2 for the names of the common alkyl groups.

The common names for alcohols are derived from the alkyl group corresponding to the parent compound. The name of the alkyl group is followed by the word *alcohol*. For some alcohols, such as ethylene glycol and glycerol, historical names are used. The following examples provide the I.U.P.A.C. and common names of several alcohols:

CH_3CHCH_3 $HOCH_2CH_2OH$ CH_3CH_2OH
$\quad|$
$\quad OH$

2-Propanol   1,2-Ethanediol   Ethanol
(isopropyl   (ethylene glycol)   (ethyl alcohol)
alcohol)   (grain alcohol)

Question 12.1

Use the I.U.P.A.C. Nomenclature System to name each of the following compounds.

a. $CH_3CHCH_2CH_2CH_2OH$
$\quad\quad|$
$\quad\quad CH_3$

b. $CH_3CHCH_2CHCH_3$
$\quad\quad|\quad\quad\quad|$
$\quad\quad OH\quad\; CH_2CH_3$

c. $CH_2—CH—CH_2$
$\quad\;|\quad\quad\;|\quad\quad\;|$
$\;\;OH\;\;OH\;\;OH$
(Common name: Glycerol)

d. $CH_3CH_2CH—CHCH_2CH_2OH$
$\quad\quad\quad\quad\quad|\quad\quad\;|$
$\quad\quad\quad\quad\;\;Cl\;\;\;CH_3$

Question 12.2

Give the common name and the I.U.P.A.C. name for each of the following compounds.

a. CH₃CH₂CH₂CH₂CH₂CH₂CH₂OH

b. CH₃CHCH₃
     |
     OH

c.   CH₃
       |
   CH₃CHCH₂OH

12.3 Medically Important Alcohols

Methanol

Methanol (methyl alcohol), CH₃OH, is a colorless and odorless liquid that is used as a solvent and as the starting material for the synthesis of methanal (formaldehyde). Methanol is often called *wood alcohol* because it can be made by heating wood in the absence of air. In fact, ancient Egyptians produced methanol by this process and, mixed with other substances, used it for embalming. It was not until 1661 that Robert Boyle first isolated pure methanol, which he called *spirit of box*, because he purified it by distillation from boxwood. Methanol is toxic and can cause blindness and perhaps death if ingested. Methanol may also be used as fuel, especially for "formula" racing cars.

Ethanol

Ethanol (ethyl alcohol), CH₃CH₂OH, is a colorless and odorless liquid and is the alcohol in alcoholic beverages. It is also widely used as a solvent and as a raw material for the preparation of other organic chemicals.

The ethanol used in alcoholic beverages comes from the **fermentation** of carbohydrates (sugars and starches). The beverage produced depends on the starting material and the fermentation process: scotch (grain), bourbon (corn), burgundy wine (grapes and grape skins), and chablis wine (grapes without red skins) (Figure 12.3). The following equation summarizes the fermentation process:

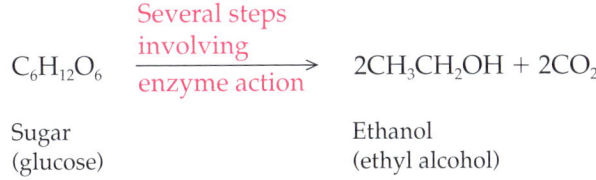

C₆H₁₂O₆ →(Several steps involving enzyme action)→ 2CH₃CH₂OH + 2CO₂

Sugar (glucose)   Ethanol (ethyl alcohol)

The alcoholic beverages listed have quite different alcohol concentrations. Wines are generally 12–13% alcohol because the yeasts that produce the ethanol are killed by ethanol concentrations of 12–13%. To produce bourbon or scotch with an alcohol concentration of 40–45% ethanol (80 or 90 proof), the original fermentation products must be distilled.

The sale and use of pure ethanol (100% ethanol) are regulated by the federal government. To prevent illegal use of pure ethanol, it is *denatured* by the addition of a denaturing agent, which makes it unfit to drink but suitable for many laboratory applications.

Figure 12.3
Champagne, a sparkling wine, results when fermentation is carried out in a sealed bottle. Under these conditions, the CO₂ produced during fermentation is trapped in the wine.

Fermentation reactions are described in detail in Section 21.4 and in A Human Perspective: Fermentations: The Good, the Bad, and the Ugly.

Distillation is the separation of compounds in a mixture based on differences in boiling points.

A Medical Perspective

Fetal Alcohol Syndrome

The first months of pregnancy are a time of great joy and anticipation but are not without moments of anxiety. On her first visit to the obstetrician, the mother-to-be is tested for previous exposure to a number of infectious diseases that could damage the fetus. She is provided with information about diet, weight gain, and drugs that could harm the baby. Among the drugs that should be avoided are alcoholic beverages.

The use of alcoholic beverages by a pregnant woman can cause *fetal alcohol syndrome (FAS)*. A *syndrome* is a set of symptoms that occur together and are characteristic of a particular disease. In this case, physicians have observed that infants born to women with chronic alcoholism showed a reproducible set of abnormalities including mental retardation, poor growth before and after birth, and facial malformations.

Mothers who report only social drinking may have children with *fetal alcohol effects*, a less severe form of fetal alcohol syndrome. This milder form is characterized by a reduced birth weight, some learning disabilities, and behavioral problems.

How does alcohol consumption cause these varied symptoms? No one is exactly sure, but it is well known that the alcohol consumed by the mother crosses the placenta and enters the bloodstream of the fetus. Within about fifteen minutes, the concentration of alcohol in the blood of the fetus is as high as that of the mother! However, the mother has enzymes to detoxify the alcohol in her blood; the fetus does not. Now consider that alcohol can cause cell division to stop or be radically altered. It is thought that even a single night on the town could be enough to cause FAS by blocking cell division during a critical developmental period.

This raises the question "How much alcohol can a pregnant woman safely drink?" As we have seen, the severity of the symptoms seems to increase with the amount of alcohol consumed by the mother. However, it is virtually impossible to do the scientific studies that would conclusively determine the risk to the fetus caused by different amounts of alcohol. There is some evidence that suggests that there is a risk associated with drinking even one ounce of absolute (100%) alcohol each day. Because of these facts and uncertainties, the American Medical Association and the U.S. Surgeon General recommend that pregnant women completely abstain from alcohol.

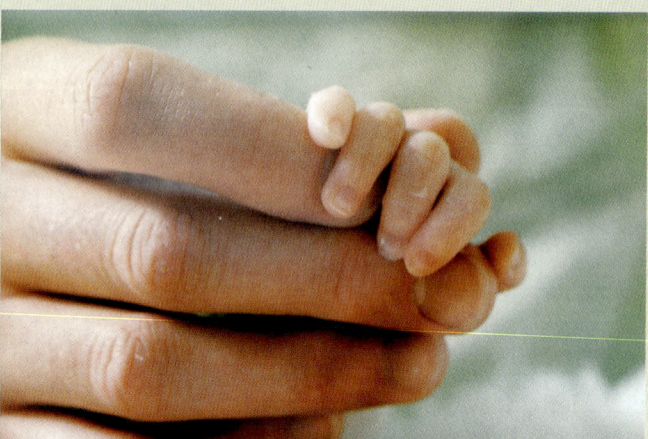

The American Medical Association recommends abstaining from alcohol during pregnancy.

For Further Understanding

In July 2004, the U.S. Centers for Disease Control issued a new document entitled *Fetal Alcohol Syndrome: Guidelines for Referral and Diagnosis*, which can be found at the following Web address: http://www.cdc.gov/ncbddd/fas/documents/FAS_guidelines_accessible.pdf. Refer to this document when answering the following questions.

What is the estimate of the number of babies born each year with fetal alcohol syndrome and why is a more accurate number so difficult to determine?

What are some of the issues, practical and ethical, involved in intervention efforts to prevent fetal alcohol syndrome?

2-Propanol

2-Propanol (isopropyl alcohol),

$$\underset{\underset{\text{OH}}{|}}{\text{CH}_3\text{CHCH}_3}$$

was commonly called *rubbing alcohol* because patients with high fevers were often given alcohol baths to reduce body temperature. Rapid evaporation of the alcohol results in skin cooling. This practice is no longer commonly used.

It is also used as a disinfectant (Figure 12.4), an astringent (skin-drying agent), an industrial solvent, and a raw material in the synthesis of organic chemicals. It is colorless, has a very slight odor, and is toxic when ingested.

1,2-Ethanediol

1,2-Ethanediol (ethylene glycol),

$$\begin{array}{c} CH_2-CH_2 \\ |\quad\ \ | \\ OH\ \ OH \end{array}$$

is used as automobile antifreeze. When added to water in the radiator, the ethylene glycol solute lowers the freezing point and raises the boiling point of the water. Ethylene glycol has a sweet taste but is extremely poisonous. For this reason, color additives are used in antifreeze to ensure that it is properly identified.

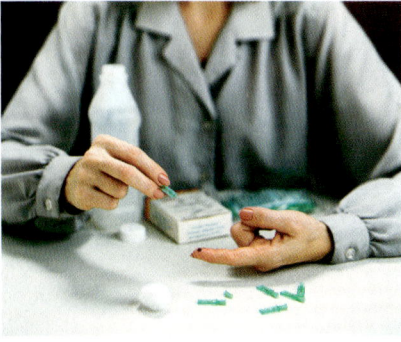

Figure 12.4
Isopropyl alcohol, or rubbing alcohol, is used as a disinfectant before and after an injection or blood test.

1,2,3-Propanetriol

1,2,3-Propanetriol (glycerol),

$$\begin{array}{c} CH_2-CH-CH_2 \\ |\quad\ \ |\quad\ \ | \\ OH\ \ OH\ \ OH \end{array}$$

is a viscous, sweet-tasting, nontoxic liquid. It is very soluble in water and is used in cosmetics, pharmaceuticals, and lubricants. Glycerol is obtained as a by-product of the hydrolysis of fats.

12.4 Classification of Alcohols

Alcohols are classified as **primary (1°), secondary (2°)**, or **tertiary (3°)**, depending on the number of alkyl groups attached to the **carbinol carbon**, the carbon bearing the hydroxyl (—OH) group. If no alkyl groups are attached, the alcohol is methyl alcohol; if there is a single alkyl group, the alcohol is a primary alcohol; an alcohol with two alkyl groups bonded to the carbon bearing the hydroxyl group is a secondary alcohol, and if three alkyl groups are attached, the alcohol is a tertiary alcohol.

 **LEARNING GOAL**

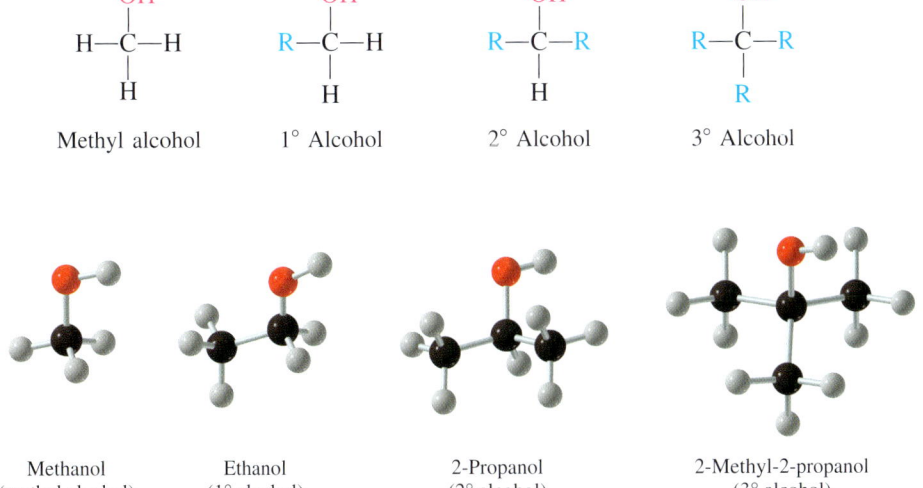

EXAMPLE 12.3 Classifying Alcohols

LEARNING GOAL 4

Classify each of the following alcohols as primary, secondary, or tertiary.

Solution

In each of the structures shown below, the carbinol carbon is shown in red:

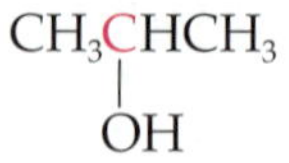

$$\text{CH}_3\text{CHCH}_3 \atop | \atop \text{OH}$$

$$\text{CH}_3 \atop | \atop \text{CH}_3\text{CCH}_3 \atop | \atop \text{OH}$$

This alcohol, 2-propanol, is a secondary alcohol because there are two alkyl groups attached to the carbinol carbon.

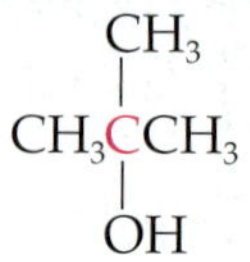

This alcohol, 2-methyl-2-propanol, is a tertiary alcohol because there are three alkyl groups attached to the carbinol carbon.

$$\text{OH} \atop | \atop \text{CH}_3\text{CH}_2\text{CHCH}_2 \atop | \atop \text{CH}_2\text{CH}_3$$

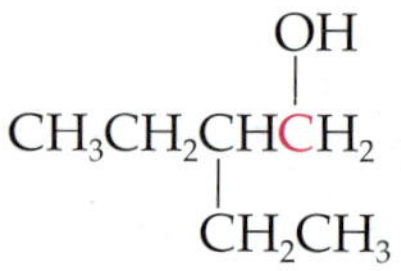

This alcohol, 2-ethyl-1-butanol, is a primary alcohol because there is only one alkyl group attached to the carbinol carbon.

Question 12.3

Classify each of the following alcohols as 1°, 2°, 3°, or aromatic (phenol).

a. $CH_3CH_2CH_2CH_2OH$

b. 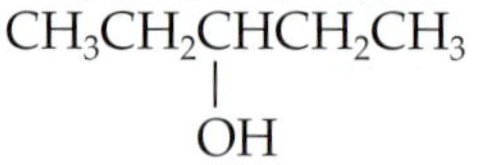 $CH_3CH_2CHCH_2CH_3$
   $|$
   OH

c. 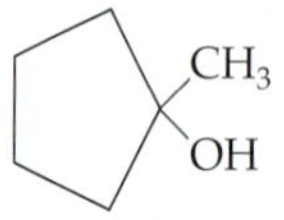 cyclopentane with CH_3 and OH on same carbon

d. 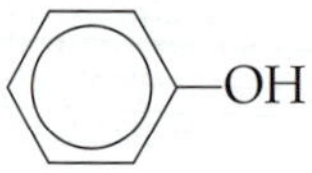 benzene ring—OH

e. 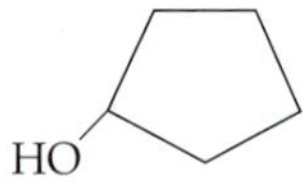 cyclopentane with HO

Question 12.4

Classify each of the following alcohols as 1°, 2°, or 3°.

a. 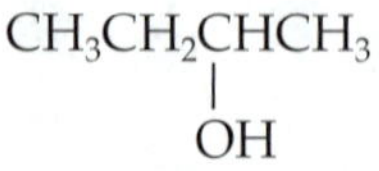 $CH_3CH_2CHCH_3$
   $|$
   OH

b. 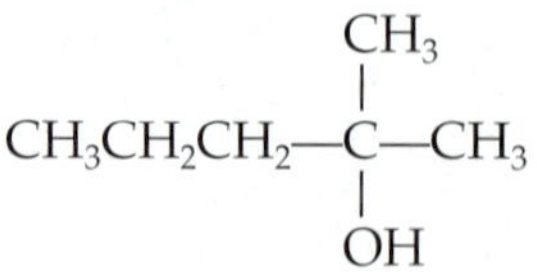 $CH_3CH_2CH_2-\underset{\underset{OH}{|}}{\overset{\overset{CH_3}{|}}{C}}-CH_3$

c. CH_3CH_2-OH

d. 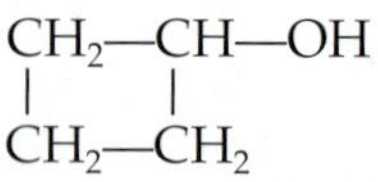 $CH_2-CH-OH$
   $||$
   CH_2-CH_2

12.5 Reactions Involving Alcohols

Preparation of Alcohols

As we saw in the last chapter, the most important reactions of alkenes are *addition reactions.* Addition of a water molecule to the carbon-carbon double bond of an alkene produces an alcohol. This reaction, called **hydration,** requires a trace of acid (H^+) as a catalyst, as shown in the following equation:

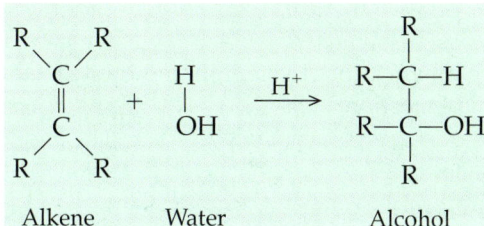

Hydration of alkenes is described in Section 11.5.

> **EXAMPLE 12.4**
>
> **Writing an Equation Representing the Preparation of an Alcohol by the Hydration of an Alkene**
>
> Write an equation representing the preparation of cyclohexanol from cyclohexene.
>
> **Solution**
>
> Begin by writing the structure of cyclohexene. Recall that cyclohexene is a six-carbon cyclic alkene. Now add the water molecule to the equation.
>
> Cyclohexene   Water
>
> You will recognize that the hydration reaction involves the addition of a water molecule to the carbon-carbon double bond. Recall that the reaction requires a trace of acid as a catalyst. Complete the equation by adding the catalyst and product, cyclohexanol.
>
> Cyclohexene   Water   Cyclohexanol

Question 12.5

Write a balanced equation showing the hydration of each of the following alkenes. Remember that Markovnikov's rule must be applied in the case of unsymmetrical alkenes.

a. $CH_3CH=CH_2$
b. $CH_2=CH_2$
c. $CH_3CH_2CH=CHCH_2CH_3$

Question 12.6

Using the I.U.P.A.C Nomenclature System, name the reactants and products for each of the reactions in Question 12.5.

Question 12.7

Classify each of the alcohols produced in the reactions of Question 12.5 as primary (1°), secondary (2°), or tertiary (3°).

Question 12.8

Which of the alkenes in Question 12.5 can be found as both *cis* and *trans* isomers?

LEARNING GOAL

Alcohols may also be prepared via the hydrogenation (reduction) of aldehydes and ketones. This reaction, summarized as follows, is discussed in Section 13.4, and is similar to the hydrogenation of alkenes.

In an aldehyde, R^1 and R^2 may be either alkyl groups or H. In ketones, R^1 and R^2 are both alkyl groups.

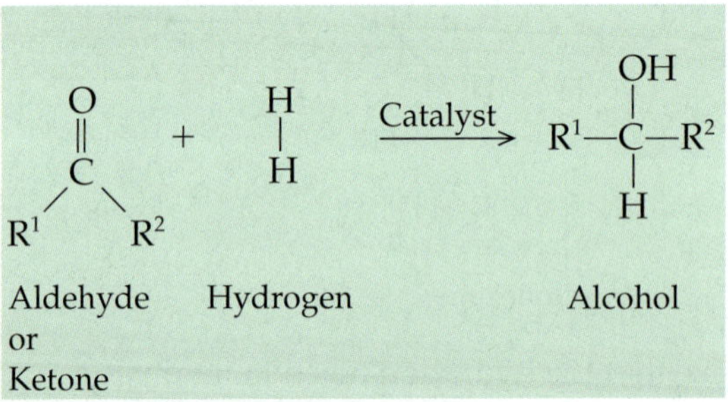

$$\underset{\substack{\text{Aldehyde} \\ \text{or} \\ \text{Ketone}}}{\overset{O}{\underset{R^1}{\overset{\|}{C}}}\underset{R^2}{\,}} + \underset{\text{Hydrogen}}{\overset{H}{\underset{H}{|}}} \xrightarrow{\text{Catalyst}} \underset{\text{Alcohol}}{\overset{OH}{\underset{H}{R^1-\overset{|}{\underset{|}{C}}-R^2}}}$$

EXAMPLE 12.5 Writing an Equation Representing the Preparation of an Alcohol by the Hydrogenation (Reduction) of an Aldehyde

Write an equation representing the preparation of 1-propanol from propanal.

Solution

Begin by writing the structure of propanal. Propanal is a three-carbon aldehyde. Aldehydes are characterized by the presence of a carbonyl group (—C=O) attached to the end of the carbon chain of the molecule. After you have drawn the structure of propanal, add diatomic hydrogen to the equation.

Continued—

12.5 Reactions Involving Alcohols

EXAMPLE 12.5 —Continued

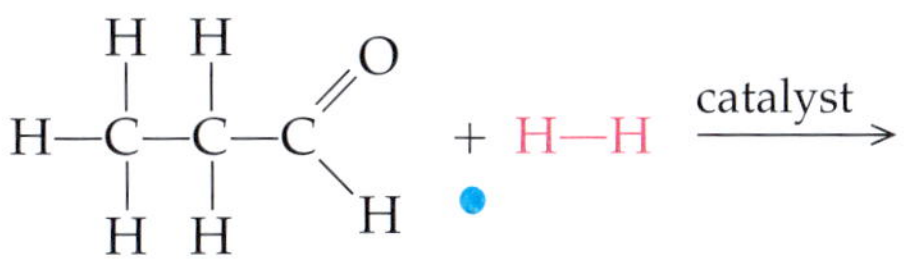

$$\text{CH}_3\text{-CH}_2\text{-CHO} + \text{H-H} \xrightarrow{\text{catalyst}}$$

Propanal · · · · · · Hydrogen

Notice that the general equation reveals this reaction to be another example of a hydrogenation reaction. As the hydrogens are added to the carbon-oxygen double bond, it is converted to a carbon-oxygen single bond, as the carbonyl oxygen becomes a hydroxyl group.

$$\text{CH}_3\text{-CH}_2\text{-CHO} + \text{H-H} \xrightarrow{\text{catalyst}} \text{CH}_3\text{-CH}_2\text{-CH}_2\text{-OH}$$

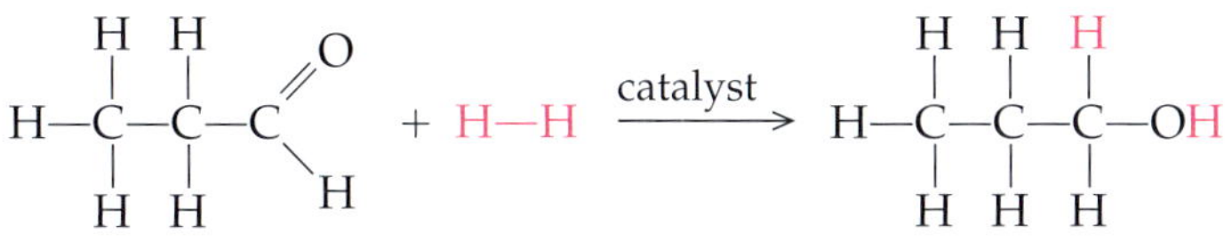

Propanal · · · · · Hydrogen · · · · · 1-Propanol

EXAMPLE 12.6

Writing an Equation Representing the Preparation of an Alcohol by the Hydrogenation (Reduction) of a Ketone

LEARNING GOAL 6

Write an equation representing the preparation of 2-propanol from propanone.

Solution

Begin by writing the structure of propanone. Propanone is a three-carbon ketone. Ketones are characterized by the presence of a carbonyl group (—C=O) located anywhere within the carbon chain of the molecule. In the structure of propanone, the carbonyl group must be associated with the center carbon. After you have drawn the structure of propanone, add diatomic hydrogen to the equation.

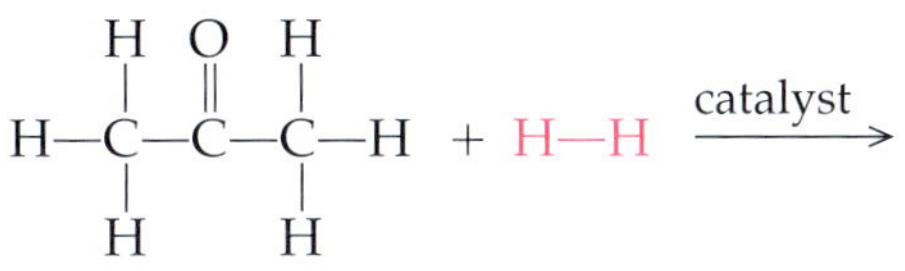

$$\text{CH}_3\text{-CO-CH}_3 + \text{H-H} \xrightarrow{\text{catalyst}}$$

Propanone · · · · · Hydrogen

Notice that this reaction is another example of a hydrogenation reaction. As the hydrogens are added to the carbon-oxygen double bond, it is converted to a carbon-oxygen single bond, as the carbonyl oxygen becomes a hydroxyl group.

$$\text{CH}_3\text{-CO-CH}_3 + \text{H-H} \xrightarrow{\text{catalyst}} \text{CH}_3\text{-CH(OH)-CH}_3$$

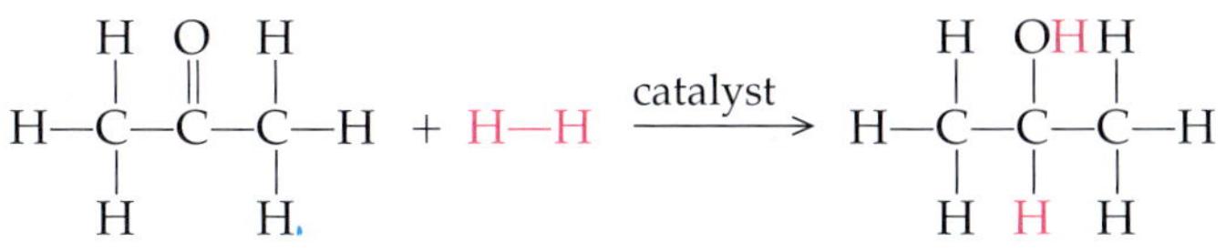

Propanone · · · · · Hydrogen · · · · · 2-Propanol

Question 12.9

Write an equation representing the reduction of butanone. *Hint:* Butanone is a four-carbon ketone.

Question 12.10

Classify the alcohol product of the reaction in Question 12.9 as a primary (1°), secondary (2°), or tertiary (3°) alcohol, and provide its I.U.P.A.C. name.

Question 12.11

Write an equation representing the reduction of butanal. Provide the structures and names for the reactants and products. *Hint:* Butanal is a four-carbon aldehyde.

Question 12.12

Classify the alcohol product of the reaction in Question 12.11 as a primary (1°), secondary (2°), or tertiary (3°) alcohol, and provide its I.U.P.A.C. name.

Dehydration of Alcohols

LEARNING GOAL 7

Alcohols undergo **dehydration** (lose water) when heated with concentrated sulfuric acid (H_2SO_4) or phosphoric acid (H_3PO_4). Dehydration is an example of an **elimination reaction**, that is, a reaction in which a molecule loses atoms or ions from its structure. In this case, the —OH and —H are "eliminated" from adjacent carbons in the alcohol to produce an alkene and water. We have just seen that alkenes can be hydrated to give alcohols. Dehydration is simply the reverse process: the conversion of an alcohol back to an alkene. This is seen in the following general reaction and the examples that follow:

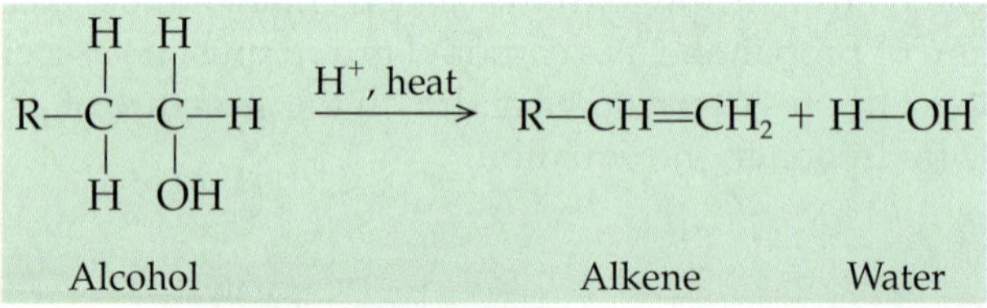

$$R-\underset{\underset{H}{|}}{\overset{\overset{H}{|}}{C}}-\underset{\underset{OH}{|}}{\overset{\overset{H}{|}}{C}}-H \xrightarrow{H^+, \text{ heat}} R-CH=CH_2 + H-OH$$

Alcohol → Alkene + Water

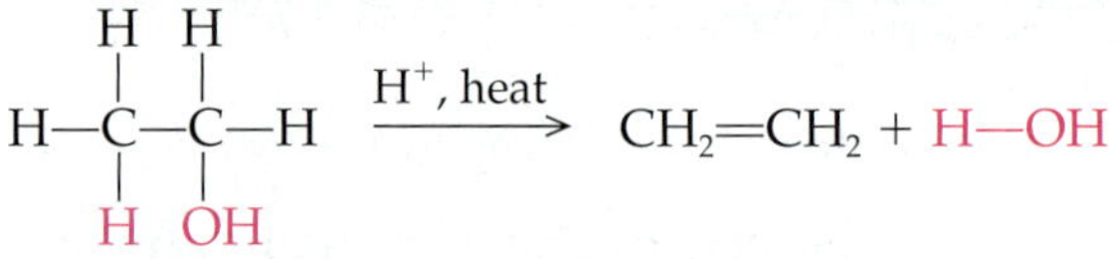

$$H-\underset{\underset{H}{|}}{\overset{\overset{H}{|}}{C}}-\underset{\underset{OH}{|}}{\overset{\overset{H}{|}}{C}}-H \xrightarrow{H^+, \text{ heat}} CH_2=CH_2 + H-OH$$

Ethanol (ethyl alcohol) → Ethene (ethylene)

$$CH_3CH_2CH_2OH \xrightarrow{H^+, \text{ heat}} CH_3CH=CH_2 + H_2O$$

1-Propanol (propyl alcohol) → Propene (propylene)

In some cases, dehydration of alcohols produces a mixture of products, as seen in the following example:

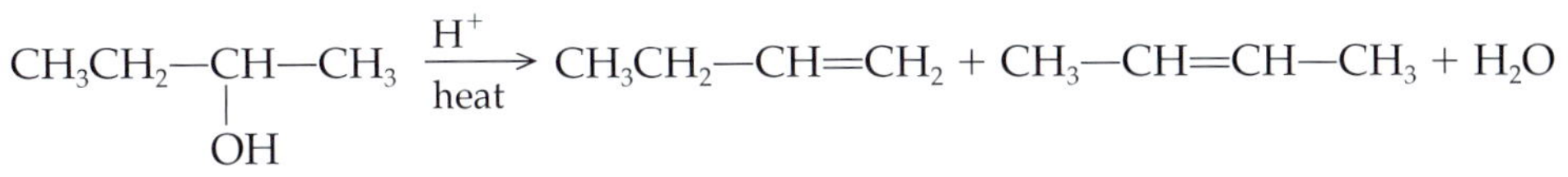

12.5 Reactions Involving Alcohols

$$CH_3CH_2-\underset{\underset{OH}{|}}{CH}-CH_3 \xrightarrow[\text{heat}]{H^+} CH_3CH_2-CH=CH_2 + CH_3-CH=CH-CH_3 + H_2O$$

2-Butanol   1-Butene   2-Butene
                   (minor product)   (major product)

Notice in the equation shown above and in Example 12.7, the major product is the more highly substituted alkene. In 1875 the Russian chemist Alexander Zaitsev developed a rule to describe such reactions. **Zaitsev's rule** states that in an elimination reaction, the alkene with the greatest number of alkyl groups on the double bonded carbons (the more highly substituted alkene) is the major product of the reaction.

EXAMPLE 12.7 Predicting the Products of Alcohol Dehydration

Predict the products of the dehydration of 3-methyl-2-butanol.

LEARNING GOAL 7

Solution

Assuming that no rearrangement occurs, the product(s) of a dehydration of an alcohol will contain a double bond in which one of the carbons was the original carbinol carbon—the carbon to which the hydroxyl group is attached. Consider the following reaction:

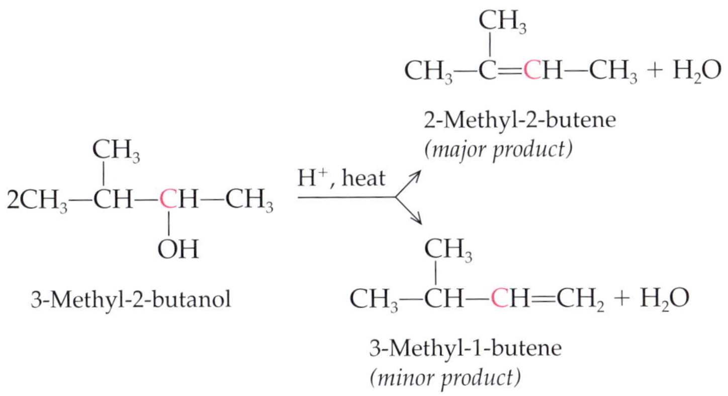

It is clear that both the major and minor products have a double bond to carbon number 2 in the original alcohol (this carbon is set off in color). Zaitsev's rule tells us that in dehydration reactions with more than one product possible, the more highly branched alkene predominates. In the reaction shown, 2-methyl-2-butene has three alkyl groups at the double bond, whereas 3-methyl-1-butene has only two alkyl groups at the double bond. The more highly branched alkene is more stable and thus is the major product.

Question 12.13

Predict the products obtained on reacting each of the following alkenes with water and a trace of acid:

a. Ethene
b. Propene
c. 2-Butene

Question 12.14

Name the products of the reactions in Question 12.13 using the I.U.P.A.C. Nomenclature System, and classify each of the product alcohols as primary (1°), secondary (2°), or tertiary (3°).

Question 12.15

Predict the products obtained on reacting each of the following alkenes with water and a trace of acid:

a. 1-Butene
b. 2-Methylpropene

Question 12.16

Name the products of the reactions in Question 12.15 using the I.U.P.A.C. Nomenclature System, and classify each of the product alcohols as primary (1°), secondary (2°), or tertiary (3°).

Question 12.17

Draw the alkene products that would be produced on dehydration of each of the following alcohols:

a. 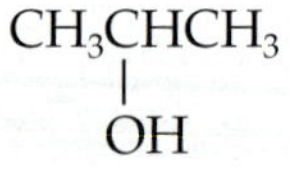 CH_3CHCH_3
      |
      OH

b. 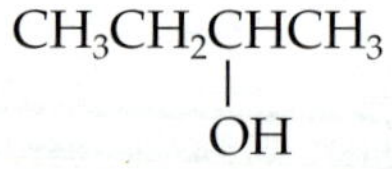 $CH_3CH_2CHCH_3$
              |
              OH

Question 12.18

Draw the alkene products that would be produced on dehydration of each of the following alcohols:

a. 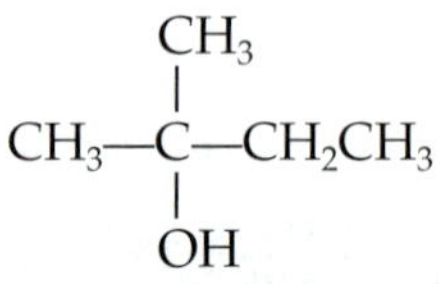 CH_3
             |
    $CH_3-C-CH_2CH_3$
             |
            OH

b. 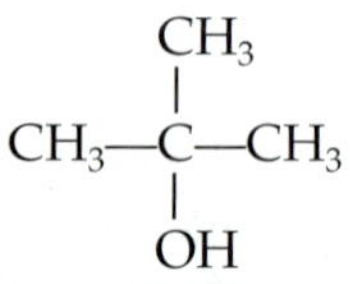 CH_3
             |
    CH_3-C-CH_3
             |
            OH

This reaction, and the other reactions of glycolysis, are considered in Section 21.3.

The dehydration of 2-phosphoglycerate to phosphoenolpyruvate is a critical step in the metabolism of the sugar glucose. In the following structures, the circled P represents a phosphoryl group (PO_4^{2-}).

$$\begin{array}{c} O^- \\ | \\ C=O \\ | \\ CH-\text{\textcircled{P}} \\ | \\ CH_2-OH \end{array} \longrightarrow \begin{array}{c} O^- \\ | \\ C=O \\ | \\ C\sim\text{\textcircled{P}} \\ || \\ CH_2 \end{array} + H_2O$$

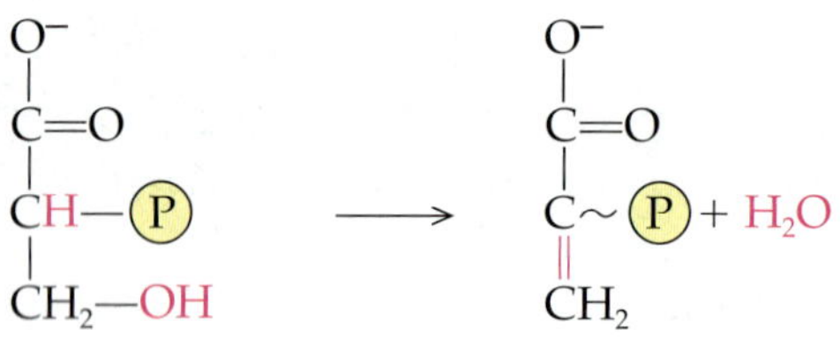

2-Phosphoglycerate → Phosphoenolpyruvate

The squiggle (~) represents a high energy bond.

Oxidation Reactions

Alcohols may be oxidized with a variety of oxidizing agents to aldehydes, ketones, and carboxylic acids. The most commonly used oxidizing agents are solutions of basic potassium permanganate (KMnO$_4$/OH$^-$) and chromic acid (H$_2$CrO$_4$). The symbol [O] over the reaction arrow is used throughout this book to designate any general oxidizing agent, as in the following reactions:

Oxidation of methanol produces the aldehyde methanal:

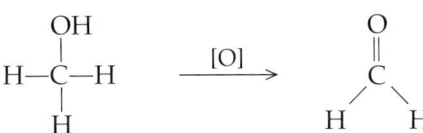

Methanol
(methyl alcohol)
An alcohol

Methanal
(formaldehyde)
An aldehyde

Oxidation of a primary alcohol produces an aldehyde:

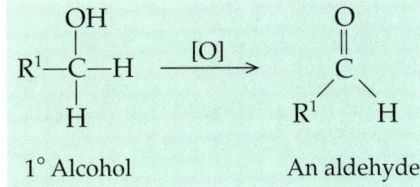

1° Alcohol

An aldehyde

 LEARNING GOAL

An *oxidation* reaction involves a gain of oxygen or the loss of hydrogen. A *reduction* reaction involves the loss of oxygen or gain of hydrogen. If two hydrogens are gained or lost for every oxygen gained or lost, the reaction is neither an oxidation nor a reduction.

Note that the symbol [O] is used throughout this book to designate any oxidizing agent.

As we will see in Section 13.4, aldehydes can undergo further oxidation to produce carboxylic acids.

EXAMPLE 12.8 Writing an Equation Representing the Oxidation of a Primary Alcohol

Write an equation showing the oxidation of 2,2-dimethylpropanol to produce 2,2-dimethylpropanal.

Solution

Begin by writing the structure of the reactant, 2,2-dimethylpropanol and indicate the need for an oxidizing agent by placing the designation [O] over the reaction arrow:

$$\text{CH}_3-\underset{\underset{\text{CH}_3}{|}}{\overset{\overset{\text{CH}_3}{|}}{\text{C}}}-\underset{\underset{\text{H}}{|}}{\overset{\overset{\text{H}}{|}}{\text{C}}}-\text{OH} \xrightarrow{[O]}$$

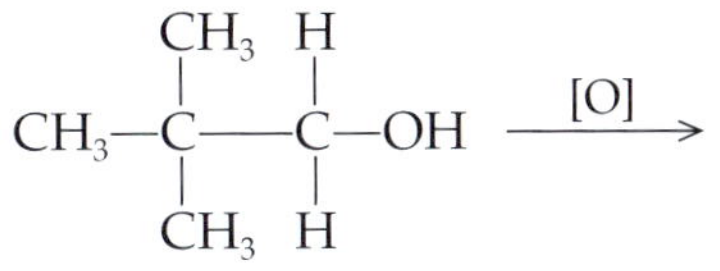

2,2-Dimethylpropanol

Now show the oxidation of the hydroxyl group to the aldehyde carbonyl group.

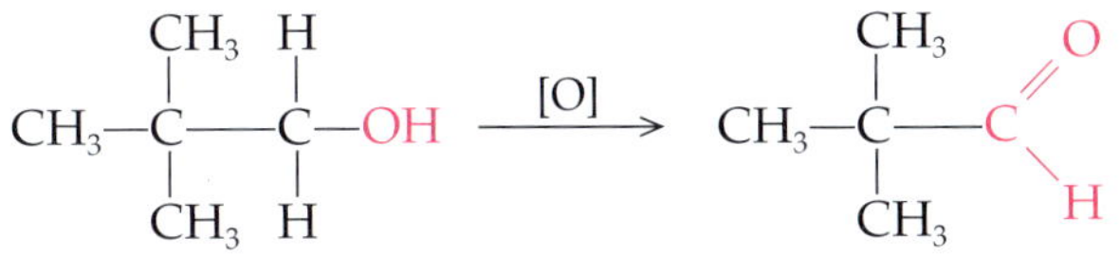

$$\text{CH}_3-\underset{\underset{\text{CH}_3}{|}}{\overset{\overset{\text{CH}_3}{|}}{\text{C}}}-\underset{\underset{\text{H}}{|}}{\overset{\overset{\text{H}}{|}}{\text{C}}}-\text{OH} \xrightarrow{[O]} \text{CH}_3-\underset{\underset{\text{CH}_3}{|}}{\overset{\overset{\text{CH}_3}{|}}{\text{C}}}-\overset{\overset{\text{O}}{\|}}{\text{C}}-\text{H}$$

2,2-Dimethylpropanol

2,2-Dimethylpropanal

Question 12.19

Write an equation showing the oxidation of the following primary alcohols:

a. 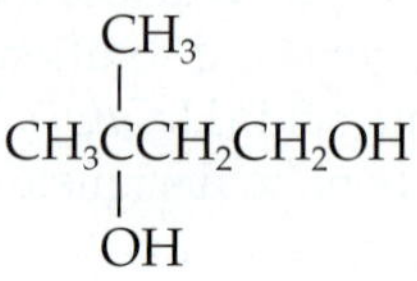
$$CH_3\text{-}\underset{\underset{OH}{|}}{\overset{\overset{CH_3}{|}}{C}}\text{-}CH_2\text{-}CH_2OH$$

b. CH_3CH_2OH

Question 12.20

Name each of the reactant alcohols and product aldehydes in Question 12.19 using the I.U.P.A.C. Nomenclature System. *Hint:* Refer to the example above, as well as to Section 13.2 to name the aldehyde products.

Oxidation of a secondary alcohol produces a ketone:

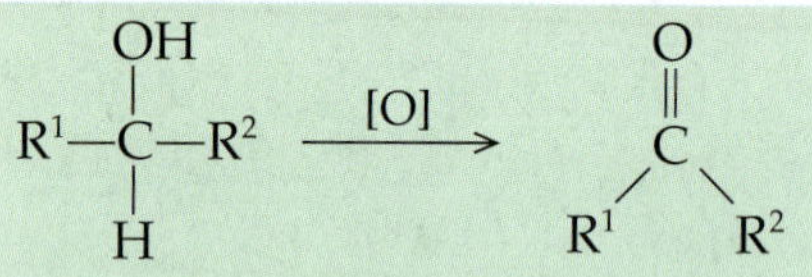

$$\underset{\text{2° Alcohol}}{R^1\text{-}\underset{\underset{H}{|}}{\overset{\overset{OH}{|}}{C}}\text{-}R^2} \xrightarrow{[O]} \underset{\text{A ketone}}{\overset{O}{\underset{R^1 \quad R^2}{\overset{\|}{C}}}}$$

EXAMPLE 12.9 Writing an Equation Representing the Oxidation of a Secondary Alcohol

LEARNING GOAL 8

Write an equation showing the oxidation of 2-propanol to produce propanone.

Solution

Begin by writing the structure of the reactant, 2-propanol, and indicate the need for an oxidizing agent by placing the designation [O] over the reaction arrow:

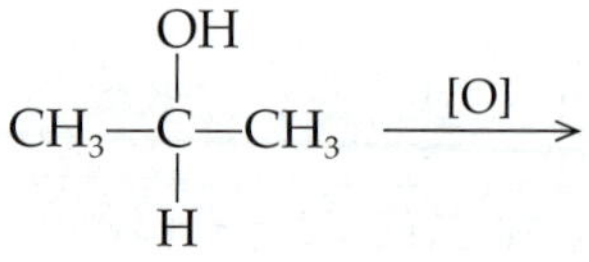

$$\underset{\text{2-Propanol}}{CH_3\text{-}\underset{\underset{H}{|}}{\overset{\overset{OH}{|}}{C}}\text{-}CH_3} \xrightarrow{[O]}$$

Now show the oxidation of the hydroxyl group to the ketone carbonyl group.

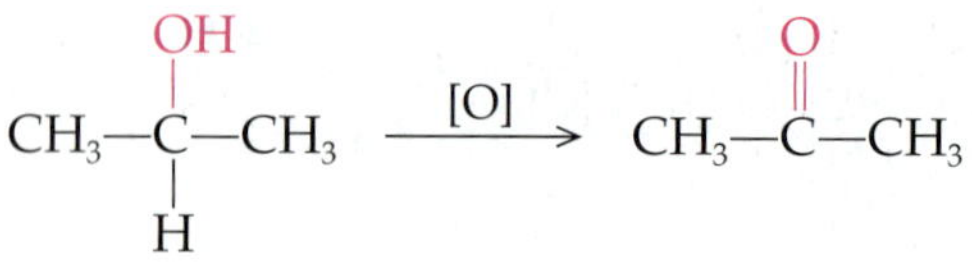

$$\underset{\text{2-Propanol}}{CH_3\text{-}\underset{\underset{H}{|}}{\overset{\overset{OH}{|}}{C}}\text{-}CH_3} \xrightarrow{[O]} \underset{\text{Propanone}}{CH_3\text{-}\overset{\overset{O}{\|}}{C}\text{-}CH_3}$$

12.5 Reactions Involving Alcohols

Question 12.21

Write an equation showing the oxidation of the following secondary alcohols:

a. 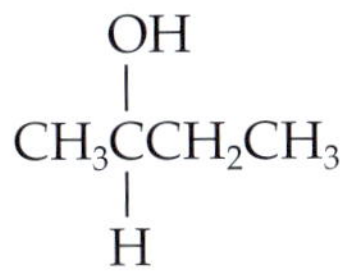
$$CH_3\underset{\underset{H}{|}}{\overset{\overset{OH}{|}}{C}}CH_2CH_3$$

b. 
$$CH_3\overset{\overset{OH}{|}}{C}HCH_2CH_2CH_3$$

Question 12.22

Name each of the reactant alcohols and product ketones in Question 12.21 using the I.U.P.A.C. Nomenclature System. *Hint:* Refer to the example above, as well as to Section 13.2, to name the ketone products.

Tertiary alcohols cannot be oxidized:

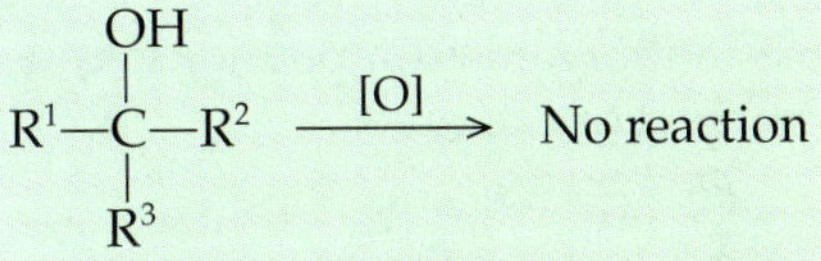

$$R^1-\underset{\underset{R^3}{|}}{\overset{\overset{OH}{|}}{C}}-R^2 \xrightarrow{[O]} \text{No reaction}$$

3° Alcohol

For the oxidation reaction to occur, the carbon bearing the hydroxyl group must contain at least one C—H bond. Because tertiary alcohols contain three C—C bonds to the carbinol carbon, they cannot undergo oxidation.

EXAMPLE 12.10

Writing an Equation Representing the Oxidation of a Tertiary Alcohol

Write an equation showing the oxidation of 2-methyl-2-propanol.

Solution

Begin by writing the structure of the reactant, 2-methyl-2-propanol and indicate the need for an oxidizing agent by placing the designation [O] over the reaction arrow:

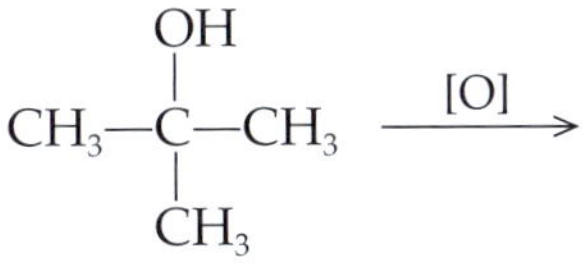

$$CH_3-\underset{\underset{CH_3}{|}}{\overset{\overset{OH}{|}}{C}}-CH_3 \xrightarrow{[O]}$$

2-Methyl-2-propanol

The structure of 2-methyl-2-propanol reveals that it is a tertiary alcohol. Therefore, no oxidation reaction can occur because the carbon bearing the hydroxyl group is bonded to three other carbon atoms, not to a hydrogen atom.

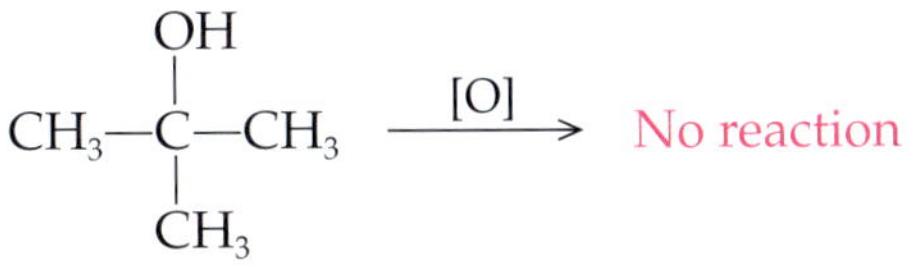

$$CH_3-\underset{\underset{CH_3}{|}}{\overset{\overset{OH}{|}}{C}}-CH_3 \xrightarrow{[O]} \text{No reaction}$$

2-Methyl-2-propanol

When ethanol is metabolized in the liver, it is oxidized to ethanal (acetaldehyde). If too much ethanol is present in the body, an overabundance of ethanal is formed, which causes many of the adverse effects of the "morning-after hangover." Continued oxidation of ethanal produces ethanoic acid (acetic acid), which is used as an energy source by the cell and eventually oxidized to CO_2 and H_2O. These reactions, summarized as follows, are catalyzed by liver enzymes.

$$CH_3CH_2-OH \longrightarrow CH_3\overset{\overset{O}{\|}}{C}-H \longrightarrow CH_3\overset{\overset{O}{\|}}{C}-OH \longrightarrow CO_2 + H_2O$$

Ethanol (ethyl alcohol)   Ethanal (acetaldehyde)   Ethanoic acid (acetic acid)

12.6 Oxidation and Reduction in Living Systems

LEARNING GOAL 9

Before beginning a discussion of oxidation and reduction in living systems, we must understand how to recognize **oxidation** (loss of electrons) and **reduction** (gain of electrons) in organic compounds. It is easy to determine when an oxidation or a reduction occurs in inorganic compounds because the process is accompanied by a change in charge. For example,

$$Ag^0 \longrightarrow Ag^+ + 1e^-$$

With the loss of an electron, the neutral atom is converted to a positive ion, which is oxidation. In contrast,

$$:\ddot{Br}\cdot + e^- \longrightarrow :\ddot{Br}:^-$$

With the gain of one electron, the bromine atom is converted to a negative ion, which is reduction.

When organic compounds are involved, however, there may be no change in charge, and it is often difficult to determine whether oxidation or reduction has occurred. The following simplified view may help.

In organic systems, *oxidation* may be recognized as a gain of oxygen or a loss of hydrogen. A *reduction* reaction may involve a loss of oxygen or gain of hydrogen.

Consider the following compounds. An alkane may be oxidized to an alcohol by gaining an oxygen. A primary or secondary alcohol may be oxidized to an aldehyde or ketone, respectively, by the loss of hydrogen. Finally, an aldehyde may be oxidized to a carboxylic acid by gaining an oxygen.

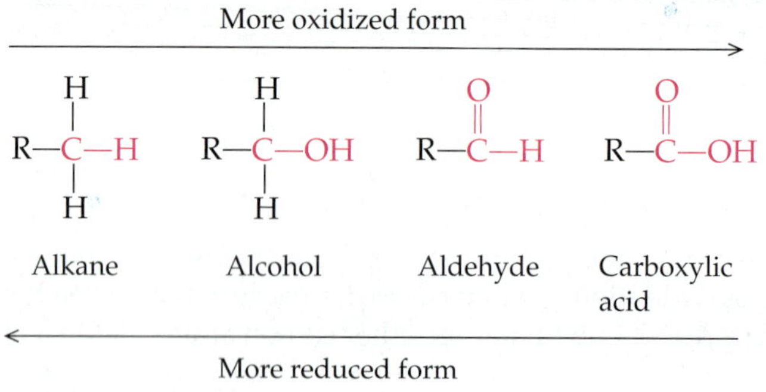

More oxidized form ⟶

$$R-\underset{\underset{H}{|}}{\overset{\overset{H}{|}}{C}}-H \quad R-\underset{\underset{H}{|}}{\overset{\overset{H}{|}}{C}}-OH \quad R-\overset{\overset{O}{\|}}{C}-H \quad R-\overset{\overset{O}{\|}}{C}-OH$$

Alkane   Alcohol   Aldehyde   Carboxylic acid

⟵ More reduced form

Thus, the conversion of an alkane to an alcohol, an alcohol to a carbonyl compound, and a carbonyl compound (aldehyde) to a carboxylic acid are all examples of oxidations. Conversions in the opposite direction are reductions.

Oxidation and reduction reactions also play an important role in the chemistry of living systems. In living systems these reactions are catalyzed by the action of various enzymes called *oxidoreductases*. These enzymes require compounds called *coenzymes* to accept or donate hydrogen in the reactions that they catalyze.

12.6 Oxidation and Reduction in Living Systems

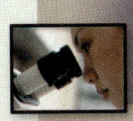

A Human Perspective

Alcohol Consumption and the Breathalyzer Test

Ethanol has been used widely as a beverage, a medicinal, and a solvent in numerous pharmaceutical preparations. Such common usage often overshadows the fact that ethanol is a toxic substance. Ethanol consumption is associated with a variety of long-term effects, including cirrhosis of the liver, death of brain cells, and alcoholism. Alcohol consumed by the mother can even affect the normal development of her unborn child and result in fetal alcohol syndrome. For these reasons, over-the-counter cough and cold medications that were once prepared in ethanol are now manufactured in alcohol-free form.

Short-term effects, linked to the social use of ethanol, center on its effects on behavior, reflexes, and coordination. Blood alcohol levels of 0.05–0.15% seriously inhibit coordination. Blood levels in excess of 0.10% are considered evidence of intoxication in most states. Blood alcohol levels in the range of 0.30–0.50% produce unconsciousness and the risk of death.

The loss of some coordination and reflex action is particularly serious when the affected individual attempts to drive a car. Law enforcement has come to rely on the "breathalyzer" test to screen for individuals suspected of driving while intoxicated. Those with a positive breathalyzer test are then given a more accurate blood test to establish their guilt or innocence.

The suspect is required to exhale into a solution that will react with the unmetabolized alcohol in the breath. The partial pressure of the alcohol in the exhaled air has been demonstrated to be proportional to the blood alcohol level. The solution is an acidic solution of dichromate ion, which is yellow-orange. The alcohol reduces the chromium in the dichromate ion from +6 to +3, the Cr^{3+} ion, which is green. The intensity of the green color is measured, and it is proportional to the amount of ethanol that was oxidized. The reaction is:

$$16H^+ + 2Cr_2O_7^{2-} + 3CH_3CH_2OH \longrightarrow$$
Yellow-orange

$$3CH_3COOH + 4Cr^{3+} + 11H_2O$$
Green

The breathalyzer test is a technological development based on a scientific understanding of the chemical reactions that ethanol may undergo—a further example of the dependence of technology on science.

For Further Understanding

Explain why the intensity of the green color in the reaction solution is proportional to the level of alcohol in the breath.

Draw a graph that represents this relationship.

Driving under the influence of alcohol impairs coordination and reflexes.

Nicotinamide adenine dinucleotide, NAD^+, is a coenzyme commonly involved in biological oxidation–reduction reactions (Figure 12.5). Now consider the final reaction of the citric acid cycle, an energy-harvesting pathway essential to life. In this reaction, catalyzed by the enzyme malate dehydrogenase, malate is oxidized to produce oxaloacetate:

Figure 12.5
Nicotinamide adenine dinucleotide.

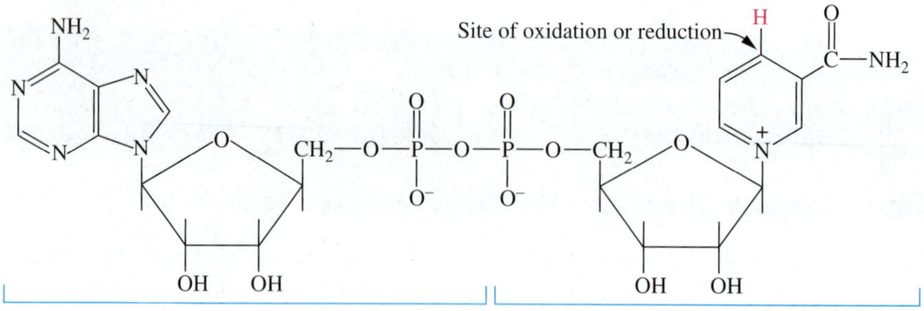

Adenine nucleotide Nicotinamide nucleotide

Nicotinamide adenine dinucleotide (NAD^+)

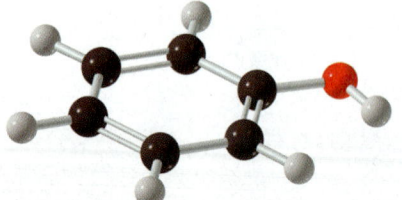

NAD^+ actually accepts a hydride anion, H^-, hydrogen with two electrons.

NAD^+ participates by accepting hydrogen from the malate. As malate is oxidized, NAD^+ is reduced to NADH.

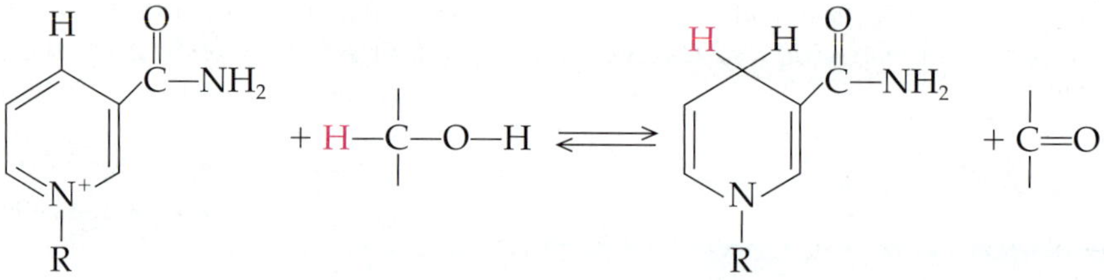

NAD^+
Oxidized form

NADH
Reduced form

We will study many other biologically important oxidation-reduction reactions in upcoming chapters.

LEARNING GOAL 10

12.7 Phenols

Figure 12.6
Ball-and-stick model of phenol. Keep in mind that this model is not completely accurate because it cannot show the cloud of shared electrons above and below the benzene ring. Review Section 11.6 for a more accurate description of the benzene ring.

Phenols are compounds in which the hydroxyl group is attached to a benzene ring (Figure 12.6). Like alcohols, they are polar compounds because of the polar hydroxyl group. Thus, the simpler phenols are somewhat soluble in water. They are found in flavorings and fragrances (mint and savory) and are used as preservatives (butylated hydroxytoluene, BHT). Examples include:

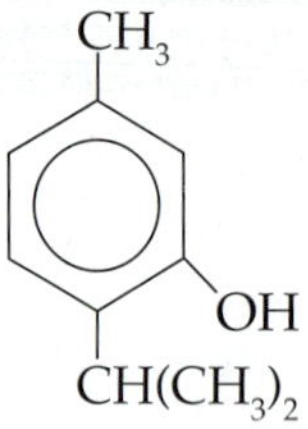

Thymol (mint)

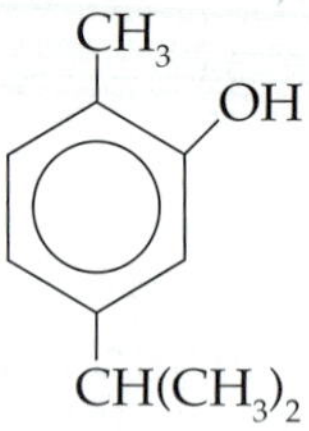

Carvacrol (savory)

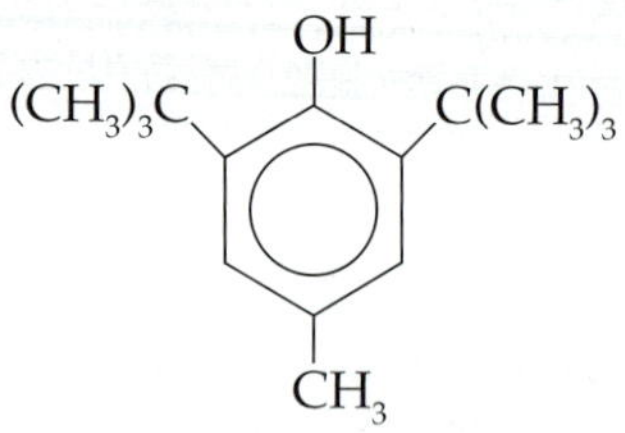

Butylated hydroxytoluene, BHT (food preservative)

A dilute solution of phenol must be used because concentrated phenol causes severe burns and because phenol is not highly soluble in water.

Phenols are also widely used in health care as germicides. In fact, carbolic acid, a dilute solution of phenol, was used as an antiseptic and disinfectant by Joseph Lister in his early work to decrease postsurgical infections. He used carbolic acid to bathe surgical wounds and to "sterilize" his instruments. Other derivatives of phenol that are used as antiseptics and disinfectants include hexachlorophene, hexylresorcinol, and *o*-phenylphenol. The structures of these compounds are shown below:

Phenol
(carbolic acid;
phenol dissolved
in water;
antiseptic)

Hexachlorophene
(antiseptic)

Hexylresorcinol
(antiseptic)

o-Phenylphenol
(antiseptic)

12.8 Ethers

Ethers have the general formula R—O—R, and thus they are structurally related to alcohols (R—O—H). The C—O bonds of ethers are polar, so ether molecules are polar (Figure 12.7). However, ethers do not form hydrogen bonds to one another because there is no —OH group. Therefore, they have much lower boiling points than alcohols of similar molecular weight but higher boiling points than alkanes of similar molecular weight. Compare the following examples:

$CH_3CH_2CH_2CH_3$

Butane
(butane)
M.W. = 58
b.p. = −0.5°C

CH_3—O—CH_2CH_3

Methoxyethane
(ethyl methyl ether)
M.W. = 60
b.p. = 7.9°C

$CH_3CH_2CH_2OH$

1-Propanol
(propyl alcohol)
M.W. = 60
b.p. = 97.2°C

In the I.U.P.A.C. system of naming ethers, the —OR substituent is named as an alkoxy group. This is analogous to the name *hydroxy* for the —OH group. Thus, CH_3—O— is methoxy, CH_3CH_2—O— is ethoxy, and so on.

Figure 12.7
Ball-and-stick model of the ether, methoxymethane (dimethyl ether).

An alkoxy group is an alkyl group bonded to an oxygen atom (—OR).

EXAMPLE 12.11 Using I.U.P.A.C. Nomenclature to Name an Ether

Name the following ether using I.U.P.A.C. nomenclature.

Solution

$$\underset{1\ \ \ 2\ \ \ 3\ \ \ 4\ \ \ 5\ \ \ 6\ \ \ 7\ \ \ 8\ \ \ 9}{CH_3CH_2\overset{\overset{\displaystyle O-CH_3}{|}}{C}HCH_2CH_2CH_2CH_2CH_2CH_3}$$

Parent compound: nonane
Position of alkoxy group: carbon-3 (*not* carbon-7)
Substituents: 3-methoxy
Name: 3-Methoxynonane

Question 12.23

Name the following ethers using I.U.P.A.C. nomenclature:

a. CH_3CH_2—O—$CH_2CH_2CH_3$
b. CH_3—O—$CH_2CH_2CH_3$

Question 12.24

Name the following ethers using I.U.P.A.C. nomenclature:

a. $CH_3CH_2-O-CH_2CH_2CH_2CH_3$
b. $CH_3CH_2CH_2CH_2-O-CH_2CH_2CH_3$

In the common system of nomenclature, ethers are named by placing the names of the two alkyl groups attached to the ether oxygen as prefixes in front of the word *ether*. The names of the two groups can be placed either alphabetically or by size (smaller to larger), as seen in the following examples:

CH_3-O-CH_3	$CH_3-O-CH_2CH_3$	$CH_3CH_2-O-CH(CH_3)_2$
Dimethyl ether or methyl ether	Ethyl methyl ether or methyl ethyl ether	Ethyl isopropyl ether

EXAMPLE 12.12 Naming Ethers Using the Common Nomenclature System

LEARNING GOAL 11

Write the common name for each of the following ethers.

Solution

$$CH_3CH_2-O-CH_2CH_3 \qquad CH_3-O-CH_2CH_2CH_3$$

Alkyl Groups:	two ethyl groups	methyl and propyl
Name:	Diethyl ether	Methyl propyl ether

Notice that there is only one correct name for methyl propyl ether because the methyl group is smaller than the propyl group and it would be first in an alphabetical listing also.

Question 12.25

Write the common name for each of the following ethers:

a. $CH_3CH_2-O-CH_2CH_2CH_3$
b. $CH_3-O-CH_2CH_2CH_3$

Question 12.26

Write the common name for each of the following ethers:

a. $CH_3CH_2-O-CH_2CH_2CH_2CH_2CH_3$
b. $CH_3CH_2CH_2CH_2-O-CH_2CH_2CH_3$

Chemically, ethers are moderately inert. They do not react with reducing agents or bases under normal conditions. However, they are extremely volatile and highly flammable (easily oxidized in air) and hence must always be treated with great care.

Ethers may be prepared by a dehydration reaction (removal of water) between two alcohol molecules, as shown in the following general reaction. The reaction requires heat and acid.

$$R^1-OH + R^2-OH \xrightarrow[\text{heat}]{H^+} R^1-O-R^2 + H_2O$$

Alcohol   Alcohol   Ether   Water

EXAMPLE 12.13

Writing an Equation Representing the Synthesis of an Ether via a Dehydration Reaction

LEARNING GOAL 12

Write an equation showing the synthesis of dimethyl ether.

Solution

The alkyl substituents of this ether are two methyl groups. Thus, the alcohol that must undergo dehydration to produce dimethyl ether is methanol.

$$CH_3OH + CH_3OH \xrightarrow{H^+} CH_3-O-CH_3 + H_2O$$

Methanol   Methanol   Dimethyl ether   Water

Question 12.27

Write an equation showing the dehydration reaction that would produce diethyl ether. Provide structures and names for all reactants and products.

Question 12.28

Write an equation showing the dehydration reaction between two molecules of 2-propanol. Provide structures and names for all reactants and products.

Diethyl ether was the first general anesthetic used. The dentist Dr. William Morton is credited with its introduction in the 1800s. Diethyl ether functions as an anesthetic by interacting with the central nervous system. It appears that diethyl ether (and many other general anesthetics) functions by accumulating in the lipid material of the nerve cells, thereby interfering with nerve impulse transmission. This results in analgesia, a lessened perception of pain.

Halogenated ethers are also routinely used as general anesthetics (Figure 12.8). They are less flammable than diethyl ether and are therefore safer to store and work with. *Penthrane* and *Enthrane* (trade names) are two of the more commonly used members of this family:

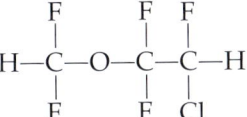

Penthrane   Enthrane

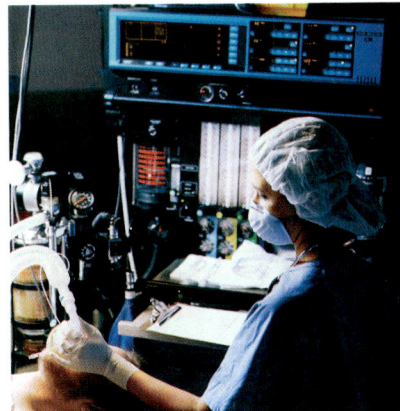

Figure 12.8
An anesthesiologist administers Penthrane to a surgical patient.

12.9 Thiols

Compounds that contain the sulfhydryl group (—SH) are called **thiols.** They are similar to alcohols in structure, but the sulfur atom replaces the oxygen atom.

Thiols and many other sulfur compounds have nauseating aromas. They are found in substances as different as the defensive spray of the North American striped skunk, onions, and garlic. The structures of the two most common compounds in the defense spray of the striped skunk, *trans*-2-butene-1-thiol and 3-methyl-1-butanethiol, are seen in Figure 12.9. These structures are contrasted with the structures of the two molecules that make up the far more pleasant scent of roses: geraniol, an unsaturated alcohol, and 2-phenylethanol, an aromatic alcohol.

Figure 12.9
This skunk on a bed of roses is surrounded by scent molecules. The two most common compounds in the defense spray of the striped skunk are the thiols *trans*-2-butene-1-thiol and 3-methyl-1-butanethiol. Two alcohols, 2-phenylethanol and geraniol, are the major components of the scent of roses.

The I.U.P.A.C. rules for naming thiols are similar to those for naming alcohols, except that the full name of the alkane is retained. The suffix *-thiol* follows the name of the parent compound.

EXAMPLE 12.14 Naming Thiols Using the I.U.P.A.C. Nomenclature System

Write the I.U.P.A.C. name for the thiols shown below.

Solution

Retain the full name of the parent compound and add the suffix *-thiol*.

CH_3CH_2—SH   HS—CH_2CH_2—SH

Parent compound:	ethane	ethane
Position of —SH:	carbon-1 (must be)	carbon-1 and carbon-2
Name:	Ethanethiol	1,2-Ethanedithiol

$$CH_3CHCH_2CH_2\text{—SH} \quad (\text{with } CH_3 \text{ on C2})$$
$$CH_3CHCH_2CH_2CH_2\text{—SH} \quad (\text{with SH on C2})$$

Parent Compound:	butane	pentane
Position of —SH:	carbon-1	carbon-1 and carbon-4
Substituent:	3-methyl	
Name:	3-Methyl-1-butanethiol	1,4-Pentanedithiol

12.9 Thiols

The amino acid cysteine is a thiol that has an important role to play in the structure and shape of many proteins. Two cysteine molecules can undergo oxidation to form cystine. The new bond formed is called a **disulfide** bond (—S—S—) bond.

Amino acids are the subunits from which proteins are made. A protein is a long polymer, or chain, of many amino acids bonded to one another.

$$
\begin{array}{c}
\text{H} \\
| \\
{}^+\text{NH}_3\text{—C—COO}^- \\
| \\
\text{CH}_2 \\
| \\
\text{S} \\
| \\
\text{H} \\
+ \\
\text{H} \\
| \\
\text{S} \\
| \\
\text{CH}_2 \\
| \\
{}^-\text{OOC—C—NH}_3{}^+ \\
| \\
\text{H}
\end{array}
\quad
\underset{\text{Reduction}}{\overset{\text{Oxidation}}{\rightleftarrows}}
\quad
\begin{array}{c}
\text{H} \\
| \\
{}^+\text{NH}_3\text{—C—COO}^- \\
| \\
\text{CH}_2 \\
| \\
\text{S} \\
| \\
\text{S} \\
| \\
\text{CH}_2 \\
| \\
{}^-\text{OOC—C—NH}_3{}^+ \\
| \\
\text{H}
\end{array}
\quad 2\text{H}^+ + 2e^-
$$

Disulfide Bond

2 Cysteine   Cystine

If the two cysteines are in different protein chains, the disulfide bond between them forms a bridge joining them together (Figure 12.10). If the two cysteines are in the same protein chain, a loop is formed. An example of the importance of disulfide bonds is seen in the production and structure of the protein hormone insulin, which controls blood sugar levels in the body. Insulin is initially produced as a protein called preproinsulin. Enzymatic removal of 24 amino acids and formation of disulfide bonds between cysteine amino acids produces proinsulin (Figure 12.10). Regions of the protein are identified as the A, B, and C chains. Notice that the A and B chains are covalently bonded to one another by two disulfide bonds. A third disulfide bond produces a hairpin loop in the A chain of the molecule. The molecule is now ready for the final stage of insulin synthesis in which the C chain is removed by the action of protein degrading enzymes (proteases). The active hormone, shown at the bottom of Figure 12.10, consists of the 21 amino acid A chain bonded to the 30 amino acid B chain by two disulfide bonds. Without these disulfide bonds, functional insulin molecules could not exist because there would be no way to keep the two chains together in the proper shape.

British Anti-Lewisite (BAL) is a dithiol used as an antidote in mercury poisoning. It was originally developed as an antidote to a mustard-gas-like chemical warfare agent called Lewisite, which was developed near the end of World War I and never used. By the onset of World War II, Lewisite was considered to be obsolete because of the discovery of BAL, an effective, inexpensive antidote. The two thiol groups of BAL form a water-soluble complex with mercury (or with the arsenic in Lewisite) that is excreted from the body in the urine.

$$
\begin{array}{c}
\text{CH}_2\text{—CH—CH}_2 \\
|\quad\quad|\quad\quad| \\
\text{OH}\quad\text{SH}\quad\text{SH}
\end{array}
$$

BAL

Coenzyme A is a thiol that serves as a "carrier" of acetyl groups (CH$_3$CO—) in biochemical reactions. It plays a central role in metabolism by shuttling acetyl groups from one reaction to another. When the two-carbon acetate group is attached to coenzyme A, the product is acetyl coenzyme A (acetyl CoA). The bond between coenzyme A and the acetyl group is a high-energy *thioester bond*. In effect, formation of the high-energy thioester bond "energizes" the acetyl group so that it can participate in other biochemical reactions.

The reactions involving coenzyme A are discussed in detail in Chapters 21, 22, and 23.

A high-energy bond is one that releases a great deal of energy when it is broken.

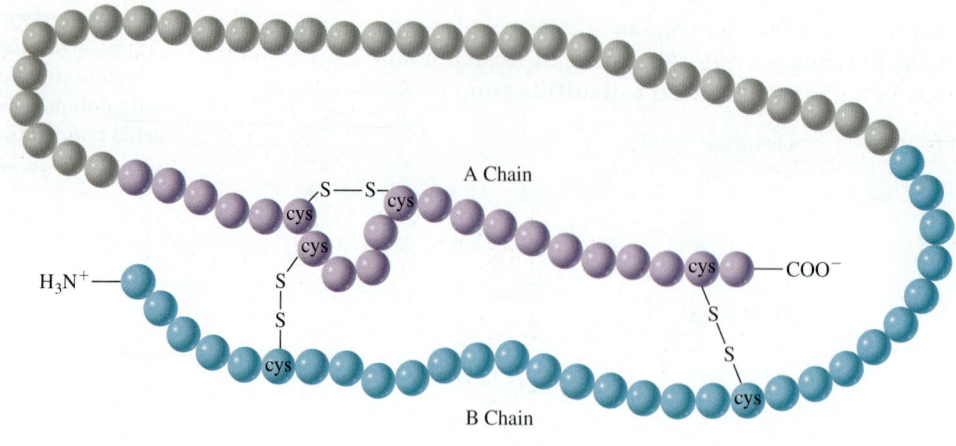

Figure 12.10
Two steps in the synthesis of insulin. Disulfide bonds hold the A and B chains together and form a hairpin loop in the A chain.

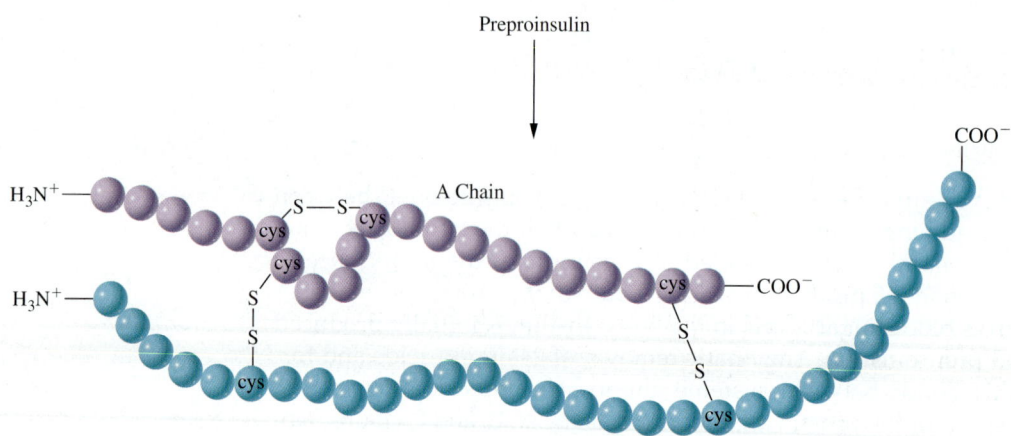

Acetyl coenzyme A
(acetyl CoA)

Acetyl CoA is made and used in the energy-producing reactions that provide most of the energy for life processes. It is also required for the biosynthesis of many biological molecules.

Summary of Reactions

Preparation of Alcohols

Hydration of alkenes:

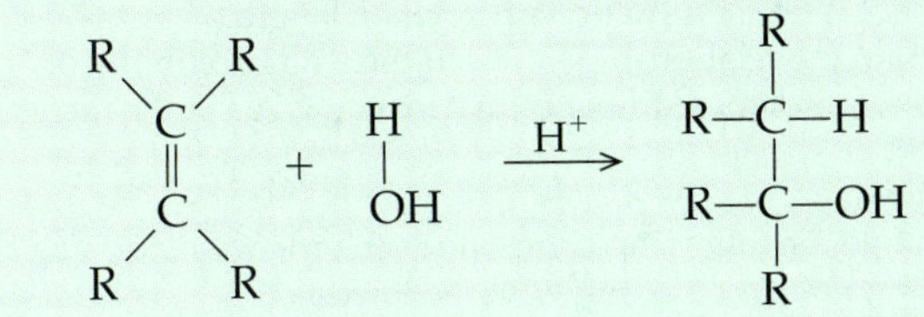

$$\underset{\text{Alkene}}{\overset{R}{\underset{R}{C}}=\overset{R}{\underset{R}{C}}} + \underset{\text{Water}}{H-OH} \xrightarrow{H^+} \underset{\text{Alcohol}}{\overset{R}{\underset{R}{R-C-H}}\ \overset{}{\underset{}{R-C-OH}}}$$

Reduction of an aldehyde or ketone:

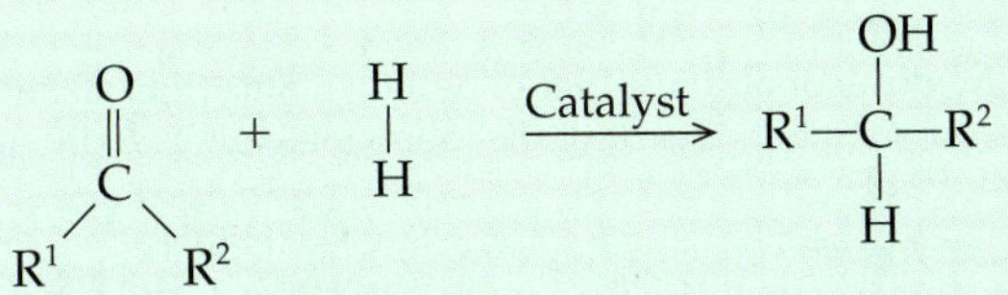

$$\underset{\substack{\text{Aldehyde}\\\text{or}\\\text{ketone}}}{\overset{O}{\underset{R^1\ \ R^2}{\parallel}}} + \underset{\text{Hydrogen}}{H-H} \xrightarrow{\text{Catalyst}} \underset{\text{Alcohol}}{R^1-\overset{OH}{\underset{H}{C}}-R^2}$$

Dehydration of Alcohols

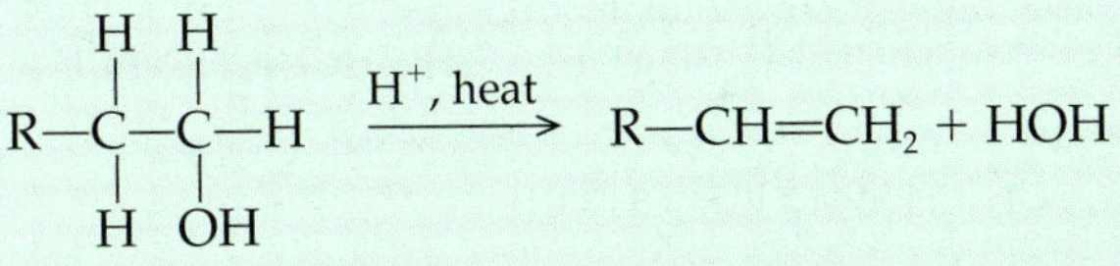

$$\underset{\text{Alcohol}}{R-\overset{H}{\underset{H}{C}}-\overset{H}{\underset{OH}{C}}-H} \xrightarrow{H^+, \text{heat}} \underset{\text{Alkene}}{R-CH=CH_2} + \underset{\text{Water}}{HOH}$$

Oxidation Reactions

Oxidation of a primary alcohol:

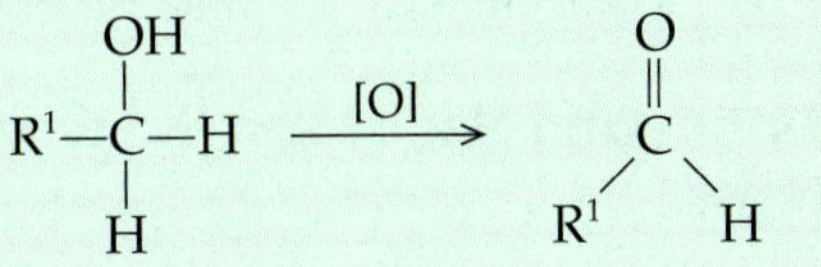

$$\underset{\text{1° Alcohol}}{R^1-\overset{OH}{\underset{H}{C}}-H} \xrightarrow{[O]} \underset{\text{An aldehyde}}{\overset{O}{\underset{R^1\ \ H}{\parallel}}}$$

Oxidation of a secondary alcohol:

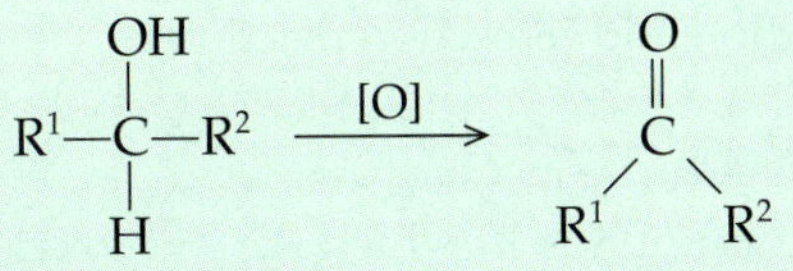

$$\underset{\text{2° Alcohol}}{R^1-\overset{OH}{\underset{H}{C}}-R^2} \xrightarrow{[O]} \underset{\text{A ketone}}{\overset{O}{\underset{R^1\ \ R^2}{\parallel}}}$$

Oxidation of a tertiary alcohol:

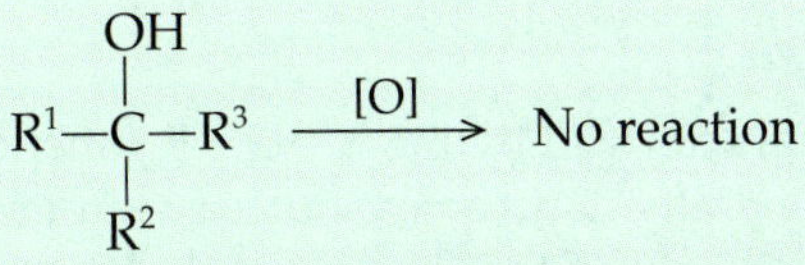

$$\underset{\text{3° Alcohol}}{R^1-\overset{OH}{\underset{R^2}{C}}-R^3} \xrightarrow{[O]} \text{No reaction}$$

Dehydration Synthesis of an Ether

$$\underset{\text{Alcohol}}{R^1-OH} + \underset{\text{Alcohol}}{R^2-OH} \xrightarrow[\text{heat}]{H^+} \underset{\text{Ether}}{R^1-O-R^2} + \underset{\text{Water}}{H_2O}$$

SUMMARY

12.1 Alcohols: Structure and Physical Properties

Alcohols are characterized by the *hydroxyl group (—OH)* and have the general formula R—OH. They are very polar, owing to the polar hydroxyl group, and are able to form intermolecular hydrogen bonds. Because of hydrogen bonding between alcohol molecules, they have higher boiling points than hydrocarbons of comparable molecular weight. The smaller alcohols are very water soluble.

12.2 Alcohols: Nomenclature

In the I.U.P.A.C. system, alcohols are named by determining the parent compound and replacing the *-e* ending with *-ol*. The chain is numbered to give the hydroxyl group the lowest possible number. Common names are derived from the alkyl group corresponding to the parent compound.

12.3 Medically Important Alcohols

Methanol is a toxic alcohol that is used as a solvent. Ethanol is the alcohol consumed in beer, wine, and distilled liquors. Isopropanol is used as a disinfectant. Ethylene glycol (1,2-ethanediol) is used as antifreeze, and glycerol (1,2,3-propanetriol) is used in cosmetics and pharmaceuticals.

12.4 Classification of Alcohols

Alcohols may be classified as *primary, secondary,* or *tertiary,* depending on the number of alkyl groups attached to the *carbinol carbon,* the carbon bearing the hydroxyl group. A primary alcohol has a single alkyl group bonded to the *carbinol carbon.* Secondary and tertiary alcohols have two and three alkyl groups, respectively.

12.5 Reactions Involving Alcohols

Alcohols can be prepared by the *hydration* of alkenes or reduction of aldehydes and ketones. Alcohols can undergo *dehydration* to yield alkenes. Primary and secondary alcohols undergo oxidation reactions to yield aldehydes and ketones, respectively. Tertiary alcohols do not undergo oxidation.

12.6 Oxidation and Reduction in Living Systems

In organic and biological systems, *oxidation* involves the gain of oxygen or loss of hydrogen. *Reduction* involves the loss of oxygen or gain of hydrogen. Nicotinamide adenine dinucleotide, NAD^+, is a coenzyme involved in many biological oxidation and reduction reactions.

12.7 Phenols

Phenols are compounds in which the hydroxyl group is attached to a benzene ring; they have the general formula Ar—OH. Many phenols are important as antiseptics and disinfectants.

12.8 Ethers

Ethers are characterized by the R—O—R functional group. Ethers are generally nonreactive but are extremely flammable. Diethyl ether was the first general anesthetic used in medical practice. It has since been replaced by Penthrane and Enthrane, which are less flammable.

12.9 Thiols

Thiols are characterized by the sulfhydryl group (—SH). The amino acid cysteine is a thiol that is extremely important for maintaining the correct shapes of proteins. Coenzyme A is a thiol that serves as a "carrier" of acetyl groups in biochemical reactions.

KEY TERMS

alcohol (12.1)
carbinol carbon (12.4)
dehydration (12.5)
disulfide (12.9)
elimination reaction (12.5)
ether (12.8)
fermentation (12.3)
hydration (12.5)
hydroxyl group (12.1)
oxidation (12.6)
phenol (12.7)
primary (1°) alcohol (12.4)
reduction (12.6)
secondary (2°) alcohol (12.4)
tertiary (3°) alcohol (12.4)
thiol (12.9)
Zaitsev's rule (12.5)

QUESTIONS AND PROBLEMS

Alcohols: Structure and Physical Properties

Foundations

12.29 Describe the relationship between the water solubility of alcohols and their hydrocarbon chain length.

12.30 Explain the relationship between the water solubility of alcohols and the number of hydroxyl groups in the molecule.

Applications

12.31 Arrange the following compounds in order of increasing boiling point, beginning with the lowest:
  a. $CH_3CH_2CH_2CH_2CH_3$
  b. $CH_3CHCH_2CHCH_3$ with OH, OH
  c. $CH_3CHCH_2CH_2CH_3$ with OH
  d. $CH_3CH_2CH_2-O-CH_2CH_3$

12.32 Why do alcohols have higher boiling points than alkanes?

12.33 Which member of each of the following pairs is more soluble in water?
  a. CH_3CH_2OH or $CH_3CH_2CH_2CH_2OH$
  b. $CH_3CH_2CH_2CH_3$ or $CH_3CH_2CH_2CH_2-OH$
  c. cyclopentanol or CH_3CHCH_3 with OH

12.34 Arrange the three alcohols in each of the following sets in order of increasing solubility in water:
  a. $CH_3CH_2CH_2CH_2CH_2OH$ $CH_3CHCH_2CHCH_2CH_3$ with OH, OH
     $CH_3CHCH_2CHCH_2CH_2OH$ with OH, OH
  b. Pentyl alcohol   1-Hexanol   Ethylene glycol

Alcohols: Nomenclature

Foundations

12.35 Briefly describe the I.U.P.A.C. rules for naming alcohols.
12.36 Briefly describe the rules for determining the common names for alcohols.

Applications

12.37 Give the I.U.P.A.C. name for each of the following compounds:
  a. $CH_3CH_2CH_2CH_2CH_2CH_2CH_2OH$
  b. CH_3CHCH_3 with OH
  c. $CH_3-C(CH_3)(CH_2OH)-CH_3$

12.38 Give the I.U.P.A.C. name for each of the following compounds:
  a. $CH_3CH_2CHCH_2CH_2OH$ with Br
  b. $CH_3CH-CCH_2CH_3$ with OH, CH_3, and $CH_2CH_2CH_3$
  c. $CH_3CH_2CCH_2CH_3$ with OH

12.39 Draw each of the following, using complete structural formulas:
  a. 3-Hexanol
  b. 1,2,3-Pentanetriol
  c. 2-Methyl-2-pentanol

12.40 Draw each of the following using condensed structural formulas:
  a. Cyclohexanol
  b. 3,4-Dimethyl-3-heptanol

12.41 Give the I.U.P.A.C. name for each of the following compounds:

  a. 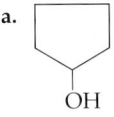 b. 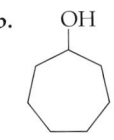 c.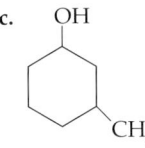

12.42 Draw each of the following alcohols:
  a. 1-Iodo-2-butanol
  b. 1,2-Butanediol
  c. Cyclobutanol

12.43 Give the common name for each of the following compounds:
  a. CH_3OH
  b. CH_3CH_2OH
  c. $CH_2\text{—}CH_2$
     $\quad\ |\quad\ \ |$
     $\quad OH\ \ OH$
  d. $CH_3CH_2CH_2OH$

12.44 Draw the structure of each of the following compounds:
  a. Pentyl alcohol
  b. Isopropyl alcohol
  c. Octyl alcohol
  d. Propyl alcohol

12.45 Draw a complete structural formula for each of the following compounds:
  a. 4-Methyl-2-hexanol
  b. Isobutyl alcohol
  c. 1,5-Pentanediol
  d. 2-Nonanol
  e. 1,3,5-Cyclohexanetriol

12.46 Name each of the following alcohols using the I.U.P.A.C. Nomenclature System:

  a. [cyclohexane with OH and CH_3]

  b. [cyclobutane with OH and Br]

  c. $CH_3CHCH_2CHCH_2CHCH_3$
     $\quad\ \ |\quad\quad\ |\quad\quad\ |$
     $\quad OH\quad OH\quad OH$

  d. $CH_3CH_2CHCHCHCH_3$
     $\quad\quad\quad\ |\ \ |$
     $\quad\quad\quad OH\ OH$

Medically Important Alcohols

12.47 What is denatured alcohol? Why is alcohol denatured?
12.48 What are the principal uses of methanol, ethanol, and isopropyl alcohol?
12.49 What is fermentation?

12.50 Why do wines typically have an alcohol concentration of 12–13%?
12.51 Why must fermentation products be distilled to produce liquors such as scotch?
12.52 If a bottle of distilled alcoholic spirits—for example, scotch whiskey—is labeled as 80 proof, what is the percentage of alcohol in the scotch?

Classification of Alcohols

Foundations

12.53 Define the term *carbinol carbon*.
12.54 Define the terms *primary, secondary,* and *tertiary alcohol,* and draw a general structure for each.

Applications

12.55 Classify each of the following as a 1°, 2°, or 3° alcohol:
  a. 3-Methyl-1-butanol
  b. 2-Methylcyclopentanol
  c. *t*-Butyl alcohol
  d. 1-Methylcyclopentanol
  e. 2-Methyl-2-pentanol

12.56 Classify each of the following as a 1°, 2°, or 3° alcohol:
  a. $CH_3CH_2CH_2CH_2CH_2CH_2CH_2OH$
  b. CH_3CHCH_3
     $\quad\ \ |$
     $\quad OH$

  c. $CH_3\text{—}\underset{\underset{CH_2OH}{|}}{\overset{\overset{CH_3}{|}}{C}}\text{—}CH_3$

  d. $CH_3CH_2\underset{\underset{Br}{|}}{C}HCH_2CH_2OH$

  e. $CH_3CH\text{—}\underset{\underset{CH_3}{|}}{\overset{\overset{CH_3}{|}}{C}}CH_2CH_3$
     $\quad\ \ |$
     $\quad OH$

12.57 Classify each of the following as a primary, secondary, or tertiary alcohol:

  a. $CH_3CH_2\underset{\underset{CH_3}{|}}{\overset{\overset{CH_2CH_3}{|}}{C}}\text{—}OH$

  b. $CH_3CHCH_2CHCH_3$
     $\quad\ \ |\quad\quad\ |$
     $\quad OH\quad Br$

  c. $CH_3CH_2CH_2OH$

  d. $CH_3\underset{\underset{OH}{|}}{\overset{\overset{CH_3}{|}}{C}}CH_2CH_2CH_3$

12.58 Classify each of the following as a primary, secondary, or tertiary alcohol:
  a. 2-Methyl-2-butanol
  b. 1,2-Dimethylcyclohexanol
  c. 2,3,4-Trimethylcyclopentanol
  d. 3,3-Dimethyl-2-pentanol

Reactions Involving Alcohols

Foundations

12.59 Write a general equation representing the preparation of an alcohol by hydration of an alkene.
12.60 Write a general equation representing the preparation of an alcohol by hydrogenation of an aldehyde or a ketone.

12.61 Write a general equation representing the dehydration of an alcohol.
12.62 Write a general equation representing the oxidation of a 1° alcohol.
12.63 Write a general equation representing the oxidation of a 2° alcohol.
12.64 Write a general equation representing the oxidation of a 3° alcohol.

Applications
12.65 Predict the products formed by the hydration of the following alkenes:
   a. 1-Pentene
   b. 2-Pentene
   c. 3-Methyl-1-butene
   d. 3,3-Dimethyl-1-butene
12.66 Draw the alkene products of the dehydration of the following alcohols:
   a. 2-Pentanol
   b. 3-Methyl-1-pentanol
   c. 2-Butanol
   d. 4-Chloro-2-pentanol
   e. 1-Propanol
12.67 Write an equation showing the hydration of each of the following alkenes. Name each of the products using the I.U.P.A.C. Nomenclature System.
   a. 2-Hexene
   b. Cyclopentene
   c. 1-Octene
   d. 1-Methylcyclohexene
12.68 Write an equation showing the dehydration of each of the following alcohols. Name each of the reactants and products using the I.U.P.A.C. Nomenclature System.
   a. CH₃CHCH₂CH₃
              |
              OH
   b.
12.69 What product(s) would result from the oxidation of each of the following alcohols with, for example, potassium permanganate? If no reaction occurs, write N.R.
   a. 2-Butanol
   b. 2-Methyl-2-hexanol
   c. Cyclohexanol
   d. 1-Methyl-1-cyclopentanol
12.70 We have seen that ethanol is metabolized to ethanal (acetaldehyde) in the liver. What would be the product formed, under the same conditions, from each of the following alcohols?
   a. CH₃OH
   b. CH₃CH₂CH₂OH
   c. CH₃CH₂CH₂CH₂OH
12.71 Give the oxidation products of the following alcohols. If no reaction occurs, write N.R.

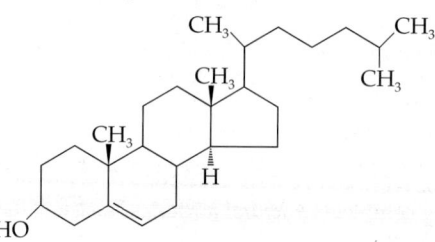

12.72 Write an equation, using complete structural formulas, demonstrating each of the following chemical transformations:
   a. Oxidation of an alcohol to an aldehyde
   b. Oxidation of an alcohol to a ketone
   c. Dehydration of a cyclic alcohol to a cycloalkene
   d. Hydrogenation of an alkene to an alkane
12.73 Write the reaction, occurring in the liver, that causes the oxidation of ethanol. What is the product of this reaction and what symptoms are caused by the product?
12.74 Write the reaction, occurring in the liver, that causes the oxidation of methanol. What is the product of this reaction and what is the possible result of the accumulation of the product in the body?
12.75 Write an equation for the preparation of 2-butanol from 1-butene. What type of reaction is involved?
12.76 Write a general equation for the preparation of an alcohol from an aldehyde or ketone. What type of reaction is involved?
12.77 Show how acetone can be prepared from propene.

$$\text{CH}_3\text{CCH}_3$$ (with =O)
Acetone

12.78 Show how bromocyclopentane can be prepared from cyclopentanol.
12.79 Give the oxidation product for cholesterol.

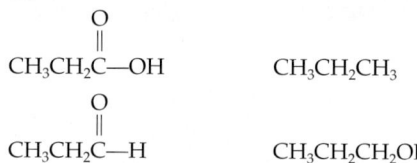
Cholesterol

12.80 Why is a tertiary alcohol not oxidized?

Oxidation and Reduction in Living Systems
Foundations
12.81 Define the terms *oxidation* and *reduction*.
12.82 How do we recognize oxidation and reduction in organic compounds?

Applications
12.83 Arrange the following compounds from the most reduced to the most oxidized:

$$\text{CH}_3\text{CH}_2\text{C}-\text{OH} \quad\quad \text{CH}_3\text{CH}_2\text{CH}_3$$

$$\text{CH}_3\text{CH}_2\text{C}-\text{H} \quad\quad \text{CH}_3\text{CH}_2\text{CH}_2\text{OH}$$

12.84 What is the role of the coenzyme nicotinamide adenine dinucleotide (NAD⁺) in enzyme-catalyzed oxidation–reduction reactions?

Phenols
Foundations
12.85 What are phenols?
12.86 Describe the water solubility of phenols.

Applications
12.87 2,4,6-Trinitrophenol is known by the common name *picric acid*. Picric acid is a solid but is readily soluble in water. In solution it is used as a biological tissue stain. As a solid, it is also known

to be unstable and may explode. In this way it is similar to 2,4,6-trinitrotoluene (TNT). Draw the structures of picric acid and TNT. Why is picric acid readily soluble in water whereas TNT is not?

12.88 Name the following aromatic compounds using the I.U.P.A.C. system:

a. [phenol with OH and NO$_2$ substituents]
b. [benzene with CH(CH$_3$)$_2$ group and OH (para)]
c. [benzene with HO, Br, and Cl substituents]
d. [benzene with OH, CH$_3$, and Br substituents]

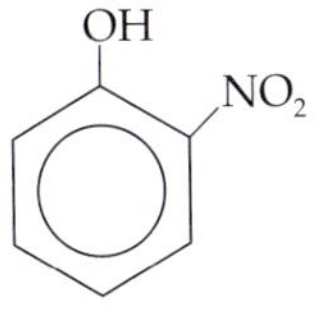

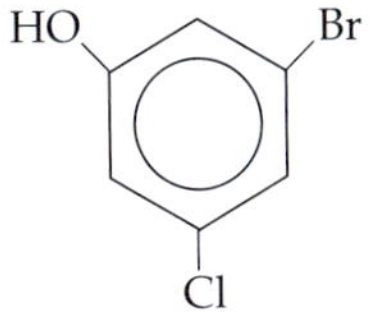

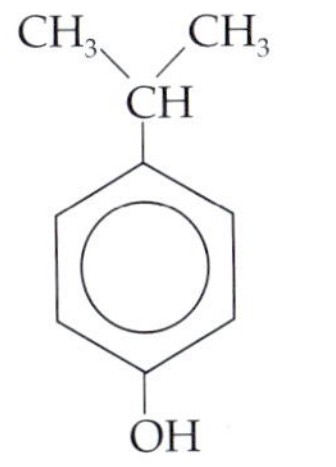

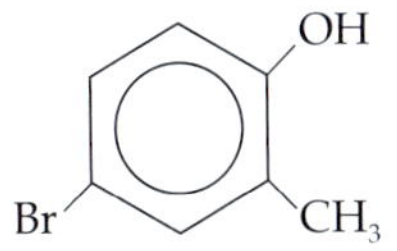

12.89 List some phenol compounds that are commonly used as antiseptics or disinfectants.

12.90 Why must a dilute solution of phenol be used for disinfecting environmental surfaces?

Ethers

12.91 Draw all of the alcohols and ethers of molecular formula $C_4H_{10}O$.

12.92 Name each of the isomers drawn for Problem 12.91.

12.93 Give the I.U.P.A.C. names for Penthrane and Enthrane (see Section 12.8).

12.94 Why have Penthrane and Enthrane replaced diethyl ether as a general anesthetic?

12.95 Ethers may be prepared by the removal of water (dehydration) between two alcohols, as shown. Give the structure(s) of the ethers formed by the reaction of the following alcohol(s) under acidic conditions with heat.

Example: $CH_3OH + HOCH_3 \xrightarrow[\text{Heat}]{H^+} CH_3OCH_3 + H_2O$

a. $2CH_3CH_2OH \longrightarrow ?$
b. $CH_3OH + CH_3CH_2OH \longrightarrow ?$
c. $(CH_3)_2CHOH + CH_3OH \longrightarrow ?$
d. $2 \text{[cyclopentyl]}-CH_2OH \longrightarrow ?$

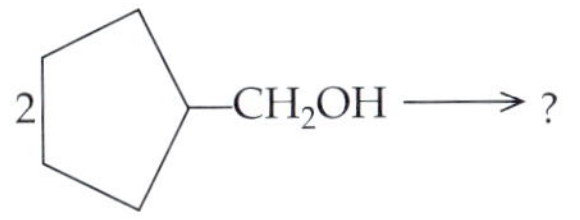

12.96 We have seen that alcohols are capable of hydrogen bonding to each other. Hydrogen bonding is also possible between alcohol molecules and water molecules or between alcohol molecules and ether molecules. Ether molecules *do not* hydrogen bond to each other, however. Explain.

12.97 Name each of the following ethers using the I.U.P.A.C. Nomenclature System:

a. $CH_3CHCH_2CH_2CH_3$
      |
      OCH_2CH_3

b. $CH_3CH_2CHCH_3$
         |
         OCH_3

c. $CH_3CH_2CH_2CH$
            |
            OCH_2CH_3

d. [cyclopentyl]—OCH_3

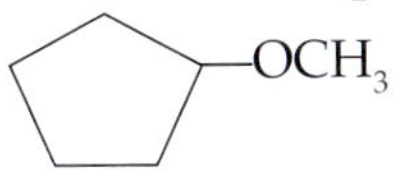

12.98 Draw the structural formula for each of the following ethers:
a. Methyl propyl ether
b. 2-Methoxyoctane
c. Diisopropyl ether
d. 3-Ethoxypentane

Thiols

12.99 Cystine is an amino acid formed from the oxidation of two cysteine molecules to form a disulfide bond. The molecular formula of cystine is $C_6H_{12}O_4N_2S_2$. Draw the structural formula of cystine. (*Hint:* For the structure of cysteine, see page 407.)

12.100 Explain the way in which British Anti-Lewisite acts as an antidote for mercury poisoning.

12.101 Give the I.U.P.A.C. name for each of the following thiols.
a. $CH_3CH_2CH_2—SH$
b. $CH_3CHCH_2CH_3$
         |
         SH

c. $CH_3-\underset{\underset{SH}{|}}{\overset{\overset{CH_2CH_3}{|}}{C}}-CH_3$

d. $HS-$[cyclohexyl]$-SH$

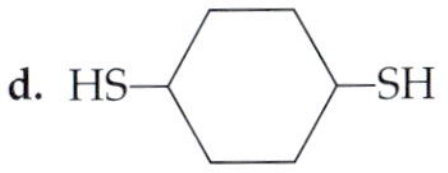

12.102 Give the I.U.P.A.C. name for each of the following thiols.
a. CH_2CHCH_3
    | |
    SH SH

b. [benzene]—SH

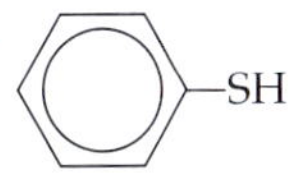

c. $CH_3CHCH_2CH_2CH_3$
         |
         SH

d. $CH_3CH_2CH_2CH_2CH_2CH_2CH_2SH$

CRITICAL THINKING PROBLEMS

1. You are provided with two solvents: water (H_2O) and hexane ($CH_3CH_2CH_2CH_2CH_2CH_3$). You are also provided with two biological molecules whose structures are shown here:

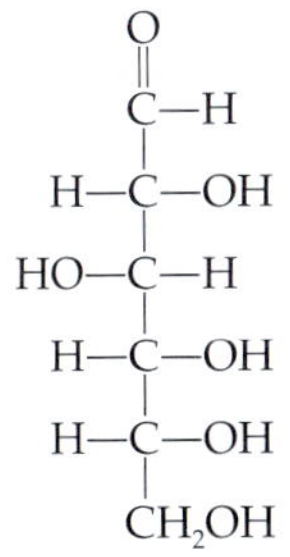

$$\begin{array}{c} O \\ \| \\ C-H \\ H-C-OH \\ HO-C-H \\ H-C-OH \\ H-C-OH \\ CH_2OH \end{array}$$

$$\begin{array}{c} H \quad O \\ | \quad \| \\ H-C-O-C-CH_2CH_2CH_2CH_2CH_2CH_2CH_2CH_2CH_2CH_2CH_2CH_2CH_3 \\ | \quad O \\ | \quad \| \\ H-C-O-C-CH_2CH_2CH_2CH_2CH_2CH_2CH_2CH_2CH_2CH_2CH_2CH_2CH_3 \\ | \quad O \\ | \quad \| \\ H-C-O-C-CH_2CH_2CH_2CH_2CH_2CH_2CH_2CH_2CH_2CH_2CH_2CH_2CH_3 \\ | \\ H \end{array}$$

Predict which biological molecule would be more soluble in water and which would be more soluble in hexane. Defend your prediction. Design a careful experiment to test your hypothesis.

Consider the digestion of dietary molecules in the digestive tract. Which of the two biological molecules shown in this problem would be more easily digested under the conditions present in the digestive tract?

2. Cholesterol is an alcohol and a steroid (Chapter 17). Diets that contain large amounts of cholesterol have been linked to heart disease and atherosclerosis, hardening of the arteries. The narrowing of the artery, caused by plaque buildup, is very apparent. Cholesterol is directly involved in this buildup. Describe the various functional groups and principal structural features of the cholesterol molecule. Would you use a polar or nonpolar solvent to dissolve cholesterol? Explain your reasoning.

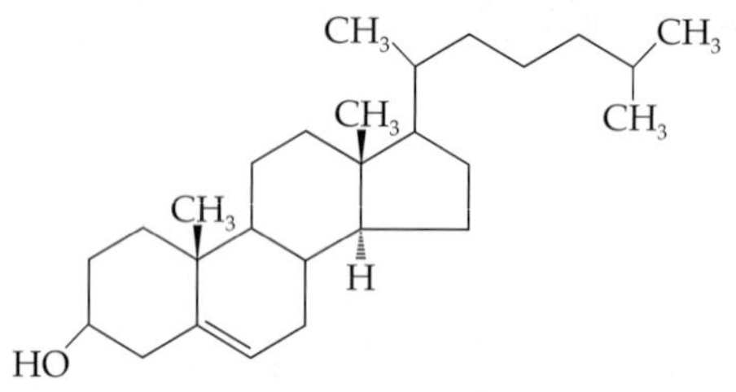

Cholesterol

3. An unknown compound A is known to be an alcohol with the molecular formula $C_4H_{10}O$. When dehydrated, compound A gave only one alkene product, C_4H_8, compound B. Compound A could not be oxidized. What are the identities of compound A and compound B?

4. Sulfides are the sulfur analogs of ethers, that is, ethers in which oxygen has been substituted by a sulfur atom. They are named in an analogous manner to the ethers with the term *sulfide* replacing *ether*. For example, CH_3—S—CH_3 is dimethyl sulfide. Draw the sulfides that correspond to the following ethers and name them:
   a. diethyl ether
   b. methyl propyl ether
   c. dibutyl ether
   d. ethyl phenyl ether

5. Dimethyl sulfoxide (DMSO) has been used by many sports enthusiasts as a linament for sore joints; it acts as an anti-inflammatory agent and a mild analgesic (pain killer). However, it is no longer recommended for this purpose because it carries toxic impurities into the blood. DMSO is a sulfoxide—it contains the S=O functional group. DMSO is prepared from dimethyl sulfide by mild oxidation, and it has the molecular formula C_2H_6SO. Draw the structure of DMSO.

ORGANIC CHEMISTRY

13

Aldehydes and Ketones

Learning Goals

1. Draw the structures and discuss the physical properties of aldehydes and ketones.

2. From the structures, write the common and I.U.P.A.C. names of aldehydes and ketones.

3. List several aldehydes and ketones that are of natural, commercial, health, and environmental interest and describe their significance.

4. Write equations for the preparation of aldehydes and ketones by the oxidation of alcohols.

5. Write equations representing the oxidation of carbonyl compounds.

6. Write equations representing the reduction of carbonyl compounds.

7. Write equations for the preparation of hemiacetals, hemiketals, acetals, and ketals.

8. Draw the keto and enol forms of aldehydes and ketones.

9. Write equations showing the aldol condensation.

Vanilla plant blossom.

Outline

Chemistry Connection:
Genetic Complexity from Simple Molecules

13.1 Structure and Physical Properties

13.2 I.U.P.A.C. Nomenclature and Common Names

13.3 Important Aldehydes and Ketones

13.4 Reactions Involving Aldehydes and Ketones

A Medical Perspective:
Formaldehyde and Methanol Poisoning

A Human Perspective:
Alcohol Abuse and Antabuse

A Medical Perspective:
That Golden Tan Without the Fear of Skin Cancer

A Human Perspective:
The Chemistry of Vision

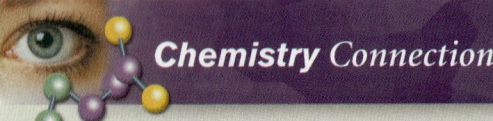

Chemistry Connection

Genetic Complexity from Simple Molecules

Examine any cellular life-form and you will find the same basic genetic system. One or more deoxyribonucleic acid (DNA) molecules carry the genetic code for all the proteins needed by the cell as enzymes—biological catalysts—as essential structural elements, and much more. But DNA cannot be read directly to produce these critical proteins. Instead, the genetic information carried by the DNA is copied to produce a variety of ribonucleic acid (RNA) molecules. These RNA molecules work together, along with other molecules, to produce the proteins.

For decades scientists have been trying to figure out how this amazing genetic system could have evolved from the small, simple molecules that were found in the shallow seas and atmosphere of earth perhaps four billion years ago. Which molecule might have formed first? After all, for the system to work all three molecules are needed: DNA to carry the information, RNA to carry and interpret it, and proteins to do all the cellular chores.

A startling discovery in the 1980s suggested an answer. It was discovered that some RNA molecules could act as biological catalysts. In other words, these RNA molecules could do two jobs: carry genetic information and catalyze chemical reactions. A hypothesis was developed: that RNA was the first biological molecule and that our genetic system evolved from it.

Could RNA have evolved from simple molecules in the "primordial soup"? In the 1960s two of the components of RNA, adenine and guanine, were synthesized in the laboratory from simple molecules and energy sources thought to be present on early earth. In 1995 researchers discovered that, by adding the carbonyl-group-containing molecule urea to their mixture, they could make large amounts of two other components of RNA, uracil and cytosine.

The remaining requirements for making an RNA molecule are phosphate groups and the sugar ribose. Phosphate would have been readily available, but what about ribose? Ribose, it turned out, could easily be produced from the simplest *aldehyde*, formaldehyde.

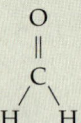

Formaldehyde would have been found in the shallow seas. We can find it today in comets, and many comets struck the early earth. In the laboratory, under conditions designed to imitate those on earth four billion years ago, formaldehyde molecules form chains. These chains twist into cyclic ring structures, including the sugar ribose.

These experiments suggest that all the precursors needed to make RNA could have formed spontaneously and thus that RNA may have been the information-carrying molecule from which our genetic system evolved. But other researchers argue that RNA is too fragile to have survived the conditions on early earth.

We may never know exactly how the first self-replicating genetic system formed. But it is intriguing to speculate about the origin of life and to consider the organic molecules and reactions that may have been involved. In this chapter we consider the aldehydes and ketones, two families of organic molecules containing the carbonyl group. As we will see in this and later chapters, the carbonyl group is a functional group that characterizes many biological molecules and affects their properties and reactivity.

Introduction

The **carbonyl group** consists of a carbon atom bonded to an oxygen atom by a double bond.

Carbonyl group

Compounds containing a carbonyl group are called carbonyl compounds. This group includes the aldehydes and ketones covered in this chapter, as well as the carboxylic acids and amides discussed in Chapters 14 and 15.

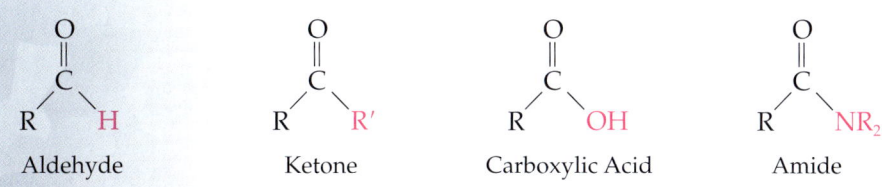

Aldehyde — Ketone — Carboxylic Acid — Amide

R=H, alkyl, or aryl group   R'=alkyl or aryl group

We will study the aldehydes and ketones together because of their similar chemical and physical properties. They are distinguished by the location of the carbonyl group within the carbon chain. In aldehydes the carbonyl group is always located at the end of the carbon chain (carbon-1). In ketones the carbonyl group is located within the carbon chain of the molecule. Thus, in ketones the carbonyl carbon is attached to two other carbon atoms. However, in aldehydes the carbonyl carbon is attached to at least one hydrogen atom; the second atom attached to the carbonyl carbon of an aldehyde may be another hydrogen or a carbon atom (Figure 13.1).

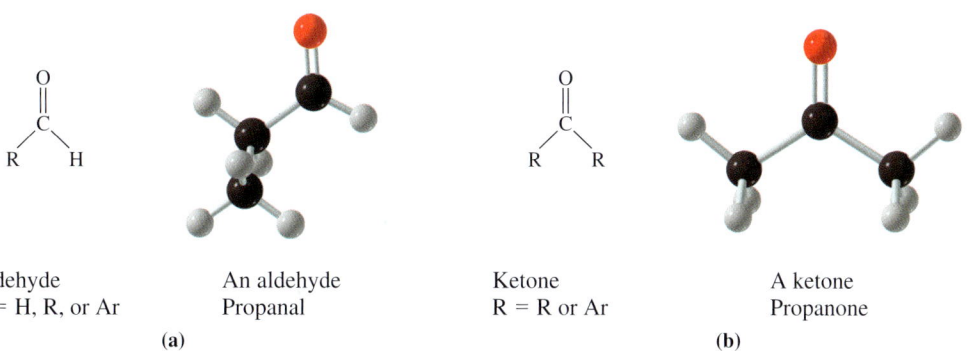

Aldehyde — An aldehyde — Ketone — A ketone
R = H, R, or Ar — Propanal — R = R or Ar — Propanone
(a) (b)

Figure 13.1
The structures of aldehydes and ketones. (a) The general structure of an aldehyde and a ball-and-stick model of the aldehyde propanal. (b) The general structure of a ketone and a ball-and-stick model of the ketone propanone.

13.1 Structure and Physical Properties

Aldehydes and **ketones** are polar compounds because of the polar carbonyl group.

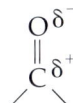

Because of the dipole-dipole attractions between molecules, they boil at higher temperatures than hydrocarbons or ethers that have the same number of carbon atoms or are of equivalent molecular weight. Because they cannot form intermolecular hydrogen bonds, their boiling points are lower than those of alcohols of comparable molecular weight. These trends are clearly demonstrated in the following examples:

1 LEARNING GOAL

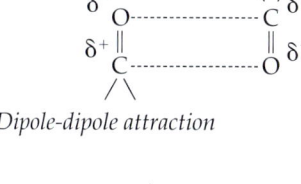
Dipole-dipole attraction

$CH_3CH_2CH_2CH_3$	$CH_3-O-CH_2CH_3$	$CH_3CH_2CH_2-OH$	$CH_3CH_2-\overset{\overset{O}{\|\|}}{C}-H$	$CH_3-\overset{\overset{O}{\|\|}}{C}-CH_3$
Butane (butane)	Methoxyethane (ethyl methyl ether)	1-Propanol (propyl alcohol)	Propanal (propionaldehyde)	Propanone (acetone)
M.W. = 58	M.W. = 60	M.W. = 60	M.W. = 58	M.W. = 58
b.p. −0.5 °C	b.p. 7.0 °C	b.p. 97.2 °C	b.p. 49 °C	b.p. 56 °C

Figure 13.2
(a) Hydrogen bonding between the carbonyl group of an aldehyde or ketone and water. (b) Polar interactions between carbonyl groups of aldehydes or ketones.

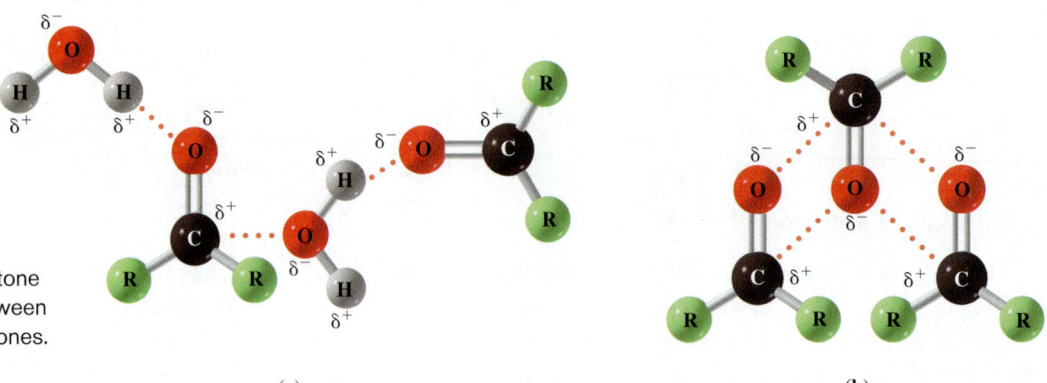

(a) (b)

Aldehydes and ketones can form intermolecular hydrogen bonds with water (Figure 13.2). As a result, the smaller members of the two families (five or fewer carbon atoms) are reasonably soluble in water. However, as the carbon chain length increases, the compounds become less polar and more hydrocarbonlike. These larger compounds are soluble in nonpolar organic solvents.

Question 13.1

Which member in each of the following pairs will be more water soluble?

a. $CH_3(CH_2)_2CH_3$ or $CH_3-\overset{\overset{O}{\|}}{C}-CH_3$

b. $CH_3-\overset{\overset{}{\underset{\underset{O}{\|}}{}}}{C}-CH_2CH_2CH_3$ or $CH_3-\underset{\underset{OH}{|}}{CH}-CH_2CH_2CH_3$

Question 13.2

Which member in each of the following pairs will be more water soluble?

a. cyclopentane-CH_3 or   cyclopentane-CHO

b. $\underset{\underset{OH}{|}}{CH_2}-\underset{\underset{OH}{|}}{CH_2}$ or $H-\overset{\overset{O}{\|}}{C}-\overset{\overset{O}{\|}}{C}-H$

Question 13.3

Which member in each of the following pairs would have a higher boiling point?

a. $CH_3CH_2\overset{\overset{O}{\|}}{C}-OH$ or $CH_3CH_2\overset{\overset{O}{\|}}{C}-H$

b. $CH_3\overset{\overset{O}{\|}}{C}-OH$ or $CH_3\overset{\overset{O}{\|}}{C}-CH_3$

Question 13.4

Which member in each of the following pairs would have a higher boiling point?

a. CH_3CH_2OH or $CH_3C(=O){-}H$
b. $CH_3(CH_2)_6CH_3$ or $CH_3(CH_2)_5C(=O){-}H$

13.2 I.U.P.A.C. Nomenclature and Common Names

Naming Aldehydes

2 LEARNING GOAL

In the I.U.P.A.C. system, aldehydes are named according to the following set of rules:

- Determine the parent compound, that is, the longest continuous carbon chain containing the carbonyl group.
- Replace the final *-e* of the parent alkane with *-al.*
- Number the chain beginning with the carbonyl carbon (or aldehyde group) as carbon-1.
- Number and name all substituents as usual. No number is used for the position of the carbonyl group because it is always at the end of the parent chain. Therefore, it must be carbon-1.

Several examples are provided here with common names given in parentheses:

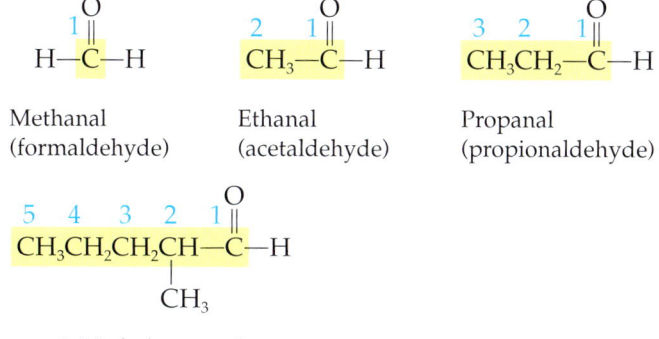

Methanal
(formaldehyde)

Ethanal
(acetaldehyde)

Propanal
(propionaldehyde)

2-Methylpentanal

Using the I.U.P.A.C. Nomenclature System to Name Aldehydes **EXAMPLE 13.1**

2 LEARNING GOAL

Name the aldehydes represented by the following condensed formulas.

Solution

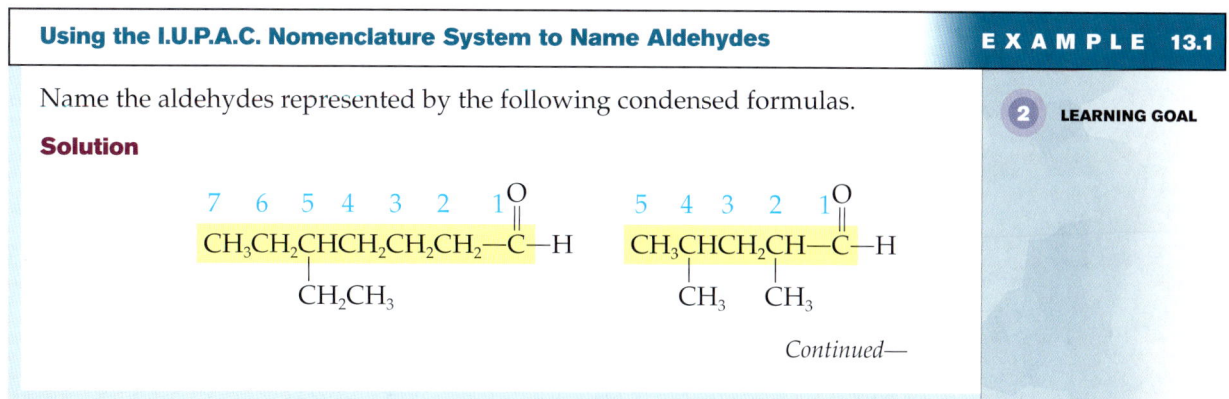

Continued—

EXAMPLE 13.1 —Continued

Parent compound:	heptane (becomes heptanal)	pentane (becomes pentanal)
Position of carbonyl group:	carbon-1	carbon-1
Substituents:	5-ethyl	2,4-dimethyl
Name:	5-Ethylheptanal	2,4-Dimethylpentanal

Notice that the position of the carbonyl group is not indicated by a number. By definition, the carbonyl group is located at the end of the carbon chain of an aldehyde. The carbonyl carbon is defined to be carbon-1; thus, it is not necessary to include the position of the carbonyl group in the name of the compound.

Question 13.5

Name each of the following aldehydes using the I.U.P.A.C. Nomenclature System.

a.  $$\text{CH}_3\text{CH}_2\text{CH}_2\overset{\overset{\text{O}}{\|}}{\text{C}}-\text{H}$$

b. 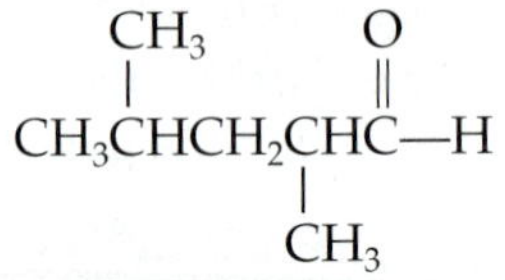 $$\text{CH}_3\overset{\overset{\text{CH}_3}{|}}{\text{CH}}\text{CH}_2\overset{\overset{\text{O}}{\|}}{\underset{\underset{\text{CH}_3}{|}}{\text{CH}}}\text{C}-\text{H}$$

Question 13.6

From the I.U.P.A.C. names, draw the structural formula for each of the following aldehydes.

a. 2,3-Dichloropentanal   b. 2-Bromobutanal   c. 4-Methylhexanal

---

Carboxylic acid nomenclature is described in Section 14.1.

The common names of the aldehydes are derived from the same Latin roots as the corresponding carboxylic acids. The common names of the first five aldehydes are presented in Table 13.1.

In the common system of nomenclature, substituted aldehydes are named as derivatives of the straight-chain parent compound (see Table 13.1). Greek letters are used to indicate the position of the substituents. The carbon atom bonded to the carbonyl group is the α-carbon, the next is the β-carbon, and so on.

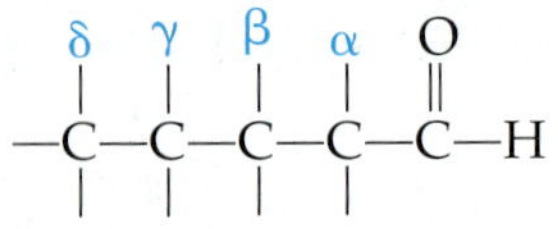

$$\overset{\delta}{\text{C}}-\overset{\gamma}{\text{C}}-\overset{\beta}{\text{C}}-\overset{\alpha}{\text{C}}-\overset{\overset{\text{O}}{\|}}{\text{C}}-\text{H}$$

Consider the following examples:

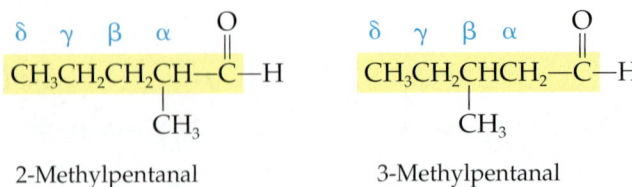

2-Methylpentanal (α-methylvaleraldehyde)

3-Methylpentanal (β-methylvaleraldehyde)

13.2 I.U.P.A.C. Nomenclature and Common Names

TABLE 13.1 — I.U.P.A.C. and Common Names and Formulas for Several Aldehydes

I.U.P.A.C. Name	Common Name	Formula
Methanal	Formaldehyde	$H-\overset{\overset{O}{\|\|}}{C}-H$
Ethanal	Acetaldehyde	$CH_3\overset{\overset{O}{\|\|}}{C}-H$
Propanal	Propionaldehyde	$CH_3CH_2\overset{\overset{O}{\|\|}}{C}-H$
Butanal	Butyraldehyde	$CH_3CH_2CH_2\overset{\overset{O}{\|\|}}{C}-H$
Pentanal	Valeraldehyde	$CH_3CH_2CH_2CH_2\overset{\overset{O}{\|\|}}{C}-H$

EXAMPLE 13.2 — Using the Common Nomenclature System to Name Aldehydes

LEARNING GOAL 2

Name the aldehydes represented by the following condensed formulas.

Solution

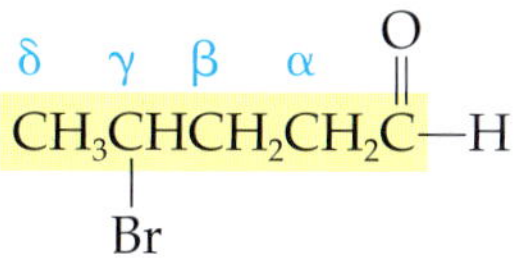

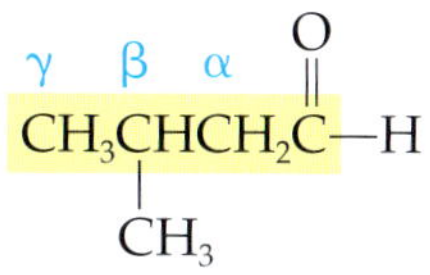

$$\overset{\delta}{C}H_3\overset{\gamma}{C}H\overset{\beta}{C}H_2\overset{\alpha}{C}H_2\overset{\overset{O}{\|\|}}{C}-H \qquad \overset{\gamma}{C}H_3\overset{\beta}{C}H\overset{\alpha}{C}H_2\overset{\overset{O}{\|\|}}{C}-H$$
$$\quad\;\;|\qquad\qquad\qquad\qquad\quad\;\;|$$
$$\quad\;\;Br\qquad\qquad\qquad\qquad\;\;CH_3$$

Parent compound:	pentane	butane
	(becomes valeraldehyde)	(becomes butyraldehyde)
Position of carbonyl group:	carbon-1 (must be!)	carbon-1 (must be!)
Substituents:	γ-bromo	β-methyl
Name:	γ-Bromovaleraldehyde	β-Methylbutyraldehyde

Notice that the substituents are designated by Greek letters, rather than by Arabic numbers. In the common system of nomenclature for aldehydes, the carbon atom bonded to the carbonyl group is called the α-carbon, the next is the β-carbon, etc. Remember to use these Greek letters to indicate the position of the substituents when naming aldehydes using the common system of nomenclature.

Also remember that by definition, the carbonyl group is located at the beginning of the carbon chain of an aldehyde. Thus, it is not necessary to include the position of the carbonyl group in the name of the compound.

Question 13.7

Use the I.U.P.A.C. and common nomenclature systems to name each of the following compounds.

a. 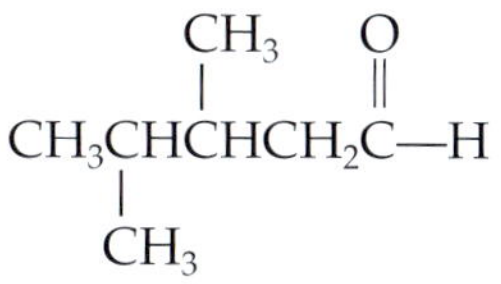 $CH_3\overset{\overset{CH_3}{\|}}{C}H\overset{}{C}H\overset{\overset{O}{\|\|}}{C}H_2C-H$
    $\qquad\;\;|$
    $\qquad CH_3$

b. 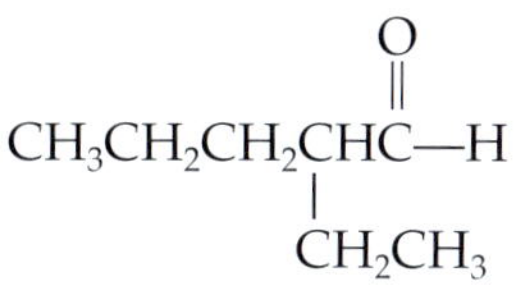 $CH_3CH_2CH_2\overset{}{C}H\overset{\overset{O}{\|\|}}{C}-H$
    $\qquad\qquad\quad\;|$
    $\qquad\qquad\;CH_2CH_3$

Question 13.8

Use the I.U.P.A.C. and common nomenclature systems to name each of the following compounds.

a. 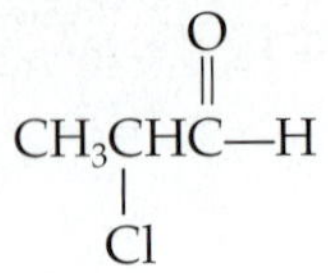 $\text{CH}_3\text{CHC}(=\text{O})\text{—H}$ with Cl on the middle C

b. 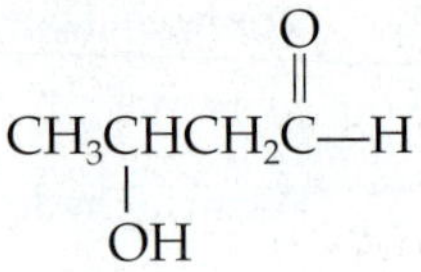 $\text{CH}_3\text{CHCH}_2\text{C}(=\text{O})\text{—H}$ with OH on the second C

Question 13.9

Write the condensed structural formula for each of the following compounds.

a. 3-Methylnonanal
b. β-Bromovaleraldehyde

Question 13.10

Write the condensed structural formula for each of the following compounds.

a. 4-Fluorohexanal
b. α,β-Dimethylbutyraldehyde

Naming Ketones

LEARNING GOAL 2

The rules for naming ketones in the I.U.P.A.C. Nomenclature System are directly analogous to those for naming aldehydes. In ketones, however, the -*e* ending of the parent alkane is replaced with the -*one* suffix of the ketone family, and the location of the carbonyl carbon is indicated with a number. The longest carbon chain is numbered to give the carbonyl carbon the lowest possible number. For example,

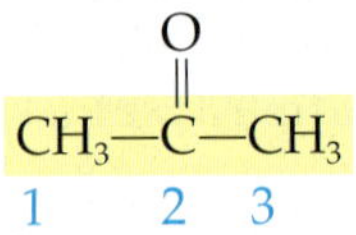

$$\underset{1\quad 2\quad 3}{\text{CH}_3\text{—C}(=\text{O})\text{—CH}_3} \qquad \underset{4\quad 3\quad 2\quad 1}{\text{CH}_3\text{CH}_2\text{—C}(=\text{O})\text{—CH}_3} \qquad \underset{8\ 7\ 6\ 5\ 4\ 3\ 2\ 1}{\text{CH}_3\text{CH}_2\text{CH}_2\text{CH}_2\text{—C}(=\text{O})\text{—CH}_2\text{CH}_2\text{CH}_3}$$

Propanone
(no number necessary)
(acetone)

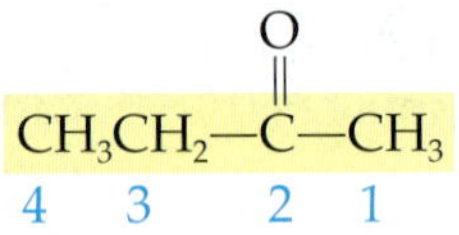

Butanone
(no number necessary)
(methyl ethyl ketone)

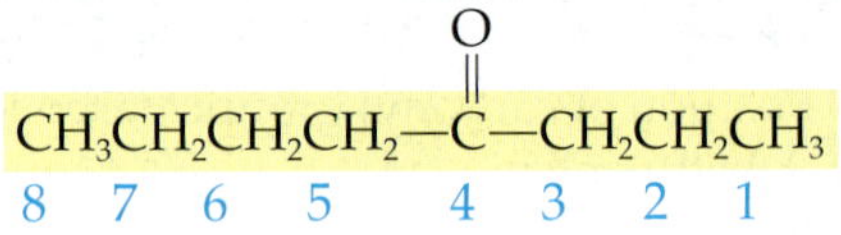

4-Octanone
(*not* 5-octanone)
(butyl propyl ketone)

EXAMPLE 13.3 Using the I.U.P.A.C Nomenclature System to Name Ketones

Name the ketones represented by the following condensed formulas.

Solution

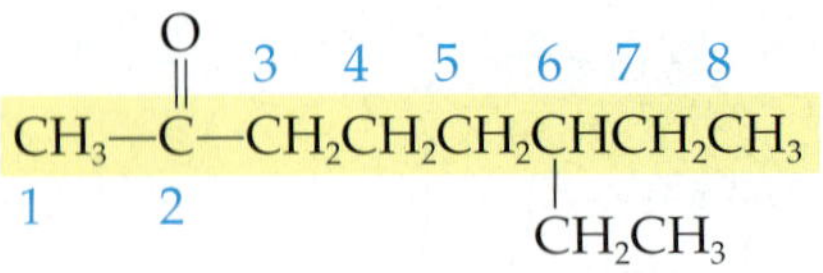

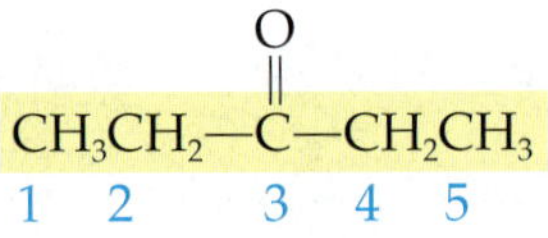

$$\underset{1\quad 2}{\text{CH}_3\text{—C}(=\text{O})\text{—}}\underset{3\ 4\ 5\ 6\ 7\ 8}{\text{CH}_2\text{CH}_2\text{CH}_2\text{CHCH}_2\text{CH}_3}$$
                with CH_2CH_3 branch

$$\underset{1\quad 2}{\text{CH}_3\text{CH}_2\text{—C}(=\text{O})\text{—}}\underset{3\ 4\ 5}{\text{CH}_2\text{CH}_3}$$

Parent compound:	octane (becomes octanone)	pentane (becomes pentanone)
Position of carbonyl group:	carbon-2 (not carbon-7)	carbon-3
Substituents:	6-ethyl	none
Name:	6-Ethyl-2-octanone	3-Pentanone

13.2 I.U.P.A.C. Nomenclature and Common Names

Question 13.11

Use the I.U.P.A.C. Nomenclature System to name each of the following compounds.

a. 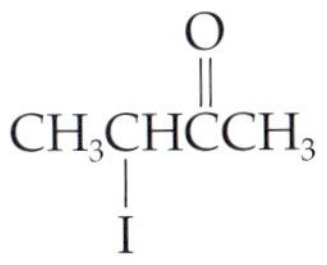 $\text{CH}_3\text{CHCCH}_3$ with O double-bonded to the third carbon and I substituent on the second carbon

b. 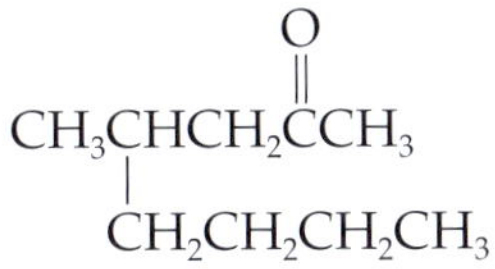 $\text{CH}_3\text{CHCH}_2\text{CCH}_3$ with O double-bonded to the fourth carbon and $\text{CH}_2\text{CH}_2\text{CH}_2\text{CH}_3$ substituent on the second carbon

Question 13.12

Use the I.U.P.A.C. Nomenclature System to name each of the following compounds.

a. 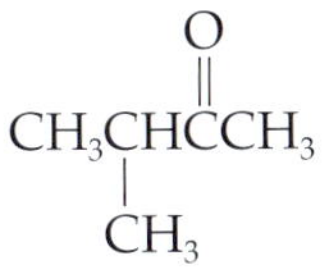 $\text{CH}_3\text{CHCCH}_3$ with O double-bonded to the third carbon and CH_3 substituent on the second carbon

b. 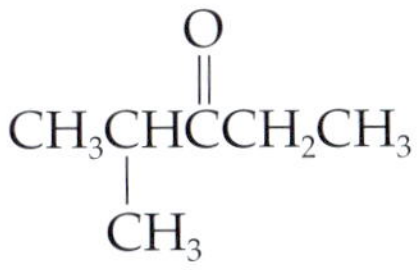 $\text{CH}_3\text{CHCCH}_2\text{CH}_3$ with O double-bonded to the third carbon and CH_3 substituent on the second carbon

c. 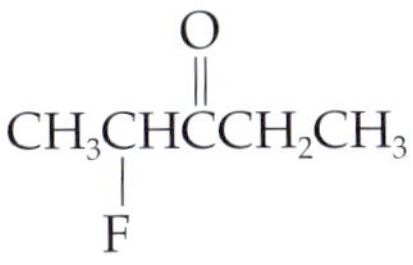 $\text{CH}_3\text{CHCCH}_2\text{CH}_3$ with O double-bonded to the third carbon and F substituent on the second carbon

---

The common names of ketones are derived by naming the alkyl groups that are bonded to the carbonyl carbon. These are used as prefixes followed by the word *ketone*. The alkyl groups may be arranged alphabetically or by size (smaller to larger).

EXAMPLE 13.4 Using the Common Nomenclature System to Name Ketones

Name the ketones represented by the following condensed formulas.

Solution

Identify the alkyl groups that are bonded to the carbonyl carbon.

$$\text{CH}_3\text{CH}_2\text{CH}_2-\overset{\overset{\displaystyle O}{\|}}{C}-\text{CH}_3 \qquad \text{CH}_3\text{CH}_2-\overset{\overset{\displaystyle O}{\|}}{C}-\text{CH}_2\text{CH}_2\text{CH}_2\text{CH}_2\text{CH}_3$$

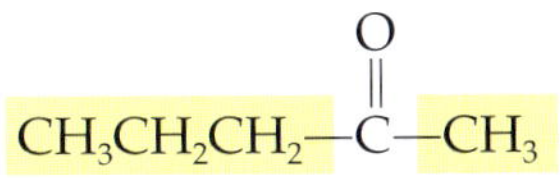

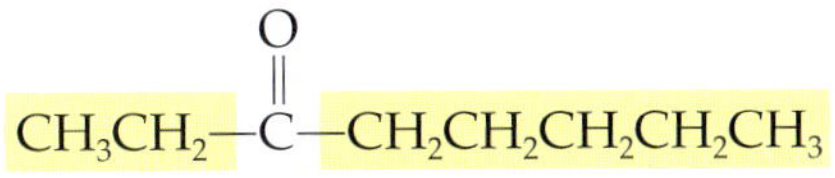

Alkyl groups:	propyl and methyl	ethyl and pentyl
Name:	Methyl propyl ketone	Ethyl pentyl ketone

Because the two *groups* bonded to the carbonyl carbon are named, a ketone is actually one carbon longer than an aldehyde with a *similar common name*. For example, methyl butyl ketone has six carbons, but β-methylbutyraldehyde has only five.

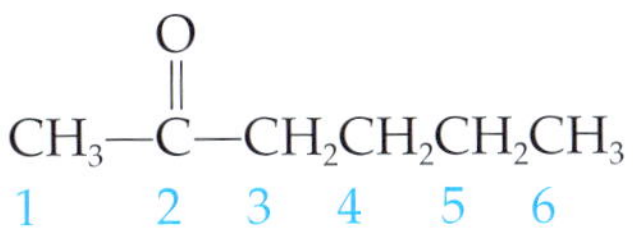

$$\underset{\substack{\text{Methyl butyl ketone}}}{\underset{1 \quad 2 \ 3 \ 4 \ 5 \ 6}{\text{CH}_3-\overset{\overset{\displaystyle O}{\|}}{C}-\text{CH}_2\text{CH}_2\text{CH}_2\text{CH}_3}} \qquad \underset{\substack{\text{β-Methylbutyraldehyde}}}{\underset{\underset{5}{\text{CH}_3}}{\overset{4 \quad 3 \quad 2}{\text{CH}_3\text{CHCH}_2}-\overset{\overset{\displaystyle O}{\|}}{\underset{1}{C}}-\text{H}}}$$

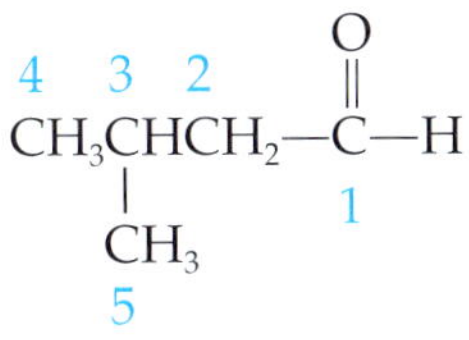

This is because the aldehyde carbonyl carbon is included in the name of the parent chain, butyraldehyde. The carbonyl carbon of the ketone is not included in the common name. It is treated only as the carbon to which the two alkyl or aryl groups are attached.

Question 13.13

Write the condensed formula for each of the following compounds.

a. Methyl isopropyl ketone (What is the I.U.P.A.C. name for this compound?)
b. 4-Heptanone

Question 13.14

Write the condensed formula for each of the following compounds.

a. 2-Fluorocyclohexanone
b. Hexachloroacetone (What is the I.U.P.A.C. name of this compound?)

13.3 Important Aldehydes and Ketones

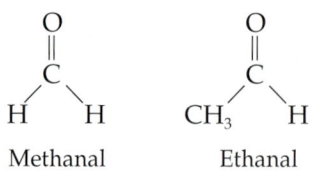

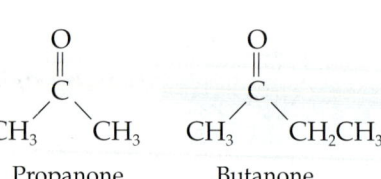

Methanal (formaldehyde) is a gas (b.p. −21°C). It is available commercially as an aqueous solution called *formalin*. Formalin has been used as a preservative for tissues and as an embalming fluid. See A Medical Perspective: Formaldehyde and Methanol Poisoning for more information on methanal.

Ethanal (acetaldehyde) is produced from ethanol in the liver. Ethanol is oxidized in this reaction, which is catalyzed by the liver enzyme alcohol dehydrogenase. The ethanal that is produced in this reaction is responsible for the symptoms of a hangover.

Propanone (acetone), the simplest ketone, is an important and versatile solvent for organic compounds. It has the ability to dissolve organic compounds and is also miscible with water. As a result, it has a number of industrial applications and is used as a solvent in adhesives, paints, cleaning solvents, nail polish, and nail polish remover. Propanone is flammable and should therefore be treated with appropriate care. *Butanone*, a four-carbon ketone, is also an important industrial solvent.

Many aldehydes and ketones are produced industrially as food and fragrance chemicals, medicinals, and agricultural chemicals. They are particularly important to the food industry, in which they are used as artificial and/or natural additives to food. Vanillin, a principal component of natural vanilla, is shown in Figure 13.3. Artificial vanilla flavoring is a dilute solution of synthetic vanillin dissolved in ethanol. Figure 13.3 also shows other examples of important aldehydes and ketones.

Question 13.15

Draw the structure of the aldehyde synthesized from ethanol in the liver.

Question 13.16

Draw the structure of a ketone that is an important, versatile solvent for organic compounds.

13.4 Reactions Involving Aldehydes and Ketones

Preparation of Aldehydes and Ketones

Aldehydes and ketones are prepared primarily by the **oxidation** of the corresponding alcohol. As we saw in Chapter 12, the oxidation of methyl alcohol gives methanal (formaldehyde). The oxidation of a primary alcohol produces an aldehyde, and the oxidation of a secondary alcohol yields a ketone. Tertiary alcohols do not undergo oxidation under the conditions normally used.

13.4 Reactions Involving Aldehydes and Ketones

Figure 13.3
Important aldehydes and ketones.

A Medical Perspective

Formaldehyde and Methanol Poisoning

Most aldehydes have irritating, unpleasant odors, and formaldehyde is no exception. Formaldehyde is also an extremely toxic substance. As an aqueous solution, called formalin, it has been used to preserve biological tissues and for embalming. It has also been used to disinfect environmental surfaces, body fluids, and feces. Under no circumstances is it used as an antiseptic on human tissue because of its toxic fumes and the skin irritations that it causes.

Formaldehyde is used in the production of some killed-virus vaccines. When a potentially deadly virus, such as polio virus, is treated with heat and formaldehyde, the genetic information (RNA) is damaged beyond repair. The proteins of the virus also react with formaldehyde. However, the shape of the proteins, which is critical for a protective immune response against the virus, is not changed. Thus, when a child is injected with the Salk killed-virus polio vaccine, the immune system recognizes viral proteins and produces antibodies that protect against polio virus infection.

Formaldehyde can also be produced in the body! As we saw in Chapter 12, methanol can be oxidized to produce formaldehyde. In the body, the liver enzyme alcohol dehydrogenase, whose function it is to detoxify alcohols by oxidizing them, catalyzes the conversion of methanol to formaldehyde (methanal). The formaldehyde then reacts with cellular macromolecules, including proteins, causing severe damage (remember, it is used as an embalming agent!). As a result, methanol poisoning can cause blindness, respiratory failure, convulsions, and death.

Clever physicians have devised a treatment for methanol poisoning that is effective if administered soon enough after ingestion. Since the same enzyme that oxidizes methanol to formaldehyde (methanal) also oxidizes ethanol to acetaldehyde (ethanal), doctors reasoned that administering an intravenous solution of ethanol to the patient could protect against the methanol poisoning. If the ethanol concentration in the body is higher than the methanol concentration, most of the alcohol dehydrogenase enzymes of the liver will be carrying out the oxidation of the ethanol. This is called *competitive inhibition* because the methanol and ethanol molecules are competing for binding to the enzymes. The molecule that is in the higher concentration will more frequently bind to the enzyme and undergo reaction. In this case, the result is that the alcohol dehydrogenase enzymes are kept busy oxidizing ethanol and producing the less-toxic (not nontoxic) product, acetaldehyde. This gives the body time to excrete the methanol before it is oxidized to the potentially deadly formaldehyde.

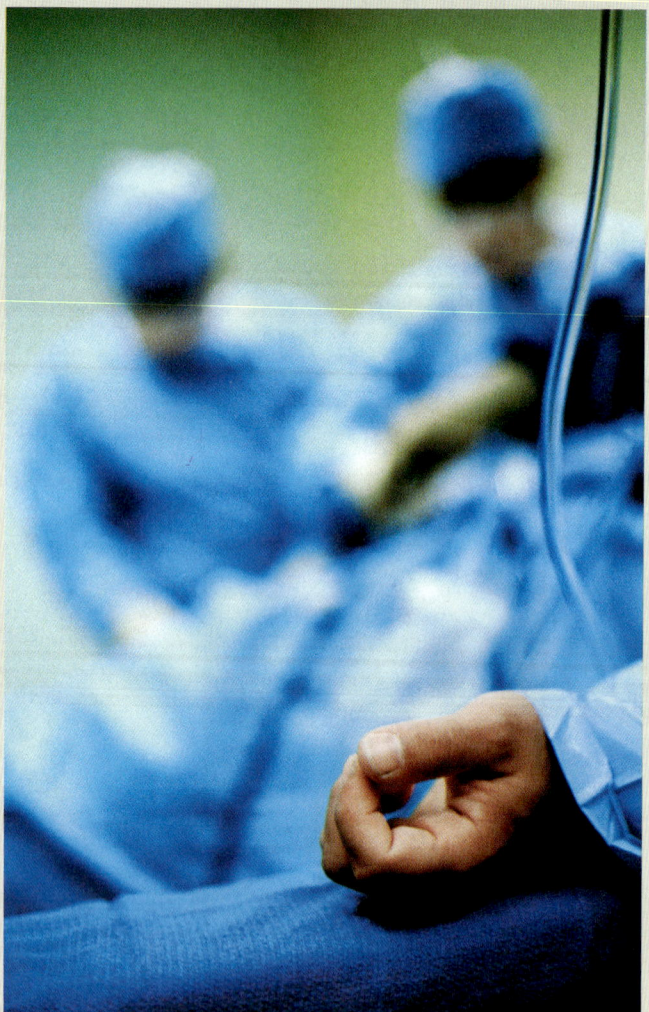

Patient receiving an intravenous solution of ethanol to treat methanol poisoning.

For Further Understanding

Acetaldehyde is described as less toxic than formaldehyde. Do some background research on the effects of these two aldehydes on biological systems.

You have studied enzymes in previous biology courses. Using what you learned in those classes with information from Sections 19.4 and 19.10 in this text, put together an explanation of the way in which competitive inhibition works. Can you think of other types of poisoning for which competitive inhibition might be used to develop an effective treatment?

13.4 Reactions Involving Aldehydes and Ketones

> **Differentiating the Oxidation of Primary, Secondary, and Tertiary Alcohols** **EXAMPLE 13.5**
>
> Use specific examples to show the oxidation of a primary, a secondary, and a tertiary alcohol.
>
> **Solution**
>
> The oxidation of a primary alcohol to an aldehyde: 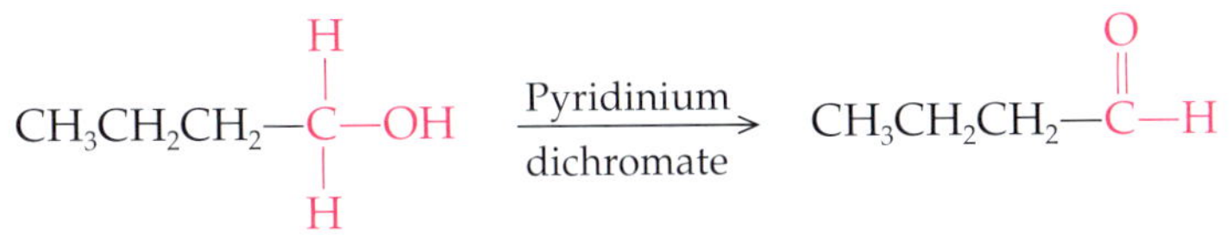
>
> $$\underset{\substack{\text{1-Butanol}\\\text{(butyl alcohol)}}}{CH_3CH_2CH_2-\underset{H}{\overset{H}{C}}-OH} \xrightarrow{\text{Pyridinium dichromate}} \underset{\substack{\text{Butanal}\\\text{(butyraldehyde)}}}{CH_3CH_2CH_2-\overset{O}{\overset{\|}{C}}-H}$$
>
> A mild oxidizing agent must be used in the oxidation of a primary alcohol to an aldehyde. Otherwise, the aldehyde will be further oxidized to a carboxylic acid.
>
> The oxidation of a secondary alcohol to a ketone: 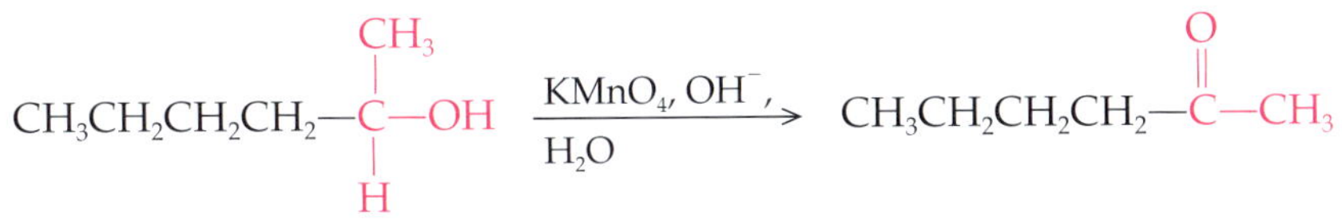
>
> $$\underset{\text{2-Hexanol}}{CH_3CH_2CH_2CH_2-\underset{H}{\overset{CH_3}{C}}-OH} \xrightarrow[H_2O]{KMnO_4,\ OH^-} \underset{\text{2-Hexanone}}{CH_3CH_2CH_2CH_2-\overset{O}{\overset{\|}{C}}-CH_3}$$
>
> Tertiary alcohols cannot undergo oxidation: 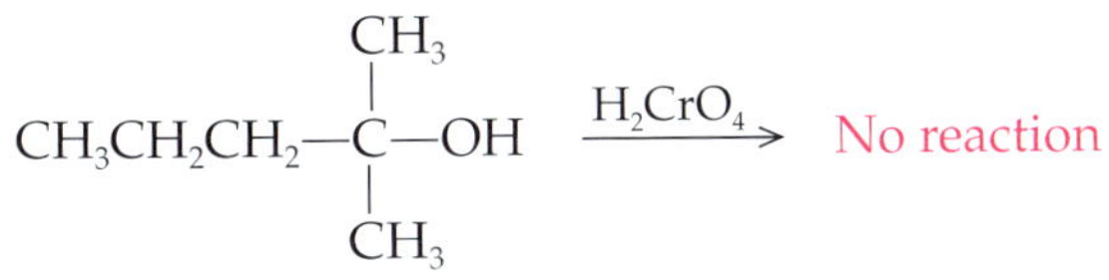
>
> $$\underset{\text{2-Methyl-2-pentanol}}{CH_3CH_2CH_2-\underset{CH_3}{\overset{CH_3}{C}}-OH} \xrightarrow{H_2CrO_4} \text{No reaction}$$

Write an equation showing the oxidation of 1-propanol. **Question 13.17**

Write an equation showing the oxidation of 2-butanol. **Question 13.18**

Oxidation Reactions

5 LEARNING GOAL

Aldehydes are easily oxidized further to carboxylic acids, whereas ketones do not generally undergo further oxidation. The reason is that an aldehydic carbon-hydrogen bond, present in the aldehyde but not in the ketone, is needed for the reaction to occur. In fact, aldehydes are so easily oxidized that it is often very difficult to prepare them because they continue to react to give the carboxylic acid rather than the desired aldehyde. Aldehydes are susceptible to air oxidation, even at room temperature, and cannot be stored for long periods. The following example shows a general equation for the oxidation of an aldehyde to a carboxylic acid:

$$\underset{\text{Aldehyde}}{R-\overset{\overset{O}{\|}}{C}-H} \xrightarrow{[O]} \underset{\text{Carboxylic acid}}{R-\overset{\overset{O}{\|}}{C}-OH}$$

In basic solution, the product is the carboxylic acid anion:

$$CH_3-\overset{\overset{O}{\|}}{C}-O^-$$

Many oxidizing agents can be used. Both basic potassium permanganate and chromic acid are good oxidizing agents, as the following specific example shows:

$$\underset{\substack{\text{Ethanal} \\ \text{(acetaldehyde)}}}{CH_3-\overset{\overset{O}{\|}}{C}-H} \xrightarrow[H_2O,OH^-]{KMnO_4} \underset{\substack{\text{Ethanoate anion} \\ \text{(acetate anion)}}}{CH_3-\overset{\overset{O}{\|}}{C}-O^-}$$

The rules for naming carboxylic acid anions are described in Section 14.1.

The oxidation of benzaldehyde to benzoic acid is an example of the conversion of an *aromatic* aldehyde to the corresponding aromatic carboxylic acid:

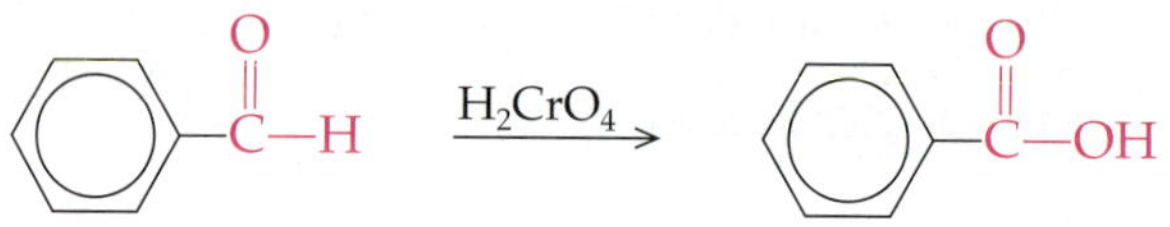

$$\underset{\text{Benzaldehyde}}{C_6H_5-\overset{\overset{O}{\|}}{C}-H} \xrightarrow{H_2CrO_4} \underset{\text{Benzoic acid}}{C_6H_5-\overset{\overset{O}{\|}}{C}-OH}$$

Silver ions are very mild oxidizing agents. They will oxidize aldehydes but not alcohols.

Aldehydes and ketones can be distinguished on the basis of differences in their reactivity. The most common laboratory test for aldehydes is the **Tollens' test**. When exposed to the Tollens' reagent, a basic solution of $Ag(NH_3)_2^+$, an aldehyde undergoes oxidation. The silver ion (Ag^+) is reduced to silver metal (Ag^0) as the aldehyde is oxidized to a carboxylic acid anion.

$$\underset{\text{Aldehyde}}{R-\overset{\overset{O}{\|}}{C}-H} + \underset{\substack{\text{Silver} \\ \text{ammonia complex—} \\ \text{Tollens' reagent}}}{Ag(NH_3)_2^+} \longrightarrow \underset{\substack{\text{Carboxylate} \\ \text{anion}}}{R-\overset{\overset{O}{\|}}{C}-O^-} + \underset{\substack{\text{Silver} \\ \text{metal} \\ \text{mirror}}}{Ag^0}$$

Silver metal precipitates from solution and coats the flask, producing a smooth silver mirror, as seen in Figure 13.4. The test is therefore often called the Tollens' silver mirror test. The commercial manufacture of silver mirrors uses a similar process. Ketones cannot be oxidized to carboxylic acids and do not react with the Tollens' reagent.

EXAMPLE 13.6 Writing Equations for the Reaction of an Aldehyde and of a Ketone with Tollens' Reagent

Write equations for the reaction of propanal and 2-pentanone with Tollens' reagent.

Solution

$$\underset{\text{Propanal}}{CH_3CH_2\overset{\overset{O}{\|}}{C}-H} + Ag(NH_3)_2^+ \longrightarrow \underset{\text{Propanoate anion}}{CH_3CH_2\overset{\overset{O}{\|}}{C}-O^-} + Ag^0$$

$$\underset{\text{2-Pentanone}}{CH_3CH_2CH_2\overset{\overset{O}{\|}}{C}CH_3} + Ag(NH_3)_2^+ \longrightarrow \text{No reaction}$$

13.4 Reactions Involving Aldehydes and Ketones

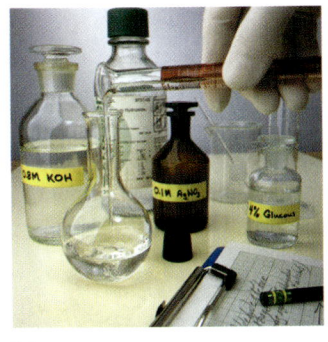
(a)

(b)

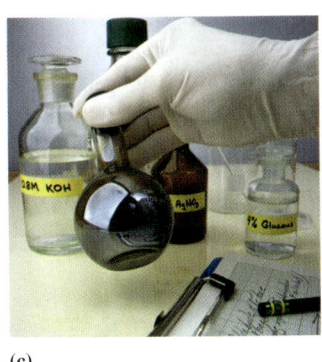
(c)

(d)

Figure 13.4
The silver precipitate produced by the Tollens' reaction is deposited on glass. The progress of the reaction is visualized in panels (a) through (d). Silver mirrors are made in a similar process.

Question 13.19
Write an equation for the reaction of ethanal with Tollens' reagent.

Question 13.20
Write an equation for the reaction of propanone with Tollens' reagent.

Another test that is used to distinguish between aldehydes and ketones is **Benedict's test.** Here, a buffered aqueous solution of copper(II) hydroxide and sodium citrate reacts to oxidize aldehydes but does not generally react with ketones. Cu^{2+} is reduced to Cu^+ in the process. Cu^{2+} is soluble and gives a blue solution, whereas the Cu^+ precipitates as the red solid copper(I) oxide, Cu_2O.

All simple sugars (monosaccharides) are either aldehydes or ketones. Glucose is an aldehyde sugar that is commonly called *blood sugar* because it is the sugar found transported in the blood and used for energy by many cells. In uncontrolled diabetes, glucose may be found in the urine. One early method used to determine the amount of glucose in the urine was to observe the color change of the Benedict's test. The amount of precipitate formed is directly proportional to the amount of glucose in the urine (Figure 13.5). The reaction of glucose with the Benedict's reagent is represented in the following equation:

$$\begin{array}{c} \text{O} \diagdown \text{H} \\ \text{C} \\ | \\ \text{H}-\text{C}-\text{OH} \\ | \\ \text{HO}-\text{C}-\text{H} \\ | \\ \text{H}-\text{C}-\text{OH} \\ | \\ \text{H}-\text{C}-\text{OH} \\ | \\ \text{CH}_2\text{OH} \\ \text{Glucose} \end{array} + 2Cu^{2+} \xrightarrow{OH^-} \begin{array}{c} \text{O} \diagdown \text{O}^- \\ \text{C} \\ | \\ \text{H}-\text{C}-\text{OH} \\ | \\ \text{HO}-\text{C}-\text{H} \\ | \\ \text{H}-\text{C}-\text{OH} \\ | \\ \text{H}-\text{C}-\text{OH} \\ | \\ \text{CH}_2\text{OH} \end{array} + Cu_2O$$

Cu(II) is an even milder oxidizing agent than silver ion.

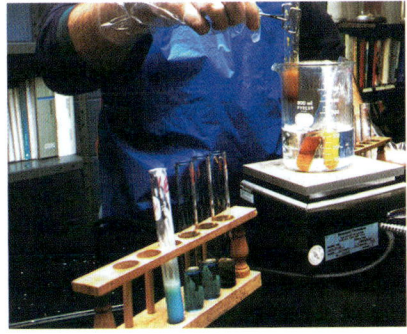

Figure 13.5
The amount of precipitate formed and thus the color change observed in the Benedict's test are directly proportional to the amount of reducing sugar in the sample.

A Human Perspective

Alcohol Abuse and Antabuse

According to a recent study carried out by the Centers for Disease Control and Prevention,[1] 75,000 Americans die each year as a result of alcohol abuse. Of these, 34,833 people died of cirrhosis of the liver, cancer, or other drinking-related diseases. The remaining 40,933 died in alcohol-related automobile accidents. Of those who died, 72% were men and 6% were under the age of 21. In fact, a separate study has estimated that 1400 college-age students die each year of alcohol-related causes.

These numbers are striking. Alcohol abuse is now the third leading cause of preventable death in the United States, outranked only by tobacco use and poor diet and exercise habits. As the study concluded, "These results emphasize the importance of adopting effective strategies to reduce excessive drinking, including increasing alcohol excise taxes and screening for alcohol misuse in clinical settings."

Tetraethylthiuram disulfide
(disulfiram)
Antabuse

One approach to treatment of alcohol abuse, the drug tetraethylthiuram disulfide or disulfiram, has been used since 1951. The activity of this drug generally known by the trade name Antabuse was discovered accidentally by a group of Danish researchers who were testing it for antiparasitic properties. They made the observation that those who had taken disulfiram became violently ill after consuming any alcoholic beverage. Further research revealed that this compound inhibits one of the liver enzymes in the pathway for the oxidation of alcohols.

In Chapter 12 we saw that ethanol is oxidized to ethanal (acetaldehyde) in the liver. This reaction is catalyzed by the enzyme alcohol dehydrogenase. Acetaldehyde, which is more toxic than ethanol, is responsible for many of the symptoms of a hangover. The enzyme acetaldehyde dehydrogenase oxidizes acetaldehyde into ethanoic acid (acetic acid), which then is used in biochemical pathways that harvest energy for cellular work or that synthesize fats.

Antabuse inhibits acetaldehyde dehydrogenase. This inhibition occurs within one to two hours of taking the drug and continues up to fourteen days. When a person who has taken Antabuse drinks an alcoholic beverage, the level of acetaldehyde quickly reaches levels that are five to ten times higher than would normally occur after a drink. Within just a few minutes, the symptoms of a severe hangover are experienced and may continue for several hours.

The drug Antabuse may be useful in treating alcohol abuse.

Experts in drug and alcohol abuse have learned that drugs such as Antabuse are generally not effective on their own. However, when used in combination with support groups and/or psychotherapy to solve underlying behavioral or psychological problems, Antabuse is an effective deterrent to alcohol abuse.

1. Alcohol-Attributable Deaths and Years of Potential Life Lost—United States, 2001, Morbidity and Mortality Weekly Report, 53 (37): 866–870, September 24, 2004, also available at http://www.cdc.gov/mmwr/preview/mmwrhtml/mm5337a2.htm.

For Further Understanding

Antabuse alone is not a cure for alcoholism. Consider some of the reasons why this is so.

Write equations showing the oxidation of ethanol to ethanoic acid as a pathway with the product of the first reaction serving as the reactant for the second. Explain the physiological effects of Antabuse in terms of these chemical reactions.

13.4 Reactions Involving Aldehydes and Ketones

We should also note that when the carbonyl group of a ketone is bonded to a —CH_2OH group, the molecule will give a positive Benedict's test. This occurs because such ketones are converted to aldehydes under basic conditions. In Chapter 16 we will see that this applies to the ketone sugars, as well. They are converted to aldehyde sugars and react with Benedict's reagent.

Reduction Reactions

Aldehydes and ketones are both readily reduced to the corresponding alcohol by a variety of reducing agents. Throughout the text the symbol [H] over the reaction arrow represents a reducing agent.

The classical method of aldehyde or ketone reduction is **hydrogenation**. The carbonyl compound is reacted with hydrogen gas and a catalyst (nickel, platinum, or palladium metal) in a pressurized reaction vessel. Heating may also be necessary. The carbon-oxygen double bond (the carbonyl group) is reduced to a carbon-oxygen single bond. This is similar to the reduction of an alkene to an alkane (the reduction of a carbon-carbon double bond to a carbon-carbon single bond). The addition of hydrogen to a carbon-oxygen double bond is shown in the following general equation:

LEARNING GOAL 6

One way to recognize reduction, particularly in organic chemistry, is the gain of hydrogen. Oxidation and reduction are discussed in Section 12.6.

Hydrogenation was first discussed in Section 11.5 for the hydrogenation of alkenes.

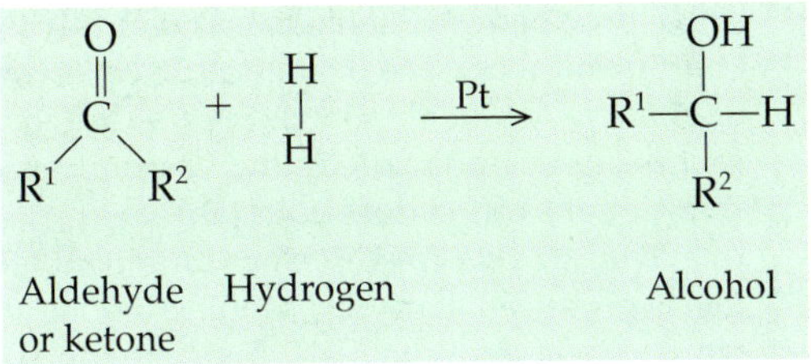

$$\underset{\substack{\text{Aldehyde} \\ \text{or ketone}}}{\overset{O}{\underset{R^1 \;\; R^2}{\|\!\!\text{C}}}} + \underset{\text{Hydrogen}}{\overset{H}{\underset{H}{|}}} \xrightarrow{\text{Pt}} \underset{\text{Alcohol}}{\overset{OH}{\underset{R^2}{R^1\!-\!\overset{|}{\underset{|}{C}}\!-\!H}}}$$

The hydrogenation (reduction) of a ketone produces a secondary alcohol, as seen in the following equation showing the reduction of the ketone, 3-octanone:

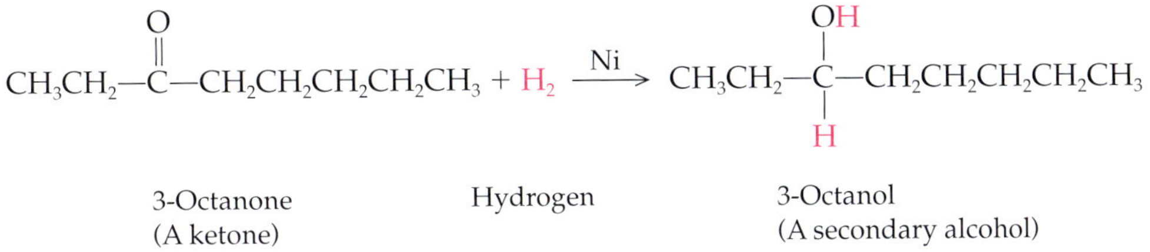

$$\underset{\substack{\text{3-Octanone} \\ \text{(A ketone)}}}{CH_3CH_2-\overset{O}{\overset{\|}{C}}-CH_2CH_2CH_2CH_2CH_3} + \underset{\text{Hydrogen}}{H_2} \xrightarrow{\text{Ni}} \underset{\substack{\text{3-Octanol} \\ \text{(A secondary alcohol)}}}{CH_3CH_2-\overset{OH}{\underset{H}{\overset{|}{\underset{|}{C}}}}-CH_2CH_2CH_2CH_2CH_3}$$

EXAMPLE 13.7 Writing an Equation Representing the Hydrogenation of a Ketone

Write an equation showing the hydrogenation of 3-pentanone.

Solution

The product of the reduction of a ketone is a secondary alcohol, in this case, 3-pentanol.

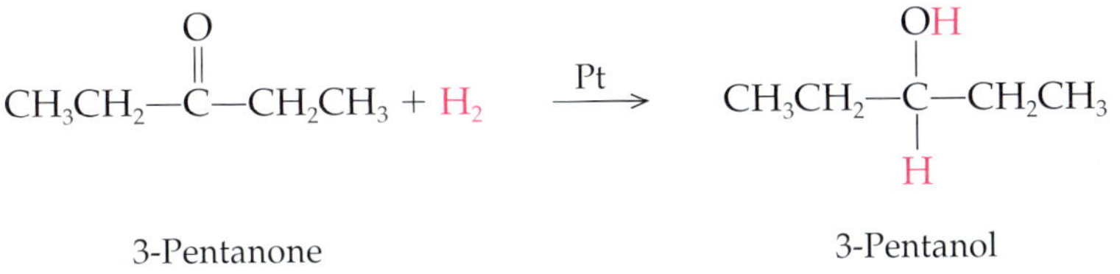

$$\underset{\text{3-Pentanone}}{CH_3CH_2-\overset{O}{\overset{\|}{C}}-CH_2CH_3} + H_2 \xrightarrow{\text{Pt}} \underset{\text{3-Pentanol}}{CH_3CH_2-\overset{OH}{\underset{H}{\overset{|}{\underset{|}{C}}}}-CH_2CH_3}$$

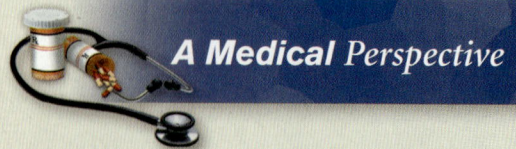

A Medical Perspective

That Golden Tan Without the Fear of Skin Cancer

Self-tanning lotions have become very popular in recent years. This seems to be the result of our growing understanding of the link between exposure to the sun and skin cancer and to improvements in the quality of the tan produced by these self-tanners.

The active ingredient in most self-tanners is dihydroxyacetone (DHA).

$$\begin{array}{c} CH_2OH \\ | \\ C=O \\ | \\ CH_2OH \end{array}$$

Dihydroxyacetone

This simple, three-carbon molecule has two hydroxyl groups (—OH) and is, thus, an alcohol. It also has a carbonyl group at the center carbon:

$$\diagdown C=O \diagup$$

Carbonyl group

This makes the compound a ketone. In fact, because it has both hydroxyl groups and a carbonyl group, DHA is a sugar, more precisely, a keto sugar. This sugar is actually a by-product of our own metabolism; we produce it in the metabolic pathway called glycolysis (Chapter 22).

A researcher by the name of Eva Wittgenstein discovered the tanning reaction while she was studying a human genetic disorder in children. These children were unable to store glycogen, a polysaccharide, or sugar polymer, which is our major energy storage molecule in the liver. She was trying to treat the disease by feeding large doses of DHA to the children. Sometimes, however, the children spit up some of the sickeningly sweet solution, which ended up on their clothes and skin. Dr. Wittgenstein noticed that the skin darkened at the site of these spills and decided to investigate the observation.

DHA works because of a reaction between its carbonyl group and a free amino group (—NH_3^+) of several amino acids in the skin protein keratin. Amino acids are the building blocks of the biological polymers called proteins (Chapter 18); keratin is just one such protein. The DHA produces brown-colored compounds called melanoids when it bonds to the keratins. These polymeric melanoids are chemically linked to cells of the stratum corneum, the dead, outermost layer of the skin. DHA does not penetrate this outer layer; so the chemical reaction that causes tanning only affects the stratum corneum. As this dead skin sloughs off, so does your tan!

Over the years research has improved the quality of the tan that is produced. Early self-tanning lotions produced an orange tan; the tans from today's lotions are much more natural. The DHA used today is in a much purer form and the other components of the lotion have been redesigned to promote greater penetration. Research has also taught us that the tanning reaction works best at acid pH; so newer formulations are buffered to pH 5. All of these changes have resulted in self-tanners that produce a longer lasting tan with a more natural, golden color.

We have also learned that it is important to exfoliate before using a self-tanner. Anywhere that the dead skin layer is thicker, there will be more keratin. From our study of chemistry, we have learned that when we begin with more reactant, we often get more product. The greater the amount of product

---

Question 13.21 Write an equation for the hydrogenation of propanone.

Question 13.22 Write an equation for the hydrogenation of butanone.

The hydrogenation of an aldehyde results in the production of a primary alcohol, as seen in the following equation showing the reduction of the aldehyde, butanal:

$$CH_3CH_2CH_2-\overset{\overset{\displaystyle O}{\|}}{C}-H + H_2 \xrightarrow{Pt} CH_3CH_2CH_2-\overset{\overset{\displaystyle OH}{|}}{\underset{\underset{\displaystyle H}{|}}{C}}-H$$

Butanal   Hydrogen   1-Butanol
(An aldehyde)   (A primary alcohol)

13.4 Reactions Involving Aldehydes and Ketones

in this case, the darker the color! The resulting tan often looked splotchy or streaky. By gently removing some of the stratum corneum by a gentle exfoliation process, the surface of the skin, and hence the tanning reaction, becomes more uniform.

More recently some companies have added a new sugar, erythrulose, to the self-tanning lotions.

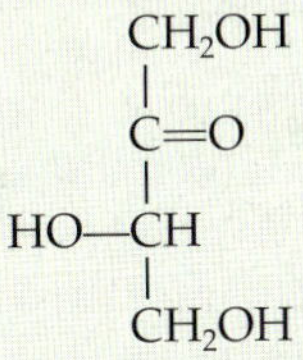

$$\begin{array}{c} CH_2OH \\ | \\ C=O \\ | \\ HO-CH \\ | \\ CH_2OH \end{array}$$

Erythrulose

Erythrulose is a four-carbon keto sugar that reacts in exactly the same way as DHA. However, since they are different compounds, they do produce different melanoids with slightly different properties, including color. The tan produced by erythrulose is less reddish in tone than that produced by DHA. However, while a DHA tan develops in two to six hours, an erythrulose tan requires two days. For this reason, erythrulose is usually not used alone, but only in combination with DHA.

People often ask whether the tan from a bottle can protect against burn, much as a natural tan does. The melanoids do absorb light of the same wavelengths absorbed by melanin (the substance formed by suntanning), so you might expect some protection against sunburn. However, the protection is minimal, rated at a sun protection factor (SPF) of only 2 or 3. Self-tanners offer an excellent substitute to a suntan. They produce the same golden tan without the danger of overexposure to the sun's harmful ultraviolet rays.

For Further Understanding

Explain in terms of a chemical reaction why using a self-tanner daily results in an increasingly darker tan.

The incidence of skin cancer in men and women has risen dramatically in recent years. Using the Internet and Chapter 20 in this book, develop a hypothesis to explain this observation.

Writing an Equation Representing the Hydrogenation of an Aldehyde

EXAMPLE 13.8

Write an equation showing the hydrogenation of 3-methylbutanal.

Solution

Recall that the reduction of an aldehyde results in the production of a primary alcohol, in this case, 3-methyl-1-butanol.

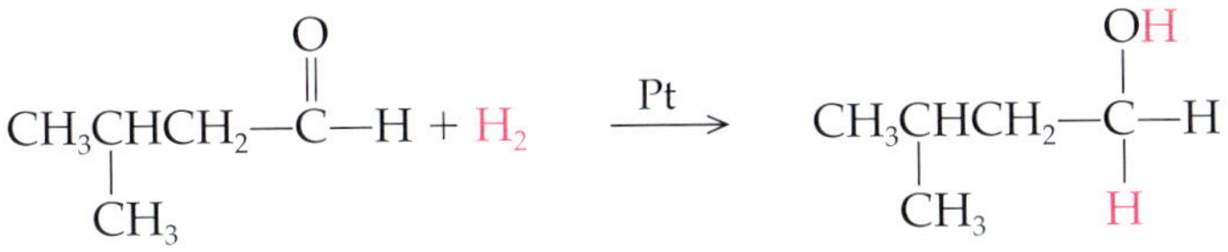

$$\underset{\text{3-Methylbutanal}}{\begin{array}{c} O \\ \| \\ CH_3CHCH_2-C-H \\ | \\ CH_3 \end{array}} + H_2 \xrightarrow{Pt} \underset{\text{3-Methyl-1-butanol}}{\begin{array}{c} OH \\ | \\ CH_3CHCH_2-C-H \\ | \quad\quad\quad | \\ CH_3 \quad\quad\; H \end{array}}$$

Question 13.23

Label each of the following as an oxidation or a reduction reaction.

a. Ethanal to ethanol
b. Benzoic acid to benzaldehyde
c. Cyclohexanone to cyclohexanol
d. 2-Propanol to propanone
e. 2,3-Butanedione (found in butter) to 2,3-butanediol

Question 13.24

Write an equation for each of the reactions in Question 13.23.

The role of the lactate fermentation in exercise is discussed in greater detail in Section 21.4.

A biological example of the reduction of a ketone occurs in the body, particularly during strenuous exercise when the lungs and circulatory system may not be able to provide enough oxygen to the muscles. Under these circumstances, the lactate fermentation begins. In this reaction, the enzyme *lactate dehydrogenase* reduces pyruvate, the product of glycolysis, a pathway for the breakdown of glucose, into lactate. The source of hydrogen ions for this reaction is nicotinamide adenine dinucleotide (NADH), which is oxidized in the course of the reaction.

$$CH_3-\underset{Pyruvate}{\overset{O}{\underset{\|}{C}}-\overset{O}{\underset{\|}{C}}-O^-} \quad \xrightarrow[NAD^+]{\underset{NADH}{Lactate\ dehydrogenase}} \quad CH_3-\underset{Lactate}{\overset{OH}{\underset{H}{C}}-\overset{O}{\underset{\|}{C}}-O^-}$$

Addition Reactions

LEARNING GOAL 7

Addition reactions of alkenes are described in detail in Section 11.5.

The principal reaction of the carbonyl group is the **addition reaction** across the polar carbon-oxygen double bond. This reaction is very similar to some that we have already studied, addition across the carbon-carbon double bond of alkenes. Such reactions require that a catalytic amount of acid be present in solution, as shown by the H$^+$ over the arrow for the reactions shown in the following examples.

An example of an addition reaction is the reaction of aldehydes with alcohols in the presence of catalytic amounts of acid. In this reaction, the hydrogen of the alcohol adds to the carbonyl oxygen. The alkoxyl group of the alcohol (—OR) adds to the carbonyl carbon. The predicted product is a **hemiacetal**.

$$R-\underset{H}{\overset{OH}{\underset{|}{\overset{|}{C}}}}-OR$$

General structure of a hemiacetal

However, this is not the product typically isolated from this reaction. Hemiacetals are quite reactive. In the presence of acid and excess alcohol, they undergo a substitution reaction in which the —OH group of the hemiacetal is exchanged for another —OR group from the alcohol. The product of this reaction is an **acetal**. Acetal formation is a reversible reaction, as the general equation shows:

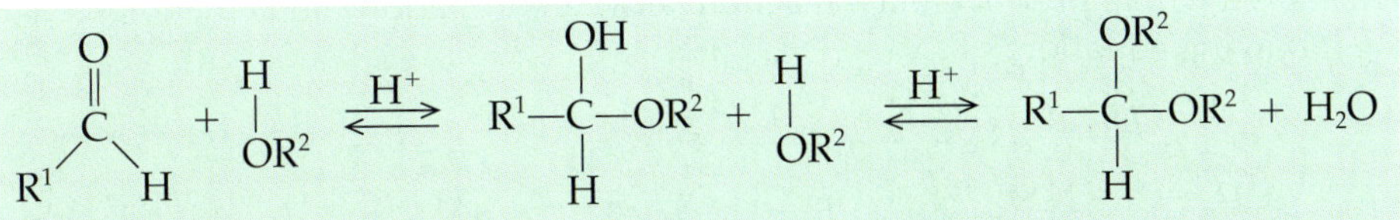

13.4 Reactions Involving Aldehydes and Ketones

$$\underset{\text{Aldehyde}}{\overset{O}{\underset{R^1}{\overset{\|}{C}}}\overset{}{\underset{H}{}}} + \underset{\text{Alcohol}}{\overset{H}{\underset{OR^2}{|}}} \underset{}{\overset{H^+}{\rightleftharpoons}} \underset{\text{Hemiacetal}}{R^1-\overset{OH}{\underset{H}{\overset{|}{C}}}-OR^2} + \overset{H}{\underset{OR^2}{|}} \overset{H^+}{\rightleftharpoons} \underset{\text{Acetal}}{R^1-\overset{OR^2}{\underset{H}{\overset{|}{C}}}-OR^2} + H_2O$$

Consider the acid-catalyzed reaction between propanal and methanol:

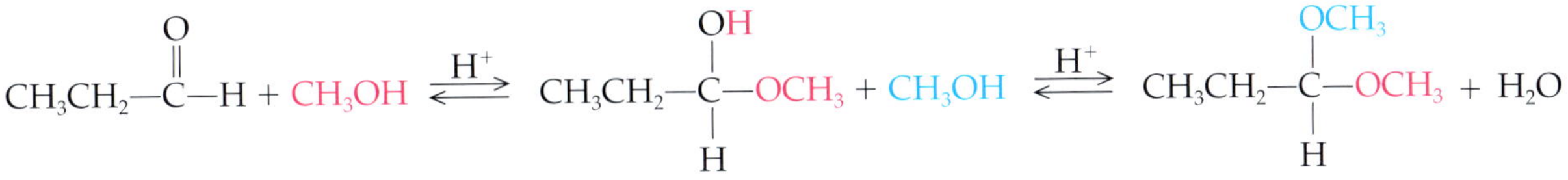

$$\underset{\text{Propanal}}{CH_3CH_2-\overset{O}{\overset{\|}{C}}-H} + \underset{\text{Methanol}}{CH_3OH} \overset{H^+}{\rightleftharpoons} \underset{\text{Hemiacetal}}{CH_3CH_2-\overset{OH}{\underset{H}{\overset{|}{C}}}-OCH_3} + CH_3OH \overset{H^+}{\rightleftharpoons} \underset{\text{Propanal dimethyl acetal}}{CH_3CH_2-\overset{OCH_3}{\underset{H}{\overset{|}{C}}}-OCH_3} + H_2O$$

Addition reactions will also occur between a ketone and an alcohol. In this case the more reactive intermediate is called a **hemiketal** and the product is called a **ketal**. The general equation for ketal formation is shown here:

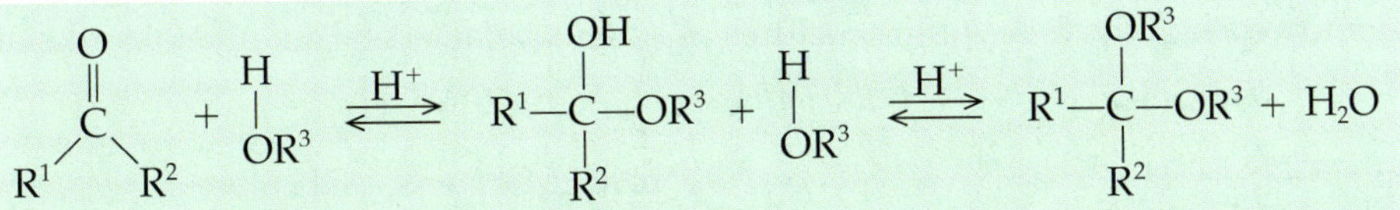

$$\underset{\text{Ketone}}{\overset{O}{\underset{R^1\ \ R^2}{\overset{\|}{C}}}} + \underset{\text{Alcohol}}{\overset{H}{\underset{OR^3}{|}}} \overset{H^+}{\rightleftharpoons} \underset{\text{Hemiketal}}{R^1-\overset{OH}{\underset{R^2}{\overset{|}{C}}}-OR^3} + \overset{H}{\underset{OR^3}{|}} \overset{H^+}{\rightleftharpoons} \underset{\text{Ketal}}{R^1-\overset{OR^3}{\underset{R^2}{\overset{|}{C}}}-OR^3} + H_2O$$

EXAMPLE 13.9

Recognizing Hemiacetals, Acetals, Hemiketals, and Ketals

Solution

A simple scheme is helpful in recognizing these four types of compounds. Begin by drawing a carbon atom with four bonds and follow the flow chart as additional groups are added that will identify the molecules.

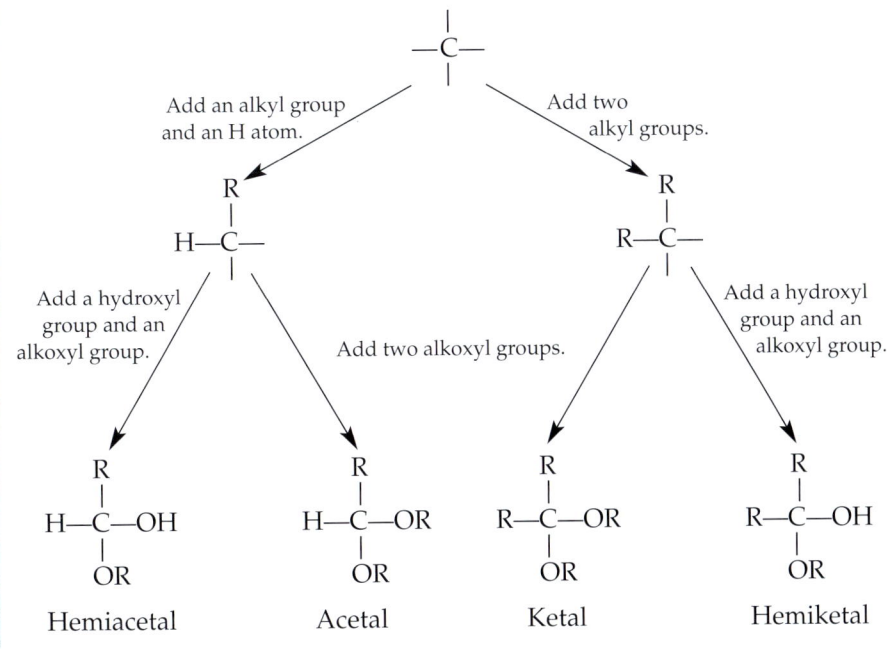

A ketal is the final product in the reaction between propanone and ethanol, seen in the following equation:

$$CH_3-\overset{O}{\underset{}{\overset{\|}{C}}}-CH_3 + CH_3CH_2OH \underset{}{\overset{H^+}{\rightleftharpoons}} CH_3-\underset{CH_3}{\overset{OH}{\underset{|}{\overset{|}{C}}}}-OCH_2CH_3 + CH_3CH_2OH \underset{}{\overset{H^+}{\rightleftharpoons}} CH_3-\underset{CH_3}{\overset{OCH_2CH_3}{\underset{|}{\overset{|}{C}}}}-OCH_2CH_3 + H_2O$$

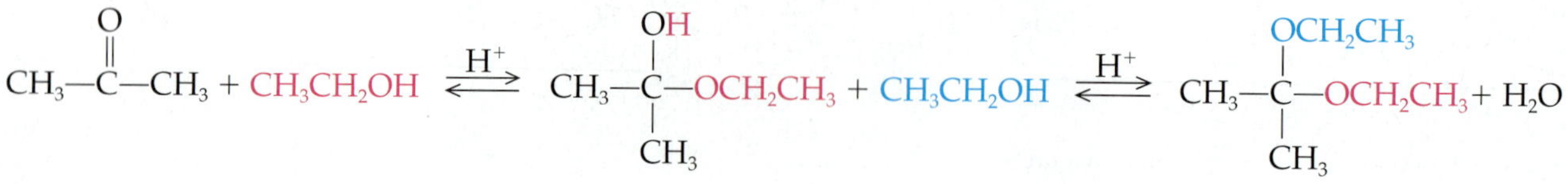

Propanone   Ethanol   Hemiketal   Ketal

Question 13.25

Identify each of the following structures as a hemiacetal, acetal, hemiketal, or ketal.

a. $H-\underset{OCH_3}{\overset{CH_3}{\underset{|}{\overset{|}{C}}}}-OH$ b. $H_3C-\underset{OCH_3}{\overset{CH_3}{\underset{|}{\overset{|}{C}}}}-OCH_3$ c. $H-\underset{OCH_3}{\overset{CH_3}{\underset{|}{\overset{|}{C}}}}-OCH_3$ d. $H_3C-\underset{OCH_3}{\overset{CH_3}{\underset{|}{\overset{|}{C}}}}-OH$

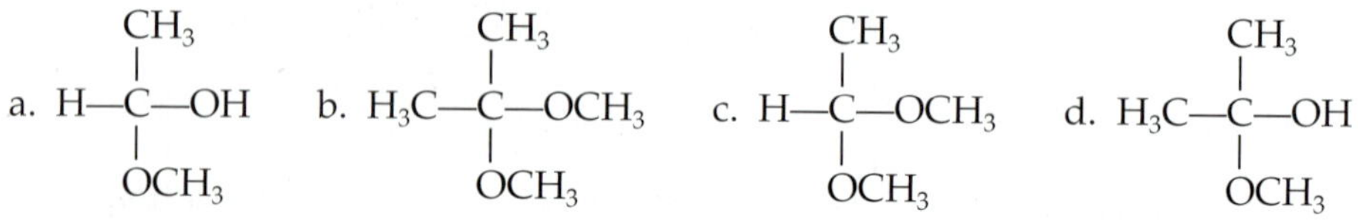

Question 13.26

Identify each of the following structures as a hemiacetal, acetal, hemiketal, or ketal.

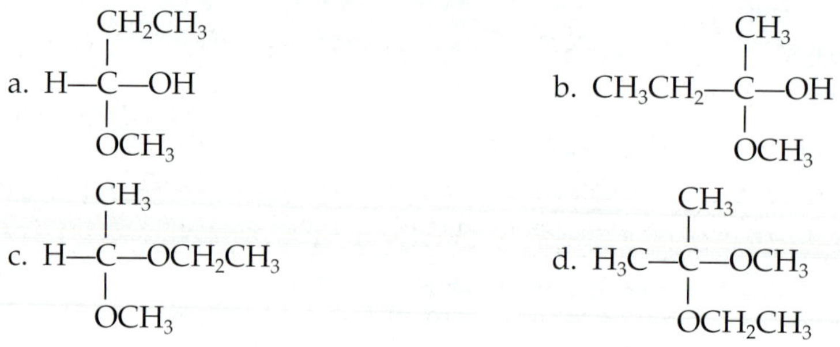

a. $H-\underset{OCH_3}{\overset{CH_2CH_3}{\underset{|}{\overset{|}{C}}}}-OH$ b. $CH_3CH_2-\underset{OCH_3}{\overset{CH_3}{\underset{|}{\overset{|}{C}}}}-OH$

c. $H-\underset{OCH_3}{\overset{CH_3}{\underset{|}{\overset{|}{C}}}}-OCH_2CH_3$ d. $H_3C-\underset{OCH_2CH_3}{\overset{CH_3}{\underset{|}{\overset{|}{C}}}}-OCH_3$

Hemiacetals and hemiketals are readily formed in carbohydrates. Monosaccharides contain several hydroxyl groups and one carbonyl group. The linear form of a monosaccharide quickly undergoes an intramolecular reaction in solution to give a cyclic hemiacetal or hemiketal.

Earlier we noted that hemiacetals and hemiketals formed in *intermolecular* reactions were unstable and continued to react, forming acetals and ketals. This is not the case with the *intramolecular* reactions involving five- or six-carbon sugars. In these reactions the cyclic or ring form of the molecule is more stable than the linear form. This reaction is shown for the sugar glucose (blood sugar) in Figure 13.6 and is discussed in detail in Section 16.2.

When the hemiacetal or hemiketal of one monosaccharide reacts with a hydroxyl group of another monosaccharide, the product is an acetal or a ketal. A sugar molecule made up of two monosaccharides is called a *disaccharide*. The C—O—C bond between the two monosaccharides is called a *glycosidic bond* (Figure 13.7).

Keto-Enol Tautomers

LEARNING GOAL 8

Many aldehydes and ketones may exist in an equilibrium mixture of two constitutional or structural isomers called *tautomers*. Tautomers differ from one another in the placement of a hydrogen atom and a double bond. One tautomer is the *keto form* (on the left in the following equation). The keto form has the structure typical of an aldehyde or ketone. The other form is called the *enol form* (on the right in the following equation). The enol form has a structure containing a carbon-carbon double bond (*en*) and a hydroxyl group, the functional group characteristic of alcohols (*ol*).

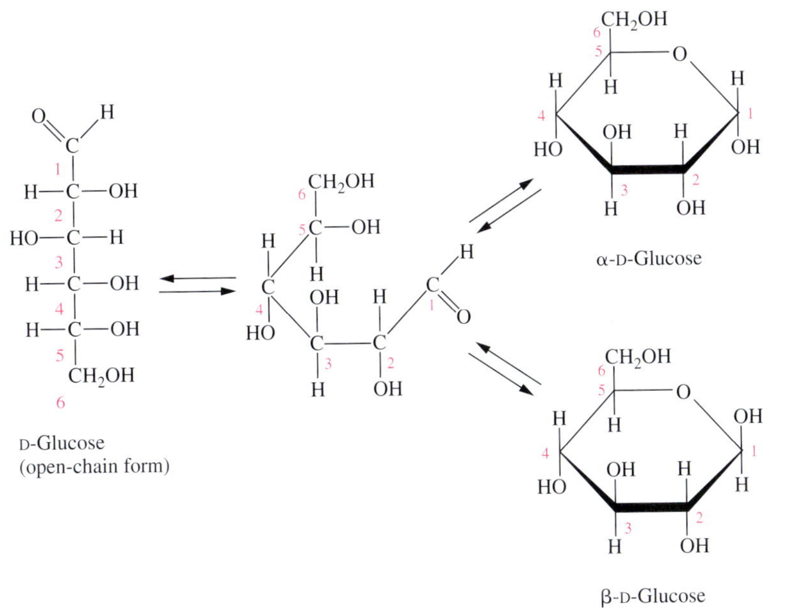

13.4 Reactions Involving Aldehydes and Ketones

Figure 13.6
Hemiacetal formation in sugars, shown for the intramolecular reaction of D-glucose.

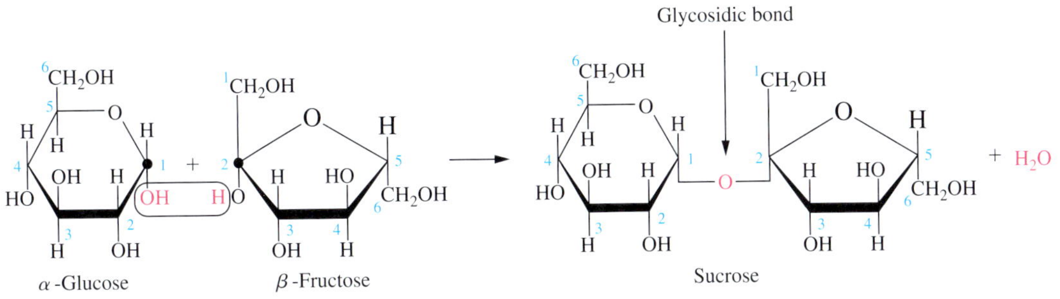

Figure 13.7
Acetal formation, demonstrated in the formation of the disaccharide sucrose, common table sugar. The reaction between the hydroxyl groups of the monosaccharides glucose and fructose produces the acetal sucrose. The bond between the two sugars is a glycosidic bond.

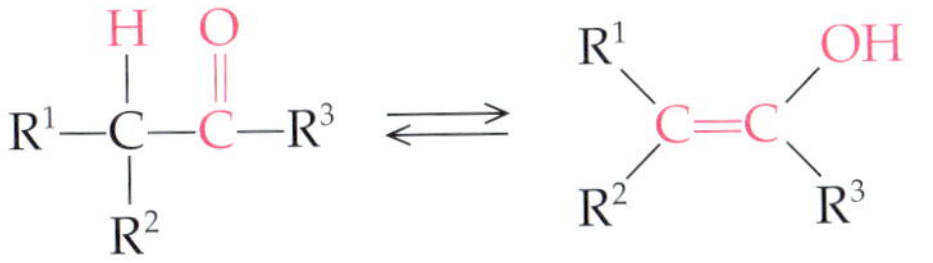

Because the keto form of most simple aldehydes and ketones is more stable, they exist mainly in that form.

EXAMPLE 13.10

Writing an Equation Representing the Equilibrium Between the Keto and Enol Forms of a Simple Aldehyde

Draw the enol form of ethanal and write an equation representing the equilibrium between the keto and enol forms of this molecule.

Solution

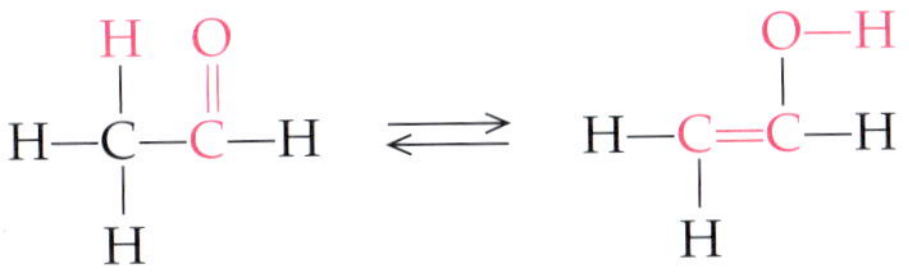

Ethanal
Keto form
More stable

Enol form
Less stable

Question 13.27

Draw the keto and enol forms of propanal.

Question 13.28

Draw the keto and enol forms of 3-pentanone.

Phosphoenolpyruvate is a biologically important enol. In fact, it is the highest energy phosphorylated compound in living systems.

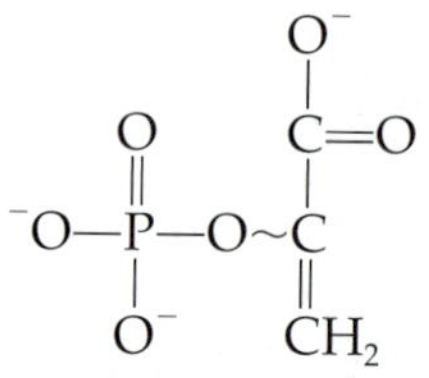

$$\begin{array}{c} O^- \\ | \\ C=O \\ O-P-O\sim C \\ | \| \\ O^- CH_2 \end{array}$$

Phosphoenolpyruvate

Phosphoenolpyruvate is produced in the next-to-last step in the metabolic pathway called *glycolysis,* which is the first stage of carbohydrate breakdown. In the final reaction of glycolysis, the phosphoryl group from phosphoenolpyruvate is transferred to adenosine diphosphate (ADP). The reaction produces ATP, the major energy currency of the cell.

The glycolysis pathway is discussed in detail in Chapter 21.

Aldol Condensation

LEARNING GOAL 9

The **aldol condensation** is a reaction in which aldehydes or ketones react to form larger molecules. A new carbon-carbon bond is formed in the process:

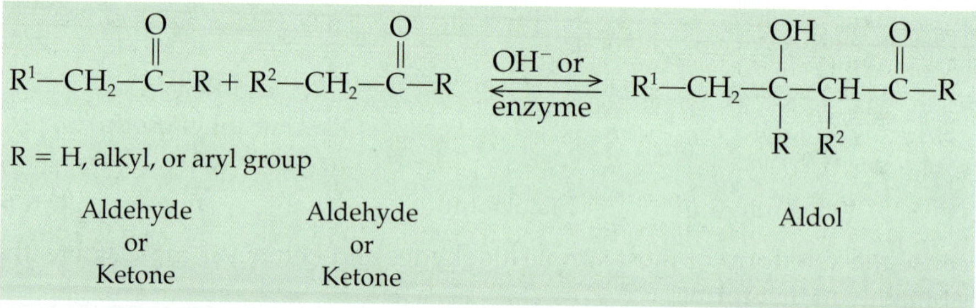

$$R^1-CH_2-\overset{\overset{\displaystyle O}{\|}}{C}-R + R^2-CH_2-\overset{\overset{\displaystyle O}{\|}}{C}-R \xrightleftharpoons[\text{enzyme}]{OH^- \text{ or}} R^1-CH_2-\overset{\overset{\displaystyle OH}{|}}{\underset{\underset{\displaystyle R}{|}}{C}}-\overset{\overset{\displaystyle }{|}}{\underset{\underset{\displaystyle R^2}{|}}{CH}}-\overset{\overset{\displaystyle O}{\|}}{C}-R$$

R = H, alkyl, or aryl group

Aldehyde or Ketone   Aldehyde or Ketone   Aldol

This is actually a very complex reaction that occurs in multiple steps. Here we focus on the end results of the reaction, using the example of the reaction between two molecules of ethanal. As shown in the equation below, the α-carbon (carbon-2) of one aldehyde forms a bond with the carbonyl carbon of a second aldehyde (shown in blue). A bond also forms between a hydrogen atom on that same α-carbon and the carbonyl oxygen (shown in red).

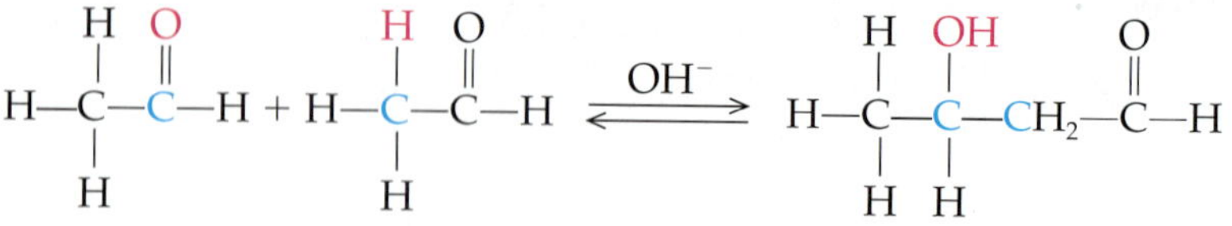

$$\underset{\text{Ethanal}}{H-\overset{\overset{\displaystyle H}{|}}{\underset{\underset{\displaystyle H}{|}}{C}}-\overset{\overset{\displaystyle O}{\|}}{C}-H} + \underset{\text{Ethanal}}{H-\overset{\overset{\displaystyle H}{|}}{\underset{\underset{\displaystyle H}{|}}{C}}-\overset{\overset{\displaystyle O}{\|}}{C}-H} \xrightleftharpoons{OH^-} \underset{\substack{\text{3-Hydroxybutanal} \\ (\beta\text{-hydroxybutyraldehyde})}}{H-\overset{\overset{\displaystyle H}{|}}{\underset{\underset{\displaystyle H}{|}}{C}}-\overset{\overset{\displaystyle OH}{|}}{\underset{\underset{\displaystyle H}{|}}{C}}-CH_2-\overset{\overset{\displaystyle O}{\|}}{C}-H}$$

13.4 Reactions Involving Aldehydes and Ketones

The result is similar when two ketones react:

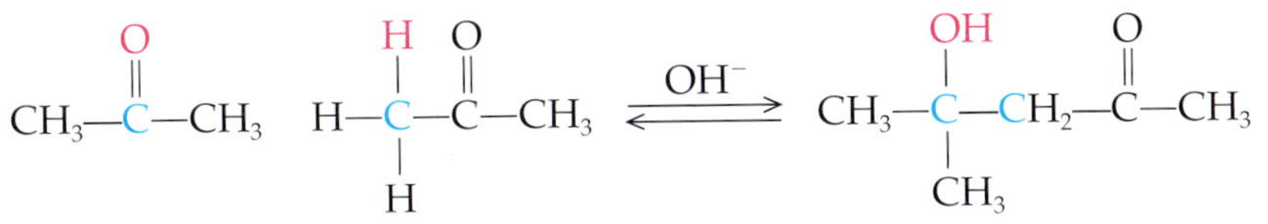

$$CH_3-\overset{O}{\underset{\|}{C}}-CH_3 \quad H-\overset{H}{\underset{H}{\overset{|}{C}}}-\overset{O}{\underset{\|}{C}}-CH_3 \quad \underset{}{\overset{OH^-}{\rightleftharpoons}} \quad CH_3-\overset{OH}{\underset{CH_3}{\overset{|}{C}}}-CH_2-\overset{O}{\underset{\|}{C}}-CH_3$$

Propanone   Propanone   4-Hydroxy-4-methyl-2-pentanone

In the laboratory, the aldol condensation is catalyzed by dilute base. But the same reaction occurs in our cells, where it is catalyzed by an enzyme. This reaction is one of many in a pathway that makes the sugar glucose from smaller molecules. This pathway is called gluconeogenesis (*gluco-* [sugar], *neo-* [new], *genesis* [beginnings]), which simply means origin of new sugar. This pathway is critical during starvation or following strenuous exercise. Under those conditions, blood glucose concentrations may fall dangerously low. Because the brain can use only glucose as an energy source, it is essential that the body be able to produce it quickly.

One of the steps in the pathway is an aldol condensation between the ketone dihydroxyacetone phosphate and the aldehyde glyceraldehyde-3-phosphate.

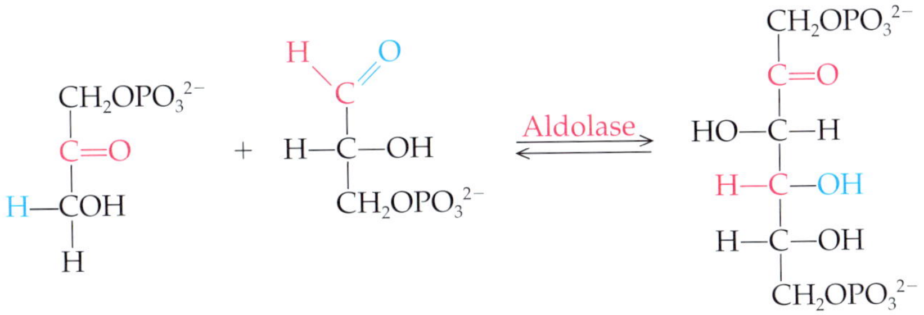

Gluconeogenesis is described in Chapter 21.

Dihydroxyacetone phosphate + Glyceraldehyde-3-phosphate $\overset{\text{Aldolase}}{\rightleftharpoons}$ Fructose-1,6-bisphosphate

Dihydroxyacetonephosphate is a phosphorylated form of dihydroxyacetone (DHA), the active ingredient in self-tanning lotions. See *A Medical Perspective: That Golden Tan Without the Fear of Skin Cancer* on page 432.

The speed and specificity of this reaction are ensured by the enzyme *aldolase*. The product is the sugar fructose-1,6-bisphosphate, which is converted by another enzyme into glucose-1,6-bisphosphate. Removal of the two phosphoryl groups results in a new molecule of glucose for use by the body as an energy source.

Gluconeogenesis occurs under starvation conditions to provide a supply of blood glucose to nourish the brain. However, when glucose is plentiful, it is broken down to provide ATP energy for the cell. The pathway for glucose degradation is called *glycolysis*. In glycolysis, the reaction just shown is reversed. In general, aldol condensation reactions are reversible. These reactions are called *reverse aldols*.

ATP, the universal energy currency, is discussed in Section 21.1.

Question 13.29

Write an equation for the aldol condensation of two molecules of propanal.

Question 13.30

Write an equation for the aldol condensation of two molecules of butanal.

A Human Perspective

The Chemistry of Vision

Sight is a complex physiological process that is dependent on two types of cells in the retina of the eye: rods and cones. Rods are primarily responsible for vision in dim light. Cones are responsible for vision in bright light.

Vision in rod cells requires an unsaturated aldehyde, 11-*cis*-retinal (see Figure below), which is produced from vitamin A. Vitamin A may be obtained directly in the diet, but it is also produced by the cleavage of β-carotene obtained in the diet.

In the retina, a protein called opsin combines with 11-*cis*-retinal to form a complex protein called rhodopsin. When light strikes the rods, the light energy is absorbed by 11-*cis*-retinal, which is then photochemically converted to 11-*trans*-retinal. This causes a change in the shape in the rhodopsin complex and results in the dissociation of 11-*trans*-retinal from the protein. This, in turn, causes ions to flow more freely into the rod cells. The influx of ions stimulates nerve cells that send signals to the brain. Interpretation of those signals produces the visual image.

Following the initial light stimulus, retinal returns to the *cis*-isomer and reassociates with opsin. The system is then

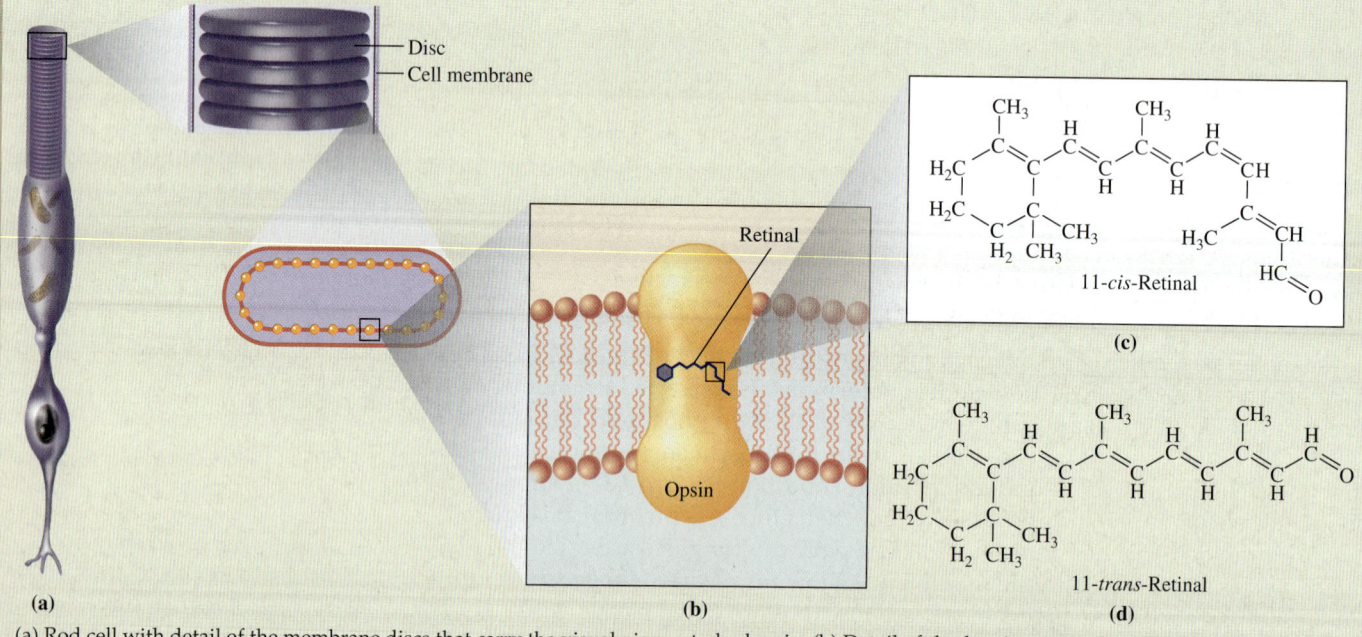

(a) Rod cell with detail of the membrane discs that carry the visual pigment, rhodopsin. (b) Detail of rhodopsin, a complex of 11-*cis*-retinal and the protein opsin, embedded in the disc membrane. (c) 11-*cis*-retinal and (d) 11-*trans*-retinal.

Summary of Reactions

Aldehydes and Ketones

Oxidation of an Aldehyde

$$\underset{\text{Aldehyde}}{\text{R}-\overset{\overset{\displaystyle O}{\|}}{\text{C}}-\text{H}} \xrightarrow{[O]} \underset{\text{Carboxylic acid}}{\text{R}-\overset{\overset{\displaystyle O}{\|}}{\text{C}}-\text{OH}}$$

Reduction of Aldehydes and Ketones

$$\underset{\substack{\text{Aldehyde} \\ \text{or Ketone}}}{\text{R}^1-\overset{\overset{\displaystyle O}{\|}}{\text{C}}-\text{R}^2} + \underset{\text{Hydrogen}}{\text{H}-\text{H}} \xrightarrow{\text{Pt}} \underset{\text{Alcohol}}{\text{R}^1-\overset{\overset{\displaystyle OH}{|}}{\underset{\underset{\displaystyle R^2}{|}}{\text{C}}}-\text{H}}$$

Summary of Reactions

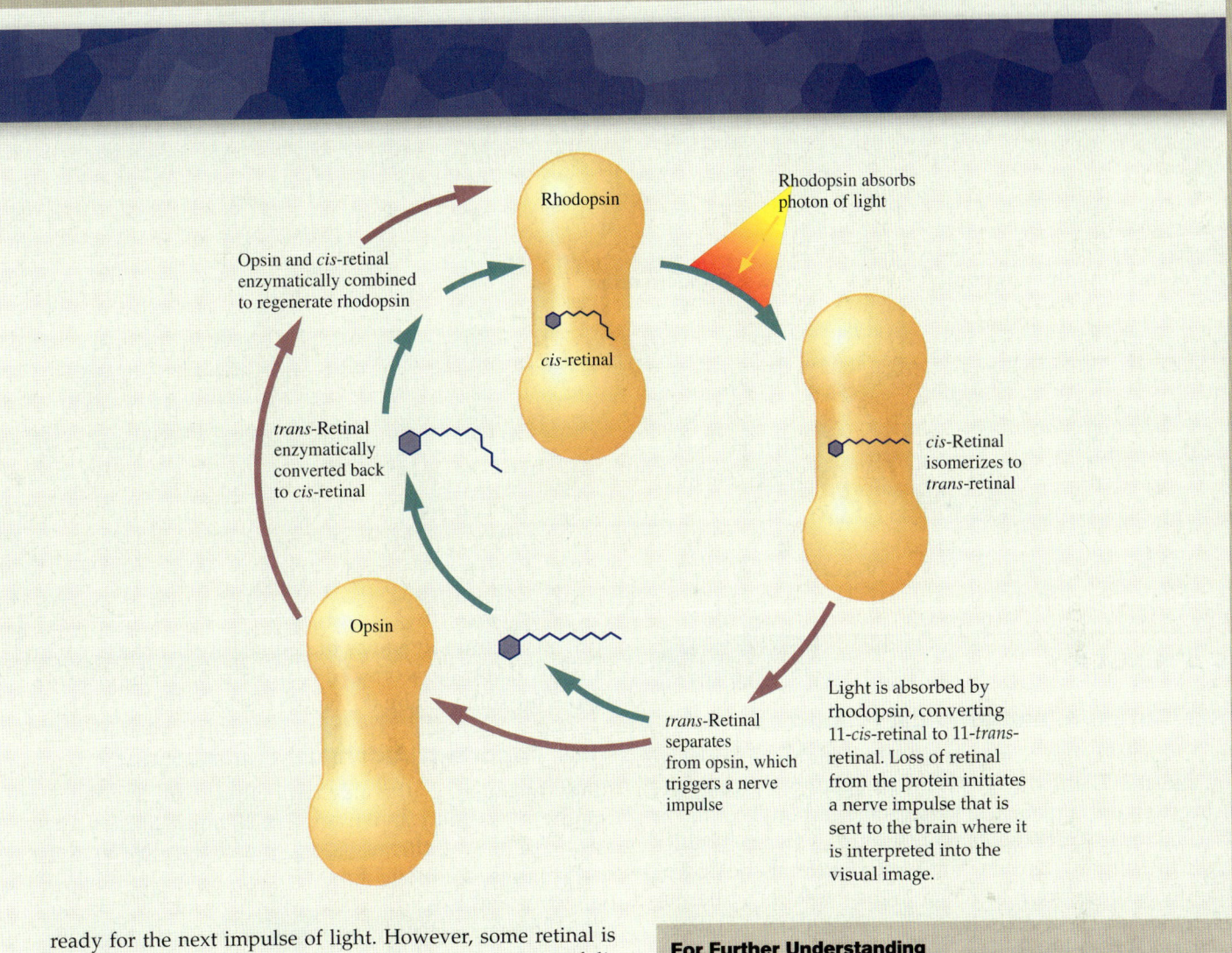

ready for the next impulse of light. However, some retinal is lost in the process and must be replaced by conversion of dietary vitamin A or β-carotene to retinal. As you might expect, a deficiency of vitamin A can have terrible consequences. In children, lack of vitamin A causes xerophthalmia, an eye disease that results first in night blindness and eventually in total blindness. This can easily be prevented by a diet rich in vitamin A or β-carotene.

For Further Understanding
What are some sources of vitamin A or β-carotene in the diet?

Xerophthalmia causes blindness in approximately 200,000 children in third world countries each year. Investigate the measures that agencies such as the World Health Organization are using to save the sight of these children.

Addition Reactions

Addition of an alcohol to a ketone—ketal formation:

$$\underset{\text{Ketone}}{\overset{\text{O}}{\underset{R^1\quad R^2}{\|}}\!\!\!C} + \underset{\text{Alcohol}}{\overset{H}{\underset{OR^3}{|}}} \overset{H^+}{\rightleftharpoons} \underset{\text{Hemiketal}}{\overset{OH}{\underset{R^2}{\underset{|}{R^1-C-OR^3}}}} + \overset{H}{\underset{OR^3}{|}} \overset{H^+}{\rightleftharpoons} \underset{\text{Ketal}}{\overset{OR^3}{\underset{R^2}{\underset{|}{R^1-C-OR^3}}}} + H_2O$$

Addition of an alcohol to an aldehyde—acetal formation:

$$R^1-\underset{H}{\underset{|}{\overset{O}{\overset{\|}{C}}}} + \underset{OR^2}{\overset{H}{\overset{|}{O}}} \underset{}{\overset{H^+}{\rightleftharpoons}} R^1-\underset{H}{\overset{OH}{\underset{|}{\overset{|}{C}}}}-OR^2 + \underset{OR^2}{\overset{H}{\overset{|}{O}}} \overset{H^+}{\rightleftharpoons} R^1-\underset{H}{\overset{OR^2}{\underset{|}{\overset{|}{C}}}}-OR^2 + H_2O$$

Aldehyde   Alcohol   Hemiacetal   Acetal

Keto-enol Tautomerization

$$R^1-\underset{R^2}{\underset{|}{\overset{H}{\overset{|}{C}}}}-\overset{O}{\overset{\|}{C}}-R^3 \rightleftharpoons \underset{R^2}{\overset{R^1}{C}}=\underset{R^3}{\overset{OH}{C}}$$

Keto form   Enol form

Aldol Condensation

$$R^1-CH_2-\overset{O}{\overset{\|}{C}}-R + R^2-CH_2-\overset{O}{\overset{\|}{C}}-R \underset{enzyme}{\overset{OH^- \text{ or}}{\rightleftharpoons}} R^1-CH_2-\underset{R}{\underset{|}{\overset{OH}{\overset{|}{C}}}}-\underset{R^2}{\underset{|}{CH}}-\overset{O}{\overset{\|}{C}}-R$$

Aldehyde   Aldehyde   Aldol

R = H, alkyl, or aryl group

SUMMARY

13.1 Structure and Physical Properties

The *carbonyl group* ($>$C=O) is characteristic of the *aldehydes* and *ketones*. The carbonyl group and the two groups attached to it are coplanar. In ketones the carbonyl carbon is attached to two carbon-containing groups, whereas in aldehydes the carbonyl carbon is attached to at least one hydrogen; the second group attached to the carbonyl carbon in aldehydes may be another hydrogen or a carbon atom. Owing to the polar carbonyl group, aldehydes and ketones are polar compounds. Their boiling points are higher than those of comparable hydrocarbons but lower than those of comparable alcohols. Small aldehydes and ketones are reasonably soluble in water because of the hydrogen bonding between the carbonyl group and water molecules. Larger carbonyl-containing compounds are less polar and thus are more soluble in nonpolar organic solvents.

13.2 I.U.P.A.C. Nomenclature and Common Names

In the I.U.P.A.C. Nomenclature System, aldehydes are named by determining the parent compound and replacing the final -e of the parent alkane with -al. The chain is numbered beginning with the carbonyl carbon as carbon-1. Ketones are named by determining the parent compound and replacing the -e ending of the parent alkane with the -one suffix of the ketone family. The longest carbon chain is numbered to give the carbonyl carbon the lowest possible number. In the common system of nomenclature, substituted aldehydes are named as derivatives of the parent compound. Greek letters indicate the position of substituents. Common names of ketones are derived by naming the alkyl groups bonded to the carbonyl carbon. These names are followed by the word *ketone*.

13.3 Important Aldehydes and Ketones

Many members of the aldehyde and ketone families are important as food and fragrance chemicals, medicinals, and agricultural chemicals. Methanal (formaldehyde) is used to preserve tissue. Ethanal causes the symptoms of a hangover and is oxidized to produce acetic acid commercially. Propanone is a useful and versatile solvent for organic compounds.

13.4 Reactions Involving Aldehydes and Ketones

In the laboratory, aldehydes and ketones are prepared by the oxidation of alcohols. Oxidation of a primary alcohol produces an aldehyde; oxidation of a secondary alcohol yields a ketone. Tertiary alcohols do not react under these conditions. Aldehydes and ketones can be distinguished from one another on the basis of their ability to undergo oxidation reactions. The *Tollens' test* and *Benedict's test* are the most common such tests. Aldehydes are easily oxidized to carboxylic acids. Ketones do not undergo further oxidation reactions. Aldehydes and ketones are readily reduced to alcohols by *hydrogenation*. The most common reaction of the carbonyl group is *addition* across the highly polar carbon-oxygen double bond. The addition of an alcohol to an aldehyde produces a *hemiacetal*. The hemiacetal reacts with a second alcohol molecule to form an *acetal*. The reaction of a ketone with an alcohol produces a *hemiketal*. A hemiketal reacts with a second alcohol molecule to form a *ketal*. Hemiacetals and hemiketals are readily formed in carbohydrates.

Aldol condensation is a reaction in which aldehydes and ketones form larger molecules. Aldehydes and ketones may exist as an equilibrium mixture of keto and enol tautomers.

KEY TERMS

acetal (13.4)
addition reaction (13.4)
aldehyde (13.1)
aldol condensation (13.4)
Benedict's test (13.4)
carbonyl group (Intro)
hemiacetal (13.4)
hemiketal (13.4)
hydrogenation (13.4)
ketal (13.4)
ketone (13.1)
oxidation (13.4)
Tollens' test (13.4)

QUESTIONS AND PROBLEMS

Structure and Physical Properties

Foundations

13.31 Explain the relationship between carbon chain length and water solubility of aldehydes or ketones.

13.32 Explain the dipole-dipole interactions that occur between molecules containing carbonyl groups.

Applications

13.33 Simple ketones (for example, acetone) are often used as industrial solvents for many organically based products such as adhesives and paints. They are often considered "universal solvents," because they dissolve so many diverse materials. Why are these chemicals such good solvents?

13.34 Explain briefly why simple (containing fewer than five carbon atoms) aldehydes and ketones exhibit appreciable solubility in water.

13.35 Draw intermolecular hydrogen bonding between ethanal and water.

13.36 Draw the polar interactions that occur between acetone molecules.

13.37 Why do alcohols have higher boiling points than aldehydes or ketones of comparable molecular weight?

13.38 Why do hydrocarbons have lower boiling points than aldehydes or ketones of comparable molecular weight?

Nomenclature

Foundations

13.39 Briefly describe the rules of the I.U.P.A.C. Nomenclature System for naming aldehydes.

13.40 Briefly describe the rules of the I.U.P.A.C. Nomenclature System for naming ketones.

13.41 Briefly describe how to determine the common name of an aldehyde.

13.42 Briefly describe how to determine the common name of a ketone.

Applications

13.43 Draw each of the following using complete structural formulas:
 a. Methanal
 b. 7,8-Dibromooctanal

13.44 Draw each of the following using condensed structural formulas:
 a. Acetone
 b. Hydroxyethanal

13.45 Draw the structure of each of the following compounds:
 a. 3-Chloro-2-pentanone
 b. Benzaldehyde

13.46 Draw the structure of each of the following compounds:
 a. 4-Bromo-3-hexanone
 b. 2-Chlorocyclohexanone

13.47 Use the I.U.P.A.C. Nomenclature System to name each of the following compounds:

 a. $CH_3\overset{\overset{O}{\|}}{C}CH_2CH_3$

 b. $H\overset{\overset{O}{\|}}{C}\underset{\underset{CH_2CH_2CH_2CH_3}{|}}{C}HCH_2CH_3$

13.48 Name each of the following using the I.U.P.A.C. Nomenclature System:

 a. $Cl\underset{\underset{Cl}{|}}{\overset{\overset{Cl}{|}}{C}}\overset{\overset{O}{\|}}{C}CH_3$

 b.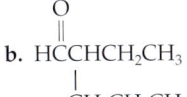

13.49 Name each of the following using the I.U.P.A.C. Nomenclature System:

 a.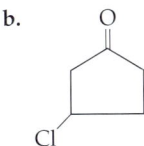

 b.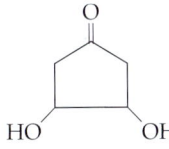

13.50 Name each of the following using the I.U.P.A.C. Nomenclature System:

 a. $CH_3CH_2CH_2\overset{\overset{O}{\|}}{C}H$

 b. $CH_3\underset{\underset{CH_3}{|}}{\overset{\overset{Br}{|}}{C}}CH_2CH_2\overset{\overset{O}{\|}}{C}H$

13.51 Give the I.U.P.A.C. name for each of the following compounds:

 a. $CH_3\underset{\underset{Br}{|}}{C}HCH_2\overset{\overset{O}{\|}}{C}H$

 b. $CH_3\underset{\underset{Cl}{|}}{\overset{\overset{CH_3}{|}}{C}}CH_2\overset{\overset{O}{\|}}{C}CH_2CH_3$

13.52 Give the I.U.P.A.C. name for each of the following compounds:

 a. $CH_3\overset{\overset{O}{\|}}{C}CH_2\underset{\underset{CH_2CH_3}{|}}{\overset{\overset{CH_2CH_3}{|}}{C}}CH_2CH_3$

 b. $CH_3\overset{\overset{O}{\|}}{C}CH_2\underset{\underset{Cl}{|}}{C}HCH_2CH_3$

13.53 Give the I.U.P.A.C. name for each of the following compounds:

 a. $CH_3\underset{\underset{CH_3}{|}}{C}HCH_2\underset{\underset{CH_3}{|}}{C}H\overset{\overset{O}{\|}}{C}CH_2CH_3$

 b.

13.54 Give the I.U.P.A.C. name for each of the following compounds:

a.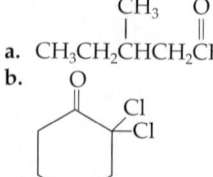

b.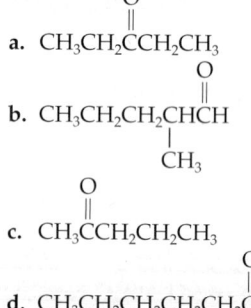

13.55 Give the common name for each of the following compounds:

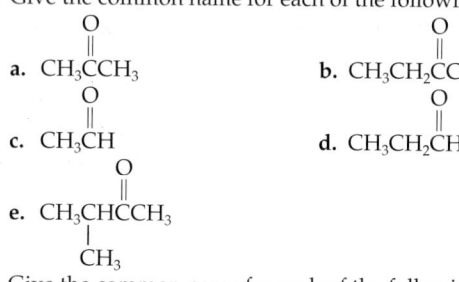

e. CH₃CHCCH₃
    |
    CH₃ (with C=O between CH and CCH₃)

13.56 Give the common name for each of the following compounds:
a. CH₃CH₂CCH₂CH₃ (with C=O)
b. CH₃CH₂CH₂CHCH (with C=O), CH₃ branch
c. CH₃CCH₂CH₂CH₃ (with C=O)
d. CH₃CH₂CH₂CH₂CH₂CH (with C=O)

13.57 Draw the structure of each of the following compounds:
a. 3-Hydroxybutanal
b. 2-Methylpentanal
c. 4-Bromohexanal
d. 3-Iodopentanal
e. 2-Hydroxy-3-methylheptanal

13.58 Draw the structure of each of the following compounds:
a. Dimethyl ketone
b. Methyl propyl ketone
c. Ethyl butyl ketone
d. Diisopropyl ketone

Important Aldehydes and Ketones

13.59 Why is acetone a good solvent for many organic compounds?
13.60 List several uses for formaldehyde.
13.61 Ethanal is produced by the oxidation of ethanol. Where does this reaction occur in the body?
13.62 List several aldehydes and ketones that are used as food or fragrance chemicals.

Reactions Involving Aldehydes and Ketones
Foundations

13.63 Explain what is meant by oxidation in organic molecules and provide an example of an oxidation reaction involving an aldehyde.
13.64 Explain what is meant by reduction in organic reactions and provide an example of a reduction reaction involving an aldehyde or ketone.
13.65 Define the term *addition reaction*. Provide an example of an addition reaction involving an aldehyde or ketone.
13.66 Define the term *aldol condensation*. Provide an example of an aldol condensation using an aldehyde or ketone.

Applications

13.67 Draw the structures of each of the following compounds. Then draw and name the product that you would expect to produce by oxidizing each of these alcohols:
a. 2-Butanol
b. 2-Methyl-1-propanol
c. Cyclopentanol

13.68 Draw the structures of each of the following compounds. Then draw and name the product that you would expect to produce by oxidizing each of these alcohols:
a. 2-Methyl-2-propanol
b. 2-Nonanol
c. 1-Decanol

13.69 Draw the generalized equation for the oxidation of a primary alcohol.

13.70 Draw the generalized equation for the oxidation of a secondary alcohol.

13.71 Draw the structures of the reactants and products for each of the following reactions. Label each as an oxidation or a reduction reaction:
a. Ethanal to ethanol
b. Cyclohexanone to cyclohexanol
c. 2-Propanol to propanone

13.72 An unknown has been determined to be one of the following three compounds:

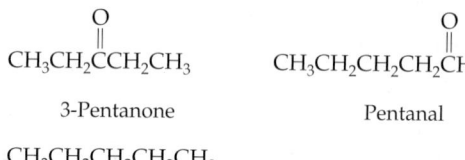

3-Pentanone   Pentanal

CH₃CH₂CH₂CH₂CH₃

Pentane

The unknown is fairly soluble in water and produces a silver mirror when treated with the silver ammonia complex. A red precipitate appears when it is treated with the Benedict's reagent. Which of the compounds is the correct structure for the unknown? Explain your reasoning.

13.73 Which of the following compounds would be expected to give a positive Tollens' test?
a. 3-Pentanone   d. Cyclopentanol
b. Cyclohexanone   e. 2,2-Dimethyl-1-pentanol
c. 3-Methylbutanal   f. Acetaldehyde

13.74 Write an equation representing the reaction of glucose with the Benedict's reagent. How was this test used in medicine?

13.75 Write an equation for the addition of one ethanol molecule to each of the following aldehydes and ketones:

a. CH₃CCH₃   b. CH₃CH

13.76 Write an equation for the addition of one ethanol molecule to each of the following aldehydes and ketones:

a. CH₃CH₂CH (with C=O)   b. CH₃CCH₂CH₂CH₃ (with C=O)

13.77 What is the general name for the product that is formed when an aldehyde reacts with one molecule of alcohol?
13.78 What is the general name of the product that is formed when a ketone reacts with one molecule of alcohol?
13.79 What is the general name for the product that is formed when an aldehyde reacts with two molecules of alcohol?
13.80 What is the general name of the product that is formed when a ketone reacts with two molecules of alcohol?
13.81 Write an equation for the addition of two methanol molecules to each of the following aldehydes and ketones:

a. CH₃CCH₃   b. CH₃CH

Critical Thinking Problems

13.82 Write an equation for the addition of two methanol molecules to each of the following aldehydes and ketones:

a. $CH_3CH_2\overset{\overset{O}{\|}}{C}H$

b. $CH_3\overset{\overset{O}{\|}}{C}CH_2CH_2CH_3$

13.83 An aldehyde can be oxidized to produce a carboxylic acid. Draw the carboxylic acid that would be produced by the oxidation of each of the following aldehydes:
a. Methanal
b. Ethanal

13.84 An aldehyde can be oxidized to produce a carboxylic acid. Draw the carboxylic acid that would be produced by the oxidation of each of the following aldehydes:
a. Propanal
b. Butanal

13.85 An alcohol can be oxidized to produce an aldehyde or a ketone. What aldehyde or ketone is produced by the oxidation of each of the following alcohols?
a. Methanol
b. 1-Propanol

13.86 An alcohol can be oxidized to produce an aldehyde or a ketone. What aldehyde or ketone is produced by the oxidation of each of the following alcohols?
a. 3-Pentanol
b. 2-Methyl-2-butanol

13.87 Indicate whether each of the following statements is true or false.
a. Aldehydes and ketones can be oxidized to produce carboxylic acids.
b. Oxidation of a primary alcohol produces an aldehyde.
c. Oxidation of a tertiary alcohol produces a ketone.
d. Alcohols can be produced by the oxidation of an aldehyde or ketone.

13.88 Indicate whether each of the following statements is true or false.
a. Ketones, but not aldehydes, react in the Tollens' silver mirror test.
b. Addition of one alcohol molecule to an aldehyde results in formation of a hemiacetal.
c. The cyclic forms of monosaccharides are intramolecular hemiacetals or intramolecular hemiketals.
d. Disaccharides (sugars composed of two covalently joined monosaccharides) are acetals, ketals, or both.

13.89 Write an equation for the aldol condensation of two molecules of ethanal.

13.90 Write an equation for the aldol condensation of two molecules of hexanal.

13.91 Draw the keto and enol forms of propanone.

13.92 Draw the keto and enol forms of 2-butanone.

13.93 Draw the hemiacetal or hemiketal that results from the reaction of each of the following aldehydes or ketones with ethanol:

a. $CH_3CH_2CH_2\overset{\overset{O}{\|}}{C}CH_3$

b. 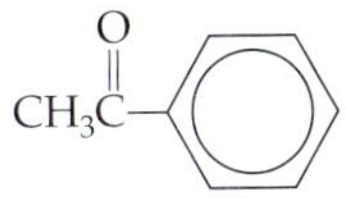 $CH_3\overset{\overset{O}{\|}}{C}-$⌬

c. (cyclopentanone)

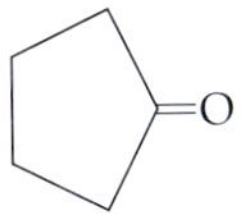

13.94 Identify each of the following compounds as a hemiacetal, hemiketal, acetal, or ketal:

a. 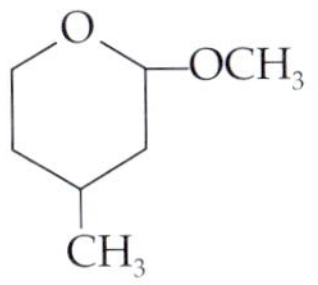 (tetrahydropyran ring with OCH₃ and CH₃ substituents)

b. 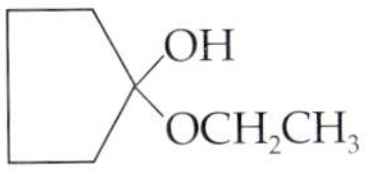 (cyclopentane with OH and OCH₂CH₃)

c. $CH_3\overset{\overset{OH}{|}}{\underset{\underset{OCH_2CH_3}{|}}{C}}CH_3$

d. 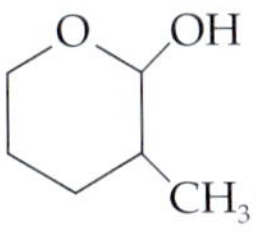 (tetrahydropyran ring with OH and CH₃)

e. $CH_3\overset{\overset{OCH_3}{|}}{\underset{\underset{OCH_2CH_3}{|}}{C}}CH_3$

f. $CH_3CH=CH\overset{\overset{OCH_3}{|}}{\underset{\underset{OH}{|}}{C}}CH_3$

13.95 Complete the following synthesis by supplying the missing reactant(s), reagent(s), or product(s) indicated by the question marks:

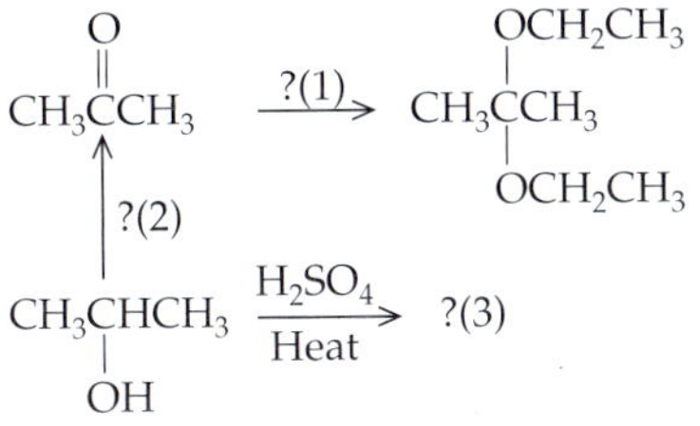

$CH_3\overset{\overset{O}{\|}}{C}CH_3 \xrightarrow{?(1)} CH_3\overset{\overset{OCH_2CH_3}{|}}{\underset{\underset{OCH_2CH_3}{|}}{C}}CH_3$

$\uparrow ?(2)$

$CH_3\overset{\overset{}{\underset{\underset{OH}{|}}{C}}}{}HCH_3 \xrightarrow[\text{Heat}]{H_2SO_4} ?(3)$

13.96 Which alcohol would you oxidize to produce each of the following compounds?

a. $CH_3\overset{\overset{CH_3}{|}}{C}HCH_2\overset{\overset{O}{\|}}{C}CH_3$

b. $H\overset{\overset{O}{\|}}{C}CH_2CH_2\overset{\overset{O}{\|}}{C}H$

c. 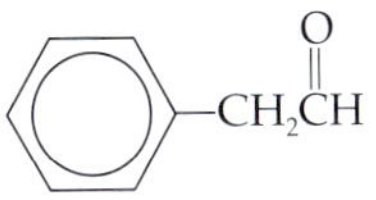 ⌬$-CH_2\overset{\overset{O}{\|}}{C}H$

d. $H\overset{\overset{O}{\|}}{C}CH_2\overset{\overset{O}{\|}}{C}CH_3$

e. $CH_3\overset{\overset{CH_3}{|}}{\underset{\underset{CH_3}{|}}{C}}CH_2CH_2\overset{\overset{O}{\|}}{C}H$

f. 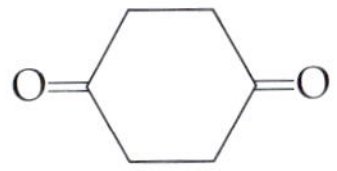 O=⌬=O

CRITICAL THINKING PROBLEMS

1. Review the material on the chemistry of vision and, with respect to the isomers of retinal, discuss the changes in structure that occur as the nerve impulses (that result in vision) are produced. Provide complete structural formulas of the retinal isomers that you discuss.

2. Classify the structure of β-D-fructose as a hemiacetal, hemiketal, acetal, or ketal. Explain your choice.

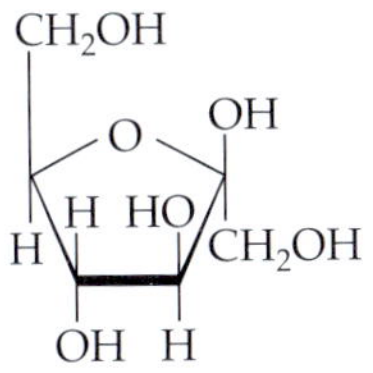

(structure of β-D-fructose shown)

3. Design a synthesis for each of the following compounds, using any inorganic reagent of your choice and any hydrocarbon or alkyl halide of your choice:
   a. Octanal
   b. Cyclohexanone
   c. 2-Phenylethanoic acid

4. When alkenes react with ozone, O_3, the double bond is cleaved, and an aldehyde and/or a ketone is produced. The reaction, called *ozonolysis*, is shown in general as:

   $$\text{C=C} + O_3 \longrightarrow \text{C=O} + \text{O=C}$$

   Predict the ozonolysis products that are formed when each of the following alkenes is reacted with ozone:
   a. 1-Butene
   b. 2-Hexene
   c. *cis*-3,6-Dimethyl-3-heptene

5. Lactose is the major sugar found in mammalian milk. It is a disaccharide composed of the monosaccharides glucose and galactose:

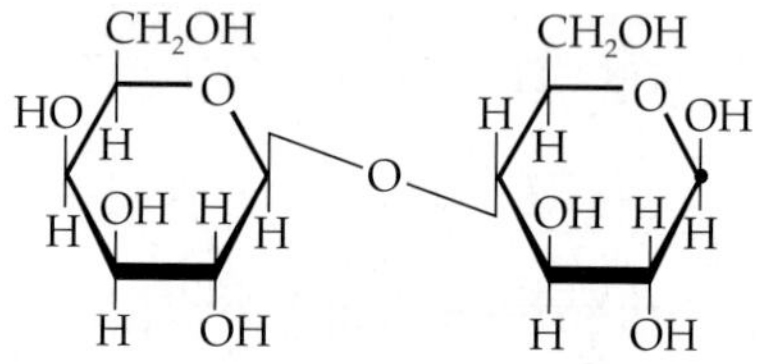

   Is lactose a hemiacetal, hemiketal, acetal, or ketal? Explain your choice or choices.

6. The following are the keto and enol tautomers of phenol:

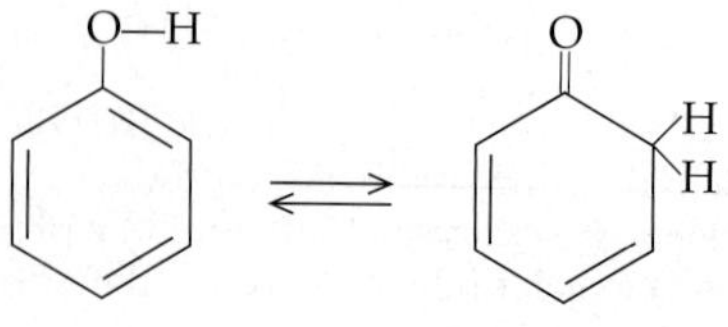

   Enol form of phenol   Keto form of phenol

   We have seen that most simple aldehydes and ketones exist mainly in the keto form because it is more stable. Phenol is an exception, existing primarily in the enol form. Propose a hypothesis to explain this.

ORGANIC CHEMISTRY

14

Carboxylic Acids and Carboxylic Acid Derivatives

Natural fruit flavors are complex mixtures of esters and other organic compounds.

Learning Goals

1. Write structures and describe the physical properties of carboxylic acids.
2. Determine the common and I.U.P.A.C. names of carboxylic acids.
3. Describe the biological, medical, or environmental significance of several carboxylic acids.
4. Write equations that show the synthesis of a carboxylic acid.
5. Write equations representing acid–base reactions of carboxylic acids.
6. Write equations representing the preparation of an ester.
7. Write structures and describe the physical properties of esters.
8. Determine the common and I.U.P.A.C. names of esters.
9. Write equations representing the hydrolysis of an ester.
10. Define the term *saponification* and describe how soap works in the emulsification of grease and oil.
11. Determine the common and I.U.P.A.C. names of acid chlorides.
12. Write equations representing the synthesis of acid chlorides.
13. Determine the common and I.U.P.A.C. names of acid anhydrides.
14. Write equations representing the synthesis of acid anhydrides.
15. Discuss the significance of thioesters and phosphoesters in biological systems.

Outline

Chemistry Connection:
Wake Up, Sleeping Gene

14.1 Carboxylic Acids

An Environmental Perspective:
Garbage Bags from Potato Peels

14.2 Esters

A Human Perspective:
The Chemistry of Flavor and Fragrance

14.3 Acid Chlorides and Acid Anhydrides

14.4 Nature's High-Energy Compounds: Phosphoesters and Thioesters

A Human Perspective:
Carboxylic Acid Derivatives of Special Interest

447

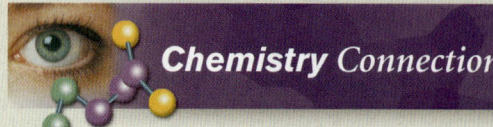

Chemistry Connection

Wake Up, Sleeping Gene

A common carboxylic acid, butyric acid, holds the promise of being an effective treatment for two age-old human genetic diseases. Sickle cell anemia and β-thalassemia are human genetic diseases of the β-globin portion of hemoglobin, the protein that carries oxygen from the lungs to tissues throughout the body. Normal hemoglobin consists of two α-globin proteins, two β-globin proteins, and four heme groups. In sickle cell anemia, the faulty β-globin gene calls for the synthesis of a sticky form of hemoglobin that forms long polymers. This distorts the red blood cells into elongated, sickled shapes that get stuck in capillaries and cannot provide the oxygen needed by the tissues. In β-thalassemia, there may be no β-globin produced at all. This results in short-lived red blood cells and severe anemia.

These two genetic diseases do not affect the fetus because before birth and for several weeks after birth, a fetal globin is made, rather than the adult β-globin. Fetal hemoglobin has a stronger affinity for oxygen than the adult form, ensuring that the fetus gets enough oxygen from the mother's blood through the placenta.

Two observations have led to a possible treatment of these diseases. First, physicians found some sickle cell anemia patients who suffered only mild symptoms because they continued to make high levels of fetal hemoglobin. Second was the observation that some babies born to diabetic mothers continued to produce fetal hemoglobin for an unusually long time after birth. Coincidentally, there was an unexpectedly high concentration of aminobutyric acid, a modified carboxylic acid, in the blood of these infants.

Susan Perrine of the Children's Hospital Oakland Research Center decided to try to reawaken the dormant fetal globin gene. She and her colleagues injected a sodium butyrate solution (the sodium salt of butyric acid) into three sickle cell patients and three β-thalassemia patients. As a result of the two- to three-week treatment, fetal hemoglobin production was boosted as much as 45% in these individuals. One β-thalassemia patient even experienced a complete reversal of the symptoms. Moreover, this treatment had few adverse side effects.

Longer studies with larger numbers of patients will be needed before this treatment can be declared a total success. However, Perrine's results hold the promise of a full and active life for individuals who were previously limited in activity and expected a short life span.

In this chapter we study the properties and reactions of the carboxylic acids; their salts, such as the sodium butyrate used to treat hemoglobin disorders; and their derivatives, the esters. We will focus on the importance of these molecules in biological systems, medicine, and the food industry.

Introduction

Carboxylic acids (Figure 14.1a) have the following general structure:

$$\text{Ar}-\overset{\overset{\displaystyle O}{\|}}{C}-\text{OH} \qquad \text{R}-\overset{\overset{\displaystyle O}{\|}}{C}-\text{OH}$$

Aromatic carboxylic acid   Aliphatic carboxylic acid

They are characterized by the carboxyl group, shown in red, which may also be written in condensed form as —COOH or —CO_2H. The name carboxylic acid describes this family of compounds quite well. The term carboxylic is taken from the terms carbonyl and hydroxyl, the two structural units that make up the carboxyl group. The word acid in the name tells us one of the more important properties of these molecules: they dissociate in water to release protons. Thus they are acids.

In this chapter we will also study the esters (Figure 14.1b), which have the following general structures:

$$\text{R}-\overset{\overset{\displaystyle O}{\|}}{C}-\text{O}-\text{R} \qquad \text{Ar}-\overset{\overset{\displaystyle O}{\|}}{C}-\text{O}-\text{Ar} \qquad \text{Ar}-\overset{\overset{\displaystyle O}{\|}}{C}-\text{O}-\text{R}$$

Examples of aliphatic and aromatic esters

> In fact, carboxylic acids are weak acids because they partially dissociate in water.

*The group shown in red is called the **acyl group**. The acyl group is part of the functional group of the carboxylic acid derivatives, including the esters, acid chlorides, acid anhydrides, and amides.*

Amides are discussed in Chapter 15.

14.1 Carboxylic Acids

Structure and Physical Properties

The **carboxyl group** consists of two very polar functional groups, the carbonyl group and the hydroxyl group. Thus **carboxylic acids** are very polar compounds. In addition, they can hydrogen bond to one another and to molecules of a polar solvent such as water. As a result of intermolecular hydrogen bonding, they boil at higher temperatures than aldehydes, ketones, or alcohols of comparable molecular weight. A comparison of the boiling points of an alkane, alcohol, ether, aldehyde, ketone, and carboxylic acid of comparable molecular weight is shown below:

LEARNING GOAL 1

$CH_3CH_2CH_2CH_3$

Butane
(butane)
M.W. = 58
b.p. −0.5°C

$CH_3-O-CH_2CH_3$

Methoxyethane
(ethyl methyl ether)
M.W. = 60
b.p. 7.0°C

$CH_3CH_2CH_2-OH$

1-Propanol
(propyl alcohol)
M.W. = 60
b.p. 97.2°C

$$CH_3CH_2\overset{O}{\overset{\|}{C}}-H$$

Propanal
(propionaldehyde)
M.W. = 58
b.p. 49°C

$$CH_3\overset{O}{\overset{\|}{C}}-CH_3$$

Propanone
(acetone)
M.W. = 58
b.p. 56°C

$$CH_3\overset{O}{\overset{\|}{C}}-OH$$

Ethanoic acid
(acetic acid)
M.W. = 60
b.p. 118°C

As with alcohols, the smaller carboxylic acids are soluble in water (Figure 14.2). However, solubility falls off dramatically as the carbon content of the carboxylic acid increases because the molecules become more hydrocarbonlike and less polar. For example, acetic acid (the two-carbon carboxylic acid found in vinegar) is completely soluble in water, but hexadecanoic acid (a sixteen-carbon carboxylic acid found in palm oil) is insoluble in water.

The lower-molecular-weight carboxylic acids have sharp, sour tastes and unpleasant aromas. Formic acid, HCOOH, is used as a chemical defense by ants and causes the burning sensation of the ant bite. Acetic acid, CH_3COOH, is found in vinegar; propionic acid, CH_3CH_2COOH, is responsible for the tangy flavor of Swiss cheese; and butyric acid, $CH_3CH_2CH_2COOH$, causes the stench associated with rancid butter and gas gangrene.

The longer-chain carboxylic acids are generally called **fatty acids** and are important components of biological membranes and triglycerides, the major lipid storage form in the body.

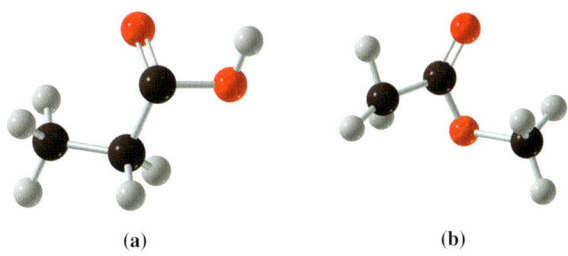

(a) (b)

Figure 14.1
Ball-and-stick models of (a) a carboxylic acid, propanoic acid, and (b) an ester, methyl ethanoate.

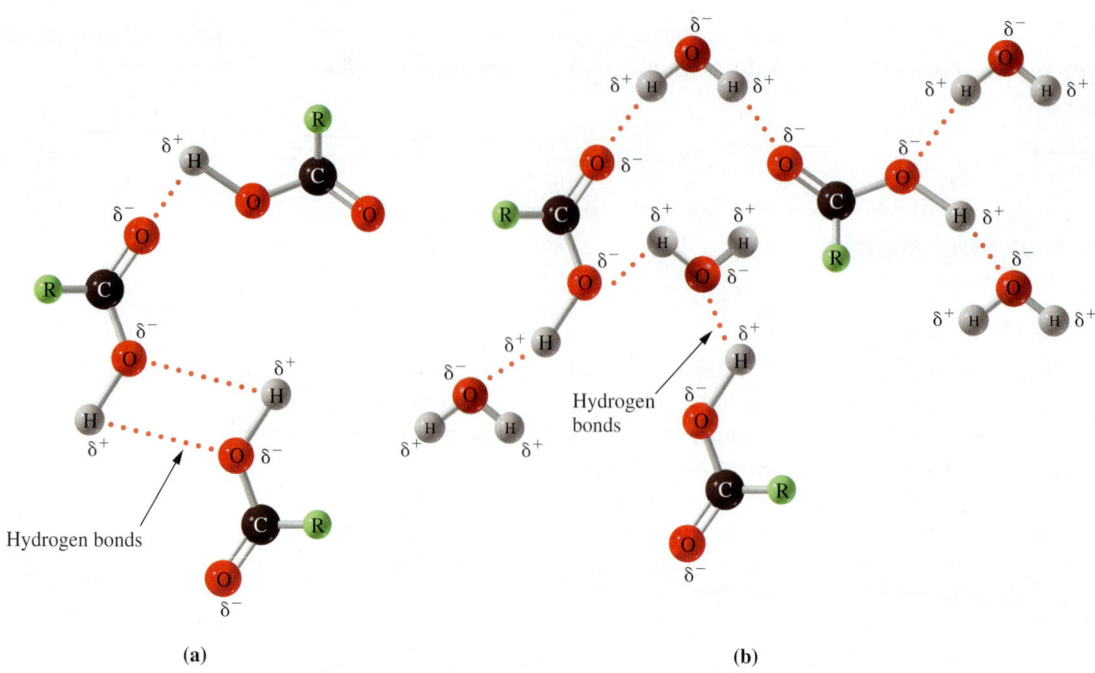

Figure 14.2
Hydrogen bonding (a) in carboxylic acids and (b) between carboxylic acids and water.

Question 14.1

Assuming that each of the following pairs of molecules has the same carbon chain length, which member of each of the following pairs has the lower boiling point?

  a. a carboxylic acid or a ketone
  b. a ketone or an alcohol
  c. an alcohol or an alkane

Question 14.2

Assuming that each of the following pairs of molecules has the same carbon chain length, which member of each of the following pairs has the lower boiling point?

  a. an ether or an aldehyde
  b. an aldehyde or a carboxylic acid
  c. an ether or an alcohol

Question 14.3

Why would you predict that a carboxylic acid would be more polar and have a higher boiling point than an aldehyde of comparable molecular weight?

Question 14.4

Why would you predict that a carboxylic acid would be more polar and have a higher boiling point than an alcohol of comparable molecular weight?

14.1 Carboxylic Acids

Nomenclature

In the I.U.P.A.C. Nomenclature System, carboxylic acids are named according to the following set of rules:

LEARNING GOAL 2

- Determine the parent compound, the longest continuous carbon chain bearing the carboxyl group.
- Number the chain so that the carboxyl carbon is carbon-1.
- Replace the -e ending of the parent alkane with the suffix -oic acid. If there are two carboxyl groups, the suffix -dioic acid is used.
- Name and number substituents in the usual way.

The following examples illustrate the naming of carboxylic acids with one carboxyl group:

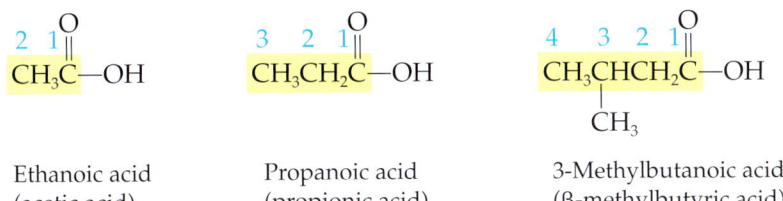

Ethanoic acid   Propanoic acid   3-Methylbutanoic acid
(acetic acid)   (propionic acid) (β-methylbutyric acid)

The following examples illustrate naming carboxylic acids with two carboxyl groups:

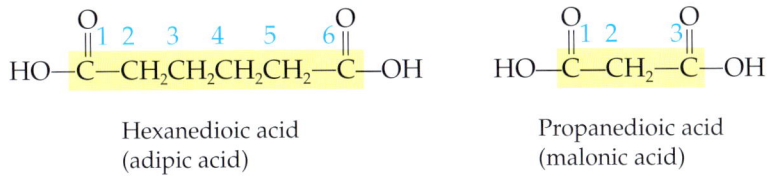

Hexanedioic acid   Propanedioic acid
(adipic acid)   (malonic acid)

EXAMPLE 14.1 Use the I.U.P.A.C. Nomenclature System to Name a Carboxylic Acid

Name the following carboxylic acids using the I.U.P.A.C. Nomenclature System.

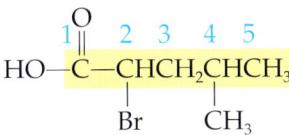

Solution

Parent compound: pentane (becomes pentanoic acid)
Position of —COOH: carbon-1 (Must be!)
Substituents: 2-bromo and 4-methyl
Name: 2-Bromo-4-methylpentanoic acid

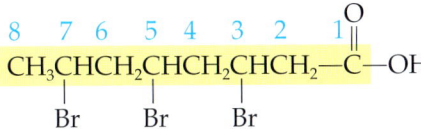

Solution

Parent compound: octane (becomes octanoic acid)
Position of —COOH: carbon-1 (Must be!)
Substituents: 3,5,7-bromo
Name: 3,5,7-Tribromooctanoic acid

The carboxylic acid derivatives of cycloalkanes are named by adding the suffix *carboxylic acid* to the name of the cycloalkane or substituted cycloalkane. The carboxyl group is always on carbon-1 and other substituents are named and numbered as usual.

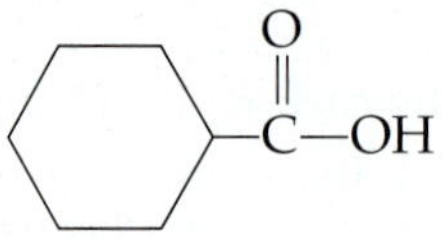

Cyclohexanecarboxylic acid

Question 14.5

Determine the I.U.P.A.C. name for each of the following structures. Remember that —COOH is an alternative way to represent the carboxyl group.

a. 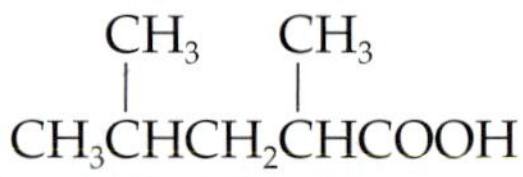 CH$_3$CHCH$_2$CHCOOH with CH$_3$ groups on the 2nd and 4th carbons

b. 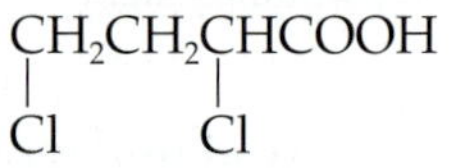 CH$_2$CH$_2$CHCOOH with Cl substituents

Question 14.6

Determine the I.U.P.A.C. name for each of the following structures.

a. 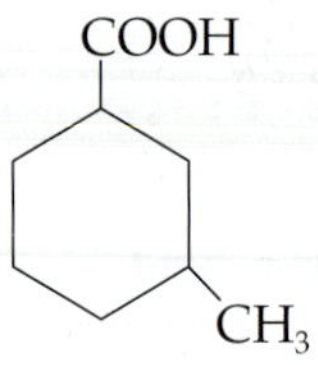 Cyclohexane with COOH and CH$_3$ substituents

b. 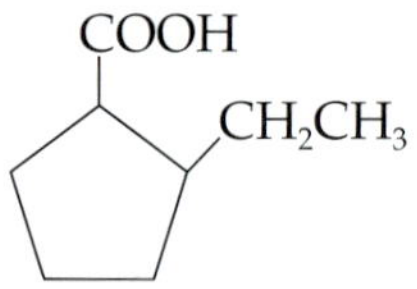 Cyclopentane with COOH and CH$_2$CH$_3$ substituents

Question 14.7

Write the structure for each of the following carboxylic acids.

a. 2,3-Dihydroxybutanoic acid
b. 2-Bromo-3-chloro-4-methylhexanoic acid

Question 14.8

Write the structure for each of the following carboxylic acids.

a. 1,4-Cyclohexanedicarboxylic acid
b. 4-Hydroxycyclohexanecarboxylic acid

LEARNING GOAL

As we have seen so often, the use of common names, rather than systematic names, still persists. Often these names have evolved from the source of a given compound. This is certainly true of the carboxylic acids. Table 14.1 shows the I.U.P.A.C. and common names of several carboxylic acids, as well as their sources

14.1 Carboxylic Acids

TABLE 14.1 **Names and Sources of Some Common Carboxylic Acids**

Name	Structure	Source	Root
Formic acid (methanoic acid)	HCOOH	Ants	L: *formica*, ant
Acetic acid (ethanoic acid)	CH_3COOH	Vinegar	L: *acetum*, vinegar
Propionic acid (propanoic acid)	CH_3CH_2COOH	Swiss cheese	Gk: *protos*, first; *pion*, fat
Butyric acid (butanoic acid)	$CH_3(CH_2)_2COOH$	Rancid butter	L: *butyrum*, butter
Valeric acid (pentanoic acid)	$CH_3(CH_2)_3COOH$	Valerian root	
Caproic acid (hexanoic acid)	$CH_3(CH_2)_4COOH$	Goat fat	L: *caper*, goat
Caprylic acid (octanoic acid)	$CH_3(CH_2)_6COOH$	Goat fat	L: *caper*, goat
Capric acid (decanoic acid)	$CH_3(CH_2)_8COOH$	Goat fat	L: *caper*, goat
Palmitic acid (hexadecanoic acid)	$CH_3(CH_2)_{14}COOH$	Palm oil	
Stearic acid (octadecanoic acid)	$CH_3(CH_2)_{16}COOH$	Tallow (beef fat)	Gk: *stear*, tallow

Note: I.U.P.A.C. names are shown in parentheses.

and the Latin or Greek words that gave rise to the common names. Not only are the prefixes different than those used in the I.U.P.A.C. system, the suffix is different as well. Common names end in *-ic acid* rather than *-oic acid*.

In the common system of nomenclature, substituted carboxylic acids are named as derivatives of the parent compound (see Table 14.1). Greek letters are used to indicate the position of the substituent. The carbon atom bonded to the carboxyl group is the α-carbon, the next is the β-carbon, and so on.

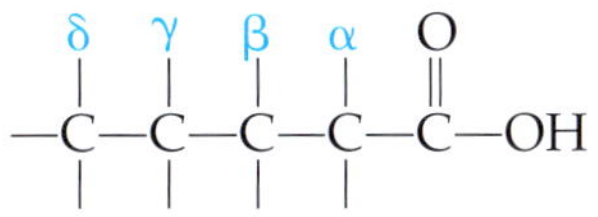

$$\overset{\delta}{-C}-\overset{\gamma}{C}-\overset{\beta}{C}-\overset{\alpha}{C}-\overset{O}{\underset{\|}{C}}-OH$$

Some examples of common names are

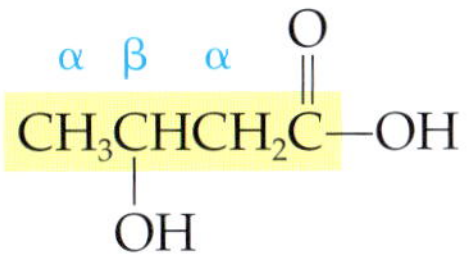

$$\underset{\underset{OH}{|}}{CH_3\overset{\alpha}{C}H}\overset{\beta}{C}H_2\overset{\alpha}{C}-OH \qquad \underset{\underset{OH}{|}}{CH_3\overset{\beta}{C}H}\overset{\alpha}{C}-OH$$

β-Hydroxybutyric acid 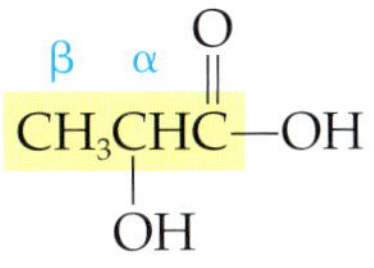   α-Hydroxypropionic acid

---

EXAMPLE 14.2 Naming Carboxylic Acids Using the Common System of Nomenclature

Write the common name for each of the following carboxylic acids.

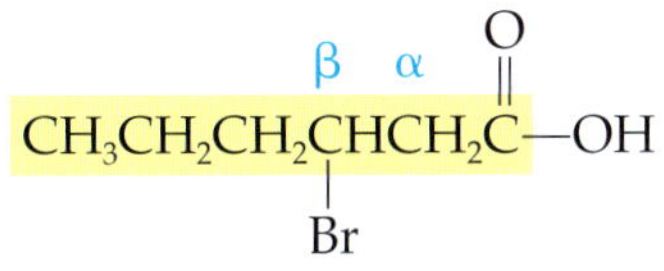

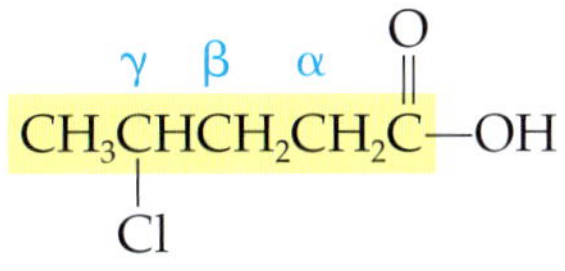

$$\underset{\underset{Br}{|}}{CH_3CH_2CH_2\overset{\beta}{C}H\overset{\alpha}{C}H_2\overset{\|}{C}-OH} \qquad \underset{\underset{Cl}{|}}{CH_3\overset{\gamma}{C}H\overset{\beta}{C}H_2\overset{\alpha}{C}H_2\overset{\|}{C}-OH}$$

Solution

Parent compound:	caproic acid	valeric acid
Substituents:	β-bromo	γ-chloro
Name:	β-Bromocaproic acid	γ-Chlorovaleric acid

Question 14.9

Provide the common name for each of the following molecules. Keep in mind that the carboxyl group can be represented as —COOH.

a. 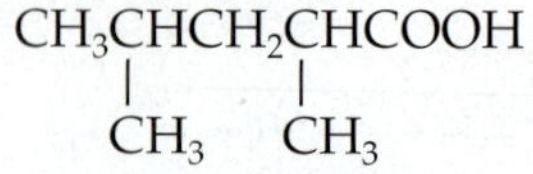 CH₃CHCH₂CHCOOH
       | |
      CH₃   CH₃

b. 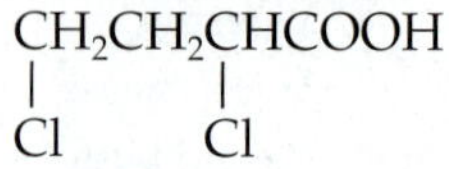 CH₂CH₂CHCOOH
    | |
    Cl   Cl

Question 14.10

Provide the common name for each of the following molecules.

a.  CH₃CHCHCH₂COOH
       | |
      Br Br

b.  CH₃CH₂CHCH₂CH₂COOH
          |
         OH

Benzoic acid is the simplest aromatic carboxylic acid.

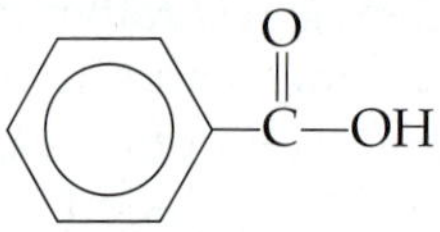

Benzoic acid

Nomenclature of aromatic compounds is described in Section 11.6.

In many cases the aromatic carboxylic acids are named, in either system, as derivatives of benzoic acid. Generally, the *-oic acid* or *-ic acid* suffix is attached to the appropriate prefix. However, "common names" of substituted benzoic acids (for example, toluic acid and phthalic acid) are frequently used.

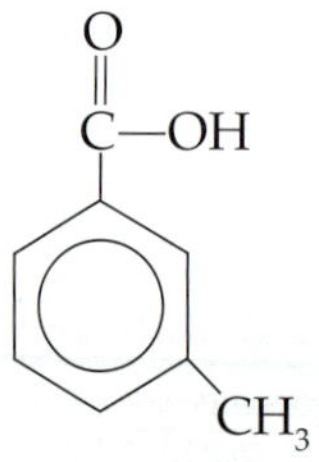

m-Toluic acid 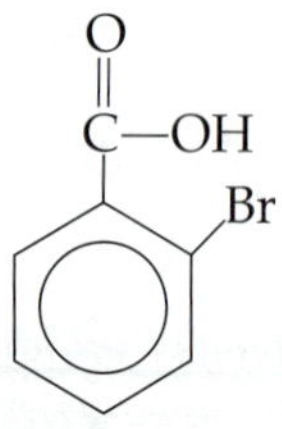   *o-Bromobenzoic acid* 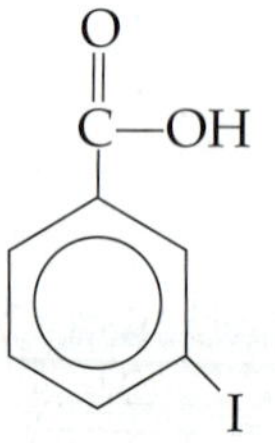   *m-Iodobenzoic acid* 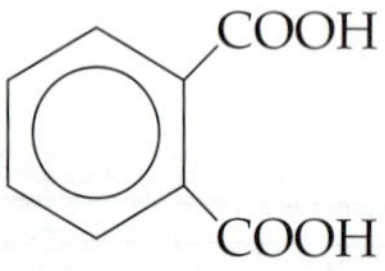   *Phthalic acid*

The phenyl group is benzene with one hydrogen removed.

Often the phenyl group is treated as a substituent, and the name is derived from the appropriate alkanoic acid parent chain. For example:

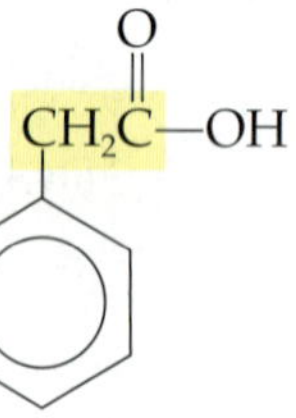

2-Phenylethanoic acid   3-Phenylpropanoic acid
(α-phenylacetic acid)   (β-phenylpropionic acid)

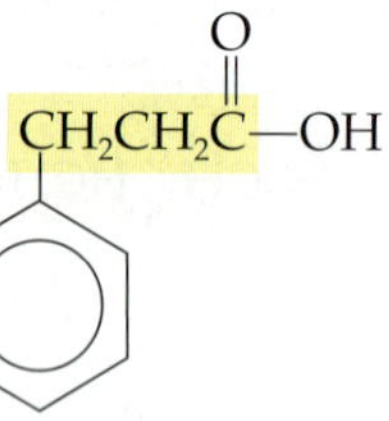

EXAMPLE 14.3

Naming Aromatic Carboxylic Acids

Name the following aromatic carboxylic acids.

(Structure: benzene ring with C(=O)-OH at top and Cl at bottom, para position)

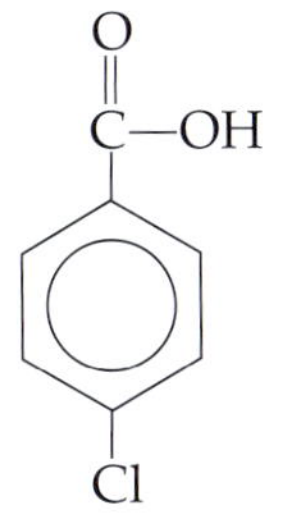

Solution

It is simplest to name the compound as a derivative of benzoic acid. The substituent, Cl, is attached to carbon-4 of the benzene ring. This compound is 4-chlorobenzoic acid or *p*-chlorobenzoic acid.

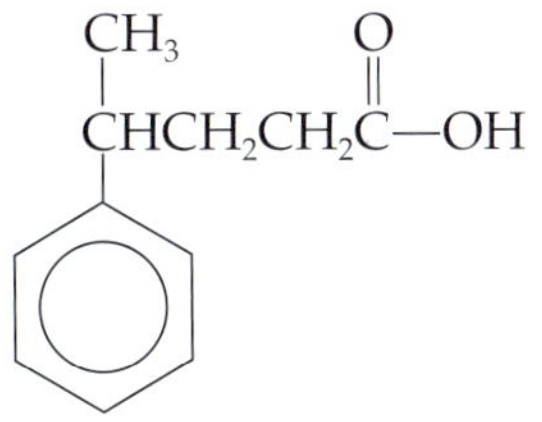

(Structure: CH₃-CH(C₆H₅)-CH₂-CH₂-C(=O)-OH)

Solution

This compound is most easily named by treating the phenyl group as a substituent. The phenyl group is bonded to carbon-4 (or the γ-carbon, in the common system of nomenclature). The parent compound is pentanoic acid (valeric acid in the common system). Hence the name of this compound is 4-phenylpentanoic acid or γ-phenylvaleric acid.

Question 14.11

Draw structures for each of the following compounds.

a. *o*-Toluic acid
b. 2,4,6-Tribromobenzoic acid
c. 2,2,2-Triphenylethanoic acid

Question 14.12

Draw structures for each of the following compounds.

a. *p*-Toluic acid
b. 3-Phenylhexanoic acid
c. 3-Phenylcyclohexanecarboxylic acid

Some Important Carboxylic Acids

 **LEARNING GOAL**

As Table 14.1 shows, many carboxylic acids occur in nature. Fatty acids can be isolated from a variety of sources including palm oil, coconut oil, butter, milk, lard, and tallow (beef fat). More complex carboxylic acids are also found in a variety of foodstuffs. For example, citric acid is found in citrus fruits and is often used to give the sharp taste to sour candies. It is also added to foods as a preservative and antioxidant. Adipic acid (hexanedioic acid) gives tartness to soft drinks and helps to retard spoilage.

An Environmental Perspective

Garbage Bags from Potato Peels?

One problem facing society is its enormous accumulation of trash. To try to control the mountains of garbage that we produce, many institutions and towns practice recycling of aluminum, paper, and plastics. One problem that remains, however, is the plastic trash bag. We stuff the trash bag full of biodegradable garbage and bury it in a landfill; but soil bacteria can't break down the plastic to get to the biodegradable materials inside. Imagine a twenty-fourth century archeologist excavating one of these monuments to our society!

The good news is that laboratory research and products of bacterial metabolism are providing new materials that have the properties of plastics, but are readily biodegradable. For instance, sheets of plastic can be made by making polymers of lactic acid, which is a natural carboxylic acid produced by fermentation of sugars, particularly in milk and working muscle. Because many common soil bacteria can break down polylactic acid (PLA), trash bags made from this polymer would be quickly broken down in landfill soil.

Making plastic from lactic acid requires a huge supply of this carboxylic acid. As it turns out, we can produce an enormous quantity of lactic acid from garbage. When french fries are produced, nearly half of the potato is wasted. That amounts to about ten billion pounds of potato waste each year. When cheese is made, the curds are separated from the whey, and several billion gallons of whey are poured down the drain each year. Potato waste and whey can easily be broken down to produce glucose, which, in turn, can be converted into lactic acid used to make biodegradable plastics. PLA plastics have been available since the early 1990's and have been used successfully for sutures, medical implants, and drug delivery systems.

Biodegradable plastic can be made from garbage, such as potato peels left over from making french fries.

$$n\ H_3C-\underset{\underset{OH}{|}}{\overset{\overset{H}{|}}{C}}-\overset{\overset{O}{\|}}{C}-OH \longrightarrow \left[-O-\underset{\underset{H}{|}}{\overset{\overset{CH_3}{|}}{C}}-\overset{\overset{O}{\|}}{C}-O-\underset{\underset{H}{|}}{\overset{\overset{CH_3}{|}}{C}}-\overset{\overset{O}{\|}}{C}-O- \right]_n$$

Lactic acid   Polylactic acid (PLA)

Nature has provided other biodegradable plastics that are produced by bacteria. The most common of these in nature is polyhydroxybutyrate (PHB) made by the bacterium *Alcaligenes eutrophus*. PHB is a homopolymer (made up of a single monomer) of 3-hydroxybutanoic acid (β-hydroxybutyric acid).

Remember, the carboxyl group can be represented as —COOH, CO$_2$H, or

$$-\overset{\overset{O}{\|}}{C}-OH$$

Bacteria in milk produce lactic acid as a product of fermentation of sugars. Lactic acid contributes a tangy flavor to yogurt and buttermilk. It is also used as a food preservative to lower the pH to a level that retards microbial growth that causes food spoilage. Lactic acid is produced in muscle cells when an individual is exercising strenuously. If the level of lactic acid in the muscle and bloodstream becomes high enough, the muscle can't continue to work.

```
      COOH              COOH              COOH
       |                  |                 |
    H—C—H              H—C—OH            H—C—H
       |                  |                 |
   HO—C—COOH             CH3             H—C—H
       |                                    |
    H—C—H                                H—C—H
       |                                    |
     COOH                                H—C—H
                                            |
                                          COOH

   Citric acid         Lactic acid       Adipic acid
  (citrus fruit)        (yogurt)         (beet juice)
```

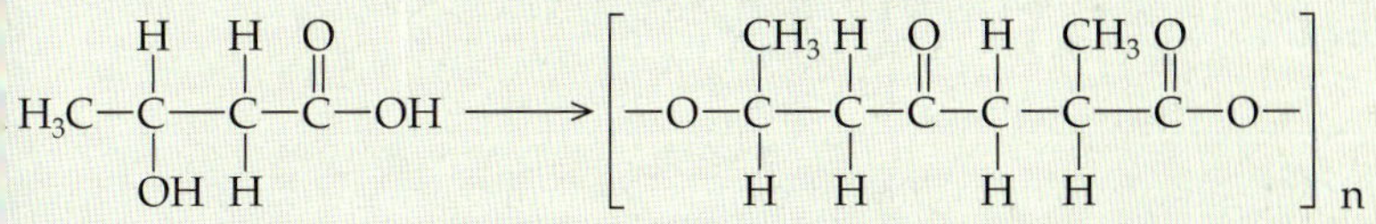

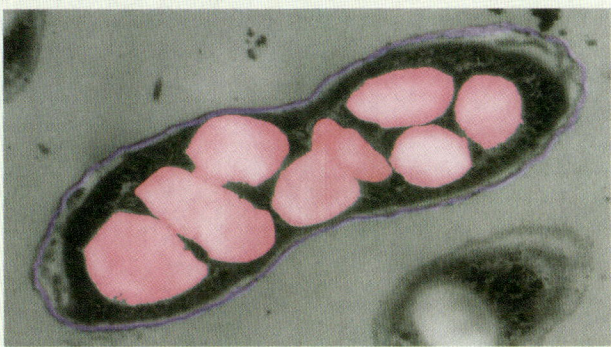

β-Hydroxybutyric acid   Polyhydroxybutyric acid (PHB)   β-Hydroxybutyric acid   β-Hydroxyvaleric acid

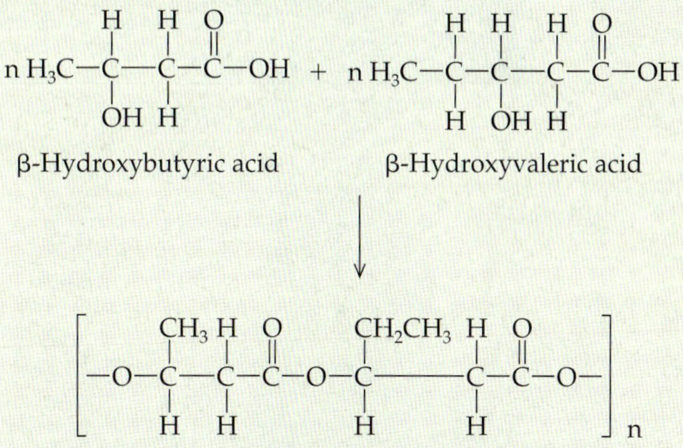

Biopol—a heteropolymer

Unfortunately, this natural polymer does not have physical properties that would allow it to be a commercially successful plastic. This led to interest in making heteropolymers, polymers composed of two or more monomers. One such polymer with useful properties is a heteropolymer of β-hydroxybutyric acid and β-hydroxyvaleric acid, which has been given the name Biopol.

Biopol has properties that make it commercially useful, and it is completely broken down into carbon dioxide and water by microorganisms in the soil. Thus, it is completely biodegradable. Because the bacteria produced a low yield of Biopol, scientists produced transgenic plants in hopes that they would produce high yields of the polymer. This approach also encountered problems.

For the time being, biodegradable plastics cannot outcompete their nonbiodegradable counterparts. Future research and development will be required to reduce the cost of commercial production and fulfill the promise of an "environmentally friendly" garbage bag.

Computer colorized inclusions of polyhydroxybutyrate in *Alcaligenes eutrophus*.

For Further Understanding

Why are biodegradable plastics useful as sutures?

Two hydrogen atoms are lost in each reaction that adds a carboxylic acid to the polymers described here. What type of chemical reaction is this?

Reactions Involving Carboxylic Acids
Preparation of Carboxylic Acids

Many of the small carboxylic acids are prepared on a commercial scale. For example, ethanoic (acetic) acid, found in vinegar, is produced commercially by the **oxidation** of either ethanol or ethanal as shown here:

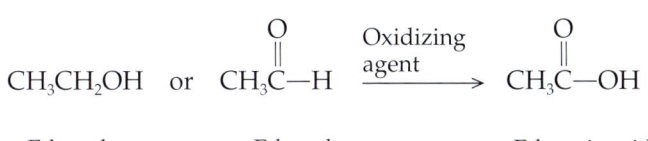

Ethanol   Ethanal   Ethanoic acid

A variety of oxidizing agents, including oxygen, can be used, and catalysts are often required to provide acceptable yields. Other simple carboxylic acids can be made by oxidation of the appropriate primary alcohol or aldehyde.

In the laboratory, carboxylic acids are prepared by the oxidation of aldehydes or primary alcohols. Most common oxidizing agents, such as chromic acid, can be used. The general reaction is

4 LEARNING GOAL

These reactions were discussed in Sections 12.5 and 13.4.

$$R-CH_2OH \xrightarrow{[O]} R-\overset{\overset{O}{\|}}{C}-H \xrightarrow{[O]} R-\overset{\overset{O}{\|}}{C}-OH$$

Primary alcohol   Aldehyde   Carboxylic acid

EXAMPLE 14.4   Writing Equations for the Oxidation of a Primary Alcohol to a Carboxylic Acid

Write an equation showing the oxidation of 1-propanol to propanoic acid.

Solution

$$CH_3CH_2\overset{\overset{H}{|}}{\underset{\underset{H}{|}}{C}}-OH \xrightarrow{H_2CrO_4} CH_3CH_2\overset{\overset{O}{\|}}{C}-H \xrightarrow[\text{oxidation}]{\text{Continued}} CH_3CH_2-\overset{\overset{O}{\|}}{C}-OH$$

1-Propanol   Propanal   Propanoic acid
(propyl alcohol)   (propionaldehyde)   (propionic acid)

Acid–Base Reactions

LEARNING GOAL 5

The carboxylic acids behave as acids because they are proton donors. They are weak acids that dissociate to form a carboxylate ion and a hydrogen ion, as shown in the following equation:

$$R-\overset{\overset{O}{\|}}{C}-OH \rightleftharpoons R-\overset{\overset{O}{\|}}{C}-O^- + H^+$$

Carboxylic acid   Carboxylate anion   Hydrogen ion

The properties of weak acids are described in Sections 8.1 and 8.2.

Carboxylic acids are weak acids because they dissociate only slightly in solution. The majority of the acid remains in solution in the undissociated form. Typically, less than 5% of the acid is ionized (approximately five carboxylate ions to every ninety-five carboxylic acid molecules).

When strong bases are added to a carboxylic acid, neutralization occurs. The acid protons are removed by the OH⁻ to form water and the carboxylate ion. The equilibrium shown in the reaction above is shifted to the right, owing to removal of H⁺. This is an illustration of LeChatelier's principle.

LeChatelier's principle is described in Section 7.4.

$$R-\overset{\overset{O}{\|}}{C}-OH + NaOH \longrightarrow R-\overset{\overset{O}{\|}}{C}-O^-Na^+ + H_2O$$

Carboxylic acid   Strong base   Carboxylic acid salt   Water

The carboxylate anion and the cation of the base form the carboxylic acid salt.

The following examples show the neutralization of acetic acid and benzoic acid in solutions of the strong base NaOH.

$$CH_3\overset{\overset{O}{\|}}{C}-OH + NaOH \longrightarrow CH_3\overset{\overset{O}{\|}}{C}-O^-Na^+ + H-O-H$$

Acetic acid   Sodium hydroxide (strong base)   Sodium acetate   Water

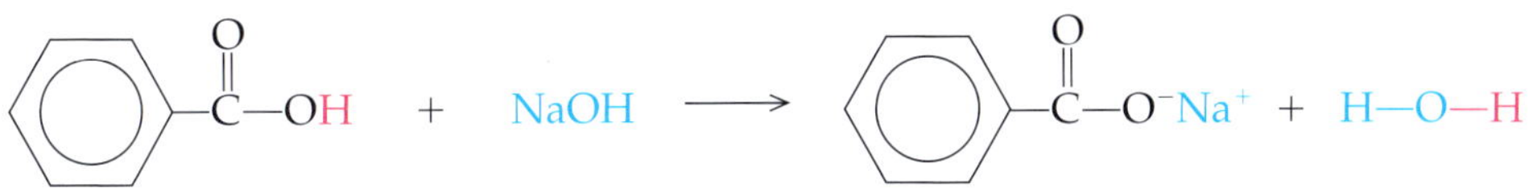

14.1 Carboxylic Acids

[Benzoic acid structure]—C(=O)—OH + NaOH ⟶ [Benzene ring]—C(=O)—O⁻Na⁺ + H—O—H

Benzoic acid   Sodium hydroxide (strong base)   Sodium benzoate

Sodium benzoate is commonly used as a food preservative.

Note that the salt of a carboxylic acid is named by replacing the *-ic acid* suffix with *-ate*. Thus acetic acid becomes acetate, and benzoic acid becomes benzoate. This name is preceded by the name of the appropriate cation, sodium in the examples above.

EXAMPLE 14.5

Writing an Equation to Show the Neutralization of a Carboxylic Acid by a Strong Base

Write an equation showing the neutralization of propanoic acid by potassium hydroxide.

Solution

The protons of the acid are removed by the OH⁻ of the base. This produces water. The cation of the base, in this case potassium ion, forms the salt of the carboxylic acid.

$$CH_3CH_2-\overset{O}{\underset{\|}{C}}-OH + KOH \longrightarrow CH_3CH_2-\overset{O}{\underset{\|}{C}}-O^-K^+ + H_2O$$

Propanoic acid   Potassium hydroxide   Potassium salt of propanoic acid   Water

---

Question 14.13

Write the formula of the organic product obtained through each of the following reactions.

a. $CH_3CH_2CH_2OH \xrightarrow{H_2CrO_4}$?

b. $H\overset{O}{\underset{\|}{C}}CH_2CH_2CH_2CH_3 \xrightarrow{H_2CrO_4}$?

Question 14.14

Write the formula of the organic product obtained through each of the following reactions.

a. $CH_3CH_2COOH + KOH \longrightarrow$?
b. $CH_3CH_2CH_2COOH + Ba(OH)_2 \longrightarrow$?

Question 14.15

Complete each of the following reactions by supplying the missing product(s).

a. $CH_3CH_2OH \xrightarrow{H_2CrO_4}$?

b. $H\overset{O}{\underset{\|}{C}}CH_2CH_2\overset{O}{\underset{\|}{C}}H \xrightarrow{H_2CrO_4}$?

Question 14.16

Complete each of the following reactions by supplying the missing products.

a. $CH_3CH_2CH_2CH_2CH_2COOH + KOH \longrightarrow$?
b. Benzoic acid + sodium hydroxide $\longrightarrow$?

EXAMPLE 14.6 Naming the Salt of a Carboxylic Acid

Write the common and I.U.P.A.C. names of the salt produced in the reaction shown in Example 14.5.

Solution

$$CH_3CH_2-\overset{\overset{O}{\|}}{C}-O^-\,K^+$$

I.U.P.A.C. name of the parent carboxylic acid:	Propanoic acid
Replace the *-ic acid* ending with *-ate*:	Propanoate
Name of the cation of the base:	Potassium
Name of the carboxylic acid salt:	Potassium propanoate
Common name of the parent carboxylic acid:	Propionic acid
Replace the *-ic acid* ending with *-ate*:	Propionate
Name of the cation of the base:	Potassium
Name of the carboxylic acid salt:	Potassium propionate

Question 14.17

Name the products of the reactions in Question 14.14.

Question 14.18

Name the products of the reactions in Question 14.16.

Soaps are made by a process called saponification, which is the base-catalyzed hydrolysis of an ester. This is described in detail in Section 14.2.

LEARNING GOAL

Carboxylic acid salts are ionic substances. As a result, they are very soluble in water. The long-chain carboxylic acid salts (fatty acid salts) are called *soaps*.

Esterification

Carboxylic acids react with alcohols to form esters and water according to the general reaction:

$$R^1-\overset{\overset{O}{\|}}{C}-OH + R^2OH \underset{}{\overset{Acid}{\rightleftarrows}} R^1-\overset{\overset{O}{\|}}{C}-OR^2 + H_2O$$

Carboxylic acid   Alcohol   Ester   Water

The details of these reactions will be examined in Section 14.2.

14.2 Esters

Structure and Physical Properties

Esters are mildly polar and have pleasant aromas. Many esters are found in natural foodstuffs; banana oil (3-methylbutyl ethanoate; common name, isoamyl acetate), pineapples (ethyl butanoate; common name, ethyl butyrate), and raspberries (isobutyl methanoate; common name, isobutyl formate) are but a few examples.

Esters boil at approximately the same temperature as aldehydes or ketones of comparable molecular weight. The simpler ones are somewhat soluble in water.

7 LEARNING GOAL

See A Human Perspective: The Chemistry of Flavor and Fragrance.

Nomenclature

Esters are **carboxylic acid derivatives,** organic compounds derived from carboxylic acids. They are formed from the reaction of a carboxylic acid with an alcohol, and both of these reactants are reflected in the naming of the ester. They are named according to the following set of rules:

- Use the *alkyl* or *aryl* portion of the alcohol name as the first name.
- The *-ic acid* ending of the name of the carboxylic acid is replaced with *-ate* and follows the first name.

8 LEARNING GOAL

For example, in the following reaction, ethanoic acid reacts with methanol to produce methyl ethanoate:

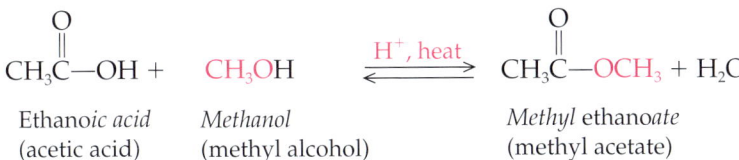

Ethan*oic acid*   Methanol   Methyl ethan*oate*
(acetic acid)   (methyl alcohol) (methyl acetate)

Similarly, *acetic acid* and *ethanol* react to produce *ethyl acetate*, and the product of the reaction between benz*oic acid* and *isopropyl* alcohol is *isopropyl* benz*oate*.

EXAMPLE 14.7   Naming Esters Using the I.U.P.A.C. and Common Nomenclature Systems

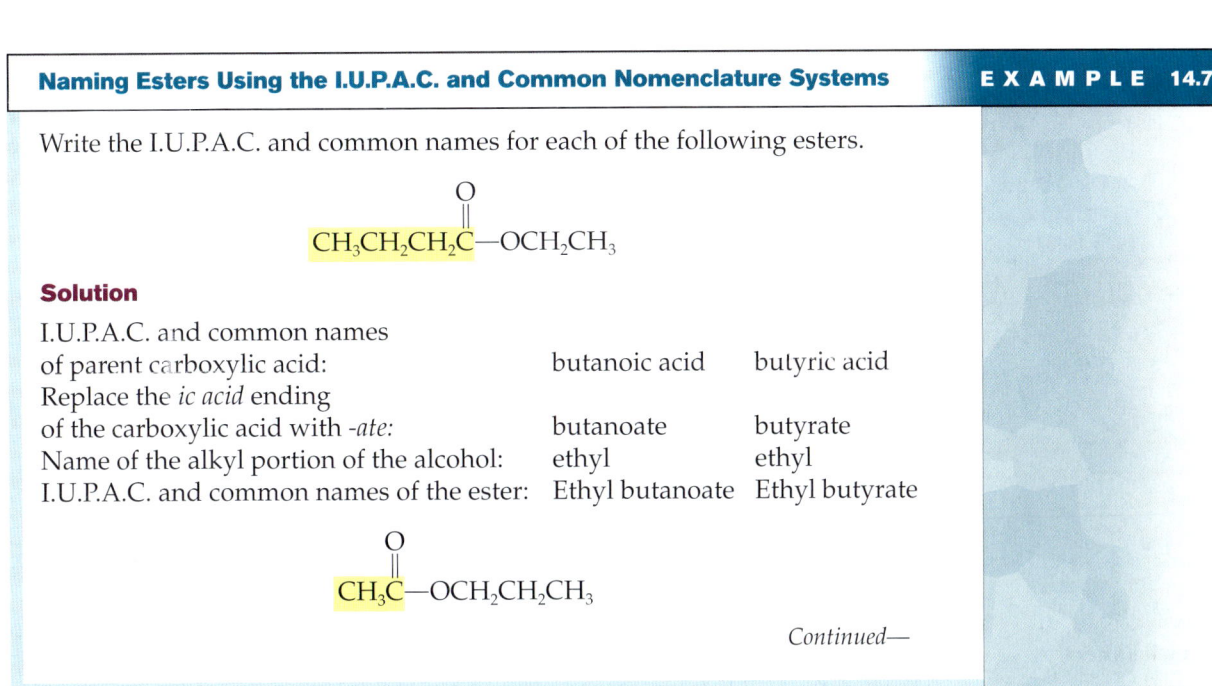

Write the I.U.P.A.C. and common names for each of the following esters.

$$CH_3CH_2CH_2\overset{O}{\underset{\|}{C}}-OCH_2CH_3$$

Solution

I.U.P.A.C. and common names of parent carboxylic acid:	butanoic acid	butyric acid
Replace the *ic acid* ending of the carboxylic acid with *-ate*:	butanoate	butyrate
Name of the alkyl portion of the alcohol:	ethyl	ethyl
I.U.P.A.C. and common names of the ester:	Ethyl butanoate	Ethyl butyrate

$$CH_3\overset{O}{\underset{\|}{C}}-OCH_2CH_2CH_3$$

Continued—

EXAMPLE 14.7 —Continued

Solution

I.U.P.A.C. and common names of parent carboxylic acid:	ethanoic acid	acetic acid
Replace the *ic acid* ending of the carboxylic acid with *-ate*:	ethanoate	acetate
Name of the alkyl portion of the alcohol:	propyl	propyl
I.U.P.A.C. and common names of the ester:	Propyl ethanoate	Propyl acetate

---

Question 14.19 Name each of the following esters using both the I.U.P.A.C. and common nomenclature systems.

a.
$$CH_3CH_2CH_2\overset{\overset{O}{\|}}{C}-OCH_2CH_2CH_3$$

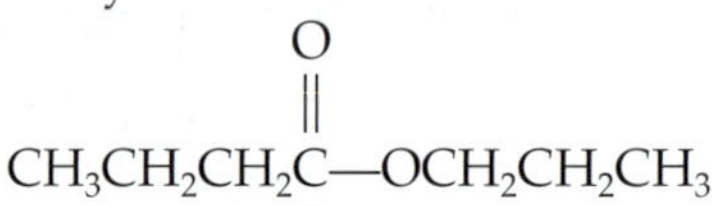

b.
$$CH_3CH_2CH_2\overset{\overset{O}{\|}}{C}-OCH_2CH_3$$

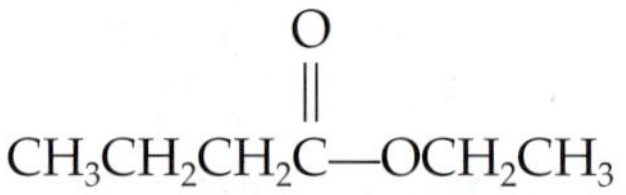

Question 14.20 Name each of the following esters using both the I.U.P.A.C. and common nomenclature systems.

a.
$$CH_3\overset{\overset{O}{\|}}{C}-OCH_2CH_3$$

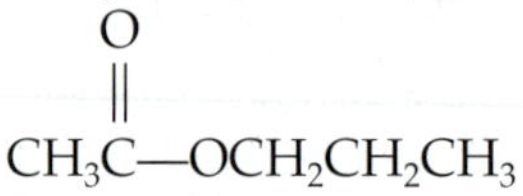

b.
$$CH_3CH_2\overset{\overset{O}{\|}}{C}-OCH_2CH_2CH_2CH_3$$

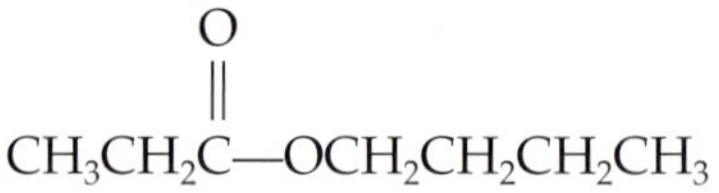

Naming esters is analogous to naming the salts of carboxylic acids. Consider the following comparison:

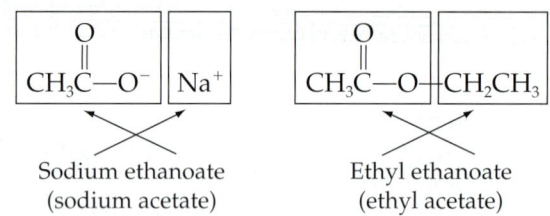

Sodium ethanoate (sodium acetate)   Ethyl ethanoate (ethyl acetate)

As shown in this example, the alkyl group of the alcohol, rather than Na$^+$, has displaced the acidic hydrogen of the carboxylic acid.

Reactions Involving Esters

Preparation of Esters

LEARNING GOAL

The conversion of a carboxylic acid to an ester requires heat and is catalyzed by a trace of acid (H$^+$). When esters are prepared directly from a carboxylic acid and an alcohol, a water molecule is lost, as in the reaction:

14.2 Esters

$$R^1-\overset{\overset{\displaystyle O}{\|}}{C}-OH + R^2OH \underset{}{\overset{H^+, \text{ heat}}{\rightleftharpoons}} R^1-\overset{\overset{\displaystyle O}{\|}}{C}-OR^2 + H_2O$$

Carboxylic acid   Alcohol   Ester   Water

> Esterification is reversible. The direction of the reaction is determined by the conditions chosen. Excess alcohol favors ester formation. The carboxylic acid is favored when excess water is present.

$$CH_3CH_2\overset{\overset{\displaystyle O}{\|}}{C}-OH + CH_3OH \overset{H^+, \text{ heat}}{\rightleftharpoons} CH_3CH_2\overset{\overset{\displaystyle O}{\|}}{C}-OCH_3 + H-O-H$$

Propanoic acid   Methanol   Methyl propanoate
(propionic acid)   (methyl alcohol)   (methyl propionate)

Esterification is a *dehydration* reaction, so called because a water molecule is eliminated during the reaction.

EXAMPLE 14.8 Writing Equations Representing Esterification Reactions

Write an equation showing the esterification reactions that would produce ethyl butanoate and propyl ethanoate.

Solution

The name, ethyl butanoate, tells us that the alcohol used in the reaction is ethanol and the carboxylic acid is butanoic acid. We must remember that a trace of acid and heat are required for the reaction and that the reaction is reversible. With this information, we can write the following equation representing the reaction:

$$CH_3CH_2CH_2\overset{\overset{\displaystyle O}{\|}}{C}-OH + CH_3CH_2OH \overset{H^+, \text{ heat}}{\rightleftharpoons} CH_3CH_2CH_2\overset{\overset{\displaystyle O}{\|}}{C}-OCH_2CH_3 + H_2O$$

Butanoic acid   Ethanol   Ethyl butanoate
(butyric acid)   (ethyl butyrate)

Similarly, the name propyl ethanoate reveals that the alcohol used in this reaction is 1-propanol and the carboxylic acid must be ethanoic acid. Knowing that we must indicate that the reaction is reversible and that heat and a trace of acid are required, we can write the following equation:

$$CH_3\overset{\overset{\displaystyle O}{\|}}{C}-OH + CH_3CH_2CH_2OH \overset{H^+, \text{ heat}}{\rightleftharpoons} CH_3\overset{\overset{\displaystyle O}{\|}}{C}-OCH_2CH_2CH_3 + H_2O$$

Ethanoic acid   1-Propanol   Propyl ethanoate
(acetic acid)   (propyl acetate)

Question 14.21 Write an equation showing the esterification reactions that would produce butyl ethanoate and ethyl propanoate.

Question 14.22 Write an equation showing the esterification reactions that would produce methyl butanoate and propyl methanoate.

A Human Perspective

The Chemistry of Flavor and Fragrance

Carboxylic acids are often foul smelling. For instance, butyric acid is one of the worst smelling compounds imaginable—the smell of rancid butter.

$$CH_3CH_2CH_2\overset{\overset{\displaystyle O}{\|}}{C}-OH$$

Butanoic acid
(butyric acid)

Butyric acid is also a product of fermentation reactions carried out by *Clostridium perfringens.* This organism is the most common cause of gas gangrene. Butyric acid contributes to the notable foul smell accompanying this infection.

By forming esters of butyric acid, one can generate compounds with pleasant smells. Ethyl butyrate is the essence of pineapple oil.

Volatile esters are often pleasant in both aroma and flavor. Natural fruit flavors are complex mixtures of many esters and other organic compounds. Chemists can isolate these mixtures and identify the chemical components. With this information they are able to synthesize artificial fruit flavors, using just a few of the esters found in the natural fruit. As a result, the artificial flavors rarely have the full-bodied flavor of nature's original blend.

Raspberries

Pineapple

$$H-\overset{\overset{\displaystyle O}{\|}}{C}-OCH_2\underset{\underset{\displaystyle CH_3}{|}}{C}HCH_3$$

Isobutyl methanoate
(isobutyl formate)

Bananas

$$CH_3CH_2CH_2\overset{\overset{\displaystyle O}{\|}}{C}-OCH_2CH_3$$

Ethyl butanoate
(ethyl butyrate)

$$CH_3\overset{\overset{\displaystyle O}{\|}}{C}-OCH_2CH_2\underset{\underset{\displaystyle CH_3}{|}}{C}HCH_3$$

3-Methylbutyl ethanoate
(isoamyl acetate)

Oranges

$$\text{CH}_3\overset{\overset{\text{O}}{\|}}{\text{C}}-\text{OCH}_2\text{CH}_2\text{CH}_2\text{CH}_2\text{CH}_2\text{CH}_2\text{CH}_3$$

Octyl ethanoate
(octyl acetate)

Apples

$$\text{CH}_3\text{CH}_2\text{CH}_2\overset{\overset{\text{O}}{\|}}{\text{C}}-\text{OCH}_3$$

Methyl butanoate
(methyl butyrate)

Apricots

$$\text{CH}_3\text{CH}_2\text{CH}_2\overset{\overset{\text{O}}{\|}}{\text{C}}-\text{OCH}_2\text{CH}_2\text{CH}_2\text{CH}_3$$

Pentyl butanoate
(pentyl butyrate)

Strawberries

$$\text{CH}_3\text{CH}_2\text{CH}_2\overset{\overset{\text{O}}{\|}}{\text{C}}-\text{SCH}_3$$

Methyl thiobutanoate
(methyl thiobutyrate)
(a thioester in which sulfur replaces oxygen)

For Further Understanding

Draw the structure of methyl salicylate, which is found in oil of wintergreen.

Write an equation for the synthesis of each of the esters shown in this Perspective.

EXAMPLE 14.9 Designing the Synthesis of an Ester

Design the synthesis of ethyl propanoate from organic alcohols.

Solution

The ease with which alcohols are oxidized to aldehydes, ketones, or carboxylic acids (depending on the alcohol that you start with and the conditions that you employ), coupled with the ready availability of alcohols, provides the pathway necessary to many successful synthetic transformations. For example, let's develop a method for synthesizing ethyl propanoate, using any inorganic reagent you wish but limiting yourself to organic alcohols that contain three or fewer carbon atoms:

$$CH_3CH_2\overset{\overset{O}{\|}}{C}-O-CH_2CH_3$$

Ethyl propanoate
(ethyl propionate)

Ethyl propanoate can be made from propanoic acid and ethanol:

$$CH_3CH_2\overset{\overset{O}{\|}}{C}-OH + CH_3CH_2OH \underset{}{\overset{H^+,\ heat}{\rightleftharpoons}} CH_3CH_2\overset{\overset{O}{\|}}{C}-O-CH_2CH_3 + H_2O$$

Propanoic acid   Ethanol   Ethyl propanoate
(propionic acid)   (ethyl alcohol)   (ethyl propionate)

Ethanol is a two-carbon alcohol that is an allowed starting material, but propanoic acid is not. Can we now make propanoic acid from an alcohol of three or fewer carbons? Yes!

$$CH_3CH_2CH_2OH \xrightarrow{[O]} CH_3CH_2\overset{\overset{O}{\|}}{C}-H \xrightarrow{[O]} CH_3CH_2\overset{\overset{O}{\|}}{C}-OH$$

Propanol   Propanoic acid
(propyl alcohol)   (propionic acid)

Propanol is a three-carbon alcohol, an allowed starting material. The synthesis is now complete. By beginning with ethanol and propanol, ethyl propanoate can be synthesized easily.

Hydrolysis of Esters

LEARNING GOAL 9

Hydrolysis, sometimes also referred to as *hydration,* refers to cleavage of any bond by the addition of a water molecule. Esters undergo hydrolysis reactions in water, as shown in the general reaction:

$$R^1-\overset{\overset{O}{\|}}{C}-OR^2 + H_2O \underset{}{\overset{H^+,\ heat}{\rightleftharpoons}} R^1-\overset{\overset{O}{\|}}{C}-OH + R^2OH$$

Ester   Water   Carboxylic   Alcohol
                                                      acid

This reaction requires heat. A small amount of acid (H$^+$) may be added to catalyze the reaction, as in the following example:

14.2 Esters

$$CH_3CH_2\overset{O}{\underset{\|}{C}}-OCH_2CH_2CH_3 + H_2O \xrightleftharpoons{H^+, heat} CH_3CH_2\overset{O}{\underset{\|}{C}}-OH + CH_3CH_2CH_2OH$$

Propyl propanoate   Propanoic acid   1-Propanol
(propyl propionate)   (propionic acid)   (propanol)

The base-catalyzed hydrolysis of an ester is called **saponification**.

$$R^1-\overset{O}{\underset{\|}{C}}-OR^2 + H_2O \xrightarrow{NaOH, heat} R^1-\overset{O}{\underset{\|}{C}}-O^-Na^+ + R^2OH$$

Ester   Water   Carboxylic   Alcohol
                                              acid salt

Under basic conditions the acid cannot exist. Thus, the reaction yields the salt of the carboxylic acid having the cation of the basic catalyst.

$$CH_3\overset{O}{\underset{\|}{C}}-OCH_2CH_2CH_2CH_3 \xrightarrow{NaOH, heat} CH_3\overset{O}{\underset{\|}{C}}-O^-Na^+ + CH_3CH_2CH_2CH_2OH$$

Butyl ethanoate   Sodium ethanoate   1-Butanol
(butyl acetate)   (sodium acetate)   (butyl alcohol)

The carboxylic acid is formed when the reaction mixture is neutralized with an acid such as HCl.

$$CH_3\overset{O}{\underset{\|}{C}}-O^-Na^+ + HCl \longrightarrow CH_3\overset{O}{\underset{\|}{C}}-OH + NaCl$$

Sodium ethanoate   Ethanoic acid
(sodium acetate)   (acetic acid)

Question 14.23

Complete each of the following reactions by drawing the structure of the missing products.

a. $CH_3\overset{O}{\underset{\|}{C}}-OCH_2CH_2CH_3 + H_2O \xrightleftharpoons{H^+, heat} ?$

b. $CH_3CH_2CH_2CH_2CH_2\overset{O}{\underset{\|}{C}}-OCH_2CH_2CH_3 + H_2O \xrightarrow{KOH, heat} ?$

c. $CH_3CH_2CH_2CH_2\overset{O}{\underset{\|}{C}}-OCH_3 + H_2O \xrightarrow{NaOH, heat} ?$

d. $CH_3CH_2CH_2CH_2CH_2\overset{O}{\underset{\|}{C}}-O\underset{\underset{CH_3}{|}}{C}HCH_2CH_2CH_3 + H_2O \xrightleftharpoons{H^+, heat} ?$

Question 14.24

Use the I.U.P.A.C. Nomenclature System to name each of the products in Question 14.23.

Triesters of glycerol are more commonly referred to as triglycerides. We know them as solid fats, generally from animal sources, and liquid oils, typically from plants. We will study triglycerides in detail in Section 17.3.

Fats and oils are triesters of the alcohol glycerol. When they are hydrolyzed by saponification, the products are **soaps,** which are the salts of long-chain carboxylic acids (fatty acid salts). According to Roman legend, soap was discovered by washerwomen following a heavy rain on Mons Sapo ("Mount Soap"). An important sacrificial altar was located on the mountain. The rain mixed with the remains of previous animal sacrifices—wood ash and animal fat—at the base of the altar. Thus, the three substances required to make soap accidentally came together—water, fat, and alkali (potassium carbonate and potassium hydroxide, called *potash*, leached from the wood ash). The soap mixture flowed down the mountain and into the Tiber River, where the washerwomen quickly realized its value.

We still use the old Roman recipe to make soap from water, a strong base, and natural fats and oils obtained from animals or plants. The carbon chain length of the fatty acid salts governs the solubility of a soap. The lower-molecular-weight carboxylic acid salts (up to twelve carbons) have greater solubility in water and give a lather containing large bubbles. The higher-molecular-weight carboxylic acid salts (fourteen to twenty carbons) are much less soluble in water and produce a lather with fine bubbles. The nature of the cation also affects the solubility of the soap. In general, the potassium salts of carboxylic acids are more soluble in water than the sodium salts. The synthesis of a soap is shown in Figure 14.3.

The role of soap in the removal of soil and grease is best understood by considering the functional groups in soap molecules and studying the way in which they interact with oil and water. The long, continuous hydrocarbon side chains of soap molecules resemble alkanes, and they dissolve other nonpolar compounds such as oils and greases ("like dissolves like"). The large nonpolar hydrocarbon part of the molecule is described as *hydrophobic,* which means "water-fearing." This part of the molecule is repelled by water. The highly polar carboxylate end of the molecule is called *hydrophilic,* which means "water-loving."

When soap is dissolved in water, the carboxylate end actually dissolves. The hydrocarbon part is repelled by the water molecules so that a thin film of soap is formed on the surface of the water with the hydrocarbon chains protruding outward. When soap solution comes in contact with oil or grease, the hydrocarbon part dissolves in the oil or grease, but the polar carboxylate group remains dissolved in water. When particles of oil or grease are surrounded by soap molecules, the resulting "units" formed are called *micelles.* A simplified view of this phenomenon is shown in Figure 14.4.

Micelles repel one another because they are surrounded on the surface by the negatively charged carboxylate ions. Mechanical action (for example, scrubbing or tumbling in a washing machine) causes oil or grease to be surrounded by soap molecules and broken into small droplets so that relatively small micelles are formed. These small micelles are then washed away. Careful examination of this solution shows that it is an *emulsion* containing suspended micelles.

A more detailed diagram of a micelle is found in Figure 23.1.

An emulsion is a suspension of very fine droplets of one liquid in another. In this case it is oil in water.

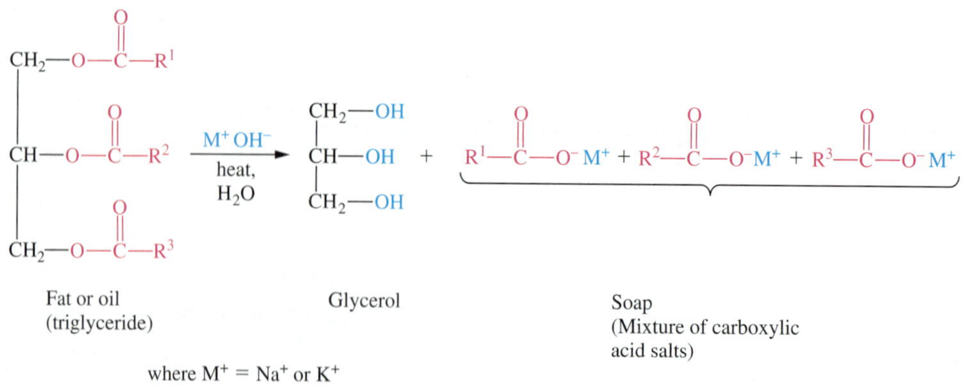

Figure 14.3
Saponification is the base-catalyzed hydrolysis of a glycerol triester.

$$CH_2-O-\overset{O}{\underset{\|}{C}}-R^1$$
$$CH-O-\overset{O}{\underset{\|}{C}}-R^2 \xrightarrow[\text{heat, } H_2O]{M^+OH^-} CH_2-OH$$
$$CH-OH + R^1-\overset{O}{\underset{\|}{C}}-O^-M^+ + R^2-\overset{O}{\underset{\|}{C}}-O^-M^+ + R^3-\overset{O}{\underset{\|}{C}}-O^-M^+$$
$$CH_2-O-\overset{O}{\underset{\|}{C}}-R^3 \qquad CH_2-OH$$

Fat or oil (triglyceride)   Glycerol   Soap (Mixture of carboxylic acid salts)

where M^+ = Na^+ or K^+

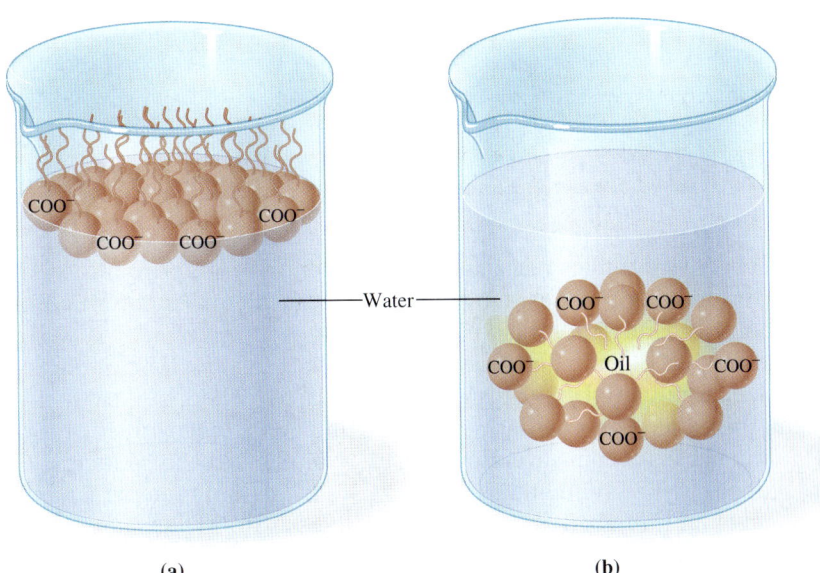

Figure 14.4
Simplified view of the action of a soap. The wiggly lines represent the long, continuous carbon chains of each soap molecule. (a) The thin film of soap molecules that forms at the water surface reduces surface tension. (b) Particles of oil and grease are surrounded by soap molecules to form a micelle.

Condensation Polymers

As we saw in Chapter 11, *polymers* are macromolecules, very large molecules. They result from the combination of many smaller molecules, usually in a repeating pattern, to give molecules whose molecular weight may be 10,000 g/mol or greater. The small molecules that make up the polymer are called *monomers*.

A polymer may be made from a single type of monomer. Such a polymer, called a homopolymer, would have the following general structure:

chain continues~A-A-A-A-A-A-A-A-A-A-A-A-A-A-A-A-A-A-A-A~chain continues

The addition polymers of alkenes that we studied in Chapter 11 are examples of this type of polymer. Alternatively, two different monomers may be copolymerized, producing a heteropolymer with the following structure:

chain continues~A-B-A-B-A-B-A-B-A-B-A-B-A-B-A-B~chain continues

Polyesters are heteropolymers. They are also known as condensation polymers. **Condensation polymers** are formed by the polymerization of monomers in a reaction that forms a small molecule such as water or an alcohol. Polyesters are synthesized by reacting a dicarboxylic acid and a dialcohol (diol). Notice that each of the combining molecules has two reactive functional groups, highlighted in red here:

n HOOC—⬡—COOH + n HOCH$_2$CH$_2$OH

Terephthalic acid   1,2-Ethanediol

H$^+$ ↓

HOCH$_2$CH$_2$O—C(=O)—⬡—COOH

Another molecule of terephthalic acid can react here.

+ H$_2$O

Another molecule of 1,2-ethanediol can react here.

Reaction continues ↓

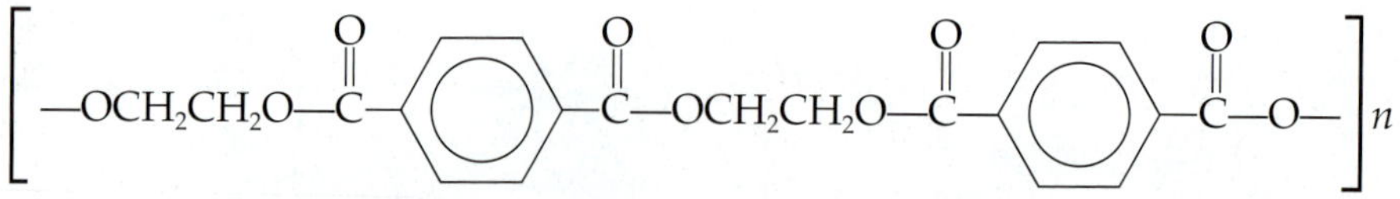

$$\left[-OCH_2CH_2O-\overset{O}{\underset{\|}{C}}-\bigcirc-\overset{O}{\underset{\|}{C}}-OCH_2CH_2O-\overset{O}{\underset{\|}{C}}-\bigcirc-\overset{O}{\underset{\|}{C}}-O- \right]_n$$

Polyethylene terephthalate
PETE

Each time a pair of molecules reacts using one functional group from each, a new molecule is formed that still has two reactive groups. The product formed in this reaction is polyethylene terephthalate, or PETE.

When formed as fibers, polyesters are used to make fabric for clothing. These polyesters were trendy in the 1970s, during the "disco" period, but lost their popularity soon thereafter. Polyester fabrics, and a number of other synthetic polymers used in clothing, have become even more fashionable since the introduction of microfiber technology. The synthetic polymers are extruded into fibers that are only half the diameter of fine silk fibers. When these fibers are used to create fabrics, the result is a fabric that drapes freely yet retains its shape. These fabrics are generally lightweight, wrinkle resistant, and remarkably strong.

Polyester can be formed into a film called Mylar. These films, coated with aluminum foil, are used to make balloons that remain inflated for long periods. They are also used as the base for recording tapes and photographic film.

PETE can be used to make shatterproof plastic bottles, such as those used for soft drinks. However, these bottles cannot be recycled and reused directly because they cannot withstand the high temperatures required to sterilize them. PETE can't be used for any foods, such as jellies, that must be packaged at high temperatures. For these uses, a new plastic, PEN, or polyethylene naphthalate, is used.

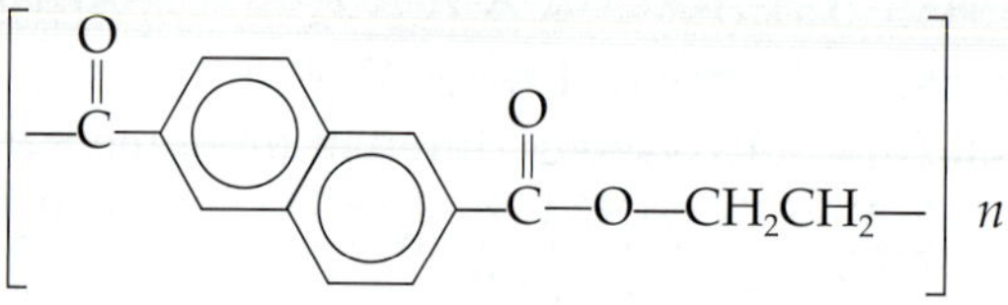

Biodegradable plastics made of both homopolymers and heteropolymers are discussed in "An Environmental Perspective: Garbage Bags from Potato Peels?" found on page 456.

$$\left[-\overset{O}{\underset{\|}{C}}-\bigcirc\bigcirc-\overset{O}{\underset{\|}{C}}-O-CH_2CH_2- \right]_n$$

Napthalate group   Ethylene group

14.3 Acid Chlorides and Acid Anhydrides

Acid Chlorides

LEARNING GOAL 11

Acid chlorides are carboxylic acid derivatives having the general formula

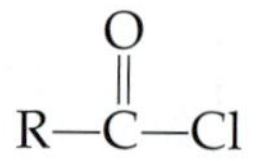

$$R-\overset{O}{\underset{\|}{C}}-Cl$$

They are named by replacing the *-ic acid* ending of the common name with *-yl chloride* or the *-oic acid* ending of the I.U.P.A.C. name of the carboxylic acid with *-oyl chloride*. For example,

$$CH_3CH_2CH_2\overset{O}{\underset{\|}{C}}-Cl \qquad CH_3\overset{O}{\underset{\|}{C}}-Cl \qquad \underset{\underset{Br}{|}}{CH_2}CH_2\overset{O}{\underset{\|}{C}}-Cl \qquad Cl-\bigcirc-\overset{O}{\underset{\|}{C}}-Cl$$

Butanoyl chloride   Ethanoyl chloride   3-Bromopropanoyl chloride   4-Chlorobenzoyl chloride
(butyryl chloride)   (acetyl chloride)   (β-bromopropionyl chloride) (*p*-chlorobenzoyl chloride)

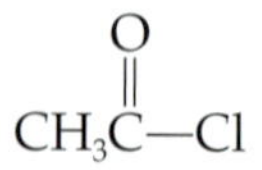

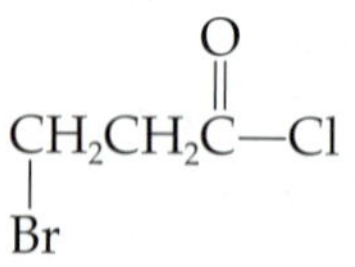

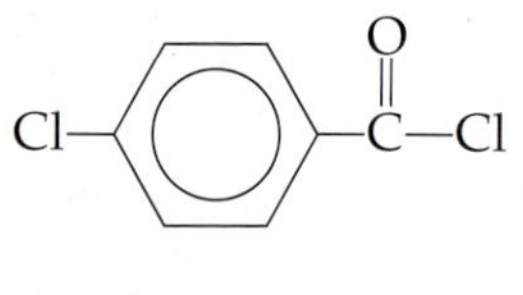

14.3 Acid Chlorides and Acid Anhydrides

Acid chlorides are noxious, irritating chemicals and must be handled with great care. They are slightly polar and boil at approximately the same temperature as the corresponding aldehyde or ketone of comparable molecular weight. They react violently with water and therefore cannot be dissolved in that solvent. Acid chlorides have little commercial value other than their utility in the synthesis of esters and amides, two of the other carboxylic acid derivatives.

Acid chlorides are prepared from the corresponding carboxylic acid by reaction with one of several inorganic acid chlorides, including PCl_3, PCl_5, or $SOCl_2$. The general reaction is summarized in the following equation:

$$\underset{\text{Carboxylic acid}}{R-\underset{\underset{O}{\|}}{C}-OH} \xrightarrow{\text{inorganic acid chloride}} \underset{\text{Acid chloride}}{R-\underset{\underset{O}{\|}}{C}-Cl} + \text{inorganic products}$$

12 LEARNING GOAL

The following equations show the synthesis of ethanoyl chloride and benzoyl chloride:

$$\underset{\substack{\text{Ethanoic acid} \\ \text{(acetic acid)}}}{CH_3\underset{\underset{O}{\|}}{C}-OH} \xrightarrow[\text{(Phosphorus trichloride)}]{PCl_3} \underset{\substack{\text{Ethanoyl chloride} \\ \text{(acetyl chloride)}}}{CH_3\underset{\underset{O}{\|}}{C}-Cl} + \text{inorganic products}$$

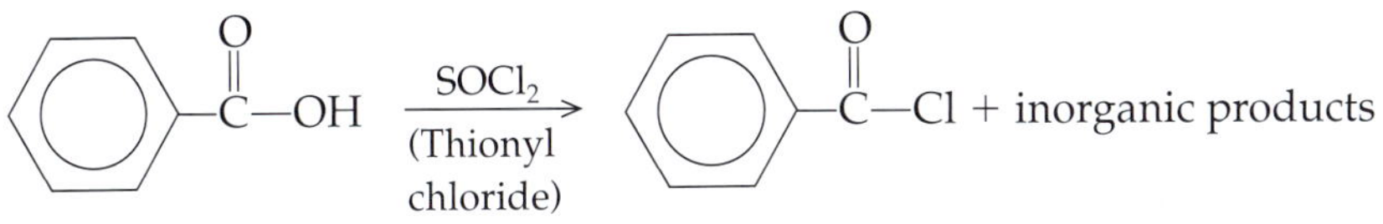

$$\underset{\text{Benzoic acid}}{C_6H_5-\underset{\underset{O}{\|}}{C}-OH} \xrightarrow[\text{(Thionyl chloride)}]{SOCl_2} \underset{\text{Benzoyl chloride}}{C_6H_5-\underset{\underset{O}{\|}}{C}-Cl} + \text{inorganic products}$$

EXAMPLE 14.10 Writing Equations Representing the Synthesis of Acid Chlorides

Write equations representing the following reactions:

a. Butanoic acid with phosphorus trichloride
b. 2-Bromobenzoic acid with thionyl chloride

Solution

a. Begin by drawing the structure of the reactant, butanoic acid. The inorganic acid chloride (phosphorus trichloride) can be represented over the reaction arrow. The product is drawn by replacing the —OH of the carboxyl group with —Cl. This gives us the following equation:

$$CH_3CH_2CH_2\underset{\underset{O}{\|}}{C}-OH \xrightarrow{PCl_3} CH_3CH_2CH_2\underset{\underset{O}{\|}}{C}-Cl + \text{inorganic products}$$

b. Begin by drawing the structure of the reactant, 2-bromobenzoic acid. The inorganic acid chloride (thionyl chloride) can be represented over the reaction arrow. The product is drawn by replacing the —OH of the carboxyl group with —Cl. This gives us the following equation:

(2-bromobenzoic acid) $\xrightarrow{SOCl_2}$ (2-bromobenzoyl chloride) + inorganic products

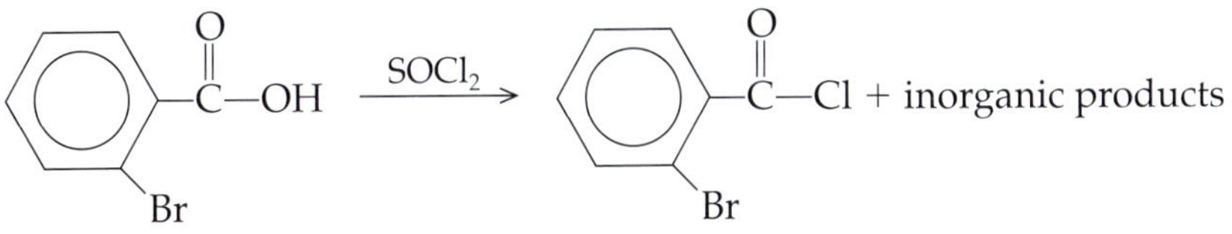

EXAMPLE 14.11 Naming Acid Chlorides

Name the products of the reactions shown in Example 14.10.

Solution

$$CH_3CH_2CH_2\overset{O}{\underset{\|}{C}}-OH \xrightarrow{PCl_3} CH_3CH_2CH_2\overset{O}{\underset{\|}{C}}-Cl + \text{inorganic products}$$

Butanoic acid
(butyric acid)

The carboxylic acid that is the reactant in this reaction is butanoic acid (butyric acid). By dropping the *-oic acid* ending of the I.U.P.A.C. name (or the *-ic acid* ending of the common name) and replacing it with *-oyl chloride* (or *-yl chloride*), we can write the I.U.P.A.C. and common names of the product of this reaction. They are butanoyl chloride and butyryl chloride, respectively.

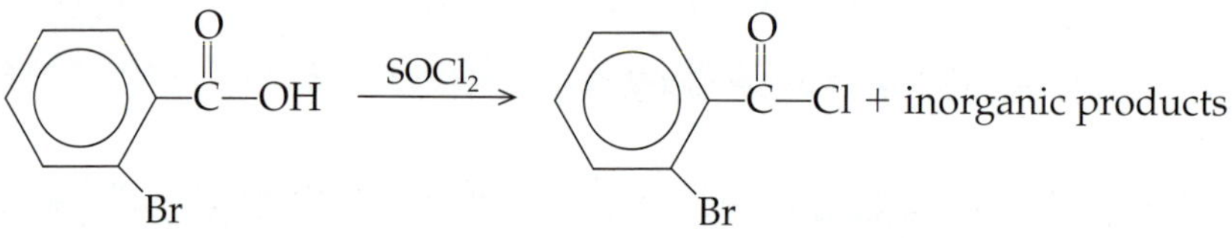

[Structure: 2-bromobenzoic acid reacting with $SOCl_2$ to give 2-bromobenzoyl chloride + inorganic products]

2-Bromobenzoic acid
(*o*-bromobenzoic acid)

The carboxylic acid in this reaction is 2-bromobenzoic acid. By dropping the *-oic acid* ending of the I.U.P.A.C. name and replacing it with *-oyl chloride*, we can name the product of this reaction: 2-bromobenzoyl chloride. It is equally correct to name this compound *o*-bromobenzoyl chloride.

The reaction that occurs when acid chlorides react violently with water is hydrolysis. The products are the carboxylic acid and hydrochloric acid.

$$CH_3\overset{O}{\underset{\|}{C}}-Cl + H_2O \longrightarrow CH_3\overset{O}{\underset{\|}{C}}-OH + HCl$$

Ethanoyl chloride   Ethanoic acid
(acetyl chloride)   (acetic acid)

Question 14.25

Write an equation showing the synthesis of each of the following acid chlorides. Provide the I.U.P.A.C. names of the carboxylic acid reactants and the acid chloride products.

a. 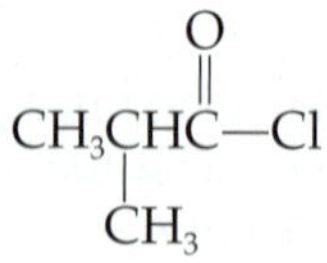 $CH_3\underset{\underset{CH_3}{|}}{CH}\overset{O}{\underset{\|}{C}}-Cl$

b. $CH_3CH_2CH_2CH_2CH_2\overset{O}{\underset{\|}{C}}-Cl$

Question 14.26

Write an equation showing the synthesis of each of the following acid chlorides. Provide the I.U.P.A.C. names of the carboxylic acid reactants and the acid chloride products.

a. 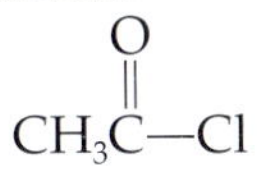 CH₃C(=O)—Cl

b. 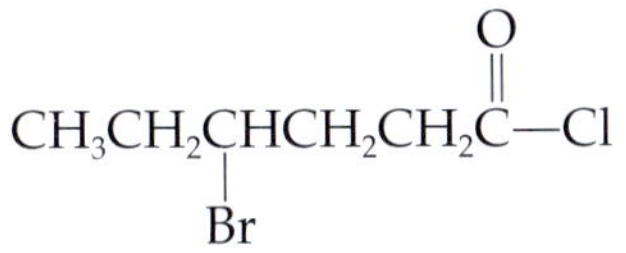 CH₃CH₂CH(Br)CH₂CH₂C(=O)—Cl

Question 14.27

Write an equation showing the synthesis of each of the following acid chlorides. Provide the common names of the carboxylic acid reactants and the acid chloride products.

a. 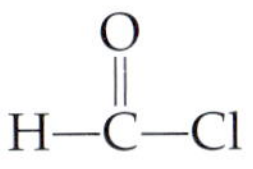 H—C(=O)—Cl

b. 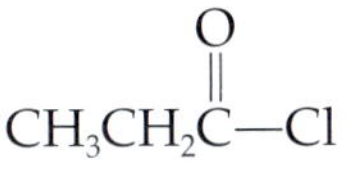 CH₃CH₂C(=O)—Cl

Question 14.28

Write an equation showing the synthesis of each of the following acid chlorides. Provide the common names of the carboxylic acid reactants and the acid chloride products.

a. 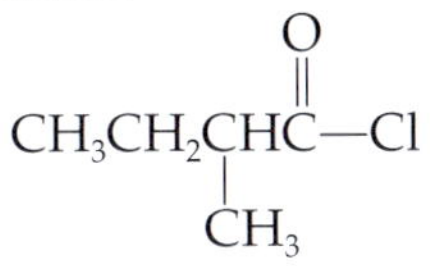 CH₃CH₂CH(CH₃)C(=O)—Cl

b. 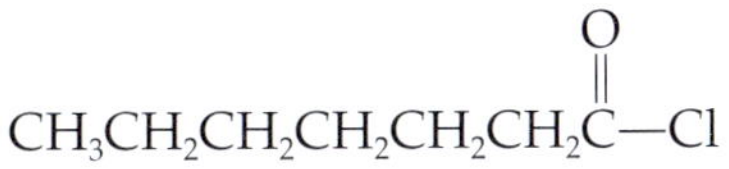 CH₃CH₂CH₂CH₂CH₂CH₂C(=O)—Cl

Acid Anhydrides

Acid anhydrides are molecules with the following general formula:

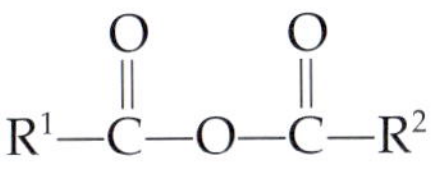

$$R^1-\overset{O}{\underset{\|}{C}}-O-\overset{O}{\underset{\|}{C}}-R^2$$

The name of the family is really quite fitting. The structure above reveals that acid anhydrides are actually two carboxylic acid molecules with a water molecule removed. The word *anhydride* means "without water."

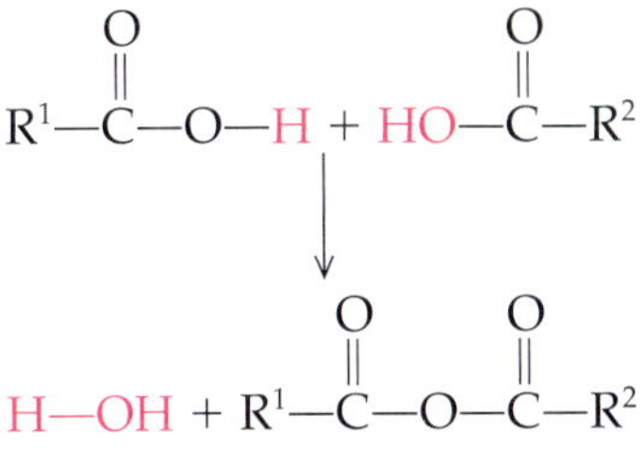

$$R^1-\overset{O}{\underset{\|}{C}}-O-H \;+\; HO-\overset{O}{\underset{\|}{C}}-R^2$$
$$\downarrow$$
$$H-OH \;+\; R^1-\overset{O}{\underset{\|}{C}}-O-\overset{O}{\underset{\|}{C}}-R^2$$

Acid anhydrides are classified as *symmetrical* if both acyl groups are the same. Symmetrical acid anhydrides are named by replacing the *acid* ending of the carboxylic acid with the word *anhydride*. For example,

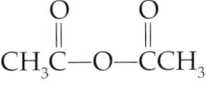

Ethanoic anhydride
(acetic anhydride)

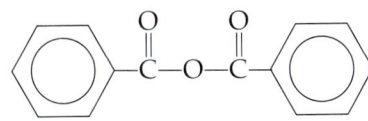

Benzoic anhydride

13 LEARNING GOAL

Chapter 14 Carboxylic Acids and Carboxylic Acid Derivatives

Unsymmetrical anhydrides are those having two different acyl groups. They are named by arranging the names of the two parent carboxylic acids and following them with the word *anhydride*. The names of the carboxylic acids may be arranged by size or alphabetically. For example:

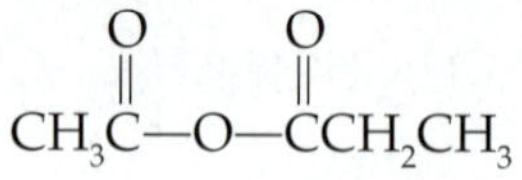

$$CH_3\overset{O}{\underset{\|}{C}}-O-\overset{O}{\underset{\|}{C}}CH_2CH_3 \qquad CH_3\overset{O}{\underset{\|}{C}}-O-\overset{O}{\underset{\|}{C}}CH_2CH_2CH_2CH_3$$

Ethanoic propanoic anhydride (acetic propionic anhydride)   Ethanoic pentanoic anhydride (acetic valeric anhydride)

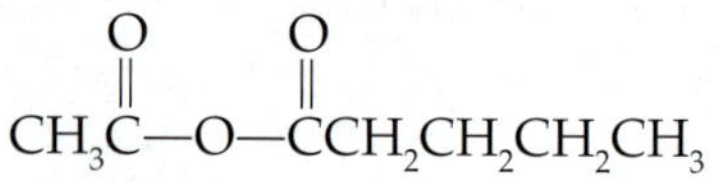

LEARNING GOAL

Most acid anhydrides cannot be formed in a reaction between the parent carboxylic acids. One typical pathway for the synthesis of an acid anhydride is the reaction between an acid chloride and a carboxylate anion. This general reaction is seen in the equation below:

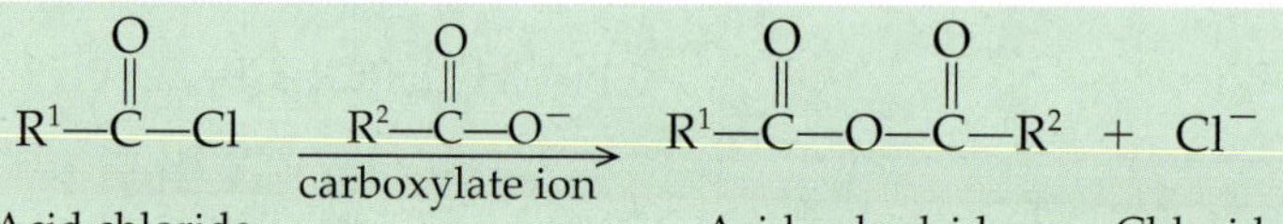

$$R^1-\overset{O}{\underset{\|}{C}}-Cl + R^2-\overset{O}{\underset{\|}{C}}-O^- \longrightarrow R^1-\overset{O}{\underset{\|}{C}}-O-\overset{O}{\underset{\|}{C}}-R^2 + Cl^-$$

Acid chloride   carboxylate ion   Acid anhydride   Chloride ion

The following equation shows the synthesis of ethanoic anhydride from ethanoic acid:

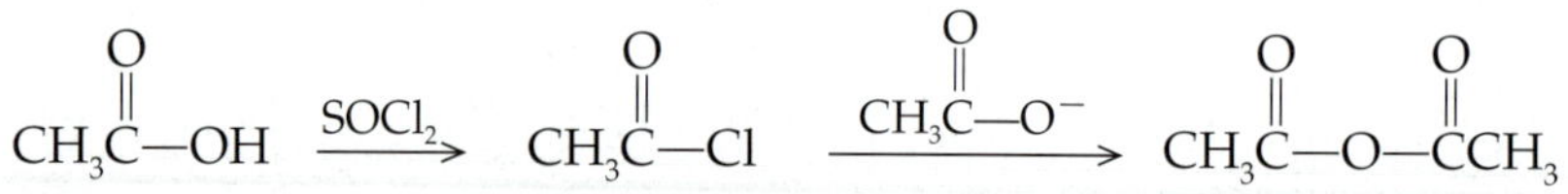

$$CH_3\overset{O}{\underset{\|}{C}}-OH \xrightarrow{SOCl_2} CH_3\overset{O}{\underset{\|}{C}}-Cl \xrightarrow{CH_3\overset{O}{\underset{\|}{C}}-O^-} CH_3\overset{O}{\underset{\|}{C}}-O-\overset{O}{\underset{\|}{C}}CH_3$$

Ethanoic acid (acetic acid)   Ethanoyl chloride (acetyl chloride)   Ethanoic anhydride (acetic anhydride)

Acid anhydrides readily undergo hydrolysis. The rate of the hydrolysis reaction may be increased by the addition of a trace of acid or hydroxide base to the solution.

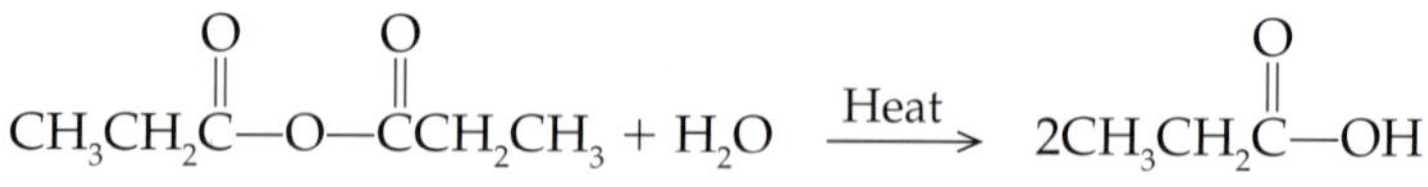

$$CH_3CH_2\overset{O}{\underset{\|}{C}}-O-\overset{O}{\underset{\|}{C}}CH_2CH_3 + H_2O \xrightarrow{Heat} 2CH_3CH_2\overset{O}{\underset{\|}{C}}-OH$$

Propanoic anhydride (propionic anhydride)   Propanoic acid (propionic acid)

EXAMPLE 14.12 Writing Equations Representing the Synthesis of Acid Anhydrides

Write an equation representing the synthesis of propanoic anhydride.

Solution

Propanoic anhydride can be synthesized in a reaction between propanoyl chloride and the propanoate anion. This gives us the following equation:

$$CH_3CH_2\overset{O}{\underset{\|}{C}}-Cl + CH_3CH_2\overset{O}{\underset{\|}{C}}-O^- \longrightarrow CH_3CH_2\overset{O}{\underset{\|}{C}}-O-\overset{O}{\underset{\|}{C}}CH_2CH_3 + Cl^-$$

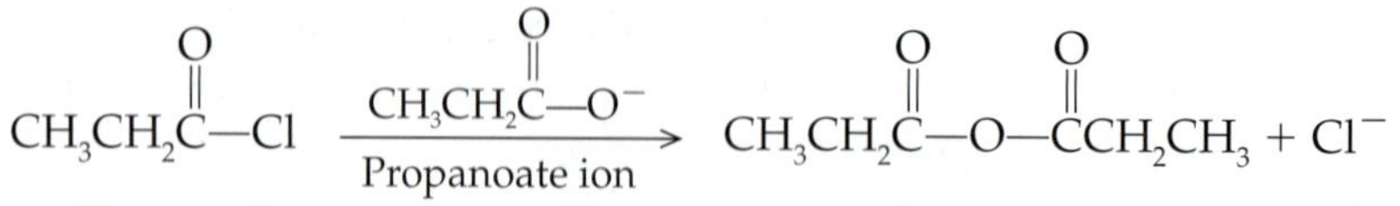

Propanoyl chloride   Propanoate ion   Propanoic anhydride   Chloride ion

14.3 Acid Chlorides and Acid Anhydrides

EXAMPLE 14.13

Naming Acid Anhydrides

Write the I.U.P.A.C. and common names for each of the following acid anhydrides.

$$\text{CH}_3\text{CH}_2\text{CH}_2\overset{\overset{\text{O}}{\|}}{\text{C}}-\text{O}-\overset{\overset{\text{O}}{\|}}{\text{C}}\text{CH}_2\text{CH}_2\text{CH}_3$$

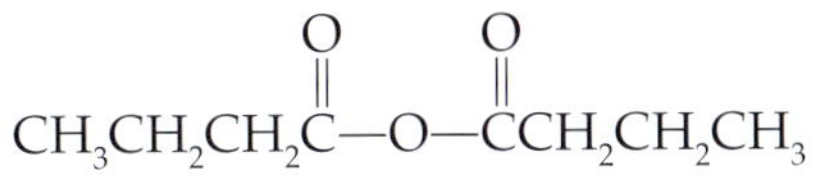

Solution

This is a symmetrical acid anhydride. The I.U.P.A.C. name of the four-carbon parent carboxylic acid is butanoic acid (common name butyric acid). To name the anhydride, simply replace the word *acid* with the word *anhydride*. The I.U.P.A.C. name of this compound is butanoic anhydride (common name butyric anhydride).

$$\text{CH}_3\overset{\overset{\text{O}}{\|}}{\text{C}}-\text{O}-\overset{\overset{\text{O}}{\|}}{\text{C}}\text{CH}_2\text{CH}_2\text{CH}_2\text{CH}_2\text{CH}_3$$

Solution

This is an unsymmetrical anhydride. The I.U.P.A.C. names of the two parent carboxylic acids are ethanoic acid (two-carbon) and hexanoic acid (six-carbon). To name an unsymmetrical anhydride, the term *anhydride* is preceded by the names of the two parent acids. The I.U.P.A.C. name of this compound is ethanoic hexanoic anhydride. The common names of the two parent carboxylic acids are acetic acid and caproic acid. Thus, the common name of this compound is acetic caproic anhydride.

Question 14.29

Write an equation showing the synthesis of each of the following acid anhydrides. Provide the I.U.P.A.C. names of the acid chloride and carboxylate anion reactants and the acid anhydride product.

a. 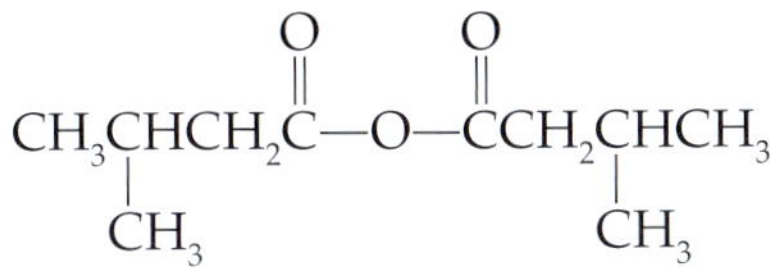 $\text{CH}_3\text{CHCH}_2\overset{\overset{\text{O}}{\|}}{\text{C}}-\text{O}-\overset{\overset{\text{O}}{\|}}{\text{C}}\text{CH}_2\text{CHCH}_3$
   $\quad\quad\;\;|\quad\quad\quad\quad\quad\quad\;\;|$
   $\quad\quad\text{CH}_3\quad\quad\quad\quad\quad\text{CH}_3$

b. 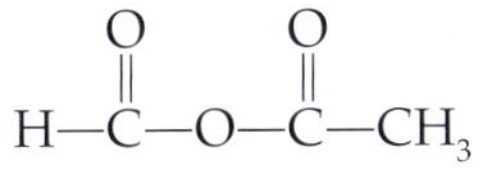 $\text{H}-\overset{\overset{\text{O}}{\|}}{\text{C}}-\text{O}-\overset{\overset{\text{O}}{\|}}{\text{C}}-\text{CH}_3$

Question 14.30

Write an equation showing the synthesis of each of the following acid anhydrides. Provide the common names of the acid chloride and carboxylate anion reactants and the acid anhydride product.

a. 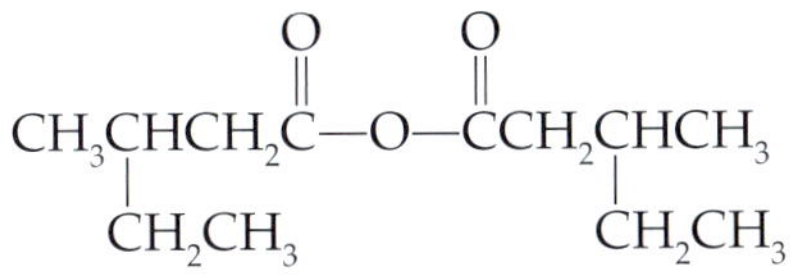 $\text{CH}_3\text{CHCH}_2\overset{\overset{\text{O}}{\|}}{\text{C}}-\text{O}-\overset{\overset{\text{O}}{\|}}{\text{C}}\text{CH}_2\text{CHCH}_3$
   $\quad\quad\;|\quad\quad\quad\quad\quad\quad\;\;|$
   $\quad\text{CH}_2\text{CH}_3\quad\quad\quad\text{CH}_2\text{CH}_3$

b. 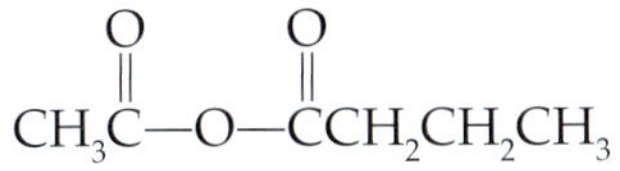 $\text{CH}_3\overset{\overset{\text{O}}{\|}}{\text{C}}-\text{O}-\overset{\overset{\text{O}}{\|}}{\text{C}}\text{CH}_2\text{CH}_2\text{CH}_3$

13 **LEARNING GOAL**

Acid anhydrides can also react with an alcohol. This reaction produces an ester and a carboxylic acid. This is an example of an acyl group transfer reaction, as shown in the following general reaction:

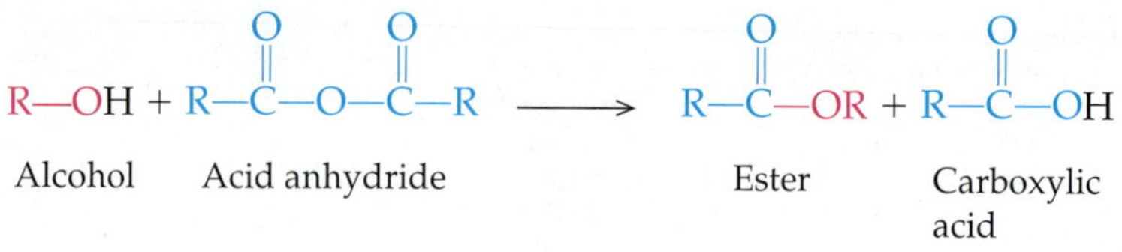

$$\text{R—OH} + \text{R—}\underset{\underset{\text{Acid anhydride}}{}}{\overset{\overset{O}{\|}}{C}}\text{—O—}\overset{\overset{O}{\|}}{C}\text{—R} \longrightarrow \text{R—}\overset{\overset{O}{\|}}{C}\text{—OR} + \text{R—}\overset{\overset{O}{\|}}{C}\text{—OH}$$

Alcohol   Acid anhydride   Ester   Carboxylic acid

The acyl group of the acid anhydride is transferred to the oxygen of the alcohol in this reaction. The alcohol and anhydride reactants and ester product are described below. The carboxylic acid product is omitted.

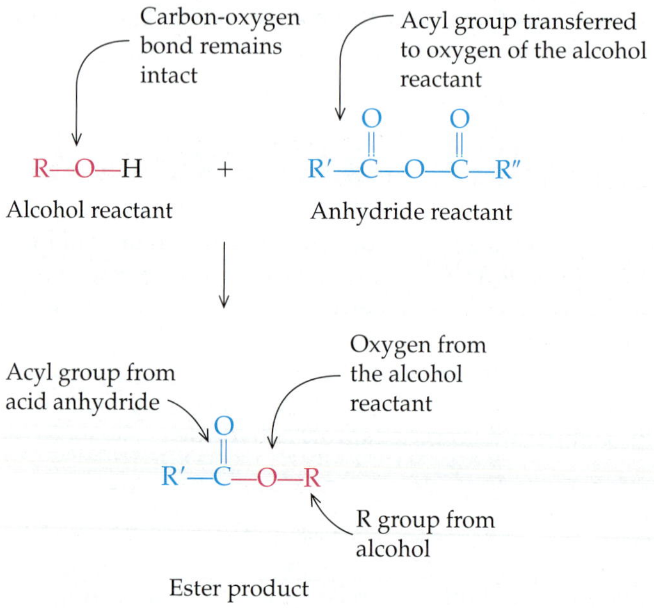

Carbon-oxygen bond remains intact — Acyl group transferred to oxygen of the alcohol reactant

R—O—H + R′—C(=O)—O—C(=O)—R″

Alcohol reactant   Anhydride reactant

Acyl group from acid anhydride — Oxygen from the alcohol reactant

R′—C(=O)—O—R

R group from alcohol

Ester product

Other acyl group donors include thioesters and esters. As we will see in the final section of this chapter, acyl transfer reactions are very important in nature, particularly in the pathways responsible for breakdown of food molecules and harvesting cellular energy.

14.4 Nature's High-Energy Compounds: Phosphoesters and Thioesters

LEARNING GOAL 15

An alcohol can react with phosphoric acid to produce a phosphate ester, or **phosphoester**, as in

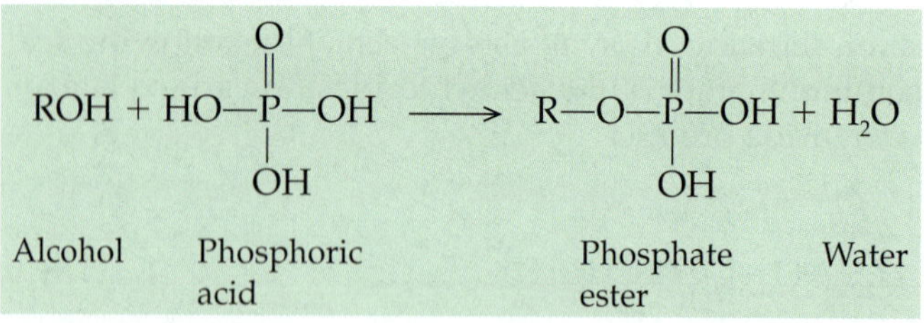

$$\text{ROH} + \text{HO—}\underset{\underset{\text{OH}}{|}}{\overset{\overset{O}{\|}}{P}}\text{—OH} \longrightarrow \text{R—O—}\underset{\underset{\text{OH}}{|}}{\overset{\overset{O}{\|}}{P}}\text{—OH} + \text{H}_2\text{O}$$

Alcohol   Phosphoric acid   Phosphate ester   Water

The many phosphorylated intermediates in the metabolism of sugars will be discussed in Chapter 21.

Phosphoesters of simple sugars or monosaccharides are very important in the energy-harvesting biochemical pathways that provide energy for all life functions. One such pathway is *glycolysis*. This pathway is the first stage in the breakdown of sugars. The first reaction in this pathway is the formation of a phosphoester of the

14.4 Nature's High-Energy Compounds: Phosphoesters and Thioesters

six-carbon sugar, glucose. The phosphorylation of glucose to produce glucose-6-phosphate is represented in the following equation:

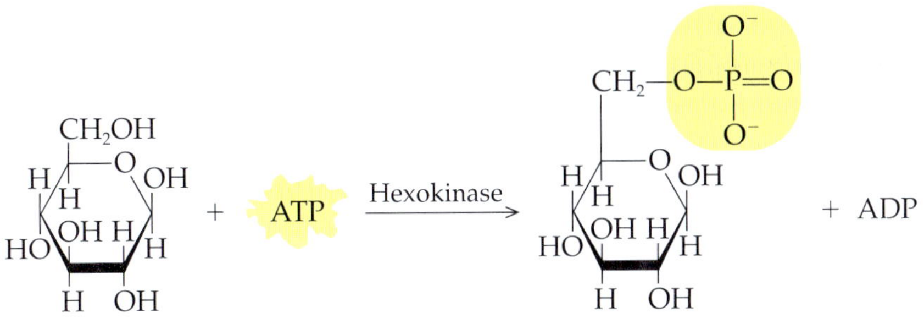

β-D-Glucose + ATP →[Hexokinase] β-D-Glucose-6-phosphate + ADP

In this reaction the source of the phosphoryl group is **adenosine triphosphate (ATP)**, which is the universal energy currency for all living organisms. As such, ATP is used to store energy released in cellular metabolic reactions and to provide the energy required for most of the reactions that occur in the cell. The transfer of a phosphoryl group from ATP to glucose "energizes" the glucose molecule in preparation for other reactions of the pathway.

ATP consists of a nitrogenous base (adenine) and a phosphate ester of the five-carbon sugar ribose (Figure 14.5). The triphosphate group attached to ribose is made up of three phosphate groups bonded to one another by phosphoanhydride bonds. When two phosphate groups react with one another, a water molecule is lost. Because water is lost, the resulting bond is called a **phosphoanhydride,** bond.

$$\text{RO}-\overset{\text{O}}{\underset{\text{OH}}{\overset{\|}{\text{P}}}}-\text{OH} + \text{HO}-\overset{\text{O}}{\underset{\text{OH}}{\overset{\|}{\text{P}}}}-\text{OH} \longrightarrow \text{RO}-\overset{\text{O}}{\underset{\text{OH}}{\overset{\|}{\text{P}}}}-\text{O}-\overset{\text{O}}{\underset{\text{OH}}{\overset{\|}{\text{P}}}}-\text{OH} + \text{H}_2\text{O}$$

Phosphate ester | Phosphate group | Phosphoanhydride bond

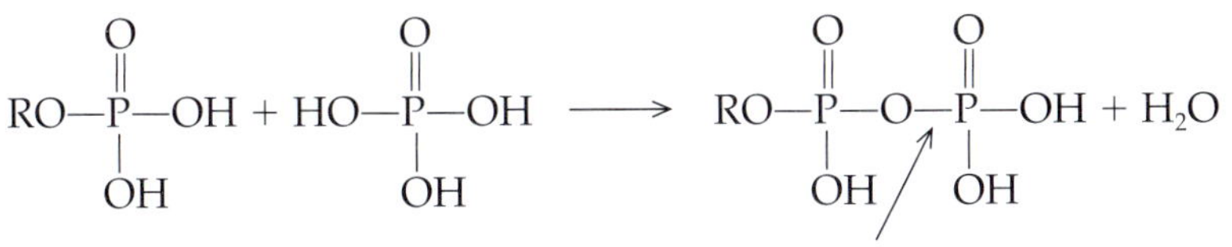

The energy of ATP is made available through hydrolysis of either of the two phosphoanhydride bonds, as shown in Figure 14.5. This is an exothermic process; that is, energy is given off. When the phosphoryl group is transferred to another

ATP → [H₂O] ADP + H₂PO₄⁻ + **energy**

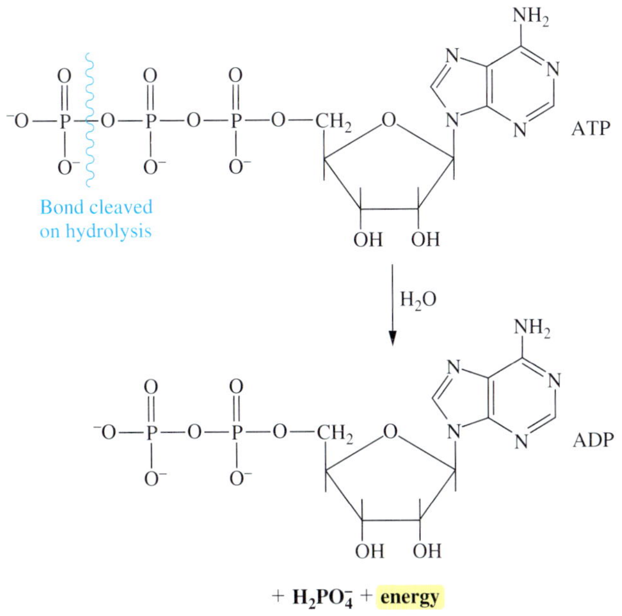

Figure 14.5
The hydrolysis of the phosphoanhydride bond of ATP is accompanied by the release of energy that is used for biochemical reactions in the cell.

In fact the word *glycolysis* comes from two Greek words that mean "splitting sugars" (*glykos,* "sweet," and *lysis,* "to split"). In this pathway, the six-carbon sugar glucose is split, and then oxidized, to produce two three-carbon molecules, called *pyruvate*.

Phosphoryl is the term used to describe the functional group derived from phosphoric acid that is part of another molecule.

The functions and properties of ATP in energy metabolism are discussed in Section 21.1.

A Human Perspective

Carboxylic Acid Derivatives of Special Interest

Analgesics (pain killers) and antipyretics (fever reducers)

Aspirin *(acetylsalicylic acid)*, derived from the bark of the willow tree (see the cover of this book), is the most widely used drug in the world. Hundreds of millions of dollars are spent annually on this compound. It is used primarily as a pain reliever (analgesic) and in the reduction of fever (antipyretic). Aspirin is among the drugs often referred to as NSAIDS, or nonsteroidal anti-inflammatory drugs. These drugs inhibit the inflammatory response by inhibiting an enzyme called cyclooxygenase, which is the first enzyme in the pathway for the synthesis of prostaglandins. Prostaglandins are responsible, in part, for pain and fever. Thus, aspirin and other NSAIDS reduce pain and fever by decreasing prostaglandin synthesis. Aspirin's side effects are a problem for some individuals. Because aspirin inhibits clotting, its use is not recommended during pregnancy, nor should it be used by individuals with ulcers. In those instances, *acetaminophen*, found in the over-the-counter pain-reliever Tylenol, is often prescribed.

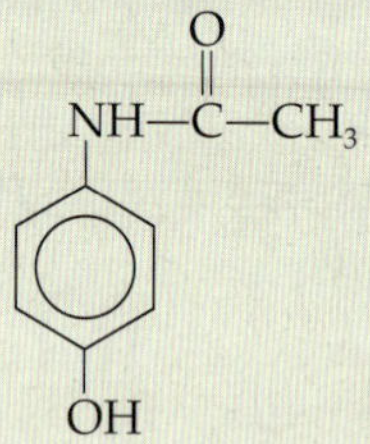

The search for NSAIDS that are more effective and yet gentler on the stomach has provided two new analgesics for the over-the-counter market. These are ibuprofen (sold as Motrin, Advil, Nuprin) and naproxen (sold as Naprosyn, Naprelan, Anaprox, and Aleve).

Many types of over-the-counter pain relievers are available.

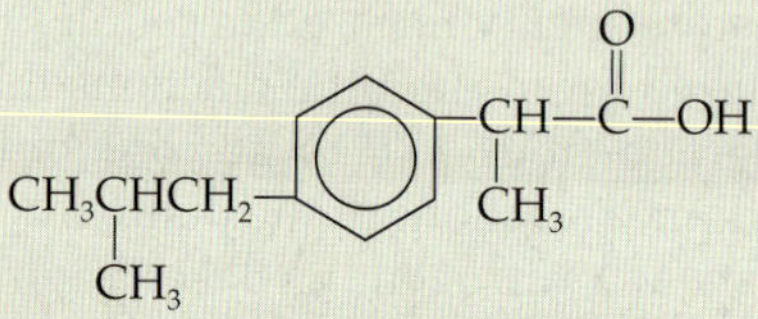

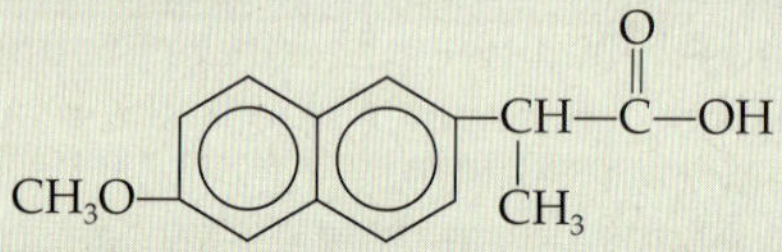

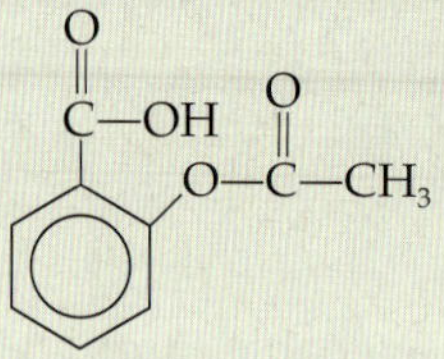

Some common analgesics.

Thiols are described in Section 12.9.

molecule—for instance, glucose—some of that energy resides in the phosphorylated sugar, thereby "energizing" it. The importance of ATP as an energy source becomes apparent when we realize that we synthesize and break down an amount of ATP equivalent to our body weight each day.

Cellular enzymes can carry out a reaction between a thiol and a carboxylic acid to produce a **thioester**:

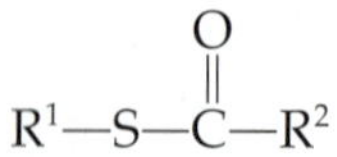

$$R^1-S-\overset{\overset{\displaystyle O}{\|}}{C}-R^2$$

Thioester

Pheromones

Pheromones, chemicals secreted by animals, influence the behavior of other members of the same species. They often represent the major means of communication among simpler animals. The term *pheromone* literally means "to carry" and "to excite" (Greek, *pherein*, to carry; Greek, *horman*, to excite). They are chemicals carried or shed by one member of the species and used to alert other members of the species.

Pheromones may be involved in sexual attraction, trail marking, aggregation or recruitment, territorial marking, or signaling alarm. Others may be involved in defense or in species socialization—for example, designating various classes within the species as a whole. Among all of the pheromones, insect pheromones have been the most intensely studied. Many of the insect pheromones are carboxylic acids or acid derivatives, as seen here:

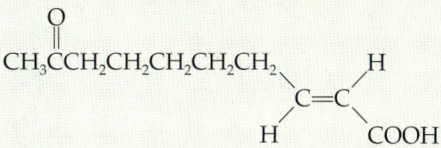

9-Keto-*trans*-2-decenoic acid
(queen bee socializing/royalty pheromone)

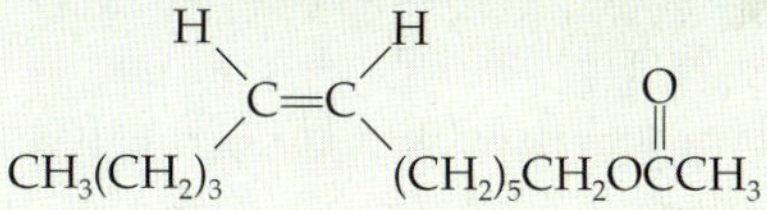

cis-7-Dodecenyl acetate
(cabbage looper sex pheromone)

$CH_3CH_2CH=CH(CH_2)_9CH_2OCCH_3$

Tetracecenyl acetate
(European corn borer sex pheromone)

For Further Understanding

A nonsteroidal anti-inflammatory drug can cause side effects if used over a long period of time. Do some research on the physiological roles of prostaglandins to develop hypotheses concerning these side effects.

How might you make use of pheromones to control pests such as the corn borer?

The reactions that produce thioesters are essential in energy-harvesting pathways as a means of "activating" acyl groups for subsequent breakdown reactions. The complex thiol coenzyme A is the most important acyl group activator in the cell. The detailed structure of coenzyme A appears in Section 12.9, but it is generally abbreviated CoA—SH to emphasize the importance of the sulfhydryl group. The most common thioester is the acetyl ester, called **acetyl coenzyme A** (acetyl CoA).

Acetyl CoA carries the acetyl group from glycolysis or β-oxidation of a fatty acid to an intermediate of the citric acid cycle. This reaction is an example of an acyl group transfer reaction. In this case, the acyl group donor is a thioester—acetyl CoA. The acyl group being transferred is the acetyl group, which is transferred to the carbonyl carbon of oxaloacetate. This reaction follows:

β-Oxidation is the pathway for the breakdown of fatty acids. Like glycolysis, it is an energy-harvesting pathway.

The acyl group of a carboxylic is named by replacing the *-oic acid* or *-ic* suffix with *-yl*. For instance, the acyl group of acetic acid is the acetyl group:

$$\text{CoA—S—}\overset{\overset{\text{O}}{\|}}{\text{C}}\text{—CH}_3$$

Acetyl coenzyme A
(acetyl CoA)

Glycolysis, β-oxidation, and the citric acid cycle are cellular energy-harvesting pathways that we will study in Chapters 21, 22, and 23.

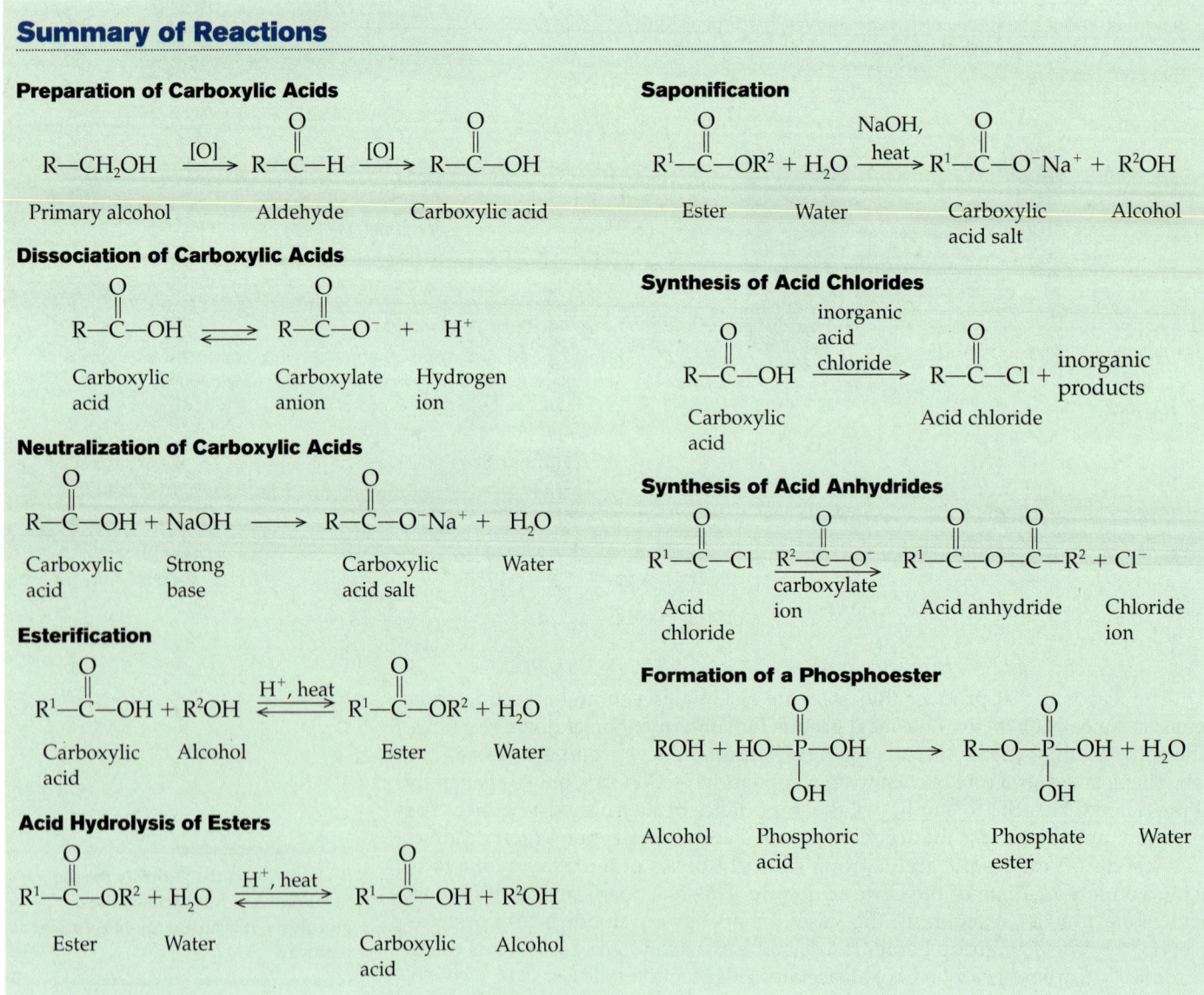

Acetyl CoA   Oxaloacetate   Citrate   Coenzyme A

As we will see in Chapter 22, the citric acid cycle is an energy-harvesting pathway that completely oxidizes the acetyl group to two CO_2 molecules. The electrons that are harvested in the process are used to produce large amounts of ATP. Coenzyme A also serves to activate the acyl group of fatty acids during β-oxidation, the pathway by which fatty acids are oxidized to produce ATP.

Summary of Reactions

Preparation of Carboxylic Acids

$$R\text{—CH}_2\text{OH} \xrightarrow{[O]} R\text{—}\overset{\overset{\text{O}}{\|}}{\text{C}}\text{—H} \xrightarrow{[O]} R\text{—}\overset{\overset{\text{O}}{\|}}{\text{C}}\text{—OH}$$

Primary alcohol   Aldehyde   Carboxylic acid

Dissociation of Carboxylic Acids

$$R\text{—}\overset{\overset{\text{O}}{\|}}{\text{C}}\text{—OH} \rightleftharpoons R\text{—}\overset{\overset{\text{O}}{\|}}{\text{C}}\text{—O}^- + \text{H}^+$$

Carboxylic acid   Carboxylate anion   Hydrogen ion

Neutralization of Carboxylic Acids

$$R\text{—}\overset{\overset{\text{O}}{\|}}{\text{C}}\text{—OH} + \text{NaOH} \longrightarrow R\text{—}\overset{\overset{\text{O}}{\|}}{\text{C}}\text{—O}^-\text{Na}^+ + \text{H}_2\text{O}$$

Carboxylic acid   Strong base   Carboxylic acid salt   Water

Esterification

$$R^1\text{—}\overset{\overset{\text{O}}{\|}}{\text{C}}\text{—OH} + R^2\text{OH} \xrightleftharpoons[]{\text{H}^+,\text{ heat}} R^1\text{—}\overset{\overset{\text{O}}{\|}}{\text{C}}\text{—O}R^2 + \text{H}_2\text{O}$$

Carboxylic acid   Alcohol   Ester   Water

Acid Hydrolysis of Esters

$$R^1\text{—}\overset{\overset{\text{O}}{\|}}{\text{C}}\text{—O}R^2 + \text{H}_2\text{O} \xrightleftharpoons[]{\text{H}^+,\text{ heat}} R^1\text{—}\overset{\overset{\text{O}}{\|}}{\text{C}}\text{—OH} + R^2\text{OH}$$

Ester   Water   Carboxylic acid   Alcohol

Saponification

$$R^1\text{—}\overset{\overset{\text{O}}{\|}}{\text{C}}\text{—O}R^2 + \text{H}_2\text{O} \xrightarrow{\text{NaOH, heat}} R^1\text{—}\overset{\overset{\text{O}}{\|}}{\text{C}}\text{—O}^-\text{Na}^+ + R^2\text{OH}$$

Ester   Water   Carboxylic acid salt   Alcohol

Synthesis of Acid Chlorides

$$R\text{—}\overset{\overset{\text{O}}{\|}}{\text{C}}\text{—OH} \xrightarrow{\text{inorganic acid chloride}} R\text{—}\overset{\overset{\text{O}}{\|}}{\text{C}}\text{—Cl} + \text{inorganic products}$$

Carboxylic acid   Acid chloride

Synthesis of Acid Anhydrides

$$R^1\text{—}\overset{\overset{\text{O}}{\|}}{\text{C}}\text{—Cl} + R^2\text{—}\overset{\overset{\text{O}}{\|}}{\text{C}}\text{—O}^- \xrightarrow{\text{carboxylate}} R^1\text{—}\overset{\overset{\text{O}}{\|}}{\text{C}}\text{—O—}\overset{\overset{\text{O}}{\|}}{\text{C}}\text{—}R^2 + \text{Cl}^-$$

Acid chloride   ion   Acid anhydride   Chloride ion

Formation of a Phosphoester

$$\text{ROH} + \text{HO}\text{—}\overset{\overset{\text{O}}{\|}}{\underset{\underset{\text{OH}}{|}}{\text{P}}}\text{—OH} \longrightarrow \text{R—O}\text{—}\overset{\overset{\text{O}}{\|}}{\underset{\underset{\text{OH}}{|}}{\text{P}}}\text{—OH} + \text{H}_2\text{O}$$

Alcohol   Phosphoric acid   Phosphate ester   Water

SUMMARY

14.1 Carboxylic Acids

The functional group of the *carboxylic acids* is the *carboxyl group* (—COOH). Because the carboxyl group is extremely polar and carboxylic acids can form intermolecular hydrogen bonds, they have higher boiling points and melting points than alcohols. The lower-molecular-weight carboxylic acids are water soluble and tend to taste sour and have unpleasant aromas. The longer-chain carboxylic acids are called *fatty acids*. Carboxylic acids are named (I.U.P.A.C.) by replacing the *-e* ending of the parent compound with *-oic acid*. Common names are often derived from the source of the carboxylic acid. They are synthesized by the oxidation of primary alcohols or aldehydes. Carboxylic acids are weak acids. They are neutralized by strong bases to form salts. *Soaps* are salts of long-chain carboxylic acids (fatty acids).

14.2 Esters

Esters are mildly polar and have pleasant aromas. The boiling points and melting points of esters are comparable to those of aldehydes and ketones. Esters are formed from the reaction between a carboxylic acid and an alcohol. They can undergo hydrolysis back to the parent carboxylic acid and alcohol. The base-catalyzed hydrolysis of an ester is called *saponification*.

14.3 Acid Chlorides and Acid Anhydrides

Acid chlorides are noxious chemicals formed in the reaction of a carboxylic acid and reagents such as PCl_3 or $SOCl_2$. *Acid anhydrides* are formed by the combination of an acid chloride and a carboxylate anion. Acid anhydrides can react with an alcohol to produce an ester and a carboxylic acid. This is an example of an acyl group transfer reaction.

14.4 Nature's High-Energy Compounds: Phosphoesters and Thioesters

An alcohol can react with phosphoric acid to produce a phosphate ester *(phosphoester)*. When two phosphate groups are joined, the resulting bond is a *phosphoanhydride* bond. These two functional groups are important to the structure and function of *adenosine triphosphate (ATP)*, the universal energy currency of all cells. Cellular enzymes can carry out a reaction between a thiol and a carboxylic acid to produce a *thioester*. This reaction is essential for the activation of acyl groups in carbohydrate and fatty acid metabolism. Coenzyme A is the most important thiol involved in these pathways.

KEY TERMS

acetyl coenzyme A (14.4)
acid anhydride (14.3)
acid chloride (14.3)
acyl group (Intro)
adenosine triphosphate (ATP) (14.4)
carboxyl group (14.1)
carboxylic acid (14.1)
carboxylic acid derivative (14.2)
condensation polymer (14.2)
ester (14.2)
fatty acid (14.1)
hydrolysis (14.2)
oxidation (14.1)
phosphoanhydride (14.4)
phosphoester (14.4)
saponification (14.2)
soap (14.2)
thioester (14.4)

QUESTIONS AND PROBLEMS

Carboxylic Acids: Structure and Properties

Foundations

14.31 The functional group is largely responsible for the physical and chemical properties of the various chemical families. Explain why a carboxylic acid is more polar and has a higher boiling point than an alcohol or an aldehyde of comparable molecular weight.

14.32 Explain why carboxylic acids are weak acids.

Applications

14.33 Which member of each of the following pairs has the lower boiling point?
   a. Hexanoic acid or 3-hexanone
   b. 3-Hexanone or 3-hexanol
   c. 3-Hexanol or hexane

14.34 Which member of each of the following pairs has the lower boiling point?
   a. Dipropyl ether or hexanal
   b. Hexanal or hexanoic acid
   c. Ethanol or ethanoic acid

14.35 Arrange the following from highest to lowest melting points:

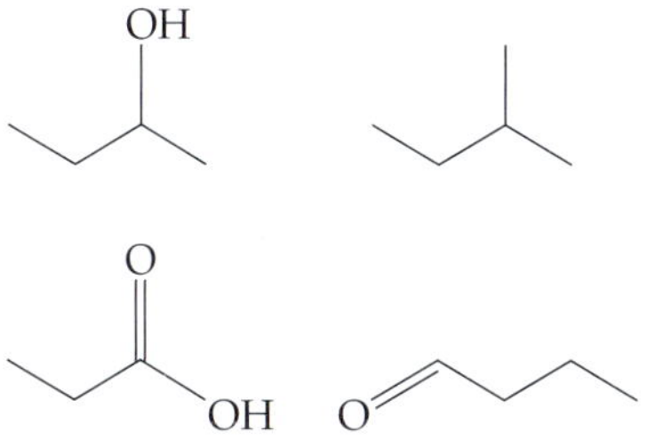

14.36 Draw the condensed structural formula for each of the line drawings in Question 14.35 and provide the I.U.P.A.C. name for each.

14.37 Which member in each of the following pairs has the higher boiling point?
   a. Heptanoic acid or 1-heptanol
   b. Propanal or 1-propanol
   c. Methyl pentanoate or pentanoic acid
   d. 1-Butanol or butanoic acid

14.38 Which member in each of the following pairs is more soluble in water?

a. 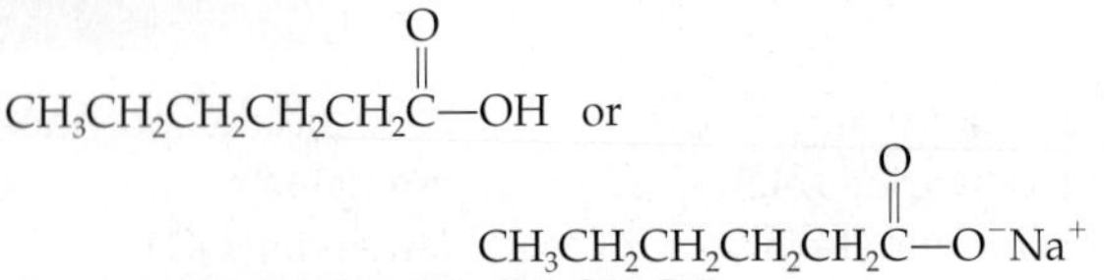 CH₃CH₂CH₂CH₂CH₂C(=O)—OH or

b. CH₃CH₂CH₂CH₂CH₂CH₂CH₂CH₂CH₃ or
CH₃CH₂CH₂CH₂CH₂C(=O)—O⁻ Na⁺
CH₃CH₂CH₂CH₂CH₂CH₂CH₂CH₂OH

c. CH₃CH₂—O—CH₂CH₃ or CH₃CH₂C(=O)—OCH₃
d. CH₃CH₂—O—CH₂CH₃ or CH₃CH₂CH₂CH₂CH₂CH₃
e. Decanoic acid or ethanoic acid

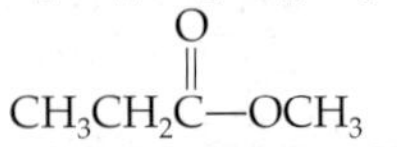

14.39 Describe the properties of low-molecular-weight carboxylic acids.

14.40 What are some of the biological functions of the long chain carboxylic acids called fatty acids?

14.41 Why are citric acid and adipic acid added to some food products?

14.42 What is the function of lactic acid in food products? Of what significance is lactic acid in muscle metabolism?

Carboxylic Acids: Structure and Nomenclature

Foundations

14.43 Summarize the I.U.P.A.C. nomenclature rules for naming carboxylic acids.

14.44 Describe the rules for determining the common names of carboxylic acids.

Applications

14.45 Write the complete structural formulas for each of the following carboxylic acids:
  a. 2-Bromopentanoic acid
  b. 2-Bromo-3-methylbutanoic acid
  c. 2-Bromocyclohexanecarboxylic acid

14.46 Write the complete structural formulas for each of the following carboxylic acids:
  a. 2,6-Dichlorocyclohexanecarboxylic acid
  b. 2,4,6-Trimethylstearic acid
  c. Propenoic acid

14.47 Name each of the following carboxylic acids, using both the common and the I.U.P.A.C. Nomenclature Systems:

  a. 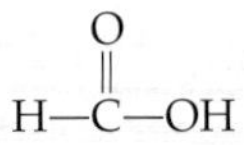 H—C(=O)—OH

  b. CH₃CH(CH₃)CH₂C(=O)—OH

  c. cyclopentyl—C(=O)—OH

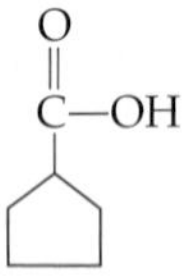

14.48 Name each of the following carboxylic acids, using both the common and I.U.P.A.C. Nomenclature Systems:

  a. 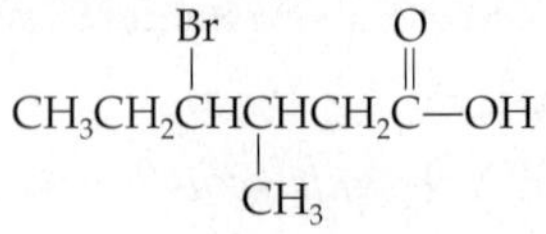 CH₃CH₂CH(Br)CH(CH₃)CH₂C(=O)—OH

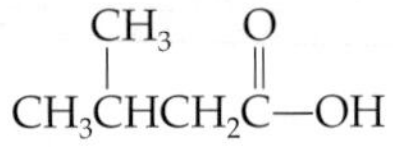

  b. 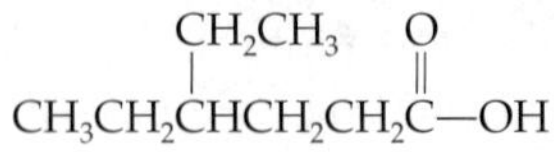 CH₃CH₂CH(CH₂CH₃)CH₂CH₂C(=O)—OH

  c. 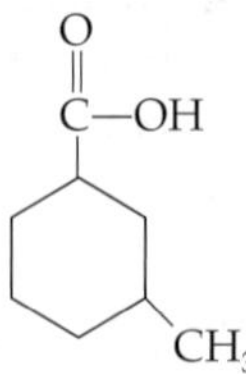 3-methylcyclohexyl—C(=O)—OH

14.49 Write a complete structural formula and determine the I.U.P.A.C. name for each of the carboxylic acids of molecular formula C₄H₈O₂.

14.50 Write the general structure of an aldehyde, a ketone, a carboxylic acid, and an ester. What similarities exist among these structures?

14.51 Write the condensed structure of each of the following carboxylic acids:
  a. 4,4-Dimethylhexanoic acid
  b. 3-Bromo-4-methylpentanoic acid
  c. 2,3-Dinitrobenzoic acid
  d. 3-Methylcyclohexanecarboxylic acid

14.52 Use I.U.P.A.C. nomenclature to write the names for each of the following carboxylic acids:

  a. 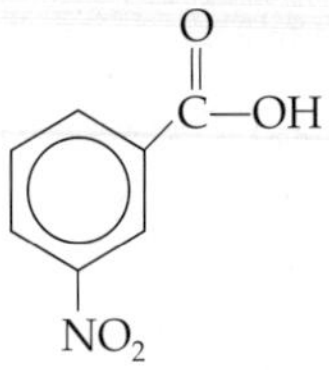 3-nitrobenzoic acid structure
  b. 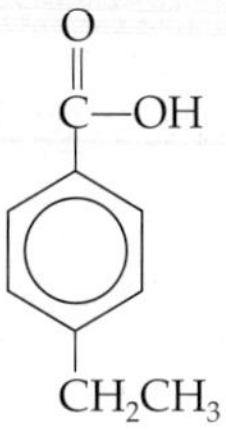 4-ethylbenzoic acid structure
  c. 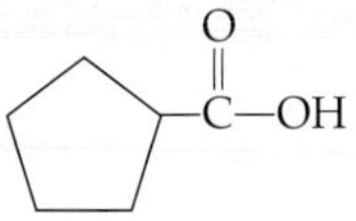 cyclopentanecarboxylic acid structure

14.53 Provide the common and I.U.P.A.C. names for each of the following compounds:

  a. 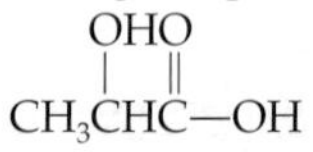 CH₃CH(OH)C(=O)—OH (with OHO labels)

  b. CH₃CH(OH)CH₂C(=O)—OH

  c. CH₃C(CH₃)(CH₃)CH₂CH₂C(=O)—OH

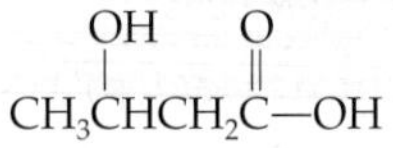

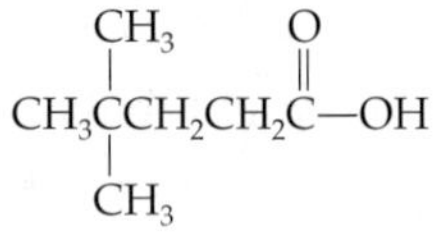

  d. 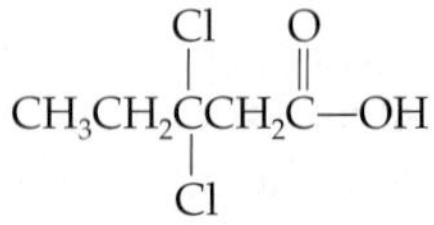 CH₃CH₂C(Cl)(Cl)CH₂C(=O)—OH

14.54 Draw the structure of each of the following carboxylic acids:
  a. β-Chlorobutyric acid
  b. α,β-Dibromovaleric acid
  c. β,γ-Dihydroxybutyric acid
  d. δ-Bromo-γ-chloro-β-methylcaproic acid

Carboxylic Acids: Reactions

Foundations

14.55 Explain what is meant by oxidation in organic molecules and provide an example of an oxidation reaction involving an aldehyde or an alcohol.

14.56 Write a general equation showing the preparation of a carboxylic acid from an alcohol.

14.57 Write a general equation showing the dissociation of a carboxylic acid in water.

14.58 Carboxylic acids are described as weak acids. To what extent do carboxylic acids generally dissociate?

14.59 What reaction occurs when a strong base is added to a carboxylic acid?

14.60 Write a general equation showing the reaction of a strong base with a carboxylic acid.

14.61 How is a soap prepared?

14.62 How do soaps assist in the removal of oil and grease from clothing?

Applications

14.63 Complete each of the following reactions by supplying the missing portion indicated by a question mark:

a. $CH_3C(=O)-H \xrightarrow{H_2CrO_4}$?

b. $CH_3CH_2CH_2C(=O)-OH + CH_3OH \xrightarrow{H^+, heat}$?

c. 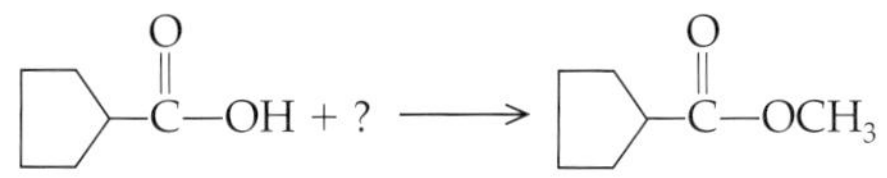 (cyclopentyl)$-C(=O)-OH$ + ? $\longrightarrow$ (cyclopentyl)$-C(=O)-OCH_3$

14.64 Complete each of the following reactions by supplying the missing part(s) indicated by the question mark(s):

a. $CH_3CH_2CH_2OH \xrightarrow{?(1)} CH_3CH_2C(=O)-OH \xrightleftharpoons[?(3)]{NaOH}$?(4)

   $\downarrow ?(2)$

   $CH_3CH_2C(=O)-OCH(CH_3)CH_3$

b. $CH_3COOH + NaOH \longrightarrow$?

c. $CH_3CH_2CH_2CH_2CH_2COOH + NaOH \longrightarrow$?

d. ? + $CH_3CH_2CH(CH_3)OH \xrightarrow{H^+} CH_3C(=O)-OCH(CH_3)CH_2CH_3$

14.65 How might $CH_3CH_2CH_2CH_2CH_2OH$ be converted to each of the following products?
a. $CH_3CH_2CH_2CH_2CHO$
b. $CH_3CH_2CH_2CH_2COOH$

14.66 Which of the following alcohols can be oxidized to a carboxylic acid? Name the carboxylic acid produced. For those alcohols that cannot be oxidized to a carboxylic acid, name the final product.
a. Ethanol
b. 2-Propanol
c. 1-Propanol
d. 3-Pentanol

Esters: Structure, Physical Properties, and Nomenclature

Foundations

14.67 Explain why esters are described as mildly polar.

14.68 Compare the boiling points of esters to those of aldehydes or ketones of similar molecular weight.

14.69 Briefly summarize the I.U.P.A.C. rules for naming esters.

14.70 How are the common names of esters derived?

Applications

14.71 Write each of the following, using condensed formulas:
a. Methyl benzoate
b. Butyl decanoate
c. Methyl propionate
d. Ethyl propionate

14.72 Write each of the following using condensed formulas:
a. Ethyl *m*-nitrobenzoate
b. Isopropyl acetate
c. Methyl butyrate

14.73 Use the I.U.P.A.C. Nomenclature System to name each of the following esters:

a. $CH_3C(=O)-OCH_2CH_3$

b. $CH_3CH_2C(=O)-OCH_3$

c. $CH_3CH(CH_3)CH_2C(=O)-OCH_3$

d. 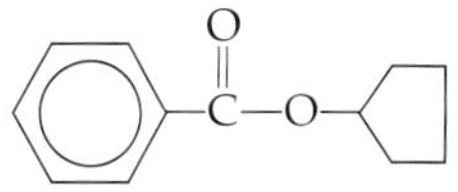 (phenyl)$-C(=O)-O-$(cyclopentyl)

14.74 Use the I.U.P.A.C. Nomenclature System to name each of the following:

a. 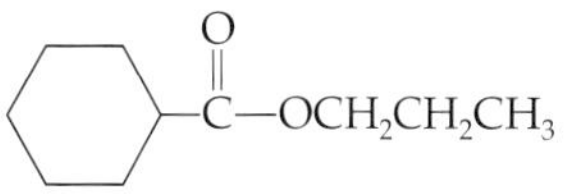 (cyclohexyl)$-C(=O)-OCH_2CH_2CH_3$

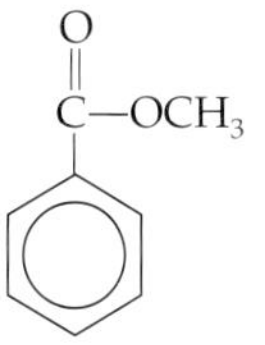

b. (phenyl)$-C(=O)-OCH_3$

c. $CH_2(Br)CH(Br)CH_2CH_2C(=O)-OCH_2CH_3$

Esters: Reactions

Foundations

14.75 Write a general reaction showing the preparation of an ester.

14.76 Why is preparation of an ester referred to as a dehydration reaction?

14.77 Write a general reaction showing the hydrolysis of an ester using an acid catalyst.

14.78 Write a general reaction showing the base-catalyzed hydrolysis of an ester.

14.79 What is meant by a hydrolysis reaction?

14.80 Why is the salt of a carboxylic acid produced in a base-catalyzed hydrolysis of an ester?

Applications

14.81 Complete each of the following reactions by supplying the missing portion indicated with a question mark:

a. $CH_3CH_2CH_2\overset{O}{\underset{\|}{C}}-OH + CH_3CH_2OH \xrightarrow{H^+, \text{heat}} ?$

b. $CH_3CH_2\overset{O}{\underset{\|}{C}}-OCH_2CH_3 + H_2O \xrightarrow{H^+, \text{heat}} ?$

c. $CH_3\overset{CH_3}{\underset{|}{CH}}CH_2CH_2\overset{O}{\underset{\|}{C}}-OH + ? \xrightarrow{H^+, \text{heat}}$

$CH_3\overset{CH_3}{\underset{|}{CH}}CH_2CH_2\overset{O}{\underset{\|}{C}}-OCH_2CH_3$

d. $CH_3CH_2\overset{Br}{\underset{|}{CH}}CH_2\overset{O}{\underset{\|}{C}}-OCH_2CH_3 + H_2O \xrightarrow{OH^-, \text{heat}} ?$

14.82 Complete each of the following reactions by supplying the missing portion indicated with a question mark:

a. $? + CH_3\overset{CH_3}{\underset{\underset{CH_3}{|}}{\overset{|}{C}}}-OH \xrightarrow{?} CH_3CH_2\overset{O}{\underset{\|}{C}}-O-\overset{CH_3}{\underset{\underset{CH_3}{|}}{\overset{|}{C}}}-CH_3$

b. $CH_3CH_2CH_2CH_2COOH + CH_3CH_2CH_2CH_2OH \xrightarrow{H^+, \text{heat}} ?$

c. $CH_3\overset{CH_3}{\underset{\underset{CH_3}{|}}{\overset{|}{C}}}CH_2\overset{O}{\underset{\|}{C}}-OCH_2CH_2\overset{CH_3}{\underset{|}{C}}CH_3 + H_2O \xrightarrow{H^+, \text{heat}} ?$

d. $CH_3CH_2\overset{O}{\underset{\|}{C}}-OCH_3 + H_2O \xrightarrow{OH^-, \text{heat}} ?$

14.83 What is saponification? Give an example using specific molecules.

14.84 When the methyl ester of hexanoic acid is hydrolyzed in aqueous sodium hydroxide in the presence of heat, a homogeneous solution results. When the solution is acidified with dilute aqueous hydrochloric acid, a new product forms. What is the new product? Draw its structure.

14.85 The structure of salicylic acid is shown. If this acid reacts with methanol, the product is an ester, methyl salicylate. Methyl salicylate is known as oil of wintergreen and is often used as a flavoring agent. Draw the structure of the product of this reaction.

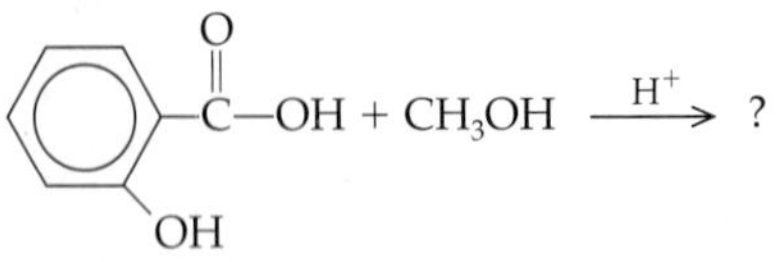

[benzene ring with -C(=O)-OH and -OH substituents] $+ CH_3OH \xrightarrow{H^+} ?$

14.86 When salicylic acid reacts with acetic anhydride, one of the products is an ester, acetylsalicylic acid. Acetylsalicylic acid is the active ingredient in aspirin. Complete the equation below by drawing the structure of acetylsalicylic acid. (*Hint:* Acid anhydrides are hydrolyzed by water.)

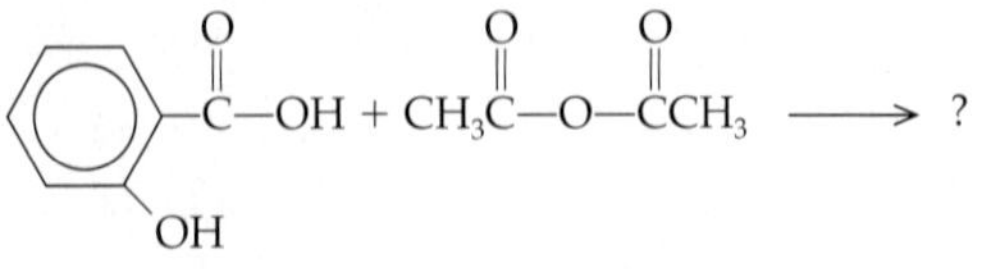

[salicylic acid structure] $-OH + CH_3\overset{O}{\underset{\|}{C}}-O-\overset{O}{\underset{\|}{C}}CH_3 \longrightarrow ?$

14.87 Compound A ($C_6H_{12}O_2$) reacts with water, acid, and heat to yield compound B ($C_5H_{10}O_2$) and compound C (CH_4O). Compound B is acidic. Deduce possible structures of compounds A, B, and C.

14.88 What products are formed when methyl *o*-bromobenzoate reacts with each of the following?
   a. Aqueous acid and heat
   b. Aqueous base and heat

14.89 Write an equation for the acid-catalyzed hydrolysis of each of the following esters:
   a. Propyl propanoate
   b. Butyl methanoate
   c. Ethyl methanoate
   d. Methyl pentanoate

14.90 Write an equation for the base-catalyzed hydrolysis of each of the following esters:
   a. Pentyl methanoate
   b. Hexyl propanoate
   c. Butyl hexanoate
   d. Methyl benzoate

Acid Chlorides and Acid Anhydrides

14.91 Supply the missing reagents (indicated by the question marks) necessary to complete each of the following transformations:

a. $CH_3\overset{O}{\underset{\|}{C}}-OH \xrightarrow{?} CH_3\overset{O}{\underset{\|}{C}}-Cl$

b. 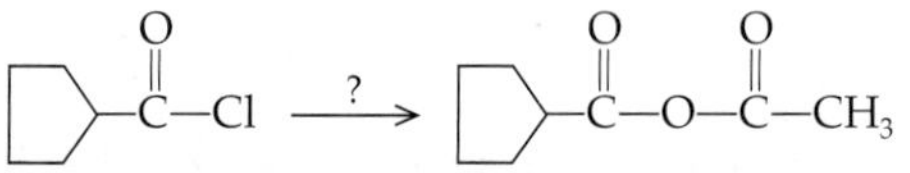 [cyclopentane]$-\overset{O}{\underset{\|}{C}}-Cl \xrightarrow{?}$ [cyclopentane]$-\overset{O}{\underset{\|}{C}}-O-\overset{O}{\underset{\|}{C}}-CH_3$

c. [benzene]$-\overset{O}{\underset{\|}{C}}-Cl \xrightarrow{?}$ [benzene]$-\overset{O}{\underset{\|}{C}}-O-\overset{O}{\underset{\|}{C}}-$[benzene]

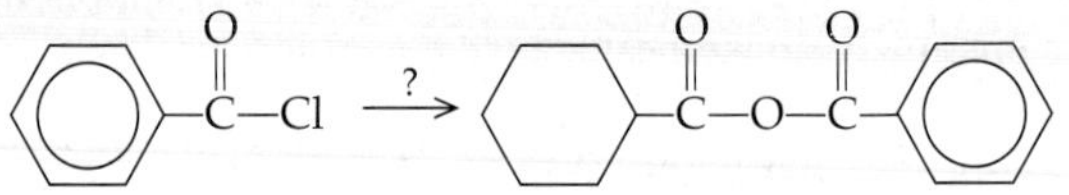

14.92 Supply the missing reagents (indicated by the question marks) necessary to complete each of the following transformations. All of the reactions may require more than one step to complete.

a. $CH_3CH_2CH_2CH_2-OH \xrightarrow{?} CH_3CH_2CH_2\overset{O}{\underset{\|}{C}}-Cl$

b. $CH_3CH_2OH \xrightarrow{?} CH_3\overset{O}{\underset{\|}{C}}-O-\overset{O}{\underset{\|}{C}}CH_2CH_3$

c. Ethanol $\xrightarrow{?}$ ethanoic anhydride

14.93 Complete each of the following reactions by supplying the missing product:

a. 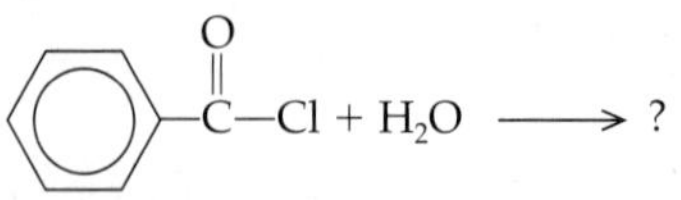 [benzene]$-\overset{O}{\underset{\|}{C}}-Cl + H_2O \longrightarrow ?$

b. $CH_3\overset{O}{\underset{\|}{C}}-O-\overset{O}{\underset{\|}{C}}CH_3 + H_2O \xrightarrow{\text{Heat}} ?$

14.94 Use the I.U.P.A.C. Nomenclature System to name the products and reactants in Problem 14.93.

14.95 Write the condensed formula for each of the following compounds:
   a. Decanoic anhydride
   b. Acetic anhydride
   c. Valeric anhydride
   d. Benzoyl chloride

14.96 Write a condensed formula for each of the following compounds:
  a. Propanoyl chloride
  b. Heptanoyl chloride
  c. Pentanoyl chloride

14.97 Describe the physical properties of acid chlorides.

14.98 Describe the physical properties of acid anhydrides.

14.99 Write an equation for the reaction of each of the following acid anhydrides with ethanol.
  a. Propanoic anhydride
  b. Ethanoic anhydride
  c. Methanoic anhydride

14.100 Write an equation for the reaction of each of the following acid anhydrides with propanol. Name each of the products using the I.U.P.A.C. Nomenclature System.
  a. Butanoic anhydride
  b. Pentanoic anhydride
  c. Methanoic anhydride

Phosphoesters and Thioesters

14.101 By reacting phosphoric acid with an excess of ethanol, it is possible to obtain the mono-, di-, and triesters of phosphoric acid. Draw all three of these products.

14.102 What is meant by a phosphoanhydride bond?

14.103 We have described the molecule ATP as the body's energy storehouse. What do we mean by this designation? How does ATP actually store energy and provide it to the body as needed?

14.104 Write an equation for each of the following reactions:
  a. Ribose + phosphoric acid
  b. Methanol + phosphoric acid
  c. Adenosine diphosphate + phosphoric acid

14.105 Draw the thioester bond between the acetyl group and coenzyme A.

14.106 Explain the significance of thioester formation in the metabolic pathways involved in fatty acid and carbohydrate breakdown.

14.107 It is also possible to form esters of other inorganic acids such as sulfuric acid and nitric acid. One particularly noteworthy product is nitroglycerine, which is both highly unstable (explosive) and widely used in the treatment of the heart condition known as angina, a constricting pain in the chest usually resulting from coronary heart disease. In the latter case, its function is to alleviate the pain associated with angina. Nitroglycerine may be administered as a tablet (usually placed just beneath the tongue when needed) or as a salve or paste that can be applied to and absorbed through the skin. Nitroglycerine is the trinitroester of glycerol. Draw the structure of nitroglycerine, using the structure of glycerol.

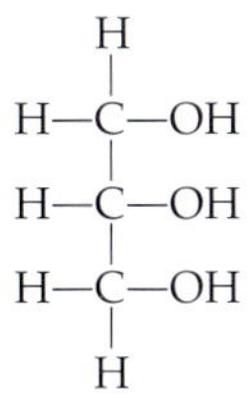

$$\begin{array}{c} H \\ | \\ H-C-OH \\ | \\ H-C-OH \\ | \\ H-C-OH \\ | \\ H \end{array}$$

Glycerol

14.108 Show the structure of the thioester that would be formed between coenzyme A and stearic acid.

CRITICAL THINKING PROBLEMS

1. Radioactive isotopes of an element behave chemically in exactly the same manner as the nonradioactive isotopes. As a result, they can be used as tracers to investigate the details of chemical reactions. A scientist is curious about the origin of the bridging oxygen atom in an ester molecule. She has chosen to use the radioactive isotope oxygen-18 to study the following reaction:

$$CH_3CH_2OH + CH_3\overset{O}{\underset{\|}{C}}-OH \xrightarrow{H^+, \text{heat}} CH_3\overset{O}{\underset{\|}{C}}-O-CH_2CH_3 + H_2O$$

   Design experiments using oxygen-18 that will demonstrate whether the oxygen in the water molecule came from the —OH of the alcohol or the —OH of the carboxylic acid.

2. Triglycerides are the major lipid storage form in the human body. They are formed in an esterification reaction between glycerol (1,2,3-propanetriol) and three fatty acids (long chain carboxylic acids). Write a balanced equation for the formation of a triglyceride formed in a reaction between glycerol and three molecules of decanoic acid.

3. Chloramphenicol is a very potent, broad-spectrum antibiotic. It is reserved for life-threatening bacterial infections because it is quite toxic. It is also a very bitter tasting chemical. As a result, children had great difficulty taking the antibiotic. A clever chemist found that the taste could be improved considerably by producing the palmitate ester. Intestinal enzymes hydrolyze the ester, producing chloramphenicol, which can then be absorbed. The following structure is the palmitate ester of chloramphenicol. Draw the structure of chloramphenicol.

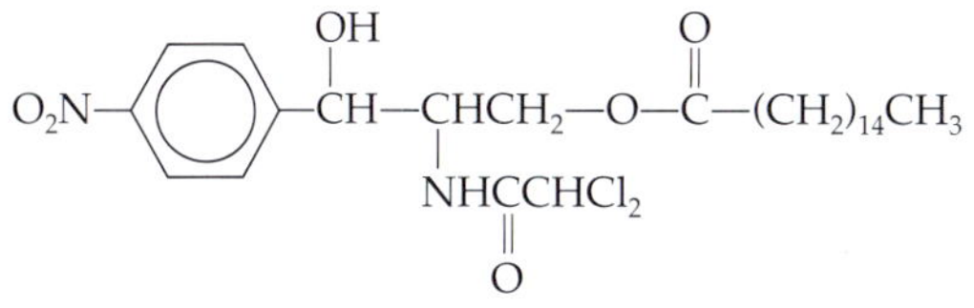

$$O_2N-\underset{}{\underset{}{\bigcirc}}-\underset{\underset{NHCCHCl_2}{\underset{\|}{|}}}{\overset{OH}{\underset{|}{C}H}}-CHCH_2-O-\overset{O}{\underset{\|}{C}}-(CH_2)_{14}CH_3$$

Chloramphenicol palmitate

4. Acetyl coenzyme A (acetyl CoA) can serve as a donor of acetate groups in biochemical reactions. One such reaction is the formation of acetylcholine, an important neurotransmitter involved in nerve signal transmission at neuromuscular junctions. The structure of choline is shown below. Draw the structure of acetylcholine.

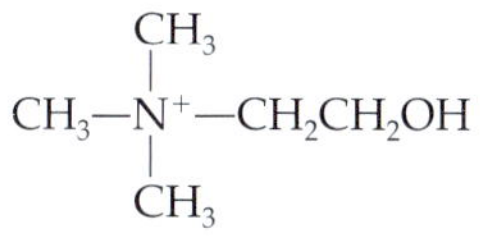

$$\begin{array}{c} CH_3 \\ | \\ CH_3-N^+-CH_2CH_2OH \\ | \\ CH_3 \end{array}$$

Choline

5. Hormones are chemical messengers that are produced in a specialized tissue of the body and travel through the bloodstream to reach receptors on cells of their target tissues. This specific binding to target tissues often stimulates a cascade of enzymatic reactions in the target cells. The work of Earl Sutherland and others led to the realization that there is a *second messenger* within the target cells. Binding of the hormone to the hormone receptor in the cell membrane triggers the enzyme adenyl cyclase to produce adenosine-3′,5′-monophosphate, which is also called *cyclic AMP*, from ATP. The reaction is summarized as follows:

$$ATP \xrightarrow{Mg^{+2}, \text{ adenyl cyclase}} \text{cyclic AMP} + PP_i + H^+$$

PP_i is the abbreviation for a pyrophosphate group, shown here:

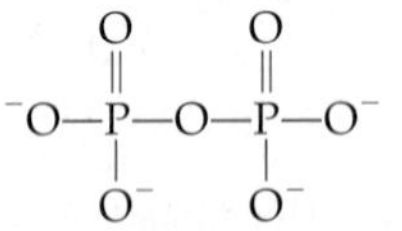

The structure of ATP is shown here with the carbon atoms of the sugar ribose numbered according to the convention used for nucleotides:

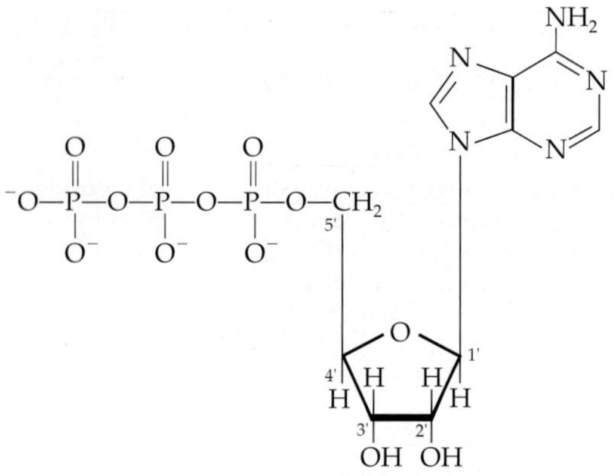

Adenosine-5′-triphosphate

Draw the structure of adenosine-3′,5′-monophosphate.

ORGANIC CHEMISTRY

15

Amines and Amides

An Amazon lily. Ethnobotanists continue to search for medically active compounds from the rain forest.

Learning Goals

1. Classify amines as primary, secondary, or tertiary.
2. Describe the physical properties of amines.
3. Draw and name simple amines using systematic and common nomenclature systems.
4. Write equations representing the synthesis of amines.
5. Write equations showing the basicity and neutralization of amines.
6. Describe the structure of quaternary ammonium salts and discuss their use as antiseptics and disinfectants.
7. Discuss the biological significance of heterocyclic amines.
8. Describe the physical properties of amides.
9. Draw the structure and write the common and I.U.P.A.C. names of amides.
10. Write equations representing the preparation of amides.
11. Write equations showing the hydrolysis of amides.
12. Draw the general structure of an amino acid.
13. Draw and discuss the structure of a peptide bond.
14. Describe the function of neurotransmitters.

Outline

Chemistry Connection:
The Nicotine Patch

15.1 Amines

A Human Perspective:
Methamphetamine

15.2 Heterocyclic Amines

15.3 Amides

A Medical Perspective:
Semisynthetic Penicillins

15.4 A Preview of Amino Acids, Proteins, and Protein Synthesis

15.5 Neurotransmitters

A Medical Perspective:
Opiate Biosynthesis and the Mutant Poppy

Chemistry Connection

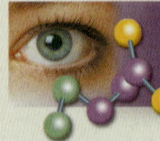

The Nicotine Patch

Smoking cigarettes is one of the most difficult habits to break—so difficult, in fact, that physicians now suspect that smoking is more than a habit: It's an addiction. The addictive drug in tobacco is *nicotine*.

Nicotine, one of the *heterocyclic amines* that we will study in this chapter (see Figure 15.4), is a highly toxic compound. In fact, it has been used as an insecticide! Small doses from cigarette smoking initially stimulate the autonomic (involuntary) nervous system. However, repeated small doses of nicotine obtained by smokers eventually depress the involuntary nervous system. As a result, the smoker *needs* another cigarette.

Some people have been able to quit smoking through behavioral modification programs, hypnosis, or sheer willpower. Others quit only after smoking has contributed to life-threatening illness, such as emphysema or a heart attack. Yet there are people who cannot quit even after a diagnosis of lung cancer.

One promising advance to help people quit smoking is the nicotine patch. The patch is applied to the smoker's skin, and nicotine from the patch slowly diffuses through the skin and into the bloodstream. Because the body receives a constant small dose of nicotine, the smoker no longer craves a cigarette. Of course, the long-range goal is to completely cure the addiction to nicotine. This is done by decreasing doses of nicotine in the patches as the treatment period continues. Eventually, after a period of about three months, the former smoker no longer needs the patches.

There are those who criticize this therapy because nicotine is a toxic chemical. However, the benefits seem to outweigh any negative aspects. A smoker inhales not only nicotine, but also dozens of other substances that have been shown to cause cancer. When someone successfully quits smoking, the body no longer suffers the risks associated with the substances in cigarette smoke.

In this chapter we study the structure and properties of amines and their derivatives, the amides. We will see that several are important pain killers, decongestants, and antibiotics, and others are addictive drugs and carcinogens.

Introduction

In this chapter we introduce an additional element into the structure of organic molecules. That element is nitrogen, the fourth most common atom in living systems. It is an important component of the structure of the nucleic acids, DNA and RNA, which are the molecules that carry the genetic information for living systems. It is also essential to the structure and function of proteins, molecules that carry out the majority of the work in biological systems. Some proteins serve as enzymes that catalyze the chemical reactions that allow life to exist. Other proteins, the antibodies, protect us against infection by a variety of infectious agents. Proteins are also structural components of the cell and of the body.

Excess nitrogen is removed from the body as urea, first synthesized by the father of organic chemistry, Friederich Wöhler (Chapter 10). Urea synthesis in the body is described in Chapter 22.

One class of organic molecules containing nitrogen is the amines. Amines are characterized by the presence of an amino group (—NH_2).

General structure of an amine

The nitrogen atom of the amino group may have one or more of its hydrogen atoms replaced by an organic group. General structures of these types of amines are shown below:

Amines are very common in biological systems and exhibit important physiological activity. Consider histamine. Histamine contributes to the inflammatory response that causes the symptoms of colds and allergies, including swollen mucous membranes, congestion, and excessive nasal secretions. We take antihistamines to help relieve these symptoms. Ephedrine is an antihistamine that has been extracted from the leaves of the ma-huang plant in China for over two thousand years. Today it is one of the antihistamines found in over-the-counter cold medications. This decongestant helps to shrink swollen mucous membranes and reduce nasal secretions. The structures of histamine and ephedrine are shown below.

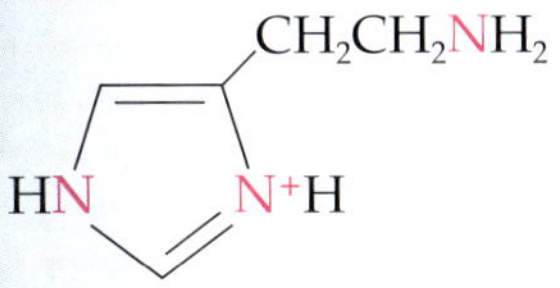

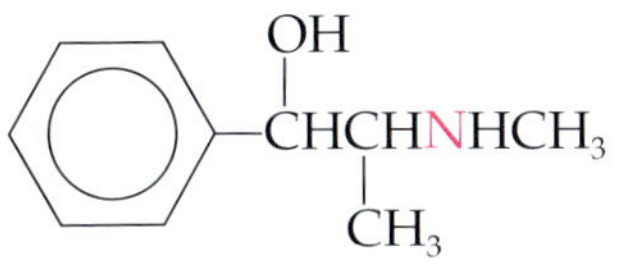

The other group of nitrogen-containing organic compounds we will investigate in this chapter is the amides. Amides are the products of a reaction between an amine and a carboxylic acid derivative. They have the following general structure:

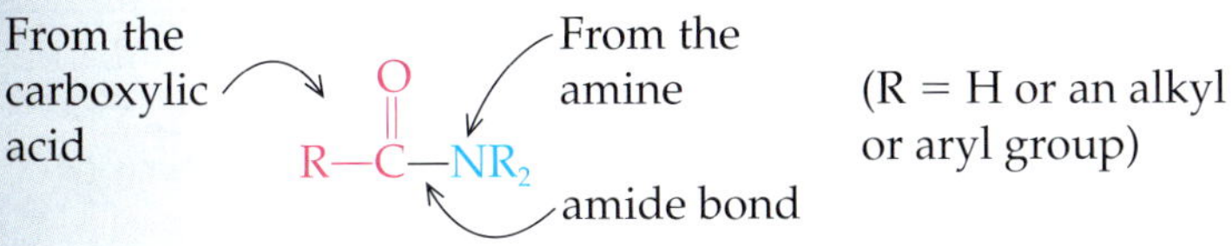

General structure of an amide

The general structure of an amino acid is

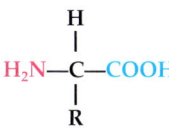

The amino group is highlighted in red and the carboxyl group in blue. Amino acids are the basic subunits of all proteins.

The amino acids are the subunits from which proteins are built. They are characterized by the presence of both an amino group and a carboxyl group. When amino acids are bonded to one another to produce a protein chain, the amino group of one amino acid reacts with the carboxyl group of another amino acid. The amide bond that results is called a peptide bond.

In this chapter we will explore the chemistry of the organic molecules that contain nitrogen. In upcoming chapters we will investigate the structure and properties of the nitrogen-containing biological molecules.

15.1 Amines

Structure and Physical Properties

Amines are organic derivatives of ammonia and, like ammonia, they are basic. In fact, amines are the most important type of organic base found in nature. We can think of them as substituted ammonia molecules in which one, two, or three of the ammonia hydrogens have been replaced by an organic group:

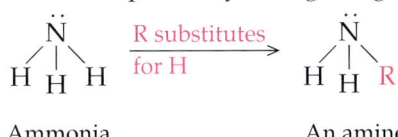

The structures drawn above and in Figure 15.1 reveal that like ammonia, amines are pyramidal. The nitrogen atom has three groups bonded to it and has a nonbonding pair of electrons.

The geometry of ammonia is described in Section 3.4.

Chapter 15 Amines and Amides

Amines are classified according to the number of alkyl or aryl groups attached to the nitrogen. In a **primary (1°) amine**, one of the hydrogens is replaced by an organic group. In a **secondary (2°) amine**, two hydrogens are replaced. In a **tertiary (3°) amine**, three organic groups replace the hydrogens:

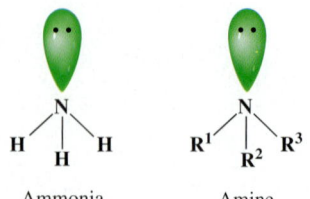

Figure 15.1
The pyramidal structure of amines. Note the similarities in structure between an amine and the ammonia molecule.

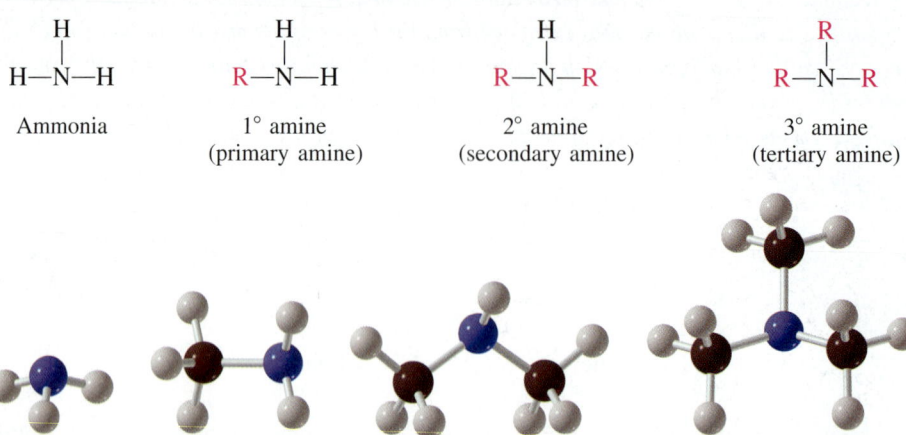

EXAMPLE 15.1 Classifying Amines as Primary, Secondary, or Tertiary

Classify each of the following compounds as a primary, secondary, or tertiary amine.

Solution

Compare the structure of the amine with that of ammonia.

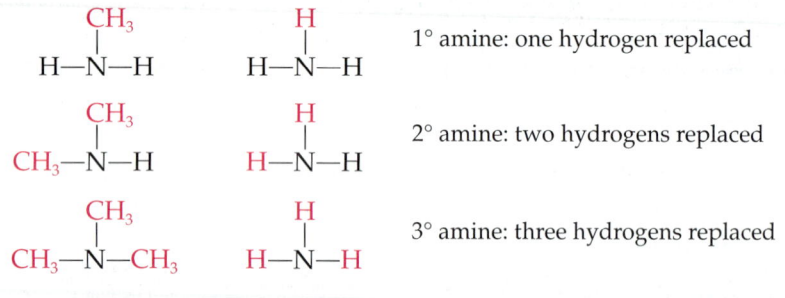

1° amine: one hydrogen replaced

2° amine: two hydrogens replaced

3° amine: three hydrogens replaced

Question 15.1

Determine whether each of the following amines is primary, secondary, or tertiary.

a. $CH_3CH_2\overset{\overset{\displaystyle CH_2CH_3}{|}}{N}CH_3$

b. $CH_3CH_2CH_2NH_2$

c. $CH_3\overset{\overset{\displaystyle H}{|}}{N}CH_3$

Question 15.2

Classify each of the following amines as primary, secondary, or tertiary.

a. CH_3NH_2

b. $CH_3CH_2CH_2CH_2\overset{\overset{\displaystyle CH_3}{|}}{N}CH_2CH_3$

c. $H\overset{\overset{\displaystyle CH_2CH_2CH_3}{|}}{N}CH_3$

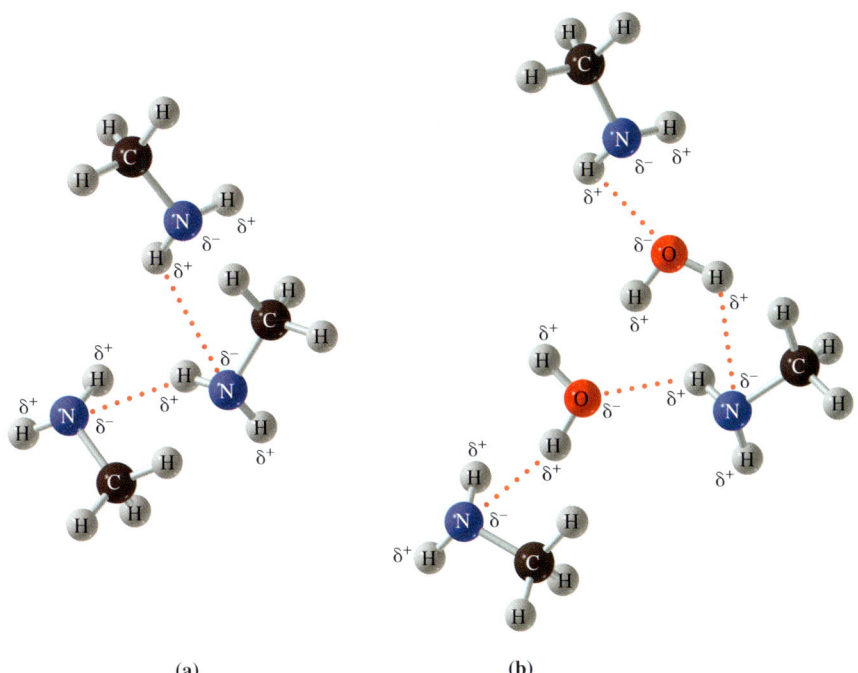

Figure 15.2
Hydrogen bonding (a) in methylamine and (b) between methylamine and water. Dotted lines represent hydrogen bonds.

The nitrogen atom is more electronegative than the hydrogen atoms in amines. As a result, the N—H bond is polar, and hydrogen bonding between amine molecules or between amine molecules and water can occur (Figure 15.2).

Hydrogen bonding is described in Section 5.2.

Question 15.3

Refer to Figure 15.2 and draw a similar figure showing the hydrogen bonding that occurs between water and a 2° amine.

Question 15.4

Refer to Figure 15.2 and draw hydrogen bonding between two primary amines.

 **LEARNING GOAL**

The ability of primary and secondary amines to form N—H···N hydrogen bonds is reflected in their boiling points (Table 15.1). Primary amines have boiling points well above those of alkanes of similar molecular weight but considerably lower than those of comparable alcohols. Consider the following examples:

$CH_3CH_2CH_3$ $CH_3CH_2NH_2$ CH_3CH_2OH

Propane   Ethanamine   Ethanol
M.W. = 44 g/mol   M.W. = 45 g/mol   M.W. = 46 g/mol
b.p. = −42.2° C   b.p. = 16.6° C   b.p. = 78.5° C

Tertiary amines do not have an N—H bond. As a result they cannot form intermolecular hydrogen bonds with other tertiary amines. Consequently, their boiling points are lower than those of primary or secondary amines of comparable molecular weight. This is seen in a comparison of the boiling points of propanamine (propylamine; M.W. = 59) and N,N-dimethylmethanamine (trimethylamine; M.W. = 59). Trimethylamine, the tertiary amine, has a boiling point of 2.9°C, whereas propylamine, the primary amine, has a boiling point of 48.7°C. Clearly the inability of trimethylamine molecules to form intermolecular hydrogen bonds results in a much lower boiling point.

TABLE 15.1 Boiling Points of Amines

Systematic Name	Common Name	Structure	Boiling Point (°C)
	Ammonia	NH_3	−33.4
Methanamine	Methylamine	CH_3NH_2	−6.3
N-Methylmethanamine	Dimethylamine	$(CH_3)_2NH$	7.4
N,N-Dimethylmethanamine	Trimethylamine	$(CH_3)_3N$	2.9
Ethanamine	Ethylamine	$CH_3CH_2NH_2$	16.6
Propanamine	Propylamine	$CH_3CH_2CH_2NH_2$	48.7
Butanamine	Butylamine	$CH_3CH_2CH_2CH_2NH_2$	77.8

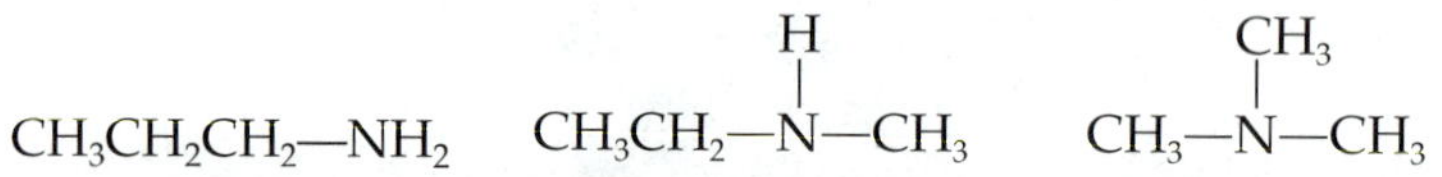

$CH_3CH_2CH_2-NH_2$ $CH_3CH_2-\overset{\overset{H}{|}}{N}-CH_3$ $CH_3-\overset{\overset{CH_3}{|}}{N}-CH_3$

Propanamine (propylamine)
M.W. = 59 g/mol
b.p. = 48.7° C

N-Methylethanamine (ethylmethylamine)
M.W. = 59 g/mol
b.p. = 36.7° C

N,N-Dimethylmethanamine (trimethylamine)
M.W. = 59 g/mol
b.p. = 2.9° C

The intermolecular hydrogen bonds formed by primary and secondary amines are not as strong as the hydrogen bonds formed by alcohols because nitrogen is not as electronegative as oxygen. For this reason primary and secondary amines have lower boiling points than alcohols (Table 15.2).

All amines can form intermolecular hydrogen bonds with water (O—H⋯N). As a result, small amines (six or fewer carbons) are soluble in water. As we have noted for other families of organic molecules, water solubility decreases as the length of the hydrocarbon (hydrophobic) portion of the molecule increases.

EXAMPLE 15.2 Predicting the Physical Properties of Amines

Which member of each of the following pairs of molecules has the higher boiling point?

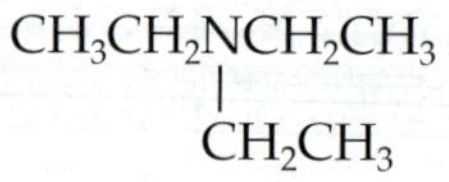

$CH_3CH_2NCH_2CH_3$
$\quad\quad\quad |$
$\quad\quad CH_2CH_3$

or $CH_3CH_2CH_2CH_2CH_2CH_2NH_2$

Solution

The molecule on the right, hexanamine, has a higher boiling point than the molecule on the left, N,N-diethylethanamine (triethylamine). Triethylamine is a tertiary amine; therefore, it has no N—H bond and cannot form intermolecular hydrogen bonds with other triethylamine molecules.

$CH_3CH_2CH_2CH_2OH$ or $CH_3CH_2CH_2CH_2NH_2$

Solution

The molecule on the left, butanol, has a higher boiling point than the molecule on the right, butanamine. Nitrogen is not as electronegative as oxygen, thus the hydroxyl group is more polar than the amino group and forms stronger hydrogen bonds.

Amine nomenclature will be studied in the next section.

15.1 Amines

TABLE 15.2 Comparison of the Boiling Points of Selected Alcohols and Amines

Name	Molecular Weight (g/mol)	Boiling Point (°C)
Methanol	32	64.5
Methanamine	31	−6.3
Ethanol	46	78.5
Ethanamine	45	16.6
Propanol	60	97.2
Propanamine	59	48.7

Question 15.5

Which compound in each of the following pairs would you predict to have a higher boiling point? Explain your reasoning.

a. Methanol or methylamine
b. Dimethylamine or water
c. Methylamine or ethylamine
d. Propylamine or butane

Question 15.6

Compare the boiling points of methylamine, dimethylamine, and trimethylamine. Explain why they are different.

Nomenclature

In systematic nomenclature, primary amines are named according to the following rules:

- Determine the name of the *parent compound*, the longest continuous carbon chain containing the amine group.
- Replace the *–e* ending of the alkane chain with *–amine*. Following this pattern, the alkane becomes an alkan*amine*; for instance, ethan*e* becomes ethan*amine*.
- Number the parent chain to give the carbon bearing the amine group the lowest possible number.
- Name and number all substituents, and add them as prefixes to the "alkanamine" name.

For instance,

CH$_3$—NH$_2$ CH$_3$CH$_2$CH$_2$—NH$_2$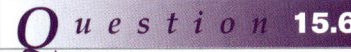

Methanamine   1-Propanamine   2-Pentanamine

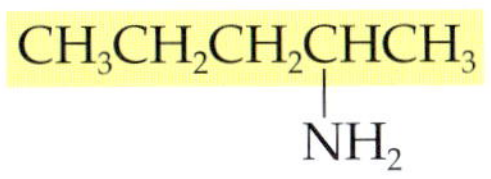

For secondary or tertiary amines the prefix *N*-alkyl is added to the name of the parent compound. For example,

CH$_3$—NH—CH$_2$CH$_3$

N-Methylethanamine   *N*,*N*-Dimethylmethanamine

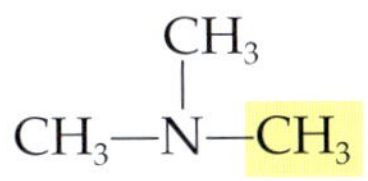

LEARNING GOAL 3

Several aromatic amines have special names that have also been approved for use by I.U.P.A.C. For example, the amine of benzene is given the name *aniline*. The systematic name for aniline is *benzenamine*.

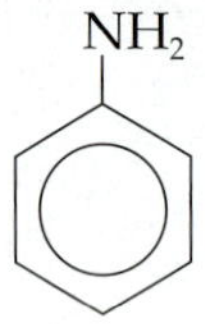

Aniline or benzenamine

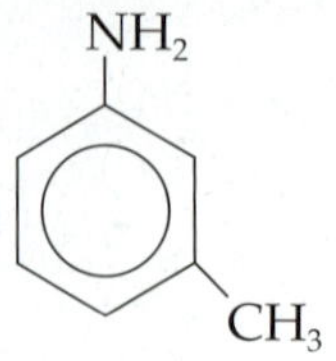

m-Toluidine or *meta*-toluidine

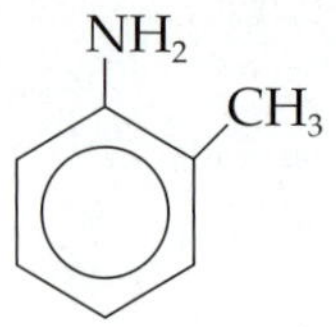

o-Toluidine or *ortho*-toluidine

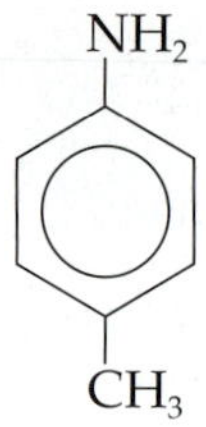

p-Toluidine or *para*-toluidine

If additional groups are attached to the nitrogen of an aromatic amine, they are indicated with the letter *N*- followed by the name of the group.

EXAMPLE 15.3 Writing the Systematic Name for an Amine

Name the following amine.

$$CH_3CH_2CH_2-NH-CH_3$$

Solution

Parent compound: propane (becomes propanamine)
Additional group on N: methyl (becomes *N*-methyl)
Name: *N*-Methylpropanamine

Common names are often used for the simple amines. The common names of the alkyl groups bonded to the amine nitrogen are followed by the ending -*amine*. Each group is listed alphabetically in one continuous word followed by the suffix -*amine*:

CH_3-NH_2 Methylamine

$CH_3-NH-CH_3$ Dimethylamine

$CH_3-N(CH_3)-CH_3$ Trimethylamine

$CH_3CH_2-NH_2$ Ethylamine

$CH_3CH_2-NH-CH_3$ Ethylmethylamine

Table 15.3 compares these systems of nomenclature for a number of simple amines.

Question 15.7

Use the structure of aniline provided and draw the complete structural formula for each of the following amines.

a. *N*-Methylaniline
b. *N,N*-Dimethylaniline
c. *N*-Ethylaniline
d. *N*-Isopropylaniline

15.1 Amines

TABLE 15.3 Systematic and Common Names of Amines

Compound	Systematic Name	Common Name
R—NH_2	Alkanamine	Alkylamine
CH_3—NH_2	Methanamine	Methylamine
CH_3CH_2—NH_2	Ethanamine	Ethylamine
$CH_3CH_2CH_2$—NH_2	1-Propanamine	Propylamine
CH_3—NH—CH_3	N-Methylmethanamine	Dimethylamine
CH_3—NH—CH_2CH_3	N-Methylethanamine	Ethylmethylamine
CH_3—N(CH_3)—CH_3	N,N-Dimethylmethanamine	Trimethylamine

Question 15.8

Name each of the following amines using the systematic and common nomenclature systems.

a.  $CH_3\underset{\underset{NH_2}{\vert}}{CH}CH_2CH_3$

b. 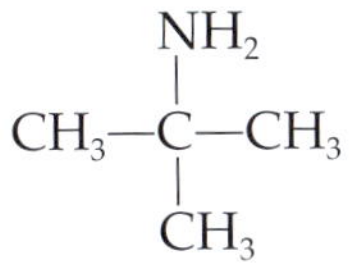 $CH_3-\underset{\underset{CH_3}{\vert}}{\overset{\overset{NH_2}{\vert}}{C}}-CH_3$

c. $CH_3\underset{\underset{NH-CH_3}{\vert}}{CH}CH_2CH_3$

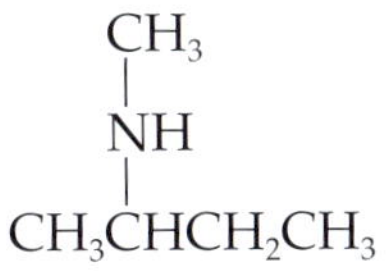

d. 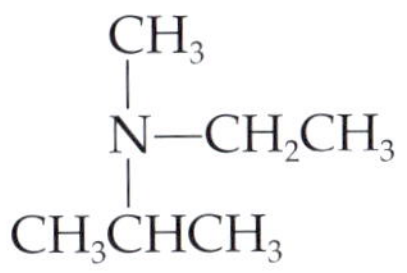 $CH_3\underset{\underset{N(CH_3)-CH_2CH_3}{\vert}}{CH}CH_3$

Question 15.9

Draw the complete structural formula for each of the following compounds.

a. 2-Propanamine
b. 3-Octanamine
c. N-Ethyl-2-heptanamine
d. 2-Methyl-2-pentanamine
e. 4-Chloro-5-iodo-1-nonanamine
f. N,N-Diethyl-1-pentanamine

Question 15.10

Draw the condensed formula for each of the following compounds.

a. Diethylmethylamine
b. 4-Methylpentylamine
c. N-Methylaniline
d. Triisopropylamine
e. Methyl-t-butylamine
f. Ethylhexylamine

Medically Important Amines

Although amines play many different roles in our day-to-day lives, one important use is in medicine. A host of drugs derived from amines is responsible for improving the quality of life, whereas others, such as cocaine and heroin, are highly addictive.

Amphetamines, such as benzedrine and methedrine, stimulate the central nervous system. They elevate blood pressure and pulse rate and are often used to decrease fatigue. Medically, they have been used to treat depression and epilepsy. Amphetamines have also been prescribed as diet pills because they decrease the appetite. Their use is controlled by federal law because excess use of amphetamines can cause paranoia and mental illness.

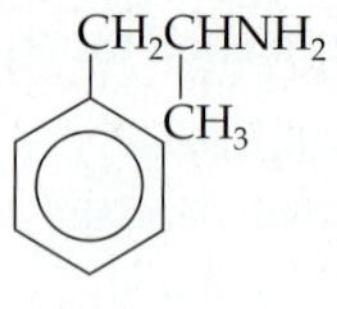

1-Phenyl-2-propanamine (Amphetamine)
Benzedrine

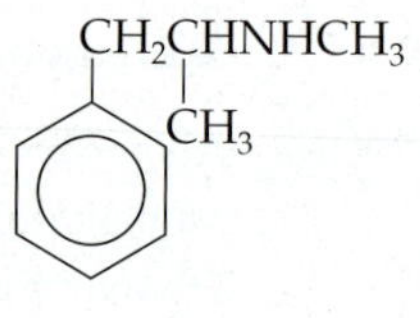

N-Methyl-1-phenyl-2-propanamine (Methamphetamine)
Methedrine

Many of the medicinal amines are *analgesics* (pain relievers) or *anesthetics* (pain blockers). Novocaine and related compounds, for example, are used as local anesthetics. Demerol is a very strong pain reliever.

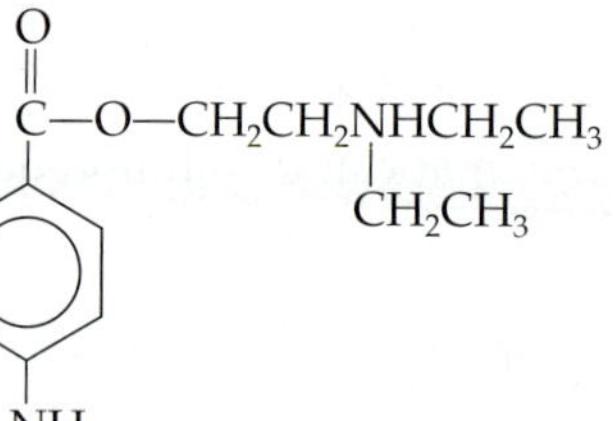

Novocaine

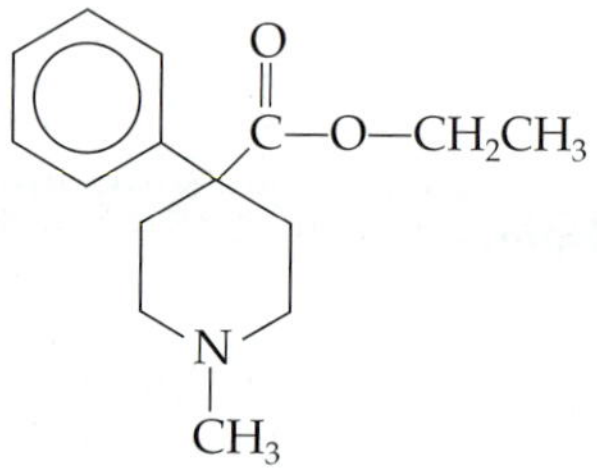

Demerol

Ephedrine, its stereoisomer pseudoephedrine, and phenylephrine (also called neosynephrine) are used as decongestants in cough syrups and nasal sprays. By shrinking of the membranes that line the nasal passages, they relieve the symptoms of congestion and stuffy nose. These compounds are very closely related to L-dopa and dopamine, which are key compounds in the function of the central nervous system.

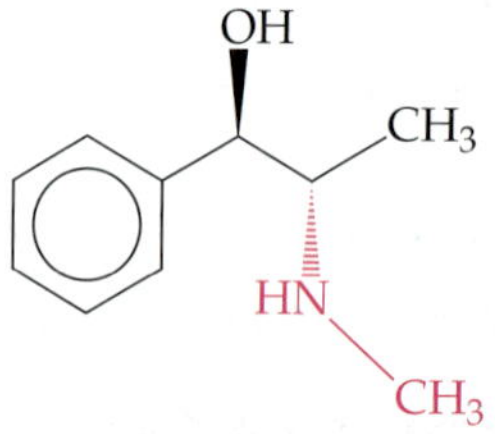

L-Dopa, dopamine, and other key neurotransmitters are described in detail in Section 15.5.

Ephedrine

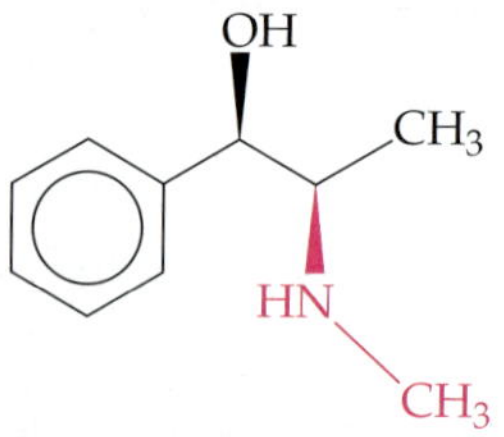

Pseudoephedrine

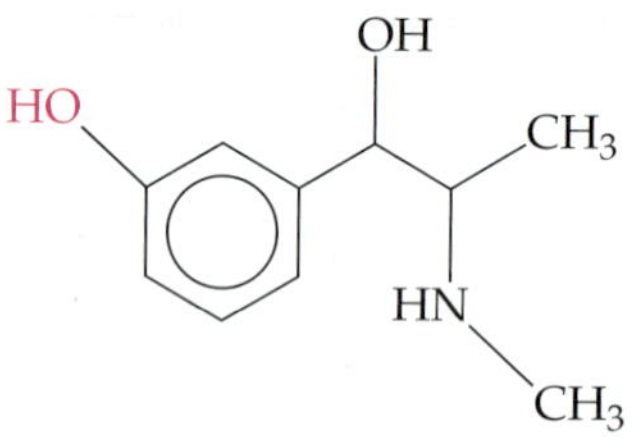

Phenylephrine (neosynephrine)

Methamphetamine use and abuse are discussed in "A Human Perspective: Methamphetamine" found on page 500.

Recently, many states have restricted the sales of products containing ephedrine and pseudoephedrine and many drug store chains have moved these products behind the counter. The reason for these precautions is that either ephedrine or pseudoephedrine can be used as the starting material in the synthesis of methamphetamines. In response to this problem, pharmaceutical companies are replacing ephedrine and pseudoephedrine in these decongestants with phenylephrine, which cannot be used as a reactant in the synthesis of methamphetamine.

Ephedrine and pseudoephedrine are also the primary active ingredients in ephedra, a plant found in the deserts of central Asia. Ephedra is used as a stimulant in a variety of products that are sold over-the-counter as aids to boost energy, promote weight loss, and enhance athletic performance. In 2004 the Food and Drug Administration banned the use of ephedra in these over-the-counter formulations after reviewing 16,000 reports of adverse side effects including nervousness, heart irregularities, seizures, heart attacks, and 80 deaths, including that of

Baltimore Orioles pitcher Steve Bechler, age 23. However, in April, 2005 a Federal Judge in Texas ruled that the FDA had failed to prove that ephedra is dangerous at doses of 10 mg or lower, opening the way for the sale of ephedra-containing herbal remedies.

The *sulfa drugs*, the first chemicals used to fight bacterial infections, are synthesized from amines.

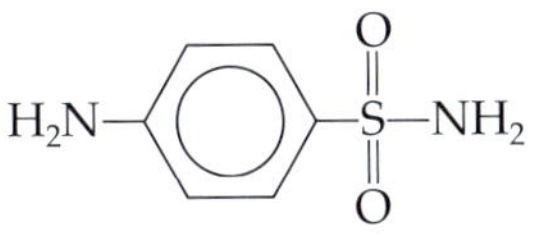

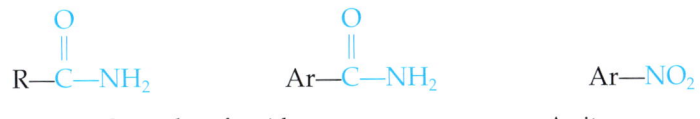

Sulfanilamide—a sulfa drug

Reactions Involving Amines
Preparation of Amines

In the laboratory, amines are prepared by the reduction of amides and nitro compounds.

 **LEARNING GOAL**

$$\underset{\text{Examples of amides}}{\underset{\|}{R-\overset{O}{C}-NH_2} \quad \underset{\|}{Ar-\overset{O}{C}-NH_2}} \quad \underset{\text{A nitro compound}}{Ar-NO_2}$$

As we will see in Section 15.3, amides are neutral nitrogen compounds that produce an amine and a carboxylic acid when hydrolyzed. Nitro compounds are prepared by the nitration of an aromatic compound.

Primary amines are readily produced by reduction of a nitro compound, as in the following reaction:

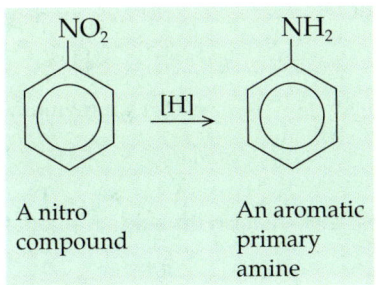

We are using the general symbol [H] to represent any reducing agent just as we used [O] to represent an oxidizing agent in previous chapters. Several different reducing agents may be used to effect the changes shown here; for example, metallic iron and acid may be used to reduce aromatic nitro compounds and LiAlH$_4$ in ether reduces amides.

In this reaction the nitro compound is nitrobenzene and the product is aniline.

Amides may also be reduced to produce primary, secondary, or tertiary amines.

$$\underset{\text{Amide}}{R^1-\overset{\overset{O}{\|}}{C}-N\overset{R^2}{\underset{R^3}{\diagdown}}} \xrightarrow{[H]} \underset{\text{Amine}}{R^1CH_2N\overset{R^2}{\underset{R^3}{\diagdown}}} \quad (R^2 \text{ and } R^3 \text{ may be a hydrogen atom or an organic group.})$$

If R^2 and R^3 are hydrogen atoms, the product will be a primary amine:

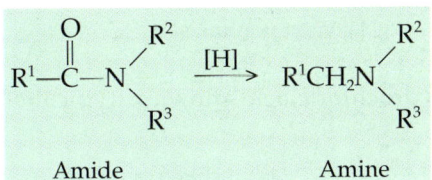

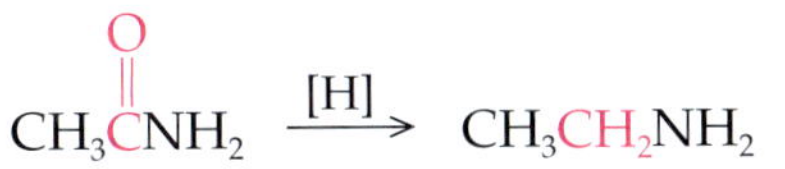

Ethanamide → Ethanamine (ethylamine)

If either R² or R³ is an organic group, the product will be a secondary amine:

$$CH_3CH_2\overset{\overset{O}{\|}}{C}NHCH_3 \xrightarrow{[H]} CH_3CH_2CH_2NHCH_3$$

N-Methylpropanamide   N-Methylpropanamine (methylpropylamine)

If both R² and R³ are organic groups, the product will be a tertiary amine:

$$CH_3\overset{\overset{O}{\|}}{C}-\overset{\overset{CH_3}{|}}{N}CH_3 \xrightarrow{[H]} CH_3CH_2\overset{\overset{CH_3}{|}}{N}CH_3$$

N,N-Dimethylethanamide   N,N-Dimethylethanamine

Basicity

LEARNING GOAL

Amines behave as weak bases, accepting H⁺, when dissolved in water. The non-bonding pair (lone pair) of electrons of the nitrogen atom can be shared with a proton (H⁺) from a water molecule, producing an **alkylammonium ion.** Hydroxide ions are also formed, so the resulting solution is basic.

$$R-\overset{\overset{H}{|}}{\underset{\underset{H}{|}}{N}}: + H-OH \rightleftharpoons R-\overset{\overset{H}{|}}{\underset{\underset{H}{|}}{N^+}}-H + OH^-$$

Amine   Water   Alkylammonium ion   Hydroxide ion

$$CH_3-\overset{\overset{H}{|}}{\underset{\underset{H}{|}}{N}}: + H-OH \rightleftharpoons CH_3-\overset{\overset{H}{|}}{\underset{\underset{H}{|}}{N^+}}-H + OH^-$$

Methylamine   Water   Methylammonium ion   Hydroxide ion

Neutralization

Because amines are bases, they react with acids to form alkylammonium salts.

$$R-\overset{\overset{H}{|}}{\underset{\underset{H}{|}}{N}}: + HCl \longrightarrow R-\overset{\overset{H}{|}}{\underset{\underset{H}{|}}{N^+}}-H\ Cl^-$$

Amine   Acid   Alkylammonium salt

Recall that the reaction of an acid and a base gives a salt (Section 8.3).

The reaction of methylamine with hydrochloric acid shown is typical of these reactions.

$$CH_3-\overset{\overset{H}{|}}{\underset{\underset{H}{|}}{N}}: + HCl \longrightarrow CH_3-\overset{\overset{H}{|}}{\underset{\underset{H}{|}}{N^+}}-H\ Cl^-$$

Methylamine   Hydrochloric acid   Methylammonium chloride

Alkylammonium salts are named by replacing the suffix *-amine* with *ammonium*. This is then followed by the name of the anion. The salts are ionic and hence are quite soluble in water.

15.1 Amines

A variety of important drugs are amines. They are usually administered as alkylammonium salts because the salts are much more soluble in aqueous solutions and in body fluids.

Question 15.11

Complete each of the following reactions by supplying the missing product(s).

a. 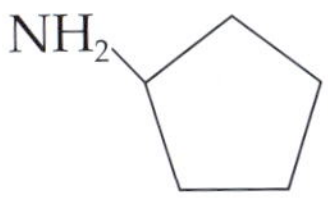 cyclopentyl-NH$_2$ + HBr ⟶ ?

b. $CH_3CH_2NHCH_3 + H_2O ⟶$?
c. $CH_3NH_2 + H_2O ⟶$?

Question 15.12

Complete each of the following reactions by supplying the missing product(s).

a. $CH_3NH_2 + HI ⟶$?
b. $CH_3CH_2NH_2 + HBr ⟶$?
c. $(CH_3CH_2)_2NH + HCl ⟶$?

Alkylammonium salts can neutralize hydroxide ions. In this reaction, water is formed and the protonated amine cation is converted into an amine.

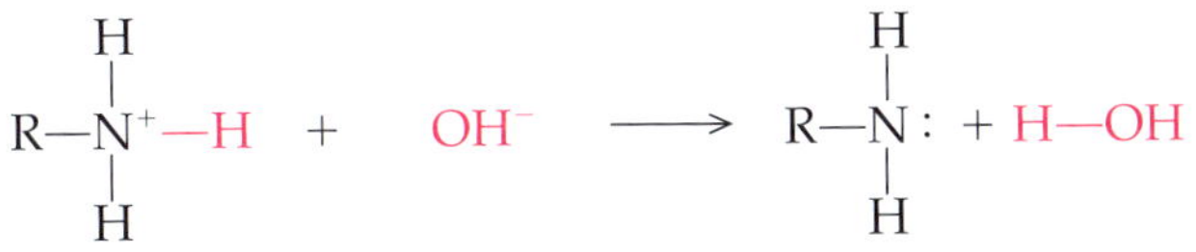

$$\underset{\substack{\text{Alkylammonium}\\\text{salt}}}{R-\overset{H}{\underset{H}{N^+}}-H} + \underset{\substack{\text{Hydroxide}\\\text{ion}}}{OH^-} \longrightarrow \underset{\text{Amine}}{R-\overset{H}{\underset{H}{N}}:} + \underset{\text{Water}}{H-OH}$$

Thus, by adding a strong acid to a water-insoluble amine, a water soluble alkylammonium salt can be formed. The salt can just as easily be converted back to an amine by the addition of a strong base. The ability to manipulate the solubility of physiologically active amines through interconversion of the amine and its corresponding salt is extremely important in the development, manufacture, and administration of many important drugs.

Pure cocaine is an amine and a base (see Figure 15.4 and structure below). This form of cocaine, referred to as "crack" or "free base" cocaine, is generally found in the form of relatively large crystals (Figure 15.3a) that may vary in color from white to dark brown or black. As we have just seen, when an amine reacts with an acid, an alkylammonium salt is formed. When cocaine reacts with HCl, the product is cocaine hydrochloride:

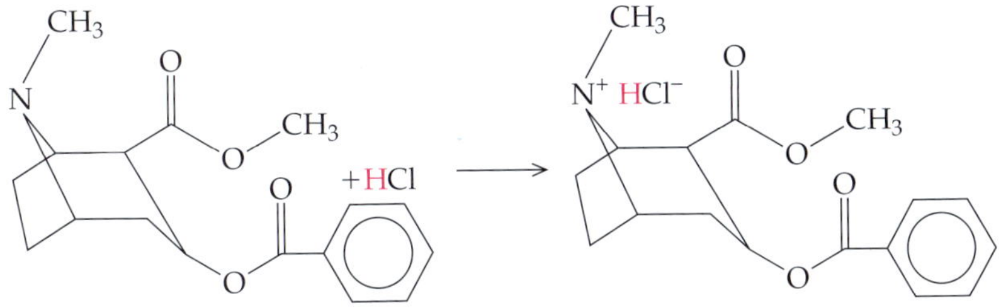

"Crack" cocaine (a base) + HCl ⟶ Cocaine hydrochloride (a salt)

The local anesthetic novocaine, which is often used in dentistry and for minor surgery, is injected as an amine salt. See Medically Important Amines earlier in this section.

A Human Perspective

Methamphetamine

Methamphetamine is an addictive drug known by many names, including "speed," "crystal," "crank," "ice," and "glass." A bitter-tasting, odorless, crystalline powder, it is easily dissolved in either water or alcohol. Methamphetamine was developed early in the twentieth century and used as a decongestant in nasal and bronchial inhalers. Now it is rarely used for medical purposes.

A 2002 Health and Human Services survey revealed that twelve million Americans age twelve and older had used methamphetamine. Use of methamphetamine was once associated with white, male, blue-collar workers; but a much more diverse audience now uses the drug. Although it is still used by people in jobs such as long-distance trucking which require long hours and mental and physical alertness, it is disturbing that use of methamphetamine is becoming increasingly associated with sexual activity, teenagers attending "raves," homeless people, and runaway youths.

Methamphetamine can be smoked, taken orally, snorted, or injected, depending on the form of the drug; and it alters the mood differently depending on how it is taken. Smoking or injecting intravenously results in a "flash" or intense rush that lasts only a few minutes. This may be followed by a high that lasts several hours. Snorting and oral ingestion result in a euphoria lasting three to five minutes in the case of snorting and up to twenty minutes in the case of oral ingestion. Because the pleasurable effects are so short-lived, methamphetamine users tend to binge to try to sustain the high.

Both the intense rush and the longer-lasting euphoria are thought to result from a release of dopamine and norepinephrine into regions of the brain that control feelings of pleasure.

Once inside nerve cells (neurons), methamphetamine causes the release of dopamine and norepinephrine. At the same time, it inhibits enzymes that would normally destroy these two neurotransmitters and the excess is transported out of neurons, causing the sensations of pleasure and euphoria. The excess norepinephrine may be responsible for the increased attention and decreased fatigue association with methamphetamine use.

Symptoms of long-term use include addiction, anxiety, violent behavior, confusion, as well as psychotic symptoms of

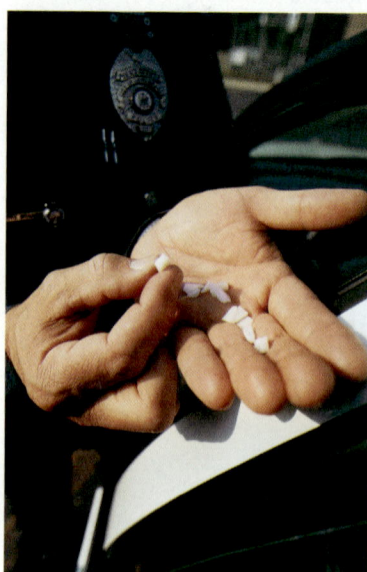
(a)

(c)

(b)

Figure 15.3
(a) Crack cocaine is a non-water-soluble base with a low melting point and a crystalline structure. (b) Powdered cocaine is a water-soluble salt of cocaine base.
(c) Cocaine is extracted from the leaves of the coca plant.

paranoia, hallucinations, and delusions. In severe cases, paranoia causes homicidal and/or suicidal feelings. Methamphetamine also increases heart rate and blood pressure. It can cause strokes, which result from irreversible damage to blood vessels in the brain, as well as respiratory problems, irregular heartbeat, and extreme anorexia. In extreme situations, it can cause cardiovascular collapse and death.

No physical symptoms accompany withdrawal from the drug, but psychological symptoms such as depression, anxiety, aggression, and intense craving are common. Of greatest concern is the brain damage that occurs in long-term users.

Dopamine release may be the cause of the drug's long-term toxic effects. Compare the structure of the neurotransmitters dopamine (Figure 15.7) with that of methamphetamine shown below. Research in humans has shown that even three years after chronic methamphetamine use, the former user continues to have a reduced ability to transport dopamine back into nerve cells. Parkinson's disease is characterized by a decrease in the dopamine-producing neurons in the brain; so it was logical to look for similarities between methamphetamine users and those suffering from Parkinson's. In fact, the brains of methamphetamine users showed damage similar to, but not as severe as, that in Parkinson's disease. Research in animals demonstrated that up to 50% of the dopamine-producing cells in parts of the brain may be destroyed by prolonged exposure, and that serotonin-containing neurons may sustain even worse damage.

Methamphetamine use continues to rise, in part, because it is easily synthesized, or "cooked," using "recipes" that are available from many sources, including the Internet. Ephedrine, an over-the-counter decongestant drug, is the starting material. (Pseudoephedrine, a stereoisomer of ephedrine, can also be used.) As shown in the equation below, ephedrine is simply reduced to produce methamphetamine.

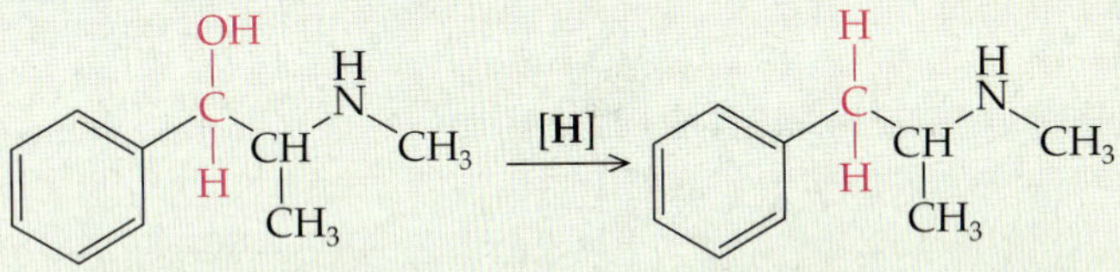

Ephedrine → [H] → Methamphetamine

While the synthesis involves a variety of dangerous chemicals, including anhydrous ammonia, anhydrous hydrochloric acid, sodium, and sodium hydroxide, most "meth cooks" do not have formal laboratory training. "Meth lab" fires are common, and the synthesis produces toxic wastes. The cleanup that followed the seizure of a major "meth lab" in 2003 took eight days and required fifty people. Over four million pounds of toxic soil and 133 drums of hazardous waste were removed from the site, at a cost of $226,000.

For Further Understanding

Compare the structures of methamphetamine and dopamine. Develop a hypothesis to explain why dopamine receptors also bind to and transport methamphetamine into neurons. (Hint: Receptors are proteins in cell membranes that have a pocket into which a specific molecule can fit.)

Explain why methamphetamine is soluble in alcohols and in water.

The salt of cocaine is a powder (Figure 15.3b) and is soluble in water. Since it is a powder, it can be snorted, and because it is water soluble, it dissolves in the fluids of the nasal mucous membranes and is absorbed into the bloodstream. This is a common form of cocaine because it is the direct product of the preparation from coca leaves (Figure 15.3c). A coca paste is made from the leaves and is mixed with HCl and water. After additional processing, the product is the salt of cocaine.

Cocaine hydrochloride salt can be converted into its base form by a process called "free basing." Although the chemistry is simple, the process is dangerous because it requires highly flammable solvents. This pure cocaine is not soluble in water. It has a relatively low melting point, however, and can be smoked. Crack, so called because of the crackling noise it makes when smoked, is absorbed into the body more quickly than the snorted powder and results in a more immediate high.

Quaternary Ammonium Salts

Quaternary ammonium salts are ammonium salts that have four organic groups bonded to the nitrogen. They have the following general structure:

 LEARNING GOAL

$R_4N^+X^-$ (R = any alkyl or aryl group;
            X^- = a halide anion, most commonly Cl^-)

Chapter 15 Amines and Amides

Quaternary ammonium salts that have a very long carbon chain, sometimes called "quats," are used as disinfectants and antiseptics because they have detergent activity. Two popular quats are benzalkonium chloride (Zephiran) and cetylpyridinium chloride, found in the mouthwash Scope.

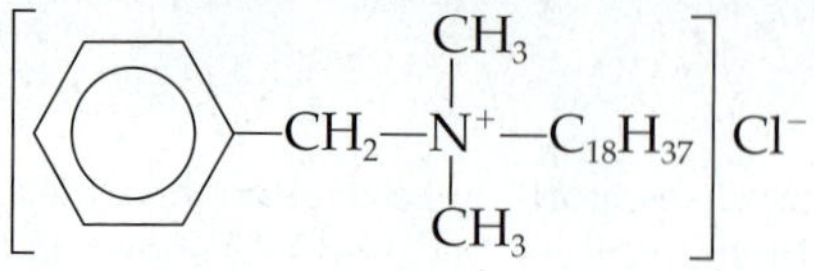

Benzalkonium chloride

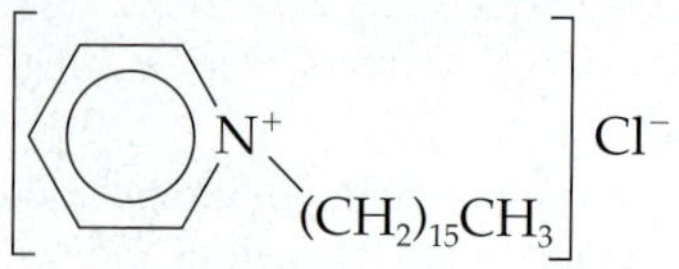

Cetylpyridinium chloride

Choline is an important quaternary ammonium salt in the body. It is part of the hydrophilic "head" of the membrane phospholipid lecithin. Choline is also a precursor for the synthesis of the neurotransmitter acetylcholine.

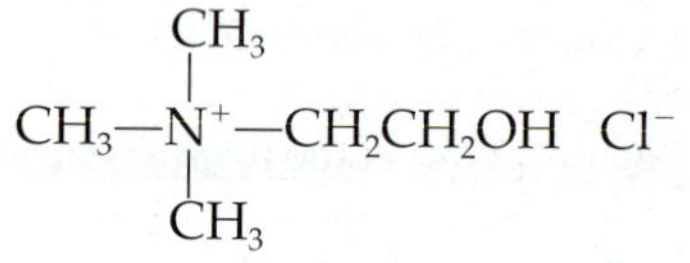

Choline

Phospholipids and biological membranes are discussed in Sections 17.3 and 17.6.

The function of acetylcholine is described in greater detail in Section 15.5.

15.2 Heterocyclic Amines

LEARNING GOAL 7

Heterocyclic amines are cyclic compounds that have at least one nitrogen atom in the ring structure. The structures and common names of several heterocyclic amines important in nature are shown here. They are represented by their structural formulas and by abbreviated line formulas.

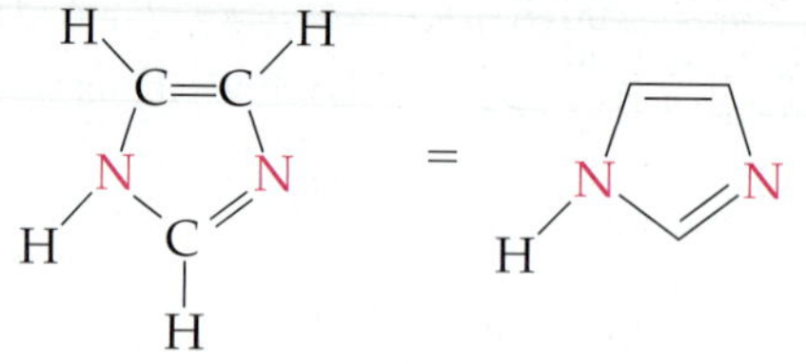

Imidazole

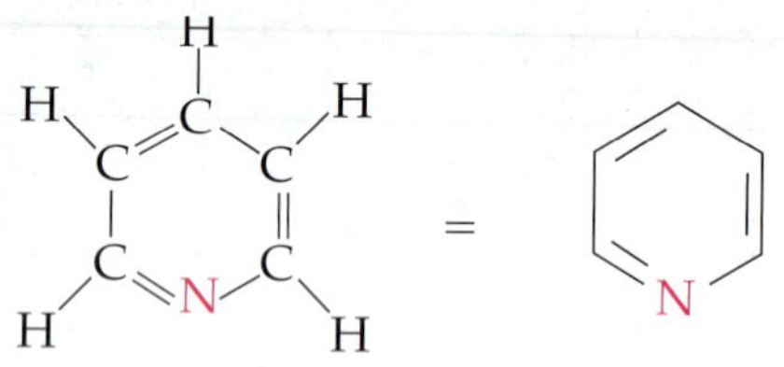

Pyridine

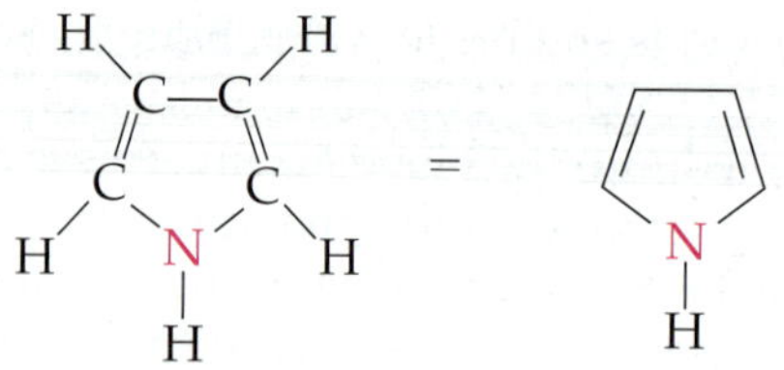

Pyrrole

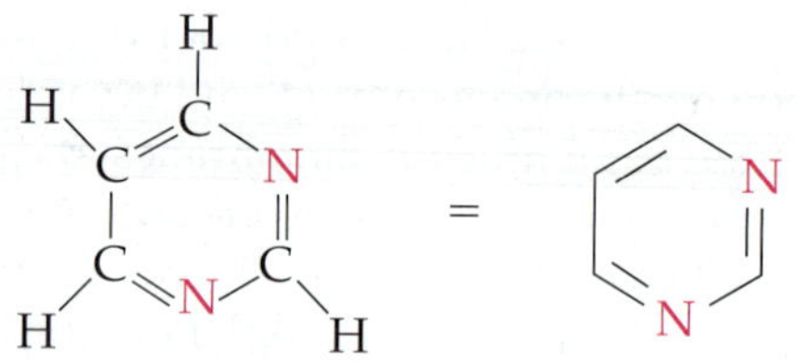

Pyrimidine

The heterocyclic amines shown below are examples of fused ring structures. Each ring pair shares two carbon atoms in common. Thus, two fused rings share one or more common bonds as part of their ring backbones. Consider the structures of a purine, indole, and porphyrin, which are shown as structural formulas and as line diagrams.

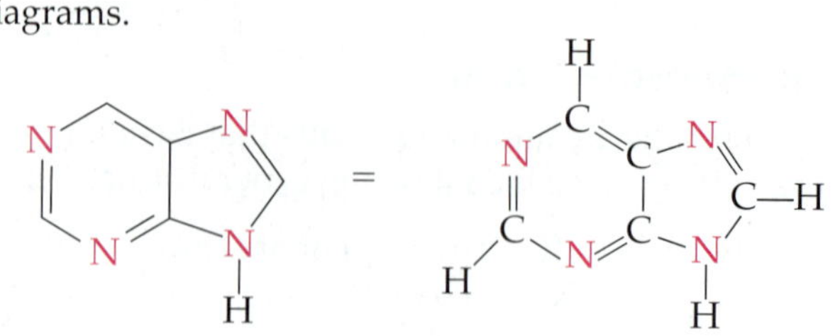

Purine

15.2 Heterocyclic Amines

Indole

Porphyrin

M$^+$ = metal ion

The pyrimidine and purine rings are found in DNA and RNA. The porphyrin ring structure is found in hemoglobin (an oxygen-carrying blood protein), myoglobin (an oxygen-carrying protein found in muscle tissue), and chlorophyll (a photosynthetic plant pigment). The indole and pyridine rings are found in many **alkaloids,** which are naturally occurring compounds with one or more nitrogen-containing heterocyclic rings. The alkaloids include cocaine, nicotine, quinine, morphine, heroin, and LSD (Figure 15.4).

The structures of purines and pyrimidines are presented in Section 20.1.

The structure of the heme group found in hemoglobin and myoglobin is presented in Section 18.9.

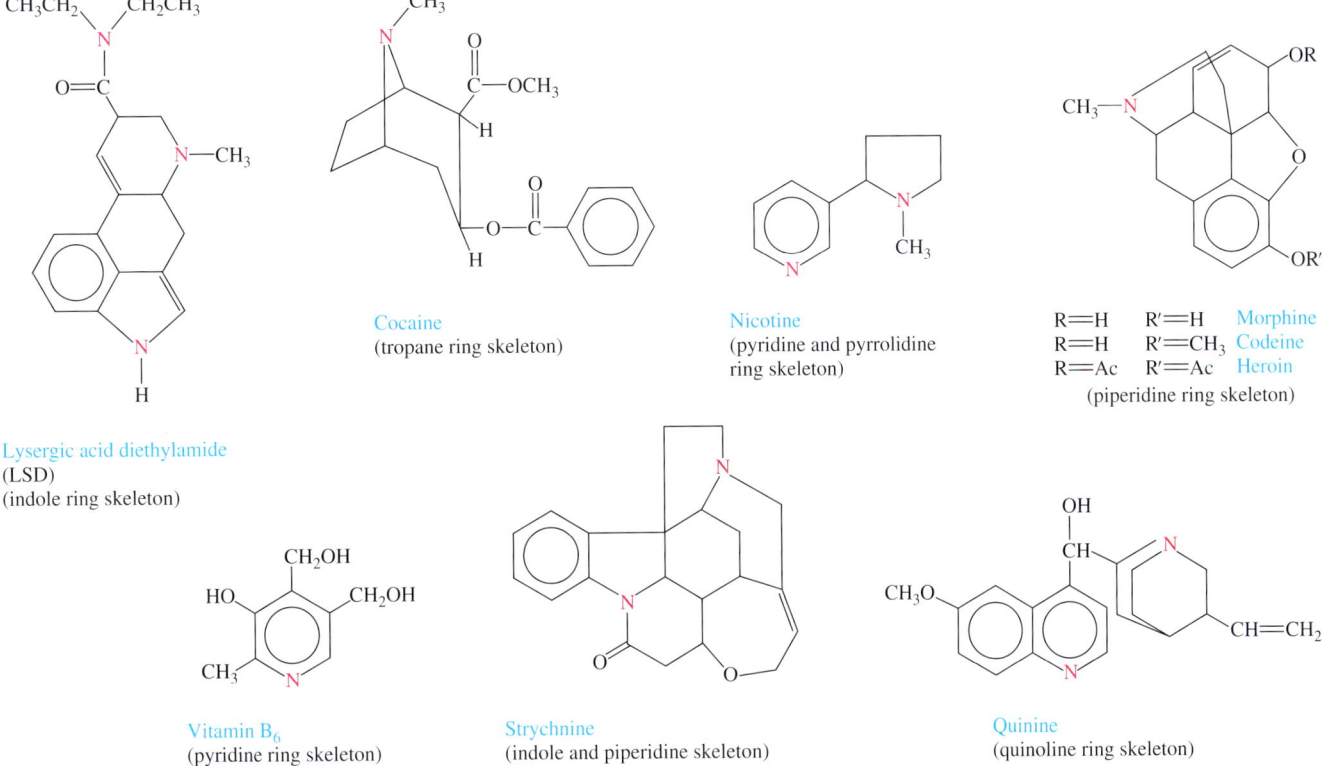

Lysergic acid diethylamide (LSD)
(indole ring skeleton)

Cocaine
(tropane ring skeleton)

Nicotine
(pyridine and pyrrolidine ring skeleton)

R=H   R'=H   Morphine
R=H   R'=CH$_3$ Codeine
R=Ac R'=Ac   Heroin
(piperidine ring skeleton)

Vitamin B$_6$
(pyridine ring skeleton)

Strychnine
(indole and piperidine skeleton)

Quinine
(quinoline ring skeleton)

Figure 15.4
Structures of several heterocyclic amines with biological activity.

Lysergic acid diethylamide (LSD) is a hallucinogenic compound that may cause severe mental disorders. Cocaine is produced by the coca plant. In small doses it is used as an anesthetic for the sinuses and eyes. An **anesthetic** is a drug that causes a lack of sensation in any part of the body (local anesthetic) or causes unconsciousness (general anesthetic). In higher doses, cocaine causes an intense feeling of euphoria followed by a deep depression. Cocaine is addictive because the user needs larger and larger amounts to overcome the depression. Nicotine is one of the simplest heterocyclic amines and appears to be the addictive component of cigarette smoke.

Morphine was the first alkaloid to be isolated from the sap of the opium poppy. Morphine is a strong **analgesic,** a drug that acts as a pain killer. However, it is a powerful and addictive narcotic. Codeine, also produced by the opium poppy, is a less powerful analgesic than morphine, but it is one of the most effective cough suppressants known. Heroin is produced in the laboratory by adding two acetyl groups to morphine. It was initially made in the hopes of producing a compound with the benefits of morphine but lacking the addictive qualities. However, heroin is even more addictive than morphine.

Strychnine is found in the seeds of an Asiatic tree. It is extremely toxic and was commonly used as a rat poison at one time. Quinine, isolated from the bark of South American trees, was the first effective treatment for malaria. Vitamin B_6 is one of the water-soluble vitamins required by the body.

15.3 Amides

Amides are the products formed in a reaction between a carboxylic acid derivative and ammonia or an amine. The general structure of an amide is shown here.

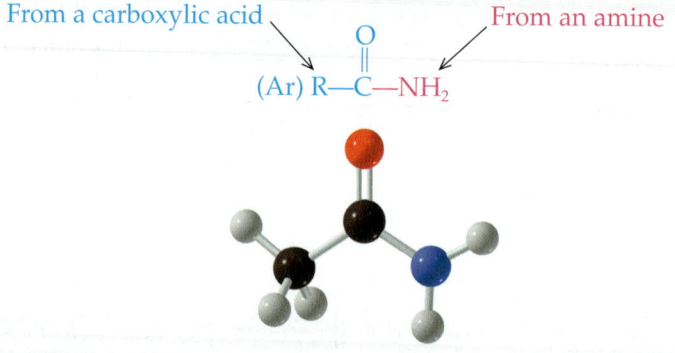

Ethanamide

The amide group is composed of two portions: the carbonyl group from a carboxylic acid and the amino group from ammonia or an amine. The bond between the carbonyl carbon and the nitrogen of the amine or ammonia is called the **amide bond.**

Structure and Physical Properties

LEARNING GOAL

Most amides are solids at room temperature. They have very high boiling points, and the simpler ones are quite soluble in water. Both of these properties are a result of strong intermolecular hydrogen bonding between the N—H bond of one amide and the C=O group of a second amide, as shown in Figure 15.5.

Unlike amines, amides are not bases (proton acceptors). The reason is that the highly electronegative oxygen atom of the carbonyl group causes a very strong attraction between the lone pair of nitrogen electrons and the carbonyl group. As a result, the unshared pair of electrons cannot "hold" a proton.

15.3 Amides

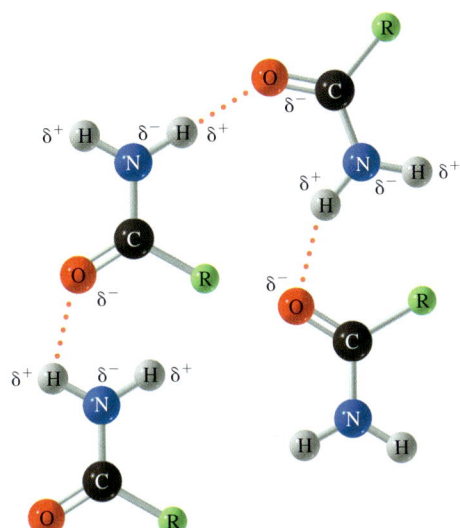

Figure 15.5
Hydrogen bonding in amides.

Because of the attraction of the carbonyl group for the lone pair of nitrogen electrons, the structure of the C—N bond of an amide is a *resonance hybrid*.

Resonance hybrids are discussed in Section 3.4.

Nomenclature

The common and I.U.P.A.C. names of the amides are derived from the common and I.U.P.A.C. names of the carboxylic acids from which they were made. Remove the *-ic acid* ending of the common name or the *-oic acid* ending of the I.U.P.A.C. name of the carboxylic acid, and replace it with the ending *-amide*. Several examples of the common and I.U.P.A.C. nomenclature are provided in Table 15.4 and in the following structures:

LEARNING GOAL 9

Nomenclature of carboxylic acids is described in Section 14.1.

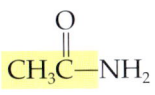

Ethanoic acid → Ethanamide
or
Acetic acid → Acetamide

Propanoic acid → Propanamide
or
Propionic acid → Propionamide

Substituents on the nitrogen are placed as prefixes and are indicated by *N-* followed by the name of the substituent. There are no spaces between the prefix and the amide name. For example:

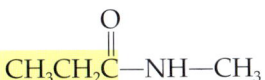

N-Methylpropanamide

N-Propylhexanamide

Medically Important Amides

Barbiturates, often called "downers," are derived from amides and are used as sedatives. They are also used as anticonvulsants for epileptics and for people suffering from a variety of brain disorders that manifest themselves in neurosis, anxiety, and tension.

TABLE 15.4 I.U.P.A.C. and Common Names of Simple Amides

Compound	I.U.P.A.C. Name	Common Name
R—C(=O)—NH$_2$	Alkan*amide* (-*amide* replaces the -*oic acid* ending of the I.U.P.A.C. name of carboxylic acid)	Alkan*amide* (-*amide* replaces the -*ic acid* ending of the common name of carboxylic acid)
H—C(=O)—NH$_2$	Methanamide	Formamide
CH$_3$—C(=O)—NH$_2$	Ethanamide	Acetamide
CH$_3$CH$_2$—C(=O)—NH$_2$	Propanamide	Propionamide
H—C(=O)—NHCH$_3$	N-Methylmethanamide	N-Methylformamide
CH$_3$—C(=O)—NHCH$_3$	N-Methylethanamide	N-Methylacetamide

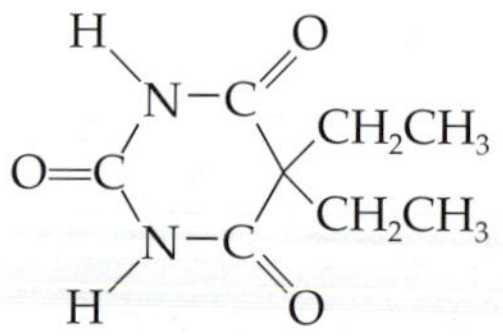

Barbital—a barbiturate

Phenacetin and acetaminophen are also amides. Acetaminophen is an aromatic amide that is commonly used in place of aspirin, particularly by people who are allergic to aspirin or who suffer stomach bleeding from the use of aspirin. It was first synthesized in 1893 and is the active ingredient in Tylenol and Datril. Like aspirin, acetaminophen relieves pain and reduces fever. However, unlike aspirin, it is not an anti-inflammatory drug.

Phenacetin was synthesized in 1887 and used as an analgesic for almost a century. Its structure and properties are similar to those of acetaminophen. However, it was banned by the U.S. Food and Drug Administration in 1983 because of the kidney damage and blood disorders that it causes.

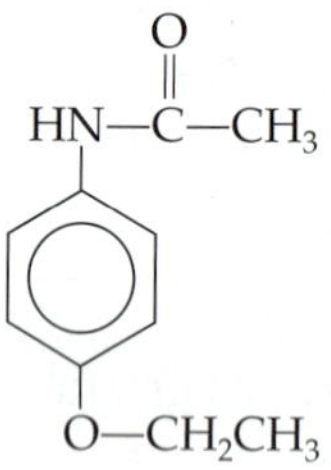

Phenacetin 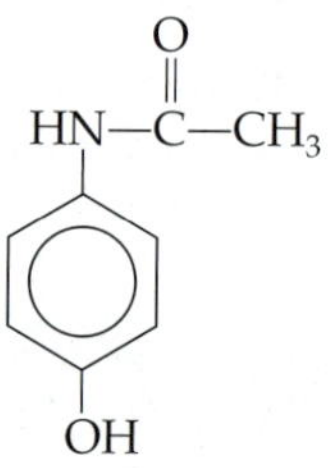   Acetaminophen

Reactions Involving Amides

Preparation of Amides

LEARNING GOAL

Amides are prepared from carboxylic acid derivatives, either acid chlorides or acid anhydrides. Recall that acid chlorides are made from carboxylic acids by reaction with reagents such as PCl$_5$.

A Medical Perspective

Semisynthetic Penicillins

The antibacterial properties of penicillin were discovered by Alexander Fleming in 1929. These natural penicillins produced by several species of the mold *Penicillium*, had a number of drawbacks. They were effective only against a type of bacteria referred to as Gram positive because of a staining reaction based on their cell wall structure. They were also very susceptible to destruction by bacterial enzymes called β-lactamases, and some were destroyed by stomach acid and had to be administered by injection.

To overcome these problems, chemists have produced semisynthetic penicillins by modifying the core structure. The core of penicillins is 6-aminopenicillanic acid, which consists of a thiazolidine ring fused to a β-lactam ring. In addition, there is an R group bonded via an amide bond to the core structure.

6-Aminopenicillanic acid

The β-lactam ring confers the antimicrobial properties. However, the R group determines the degree of antibacterial activity, the pharmacological properties, including the types of bacteria against which it is active, and the degree of resistance to the β-lactamases exhibited by any particular penicillin antibiotic. These are the properties that must be modified to produce penicillins that are acid resistant, effective with a broad spectrum of bacteria, and β-lactamase resistant.

Chemists simply remove the natural R group by cleaving the amide bond with an enzyme called an amidase. They then replace the R group and test the properties of the "new" antibiotic. Among the resulting semisynthetic penicillins are ampicillin, methicillin, and oxacillin.

Ampicillin

Methicillin

Oxacillin

For Further Understanding

Using the Internet and other resources, investigate and describe the properties of some new penicillins and the bacteria against which they are effective.

Why does changing the R group of a penicillin result in altered chemical and physiological properties?

$$\underset{\text{Carboxylic acid}}{\text{R}-\overset{\overset{\text{O}}{\|}}{\text{C}}-\text{OH}} \xrightarrow{\text{PCl}_5} \underset{\text{Acid chloride}}{\text{R}-\overset{\overset{\text{O}}{\|}}{\text{C}}-\text{Cl}} + \text{inorganic products}$$

Formation of acid chlorides is described in Section 14.3.

These acid chlorides rapidly react with either ammonia or amines, as in:

$$\underset{\substack{\text{Acid} \\ \text{chloride}}}{\text{R}-\overset{\overset{\text{O}}{\|}}{\text{C}}-\text{Cl}} + \underset{\substack{\text{Ammonia} \\ \text{or} \\ \text{amine}}}{2\text{NH}_3} \longrightarrow \underset{\text{Amide}}{\text{R}-\overset{\overset{\text{O}}{\|}}{\text{C}}-\text{NH}_2} + \underset{\substack{\text{Ammonium chloride} \\ \text{or} \\ \text{alkylammonium chloride}}}{\text{NH}_4^+\text{Cl}^-}$$

Chapter 15 Amines and Amides

Note that two molar equivalents of ammonia or amine are required in this reaction and that this is an acyl group transfer reaction. The **acyl group**

$$R-\overset{\overset{O}{\|}}{C}-$$

of the acid chloride is transferred from the Cl atom to the N atom of one of the ammonia or amine molecules. The second ammonia (or amine) reacts with the HCl formed in the transfer reaction to produce ammonium chloride or alkylammonium chloride.

The reaction between butanoyl chloride and methanamine to produce *N*-methylbutanamide is an example of an acyl group transfer reaction.

$$CH_3CH_2CH_2\overset{\overset{O}{\|}}{C}-Cl + 2CH_3NH_2 \longrightarrow$$

Butanoyl   Methanamine
chloride

$$CH_3CH_2CH_2\overset{\overset{O}{\|}}{C}-NH-CH_3 + CH_3NH_3{}^+Cl^-$$

N-Methylbutanamide   Methylammonium chloride

The reaction between an amine and an acid anhydride is also an acyl group transfer. The general equation for the synthesis of an amide in the reaction between an acid anhydride and ammonia or an amine is

$$R-\overset{\overset{O}{\|}}{C}-O-\overset{\overset{O}{\|}}{C}-R + 2NH_3 \longrightarrow R-\overset{\overset{O}{\|}}{C}-NH_2 + R-\overset{\overset{O}{\|}}{C}-O^-NH_4{}^+$$

Acid anhydride   Ammonia   Amide   Carboxylic acid
                 or   salt
                 amine

When subjected to heat, the ammonium salt loses a water molecule to produce a second amide molecule.

A well-known commercial amide is the artificial sweetener aspartame or NutraSweet. Although the name suggests that it is a sugar, it is not a sugar at all. In fact, it is the methylester of a molecule composed of two amino acids, aspartic acid and phenylalanine, joined by an amide bond (Figure 15.6a).

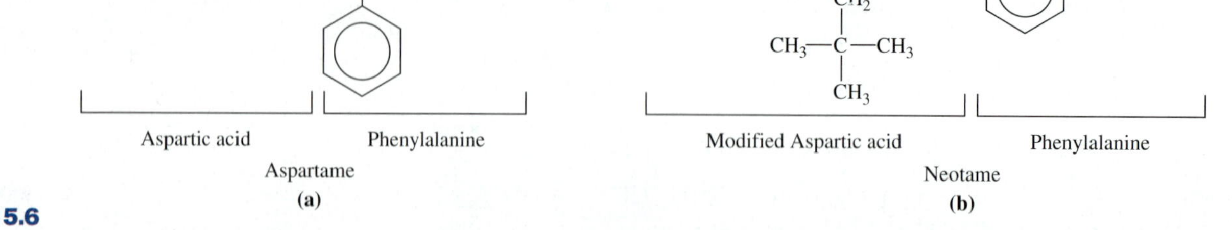

Amino acids have both a carboxyl group and an amino group and are discussed in detail in Sections 15.4 and 18.1.

Figure 15.6
The amide bond. (a) NutraSweet, the dipeptide aspartame, is a molecule composed of two amino acids joined by an amide (peptide) bond. (b) Neotame, a newly approved sweetener, is also a dipeptide. One of the amino acids has been modified so that it is safe for use by phenylketonurics.

Packages of aspartame carry the warning: "Phenylketonurics: Contains Phenylalanine." Digestion of aspartame and heating to high temperatures during cooking break both the ester bond and the amide bond, which releases the amino acid phenylalanine. People with the genetic disorder phenylketonuria (PKU) cannot metabolize this amino acid. As a result, it builds up to toxic levels that can cause mental retardation in an infant born with the condition. This no longer occurs because every child is tested for PKU at the time of birth and each is treated with a diet that limits the amount of phenylalanine to only the amount required for normal growth.

In July 2002 the Food and Drug Administration approved a new artificial sweetener that is related to aspartame. Called neotame, it has the same core structure as aspartame, but a 3,3-dimethylbutyl group has been added to the aspartic acid (Figure 15.6b). Digestion and heating still cause breakage of the ester bond, but the bulky 3,3-dimethylbutyl group blocks the breakage of the amide bond. Neotame can be used without risk by people with PKU and also retains its sweetness during cooking.

Question 15.13

What is the structure of the amine that, on reaction with the acid chlorides shown, will give each of the following products?

a. ? + $CH_3\overset{O}{\overset{\|}{C}}-Cl \longrightarrow CH_3\overset{O}{\overset{\|}{C}}NHCH_3 + CH_3NH_3{}^+Cl^-$

b. ? + $CH_3CH_2CH_2CH_2\underset{CH_2CH_3}{\overset{\overset{O}{\|}}{C}H\overset{}{C}}-Cl \longrightarrow (CH_3)_2N\overset{O}{\overset{\|}{C}}CH\underset{CH_2CH_3}{}CH_2CH_2CH_3 + (CH_3)_2NH_2{}^+Cl^-$

Question 15.14

What are the structures of the acid chlorides and the amines that will react to give each of the following products?

a. *N*-Ethylhexanamide
b. *N*-Propylbutanamide

Hydrolysis of Amides

11 LEARNING GOAL

Hydrolysis of an amide results in breaking the amide bond to produce a carboxylic acid and ammonia or an amine. It is very difficult to hydrolyze the amide bond. In fact, the reaction requires heating the amide in the presence of a strong acid or base.

$$R-\overset{O}{\overset{\|}{C}}-NH-R^1 + H_3O^+ \longrightarrow R-\overset{O}{\overset{\|}{C}}-OH + R^1-N^+H_3$$

Amide — Strong acid — Carboxylic acid — Alkylammonium ion or ammonium ion

$$CH_3CH_2CH_2\overset{O}{\overset{\|}{C}}-NH_2 + H_3O^+ \longrightarrow CH_3CH_2CH_2\overset{O}{\overset{\|}{C}}-OH + N^+H_4$$

Butanamide (butyramide) — Butanoic acid (butyric acid)

If a strong base is used, the products are the amine and the salt of the carboxylic acid:

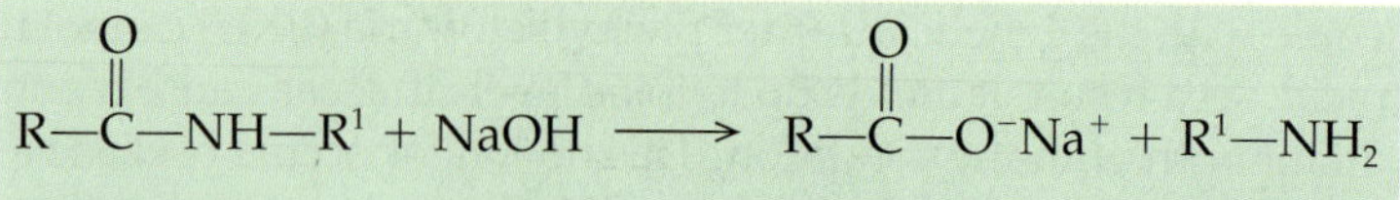

$$\underset{\text{Amide}}{R-\overset{O}{\underset{\|}{C}}-NH-R^1} + \underset{\text{Strong base}}{NaOH} \longrightarrow \underset{\text{Carboxylic acid salt}}{R-\overset{O}{\underset{\|}{C}}-O^-Na^+} + \underset{\substack{\text{Amine} \\ \text{or} \\ \text{ammonia}}}{R^1-NH_2}$$

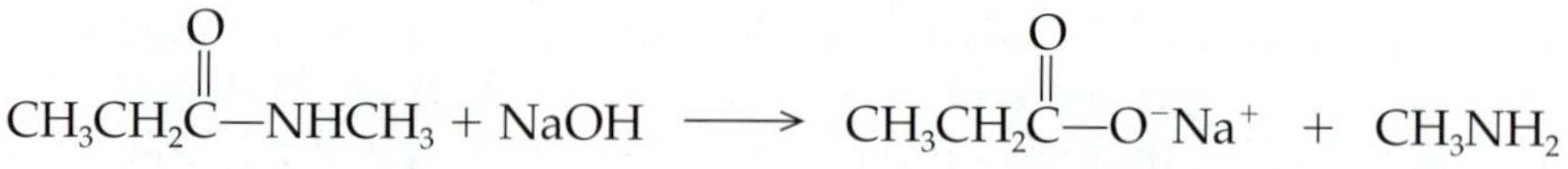

$$\underset{\substack{N\text{-Methylpropanamide} \\ (N\text{-methylpropionamide})}}{CH_3CH_2\overset{O}{\underset{\|}{C}}-NHCH_3} + NaOH \longrightarrow \underset{\substack{\text{Sodium propanoate} \\ \text{(sodium propionate)}}}{CH_3CH_2\overset{O}{\underset{\|}{C}}-O^-Na^+} + \underset{\substack{\text{Methanamine} \\ \text{(methylamine)}}}{CH_3NH_2}$$

15.4 A Preview of Amino Acids, Proteins, and Protein Synthesis

LEARNING GOAL 12

LEARNING GOAL 13

In Chapter 18 we will describe the structure of proteins, the molecules that carry out the majority of the biological processes essential to life. Proteins are polymers of amino acids. As the name suggests, amino acids have two essential functional groups, an amino group (—NH_2) and a carboxyl group (—COOH). Typically amino acids have the following general structure:

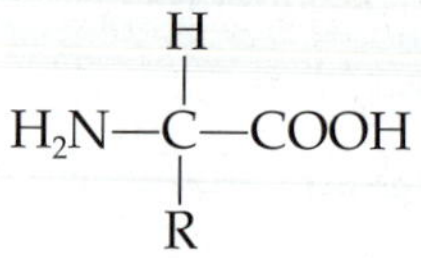

$$H_2N-\underset{R}{\overset{H}{\underset{|}{\overset{|}{C}}}}-COOH \quad \text{(R may be a hydrogen atom or an organic group.)}$$

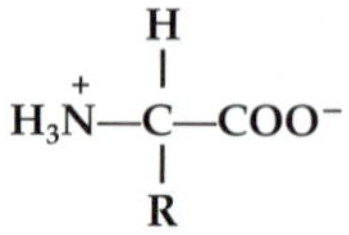

The amide bond that forms between the carboxyl group of one amino acid and the amino group of another is called the **peptide bond.**

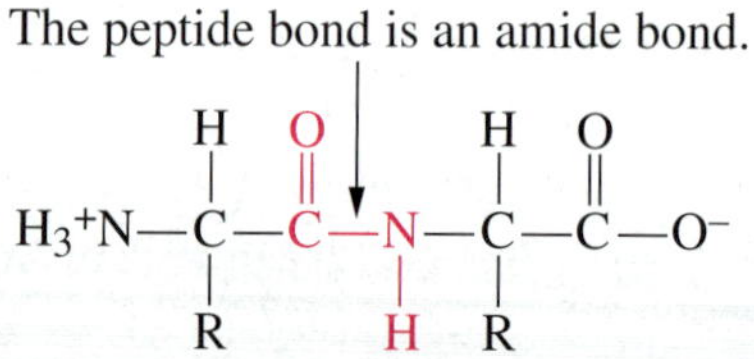

The peptide bond is an amide bond.

$$H_3{}^+N-\underset{R}{\overset{H}{\underset{|}{\overset{|}{C}}}}-\overset{O}{\underset{\|}{C}}-\underset{H}{\overset{}{\underset{|}{N}}}-\underset{R}{\overset{H}{\underset{|}{\overset{|}{C}}}}-\overset{O}{\underset{\|}{C}}-O^-$$

In the cell the amino group is usually protonated and the carboxyl group is usually ionized to the carboxylate anion. In the future we will represent an amino acid in the following way:

$$H_3{}^+N-\underset{R}{\overset{H}{\underset{|}{\overset{|}{C}}}}-COO^-$$

The joining of amino acids by amide bonds produces small *peptides* and larger *proteins*. Because protein structure and function are essential for life processes, it is fortunate indeed that the amide bonds that hold them together are not easily hydrolyzed at physiological pH and temperature.

The process of protein synthesis in the cell mimics amide formation in the laboratory; it involves acyl group transfer. There are several important differences between the chemistry in the laboratory and the chemistry in the cell. During protein synthesis, the **aminoacyl group** of the amino acid is transferred, rather than the acyl group of a carboxylic acid. In addition, the aminoacyl group is not transferred from a carboxylic acid derivative; it is transferred from a special carrier molecule called a **transfer RNA (tRNA).** When the aminoacyl group is covalently bonded to a tRNA, the resulting structure is called an *aminoacyl tRNA*:

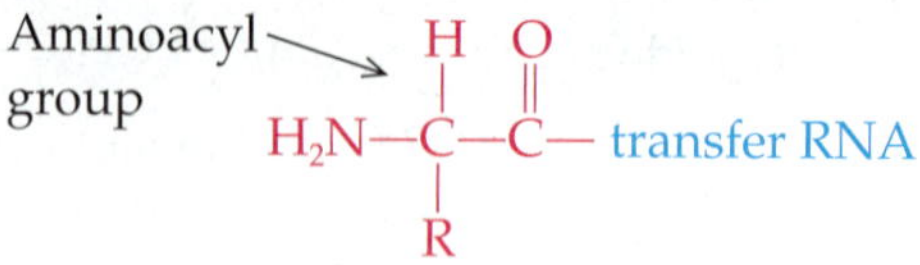

$$\underset{\text{group}}{\text{Aminoacyl}} \searrow \quad H_2N-\underset{R}{\overset{H}{\underset{|}{\overset{|}{C}}}}-\overset{O}{\underset{\|}{C}}-\text{transfer RNA}$$

The aminoacyl group of the aminoacyl tRNA is transferred to the amino group nitrogen to form a peptide bond. The transfer RNA is recycled by binding to another of the same kind of aminoacyl group.

More than one hundred kinds of proteins, nucleotides, and RNA molecules participate in the incredibly intricate process of protein synthesis. In Chapter 18 we will study protein structure and learn about the many functions of proteins in the life of the cell. In Chapter 20 we will study the details of protein synthesis to see how these aminoacyl transfer reactions make us the individuals that we are.

15.5 Neurotransmitters

Neurotransmitters are chemicals that carry messages, or signals, from a nerve cell to a target cell, which may be another nerve cell or a muscle cell. Neurotransmitters are classified as being *excitatory*, stimulating their target cell, or *inhibitory*, decreasing activity of the target cell. One feature shared by the neurotransmitters is that they are all nitrogen-containing compounds. Some of them have rather complex structures and one, nitric oxide (NO), consists of only two atoms. Several important neurotransmitters are discussed in the following sections.

Catecholamines

All of the catecholamine neurotransmitters, including dopamine, epinephrine, and norepinephrine, are synthesized from the amino acid tyrosine (Figure 15.7). *Dopamine* is critical to good health. A deficiency in this neurotransmitter, for example, results in Parkinson's disease, a disorder characterized by tremors, monotonous speech, loss of memory and problem-solving ability, and loss of motor function. In the brain, dopamine is synthesized from L-dopa; so it would seem logical to treat Parkinson's disease with dopamine. Unfortunately, dopamine cannot cross the blood-brain barrier to enter brain cells. As a result, L-dopa, which is converted to dopamine in brain cells, is used to treat this disorder.

Just as too little dopamine causes Parkinson's disease, an excess is associated with schizophrenia. Dopamine also appears to play a role in addictive behavior. In proper amounts, it causes a pleasant, satisfied feeling. The greater the amount of dopamine, the more intense the sensation, the "high." Several drugs have been shown to increase the levels of dopamine. Among these are cocaine, heroin, amphetamines, alcohol, and nicotine. Marijuana also causes an increase in brain dopamine, raising the possibility that it, too, has the potential to produce addiction.

Both *epinephrine* (adrenaline) and *norepinephrine* are involved in the "fight or flight" response. Epinephrine stimulates the breakdown of glycogen to produce glucose, which is then metabolized to provide energy for the body. Norepinephrine is involved with the central nervous system in the stimulation of other glands and the constriction of blood vessels. All of these responses prepare the body to meet the stressful situation.

Serotonin

Serotonin is synthesized from the amino acid tryptophan (Figure 15.8). A deficiency of serotonin has been associated with depression. It is also thought to be involved in bulimia and anorexia nervosa, as well as the carbohydrate-cravings that characterize seasonal affective disorder (SAD), a depression caused by a decrease in daylight during autumn and winter.

Serotonin also affects the perception of pain, thermoregulation, and sleep. There are those who believe that a glass of warm milk will help you fall asleep. We have all noticed how sleepy we become after that big Thanksgiving turkey dinner.

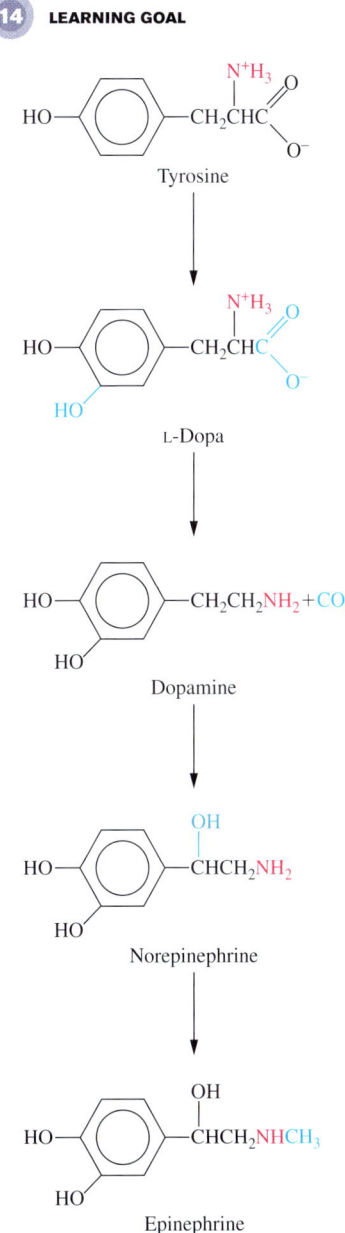

Figure 15.7
The pathway for synthesis of dopamine, epinephrine, and norepinephrine.

A Medical Perspective

Opiate Biosynthesis and the Mutant Poppy

Hippocrates, the Father of Medicine, left us the first record of the therapeutic use of opium (460 B.C.). Although not recorded, it is probably true that the addictive properties of opium were recognized soon thereafter!

The opium poppy *(Papaver somniferum)* is cultivated, legally and illegally, in many parts of the world. The flowers vary in color from white to deep red, but it is the seed pod that is sought after. In the seed pod is a milky fluid that contains morphine and codeine, and a small amount of an opioid called thebaine. The juice is extracted from the unripe seed pods and dried, and the opium alkaloids are extracted and purified.

In the legal pharmaceutical world, morphine and codeine are used to ease pain and spasmodic coughing. Thebaine is used as a reactant in pharmaceutical synthesis to produce a number of synthetic opioid compounds with a variety of biological effects. These include the analgesics oxycodone (brand name OxyContin), oxymorphone, and nalbuphine; naloxone, which is used to treat opioid overdosage; naltrexone, which is useful in helping people with narcotic or alcohol addictions to remain drug free; and buprenorphine, which is useful in the treatment of opiate addiction because it prevents withdrawal symptoms.

Approximately 40% of the world's legal opium poppies are grown on the Australian island state of Tasmania. This is big business, and the industry has developed an active research program to study the biochemical pathway for the synthesis of morphine and codeine. That pathway begins with the amino acid tyrosine, the same amino acid that is the initial reactant for the synthesis of dopamine and epinephrine (Section 15.5).

    7 steps   6 steps   3 steps   1 step

Tyrosine → Reticuline → Thebaine → Codeine → Morphine

Through seven chemical reactions, tyrosine is converted to reticuline. Another six reactions convert reticuline to thebaine. Three chemical modifications convert thebaine to codeine, which undergoes an ester hydrolysis to produce morphine.

In the course of their studies, researchers produced a mutant strain of poppy that cannot make morphine or codeine, but does produce high levels of thebaine. "Norman," for "No Morphine," has been the most common strain of poppy grown in Tasmania since 1997. The mutation that causes Norman to produce high levels of thebaine is an alteration in one of the enzymes that catalyzes the conversion of thebaine to codeine. Since it can't be converted into codeine, large amounts of thebaine accumulate in the seed pods.

Synthetic opioids, such as naloxone and buprenorphine and the others mentioned above, have become much more important commercially than codeine and morphine. The economic value of Norman is that it produces large amounts of the starting material for the synthesis of these synthetic opioids, as well as the experimental synthesis of new drugs with unknown potential.

(a)

(b)

(a) Opium poppies are the source of morphine and codeine. (b) The sap of an opium poppy is white. The sap of the no-morphine mutant poppy is red.

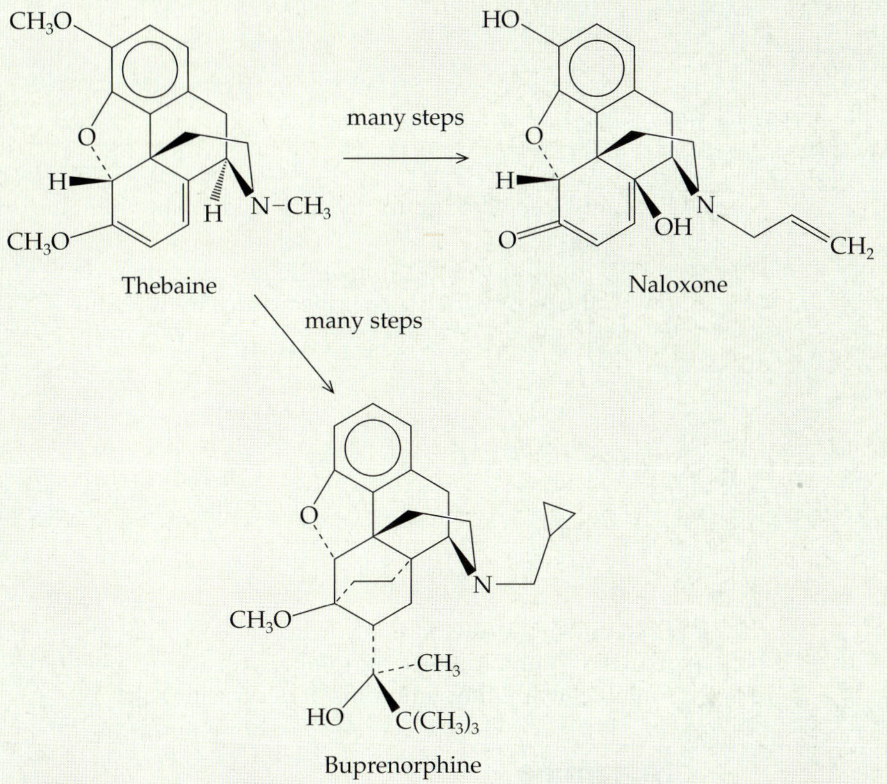

15.5 Neurotransmitters

[Structures of Thebaine, Naloxone, and Buprenorphine shown, with "many steps" arrows from Thebaine to both Naloxone and Buprenorphine.]

Consider the drug formulation Suboxone, a combination of buprenorphine and naloxone, approved for use in the United States in 2003 and produced by the British company Reckitt Benckiser. This combination of synthetic opiates calms the addict's craving for opiates and yet poses little risk of being abused.

Buprenorphine or "bupe" works by binding to the same receptors in the brain to which heroin binds; but the drug is only a partial heroin agonist, so there is no high. As a result, buprenorphine is much less addictive than drugs such a methadone, leaves patients much more clearheaded, and makes it easier for them to withdraw from the drug after a few months.

Because addicts don't get high from Suboxone, it is much less likely than methadone to be stolen and sold illegally. It has the added advantage that its effects are longer lasting than those of methadone; so addicts need only one pill every two or three days. In addition, because naloxone is an opioid antagonist, it causes instant withdrawal symptoms if an addict tries to inject Suboxone for a high. As a result, Suboxone can be given to addicts to be taken at home, rather than being dispensed only at clinics, as methadone is. With all of these features, Suboxone begins to sound like a miracle drug; but as with any addiction treatment, it will only help those who want to quit and are willing to work with counselors and support groups to resolve the underlying problems that caused the addiction in the first place.

For Further Understanding

In 1998 researchers in England reported that some individuals who had eaten poppy seed rolls or cake tested positive in an opiate drug screen. Using Internet or other resources, investigate the "poppy seed defense" and suggest guidelines for opiate testing that would protect the innocent.

OxyContin is the brand name for a formulation of oxycodone in a timed-release tablet. It is prescribed to provide up to twelve hours of relief from chronic pain. Recently, OxyContin has become a commonly abused drug and is thought to be responsible for a number of deaths. Abusers crush the time-release tablets and ingest or snort the drug, which results in a rapid and powerful high often compared to the euphoria experienced from taking heroin. Use the Internet or other resources to explain why abuse of this prescription medication has overshadowed heroin use in some areas. Consider ways to prevent such abuse.

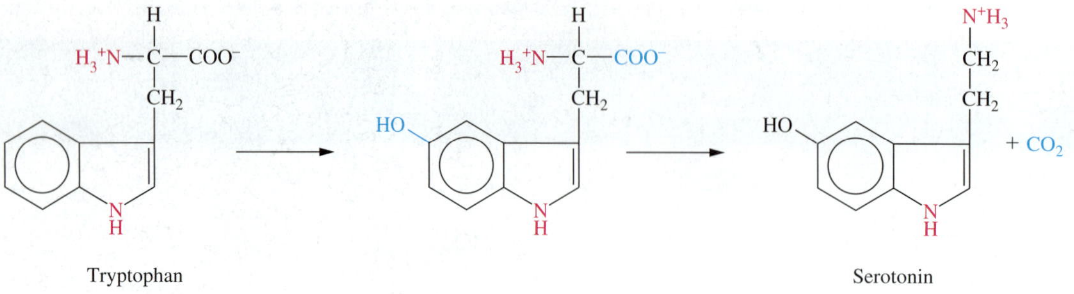

Figure 15.8
Synthesis of serotonin from the amino acid tryptophan.

Tryptophan → (intermediate) → Serotonin + CO_2

Both milk protein and turkey are exceptionally high in tryptophan, the precursor of serotonin!

Prozac (fluoxetine), one of the newest generation of antidepressant drugs, is one of the most widely prescribed drugs in the United States.

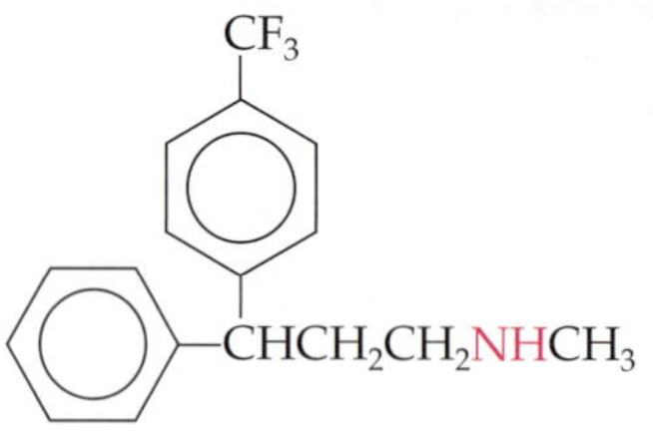

Prozac (fluoxetine)

It is a member of a class of drugs called selective serotonin reuptake inhibitors (SSRI). By inhibiting the reuptake, Prozac effectively increases the level of serotonin, relieving the symptoms of depression.

Histamine

Histamine is a neurotransmitter that is synthesized in many tissues by removing the carboxyl group from the amino acid histidine (Figure 15.9). It has many, often annoying, physiological roles. Histamine is released during the allergic response. It causes the itchy skin rash associated with poison ivy or insect bites. It also promotes the red, watery eyes and respiratory symptoms of hay fever.

Many antihistamines are available to counteract the symptoms of histamine release. These act by competing with histamine for binding to target cells. If histamine cannot bind to these target cells, the allergic response stops.

Benadryl is an antihistamine that is available as an ointment to inhibit the itchy rash response to allergens. It is also available as an oral medication to block the symptoms of systemic allergies. You need only visit the "colds and allergies" aisle of your grocery store to find dozens of medications containing antihistamines.

Histamines also stimulate secretion of stomach acid. When this response occurs frequently, the result can be chronic heartburn. The reflux of stomach acid into

Figure 15.9
Synthesis of histamine from the amino acid histidine.

Histidine → Histamine + CO_2

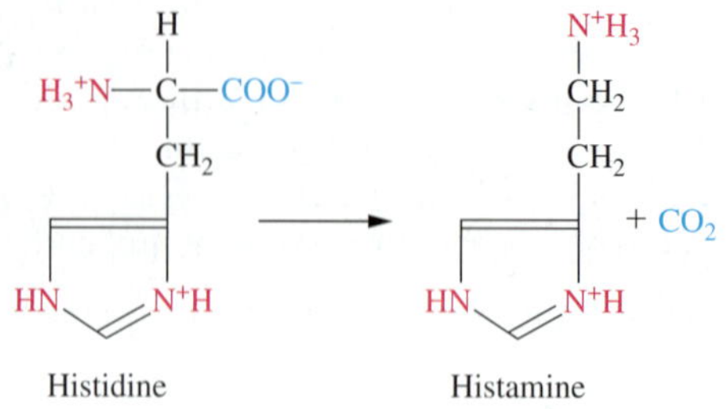

the esophagus can result in erosion of tissue and ulceration. The excess stomach acid may also contribute to development of stomach ulcers. The drug marketed as Tagamet (cimetidine) has proven to be an effective inhibitor of this histamine response, providing relief from chronic heartburn.

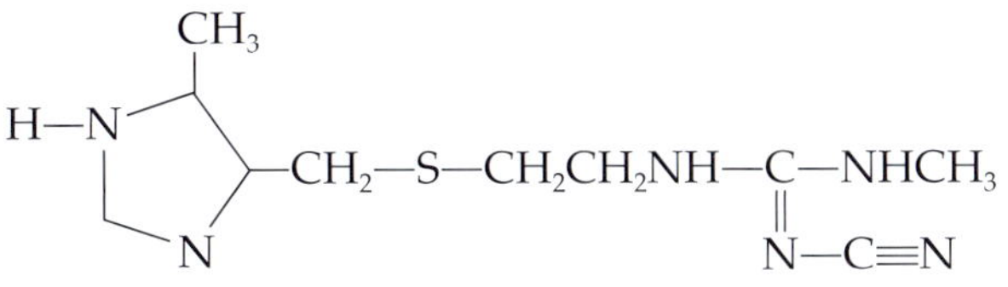

Tagamet (cimetidine)

γ-Aminobutyric Acid and Glycine

γ-Aminobutyric acid (GABA) is produced by removal of a carboxyl group from the amino acid glutamate (Figure 15.10). Both GABA and the amino acid *glycine*

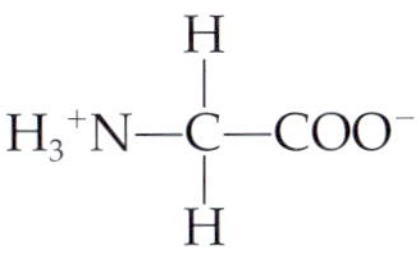

Glycine

are inhibitory neurotransmitters acting in the central nervous system. One class of tranquilizers, the benzodiazopines, relieves aggressive behavior and anxiety. These drugs have been shown to enhance the inhibitory activity of GABA, suggesting one of the roles played by this neurotransmitter.

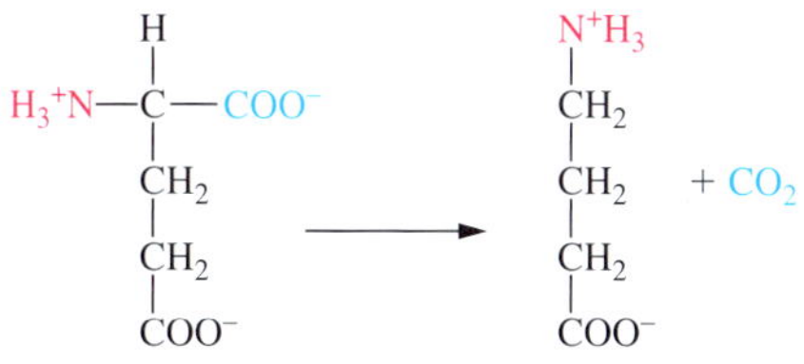

Figure 15.10
Synthesis of GABA from the amino acid glutamate.

Acetylcholine

Acetylcholine is a neurotransmitter that functions at the neuromuscular junction, carrying signals from the nerve to the muscle. It is synthesized in a reaction between the quaternary ammonium ion choline and acetyl coenzyme A (Figure 15.11). When it is released from the nerve cell, acetylcholine binds to receptors on the surface of muscle cells. This binding stimulates the muscle cell to contract. Acetylcholine is then broken down to choline and acetate ion.

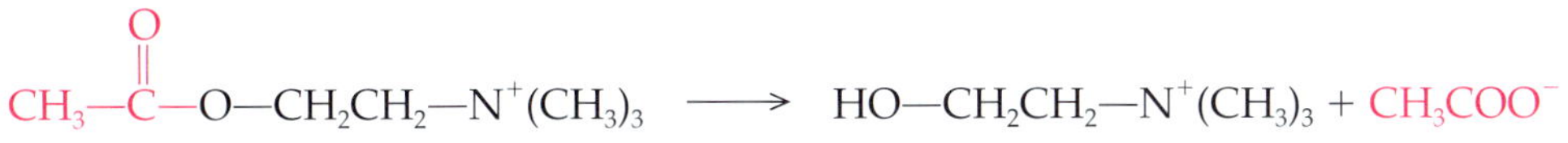

$$CH_3-\overset{O}{\overset{\|}{C}}-O-CH_2CH_2-N^+(CH_3)_3 \longrightarrow HO-CH_2CH_2-N^+(CH_3)_3 + CH_3COO^-$$

Acetylcholine   Choline   Acetate

These molecules are essentially recycled. They are taken up by the nerve cell where they are used to resynthesize acetylcholine, which is stored in the nerve cell until it is needed.

Nicotine is an agonist of acetylcholine. An agonist is a compound that binds to the receptor for another compound and causes or enhances the biological response.

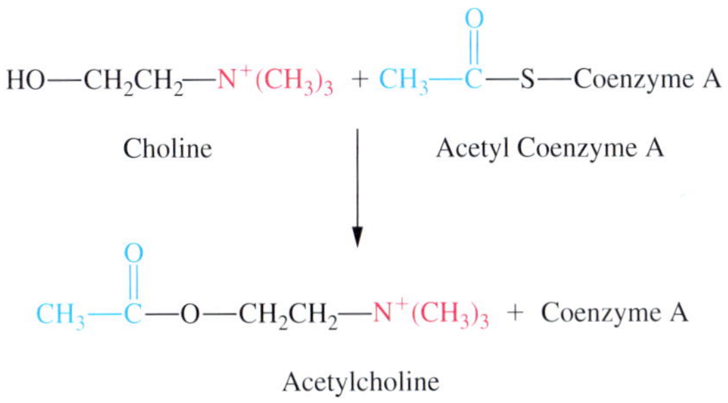

Figure 15.11
Synthesis of acetylcholine.

By binding to acetylcholine receptors, nicotine causes the sense of alertness and calm many smokers experience. Nerve cells that respond to nicotine may also signal nerve cells that produce dopamine. As noted above, the dopamine may be responsible for the addictive property of nicotine.

Inhibitors of acetylcholinesterase, the enzyme that catalyzes the breakdown of acetylcholine, are used both as poisons and as drugs. Among the most important poisons of acetylcholinesterase are a class of compounds known as organophosphates. One of these is *diisopropyl fluorophosphate* (DIFP). This molecule forms a covalently bonded intermediate with the enzyme, irreversibly inhibiting its activity.

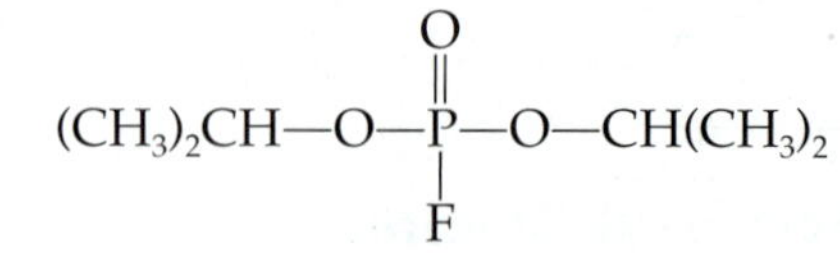

$$(CH_3)_2CH-O-\underset{F}{\overset{\overset{O}{\|}}{P}}-O-CH(CH_3)_2$$

Diisopropyl fluorophosphate (DIFP)

Thus, it is unable to break down the acetylcholine and nerve transmission continues, resulting in muscle spasm. Death may occur as a result of laryngeal spasm. Antidotes for poisoning by organophosphates, which include many insecticides and nerve gases, have been developed. The antidotes work by reversing the effects of the inhibitor. One of these antidotes is *pyridine aldoxime methiodide* (PAM). This molecule displaces the organophosphate group from the active site of the enzyme, alleviating the effects of the poison.

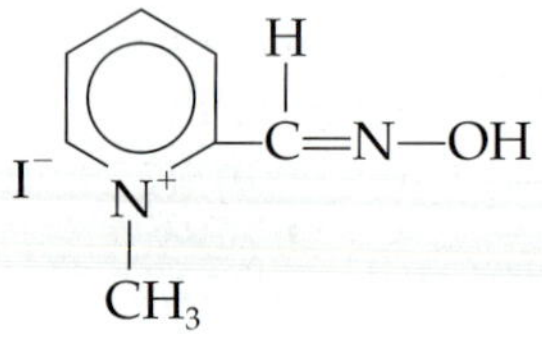

Pyridine aldoxime methiodide

Succinylcholine is a competitive inhibitor of acetylcholine which is used as a muscle relaxant in surgical procedures. Competitive inhibition occurs because the two molecules have structures so similar that both can bind to the acetylcholine receptor (compare the structures below). When administered to a patient, there is more succinylcholine than acetylcholine in the synapse, and thus more succinylcholine binding to the receptor. Because it cannot stimulate muscle contraction, succinylcholine causes muscles to relax. Normal muscle contraction resumes when the drug is no longer administered.

$$CH_3\overset{\overset{O}{\|}}{C}-O-CH_2CH_2-N^+(CH_3)_3 \qquad (CH_3)_3{}^+N-CH_2CH_2-O-\overset{\overset{O}{\|}}{C}CH_2CH_2\overset{\overset{O}{\|}}{C}-O-CH_2CH_2-N^+(CH_3)_3$$

Acetylcholine 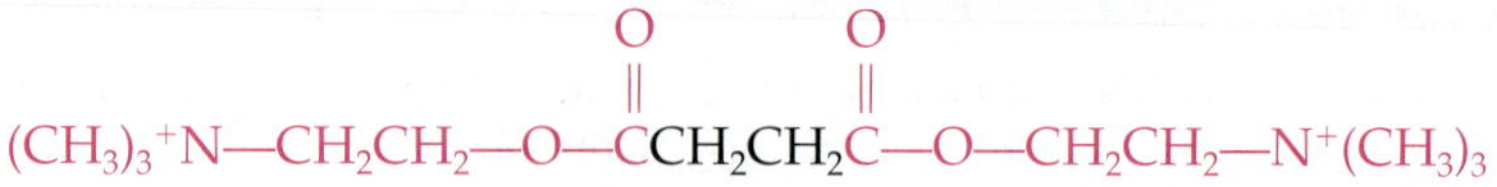 · Succinylcholine

Acetylcholine nerve transmission is discussed in further detail in A Clinical Perspective: Enzymes, Nerve Transmission, and Nerve Agents in Chapter 19.

Nitric Oxide and Glutamate

Nitric oxide (NO) is an amazing little molecule that has been shown to have many physiological functions. Among these is its ability to act as a neurotransmitter. NO is synthesized in many areas of the brain from the amino acid arginine. Research has suggested that NO works in conjunction with another neurotransmitter, the amino acid glutamate (see the structure of glutamate in Figure 15.10). Glutamate released from one nerve cell binds to receptors on its target cell. This triggers the target cell to produce NO, which then diffuses back to the original nerve cell. The NO signals the cell to release more glutamate, thus stimulating this neural pathway even further. This is a kind of positive feedback loop. It is thought that this NO-glutamate mechanism is involved in learning and the formation of memories.

Summary of Reactions

Preparation of Amines

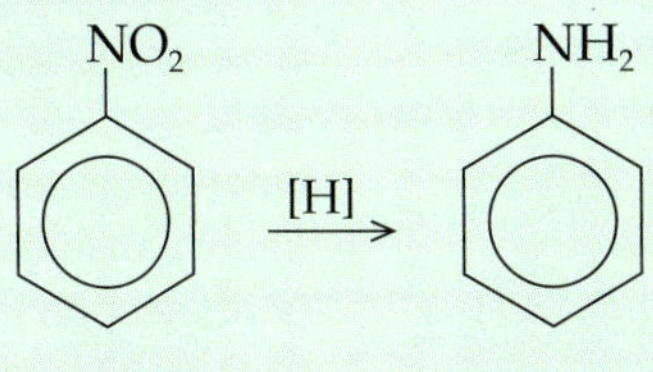

$$\text{Ar-NO}_2 \xrightarrow{[H]} \text{Ar-NH}_2$$

A nitro compound → An aromatic primary amine

$$R^1-\underset{\underset{R^3}{|}}{\overset{\overset{O}{\|}}{C}}-N{\overset{R^2}{\diagdown}} \xrightarrow{[H]} R^1CH_2N{\overset{R^2}{\diagdown}}_{R^3}$$

Amide → Amine

Basicity of Amines

$$R-NH_2 + H-OH \rightleftharpoons R-\underset{\underset{H}{|}}{\overset{\overset{H}{|}}{N^+}}-H + OH^-$$

Amine Water Alkylammonium ion Hydroxide ion

Neutralization of Amines

$$R-NH_2 + HCl \longrightarrow R-\underset{\underset{H}{|}}{\overset{\overset{H}{|}}{N^+}}-H\ Cl^-$$

Amine Acid Alkylammonium salt

Preparation of Amides

$$R-\overset{\overset{O}{\|}}{C}-Cl + 2NH_3 \longrightarrow$$

Acid chloride Ammonia or amine

$$R-\overset{\overset{O}{\|}}{C}-NH_2 + NH_4^+Cl^-$$

Amide Ammonium chloride or alkylammonium chloride

$$R-\overset{\overset{O}{\|}}{C}-O-\overset{\overset{O}{\|}}{C}-R + 2NH_3 \longrightarrow$$

Acid anhydride Ammonia or amine

$$R-\overset{\overset{O}{\|}}{C}-NH_2 + R-\overset{\overset{O}{\|}}{C}-O^-NH_4^+$$

Amide Carboxylic acid salt

Hydrolysis of Amides

$$R-\overset{\overset{O}{\|}}{C}-NH-R^1 + H_3O^+ \longrightarrow R-\overset{\overset{O}{\|}}{C}-OH + R^1-N^+H_3$$

Amide Strong acid Carboxylic acid Alkylammonium ion

$$R-\overset{\overset{O}{\|}}{C}-NH-R^1 + NaOH \longrightarrow R-\overset{\overset{O}{\|}}{C}-O^-Na^+ + R^1-NH_2$$

Amide Strong base Carboxylic acid salt Amine or ammonia

SUMMARY

15.1 Amines

Amines are a family of organic compounds that contain an amino group or substituted amino group. A *primary amine* has the general formula RNH_2; a *secondary amine* has the general formula R_2NH; and a *tertiary amine* has the general formula R_3N. In the systematic nomenclature system, amines are named as *alkanamines*. In the common system they are named as *alkylamines*. Amines behave as weak bases, forming *alkylammonium ions* in water and alkylammonium salts when they react with acids. *Quaternary ammonium salts* are ammonium salts that have four organic groups bonded to the nitrogen atom.

15.2 Heterocyclic Amines

Heterocyclic amines are cyclic compounds having at least one nitrogen atom in the ring structure. *Alkaloids* are natural plant products that contain at least one heterocyclic ring. Many alkaloids have powerful biological effects.

15.3 Amides

Amides are formed in a reaction between a carboxylic acid derivative and an amine (or ammonia). The *amide bond* is the bond between the carbonyl carbon of the *acyl group* and the nitrogen of the amine. In the I.U.P.A.C. Nomenclature System, they are named by replacing the *-oic acid* ending of the carboxylic acid with the *-amide* ending. In the common system of nomenclature the *-ic acid* ending of the carboxylic acid is replaced by the *-amide* ending. Hydrolysis of an amide produces a carboxylic acid and an amine (or ammonia).

15.4 A Preview of Amino Acids, Proteins, and Protein Synthesis

Proteins are polymers of amino acids joined to one another by *amide bonds* called *peptide bonds*. During protein synthesis, the *aminoacyl group* of one amino acid is transferred from a carrier molecule called a *transfer RNA* to the amino group nitrogen of another amino acid.

15.5 Neurotransmitters

Neurotransmitters are chemicals that carry messages, or signals, from a nerve cell to a target cell, which may be another nerve cell or a muscle cell. They may be inhibitory or excitatory and all are nitrogen-containing compounds. The catecholamines include dopamine, norepinephrine, and epinephrine. Too little dopamine results in Parkinson's disease. Too much is associated with schizophrenia. Dopamine is also associated with addictive behavior. A deficiency of serotonin is associated with depression and eating disorders. Serotonin is involved in pain perception, regulation of body temperature, and sleep. Histamine contributes to allergy symptoms. Antihistamines block histamines and provide relief from allergies. γ-Aminobutyric acid (GABA) and glycine are inhibitory neurotransmitters. It is believed that GABA is involved in control of aggressive behavior. Acetylcholine is a neurotransmitter that functions at the neuromuscular junction, carrying signals from the nerve to the muscle. Nitric oxide and glutamate function in a positive feedback loop that is thought to be involved in learning and the formation of memories.

KEY TERMS

acyl group (15.3)
alkaloid (15.2)
alkylammonium ion (15.1)
amide (15.3)
amide bond (15.3)
amine (15.1)
aminoacyl group (15.4)
analgesic (15.2)
anesthetic (15.2)
heterocyclic amine (15.2)
neurotransmitter (15.5)
peptide bond (15.4)
primary (1°) amine (15.1)
quaternary ammonium salt (15.1)
secondary (2°) amine (15.1)
tertiary (3°) amine (15.1)
transfer RNA (tRNA) (15.4)

QUESTIONS AND PROBLEMS

Amines

Foundations

15.15 Compare the boiling points of amines, alkanes, and alcohols of the same molecular weight. Explain these differences in boiling points.

15.16 Describe the water solubility of amines in relation to their carbon chain length.

15.17 Describe the systematic rules for naming amines.

15.18 How are the common names of amines derived?

15.19 Describe the physiological effects of amphetamines.

15.20 Define the terms *analgesic* and *anesthetic* and list some amines that have these activities.

Applications

15.21 For each pair of compounds predict which would have greater solubility in water. Explain your reasoning.
   a. Pentane or 1-butanamine
   b. Cyclohexane or 2-pentanamine

15.22 For each pair of compounds predict which would have the higher boiling point. Explain your reasoning.
   a. Ethanamine or ethanol
   b. Butane or 1-propanamine
   c. Methanamine or water
   d. Ethylmethylamine or butane

15.23 Explain why a tertiary amine such as triethylamine has a significantly lower boiling point than its primary amine isomer, 1-hexanamine.

15.24 Draw a diagram to illustrate your answer to Problem 15.23.

15.25 Use systematic nomenclature to name each of the following amines:
a. CH$_3$CH$_2$CHNH$_2$
         |
         CH$_3$
b. CH$_3$CH$_2$CH$_2$CHCH$_2$CH$_3$
                   |
                   NH$_2$
c. 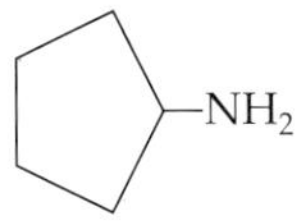 cyclopentyl—NH$_2$
d. (CH$_3$)$_3$C—NH$_2$

15.26 Use systematic and common nomenclature to name each of the following amines:
a. CH$_3$CH$_2$CH$_2$CH$_2$CH$_2$CH$_2$CH$_2$CH$_2$NH$_2$
b. 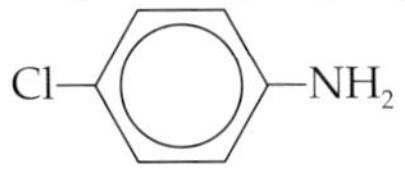 Cl—C$_6$H$_4$—NH$_2$
c. CH$_3$CHCH$_2$CH$_3$
      |
      NH$_2$
      CH$_3$
      |
d. CH$_3$NCH$_2$CH$_3$

15.27 Draw the structure of each of the following compounds:
a. Diethylamine
b. Butylamine
c. 3-Decanamine
d. 3-Bromo-2-pentanamine
e. Triphenylamine

15.28 Draw the structure of each of the following compounds:
a. N,N-Dipropylaniline
b. Cyclohexanamine
c. 2-Bromocyclopentanamine
d. Tetraethylammonium iodide
e. 3-Bromobenzenamine

15.29 Draw each of the following compounds with condensed formulas:
a. 2-Pentanamine
b. 2-Bromo-1-butanamine
c. Ethylisopropylamine
d. Cyclopentanamine

15.30 Draw each of the following compounds with condensed formulas:
a. Dipentylamine
b. 3,4-Dinitroaniline
c. 4-Methyl-3-heptanamine
d. t-Butylpentylamine
e. 3-Methyl-3-hexanamine
f. Trimethylammonium iodide

15.31 Draw structural formulas for the eight isomeric amines that have the molecular formula C$_4$H$_{11}$N. Name each of the isomers using the systematic names and determine whether each isomer is a 1°, 2°, or 3° amine.

15.32 Draw all of the isomeric amines of molecular formula C$_3$H$_9$N. Name each of the isomers, using the systematic names and determine whether each isomer is a primary, secondary, or tertiary amine.

15.33 Classify each of the following amines as 1°, 2°, or 3°:
a. Cyclohexanamine
b. Dibutylamine
c. 2-Methyl-2-heptanamine
d. Tripentylamine

15.34 Classify each of the following amines as primary, secondary, or tertiary:
a. Benzenamine
b. N-Ethyl-2-pentanamine
c. Ethylmethylamine
d. Tripropylamine
e. m-Chloroaniline

15.35 Write an equation to show a reaction that would produce each of the following products:

a.

b. 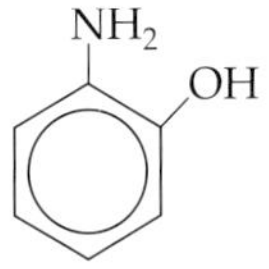 2-aminophenol (benzene ring with NH$_2$ and OH)

c. aniline (C$_6$H$_5$NH$_2$)

d. 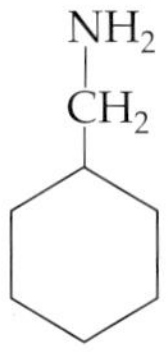 NH$_2$—CH$_2$—cyclohexyl

15.36 Write an equation to show a reaction that would produce each of the following amines:
a. 1-Pentanamine
b. N,N-Dimethylethanamine
c. N-Ethylpropanamine

15.37 Complete each of the following reactions by supplying the missing reactant or product indicated by a question mark:

a. CH$_3$NH(CH$_3$) + ? ⟶ CH$_3$N$^+$H(CH$_3$)(H) + OH$^-$

b. CH$_3$CH$_2$N(CH$_3$)(CH$_2$CH$_3$) + ? ⟶ CH$_3$CH$_2$N$^+$H(CH$_3$)(CH$_2$CH$_3$) Br$^-$

c. CH$_3$CH$_2$CH$_2$NH$_2$ + H$_2$O ⟶ ? + OH$^-$

d. CH$_3$CH$_2$NH(CH$_2$CH$_3$) + HCl ⟶ ?

15.38 Complete each of the following reactions by supplying the missing reactant or product indicated by a question mark:

a. CH$_3$CH$_2$NH$_2$ + H$_2$O ⟶ ? + OH$^-$

b. ? + HCl ⟶ CH$_3$CH$_2$CH$_2$N$^+$H(CH$_2$CH$_2$CH$_3$)(H) Cl$^-$

c. CH$_3$CHNH(CH$_3$) + H$_2$O ⟶ ? + ?
     |
     CH$_3$

d. NH$_3$ + HBr ⟶ ?

15.39 Briefly explain why the lower-molecular-weight amines (fewer than five carbons) exhibit appreciable solubility in water.

15.40 Why is the salt of an amine appreciably more soluble in water than the amine from which it was formed?

15.41 Most drugs containing amine groups are not administered as the amine but rather as the ammonium salt. Can you suggest a reason why?

15.42 Why does aspirin upset the stomach, whereas acetaminophen (Tylenol) does not?

15.43 Putrescine and cadaverine are two odoriferous amines that are produced by decaying flesh. Putrescine is 1,4-butanediamine, and cadaverine is 1,5-pentanediamine. Draw the structures of these two compounds.

15.44 How would you quickly convert an alkylammonium salt into a water-insoluble amine? Explain the rationale for your answer.

Heterocyclic Amines

15.45 Indole and pyridine rings are found in alkaloids.
  a. Sketch each ring.
  b. Name one compound containing each of the ring structures and indicate its use.

15.46 What is an alkaloid?

15.47 List some heterocyclic amines that are used in medicine.

15.48 Distinguish between the terms *analgesic* and *anesthetic*.

Amides

Foundations

15.49 Why do amides have very high boiling points?

15.50 Describe the water solubility of amides in relation to their carbon chain length.

15.51 Explain the I.U.P.A.C. nomenclature rules for naming amides.

15.52 How are the common names of amides derived?

15.53 Describe the physiological effects of barbiturates.

15.54 Why is acetaminophen often recommended in place of aspirin?

Applications

15.55 Use the I.U.P.A.C. and common systems of nomenclature to name the following amides:
  a. $CH_3CH_2CNH_2$ (with C=O)

  b. $CH_3CH_2CH_2CH_2CNH_2$ (with C=O)

  c. $CH_3CN(CH_3)_2$ (with C=O)

15.56 Use the I.U.P.A.C. Nomenclature System to name each of the following amides:
  a. $CH_3CH_2CHCH_2CNH_2$ with Br substituent (with C=O)

  b. 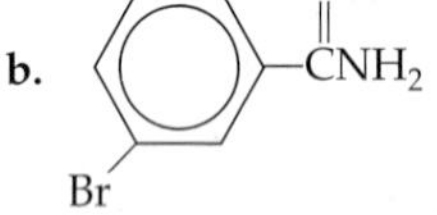 benzene ring with –CNH$_2$ (C=O) and Br substituent

  c. CH_3CHCNH_2 with CH_3 substituent (with C=O)

15.57 Draw the condensed structural formula of each of the following amides:
  a. Ethanamide
  b. N-Methylpropanamide
  c. N,N-Diethylbenzamide
  d. 3-Bromo-4-methylhexanamide
  e. N,N-Dimethylacetamide

15.58 Draw the condensed formula of each of the following amides:
  a. Acetamide
  b. 4-Methylpentanamide
  c. N,N-Dimethylpropanamide
  d. Formamide
  e. N-Ethylpropionamide

15.59 The active ingredient in many insect repellents is N,N-diethyl-m-toluamide. Draw the structure of this compound. Which carboxylic acid and amine would be released by hydrolysis of this compound?

15.60 When an acid anhydride and an amine are combined, an amide is formed. This approach may be used to synthesize acetaminophen, the active ingredient in Tylenol. Complete the following reaction to determine the structure of acetaminophen:

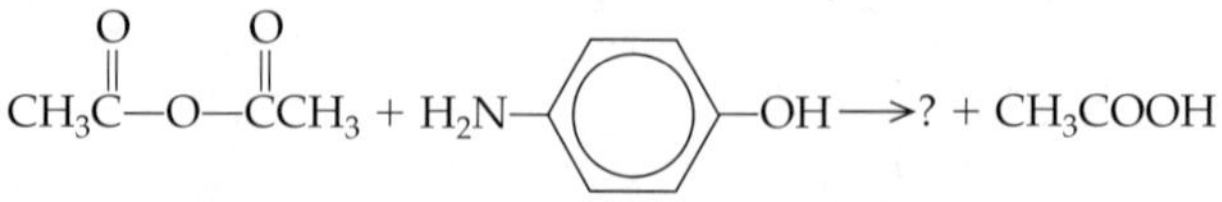

$$CH_3C-O-CCH_3 + H_2N-\text{(benzene)}-OH \longrightarrow ? + CH_3COOH$$

15.61 Explain why amides are neutral in the acid-base sense.

15.62 The amide bond is stabilized by resonance. Draw the contributing resonance forms of the amide bond.

15.63 Lidocaine is often used as a local anesthetic. For medicinal purposes it is often used in the form of its hydrochloride salt because the salt is water soluble. In the structure of lidocaine hydrochloride shown, locate the amide functional group.

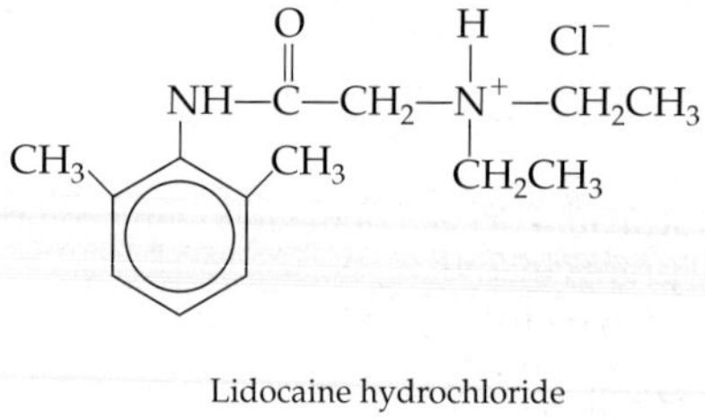

Lidocaine hydrochloride

15.64 Locate the amine functional group in the structure of lidocaine. Is lidocaine a primary, secondary, or tertiary amine?

15.65 The antibiotic penicillin BT contains functional groups discussed in this chapter. In the structure of penicillin BT shown, locate and name as many functional groups as you can.

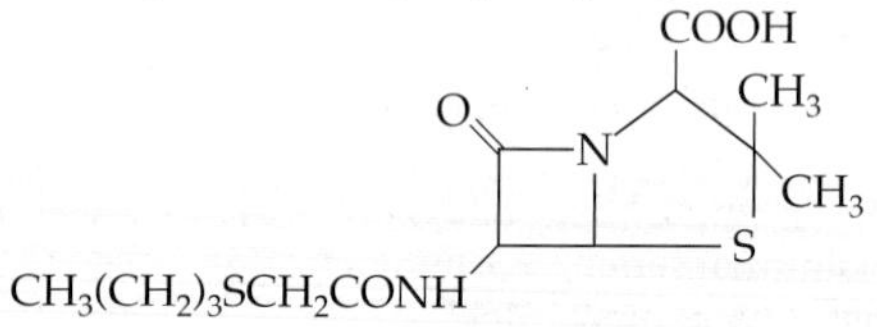

Penicillin BT

15.66 The structure of saccharin, an artificial sweetener, is shown. Circle the amide group.

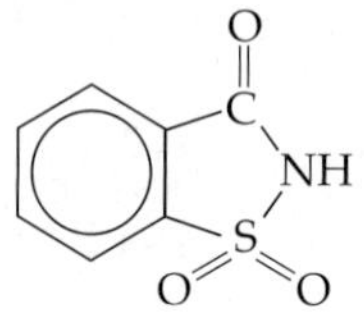

Saccharin

15.67 Complete each of the following reactions by supplying the missing reactant(s) or product(s) indicated by a question mark. Provide the systematic name for all the reactants and products.

  a. $CH_3CNHCH_3 + H_3O^+ \longrightarrow ? + ?$ (with C=O)

b. $? + H_3O^+ \longrightarrow CH_3CH_2CH_2\overset{O}{\underset{\|}{C}}-OH + CH_3N^+H_3$

c. $CH_3\underset{\underset{CH_3}{|}}{CH}CH_2\overset{O}{\underset{\|}{C}}NHCH_2CH_3 + ? \longrightarrow$

$CH_3\underset{\underset{CH_3}{|}}{CH}CH_2\overset{O}{\underset{\|}{C}}-OH + ?$

15.68 Complete each of the following by supplying the missing reagents. Draw the structures of each of the reactants and products.
  a. N-Methylpropanamide + ? $\longrightarrow$ propanoic acid + ?
  b. N,N-Dimethylacetamide + strong acid $\longrightarrow$? + ?
  c. Formamide + strong acid $\longrightarrow$? + ?

15.69 Complete each of the following reactions by supplying the missing reactant(s) or product(s) indicated by a question mark.
  a. $? + 2CH_3CH_2CH_2NH_2 \longrightarrow$

  $CH_3CH_2CH_2NH\overset{O}{\underset{\|}{C}}CH_2CH_3 +$

  $CH_3CH_2\overset{O}{\underset{\|}{C}}-O^-\ N^+H_3-CH_2CH_3$

  b. $CH_3CH_2\overset{O}{\underset{\|}{C}}-Cl + 2NH_3 \longrightarrow ? + ?$

  c. $? + ? \longrightarrow CH_3CH_2CH_2\overset{O}{\underset{\|}{C}}NHCH_2CH_3 +$

  $CH_3CH_2-NH_3^+Cl^-$

15.70 Write two equations for the synthesis of each of the following amides. In one equation use an acid chloride as a reactant. In the second equation use an acid anhydride.
  a. Ethanamide
  b. N-Propylpentanamide
  c. Propionamide

A Preview of Amino Acids, Proteins, and Protein Synthesis

Foundations
15.71 Draw the general structure of an amino acid.
15.72 What is the name of the amide bond formed between two amino acids?

Applications
15.73 The amino acid glycine has a hydrogen atom as its R group, and the amino acid alanine has a methyl group. Draw these two amino acids.
15.74 Draw a dipeptide composed of glycine and alanine. Begin by drawing glycine with its amino group on the left. Circle the amide bond.
15.75 Draw the amino acid alanine (see Problem 15.73). Place a star by the chiral carbon.
15.76 Does glycine have a chiral carbon? Explain your reasoning.
15.77 Describe acyl group transfer.
15.78 Describe the relationship between acyl group transfer and the process of protein synthesis.

Neurotransmitters

Foundations
15.79 Define the term *neurotransmitter*.
15.80 What are the two general classes of neurotransmitters? What distinguishes them from one another?

Applications
15.81 a. What symptoms result from a deficiency of dopamine?
  b. What is the name of this condition?
  c. What symptoms result from an excess of dopamine?
15.82 What is the starting material in the synthesis of dopamine, epinephrine, and norepinephrine?
15.83 Explain the connection between addictive behavior and dopamine.
15.84 Why is L-dopa used to treat Parkinson's disease rather than dopamine?
15.85 What is the function of epinephrine?
15.86 What is the function of norepinephrine?
15.87 What is the starting material from which serotonin is made?
15.88 What symptoms are associated with a deficiency of serotonin?
15.89 What physiological processes are affected by serotonin?
15.90 How does Prozac relieve the symptoms of depression?
15.91 What are the physiological roles of histamine?
15.92 How do antihistamines function to control the allergic response?
15.93 What type of neurotransmitters are γ-aminobutyric acid and glycine?
15.94 Explain the evidence for a relationship between γ-aminobutyric acid and aggressive behavior.
15.95 Explain the function of acetylcholine at the neuromuscular junction.
15.96 Explain why organophosphates are considered to be poisons.
15.97 How does pyridine aldoxime methiodide function as an antidote for organophosphate poisoning?
15.98 Explain the mechanism by which glutamate and NO may function to promote development of memories and learning.

CRITICAL THINKING PROBLEMS

1. Histamine is made and stored in blood cells called *mast cells*. Mast cells are involved in the allergic response. Release of histamine in response to an allergen causes dilation of capillaries. This, in turn, allows fluid to leak out of the capillary resulting in local swelling. It also causes an increase in the volume of the vascular system. If this increase is great enough, a severe drop in blood pressure may cause shock. Histamine is produced by decarboxylation (removal of the carboxylate group as CO_2) of the amino acid histidine shown below. Draw the structure of histamine.

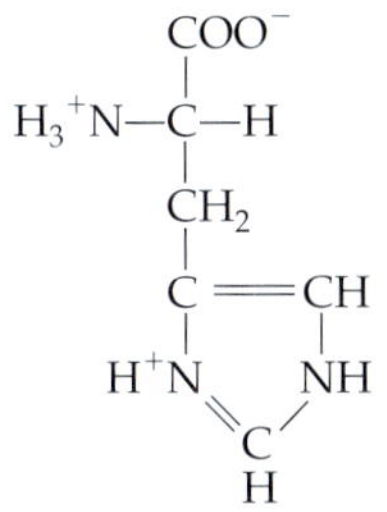

2. Carnitine tablets are sold in health food stores. It is claimed that carnitine will enhance the breakdown of body fat. Carnitine is a tertiary amine found in mitochondria, cell organelles in which food molecules are completely oxidized and ATP is produced. Carnitine is involved in transporting the acyl groups of fatty acids from the cytoplasm into the mitochondria. The fatty acyl group is transferred from a fatty acyl CoA molecule and esterified to carnitine. Inside the mitochondria the reaction is reversed and the fatty acid is completely oxidized. The structure of carnitine is shown here:

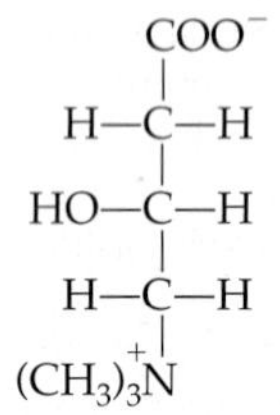

$$\begin{array}{c} COO^- \\ | \\ H-C-H \\ | \\ HO-C-H \\ | \\ H-C-H \\ | \\ (CH_3)_3\overset{+}{N} \end{array}$$

Draw the acyl carnitine molecule that is formed by esterification of palmitic acid with carnitine.

3. The amino acid proline has a structure that is unusual among amino acids. Compare the general structure of an amino acid with that of proline, shown here:

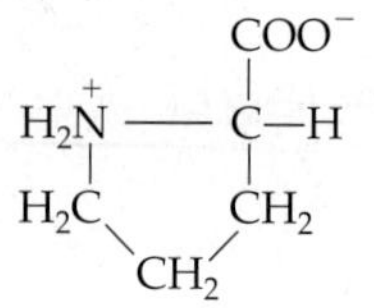

$$\begin{array}{c} COO^- \\ | \\ H_2\overset{+}{N}\text{——}C-H \\ | \quad\quad\quad | \\ H_2C \quad\quad CH_2 \\ \diagdown\;\diagup \\ CH_2 \end{array}$$

What is the major difference between proline and the other amino acids? Draw the structure of a dipeptide in which the amino group of proline forms a peptide bond with the carboxyl group of alanine.

4. Bulletproof vests are made of the polymer called Kevlar. It is produced by the copolymerization of the following two monomers:

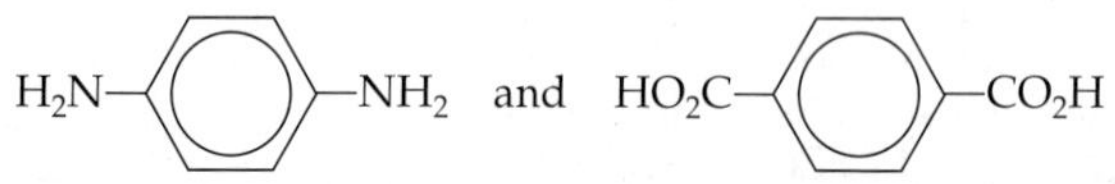

$$H_2N\text{—}\bigcirc\text{—}NH_2 \quad \text{and} \quad HO_2C\text{—}\bigcirc\text{—}CO_2H$$

Draw the structure of a portion of Kevlar polymer.

BIOCHEMISTRY
Carbohydrates 16

Learning Goals

1. Explain the difference between complex and simple carbohydrates and know the amounts of each recommended in the daily diet.

2. Apply the systems of classifying and naming monosaccharides according to the functional group and number of carbons in the chain.

3. Determine whether a molecule has a chiral center.

4. Explain stereoisomerism.

5. Identify monosaccharides as either D- or L-.

6. Draw and name the common monosaccharides using structural formulas.

7. Given the linear structure of a monosaccharide, draw the Haworth projection of its α- and β-cyclic forms and vice versa.

8. By inspection of the structure, predict whether a sugar is a reducing or a nonreducing sugar.

9. Discuss the use of the Benedict's reagent to measure the level of glucose in urine.

10. Draw and name the common disaccharides and discuss their significance in biological systems.

11. Describe the difference between galactosemia and lactose intolerance.

12. Discuss the structural, chemical, and biochemical properties of starch, glycogen, and cellulose.

Would "looking-glass milk" be nutritious?

Outline

Chemistry Connection:
Chemistry Through the Looking Glass

16.1 Types of Carbohydrates

A Human Perspective:
Tooth Decay and Simple Sugars

16.2 Monosaccharides

16.3 Stereoisomers and Stereochemistry

16.4 Biologically Important Monosaccharides

16.5 Biologically Important Disaccharides

A Human Perspective:
Blood Transfusions and the Blood Group Antigens

16.6 Polysaccharides

A Medical Perspective:
Monosaccharide Derivatives and Heteropolysaccharides of Medical Interest

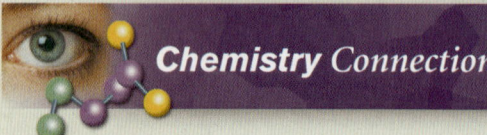

Chemistry Connection

Chemistry Through the Looking Glass

In his children's story *Through the Looking Glass*, Lewis Carroll's heroine Alice wonders whether "looking-glass milk" would be good to drink. As we will see in this chapter, many biological molecules, such as the sugars, exist as two stereoisomers, *enantiomers*, that are mirror images of one another. Because two mirror-image forms occur, it is rather remarkable that in our bodies, and in most of the biological world, only one of the two is found. For instance, the common sugars are members of the D-family, whereas all the common amino acids that make up our proteins are members of the L-family. It is not too surprising, then, that the enzymes in our bodies that break down the sugars and proteins we eat are *stereospecific*, that is, they recognize only one mirror-image isomer. Knowing this, we can make an educated guess that "looking-glass milk" could not be digested by our enzymes and therefore would not be a good source of food for us. It is even possible that it might be toxic to us!

Pharmaceutical chemists are becoming more and more concerned with the stereochemical purity of the drugs that we take. Consider a few examples. In 1960 the drug thalidomide was commonly prescribed in Europe as a sedative. However, during that year, hundreds of women who took thalidomide during pregnancy gave birth to babies with severe birth defects. Thalidomide, it turned out, was a mixture of two enantiomers. One is a sedative; the other is a teratogen, a chemical that causes birth defects.

One of the common side effects of taking antihistamines for colds or allergies is drowsiness. Again, this is the result of the fact that antihistamines are mixtures of enantiomers. One causes drowsiness; the other is a good decongestant.

One enantiomer of the compound carvone is associated with the smell of spearmint; the other produces the aroma of caraway seeds or dill. One mirror-image form of limonene smells like lemons; the other has the aroma of oranges.

The pain reliever ibuprofen is currently sold as a mixture of enantiomers, but one is a much more effective analgesic than the other.

Taste, smell, and the biological effects of drugs in the body all depend on the stereochemical form of compounds and their interactions with cellular enzymes or receptors. As a result, chemists are actively working to devise methods of separating the isomers in pure form. Alternatively, methods of conducting stereospecific syntheses that produce only one stereoisomer are being sought. By preparing pure stereoisomers, the biological activity of a compound can be much more carefully controlled. This will lead to safer medications.

In this chapter we will begin our study of stereochemistry, the spatial arrangement of atoms in molecules, with the carbohydrates. Later, we will examine the stereochemistry of the amino acids that make up our proteins and consider the stereochemical specificity of the metabolic reactions that are essential to life. A more complete treatment of stereochemistry is found online at www.mhhe.com/denniston5e, in Stereochemistry and Stereoisomers Revisited.

Introduction

Emil Fischer's father, a wealthy businessman, once said that Emil was too stupid to be a businessman and had better be a student. Lucky for the field of biochemistry, Emil did just that. Originally he wanted to study physics, but his cousin Otto Fischer convinced him to study chemistry. Beginning as an organic chemist, Fischer launched a career that eventually led to groundbreaking research in biochemistry. By following his career, we get a glimpse of the entire field.

Early on, Fischer discovered the active ingredients in coffee and tea, caffeine and theobromine. Eventually he discovered their structures and synthesized them in the laboratory. In work that he carried out between 1882 and 1906, Fischer demonstrated that adenine and guanine, along with some other compounds found in plants and animals, all belonged to one family of compounds. He called these the purines. All the purines have the same core structure and differ from one another by the functional groups attached to the ring.

Introduction

Purine core structure

We will study the purines in Chapter 20 where we will learn that these are two of the essential components of the genetic molecules DNA and RNA. We will see that DNA is a double helix. Each strand of the helix is made up of a backbone of alternating sugars (ribose in RNA and deoxyribose in DNA) and phosphoryl groups. The purines are one of the two types of nitrogenous bases that project into the helix. By reading the order of these nitrogenous bases, we can decipher the genetic code of an organism.

In 1884 Fischer began his monumental work on sugars. In 1890 he established the stereochemical nature of all sugars, and between 1891 and 1894 he worked out the stereochemical configuration of all the known sugars and predicted all the possible stereoisomers. Stereochemistry is the study of molecules that have two mirror-image isomers. We will find, as Fischer did, that nature has "selected" only one of the two mirror-image forms for common use in biological systems. Fischer studied virtually all the sugars, but one of his greatest successes was the synthesis of glucose, fructose, and mannose, three six-carbon sugars.

Between 1899 and 1908, Fischer turned his attention to proteins. He developed methods to separate and identify individual amino acids and discovered an entirely new class, the cyclic amino acids (those with ring structures). Fischer also worked on synthesis of proteins from amino acids. He demonstrated the nature of the peptide bond and discovered how to make small peptides in the laboratory.

As we will learn in Chapter 18, amino acids are all characterized by a common core structure, having both a carboxyl group and an amino group.

Amino acid core structure

The peptide bond that forms between amino acids is actually an amide bond between these two groups.

While each of the amino acids has a common core, they differ from one another in the nature of a side chain, or R group. The amino acids are classified based on the properties they acquire from these R groups.

Quite late in his career, Fischer even got around to studying the fats, a heterogenous group of substances characterized by their hydrophobic nature. In Chapter 17, we will study the incredibly diverse family of fats, or lipids.

Fischer's personal life was not as happy as his professional life. His wife died after only seven years of marriage, leaving Fischer with three sons. One son died in World War I and another committed suicide at the age of twenty-five. However, his third son, Hermann Otto Laurenz Fischer, followed in his father's footsteps, becoming a professor of biochemistry at the University of California at Berkeley.

In 1902, Fischer was awarded the Nobel Prize for his work on sugar and purine synthesis. But a glance at all of his accomplishments helps us to realize that this great man, who began his career as an organic chemist, established the field of biochemistry through his extraordinary studies of the molecules of life.

The structure of the purines and pyrimidines was first described in Section 15.2. It is discussed in much more detail in Sections 20.1 and 20.2.

The structure of simple sugars and the function in biological systems are discussed in Sections 16.2 and 16.4. Stereochemistry is discussed in Section 16.3.

We will study the amino acids and the structure of proteins in Chapter 18.

See also Section 15.4, A Preview of Amino Acids, Proteins, and Protein Synthesis.

LEARNING GOAL

A kilocalorie is the same as the calorie referred to in the "count-your-calories" books and on nutrition labels.

See a Human Perspective; Tooth Decay and Simple Sugars on page 528.

16.1 Types of Carbohydrates

We begin our study of biochemistry with the carbohydrates. Carbohydrates are produced in plants by photosynthesis (Figure 16.1). Natural carbohydrate sources such as grains and cereals, breads, sugar cane, fruits, milk, and honey are an important source of energy for animals. **Carbohydrates** include simple sugars having the general formula $(CH_2O)_n$, as well as long polymers of these simple sugars, for instance potato starch, and a variety of molecules of intermediate size. The simple sugar glucose, $C_6H_{12}O_6$, is the primary energy source for the brain and nervous system and can be used by many other tissues. When "burned" by cells for energy, each gram of carbohydrate releases approximately four kilocalories of energy.

A healthy diet should contain both complex carbohydrates, such as starches and cellulose, and simple sugars, such as fructose and sucrose (Figure 16.2). However, the quantity of simple sugars, especially sucrose, should be minimized because large quantities of sucrose in the diet promote obesity and tooth decay.

Complex carbohydrates are better for us than the simple sugars. Starch, found in rice, potatoes, breads, and cereals, is an excellent energy source. In addition, the complex carbohydrates, such as cellulose, provide us with an important supply of dietary fiber.

It is hard to determine exactly what percentage of the daily diet *should* consist of carbohydrates. The *actual* percentage varies widely throughout the world, from 80% in the Far East, where rice is the main component of the diet, to 40–50% in the United States. Currently, it is recommended that 45–65% of the calories in the diet should come from carbohydrates and that no more than 10% of the daily caloric intake should be sucrose.

Figure 16.1
Carbohydrates are produced by plants such as this potato in the process of photosynthesis, which uses the energy of sunlight to produce hexoses from CO_2 and H_2O.

Figure 16.2
Carbohydrates from a variety of foods are an essential component of the diet.

> **Question 16.1**
>
> What is the current recommendation for the amount of carbohydrates that should be included in the diet? Of the daily intake of carbohydrates, what percentage should be simple sugar?

> **Question 16.2**
>
> Distinguish between simple and complex sugars. What are some sources of complex carbohydrates?

Monosaccharides such as glucose and fructose are the simplest carbohydrates because they contain a single *(mono-)* sugar *(saccharide)* unit. **Disaccharides,** including sucrose and lactose, consist of two monosaccharide units joined through bridging oxygen atoms. Such a bond is called a **glycosidic bond. Oligosaccharides** consist of three to ten monosaccharide units joined by glycosidic bonds. The largest and most complex carbohydrates are the **polysaccharides,** which are long, often highly branched, chains of monosaccharides. Starch, glycogen, and cellulose are all examples of polysaccharides.

16.2 Monosaccharides

Monosaccharides are composed of carbon, hydrogen, and oxygen, and most are characterized by the general formula $(CH_2O)_n$, in which n is any integer from 3 to 7. As we will see, this general formula is an oversimplification because several biologically important monosaccharides are chemically modified. For instance, several blood group antigen and bacterial cell wall monosaccharides are substituted with amino groups. Many of the intermediates in carbohydrate metabolism carry phosphate groups. Deoxyribose, the monosaccharide found in DNA, has one fewer oxygen atom than the general formula above would predict.

The importance of phosphorylated sugars in metabolic reactions is discussed in Sections 14.4 and 21.3.

 LEARNING GOAL

Monosaccharides can be named on the basis of the functional groups they contain. A monosaccharide with a ketone (carbonyl) group is a **ketose.** If an aldehyde (carbonyl) group is present, it is called an **aldose.** Because monosaccharides also contain many hydroxyl groups, they are sometimes called *polyhydroxyaldehydes* or *polyhydroxyketones*.

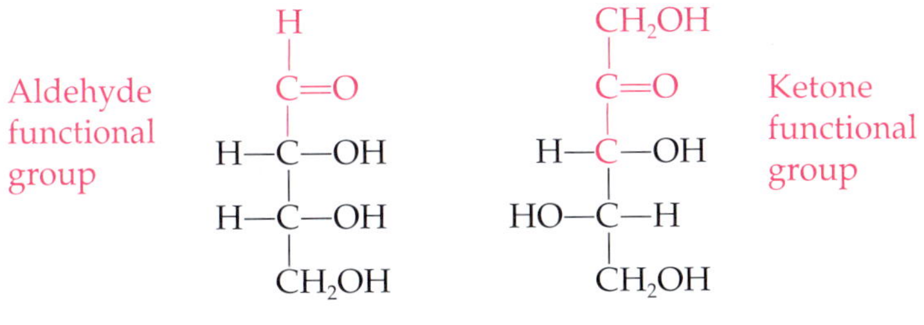

```
                    H                        CH₂OH
                    |                        |
 Aldehyde           C=O                      C=O         Ketone
 functional         |                        |           functional
 group          H—C—OH                   H—C—OH          group
                    |                        |
                H—C—OH                  HO—C—H
                    |                        |
                 CH₂OH                    CH₂OH

               An aldose                 A ketose
```

Another system of nomenclature tells us the number of carbon atoms in the main skeleton. A three-carbon monosaccharide is a *triose,* a four-carbon sugar is a *tetrose,* a five-carbon sugar is a *pentose,* a six-carbon sugar is a *hexose,* and so on. Combining the two naming systems gives even more information about the structure and composition of a sugar. For example, an aldotetrose is a four-carbon sugar that is also an aldehyde.

In addition to these general names, each monosaccharide has a unique name. These names are shown in blue for the following structures. Because the monosaccharides can exist in several different isomeric forms, it is important to provide the

A Human Perspective

Tooth Decay and Simple Sugars

How many times have you heard the lecture from parents or your dentist about brushing your teeth after a sugary snack? Annoying as this lecture might be, it is based on sound scientific data that demonstrate that the cause of tooth decay is plaque and acid formed by the bacterium *Streptococcus mutans* using sucrose as its substrate.

Saliva is teeming with bacteria in concentrations up to one hundred million (10^8) per milliliter of saliva! Within minutes after you brush your teeth, sticky glycoproteins in the saliva adhere to tooth surfaces. Then millions of oral bacteria immediately bind to this surface.

Although many oral bacteria stick to the tooth surface, as the diagram below shows, only *S. mutans* causes cavities. The reason for this is that this organism alone can make the enzyme *glucosyl transferase*. This enzyme acts only on the disaccharide sucrose, breaking it down into glucose and fructose. The glucose is immediately added to a growing polysaccharide called *dextran*, the glue that allows the bacteria to adhere to the tooth surface, contributing to the formation of plaque.

Now the bacteria embedded in the dextran take in the fructose and use it in the lactic acid fermentation. The lactic acid that is produced lowers the pH on the tooth surface and begins to dissolve calcium from the tooth enamel. Even though we produce about one liter of saliva each day, the acid cannot be washed away from the tooth surface because the dextran plaque is not permeable to saliva.

So what can we do to prevent tooth decay? Of course, brushing after each meal and flossing regularly reduce plaque buildup. Eating a diet rich in calcium also helps build strong tooth enamel. Foods rich in complex carbohydrates, such as fruits and vegetables, help prevent cavities in two ways. Glucosyl transferase can't use complex carbohydrates in its cavity-causing chemistry, and eating fruits and vegetables helps to mechanically remove plaque.

Perhaps the most effective way to prevent tooth decay is to avoid sucrose-containing snacks between meals. Studies have shown that eating sucrose-rich foods doesn't cause much tooth decay if followed immediately by brushing. However, even small amounts of sugar eaten between meals actively promote cavity formation.

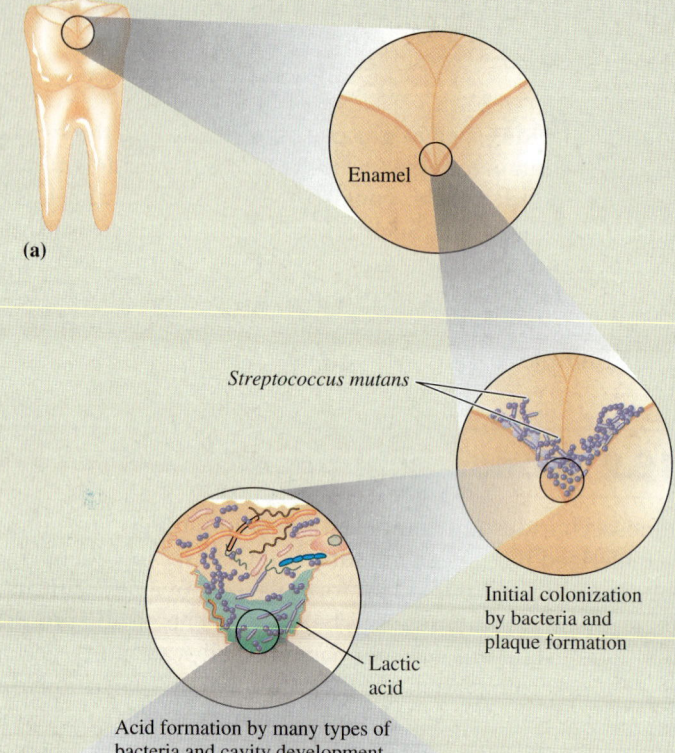

(a) The complex process of tooth decay. (b) Electron micrograph of dental plaque.

For Further Understanding

It has been suggested that tooth decay could be prevented by a vaccine that would rid the mouth of *Streptococcus mutans*. Explain this from the point of view of the chemical reactions that are described above.

What steps could you take following a sugary snack to help prevent tooth decay, even when it is not possible to brush your teeth?

complete name. Thus the complete names of the following structures are D-glyceraldehyde, D-glucose, and D-fructose. These names tell us that the structure represents one particular sugar and also identifies the sugar as one of two possible isomeric forms (D- or L-).

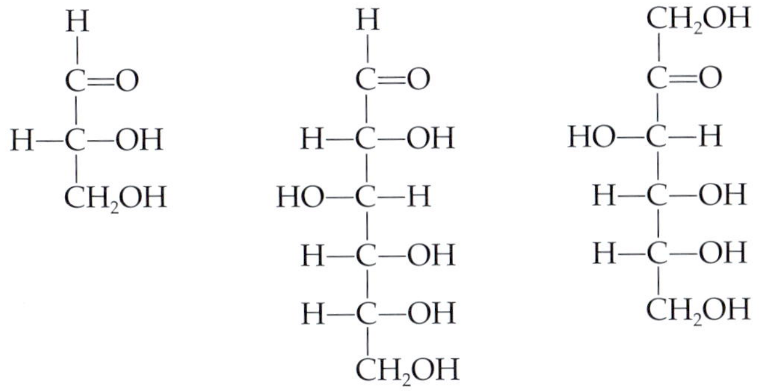

```
      H                  H                 CH₂OH
      |                  |                   |
      C=O                C=O                 C=O
      |                  |                   |
   H—C—OH             H—C—OH              HO—C—H
      |                  |                   |
     CH₂OH            HO—C—H              H—C—OH
                        |                   |
                     H—C—OH              H—C—OH
                        |                   |
                     H—C—OH               CH₂OH
                        |
                      CH₂OH
```

Aldose	Aldose	Ketose
Triose	Hexose	Hexose
Aldotriose	Aldohexose	Ketohexose
D-Glyceraldehyde	D-Glucose	D-Fructose

Question 16.3

What is the structural difference between an aldose and a ketose?

Question 16.4

Explain the difference between:

a. A ketohexose and an aldohexose
b. A triose and a pentose

16.3 Stereoisomers and Stereochemistry

Stereoisomers

The prefixes D- and L- found in the complete name of a monosaccharide are used to identify one of two possible isomeric forms called **stereoisomers**. By definition, each member of a pair of stereoisomers must have the same molecular formula and the same bonding. How then do isomers of the D-family differ from those of the L-family? D- and L-isomers differ in the spatial arrangements of atoms in the molecule.

Stereochemistry is the study of the different spatial arrangements of atoms. A general example of a pair of stereoisomers is shown in Figure 16.3. In this example the general molecule C-abcd is formed from the bonding of a central carbon to four different groups: a, b, c, and d. This results in two molecules rather than one. Each isomer is bonded together through the exact *same* bonding pattern, yet they are *not* identical. If they were identical, they would be superimposable one upon the other; *they are not*. They are therefore stereoisomers. These two stereoisomers have a mirror-image relationship that is analogous to the mirror-image relationship of the left and right hands (see Figure 16.3b).

Two stereoisomers that are nonsuperimposable mirror images of one another are called a pair of **enantiomers**. Molecules that can exist in enantiomeric forms are called **chiral molecules**. The term simply means that as a result of different

3 LEARNING GOAL

4 LEARNING GOAL

5 LEARNING GOAL

Build models of these compounds using toothpicks and gumdrops of five different colors to prove this to yourself.

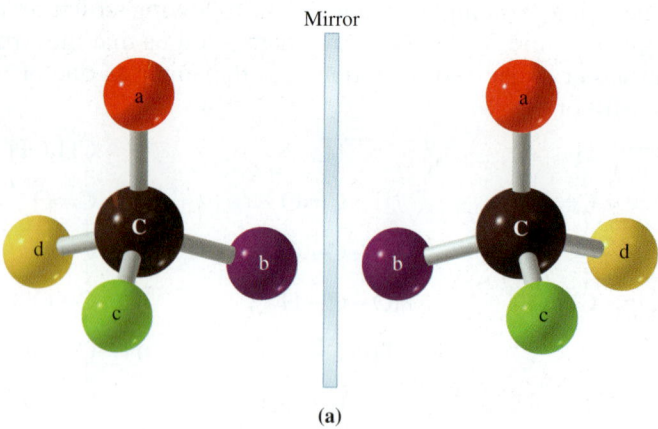

Figure 16.3
(a) A pair of enantiomers for the general molecule C-abcd. (b) Mirror-image right and left hands.

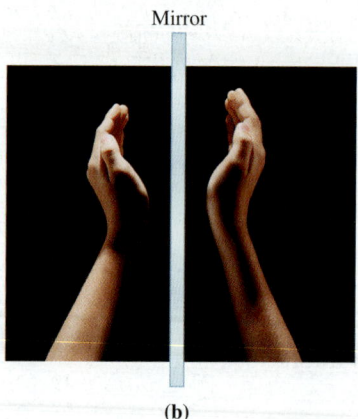

You can find further information online at www.mhhe.com/denniston5e in Stereochemistry and Stereoisomers Revisited.

three-dimensional arrangements of atoms, the molecule can exist in two mirror-image forms. For any pair of nonsuperimposable mirror-image forms (enantiomers), one is always designated D- and the other L-.

A carbon atom that has four different groups bonded to it is called a **chiral carbon** atom. Any molecule containing a chiral carbon can exist as a pair of enantiomers. Consider the simplest carbohydrate, **glyceraldehyde,** which is shown in Figure 16.4. Note that the second carbon is bonded to four different groups. It is therefore a chiral carbon. As a result, we can draw two enantiomers of glyceraldehyde that are nonsuperimposable mirror images of one another. Larger biological molecules typically have more than one chiral carbon.

Rotation of Plane-Polarized Light

Stereoisomers can be distinguished from one another by their different optical properties. Each member of a pair of stereoisomers will rotate plane-polarized light in different directions.

The polarimeter, measurement of the rotation of plane-polarized light, and the calculation of specific rotation are discussed in detail online at www.mhhe.com/denniston5e in Stereochemistry and Stereoisomers Revisited.

As we learned in Chapter 2, white light is a form of electromagnetic radiation that consists of many different wavelengths (colors) vibrating in *planes* that are all perpendicular to the direction of the light beam (Figure 16.5). To measure optical properties of enantiomers, scientists use special light sources to produce *monochromatic light,* that is, light of a single wavelength. The monochromatic light is passed through a polarizing material, like a Polaroid lens, so that only waves in one plane can pass through. The light that emerges from the lens is *plane-polarized light* (Figure 16.5).

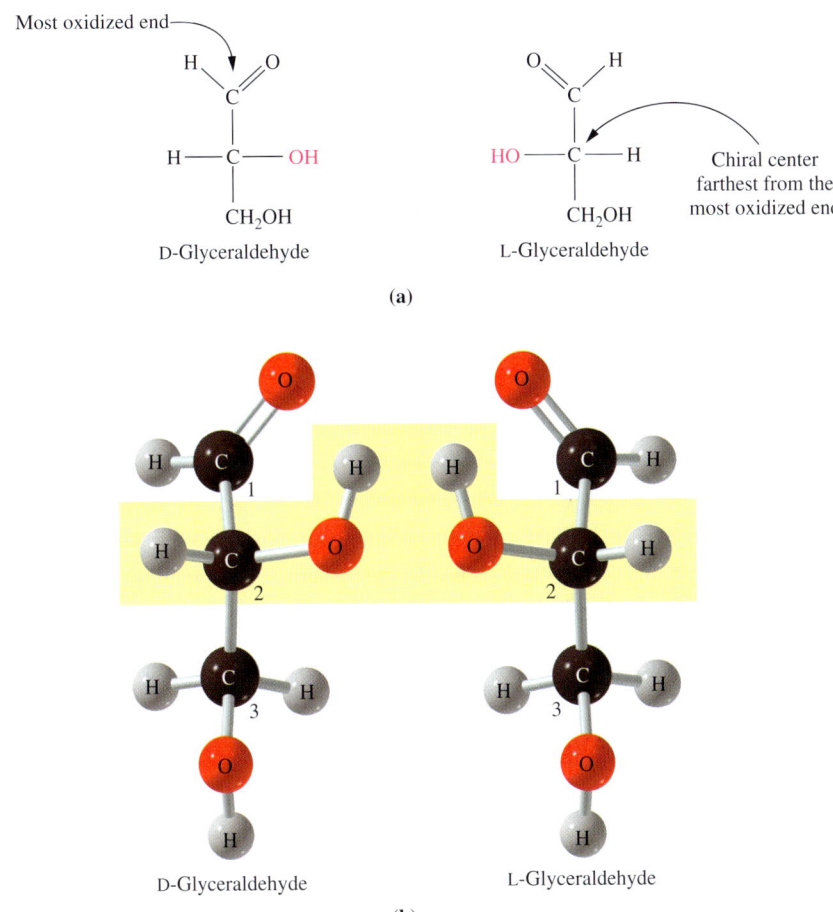

Figure 16.4
(a) Structural formulas of D- and L-glyceraldehyde. The end of the molecule with the carbonyl group is the most oxidized end. The D- or L-configuration of a monosaccharide is determined by the orientation of the functional groups attached to the chiral carbon farthest from the oxidized end. In the D-enantiomer, the —OH is to the right. In the L-enantiomer, the —OH is to the left. (b) A three-dimensional representation of D- and L-glyceraldehyde.

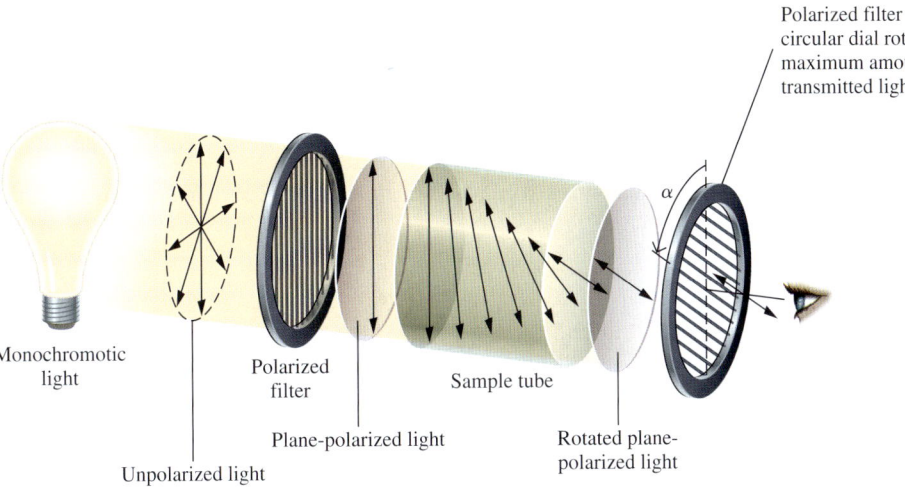

Figure 16.5
Schematic drawing of a polarimeter.

Applying these principles, scientists have developed the *polarimeter* to measure the ability of a compound to change the angle of the plane of plane-polarized light (Figure 16.5). The polarimeter allows the determination of the specific rotation of a compound, that is, the measure of its ability to rotate plane-polarized light.

Some compounds rotate light in a clockwise direction. These are said to be *dextrorotatory* and are designated by a plus sign (+) before the specific rotation value. Other substances rotate light in a counterclockwise direction. These are called *levorotatory* and are indicated by a minus sign (−) before the specific rotation value.

The Relationship Between Molecular Structure and Optical Activity

In 1848, Louis Pasteur was the first to see a relationship between the structure of a compound and the effect of that compound on plane-polarized light. In his studies of winemaking, Pasteur noticed that salts of tartaric acid were formed as a byproduct. It is a tribute to his extraordinary powers of observation that he noticed that two types of crystals were formed and that they were mirror images of one another. Using a magnifying glass and forceps, Pasteur separated the left-handed and right-handed crystals into separate piles. When he measured the optical activity of each of the mirror-image forms and of the original mixed sample, he obtained the following results:

- A solution of the original mixture of crystals was optically inactive.
- But both of the mirror-image crystals were optically active. In fact, the specific rotation produced by each was identical in magnitude but was of opposite sign.

Although Pasteur's work opened the door to understanding the relationship between structure and optical activity, it was not until 1874 that the Dutch chemist van't Hoff and the French chemist LeBel independently came up with a basis for the observed optical activity: tetrahedral carbon atoms bonded to four different atoms or groups of atoms. Thus, two enantiomers, which are identical to one another in all other chemical and physical properties, will rotate plane-polarized light to the same degree, but in opposite directions.

Fischer Projection Formulas

Emil Fischer devised a simple way to represent the structure of stereoisomers. The **Fischer Projection** is a two-dimensional drawing of a molecule that shows a chiral carbon at the intersection of two lines. The horizontal lines represent bonds projecting out of the page, and the vertical lines represent bonds that project into the page. Figure 16.6 demonstrates how to draw the Fischer Projections for the stereoisomers of bromochlorofluoromethane. In Figure 16.6a, the two isomers are represented using ball-and-stick models. The molecules are reinterpreted using the wedge-and-dash representations in Figure 16.6b. In the Fischer Projections shown in Figure 16.6c, the point at which two lines cross represents the chiral carbon. Horizontal lines replace the solid wedges indicating that the bonds are projecting toward the reader. Vertical lines replace the dashed wedges, indicating that the

Figure 16.6
Drawing a Fischer Projection. (a) The ball-and-stick models for the stereoisomers of bromochlorofluoromethane. (b) The wedge-and-dash and (c) Fischer Projections of these molecules.

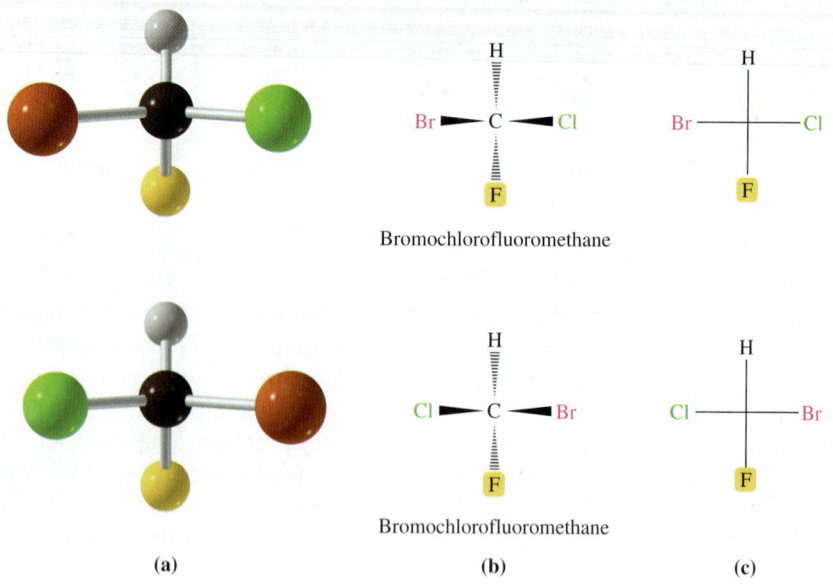

bonds are projecting away from the reader. For sugars, the aldehyde or ketone group, the most oxidized carbon, is always represented at the "top."

EXAMPLE 16.1 Drawing Fischer Projections for a Sugar

Draw the Fischer Projections for the stereoisomers of glyceraldehyde.

Solution

Review the structures of the two stereoisomers of glyceraldehydes (Figure 16.4b). The ball-and-stick models can be represented using three-dimensional wedge drawings. Remember that for sugars the most oxidized carbon (the aldehyde or ketone group) is always drawn at the top of the structure.

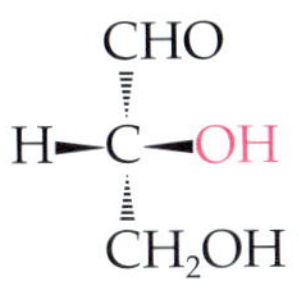

```
        CHO                    CHO
         |                      |
     H—C—OH                HO—C—H
         |                      |
       CH₂OH                  CH₂OH
   D-Glyceraldehyde       L-Glyceraldehyde
```

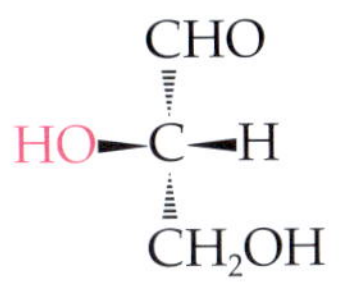

Remember that in the wedge diagram, the solid wedges represent bonds directed toward the reader. The dashed wedges represent bonds directed away from the reader and into the page. In these molecules, the center carbon is the only chiral carbon in the structure. To convert these wedge representations to a Fischer Projection, simply use a horizontal line in place of each solid wedge and use a vertical line to represent each dashed wedge. The chiral carbon is represented by the point at which the vertical and horizontal lines cross, as shown below.

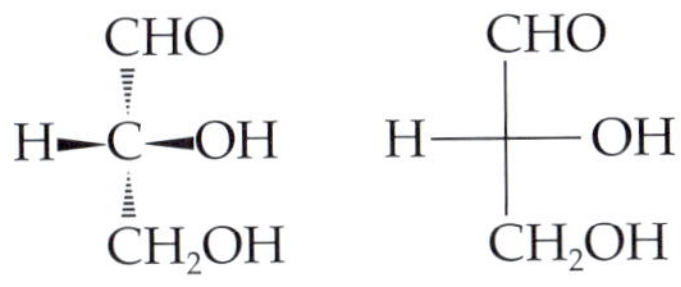

```
   CHO          CHO              CHO           CHO
    |            |                |             |
H—C—OH      H—|—OH          HO—C—H        HO—|—H
    |            |                |             |
  CH₂OH        CH₂OH            CH₂OH         CH₂OH
       D-Glyceraldehyde              L-Glyceraldehyde
```

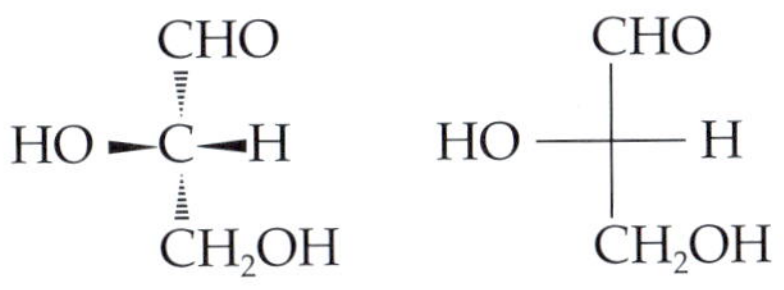

Question 16.5

Indicate whether each of the following molecules is an aldose or a ketose.

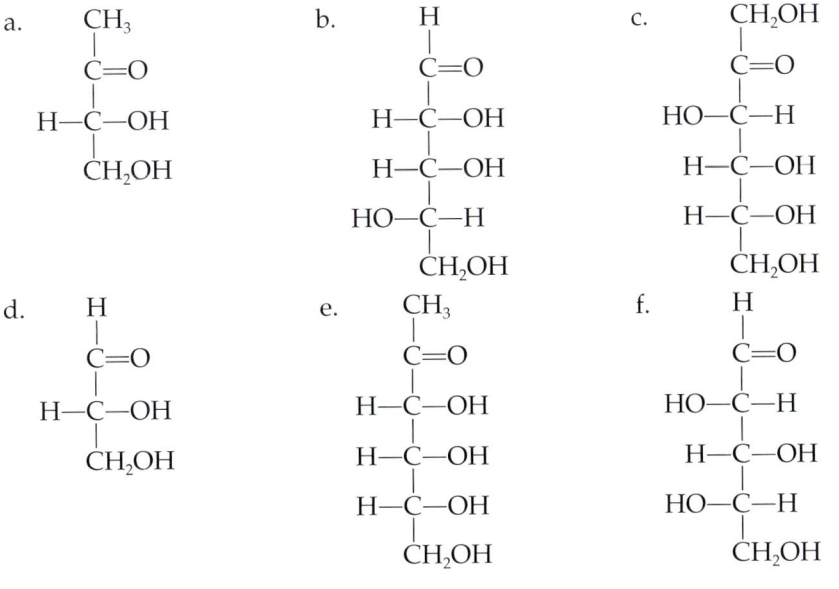

Question 16.6
Draw the Fischer Projection for each molecule in Question 16.5.

Question 16.7
Place an asterisk beside each chiral carbon in the Fischer Projections you drew for Question 16.6.

Question 16.8
In the Fischer Projections you drew for Question 16.6, indicate which bonds project toward you and which project into the page.

The D- and L- System of Nomenclature

In 1891 Emil Fischer devised a nomenclature system that would allow scientists to distinguish between enantiomers. Fischer knew that there are two enantiomers of glyceraldehyde that rotated plane-polarized light in opposite directions. He did not have the sophisticated tools needed to make an absolute connection between the structure and the direction of rotation of plane-polarized light. He simply decided that the (+) enantiomer would be the one with the hydroxyl group of the chiral carbon on the right. The enantiomer that rotated plane-polarized light in the (−) or levorotatory direction, he called L-glyceraldehyde (Figure 16.4).

While specific rotation is an experimental value that must be measured, the D- and L- designations of all other monosaccharides are determined by comparison of their structures with D- and L-glyceraldehyde. Sugars with more than three carbons will have more than one chiral carbon. *By convention*, it is the position of the hydroxyl group on the chiral carbon farthest from the carbonyl group (the most oxidized end of the molecule) that determines whether a monosaccharide is in the D- or L- configuration. If the —OH group is on the right, the molecule is in the D-configuration. If the —OH group is on the left, the molecule is in the L-configuration. Almost all carbohydrates in living systems are members of the D-family.

It was not until 1952 that researchers were able to demonstrate that Fischer had guessed correctly when he proposed the structures of the (+) and (−) enantiomers of glyceraldehyde.

The structures and designations of D- and L-glyceraldehyde are defined by convention. In fact, the D- and L- terminology is generally applied only to carbohydrates and amino acids. For organic molecules, the D- and L- convention has been replaced by a new system that provides the absolute configuration of a chiral carbon. This system, called the (R) and (S) system, is described in Stereochemistry and Stereoisomers Revisited, online at www.mhhe.com/denniston5e.

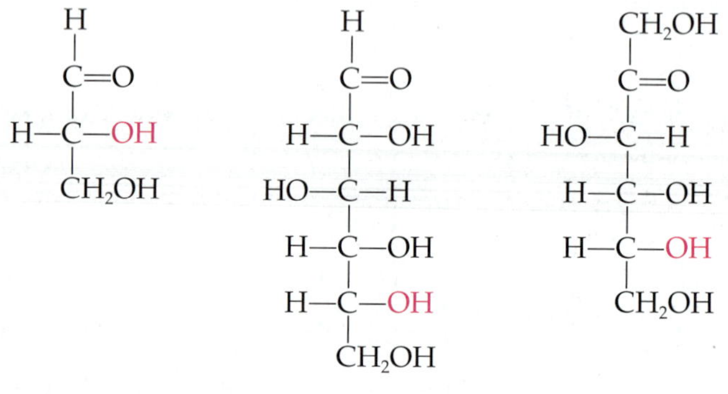

D-Glyceraldehyde, D-Glucose, D-Fructose (Fischer projections)

Question 16.9
Determine the configuration (D- or L-) for each of the molecules in Question 16.5.

Question 16.10
Explain the difference between the D- and L- designation and the (+) and (−) designation.

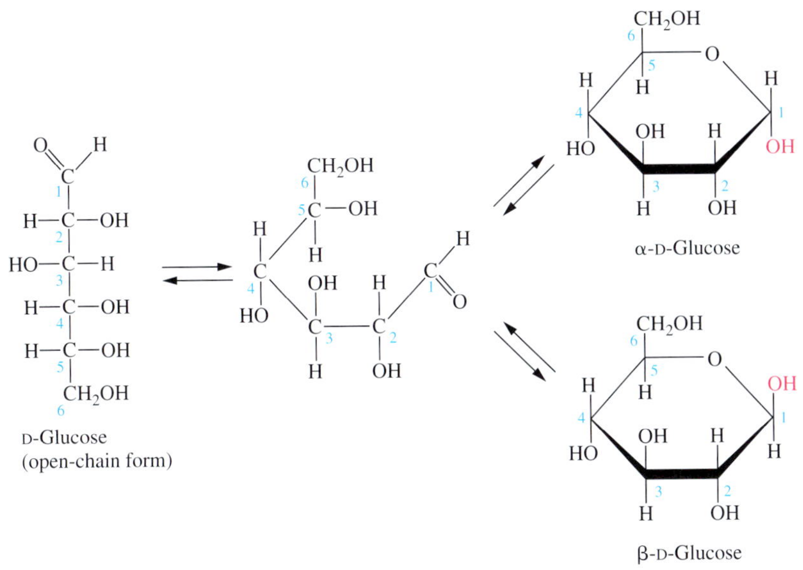

16.4 Biologically Important Monosaccharides

Figure 16.7 Cyclization of glucose to give α- and β-D-glucose. Note that the carbonyl carbon (C-1) becomes chiral in this process, yielding the α- and β- forms of glucose.

Monosaccharides, the simplest carbohydrates, have backbones of from three to seven carbons. There are many monosaccharides, but we will focus on those that are most common in biological systems. These include the five- and six-carbon sugars: glucose, fructose, galactose, ribose, and deoxyribose.

Glucose

Glucose is the most important sugar in the human body. It is found in numerous foods and has several common names, including dextrose, grape sugar, and blood sugar. Glucose is broken down in glycolysis and other pathways to release energy for body functions.

The concentration of glucose in the blood is critical to normal body function. As a result, it is carefully controlled by the hormones insulin and glucagon. Normal blood glucose levels are 100–120 mg/100 mL, with the highest concentrations appearing after a meal. Insulin stimulates the uptake of the excess glucose by most cells of the body, and after one to two hours, levels return to normal. If blood glucose concentrations drop too low, the individual feels lightheaded and shaky. When this happens, glucagon stimulates the liver to release glucose into the blood, reestablishing normal levels. We will take a closer look at this delicate balancing act in Section 23.6.

The molecular formula of glucose, an aldohexose, is $C_6H_{12}O_6$. The structure of glucose is shown in Figure 16.7, and the method used to draw this structure is described in Example 16.2.

EXAMPLE 16.2 Drawing the Structure of a Monosaccharide

Draw the structure for D-glucose.

Solution

Glucose is an aldohexose.

Continued—

EXAMPLE 16.2 —Continued

Step 1. Draw six carbons in a straight vertical line; each carbon is separated from the ones above and below it by a bond:

```
1 C
  |
2 C
  |
3 C
  |
4 C
  |
5 C
  |
6 C
```

Step 2. The most highly oxidized carbon is, by convention, drawn as the uppermost carbon (carbon-1). In this case, carbon-1 is an aldehyde carbon:

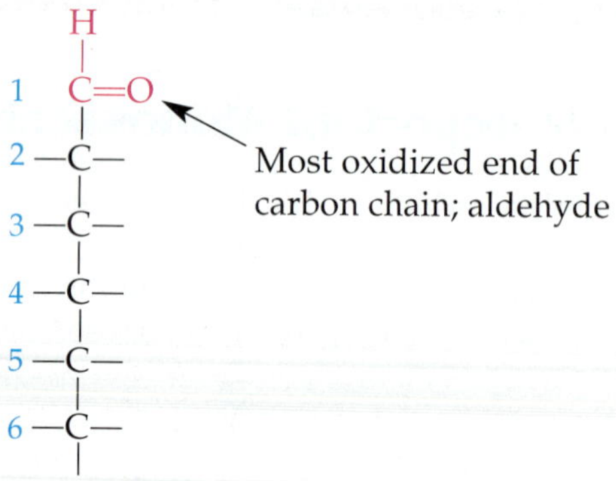

```
      H
      |
  1   C=O    ← Most oxidized end of
      |        carbon chain; aldehyde
  2 —C—
      |
  3 —C—
      |
  4 —C—
      |
  5 —C—
      |
  6 —C—
      |
```

Step 3. The atoms are added to the next to the last carbon atom, at the bottom of the chain, to give either the D- or L-configuration as desired. Remember, when the —OH group is to the right, you have D-glucose. When in doubt, compare your structure to D-glyceraldehyde!

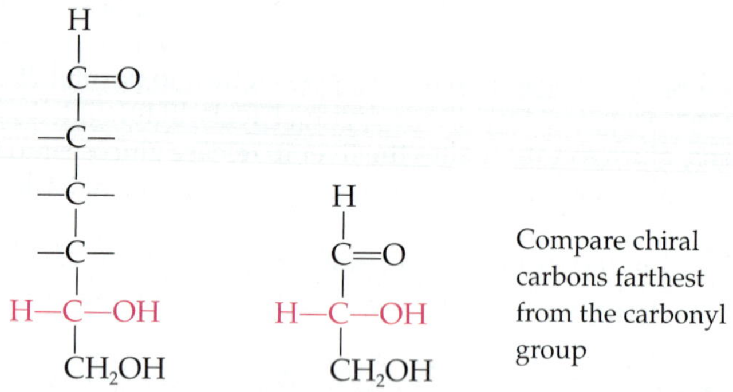

```
      H                    
      |                    
      C=O                  
      |                    
    —C—                    
      |                    
    —C—            H       
      |           |        
    —C—           C=O         Compare chiral
      |           |           carbons farthest
   H—C—OH      H—C—OH         from the carbonyl
      |           |           group
     CH₂OH       CH₂OH

    D-Isomer    D-Glyceraldehyde
```

Step 4. All the remaining atoms are then added to give the desired carbohydrate. For example, one would draw the following structure for D-glucose.

Continued—

16.4 Biologically Important Monosaccharides

EXAMPLE 16.2 —Continued

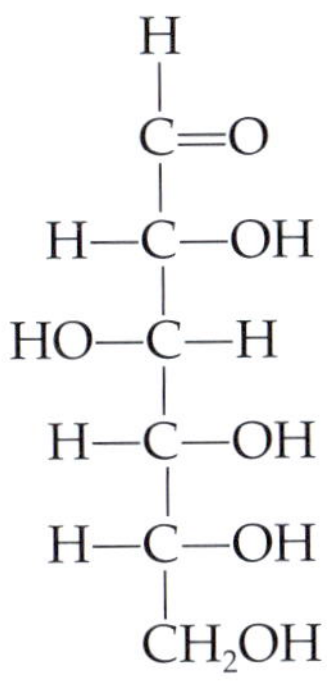

$$
\begin{array}{c}
H \\
| \\
C=O \\
| \\
H-C-OH \\
| \\
HO-C-H \\
| \\
H-C-OH \\
| \\
H-C-OH \\
| \\
CH_2OH
\end{array}
$$

D-Glucose

The positions for the hydrogen atoms and the hydroxyl groups on the remaining carbons must be learned for each sugar.

Question 16.11

Draw the structure of D-ribose. (Information on the structure of D-ribose is found later in this chapter.)

Question 16.12

Draw the structure of D-galactose. (Information on the structure of D-galactose is found later in this chapter.)

Question 16.13

Draw the structure of L-ribose.

Question 16.14

Draw the structure of L-galactose.

In actuality the open-chain form of glucose is present in very small concentrations in cells. It exists in cyclic form under physiological conditions because the carbonyl group at C-1 of glucose reacts with the hydroxyl group at C-5 to give a six-membered ring. In the discussion of aldehydes, we noted that the reaction between an aldehyde and an alcohol yields a **hemiacetal**. When the aldehyde portion of the glucose molecule reacts with the C-5 hydroxyl group, the product is a cyclic intramolecular hemiacetal. For D-glucose, two isomers can be formed in this reaction (see Figure 16.7). These isomers are called α- and β-D-glucose. The two isomers formed differ from one another in the location of the —OH attached to the hemiacetal carbon, C-1. Such isomers, differing in the arrangement of bonds around the hemiacetal carbon, are called **anomers**. In the α-anomers, the C-1 (anomeric carbon) hydroxyl group is below the ring, and in the β-anomers, the C-1 hydroxyl group is above the ring. Like the stereoisomers discussed previously, the α and β forms can be distinguished from one another because they rotate plane-polarized light differently.

Hemiacetal structure,

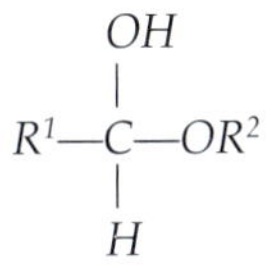

$$
\begin{array}{c}
OH \\
| \\
R^1-C-OR^2 \\
| \\
H
\end{array}
$$

and formation are described in Section 13.4.

The term *intramolecular* tells us that the reacting carbonyl and hydroxyl groups are part of the same molecule.

Chapter 16 Carbohydrates

In Figure 16.7 a new type of structural formula, called a **Haworth projection,** is presented. Although on first inspection it appears complicated, it is quite simple to derive a Haworth projection from a structural formula, as Example 16.3 shows.

EXAMPLE 16.3 — **Drawing the Haworth Projection of a Monosaccharide from the Structural Formula**

LEARNING GOAL 7

Draw the Haworth projections of α- and β-D-glucose.

Solution

1. Before attempting to draw a Haworth projection, look at the first steps of ring formation shown here:

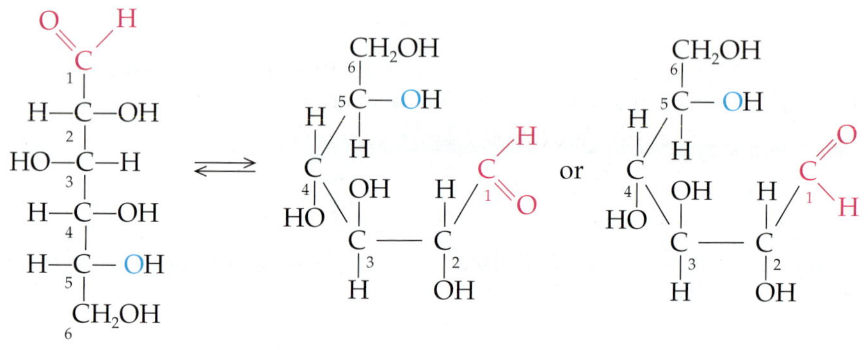

   Glucose (open chain) ⇌ Glucose (intermediates in ring formation)

   Try to imagine that you are seeing the molecules shown above in three dimensions. Some of the substituent groups on the molecule will be above the ring, and some will be beneath it. The question then becomes: How do you determine which groups to place above the ring and which to place beneath the ring?

2. Look at the two-dimensional structural formula. Note the groups (drawn in blue) to the left of the carbon chain. These are placed above the ring in the Haworth projection.

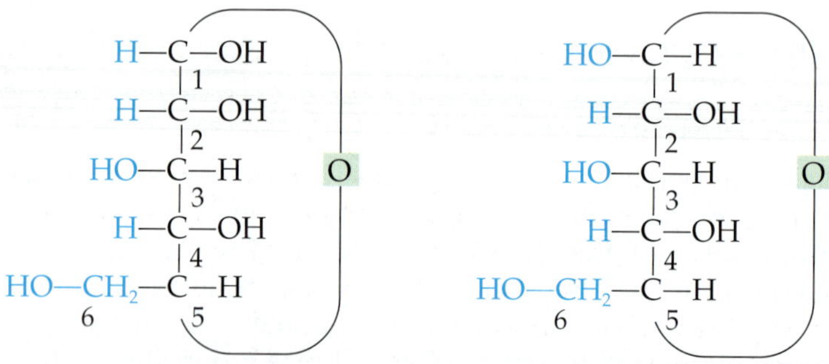

   α-D-Glucose   β-D-Glucose

3. Now note the groups (drawn in red) to the right of the carbon chain. These will be located beneath the carbon ring in the Haworth projection.

Continued—

EXAMPLE 16.3 —Continued

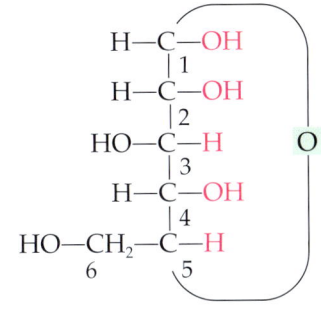

α-D-Glucose 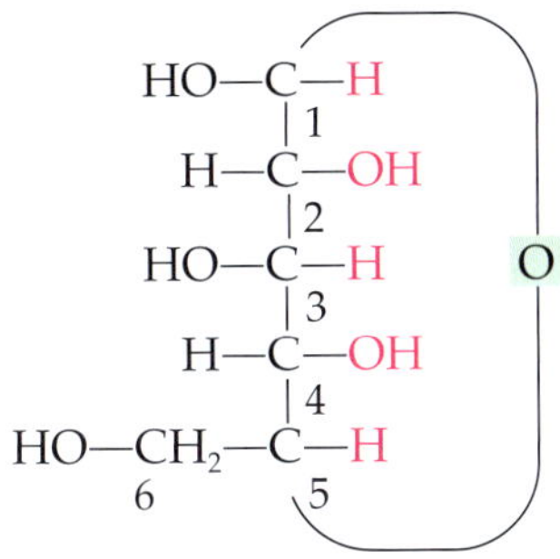   β-D-Glucose

4. Thus in the Haworth projection of the cyclic form of any D-sugar the —CH₂OH group is always "up." When the —OH group at C-1 is also "up," *cis* to the —CH₂OH group, the sugar is β-D-glucose. When the —OH group at C-1 is "down," *trans* to the —CH₂OH group, the sugar is α-D-glucose.

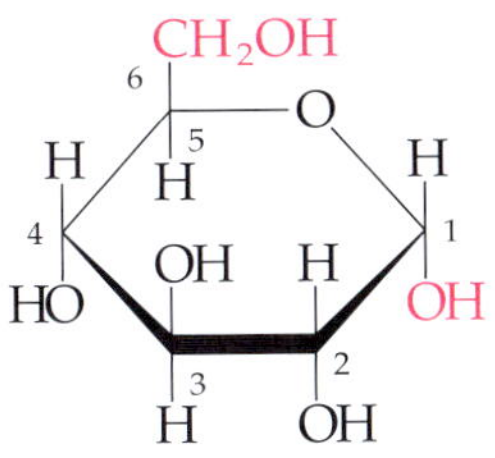

Haworth projection
α-D-Glucose

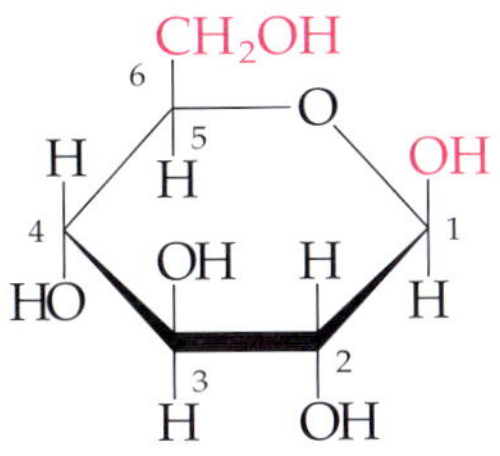

Haworth projection
β-D-Glucose

Question 16.15

Refer to the linear structure of D-galactose. Draw the Haworth projections of α- and β-D-galactose.

Question 16.16

Refer to the linear structure of D-ribose. Draw the Haworth projections of α- and β-D-ribose. Note that D-ribose is a pentose.

Fructose

Fructose, also called levulose and fruit sugar, is the sweetest of all sugars. It is found in large amounts in honey, corn syrup, and sweet fruits. The structure of fructose is similar to that of glucose. When there is a —CH₂OH group instead of a —CHO group at carbon-1 and a —C=O group instead of CHOH at carbon-2, the sugar is a ketose. In this case it is D-fructose.

Cyclization of fructose produces α- and β-D-fructose:

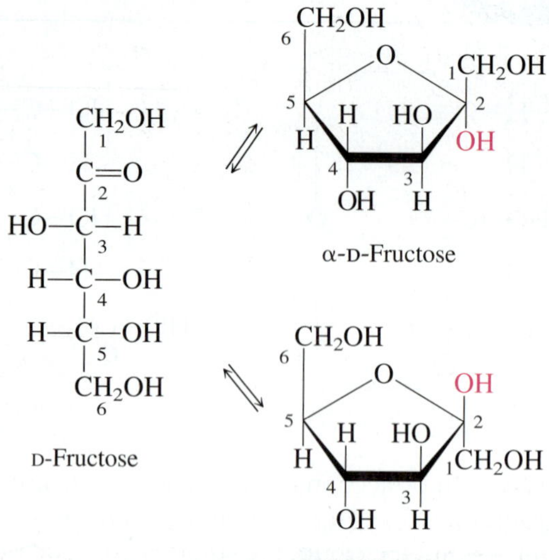

[Structure: D-Fructose linear form with C1 CH₂OH, C2 C=O, C3 HO—C—H, C4 H—C—OH, C5 H—C—OH, C6 CH₂OH]

[Structure: α-D-Fructose Haworth projection]

[Structure: β-D-Fructose Haworth projection]

Hemiketal formation is described in Section 13.4.

Fructose is a ketose, or ketone sugar. Recall that the reaction between an alcohol and a ketone yields a **hemiketal**. Thus the reaction between the C-2 keto group and the C-5 hydroxyl group in the fructose molecule produces an *intramolecular hemiketal*. Fructose forms a five-membered ring structure.

Galactose

Another important hexose is **galactose**. The linear structure of D-galactose and the Haworth projections of α-D-galactose and β-D-galactose are shown here:

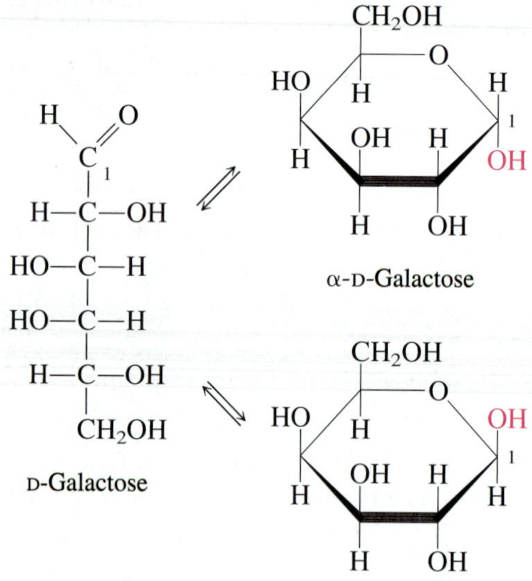

[Structure: D-Galactose linear form with C1 CHO, C2 H—C—OH, C3 HO—C—H, C4 HO—C—H, C5 H—C—OH, C6 CH₂OH]

[Structure: α-D-Galactose Haworth projection]

[Structure: β-D-Galactose Haworth projection]

Galactose is found in biological systems as a component of the disaccharide lactose, or milk sugar. This is the principal sugar found in the milk of most mammals. β-D-Galactose and a modified form, β-D-N-acetylgalactosamine, are also components of the blood group antigens.

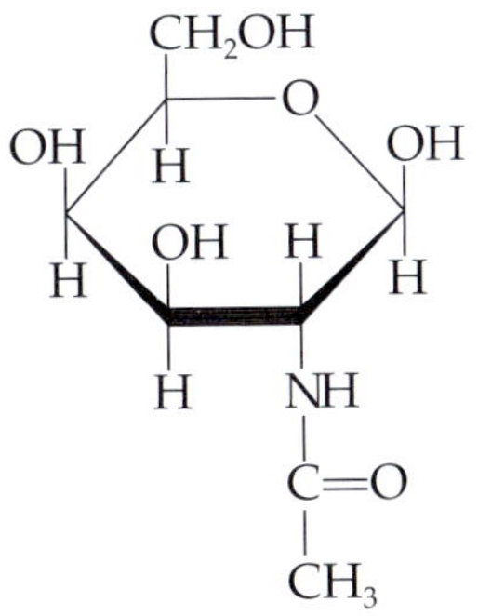

16.4 Biologically Important Monosaccharides

β-D-N-Acetylgalactosamine

Ribose and Deoxyribose, Five-Carbon Sugars

Ribose is a component of many biologically important molecules, including RNA and various coenzymes that are required by many of the enzymes that carry out biochemical reactions in the body. The structure of the five-carbon sugar D-ribose is shown in its open-chain form and in the α- and β-cyclic forms.

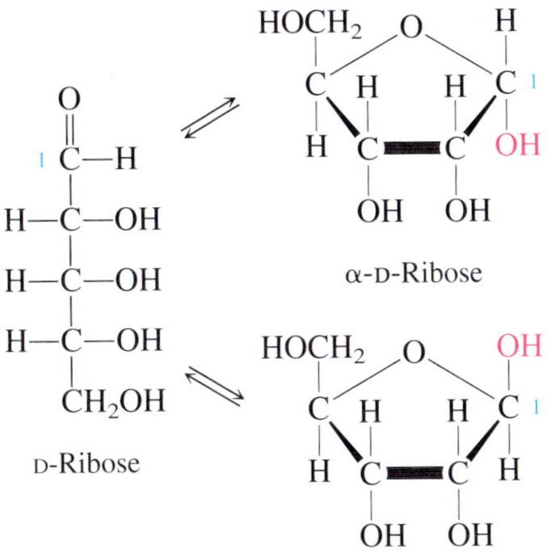

DNA, the molecule that carries the genetic information of the cell, contains 2-deoxyribose. In this molecule the —OH group at C-2 has been replaced by a hydrogen, hence the designation "2-deoxy," indicating the absence of an oxygen.

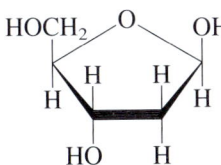

β-D-2-Deoxyribose

Reducing Sugars

The aldehyde group of aldoses is readily oxidized by the Benedict's reagent. Recall that the **Benedict's reagent** is a basic buffer solution that contains Cu^{2+} ions. The Cu^{2+} ions are reduced to Cu^+ ions, which, in basic solution, precipitate as brick-red Cu_2O. The aldehyde group of the aldose is oxidized to a carboxylic acid, which undergoes an acid-base reaction to produce a carboxylate anion.

8 LEARNING GOAL

9 LEARNING GOAL

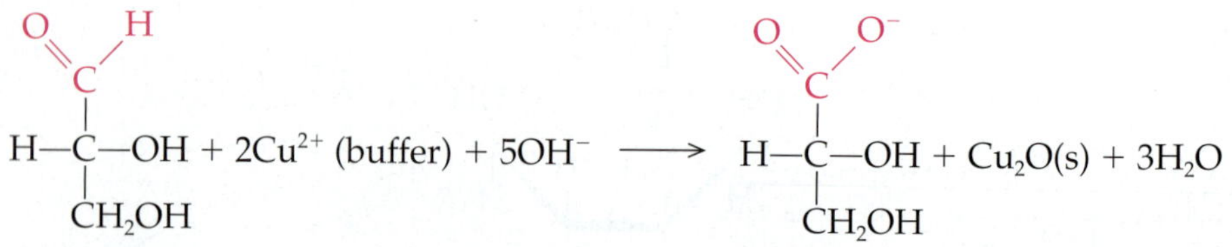

$$\underset{\underset{CH_2OH}{|}}{\overset{\overset{O\;\;\;H}{\diagdown\!/}}{\underset{|}{C}}} + 2Cu^{2+} \text{(buffer)} + 5OH^- \longrightarrow \underset{\underset{CH_2OH}{|}}{\overset{\overset{O\;\;\;O^-}{\diagdown\!\!/}}{\underset{|}{C}}} + Cu_2O(s) + 3H_2O$$
$$\text{H—C—OH} \hspace{4em} \text{H—C—OH}$$

Although ketones generally are not easily oxidized, ketoses are an exception to that rule. Because of the —OH group on the carbon next to the carbonyl group, ketoses can be converted to aldoses, under basic conditions, via an *enediol reaction*:

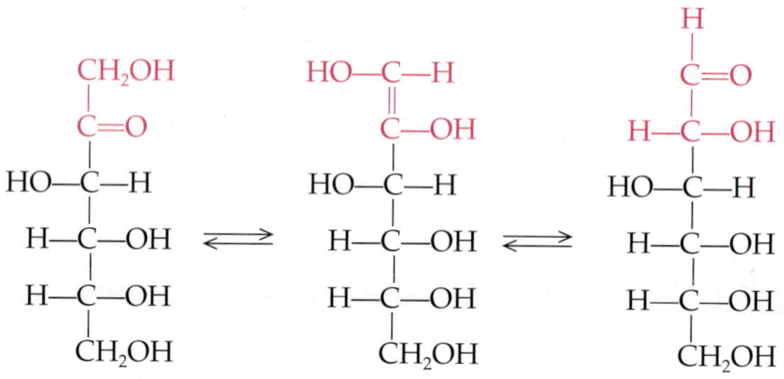

[Structures of D-Fructose ⇌ Enediol ⇌ D-Glucose]

The name of the enediol reaction is derived from the structure of the intermediate through which the ketose is converted to the aldose: It has a double bond (ene), and it has two hydroxyl groups (diol). Because of this enediol reaction, ketoses are also able to react with Benedict's reagent, which is basic. Because the metal ions in the solution are reduced, the sugars are serving as reducing agents and are called **reducing sugars.** All monosaccharides and all the common disaccharides, except sucrose, are reducing sugars.

See A Medical Perspective: Diabetes Mellitus and Ketone Bodies in Chapter 23.

For many years the Benedict's reagent was used to test for *glucosuria*, the presence of excess glucose in the urine. Individuals suffering from *Type I insulin-dependent diabetes mellitus* do not produce the hormone insulin, which controls the uptake of glucose from the blood. When the blood glucose level rises above 160 mg/100 mL, the kidney is unable to reabsorb the excess, and glucose is found in the urine. Although the level of blood glucose could be controlled by the injection of insulin, urine glucose levels were monitored to ensure that the amount of insulin injected was correct. The Benedict's reagent was a useful tool because the amount of Cu_2O formed, and hence the degree of color change in the reaction, is directly proportional to the amount of reducing sugar in the urine. A brick-red color indicates a very high concentration of glucose in the urine. Yellow, green, and blue-green solutions indicate decreasing amounts of glucose in the urine, and a blue solution indicates an insignificant concentration.

Use of Benedict's reagent to test urine glucose levels has largely been replaced by chemical tests that provide more accurate results. The most common technology is based on a test strip that is impregnated with the enzyme glucose oxidase and other agents that will cause a measurable color change. In one such kit, the compounds that result in color development include the enzyme peroxidase, a compound called orthotolidine, and a yellow dye. When a drop of urine is placed on the strip, the glucose oxidase catalyzes the conversion of glucose into gluconic acid and hydrogen peroxide.

16.5 Biologically Important Disaccharides

$$\begin{array}{c}\text{C—H} \\ \| \\ \text{O}\end{array} \quad\quad \begin{array}{c}\text{C—OH} \\ \| \\ \text{O}\end{array}$$

H—C—OH   H—C—OH
HO—C—H + O₂ →(Glucose oxidase)→ HO—C—H + H₂O₂
H—C—OH   H—C—OH
H—C—OH   H—C—OH
CH₂OH   CH₂OH

D-Glucose   D-Gluconic acid

The enzyme peroxidase catalyzes a reaction between the hydrogen peroxide and orthotolidine. This produces a blue product. The yellow dye on the test strip simply serves to "dilute" the blue end product, thereby allowing greater accuracy of the test over a wider range of glucose concentrations. The test strip remains yellow if there is no glucose in the sample. It will vary from a pale green to a dark blue, depending on the concentration of glucose in the urine sample.

Frequently, doctors recommend that diabetics monitor their *blood* glucose levels multiple times each day because this provides a more accurate indication of how well the diabetic is controlling his or her diet. Many small, inexpensive glucose meters are available that couple the oxidation of glucose by glucose oxidase with an appropriate color change system. As with the urine test, the intensity of the color change is proportional to the amount of glucose in the blood. A photometer within the device reads the color change and displays the glucose concentration. An even newer technology uses a device that detects the electrical charge generated by the oxidation of glucose. In this case, it is the amount of electrical charge that is proportional to the glucose concentration.

> Actually, glucose oxidase can only oxidize β-D-glucose. However, in the blood there is an equilibrium mixture of the α and β anomers of glucose. Fortunately, α-D-glucose is very quickly converted to β-D-glucose.

16.5 Biologically Important Disaccharides

Recall that disaccharides consist of two monosaccharides joined through an "oxygen bridge." In biological systems, monosaccharides exist in the cyclic form and, as we have seen, they are actually hemiacetals or hemiketals. Recall that when a hemiacetal reacts with an alcohol, the product is an *acetal*, and when a hemiketal reacts with an alcohol, the product is a *ketal*. In the case of disaccharides, the alcohol comes from a second monosaccharide. The acetals or ketals formed are given the general name *glycosides*, and the carbon-oxygen bonds are called *glycosidic bonds*.

Glycosidic bond formation is nonspecific; that is, it can occur between a hemiacetal or hemiketal and any of the hydroxyl groups on the second monosaccharide. However, in biological systems, we commonly see only particular disaccharides, such as maltose (Figure 16.8), lactose (see Figure 16.10), or sucrose (see Figure 16.11). These specific disaccharides are produced in cells because the reactions are catalyzed by enzymes. Each enzyme catalyzes the synthesis of one specific disaccharide, ensuring that one particular pair of hydroxyl groups on the reacting monosaccharides participates in glycosidic bond formation.

10 LEARNING GOAL

Acetals and ketals are described in Section 13.4:

$$\begin{array}{cc} \text{OR} & \text{OR} \\ | & | \\ \text{R—C—OR} & \text{R—C—OR} \\ | & | \\ \text{H} & \text{R} \\ \textit{Acetal} & \textit{Ketal} \end{array}$$

Maltose

If an α-D-glucose and a second glucose are linked, as shown in Figure 16.8, the disaccharide is **maltose**, or malt sugar. This is one of the intermediates in the hydrolysis of starch. Because the C-1 hydroxyl group of α-D-glucose is attached to C-4 of another glucose molecule, the disaccharide is linked by an α (1 → 4) glycosidic bond.

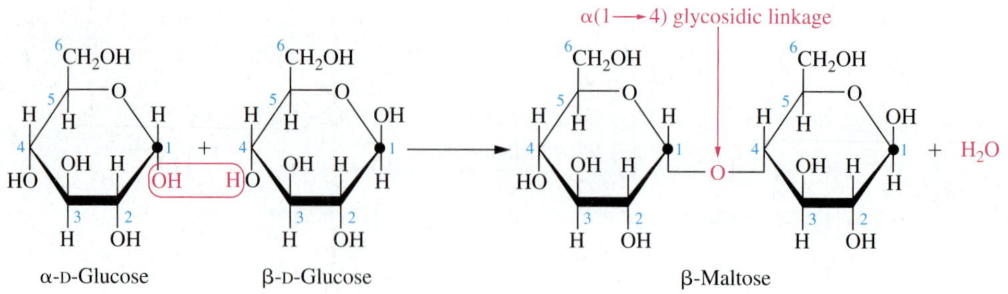

Figure 16.8
Glycosidic bond formed between the C-1 hydroxyl group of α-D-glucose and the C-4 hydroxyl group of β-D-glucose. The disaccharide is called β-maltose because the hydroxyl group at the reducing end of the disaccharide has the β-configuration.

Maltose is a reducing sugar. Any disaccharide that has a hemiacetal hydroxyl group (a free —OH group at C-1) is a reducing sugar. This is because the cyclic structure can open at this position to form a free aldehyde. Disaccharides that do not contain a hemiacetal group on C-1 do not react with the Benedict's reagent and are called **nonreducing sugars**.

Lactose

LEARNING GOAL 11

Milk sugar, or **lactose,** is a disaccharide made up of one molecule of β-D-galactose and one of either α- or β-D-glucose. Galactose differs from glucose only in the configuration of the hydroxyl group at C-4 (Figure 16.9). In the cyclic form of glucose, the C-4 hydroxyl group is "down," and in galactose it is "up." In lactose the C-1 hydroxyl group of β-D-galactose is bonded to the C-4 hydroxyl group of either an α- or β-glucose. The bond between the two monosaccharides is therefore a β(1 → 4) glycosidic bond (Figure 16.10).

Lactose is the principal sugar in the milk of most mammals. To be used by the body as an energy source, lactose must be hydrolyzed to produce glucose and galactose. Note that this is simply the reverse of the reaction shown in Figure 16.10. Glucose liberated by the hydrolysis of lactose is used directly in the energy-harvesting reactions of glycolysis. However, a series of reactions is necessary to convert galactose into a phosphorylated form of glucose that can be used in cellular metabolic reactions. In humans the genetic disease **galactosemia** is caused by the absence of one or more of the enzymes needed for this conversion. A toxic

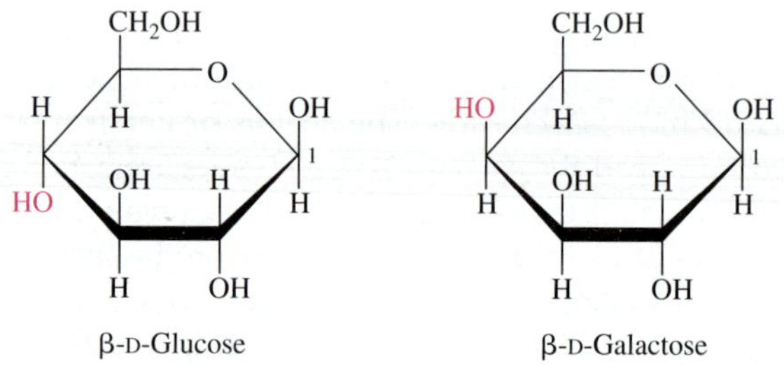

Glycolysis is discussed in Chapter 21.

Figure 16.9
Comparison of the cyclic forms of glucose and galactose. Note that galactose is identical to glucose except in the position of the C-4 hydroxyl group.

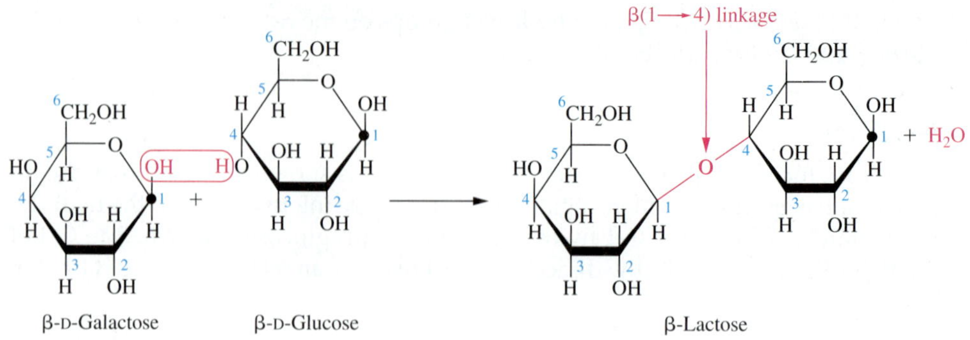

Figure 16.10
Glycosidic bond formed between the C-1 hydroxyl group of β-D-galactose and the C-4 hydroxyl group of β-D-glucose. The disaccharide is called β-lactose because the hydroxyl group at the reducing end of the disaccharide has the β-configuration.

compound formed from galactose accumulates in people who suffer from galactosemia. If the condition is not treated, galactosemia leads to severe mental retardation, cataracts, and early death. However, the effects of this disease can be avoided entirely by providing galactosemic infants with a diet that does not contain galactose. Such a diet, of course, cannot contain lactose and therefore must contain no milk or milk products.

Many adults, and some children, are unable to hydrolyze lactose because they do not make the enzyme *lactase*. This condition, which affects 20% of the population of the United States, is known as **lactose intolerance.** Undigested lactose remains in the intestinal tract and causes cramping and diarrhea that can eventually lead to dehydration. Some of the lactose is metabolized by intestinal bacteria that release organic acids and CO_2 gas into the intestines, causing further discomfort. Lactose intolerance is unpleasant, but its effects can be avoided by a diet that excludes milk and milk products. Alternatively, the enzyme that hydrolyzes lactose is available in tablet form. When ingested with dairy products, it breaks down the lactose, preventing symptoms.

Sucrose

Sucrose is also called table sugar, cane sugar, or beet sugar. Sucrose is an important carbohydrate in plants. It is water soluble and can easily be transported through the circulatory system of the plant. It cannot be synthesized by animals. High concentrations of sucrose produce a high osmotic pressure, which inhibits the growth of microorganisms, so it is used as a preservative. Of course, it is also widely used as a sweetener. In fact, it is estimated that the average American consumes 100–125 pounds of sucrose each year. It has been suggested that sucrose in the diet is undesirable because it represents a source of empty calories; that is, it contains no vitamins or minerals. However, the only negative association that has been scientifically verified is the link between sucrose in the diet and dental caries, or cavities (see A Human Perspective: Tooth Decay and Simple Sugars on p. 528).

Sucrose is a disaccharide of α-D-glucose joined to β-D-fructose (Figure 16.11). The glycosidic linkage between α-D-glucose and β-D-fructose is quite different from those that we have examined for lactose and maltose. This bond involves the anomeric carbons of *both* sugars! This bond is called an (α1 → β2) glycosidic linkage, since it involves the C-1 anomeric carbon of glucose and the C-2 anomeric carbon of fructose (noted in red in Figure 16.11). Because the (α1 → β2) glycosidic bond joins both anomeric carbons, there is no hemiacetal group. As a result, sucrose will not react with Benedict's reagent and is not a reducing sugar.

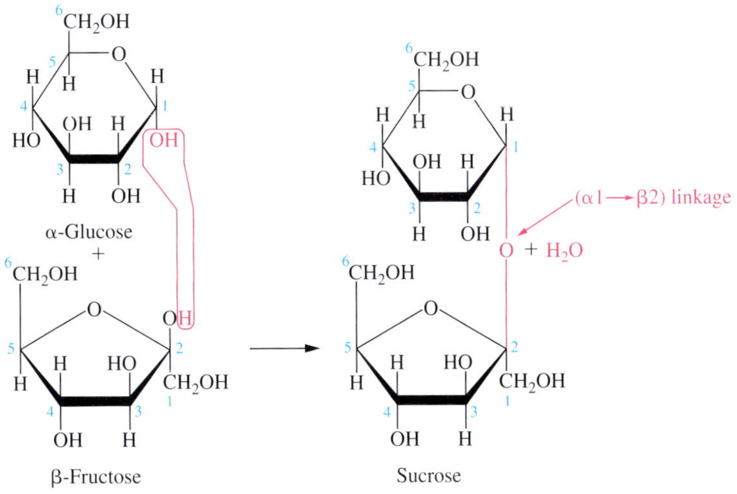

Figure 16.11
Glycosidic bond formed between the C-1 hydroxyl of α-D-glucose and the C-2 hydroxyl of β-D-fructose. This bond is called an (α1 → β2) glycosidic linkage. The disaccharide formed in this reaction is sucrose.

A Human Perspective

Blood Transfusions and the Blood Group Antigens

The first blood transfusions were tried in the seventeenth century, when physicians used animal blood to replace human blood lost by hemorrhages. Unfortunately, many people died as a result of this attempted cure, and transfusions were banned in much of Europe. Transfusions from human donors were somewhat less lethal, but violent reactions often led to the death of the recipient, and by the nineteenth century, transfusions had been abandoned as a medical failure.

In 1904, Dr. Karl Landsteiner performed a series of experiments on the blood of workers in his laboratory. His results explained the mysterious transfusion fatalities, and blood transfusions were reinstated as a lifesaving clinical tool. Landsteiner took blood samples from his coworkers. He separated the blood cells from the serum, the liquid component of the blood, and mixed these samples in test tubes. When he mixed serum from one individual with blood cells of another, Landsteiner observed that, in some instances, the serum samples caused clumping, or *agglutination*, of red blood cells (RBC). (See figure below.) The agglutination reaction always indicated that the two bloods were incompatible and transfusion could lead to life-threatening reactions. As a result of many such experiments, Landsteiner showed that there are four human blood groups, designated A, B, AB, and O.

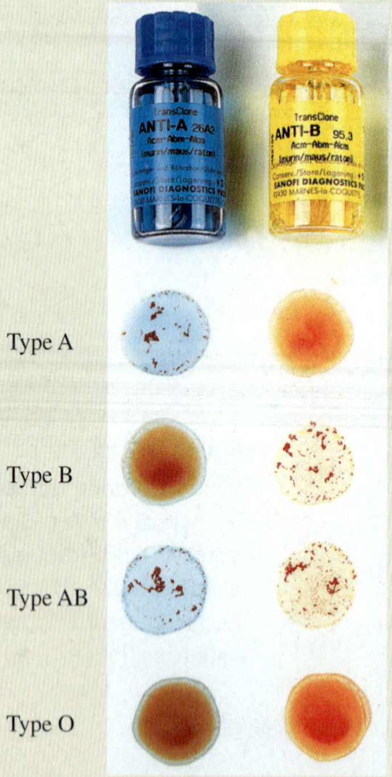

We now know that differences among blood groups reflect differences among oligosaccharides attached to the proteins and lipids of the RBC membranes. The oligosaccharides on the RBC surface have a common core, as shown in the accompanying figure, consisting of β-D-N-acetylgalactosamine, galactose, N-acetylneuraminic acid (sialic acid), and L-fucose. It is the terminal monosaccharide of this oligosaccharide that distinguishes the cells and governs the compatibility of the blood types.

The A blood group antigen has β-D-N-acetylgalactosamine at its end, whereas the B blood group antigen has α-D-galactose. In type O blood, neither of these sugars is found on the cell surface; only the core oligosaccharide is present. Some of the oligosaccharides on type AB blood cells have a terminal β-D-N-acetylgalactosamine, whereas others have a terminal α-D-galactose.

Why does agglutination occur? The clumping reaction that occurs when incompatible bloods are mixed is an antigen-antibody reaction. Antigens are large molecules, often portions of bacteria or viruses, that stimulate the immune defenses of the body to produce protective antibodies. Antibodies bind to the foreign antigens and help to destroy them.

People with type A blood also have antibodies against type B blood (anti-B antibodies) in the blood serum. If the person with type A blood receives a transfusion of type B blood, the anti-B antibodies bind to the type B blood cells, causing clumping and destruction of those cells that can result in death. Individuals with type B blood also produce anti-A antibodies and therefore cannot receive a transfusion from a type A individual. Those with type AB blood have neither anti-A nor anti-B antibodies in their blood. (If they did, they would destroy their own red blood cells!) Type O blood has no A or B antigens on the RBC but has both anti-A and anti-B antibodies. Because of the presence of both types of antibodies, type O individuals can receive transfusions only from a person who is also type O.

ABO blood typing kit. When antibodies bind to antigens on the cell surface, clumping occurs.

For Further Understanding

People with type AB blood are referred to as universal recipients. Explain why these people can receive blood of any of the four ABO types.

People with type O blood are referred to as universal donors. Explain why type O blood can be safely transfused into patients of any ABO blood type.

16.5 Biologically Important Disaccharides

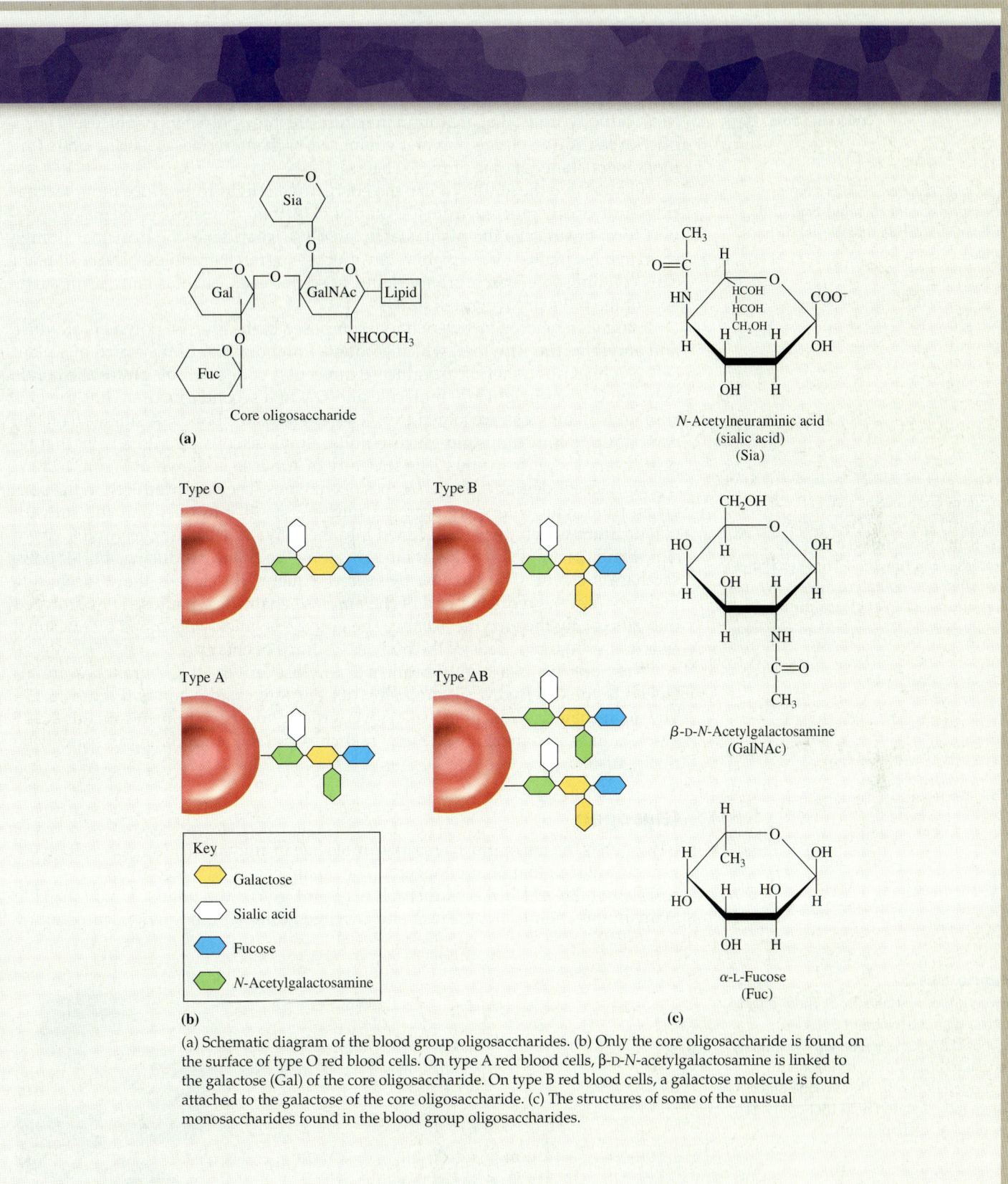

(a) Schematic diagram of the blood group oligosaccharides. (b) Only the core oligosaccharide is found on the surface of type O red blood cells. On type A red blood cells, β-D-N-acetylgalactosamine is linked to the galactose (Gal) of the core oligosaccharide. On type B red blood cells, a galactose molecule is found attached to the galactose of the core oligosaccharide. (c) The structures of some of the unusual monosaccharides found in the blood group oligosaccharides.

16.6 Polysaccharides

Starch

Most carbohydrates that are found in nature are large polymers of glucose. Thus a polysaccharide is a large polymer composed of many monosaccharide units (the monomers) joined in one or more chains.

Plants have the ability to use the energy of sunlight to produce monosaccharides, principally glucose, from CO_2 and H_2O. Although sucrose is the major transport form of sugar in the plant, starch (a polysaccharide) is the principal storage form in most plants. These plants store glucose in starch granules. Nearly all plant cells contain some starch granules, but in some seeds, such as corn, as much as 80% of the cell's dry weight is starch.

Starch is a heterogeneous material composed of the glucose polymers **amylose** and **amylopectin**. Amylose, which accounts for about 80% of the starch of a plant cell, is a linear polymer of α-D-glucose molecules connected by glycosidic bonds between C-1 of one glucose molecule and C-4 of a second glucose. Thus the glucose units in amylose are joined by α(1 → 4) glycosidic bonds. A single chain can contain up to four thousand glucose units. Amylose coils up into a helix that repeats every six glucose units. The structure of amylose is shown in Figure 16.12.

Amylose is degraded by two types of enzymes. They are produced in the pancreas, from which they are secreted into the small intestine, and the salivary glands, from which they are secreted into the saliva. α-Amylase cleaves the glycosidic bonds of amylose chains at random along the chain, producing shorter polysaccharide chains. The enzyme β-*amylase* sequentially cleaves the disaccharide maltose from the reducing end of the amylose chain. The maltose is hydrolyzed into glucose by the enzyme *maltase*. The glucose is quickly absorbed by intestinal cells and used by the cells of the body as a source of energy.

Amylopectin is a highly branched amylose in which the branches are attached to the C-6 hydroxyl groups by α(1 → 6) glycosidic bonds (Figure 16.13). The main chains consist of α(1 → 4) glycosidic bonds. Each branch contains 20–25 glucose units, and there are so many branches that the main chain can scarcely be distinguished.

Glycogen

Glycogen is the major glucose storage molecule in animals. The structure of glycogen is similar to that of amylopectin. The "main chain" is linked by α(1 → 4) glycosidic bonds, and it has numerous α(1 → 6) glycosidic bonds, which provide many branch points along the chain. Glycogen differs from amylopectin only by

A polymer (Section 11.5) is a large molecule made up of many small units, the monomers, held together by chemical bonds.

Enzymes are proteins that serve as biological catalysts. They speed up biochemical reactions so that life processes can function. These enzymes are called α(1 → 4) glycosidases because they cleave α(1 → 4) glycosidic bonds.

Figure 16.12
Structure of amylose. (a) A linear chain of α-D-glucose joined in α(1 → 4) glycosidic linkage makes up the primary structure of amylose. (b) Owing to hydrogen bonding, the amylose chain forms a left-handed helix that contains six glucose units per turn.

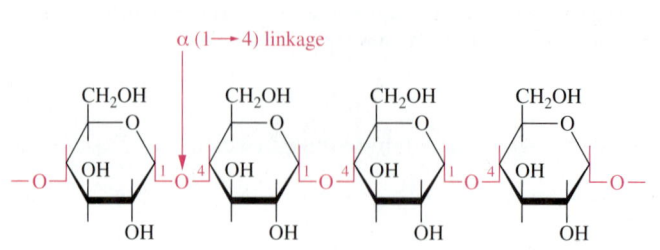

(a)

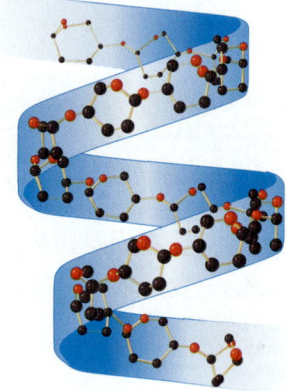

(b)

16.6 Polysaccharides

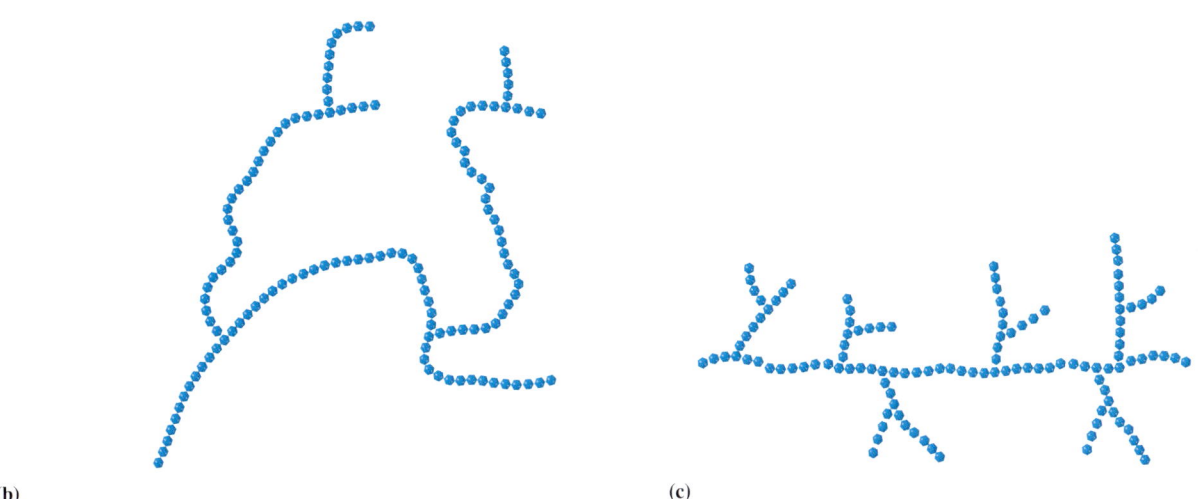

Figure 16.13
Structure of amylopectin and glycogen. (a) Both amylopectin and glycogen consist of chains of α-D-glucose molecules joined in α(1 → 4) glycosidic linkages. Branching from these chains are other chains of the same structure. Branching occurs by formation of α(1 → 6) glycosidic bonds between glucose units. (b) A representation of the branched-chain structure of amylopectin. (c) A representation of the branched-chain structure of glycogen. Glycogen differs from amylopectin only in that the branches are shorter and there are more of them.

having more and shorter branches. Otherwise, the two molecules are virtually identical. The structure of glycogen is shown in Figure 16.13.

Glycogen is stored in the liver and skeletal muscle. Glycogen synthesis and degradation in the liver are carefully regulated. As we will see in Section 21.7, these two processes are intimately involved in keeping blood glucose levels constant.

Cellulose

The most abundant polysaccharide, indeed the most abundant organic molecule in the world, is **cellulose,** a polymer of β-D-glucose units linked by β(1 → 4) glycosidic bonds (Figure 16.14). A molecule of cellulose typically contains about 3000 glucose units, but the largest known cellulose, produced by the alga *Valonia*, contains 26,000 glucose molecules.

Cellulose is a structural component of the plant cell wall. The unbranched structure of the cellulose polymer and the β(1 → 4) glycosidic linkages allow cellulose molecules to form long, straight chains of parallel cellulose molecules called *fibrils*. These fibrils are quite rigid and are held together tightly by hydrogen bonds; thus it is not surprising that cellulose is a cell wall structural element.

In contrast to glycogen, amylose, and amylopectin, cellulose *cannot* be digested by humans. The reason is that we cannot synthesize the enzyme *cellulase*, which can hydrolyze the β(1 → 4) glycosidic linkages of the cellulose polymer. Indeed, only a few animals, such as termites, cows, and goats, are able to digest cellulose. These animals have, within their digestive tracts, microorganisms that

A Medical Perspective

Monosaccharide Derivatives and Heteropolysaccharides of Medical Interest

Many of the carbohydrates with important functions in the human body are either derivatives of simple monosaccharides or are complex polymers of monosaccharide derivatives. One type of monosaccharide derivatives, the uronates, is formed when the terminal—CH_2OH group of a monosaccharide is oxidized to a carboxylate group. α-D-Glucuronate is a uronate of glucose:

α-D-Glucuronate

In liver cells, α-D-glucuronate is bonded to hydrophobic molecules, such as steroids, to increase their solubility in water. When bonded to the modified sugar, steroids are more readily removed from the body.

Amino sugars are a second important group of monosaccharide derivatives. In amino sugars one of the hydroxyl groups (usually on carbon-2) is replaced by an amino group. Often these are found in complex oligosaccharides that are attached to cellular proteins and lipids. The most common amino sugars, D-glucosamine and D-galactosamine, are often found in the N-acetyl form. N-acetylglucosamine is a component of bacterial cell walls and N-acetylgalactosamine is a component of the A, B, O blood group antigens (see preceding, A Human Perspective: Blood Transfusions and the Blood Group Antigens).

α-D-Glucosamine

α-D-N-Acetylglucosamine

Heteropolysaccharides are long-chain polymers that contain more than one type of monosaccharide, many of which are amino sugars. These *glycosaminoglycans* include chondroitin sulfate, hyaluronic acid, and heparin. Hyaluronic acid is abundant in the fluid of joints and in the vitreous humor of the eye. Chondroitin sulfate is an important component of cartilage; and heparin has anticoagulant function. The structures of the repeat units of these polymers are shown below.

Repeat unit of chondroitin sulfate

Figure 16.14
The structure of cellulose.

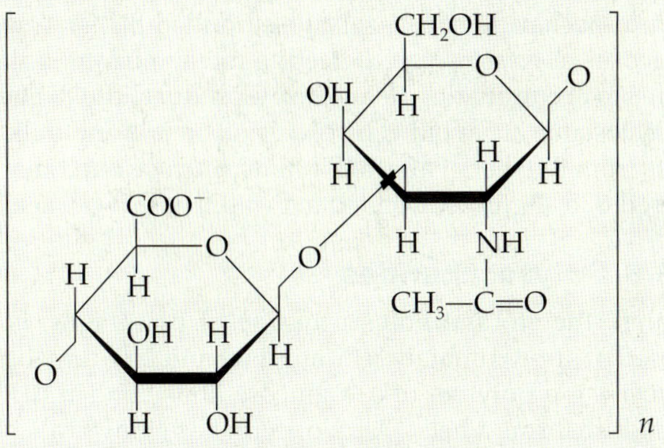

16.6 Polysaccharides

onset and progression of arthritis. It has been suggested that ingestion of D-glucosamine can actually "jump-start" production of cartilage and help repair eroded cartilage in arthritic joints.

It has also been suggested that chondroitin sulfate can protect existing cartilage from premature breakdown. It absorbs large amounts of water, which is thought to facilitate diffusion of nutrients into the cartilage, providing precursors for the synthesis of new cartilage. The increased fluid also acts as a shock absorber.

Studies continue on the effects that D-glucosamine and chondroitin sulfate have on degenerative joint disease. To date the studies are inconclusive because a large placebo effect is observed with sufferers of osteoarthritis. Many people in the control groups of these studies also experience relief of symptoms when they receive treatment with a placebo, such as a sugar pill.

Capsules containing D-glucosamine and chondroitin sulfate are available over the counter, and many sufferers of osteoarthritis prefer to take this nutritional supplement as an alternative to any nonsteroidal anti-inflammatory drugs (NSAID), such as ibuprofen. Although NSAIDs can reduce inflammation and pain, long-term use of NSAIDs can result in stomach ulcers, damage to auditory nerves, and kidney damage.

Repeat unit of hyaluronic acid

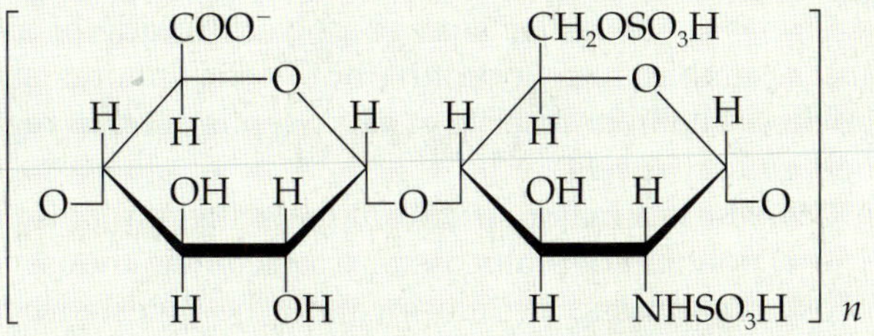

Repeat unit of heparin

Two of these molecules have been studied as potential treatments for osteoarthritis, a painful, degenerative disease of the joints. The amino sugar D-glucosamine is thought to stimulate the production of collagen. Collagen is one of the main components of articular cartilage, which is the shock-absorbing cushion within the joints. With aging, some of the D-glucosamine is lost, leading to a reduced cartilage layer and to the

For Further Understanding

In Chapter 15 we learned that nonsteroidal anti-inflammatory drugs (NSAIDs), such as ibuprofen, are analgesics used to treat pain, such as that associated with osteoarthritis. Why do many people prefer to treat osteoarthritis with D-glucosamine and chondroitin sulfate rather than NSAIDs?

Explain why attaching a molecule such as α-D-glucuronate to a steroid molecule would increase its water solubility.

---

produce the enzyme cellulase. The sugars released by this microbial digestion can then be absorbed and used by these animals. In humans, cellulose from fruits and vegetables serves as fiber in the diet.

Question 16.17 What chemical reactions are catalyzed by α-amylase and β-amylase?

Question 16.18 What is the function of cellulose in the human diet? How does this relate to the structure of cellulose?

SUMMARY

16.1 Types of Carbohydrates

Carbohydrates are found in a wide variety of naturally occurring substances and serve as principal energy sources for the body. Dietary carbohydrates include complex carbohydrates, such as starch in potatoes, and simple carbohydrates, such as sucrose.

Carbohydrates are classified as *monosaccharides* (one sugar unit), *disaccharides* (two sugar units), *oligosaccharides* (three to ten sugar units), or *polysaccharides* (many sugar units).

16.2 Monosaccharides

Monosaccharides that have an aldehyde as their most oxidized functional group are *aldoses*, and those having a ketone group as their most oxidized functional group are *ketoses*. They may be classified as *trioses, tetroses, pentoses,* and so forth, depending on the number of carbon atoms in the carbohydrate.

16.3 Stereoisomers and Stereochemistry

Stereoisomers of monosaccharides exist because of the presence of *chiral carbon* atoms. They are classified as D- or L- depending on the arrangement of the atoms on the chiral carbon farthest from the aldehyde or ketone group. If the —OH on this carbon is to the right, the stereoisomer is of the D-family. If the —OH group is to the left, the stereoisomer is of the L-family.

Each member of a pair of stereoisomers will rotate plane-polarized light in different directions. A polarimeter is used to measure the direction of rotation of plane-polarized light. Compounds that rotate light in a clockwise direction are termed dextrorotatory and are designated by a plus sign (+). Compounds that rotate light in a counterclockwise direction are called levorotatory and are indicated by a minus sign (−).

The *Fischer Projection* is a two-dimensional drawing of a molecule that shows a chiral carbon at the intersection of two lines. Horizontal lines represent bonds projecting out of the page and vertical lines represent bonds that project into the page. The most oxidized carbon is always represented at the "top" of the structure.

16.4 Biologically Important Monosaccharides

Important monosaccharides include *glyceraldehyde, glucose, fructose,* and *ribose*. Monosaccharides containing five or six carbon atoms can exist as five-membered or six-membered rings. Formation of a ring produces a new chiral carbon at the original carbonyl carbon, which is designated either α or β depending on the orientation of the groups. The cyclization of an aldose produces an intramolecular *hemiacetal*, and the cyclization of a ketose yields an intramolecular *hemiketal*.

Reducing sugars are oxidized by the *Benedict's reagent*. All monosaccharides and all common disaccharides, except sucrose, are reducing sugars. At one time Benedict's reagent was used to determine the concentration of glucose in urine.

16.5 Biologically Important Disaccharides

Important disaccharides include *lactose* and *sucrose*. Lactose is a disaccharide of β-D-galactose bonded β(1 → 4) with D-glucose. In *galactosemia*, defective metabolism of galactose leads to accumulation of a toxic by-product. The ill effects of galactosemia are avoided by exclusion of milk and milk products from the diet of affected infants. Sucrose is a dimer composed of α-D-glucose bonded (α1 → β2) with β-D-fructose.

16.6 Polysaccharides

Starch, the storage polysaccharide of plant cells, is composed of approximately 80% amylose and 20% amylopectin. *Amylose* is a polymer of α-D-glucose units bonded α(1 → 4). Amylose forms a helix. Amylopectin has many branches. Its main chain consists of α-D-glucose units bonded α(1 → 4). The branches are connected by α(1 → 6) glycosidic bonds.

Glycogen, the major storage polysaccharide of animal cells, resembles amylopectin, but it has more, shorter branches. The liver reserve of glycogen is used to regulate blood glucose levels.

Cellulose is a major structural molecule of plants. It is a β(1 → 4) polymer of D-glucose that can contain thousands of glucose monomers. Cellulose cannot be digested by animals because they do not produce an enzyme capable of cleaving the β(1 → 4) glycosidic linkage.

KEY TERMS

aldose (16.2)
amylopectin (16.6)
amylose (16.6)
anomer (16.4)
Benedict's reagent (16.4)
carbohydrate (16.1)
cellulose (16.6)
chiral carbon (16.3)
chiral molecule (16.3)
disaccharide (16.1)
enantiomers (16.3)
Fischer Projection (16.3)
fructose (16.4)
galactose (16.4)
galactosemia (16.5)
glucose (16.4)
glyceraldehyde (16.3)
glycogen (16.6)
glycosidic bond (16.1)
Haworth projection (16.4)

hemiacetal (16.4)
hemiketal (16.4)
hexose (16.2)
ketose (16.2)
lactose (16.5)
lactose intolerance (16.5)
maltose (16.5)
monosaccharide (16.1)
nonreducing sugar (16.5)
oligosaccharide (16.1)
pentose (16.2)
polysaccharide (16.1)
reducing sugar (16.4)
ribose (16.4)
saccharide (16.1)
stereochemistry (16.3)
stereoisomers (16.3)
sucrose (16.5)
tetrose (16.2)
triose (16.2)

QUESTIONS AND PROBLEMS

Types of Carbohydrates

Foundations
16.19 What is the difference between a monosaccharide and a disaccharide?
16.20 What is a polysaccharide?
16.21 What is the general molecular formula for a simple sugar?
16.22 What biochemical process is ultimately responsible for the synthesis of sugars?

Applications
16.23 Read the labels on some of the foods in your kitchen, and see how many products you can find that list one or more carbohydrates among the ingredients in the package. Make a list of these compounds, and attempt to classify them as monosaccharides, disaccharides, or polysaccharides.
16.24 Some disaccharides are often referred to by their common names. What are the chemical names of (a) milk sugar, (b) beet sugar, and (c) cane sugar?
16.25 How many kilocalories of energy are released when 1 g of carbohydrate is "burned" or oxidized?
16.26 List some natural sources of carbohydrates.
16.27 Draw and provide the names of an aldohexose and a ketohexose.
16.28 Draw and provide the name of an aldotriose.

Monosaccharides

Foundations
16.29 Define the term *aldose*.
16.30 Define the term *ketose*.
16.31 What is a tetrose?
16.32 What is a hexose?
16.33 What is a ketopentose?
16.34 What is an aldotriose?

Applications
16.35 Identify each of the following sugars. Label each as either a hemiacetal or a hemiketal:

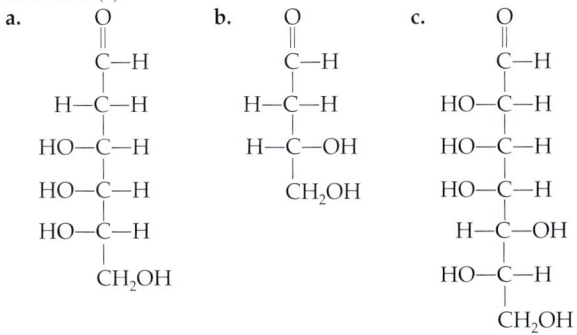

16.36 Draw the open-chain form of the sugars in Problem 16.35.
16.37 Draw all of the different possible aldotrioses of molecular formula $C_3H_6O_3$.
16.38 Draw all of the different possible aldotetroses of molecular formula $C_4H_8O_4$.

Stereoisomers and Stereochemistry

Foundations
16.39 Define the term *stereoisomer*.
16.40 Define the term *enantiomer*.
16.41 Define the term *chiral carbon*.

16.42 Draw an aldotetrose. Note each chiral carbon with an asterisk (*).
16.43 Explain how a polarimeter works.
16.44 What is plane-polarized light?
16.45 What is a Fischer Projection?
16.46 How would you produce a Fischer Projection beginning with a three-dimensional model of a sugar?

Applications
16.47 Is there any difference between dextrose and D-glucose?
16.48 The linear structure of D-glucose is shown in Figure 16.7. Draw its mirror image.
16.49 How are D- and L-glyceraldehyde related?
16.50 Determine whether each of the following is a D- or L-sugar:

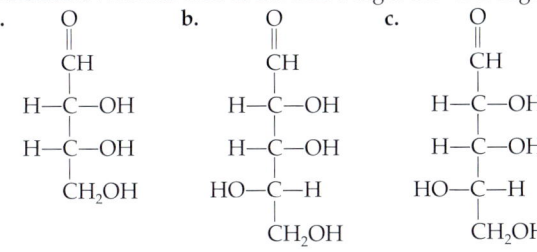

16.51 Draw a Fischer Projection formula for each of the following compounds. Indicate each of the chiral carbons with asterisks (*).

a.
$$\begin{array}{c} O \\ \| \\ C-H \\ HO-C-H \\ H-C-OH \\ HO-C-H \\ HO-C-H \\ CH_2OH \end{array}$$

b.
$$\begin{array}{c} O \\ \| \\ C-H \\ H-C-OH \\ H-C-OH \\ CH_2OH \end{array}$$

c.
$$\begin{array}{c} O \\ \| \\ C-H \\ HO-C-H \\ H-C-OH \\ HO-C-H \\ H-C-OH \\ HO-C-H \\ CH_2OH \end{array}$$

16.52 Draw a Fischer Projection formula for each of the following compounds. Indicate each of the chiral carbons with asterisks (*).

a.
$$\begin{array}{c} O \\ \| \\ C-H \\ H-C-H \\ HO-C-H \\ HO-C-H \\ HO-C-H \\ CH_2OH \end{array}$$

b.
$$\begin{array}{c} O \\ \| \\ C-H \\ H-C-H \\ H-C-OH \\ CH_2OH \end{array}$$

c.
$$\begin{array}{c} O \\ \| \\ C-H \\ HO-C-H \\ HO-C-H \\ HO-C-H \\ H-C-OH \\ HO-C-H \\ CH_2OH \end{array}$$

Biologically Important Monosaccharides

Foundations
16.53 Define the term *anomer*.
16.54 What is a Haworth projection?
16.55 What is a hemiacetal?
16.56 What is a hemiketal?
16.57 Explain why the cyclization of D-glucose forms a hemiacetal.
16.58 Explain why cyclization of D-fructose forms a hemiketal.

Applications

16.59 Why does cyclization of D-glucose give two isomers, α- and β-D-glucose?

16.60 Draw the structure of the open-chain form of D-fructose, and show how it cyclizes to form α- and β-D-fructose.

16.61 Which of the following would give a positive Benedict's test?
   a. Sucrose
   b. Glycogen
   c. β-Maltose
   d. α-Lactose

16.62 Why was the Benedict's reagent useful for determining the amount of glucose in the urine?

16.63 Describe what is meant by a pair of enantiomers. Draw an example of a pair of enantiomers.

16.64 What is a chiral carbon atom?

16.65 When discussing sugars, what do we mean by an intramolecular hemiacetal?

16.66 When discussing sugars, what do we mean by an intramolecular hemiketal?

Biologically Important Disaccharides

Foundations

16.67 What is a ketal?

16.68 What is an acetal?

16.69 What is a glycosidic bond?

16.70 Why are glycosidic bonds either acetals or ketals?

Applications

16.71 Maltose is a disaccharide isolated from amylose that consists of two glucose units linked α(1 → 4). Draw the structure of this molecule.

16.72 Sucrose is a disaccharide formed by linking α-D-glucose and β-D-fructose by an (α1 → β2) bond. Draw the structure of this disaccharide.

16.73 What is the major biological source of lactose?

16.74 What metabolic defect causes galactosemia?

16.75 What simple treatment prevents most of the ill effects of galactosemia?

16.76 What are the major physiological effects of galactosemia?

16.77 What is lactose intolerance?

16.78 What is the difference between lactose intolerance and galactosemia?

Polysaccharides

Foundations

16.79 What is a polymer?

16.80 What form of sugar is used as the major transport sugar in a plant?

16.81 What is the major storage form of sugar in a plant?

16.82 What is the major structural form of sugar in a plant?

Applications

16.83 What is the difference between the structure of cellulose and the structure of amylose?

16.84 How does the structure of amylose differ from that of amylopectin and glycogen?

16.85 What is the major physiological purpose of glycogen?

16.86 Where in the body do you find glycogen stored?

16.87 Where are α-amylase and β-amylase produced?

16.88 Where do α-amylase and β-amylase carry out their enzymatic functions?

CRITICAL THINKING PROBLEMS

1. The six-member glucose ring structure is not a flat ring. Like cyclohexane, it can exist in the chair conformation. Build models of the chair conformation of α- and β-D-glucose. Draw each of these structures. Which would you predict to be the more stable isomer? Explain your reasoning.

2. The following is the structure of salicin, a bitter-tasting compound found in the bark of willow trees:

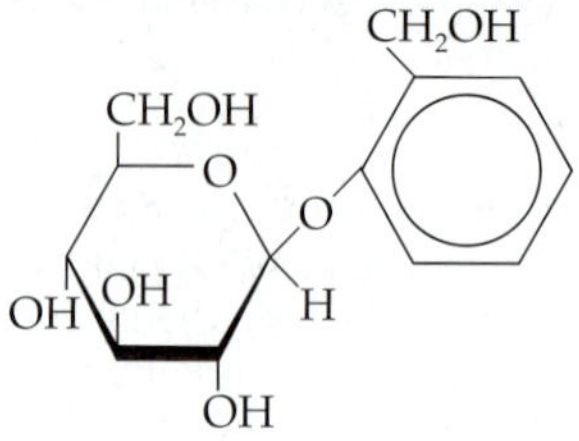

   Salicin

   The aromatic ring portion of this structure is quite insoluble in water. How would forming a glycosidic bond between the aromatic ring and β-D-glucose alter the solubility? Explain your answer.

3. Ancient peoples used salicin to reduce fevers. Write an equation for the acid-catalyzed hydrolysis of the glycosidic bond of salicin. Compare the aromatic product with the structure of acetylsalicylic acid (aspirin). Use this information to develop a hypothesis explaining why ancient peoples used salicin to reduce fevers.

4. Chitin is a modified cellulose in which the C-2 hydroxyl group of each glucose is replaced by

   This nitrogen-containing polysaccharide makes up the shells of lobsters, crabs, and the exoskeletons of insects. Draw a portion of a chitin polymer consisting of four monomers.

5. Pectins are polysaccharides obtained from fruits and berries and used to thicken jellies and jams. Pectins are α(1 → 4) linked D-galacturonic acid. D-Galacturonic acid is D-galactose in which the C-6 hydroxyl group has been oxidized to a carboxyl group. Draw a portion of a pectin polymer consisting of four monomers.

6. Peonin is a red pigment found in the petals of peony flowers. Consider the structure of peonin:

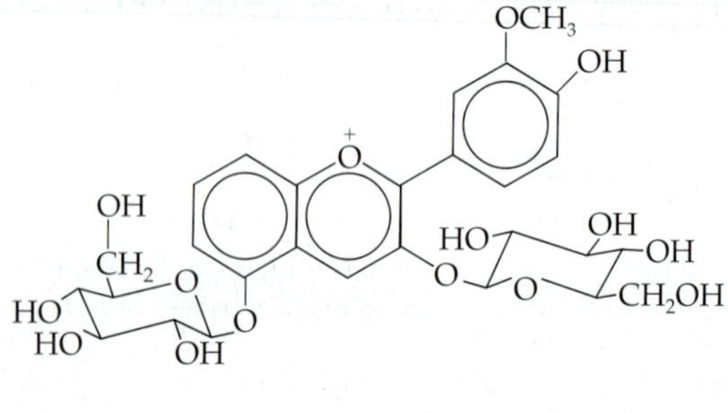

   Why do you think peonin is bonded to two hexoses? What monosaccharide(s) would be produced by acid-catalyzed hydrolysis of peonin?

BIOCHEMISTRY

17

Lipids and Their Functions in Biochemical Systems

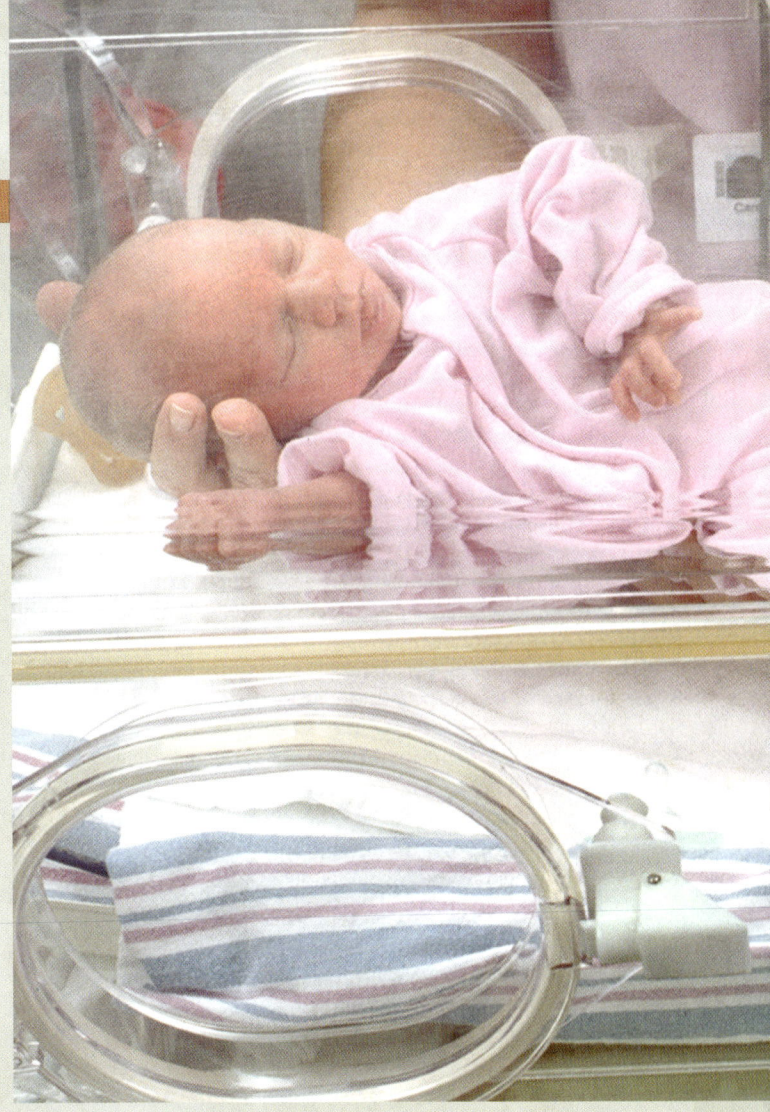

A premature baby in an incubator.

Learning Goals

1. Discuss the physical and chemical properties and biological function of each of the families of lipids.
2. Write the structures of saturated and unsaturated fatty acids.
3. Compare and contrast the structure and properties of saturated and unsaturated fatty acids.
4. Write equations representing the reactions that fatty acids undergo.
5. Describe the functions of prostaglandins.
6. Discuss the mechanism by which aspirin reduces pain.
7. Draw the structure of a phospholipid and discuss its amphipathic nature.
8. Discuss the general classes of sphingolipids and their functions.
9. Draw the structure of the steroid nucleus and discuss the functions of steroid hormones.
10. Describe the function of lipoproteins in triglyceride and cholesterol transport in the body.
11. Draw the structure of the cell membrane and discuss its functions.
12. Discuss passive and facilitated diffusion of materials through a cell membrane.
13. Explain the process of osmosis.
14. Describe the mechanism of action of a Na^+-K^+ ATPase.

Outline

Chemistry Connection:
Lifesaving Lipids

17.1 Biological Functions of Lipids
17.2 Fatty Acids

A Human Perspective:
Mummies Made of Soap

17.3 Glycerides
17.4 Nonglyceride Lipids

A Medical Perspective:
Disorders of Sphingolipid Metabolism

A Medical Perspective:
Steroids and the Treatment of Heart Disease

17.5 Complex Lipids
17.6 The Structure of Biological Membranes

A Medical Perspective:
Liposome Delivery Systems

A Medical Perspective:
Antibiotics That Destroy Membrane Integrity

555

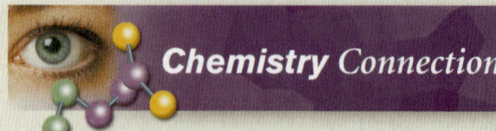

Chemistry Connection

Lifesaving Lipids

In the intensive-care nursery the premature infant struggles for life. Born three and a half months early, the baby weighs only 1.6 pounds, and the lungs labor to provide enough oxygen to keep the tiny body alive. Premature infants often have respiratory difficulties because they have not yet begun to produce *pulmonary surfactant*.

Pulmonary surfactant is a combination of phospholipids and proteins that reduces surface tension in the alveoli of the lungs. (Alveoli are the small, thin-walled air sacs in the lungs.) This allows efficient gas exchange across the membranes of the alveolar cells; oxygen can more easily diffuse from the air into the tissues and carbon dioxide can easily diffuse from the tissues into the air. Without pulmonary surfactant, gas exchange in the lungs is very poor.

Pulmonary surfactant is not produced until early in the sixth month of pregnancy. Premature babies born before they have begun secretion of natural surfactant suffer from *respiratory distress syndrome (RDS)*, which is caused by the severe difficulty they have obtaining enough oxygen from the air that they breathe.

Until recently, RDS was a major cause of death among premature infants, but now a lifesaving treatment is available. A fine aerosol of an artificial surfactant is administered directly into the trachea. The Glaxo-Wellcome Company product EXOSURF Neonatal contains the phospholipid lecithin to reduce surface tension; 1-hexadecanol, which spreads the lecithin; and a polymer called tyloxapol, which disperses both the lecithin and the 1-hexadecanol.

Artificial pulmonary surfactant therapy has dramatically reduced premature infant death caused by RDS and appears to have reduced overall mortality for all babies born weighing less than 700 g (about 1.5 pounds). Advances such as this have come about as a result of research on the makeup of body tissues and secretions in both healthy and diseased individuals. Often, such basic research provides the information needed to develop effective therapies.

In this chapter we will study the chemistry of lipids with a wide variety of structures and biological functions. Among these are the triglycerides that stock our adipose tissue, pain-producing prostaglandins, and steroids that determine our secondary sexual characteristics.

Introduction

Lipids seem to be the most controversial group of biological molecules, particularly in the fields of medicine and nutrition. We are concerned about the use of anabolic steroids by athletes. Although these hormones increase muscle mass and enhance performance, we are just beginning to understand the damage they cause to the body.

We worry about what types of dietary fat we should consume. We hear frequently about the amounts of saturated fats and cholesterol in our diets because a strong correlation has been found between these lipids and heart disease. Large quantities of dietary saturated fats may also predispose an individual to colon, esophageal, stomach, and breast cancers. As a result, we are advised to reduce our intake of cholesterol and saturated fats.

Nonetheless, lipids serve a wide variety of functions essential to living systems and are required in our diet. Standards of fat intake have not been experimentally determined. However, the most recent U.S. Dietary Guidelines recommend that dietary fat not exceed 35% of the daily caloric intake, and no more than 10% should be saturated fats. Dietary cholesterol should be no more than 300 mg/day.

17.1 Biological Functions of Lipids

The term **lipids** actually refers to a collection of organic molecules of varying chemical composition. They are grouped together on the basis of their solubility in nonpolar solvents. Lipids are commonly subdivided into four main groups:

1. *Fatty acids* (saturated and unsaturated)
2. *Glycerides* (glycerol-containing lipids)
3. *Nonglyceride lipids* (sphingolipids, steroids, waxes)
4. *Complex lipids* (lipoproteins)

In this chapter we examine the structure, properties, chemical reactions, and biological functions of each of the lipid groups shown in Figure 17.1.

As a result of differences in their structures, lipids serve many different functions in the human body. The following brief list will give you an idea of the importance of lipids in biological processes:

- *Energy source.* Like carbohydrates, lipids are an excellent source of energy for the body. When oxidized, each gram of fat releases 9 kcal of energy, or more than twice the energy released by oxidation of a gram of carbohydrate.
- *Energy storage.* Most of the energy stored in the body is in the form of lipids (triglycerides). Stored in fat cells called *adipocytes*, these fats are a particularly rich source of energy for the body.
- *Cell membrane structural components.* Phosphoglycerides, sphingolipids, and steroids make up the basic structure of all cell membranes. These membranes control the flow of molecules into and out of cells and allow cell-to-cell communication.
- *Hormones.* The steroid hormones are critical chemical messengers that allow tissues of the body to communicate with one another. The hormonelike prostaglandins exert strong biological effects on both the cells that produce them and other cells of the body.
- *Vitamins.* The lipid-soluble vitamins, A, D, E, and K, play a major role in the regulation of several critical biological processes, including blood clotting and vision.
- *Vitamin absorption.* Dietary fat serves as a carrier of the lipid-soluble vitamins. All are transported into cells of the small intestine in association with fat molecules. Therefore a diet that is too low in fat (less than 20% of calories) can result in a deficiency of these four vitamins.
- *Protection.* Fats serve as a shock absorber, or protective layer, for the vital organs. About 4% of the total body fat is reserved for this critical function.
- *Insulation.* Fat stored beneath the skin (subcutaneous fat) serves to insulate the body from extremes of cold temperatures.

LEARNING GOAL

Lipid-soluble vitamins are discussed in detail online at www.mhhe.com/denniston5e in Lipid-Soluble Vitamins.

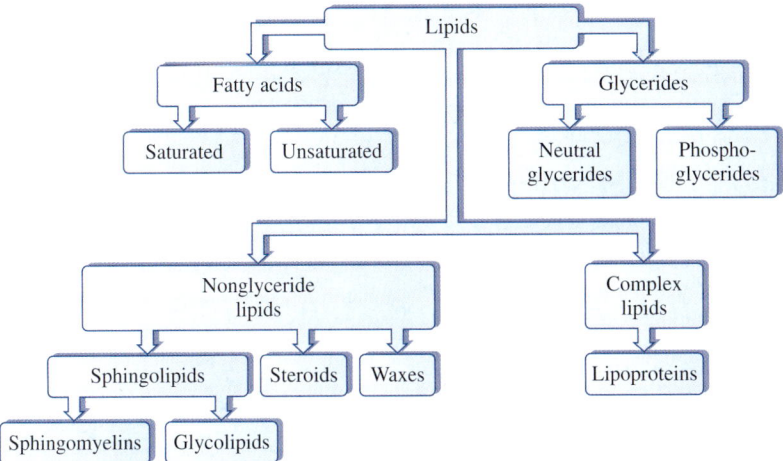

Figure 17.1
The classification of lipids.

17.2 Fatty Acids

Structure and Properties

LEARNING GOAL

Fatty acids are long-chain monocarboxylic acids. As a consequence of their biosynthesis, fatty acids generally contain an *even number* of carbon atoms. The general formula for a **saturated fatty acid** is $CH_3(CH_2)_n COOH$, in which n in biological systems is an even integer. If $n = 16$, the result is an 18-carbon saturated fatty acid, stearic acid, having the following structural formula:

$$H-\underset{H}{\overset{H}{C}}-\underset{H}{\overset{H}{C}}-\underset{H}{\overset{H}{C}}-\underset{H}{\overset{H}{C}}-\underset{H}{\overset{H}{C}}-\underset{H}{\overset{H}{C}}-\underset{H}{\overset{H}{C}}-\underset{H}{\overset{H}{C}}-\underset{H}{\overset{H}{C}}-\underset{H}{\overset{H}{C}}-\underset{H}{\overset{H}{C}}-\underset{H}{\overset{H}{C}}-\underset{H}{\overset{H}{C}}-\underset{H}{\overset{H}{C}}-\underset{H}{\overset{H}{C}}-\underset{H}{\overset{H}{C}}-\underset{H}{\overset{H}{C}}-\overset{O}{\overset{\|}{C}}-OH$$

The saturated fatty acids may be thought of as derivatives of alkanes, the saturated hydrocarbons described in Chapter 10.

Note that each of the carbons in the chain is bonded to the maximum number of hydrogen atoms. To help remember the structure of a saturated fatty acid, you might think of each carbon in the chain being "saturated" with hydrogen atoms. Examples of common saturated fatty acids are given in Table 17.1. An example of an **unsaturated fatty acid** is the 18-carbon unsaturated fatty acid oleic acid, which has the following structural formula:

$$H-\underset{H}{\overset{H}{C}}-\underset{H}{\overset{H}{C}}-\underset{H}{\overset{H}{C}}-\underset{H}{\overset{H}{C}}-\underset{H}{\overset{H}{C}}-\underset{H}{\overset{H}{C}}-\underset{H}{\overset{H}{C}}-\underset{H}{\overset{H}{C}}\underset{\underset{H}{C}=\underset{H}{C}}{}\underset{H}{\overset{H}{C}}-\underset{H}{\overset{H}{C}}-\underset{H}{\overset{H}{C}}-\underset{H}{\overset{H}{C}}-\underset{H}{\overset{H}{C}}-\underset{H}{\overset{H}{C}}-\underset{H}{\overset{H}{C}}-\overset{O}{\overset{\|}{C}}-OH$$

The unsaturated fatty acids may be thought of as derivatives of the alkenes, the unsaturated hydrocarbons discussed in Chapter 11.

A discussion of trans-*fatty acids is found in Section 11.3.*

In the case of unsaturated fatty acids, there is at least one carbon-to-carbon double bond. Because of the double bonds, the carbon atoms involved in these bonds are not "saturated" with hydrogen atoms. The double bonds found in almost all naturally occurring unsaturated fatty acids are in the *cis* configuration. In addition, the double bonds are not randomly located in the hydrocarbon chain. Both the placement and the geometric configuration of the double bonds are dictated by the enzymes that catalyze the biosynthesis of unsaturated fatty acids. Examples of common unsaturated fatty acids are also given in Table 17.1.

EXAMPLE 17.1 **Writing the Structural Formula of an Unsaturated Fatty Acid**

Draw the structural formula for palmitoleic acid.

Solution

The I.U.P.A.C. name of palmitoleic acid is *cis*-9-hexadecenoic acid. The name tells us that this is a 16-carbon fatty acid having a carbon-to-carbon double bond between carbons 9 and 10. The name also reveals that this is the *cis* isomer.

$$H-\underset{H}{\overset{H}{C}}-\underset{H}{\overset{H}{C}}-\underset{H}{\overset{H}{C}}-\underset{H}{\overset{H}{C}}-\underset{H}{\overset{H}{C}}-\underset{H}{\overset{H}{C}}\underset{\underset{H}{C}=\underset{H}{C}}{}\underset{H}{\overset{H}{C}}-\underset{H}{\overset{H}{C}}-\underset{H}{\overset{H}{C}}-\underset{H}{\overset{H}{C}}-\underset{H}{\overset{H}{C}}-\underset{H}{\overset{H}{C}}-\underset{H}{\overset{H}{C}}-COOH$$

16 15 14 13 12 11 10 9   8 7 6 5 4 3 2 1

TABLE 17.1 Common Saturated and Unsaturated Fatty Acids

Common Saturated Fatty Acids

Common Name	I.U.P.A.C. Name	Melting Point (°C)	RCOOH Formula	Condensed Formula
Capric	Decanoic	32	$C_9H_{19}COOH$	$CH_3(CH_2)_8COOH$
Lauric	Dodecanoic	44	$C_{11}H_{23}COOH$	$CH_3(CH_2)_{10}COOH$
Myristic	Tetradecanoic	54	$C_{13}H_{27}COOH$	$CH_3(CH_2)_{12}COOH$
Palmitic	Hexadecanoic	63	$C_{15}H_{31}COOH$	$CH_3(CH_2)_{14}COOH$
Stearic	Octadecanoic	70	$C_{17}H_{35}COOH$	$CH_3(CH_2)_{16}COOH$
Arachidic	Eicosanoic	77	$C_{19}H_{39}COOH$	$CH_3(CH_2)_{18}COOH$

Common Unsaturated Fatty Acids

Common Name	I.U.P.A.C. Name	Melting Point (°C)	RCOOH Formula	Number of Double Bonds	Position of Double Bonds
Palmitoleic	cis-9-Hexadecenoic	0	$C_{15}H_{29}COOH$	1	9
Oleic	cis-9-Octadecenoic	16	$C_{17}H_{33}COOH$	1	9
Linoleic	cis,cis-9,12-Octadecadienoic	5	$C_{17}H_{31}COOH$	2	9, 12
Linolenic	All cis-9,12,15-Octadecatrienoic	−11	$C_{17}H_{29}COOH$	3	9, 12, 15
Arachidonic	All cis-5,8,11,14-Eicosatetraenoic	−50	$C_{19}H_{31}COOH$	4	5, 8, 11, 14

Condensed Formula

Palmitoleic	$CH_3(CH_2)_5CH=CH(CH_2)_7COOH$
Oleic	$CH_3(CH_2)_7CH=CH(CH_2)_7COOH$
Linoleic	$CH_3(CH_2)_4CH=CH-CH_2-CH=CH(CH_2)_7COOH$
Linolenic	$CH_3CH_2CH=CH-CH_2-CH=CH-CH_2-CH=CH(CH_2)_7COOH$
Arachidonic	$CH_3(CH_2)_4CH=CH-CH_2-CH=CH-CH_2-CH=CH-CH_2-CH=CH-(CH_2)_3COOH$

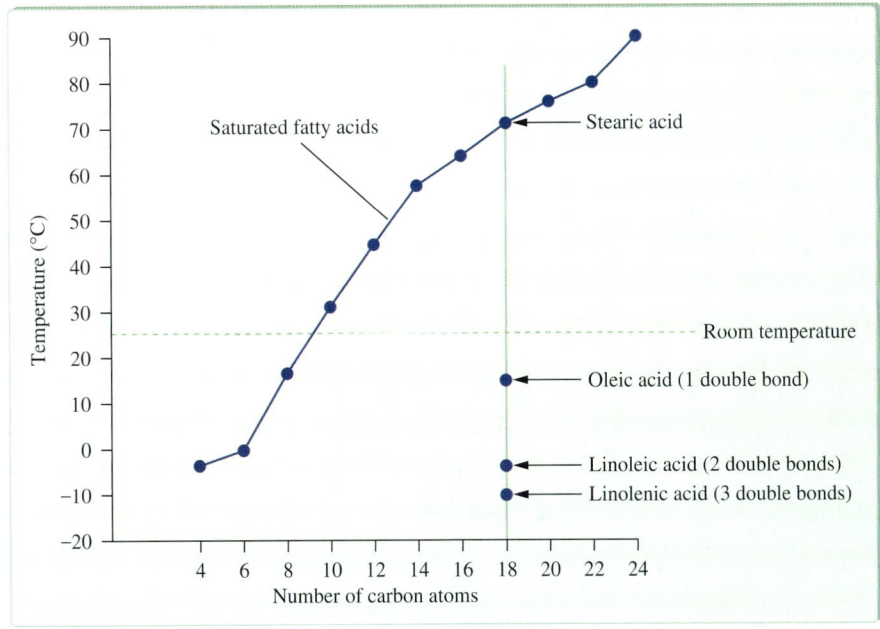

Figure 17.2
The melting points of fatty acids. Melting points of both saturated and unsaturated fatty acids increase as the number of carbon atoms in the chain increases. The melting points of unsaturated fatty acids are lower than those of the corresponding saturated fatty acid with the same number of carbon atoms. Also, as the number of double bonds in the chain increases, the melting points decrease.

Examination of Table 17.1 and Figure 17.2 reveals several interesting and important points about the physical properties of fatty acids.

3 LEARNING GOAL

The relationship between alkane chain length and melting point is described in Section 10.2.

- The melting points of saturated fatty acids increase with increasing carbon number, as is the case with alkanes. Saturated fatty acids containing ten or more carbons are solids at room temperature.
- The melting point of a saturated fatty acid is greater than that of an unsaturated fatty acid of the same chain length. The reason is that saturated fatty acid chains tend to be fully extended and to stack in a regular structure, thereby causing increased intermolecular attraction. Introduction of a *cis* double bond into the hydrocarbon chain produces a rigid 30° bend. Such "kinked" molecules cannot stack in an organized arrangement and thus have lower intermolecular attractions and lower melting points.

The relationship between alkene chain length and melting point is described in Section 11.1.

- As in the case for saturated fatty acids, the melting points of unsaturated fatty acids increase with increasing hydrocarbon chain length.

EXAMPLE 17.2 Examining the Similarities and Differences Between Saturated and Unsaturated Fatty Acids

Construct a table comparing the structure and properties of saturated and unsaturated fatty acids.

Solution

Property	Saturated Fatty Acid	Unsaturated Fatty Acid
Chemical composition	Carbon, hydrogen, oxygen	Carbon, hydrogen, oxygen
Chemical structure	Hydrocarbon chain with a terminal carboxyl group	Hydrocarbon chain with a terminal carboxyl group
Carbon-carbon bonds within the hydrocarbon chain	Only C—C single bonds	At least one C—C double bond
Hydrocarbon chains are characteristic of what group of hydrocarbons	Alkanes	Alkenes
"Shape" of hydrocarbon chain	Linear, fully extended	Bend in carbon chain at site of C—C double bond
Physical state at room temperature	Solid	Liquid
Melting point for two fatty acids of the same hydrocarbon chain length	Higher	Lower
Relationship between melting point and chain length	Longer chain length, higher melting point	Longer chain length, higher melting point

Question 17.1

Draw formulas for each of the following fatty acids:

a. Oleic acid
b. Lauric acid
c. Linoleic acid
d. Stearic acid

Question 17.2

What is the I.U.P.A.C. name for each of the fatty acids in Question 17.1? (*Hint:* Review the naming of carboxylic acids in Section 14.1 and Table 17.1.)

Chemical Reactions of Fatty Acids

The reactions of fatty acids are identical to those of short-chain carboxylic acids. The major reactions that they undergo include esterification, acid hydrolysis of esters, saponification, and addition at the double bond.

Esterification

In **esterification,** fatty acids react with alcohols to form esters and water according to the following general equation:

Esterification is described in Sections 14.1 and 14.2.

$$R^1-\underset{\text{Fatty acid}}{\underset{\|}{\overset{O}{C}}}-OH + \underset{\text{Alcohol}}{HOR^2} \xrightarrow{H^+, \text{heat}} R^1-\underset{\text{Ester}}{\underset{\|}{\overset{O}{C}}}-OR^2 + \underset{\text{Water}}{H-OH}$$

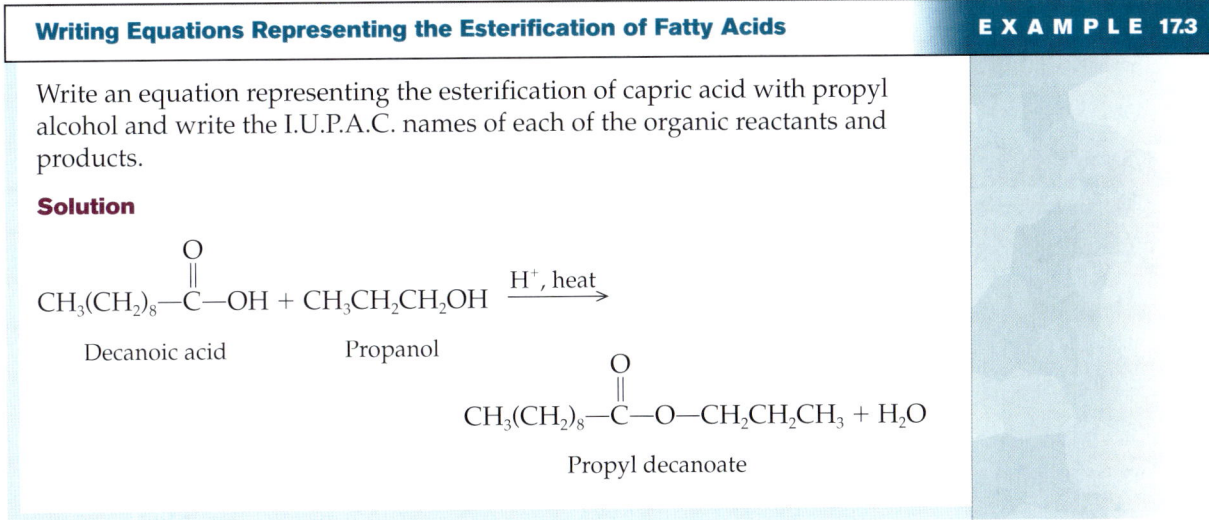

EXAMPLE 17.3 Writing Equations Representing the Esterification of Fatty Acids

Write an equation representing the esterification of capric acid with propyl alcohol and write the I.U.P.A.C. names of each of the organic reactants and products.

Solution

$$\underset{\text{Decanoic acid}}{CH_3(CH_2)_8-\underset{\|}{\overset{O}{C}}-OH} + \underset{\text{Propanol}}{CH_3CH_2CH_2OH} \xrightarrow{H^+, \text{heat}}$$

$$\underset{\text{Propyl decanoate}}{CH_3(CH_2)_8-\underset{\|}{\overset{O}{C}}-O-CH_2CH_2CH_3} + H_2O$$

Question 17.3

Write the complete equation for the esterification of lauric acid and ethyl alcohol. Write the I.U.P.A.C. name for each of the organic reactants and products.

Question 17.4

Write the complete equation for the esterification of capric acid and 1-pentanol. Write the I.U.P.A.C. name for each of the organic reactants and products.

Acid Hydrolysis

Recall that hydrolysis is the reverse of esterification, producing fatty acids from esters:

Acid hydrolysis is discussed in Section 14.2.

$$R^1-\underset{\text{Ester}}{\underset{\|}{\overset{O}{C}}}-OR^2 + \underset{\text{Water}}{HO-H} \xrightarrow{H^+, \text{heat}} R^1-\underset{\text{Fatty acid}}{\underset{\|}{\overset{O}{C}}}-OH + \underset{\text{Alcohol}}{R^2OH}$$

EXAMPLE 17.4 Writing Equations Representing the Acid Hydrolysis of a Fatty Acid Ester

Write an equation representing the acid hydrolysis of methyl decanoate and write the I.U.P.A.C. names of each of the organic reactants and products.

Solution

$$CH_3(CH_2)_8-\overset{O}{\underset{\|}{C}}-O-CH_3 + H_2O \xrightarrow{H^+, \text{heat}} CH_3(CH_2)_8-\overset{O}{\underset{\|}{C}}-OH + CH_3OH$$

Methyl decanoate → Decanoic acid + Methanol

Question 17.5

Write a complete equation for the acid hydrolysis of butyl propionate. Write the I.U.P.A.C. name for each of the organic reactants and products.

Question 17.6

Write a complete equation for the acid hydrolysis of ethyl butyrate. Write the I.U.P.A.C. name for each of the organic reactants and products.

Saponification

Saponification is described in Section 14.2.

Saponification is the base-catalyzed hydrolysis of an ester:

$$R^1-\overset{O}{\underset{\|}{C}}-OR^2 + NaOH \longrightarrow R^1-\overset{O}{\underset{\|}{C}}-O^-Na^+ + R^2OH$$

Ester + Base → Salt + Alcohol

The role of soaps in removal of dirt and grease is described in Section 14.2.

Examples of micelles are shown in Figures 14.4 and 23.1.

The product of this reaction, an ionized salt, is a soap. Because soaps have a long uncharged hydrocarbon tail and a negatively charged terminus (the carboxylate group), they form micelles that dissolve oil and dirt particles. Thus the dirt is emulsified and broken into small particles, and can be rinsed away.

EXAMPLE 17.5 Writing Equations Representing the Base-Catalyzed Hydrolysis of a Fatty Acid Ester

Write an equation representing the base-catalyzed hydrolysis of ethyl dodecanoate and write the I.U.P.A.C. names of each of the organic reactants and products.

Solution

$$CH_3(CH_2)_{10}-\overset{O}{\underset{\|}{C}}-O-CH_2CH_3 + NaOH \longrightarrow$$

Ethyl dodecanoate

$$CH_3(CH_2)_{10}-\overset{O}{\underset{\|}{C}}-O^-Na^+ + CH_3CH_2OH$$

Sodium dodecanoate + Ethanol

A Human Perspective

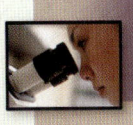

Mummies Made of Soap

In the Smithsonian Museum of Natural History in Washington, D.C., one can find a great many wonders of the natural world. None is quite as macabre as the corpse made of soap. The man in question died of yellow fever in the eighteenth century and was buried near Boston. Actually, he was buried alongside a woman, perhaps the love of his life, who has been dubbed "Soap Woman." She, too, died of yellow fever. However, the couple has been separated for quite some time, because Soap Woman has been on display at the Mutter Museum at the College of Physicians in Philadelphia since 1874.

Recently Soap Woman became a television celebrity when a CT Scan done to examine the body was filmed for "The Mummy Road Show," a presentation of the National Geographic Channel. One reason for the examination was to try to understand the conditions that caused this chemical conversion. At the present time, no one is precisely sure how these two people turned to soap. One clue resides in the environment of the burial site. Apparently the groundwater running through the graves was very basic. Another clue to the mystery is that our soap couple was overweight. Certainly these two factors contributed to the saponification reactions that converted this chubby couple into blocks of soap.

For Further Understanding

Adipocere is the technical term for "soap mummification." It comes from the Latin words *adipis* or fat, as in adipose tissue, and *cera*, which means wax. Draw a triglyceride composed of the fatty acids myristic acid, stearic acid, and oleic acid. Write a balanced equation showing a possible reaction that would lead to the formation of adipocere.

Forensic scientists are studying adipocere formation as a possible source of information to determine postmortem interval (length of time since death) of bodies of murder or accident victims. Among the factors being studied are the type of soil, including pH, moisture, temperature, and presence or absence of lime. How might each of these factors influence the rate of adipocere formation and hence the determination of the postmortem interval?

Soap woman.

Question 17.7

Write a complete equation for the reaction of butyl propionate with KOH. Write the I.U.P.A.C. name for each of the organic reactants and products.

Question 17.8

Write a complete equation for the reaction of ethyl butyrate with NaOH. Write the I.U.P.A.C. name for each of the organic reactants and products.

Problems can arise when "hard" water is used for cleaning because the high concentrations of Ca^{2+} and Mg^{2+} in such water cause fatty acid salts to precipitate. Not only does this interfere with the emulsifying action of the soap, it also leaves a hard scum on the surface of sinks and tubs.

$$2R-\overset{O}{\underset{\|}{C}}-O^- + Ca^{2+} \longrightarrow (R-\overset{O}{\underset{\|}{C}}-O^-)_2 Ca^{2+}(s)$$

Reaction at the Double Bond (Unsaturated Fatty Acids)

Hydrogenation is described in Section 11.5.

Hydrogenation is an example of an addition reaction. The following is a typical example of the addition of hydrogen to the double bonds of a fatty acid:

$$CH_3(CH_2)_4CH=CHCH_2CH=CH(CH_2)_7COOH \xrightarrow{2H_2,\ Ni} CH_3(CH_2)_{16}COOH$$

Linoleic acid → Stearic acid

EXAMPLE 17.6 — Writing Equations for the Hydrogenation of a Fatty Acid

Write an equation representing the hydrogenation of oleic acid and write the I.U.P.A.C. names of each of the organic reactants and products.

Solution

$$CH_3(CH_2)_7CH=CH(CH_2)_7-\overset{O}{\underset{\|}{C}}-OH \xrightarrow{H_2,\ Ni} CH_3(CH_2)_{16}-\overset{O}{\underset{\|}{C}}-OH$$

cis-9-Octadecenoic acid → Octadecanoic acid

Question 17.9

Write a balanced equation showing the hydrogenation of *cis*-9-hexadecenoic acid.

Question 17.10

Write a balanced equation showing the hydrogenation of arachidonic acid.

Hydrogenation is used in the food industry to convert polyunsaturated vegetable oils into saturated solid fats. *Partial hydrogenation* is carried out to add hydrogen to some, but not all, double bonds in polyunsaturated oils. In this way liquid vegetable oils are converted into solid form. Crisco is one example of a hydrogenated vegetable oil.

Margarine is also produced by partial hydrogenation of vegetable oils, such as corn oil or soybean oil. The extent of hydrogenation is carefully controlled so that the solid fat will be spreadable and have the consistency of butter when eaten. If too many double bonds were hydrogenated, the resulting product would have the undesirable consistency of animal fat. Artificial color is added to the product, and it may be mixed with milk to produce a butterlike appearance and flavor.

Hydrogenation of vegetable oils produces a mixture of cis and trans unsaturated fatty acids. The trans unsaturated fatty acids are thought to contribute to atherosclerosis (hardening of the arteries).

Eicosanoids: Prostaglandins, Leukotrienes, and Thromboxanes

Some of the unsaturated fatty acids containing more than one double bond cannot be synthesized by the body. For many years it has been known that linolenic acid and linoleic acid, called the **essential fatty acids,** are necessary for specific biochemical functions and must be supplied in the diet (see Table 17.1). The function of linoleic acid became clear in the 1960s when it was discovered that linoleic acid is required for the biosynthesis of **arachidonic acid,** the precursor of a class of hormonelike molecules known as **eicosanoids.** The name is derived from the Greek word *eikos,* meaning "twenty," because they are all derivatives of twenty-carbon fatty acids. The eicosanoids include three groups of structurally related compounds: the prostaglandins, the leukotrienes, and the thromboxanes.

A hormone is a chemical signal that is produced by a specialized tissue and is carried by the bloodstream to target tissues. Eicosanoids are referred to as hormonelike because they affect the cells that produce them, as well as other target tissues.

Prostaglandins are extremely potent biological molecules with hormonelike activity. They got their name because they were originally isolated from seminal fluid produced in the prostate gland. More recently they also have been isolated from most animal tissues. Prostaglandins are unsaturated carboxylic acids consisting of a twenty-carbon skeleton that contains a five-carbon ring.

Several general classes of prostaglandins are grouped under the designations A, B, E, and F, among others. The nomenclature of prostaglandins is based on the arrangement of the carbon skeleton and the number and orientation of double bonds, hydroxyl groups, and ketone groups. For example, in the name PGF_2, PG stands for prostaglandin, F indicates a particular group of prostaglandins with a hydroxyl group bonded to carbon-9, and 2 indicates that there are two carbon-carbon double bonds in the compound. The examples in Figure 17.3 illustrate the general structure of prostaglandins and the current nomenclature system.

Prostaglandins are made in most tissues, and exert their biological effects on the cells that produce them and on other cells in the immediate vicinity. Because the prostaglandins and the closely related leukotrienes and thromboxanes affect so many body processes and because they often cause opposing effects in different tissues, it can be difficult to keep track of their many regulatory functions. The following is a brief summary of some of the biological processes that are thought to be regulated by the prostaglandins, leukotrienes, and thromboxanes.

1. **Blood clotting.** Blood clots form when a blood vessel is damaged, yet such clotting along the walls of undamaged vessels could result in heart attack or stroke. *Thromboxane A_2* (Figure 17.4) is produced by platelets in the blood and stimulates constriction of the blood vessels and aggregation of the platelets. Conversely, PGI_2 (prostacyclin) is produced by the cells lining the blood vessels and has precisely the opposite effect of thromboxane A_2. Prostacyclin inhibits platelet aggregation and causes dilation of blood vessels and thus prevents the untimely production of blood clots.
2. **The inflammatory response.** The inflammatory response is another of the body's protective mechanisms. When tissue is damaged by mechanical injury, burns, or invasion by microorganisms, a variety of white blood cells descend on the damaged site to try to minimize the tissue destruction. The result of this response is swelling, redness, fever, and pain. Prostaglandins are thought to promote certain aspects of the inflammatory response, especially pain and fever. Drugs such as aspirin block prostaglandin synthesis and help to relieve the symptoms. We will examine the mechanism of action of these drugs later in this section.
3. **Reproductive system.** PGE_2 stimulates smooth muscle contraction, particularly uterine contractions. An increase in the level of prostaglandins has been noted immediately before the onset of labor. PGE_2 has also been used to induce second trimester abortions. There is strong evidence that dysmenorrhea (painful menstruation) suffered by many women may be the result of an excess of two prostaglandins. Indeed, drugs, such as ibuprofen, that inhibit prostaglandin synthesis have been approved by the FDA and are found to provide relief from these symptoms.
4. **Gastrointestinal tract.** Prostaglandins have been shown to both inhibit the secretion of acid and increase the secretion of a protective mucus layer into the stomach. In this way, prostaglandins help to protect the stomach lining. Consider for a moment the possible side effect that prolonged use of a drug such as aspirin might have on the stomach—ulceration of the stomach lining. Because aspirin inhibits prostaglandin synthesis, it may actually encourage stomach ulcers by inhibiting the formation of the normal protective mucus layer, while simultaneously allowing increased secretion of stomach acid.

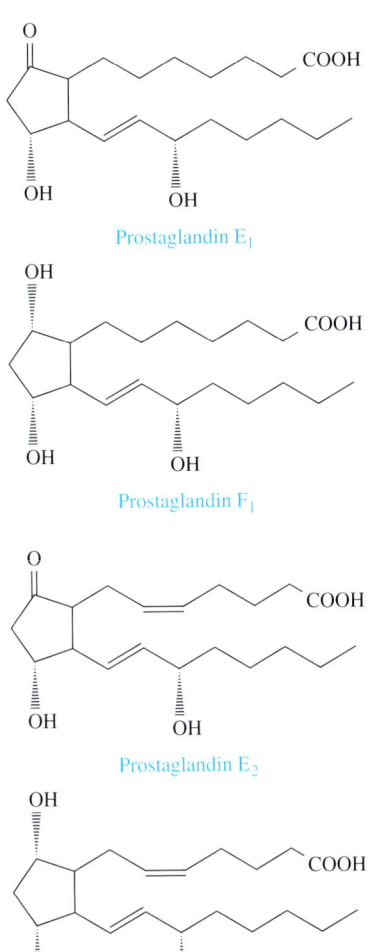

Figure 17.3
The structures of four prostaglandins.

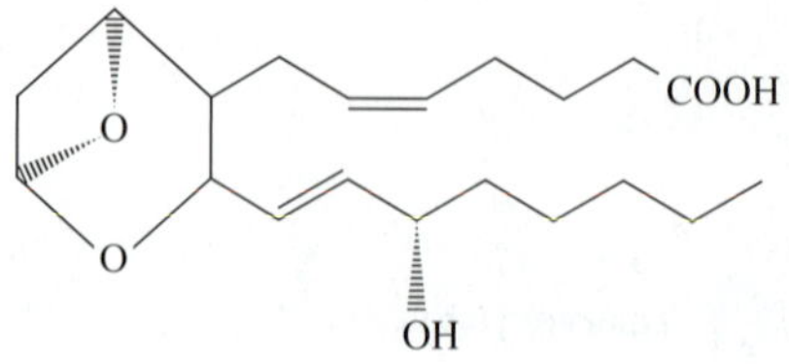

Figure 17.4
The structures of thromboxane A₂ and leukotriene B₄.

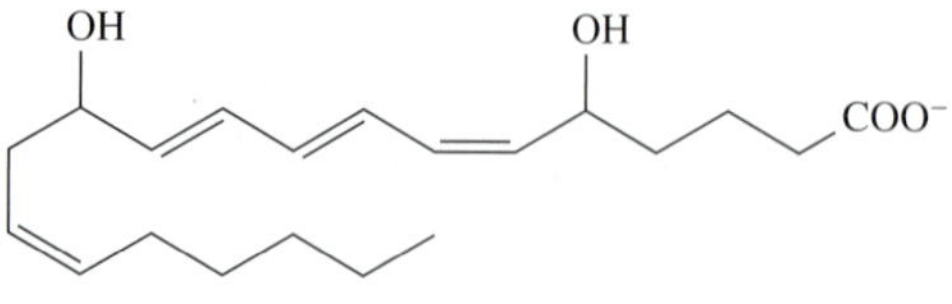

5. **Kidneys.** Prostaglandins produced in the kidneys cause the renal blood vessels to dilate. The greater flow of blood through the kidney results in increased water and electrolyte excretion.
6. **Respiratory tract.** Eicosanoids produced by certain white blood cells, the *leukotrienes* (see Figure 17.4), promote the constriction of the bronchi associated with asthma. Other prostaglandins promote bronchodilation.

As this brief survey suggests, the prostaglandins have numerous, often antagonistic effects. Although they do not fit the formal definition of a hormone (a substance produced in a specialized tissue and transported by the circulatory system to target tissues *elsewhere* in the body), the prostaglandins are clearly strong biological regulators with far-reaching effects.

As mentioned, prostaglandins stimulate the inflammatory response and, as a result, are partially responsible for the cascade of events that cause pain. Aspirin has long been known to alleviate such pain, and we now know that it does so by inhibiting the synthesis of prostaglandins (Figure 17.5).

The first two steps of prostaglandin synthesis (Figure 17.6), the release of arachidonic acid from the membrane and its conversion to PGH$_2$ by the enzyme cyclooxygenase, occur in all tissues that are able to produce prostaglandins. The conversion of PGH$_2$ into the other biologically active forms is tissue specific and requires the appropriate enzymes, which are found only in certain tissues.

Figure 17.5
Aspirin inhibits the synthesis of prostaglandins by acetylating the enzyme cyclooxygenase. The acetylated enzyme is no longer functional.

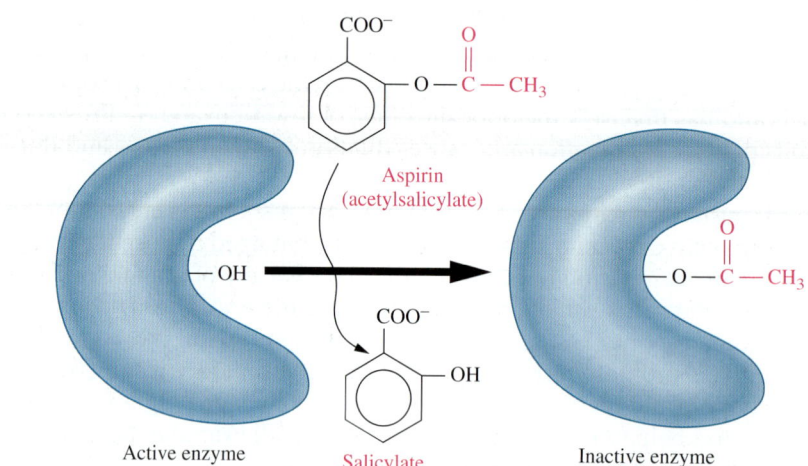

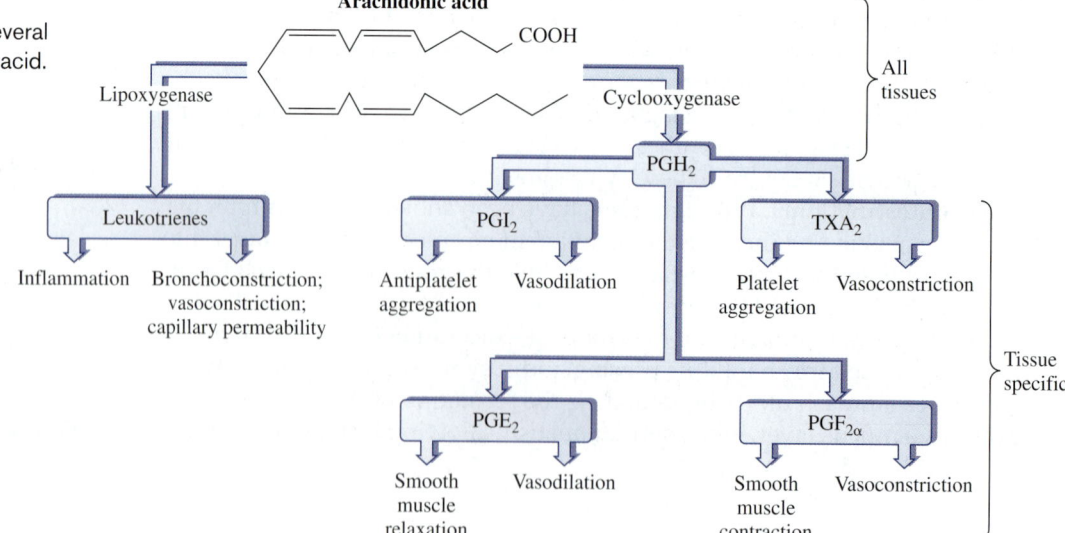

Figure 17.6
A summary of the synthesis of several prostaglandins from arachidonic acid.

Aspirin works by inhibiting the cyclooxygenase, which catalyzes the first step in the pathway leading from arachidonic acid to PGH_2. The acetyl group of aspirin becomes covalently bound to the enzyme, thereby inactivating it (Figure 17.5). Because the reaction catalyzed by cyclooxygenase occurs in all cells, aspirin effectively inhibits synthesis of all of the prostaglandins.

17.3 Glycerides

Neutral Glycerides

Glycerides are lipid esters that contain the glycerol molecule and fatty acids. They may be subdivided into two classes: neutral glycerides and phosphoglycerides. Neutral glycerides are nonionic and nonpolar. Phosphoglyceride molecules have a polar region, the phosphoryl group, in addition to the nonpolar fatty acid tails. The structures of each of these types of glycerides are critical to their function.

The esterification of glycerol with a fatty acid produces a **neutral glyceride.** Esterification may occur at one, two, or all three positions, producing **monoglycerides, diglycerides,** or **triglycerides.** You will also see these referred to as *mono-, di-,* or *triacylglycerols.*

Writing an Equation for the Synthesis of a Monoglyceride

EXAMPLE 17.7

Write a general equation for the esterification of glycerol and one fatty acid.

Solution

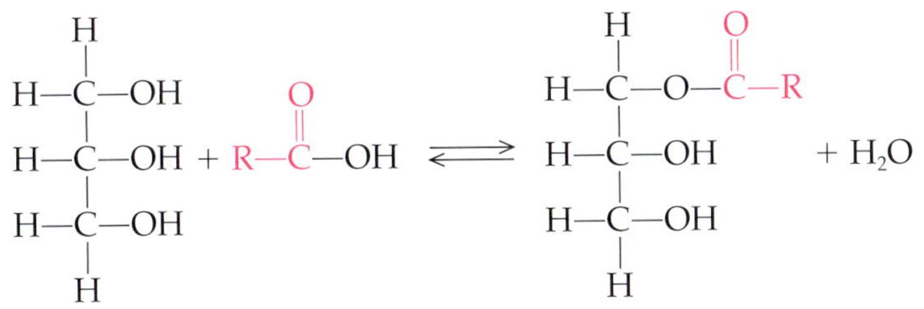

$$\begin{array}{c}\text{H} \\ | \\ \text{H—C—OH} \\ | \\ \text{H—C—OH} \\ | \\ \text{H—C—OH} \\ | \\ \text{H} \end{array} + \text{R—}\overset{\text{O}}{\overset{\|}{\text{C}}}\text{—OH} \rightleftharpoons \begin{array}{c}\text{H} \\ | \\ \text{H—C—O—}\overset{\text{O}}{\overset{\|}{\text{C}}}\text{—R} \\ | \\ \text{H—C—OH} \\ | \\ \text{H—C—OH} \\ | \\ \text{H} \end{array} + \text{H}_2\text{O}$$

Glycerol   Fatty acid   Monoglyceride   Water

Question 17.11

Write an equation for the esterification of glycerol with two molecules of stearic acid.

Question 17.12

Write a balanced equation for the esterification of glycerol with one molecule of myristic acid.

Although monoglycerides and diglycerides are present in nature, the most important neutral glycerides are the triglycerides, the major component of fat cells. The triglyceride consists of a glycerol backbone (shown in black) joined to three fatty acid units through ester bonds (shown in red). The formation of a triglyceride is shown in the following equation:

$$\underset{\text{Glycerol}}{\begin{array}{c} H \\ | \\ H-C-OH \\ | \\ H-C-OH \\ | \\ H-C-OH \\ | \\ H \end{array}} + \underset{\text{Fatty acids}}{3R-\overset{\overset{O}{\|}}{C}-OH} \rightleftarrows \underset{\text{Triglyceride}}{\begin{array}{c} H \\ | \\ H-C-O-\overset{\overset{O}{\|}}{C}-R \\ | \\ H-C-O-\overset{\overset{O}{\|}}{C}-R \\ | \\ H-C-O-\overset{\overset{O}{\|}}{C}-R \\ | \\ H \end{array}} + \underset{\text{Water}}{3H_2O}$$

Because there are no charges (+ or −) on these molecules, they are called *neutral glycerides*. These long molecules readily stack with one another and constitute the majority of the lipids stored in the body's fat cells.

The principal function of triglycerides in biochemical systems is the storage of energy. If more energy-rich nutrients are consumed than are required for metabolic processes, much of the excess is converted to neutral glycerides and stored as triglycerides in fat cells of *adipose tissue*. When energy is needed, the triglycerides are metabolized by the body, and energy is released. For this reason, exercise, along with moderate reduction in caloric intake, is recommended for overweight individuals. Exercise, an energy-demanding process, increases the rate of metabolism of fats and results in weight loss.

Phosphoglycerides

Phospholipids are a group of lipids that are phosphate esters. The presence of the phosphoryl group results in a molecule with a polar head (the phosphoryl group) and a nonpolar tail (the alkyl chain of the fatty acid). Because the phosphoryl group ionizes in solution, a charged lipid results.

The most abundant membrane lipids are derived from glycerol-3-phosphate and are known as **phosphoglycerides.** Phosphoglycerides contain acyl groups derived from long-chain fatty acids at C-1 and C-2 of glycerol-3-phosphate. At C-3 the phosphoryl group is joined to glycerol by a phosphoester bond. The simplest phosphoglyceride contains a free phosphoryl group and is known as a **phosphatidate** (Figure 17.7). When the phosphoryl group is attached to another hydrophilic molecule, a more complex phosphoglyceride is formed. For example, *phosphatidylcholine (lecithin)* and *phosphatidylethanolamine (cephalin)* are found in the membranes of most cells (Figure 17.7).

Lecithin possesses a polar "head" and a nonpolar "tail." Thus, it is an *amphipathic* molecule. This structure is similar to that of soap and detergent molecules, discussed earlier. The ionic "head" is hydrophilic and interacts with water molecules, whereas the nonpolar "tail" is hydrophobic and interacts with nonpolar molecules. This amphipathic nature is central to the structure and function of cell membranes.

In addition to being a component of cell membranes, lecithin is the major phospholipid in pulmonary surfactant. It is also found in egg yolks and soybeans and is used as an emulsifying agent in ice cream. An **emulsifying agent** aids in the suspension of triglycerides in water. The amphipathic lecithin serves as a bridge, holding together the highly polar water molecules and the nonpolar triglycerides. Emulsification occurs because the hydrophilic head of lecithin dissolves in water and its hydrophobic tail dissolves in the triglycerides.

Cephalin is similar in general structure to lecithin; the amine group bonded to the phosphoryl group is the only difference.

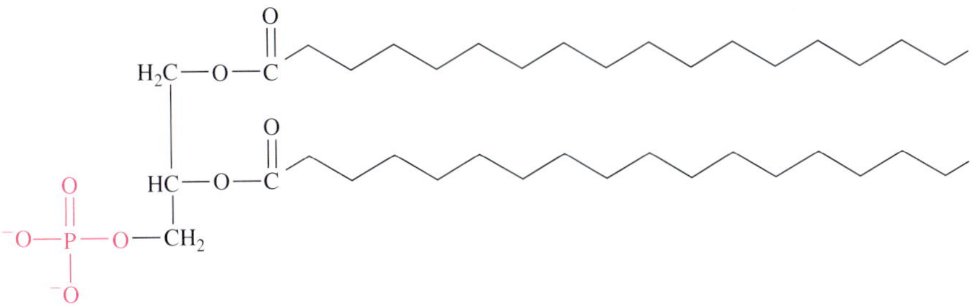

Phosphatidate

(a)

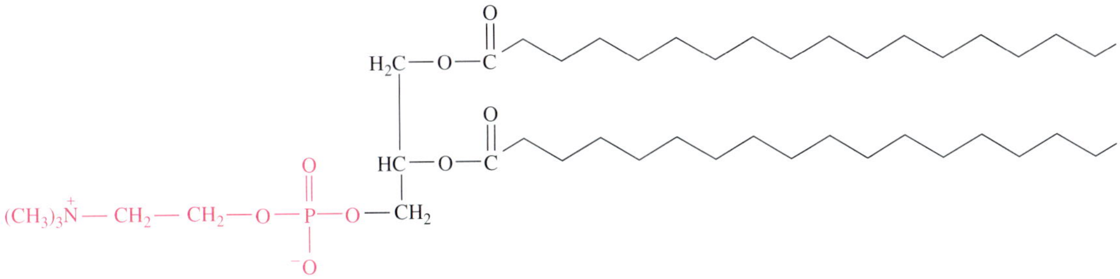

Phosphatidylcholine (lecithin)

(b)

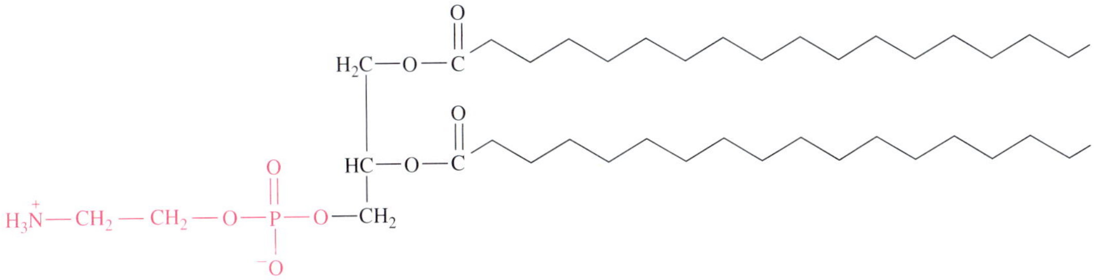

Phosphatidylethanolamine (cephalin)

(c)

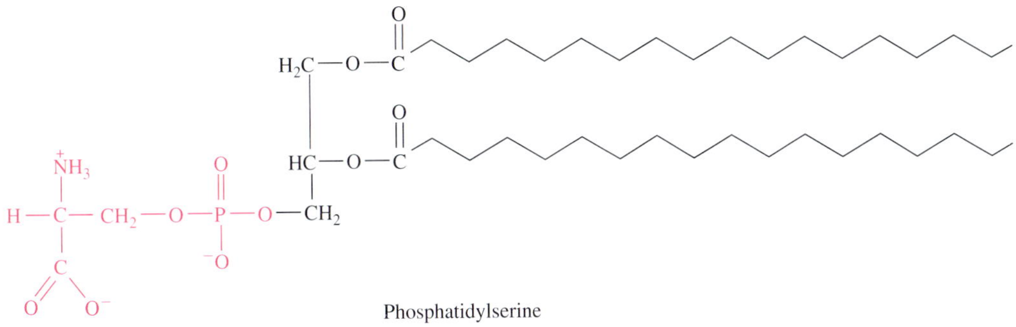

Phosphatidylserine

(d)

Figure 17.7

The structures of (a) phosphatidate and the common membrane phospholipids, (b) phosphatidylcholine (lecithin), (c) phosphatidylethanolamine (cephalin), and (d) phosphatidylserine.

Question 17.13

Using condensed formulas, draw the mono-, di-, and triglycerides that would result from the esterification of glycerol with each of the following fatty acids.

a. Oleic acid
b. Capric acid

Question 17.14

Using condensed formulas, draw the mono-, di-, and triglycerides that would result from the esterification of glycerol with each of the following fatty acids.

a. Palmitic acid
b. Lauric acid

17.4 Nonglyceride Lipids

Sphingolipids

LEARNING GOAL 8

Sphingolipids are lipids that are not derived from glycerol. Like phospholipids, sphingolipids are amphipathic, having a polar head group and two nonpolar fatty acid tails, and are structural components of cellular membranes. They are derived from sphingosine, a long-chain, nitrogen-containing (amino) alcohol:

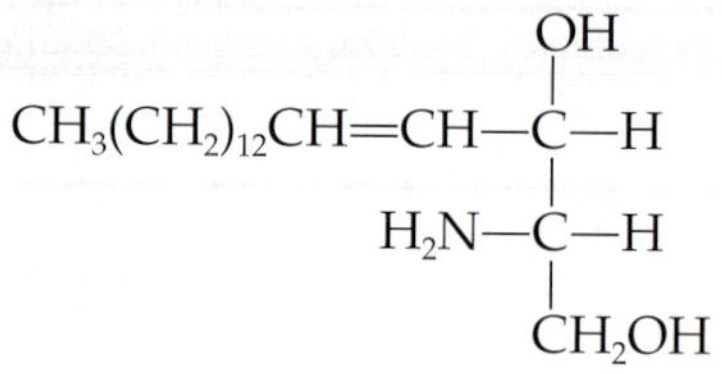

$$CH_3(CH_2)_{12}CH=CH-\underset{\underset{CH_2OH}{|}}{\underset{H_2N-C-H}{|}}\overset{OH}{\underset{|}{C}}-H$$

Sphingosine

The sphingolipids include the sphingomyelins and the glycosphingolipids. The **sphingomyelins** are the only class of sphingolipids that are also phospholipids:

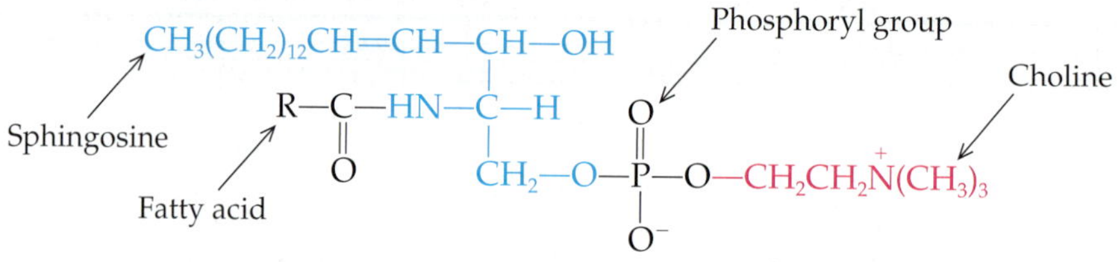

Sphingomyelin

Sphingomyelins are located throughout the body, but are particularly important structural lipid components of nerve cell membranes. They are found in abundance in the myelin sheath that surrounds and insulates cells of the central nervous system. In humans, about 25% of the lipids of the myelin sheath are sphingomyelins. Their role is essential to proper cerebral function and nerve transmission.

17.4 Nonglyceride Lipids

Glycosphingolipids, or *glycolipids*, include the cerebrosides, sulfatides, and gangliosides and are built on a ceramide backbone structure, which is a fatty acid amide derivative of sphingosine:

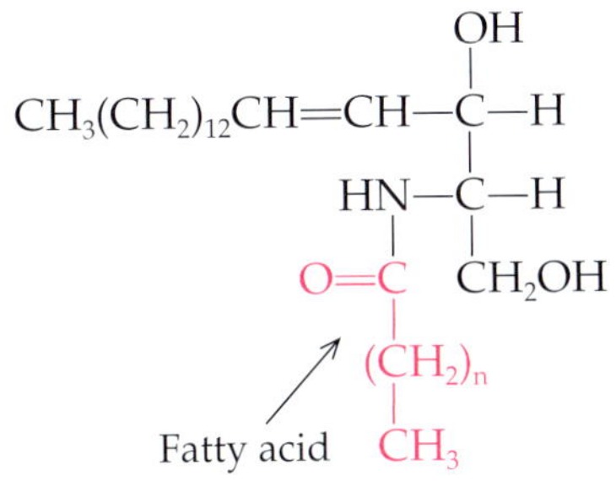

$$CH_3(CH_2)_{12}CH=CH-\underset{\underset{\underset{\underset{CH_3}{|}}{\underset{(CH_2)_n}{|}}}{\underset{O=C}{|}}}{\overset{\overset{OH}{|}}{C}}-H$$

$$\underset{}{\overset{}{\underset{}{HN-C-H}}}$$

$$\underset{}{\overset{}{CH_2OH}}$$

Fatty acid

Ceramide

The *cerebrosides* are characterized by the presence of a single monosaccharide head group. Two common cerebrosides are glucocerebroside, found in the membranes of macrophages (cells that protect the body by ingesting and destroying foreign microorganisms) and galactocerebroside, found almost exclusively in the membranes of brain cells. Glucocerebroside consists of ceramide bonded to the hexose glucose; galactocerebroside consists of ceramide joined to the monosaccharide galactose.

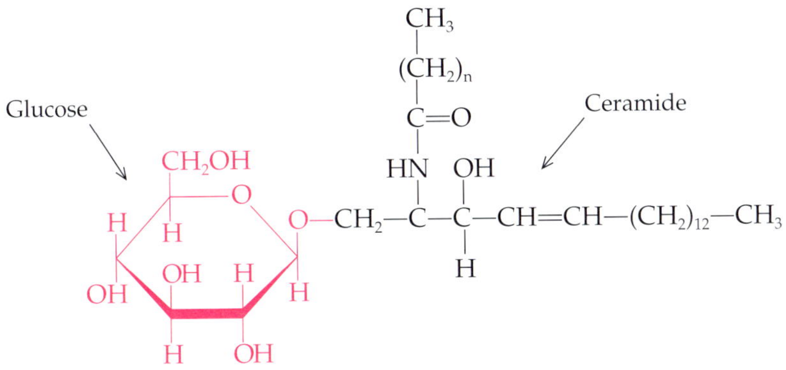

Glucocerebroside

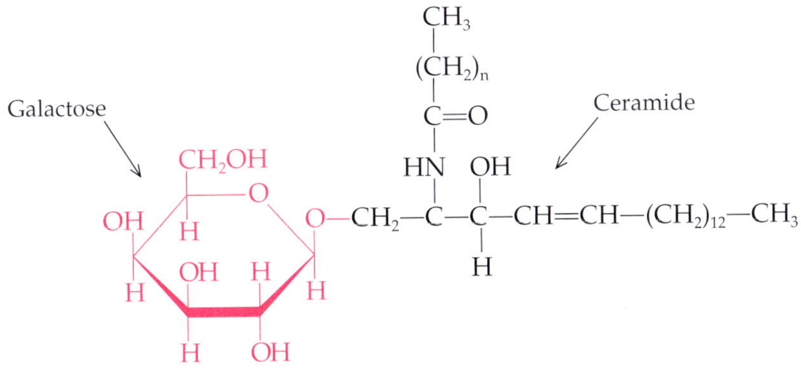

Galactocerebroside

Sulfatides are derivatives of galactocerebroside that contain a sulfate group. Notice that they carry a negative charge at physiological pH.

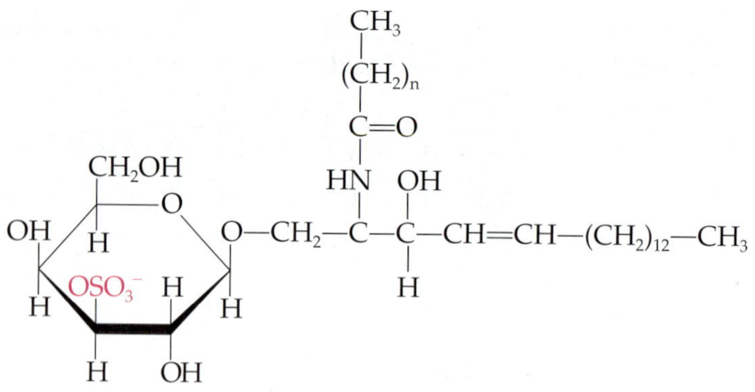

A sulfatide of galactocerebroside

Gangliosides are glycolipids that possess oligosaccharide groups, including one or more molecules of N-acetylneuraminic acid (sialic acid). First isolated from membranes of nerve tissue, gangliosides are found in most tissues of the body.

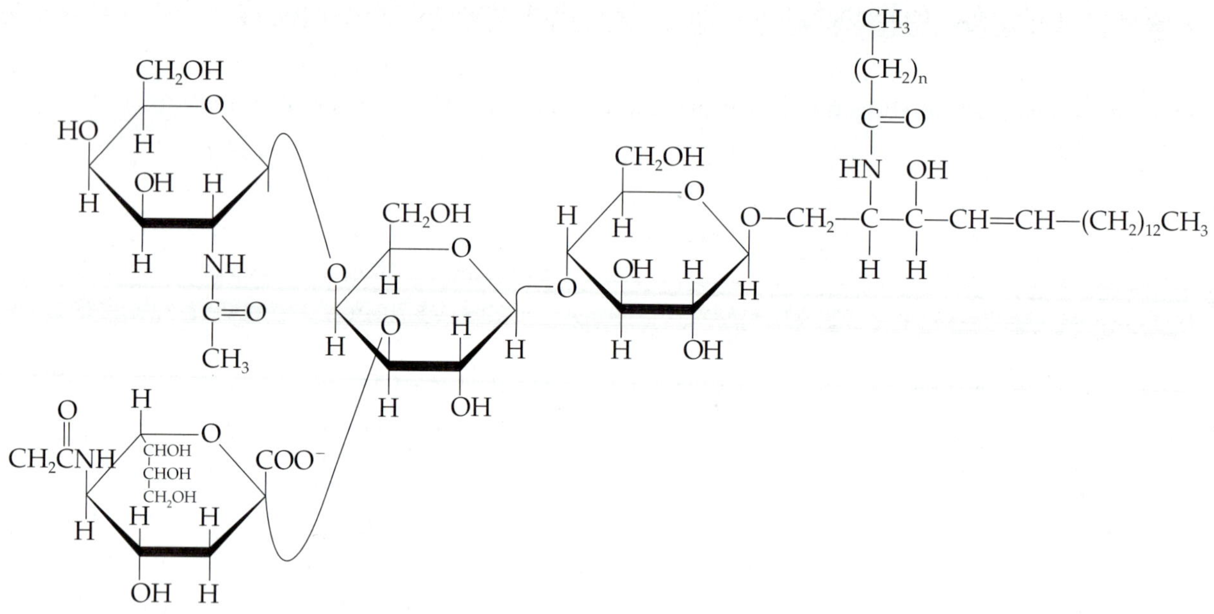

A ganglioside associated with Tay-Sachs disease

Steroids

LEARNING GOAL 9

Lipid digestion is described in Section 23.1.

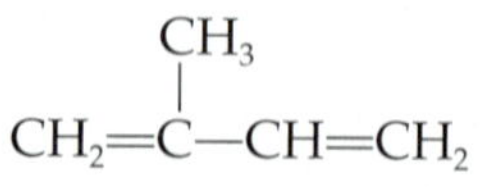

Isoprene

Steroids are a naturally occurring family of organic molecules of biochemical and medical interest. A great deal of controversy has surrounded various steroids. We worry about the amount of cholesterol in the diet and the possible health effects. We are concerned about the use of anabolic steroids by athletes wishing to build muscle mass and improve their performance. However, members of this family of molecules derived from cholesterol have many important functions in the body. The bile salts that aid in the emulsification and digestion of lipids are steroid molecules, as are the sex hormones testosterone and estrone.

The steroids are members of a large, diverse collection of lipids called the *isoprenoids*. All of these compounds are built from one or more five-carbon units called *isoprene*.

Terpene is the general term for lipids that are synthesized from isoprene units. Examples of terpenes include the steroids and bile salts, the lipid-soluble vitamins, chlorophyll, and certain plant hormones.

A Medical Perspective

Disorders of Sphingolipid Metabolism

There are a number of human genetic disorders that are caused by a deficiency in one of the enzymes responsible for the breakdown of sphingolipids. In general, the symptoms are caused by the accumulation of abnormally large amounts of these lipids within particular cells. It is interesting to note that three of these diseases, Niemann-Pick disease, Gaucher's disease, and Tay-Sachs disease are found much more frequently among Ashkenazi Jews of Northern European heritage than among other ethnic groups.

Of the four subtypes of Niemann-Pick disease, type A is the most severe. It is inherited as a recessive disorder (i.e., a defective copy of the gene must be inherited from each parent) that results in an absence of the enzyme sphingomyelinase. The absence of this enzyme causes the storage of large amounts of sphingomyelin and cholesterol in the brain, bone marrow, liver, and spleen.

Symptoms may begin when a baby is only a few months old. The parents may notice a delay in motor development and/or problems with feeding. Although the infants may develop some motor skills, they quickly begin to regress as they lose muscle strength and tone, as well as vision and hearing. The disease progresses rapidly and the children typically die within the first few years of life.

Tay-Sachs disease is a lipid storage disease caused by an absence of the enzyme hexosaminidase, which functions in ganglioside metabolism. As a result of the enzyme deficiency, the ganglioside, shown in Section 17.4, accumulates in the cells of the brain causing neurological deterioration. Like Niemann-Pick disease, it is an autosomal recessive genetic trait that becomes apparent in the first few months of the life of an infant and rapidly progresses to death within a few years. Symptoms include listlessness, irritability, seizures, paralysis, loss of muscle tone and function, blindness, deafness, and delayed mental and social skills.

Gaucher's disease is an autosomal recessive genetic disorder resulting in a deficiency of the enzyme glucocerebrosidase. In the normal situation, this enzyme breaks down glucocerebroside, which is an intermediate in the synthesis and degradation of complex glycosphingolipids found in cellular membranes. In Gaucher's disease, glucocerebroside builds up in macrophages found in the liver, spleen, and bone marrow. These cells become engorged with excess lipid and displace healthy, normal cells in bone marrow. The symptoms of Gaucher's disease include severe anemia, thrombocytopenia (reduction in the number of platelets), and hepatosplenomegaly (enlargement of the spleen and liver). There can also be skeletal problems including bone deterioration and secondary fractures.

Fabry's disease is an X-linked inherited disorder caused by the deficiency of the enzyme α-galactosidase A. This disease afflicts as many as fifty thousand people worldwide. Typically, symptoms, including pain in the fingers and toes and a red rash around the waist, begin to appear when individuals reach their early twenties. A preliminary diagnosis can be confirmed by determining the concentration of the enzyme α-galactosidase A. Patients with Fabry's disease have an increased risk of kidney and heart disease, and a reduced life expectancy. Because this is an X-linked disorder, it is more common among males than females.

For Further Understanding

A defect in the enzyme sphingomyelinase is the cause of Niemann-Pick disease. Write a chemical equation to represent the reaction catalyzed by the enzyme sphingomyelinase, the cleavage of sphingomyelin to produce phosphorylcholine and ceramide. Show the structural formulas for the reactant and products of this reaction.

A defect in the enzyme glucocerebrosidase is the cause of Gaucher's disease. Write a chemical equation to represent the reaction catalyzed by the enzyme glucocerebrosidase, the cleavage of glucocerebroside to produce glucose and ceramide. Show the structural formulas for the reactant and products of this reaction.

All steroids contain the steroid nucleus (steroid carbon skeleton) as shown here:

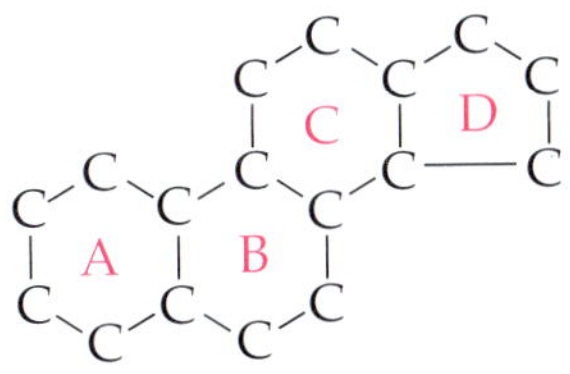

Carbon skeleton of the steroid nucleus

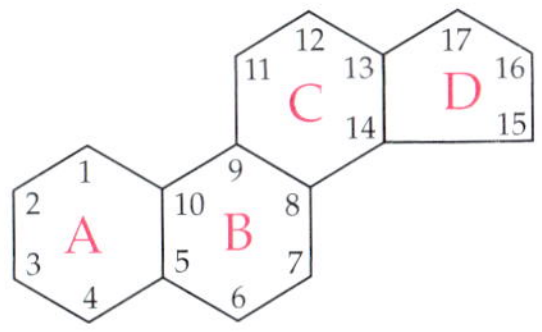

Steroid nucleus

The structure and function of the lipid-soluble vitamins are found online at www.mhhe.com/denniston5e in Lipid-Soluble Vitamins.

A Medical Perspective

Steroids and the Treatment of Heart Disease

The foxglove plant *(Digitalis purpurea)* is an herb that produces one of the most powerful known stimulants of heart muscle. The active ingredients of the foxglove plant (digitalis) are the so-called cardiac glycosides, or *cardiotonic steroids*, which include digitoxin, digosin, and gitalin.

Digitalis purpurea, the foxglove plant.

Digitoxin

The structure of digitoxin, one of the cardiotonic steroids produced by the foxglove plant.

These drugs are used clinically in the treatment of congestive heart failure, which results when the heart is not beating with strong, efficient strokes. When the blood is not propelled through the cardiovascular system efficiently, fluid builds up in the lungs and lower extremities (edema). The major symptoms of congestive heart failure are an enlarged heart, weakness, edema, shortness of breath, and fluid accumulation in the lungs.

This condition was originally described in 1785 by a physician, William Withering, who found a peasant woman whose folk medicine was famous as a treatment for chronic heart problems. Her potion contained a mixture of more than twenty herbs, but Dr. Withering, a botanist as well as physician, quickly discovered that foxglove was the active ingredient in the mixture. Withering used *Digitalis purpurea* successfully to treat congestive heart failure and even described some cautions in its use.

The cardiotonic steroids are extremely strong heart stimulants. A dose as low as 1 mg increases the stroke volume of the heart (volume of blood per contraction), increases the strength of the contraction, and reduces the heart rate. When the heart is pumping more efficiently because of stimulation by digitalis, the edema disappears.

Digitalis can be used to control congestive heart failure, but the dose must be carefully determined and monitored because the therapeutic dose is close to the dose that causes toxicity. The symptoms that result from high body levels of cardiotonic steroids include vomiting, blurred vision and lightheadedness, increased water loss, convulsions, and death. Only a physician can determine the initial dose and maintenance schedule for an individual to control congestive heart failure and yet avoid the toxic side effects.

For Further Understanding

Foxglove is a perennial plant, that is, it is a plant that will grow back each year for at least three years. Occasionally, foxglove first-year growth has been mistaken for comfrey, another plant with medical applications. Greeks and Romans used comfrey to treat wounds and to stop heavy bleeding, as well as for bronchial problems. Explain why the use of foxglove in place of comfrey might have fatal consequences.

Drugs such as digitalis are referred to as cardiac glycosides and as cardiotonic steroids. Explain why both these names are valid.

The steroid carbon skeleton consists of four fused rings. Each ring pair has two carbons in common. Thus two fused rings share one or more common bonds as part of their ring backbones. For example, rings A and B, B and C, and C and D are all fused in the preceding structure. Many steroids have methyl groups attached to carbons 10 and 13, as well as an alkyl, alcohol, or ketone group attached to carbon-17.

17.4 Nonglyceride Lipids

Cholesterol, a common steroid, is found in the membranes of most animal cells. It is an amphipathic molecule and is readily soluble in the hydrophobic region of membranes. It is involved in regulation of the fluidity of the membrane as a result of the nonpolar fused ring. However, the hydroxyl group is polar and functions like the polar heads of sphingolipids and phospholipids. There is a strong correlation between the concentration of cholesterol found in the blood plasma and heart disease, particularly **atherosclerosis** (hardening of the arteries). Cholesterol, in combination with other substances, contributes to a narrowing of the artery passageway. As narrowing increases, more pressure is necessary to ensure adequate blood flow, and high blood pressure *(hypertension)* develops. Hypertension is also linked to heart disease.

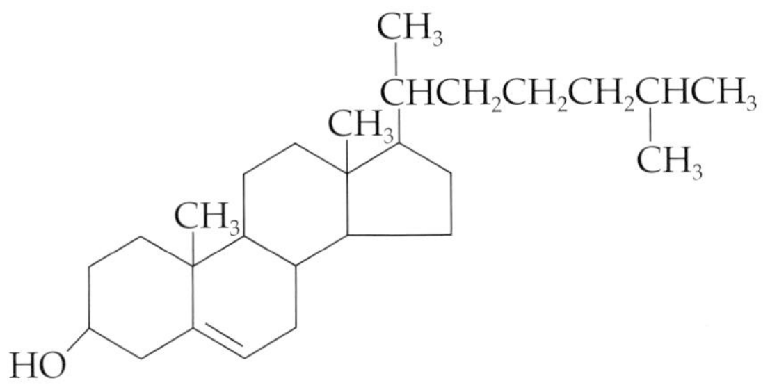

Cholesterol

Egg yolks contain a high concentration of cholesterol, as do many dairy products and animal fats. As a result, it has been recommended that the amounts of these products in the diet be regulated to moderate the dietary intake of cholesterol.

Bile salts are amphipathic derivatives of cholesterol that are synthesized in the liver and stored in the gallbladder. The principal bile salts in humans are cholate and chenodeoxycholate.

Bile salts are described in greater detail in Section 23.1.

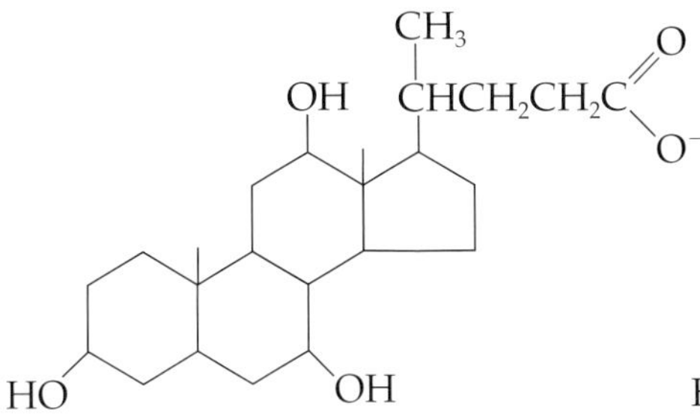

Cholate 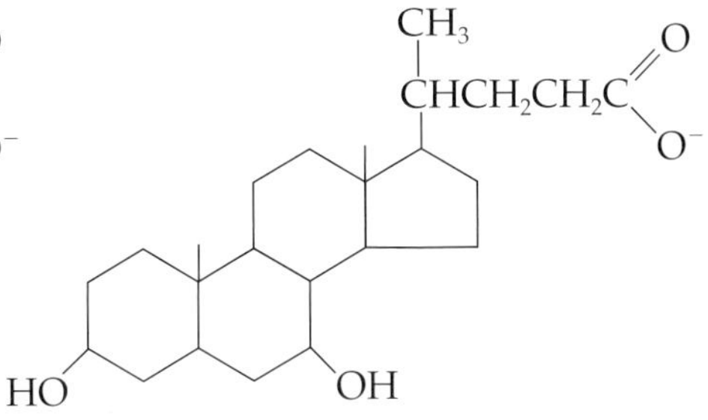   Chenodeoxycholate

Bile salts are emulsifying agents whose polar hydroxyl groups interact with water and whose hydrophobic regions bind to lipids. Following a meal, bile flows from the gallbladder to the duodenum (the uppermost region of the small intestine). Here the bile salts emulsify dietary fats into small droplets that can be more readily digested by lipases (lipid digesting enzymes) also found in the small intestine.

Steroids play a role in the reproductive cycle. In a series of chemical reactions, cholesterol is converted to the steroid *progesterone,* the most important hormone associated with pregnancy. Produced in the ovaries and in the placenta, progesterone is responsible for both the successful initiation and the successful completion of pregnancy. It prepares the lining of the uterus to accept the fertilized egg. Once the egg is attached, progesterone is involved in the development of the fetus and plays a role in the suppression of further ovulation during pregnancy.

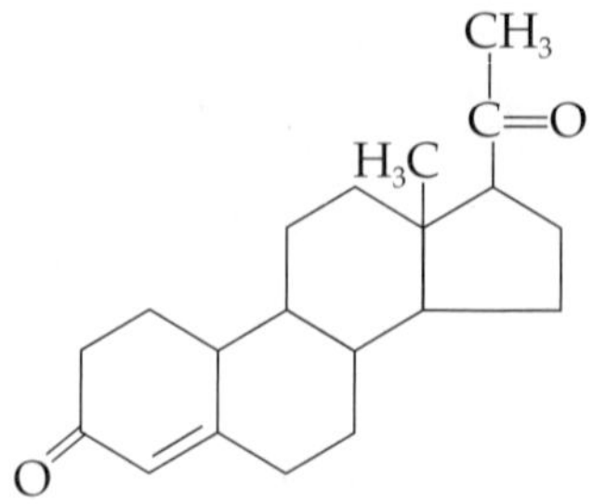

Chapter 17 Lipids and Their Functions in Biochemical Systems

19-Norprogesterone

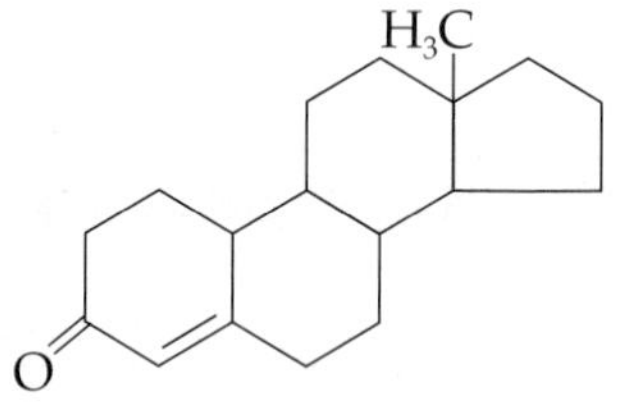

Norlutin

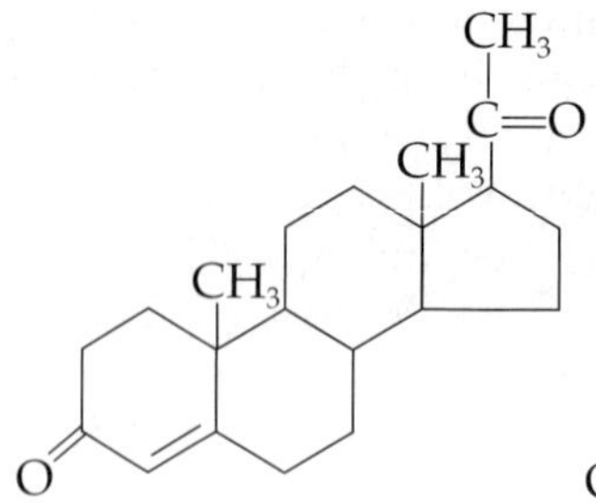

Progesterone

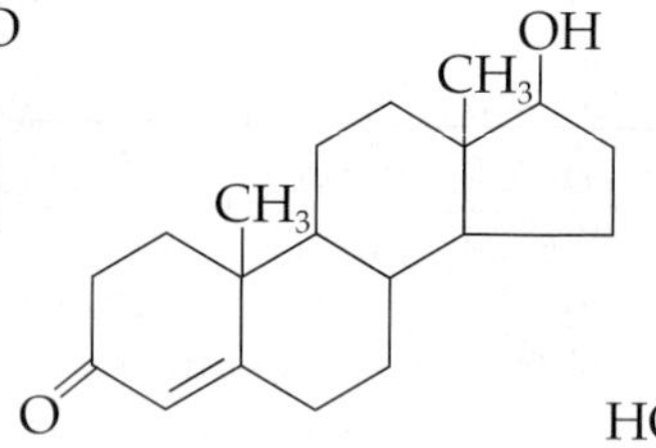

Testosterone

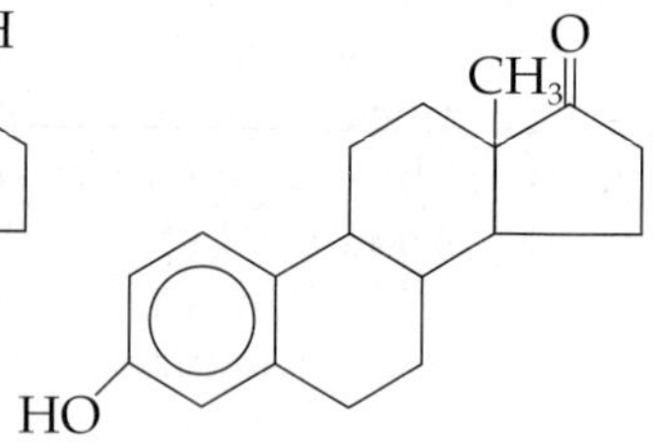

Estrone

Testosterone, a male sex hormone found in the testes, and *estrone*, a female sex hormone, are both produced by the chemical modification of progesterone. These hormones are involved in the development of male and female sex characteristics.

Many steroids, including progesterone, have played important roles in the development of birth control agents. 19-Norprogesterone was one of the first synthetic birth control agents. It is approximately ten times as effective as progesterone in providing birth control. However, its utility was severely limited because this compound could not be administered orally and had to be taken by injection. A related compound, norlutin (chemical name: 17-α-ethynyl-19-nortestosterone), was found to provide both the strength and the effectiveness of 19-norprogesterone and could be taken orally.

Currently "combination" oral contraceptives are prescribed most frequently. These include a progesterone and an estrogen. These newer products confer better contraceptive protection than either agent administered individually. They are also used to regulate menstruation in patients with heavy menstrual bleeding. First investigated in the late 1950s and approved by the FDA in 1961, there are at least thirty combination pills currently available. In addition, a transdermal patch for the treatment of postmenopausal osteoporosis is being investigated.

All of these compounds act by inducing a false pregnancy, which prevents ovulation. When oral contraception is discontinued, ovulation usually returns within three menstrual cycles. Although there have been problems associated with "the pill," it appears to be an effective and safe method of family planning for much of the population.

Cortisone is also important to the proper regulation of a number of biochemical processes. For example, it is involved in the metabolism of carbohydrates. Cortisone is also used in the treatment of rheumatoid arthritis, asthma, gastrointestinal disorders, many skin conditions, and a variety of other diseases. However, treatment with cortisone is not without risk. Some of the possible side effects of cortisone therapy include fluid retention, sodium retention, and potassium loss that can lead to congestive heart failure. Other side effects include muscle weakness, osteoporosis, gastrointestinal upsets including peptic ulcers, and neurological symptoms, including vertigo, headaches, and convulsions.

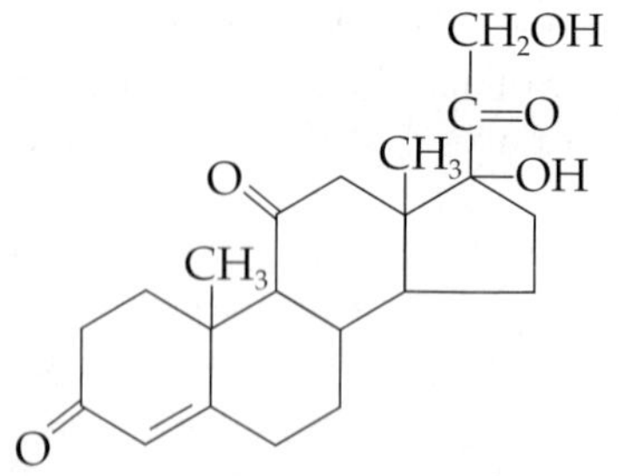

Cortisone

Aldosterone is a steroid hormone produced by the adrenal cortex and secreted into the bloodstream when blood sodium ion levels are too low. Upon reaching its target tissues in the kidney, aldosterone activates a set of reactions that cause sodium ions and water to be returned to the blood. If sodium levels are elevated, aldosterone is not secreted from the adrenal cortex and the sodium ions filtered out of the blood by the kidney will be excreted.

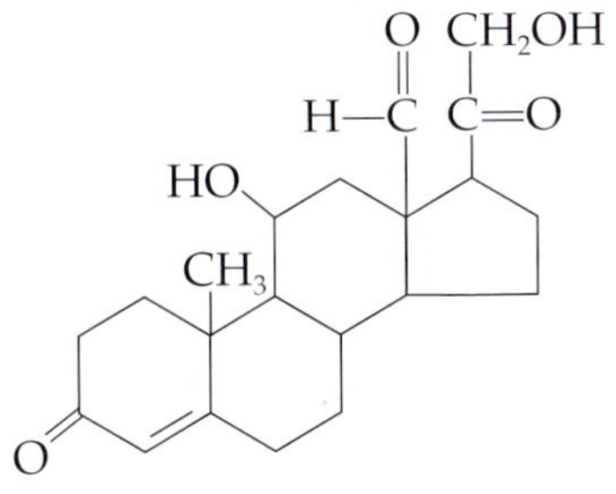

Aldosterone

Question 17.15

Draw structure of the steroid nucleus. Note the locations of the A, B, C, and D steroid rings.

Question 17.16

What is meant by the term *fused ring*?

Waxes

Waxes are derived from many different sources and have a variety of chemical compositions, depending on the source. Paraffin wax, for example, is composed of a mixture of solid hydrocarbons (usually straight-chain compounds). The natural waxes generally are composed of a long-chain fatty acid esterified to a long-chain alcohol. Because the long hydrocarbon tails are extremely hydrophobic, waxes are completely insoluble in water. Waxes are also solid at room temperature, owing to their high molecular weights. Two examples of waxes are myricyl palmitate, a major component of beeswax, and whale oil (spermaceti wax), from the head of the sperm whale, which is composed of cetyl palmitate.

Naturally occurring waxes have a variety of uses. Lanolin, which serves as a protective coating for hair and skin, is used in skin creams and ointments. Carnauba wax is used in automobile polish. Whale oil was once used as a fuel, in ointments, and in candles. However, synthetic waxes have replaced whale oil to a large extent, because of efforts to ban the hunting of whales.

$$CH_3(CH_2)_{14}-\overset{O}{\underset{\|}{C}}-O-(CH_2)_{29}CH_3$$

Myricyl palmitate (beeswax)

$$CH_3(CH_2)_{14}-\overset{O}{\underset{\|}{C}}-O-(CH_2)_{15}CH_3$$

Cetyl palmitate (whale oil)

17.5 Complex Lipids

Complex lipids are lipids that are bonded to other types of molecules. The most common and important complex lipids are plasma lipoproteins, which are responsible for the transport of other lipids in the body.

Lipids are only sparingly soluble in water, and the movement of lipids from one organ to another through the bloodstream requires a transport system that uses **plasma lipoproteins.** Lipoprotein particles consist of a core of hydrophobic lipids surrounded by amphipathic proteins, phospholipids, and cholesterol (Figure 17.8).

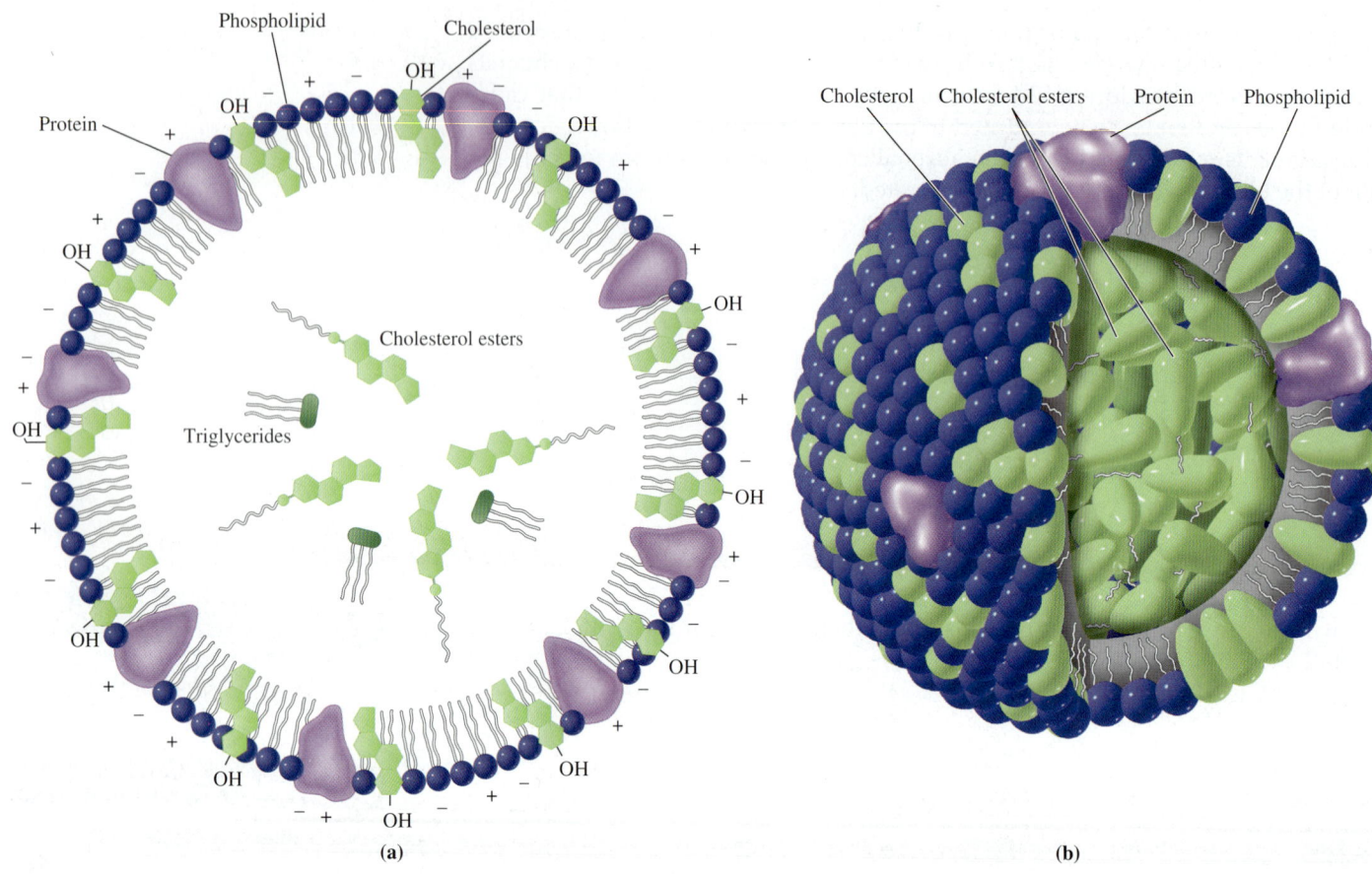

Figure 17.8
A model for the structure of a plasma lipoprotein. The various lipoproteins are composed of a shell of protein, cholesterol, and phospholipids surrounding more hydrophobic molecules such as triglycerides or cholesterol esters (cholesterol esterified to fatty acids). (a) Cross section, (b) three-dimensional view.

There are four major classes of human plasma lipoproteins:

- **Chylomicrons,** which have a density of less than 0.95 g/mL, carry dietary triglycerides from the intestine to other tissues. The remaining lipoproteins are classified by their densities.
- **Very low density lipoproteins (VLDL)** have a density of 0.95–1.019 g/mL. They bind triglycerides synthesized in the liver and carry them to adipose and other tissues for storage.
- **Low-density lipoproteins (LDL)** are characterized by a density of 1.019–1.063 g/mL. They carry cholesterol to peripheral tissues and help regulate cholesterol levels in those tissues. These are richest in cholesterol, frequently carrying 40% of the plasma cholesterol.
- **High-density lipoproteins (HDL)** have a density of 1.063–1.210 g/mL. They are bound to plasma cholesterol; however, they transport cholesterol from peripheral tissues to the liver.

A summary of the composition of each of the plasma lipoproteins is presented in Figure 17.9.

Chylomicrons are aggregates of triglycerides and protein that transport dietary triglycerides to cells throughout the body. Not all lipids in the blood are derived directly from the diet. Triglycerides and cholesterol are also synthesized in the liver and also are transported through the blood in lipoprotein packages. Triglycerides are assembled into VLDL particles that carry the energy-rich lipid molecules either to tissues requiring an energy source or to adipose tissue for stor-

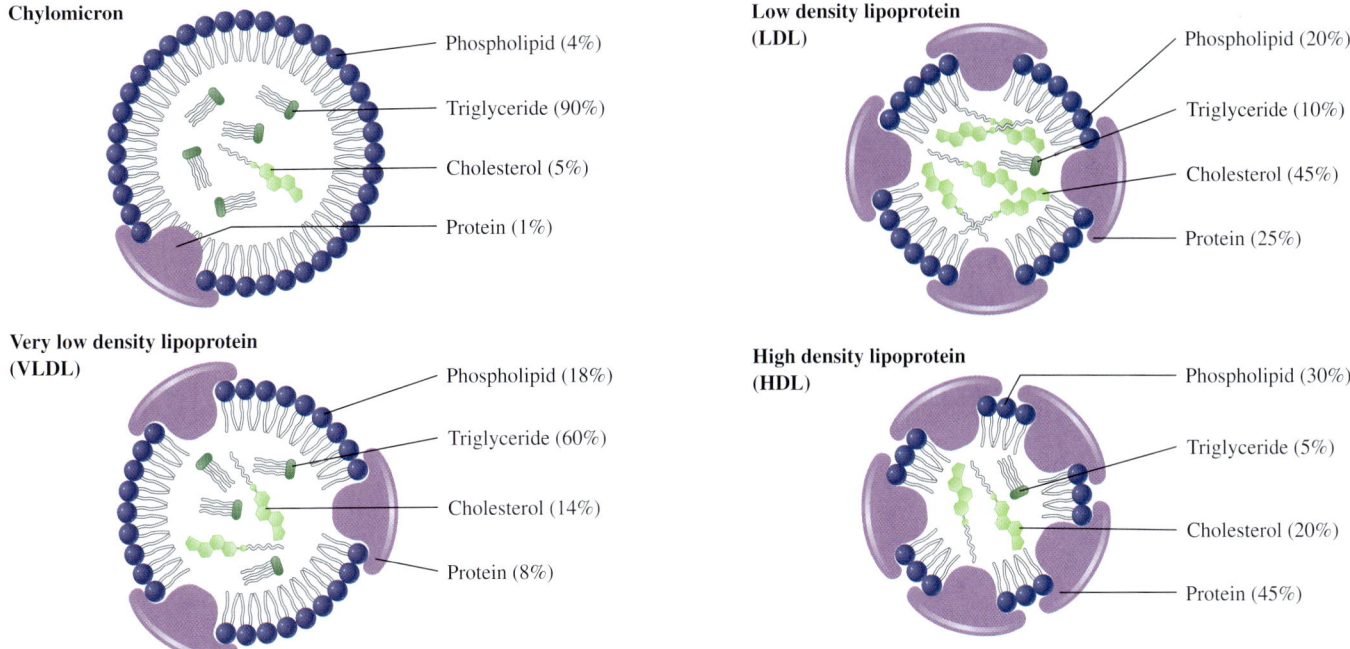

Figure 17.9
A summary of the relative amounts of cholesterol, phospholipid, protein, and triglycerides in the four classes of lipoproteins.

age. Similarly, cholesterol is assembled into LDL particles for transport from the liver to peripheral tissues.

Entry of LDL particles into the cell is dependent on a specific recognition event and binding between the LDL particle and a protein receptor embedded within the membrane. Low-density lipoprotein receptors (LDL receptors) are found in the membranes of cells outside the liver and are responsible for the uptake of cholesterol by the cells of various tissues. LDL (lipoprotein bound to cholesterol) binds specifically to the LDL receptor, and the complex is taken into the cell by a process called *receptor-mediated endocytosis* (Figure 17.10). The membrane begins to be pulled into the cell at the site of the LDL receptor complexes. This draws the entire LDL particle into the cell. Eventually, the portion of the membrane surrounding the LDL particles pinches away from the cell membrane and forms a membrane around the LDL particles. As we will see in Section 17.6, membranes are fluid and readily flow. Thus, they can form a vesicle or endosome containing the LDL particles.

Cellular digestive organelles known as *lysosomes* fuse with the endosomes. This fusion is accomplished when the membranes of the endosome and the lysosome flow together to create one larger membrane-bound body or vesicle. Hydrolytic enzymes from the lysosome then digest the entire complex to release cholesterol into the cytoplasm of the cell. There, cholesterol inhibits its own biosynthesis and activates an enzyme that stores cholesterol in cholesterol ester droplets. High concentrations of cholesterol inside the cell also inhibit the synthesis of LDL receptors to ensure that the cell will not take up too much cholesterol. People who have a genetic defect in the gene coding for the LDL receptor do not take up as much cholesterol. As a result they accumulate LDL cholesterol in the plasma. This excess plasma cholesterol is then deposited on the artery walls, causing atherosclerosis. This disease is called *hypercholesterolemia*.

Liver lipoprotein receptors enable large amounts of cholesterol to be removed from the blood, thus ensuring low concentrations of cholesterol in the blood plasma. Other factors being equal, the person with the most lipoprotein receptors will be the least vulnerable to a high-cholesterol diet and have the least likelihood of developing atherosclerosis.

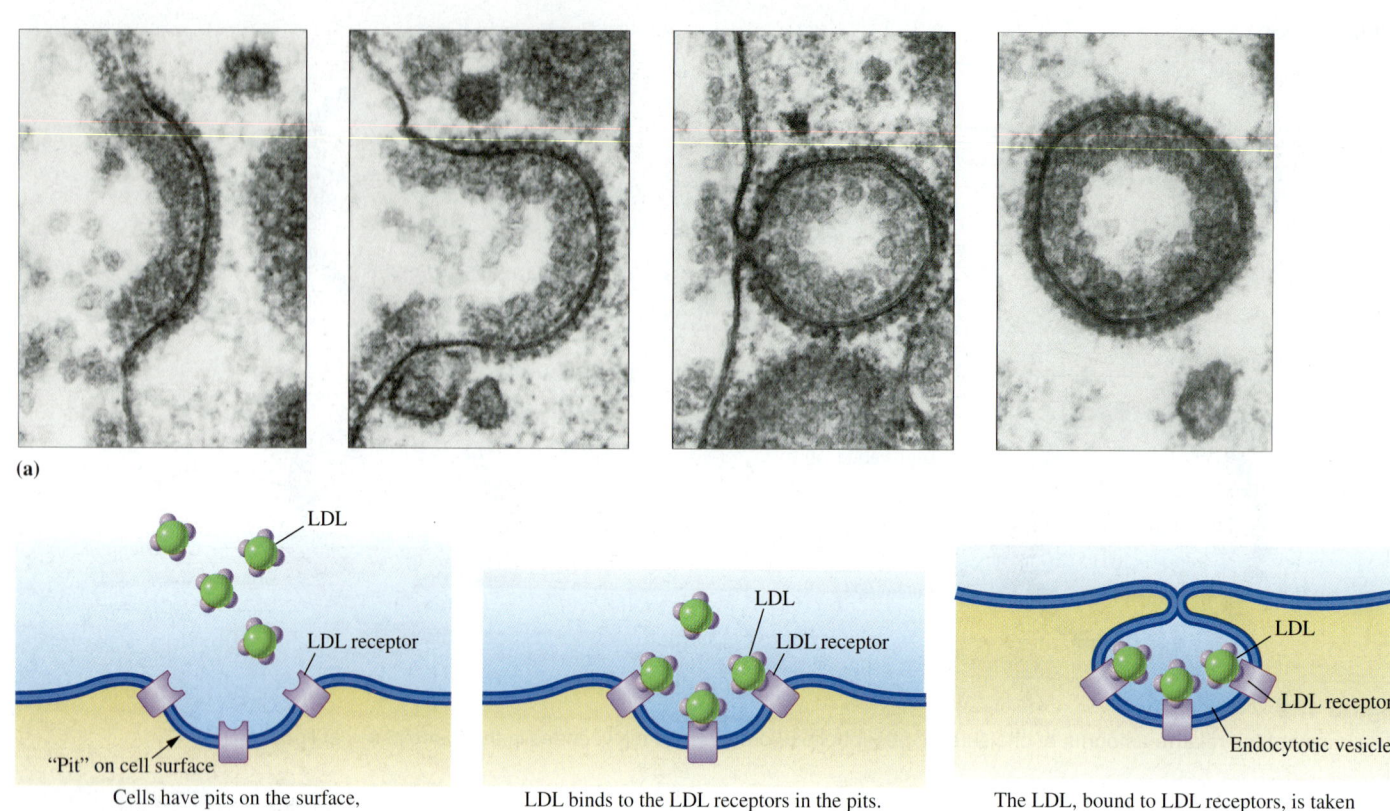

Figure 17.10
Receptor-mediated endocytosis. (a) Electron micrographs of the process of receptor-mediated endocytosis. (b) Summary of the events of receptor-mediated endocytosis of LDL.

There is also evidence that high levels of HDL in the blood help reduce the incidence of atherosclerosis, perhaps because HDL carries cholesterol from the peripheral tissues back to the liver. In the liver, some of the cholesterol is used for bile synthesis and secreted into the intestines, from which it is excreted.

A final correlation has been made between diet and atherosclerosis. People whose diet is high in saturated fats tend to have high levels of cholesterol in the blood. Although the relationship between saturated fatty acids and cholesterol metabolism is unclear, it is known that a diet rich in unsaturated fats results in decreased cholesterol levels. In fact, the use of unsaturated fat in the diet results in a decrease in the level of LDL and an increase in the level of HDL. With the positive correlation between heart disease and high cholesterol levels, the current dietary recommendations include a diet that is low in fat and the substitution of unsaturated fats (vegetable oils) for saturated fats (animal fats).

Recently an inflammatory protein, the C-reactive protein (CRP), has been implicated in atherosclerosis. A test for the level of this protein in the blood is being suggested as a way to predict the risk of heart attack. A high sensitivity CRP test is now widely available.

Question 17.17 What is the mechanism of uptake of cholesterol from plasma?

Question 17.18 What is the role of lysosomes in the metabolism of plasma lipoproteins?

17.6 The Structure of Biological Membranes

Biological membranes are *lipid bilayers* in which the hydrophobic hydrocarbon tails are packed in the center of the bilayer and the ionic head groups are exposed on the surface to interact with water (Figure 17.11). The hydrocarbon tails of membrane phospholipids provide a thin shell of nonpolar material that prevents mixing of molecules on either side. The nonpolar tails of membrane phospholipids thus provide a barrier between the interior of the cell and its surroundings. The polar heads of lipids are exposed to water, and they are highly solvated.

The two layers of the phospholipid bilayer membrane are not identical in composition. For instance, in human red blood cells, approximately 80% of the phospholipids in the outer layer of the membrane are phosphatidylcholine and sphingomyelin; whereas phosphatidylethanolamine and phosphatidylserine make up approximately 80% of the inner layer. In addition, carbohydrate groups are found attached only to those phospholipids found on the outer layer of a membrane. Here they participate in receptor and recognition functions.

LEARNING GOAL 11

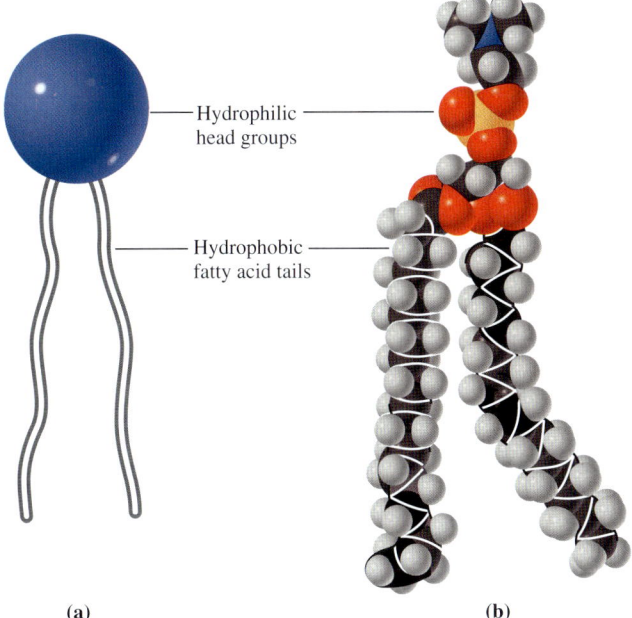

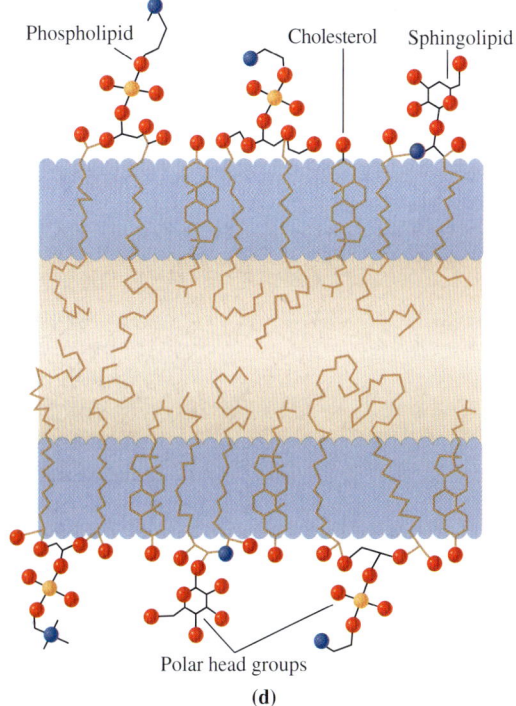

Figure 17.11
(a) Representation of a phospholipid.
(b) Space-filling model of a phospholipid.
(c) Representation of a phospholipid bilayer membrane. (d) Line formula structure of a bilayer membrane composed of phospholipids, cholesterol, and sphingolipids.

Fluid Mosaic Structure of Biological Membranes

As we have just noted, membranes are not static; they are composed of molecules in motion. The fluidity of biological membranes is determined by the proportions of saturated and unsaturated fatty acid groups in the membrane phospholipids. About half of the fatty acids that are isolated from membrane lipids from all sources are unsaturated.

The unsaturated fatty acid tails of the phospholipids contribute to membrane fluidity because of the bends introduced into the hydrocarbon chain by the double bonds. Because of these "kinks," the fatty acid tails do not pack together tightly.

We also find that the percentage of unsaturated fatty acid groups in membrane lipids is inversely proportional to the temperature of the environment. Bacteria, for example, have different ratios of saturated and unsaturated fatty acids in their membrane lipids, depending on the temperatures of their surroundings. For instance, the membranes of bacteria that grow in the Arctic Ocean have high levels of unsaturated fatty acids so that their membranes remain fluid even at these frigid temperatures. Conversely, the organisms that live in the hot springs of Yellowstone National Park, with temperatures near the boiling point of water, have membranes with high levels of saturated fatty acids. This flexibility in fatty acid content enables the bacteria to maintain the same membrane fluidity over a temperature range of almost 100°C.

Generally, the body temperatures of mammals are quite constant, and the fatty acid composition of their membrane lipids is therefore usually very uniform. One interesting exception is the reindeer. Much of the year the reindeer must travel through ice and snow. Thus the hooves and lower legs must function at much colder temperatures than the rest of the body. Because of this, the percentage of unsaturation in the membranes varies along the length of the reindeer leg. We find that the proportion of unsaturated fatty acids increases closer to the hoof, permitting the membranes to function in the low temperatures of ice and snow to which the lower leg is exposed.

Thus, membranes are fluid, regardless of the environmental temperature conditions. In fact, it has been estimated that membranes have the consistency of olive oil.

Although the hydrophobic barrier created by the fluid lipid bilayer is an important feature of membranes, the proteins embedded within the lipid bilayer are equally important and are responsible for critical cellular functions. The presence of these membrane proteins was revealed by an electron microscopic technique called *freeze-fracture*. Cells are frozen to very cold temperatures and then fractured with a very fine diamond knife. Some of the cells are fractured between the two layers of the lipid bilayer. When viewed with the electron microscope, the membrane appeared to be a mosaic, studded with proteins. Because of the fluidity of membranes and the appearance of the proteins seen by electron microscopy, our concept of membrane structure is called the **fluid mosaic model** (Figure 17.12).

Some of the observed proteins, called **peripheral membrane proteins,** are bound only to one of the surfaces of the membrane by interactions between ionic head groups of the membrane lipids and ionic amino acids on the surface of the peripheral protein. Other membrane proteins, called **transmembrane proteins,** are embedded within the membrane and extend completely through it, being exposed both inside and outside the cell.

Just as the phospholipid composition of the membrane is asymmetric, so too is the orientation of transmembrane proteins. Each transmembrane protein has hydrophobic regions that associate with the fatty acid tails of membrane phospholipids. Each also has a unique hydrophilic domain that is always found associated with the outer layer of the membrane and is located on the outside of the cell. This region of the protein typically has oligosaccharides covalently attached. Hence

17.6 The Structure of Biological Membranes

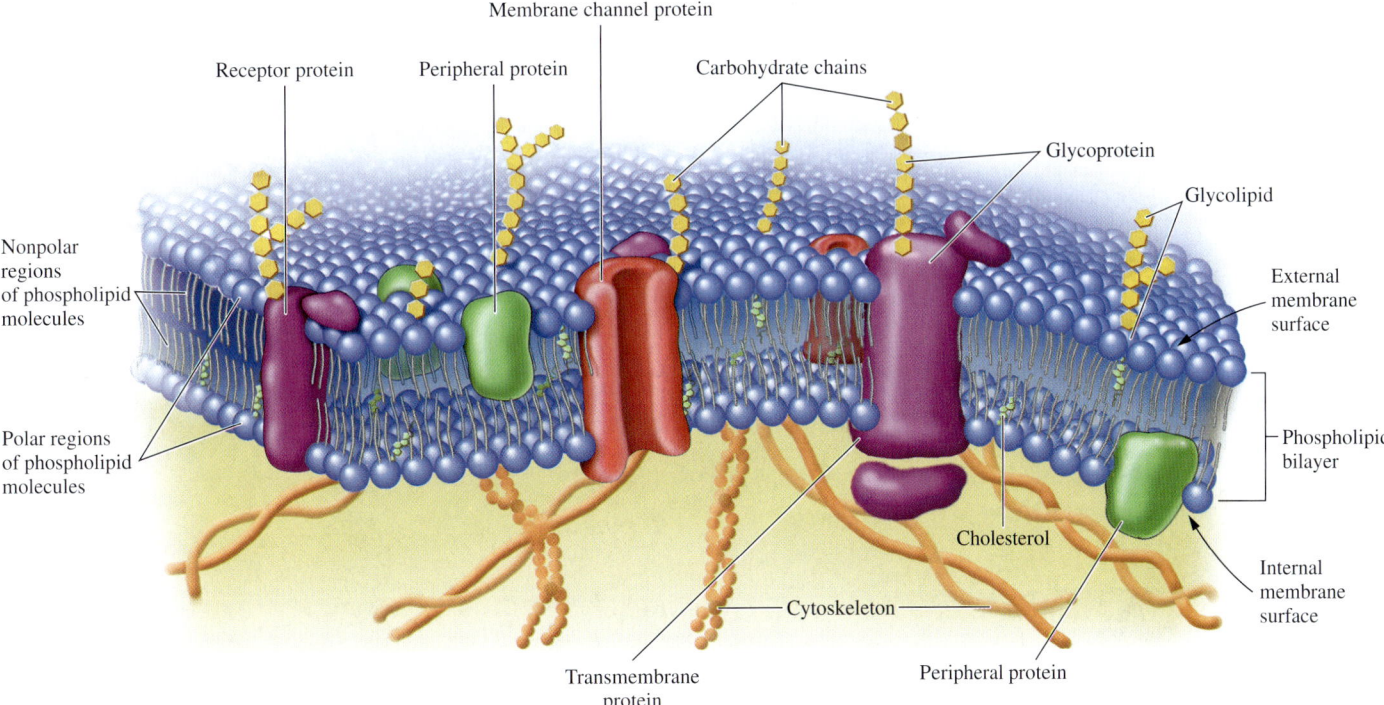

Figure 17.12
The fluid mosaic model of membrane structure.

these proteins are *glycoproteins.* Similarly, each transmembrane protein has a second hydrophilic domain that is always found associated with the inner layer of the membrane and projects into the cytoplasm of the cell. Typically this region of the transmembrane protein is attached to filaments of the cytoplasmic skeleton.

Membranes are dynamic structures. The mobility of proteins embedded in biological membranes was studied by labeling certain proteins in human and mouse cell membranes with red and green fluorescent dyes. The human and mouse cells were fused; in other words, special techniques were used to cause the membranes of the mouse and human cell to flow together to create a single cell. The new cell was observed through a special ultraviolet or fluorescence microscope. The red and green patches were localized within regions of their original cell membranes when the experiment began. Forty minutes later the color patches were uniformly distributed in the fused cellular membrane (Figure 17.13). This experiment suggests that we can think of the fluid mosaic membrane as an ocean filled with mobile, floating icebergs.

Membrane Transport

The cell membrane mediates the interaction of the cell with its environment and is responsible for the controlled passage of material into and out of the cell. The external cell membrane controls the entrance of fuel and the exit of waste products. Most of these transport processes are controlled by transmembrane transport proteins. These transport proteins are the cellular gatekeepers, whose function in membrane transport is analogous to the function of enzymes in carrying out cellular chemical reactions. However, some molecules pass through the membranes unassisted, by the **passive transport** processes of diffusion and osmosis. These are referred to as passive processes because they do not require any energy expenditure by the cell.

 **LEARNING GOAL**

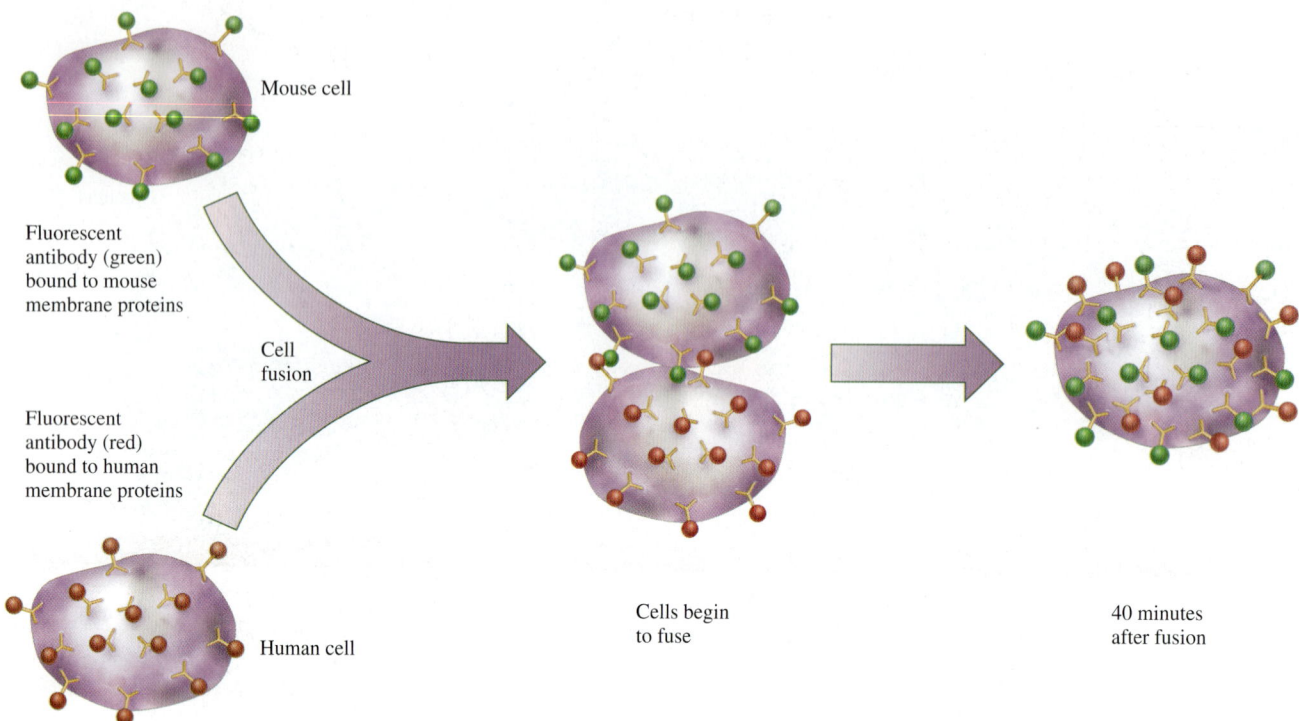

Figure 17.13
Demonstration that membranes are fluid and that proteins move freely in the plane of the lipid bilayer.

Passive Diffusion: The Simplest Form of Membrane Transport

If we put a teaspoon of instant iced tea on the surface of a glass of water, the molecules soon spread throughout the solution. The molecules of both the solute (tea) and the solvent (water) are propelled by random molecular motion. The initially concentrated tea becomes more and more dilute. This process of the *net* movement of a solute with the gradient (from an area of high concentration to an area of low concentration) is called *diffusion* (Figure 17.14).

Diffusion is one means of passive transport across membranes. Let's now suppose that a biological membrane is present and that a substance is found at one concentration outside the membrane and at half that concentration inside the membrane. If the solute can pass through the membrane, diffusion will occur with net transport of material from the region of initial high concentration to the region of initial low concentration, and the substance will equilibrate across the cell membrane (Figures 17.15 and 17.16). After a while, the concentration of the substance will be the same on both sides of the membrane; the system will be at equilibrium, and no more net change will occur.

Of what practical value is the process of diffusion to the cell? Certainly, diffusion is able to distribute metabolites effectively throughout the interior of the cell. But what about the movement of molecules through the membrane? Because of

Figure 17.14
Diffusion results in the *net* movement of sugar and water molecules from the area of high concentration to the area of low concentration. Eventually, the concentrations of sugar and water throughout the beaker will be equal.

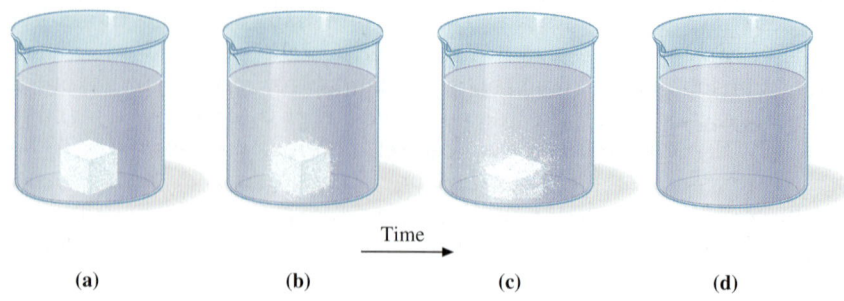

17.6 The Structure of Biological Membranes

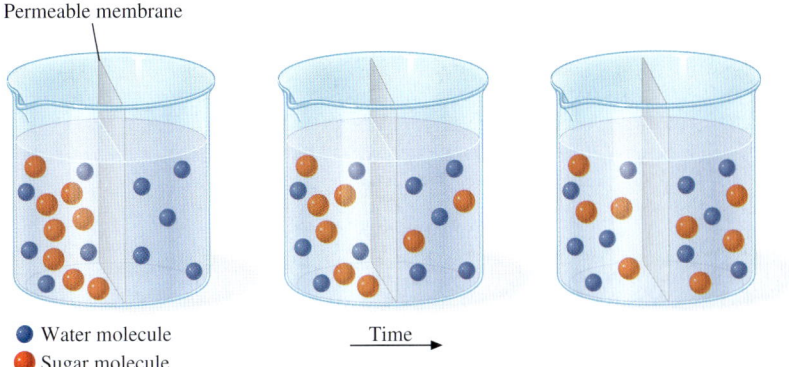

Figure 17.15
Diffusion of a solute through a membrane.

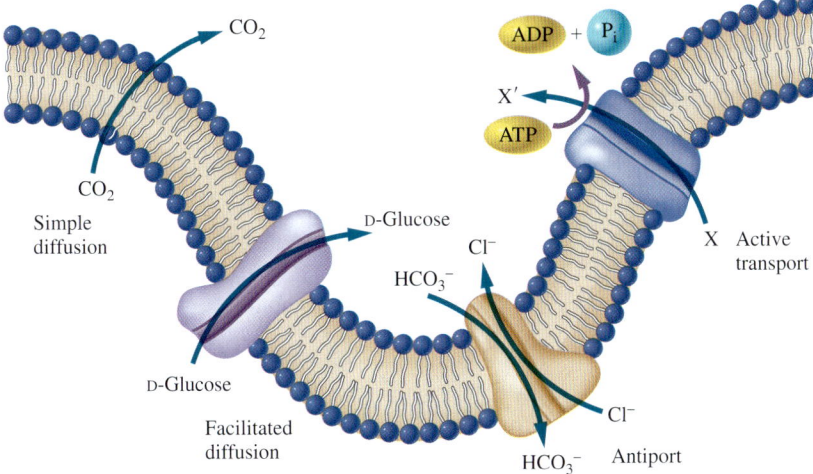

Figure 17.16
Transport across a cell membrane may occur by simple diffusion or facilitated diffusion. Transport of Cl^- and HCO_3^- occurs in opposite directions across cell membranes, a process called antiport. Energy requiring active transport mechanisms are required for many important substances.

the lipid bilayer structure of the membrane, only a few molecules are able to diffuse freely across a membrane. These include small molecules such as O_2 and CO_2. Any large or highly charged molecules or ions are not able to pass through the lipid bilayer directly. Such molecules require an assist from cell membrane proteins. Any membrane that allows the diffusion of some molecules but not others is said to be *selectively permeable*.

Facilitated Diffusion: Specificity of Molecular Transport

Most molecules are transported across biological membranes by specific protein carriers known as *permeases*. When a solute diffuses through a membrane from an area of high concentration to an area of low concentration by passing through a channel within a permease, the process is known as **facilitated diffusion**. No energy is consumed by facilitated diffusion; thus it is another means of passive transport, and the direction of transport depends upon the concentrations of metabolite on each side of the membrane.

Transport across cell membranes by facilitated diffusion occurs through pores within the permease that have conformations, or shapes, that are complementary to those of the transported molecules. The charge and conformation of the pore define the specificity of the carrier (Figure 17.16). Only molecules that have the correct shape can enter the pore. As a result, the rate of diffusion for any molecule is limited by the number of carrier permease molecules in the membrane that are responsible for the passage of that molecule.

A Medical Perspective

Liposome Delivery Systems

Liposomes were discovered by Dr. Alec Bangham in 1961. During his studies on phospholipids and blood clotting, he found that if he mixed phospholipids and water, tiny phospholipid bilayer sacs, called liposomes, would form spontaneously. Since that first observation, liposomes have been developed as efficient delivery systems for everything from antitumor and antiviral drugs, to the hair-loss therapy minoxidil!

If a drug is included in the solution during formation of liposomes, the phospholipids will form a sac around the solution. In this way the drug becomes encapsulated within the phospholipid sphere. These liposomes can be injected intravenously or applied to body surfaces. Sometimes scientists include hydrophilic molecules in the surface of the liposome. This increases the length of time that they will remain in circulation in the bloodstream. These so-called stealth liposomes are being used to carry anticancer drugs, such as doxorubicin and mitoxantrone. Liposomes are also being used as carriers for the antiviral drugs, such as AZT and ddC, that are used to treat human immunodeficiency virus infection.

A clever trick to help target the drug-carrying liposome is to include an antibody on the surface of the liposome. These antibodies are proteins designed to bind specifically to the surface of a tumor cell. Upon attaching to the surface of the tumor cell, the liposome "membrane" fuses with the cell membrane. In this way the deadly chemicals are delivered only to those cells targeted for destruction. This helps to avoid many of the unpleasant side effects of chemotherapy treatment that occur when normal healthy cells are killed by the drug.

Another application of liposomes is in the cosmetics industry. Liposomes can be formed that encapsulate a vitamin, herbal agent, or other nutritional element. When applied to the skin, the liposomes pass easily through the outer layer of dead skin, delivering their contents to the living skin cells beneath. As with the pharmaceutical liposomes, these liposomes, sometimes called cosmeceuticals, fuse with skin cells. Thus, they directly deliver the beneficial cosmetic agent directly to the cells that can benefit the most.

Since their accidental discovery forty years ago, much has been learned about the formation of liposomes and ways to engineer them for more efficient delivery of their contents. This is another example of the marriage of serendipity, an accidental discovery, with scientific research and technological application. As the development of new types of liposomes continues, we can expect that even more ways will be found to improve the human condition.

For Further Understanding

From what you know of the structure of phospholipids, explain the molecular interactions that cause liposomes to form.

Could you use liposome technology to deliver a hydrophobic drug? Explain your answer.

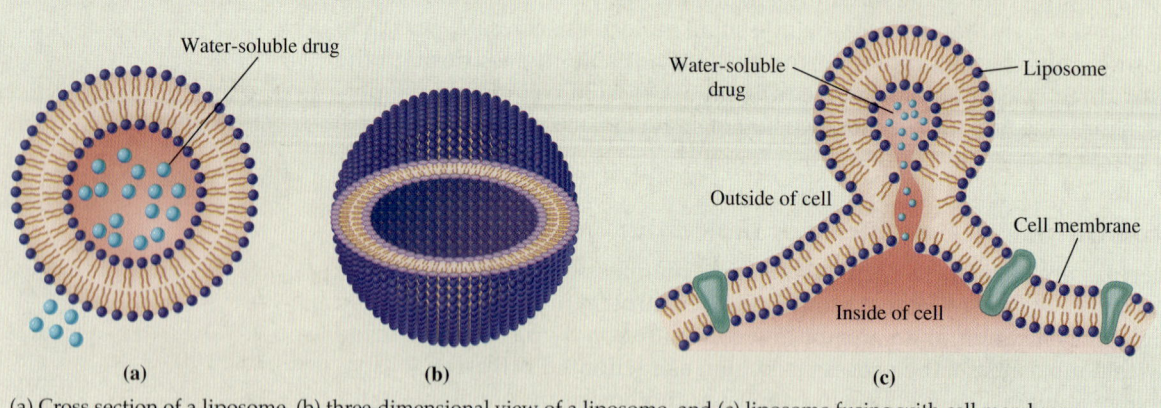

(a) Cross section of a liposome, (b) three-dimensional view of a liposome, and (c) liposome fusing with cell membrane.

The transport of glucose illustrates the specificity of carrier permease proteins. D-Glucose is transported by the glucose carrier, but its enantiomer, L-glucose, is not. Thus the glucose permease exhibits stereospecificity. In other words, the solute to be brought into the cell must "fit" precisely, like a hand in a glove, into a recognition site within the structure of the permease. In Chapter 19 we will see that

17.6 The Structure of Biological Membranes

enzymes that catalyze the biochemical reactions within cells show this same type of specificity.

The rate of transport of metabolites into the cell has profound effects on the net metabolic rate of many cells. Insulin, a polypeptide hormone synthesized by the islet β-cells of the pancreas, increases the maximum rate of glucose transport by a factor of three to four. The result is that the metabolic activity of the cell is greatly increased.

Red blood cells use the same anion channel to transport Cl^- into the cell in exchange for HCO_3^- ions (Figure 17.16). This two-way transport is known as *antiport*. This transport occurs by facilitated diffusion so that Cl^- flows from a high exterior concentration to a low interior one, and bicarbonate flows from a high interior concentration to a low exterior one.

Osmosis: Passive Movement of a Solvent Across a Membrane

Because a cell membrane is selectively permeable, it is not always possible for solutes to pass through it in response to a concentration gradient. In such cases the solvent diffuses through the membrane. Such membranes, permeable to solvent but not to solute, are specifically called **semipermeable membranes. Osmosis** is the diffusion of a solvent (water in biological systems) through a semipermeable membrane in response to a water concentration gradient.

Suppose that we place a 0.5 M glucose solution in a dialysis bag that is composed of a membrane with pores that allow the passage of water molecules but not glucose molecules. Consider what will happen when we place this bag into a beaker of pure water. We have created a gradient in which there is a higher concentration of glucose inside the bag than outside, but the glucose cannot diffuse through the bag to achieve equal concentration on both sides of the membrane.

Now let's think about this situation in another way. We have a higher concentration of water molecules outside the bag (where there is only pure water) than inside the bag (where some of the water molecules are occupied in the hydration of solute particles and are consequently unable to move freely in the system). Because water can diffuse through the membrane, a net diffusion of water will occur through the membrane into the bag. This is the process of osmosis (Figure 17.17).

As you have probably already guessed, this system can never reach equilibrium (equal concentrations inside and outside the bag) because regardless of how much water diffuses into the bag, diluting the glucose solution, the concentration of glucose will always be higher inside the bag (and the accompanying free water concentration will always be lower).

What happens when the bag has taken in as much water as it can, when it has expanded as much as possible? Now the walls of the bag exert a force that will stop the *net* flow of water into the bag. **Osmotic pressure** is the pressure that must be exerted to stop the flow of water across a selectively permeable membrane by

Enzyme specificity is discussed in Section 19.5.

The metabolic effects of insulin are described in Sections 21.7 and 23.6.

 **LEARNING GOAL**

Recall that there is an inverse relationship between the osmotic (solute) concentration of a solution and the water concentration of that solution, as discussed in Chapter 6.

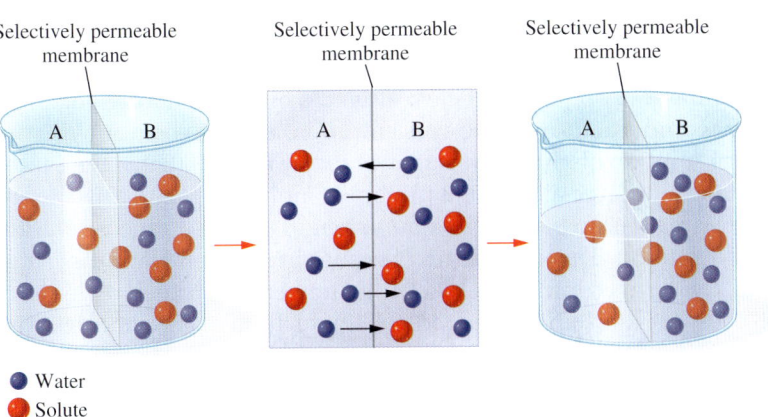

Figure 17.17
Osmosis across a membrane. The solvent, water, diffuses from an area of lower solute concentration (side A) to an area of higher solute concentration (side B).

A Medical Perspective

Antibiotics That Destroy Membrane Integrity

The "age of antibiotics" began in 1927 when Alexander Fleming discovered, quite by accident, that a product of the mold *Penicillium* can kill susceptible bacteria. We now know that penicillin inhibits bacterial growth by interfering with cell wall synthesis. Since Fleming's time, hundreds of antibiotics, which are microbial products that either kill or inhibit the growth of susceptible bacteria or fungi, have been discovered. The key to antibiotic therapy is to find a "target" in the microbe, a metabolic process or structure that the human does not have. In this way the antibiotic will selectively inhibit the disease-causing organism without harming the patient.

Many antibiotics disrupt cell membranes. The cell membrane is not an ideal target for antibiotic therapy because all cells, human and bacterial, have membranes. Therefore both types of cells are damaged. Because these antibiotics exhibit a wide range of toxic side effects when ingested, they are usually used to combat infections topically (on body surfaces). In this way, damage to the host is minimized but the inhibitory effect on the microbe is maximized.

Polymyxins are antibiotics produced by the bacterium *Bacillus polymyxa*. They are protein derivatives having one end that is hydrophobic because of an attached fatty acid. The opposite end is hydrophilic. Because of these properties, the polymyxins bind to membranes with the hydrophobic end embedded within the membrane, while the hydrophilic end remains outside the cell. As a result, the integrity of the membrane is disrupted, and leakage of cellular constituents occurs, causing cell death.

Although the polymyxins have been found to be useful in treating some urinary tract infections, pneumonias, and infec-

The structures of amphotericin B and nystatin, two antifungal antibiotics.

osmosis. Stated more precisely, the osmotic pressure of a solution is the net pressure with which water enters it by osmosis from a pure water compartment when the two compartments are separated by a semipermeable membrane.

Osmotic concentration or *osmolarity* is the term used to describe the osmotic strength of a solution. It depends only on the ratio of the number of solute particles to the number of solvent particles. Thus the chemical nature and size of the solute are not important, only the concentration, expressed in molarity. For instance, a 2 M solution of glucose (a sugar of molar mass 180) has the same osmolarity as a 2 M solution of albumin (a protein of molar mass 60,000).

Blood plasma has an osmolarity equivalent to a 0.30 M glucose solution or a 0.15 M NaCl solution. This latter is true because in solution NaCl dissociates into Na^+ and Cl^- and thus contributes twice the number of solute particles as a molecule that does not ionize. If red blood cells, which have an osmolarity equal to blood plasma, are placed in a 0.30 M glucose solution, no net osmosis will occur because the osmolarity and water concentration inside the red blood cell are equal to those of the 0.30 M glucose solution. The solutions inside and outside the red

17.6 The Structure of Biological Membranes

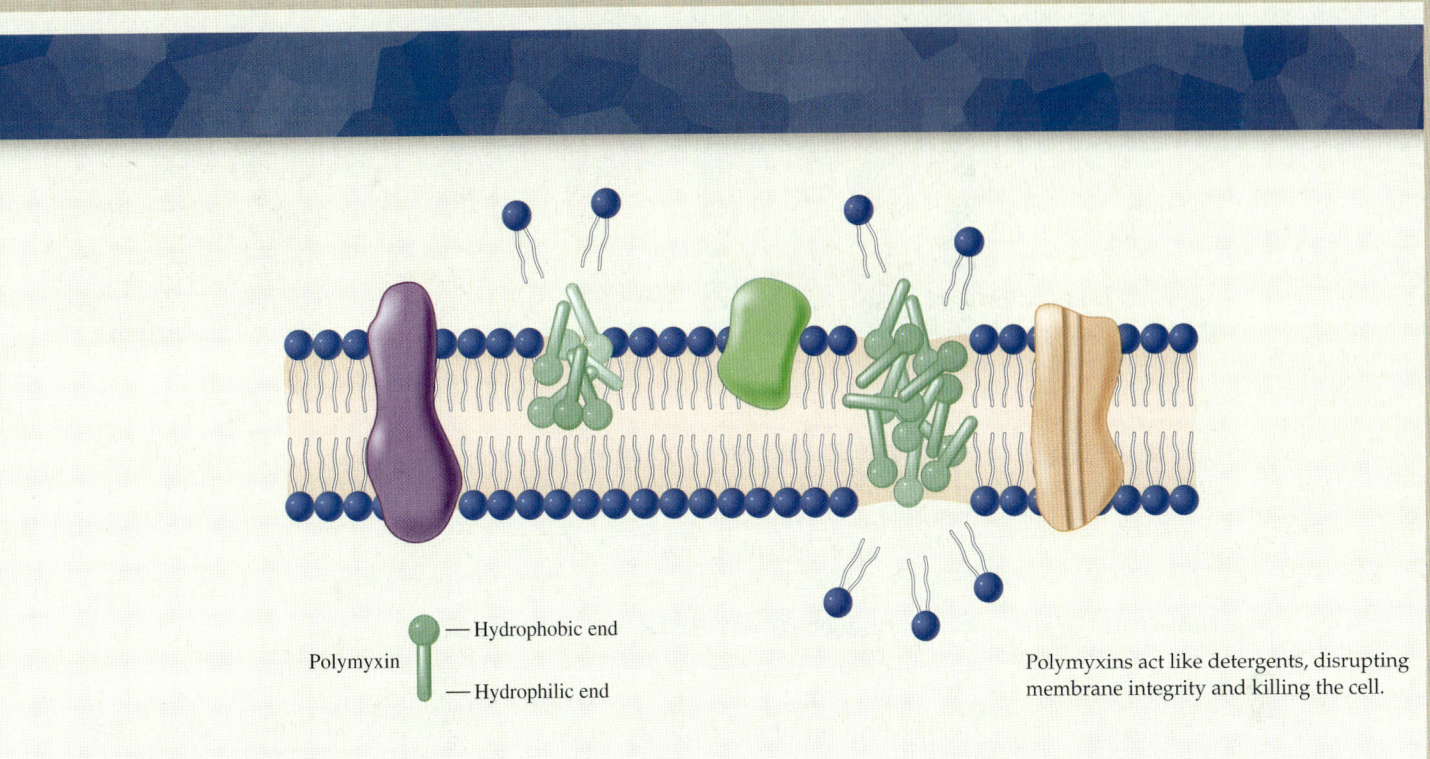

Polymyxin — Hydrophobic end / Hydrophilic end

Polymyxins act like detergents, disrupting membrane integrity and killing the cell.

tions of burn patients, other antibiotics are now favored because of the toxic effects of the polymyxins on the kidney and central nervous system. Polymyxin B is still used topically and is available as an over-the-counter ointment in combination with two other antibiotics, neomycin and bacitracin.

Two other antibiotics that destroy membranes, amphotericin B and nystatin, are large ring structures that are used in treating serious systemic fungal infections. These antibiotics form complexes with ergosterol in the fungal cell membrane, and they disrupt the membrane permeability and cause leakage of cellular constituents. Neither is useful in treating bacterial infections because most bacteria have no ergosterol in their membranes. Both amphotericin B and nystatin are extremely toxic and cause symptoms that include nausea and vomiting, fever and chills, anemia, and renal failure. It is easy to understand why the use of these drugs is restricted to treatment of life-threatening fungal diseases.

For Further Understanding
Why are drugs such as amphotericin B not administered orally?

Draw the structure of amphotericin B and identify the hydrophobic and hydrophilic regions of the molecule.

---

blood cell are said to be **isotonic** (*iso* means "same," and *tonic* means "strength") **solutions.** Because the osmolarity is the same inside and outside, the red blood cell will remain the same size (Figure 17.18b).

What happens if we now place the red blood cells into a **hypotonic solution,** in other words, a solution having a lower osmolarity than the cytoplasm of the cell? In this situation there will be a net movement of water into the cell as water diffuses down its concentration gradient. The membrane of the red blood cell does not have the strength to exert a sufficient pressure to stop this flow of water, and the cell will swell and burst (Figure 17.18c). Alternatively, if we place the red blood cells into a **hypertonic solution** (one with a greater osmolarity than the cell), water will pass out of the cells, and they will shrink dramatically in size (Figure 17.18a).

These principles have important applications in the delivery of intravenous (IV) solutions into an individual. Normally, any fluids infused intravenously must have the correct osmolarity; they must be isotonic with the blood cells and the blood plasma. Such infusions are frequently either 5.5% dextrose (glucose) or "normal saline." The first solution is composed of 5.5 g of glucose per 100 mL of

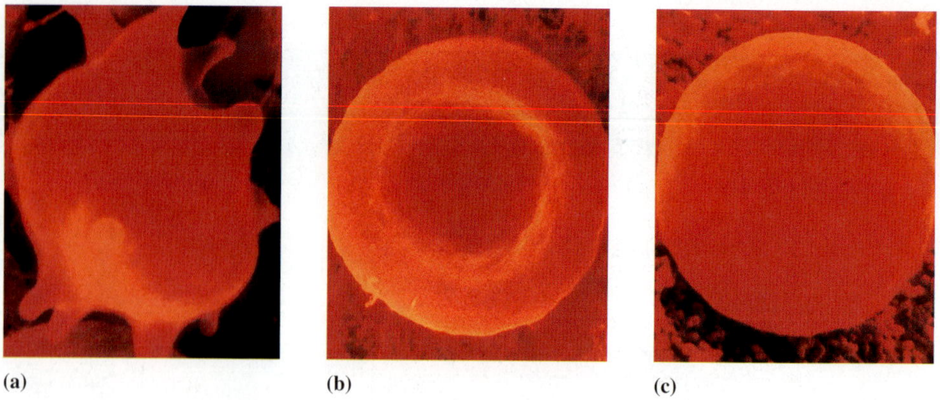

Figure 17.18
Scanning electron micrographs of red blood cells exposed to (a) hypertonic, (b) isotonic, and (c) hypotonic solutions.

(a)   (b)   (c)

solution (0.30 M), and the latter of 9.0 g of NaCl per 100 mL of solution (0.15 M). In either case, they have the same osmotic pressure and osmolarity as the plasma and blood cells and can therefore be safely administered without upsetting the osmotic balance between the blood and the blood cells.

Energy Requirements for Transport

Simple diffusion, facilitated diffusion, and osmosis involve the spontaneous flow of materials from a region of higher concentration to an area of lower concentration (a concentration gradient). To survive, cells often must move substances "uphill," against a concentration gradient. This phenomenon, called **active transport**, requires energy (Figure 17.19). Many ions and food molecules are imported through the cell membrane by active transport. The energy used for this process may consume more than half of the total energy harvested by cellular metabolism.

A good example of active transport is the Na^+-K^+ ATPase, which moves these ions into and out of the cell against their gradients (Figure 17.19). Cells must maintain a high concentration of Na^+ outside the cell and a high concentration of K^+ inside the cell. This requires a continuous supply of cellular energy in the form of adenosine triphosphate (ATP). Over one-third of the total ATP produced by the cell is used to maintain these Na^+ and K^+ concentration gradients across the cell membrane. Thus, the name *Na^+-K^+ ATPase* refers to the enzymatic activity that hydrolyzes ATP. The hydrolysis of ATP releases the energy needed to move Na^+ and K^+ ions across the cell membrane. For each ATP molecule hydrolyzed, three Na^+ are moved out of the cell and two K^+ are transported into the cell.

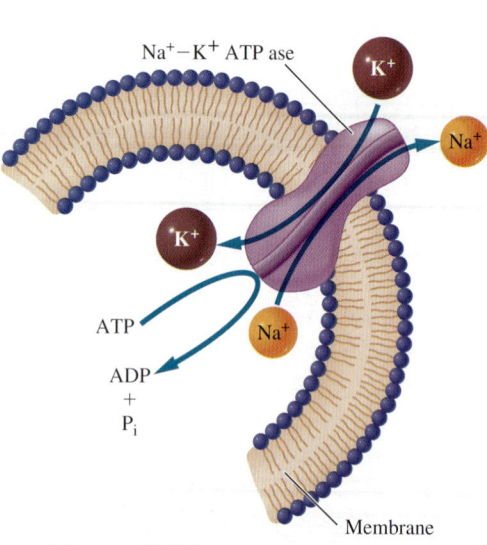

Figure 17.19
Schematic diagram of the active transport mechanism of the Na^+-K^+ ATPase.

How does membrane transport resemble enzyme catalysis?

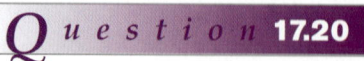

Why is D-glucose transported by the glucose transport protein whereas L-glucose is not?

SUMMARY

17.1 Biological Functions of Lipids

Lipids are organic molecules characterized by their solubility in nonpolar solvents. Lipids are subdivided into classes based on structural characteristics: *fatty acids, glycerides, nonglycerides,* and *complex lipids.* Lipids serve many functions in the body, including energy storage, protection of organs, insulation, and absorption of vitamins. Other lipids are energy sources, hormones, or vitamins. Cells store chemical energy in the form of lipids, and the cell membrane is a lipid bilayer.

17.2 Fatty Acids

Fatty acids are *saturated* and *unsaturated* carboxylic acids containing between twelve and twenty-four carbon atoms. Fatty acids with even numbers of carbon atoms occur most frequently in nature. The reactions of fatty acids are identical to those of carboxylic acids. They include esterification, production by acid hydrolysis of esters, saponification, and addition at the double bond. *Prostaglandins,* thromboxanes, and leukotrienes are derivatives of twenty-carbon fatty acids that have a variety of physiological effects.

17.3 Glycerides

Glycerides are the most abundant lipids. The triesters of glycerol *(triglycerides)* are of greatest importance. Neutral triglycerides are important because of their ability to store energy. The ionic *phospholipids* are important components of all biological membranes.

17.4 Nonglyceride Lipids

Nonglyceride lipids consist of *sphingolipids, steroids,* and *waxes. Sphingomyelin* is a component of the myelin sheath around cells of the central nervous system. The *steroids* are important for many biochemical functions: *cholesterol* is a membrane component; testosterone, progesterone, and estrone are sex hormones; and cortisone is an anti-inflammatory steroid that is important in the regulation of many biochemical pathways.

17.5 Complex Lipids

Plasma lipoproteins are complex lipids that transport other lipids through the bloodstream. *Chylomicrons* carry dietary triglycerides from the intestine to other tissues. *Very low density lipoproteins* carry triglycerides synthesized in the liver to other tissues for storage. *Low-density lipoproteins* carry cholesterol to peripheral tissues and help regulate blood cholesterol levels. *High-density lipoproteins* transport cholesterol from peripheral tissues to the liver.

17.6 The Structure of Biological Membranes

The *fluid mosaic model* of membrane structure pictures biological membranes that are composed of lipid bilayers in which proteins are embedded. Membrane lipids contain polar head groups and nonpolar hydrocarbon tails. The hydrocarbon tails of phospholipids are derived from saturated and unsaturated long-chain fatty acids containing an even number of carbon atoms. The lipids and proteins diffuse rapidly in the lipid bilayer but seldom cross from one side to the other.

The simplest type of membrane transport is passive diffusion of a substance across the lipid bilayer from the region of higher concentration to that of lower concentration. Many metabolites are transported across biological membranes by permeases that form pores through the membrane. The conformation of the pore is complementary to that of the substrate to be transported. Cells use energy to transport molecules across the plasma membrane against their concentration gradients, a process known as *active transport.*

The Na^+-K^+ ATPase hydrolyzes one molecule of ATP to provide the driving force for pumping three Na^+ out of the cell in exchange for two K^+.

KEY TERMS

active transport (17.6)
arachidonic acid (17.2)
atherosclerosis (17.4)
cholesterol (17.4)
chylomicron (17.5)
complex lipid (17.5)
diglyceride (17.3)
eicosanoid (17.2)
emulsifying agent (17.3)
essential fatty acid (17.2)
esterification (17.2)
facilitated diffusion (17.6)
fatty acid (17.2)
fluid mosaic model (17.6)
glyceride (17.3)
high-density lipoprotein (HDL) (17.5)
hydrogenation (17.2)
hypertonic solution (17.6)
hypotonic solution (17.6)
isotonic solution (17.6)
lipid (17.1)
low-density lipoprotein (LDL) (17.5)
monoglyceride (17.3)
neutral glyceride (17.3)
osmosis (17.6)
osmotic pressure (17.6)
passive transport (17.6)
peripheral membrane protein (17.6)
phosphatidate (17.3)
phosphoglyceride (17.3)
phospholipid (17.3)
plasma lipoprotein (17.5)
prostaglandin (17.2)
saponification (17.2)
saturated fatty acid (17.2)
semipermeable membrane (17.6)
sphingolipid (17.4)
sphingomyelin (17.4)
steroid (17.4)
terpene (17.4)
transmembrane protein (17.6)
triglyceride (17.3)
unsaturated fatty acid (17.2)
very low density lipoprotein (VLDL) (17.5)
wax (17.4)

QUESTIONS AND PROBLEMS

Biological Functions of Lipids

Foundations
17.21 List the four main groups of lipids.
17.22 List the biological functions of lipids.

Applications
17.23 In terms of solubility, explain why a diet that contains no lipids can lead to a deficiency of the lipid-soluble vitamins.
17.24 Why are lipids (triglycerides) such an efficient molecule for the storage of energy in the body?

Fatty Acids

Foundations
17.25 What is the difference between a saturated and an unsaturated fatty acid?
17.26 Write the structures for a saturated and an unsaturated fatty acid.
17.27 As the length of the hydrocarbon chain of saturated fatty acids increases, what is the effect on the melting points?
17.28 As the number of carbon-carbon double bonds in fatty acids increases, what is the effect on the melting points?

Applications
17.29 Explain the relationship between fatty-acid-chain length and melting points that you described in answer to Question 17.27.
17.30 Explain the relationship that you described in answer to Question 17.28 for the effect of the number of carbon-carbon bonds in fatty acids on their melting points.
17.31 Draw the structures of each of the following fatty acids:
  a. Decanoic acid
  b. Stearic acid
17.32 Draw the structures of each of the following fatty acids:
  a. *trans*-5-Decenoic acid
  b. *cis*-5-Decenoic acid
17.33 What are the common and I.U.P.A.C. names of each of the following fatty acids?
  a. $C_{15}H_{31}COOH$
  b. $C_{11}H_{23}COOH$
17.34 What are the common and I.U.P.A.C. names of each of the following fatty acids?
  a. $CH_3(CH_2)_5CH=CH(CH_2)_7COOH$
  b. $CH_3(CH_2)_7CH=CH(CH_2)_7COOH$
17.35 Write an equation for each of the following reactions:
  a. Esterification of glycerol with three molecules of myristic acid
  b. Acid hydrolysis of a triglyceride containing three stearic acid molecules
  c. Reaction of decanoic acid with KOH
  d. Hydrogenation of linoleic acid
17.36 Write an equation for each of the following reactions:
  a. Esterification of glycerol with three molecules of palmitic acid
  b. Acid hydrolysis of a triglyceride containing three oleic acid molecules
  c. Reaction of stearic acid with KOH
  d. Hydrogenation of oleic acid
17.37 What is the function of the essential fatty acids?
17.38 What molecules are formed from arachidonic acid?
17.39 What is the biochemical basis for the effectiveness of aspirin in decreasing the inflammatory response?
17.40 What is the role of prostaglandins in the inflammatory response?
17.41 List four effects of prostaglandins.
17.42 What are the functions of thromboxane A_2 and leukotrienes?

Glycerides

Foundations
17.43 Define the term *glyceride*.
17.44 Define the term *phosphatidate*.
17.45 What are emulsifying agents and what are their practical uses?
17.46 Why are triglycerides also referred to as triacylglycerols?

Applications
17.47 What do you predict the physical state would be of a triglyceride with three saturated fatty acid tails? Explain your reasoning.
17.48 What do you predict the physical state would be of a triglyceride with three unsaturated fatty acid tails? Explain your reasoning.
17.49 Draw the structure of the triglyceride molecule formed by esterification at C-1, C-2, and C-3 with hexadecanoic acid, *trans*-9-hexadecenoic acid, and *cis*-9-hexadecenoic acid, respectively.
17.50 Draw one possible structure of a triglyceride that contains the three fatty acids stearic acid, palmitic acid, and oleic acid.
17.51 Draw the structure of the phosphatidate formed between glycerol-3-phosphate that is esterified at C-1 and C-2 with capric and lauric acids, respectively.
17.52 Draw the structure of a lecithin molecule in which the fatty acyl groups are derived from stearic acid.
17.53 What are the structural differences between triglycerides (triacylglycerols) and phospholipids?
17.54 How are the structural differences between triglycerides and phospholipids reflected in their different biological functions?

Nonglyceride Lipids

Foundations
17.55 Define the term *sphingolipid*.
17.56 What are the two major types of sphingolipids?
17.57 Define the term *glycosphingolipid*.
17.58 Distinguish among the three types of glycosphingolipids, cerebrosides, sulfatides, and gangliosides.

Applications
17.59 What is the biological function of sphingomyelin?
17.60 Why are sphingomyelins amphipathic?
17.61 What is the role of cholesterol in biological membranes?
17.62 How does cholesterol contribute to atherosclerosis?
17.63 What are the biological functions of progesterone, testosterone, and estrone?
17.64 How has our understanding of the steroid sex hormones contributed to the development of oral contraceptives?
17.65 What is the medical application of cortisone?
17.66 What are the possible side effects of cortisone treatment?
17.67 A wax found in beeswax is myricyl palmitate. What fatty acid and what alcohol are used to form this compound?
17.68 A wax found in the head of sperm whales is cetyl palmitate. What fatty acid and what alcohol are used to form this compound?
17.69 What are isoprenoids?
17.70 What is a terpene?
17.71 List some important biological molecules that are terpenes.
17.72 Draw the five-carbon isoprene unit.

Complex Lipids

Foundations
17.73 What are the four major types of plasma lipoproteins?
17.74 What is the function of each of the four types of plasma lipoproteins?

Applications
17.75 There is a single, unique structure for the cholesterol molecule. What is meant by the terms *good* and *bad* cholesterol?
17.76 Distinguish among the four plasma lipoproteins in terms of their composition and their function.
17.77 What is the relationship between atherosclerosis and high blood pressure?
17.78 How is LDL taken into cells?
17.79 How does a genetic defect in the LDL receptor contribute to atherosclerosis?
17.80 What is the correlation between saturated fats in the diet and atherosclerosis?

The Structure of Biological Membranes

Foundations
17.81 What is the basic structure of a biological membrane?
17.82 Describe the fluid mosaic model of membrane structure.
17.83 Describe peripheral membrane proteins.
17.84 Describe transmembrane proteins and list some of their functions.

Applications
17.85 What is the major effect of cholesterol on the properties of biological membranes?
17.86 Why do the hydrocarbon tails of membrane phospholipids provide a barrier between the inside and outside of the cell?
17.87 What experimental observation shows that proteins diffuse within the lipid bilayers of biological membranes?
17.88 Why don't proteins turn around in biological membranes like revolving doors?
17.89 How will the properties of a biological membrane change if the fatty acid tails of the phospholipids are converted from saturated to unsaturated chains?
17.90 What is the function of unsaturation in the hydrocarbon tails of membrane lipids?

Biological Membranes: Transport

Foundations
17.91 Explain the difference between simple diffusion across a membrane and facilitated diffusion.
17.92 Explain what would happen to a red blood cell placed in each of the following solutions:
   a. hypotonic   c. hypertonic
   b. isotonic

Applications
17.93 How does active transport differ from facilitated diffusion?
17.94 By what mechanism are Cl^- and HCO_3^- ions transported across the red blood cell membrane?
17.95 What is the meaning of the term *antiport*?
17.96 How does insulin affect the transport of glucose?
17.97 What properties of a transport protein (permease) determine its specificity?
17.98 Why is the function of the Na^+-K^+ ATPase an example of active transport?
17.99 What is the stoichiometry of the Na^+-K^+ ATPase?
17.100 How will the Na^+ and K^+ concentrations of a cell change if the Na^+-K^+ ATPase is inhibited?

CRITICAL THINKING PROBLEMS

1. Olestra is a fat substitute that provides no calories, yet has all the properties of a naturally occurring fat. It has a creamy, tongue-pleasing consistency. Unlike other fat substitutes, olestra can withstand heating. Thus, it can be used to prepare foods such as potato chips and crackers. Olestra is a sucrose polyester and is produced by esterification of six, seven, or eight fatty acids to molecules of sucrose. Draw the structure of one such molecule having eight stearic acid acyl groups attached.
2. Liposomes can be made by vigorously mixing phospholipids (like phosphatidylcholine) in water. When the mixture is allowed to settle, spherical vesicles form that are surrounded by a phospholipid bilayer "membrane." Pharmaceutical chemists are trying to develop liposomes as a targeted drug delivery system. By adding the drug of choice to the mixture described above, liposomes form around the solution of drug. Specific proteins can be incorporated into the mixture that will end up within the phospholipid bilayers of the liposomes. These proteins are able to bind to targets on the surface of particular kinds of cells in the body. Explain why injection of liposome encapsulated pharmaceuticals might be a good drug delivery system.
3. "Cholesterol is bad and should be eliminated from the diet." Do you agree or disagree? Defend your answer.
4. Why would a phospholipid such as lecithin be a good emulsifying agent for ice cream?
5. When a plant becomes cold-adapted, the composition of the membranes changes. What changes in fatty acid and cholesterol composition would you predict? Explain your reasoning.
6. In terms of osmosis, explain why it would be preferable for a cell to store 10,000 molecules of glycogen each composed of 10^5 molecules of glucose rather than to store 10^9 individual molecules of glucose.

BIOCHEMISTRY

Protein Structure and Function

18

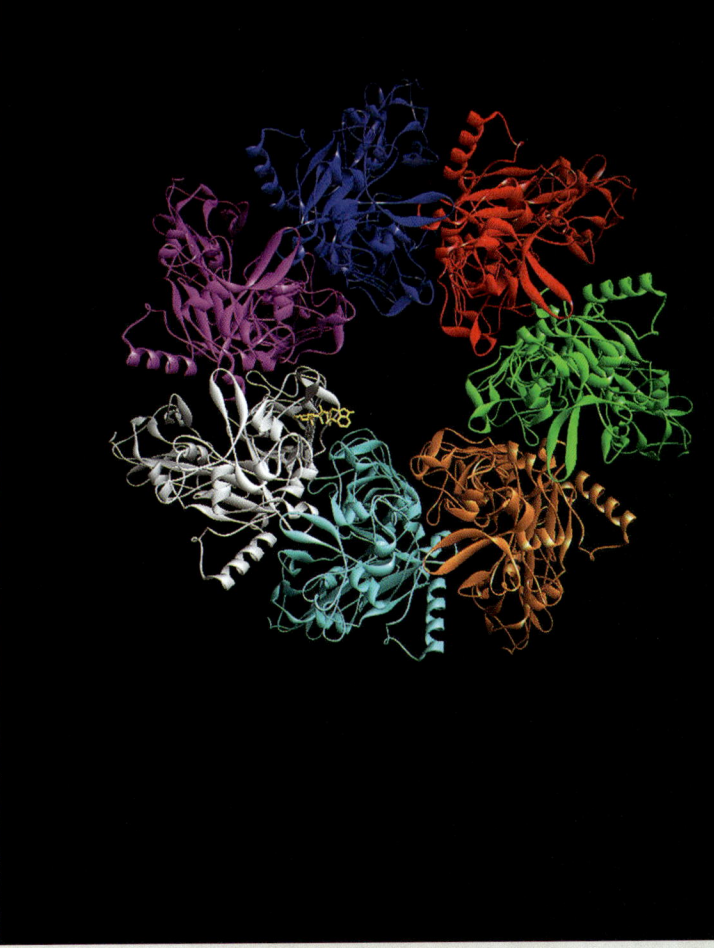

The cholera toxin B subunit.

Learning Goals

1. List the functions of proteins.
2. Draw the general structure of an amino acid and classify amino acids based on their R groups.
3. Describe the primary structure of proteins and draw the structure of the peptide bond.
4. Draw the structures of small peptides and name them.
5. Describe the types of secondary structure of a protein.
6. Discuss the forces that maintain secondary structure.
7. Describe the structure and functions of fibrous proteins.
8. Describe the tertiary and quaternary structure of a protein.
9. List the R group interactions that maintain protein conformation.
10. List examples of proteins that require prosthetic groups and explain the way in which they function.
11. Discuss the importance of the three-dimensional structure of a protein to its function.
12. Describe the roles of hemoglobin and myoglobin.
13. Describe how extremes of pH and temperature cause denaturation of proteins.
14. Explain the difference between essential and nonessential amino acids.

Outline

Chemistry Connection:
Angiogenesis Inhibitors: Proteins That Inhibit Tumor Growth

18.1 Cellular Functions of Proteins

A Medical Perspective:
Proteins in the Blood

18.2 The α-Amino Acids
18.3 The Peptide Bond

A Human Perspective:
The Opium Poppy and Peptides in the Brain

18.4 The Primary Structure of Proteins
18.5 The Secondary Structure of Proteins
18.6 The Tertiary Structure of Proteins

A Human Perspective:
Collagen: A Protein That Holds Us Together

18.7 The Quaternary Structure of Proteins
18.8 An Overview of Protein Structure and Function
18.9 Myoglobin and Hemoglobin

A Medical Perspective:
Immunoglobulins: Proteins That Defend the Body

18.10 Denaturation of Proteins
18.11 Dietary Protein and Protein Digestion

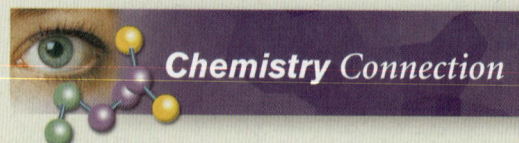

Chemistry Connection

Angiogenesis Inhibitors: Proteins That Inhibit Tumor Growth

Cancer researchers have long known that solid tumors cannot grow larger than the size of a pinhead unless they stimulate the formation of new blood vessels that provide the growing tumor with nutrients and oxygen and remove the waste products of cellular metabolism. Studies of *angiogenesis,* the formation of new blood vessels, in normal tissues have provided new weapons in the arsenal of anticancer drugs.

Angiogenesis occurs through a carefully controlled sequence of steps. Consider the process of tissue repair. One of several protein growth factors stimulates the endothelial cells that form the lining of an existing blood vessel to begin growing, dividing, and migrating into the tissue to be repaired. Threads of new endothelial cells organize themselves into hollow cylinders, or tubules. These tubules become a new network of blood vessels throughout the damaged tissue. These new blood vessels bring the needed nutrients, oxygen, and other factors to the site of damage, allowing the tissue to be repaired and healing to occur.

In addition to the growth factors that stimulate this process, there are several other proteins that inhibit the formation of new blood vessels. In fact, the normal process of angiogenesis is dependent on the appropriate balance of the stimulatory growth factors and the inhibitory proteins.

The normal events of angiogenesis are duplicated at a critical moment in the growth of a tumor. Cells of the tumor secrete one or more of the growth factors known to stimulate angiogenesis. The newly formed blood vessels provide the cells of the growing tumor with everything needed to continue growing and dividing.

Metastasis, the spreading of tumor cells to other sites in the body, also requires angiogenesis. Typically, those tumors having more blood vessels are more likely to metastasize. Clinically, treatment of these tumors has a poorer outcome.

Researchers considered all of this information known about angiogenesis and its impact on tumor formation and metastasis. They developed the hypothesis that proteins that inhibit blood vessel formation might be effective weapons against developing tumors. If this hypothesis turned out to be supported by experimental data, there would be a number of advantages to the use of angiogenesis inhibitors. Because these proteins are normally produced by the human body, they should not have the toxic side effects caused by so many anticancer drugs. In addition, angiogenesis inhibitors can overcome the problem of cancer cell drug resistance. Most cancer cells are prone to mutations and mutant cells resistant to the anticancer drugs develop. The angiogenesis inhibitors target normal endothelial cells, which are genetically stable. As a result, drug resistance is much less likely to occur.

Endostatin is one of the anti-angiogenesis proteins. Discovered in 1997, it was found to be a protein of 20,000 g/mol, which is a fragment of the C-terminus of collagen XVIII. Experimentally, endostatin is a potent inhibitor of tumor growth. It binds to the heparin sulfate proteoglycans of the cell surface and interferes with growth factor signaling. As a result, the growth and division of endothelial cells is inhibited and new blood vessels are not formed.

Angiostatin is another anti-angiogenesis protein normally found in the human body. Discovered in 1994, it is a protein fragment of human plasminogen and has a molar mass of 50,000 g/mol. The role of angiostatin in the human body is to block the growth of diseased tissue by inhibiting the formation of blood vessels. Like endostatin, it is hoped that angiostatin will block the growth of tumors by depriving them of their blood supply.

Currently, there are about twenty angiogenesis inhibitors being tested in clinical trials involving humans. Most are in phase I or II trials, which allow scientists to determine a safe dosage and assess the severity of any side effects. Only a small number of people are involved in phase I or II trials. In phase III trials, a large number of patients are divided into two groups. One group receives standard anticancer treatment plus a placebo. The other group receives standard treatment and the new drug.

As we await the results of the clinical trials involving the proteins endostatin and angiostatin, scientists explore alternative methods to attack cancer cells. Some of these involve a class of proteins called *antibodies* that can bind specifically to cancer cells and help to inhibit or destroy them.

As we will discover, there are many different classes of proteins that carry out a variety of functions for the body. Endostatin and angiostatin serve as regulatory proteins; the antibodies serve as the body's defense system against infectious diseases. These and many other proteins are the focus of this chapter.

Introduction

*In the 1800s, Johannes Mulder came up with the name **protein,** a term derived from a Greek word that means "of first importance." Indeed, proteins are a very important class of food molecules because they provide an organism not only with carbon and hydrogen, but also with nitrogen and sulfur. These latter two elements are unavailable from fats and carbohydrates, the other major classes of food molecules.*

In addition to their dietary importance, the proteins are the most abundant macromolecules in the cell, and they carry out most of the work in a cell. Protection of the body from infection, mechanical support and strength, and catalysis of metabolic reactions—all are functions of proteins that are essential to life.

18.1 Cellular Functions of Proteins

Proteins have many biological functions, as the following short list suggests.

 LEARNING GOAL

- **Enzymes** are biological catalysts. The majority of the enzymes that have been studied are proteins. Reactions that would take days or weeks or require extremely high temperatures without enzymes are completed in an instant. For example, the digestive enzymes *pepsin, trypsin,* and *chymotrypsin* break down proteins in our diet so that subunits can be absorbed for use by our cells.
- **Defense proteins** include **antibodies** (also called *immunoglobulins*) which are specific protein molecules produced by specialized cells of the immune system in response to foreign **antigens.** These foreign invaders include bacteria and viruses that infect the body. Each antibody has regions that precisely fit and bind to a single antigen. It helps to end the infection by binding to the antigen and helping to destroy it or remove it from the body.

In the broadest sense, an antigen is any substance that stimulates an immune response.

- **Transport proteins** carry materials from one place to another in the body. The protein *transferrin* transports iron from the liver to the bone marrow, where it is used to synthesize the heme group for hemoglobin. The proteins *hemoglobin* and *myoglobin* are responsible for transport and storage of oxygen in higher organisms, respectively.
- **Regulatory proteins** control many aspects of cell function, including metabolism and reproduction. We can function only within a limited set of conditions. For life to exist, body temperature, the pH of the blood, and blood glucose levels must be carefully regulated. Many of the hormones that regulate body function, such as *insulin* and *glucagon,* are proteins.
- **Structural proteins** provide mechanical support to large animals and provide them with their outer coverings. Our hair and fingernails are largely composed of the protein *keratin.* Other proteins provide mechanical strength for our bones, tendons, and skin. Without such support, large, multicellular organisms like ourselves could not exist.
- **Movement proteins** are necessary for all forms of movement. Our muscles, including that most important muscle, the heart, contract and expand through the interaction of *actin* and *myosin* proteins. Sperm can swim because they have long flagella made up of proteins.
- **Nutrient proteins** serve as sources of amino acids for embryos or infants. Egg *albumin* and *casein* in milk are examples of nutrient storage proteins.

18.2 The α-Amino Acids

Structure of Amino Acids

The proteins of the body are made up of some combination of twenty different subunits called **α-amino acids.** The general structure of an α-amino acid is shown in Figure 18.1. We find that nineteen of the twenty amino acids that are commonly isolated from proteins have this same general structure; they are primary amines on the α-carbon. The remaining amino acid, proline, is a secondary amine.

 LEARNING GOAL

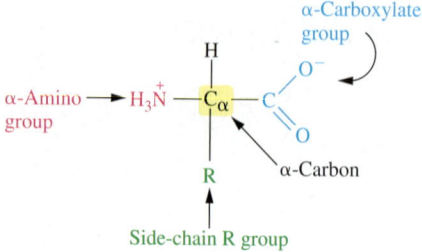

Figure 18.1
General structure of an α-amino acid. All amino acids isolated from proteins, with the exception of proline, have this general structure.

Conjugate acids and bases are described in detail in Section 8.1.

Stereochemistry is discussed in Section 16.3 and online at www.mhhe.com/denniston5e in Stereochemistry and Stereoisomers Revisited.

Figure 18.2
(a) Structure of D- and L-glyceraldehyde and their relationship to D- and L-alanine. (b) Ball-and-stick models of D- and L-alanine.

Notice that the α-carbon in the general structure is attached to a carboxylate group (a carboxyl group that has lost a proton, —COO⁻) and a protonated amino group (an amino group that has gained a proton, —NH$_3^+$). At pH 7, conditions required for life functions, you will not find amino acids in which the carboxylate group is protonated (—COOH) and the amino group is unprotonated (—NH$_2$). Under these conditions, the carboxyl group is in the conjugate base form (—COO⁻), and the amino group is in its conjugate acid form (—NH$_3^+$). Any neutral molecule with equal numbers of positive and negative charges is called a *zwitterion*. Thus, amino acids in water exist as dipolar ions called zwitterions.

The α-carbon of each amino acid is also bonded to a hydrogen atom and a side chain, or R group. In a protein, the R groups interact with one another through a variety of weak attractive forces. These interactions participate in folding the protein chain into a precise three-dimensional shape that determines its ultimate function. They also serve to maintain that three-dimensional conformation.

Stereoisomers of Amino Acids

The α-carbon is attached to four different groups in all amino acids except glycine. The α-carbon of most α-amino acids is therefore chiral, allowing mirror-image forms, enantiomers, to exist. Glycine has two hydrogen atoms attached to the α-carbon and is the only amino acid commonly found in proteins that is not chiral.

The configuration of α-amino acids isolated from proteins is L-. This is based on comparison of amino acids with D-glyceraldehyde (Figure 18.2). In Figure 18.2a we see a comparison of D- and L-glyceraldehyde with D- and L-alanine. Notice that the most oxidized end of the molecule, in each case the carbonyl group, is drawn at the top of the molecule. In the D-isomer of glyceraldehyde, the —OH group is on

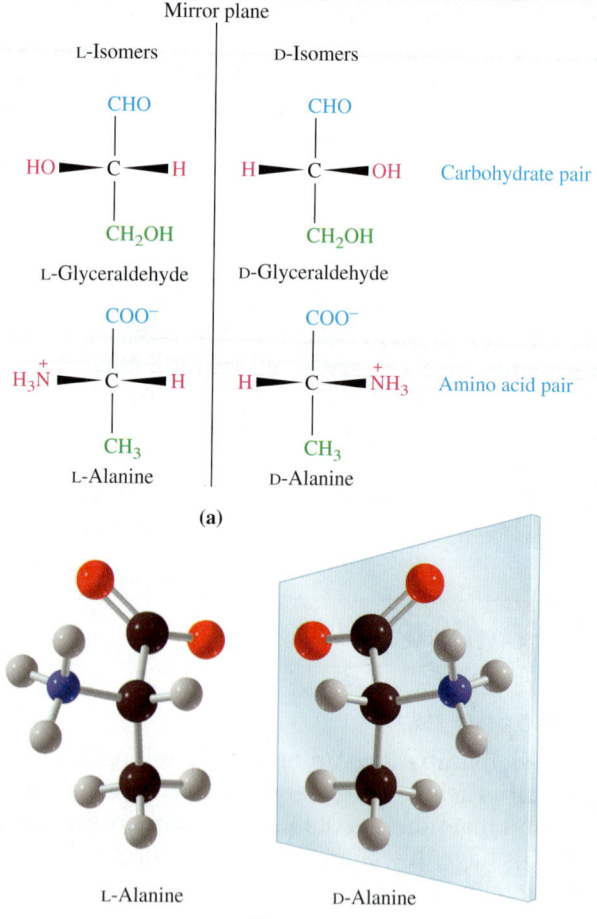

A Medical Perspective

Proteins in the Blood

The blood plasma of a healthy individual typically contains 60–80 g/L of protein. This protein can be separated into five classes designated α through γ. The separation is based on the overall surface charge on each of the types of protein. (See A Medical Perspective: Enzymes, Isoenzymes, and Myocardial Infarction in Chapter 19 for a discussion of the separation of proteins based on surface charge.)

The most abundant protein in the blood is albumin, making up about 55% of the blood protein. Albumin contributes to the osmotic pressure of the blood simply because it is a dissolved molecule. It also serves as a nonspecific transport molecule for important metabolites that are otherwise poorly soluble in water. Among the molecules transported through the blood by albumin are bilirubin (a waste product of the breakdown of hemoglobin), Ca^{2+}, and fatty acids (organic anions).

The α-globulins ($α_1$ and $α_2$) make up 13% of the plasma proteins. They include glycoproteins (proteins with sugar groups attached), high-density lipoproteins, haptoglobin (a transport protein for free hemoglobin), ceruloplasmin (a copper transport protein), prothrombin (a protein involved in blood clotting), and very low density lipoproteins. The most abundant is $α_1$-globulin $α_1$-antitrypsin. Although the name leads us to believe that this protein inhibits a digestive enzyme, trypsin, the primary function of $α_1$-antitrypsin is the inactivation of an enzyme that causes damage in the lungs (see also, A Medical Perspective: $α_1$-Antitrypsin and Familial Emphysema in Chapter 19). $α_1$-Antichymotrypsin is another inhibitor found in the bloodstream. This protein, along with amyloid proteins, is found in the amyloid plaques characteristic of Alzheimer's disease (AD). As a result, it has been suggested that an overproduction of this protein may contribute to AD. In the blood, $α_1$-antichymotrypsin is also found complexed to prostate specific antigen (PSA), the protein antigen that is measured as an indicator of prostate cancer. Elevated PSA levels are observed in those with the disease. It is interesting to note that PSA is a chymotrypsin-like proteolytic enzyme.

The β-globulins represent 13% of the blood plasma proteins and include transferrin (an iron transport protein) and low-density lipoprotein. Fibrinogen, a protein involved in coagulation of blood, comprises 7% of the plasma protein. Finally, the γ-globulins, IgG, IgM, IgA, IgD, and IgE, make up the remaining 11% of the plasma proteins. The γ-globulins are synthesized by B lymphocytes, but most of the remaining plasma proteins are synthesized in the liver. In fact, a frequent hallmark of liver disease is reduced amounts of one or more of the plasma proteins.

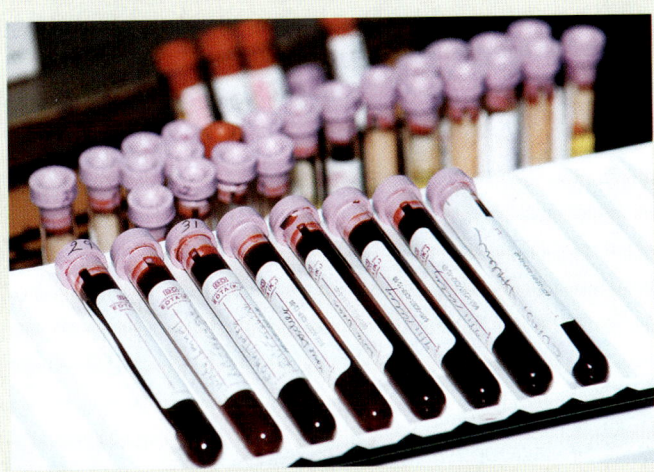

Blood samples drawn from patients.

For Further Understanding

Develop a hypothesis to explain why albumin in the blood can serve as a nonspecific carrier for such diverse substances as bilirubin, Ca^{2+}, and fatty acids. (Hint: Consider what you know about the structures of amino acid R groups.)

Fibrinogen and prothrombin are both involved in formation of blood clots when they are converted into proteolytic enzymes. However, they are normally found in the blood in an inactive form. Develop an explanation for this observation.

the right. Similarly, in the D-isomer of alanine, the —N^+H_3 is on the right. In the L-isomers of the two compounds, the —OH and —N^+H_3 groups are on the left. By this comparison with the enantiomers of glyceraldehyde, we can define the D- and L-enantiomers of the amino acids. Figure 18.2b shows the ball-and-stick models of the D- and L-isomers of alanine.

In Chapter 16 we learned that almost all of the monosaccharides found in nature are in the D-family. Just the opposite is true of the α-amino acids. Almost all of the α-amino acids isolated from proteins in nature are members of the L-family. In other words, the orientation of the four groups around the chiral carbon of these α-amino acids resembles the orientation of the four groups around the chiral carbon of L-glyceraldehyde.

Classes of Amino Acids

Because all of the amino acids have a carboxyl group and an amino group, all differences between amino acids depend upon their side-chain R groups. The amino acids are grouped in Figure 18.3 according to the polarity of their side chains.

The side chains of some amino acids are nonpolar. They prefer contact with one another over contact with water and are said to be **hydrophobic** ("water-fearing") **amino acids.** They are generally found buried in the interior of proteins, where they can associate with one another and remain isolated from water. Nine amino acids fall into this category: alanine, valine, leucine, isoleucine, proline, glycine, methionine, phenylalanine, and tryptophan. The R group of proline is unique; it is actually bonded to the α-amino group, forming a secondary amine.

The side chains of the remaining amino acids are polar. Because they are attracted to polar water molecules, they are said to be **hydrophilic** ("water-loving") **amino acids.** The hydrophilic side chains are often found on the surfaces of proteins. The polar amino acids can be subdivided into three classes.

- *Polar, neutral amino acids* have R groups that have a high affinity for water but that are not ionic at pH 7. Serine, threonine, tyrosine, cysteine, asparagine, and glutamine fall into this category. Most of these amino acids associate with one another by hydrogen bonding; but cysteine molecules form disulfide bonds with one another, as we will discuss in Section 18.6.
- *Negatively charged amino acids* have ionized carboxyl groups in their side chains. At pH 7 these amino acids have a net charge of -1. Aspartate and glutamate are the two amino acids in this category. They are acidic amino acids because ionization of the carboxylic acid releases a proton.
- *Positively charged amino acids.* At pH 7, lysine, arginine, and histidine have a net positive charge because their side chains contain positive groups. These amino groups are basic because the side chain reacts with water, picking up a proton and releasing a hydroxide anion.

The names of the amino acids can be abbreviated by a three-letter code. These abbreviations are shown in Table 18.1.

The hydrophobic interaction between nonpolar R groups is one of the forces that helps maintain the proper three-dimensional shape of a protein.

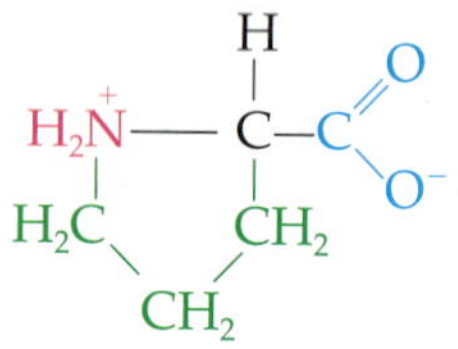

Proline (Pro)

Hydrogen bonding (Section 5.2) is another weak interaction that helps maintain the proper three-dimensional structure of a protein. The positively and negatively charged amino acids within a protein interact with one another to form ionic bridges that also help to keep the protein chain folded in a precise way.

TABLE 18.1 Names and Three-Letter Abbreviations of the α-Amino Acids

Amino Acid	Three-Letter Abbreviation
Alanine	ala
Arginine	arg
Asparagine	asn
Aspartate	asp
Cysteine	cys
Glutamic acid	glu
Glutamine	gln
Glycine	gly
Histidine	his
Isoleucine	ile
Leucine	leu
Lysine	lys
Methionine	met
Phenylalanine	phe
Proline	pro
Serine	ser
Threonine	thr
Tryptophan	trp
Tyrosine	tyr
Valine	val

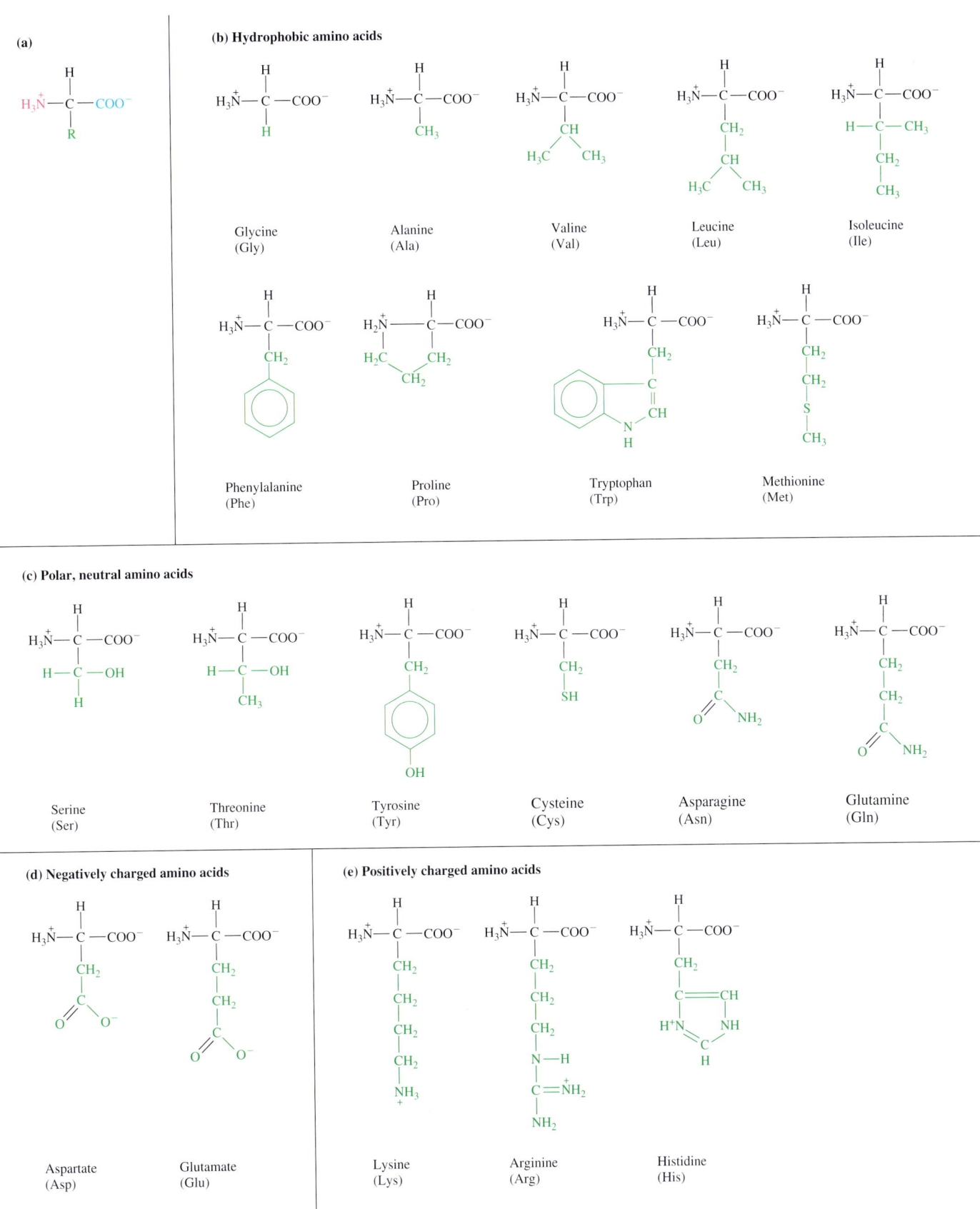

Figure 18.3
Structures of the amino acids at pH 7.0. (a) The general structure of an amino acid. Structures of (b) the hydrophobic; (c) polar, neutral; (d) negatively charged; and (e) positively charged amino acids.

Question 18.1

Write the three-letter abbreviation and draw the structure of each of the following amino acids.

a. Glycine
b. Proline
c. Threonine
d. Aspartate
e. Lysine

Question 18.2

Indicate whether each of the amino acids listed in Question 18.1 is polar, nonpolar, basic, or acidic.

18.3 The Peptide Bond

LEARNING GOAL

Proteins are linear polymers of L-α-amino acids in which the carboxyl group of one amino acid is linked to the amino group of another amino acid. The **peptide bond** is an *amide bond* formed between the —COO⁻ group of one amino acid and the α-N⁺H₃ group of another amino acid. The reaction, shown below for the amino acids glycine and alanine, is a dehydration reaction, because a water molecule is lost as the amide bond is formed.

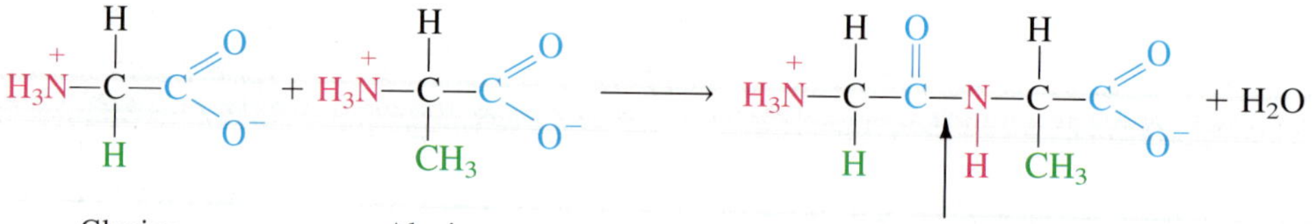

$$H_3\overset{+}{N}-\underset{H}{\overset{H}{C}}-C\overset{O}{\underset{O^-}{\diagup}} + H_3\overset{+}{N}-\underset{CH_3}{\overset{H}{C}}-C\overset{O}{\underset{O^-}{\diagup}} \longrightarrow H_3\overset{+}{N}-\underset{H}{\overset{H}{C}}-\overset{O}{\overset{\|}{C}}-\underset{H}{\overset{H}{N}}-\underset{CH_3}{\overset{H}{C}}-C\overset{O}{\underset{O^-}{\diagup}} + H_2O$$

Glycine   Alanine   Peptide bond (amide bond)

Glycyl-alanine

To understand why the N-terminal amino acid is placed first and the C-terminal amino acid is placed last, we need to look at the process of protein synthesis. As we will see in Section 20.6, the N-terminal amino acid is the first amino acid of the protein. It forms a peptide bond involving its carboxyl group and the amino group of the second amino acid in the protein. Thus, a free amino group literally projects from the "left" end of the protein. Similarly, the C-terminal amino acid is the last amino acid added to the protein during protein synthesis. Because the peptide bond is formed between the amino group of this amino acid and the carboxyl group of the previous amino acid, a free carboxyl group projects from the "right" end of the protein chain.

The molecule formed by condensing two amino acids is called a *dipeptide*. The amino acid with a free α-N⁺H₃ group is known as the amino terminal, or simply the **N-terminal amino acid,** and the amino acid with a free —COO⁻ group is known as the carboxyl, or **C-terminal amino acid.** Structures of proteins are conventionally written with their N-terminal amino acid on the left.

The number of amino acids in small peptides is indicated by the prefixes *di-* (two units), *tri-* (three units), *tetra-* (four units), and so forth. Peptides are named as derivatives of the C-terminal amino acid, which receives its entire name. For all other amino acids, the ending *-ine* is changed to *-yl*. Thus, the dipeptide alanyl-glycine has glycine as its C-terminal amino acid, as indicated by its full name, *glycine:*

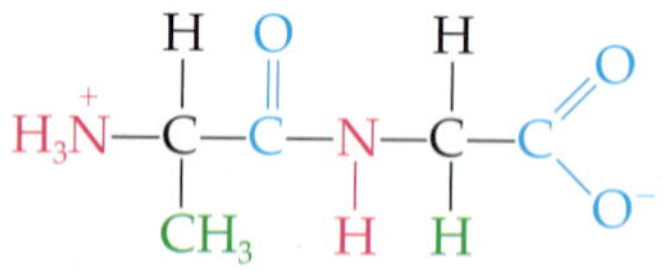

$$H_3\overset{+}{N}-\underset{CH_3}{\overset{H}{C}}-\overset{O}{\overset{\|}{C}}-\underset{H}{\overset{H}{N}}-\underset{H}{\overset{H}{C}}-C\overset{O}{\underset{O^-}{\diagup}}$$

Alanyl-glycine
(ala-gly)

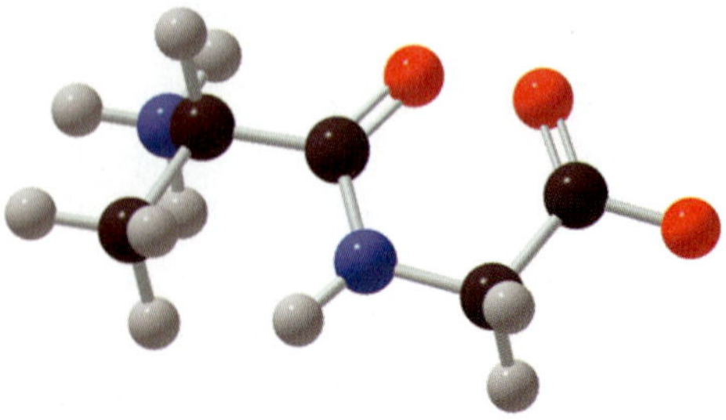

Alanyl-glycine

The dipeptide formed from alanine and glycine that has alanine as its C-terminal amino acid, glycyl-alanine, is the product of the reaction shown above. These two dipeptides have the same amino acid composition, but different amino acid sequences.

The structures of small peptides can easily be drawn with practice if certain rules are followed. First note that the backbone of the peptide contains the repeating sequence

$$\text{N—C—C—N—C—C—N—C—C}$$
$$\phantom{\text{N—}}1\phantom{\text{—}}2\phantom{\text{—N—}}1\phantom{\text{—}}2\phantom{\text{—N—}}1\phantom{\text{—}}2$$

in which N is the α-amino group, carbon-1 is the α-carbon, and carbon-2 is the carboxyl group. Carbon-1 is always bonded to a hydrogen atom and to the R group side chain that is unique to each amino acid. Continue drawing as outlined in Example 18.1.

EXAMPLE 18.1 Writing the Structure of a Tripeptide

Draw the structure of the tripeptide alanyl-glycyl-valine.

Solution

Step 1. Write the backbone for a tripeptide. It will contain three sets of three atoms, or nine atoms in all. Remember that the N-terminal amino acid is written to the left.

$$\underbrace{\text{N—C—C}}_{\text{Set 1}} \quad \underbrace{\text{N—C—C}}_{\text{Set 2}} \quad \underbrace{\text{N—C—C}}_{\text{Set 3}}$$

Step 2. Add oxygens to the carboxyl carbons and hydrogens to the amino nitrogens:

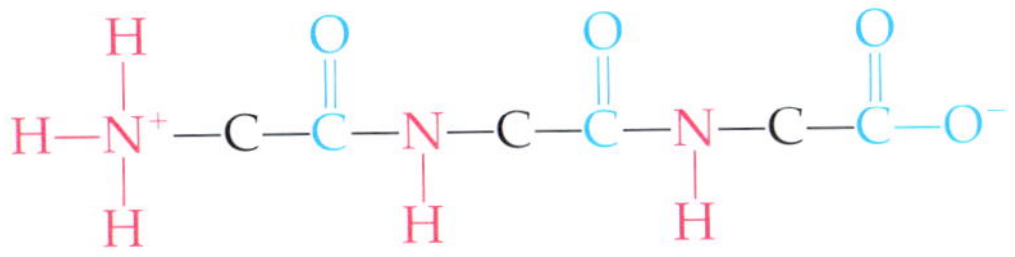

$$\text{H—N}^+\text{—C—C—N—C—C—N—C—C—O}^-$$

Step 3. Add hydrogens to the α-carbons:

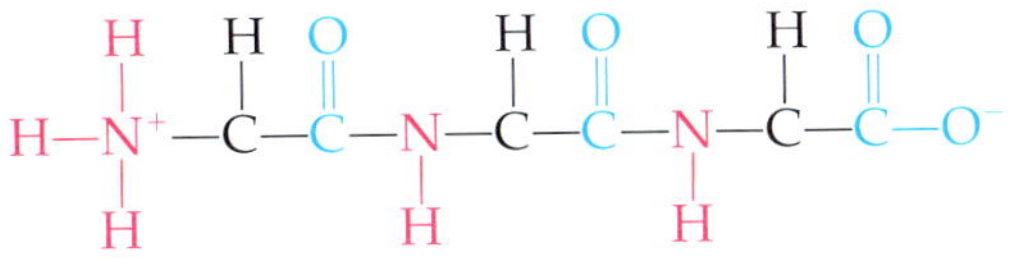

$$\text{H—N}^+\text{—C—C—N—C—C—N—C—C—O}^-$$

Step 4. Add the side chains. In this example (ala-gly-val) they are, from left to right, —CH$_3$, H, and —CH(CH$_3$)$_2$:

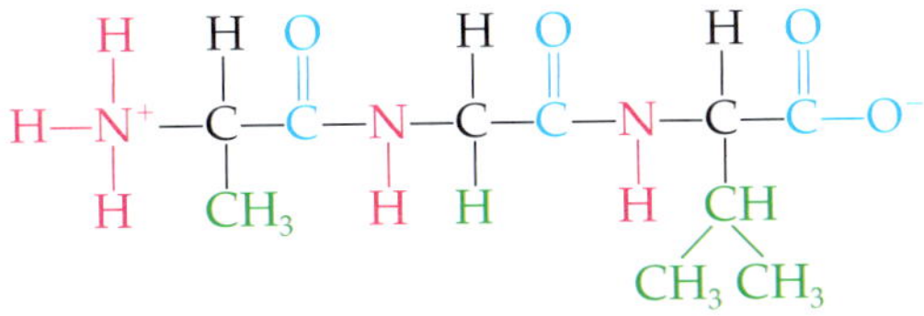

$$\text{H—N}^+\text{—C(CH}_3\text{)—C—N—C(H,H)—C—N—C(H, CH(CH}_3\text{)}_2\text{)—C—O}^-$$

Question 18.3

Write the structure of each of the following peptides at pH 7:

a. Alanyl-phenylalanine
b. Lysyl-alanine
c. Phenylalanyl-tyrosyl-leucine

A Human Perspective

The Opium Poppy and Peptides in the Brain

The seeds of the oriental poppy contain morphine. *Morphine* is a narcotic that has a variety of effects on the body and the brain, including drowsiness, euphoria, mental confusion, and chronic constipation. Although morphine was first isolated in 1805, not until the 1850s and the advent of the hypodermic was it effectively used as a painkiller. During the American Civil War, morphine was used extensively to relieve the pain of wounds and amputations. It was at this time that the addictive properties were noticed. By the end of the Civil War, over 100,000 soldiers were addicted to morphine.

As a result of the Harrison Act (1914), morphine came under government control and was made available only by prescription. Although morphine is addictive, *heroin*, a derivative of morphine, is much more addictive and induces a greater sense of euphoria that lasts for a longer time.

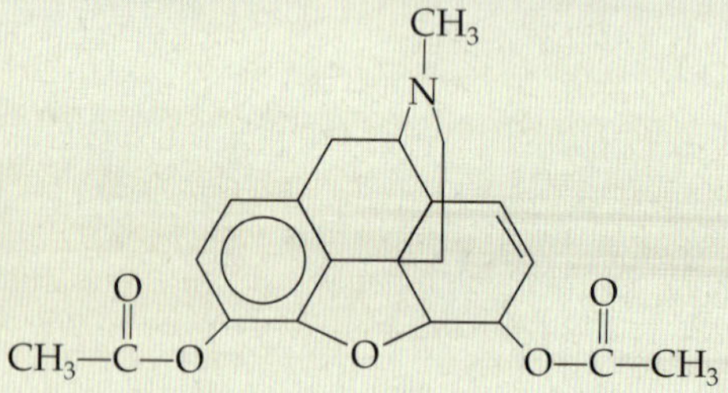

The structures of heroin and morphine.

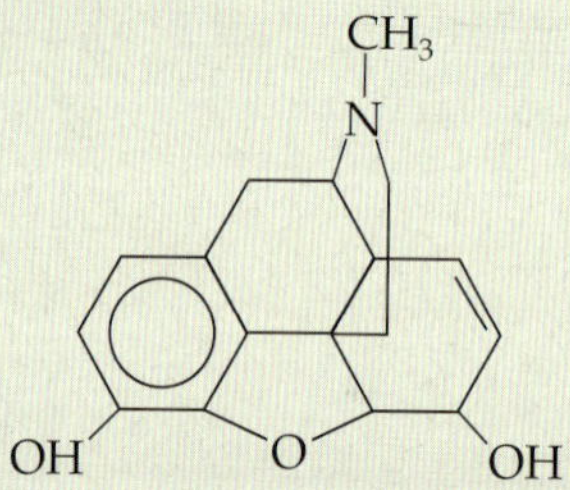

Why do heroin and morphine have such powerful effects on the brain? Both drugs have been found to bind to *receptors* on the surface of the cells of the brain. The function of these receptors is to bind specific chemical signals and to direct the brain cells to respond. Yet it seemed odd that the cells of our brain should have receptors for a plant chemical. This mystery was solved in 1975, when John Hughes discovered that the brain itself synthesizes small peptide hormones with a morphinelike structure. Two of these opiate peptides are called *methionine enkephalin*, or met-enkephalin, and *leucine enkephalin*, or leu-enkephalin.

These neuropeptide hormones have a variety of effects. They inhibit intestinal motility and blood flow to the gastrointestinal tract. This explains the chronic constipation of morphine users. In addition, it is thought that these *enkephalins* play a role in pain perception, perhaps serving as a pain blockade. This is supported by the observation that they are found in higher concentrations in the bloodstream following painful stimulation. It is further suspected that they may play a role in mood and mental health. The so-called runner's high is thought to be a euphoria brought about by an excessively long or strenuous run!

Unlike morphine, the action of enkephalins is short-lived. They bind to the cellular receptor and thereby induce the cells to respond. Then they are quickly destroyed by enzymes in the brain that hydrolyze the peptide bonds of the enkephalin. Once destroyed, they are no longer able to elicit a cellular response. Morphine and heroin bind to these same receptors and induce the cells to respond. However, these drugs are not destroyed and therefore persist in the brain for long periods at concentrations high enough to continue to cause biological effects.

Many researchers are working to understand why drugs like morphine and heroin are addictive. Studies with cells in culture have suggested one mechanism for morphine tolerance and addiction. Normally, when the cell receptors bind to enkephalins, this signals the cell to decrease the production of a chemical messenger called *cyclic AMP*, or simply cAMP. (This compound is very closely related to the nucleotide adenosine-

Poppies seen here growing in the wild are the source of the natural opiate drugs morphine and codeine.

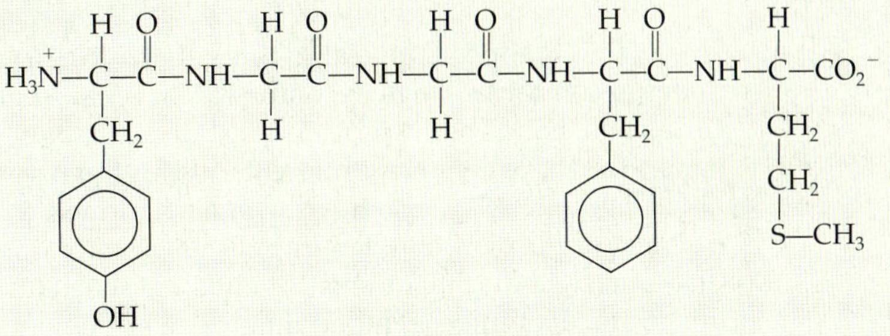

Tyr-Gly-Gly-Phe-Met
Methionine enkephalin

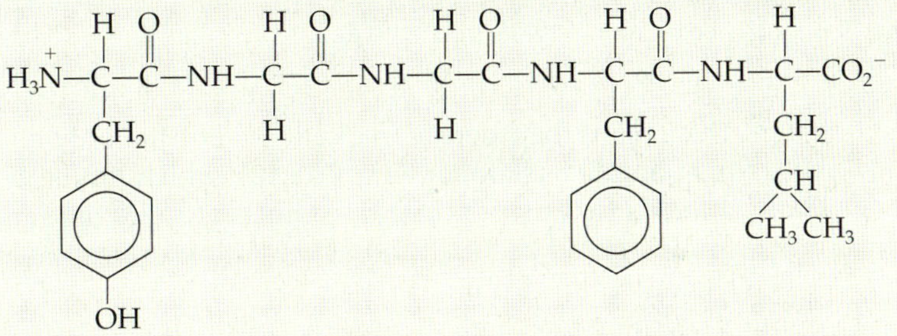

Tyr-Gly-Gly-Phe-Leu
Leucine enkephalin

Structures of the peptide opiates leucine enkephalin and methionine enkephalin. These are the body's own opiates.

5'-monophosphate.) The decrease in cAMP level helps to block pain and elevate one's mood. When morphine is applied to these cells they initially respond by decreasing cAMP levels. However, with chronic use of morphine the cells become desensitized; that is, they do not decrease cAMP production and thus behave as though no morphine were present. However, a greater amount of morphine will once again cause the decrease in cAMP levels. Thus addiction and the progressive need for more of the drug seem to result from biochemical reactions in the cells.

This logic can be extended to understand withdrawal symptoms. When an addict stops using the drug, he or she exhibits withdrawal symptoms that include excessive sweating, anxiety, and tremors. The cause may be that the high levels of morphine were keeping the cAMP levels low, thus reducing pain and causing euphoria. When morphine is removed completely, the cells overreact and produce huge quantities of cAMP. The result is all of the unpleasant symptoms known collectively as the *withdrawal syndrome*.

Clearly, morphine and heroin have demonstrated the potential for misuse and are a problem for society in several respects. Nonetheless, morphine remains one of the most effective painkillers known. Certainly, for people suffering from cancer, painful burns, or serious injuries, the risk of addiction is far outweighed by the benefits of relief from excruciating pain.

For Further Understanding

Compare the structures of leucine and methionine enkephalin with those of heroin and morphine. What similarities do you see that might cause them to bind to the same receptors on the surfaces of nerve cells?

What characteristics might you look for in a nonaddictive drug that could be used both to combat heroin addiction and to treat the symptoms of withdrawal?

Question 18.4

Write the structure of each of the following peptides at pH 7:

a. Glycyl-valyl-serine
b. Threonyl-cysteine
c. Isoleucyl-methionyl-aspartate

At first it appears logical to think that a long polymer of amino acids would undergo constant change in conformation because of free rotation around the -N-C-C- single bonds of the peptide backbone. In reality, this is not the case. An explanation for this phenomenon resulted from the early X-ray diffraction studies of Linus Pauling. By interpreting the pattern formed when X-rays were diffracted by a crystal of pure protein, Pauling discovered that peptide bonds are both planar (flat) and rigid and that the N-C bonds are shorter than expected. What did all of this mean? Pauling concluded that the peptide bond has a partially double bond character because it exhibits resonance.

$$-C_\alpha\diagdown_{\displaystyle C-N}\diagup^{\displaystyle H}_{\displaystyle C_\alpha-} \quad\longleftrightarrow\quad -C_\alpha\diagdown_{\displaystyle C=N}\diagup^{\displaystyle H}_{\displaystyle C_\alpha-}$$

This means that there is free rotation around only two of the three single bonds of the peptide backbone (Figure 18.4a), which limits the number of possible conformations for any peptide. A second feature of the rigid peptide bond is that the R groups on adjacent amino acids are on opposite sides of the extended peptide chain (Figure 18.4b).

18.4 The Primary Structure of Proteins

 LEARNING GOAL 3

The genetic code and the process of protein synthesis are described in Sections 20.5 and 20.6.

The **primary structure** of a protein is the amino acid sequence of the protein chain. It results from the covalent bonding between the amino acids in the chain (peptide bonds). The primary structures of proteins are translations of information contained in genes. Each protein has a different primary structure with different amino acids in different places along the chain.

Ultimately, it is the primary structure of a protein that will determine its biologically active form. The interactions among the R groups of the amino acids in the protein chain depend on the location of those R groups along the chain. These interactions will govern how the protein chain folds, which, in turn, dictates its final three-dimensional structure and its biological function.

Mutations and their effect on protein synthesis are discussed in Section 20.7.

Genes can change by the process of mutation during the course of evolution. A mutation in a gene can result in a change in the primary amino acid sequence of a protein. Over longer periods, more of these changes will occur. If two species of organisms diverged (became new species) very recently, the differences in the amino acid sequences of their proteins will be few. On the other hand, if they diverged millions of years ago, there will be many more differences in the amino acid sequences of their proteins. As a result, we can compare evolutionary relationships between species by comparing the primary structures of proteins present in both species.

18.5 The Secondary Structure of Proteins

 LEARNING GOAL 5

 LEARNING GOAL 6

The primary sequence of a protein, the chain of covalently linked amino acids, folds into regularly repeating structures that resemble designs in a tapestry. These repeating structures define the **secondary structure** of the protein. The secondary

18.5 The Secondary Structure of Proteins

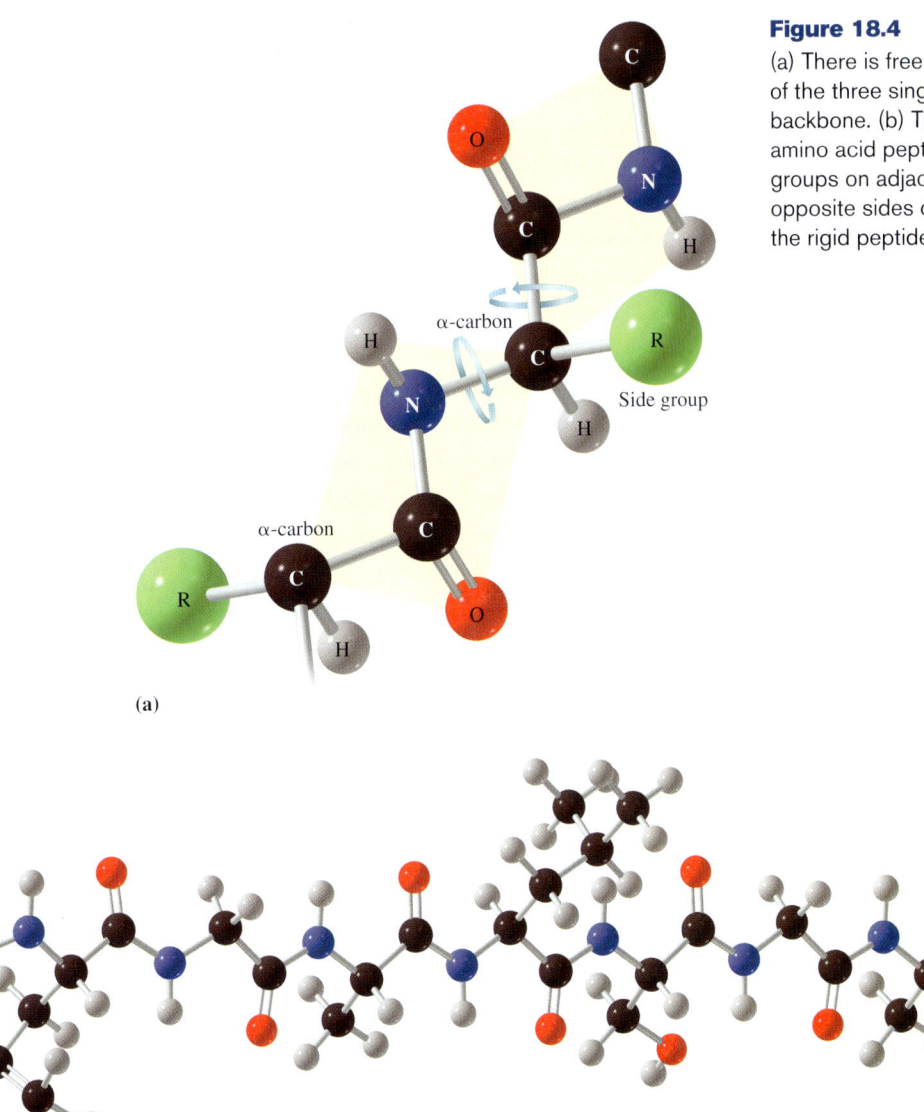

Figure 18.4
(a) There is free rotation around only two of the three single bonds of a peptide backbone. (b) This model of an eight amino acid peptide shows that the R groups on adjacent amino acids are on opposite sides of the chain because of the rigid peptide bond.

structure is the result of hydrogen bonding between the amide hydrogens and carbonyl oxygens of the peptide bonds. Many hydrogen bonds are needed to maintain the secondary structure and thereby the overall structure of the protein. Different regions of a protein chain may have different types of secondary structure. Some regions of a protein chain may have a random or nonregular structure; however, the two most common types of secondary structure are the α-helix and the β-pleated sheet because they maximize hydrogen bonding in the backbone.

α-Helix

The most common type of secondary structure is a coiled, helical conformation known as the **α-helix** (Figure 18.5). The α-helix has several important features.

- Every amide hydrogen and carbonyl oxygen associated with the peptide backbone is involved in a hydrogen bond when the chain coils into an α-helix. These hydrogen bonds lock the α-helix into place.

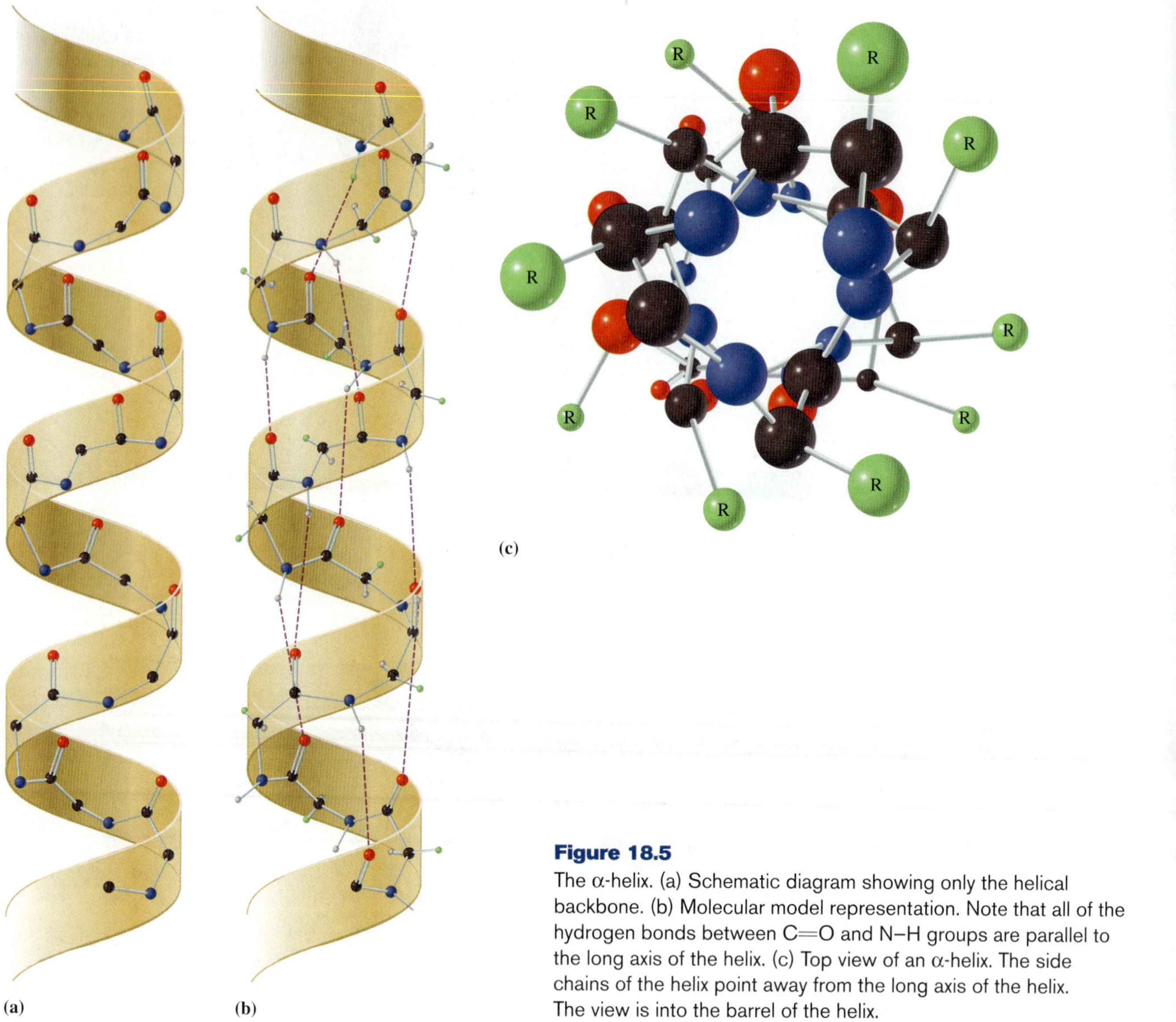

Figure 18.5
The α-helix. (a) Schematic diagram showing only the helical backbone. (b) Molecular model representation. Note that all of the hydrogen bonds between C=O and N–H groups are parallel to the long axis of the helix. (c) Top view of an α-helix. The side chains of the helix point away from the long axis of the helix. The view is into the barrel of the helix.

- Every carbonyl oxygen is hydrogen-bonded to an amide hydrogen four amino acids away in the chain.
- The hydrogen bonds of the α-helix are parallel to the long axis of the helix (see Figure 18.5).
- The polypeptide chain in an α-helix is right-handed. It is oriented like a normal screw. If you turn a screw clockwise it goes into the wall; turned counterclockwise, it comes out of the wall.
- The repeat distance of the helix, or its pitch, is 5.4 Å, and there are 3.6 amino acids per turn of the helix.

LEARNING GOAL

Fibrous proteins are structural proteins arranged in fibers or sheets that have only one type of secondary structure. The **α-keratins** are fibrous proteins that form the covering (hair, wool, nails, hooves, and fur) of most land animals. Human hair provides a typical example of the structure of the α-keratins. The proteins of hair consist almost exclusively of polypeptide chains coiled up into α-helices. A single α-helix is coiled in a bundle with two other helices to give a three-stranded superstructure called a *protofibril* that is part of an array known as a *microfibril* (Figure 18.6). These structures, which resemble "molecular pigtails," possess great mechanical strength, and they are virtually insoluble in water.

The major structural property of a coiled coil superstructure of α-helices is its great mechanical strength. This property is applied very efficiently in both the fibrous proteins of skin and those of muscle. As you can imagine, these proteins must be very strong to carry out their functions of mechanical support and muscle contraction.

β-Pleated Sheet

The second common secondary structure in proteins resembles the pleated folds of drapery and is known as **β-pleated sheet** (Figure 18.7a). All of the carbonyl oxygens and amide hydrogens in a β-pleated sheet are involved in hydrogen bonds, and the polypeptide chain is nearly completely extended. The polypeptide chains in a β-pleated sheet can have two orientations. If the N-termini are head to head, the structure is known as a *parallel* β-pleated sheet. And if the N-terminus of one chain is aligned with the C-terminus of a second chain (head to tail), the structure is known as an *antiparallel* β-pleated sheet.

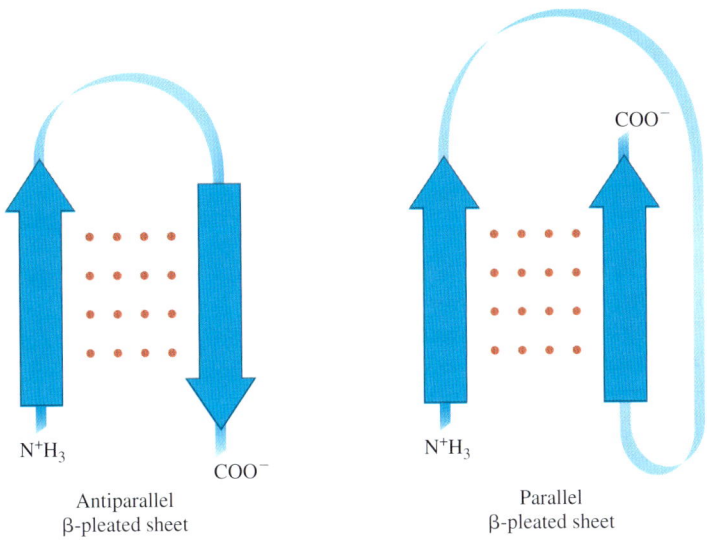

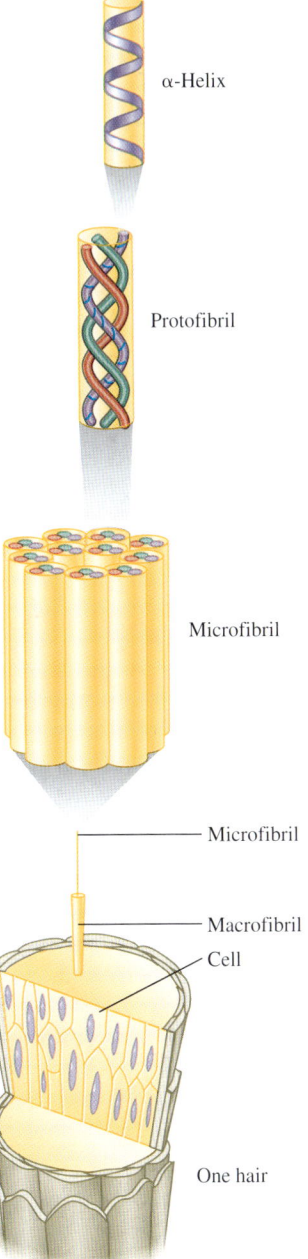

Figure 18.6
Structure of the α-keratins. These proteins are assemblies of triple-helical protofibrils that are assembled in an array known as a *microfibril*. These in turn are assembled into macrofibrils. Hair is a collection of macrofibrils and hair cells.

Some fibrous proteins are composed of β-pleated sheets. For example, the silkworm produces *silk fibroin,* a protein whose structure is an antiparallel β-pleated sheet (Figure 18.7). The polypeptide chains of a β-pleated sheet are almost completely extended, and silk does not stretch easily. Glycine accounts for nearly half of the amino acids of silk fibroin. Alanine and serine account for most of the others. The methyl groups of alanines and the hydroxymethyl groups of serines lie on opposite sides of the sheet. Thus the stacked sheets nestle comfortably, like sheets of corrugated cardboard, because the R groups are small enough to allow the stacked-sheet superstructure.

18.6 The Tertiary Structure of Proteins

Most fibrous proteins, such as silk, collagen, and the α-keratins, are almost completely insoluble in water. (Our skin would do us very little good if it dissolved in the rain.) The majority of cellular proteins, however, are soluble in the cell cytoplasm. Soluble proteins are usually **globular proteins.** Globular proteins have three-dimensional structures called the **tertiary structure** of the protein, which are distinct from their secondary structure. The polypeptide chain with its regions of secondary structure, α-helix and β-pleated sheet, further folds on itself to achieve the tertiary structure.

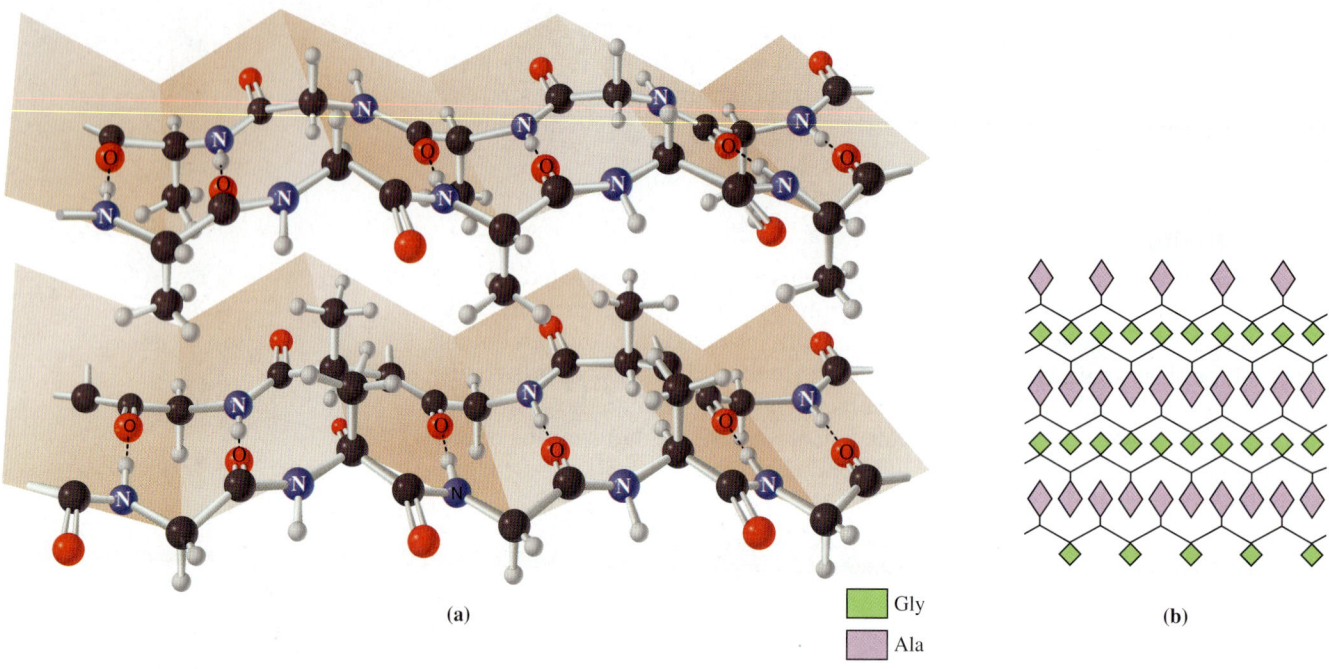

Figure 18.7
The structure of silk fibroin is almost entirely antiparallel β-pleated sheet. (a) The molecular structure of a portion of the silk fibroin protein. (b) A schematic representation of the antiparallel β-pleated sheet with the nestled R groups.

We have seen that the forces that maintain the secondary structure of a protein are hydrogen bonds between the amide hydrogen and the carbonyl oxygen of the peptide bond. What are the forces that maintain the tertiary structure of a protein? The globular tertiary structure forms spontaneously and is maintained as a result of interactions among the side chains, the R groups, of the amino acids. The structure is maintained by the following molecular interactions:

LEARNING GOAL

- Van der Waals forces between the R groups of nonpolar amino acids that are hydrophobic
- Hydrogen bonds between the polar R groups of the polar amino acids
- Ionic bonds (salt bridges) between the R groups of oppositely charged amino acids
- Covalent bonds between the thiol-containing amino acids. Two of the polar cysteines can be oxidized to a dimeric amino acid called *cystine* (Figure 18.8). The disulfide bond of cystine can be a cross-link between different proteins, or it can tie two segments within a protein together.

Figure 18.8
Oxidation of two cysteines to give the dimer cystine. This reaction occurs in cells and is readily reversible.

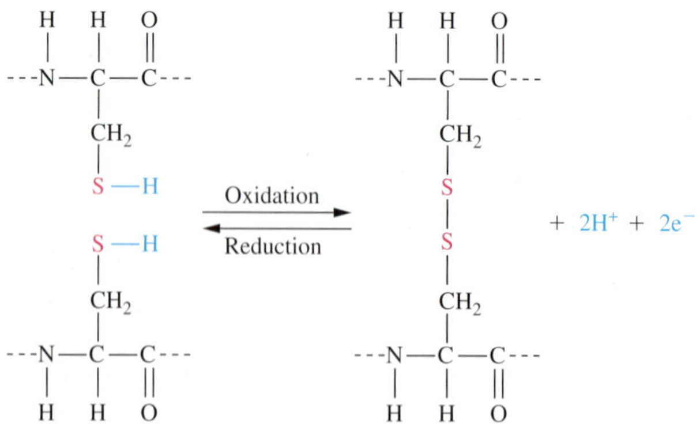

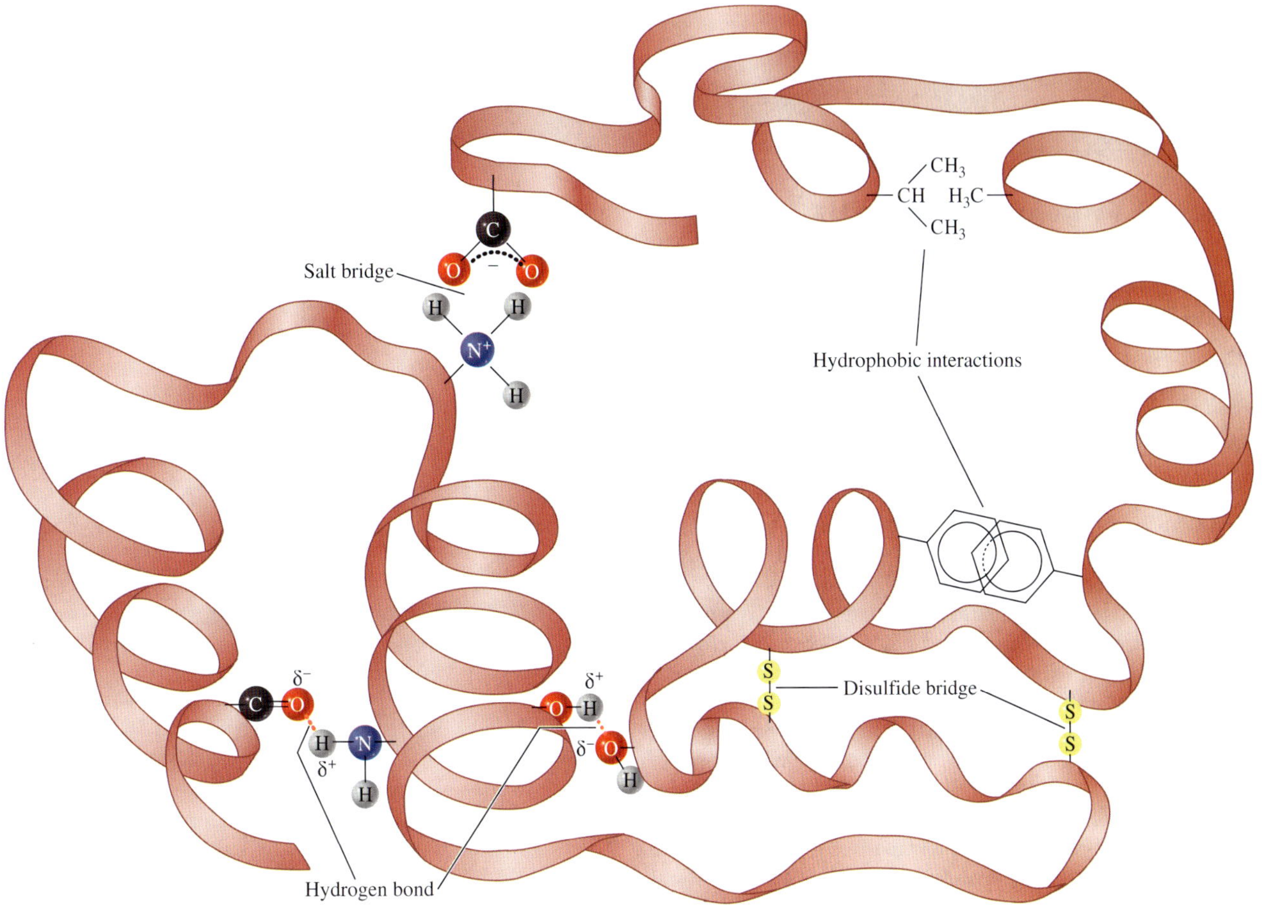

Figure 18.9
Summary of the weak interactions that help maintain the tertiary structure of a protein.

The bonds that maintain the tertiary structure of proteins are shown in Figure 18.9. The importance of these bonds becomes clear when we realize that it is the tertiary structure of the protein that defines its biological function. Most of the time, nonpolar amino acids are buried, closely packed, in the interior of a globular protein, out of contact with water. Polar and charged amino acids lie on the surfaces of globular proteins. Globular proteins are extremely compact. The tertiary structure can contain regions of α-helix and regions of β-pleated sheet. "Hinge" regions of random coil connect regions of α-helix and β-pleated sheet. Because of its cyclic structure, proline disrupts an α-helix. As a result, proline is often found in these hinge regions.

In the next section we will see that some proteins have an additional level of structure, quaternary structure, that also influences function.

18.7 The Quaternary Structure of Proteins

For many proteins the functional form is not composed of a single peptide but is rather an aggregate of smaller globular peptides. For instance, the protein hemoglobin is composed of four individual globular peptide subunits: two identical α-subunits and two identical β-subunits. Only when the four peptides are bound to one another is the protein molecule functional. The association of several polypeptides to produce a functional protein defines the **quaternary structure** of a protein.

The forces that hold the quaternary structure of a protein are the same as those that hold the tertiary structure. These include hydrogen bonds between polar amino acids, ionic bridges between oppositely charged amino acids, van der Waals forces between nonpolar amino acids, and disulfide bridges.

  LEARNING GOAL

The designations α- and β- used to describe the subunits of hemoglobin do not refer to types of secondary structure.

A Human Perspective

Collagen: A Protein That Holds Us Together

Collagen is the most abundant protein in the human body, making up about one-third of the total protein content. It provides mechanical strength to bone, tendon, skin, and blood vessels. Collagen fibers in bone provide a scaffolding around which *hydroxyapatite* (a calcium phosphate polymer) crystals are arranged. Skin contains loosely woven collagen fibers that can expand in all directions. The corneas of the eyes are composed of collagen. As we consider these tissues, we realize that they have quite different properties, ranging from tensile strength (tendons) and flexibility (blood vessels) to transparency (cornea).

How could such diverse structures be composed of a single protein? The answer lies in the fact that collagen is actually a family of twenty genetically distinct, but closely related proteins. Although the differences in the amino acid sequence of these different collagen proteins allow them to carry out a variety of functions in the body, they all have a similar three-dimensional structure. Collagen is composed of three left-handed polypeptide helices that are twisted around one another to form a "superhelix" called a *triple helix*. Each of the individual peptide chains of collagen is a left-handed helix, but they are wrapped around one another in the right-handed sense.

Every third amino acid in the collagen chain is glycine. It is important to the structure because the triple-stranded helix forms as a result of interchain hydrogen bonding involving glycine. Thus, every third amino acid on one strand is in very close contact with the other two strands. Glycine has another advantage; it is the only amino acid with an R group small enough for the space allowed by the triple-stranded structure.

Structure of the collagen triple helix.

LEARNING GOAL 10

In some cases the quaternary structure of a functional protein involves binding to a nonprotein group. This additional group is called a **prosthetic group.** For example, many of the receptor proteins on cell surfaces are **glycoproteins.** These are proteins with sugar groups covalently attached. Each of the subunits of hemoglobin is bound to an iron-containing heme group. The heme group is a large, unsaturated organic cyclic amine with an iron ion coordinated within it. As in the case of hemoglobin, the prosthetic group often determines the function of a protein. For instance, in hemoglobin it is the iron-containing heme groups that have the ability to bind reversibly to oxygen.

Question 18.5 Describe the four levels of protein structure.

Question 18.6 What are the weak interactions that maintain the tertiary structure of a protein?

Two unusual, hydroxylated amino acids account for nearly one-fourth of the amino acids in collagen. These amino acids are 4-hydroxyproline and 5-hydroxylysine.

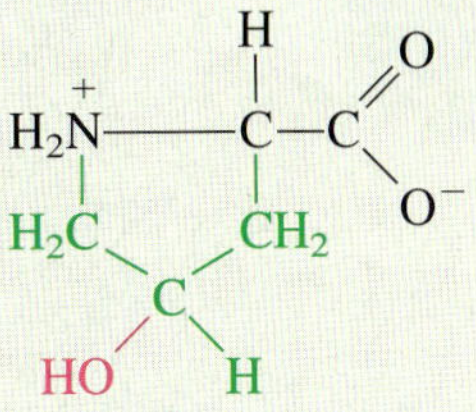

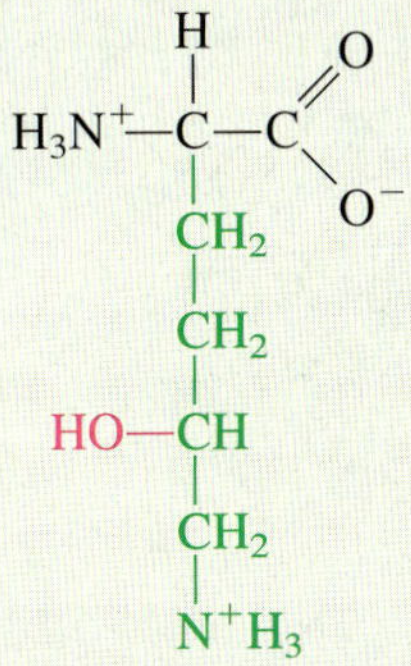

Structures of 4-hydroxyproline and 5-hydroxylysine, two amino acids found only in collagen.

These amino acids are an important component of the structure of collagen because they form covalent cross-linkages between adjacent molecules within the triple strand. They can also participate in interstrand hydrogen bonding to further strengthen the structure.

When collagen is synthesized, the amino acids proline and lysine are incorporated into the chain of amino acids. These are later modified by two enzymes to form 4-hydroxyproline and 5-hydroxylysine. Both of these enzymes require vitamin C to carry out these reactions. In fact, this is the major known physiological function of vitamin C. Without hydroxylation, hydrogen bonds cannot form and the triple helix is weak, resulting in fragile blood vessels

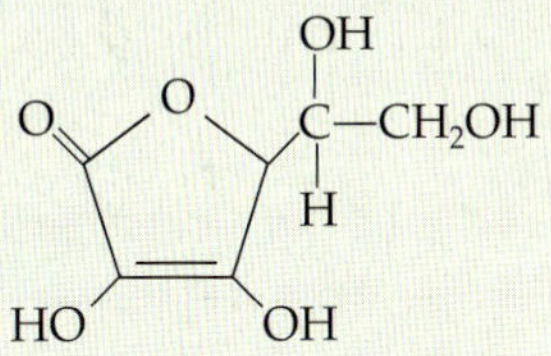

Vitamin C (ascorbic acid)

People who are deprived of vitamin C, as were sailors on long voyages before the eighteenth century, develop *scurvy*, a disease of collagen metabolism. The symptoms of scurvy include skin lesions, fragile blood vessels, and bleeding gums. The British Navy provided the antidote to scurvy by including limes, which are rich in vitamin C, in the diets of its sailors. The epithet *limey*, a slang term for *British*, entered the English language as a result.

For Further Understanding

What feature of glycine is responsible for its importance in the hydrogen bonding that maintains the helical structure of collagen?

Collagen may have great tensile strength (as in tendons) and flexibility (as in skin and blood vessels), and may even be transparent (cornea of the eye). Propose a hypothesis to explain the biochemical differences between different forms of collagen that give rise to such different properties.

18.8 An Overview of Protein Structure and Function

Let's summarize the various types of protein structure and their relationship to one another (Figure 18.10).

- *Primary Structure:* The primary structure of the protein is the amino acid sequence of the protein. The primary structure results from the formation of covalent peptide bonds between amino acids. Peptide bonds are amide bonds formed between the α-carboxylate group of one amino acid and the α-amino group of another.
- *Secondary Structure:* As the protein chain grows, numerous opportunities for noncovalent interactions in the backbone of the polypeptide chain become available. These cause the chain to fold and orient itself in a variety of conformational arrangements. The secondary level of structure includes the α-helix and the β-pleated sheet, which are the result of hydrogen bonding between the amide hydrogens and carbonyl oxygens of the peptide bonds. Different portions of the chain may be involved in different types of

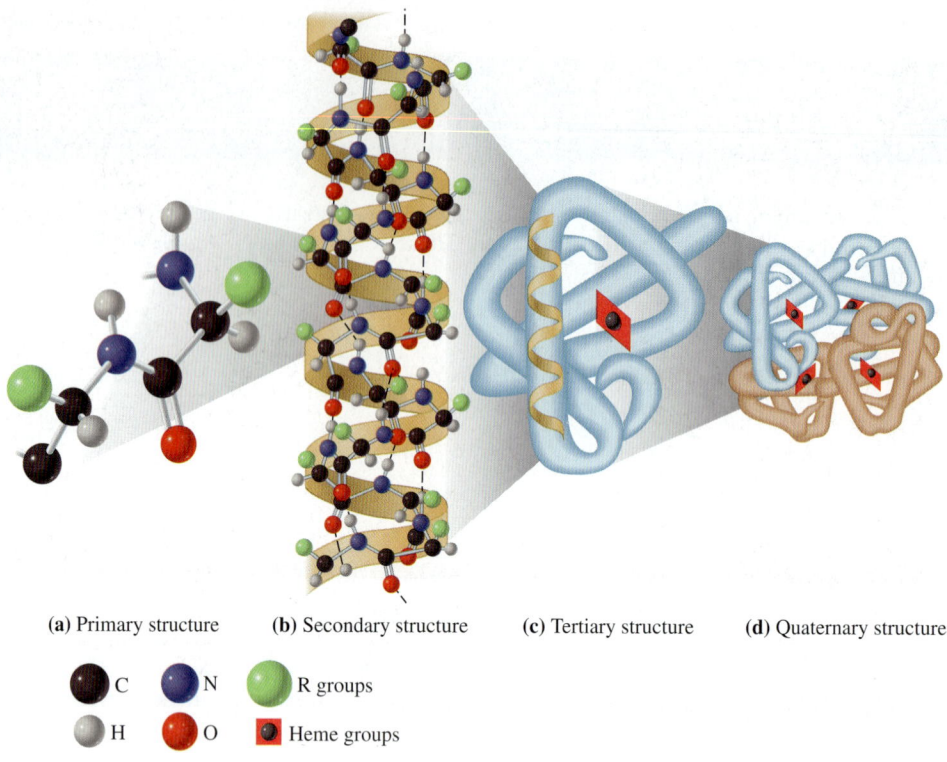

(a) Primary structure (b) Secondary structure (c) Tertiary structure (d) Quaternary structure

● C   ● N   ● R groups
○ H   ● O   ▰ Heme groups

Figure 18.10
Summary of the four levels of protein structure, using hemoglobin as an example.

The three-dimensional structure of a protein is the feature that allows it to carry out its specific biological function. However, we must always remember that it is the primary structure, the order of the R groups, that determines how the protein will fold and what the ultimate shape will be.

LEARNING GOAL

secondary structure arrangements; some regions might be α-helix and others might be a β-pleated sheet.
- *Tertiary Structure:* When we discuss tertiary structure, we are interested in the overall folding of the entire chain. In other words, we are concerned with the further folding of the secondary structure. Are the two ends of the chain close together or far apart? What general shape is involved? Both noncovalent interactions between the R groups of the amino acids and covalent disulfide bridges play a role in determining the tertiary structure. The noncovalent interactions include hydrogen bonding, ionic bonding, and van der Waals forces.
- *Quaternary Structure:* Like tertiary structure, quaternary structure is concerned with the topological, spatial arrangements of two or more peptide chains with respect to each other. How is one chain oriented with respect to another? What is the overall shape of the final functional protein?

The quaternary structure is maintained by the same forces that are responsible for the tertiary structure. It is the tertiary and quaternary structures of the protein that ultimately define its function. Some have a fibrous structure with great mechanical strength. These make up the major structural components of the cell and the organism. Often they are also responsible for the movement of the organism. Others fold into globular shapes. Most of the transport proteins, regulatory proteins, and enzymes are globular proteins. The very precise three-dimensional structure of the transport proteins allows them to recognize a particular molecule and facilitate its entry into the cell. Similarly, it is the specific three-dimensional shape of regulatory proteins that allows them to bind to their receptors on the surfaces of the target cell. In this way they can communicate with the cell, instructing it to take some course of action. In Chapter 19 we will see that the three-dimensional structure of enzyme active sites allows them to bind to their specific reactants and speed up biochemical reactions.

As we will see with the example of sickle cell hemoglobin in the next section, an alteration of just a single amino acid within the primary structure of a protein

18.9 Myoglobin and Hemoglobin

can have far-reaching implications. When an amino acid replaces another in a peptide, there is a change in the R group at that position in the protein chain. This leads to different tertiary and perhaps quaternary structure because the nature of the noncovalent interactions is altered by changing the R group that is available for that bonding. Similarly, replacement of another amino acid with proline can disrupt important regions of secondary structure. Thus changes in the primary amino acid sequence can change the three-dimensional structure of a protein in ways that cause it to be nonfunctional. In the case of sickle cell hemoglobin, this protein malfunction can lead to death.

18.9 Myoglobin and Hemoglobin

Myoglobin and Oxygen Storage

Most of the cells of our bodies are buried in the interior of the body and cannot directly get food molecules or eliminate waste. The circulatory system solves this problem by delivering nutrients and oxygen to body cells and carrying away wastes. Our cells require a steady supply of oxygen, but oxygen is only slightly soluble in aqueous solutions. To overcome this solubility problem, we have an oxygen transport protein, **hemoglobin.** Hemoglobin is found in red blood cells and is the oxygen transport protein of higher animals. **Myoglobin** is the oxygen storage protein of skeletal muscle.

The structure of myoglobin (Mb) is shown in Figure 18.11. The **heme group** (Figure 18.12) is also an essential component of this protein. The Fe^{2+} ion in the heme group is the binding site for oxygen in both myoglobin and hemoglobin. Fortunately, myoglobin has a greater attraction for oxygen than does hemoglobin, which allows efficient transfer of oxygen from the bloodstream to the cells of the body.

12 LEARNING GOAL

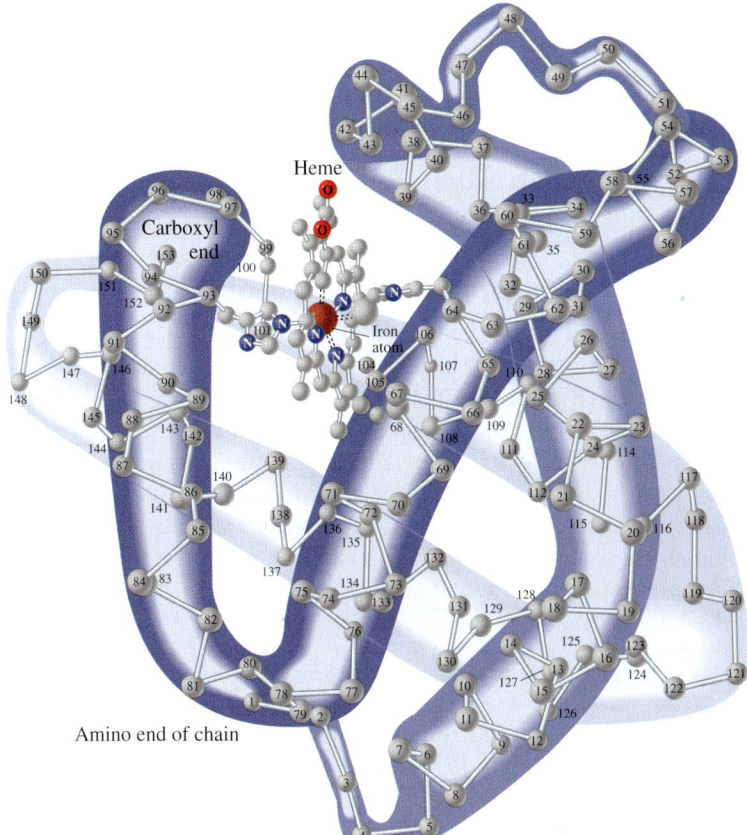

Figure 18.11
Myoglobin. The heme group has an iron atom to which oxygen binds.

Figure 18.12
Structure of the heme prosthetic group, which binds to myoglobin and hemoglobin.

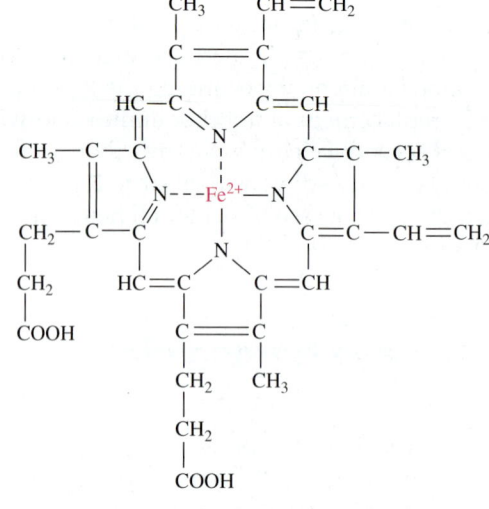

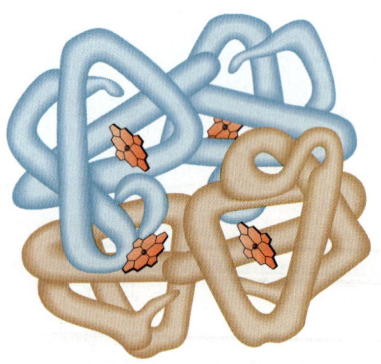

Hemoglobin
- α-chains
- β-chains
- Heme groups

Figure 18.13
Structure of hemoglobin. The protein contains four subunits, designated α and β. The α- and β-subunits face each other across a central cavity. Each subunit in the tetramer contains a heme group that binds oxygen.

See *Chemistry Connection: Wake Up, Sleeping Gene*, in Chapter 14.

Hemoglobin and Oxygen Transport

Hemoglobin (Hb) is a tetramer composed of four polypeptide subunits: two α-subunits and two β-subunits (Figure 18.13). Because each subunit of hemoglobin contains a heme group, a hemoglobin molecule can bind four molecules of oxygen:

$$\text{Hb} + 4\text{O}_2 \longrightarrow \text{Hb}(\text{O}_2)_4$$

Deoxyhemoglobin   Oxyhemoglobin

The oxygenation of hemoglobin in the lungs and the transfer of oxygen from hemoglobin to myoglobin in the tissues are very complex processes. We begin our investigation of these events with the inhalation of a breath of air.

The oxygenation of hemoglobin in the lungs is greatly favored by differences in the oxygen partial pressure (pO_2) in the lungs and in the blood. The pO_2 in the air in the lungs is approximately 100 mm Hg; the pO_2 in oxygen-depleted blood is only about 40 mm Hg. Oxygen diffuses from the region of high pO_2 in the lungs to the region of low pO_2 in the blood. There it enters red blood cells and binds to the Fe^{2+} ions of the heme groups of deoxyhemoglobin, forming oxyhemoglobin. This binding actually helps bring more O_2 into the blood.

Oxygen Transport from Mother to Fetus

A fetus receives its oxygen from its mother by simple diffusion across the placenta. If both the fetus and the mother had the same type of hemoglobin, this transfer process would not be efficient, because the hemoglobin of the fetus and the mother would have the same affinity for oxygen. The fetus, however, has a unique type of hemoglobin, called *fetal hemoglobin*. This unique hemoglobin molecule has a greater affinity for oxygen than does the mother's hemoglobin. Oxygen is therefore efficiently transported, via the circulatory system, from the lungs of the mother to the fetus. The biosynthesis of fetal hemoglobin stops shortly after birth when the genes encoding fetal hemoglobin are switched "off" and the genes coding for adult hemoglobin are switched "on."

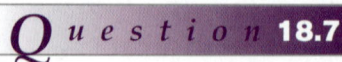

Why is oxygen efficiently transferred from hemoglobin in the blood to myoglobin in the muscles?

How is oxygen efficiently transferred from mother to fetus?

Question 18.8

Sickle Cell Anemia

Sickle cell anemia is a human genetic disease that first appeared in tropical west and central Africa. It afflicts about 0.4% of African Americans. These individuals produce a mutant hemoglobin known as sickle cell hemoglobin (Hb S). Sickle cell anemia receives its name from the sickled appearance of the red blood cells that form in this condition (Figure 18.14). The sickled cells are unable to pass through the small capillaries of the circulatory system, and circulation is hindered. This results in damage to many organs, especially bone and kidney, and can lead to death at an early age.

Sickle cell hemoglobin differs from normal hemoglobin by a single amino acid. In the β-chain of sickle cell hemoglobin, a valine (a hydrophobic amino acid) has replaced a glutamic acid (a negatively charged amino acid). This substitution provides a basis for the binding of hemoglobin S molecules to one another. When oxyhemoglobin S unloads its oxygen, individual deoxyhemoglobin S molecules bind to one another as long polymeric fibers. This occurs because the valine fits into a hydrophobic pocket on the surface of a second deoxyhemoglobin S molecule. The fibers generated in this way radically alter the shape of the red blood cell, resulting in the sickling effect.

Sickle cell anemia occurs in individuals who have inherited the gene for sickle cell hemoglobin from both parents. Afflicted individuals produce 90–100% defective β-chains. Individuals who inherit one normal gene and one defective gene produce both normal and altered β-chains. About 10% of African Americans carry a single copy of the defective gene, a condition known as *sickle cell trait*. Although not severely affected, they have a 50% chance of passing the gene to each of their children.

An interesting relationship exists between sickle cell trait and resistance to malaria. In some parts of Africa, up to 20% of the population has sickle cell trait. In those same parts of Africa, one of the leading causes of death is malaria. The presence of sickle cell trait is linked to an increased resistance to malaria because the malarial parasite cannot feed efficiently on sickled red blood cells. People who have sickle cell disease die young; those without sickle cell trait have a high probability of succumbing to malaria. Occupying the middle ground, people who have sickle cell trait do not suffer much from sickle cell anemia and simultaneously resist deadly malaria. Because those with sickle cell trait have a greater chance of survival and reproduction, the sickle cell hemoglobin gene is maintained in the population.

The genetic basis of this alteration is discussed in Chapter 20.

When hemoglobin is carrying O_2, it is called oxyhemoglobin. When it is not bound to O_2, it is called deoxyhemoglobin.

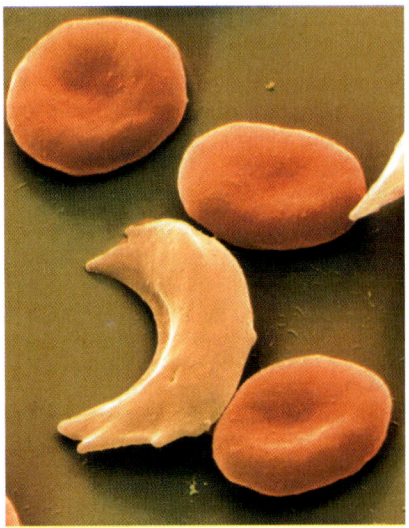

Figure 18.14
Scanning electron micrographs of normal and sickled red blood cells.

18.10 Denaturation of Proteins

We have shown that the shape of a protein is absolutely essential to its function. We have also mentioned that life can exist only within a rather narrow range of temperature and pH. How are these two concepts related? As we will see, extremes of pH or temperature have a drastic effect on protein conformation, causing the molecules to lose their characteristic three-dimensional shape. **Denaturation** occurs when the organized structures of a globular protein, the α-helix, the β-pleated sheet, and tertiary folds become completely disorganized. However, it does not alter the primary structure. Denaturation of an α-helical protein is shown in Figure 18.15.

LEARNING GOAL

A Medical Perspective

Immunoglobulins: Proteins That Defend the Body

A living organism is subjected to a constant barrage of bacterial, viral, parasitic, and fungal diseases. Without a defense against such perils we would soon perish. All vertebrates possess an *immune system*. In humans the immune system is composed of about 10^{12} cells, about as many as the brain or liver, which protect us from foreign invaders. This immune system has three important characteristics.

1. **It is highly specific.** The immune response to each infection is specific to, or directed against, only one disease organism or similar, related organisms.
2. **It has a memory.** Once the immune system has responded to an infection, the body is protected against reinfection by the same organism. This is the reason that we seldom suffer from the same disease more than once. Most of the diseases that we suffer recurrently, such as the common cold and flu, are actually caused by many different strains of the same virus. Each of these strains is "new" to the immune system.
3. **It can recognize "self" from "nonself."** When we are born, our immune system is already aware of all the antigens of our bodies. These it recognizes as "self" and will not attack. Every antigen that is not classified as "self" will be attacked by the immune system when it is encountered. Some individuals suffer from a defect of the immune response that allows it to attack the cells of one's own body. The result is an *autoimmune reaction* that can be fatal.

One facet of the immune response is the synthesis of *immunoglobulins*, or *antibodies*, that specifically bind a single macromolecule called an *antigen*. These antibodies are produced by specialized white blood cells called *B lymphocytes*. We are born with a variety of B lymphocytes that are capable of producing antibodies against perhaps a million different antigens. When a foreign antigen enters the body, it binds to the B lymphocyte that was preprogrammed to produce antibodies to destroy it. This stimulates the B cell to grow and divide. Then all of these new B cells produce antibodies that will bind to the disease agent and facilitate its destruction. Each B cell produces only one type of antibody with an absolute specificity for its target antigen. Many different B cells respond to each infection because the disease-causing agent is made up of many different antigens. Antibodies are made that bind to many of the antigens of the invader. This primary immune response is rather slow. It can take a week or two before there are enough B cells to produce a high enough level of antibodies in the blood to combat an infection.

Because the immune response has a memory, the second time we encounter a disease-causing agent the antibody response is immediate. This is why it is extremely rare to suffer from mumps, measles, or chickenpox a second time. We take advantage of this property of the immune system to protect ourselves against many diseases. In the process of *vaccination* a person can be immunized against an infectious disease by injection of a small amount of the antigens of the virus or mi-

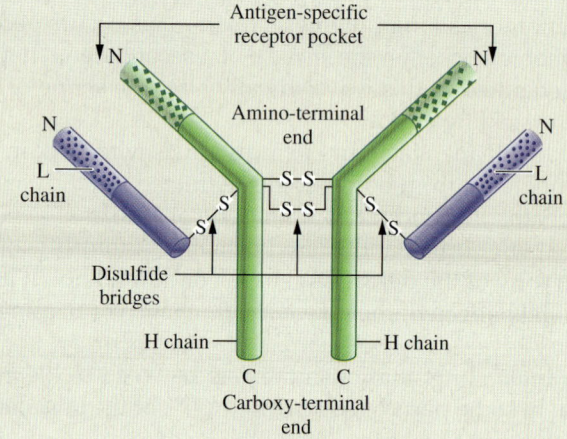

Schematic diagram of a Y-shaped immunoglobulin molecule. The binding sites for antigens are at the tips of the Y.

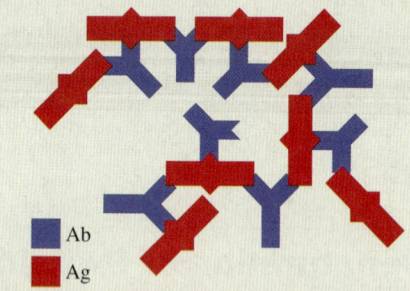

Schematic diagram of cross-linked immunoglobulin–antigen lattice.

Temperature

Consider the effect of increasing temperature on a solution of proteins—for instance, egg white. At first, increasing the temperature simply increases the rate of molecular movement, the movement of the individual molecules within the solution. Then, as the temperature continues to increase, the bonds within the proteins begin to vibrate more violently. Eventually, the weak interactions, like hydrogen

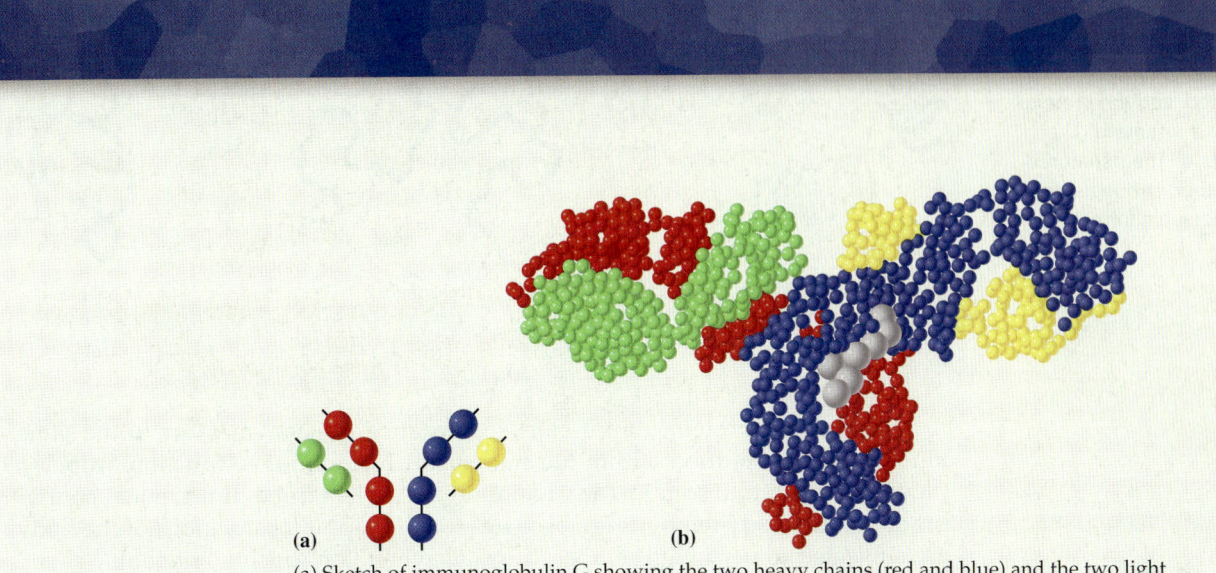

(a) Sketch of immunoglobulin G showing the two heavy chains (red and blue) and the two light chains (green and yellow). (b) Space-filling model of immunoglobulin G. The color code is the same as in (a). The gray balls represent sugar groups attached to the immunoglobulin molecule.

croorganism (the vaccine). The B lymphocytes of the body then manufacture antibodies against the antigens of the infectious agent. If the individual comes into contact with the disease-causing microorganism at some later time, the sensitized B lymphocytes "remember" the antigen and very quickly produce a large amount of specific antibody to overwhelm the microorganism or virus before it can cause overt disease.

Immunoglobulin molecules contain four peptide chains that are connected by disulfide bonds and arranged in a Y-shaped quaternary structure.

Each immunoglobulin has two identical antigen-binding sites located at the tips of the Y. Because most antigens have three or more antibody-binding sites, immunoglobulins can form large cross-linked antigen-antibody complexes that precipitate from solution.

Immunoglobulin G (IgG) is the major serum immunoglobulin. Some immunoglobulin G molecules can cross cell membranes and thus can pass between mother and fetus through the placenta, before birth. This is important because the immune system of a fetus is immature and cannot provide adequate protection from disease. Fortunately, the IgG acquired from the mother protects the fetus against most bacterial and viral infections that it might encounter before birth.

There are four additional types of antibody molecules that vary in their protein composition, but all have the same general Y shape. One of these is IgM, which is the first antibody produced in response to an infection. Secondarily, the B cell produces IgG molecules with the same antigen-binding region but a different protein composition in the rest of the molecule. IgA is the immunoglobulin responsible for protecting the body surfaces, such as the mucous membranes of the gut, the oral cavity, and the genitourinary tract. IgA is also found in mother's milk, protecting the newborn against diseases during the first few weeks of life. IgD is found in very small amounts and is thought to be involved in the regulation of antibody synthesis. The last type of immunoglobulin is IgE. For many years the function of IgE was unknown. It is found in large quantities in the blood of people suffering from allergies and is therefore thought to be responsible for this "overblown" immunological reaction to dust particles and pollen grains.

For Further Understanding

Develop a hypothesis to explain why we may suffer from dozens of cases of the common cold caused by rhinoviruses. (Hint: Think about the structure of the proteins at the surface of the virus that serve as antigens.)

Describe the kinds of interactions that you would expect to find when an antibody binds to its cognate antigen.

bonds and hydrophobic interactions, that maintain the protein structure are disrupted. The protein molecules are denatured as they lose their characteristic three-dimensional conformation and become completely disorganized. **Coagulation** occurs as the protein molecules then unfold and become entangled. At this point they are no longer in solution; they have aggregated to become a solid (see Figure 18.15). The egg white began as a viscous solution of egg albumins; but when it was cooked, the proteins were denatured and coagulated to become solid.

Figure 18.15
The denaturation of proteins by heat. (a) The α-helical proteins are in solution. (b) As heat is applied, the hydrogen bonds maintaining the secondary structure are disrupted, and the protein structure becomes disorganized. The protein is denatured. (c) The denatured proteins clump together, or coagulate, and are now in an insoluble form.

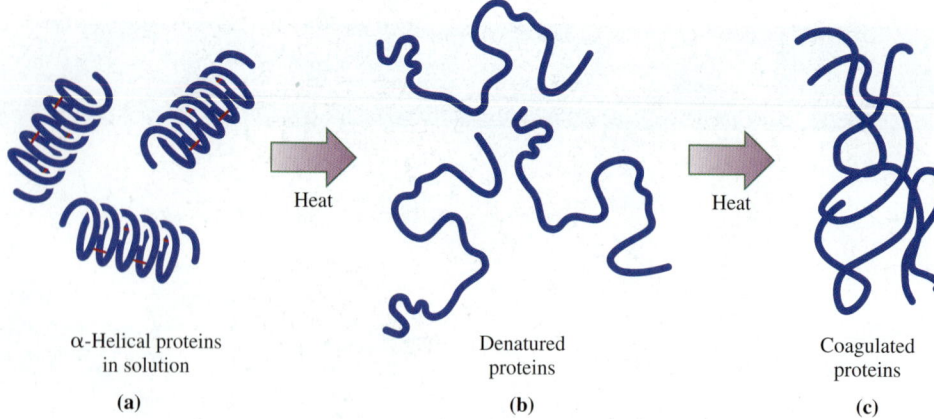

α-Helical proteins in solution
(a)

Denatured proteins
(b)

Coagulated proteins
(c)

Many of the proteins of our cells, for instance, the enzymes, are in the same kind of viscous solution within the cytoplasm. To continue to function properly, they must remain in solution and maintain the correct three-dimensional configuration. If the body temperature becomes too high, or if local regions of the body are subjected to very high temperatures, as when you touch a hot cookie sheet, cellular proteins become denatured. They lose their function, and the cell or the organism dies.

pH

Because of the R groups of the amino acids, all proteins have a characteristic electric charge. Because every protein has a different amino acid composition, each will have a characteristic net electric charge on its surface. The positively and negatively charged R groups on the surface of the molecule interact with ions and water molecules, and these interactions keep the protein in solution within the cytoplasm.

The protein shown in Figure 18.16a has a net charge of 2+ because it has two extra —N^+H_3 groups. If we add 2 moles of base, such as NaOH, the protonated amino groups lose their protons and thus become electrically neutral. Now the net charge of the protein is zero. The pH at which a protein has an equal number of positive and negative charges, that is, a net charge of zero, is called the **isoelectric point.** The protein shown in Figure 18.16b has a net charge of 2− because of two additional carboxylate groups. When 2 moles of acid are added, the carboxylate groups become protonated. They are now electrically neutral, and the net charge on the protein is zero. As in the preceding example, the protein solution is at the isoelectric point.

When the pH of a protein solution is above the isoelectric point, all the protein molecules will have a net negative surface charge. Below the isoelectric point, they will have a net positive charge. In either case, these like-charged molecules repel one another, and this repulsion helps keep these very large molecules in solution.

At the isoelectric point the protein molecules no longer have a net surface charge. As a result they no longer strongly repel one another and are at their least soluble. Under these conditions, there is a tendency for them to clump together and precipitate out of solution. In this case, proteins may coagulate even though they are not denatured.

This is a reaction that you have probably observed in your own kitchen. When milk sits in the refrigerator for a prolonged period, the bacteria in the milk begin to grow. They use the milk sugar, lactose, as an energy source in the process of fermentation and produce lactic acid as a by-product. As the bacteria continue to grow, the concentration of lactic acid increases. The additional acid results in the protonation of exposed carboxylate groups on the surface of the dissolved milk proteins. They become isoelectric and coagulate into a solid curd.

Lactate fermentation is discussed in Section 21.4.

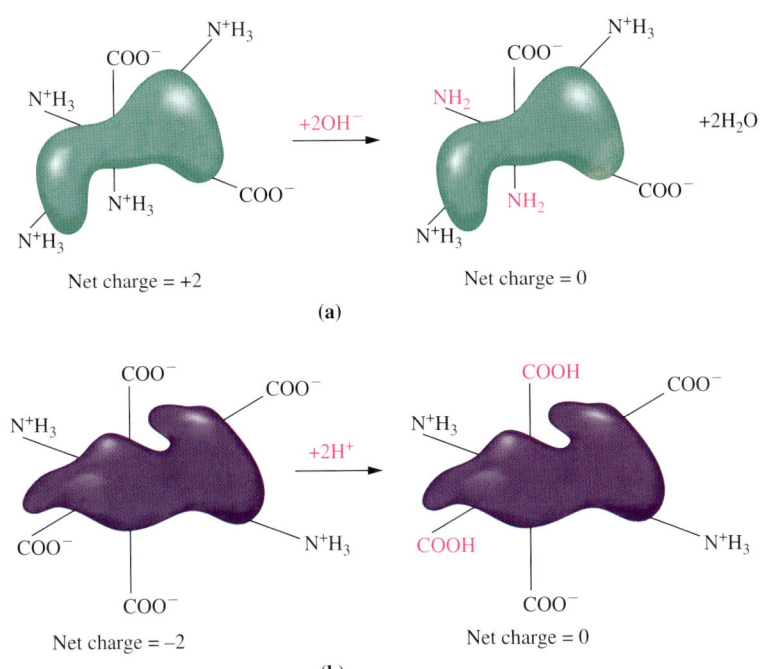

Figure 18.16
The effect of pH on proteins. (a) This protein has an overall charge of 2+. When a base is added, some of the protonated amino groups lose their protons. Now the protein is isoelectric; it has an equal number of positive and negative charges. (b) This protein has an overall charge of 2−. As acid is added, some of the carboxylate groups are protonated. The result is that the protein becomes isoelectric.

Imagine for a moment what would happen if the pH of the blood were to become too acidic or too basic. Blood is a fluid that contains water and dissolved electrolytes, a variety of cells, including the red blood cells responsible for oxygen transport, and many different proteins. These proteins include fibrinogen, which is involved in the clotting reaction; immunoglobulins, which protect us from disease; and albumins, which carry hydrophobic molecules in the blood.

When the blood pH drops too low, blood proteins become polycations. Similarly, when the blood pH rises too high, the proteins become polyanions. In either case, the proteins will unfold because of charge repulsion and loss of stabilizing ionic interactions. Under these extreme conditions, the denatured blood proteins would no longer be able to carry out their required functions. The blood cells would also die as their critical enzymes were denatured. The hemoglobin in the red blood cells would become denatured and would no longer be able to transport oxygen. Fortunately, the body has a number of mechanisms to avoid the radical changes in the blood pH that can occur as a result of metabolic or respiratory difficulties.

See A Medical Perspective: Proteins in the Blood on page 599.

Organic Solvents

Polar organic solvents, such as rubbing alcohol (2-propanol), denature proteins by disrupting hydrogen bonds within the protein, in addition to forming hydrogen bonds with the solvent, water. The nonpolar regions of these solvents interfere with hydrophobic interactions in the interior of the protein molecule, thereby disrupting the conformation. Traditionally, a 70% solution of rubbing alcohol was often used as a disinfectant or antiseptic. However, recent evidence suggests that it is not an effective agent in this capacity.

Detergents

Detergents have both a hydrophobic region (the fatty acid tail) and a polar or hydrophilic region. When detergents interact with proteins, they disrupt hydrophobic interactions, causing the protein chain to unfold.

Heavy Metals

Heavy metals such as mercury (Hg^{2+}) or lead (Pb^{2+}) may form bonds with negatively charged side chain groups. This interferes with the salt bridges formed between amino acid R groups of the protein chain, resulting in loss of conformation. Heavy metals may also bind to sulfhydryl groups of a protein. This may cause a profound change in the three-dimensional structure of the protein, accompanied by loss of function.

Mechanical Stress

Stirring, whipping, or shaking can disrupt the weak interactions that maintain protein conformation. This is the reason that whipping egg whites produces a stiff meringue.

Question 18.9 How does high temperature denature proteins?

Question 18.10 How does extremely low pH cause proteins to coagulate?

18.11 Dietary Protein and Protein Digestion

LEARNING GOAL

Proteins, as well as carbohydrates and fats, are an energy source in the diet. As do carbohydrates and fats, proteins serve several dietary purposes. They can be oxidized to provide energy. In addition, the amino acids liberated by the hydrolysis of proteins are used directly in biosynthesis. The protein synthetic machinery of the cell can incorporate amino acids, released by the digestion of dietary protein, directly into new cellular proteins. Amino acids are also used in the biosynthesis of a large number of important molecules called the *nitrogen compounds*. This group includes some hormones, the heme groups of hemoglobin and myoglobin, and the nitrogen-containing bases found in DNA and RNA.

Digestion of dietary protein begins in the stomach. The stomach enzyme *pepsin* begins the digestion by hydrolyzing some of the peptide bonds of the protein. This breaks the protein down into smaller peptides.

Production of pepsin and other proteolytic digestive enzymes must be carefully controlled because the active enzymes would digest and destroy the cell that produces them. Thus, the stomach lining cells that make pepsin actually synthesize and secrete an inactive form called *pepsinogen*. Pepsinogen has an additional forty-two amino acids in its primary structure. These are removed in the stomach to produce active pepsin.

Protein digestion continues in the small intestine where the enzymes trypsin, chymotrypsin, elastase, and others catalyze the hydrolysis of peptide bonds at different sites in the protein. For instance, chymotrypsin cleaves peptide bonds on the carbonyl side of aromatic amino acids and trypsin cleaves peptide bonds on the carbonyl side of basic amino acids. Together these proteolytic enzymes degrade large dietary proteins into amino acids that can be absorbed by cells of the small intestine.

Amino acids can be divided into two major nutritional classes. **Essential amino acids** are those that cannot be synthesized by the body and are required in the diet. **Nonessential amino acids** are those amino acids that can be synthesized

The inactive form of a proteolytic enzyme is called a proenzyme. These are discussed in Section 19.9.

The specificity of proteolytic enzymes is described in Section 19.11.

See Section 19.11 and Figure 19.12 for a more detailed picture of the action of digestive proteases.

18.11 Dietary Protein and Protein Digestion

TABLE 18.2 The Essential and Nonessential Amino Acids

Essential Amino Acids	Nonessential Amino Acids
Isoleucine	Alanine
Leucine	Arginine[1]
Lysine	Asparagine
Methionine	Aspartate
Phenylalanine	Cysteine[2]
Threonine	Glutamate
Tryptophan	Glutamine
Valine	Glycine
	Histidine[1]
	Proline
	Serine
	Tyrosine[2]

[1]Histidine and arginine are essential amino acids for infants but not for healthy adults.
[2]Cysteine and tyrosine are considered to be semiessential amino acids. They are required by premature infants and adults who are ill.

by the body and need not be included in the diet. Table 18.2 lists the essential and nonessential amino acids.

Proteins are also classified as *complete* or *incomplete*. Protein derived from animal sources is generally **complete protein.** That is, it provides all of the essential and nonessential amino acids in approximately the correct amounts for biosynthesis. In contrast, protein derived from vegetable sources is generally **incomplete protein** because it lacks a sufficient amount of one or more essential amino acids. People who want to maintain a strictly vegetarian diet or for whom animal protein is often not available have the problem that no single high-protein vegetable has all of the essential amino acids to ensure a sufficient daily intake. For example, the major protein of beans contains abundant lysine and tryptophan but very little methionine, whereas corn contains considerable methionine but very little tryptophan or lysine. A mixture of corn and beans, however, satisfies both requirements. This combination, called *succotash,* was a staple of the diet of Native Americans for centuries.

Eating a few vegetarian meals each week can provide all the required amino acids and simultaneously help reduce the amount of saturated fats in the diet. Many ethnic foods apply the principle of mixing protein sources. Mexican foods such as tortillas and refried beans, Cajun dishes of spicy beans and rice, Indian cuisine of rice and lentils, and even the traditional American peanut butter sandwich are examples of ways to mix foods to provide complete protein.

Question 18.11 Why must vegetable sources of protein be mixed to provide an adequate diet?

Question 18.12 What are some common sources of dietary protein?

SUMMARY

18.1 Cellular Functions of Proteins

Proteins serve as biological catalysts (enzymes) and protective *antibodies*. *Transport proteins* carry materials throughout the body. Protein hormones regulate conditions in the body. Proteins also provide mechanical support and are needed for movement.

18.2 The α-Amino Acids

Proteins are made from twenty different *amino acids*, each having an α-COO$^-$ group and an α-N$^+$H$_3$ group. They differ only in their side-chain R groups. All α-*amino acids* are chiral except glycine. Naturally occurring amino acids have the same chirality, designated L. The amino acids are grouped according to the polarity of their R groups.

18.3 The Peptide Bond

Amino acids are joined by *peptide bonds* to produce peptides and proteins. The peptide bond is an amide bond formed in the reaction between the carboxyl group of one amino acid and the amino group of another. The peptide bond is planar and relatively rigid.

18.4 The Primary Structure of Proteins

Proteins are linear polymers of amino acids. The linear sequence of amino acids defines the *primary structure* of the protein. Evolutionary relationships between species of organisms can be deduced by comparing the primary structures of their proteins.

18.5 The Secondary Structure of Proteins

The *secondary structure* of a protein is the folding of the primary sequence into an α-*helix* or β-*pleated sheet*. These structures are maintained by hydrogen bonds between the amide hydrogen and the carbonyl oxygen of the peptide bond. Usually *structural proteins*, such as the α-keratins and silk fibroin, are composed entirely of α-helix or β-pleated sheet.

18.6 The Tertiary Structure of Proteins

Globular proteins contain varying amounts of α-helix and β-pleated sheet folded into higher levels of structure called the *tertiary structure*. The tertiary structure of a protein is maintained by attractive forces between the R groups of amino acids. These forces include hydrophobic interactions, hydrogen bonds, ionic bridges, and disulfide bonds.

18.7 The Quaternary Structure of Proteins

Some proteins are composed of more than one peptide. They are said to have *quaternary structure*. Weak attractions between amino acid R groups hold the peptide subunits of the protein together. Some proteins require an attached, nonprotein *prosthetic group*.

18.8 An Overview of Protein Structure and Function

The primary structure of a protein dictates the way in which it folds into secondary and tertiary levels of structure. It also determines the way in which a protein may associate with other peptide subunits in the quaternary structure. An amino acid change in the primary structure may drastically affect protein folding. If the protein does not fold properly, and assume its correct three-dimensional shape, it will not be able to carry out its cellular function.

18.9 Myoglobin and Hemoglobin

Myoglobin, the oxygen storage protein of skeletal muscle, has a prosthetic group called the *heme group*. The heme group is the site of oxygen binding. *Hemoglobin* consists of four peptides. It transports oxygen from the lungs to the tissues. Myoglobin has a greater affinity for oxygen than does hemoglobin, and so oxygen is efficiently transferred from hemoglobin in the blood to myoglobin in tissues. Fetal hemoglobin has a greater affinity for oxygen than does maternal hemoglobin, and oxygen transfer occurs efficiently across the placenta from the mother to the fetus. A mutant hemoglobin is responsible for the genetic disease *sickle cell anemia*.

18.10 Denaturation of Proteins

Heat disrupts the hydrogen bonds and hydrophobic interactions that maintain protein structure. As a result, the protein unfolds and the organized structure is lost. The protein is said to be *denatured*. *Coagulation*, or clumping, occurs when the protein chains unfold and become entangled. When this occurs, proteins are no longer soluble. Changes in pH may cause proteins to become *isoelectric* (equal numbers of positive and negative charges). Isoelectric proteins coagulate because they no longer repel each other. If pH drops very low, proteins become polycations; if pH rises very high, proteins become polyanions. In either case, proteins become denatured owing to charge repulsion.

18.11 Dietary Protein and Protein Digestion

Essential amino acids must be acquired in the diet; *nonessential amino acids* can be synthesized by the body. *Complete proteins* contain all the essential and nonessential amino acids. *Incomplete proteins* are missing one or more essential amino acids. Protein digestion begins in the stomach, where proteins are degraded by the enzyme pepsin. Further digestion occurs in the small intestine by enzymes such as trypsin and chymotrypsin.

KEY TERMS

α-amino acid (18.2)
antibody (18.1)
antigen (18.1)
coagulation (18.10)
complete protein (18.11)
C-terminal amino acid (18.3)
defense proteins (18.1)
denaturation (18.10)
enzyme (18.1)
essential amino acid (18.11)
fibrous protein (18.5)
globular protein (18.6)
glycoprotein (18.7)
α-helix (18.5)
heme group (18.9)
hemoglobin (18.9)
hydrophilic amino acid (18.2)
hydrophobic amino acid (18.2)
incomplete protein (18.11)
isoelectric point (18.10)
α-keratin (18.5)
movement protein (18.1)
myoglobin (18.9)
nonessential amino acid (18.11)
N-terminal amino acid (18.3)
nutrient protein (18.1)
peptide bond (18.3)
β-pleated sheet (18.5)
primary structure (of a protein) (18.4)
prosthetic group (18.7)
protein (Intro)
quaternary structure (of a protein) (18.7)
regulatory protein (18.1)
secondary structure (of a protein) (18.5)
sickle cell anemia (18.9)
structural protein (18.1)
tertiary structure (of a protein) (18.6)
transport protein (18.1)

QUESTIONS AND PROBLEMS

Cellular Functions of Proteins

Foundations

18.13 Define the term *enzyme*.
18.14 Define the term *antibody*.
18.15 What is a transport protein?
18.16 What are the functions of structural proteins?

Applications

18.17 Of what significance are enzymes in the cell?
18.18 How do antibodies protect us against infection?
18.19 List two transport proteins and describe their significance to the organism.
18.20 What is the function of regulatory proteins?
18.21 Provide two examples of nutrient proteins.
18.22 Provide two examples of proteins that are required for movement.

The α-Amino Acids

Foundations

18.23 Write the basic general structure of an L-α-amino acid.
18.24 Draw the D- and L-isomers of serine. Which would you expect to find in nature?
18.25 What is a zwitterion?
18.26 Why are amino acids zwitterions at pH 7.0?

Applications

18.27 What is a chiral carbon?
18.28 Why are all of the α-amino acids except glycine chiral?
18.29 What is the importance of the R groups of the amino acids?
18.30 Describe the classification of the R groups of the amino acids, and provide an example of each class.
18.31 Write the structures of the nine amino acids that have hydrophobic side chains.
18.32 Write the structures of the aromatic amino acids. Indicate whether you would expect to find each on the surface or buried in a globular protein.

The Peptide Bond

Foundations

18.33 Define the term *peptide bond*.
18.34 What type of bond is the peptide bond? Explain why the peptide bond is rigid.
18.35 What observations led Linus Pauling and his colleagues to hypothesize that the peptide bond exists as a resonance hybrid?
18.36 Draw the resonance hybrids that represent the peptide bond.

Applications

18.37 Write the structure of each of the following peptides:
   a. His-trp-cys
   b. Gly-leu-ser
   c. Arg-ile-val
18.38 Write the structure of each of the following peptides:
   a. Ile-leu-phe
   b. His-arg-lys
   c. Asp-glu-ser

The Primary Structure of Proteins

Foundations

18.39 Define the *primary structure* of a protein.
18.40 What type of bond joins the amino acids to one another in the primary structure of a protein?

Applications

18.41 How does the primary structure of a protein determine its three-dimensional shape?
18.42 How does the primary structure of a protein ultimately determine its biological function?
18.43 Explain the relationship between the primary structure of a protein and the gene for that protein.
18.44 Explain how comparison of the primary structure of a protein from different organisms can be used to deduce evolutionary relationships between them.

The Secondary Structure of Proteins

Foundations

18.45 Define the secondary structure of a protein.
18.46 What are the two most common types of secondary structure?

Applications

18.47 What type of secondary structure is characteristic of:
   a. The α-keratins?
   b. Silk fibroin?
18.48 Describe the forces that maintain the two types of secondary structure: α-helix and β-pleated sheet.
18.49 Define fibrous proteins.
18.50 What is the relationship between the structure of fibrous proteins and their functions?
18.51 Describe a parallel β-pleated sheet.
18.52 Compare a parallel β-pleated sheet to an antiparallel β-pleated sheet.

The Tertiary Structure of Proteins

Foundations
18.53 Define the tertiary structure of a protein.
18.54 Use examples of specific amino acids to show the variety of weak interactions that maintain tertiary protein structure.

Applications
18.55 Write the structure of the amino acid produced by the oxidation of cysteine.
18.56 What is the role of cystine in maintaining protein structure?
18.57 Explain the relationship between the secondary and tertiary protein structures.
18.58 Why is the amino acid proline often found in the random coil hinge regions of the tertiary structure?

The Quaternary Structure of Proteins

Foundations
18.59 Describe the quaternary structure of proteins.
18.60 What weak interactions are responsible for maintaining quaternary protein structure?

Applications
18.61 What is a glycoprotein?
18.62 What is a prosthetic group?

An Overview of Protein Structure and Function

Applications
18.63 Why is hydrogen bonding so important to protein structure?
18.64 Explain why α-keratins that have many disulfide bonds between adjacent polypeptide chains are much less elastic and much harder than those without disulfide bonds.
18.65 How does the structure of the peptide bond make the structure of proteins relatively rigid?
18.66 The primary structure of a protein known as histone H4, which tightly binds DNA, is identical in all mammals and differs by only one amino acid between the calf and pea seedlings. What does this extraordinary conservation of primary structure imply about the importance of that one amino acid?
18.67 What does it mean to say that the structure of proteins is genetically determined?
18.68 Explain why genetic mutations that result in the replacement of one amino acid with another can lead to the formation of a protein that cannot carry out its biological function.

Myoglobin and Hemoglobin

Foundations
18.69 What is the function of hemoglobin?
18.70 What is the function of myoglobin?
18.71 Describe the structure of hemoglobin.
18.72 Describe the structure of myoglobin.
18.73 What is the function of heme in hemoglobin and myoglobin?
18.74 Write an equation representing the binding to and release of oxygen from hemoglobin.

Applications
18.75 Carbon monoxide binds tightly to the heme groups of hemoglobin and myoglobin. How does this affinity reflect the toxicity of carbon monoxide?
18.76 The blood of the horseshoe crab is blue because of the presence of a protein called hemocyanin. What is the function of hemocyanin?
18.77 Why does replacement of glutamic acid with valine alter hemoglobin and ultimately result in sickle cell anemia?
18.78 How do sickled red blood cells hinder circulation?
18.79 What is the difference between sickle cell disease and sickle cell trait?
18.80 How is it possible for sickle cell trait to confer a survival benefit on the person who possesses it?

Denaturation of Proteins

Foundations
18.81 Define the term *denaturation*.
18.82 What is the difference between denaturation and coagulation?

Applications
18.83 Why is heat an effective means of sterilization?
18.84 As you increase the temperature of an enzyme-catalyzed reaction, the rate of the reaction initially increases. It then reaches a maximum rate and finally dramatically declines. Keeping in mind that enzymes are proteins, how do you explain these changes in reaction rate?
18.85 Why is it important that blood have several buffering mechanisms to avoid radical pH changes?
18.86 Define the term *isoelectric*.
18.87 Why do proteins become polycations at extremely low pH?
18.88 Why do proteins become polyanions at very high pH?
18.89 Yogurt is produced from milk by the action of dairy bacteria. These bacteria produce lactic acid as a by-product of their metabolism. The pH decrease causes the milk proteins to coagulate. Why are food preservatives not required to inhibit the growth of bacteria in yogurt?
18.90 Wine is made from the juice of grapes by varieties of yeast. The yeast cells produce ethanol as a by-product of their fermentation. However, when the ethanol concentration reaches 12–13%, all the yeast die. Explain this observation.

Dietary Protein and Protein Digestion

Foundations
18.91 Define the term *essential amino acid*.
18.92 Define the term *nonessential amino acid*.
18.93 Define the term *complete protein*.
18.94 Define the term *incomplete protein*.

Applications
18.95 Write an equation representing the action of the proteolytic enzyme chymotrypsin. (Hint: In order to write the structure of a dipeptide that would be an appropriate reactant, you must consider what is known about where chymotrypsin cleaves a protein chain.)
18.96 Write an equation representing the action of the proteolytic enzyme trypsin. (Hint: In order to write the structure of a dipeptide that would be an appropriate reactant, you must consider what is known about where trypsin cleaves a protein chain.)
18.97 Why is it necessary to mix vegetable proteins to provide an adequate vegetarian diet?
18.98 Name some ethnic foods that apply the principle of mixing vegetable proteins to provide all of the essential amino acids.
18.99 Why must synthesis of digestive enzymes be carefully controlled?
18.100 What is the relationship between pepsin and pepsinogen?

CRITICAL THINKING PROBLEMS

1. Calculate the length of an α-helical polypeptide that is twenty amino acids long. Calculate the length of a region of antiparallel β-pleated sheet that is forty amino acids long.
2. Proteins involved in transport of molecules or ions into or out of cells are found in the membranes of all cells. They are classified as transmembrane proteins because some regions are embedded within the lipid bilayer, whereas other regions protrude into the cytoplasm or outside the cell. Review the classification of amino acids based on the properties of their R groups. What type of amino acids would you expect to find in the regions of the proteins embedded within the membrane? What type of amino acids would you expect to find on the surface of the regions in the cytoplasm or that protrude outside the cell?
3. A biochemist is trying to purify the enzyme hexokinase from a bacterium that normally grows in the Arctic Ocean at 5°C. In the next lab, a graduate student is trying to purify the same protein from a bacterium that grows in the vent of a volcano at 98°C. To maintain the structure of the protein from the Arctic bacterium, the first biochemist must carry out all her purification procedures at refrigerator temperatures. The second biochemist must perform all his experiments in a warm room incubator. In molecular terms, explain why the same kind of enzyme from organisms with different optimal temperatures for growth can have such different thermal properties.
4. The α-keratin of hair is rich in the amino acid cysteine. The location of these cysteines in the protein chain is genetically determined; as a result of the location of the cysteines in the protein, a person may have curly, wavy, or straight hair. How can the location of cysteines in α-keratin result in these different styles of hair? Propose a hypothesis to explain how a "perm" causes straight hair to become curly.
5. Calculate the number of different pentapeptides you can make in which the amino acids phenylalanine, glycine, serine, leucine, and histidine are each found. Imagine how many proteins could be made from the twenty amino acids commonly found in proteins.

BIOCHEMISTRY

19

Enzymes

A hot-spring-fed lake at Yellowstone National Park.

Learning Goals

1. Classify enzymes according to the type of reaction catalyzed and the type of specificity.
2. Give examples of the correlation between an enzyme's common name and its function.
3. Describe the effect that enzymes have on the activation energy of a reaction.
4. Explain the effect of substrate concentration on enzyme-catalyzed reactions.
5. Discuss the role of the active site and the importance of enzyme specificity.
6. Describe the difference between the lock-and-key model and the induced fit model of enzyme-substrate complex formation.
7. Discuss the roles of cofactors and coenzymes in enzyme activity.
8. Explain how pH and temperature affect the rate of an enzyme-catalyzed reaction.
9. Describe the mechanisms used by cells to regulate enzyme activity.
10. Discuss the mechanisms by which certain chemicals inhibit enzyme activity.
11. Discuss the role of the enzyme chymotrypsin and other serine proteases.
12. Provide examples of medical uses of enzymes.

Outline

Chemistry Connection:
Super Hot Enzymes and the Origin of Life

19.1 Nomenclature and Classification
19.2 The Effect of Enzymes on the Activation Energy of a Reaction
19.3 The Effect of Substrate Concentration on Enzyme-Catalyzed Reactions
19.4 The Enzyme-Substrate Complex
19.5 Specificity of the Enzyme-Substrate Complex
19.6 The Transition State and Product Formation

A Medical Perspective:
HIV Protease Inhibitors and Pharmaceutical Drug Design

19.7 Cofactors and Coenzymes
19.8 Environmental Effects

A Medical Perspective:
α_1-Antitrypsin and Familial Emphysema

19.9 Regulation of Enzyme Activity
19.10 Inhibition of Enzyme Activity

A Medical Perspective:
Enzymes, Nerve Transmission, and Nerve Agents

19.11 Proteolytic Enzymes

A Medical Perspective:
Enzymes, Isoenzymes, and Myocardial Infarction

19.12 Uses of Enzymes in Medicine

629

Chemistry Connection

Super Hot Enzymes and the Origin of Life

Imagine the earth about four billion years ago: it was young then, not even a billion years old. Beginning as a red-hot molten sphere, slowly the earth's surface had cooled and become solid rock. But the interior, still extremely hot, erupted through the crust spewing hot gases and lava. Eventually these eruptions produced craggy land masses and an atmosphere composed of gases like hydrogen, carbon dioxide, ammonia, and water vapor. As the water vapor cooled, it condensed into liquid water, forming ponds and shallow seas.

At the dawn of biological life, the surface of the earth was still very hot and covered with rocky peaks and hot shallow oceans. The atmosphere was not very inviting either—filled with noxious gases and containing no molecular oxygen. Yet this is the environment where life on our planet began.

Some scientists think that they have found bacteria—living fossils—that may be very closely related to the first inhabitants of earth. These bacteria thrive at temperatures higher than the boiling point of water. Some need only H_2, CO_2, and H_2O for their metabolic processes and they quickly die in the presence of molecular oxygen.

But this lifestyle raises some uncomfortable questions. For instance, how do these bacteria survive at these extreme temperatures that would cook the life-forms with which we are more familiar? Researcher Mike Adams of the University of Georgia has found some of the answers. Adams and his students have studied the structure of an enzyme, a protein that acts as a biological catalyst, from one of these extraordinary bacteria. He compared the structure of the super hot enzyme with that of the same enzyme purified from an organism that grows at "normal" temperatures. The overall three-dimensional structures of the two enzymes were very similar. This makes sense because they both catalyze the same reaction.

The question, then, is why is the super hot enzyme so stable at very high temperatures, while its low temperature counterpart is not. The answer lay in the tertiary structure of the enzyme. Adams observed that the three-dimensional structure of the super hot enzyme is held together by many more R group interactions than are found in the low-temperature version. These R group interactions, along with other differences, keep the protein stable and functional even at temperatures above 100°C.

In Chapter 18 we studied the structure and properties of proteins. We are now going to apply that knowledge to the study of a group of proteins that do the majority of the work for the cell. These special proteins, the enzymes, catalyze the biochemical reactions that break down food molecules to allow the cell to harvest energy. They also catalyze the biosynthetic reactions that produce the molecules required for cellular life. In this chapter we will study the properties of this extraordinary group of proteins and learn how they dramatically speed up biochemical reactions.

Introduction

The enzymes discussed in this chapter are proteins; however, several ribonucleic acid (RNA) molecules have been demonstrated to have the ability to catalyze biological reactions. These are called *ribozymes*.

An **enzyme** is a biological molecule that serves as a catalyst for a biochemical reaction. The majority of enzymes are proteins. Without enzymes to speed up biochemical reactions, life could not exist. The life of the cell depends on the simultaneous occurrence of hundreds of chemical reactions that must take place rapidly under mild conditions. It is possible, for example, to add water to an alkene. However, this reaction is usually carried out at a temperature of 100°C in aqueous sulfuric acid. Such conditions would kill a cell. The fragile cell must carry out its chemical reactions at body temperature (37°C) and in the absence of any strong acids or bases. How can this be accomplished? In Section 7.3 we saw that catalysts lower the energy of activation of a chemical reaction and thereby increase the rate of the reaction. This allows reactions to occur under milder conditions. The cell uses enzymes to solve the problem of chemical reactions that must occur rapidly under the mild conditions found within the cell. The enzyme facilitates a biochemical reaction, lowering the energy of activation and increasing the rate of the reaction. The efficient functioning of enzymes is essential for the life of the cell and of the organism.

The twin phenomena of high specificity and rapid reaction rates are the cornerstones of enzyme activity and the topic of this chapter. A typical cell contains thousands of different molecules, each of which is important to the chemistry of life processes. Each enzyme "recognizes" only one, or occasionally a few, of these molecules. One of the most remarkable features of enzymes is this specificity. Each can recognize and bind to a single type of substrate or reactant. The molecular size, shape, and charge distribution of both the enzyme and substrate must be compatible for this selective binding process to occur. The enzyme then transforms the substrate into the product with lightning speed. In fact, enzyme-catalyzed reactions often occur from one million to one hundred million times faster than the corresponding uncatalyzed reaction.

The enzyme catalase provides one of the most spectacular examples of the increase in reaction rates brought about by enzymes. This enzyme is required for life in an oxygen-containing environment. In this environment the process of the aerobic (oxygen-requiring) breakdown of food molecules produces hydrogen peroxide (H_2O_2). Because H_2O_2 is toxic to the cell, it must be destroyed. One molecule of catalase converts forty million molecules of hydrogen peroxide to harmless water and oxygen every second:

$$2H_2O_2 \xrightarrow[\text{(an enzyme)}]{\text{Catalase}} 2H_2O + O_2$$

Reaction occurs forty million times every second!

This is the same reaction that you witness when you pour hydrogen peroxide on a wound. The catalase released from injured cells rapidly breaks down the hydrogen peroxide. The bubbles that you see are oxygen gas released as a product of the reaction.

19.1 Nomenclature and Classification

Classification of Enzymes

Enzymes may be classified according to the type of reaction that they catalyze. The six classes are as follows.

Oxidoreductases

Oxidoreductases are enzymes that catalyze oxidation–reduction (redox) reactions. *Lactate dehydrogenase* is an oxidoreductase that removes hydrogen from a molecule of lactate. Other subclasses of the oxidoreductases include oxidases and reductases.

Recall that redox reactions involve electron transfer from one substance to another (Section 8.5).

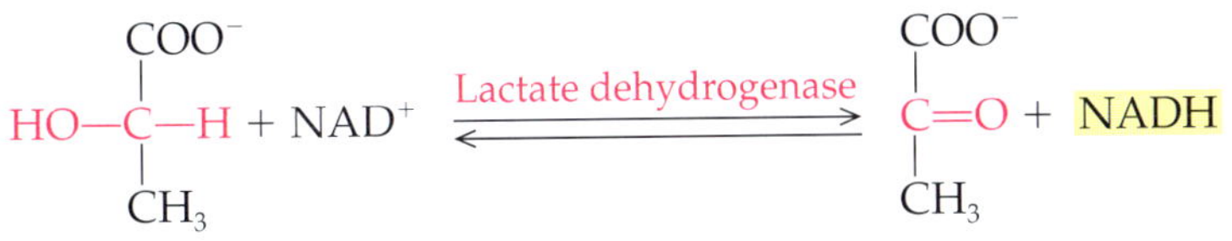

Lactate + NAD⁺ ⇌ (Lactate dehydrogenase) Pyruvate + NADH

Transferases

Transferases are enzymes that catalyze the transfer of functional groups from one molecule to another. For example, a *transaminase* catalyzes the transfer of an amino functional group, and a *kinase* catalyzes the transfer of a phosphate group. Kinases play a major role in energy-harvesting processes involving ATP. In the adrenal glands, norepinephrine is converted to epinephrine by the enzyme *phenylethanolamine-N-methyltransferase* (PNMT), a *transmethylase*.

The significance of phosphate group transfers in energy metabolism is discussed in Sections 21.1 and 21.3.

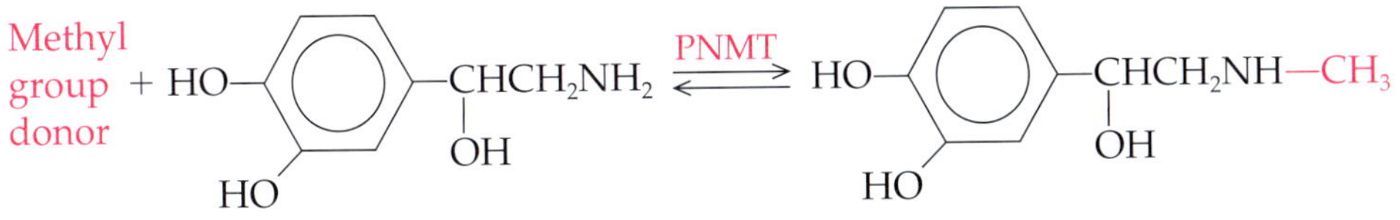

Methyl group donor + Norepinephrine ⇌ (PNMT) Epinephrine

Hydrolases

Hydrolysis of esters is described in Section 14.2. The action of lipases in digestion is discussed in Section 23.1.

Hydrolases catalyze hydrolysis reactions, that is, the addition of a water molecule to a bond resulting in bond breakage. These reactions are important in the digestive process. For example, *lipases* catalyze the hydrolysis of the ester bonds in triglycerides:

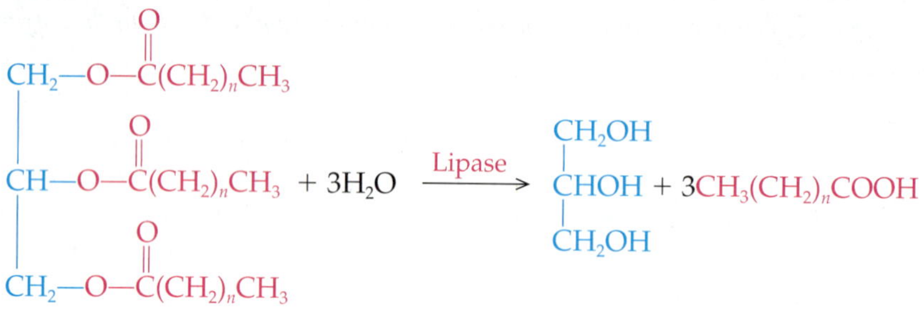

$$\begin{array}{c} CH_2-O-\overset{O}{\underset{\|}{C}}(CH_2)_nCH_3 \\ | \\ CH-O-\overset{O}{\underset{\|}{C}}(CH_2)_nCH_3 \\ | \\ CH_2-O-\overset{O}{\underset{\|}{C}}(CH_2)_nCH_3 \end{array} + 3H_2O \xrightarrow{\text{Lipase}} \begin{array}{c} CH_2OH \\ | \\ CHOH \\ | \\ CH_2OH \end{array} + 3CH_3(CH_2)_nCOOH$$

Triglyceride → Glycerol + Fatty acids

Lyases

The reactions of the citric acid cycle are described in Section 22.4.

Lyases catalyze the addition of a group to a double bond or the removal of a group to form a double bond. *Fumarase* is an example of a lyase. In the citric acid cycle, fumarase catalyzes the addition of a water molecule to the double bond of the substrate fumarate. The product is malate.

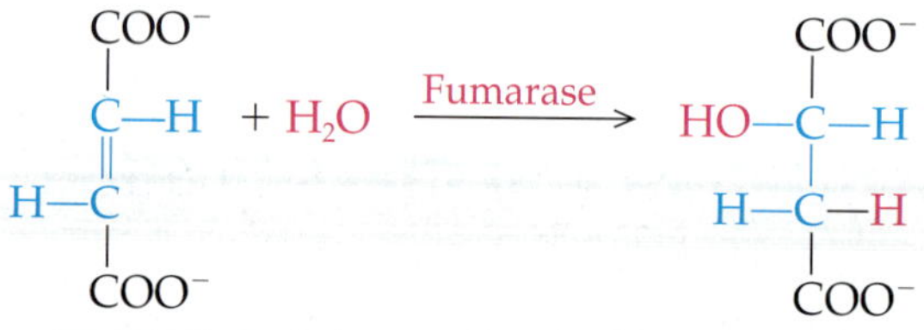

$$\begin{array}{c} COO^- \\ | \\ C-H \\ \| \\ H-C \\ | \\ COO^- \end{array} + H_2O \xrightarrow{\text{Fumarase}} \begin{array}{c} COO^- \\ | \\ HO-C-H \\ | \\ H-C-H \\ | \\ COO^- \end{array}$$

Fumarate → Malate

Citrate lyase catalyzes a far more complicated reaction in which we see the removal of a group and formation of a double bond. Specifically, citrate lyase catalyzes the removal of an acetyl group from a molecule of citrate. The products of this reaction include oxaloacetate, acetyl CoA, ADP, and an inorganic phosphate group (P_i):

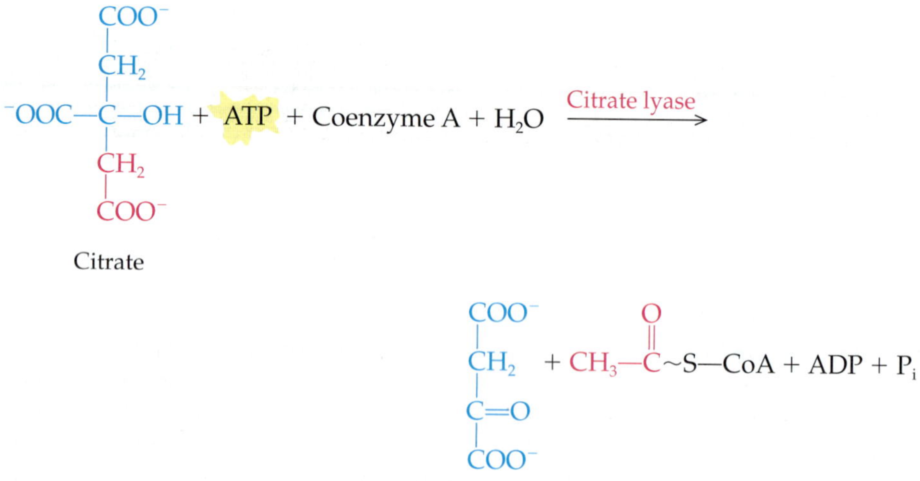

$$\begin{array}{c} COO^- \\ | \\ CH_2 \\ | \\ ^-OOC-C-OH \\ | \\ CH_2 \\ | \\ COO^- \end{array} + ATP + \text{Coenzyme A} + H_2O \xrightarrow{\text{Citrate lyase}}$$

Citrate

$$\begin{array}{c} COO^- \\ | \\ CH_2 \\ | \\ C=O \\ | \\ COO^- \end{array} + CH_3-\overset{O}{\underset{\|}{C}}{\sim}S-CoA + ADP + P_i$$

Oxaloacetate + Acetyl CoA

Recall that the squiggle (~) represents a high-energy bond.

Isomerases

Isomerases rearrange the functional groups within a molecule and catalyze the conversion of one isomer into another. For example, *phosphoglycerate mutase* converts one structural isomer, 3-phosphoglycerate, into another, 2-phosphoglycerate:

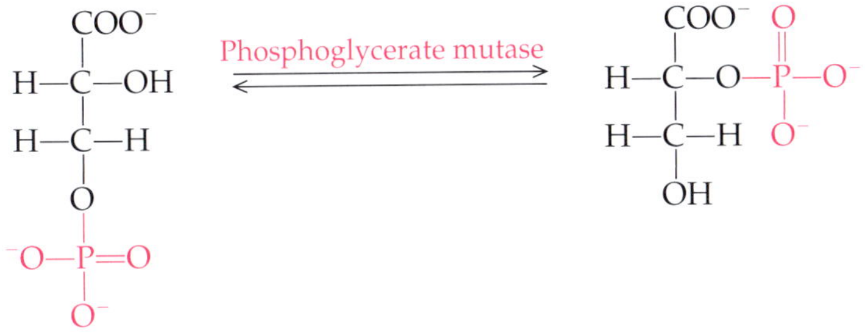

$$\text{3-Phosphoglycerate} \xrightleftharpoons{\text{Phosphoglycerate mutase}} \text{2-Phosphoglycerate}$$

Ligases

Ligases are enzymes that catalyze a reaction in which a C—C, C—S, C—O, or C—N bond is made or broken. This is accompanied by an ATP-ADP interconversion. For example, *DNA ligase* catalyzes the joining of the hydroxyl group of a nucleotide in a DNA strand with the phosphoryl group of the adjacent nucleotide to form a phosphoester bond:

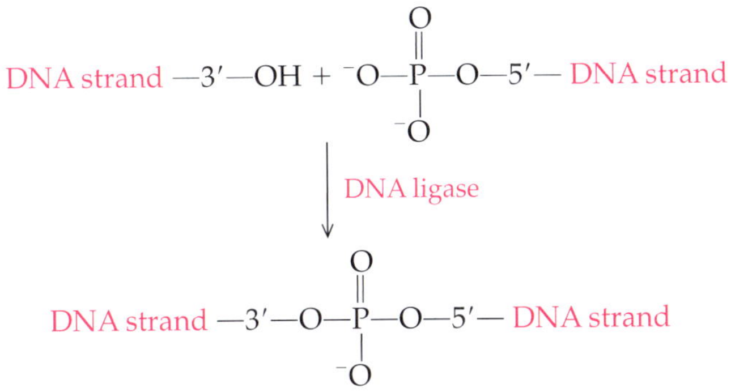

$$\text{DNA strand} - 3' - \text{OH} + {}^-\text{O}-\overset{\overset{\text{O}}{\|}}{\underset{\underset{\text{O}^-}{|}}{\text{P}}}-\text{O}-5'-\text{DNA strand} \xrightarrow{\text{DNA ligase}} \text{DNA strand}-3'-\text{O}-\overset{\overset{\text{O}}{\|}}{\underset{\underset{\text{O}^-}{|}}{\text{P}}}-\text{O}-5'-\text{DNA strand}$$

The use of DNA ligase in recombinant DNA studies is detailed in Section 20.8.

ATP, adenosine triphosphate, is the universal energy currency for all lifeforms we have studied. It is formed by the addition of a phosphoryl group to a molecule of ADP, adenosine diphosphate. Formation of ATP requires energy. Thus, when a ligase breaks a bond, the energy released is used to form ATP. When a ligase forms a bond, energy is required. That energy is provided by the hydrolysis of the terminal phosphoanhydride bond in ATP.

EXAMPLE 19.1 Classifying Enzymes According to the Type of Reaction That They Catalyze

Classify the enzyme that catalyzes each of the following reactions, and explain your reasoning.

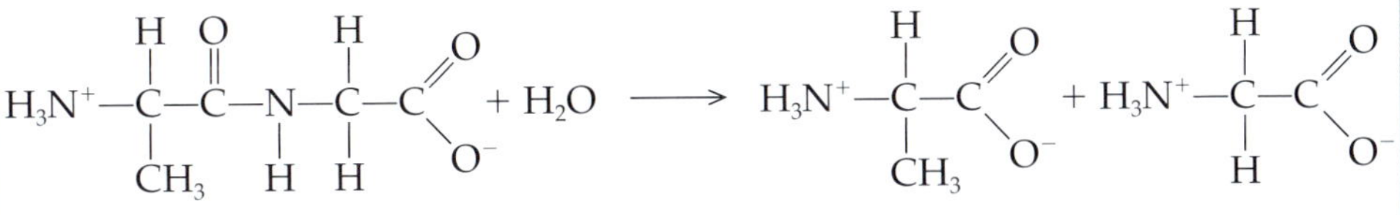

Alanyl-glycine + H₂O ⟶ Alanine + Glycine

Solution

The reaction occurring here involves breaking a bond, in this case a peptide bond, by adding a water molecule. The enzyme is classified as a *hydrolase*, specifically a *peptidase*.

Continued—

EXAMPLE 19.1 —Continued

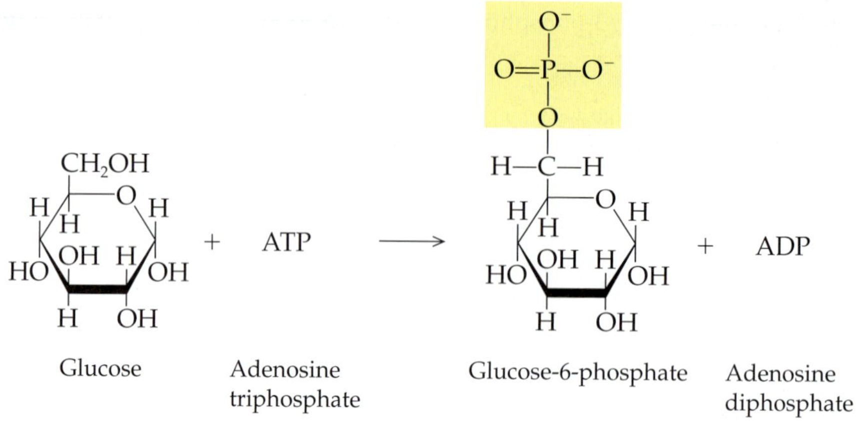

Glucose + ATP → Glucose-6-phosphate + ADP
(Adenosine triphosphate) (Adenosine diphosphate)

Solution

This is the first reaction in the biochemical pathway called *glycolysis*. A phosphoryl group is transferred from a donor molecule, adenosine triphosphate, to the recipient molecule, glucose. The products are glucose-6-phosphate and adenosine diphosphate. This enzyme, called *hexokinase*, is an example of a *transferase*.

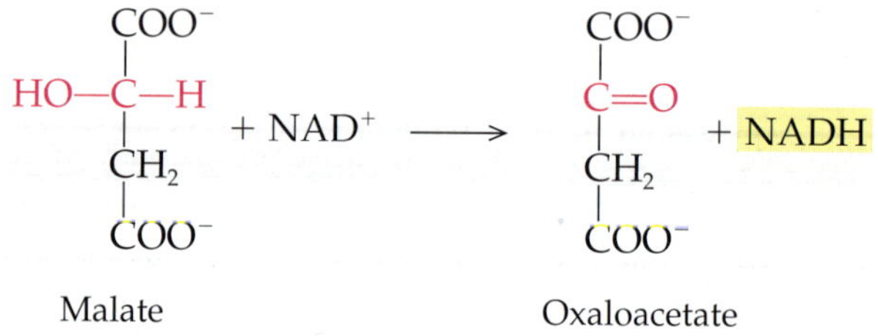

Malate + NAD$^+$ → Oxaloacetate + NADH

Solution

In this reaction, the reactant malate is oxidized and the coenzyme NAD$^+$ is reduced. The enzyme that catalyzes this reaction, *malate dehydrogenase*, is an *oxidoreductase*.

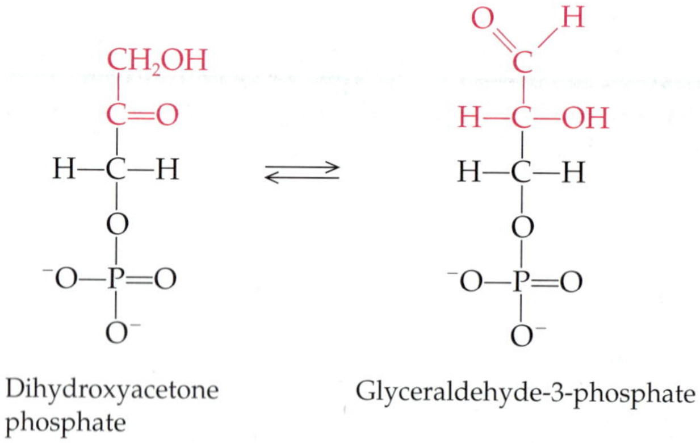

Dihydroxyacetone phosphate ⇌ Glyceraldehyde-3-phosphate

Solution

Careful inspection of the structure of the reactant and the product reveals that they each have the same number of carbon, hydrogen, oxygen, and phosphorus atoms; thus, they must be structural isomers. The enzyme must be an *isomerase*. Its name is *triose phosphate isomerase*.

Question 19.1

To which class of enzymes does each of the following belong?

a. Pyruvate kinase
b. Alanine transaminase
c. Triose phosphate isomerase
d. Pyruvate dehydrogenase
e. Lactase

Question 19.2

To which class of enzymes does each of the following belong?

a. Phosphofructokinase
b. Lipase
c. Acetoacetate decarboxylase
d. Succinate dehydrogenase

Question 19.3

Write an equation representing the reaction catalyzed by each of the enzymes listed in Question 19.1. (Hint: You may need to refer to the index of this book to learn more about the substrates and the reactions that are catalyzed.)

Question 19.4

Write an equation representing the reaction catalyzed by each of the enzymes listed in Question 19.2. (Hint: You may need to refer to the index of this book to learn more about the substrates and the reactions that are catalyzed.)

Nomenclature of Enzymes

 **LEARNING GOAL**

The common names for some enzymes are derived from the name of the **substrate**, the reactant that binds to the enzyme and is converted into product. In many cases, the name of the enzyme is simply derived by adding the suffix -*ase* to the name of the substrate. For instance, *urease* catalyzes the hydrolysis of urea and *lactase* catalyzes the hydrolysis of the disaccharide lactose.

Names of other enzymes reflect the type of reaction that they catalyze. *Dehydrogenases* catalyze the removal of hydrogen atoms from a substrate, while *decarboxylases* catalyze the removal of carboxyl groups. *Hydrogenases* and *carboxylases* carry out the opposite reaction, adding hydrogen atoms or carboxyl groups to their substrates.

Thus, the common name of an enzyme often tells us a great deal about the function of an enzyme. Yet other enzymes have historical names that have no relationship to either the substrates or the reactions that they catalyze. A few examples include catalase, trypsin, pepsin, and chymotrypsin. In these cases, the names of the enzymes and the reactions that they catalyze must simply be memorized.

The systematic names for enzymes tell us the substrate, the type of reaction that is catalyzed, and the name of any coenzyme that is required. For instance, the systematic name of the oxidoreductase lactate dehydrogenase is lactate: NAD oxidoreductase.

Coenzymes are molecules required by some enzymes to serve as donors or acceptors of electrons, hydrogen atoms, or other functional groups during a chemical reaction. Coenzymes are discussed in Section 19.7.

Question 19.5 What is the substrate for each of the following enzymes?

a. Sucrase
b. Pyruvate decarboxylase
c. Succinate dehydrogenase

Question 19.6 What chemical reaction is mediated by each of the enzymes in Question 19.5?

19.2 The Effect of Enzymes on the Activation Energy of a Reaction

LEARNING GOAL 3

How does an enzyme speed up a chemical reaction? It changes the path by which the reaction occurs, providing a lower energy route for the conversion of the substrate into the **product,** the substance that results from the enzyme-catalyzed reaction. Thus enzymes speed up reactions by lowering the activation energy of the reaction.

The activation energy (Section 7.3) of a reaction is the threshold energy that must be overcome to produce a chemical reaction.

Recall that every chemical reaction is characterized by an equilibrium constant. Consider, for example, the simple equilibrium

$$aA \rightleftharpoons bB$$

Equilibrium constants are described in Section 7.4.

The equilibrium constant for this reaction, K_{eq}, is defined as

$$K_{eq} = \frac{[B]^b}{[A]^a} = \frac{[\text{product}]^b}{[\text{reactant}]^a}$$

Energy, rate, and equilibrium are described in Chapter 7.

This equilibrium constant is actually a reflection of the difference in energy between reactants and products. It is a measure of the relative stabilities of the reactants and products. No matter how the chemical reaction occurs (which path it follows), the difference in energy between the reactants and the products is always the same. **For this reason, an enzyme cannot alter the equilibrium constant for the reaction that it catalyzes.** An enzyme does, however, change the path by which the process occurs, providing a lower energy route for the conversion of the substrate into the product. An enzyme increases the rate of a chemical reaction by lowering the activation energy for the reaction (Figure 19.1). An enzyme thus increases the rate at which the reaction it catalyzes reaches equilibrium.

Figure 19.1
Diagram of the difference in energy between the reactants (A and B) and products (C and D) for a reaction. Enzymes cannot change this energy difference but act by lowering the activation energy (E_a) for the reaction, thereby speeding up the reaction.

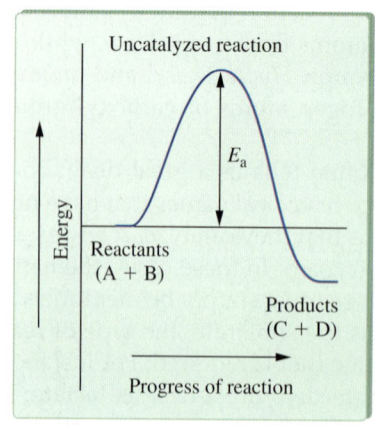
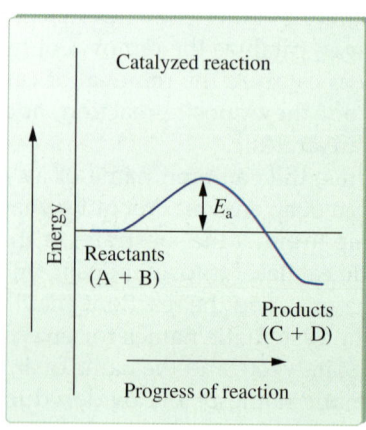

(a)   (b)

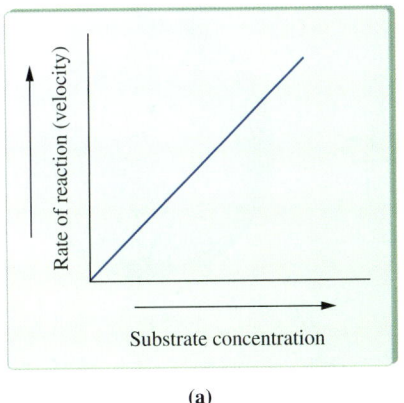
(a)

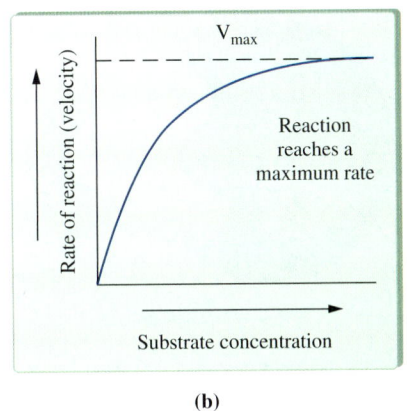
(b)

Figure 19.2
Plot of the rate or velocity, V, of a reaction versus the concentration of substrate, [S], for (a) an uncatalyzed reaction and (b) an enzyme-catalyzed reaction. For an enzyme-catalyzed reaction, the rate is at a maximum when all of the enzyme molecules are bound to the substrate. Beyond this concentration of substrate, further increases in substrate concentration have no effect on the rate of the reaction.

19.3 The Effect of Substrate Concentration on Enzyme-Catalyzed Reactions

The rates of uncatalyzed chemical reactions often double every time the substrate concentration is doubled (Figure 19.2a). Therefore as long as the substrate concentration increases, there is a direct increase in the rate of the reaction. For enzyme-catalyzed reactions, however, this is not the case. Although the rate of the reaction is initially responsive to the substrate concentration, at a certain concentration of substrate the rate of the reaction reaches a maximum value. A graph of the rate of reaction, V, versus the substrate concentration, [S], is shown in Figure 19.2b. We see that the rate of the reaction initially increases rapidly as the substrate concentration is increased but that the rate levels off at a maximum value. At its maximum rate, the active sites of all the enzyme molecules are occupied by a substrate molecule. The active site is the region of the enzyme that specifically binds the substrate and catalyzes the reaction. A new molecule of substrate cannot bind to the enzyme molecule until the substrate molecule already held in the active site is converted to product and released. Thus, it appears that the enzyme-catalyzed reaction occurs in two stages. The first, rapid step is the formation of the *enzyme-substrate complex*. The second step is slower and involves conversion of the substrate to product and the release of the product and enzyme from the resulting enzyme-product complex. It is called the *rate-limiting step* because the rate of the reaction is limited by the speed with which the substrate is converted into product and the product is released. Thus, the reaction rate is dependent on the amount of enzyme available.

4 LEARNING GOAL

19.4 The Enzyme-Substrate Complex

The following series of reversible reactions represents the steps in an enzyme-catalyzed reaction. The first step (highlighted in blue) involves the encounter of the enzyme with its substrate and the formation of an **enzyme-substrate complex**.

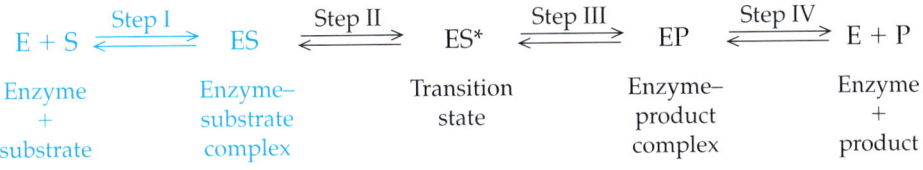

5 LEARNING GOAL

The part of the enzyme that binds with the substrate is called the **active site**. The characteristics of the active site that are crucial to enzyme function include the following:

Figure 19.3
(a) The lock-and-key model of enzyme-substrate binding assumes that the enzyme active site has a rigid structure that is precisely complementary in shape and charge distribution to the substrate. (b) The induced fit model of enzyme-substrate binding. As the enzyme binds to the substrate, the shape of the active site conforms precisely to the shape of the substrate. The shape of the substrate may also change.

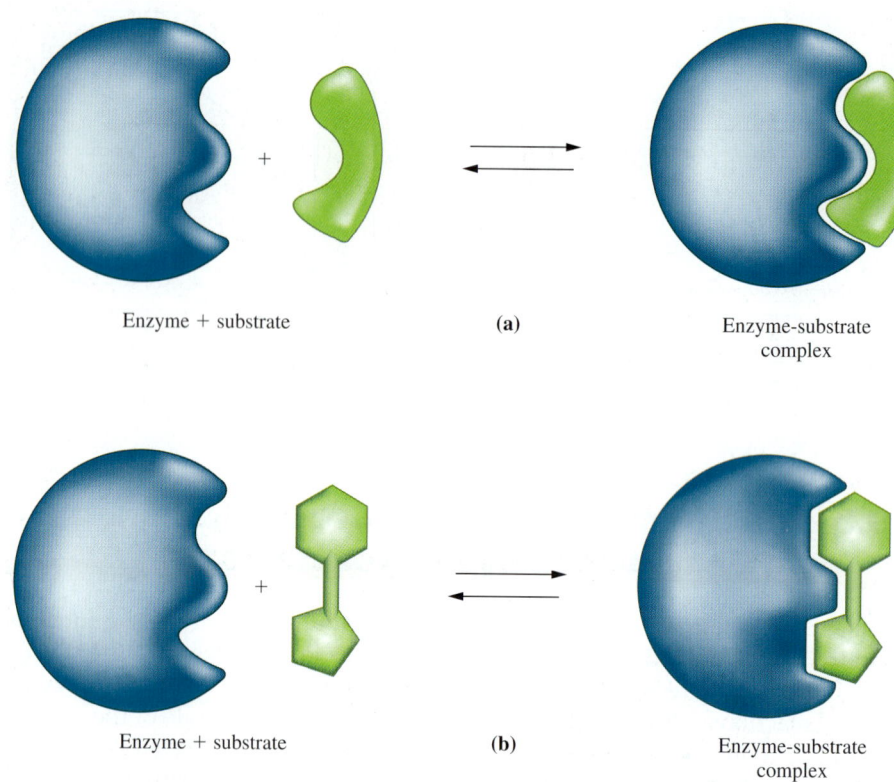

Enzyme + substrate   (a)   Enzyme-substrate complex

Enzyme + substrate   (b)   Enzyme-substrate complex

- Enzyme active sites are pockets or clefts in the surface of the enzyme. The R groups in the active site that are involved in catalysis are called *catalytic groups*.
- The shape of the active site is complementary to the shape of the substrate. That is, the substrate fits neatly into the active site of the enzyme.
- An enzyme attracts and holds its substrate by weak, noncovalent interactions. The R groups involved in substrate binding, and not necessarily catalysis, make up the *binding site*.
- The conformation of the active site determines the specificity of the enzyme because only the substrate that fits into the active site will be used in a reaction.

The **lock-and-key model** of enzyme activity, shown in Figure 19.3a, was devised by Emil Fischer in 1894. At that time it was thought that the substrate simply snapped into place like a piece of a jigsaw puzzle or a key into a lock.

Today we know that proteins are flexible molecules. This led Daniel E. Koshland, Jr., to propose a more sophisticated model of the way enzymes and substrates interact. This model, proposed in 1958, is called the **induced fit model** (Figure 19.3b). In this model, the active site of the enzyme is not a rigid pocket into which the substrate fits precisely; rather, it is a flexible pocket that *approximates* the shape of the substrate. When the substrate enters the pocket, the active site "molds" itself around the substrate. This produces the perfect enzyme-substrate "fit."

The overall shape of a protein is maintained by many weak interactions. At any time a few of these weak interactions may be broken by heat energy or a local change in pH. If only a few bonds are broken, they will re-form very quickly. The overall result is that there is a brief change in the shape of the enzyme. Thus the protein or enzyme can be viewed as a flexible molecule, changing shape slightly in response to minor local changes.

Question 19.7 Compare the lock-and-key and induced fit models of enzyme-substrate binding.

Question 19.8 What is the relationship between an enzyme active site and its substrate?

19.5 Specificity of the Enzyme-Substrate Complex

For an enzyme-substrate interaction to occur, the surfaces of the enzyme and substrate must be complementary. It is this requirement for a specific fit that determines whether an enzyme will bind to a particular substrate and carry out a chemical reaction.

Enzyme specificity is the ability of an enzyme to bind only one, or a very few, substrates and thus catalyze only a single reaction. To illustrate the specificity of enzymes, consider the following reactions.

The enzyme urease catalyzes the hydrolysis of urea to carbon dioxide and ammonia as follows:

$$H_2N-\underset{\text{Urea}}{\overset{\overset{O}{\|}}{C}}-NH_2 + H_2O \xrightarrow{\text{Urease}} CO_2 + 2NH_3$$

Methylurea, in contrast, though structurally similar to urea, is not affected by urease:

$$H_2N-\underset{\text{Methylurea}}{\overset{\overset{O}{\|}}{C}}-NHCH_3 + H_2O \xrightarrow{\text{Urease}} \text{no reaction}$$

Not all enzymes exhibit the same degree of specificity. Four classes of enzyme specificity have been observed.

- **Absolute specificity:** An enzyme that catalyzes the reaction of only one substrate has absolute specificity. Aminoacyl tRNA synthetases exhibit absolute specificity. Each must attach the correct amino acid to the correct transfer RNA molecule. If the wrong amino acid is attached to the transfer RNA, it may be mistakenly added to a peptide chain, producing a nonfunctional protein.
- **Group specificity:** An enzyme that catalyzes processes involving similar molecules containing the same functional group has group specificity. Hexokinase is a group-specific enzyme that catalyzes the addition of a phosphoryl group to the hexose sugar glucose in the first step of glycolysis. Hexokinase can also add a phosphoryl group to several other six-carbon sugars.
- **Linkage specificity:** An enzyme that catalyzes the formation or breakage of only certain bonds in a molecule has linkage specificity. Proteases, such as trypsin, chymotrypsin, and elastase, are enzymes that selectively hydrolyze peptide bonds. Thus, these enzymes are linkage specific.
- **Stereochemical specificity:** An enzyme that can distinguish one enantiomer from the other has stereochemical specificity. Most of the enzymes of the human body show stereochemical specificity. Because we use only D-sugars and L-amino acids, the enzymes involved in digestion and metabolism recognize only those particular stereoisomers.

This reaction is one that you may have observed if you have a cat. Urea is a waste product of the breakdown of proteins and is removed from the body in urine. Bacteria in kitty litter produce urease. As the urease breaks down the urea in the cat urine, the ammonia released produces the distinctive odor of an untended litter box.

Aminoacyl tRNA synthetases are discussed in Section 20.6. Aminoacyl group transfer reactions were described in Section 15.4.

Hexokinase activity is described in Section 21.3.

Proteolytic enzymes are discussed in Section 19.11.

19.6 The Transition State and Product Formation

How does enzyme-substrate binding result in a faster chemical reaction? The precise answer to this question is probably different for each enzyme-substrate pair, and, indeed, we understand the exact mechanism of catalysis for very few enzymes. Nonetheless, we can look at the general features of enzyme-substrate interactions that result in enhanced reaction rate and product formation. To do this, we must once again look at the steps of an enzyme-catalyzed reaction, focusing on the steps highlighted in blue:

$$E + S \underset{}{\overset{\text{Step I}}{\rightleftharpoons}} ES \underset{}{\overset{\text{Step II}}{\rightleftharpoons}} ES^* \underset{}{\overset{\text{Step III}}{\rightleftharpoons}} EP \underset{}{\overset{\text{Step IV}}{\rightleftharpoons}} E + P$$

Enzyme + substrate | Enzyme–substrate complex | Transition state | Enzyme–product complex | Enzyme + product

In Section 19.4 we examined the events of step I by which the enzyme and substrate interact to form the enzyme-substrate complex. In this section, we will look at the events that lead to product formation. We have already described the enzyme as a flexible molecule; the substrate also has a degree of flexibility. The continued interaction between the enzyme and substrate changes the shape or position of the substrate in such a way that the molecular configuration is no longer energetically stable (step II). In this state, the **transition state,** the shape of the substrate is altered, because of its interaction with the enzyme, into an intermediate form having features of both the substrate and the final product. This transition state, in turn, favors the conversion of the substrate into product (step III). The product remains bound to the enzyme for a very brief time, then in step IV the product and enzyme dissociate from one another, leaving the enzyme completely unchanged.

What kinds of transition state changes might occur in the substrate that would make a reaction proceed more rapidly?

1. The enzyme might put "stress" on a bond and thereby facilitate bond breakage. Consider the hydrolysis of the sugar sucrose by the enzyme sucrase. The enzyme catalyzes the hydrolysis of the disaccharide sucrose into the monosaccharides glucose and fructose. The formation of the enzyme-substrate complex (Figures 19.4a and 19.4b) results in a change in the shape of the enzyme. This, in turn, may stretch or distort one of the bonds of the substrate. Such a stress weakens the bond, allowing it to be broken much more easily than in the absence of the enzyme. This is represented in Figure 19.4c as the bending of the glycosidic bond between the fructose and the glucose. In the transition state the substrate has a molecular form resembling both the disaccharide, the original substrate, and the two monosaccharides, the eventual products. Clearly, the stress placed on the bond weakens it, and much less energy is required to break the bond to form products (Figures 19.4d and 19.4e). This has the effect of speeding up the reaction.

2. An enzyme may facilitate a reaction by bringing two reactants into close proximity and in the proper orientation for reaction to occur. Consider now the condensation reaction between glucose and fructose to produce sucrose (Figure 19.5a). Each of the sugars has five hydroxyl groups that could undergo condensation to produce a disaccharide. But the purpose is to produce sucrose, not some other disaccharide. By random molecular collision there is a one in twenty-five chance that the two molecules will collide in the proper orientation to produce sucrose. The probability that the two will react is actually much less than that because of a variety of conditions in addition to orientation that must be satisfied for the reaction to occur. For example, at body temperature, most molecular collisions will not have a sufficient amount of energy to overcome the energy of activation, even if the molecules are in the proper orientation. The enzyme can facilitate the reaction by bringing the two molecules close together in the correct alignment (Figure 19.5b), thereby forcing the desired reactive groups of the two molecules together in the transition state and greatly speeding up the reaction.

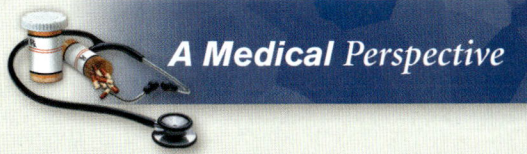

A Medical Perspective

HIV Protease Inhibitors and Pharmaceutical Drug Design

In 1981 the Centers for Disease Control in Atlanta, Georgia, recognized a new disease syndrome, acquired immune deficiency syndrome (AIDS). The syndrome is characterized by an impaired immune system, a variety of opportunistic infections and cancer, and brain damage that results in dementia. It soon became apparent that the disease was being transmitted by blood and blood products, as well as by sexual conduct.

The earliest drugs that proved effective in the treatment of HIV infections all inhibited replication of the genetic material of the virus. While these treatments were initially effective, prolonging the lives of many, it was not long before viral mutants resistant to these drugs began to appear. Clearly, a new approach was needed.

In 1989 a group of scientists revealed the three-dimensional structure of the HIV protease. This structure is shown in the accompanying figure. This enzyme is necessary for viral replication because the virus has an unusual strategy for making all of its proteins. Rather than make each protein individually, it makes large "polyproteins" that must then be cut by the HIV protease to form the final proteins required for viral replication.

Since scientists realized that this enzyme was essential for HIV replication, they decided to engineer a substance that would inhibit the enzyme by binding irreversibly to the active site, in essence plugging it up. The challenge, then, was to design a molecule that would be the plug. Researchers knew the primary structure (amino acid sequence) of the HIV protease from earlier nucleic acid sequencing studies. By 1989 they also had a very complete picture of the three-dimensional nature of the molecule, which they had obtained by X-ray crystallography.

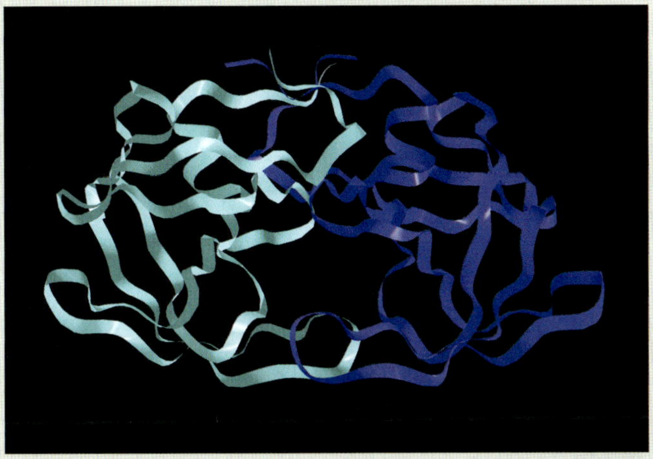

Putting all of this information into a sophisticated computer modeling program, they could look at the protease from any angle. They could see the location of each of the R groups of each of the amino acids in the active site. This kind of information allowed the scientists to design molecules that would be complementary to the shape and charge distribution of the enzyme active site—in other words, structural analogs of the normal protease substrate. It was not long before the scientists had produced several candidates for the HIV protease inhibitor.

But, there are many tests that a drug candidate must pass before it can be introduced into the market as safe and effective. Scientists had to show that the candidate drugs would bind effectively to the HIV protease and block its function, thereby inhibiting virus replication. Properties such as the solubility, the efficiency of absorption by the body, the period of activity in the body, and the toxicity of the drug candidates all had to be determined.

By 1996 there were three protease inhibitors available to combat HIV infection. There are currently seven of these drugs on the market. In many cases development and testing of a drug candidate can take up to fifteen years. In the case of the first HIV protease inhibitors, the first three drugs were on the market in less than eight years. This is a testament both to the urgent need for HIV treatments and to the technology available to attack the problem.

The human immunodeficiency virus protease.

For Further Understanding

What particular concerns would you have, as a medical researcher, about administering a drug that is a protease inhibitor?

Often a protease inhibitor is prescribed along with an inhibitor of replication of the viral genetic material. Develop a hypothesis to explain this strategy.

3. The active site of an enzyme may modify the pH of the microenvironment surrounding the substrate. To accomplish this, the enzyme may, for example, serve as a donor or an acceptor of H^+. As a result, there would be a change in the pH in the vicinity of the substrate without disturbing the normal pH elsewhere in the cell.

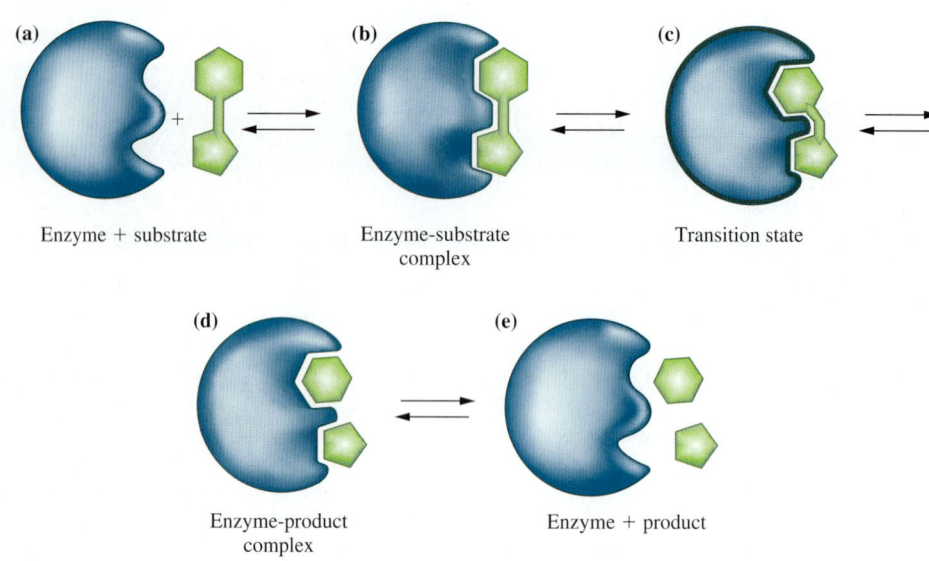

Figure 19.4
Bond breakage is facilitated by the enzyme as a result of stress on a bond. (a, b) The enzyme-substrate complex is formed. (c) In the transition state, the enzyme changes shape and thereby puts stress on the glycosidic linkage holding the two monosaccharides together. This lowers the energy of activation of this reaction. (d, e) The bond is broken, and the products are released.

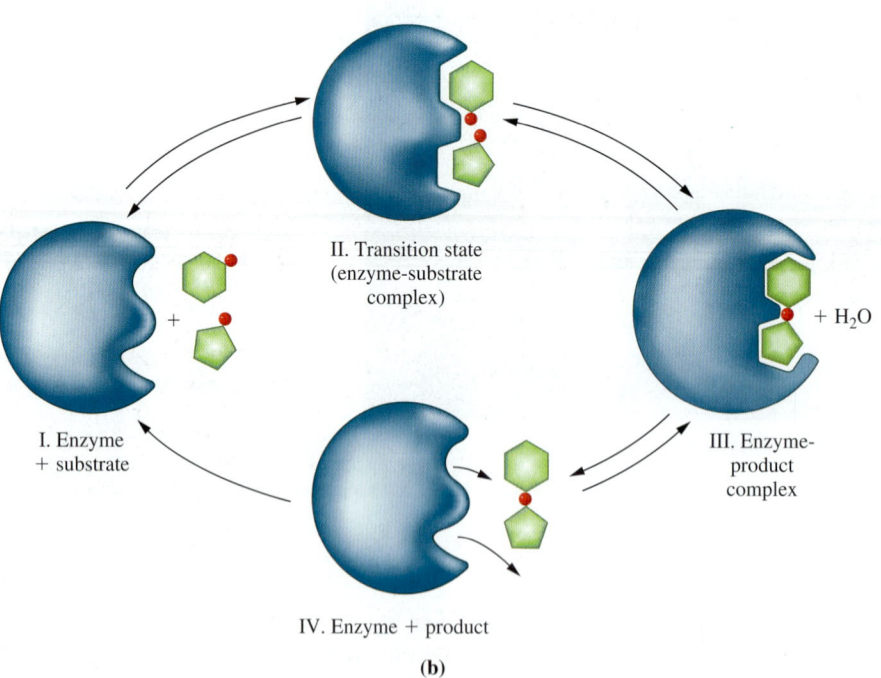

Figure 19.5
An enzyme may lower the energy of activation required for a reaction by holding the substrates in close proximity and in the correct orientation. (a) A condensation reaction in which glucose and fructose are joined in glycosidic linkage to produce sucrose. (b) The enzyme-substrate complex forms, bringing the two monosaccharides together with the correct hydroxyl groups extended toward one another.

> **Question 19.9**
> Summarize three ways in which an enzyme might lower the energy of activation of a reaction.

> **Question 19.10**
> What is the transition state in an enzyme-catalyzed reaction?

19.7 Cofactors and Coenzymes

In Section 18.7 we saw that some proteins require an additional nonprotein prosthetic group to function. The same is true of some enzymes. The polypeptide portion of such an enzyme is called the **apoenzyme,** and the nonprotein prosthetic group is called the **cofactor.** Together they form the active enzyme called the **holoenzyme.** Cofactors may be metal ions, organic compounds, or organometallic compounds. They must be bound to the enzyme to maintain the correct configuration of the enzyme active site (Figure 19.6). When the cofactor is bound and the active site is in the proper conformation, the enzyme can bind the substrate and catalyze the reaction.

Other enzymes require the temporary binding of a **coenzyme.** Such binding is generally mediated by weak interactions like hydrogen bonds. The coenzymes are organic molecules that generally serve as carriers of electrons or chemical groups. In chemical reactions, they may either donate groups to the substrate or serve as recipients of groups that are removed from the substrate. The example in Figure 19.7 shows a coenzyme accepting a functional group from one substrate and donating it to the second substrate in a reaction catalyzed by a transferase.

Often coenzymes contain modified vitamins as part of their structure. A **vitamin** is an organic substance that is required in the diet in only small amounts. Of the water-soluble vitamins, only vitamin C has not been associated with a coenzyme. Table 19.1 is a summary of some coenzymes and the water-soluble vitamins from which they are made.

7 LEARNING GOAL

Water soluble vitamins are discussed in greater detail online at www.mhhe.com/denniston5e in Water-Soluble Vitamins.

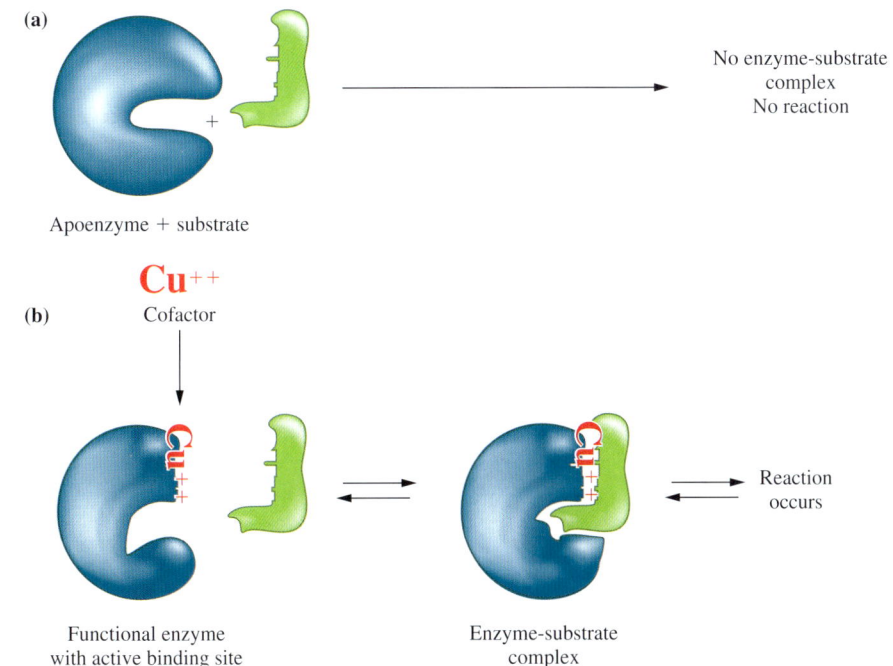

Figure 19.6
(a) The apoenzyme is unable to bind to its substrate. (b) When the required cofactor, in this case a copper ion, Cu^{2+}, is available, it binds to the apoenzyme. Now the active site takes on the correct configuration, the enzyme-substrate complex forms, and the reaction occurs.

Figure 19.7
Some enzymes require a coenzyme to facilitate the reaction.

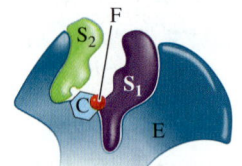

1. An enzyme with a coenzyme positioned to react with two substrates.

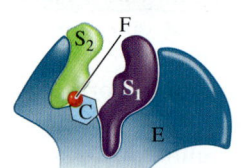

2. Coenzyme picks up a functional group from substrate 1.

3. Coenzyme transfers the functional group to substrate 2.

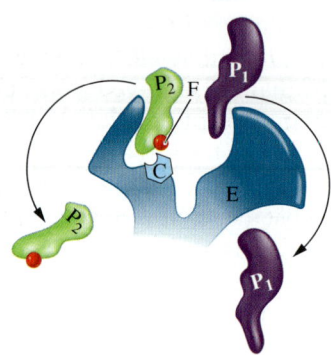

4. Products are released from enzyme.

TABLE 19.1 The Water-Soluble Vitamins and their Coenzymes

Vitamin	Coenzyme	Function
Thiamine (B_1)	Thiamine pyrophosphate	Decarboxylation reactions
Riboflavin (B_2)	Flavin mononucleotide (FMN)	Carrier of H atoms
	Flavin adenine dinucleotide (FAD)	
Niacin (B_3)	Nicotinamide adenine dinucleotide (NAD^+)	Carrier of hydride ions
	Nicotinamide adenine dinucleotide phosphate ($NADP^+$)	
Pyridoxine (B_6)	Pyridoxal phosphate	Carriers of amino and carboxyl groups
	Pyridoxamine phosphate	
Cyanocobalamin (B_{12})	Deoxyadenosyl cobalamin	Coenzyme in amino acid metabolism
Folic acid	Tetrahydrofolic acid	Coenzyme for 1-C transfer
Pantothenic acid	Coenzyme A	Acyl group carrier
Biotin	Biocytin	Coenzyme in CO_2 fixation
Ascorbic acid	Unknown	Hydroxylation of proline and lysine in collagen

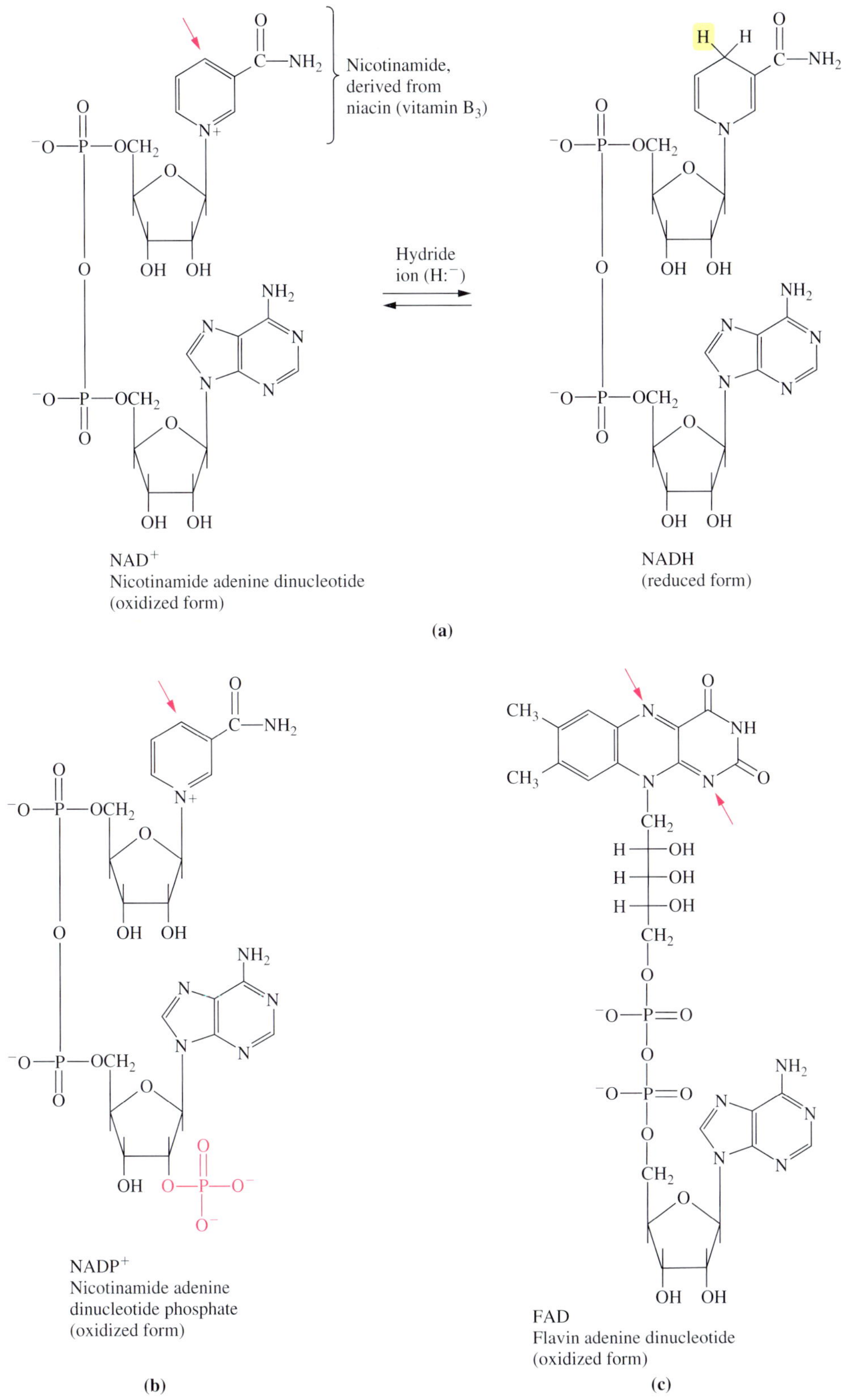

19.7 Cofactors and Coenzymes

Figure 19.8

The structure of three coenzymes. (a) The oxidized and reduced forms of nicotinamide adenine dinucleotide. (b) The oxidized form of the closely related hydride ion carrier, nicotinamide adenine dinucleotide phosphate (NADP$^+$), which accepts hydride ions at the same position as NAD$^+$ (colored arrow). (c) The oxidized form of flavin adenine dinucleotide (FAD) accepts hydrogen atoms at the positions indicated by the colored arrows.

Nicotinamide adenine dinucleotide (NAD$^+$), shown in Figure 19.8, is an example of a coenzyme that is of critical importance in the oxidation reactions of the cellular energy-harvesting processes. The NAD$^+$ molecule can accept a hydride ion, a hydrogen atom with two electrons, from the substrate of these reactions. The substrate is oxidized, and the portion of NAD$^+$ that is derived from the vitamin *niacin* is reduced to produce NADH. The NADH subsequently yields the hydride ion to the first acceptor in an electron transport chain. This regenerates the NAD$^+$ and provides electrons for the generation of ATP, the chemical energy required by the cell. Also shown in Figure 19.8 is the hydride carrier NADP$^+$ and the hydrogen atom carrier FAD. Both are used in the oxidation-reduction reactions that harvest energy for the cell. Unlike NADH and FADH$_2$, NADPH serves as "reducing power" for the cell by donating hydride ions in biochemical reactions. Like NAD$^+$, NADP$^+$ is derived from niacin. FAD is made from the vitamin *riboflavin*.

Question 19.11

Why does the body require the water-soluble vitamins?

Question 19.12

What are the coenzymes formed from each of the following vitamins? What are the functions of each of these coenzymes?

a. Pantothenic acid
b. Niacin
c. Riboflavin

19.8 Environmental Effects

Effect of pH

LEARNING GOAL 8

Most enzymes are active only within a very narrow pH range. The cellular cytoplasm has a pH of 7, and most cytoplasmic enzymes function at a maximum efficiency at this pH. A plot of the relative rate at which a typical cytoplasmic enzyme catalyzes its specific reaction versus pH is provided in Figure 19.9.

The pH at which an enzyme functions optimally is called the **pH optimum**. Making the solution more basic or more acidic sharply decreases the rate of the reaction. As was discussed in Section 18.10, at extremes of pH the enzyme actually loses its biologically active conformation and is *denatured*. This is because pH changes alter the degree of ionization of the R groups of the amino acids within the protein chain, as well as the extent to which they can hydrogen bond. Less drastic changes in the R groups of an enzyme active site can also destroy the ability to form the enzyme-substrate complex.

Although the cytoplasm of the cell and the fluids that bathe the cells have a pH that is carefully controlled so that it remains at about pH 7, there are environments within the body in which enzymes must function at a pH far from 7. Protein sequences have evolved that can maintain the proper three-dimensional structure under extreme conditions of pH. For instance, the pH of the stomach is approximately 2 as a result of the secretion of hydrochloric acid by specialized cells of the stomach lining. The proteolytic digestive enzyme *pepsin* must effectively degrade proteins at this extreme pH. In the case of pepsin the enzyme has evolved in such a way that it can maintain a stable tertiary structure at a pH of 2 and is catalytically most active in the hydrolysis of peptides that have been denatured by very low pH. Thus pepsin has a pH optimum of 2.

In a similar fashion, another proteolytic enzyme, *trypsin*, functions under the conditions of higher pH found in the intestine. Both pepsin and trypsin cleave peptide bonds by virtually identical mechanisms, yet their amino acid sequences have evolved so that they are stable and active in very different environments.

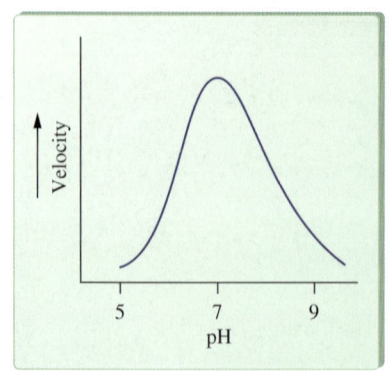

Figure 19.9
Effect of pH on the rate of an enzyme-catalyzed reaction. This enzyme functions most efficiently at pH 7. The rate of the reaction falls rapidly as the solution is made either more acidic or more basic.

The body has used adaptation of enzymes to different environments to protect itself against one of its own destructive defense mechanisms. Within the cytoplasm of a cell are organelles called *lysosomes*. Christian de Duve, who discovered lyosomes in 1956, called them "suicide bags" because they are membrane-bound vesicles containing about fifty different kinds of hydrolases which degrade large biological molecules into small molecules that are useful for energy-harvesting reactions. For instance, some of the enzymes in the lysosomes can degrade proteins to amino acids, and others hydrolyze polysaccharides into monosaccharides. Certain cells of the immune defense system engulf foreign invaders, such as bacteria and viruses. They then use the hydrolytic enzymes in the lysosomes to degrade and destroy the invaders and use the simple sugars, amino acids, and lipids that are produced as energy sources.

What would happen if the hydrolytic enzymes of the lysosome were accidentally released into the cytoplasm of the cell? Certainly, the result would be the destruction of cellular macromolecules and death of the cell. Because of this danger, the cell invests a great deal of energy in maintaining the integrity of the lysosomal membranes. An additional protective mechanism relies on the fact that lysosomal enzymes function optimally at an acid pH (pH 4.8). Should some of these enzymes leak out of the lysosome or should a lysosome accidentally rupture, the cytoplasmic pH of 7.0–7.3 renders them inactive.

Effect of Temperature

Enzymes are rapidly destroyed if the temperature of the solution rises much above 37°C, but they remain stable at much lower temperatures. It is for this reason that solutions of enzymes used for clinical assays are stored in refrigerators or freezers before use. Figure 19.10 shows the effects of temperature on enzyme-catalyzed and uncatalyzed reactions. The rate of the uncatalyzed reaction steadily increases with increasing temperature because more collisions occur with sufficient energy to overcome the energy barrier for the reaction. The rate of an enzyme-catalyzed reaction also increases with modest increases in temperature because there are increasing numbers of collisions between the enzyme and the substrate. At the **temperature optimum,** the enzyme is functioning optimally and the rate of the reaction is maximal. Above the temperature optimum, increasing temperature begins to increase the vibrational energy of the bonds within the enzyme. Eventually, so many bonds and weak interactions are disrupted that the enzyme becomes denatured, and the reaction stops.

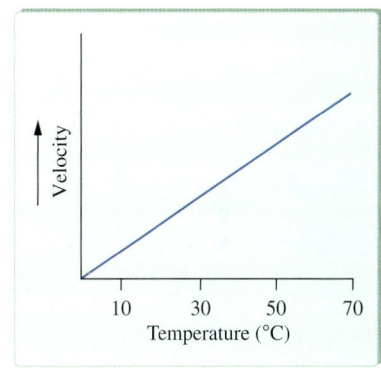

(a)

Because heating enzymes and other proteins destroys their characteristic three-dimensional structure, and hence their activity, a cell cannot survive very high temperatures. Thus, heat is an effective means of sterilizing medical instruments and solutions for transfusion or clinical tests. Although instruments can be sterilized by dry heat (160°C) applied for at least two hours in a dry air oven, autoclaving is a quicker, more reliable procedure. The autoclave works on the principle of the pressure cooker. Air is pumped out of the chamber, and steam under pressure is pumped into the chamber until a pressure of two atmospheres is achieved. The pressure causes the temperature of the steam, which would be 100°C at atmospheric pressure, to rise to 121°C. Within twenty minutes, all the bacteria and viruses are killed. This is the most effective means of destroying the very heat-resistant endospores that are formed by many bacteria of clinical interest. These bacteria include the genera *Bacillus* and *Clostridium,* which are responsible for such unpleasant and deadly diseases as anthrax, gas gangrene, tetanus, and botulism food poisoning.

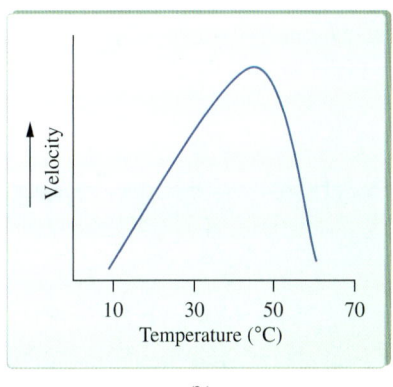

(b)

Figure 19.10
Effect of temperature on (a) uncatalyzed reactions and (b) enzyme-catalyzed reactions.

However, not all enzymes are inactivated by heating, even to rather high temperatures. Certain bacteria live in such out-of-the-way places as coal slag heaps, which are actually burning. Others live in deep vents on the ocean floor where temperatures and pressures are extremely high. Still others grow in the

A Medical Perspective

α_1-Antitrypsin and Familial Emphysema

Nearly two million people in the United States suffer from emphysema. Emphysema is a respiratory disease caused by destruction of the alveoli, the tiny, elastic air sacs of the lung. This damage results from the irreversible destruction of a protein called elastin, which is needed for the strength and flexibility of the walls of the alveoli. When elastin is destroyed, the small air passages in the lungs, called bronchioles, become narrower or may even collapse. This severely limits the flow of air into and out of the lung, causing respiratory distress, and in extreme conditions, death.

Some people have a genetic predisposition to emphysema. This is called familial emphysema. These individuals have a genetic defect in the gene that encodes the human plasma protein α_1-antitrypsin. As the name suggests, α_1-antitrypsin is an inhibitor of the proteolytic enzyme trypsin. But, as we have seen in this chapter, trypsin is just one member of a large family of proteolytic enzymes called the serine proteases. In the case of the α_1-antitrypsin activity in the lung, it is the inhibition of the enzyme elastase that is the critical event.

Elastase damages or destroys elastin, which in turn promotes the development of emphysema. People with normal levels of α_1-antitrypsin are protected from familial emphysema because their α_1-antitrypsin inhibits elastase and, thus, protects the elastin. The result is healthy alveoli in the lungs. However, individuals with a genetic predisposition to emphysema have very low levels of α_1-antitrypsin. This is due to a mutation that causes a single amino acid substitution in the protein chain. Because elastase in the lungs is not effectively controlled, severe lung damage characteristic of emphysema occurs.

Emphysema is also caused by cigarette smoking. Is there a link between these two forms of emphysema? The answer is yes; research has revealed that components of cigarette smoke cause the oxidation of a methionine near the amino terminus of α_1-antitrypsin. This chemical damage destroys α_1-antitrypsin activity. There are enzymes in the lung that reduce the methionine, converting it back to its original chemical form and restoring α_1-antitrypsin activity. However, it is obvious that over a long period, smoking seriously reduces the level of α_1-antitrypsin activity. The accumulated lung damage results in emphysema in many chronic smokers.

At the current time the standard treatment of emphysema is the use of inhaled oxygen. Studies have shown that intravenous infusion of α_1-antitrypsin isolated from human blood is both safe and effective. However, the level of α_1-antitrypsin in the blood must be maintained by repeated administration.

The α_1-antitrypsin gene has been cloned. In experiments with sheep it was shown that the protein remains stable when administered as an aerosol. It is still functional after it has passed through the pulmonary epithelium. This research offers hope of an effective treatment for this frightful disease.

For Further Understanding

Draw the structure of methionine and write an equation showing the reversible oxidation of this amino acid.

Develop a hypothesis to explain why an excess of elastase causes emphysema. What is the role of elastase in this disease?

See also the Chemistry Connection: Super Hot Enzymes and the Origin of Life at the beginning of this chapter.

hot springs of Yellowstone National Park, where they thrive at temperatures near the boiling point of water. These organisms, along with their enzymes, survive under such incredible conditions because the amino acid sequences of their proteins dictate structures that are stable at such seemingly impossible temperature extremes.

Question 19.13

How does a decrease in pH alter the activity of an enzyme?

Question 19.14

Heating is an effective mechanism for killing bacteria on surgical instruments. How does elevated temperature result in cellular death?

19.9 Regulation of Enzyme Activity

One of the major ways in which enzymes differ from nonbiological catalysts is that the activity of the enzyme is often regulated by the cell. There are many reasons for this control of enzyme function. Some involve energy considerations. If the cell runs out of chemical energy, it will die; therefore many mechanisms exist to conserve cellular energy. For instance, it is a great waste of energy to produce an enzyme if the substrate is not available. Similarly, if the product of an enzyme-catalyzed reaction is present in excess, it is a waste of energy for the enzyme to continue to catalyze the reaction, thereby producing more of the unwanted product.

Just as there are many reasons for regulation of enzyme activity, there are many mechanisms for such regulation. The simplest mechanism is to produce the enzyme only when the substrate is present. This mechanism is used by bacteria to regulate the enzymes needed to break down various sugars to yield ATP for cellular work. The bacteria have no control over their environment or over what food sources, if any, might be available. It would be an enormous waste of energy to produce all of the enzymes that are needed to break down all the possible sugars. Thus the bacteria save energy by producing the enzymes only when a specific sugar substrate is available. Other mechanisms for regulating enzyme activity include use of allosteric enzymes, feedback inhibition, production of proenzymes, and protein modification. Let's take a look at these regulatory mechanisms in some detail.

Allosteric Enzymes

One type of enzyme regulation involves enzymes that have more than a single binding site. These enzymes, called **allosteric enzymes,** meaning "other forms," are enzymes whose active sites can be altered by binding of small molecules called *effector molecules*. As shown in Figure 19.11, the effector binding to its binding site alters the shape of the active site of the enzyme. The result can be to convert the active site to an inactive configuration, **negative allosterism,** or to convert the active site to the active configuration, **positive allosterism.** In either case, binding of the effector molecule regulates enzyme activity by determining whether it will be active or inactive.

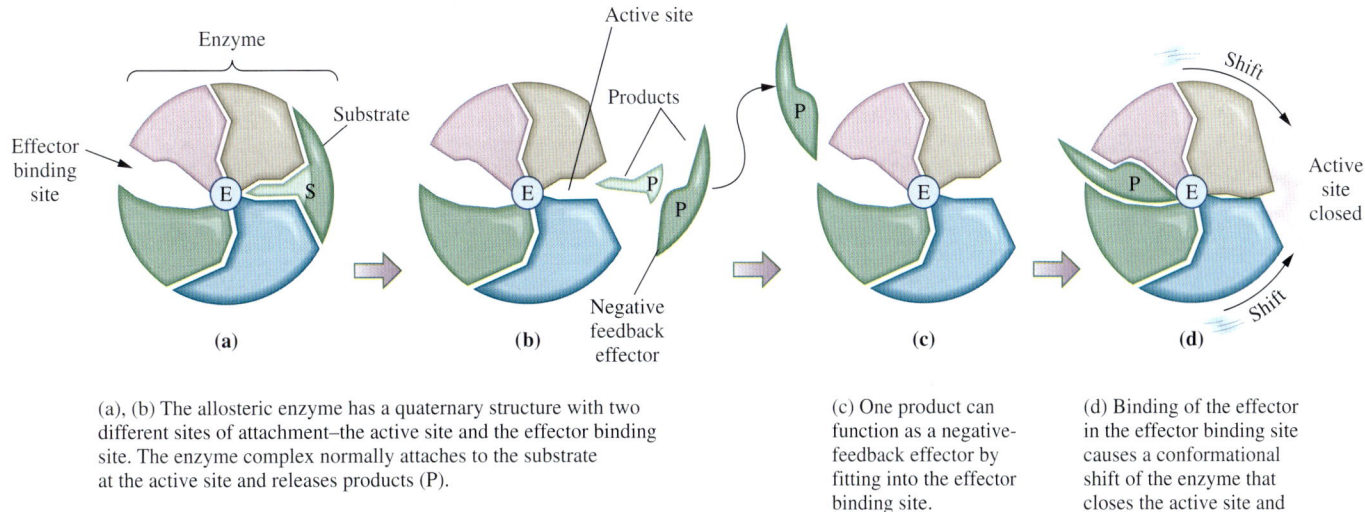

(a), (b) The allosteric enzyme has a quaternary structure with two different sites of attachment–the active site and the effector binding site. The enzyme complex normally attaches to the substrate at the active site and releases products (P).

(c) One product can function as a negative-feedback effector by fitting into the effector binding site.

(d) Binding of the effector in the effector binding site causes a conformational shift of the enzyme that closes the active site and inactivates the enzyme.

Figure 19.11
A mechanism of negative allosterism. This is an example of feedback inhibition.

In upcoming chapters we will study metabolic pathways. A metabolic pathway is a series of biochemical reactions that accomplishes the breakdown or synthesis of one or more biological molecules. One of these is glycolysis, which is the first stage of the breakdown of carbohydrates to produce ATP energy for the cell. This pathway must be responsive to the demands of the body. When more energy is required, the reactions of the pathway should occur more quickly, producing more ATP. However, if the energy demand is low, the reactions should slow down.

The third reaction in glycolysis is the transfer of a phosphoryl group from an ATP molecule to a molecule of fructose-6-phosphate. This reaction, shown here, is catalyzed by an enzyme called *phosphofructokinase*:

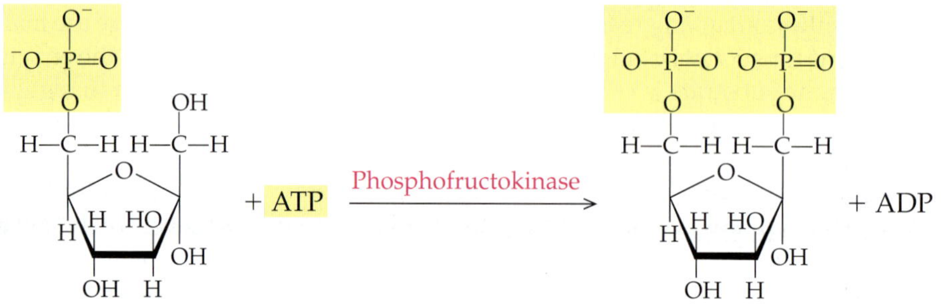

Fructose-6-phosphate + ATP $\xrightarrow{\text{Phosphofructokinase}}$ Fructose-1,6-bisphosphate + ADP

Phosphofructokinase activity is sensitive to both positive and negative allosterism. For instance, when ATP is present in abundance, a signal that the body has sufficient energy, it binds to an effector binding site on phosphofructokinase. This inhibits the activity of the enzyme and, thus, slows the entire pathway. An abundance of AMP, which is a precursor of ATP, is evidence that the body needs to make ATP to have a sufficient energy supply. When AMP binds to an effector binding site on phosphofructokinase, enzyme activity is increased, speeding up the reaction and the entire pathway.

Feedback Inhibition

Allosteric enzymes are the basis for **feedback inhibition** of biochemical pathways. This system functions on the same principle as the thermostat on your furnace. You set the thermostat at 70°F; the furnace turns on and produces heat until the sensor in the thermostat registers a room temperature of 70°F. It then signals the furnace to shut off.

Feedback inhibition usually regulates pathways of enzymes involved in the synthesis of a biological molecule. Such a pathway can be shown schematically as follows:

$$A \xrightarrow{E_1} B \xrightarrow{E_2} C \xrightarrow{E_3} D \xrightarrow{E_4} E \xrightarrow{E_5} F$$

In this pathway the starting material, A, is converted to B by the enzyme E_1. Enzyme E_2 immediately converts B to C, and so on until the final product, F, has been synthesized. If F is no longer needed, it is a waste of cellular energy to continue to produce it.

To avoid this waste of energy, the cell uses feedback inhibition, in which the product can shut off the entire pathway for its own synthesis. This is the result of the fact that the product, F, acts as a negative allosteric effector on one of the early enzymes of the pathway. For instance, enzyme E_1 may have an effector-binding site for F in addition to the active site that binds to A. When F is present in excess, it binds to the effector-binding site. This binding causes the active site to close so that it cannot bind to substrate A. Thus A is not converted to B. If no B is produced, there is no substrate for enzyme E_2, and the entire pathway ceases to operate. The product, *F,* has turned off all the steps involved in its own synthesis, just as the heat produced by the furnace is ultimately responsible for turning off the furnace itself.

When the concentration of *F* drops, it will dissociate from the effector binding site. When this occurs, the enzyme is once again active. Thus, feedback inhibition is an effective metabolic on-off switch.

Proenzymes

Another means of regulating enzyme activity involves the production of the enzyme in an inactive form called a **proenzyme**. It is then converted by proteolysis (hydrolysis of the protein), to the active form when it has reached the site of its activity. What is the purpose of this type of mechanism? On first examination it seems wasteful to add a step to the synthesis of an enzyme. But consider for a moment the very destructive nature of some of the enzymes that are necessary for life. The enzymes pepsin, trypsin, and chymotrypsin are all proteolytic enzymes of the digestive tract. They are necessary to life because they degrade dietary proteins into amino acids that are used by the cell. But what would happen to the cells that produce these enzymes if they were synthesized in active form? Those cells would be destroyed. Thus the cells of the stomach that produce pepsin actually produce an inactive proenzyme, called *pepsinogen.* Pepsinogen has an additional forty-two amino acids. In the presence of stomach acid and previously activated pepsin, the extra forty-two amino acids are cleaved off, and the proenzyme is transformed into the active enzyme. Table 19.2 lists several other proenzymes and the enzymes that convert them to active form.

Protein Modification

Protein modification is another mechanism that the cell can use to turn an enzyme on or off. This is a process in which a chemical group is covalently added to or removed from the protein. This covalent modification either activates the enzyme or turns it off.

The most common type of protein modification is phosphorylation or dephosphorylation of an enzyme. Typically, the phosphoryl group is added to (or removed from) the R group of a serine, tyrosine, or threonine in the protein chain of the enzyme. Notice that these three amino acids have a free —OH in their R group, which serves as the site for the addition of the phosphoryl group.

The covalent modification of an enzyme's structure is catalyzed by other enzymes. *Protein kinases* add phosphoryl groups to a target enzyme, while *phosphatases* remove them. For some enzymes it is the phosphorylated form that is active. For instance, in adipose tissue, phosphorylation activates the enzyme triacylglycerol lipase, an enzyme that breaks triglycerides down to fatty acids and

TABLE 19.2 Proenzymes of the Digestive Tract

Zymogen	Activator	Enzyme
Proelastase	Trypsin	Elastase
Trypsinogen	Trypsin	Trypsin
Chymotrypsinogen A	Trypsin + chymotrypsin	Chymotrypsin
Pepsinogen	Acid pH + pepsin	Pepsin
Procarboxypeptidases	Trypsin	Carboxypeptidase A, carboxypeptidase B

glycerol. Glycogen phosphorylase, an enzyme involved in the breakdown of glycogen, is also activated by the addition of a phosphoryl group. However, for some enzymes phosphorylation inactivates the enzyme. This is true for glycogen synthase, an enzyme involved in the synthesis of glycogen. When this enzyme is phosphorylated, it becomes inactive.

The convenient aspect of this type of regulation is the reversibility. An enzyme can quickly be turned on or off in response to environmental or physiological conditions.

19.10 Inhibition of Enzyme Activity

LEARNING GOAL

Many chemicals can bind to enzymes and either eliminate or drastically reduce their catalytic ability. These chemicals, called *enzyme inhibitors*, have been used for hundreds of years. When she poisoned her victims with arsenic, Lucretia Borgia was unaware that it was binding to the thiol groups of cysteine amino acids in the proteins of her victims and thus interfering with the formation of disulfide bonds needed to stabilize the tertiary structure of enzymes. However, she was well aware of the deadly toxicity of heavy metal salts like arsenic and mercury. When you take penicillin for a bacterial infection, you are taking another enzyme inhibitor. Penicillin inhibits several enzymes that are involved in the synthesis of bacterial cell walls.

Enzyme inhibitors are classified on the basis of whether the inhibition is reversible or irreversible, competitive or noncompetitive. Reversibility deals with whether the inhibitor will eventually dissociate from the enzyme, releasing it in the active form. Competition refers to whether the inhibitor is a structural analog, or look-alike, of the natural substrate. If so, the inhibitor and substrate will compete for the enzyme active site.

Irreversible Inhibitors

Irreversible enzyme inhibitors, such as arsenic, usually bind very tightly, sometimes even covalently, to the enzyme. This generally involves binding of the inhibitor to one of the R groups of an amino acid in the active site. Inhibitor binding may block the active site binding groups so that the enzyme-substrate complex cannot form. Alternatively, an inhibitor may interfere with the catalytic groups of the active site, thereby effectively eliminating catalysis. Irreversible inhibitors, which include snake venoms and nerve gases, generally inhibit many different enzymes.

Reversible, Competitive Inhibitors

See A Medical Perspective: Fooling the AIDS Virus with "Look-Alike" Nucleotides in Chapter 20.

Reversible, competitive enzyme inhibitors are often referred to as **structural analogs,** that is, they are molecules that resemble the structure and charge distribution of the natural substrate for a particular enzyme. Because of this resemblance, the inhibitor can occupy the enzyme active site. However, no reaction can occur, and enzyme activity is inhibited (Figure 19.12). This inhibition is said to be competitive because the inhibitor and the substrate compete for binding to the enzyme active site. Thus, the degree of inhibition depends on their relative concentrations. If the inhibitor is in excess or binds more strongly to the active site, it will occupy the active site more frequently, and enzyme activity will be greatly decreased. On the other hand, if the natural substrate is present in excess, it will more frequently occupy the active site, and there will be little inhibition.

The sulfa drugs, the first antimicrobics to be discovered, are **competitive inhibitors** of a bacterial enzyme needed for the synthesis of the vitamin folic acid. *Folic acid* is a vitamin required for the transfer of methyl groups in the biosynthesis of methionine and the nitrogenous bases required to make DNA and RNA.

19.10 Inhibition of Enzyme Activity

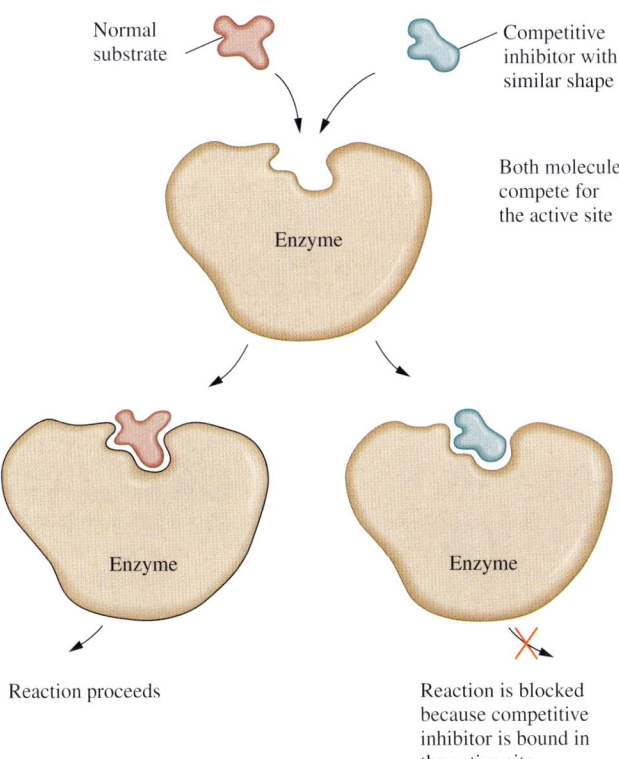

Figure 19.12
Competitive inhibition.

Humans cannot synthesize folic acid and must obtain it from the diet. Bacteria, on the other hand, must make folic acid because they cannot take it in from the environment.

para-Aminobenzoic acid (PABA) is the substrate for an early step in folic acid synthesis. The sulfa drugs, the prototype of which was discovered in the 1930s by Gerhard Domagk, are structural analogs of PABA and thus competitive inhibitors of the enzyme that uses PABA as its normal substrate.

p-Aminobenzoic acid   Sulfanilamide

If the correct substrate (PABA) is bound by the enzyme, the reaction occurs, and the bacterium lives. However, if the sulfa drug is present in excess over PABA, it binds more frequently to the active site of the enzyme. No folic acid will be produced, and the bacterial cell will die.

Because we obtain our folic acid from our diets, sulfa drugs do not harm us. However, bacteria are selectively killed. Luckily, we can capitalize on this property for the treatment of bacterial infections, and as a result, sulfa drugs have saved countless lives. Although bacterial infection was the major cause of death before the discovery of sulfa drugs and other antibiotics, death caused by bacterial infection is relatively rare at present.

> In addition to the folic acid supplied in the diet, we obtain folic acid from our intestinal bacteria.

Reversible, Noncompetitive Inhibitors

Reversible, noncompetitive enzyme inhibitors bind to R groups of amino acids or perhaps to the metal ion cofactors. Unlike the situation of irreversible inhibition, however, the binding is weak, and the enzyme activity is restored when the

A Medical Perspective

Enzymes, Nerve Transmission, and Nerve Agents

The transmission of nerve impulses at the *neuromuscular junction* involves many steps, one of which is the activity of a critical enzyme, called *acetylcholinesterase*, which catalyzes the hydrolysis of the chemical messenger, *acetylcholine*, that initiated the nerve impulse. The need for this enzyme activity becomes clear when we consider the events that begin with a message from the nerve cell and end in the appropriate response by the muscle cell. Acetylcholine is a *neurotransmitter*, that is, a chemical messenger that transmits a message from the nerve cell to the muscle cell. Acetylcholine is stored in membrane-bound bags, called *synaptic vesicles*, in the nerve cell ending.

Acetylcholinesterase comes into play in the following way. The arrival of a nerve impulse at the end plate of the nerve axon causes an influx of Ca^{2+}. This causes the acetylcholine-containing vesicles to migrate to the nerve cell membrane that is in contact with the muscle cell. This is called the *presynaptic membrane*. The vesicles fuse with the presynaptic membrane and release the neurotransmitter. The acetylcholine then diffuses across the *nerve synapse* (the space between the nerve and muscle cells) and binds to the acetylcholine receptor protein in the *postsynaptic membrane* of the muscle cell. This receptor then opens pores in the membrane through which Na^+ and K^+ ions flow into and out of the cell, respectively. This generates the nerve impulse and causes the muscle to contract. If acetylcholine remains at the neuromuscular junction, it will continue to stimulate the muscle contraction. To stop this continued stimulation, acetylcholine is hydrolyzed, and hence, destroyed by acetylcholinesterase. When this happens, nerve stimulation ceases.

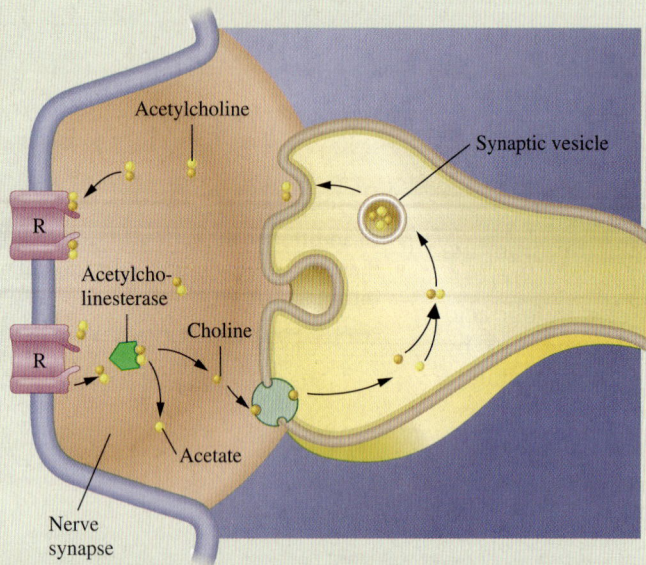

Schematic diagram of the synapse at the neuromuscular junction. The nerve impulse causes acetylcholine to be released from synaptic vesicles. Acetylcholine diffuses across the synaptic cleft and binds to a specific receptor protein (R) on the postsynaptic membrane. A channel opens that allows Na^+ ions to flow into the cell and K^+ ions to flow out of the cell. This results in muscle contraction. Any acetylcholine remaining in the synaptic cleft is destroyed by acetylcholinesterase to terminate the stimulation of the muscle cell.

$$H_3C-\overset{\overset{O}{\|}}{C}-O-CH_2CH_2-N^+-(CH_3)_3 + H_2O$$

Acetylcholine

$$\updownarrow \text{Acetylcholinesterase}$$

$$H_3C-C\overset{O}{\underset{O^-}{\lessgtr}} + HO-CH_2CH_2-N^+-(CH_3)_3 + H^+$$

Acetate   Choline

Inhibitors of acetylcholinesterase are used both as poisons and as drugs. Among the most important inhibitors of acetylcholinesterase are a class of compounds known as *organophosphates*. One of these is the nerve agent Sarin (isopropylmethylfluorophosphate). Sarin forms a covalently bonded

inhibitor dissociates from the enzyme-inhibitor complex. These inhibitors do not bind to the active site, but they do modify the shape of the active site by binding elsewhere in the protein structure. Keep in mind that the entire three-dimensional structure of an enzyme is needed to maintain the correct shape of the active site. Noncovalent interactions between the inhibitor and one or more sites on the enzyme surface can alter the shape of the active site in a fashion analogous to that of an allosteric effector. These inhibitors also inactivate a broad range of enzymes.

19.10 Inhibition of Enzyme Activity

intermediate with the active site of acetylcholinesterase. Thus, it acts as an irreversible, noncompetitive inhibitor.

[Structure: Sarin (CH₃)₂CH—O—P(=O)(F)—CH₃ reacting with HO— Serine in the acetylcholinesterase active site]

↓ (HF released)

[Structure: Sarin covalently bonded to the serine in the active site]

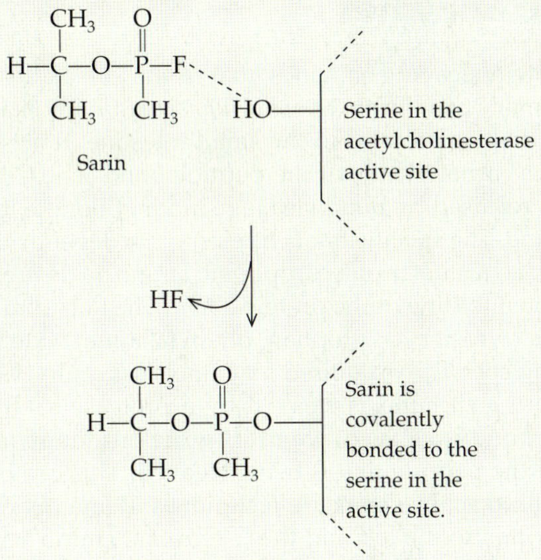

The covalent intermediate is stable, and acetylcholinesterase is therefore inactive, no longer able to break down acetylcholine. Nerve transmission continues, resulting in muscle spasm. Death may occur as a result of laryngeal spasm. Antidotes for poisoning by organophosphates, which include many insecticides and nerve gases, have been developed. The antidotes work by reversing the effects of the inhibitor. One of these antidotes is known as *PAM*, an acronym for *pyridine aldoxime methiodide*. This molecule displaces the organophosphate group from the active site of the enzyme, alleviating the effects of the poison.

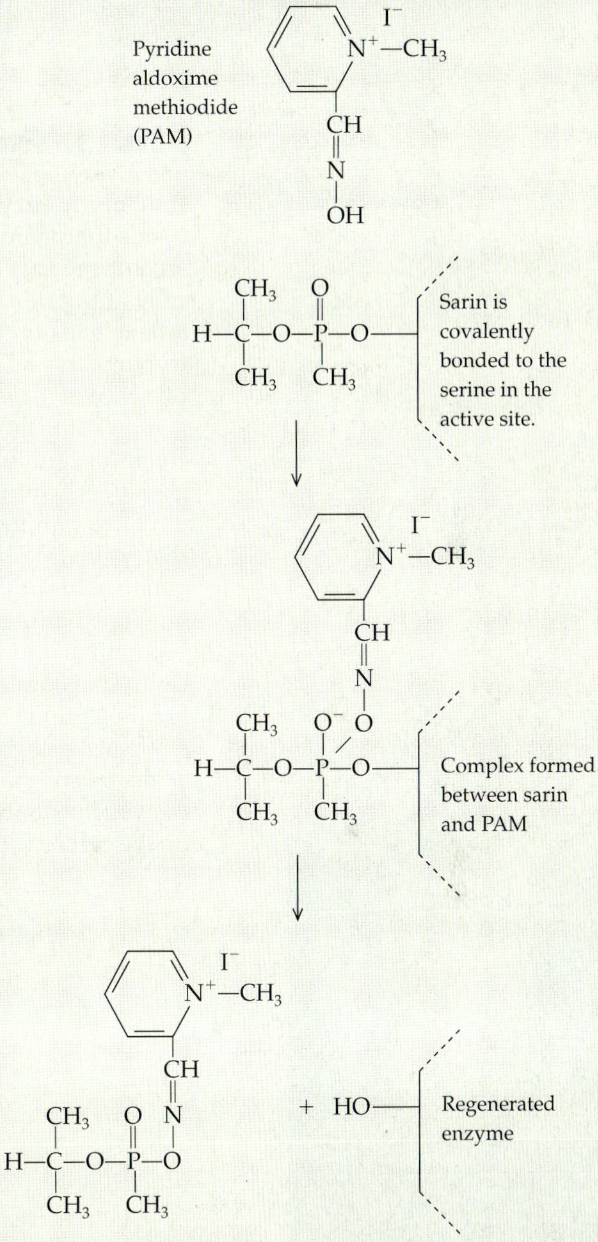

For Further Understanding

Botulinum toxin inhibits release of neurotransmitters from the presynaptic membrane. What symptoms do you predict would result from this?

Why must Na^+ and K^+ enter and exit the cell through a protein channel?

[Right column: Pyridine aldoxime methiodide (PAM) structure, followed by reaction showing Sarin covalently bonded to the serine in the active site, then complex formed between sarin and PAM, then regenerated enzyme + HO—]

Question 19.15
Why are irreversible inhibitors considered to be poisons?

Question 19.16
Explain the difference between an irreversible inhibitor and a reversible, noncompetitive inhibitor.

Question 19.17 What is a structural analog?

Question 19.18 How can structural analogs serve as enzyme inhibitors?

19.11 Proteolytic Enzymes

LEARNING GOAL

Proteolytic enzymes are protein-cleaving enzymes, that is, they break the peptide bonds that maintain the primary protein structure. *Chymotrypsin,* for example, is an enzyme that hydrolyzes dietary proteins in the small intestine. It acts specifically at peptide bonds on the carbonyl side of the peptide bond. The C-terminal amino acids of the peptides released by bond cleavage are methionine, tyrosine, tryptophan, and phenylalanine. The specificity of chymotrypsin depends upon the presence of a *hydrophobic pocket*, a cluster of hydrophobic amino acids brought together by the three-dimensional folding of the protein chain. The flat aromatic side chains of certain amino acids (tyrosine, tryptophan, phenylalanine) slide into this pocket, providing the binding specificity required for catalysis (Figure 19.13).

How can we determine which bond is cleaved by a protease such as chymotrypsin? To know which bond is cleaved, we must write out the sequence of amino acids in the region of the peptide that is being cleaved. This can be determined experimentally by amino acid sequencing techniques. Remember that the N-terminal amino acid is written to the left and the C-terminal amino acid to the right. Consider a protein having within it the sequence —Ala-Phe-Gly—. A reaction is set up in which the enzyme, chymotrypsin, is mixed with the protein substrate. After the reaction has occurred, the products are purified, and their amino acid sequences are determined. Experiments of this sort show that chymotrypsin cleaves the bond between phenylalanine and glycine, which is the peptide bond on the carbonyl side of amino acids having an aromatic side chain.

$$H_3{}^+N-\underset{CH_3}{\underset{|}{C}}\underset{H}{\overset{H}{|}}-\overset{O}{\overset{\|}{C}}-NH-\underset{CH_2-C_6H_5}{\underset{|}{C}}\underset{H}{\overset{H}{|}}-\overset{O}{\overset{\|}{C}}-NH-\underset{H}{\underset{|}{C}}\underset{H}{\overset{H}{|}}-\overset{O}{\overset{\|}{C}}-O^-$$

Peptide bond cleaved

Ala ———— Phe ———— Gly

The **pancreatic serine proteases** trypsin, chymotrypsin, and elastase all hydrolyze peptide bonds. These enzymes are the result of *divergent evolution* in which a single ancestral gene was first duplicated. Then each copy evolved individually. They have similar primary structures, similar tertiary structures (Figure 19.14), and virtually identical mechanisms of action. However, as a result of evolution, these enzymes all have different specificities:

- Chymotrypsin cleaves peptide bonds on the carbonyl side of aromatic amino acids and large, hydrophobic amino acids such as methionine.
- Trypsin cleaves peptide bonds on the carbonyl side of basic amino acids.
- Elastase cleaves peptide bonds on the carbonyl side of glycine and alanine.

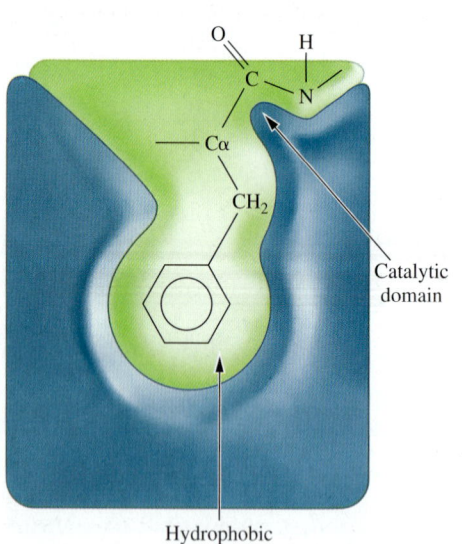

Figure 19.13
The specificity of chymotrypsin is determined by a hydrophobic pocket that holds the aromatic side chain of the substrate. This brings the peptide bond to be cleaved into the catalytic domain of the active site.

19.11 Proteolytic Enzymes

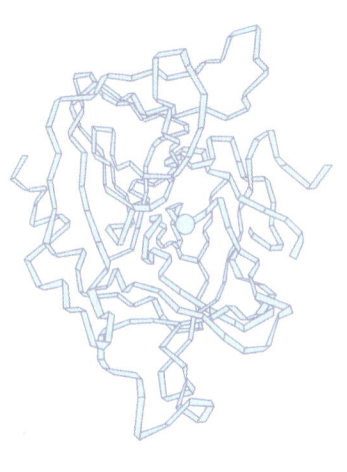
Chymotrypsin

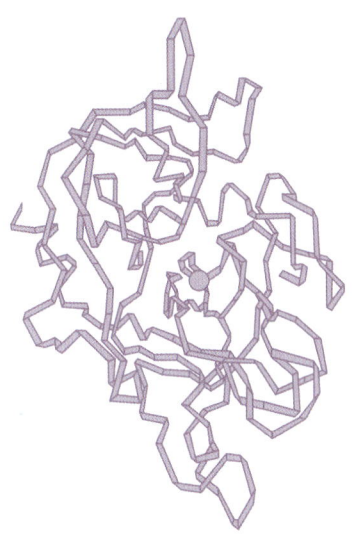
Elastase

Figure 19.14
Structures of chymotrypsin and elastase are virtually identical, suggesting that these enzymes have evolved from a common ancestral protease.

These enzymes have different pockets for the side chains of their substrates; *different keys fit different locks*. This difference manifests itself in the substrate specificity alluded to above. For example, the binding pocket of trypsin is long, narrow, and negatively charged to accommodate lysine or arginine R groups. Yet although the binding pockets have undergone divergent evolution, the catalytic sites have remained unchanged, and the mechanism of proteolytic action is the same for all the serine proteases. In each case, the mechanism involves a serine R group.

These enzymes are called *serine proteases* because they have the amino acid serine in the catalytic region of the active site that is essential for hydrolysis of the peptide bond.

Question 19.19
Draw the structural formulas of the following peptides and show which bond would be cleaved by chymotrypsin.

a. ala-phe-ala
b. tyr-ala-tyr

Question 19.20
Draw the structural formulas of the following peptides and show which bond would be cleaved by chymotrypsin.

a. trp-val-gly
b. phe-ala-pro

Question 19.21
Draw the structural formula of the peptide val-phe-ala-gly-leu. Which bond would be cleaved if this peptide were reacted with chymotrypsin? With elastase?

Question 19.22
Draw the structural formula of the peptide trp-val-lys-ala-ser. Show which bonds would be cleaved by trypsin, chymotrypsin, and elastase.

A Medical Perspective

Enzymes, Isoenzymes, and Myocardial Infarction

A patient is brought into the emergency room with acute, squeezing chest pains; shallow, irregular breathing; and pale, clammy skin. The immediate diagnosis is myocardial infarction, a heart attack. The first thoughts of the attending nurses and physicians concern the series of treatments and procedures that will save the patient's life. It is a short time later, when the patient's condition has stabilized, that the doctor begins to consider the battery of enzyme assays that will confirm the diagnosis.

Myocardial infarction (MI) occurs when the blood supply to the heart muscle is blocked for an extended time. If this lack of blood supply, called *ischemia,* is prolonged, the myocardium suffers irreversible cell damage and muscle death, or infarction. When this happens, the concentration of cardiac enzymes in the blood rises dramatically as the dead cells release their contents into the bloodstream. Although many enzymes are liberated, three are of prime importance. These three enzymes, creatine phosphokinase (CPK), lactate dehydrogenase (LDH), and aspartate aminotransferase/serum glutamate–oxaloacetate transaminase (AST/SGOT), show a characteristic sequential rise in blood serum level following myocardial infarction and then return to normal. This enzyme profile, shown in the accompanying figure, is characteristic of and the basis for the diagnosis of a heart attack.

To ensure against misdiagnosis caused by tissue damage in other organs, the levels of other serum enzymes, including alanine aminotransferase/serum glutamate–pyruvate transaminase (ALT/SGPT) and isocitrate dehydrogenase (ICD), are also measured. ALT and AST are usually determined simultaneously to differentiate between cardiac and hepatic disease. The concentration of ALT is higher in liver disease, whereas the serum concentration of AST is higher following acute myocardial infarction. ICD is found primarily in the liver, and serum levels would not be elevated after a heart attack.

The use of LDH and CPK levels alone can also lead to a misdiagnosis because these enzymes are produced by many tissues. How can a clinician diagnose heart disease with confidence when the elevated serum enzyme levels could indicate coexisting disease in another tissue? The physician is able to make such a decision because of the presence of *isoenzymes*, which provide diagnostic accuracy because they reveal the tissue of origin.

Isoenzymes are forms of the same enzyme with slightly different amino acid sequences. The binding and catalytic sites are the same, but there are differences in the scaffolding sequences of the enzyme that maintain the three-dimensional structure of the protein. Each of the cells of the body contains the genes that could direct the production of all the different forms of these enzymes, yet the expression is *tissue-specific*. This means that the genes for certain isoenzymes are expressed preferentially in different types of tissue.

It is not clear why a certain isoenzyme is "turned on" in the liver whereas another predominates in the heart, but it is known that we can distinguish among the different forms in the laboratory on the basis of their migration through a gel placed in an electric field. This process is called *gel electrophoresis*. This test is based on the fact that each protein has a characteristic surface charge resulting from the R groups of the amino acids. If these proteins are placed in a gel matrix and an electrical current is applied, the proteins migrate as a function of that charge. The figure on the next page shows the position of the five isoenzymes of LDH following electrophoresis.

Imagine a mixture of serum proteins, each with a different overall charge, subjected to an electric field. The proteins with the greatest negative charge will be most strongly attracted to the positive pole and will migrate rapidly toward it, whereas those with a lesser negative charge will migrate much more slowly. Thus, the proteins will be distributed throughout the gel based on their characteristic overall charge.

Once electrophoresis is terminated, the location of the bands of each isoenzyme must be determined and the amount

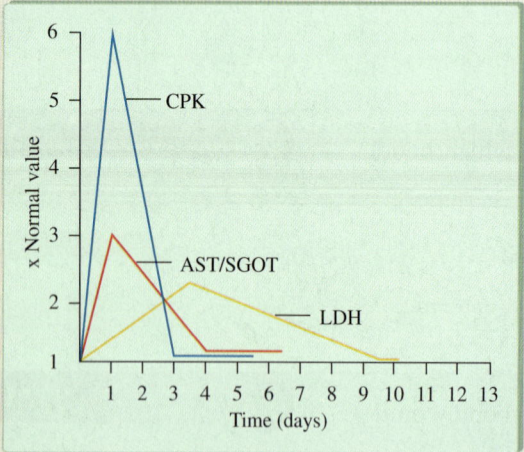

Characteristic pattern of serum cardiac enzyme concentrations following a myocardial infarction.

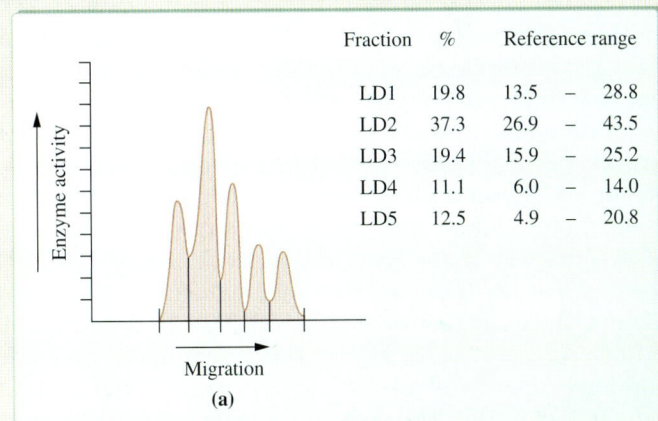
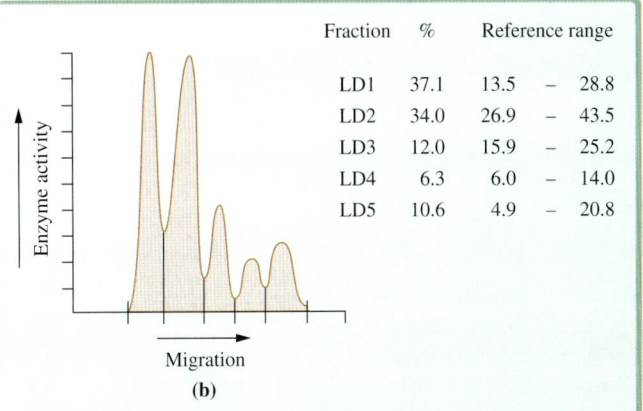

A profile of the serum isoenzymes of lactate dehydrogenase. (a) The pattern of LDH isoenzymes from a normal individual. (b) The pattern of LDH isoenzymes of an individual suffering from a myocardial infarction.

of each must be measured. To do this an enzyme assay is carried out. The substrate is added to the gel in a solution that provides proper conditions for the enzyme. The product is a colored substance that can be seen visually. By inspecting the positions of the stained bands on the gel, one can determine which tissue isoenzymes are present.

To measure the amount of each, the intensity of the color can be measured with a spectrophotometer. The intensity of the color is directly proportional to the amount of the product, and thus to the amount of enzyme in the original sample. This allows the laboratory to calculate the concentration of each isoenzyme in the sample.

These data give the clinician an accurate picture of the nature of the diseased tissue. For a heart attack victim, only creatine phosphokinase isoenzyme 2 (CPK II, the predominant heart isoenzyme) will be elevated in the three days following the heart attack. CPK I (brain) and CPK III (skeletal muscle) levels will remain unchanged. Similarly, only LDH 1, the lactate dehydrogenase isoenzyme made in heart muscle, will be elevated. The levels of LDH 2–5 will remain within normal values.

The physician also has enzymes available to treat a heart attack patient. Most MIs are the result of a *thrombus*, or clot, within a coronary blood vessel. The clot restricts blood flow to the heart muscle. One technique that shows promise for treatment following a coronary thrombosis, a heart attack caused by the formation of a clot, is destruction of the clot by intravenous or intracoronary injection of an enzyme called *streptokinase*. This enzyme, formerly purified from the pathogenic bacterium *Streptococcus pyogenes* but now available through recombinant DNA techniques, catalyzes the production of the proteolytic enzyme plasmin from its proenzyme, plasminogen. Plasmin can degrade a fibrin clot into subunits. This has the effect of dissolving the clot that is responsible for restricted blood flow to the heart, but there is an additional protective function as well. The subunits produced by plasmin degradation of fibrin clots are able to inhibit further clot formation by inhibiting thrombin.

Recombinant DNA technology has provided medical science with yet another, perhaps more promising, clot-dissolving enzyme. *Tissue-type plasminogen activator (TPA)* is a proteolytic enzyme that occurs naturally in the body as a part of the anti-clotting mechanisms. TPA converts the proenzyme, plasminogen, into the active enzyme, plasmin. Injection of TPA within two hours of the initial chest pain can significantly improve the circulation to the heart and greatly improve the patient's chances of survival.

For Further Understanding

When an enzyme assay is carried out following electrophoresis of a sample, why is the intensity of the color produced proportional to the amount of enzyme in the sample?

Aspirin has also been suggested as a treatment to enhance a patient's probability of surviving a heart attack. What mechanism can you devise to explain this suggestion?

19.12 Uses of Enzymes in Medicine

LEARNING GOAL

Analysis of blood serum for levels (concentrations) of certain enzymes can provide a wealth of information about a patient's medical condition. Often, such tests are used to confirm a preliminary diagnosis based on the disease symptoms or clinical picture.

For example, when a heart attack occurs, a lack of blood supplied to the heart muscle causes some of the heart muscle cells to die. These cells release their contents, including their enzymes, into the bloodstream. Simple tests can be done to measure the amounts of certain enzymes in the blood. Such tests, called *enzyme assays*, are very precise and specific because they are based on the specificity of the enzyme-substrate complex. If you wish to test for the enzyme lactate dehydrogenase (LDH), you need only to add the appropriate substrate, in this case pyruvate and NADH. The reaction that occurs is the oxidation of NADH to NAD^+ and the reduction of pyruvate to lactate. To measure the rate of the chemical reaction, one can measure the disappearance of the substrate or the accumulation of one of the products. In the case of LDH, spectrophotometric methods (based on the light-absorbing properties of a substrate or product) are available to measure the rate of production of NAD^+. The choice of substrate determines what enzyme activity is to be measured.

Elevated blood serum concentrations of the enzymes amylase and lipase are indications of pancreatitis, an inflammation of the pancreas. Liver diseases such as cirrhosis and hepatitis result in elevated levels of one of the isoenzymes of lactate dehydrogenase (LDH_5), and elevated levels of alanine aminotransferase/serum glutamate–pyruvate transaminase (ALT/SGPT) and aspartate aminotransferase/serum glutamate–oxaloacetate transaminase (AST/SGOT) in blood serum. In fact, these latter two enzymes also increase in concentration following heart attack, but the physician can differentiate between these two conditions by considering the relative increase in the two enzymes. If ALT/SGPT is elevated to a greater extent than AST/SGOT, it can be concluded that the problem is liver dysfunction.

Enzymes are also used as analytical reagents in the clinical laboratory owing to their specificity. They often selectively react with one substance of interest, producing a product that is easily measured. An example of this is the clinical analysis of urea in blood. The measurement of urea levels in blood is difficult because of the complexity of blood. However, if urea is converted to ammonia using the enzyme urease, the ammonia becomes an *indicator* of urea, because it is produced from urea, and it is easily measured. This test, called the *blood urea nitrogen (BUN) test*, is useful in the diagnosis of kidney malfunction and serves as one example of the utility of enzymes in clinical chemistry.

Enzyme replacement therapy can also be used in the treatment of certain diseases. One such disease, Gaucher's disease, is a genetic disorder resulting in a deficiency of the enzyme *glucocerebrosidase*. In the normal situation, this enzyme breaks down a glycolipid called *glucocerebroside,* which is an intermediate in the synthesis and degradation of complex glycosphingolipids found in cellular membranes. Glucocerebrosidase is found in the lysosomes, where it hydrolyzes glucocerebroside into glucose and ceramide.

The role of LDH and other enzymes in disease diagnosis is discussed in A Medical Perspective: Enzymes, Isoenzymes, and Myocardial Infarction.

See A Medical Perspective: Disorders of Sphingolipid Metabolism in Chapter 17.

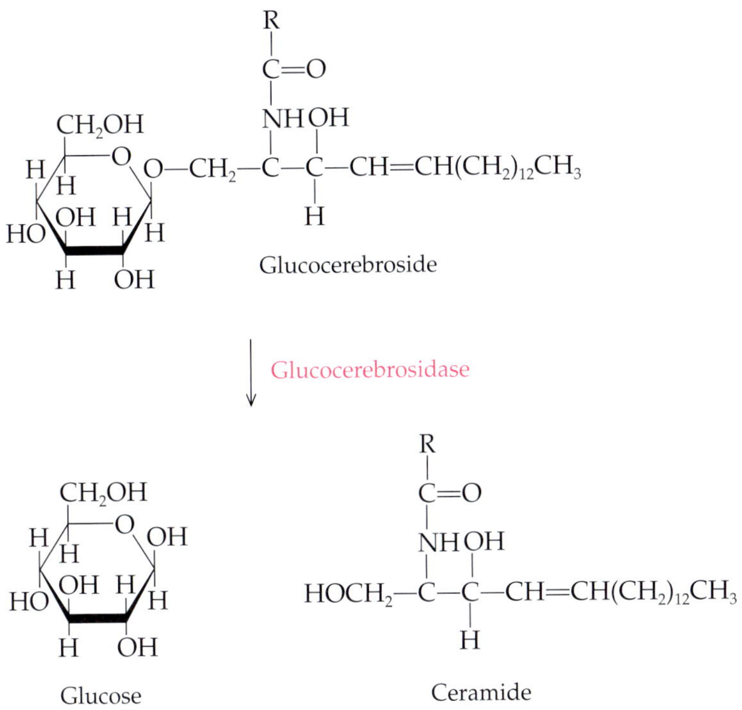

19.12 Uses of Enzymes in Medicine

[Structure of Glucocerebroside, with glucose ring linked via O–CH₂ to a ceramide portion containing R–C(=O)–NHOH, CH, CH=CH(CH₂)₁₂CH₃]

Glucocerebroside

↓ *Glucocerebrosidase*

[Structures of Glucose and Ceramide: HOCH₂–C–C–CH=CH(CH₂)₁₂CH₃ with NHOH and C(=O)R]

Glucose   Ceramide

In Gaucher's disease, the enzyme is not present and glucocerebroside builds up in macrophages found in the liver, spleen, and bone marrow. These cells become engorged with excess lipid that cannot be metabolized and then displace healthy, normal cells in bone marrow. The symptoms of Gaucher's disease include severe anemia, thrombocytopenia (reduction in the number of platelets), and hepatosplenomegaly (enlargement of the spleen and liver). There can also be skeletal problems including bone deterioration and secondary fractures.

Recombinant DNA technology has been used by the Genzyme Corporation to produce the human lysosomal enzyme β-glucocerebrosidase. Given the trade name *Cerezyme*, the enzyme hydrolyzes glucocerebroside into glucose and ceramide so that the products can be metabolized normally. Patients receive Cerezyme intravenously over the course of one to two hours. The dosage and treatment schedule can be tailored to the individual. The results of testing are very encouraging. Patients experience improved red blood cell and platelet counts and reduced hepatosplenomegaly.

SUMMARY

19.1 Nomenclature and Classification

Enzymes are most frequently named by using the common system of nomenclature. The names are useful because they are often derived from the name of the substrate and/or the reaction of the substrate that is catalyzed by the enzyme. Enzymes are classified according to function. The six general classes are *oxidoreductases, transferases, hydrolases, lyases, isomerases,* and *ligases*.

19.2 The Effect of Enzymes on the Activation Energy of a Reaction

Enzymes are the biological catalysts of cells. They lower the activation energies but do not alter the equilibrium constants of the reactions they catalyze.

19.3 The Effect of Substrate Concentration on Enzyme-Catalyzed Reactions

With uncatalyzed reactions, increases in substrate concentration result in an increase in reaction rate. For enzyme-catalyzed reactions, an increase in substrate concentration initially causes an increase in reaction rate, but at a particular concentration the reaction rate reaches a maximum. At this concentration all enzyme active sites are filled with substrate.

19.4 The Enzyme-Substrate Complex

Formation of an *enzyme-substrate complex* is the first step of an enzyme-catalyzed reaction. This involves the binding of the substrate to the active site of the enzyme. The lock-and-key model of substrate binding describes the enzyme as a rigid structure into which the substrate fits precisely. The newer induced fit model describes the enzyme as a flexible molecule. The shape of the active site approximates the shape of the substrate and then "molds" itself around the substrate.

19.5 Specificity of the Enzyme-Substrate Complex

Enzymes are also classified on the basis of their specificity. The four classifications of specificity are *absolute, group, linkage,* and *stereochemical specificity.* An enzyme with *absolute specificity* catalyzes the reaction of only a single substrate. An enzyme with group specificity catalyzes reactions involving similar substrates with the same functional group. An enzyme with *linkage specificity* catalyzes reactions involving a single kind of bond. An enzyme with *stereochemical specificity* catalyzes reactions involving only one enantiomer.

19.6 The Transition State and Product Formation

An enzyme-catalyzed reaction is mediated through an unstable *transition state*. This may involve the enzyme putting "stress" on a bond, bringing reactants into close proximity and in the correct orientation, or altering the local pH.

19.7 Cofactors and Coenzymes

Cofactors are metal ions, organic compounds, or organometallic compounds that bind to an enzyme and help maintain the correct configuration of the active site. The term *coenzyme* refers specifically to an organic group that binds transiently to the enzyme during the reaction. It accepts or donates chemical groups.

19.8 Environmental Effects

Enzymes are sensitive to pH and temperature. High temperatures or extremes of pH rapidly inactivate most enzymes by denaturing them.

19.9 Regulation of Enzyme Activity

Enzymes differ from inorganic catalysts in that they are regulated by the cell. Some of the means of enzyme regulation include allosteric regulation, *feedback inhibition*, production of inactive forms, or *proenzymes*, and *protein modification*. *Allosteric enzymes* have an effector binding site, as well as an active site. Effector binding renders the enzyme active (*positive allosterism*) or inactive (*negative allosterism*). In feedback inhibition the product of a biosynthetic pathway turns off the entire pathway via negative allosterism. In protein modification, adding or removing a covalently bound group either activates or inactivates an enzyme.

19.10 Inhibition of Enzyme Activity

Enzyme activity can be destroyed by a variety of inhibitors. *Irreversible inhibitors*, or poisons, bind tightly to enzymes and destroy their activity permanently. *Competitive inhibitors* are generally *structural analogs* of the natural substrate for the enzyme. They compete with the normal substrate for binding to the active site. When the competitive inhibitor is bound by the active site, the reaction cannot occur, and no product is produced.

19.11 Proteolytic Enzymes

Proteolytic enzymes (proteases) catalyze the hydrolysis of peptide bonds. The *pancreatic serine proteases* chymotrypsin, trypsin, and elastase have similar structures and mechanisms of action, but different substrate specificities. It is thought that they evolved from a common ancestral protease.

19.12 Uses of Enzymes in Medicine

Analysis of blood serum for unusually high levels of certain enzymes provides valuable information on a patient's condition. Such analysis is used to diagnose heart attack, liver disease, and pancreatitis. Enzymes are also used as analytical reagents, as in the blood urea nitrogen (BUN) test, and in the treatment of disease.

KEY TERMS

absolute specificity (19.5)
active site (19.4)
allosteric enzyme (19.9)
apoenzyme (19.7)
coenzyme (19.7)
cofactor (19.7)
competitive inhibitor (19.10)
enzyme (Intro)
enzyme specificity (19.5)
enzyme-substrate complex (19.4)
feedback inhibition (19.9)
group specificity (19.5)
holoenzyme (19.7)
hydrolase (19.1)
induced fit model (19.4)
irreversible enzyme inhibitor (19.10)
isomerase (19.1)
ligase (19.1)
linkage specificity (19.5)
lock-and-key model (19.4)
lyase (19.1)
negative allosterism (19.9)
oxidoreductase (19.1)
pancreatic serine protease (19.11)
pH optimum (19.8)
positive allosterism (19.9)
product (19.2)
proenzyme (19.9)
protein modification (19.9)
proteolytic enzyme (19.11)
reversible, competitive enzyme inhibitor (19.10)
reversible, noncompetitive enzyme inhibitor (19.10)
stereochemical specificity (19.5)
structural analog (19.10)
substrate (19.1)
temperature optimum (19.8)
transferase (19.1)
transition state (19.6)
vitamin (19.7)

QUESTIONS AND PROBLEMS

Nomenclature and Classification

Foundations
19.23 How are the common names of enzymes often derived?
19.24 What is the most common characteristic used to classify enzymes?

Applications
19.25 Match each of the following substrates with its corresponding enzyme:
  1. Urea
  2. Hydrogen peroxide
  3. Lipid
  4. Aspartic acid
  5. Glucose-6-phosphate
  6. Sucrose

  a. Lipase
  b. Glucose-6-phosphatase
  c. Peroxidase
  d. Sucrase
  e. Urease
  f. Aspartase

19.26 Give a systematic name for the enzyme that would act on each of the following substrates:
  a. Alanine
  b. Citrate
  c. Ampicillin
  d. Ribose
  e. Methylamine

19.27 Describe the function implied by the name of each of the following enzymes:
  a. Citrate decarboxylase
  b. Adenosine diphosphate phosphorylase
  c. Oxalate reductase
  d. Nitrite oxidase
  e. cis-trans Isomerase

19.28 List the six classes of enzymes based on the type of reaction catalyzed. Briefly describe the function of each class, and provide an example of each.

The Effect of Enzymes on the Activation Energy of a Reaction

Foundations
19.29 Define the term *substrate*.
19.30 Define the term *product*.

Applications
19.31 What is the activation energy of a reaction?
19.32 What is the effect of an enzyme on the activation energy of a reaction?
19.33 Write and explain the equation for the equilibrium constant of an enzyme-mediated reaction. Does the enzyme alter the K_{eq}?
19.34 If an enzyme does not alter the equilibrium constant of a reaction, how does it speed up the reaction?

The Effect of Substrate Concentration on Enzyme-Catalyzed Reactions

Foundations
19.35 What is the effect of doubling the substrate concentration on the rate of a chemical reaction?
19.36 Why doesn't the rate of an enzyme-catalyzed reaction increase indefinitely when the substrate concentration is made very large?

Applications
19.37 What is meant by the term *rate limiting step*?
19.38 How does the rate limiting step influence an enzyme-catalyzed reaction?
19.39 Draw a graph that describes the effect of increasing the concentration of the substrate on the rate of an enzyme-catalyzed reaction.
19.40 What does a graph of enzyme activity versus substrate concentration tell us about the nature of enzyme-catalyzed reactions?

The Enzyme-Substrate Complex

Foundations
19.41 Define the term *enzyme-substrate complex*.
19.42 Define the term *active site*.
19.43 What are catalytic groups of an enzyme active site?
19.44 What is the binding site of an enzyme active site?

Applications
19.45 Name three major properties of enzyme active sites.
19.46 If enzyme active sites are small, why are enzymes so large?
19.47 What is the lock-and-key model of enzyme-substrate binding?
19.48 Why is the induced fit model of enzyme-substrate binding a much more accurate model than the lock-and-key model?

Specificity of the Enzyme-Substrate Complex

Foundations
19.49 Define the term *enzyme specificity*.
19.50 What region of an enzyme is responsible for its specificity?
19.51 What is meant by the term *group specificity*?
19.52 What is meant by the term *linkage specificity*?
19.53 What is meant by the term *absolute specificity*?
19.54 What is meant by the term *stereochemical specificity*?

Applications
19.55 Provide an example of an enzyme with group specificity and explain the advantage of group specificity for that particular enzyme.
19.56 Provide an example of an enzyme with linkage specificity and explain the advantage of linkage specificity for that particular enzyme.
19.57 Provide an example of an enzyme with absolute specificity and explain the advantage of absolute specificity for that particular enzyme.
19.58 Provide an example of an enzyme with stereochemical specificity and explain the advantage of stereochemical specificity for that particular enzyme.

The Transition State and Product Formation

Foundations
19.59 Outline the four general stages in an enzyme-catalyzed reaction.
19.60 Describe the transition state.

Applications
19.61 What types of transition states might be envisioned that would decrease the energy of activation of an enzyme?
19.62 If an enzyme catalyzed a reaction by modifying the local pH, what kind of amino acid R groups would you expect to find in the active site?

Cofactors and Coenzymes

Foundations
19.63 What is the role of a cofactor in enzyme activity?
19.64 How does a coenzyme function in an enzyme-catalyzed reaction?

Applications
19.65 What is the function of NAD^+? What class of enzymes would require a coenzyme of this sort?
19.66 What is the function of FAD? What class of enzymes would require this coenzyme?

Environmental Effects

Foundations
19.67 List the factors that affect enzyme activity.
19.68 Define the optimum pH for enzyme activity.
19.69 How will each of the following changes in conditions alter the rate of an enzyme-catalyzed reaction?
  a. Decreasing the temperature from 37°C to 10°C
  b. Increasing the pH of the solution from 7 to 11
  c. Heating the enzyme from 37°C to 100°C
19.70 Why does an enzyme lose activity when the pH is drastically changed from optimum pH?

Applications
19.71 High temperature is an effective mechanism for killing bacteria on surgical instruments. How does high temperature result in cellular death?
19.72 An increase in temperature will increase the rate of a reaction if a nonenzymatic catalyst is used; however, an increase in temperature will eventually *decrease* the rate of a reaction when an enzyme catalyst is used. Explain the apparent contradiction of these two statements.
19.73 What is a lysosome?
19.74 Of what significance is it that lysosomal enzymes have a pH optimum of 4.8?
19.75 Why are enzymes that are used for clinical assays in hospitals stored in refrigerators?
19.76 Why do extremes of pH inactivate enzymes?

Regulation of Enzyme Activity

Foundations
19.77 a. Why is it important for cells to regulate the level of enzyme activity?
  b. Why must synthesis of digestive enzymes be carefully controlled?
19.78 What is an allosteric enzyme?
19.79 What is the difference between positive and negative allosterism?
19.80 a. Define feedback inhibition.
  b. Describe the role of allosteric enzymes in feedback inhibition.
  c. Is this positive or negative allosterism?
19.81 What is a proenzyme?
19.82 Three proenzymes that are involved in digestion of proteins in the stomach and intestines are pepsinogen, chymotrypsinogen, and trypsinogen. What is the advantage of producing these enzymes as inactive peptides?

Applications
19.83 The blood clotting mechanism consists of a set of proenzymes that act in a cascade that results in formation of a blood clot. Develop a hypothesis to explain the value of this mechanism for blood clotting.
19.84 What is the benefit for an enzyme such as triacylglycerol lipase to be regulated by covalent modification, in this case phosphorylation?

Inhibition of Enzyme Activity

Foundations
19.85 Define *competitive enzyme inhibition*.
19.86 How do the sulfa drugs selectively kill bacteria while causing no harm to humans?
19.87 Describe the structure of a structural analog.
19.88 How can structural analogs serve as enzyme inhibitors?
19.89 Define *irreversible enzyme inhibition*.
19.90 Why are irreversible enzyme inhibitors often called *poisons*?

Applications
19.91 Suppose that a certain drug company manufactured a compound that had nearly the same structure as a substrate for a certain enzyme but that could not be acted upon chemically by the enzyme. What type of interaction would the compound have with the enzyme?
19.92 The addition of phenylthiourea to a preparation of the enzyme polyphenoloxidase completely inhibits the activity of the enzyme.
  a. Knowing that phenylthiourea binds all copper ions, what conclusion can you draw about whether polyphenoloxidase requires a cofactor?
  b. What kind of inhibitor is phenylthiourea?

Proteolytic Enzymes

Applications
19.93 What do the similar structures of chymotrypsin, trypsin, and elastase suggest about their evolutionary relationship?
19.94 What properties are shared by chymotrypsin, trypsin, and elastase?
19.95 Draw the complete structural formula for the peptide tyr-lys-ala-phe. Show which bond would be broken when this peptide is reacted with chymotrypsin.
19.96 Repeat Question 19.95 for the peptide trp-pro-gly-tyr.
19.97 The sequence of a peptide that contains ten amino acids is as follows:

ala-gly-val-leu-trp-lys-ser-phe-arg-pro

Which peptide bond(s) are cleaved by elastase, trypsin, and chymotrypsin?
19.98 What structural features of trypsin, chymotrypsin, and elastase account for their different specificities?

Uses of Enzymes in Medicine

19.99 List the enzymes whose levels are elevated in blood serum following a myocardial infarction.
19.100 List the enzymes whose levels are elevated as a result of hepatitis or cirrhosis of the liver.

CRITICAL THINKING PROBLEMS

1. Ethylene glycol is a poison that causes about fifty deaths a year in the United States. Treating people who have drunk ethylene glycol with massive doses of ethanol can save their lives. Suggest a reason for the effect of ethanol.

2. Generally speaking, feedback inhibition involves regulation of the first step in a pathway. Consider the following hypothetical pathway:

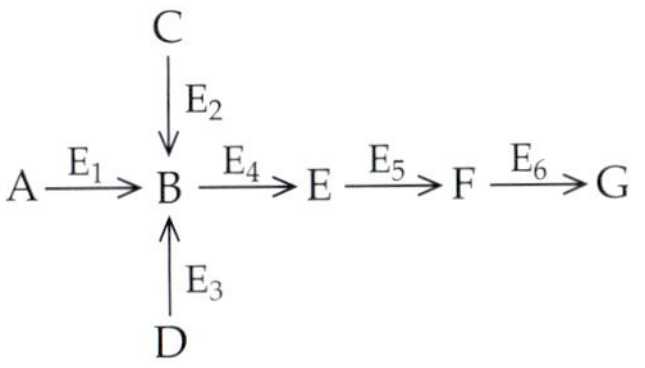

$$A \xrightarrow{E_1} B \xrightarrow{E_4} E \xrightarrow{E_5} F \xrightarrow{E_6} G$$

with $B \xrightarrow{E_2} C$ and $D \xrightarrow{E_3} B$

Which step in this pathway do you think should be regulated? Explain your reasoning.

3. In an amplification cascade, each step greatly increases the amount of substrate available for the next step, so that a very large amount of the final product is made. Consider the following hypothetical amplification cascade:

A_{active} ↓
$B_{inactive} \longrightarrow B_{active}$ ↓
$C_{inactive} \longrightarrow C_{active}$ ↓
$D_{inactive} \longrightarrow D_{active}$

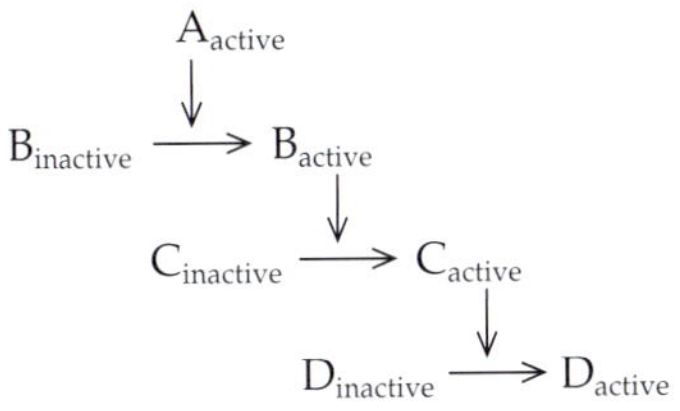

If each active enzyme in the pathway converts one hundred molecules of its substrate to active form, how many molecules of D will be produced if the pathway begins with one molecule of A?

4. L-1-(p-toluenesulfonyl)-amido-2-phenylethylchloromethyl ketone (TPCK, shown below) inhibits chymotrypsin, but not trypsin. Propose a hypothesis to explain this observation.

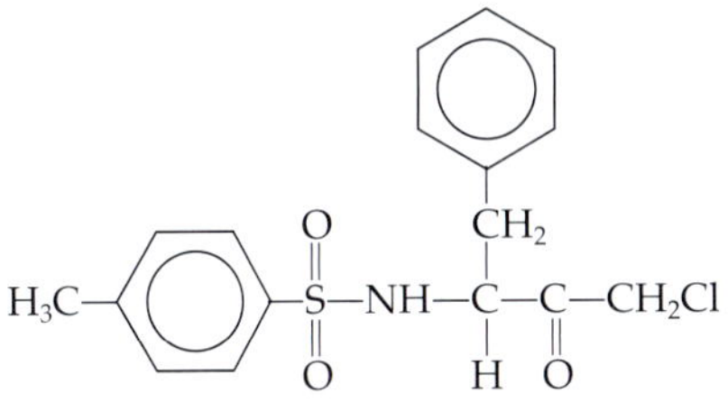

5. A graduate student is trying to make a "map" of a short peptide so that she can eventually determine the amino acid sequence. She digested the peptide with several proteases and determined the sizes of the resultant digestion products.

Enzyme	M.W. of Digestion Products
Trypsin	2000, 3000
Chymotrypsin	500, 1000, 3500
Elastase	500, 1000, 1500, 2000

Suggest experiments that would allow the student to map the order of the enzyme digestion sites along the peptide.

BIOCHEMISTRY

Introduction to Molecular Genetics

20

Identical twins.

Learning Goals

1. Draw the general structure of DNA and RNA nucleotides.
2. Describe the structure of DNA and compare it with RNA.
3. Explain DNA replication.
4. List three classes of RNA molecules and describe their functions.
5. Explain the process of transcription.
6. List and explain the three types of post-transcriptional modifications of eukaryotic mRNA.
7. Describe the essential elements of the genetic code, and develop a "feel" for its elegance.
8. Describe the process of translation.
9. Define mutation and understand how mutations cause cancer and cell death.
10. Describe the tools used in the study of DNA and in genetic engineering.
11. Describe the process of polymerase chain reaction and discuss potential uses of the process.
12. Discuss strategies for genome analysis and DNA sequencing.

Outline

Chemistry Connection:
Molecular Genetics and Detection of Human Genetic Disease

20.1 The Structure of the Nucleotide

20.2 The Structure of DNA and RNA

A Medical Perspective:
Fooling the AIDS Virus with "Look-Alike" Nucleotides

20.3 DNA Replication

20.4 Information Flow in Biological Systems

20.5 The Genetic Code

20.6 Protein Synthesis

20.7 Mutation, Ultraviolet Light, and DNA Repair

A Medical Perspective:
The Ames Test for Carcinogens

20.8 Recombinant DNA

A Human Perspective:
DNA Fingerprinting

A Medical Perspective:
A Genetic Approach to Familial Emphysema

20.9 Polymerase Chain Reaction

20.10 The Human Genome Project

667

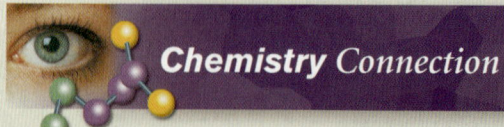

Chemistry Connection

Molecular Genetics and Detection of Human Genetic Disease

It is estimated that 3–5% of the human population suffers from a serious genetic defect. That's 200 million people! But what if genetic disease could be detected and "cured"? Two new technologies, *gene therapy* and *preimplantation diagnosis*, may help us realize this dream.

For a couple with a history of a genetic disease in the family, pregnancy is a time of anxiety. Through *genetic counseling* these couples can learn the probability that their child has the disease. For about 200 genetic diseases the uncertainty can be eliminated. *Amniocentesis* (removal of 10–20 mL of fluid from the sac around the fetus) and *chorionic villus sampling* (removal of cells from a fetal membrane) are two procedures that are used to obtain fetal cells for genetic testing. Fetal cells are cultured and tested by enzyme assays and DNA tests to look for genetic diseases. If a genetic disease is diagnosed, the parents must make a difficult decision: to abort the fetus or to carry the child to term and deal with the effects of the genetic disease.

The power of modern molecular genetics is obvious in our ability to find a "bad" gene from just a few cells. But scientists have developed an even more impressive way to test for genetic disease before the embryo implants into the uterine lining. This technique, called *preimplantation diagnosis*, involves fertilizing a human egg and allowing the resulting zygote to divide in a sterile petri dish. When the zygote consists of 8–16 cells, *one* cell is removed for genetic testing. Only genetically normal embryos are implanted in the mother. Thus, the genetic diseases that we can detect could be eliminated from the population by preimplantation diagnosis because only a zygote with "good" genes is used.

Gene therapy is a second way in which genetic diseases may one day be eliminated. Foreign genes, including growth hormone, have been introduced into fertilized mouse eggs and the zygotes implanted in female mice. The baby mice born with the foreign growth hormone gene were about three times larger than their normal littermates! One day, this kind of technology may be used to introduce normal genes into human fertilized eggs carrying a defective gene, thereby replacing the defective gene with a normal one.

In this chapter we will examine the molecules that carry and express our genetic information, DNA and RNA. Only by understanding the structure and function of these molecules have we been able to develop the amazing array of genetic tools that currently exists. We hope that as we continue to learn more about human genetics, we will be able to detect and one day correct most of the known genetic diseases.

Introduction

Look around at the students in your chemistry class. They all share many traits: upright stance, a head with two eyes, a nose, and a mouth facing forward, one ear on each side of the head, and so on. You would have no difficulty listing the similarities that define you and your classmates as Homo sapiens.

As you look more closely at the individuals you begin to notice many differences. Eye color, hair color, skin color, the shape of the nose, height, body build: all these traits, and many more, show amazing variety from one person to the next. Even within one family, in which the similarities may be more pronounced, each individual has a unique appearance. In fact, only identical twins look exactly alike—well, most of the time.

The molecule responsible for all these similarities and differences is deoxyribonucleic acid (DNA). Tightly wound up in structures called chromosomes in the nucleus of the cell, DNA carries the genetic code to produce the thousands of different proteins that make us who we are. These proteins include enzymes that are responsible for production of the pigment melanin. The more melanin we are genetically programmed to make, the darker our hair, eyes, and skin will be. Others are structural proteins. The gene for α-keratin that makes up hair determines whether our hair will be wavy, straight, or curly. Thousands of genes carry the genetic information for thousands of proteins that dictate our form and, some believe, our behavior.

Genetic traits are passed from one generation to the next. When a sperm fertilizes an ovum, a zygote is created from a single set of maternal chromosomes and a single set of paternal chromosomes. As this fertilized egg divides, each daughter cell will receive one copy of each of these chromosomes. The genes on these chromosomes will direct fetal development from that fertilized cell to a newborn with all the characteristics we recognize as human.

In this chapter we will explore the structure of DNA and the molecular events that translate the genetic information of a gene into the structure of a protein.

20.1 The Structure of the Nucleotide

Even before the philosopher Aristotle observed that "like begets like," humans were curious about the way in which family likenesses are passed from one generation to the next. In the 1860s Gregor Mendel combined astute observations, careful experimental design, and mathematical analysis to explain inheritance. Presented at a meeting in 1865 and published in 1866, Mendel's brilliant work was largely ignored by a scientific community that simply could not understand it.

At about the same time (1869), Friedrich Miescher discovered a substance in the nuclei of white blood cells recovered from pus. Chemical analysis indicated that, in addition to carbon, hydrogen, and oxygen, nuclein contained 14% nitrogen and 3% phosphorus. As microscopes improved in the last decades of the nineteenth century, biologists were able to peer into the nuclei of cells. They observed structures, later called chromosomes, which seemed to play a critical role in the process of cell division. Interestingly, egg and sperm cells were observed to have only half the chromosomes of the cells that produced them.

Chemical analysis of chromosomes indicated that they were composed of both protein and nuclein. But which of these molecules represented the genetic material? Most were convinced that the answer to this question was protein. The reasoning was that the genetic material must have a structure that would allow it to encode the enormous variation seen in the biological world. Both nuclein and protein were known to be polymers. However, proteins were polymers of twenty different subunits, the amino acids. Based on the results of Phoebus Levene, working with Emil Fischer and Albrecht Kossel, nuclein was composed of only four subunits. It appeared to lack the complexity required of a molecule responsible for the great diversity seen among plants, animals, and microbes.

In 1950, the genetic information was demonstrated to be nuclein, now called deoxyribonucleic acid, or DNA. In 1953, just over fifty years ago, James Watson and Francis Crick published a paper describing the structure of the DNA molecules.

Nucleotide Structure

From the work of Watson and Crick, as well as that of Miescher, Levene, and many others, we now know that **deoxyribonucleic acid (DNA)** and **ribonucleic acid (RNA)** are long polymers of **nucleotides.** Every nucleotide is composed of a nitrogenous base, a five-carbon sugar, and at least one phosphoryl group.

Nitrogenous bases are heterocyclic amines; that is, they are cyclic compounds with at least one nitrogen atom in the ring structure. There are two types of nitrogenous bases: purines and pyrimidines. **Purines,** which include adenine and guanine, consist of a six-member ring fused to a five-member ring. **Pyrimidines,** which include thymine, cytosine, and uracil, consist of a single six-member ring. Structures of these molecules are shown in Figure 20.1.

The five-carbon sugar in RNA is ribose, and the sugar in DNA is 2'-deoxyribose. The only difference between these two sugars is found at the 2'-carbon. Ribose has a hydroxyl group attached to this carbon, while deoxyribose has a hydrogen atom.

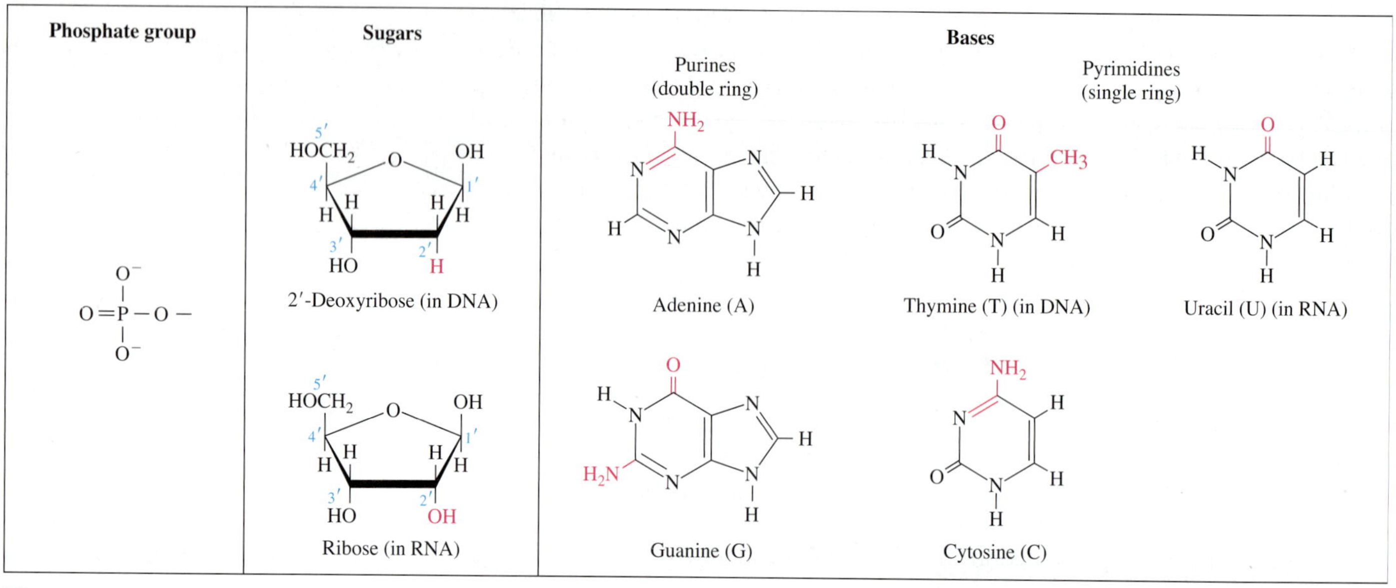

Figure 20.1
The components of nucleic acids include phosphate groups, the five-carbon sugars ribose and deoxyribose, and purine and pyrimidine nitrogenous bases. The ring positions of the sugars are designated with primes (′) to distinguish them from the ring positions of the bases.

Each nucleotide consists of either ribose or deoxyribose, one of the five nitrogenous bases, and one or more phosphoryl groups (Figure 20.2). A nucleotide with the sugar ribose is a **ribonucleotide**, and one having the sugar 2′-deoxyribose is a **deoxyribonucleotide**. Because there are two cyclic molecules in a nucleotide, the sugar, and the base, we need an easy way to describe the ring atoms of each. For this reason the ring atoms of the sugar are designated with a prime to distinguish them from the atoms of the base (Figures 20.1 and 20.2).

The covalent bond between the sugar and the phosphoryl group is a phosphoester bond formed by a condensation reaction between the 5′-OH of the sugar and an —OH of the phosphoryl group. The bond between the base and the sugar is a β-N-glycosidic linkage that joins the 1′-carbon of the sugar and a nitrogen atom of the base (N-9 of purines and N-1 of pyrimidines).

Question 20.1

Referring to the structures in Figures 20.1 and 20.2, draw the structures for nucleotides consisting of the following units.

a. Ribose, adenine, two phosphoryl groups
b. 2′-Deoxyribose, guanine, three phosphoryl groups

Question 20.2

Referring to the structures in Figures 20.1 and 20.2, draw the structures for nucleotides consisting of the following units.

a. 2′-Deoxyribose, thymine, one phosphoryl group
b. Ribose, cytosine, three phosphoryl groups
c. Ribose, uracil, one phosphoryl group

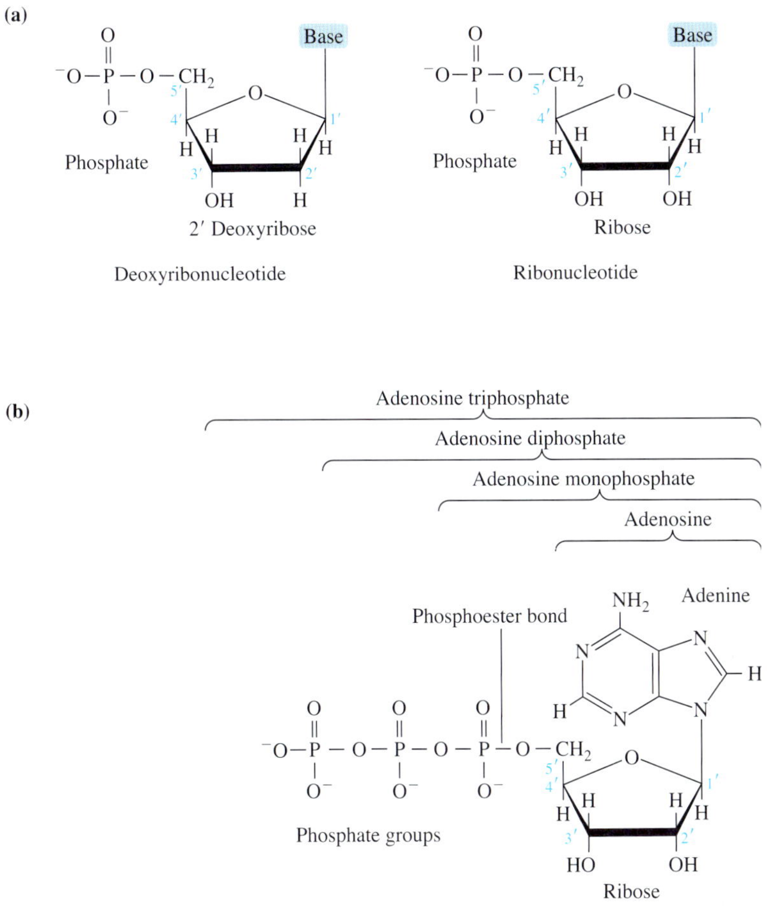

20.2 The Structure of DNA and RNA

Figure 20.2
(a) The general structures of a deoxyribonucleotide and a ribonucleotide. (b) A specific example of a ribonucleotide, adenosine triphosphate.

20.2 The Structure of DNA and RNA

A single strand of DNA is a polymer of nucleotides bonded to one another by 3′–5′ phosphodiester bonds. The backbone of the polymer is called the *sugar-phosphate backbone* because it is composed of alternating units of the five-carbon sugar 2′-deoxyribose and phosphoryl groups in phosphodiester linkage. A nitrogenous base is bonded to each sugar by an *N*-glycosidic linkage (Figure 20.3).

DNA Structure: The Double Helix

James Watson and Francis Crick were the first to describe the three-dimensional structure of DNA in 1953. They deduced the structure by building models based on the experimental results of others. Irwin Chargaff observed that the amount of adenine in any DNA molecule is equal to the amount of thymine. Similarly, he found that the amounts of cytosine and guanine are also equal. The X-ray diffraction studies of Rosalind Franklin and Maurice Wilkens revealed several repeat distances that characterize the structure of DNA: 0.34 nm, 3.4 nm, and 2 nm. (Look at the structure of DNA in Figure 20.4 to see the significance of these measurements.)

With this information, Watson and Crick concluded that DNA is a **double helix** of two strands of DNA wound around one another. The structure of the double helix is often compared to a spiral staircase. The sugar-phosphate backbones of the two strands of DNA spiral around the outside of the helix like the handrails on a spiral staircase. The nitrogenous bases extend into the center at right angles to the axis of the

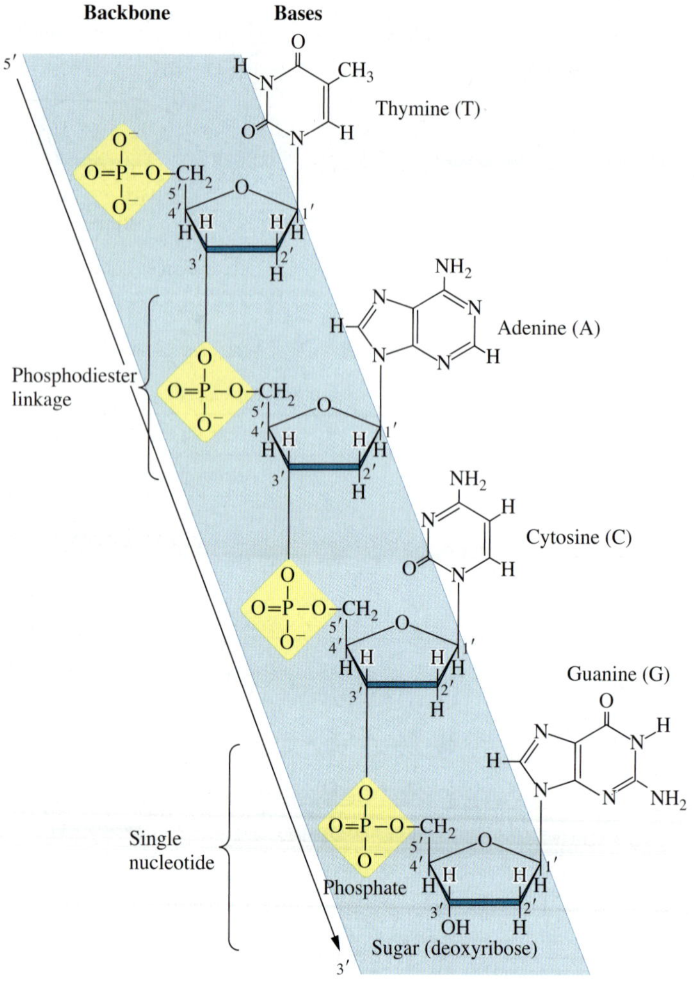

Figure 20.3
The covalent, primary structure of DNA.

helix. You can imagine the nitrogenous bases forming the steps of the staircase. The structure of this elegant molecule is shown in Figure 20.4.

One noncovalent attraction that helps maintain the double helix structure is hydrogen bonding between the nitrogenous bases in the center of the helix. Adenine forms two hydrogen bonds with thymine, and cytosine forms three hydrogen bonds with guanine (Figures 20.4). These are called **base pairs.** The two strands of DNA are **complementary strands** because the sequence of bases on one automatically determines the sequence of bases on the other. When there is an adenine on one strand, there will always be a thymine in the same location on the opposite strand.

The diameter of the double helix is 2.0 nm. This is dictated by the dimensions of the purine-pyrimidine base pairs. The helix completes one turn every ten base pairs. One complete turn is 3.4 nm. Thus, each base pair advances the helix by 0.34 nm.

One last important feature of the DNA double helix is that the two strands are **antiparallel strands,** as this example shows:

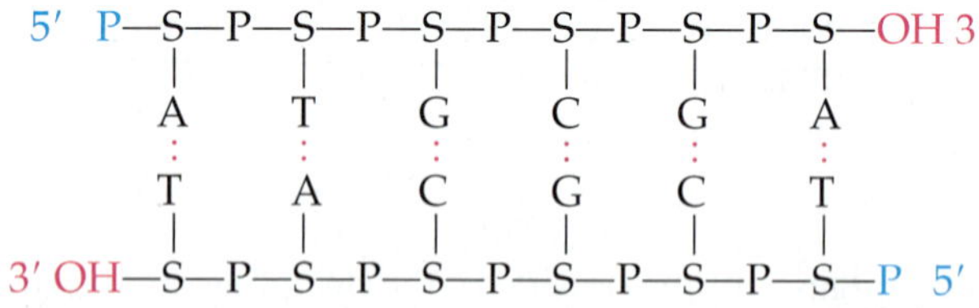

```
5'  P—S—P—S—P—S—P—S—P—S—P—S—OH 3'
        |     |     |     |     |     |
        A     T     G     C     G     A
        ⋮     ⋮     ⋮     ⋮     ⋮     ⋮
        T     A     C     G     C     T
        |     |     |     |     |     |
3' OH—S—P—S—P—S—P—S—P—S—P—S—P  5'
```

In other words, the two strands of the helix run in opposite directions (see Figure 20.4). Only when the two strands are antiparallel can the base pairs form the hydrogen bonds that hold the two strands together.

20.2 The Structure of DNA and RNA

Key Features

- Two strands of DNA form a right-handed double helix.
- The bases in opposite strands hydrogen bond according to the AT/GC rule.
- The 2 strands are antiparallel with regard to their 5' to 3' directionality.
- There are ~10.0 nucleotides in each strand per complete 360° turn of the helix.

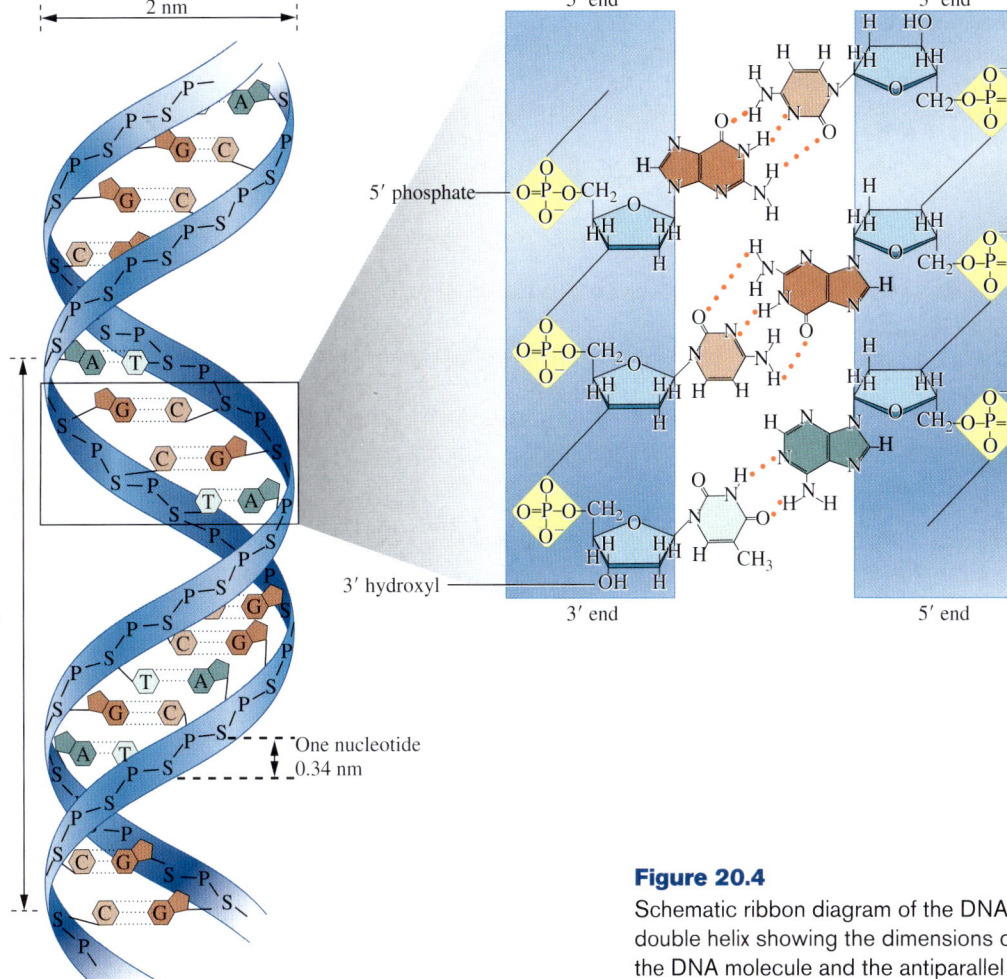

Figure 20.4
Schematic ribbon diagram of the DNA double helix showing the dimensions of the DNA molecule and the antiparallel orientation of the two strands.

Chromosomes

Chromosomes are pieces of DNA that carry the genetic instructions, or genes, of an organism. Organisms such as the prokaryotes have only a single chromosome and its structure is relatively simple. Others, the eukaryotes, have many chromosomes, each of which has many different levels of structure. The complete set of genetic information in all the chromosomes of an organism is called the **genome.**

Prokaryotes are organisms with a simple cellular structure in which there is no true nucleus surrounded by a nuclear membrane and there are no true membrane-bound organelles. This group includes all of the bacteria. In these organisms the chromosome is a circular DNA molecule that is supercoiled, which means that the helix is coiled on itself. The supercoiled DNA molecule is attached to a complex of proteins at roughly forty sites along its length, forming a series of loops. This structure, called the nucleoid, can be seen in Figure 20.5.

Eukaryotes are organisms that have cells containing a true nucleus enclosed by a nuclear membrane. They also have a variety of membrane-bound organelles that segregate different cellular functions into different compartments. As an example, the reactions of aerobic respiration are located within the mitochondria.

All animals, plants, and fungi are eukaryotes. The number and size of the chromosomes of eukaryotes vary from one species to the next. For instance, humans have 23 pairs of chromosomes, while the Adder's Tongue fern has 631 pairs of chromosomes. But the chromosome structure is the same for all those organisms that have been studied.

Figure 20.5
Structure of a bacterial nucleoid. The nucleoid is made up of the supercoiled, circular chromosome attached to a protein core.

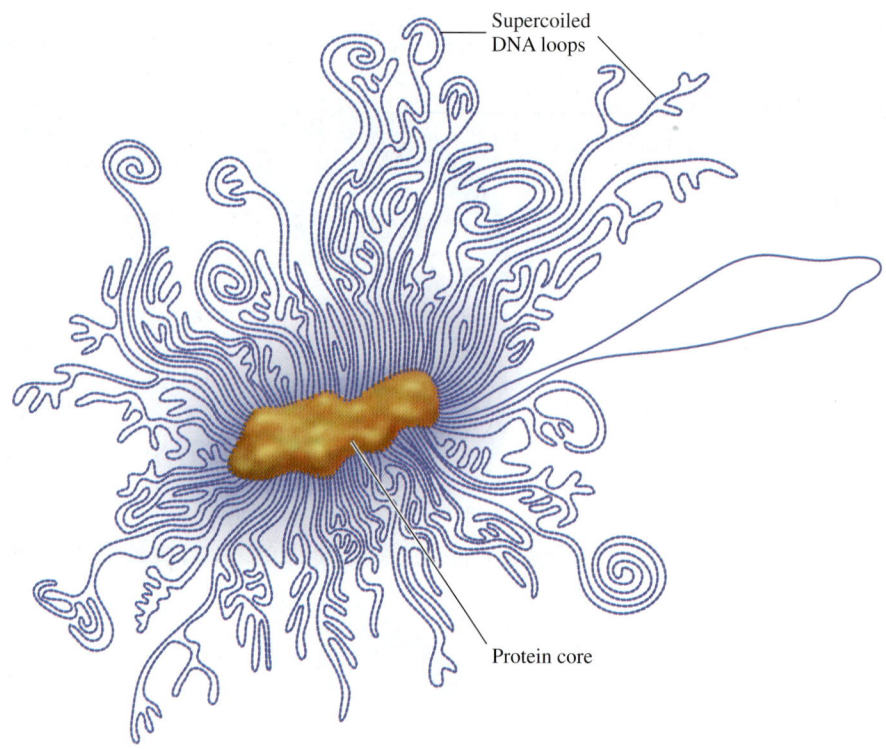

Eukaryotic chromosomes are very complex structures (Figure 20.6). The first level of structure is the **nucleosome,** which consists of a strand of DNA wrapped around a small disk made up of histone proteins. At this level the DNA looks like beads along a string. The string of beads then coils into a larger structure called the *30 nm fiber*. These, in turn, are further coiled into a *200 nm fiber*. Other proteins are probably involved in the organization of the 200 nm fiber. The full complexities of the eukaryotic chromosome are not yet understood, but there are probably many such levels of coiled structures.

RNA Structure

The sugar-phosphate backbone of RNA consists of ribonucleotides, also linked by 3′–5′ phosphodiester bonds. These phosphodiester bonds are identical to those found in DNA. However, RNA molecules differ from DNA molecules in three basic properties.

- RNA molecules are usually *single-stranded.*
- The sugar-phosphate backbone of RNA consists of *ribonucleotides* linked by 3′–5′ phosphodiester bonds. Thus the sugar *ribose* is found in place of 2′-deoxyribose.
- The nitrogenous base *uracil* (U) replaces thymine (T).

Although RNA molecules are single-stranded, base pairing between uracil and adenine and between guanine and cytosine can still occur. We will show the importance of this property as we examine the way in which RNA molecules are involved in the expression of the genetic information in DNA.

20.3 DNA Replication

LEARNING GOAL 3

DNA must be replicated before a cell divides so that each daughter cell inherits a copy of each gene. A cell that is missing a critical gene will die, just as an individual with a genetic disease, a defect in an important gene, may die early in life. Thus

20.3 DNA Replication

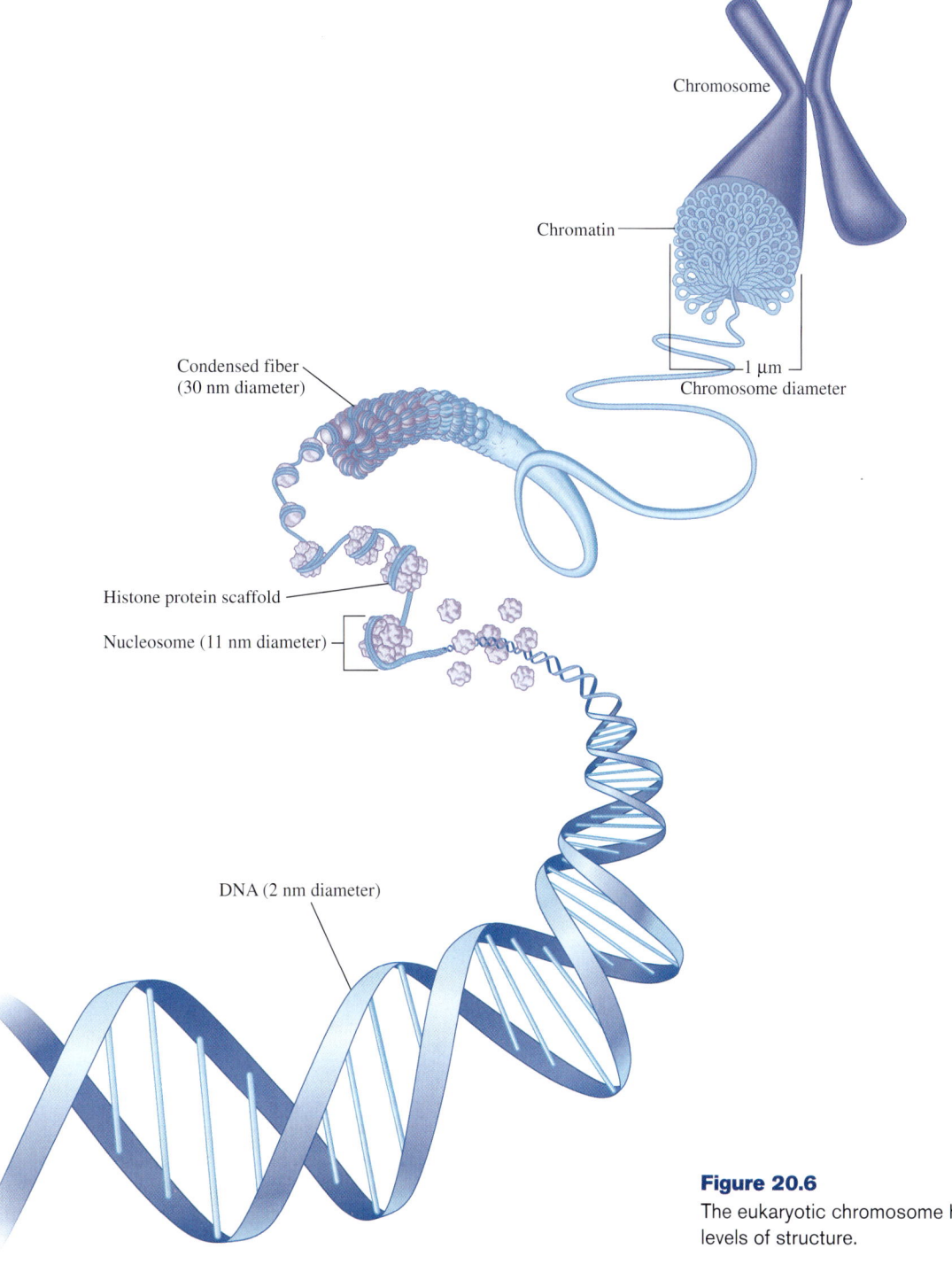

Figure 20.6
The eukaryotic chromosome has many levels of structure.

it is essential that the process of DNA replication produces an absolutely accurate copy of the original genetic information. If mistakes are made in critical genes, the result may be lethal mutations.

The structure of the DNA molecule suggested the mechanism for its accurate replication. Since adenine can base pair only with thymine and cytosine with guanine, Watson and Crick first suggested that an enzyme could "read" the nitrogenous bases on one strand of a DNA molecule and add complementary bases to a strand of DNA being synthesized. The product of this mechanism would be a new DNA molecule in which one strand is the original, or parent, strand and the

A Medical Perspective

Fooling the AIDS Virus with "Look-Alike" Nucleotides

The virus that is responsible for the acquired immune deficiency syndrome (AIDS) is called the *human immunodeficiency virus*, or *HIV*. HIV is a member of a family of viruses called *retroviruses*, all of which have single-stranded RNA as their genetic material. The RNA is copied by a viral enzyme called *reverse transcriptase* into a double-stranded DNA molecule. This process is the opposite of the central dogma, which states that the flow of genetic information is from DNA to RNA. But these viruses reverse that flow, RNA to DNA. For this reason, these viruses are called retroviruses, which literally means "backward viruses." The process of producing a DNA copy of the RNA is called *reverse transcription*.

Because our genetic information is DNA and it is expressed by the classical DNA → RNA → protein pathway, our cells have no need for a reverse transcriptase enzyme. Thus, the HIV reverse transcriptase is a good target for antiviral chemotherapy because inhibition of reverse transcription should kill the virus but have no effect on the human host. Many drugs have been tested for the ability to selectively inhibit HIV reverse transcription. Among these is the DNA chain terminator 3'-azido-2', 3'-dideoxythymidine, commonly called *AZT* or *zidovudine*.

How does AZT work? It is one of many drugs that looks like one of the normal nucleosides. These are called *nucleoside analogs*. A nucleoside is just a nucleotide without any phosphate groups attached. The analog is phosphorylated by the cell and then tricks a polymerase, in this case viral reverse transcriptase, into incorporating it into the growing DNA chain in place of the normal phosphorylated nucleoside. AZT is a nucleoside analog that looks like the nucleoside thymidine except that in the 3' position of the deoxyribose sugar there is an azido group ($-N_3$) rather than the 3'-OH group. Compare the structures of thymidine and AZT shown in the accompanying figure. The 3'-OH group is necessary for further DNA polymerization because it is there that the phosphoester linkage must be made between the growing DNA strand and the next nucleotide. If an azido group or some other group is present at the 3' position, the nucleotide analog can be incorporated into the growing DNA strand, but further chain elongation is blocked, as shown in the following figure. If the viral RNA cannot be reverse transcribed into the DNA form, the virus will not be able to replicate and can be considered to be dead.

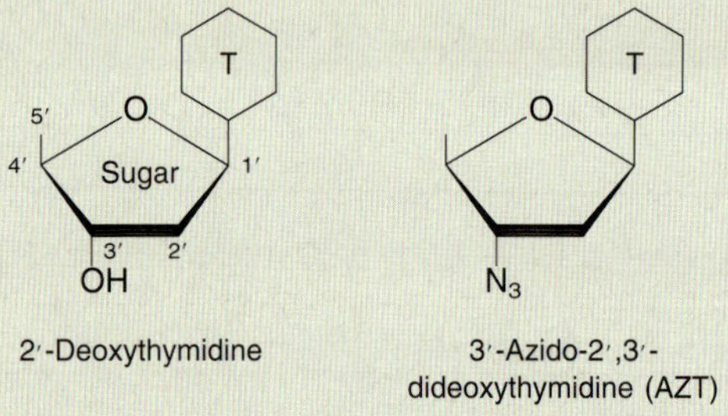

Comparison of the structures of the normal nucleoside, 2'-deoxythymidine, and the nucleoside analog, 3'-azido-2', 3'-dideoxythymidine.

AZT is particularly effective because the HIV reverse transcriptase actually prefers it over the normal nucleotide, thymidine. Nonetheless, AZT is not a cure. At best it prolongs the life of a person with AIDS for a year or two. Eventually, however, AZT has a negative effect on the body. The cells of our bone marrow are constantly dividing to produce new blood cells: red blood cells to carry oxygen to the tissues, white blood cells of the immune system, and platelets for blood clotting. For cells to divide, they must replicate their DNA. The DNA polymerases of these dividing cells also accidentally incorporate AZT into

---

second strand is a newly synthesized, or daughter, strand. This mode of DNA replication is called **semiconservative replication** (Figure 20.7).

Experimental evidence for this mechanism of DNA replication was provided by an experiment designed by Matthew Meselson and Franklin Stahl in 1958. *Escherichia coli* cells were grown in a medium in which $^{15}NH_4^+$ was the sole nitrogen source. ^{15}N is a nonradioactive, heavy isotope of nitrogen. Thus, growing the cells in this medium resulted in all of the cellular DNA containing this heavy isotope.

The cells containing only $^{15}NH_4^+$ were then added to a medium containing only the abundant isotope of nitrogen, $^{14}NH_4^+$, and were allowed to grow for one cycle of cell division. When the daughter DNA molecules were isolated and analyzed, it was found that each was made up of one strand of "heavy" DNA, the parental strand, and one strand of "light" DNA, the new daughter strand. After a second round of cell division, half of the isolated DNA contained no ^{15}N and half contained a 50/50 mixture of ^{14}N and ^{15}N-labeled DNA (Figure 20.8). This demonstrated conclusively that each parental strand of the DNA molecule serves as the

> Isotopes are atoms of the same element having the same number of protons but different numbers of neutrons and, therefore, different mass numbers.

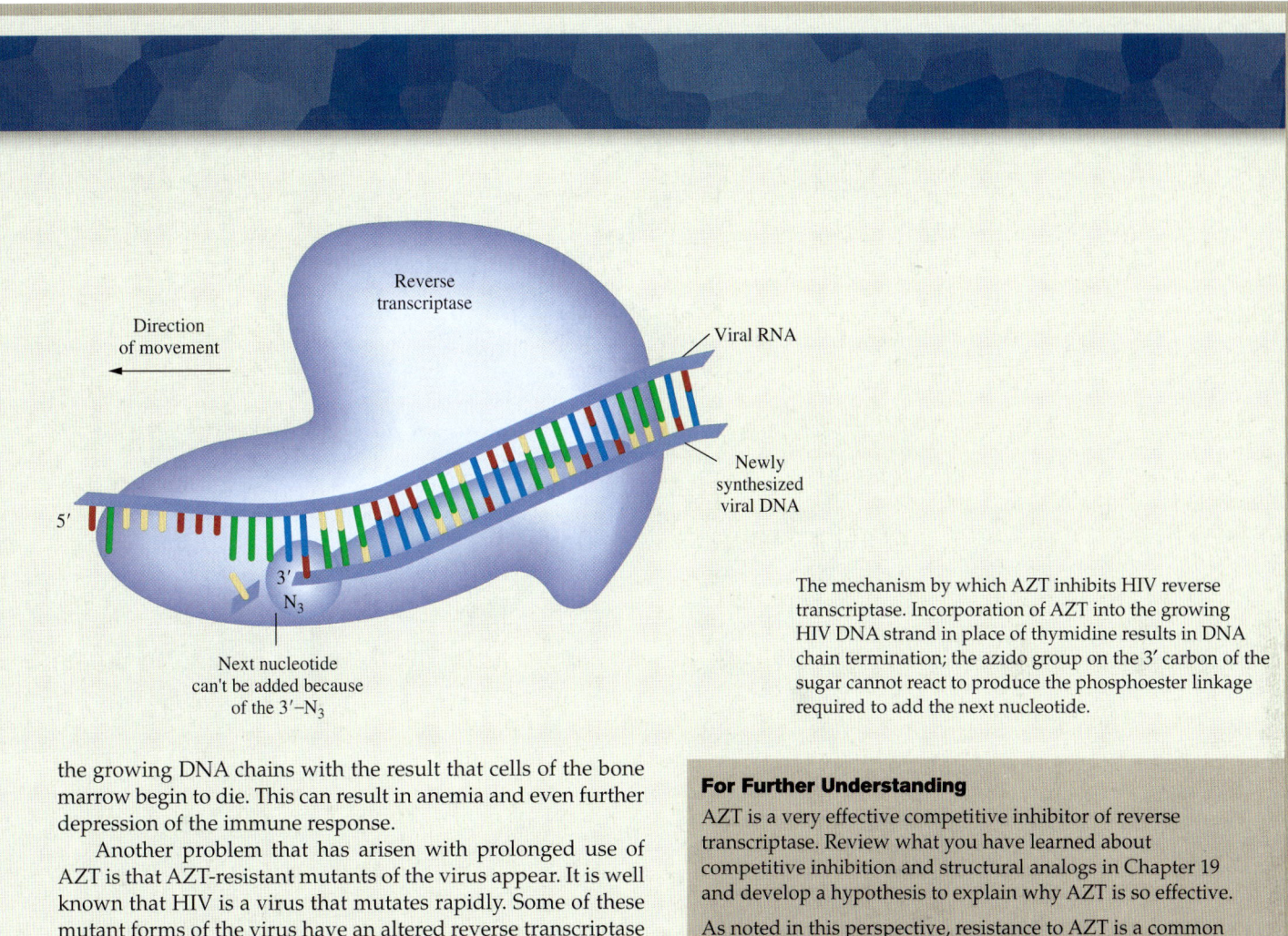

The mechanism by which AZT inhibits HIV reverse transcriptase. Incorporation of AZT into the growing HIV DNA strand in place of thymidine results in DNA chain termination; the azido group on the 3′ carbon of the sugar cannot react to produce the phosphoester linkage required to add the next nucleotide.

the growing DNA chains with the result that cells of the bone marrow begin to die. This can result in anemia and even further depression of the immune response.

Another problem that has arisen with prolonged use of AZT is that AZT-resistant mutants of the virus appear. It is well known that HIV is a virus that mutates rapidly. Some of these mutant forms of the virus have an altered reverse transcriptase that will no longer use AZT. When these mutants appear, AZT is no longer useful in treating the infection.

Fortunately, research with other nucleoside analogs, alternative types of antiviral treatments, and combinations of drugs has provided a more effective means of treating HIV infection.

For Further Understanding

AZT is a very effective competitive inhibitor of reverse transcriptase. Review what you have learned about competitive inhibition and structural analogs in Chapter 19 and develop a hypothesis to explain why AZT is so effective.

As noted in this perspective, resistance to AZT is a common problem. This has led to the simultaneous use of two or more drugs in the treatment of HIV AIDS, for instance, AZT and a protease inhibitor. Explain why multiple drug therapy reduces the problem of viral drug resistance.

template for the synthesis of a daughter strand and that each newly synthesized DNA molecule is composed of one parental strand and one newly synthesized daughter strand.

Bacterial DNA Replication

The bacterial chromosome is a circular molecule of DNA made up of about three million nucleotides. DNA replication begins at a unique sequence on the circular chromosome known as the **replication origin** (Figure 20.9). Replication occurs bidirectionally at the rate of about five hundred new nucleotides every second! The point at which the new deoxyribonucleotide is added to the growing daughter strand is called the **replication fork** (Figure 20.9). It is here that the DNA has been opened to allow binding of the various proteins and enzymes responsible for DNA replication. Since DNA synthesis occurs bidirectionally, there are two replication forks moving in opposite directions. Replication is complete when the replication forks collide approximately half way around the circular chromosome.

Figure 20.7
In semiconservative DNA replication, each parent strand serves as a template for the synthesis of a new daughter strand.

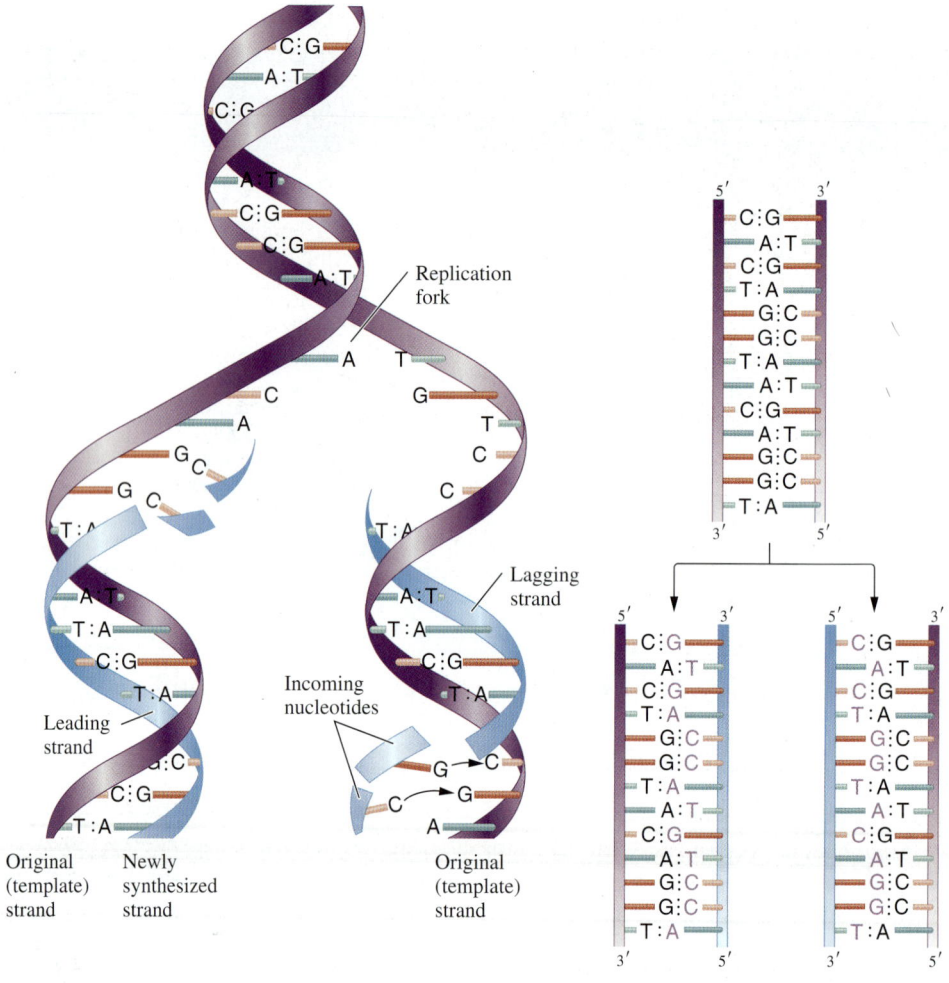

(a) The mechanism of DNA replication

(b) The products of replication

The first step in DNA replication (Figure 20.10) is the separation of the strands of DNA. The protein *helicase* does this by breaking the hydrogen bonds between the base pairs. This, in turn, causes supercoiling of the molecule. This stress is relieved by the enzyme *topoisomerase*, which travels along the DNA ahead of the replication fork. At this point, *single-strand binding protein* binds to the separated strands, preventing them from coming back together. In the next step, the enzyme *primase* catalyzes the synthesis of a small piece of RNA (ten to twelve nucleotides) called an *RNA primer* that serves to "prime" the process of DNA replication.

Now the enzyme **DNA polymerase III** "reads" each parental strand, also called the *template*, and catalyzes the polymerization of a complementary daughter strand. Deoxyribonucleotide triphosphate molecules are the precursors for DNA replication (Figure 20.11). In this reaction, a pyrophosphate group is released as a phosphoester bond is formed between the 5′-phosphoryl group of the nucleotide being added to the chain and the 3′-OH of the nucleotide on the daughter strand. This is called 5′ to 3′ synthesis.

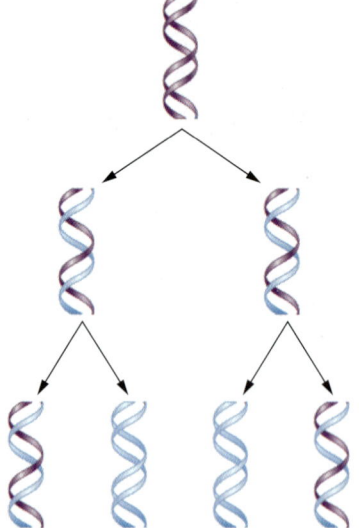

Figure 20.8
Representation of the Meselson and Stahl experiment. The DNA from cells grown in medium containing $^{15}NH_4^+$ is shown in purple. After a single generation in medium containing $^{14}NH_4^+$, the daughter DNA molecules have one ^{15}N-labeled parent strand and one ^{14}N-labeled daughter strand (blue). After a second generation in $^{14}NH_4^+$ containing medium, there are equal numbers of $^{14}N/^{15}N$ DNA molecules and $^{14}N/^{14}N$ DNA molecules.

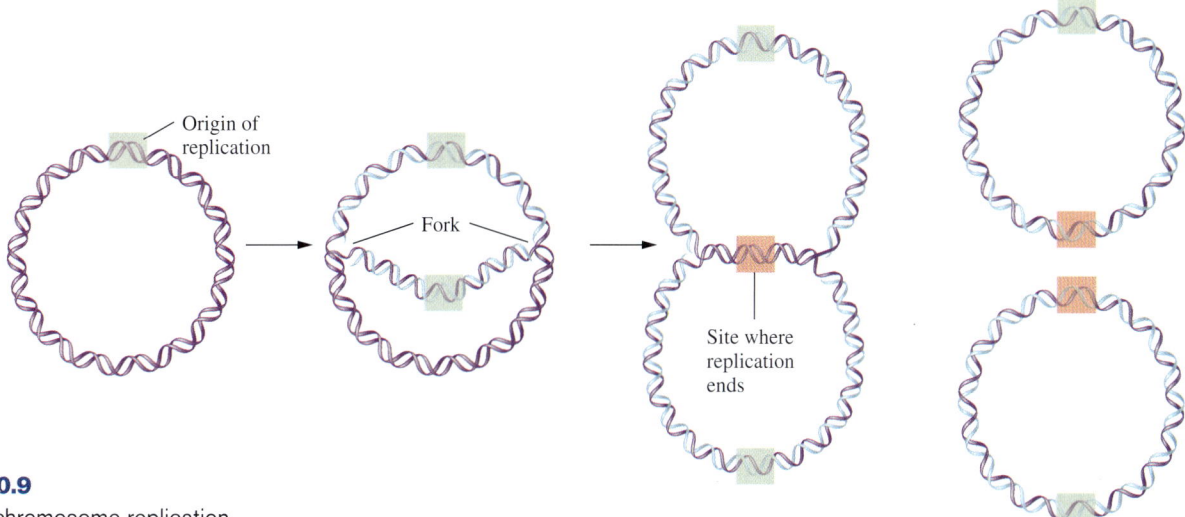

Figure 20.9
Bacterial chromosome replication.

Functions of key proteins involved with DNA replication

- **DNA helicase** breaks the hydrogen bonds between the DNA strands.
- **Topoisomerase** alleviates positive supercoiling.
- **Single-strand binding proteins** keep the parental strands apart.
- **Primase** synthesizes an RNA primer.
- **DNA polymerase III** synthesizes a daughter strand of DNA.
- **DNA polymerase I** excises the RNA primers and fills in with DNA.
- **DNA ligase** covalently links the DNA fragments together.

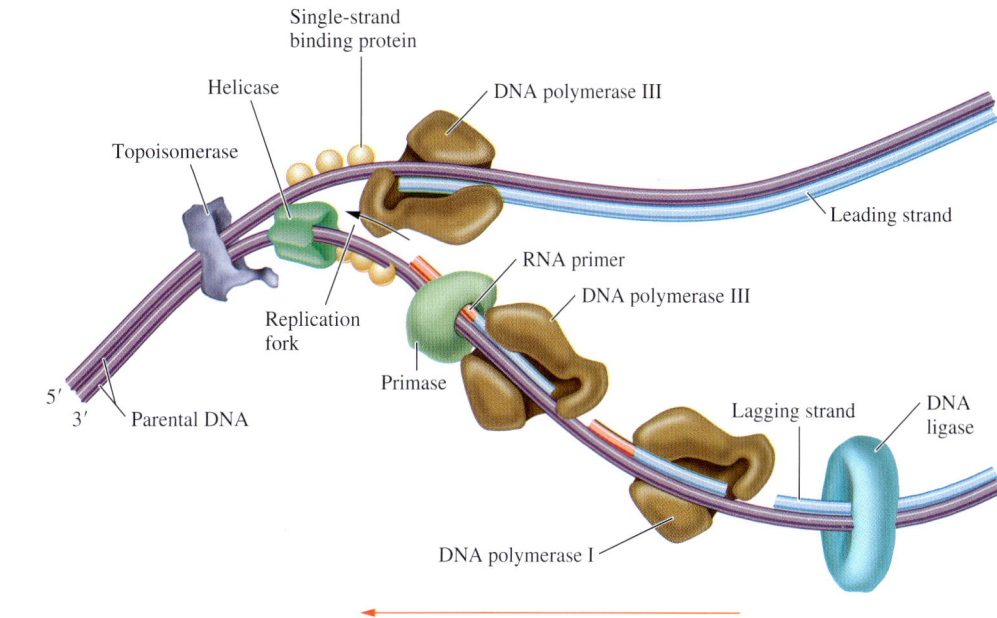

Figure 20.10
Because the two strands of DNA are antiparallel and DNA polymerase can only catalyze 5′ → 3′ replication, only one of the two DNA strands (top strand) can be read continuously to produce a daughter strand. The other must be synthesized in segments that are extended away from the direction of movement of the replication fork (bottom strand). These discontinuous segments are later covalently joined together by DNA ligase.

One complicating factor in the process of DNA replication is the fact that the two strands of DNA are antiparallel to one another. DNA polymerase III can only catalyze DNA chain elongation in the 5′ to 3′ direction, yet the replication fork proceeds in one direction, while both strands are replicated simultaneously. Another complication is the need for an RNA primer to serve as the starting point for DNA replication. As a result of these two obstacles, there are different mechanisms for replication of the two strands. One strand, called the **leading strand,** is replicated continuously. The opposite strand, called the **lagging strand,** is replicated discontinuously.

Figure 20.11
The reaction catalyzed by DNA polymerase.

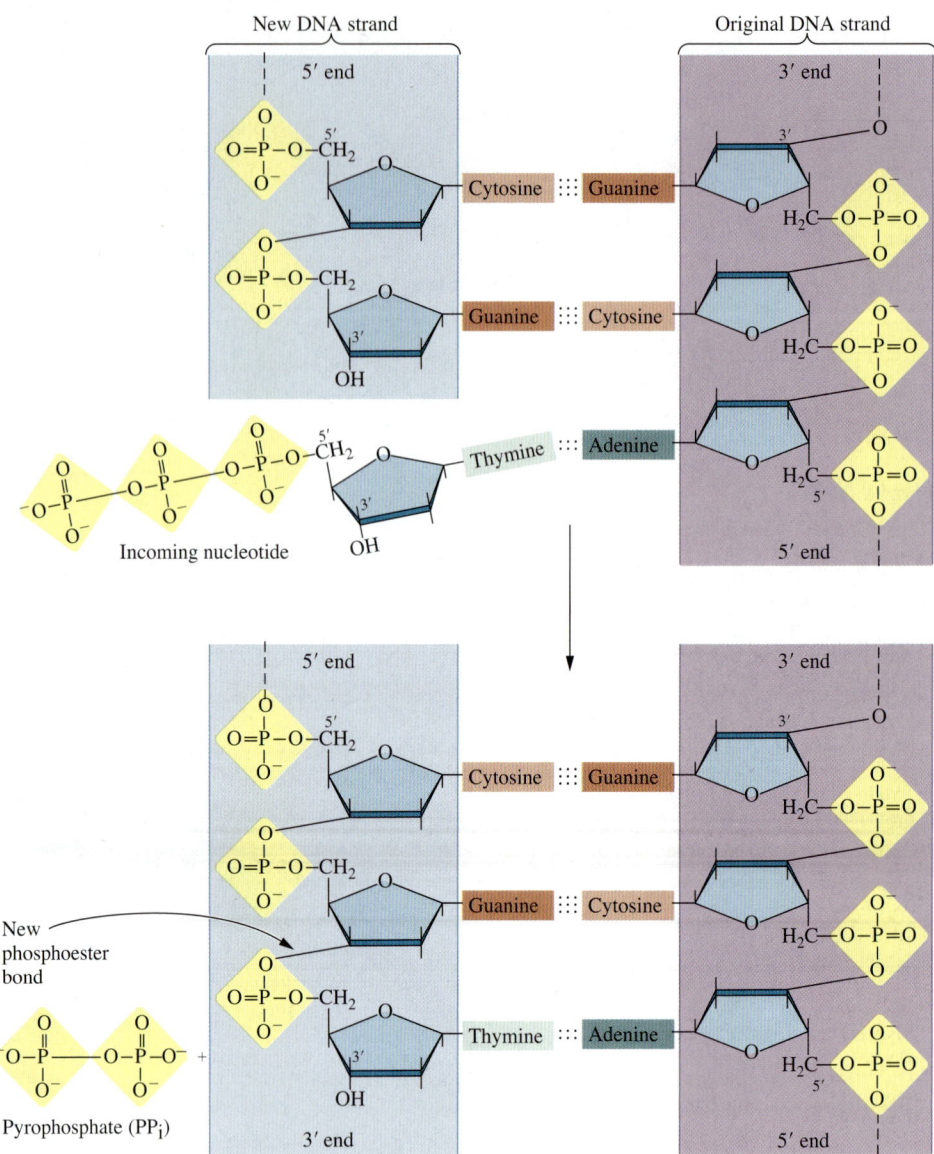

The two mechanisms are shown in Figures 20.10 and 20.12. For the leading strand, a single RNA primer is produced at the replication origin and DNA polymerase III continuously catalyzes the addition of nucleotides in the 5' to 3' direction, beginning with addition of the first nucleotide to the RNA primer.

On the lagging strand, many RNA primers are produced as the replication fork proceeds along the molecule. DNA polymerase III catalyzes DNA chain elongation from each of these primers. When the new strand "bumps" into a previous one, synthesis stops at that site. Meanwhile, at the replication fork, a new primer is being synthesized by primase. The final steps of synthesis on the lagging strand involve removal of the primers, repair of the gaps, and sealing of the fragments into an intact strand of DNA. The enzyme DNA polymerase I catalyzes the removal of the RNA primer and its replacement with DNA nucleotides. In the final step of the process, the enzyme DNA ligase catalyzes the formation of a phosphoester bond between the two adjacent fragments. It is little wonder that this is referred to as lagging strand replication! A more accurate model of the replication fork is shown in Figure 20.12.

Because it is critical to produce an accurate copy of the parental DNA, it is very important to avoid errors in the replication process. In addition to catalyzing

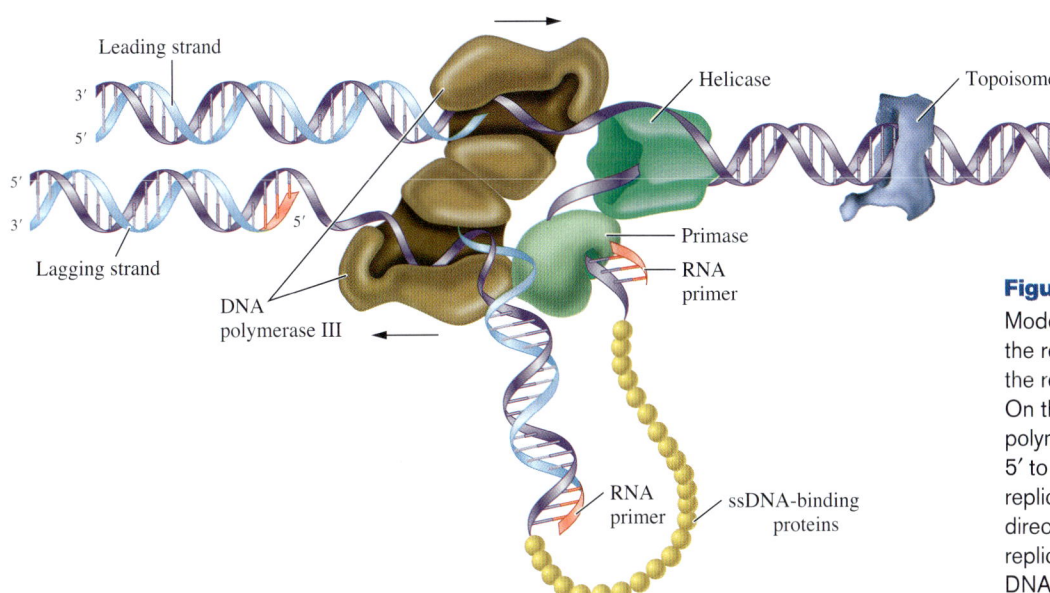

Figure 20.12
Model of the complex events occurring at the replication fork. In this representation, the replication fork is moving to the right. On the leading (top) strand, DNA polymerase III synthesizes DNA in the 5' to 3' direction continuously. Thus, replication proceeds in the same direction as the movement of the replication fork. On the lagging strand, DNA polymerase III also synthesizes DNA in the 5' to 3' direction. However, since the DNA strands are antiparallel, DNA polymerase III must read this strand in short segments (discontinuously) and in the opposite direction of the movement of the replication fork.

the replication of new DNA, DNA polymerase III is able to proofread the newly synthesized strand. If the wrong nucleotide has been added to the growing DNA strand, it is removed and replaced with the correct one. In this way, a faithful copy of the parental DNA is ensured.

Eukaryotic DNA Replication

DNA replication in eukaryotes is more complex. The human genome consists of approximately three billion nucleotide pairs. Just one chromosome may be nearly one hundred times longer than a bacterial chromosome. To accomplish this huge job, DNA replication begins at many replication origins and proceeds bidirectionally along each chromosome.

20.4 Information Flow in Biological Systems

The **central dogma** of molecular biology states that in cells the flow of genetic information contained in DNA is a one-way street that leads from DNA to RNA to protein. The process by which a single strand of DNA serves as a template for the synthesis of an RNA molecule is called **transcription**. The word *transcription* is derived from the Latin word *transcribere* and simply means "to make a copy." Thus, in this process, part of the information in the DNA is copied into a strand of RNA. The process by which the message is converted into protein is called **translation.** Unlike transcription the process of translation involves converting the information from one language to another. In this case the genetic information in the linear sequence of nucleotides is being translated into a protein, a linear sequence of amino acids. The expression of the information contained in DNA is fundamental to the growth, development, and maintenance of all organisms.

Classes of RNA Molecules

Three classes of RNA molecules are produced by transcription: messenger RNA, transfer RNA, and ribosomal RNA.

 **LEARNING GOAL**

1. **Messenger RNA (mRNA)** carries the genetic information for a protein from DNA to the ribosomes. It is a complementary RNA copy of a gene on the DNA.

Figure 20.13
Structure of tRNA. The primary structure of a tRNA is the linear sequence of ribonucleotides. Here we see the hydrogen-bonded secondary structure of a tRNA showing the three loops and the amino acid accepting end.

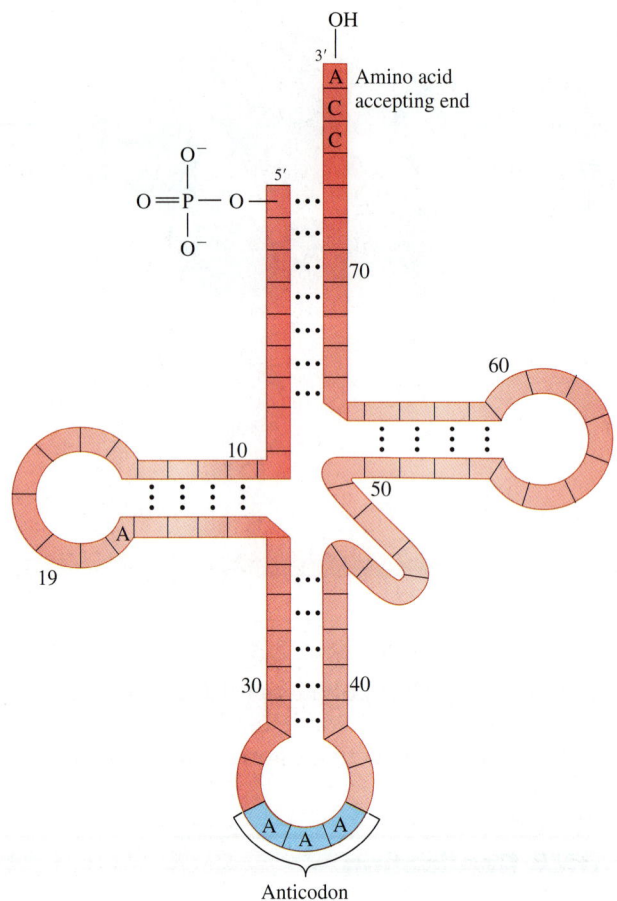

2. **Ribosomal RNA (rRNA)** is a structural and functional component of the ribosomes, which are "platforms" on which protein synthesis occurs. There are three types of rRNA molecules in bacterial ribosomes and four in the ribosomes of eukaryotes.
3. **Transfer RNA (tRNA)** translates the genetic code of the mRNA into the primary sequence of amino acids in the protein. In addition to the primary structure, tRNA molecules have a cloverleaf-shaped secondary structure resulting from base pair hydrogen bonding (A—U and G—C) and a roughly L-shaped tertiary structure (Figure 20.13). The sequence CCA is found at the 3′ end of the tRNA. The 3′–OH group of the terminal nucleotide, adenosine, can be covalently attached to an amino acid. Three nucleotides at the base of the cloverleaf structure form the **anticodon.** As we will discuss in more detail in Section 20.6, this triplet of bases forms hydrogen bonds to a **codon** (complementary sequence of bases) on a messenger RNA (mRNA) molecule on the surface of a ribosome during protein synthesis. This hydrogen bonding of codon and anticodon brings the correct amino acid to the site of protein synthesis at the appropriate location in the growing peptide chain.

Transcription

LEARNING GOAL

Transcription, shown in Figure 20.14, is catalyzed by the enzyme **RNA polymerase.** The process occurs in three stages. The first, called *initiation*, involves binding of RNA polymerase to a specific nucleotide sequence, the **promoter,** at the beginning of a gene. This interaction of RNA polymerase with specific promoter DNA se-

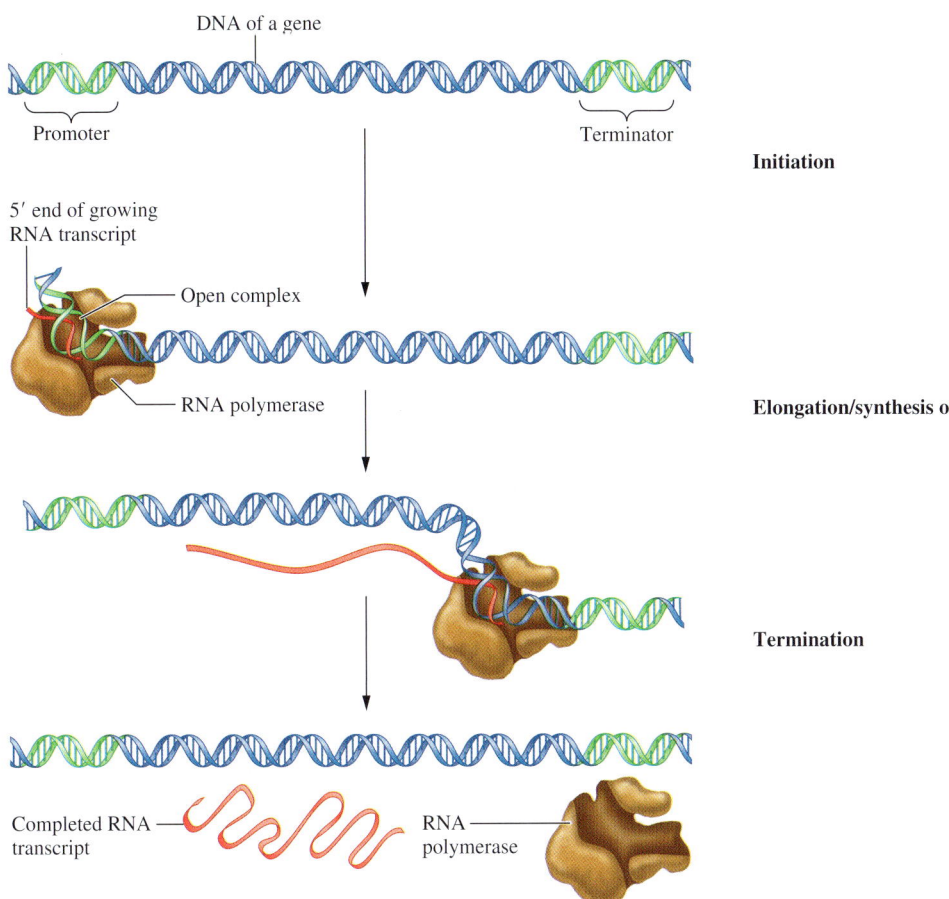

Figure 20.14
The stages of transcription.

quences allows RNA polymerase to recognize the start point for transcription. It also determines which DNA strand will be transcribed. Unlike DNA replication, transcription produces a complementary copy of only one of the two strands of DNA. As it binds to the DNA, RNA polymerase separates the two strands of DNA so that it can "read" the base sequence of the DNA.

The second stage, chain elongation, begins as the RNA polymerase "reads" the DNA template strand and catalyzes the polymerization of a complementary RNA copy. With each catalytic step, RNA polymerase transfers a complementary ribonucleotide to the end of the growing RNA chain and catalyzes the formation of a 3′–5′ phosphodiester bond between the 5′ phosphoryl group of the incoming ribonucleotide and the 3′ hydroxyl group of the last ribonucleotide of the growing RNA chain.

The final stage of transcription is termination. The RNA polymerase finds a termination sequence at the end of the gene and releases the newly formed RNA molecule.

Question 20.3 What is the function of RNA polymerase in the process of transcription?

Question 20.4 What is the function of the promoter sequence in the process of transcription?

LEARNING GOAL

Post-transcriptional Processing of RNA

In bacteria, which are prokaryotes, termination releases a mature mRNA for translation. In fact, because prokaryotes have no nuclear membrane separating the DNA from the cytoplasm, translation begins long before the mRNA is completed. In eukaryotes, transcription produces a **primary transcript** that must undergo extensive **post-transcriptional modification** before it is exported out of the nucleus for translation in the cytoplasm.

Eukaryotic primary transcripts undergo three post-transcriptional modifications. These are the addition of a 5′ cap structure and a 3′ poly(A) tail, and RNA splicing.

In the first modification, a **cap structure** is enzymatically added to the 5′ end of the primary transcript. The cap structure (Figure 20.15) consists of 7-methyl-guanosine attached to the 5′ end of the RNA by a 5′–5′ triphosphate bridge. The first two nucleotides of the mRNA are also methylated. The cap structure is required for efficient translation of the final mature mRNA.

The second modification is the enzymatic addition of a **poly(A)** tail to the 3′ end of the transcript. *Poly(A) polymerase* uses ATP and catalyzes the stepwise polymerization of one hundred to two hundred adenosine nucleotides on the 3′ end of the RNA. The poly(A) tail protects the 3′ end of the mRNA from enzymatic degradation and thus prolongs the lifetime of the mRNA.

The third modification, **RNA splicing,** involves the removal of portions of the primary transcript that are not protein coding. Bacterial genes are continuous; all the nucleotide sequences of the gene are found in the mRNA. However, study of the gene structure of eukaryotes revealed a fascinating difference. Eukaryotic genes are discontinuous; there are *extra* DNA sequences within these genes that do not encode any amino acid sequences for the protein. These sequences are called *intervening sequences* or **introns.** The primary transcript contains both the introns and the protein coding sequences, called **exons.** The presence of introns in the mRNA would make it impossible for the process of translation to synthesize the

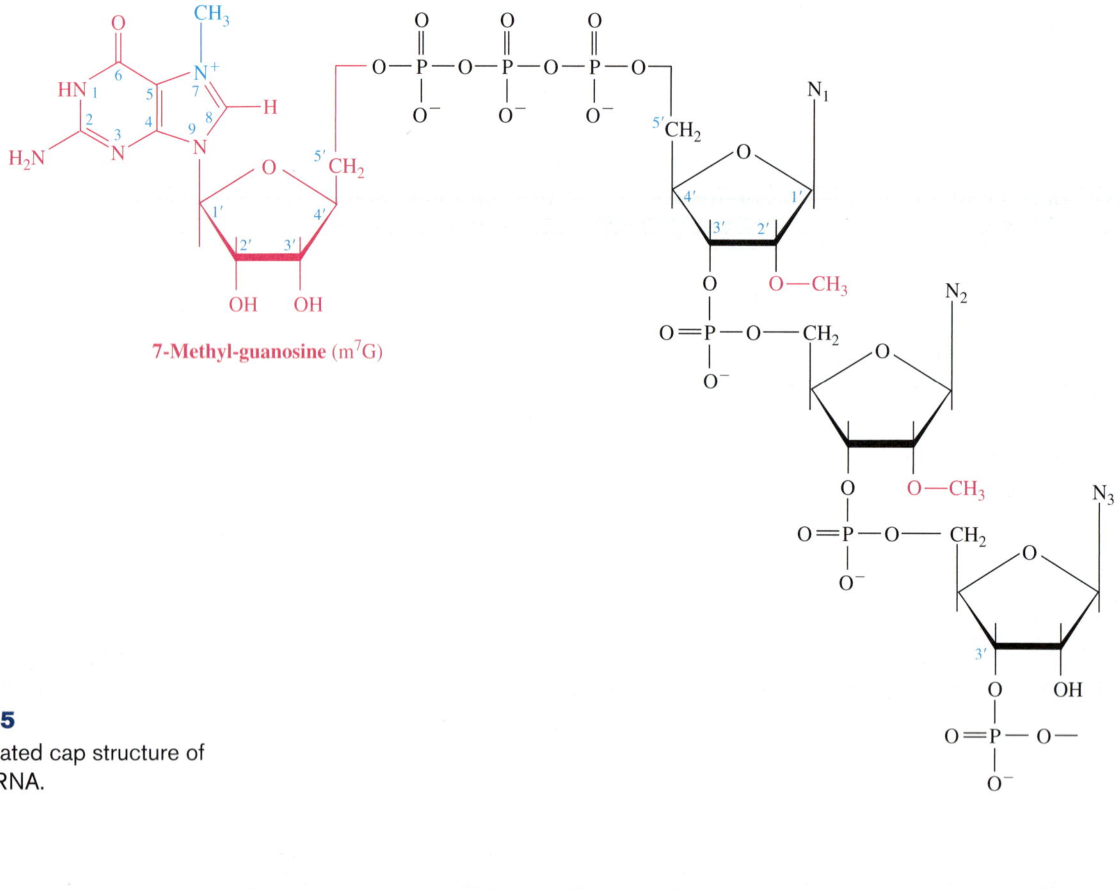

Figure 20.15
The 5′-methylated cap structure of eukaryotic mRNA.

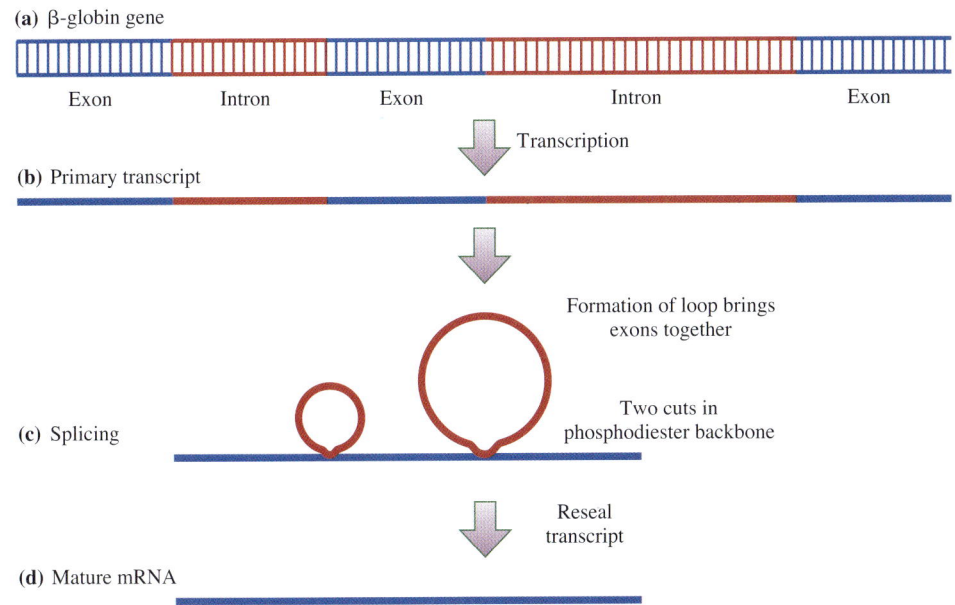

Figure 20.16
Schematic diagram of mRNA splicing. (a) The β-globin gene contains protein coding exons, as well as noncoding sequences called introns. (b) The primary transcript of the DNA carries both the introns and the exons. (c) The introns are looped out, the phosphodiester backbone of the mRNA is cut twice, and the pieces are tied together. (d) The final mature mRNA now carries only the coding sequences (exons) of the gene.

correct protein. Therefore they must be removed, which is done by the process of RNA splicing.

As you can imagine, RNA splicing must be very precise. If too much, or too little, RNA is removed, the mRNA will not carry the correct code for the protein. Thus, there are "signals" in the DNA to mark the boundaries of the introns. The sequence GpU is always found at the intron's 5′ boundary and the sequence ApG is found at the 3′ boundary.

Recognition of the splice boundaries and stabilization of the splicing complex requires the assistance of particles called *spliceosomes*. Spliceosomes are composed of a variety of *small nuclear ribonucleoproteins* (snRNPs, read "snurps"). Each snRNP consists of a small RNA and associated proteins. The RNA components of different snRNPs are complementary to different sequences involved in splicing. By hydrogen bonding to a splice boundary or intron sequences the snRNPs recognize and bring together the sequences involved in the splicing reactions.

One of the first eukaryotic genes shown to contain introns was the gene for the β subunit of adult hemoglobin (Figure 20.16). On the DNA, the gene for β-hemoglobin is 1200 nucleotides long, but only 438 nucleotides carry the genetic information for protein. The remaining sequences are found in two introns of 116 and 646 nucleotides that are removed by splicing before translation. It is interesting that the larger intron is longer than the final β-hemoglobin mRNA! In the genes that have been studied, introns have been found to range in size from 50 to 20,000 nucleotides in length, and there may be many throughout a gene. Thus a typical human gene might be 10–30 times longer than the final mRNA.

20.5 The Genetic Code

The mRNA carries the genetic code for a protein. But what is the nature of this code? In 1954, George Gamow proposed that because there are only four "letters" in the DNA alphabet (A, T, G, and C) and because there are twenty amino acids, the genetic code must contain words made of at least three letters taken from the four letters in the DNA alphabet. How did he come to this conclusion? He reasoned that a code of two-letter words constructed from any combination of the four letters has a "vocabulary" of only sixteen words (4^2). In other words, there are only sixteen different ways to put A, T, C, and G together two bases at a time (AA,

LEARNING GOAL

AT, AC, AG, TT, TA, etc.). That is not enough to encode all twenty amino acids. A code of four-letter words gives 256 words (4^4), far more than are needed. A code of three-letter words, however, has a possible vocabulary of sixty-four words (4^3), sufficient to encode the twenty amino acids but not too excessive.

A series of elegant experiments proved that Gamow was correct by demonstrating that the genetic code is, indeed, a triplet code. Mutations were introduced into the DNA of a bacterial virus. These mutations inserted (or deleted) one, two, or three nucleotides into a gene. The researchers then looked for the protein encoded by that gene. When one or two nucleotides were inserted, no protein was produced. However, when a third base was inserted, the sense of the mRNA was restored, and the protein was made. You can imagine this experiment by using a sentence composed of only three-letter words. For instance,

THE CAT RAN OUT

What happens to the "sense" of the sentence if we insert one letter?

THE FCA TRA NOU T

The reading frame of the sentence has been altered, and the sentence is now nonsense. Can we now restore the sense of the sentence by inserting a second letter?

THE FAC ATR ANO UT

No, we have not restored the sense of the sentence. Once again, we have altered the reading frame, but because our code has only three-letter words, the sentence is still nonsense. If we now insert a third letter, it should restore the correct reading frame:

THE FAT CAT RAN OUT

Indeed, by inserting three new letters we have restored the sense of the message by restoring the reading frame. This is exactly the way in which the message of the mRNA is interpreted. Each group of three nucleotides in the sequence of the mRNA is called a *codon,* and each codes for a single amino acid. If the sequence is interrupted or changed, it can change the amino acid composition of the protein that is produced or even result in the production of no protein at all.

As we noted, a three-letter genetic code contains sixty-four words, called *codons,* but there are only twenty amino acids. Thus, there are forty-four more codons than are required to specify all of the amino acids found in proteins. Three of the codons—UAA, UAG, and UGA—specify termination signals for the process of translation. But this still leaves us with forty-one additional codons. What is the function of the "extra" code words? Francis Crick (recall Watson and Crick and the double helix) proposed that the genetic code is a **degenerate code.** The term *degenerate* is used to indicate that different triplet codons may serve as code words for the same amino acid.

The complete genetic code is shown in Figure 20.17. We can make several observations about the genetic code. First, methionine and tryptophan are the only amino acids that have a single codon. All others have at least two codons, and serine and leucine have six codons each. The genetic code is also somewhat mutation-resistant. For those amino acids that have multiple codons, the first two bases are often identical and thus identify the amino acid, and only the third position is variable. Mutations—changes in the nucleotide sequence—in the third position therefore often have no effect on the amino acid that is incorporated into a protein.

Question 20.5 Why is the genetic code said to be degenerate?

Question 20.6 Why is the genetic code said to be mutation-resistant?

Figure 20.17

FIRST BASE	SECOND BASE				THIRD BASE
	U	C	A	G	
U	UUU Phenylalanine	UCU Serine	UAU Tyrosine	UGU Cysteine	U
U	UUC Phenylalanine	UCC Serine	UAC Tyrosine	UGC Cysteine	C
U	UUA Leucine	UCA Serine	UAA STOP	UGA STOP	A
U	UUG Leucine	UCG Serine	UAG STOP	UGG Tryptophan	G
C	CUU Leucine	CCU Proline	CAU Histidine	CGU Arginine	U
C	CUC Leucine	CCC Proline	CAC Histidine	CGC Arginine	C
C	CUA Leucine	CCA Proline	CAA Glutamine	CGA Arginine	A
C	CUG Leucine	CCG Proline	CAG Glutamine	CGG Arginine	G
A	AUU Isoleucine	ACU Threonine	AAU Asparagine	AGU Serine	U
A	AUC Isoleucine	ACC Threonine	AAC Asparagine	AGC Serine	C
A	AUA Isoleucine	ACA Threonine	AAA Lysine	AGA Arginine	A
A	AUG (START) Methionine	ACG Threonine	AAG Lysine	AGG Arginine	G
G	GUU Valine	GCU Alanine	GAU Aspartic acid	GGU Glycine	U
G	GUC Valine	GCC Alanine	GAC Aspartic acid	GGC Glycine	C
G	GUA Valine	GCA Alanine	GAA Glutamic acid	GGA Glycine	A
G	GUG Valine	GCG Alanine	GAG Glutamic acid	GGG Glycine	G

The genetic code. The table shows the possible codons found in mRNA. To read the universal biological language from this chart, find the first base in the column on the left, the second base from the row across the top, and the third base from the column to the right. This will direct you to one of the sixty-four squares in the matrix. Within that square you will find the codon and the amino acid that it specifies. In the cell this message is decoded by tRNA molecules like those shown to the right of the table.

20.6 Protein Synthesis

The process of protein synthesis is called *translation*. It involves translating the genetic information from the sequence of nucleotides into the sequence of amino acids in the primary structure of a protein. Figure 20.18 shows the relationship through which the nucleotide sequence of a DNA molecule is transcribed into a complementary sequence of ribonucleotides, the mRNA molecule. Each mRNA has a short untranslated region followed by the sequences that carry the information for the order of the amino acids in the protein that will be produced in the process of translation. That genetic information is the sequence of codons along the mRNA. The decoding process is carried out by tRNA molecules.

Translation is carried out on **ribosomes,** which are complexes of ribosomal RNA (rRNA) and proteins. Each ribosome is made up of two subunits: a small and a large ribosomal subunit (Figure 20.19a). In eukaryotic cells, the small ribosomal subunit contains one rRNA molecule and thirty-three different ribosomal proteins, and the large subunit contains three rRNA molecules and about forty-nine different proteins.

LEARNING GOAL 8

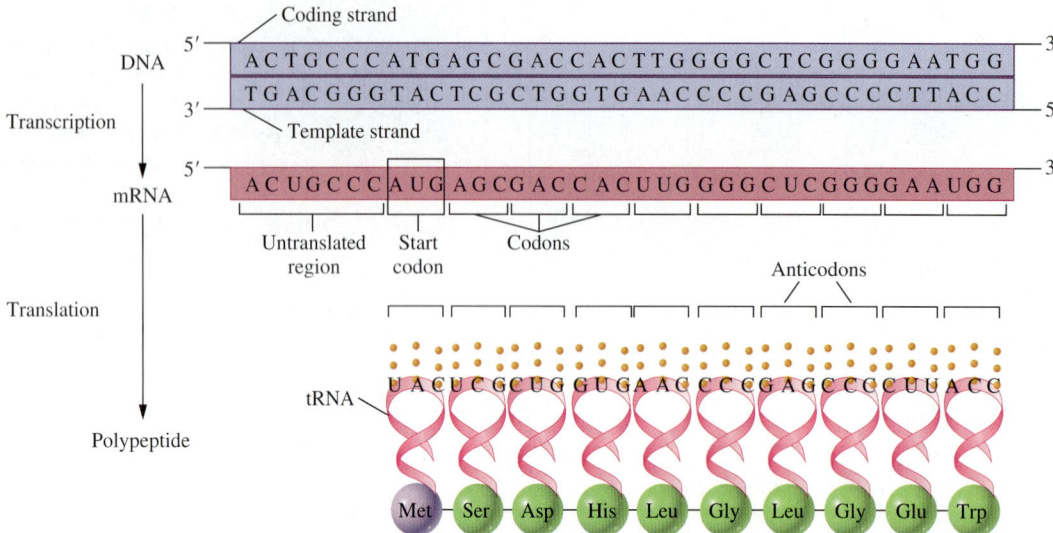

Figure 20.18
Messenger RNA (mRNA) is an RNA copy of one strand of a gene in the DNA. Each codon on the mRNA that specifies a particular amino acid is recognized by the complementary anticodon on a transfer RNA (tRNA).

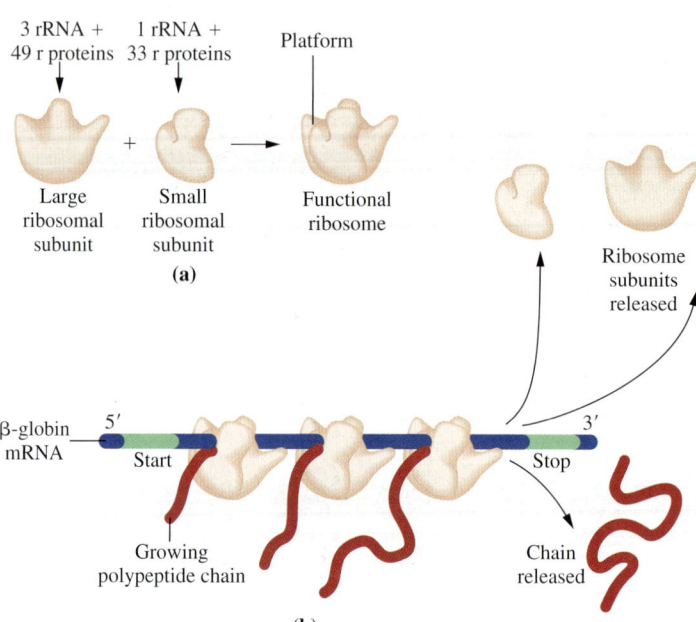

Figure 20.19
Structure of the ribosome. (a) The large and small subunits form the functional complex in association with an mRNA molecule. (b) A polyribosome translating the mRNA for a β-globin chain of hemoglobin.

Protein synthesis involves the simultaneous action of many ribosomes on a single mRNA molecule. These complexes of many ribosomes along a single mRNA are known as *polyribosomes* or **polysomes** (Figure 20.19b). Each ribosome is synthesizing one copy of the protein molecule encoded by the mRNA. Thus, many copies of a protein are simultaneously produced.

The Role of Transfer RNA

The codons of mRNA must be read if the genetic message is to be translated into protein. The molecule that decodes the information in the mRNA molecule into the primary structure of a protein is transfer RNA (tRNA). To decode the genetic message into the primary sequence of a protein, the tRNA must faithfully perform two functions.

20.6 Protein Synthesis

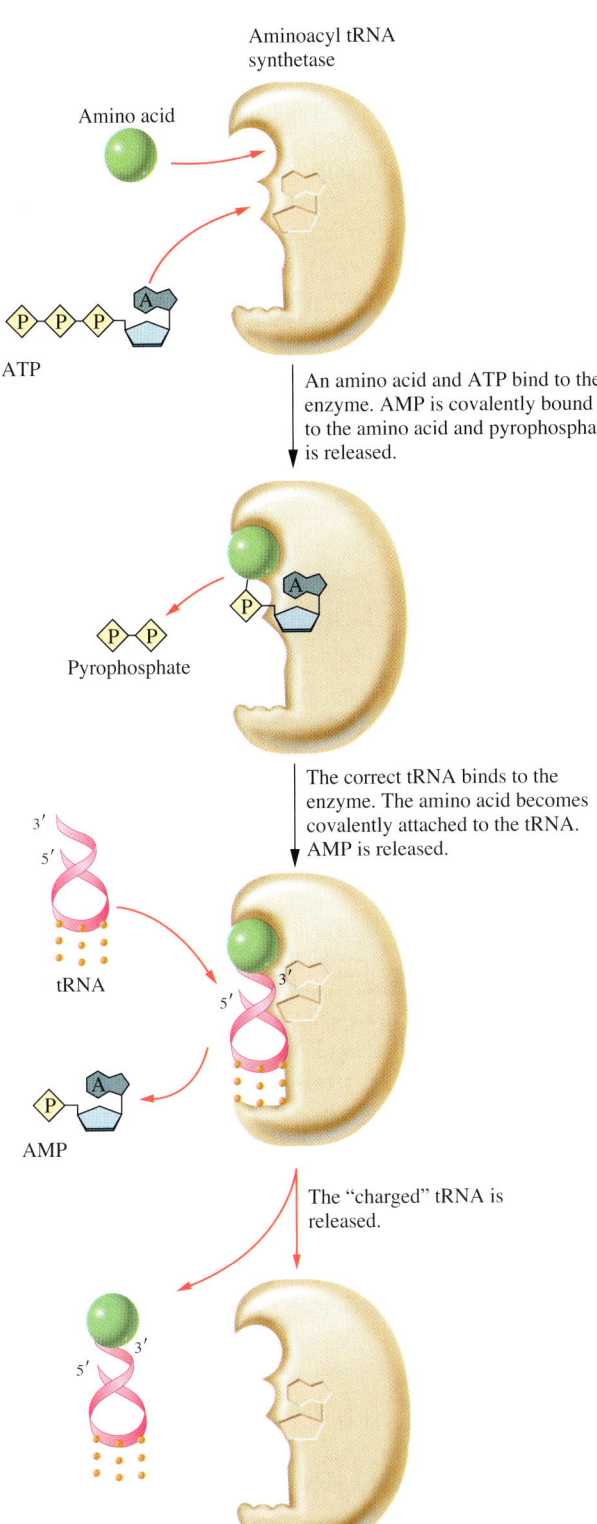

Figure 20.20
Aminoacyl tRNA synthetase binds the amino acid in one region of the active site and the appropriate tRNA in another. The acylation reaction occurs and the aminoacyl tRNA is released.

First, the tRNA must covalently bind one, and only one, specific amino acid. There is at least one transfer RNA for each amino acid. All tRNA molecules have the sequence CCA at their 3' ends. This is the site where the amino acid will be covalently attached to the tRNA molecule. Each tRNA is specifically recognized by the active site of an enzyme called an **aminoacyl tRNA synthetase**. This enzyme also recognizes the correct amino acid and covalently links the amino acid to the 3' end of the tRNA molecule (Figure 20.20). The resulting structure is called an **aminoacyl** tRNA. The covalently bound amino acid will be transferred from the tRNA to a growing polypeptide chain during protein synthesis.

Second, the tRNA must be able to recognize the appropriate codon on the mRNA that calls for that amino acid. This is mediated through a sequence of three bases called the *anticodon*, which is located at the bottom of the tRNA cloverleaf (refer to Figure 20.13). The anticodon sequence for each tRNA is complementary to the codon on the mRNA that specifies a particular amino acid. As you can see in Figure 20.18, the anticodon-codon complementary hydrogen bonding will bring the correct amino acid to the site of protein synthesis.

Question 20.7

How are codons related to anticodons?

Question 20.8

If the sequence of a codon on the mRNA is 5'-AUG-3', what will the sequence of the anticodon be? Remember that the hydrogen bonding rules require antiparallel strands. It is easiest to write the anticodon first 3' → 5' and then reverse it to the 5' → 3' order.

The Process of Translation

Initiation

LEARNING GOAL 8

The first stage of protein synthesis is *initiation*. Proteins called **initiation factors** assist in the formation of a translation complex composed of an mRNA molecule, the small and large ribosomal subunits, and the initiator tRNA. This initiator tRNA recognizes the codon AUG and carries the amino acid methionine.

The ribosome has two sites for binding tRNA molecules. The first site, called the **peptidyl tRNA binding site (P-site),** holds the peptidyl tRNA, the growing peptide bound to a tRNA molecule. The second site, called the **aminoacyl tRNA binding site (A-site),** holds the aminoacyl tRNA carrying the next amino acid to be added to the peptide chain. Each of the tRNA molecules is hydrogen bonded to the mRNA molecule by codon-anticodon complementarity. The entire complex is further stabilized by the fact that the mRNA is also bound to the ribosome. Figure 20.21a shows the series of events that result in the formation of the initiation complex. The initiator methionyl tRNA occupies the P-site in this complex.

Chain Elongation

The second stage of translation is *chain elongation*. This occurs in three steps that are repeated until protein synthesis is complete. We enter the action after a tetrapeptide has already been assembled, and a peptidyl tRNA occupies the P-site (Figure 20.21b).

The first event is binding of an aminoacyl-tRNA molecule to the empty A-site. Next, peptide bond formation occurs. This is catalyzed by an enzyme on the ribosome called *peptidyl transferase*. Now the peptide chain is shifted to the tRNA that occupies the A-site. Finally, the tRNA in the P-site falls away, and the ribosome changes positions so that the next codon on the mRNA occupies the A-site. This movement of the ribosome is called **translocation.** The process shifts the new peptidyl tRNA from the A-site to the P-site. The chain elongation stage of translation requires the hydrolysis of GTP to GDP and P_i. Several **elongation factors** are also involved in this process.

Recent evidence indicates that the peptidyl transferase is a catalytic region of the 28S ribosomal RNA.

20.6 Protein Synthesis

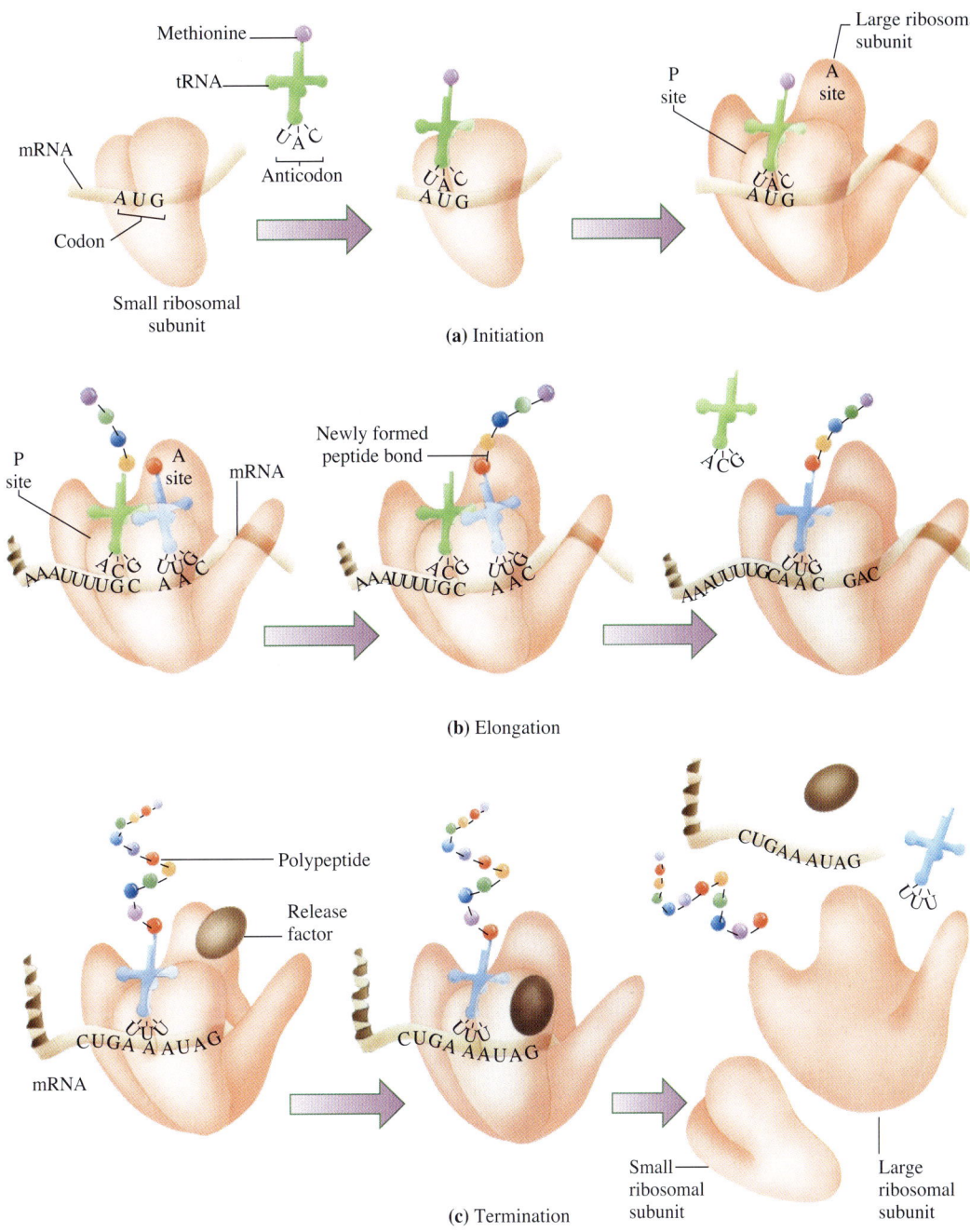

Figure 20.21
(a) Formation of an initiation complex sets protein synthesis in motion. The mRNA and proteins called initiation factors bind to the small ribosomal subunit. Next, a charged methionyl tRNA molecule binds, and finally, the initiation factors are released, and the large subunit binds. (b) The elongation phase of protein synthesis involves addition of new amino acids to the C-terminus of the growing peptide. An aminoacyl tRNA molecule binds at the empty A-site, and the peptide bond is formed. The uncharged tRNA molecule is released, and the peptidyl tRNA is shifted to the P-site as the ribosome moves along the mRNA. (c) Termination of protein synthesis occurs when a release factor binds the stop codon on mRNA. This leads to the hydrolysis of the ester bond linking the peptide to the peptidyl tRNA molecule in the P-site. The ribosome then dissociates into its two subunits, releasing the mRNA and the newly synthesized peptide.

Termination

The last stage of translation is *termination*. There are three **termination codons**—UAA, UAG, and UGA—for which there are no corresponding tRNA molecules. When one of these "stop" codons is encountered, translation is terminated. A **release factor** binds the empty A-site. The peptidyl transferase that had previously catalyzed peptide bond formation hydrolyzes the ester bond between the peptidyl

Question 20.9
What is the function of the ribosomal P-site in protein synthesis?

Question 20.10
What is the function of the ribosomal A-site in protein synthesis?

Post-translational proteolytic cleavage of digestive enzymes is discussed in Section 19.11.

The peptide that is released following translation is not necessarily in its final functional form. In some cases the peptide is proteolytically cleaved before it becomes functional. Synthesis of digestive enzymes uses this strategy. Sometimes the protein must associate with other peptides to form a functional protein, as in the case of hemoglobin. Cellular enzymes add carbohydrate or lipid groups to some proteins, especially those that will end up on the cell surface. These final modifications are specific for particular proteins and, like the sequence of the protein itself, are directed by the cellular genetic information.

The quaternary structure of hemoglobin is described in Section 18.9.

20.7 Mutation, Ultraviolet Light, and DNA Repair

The Nature of Mutations

LEARNING GOAL 9

Changes can occur in the nucleotide sequence of a DNA molecule. Such a genetic change is called a **mutation.** Mutations can arise from mistakes made by DNA polymerase during DNA replication. They also result from the action of chemicals, called **mutagens,** that damage the DNA.

Mutations are classified by the kind of change that occurs in the DNA. The substitution of a single nucleotide for another is called a **point mutation:**

        ATGGACTTC: normal DNA sequence
        ATGCACTTC: point mutation

Sometimes a single nucleotide or even large sections of DNA are lost. These are called **deletion mutations:**

        ATGGACTTC: normal DNA sequence
        ATGTTC: deletion mutation

Occasionally, one or more nucleotides are added to a DNA sequence. These are called **insertion mutations:**

        ATGGACTTC: normal DNA sequence
        ATGCTCGACTTC: insertion mutation

The Results of Mutations

Some mutations are **silent mutations;** that is, they cause no change in the protein. Often, however, a mutation has a negative effect on the health of the organism. The effect of a mutation depends on how it alters the genetic code for a protein. Consider the two codons for glutamic acid: GAA and GAG. A point mutation that alters the third nucleotide of GA<u>A</u> to GA<u>G</u> will still result in the incorporation of glutamic acid at the correct position in the protein. Similarly, a GA<u>G</u> to GA<u>A</u> mutation will also be silent.

Many mutations are not silent. There are approximately four thousand human genetic disorders that result from such mutations. These occur because the mutation in the DNA changes the codon and results in incorporation of the wrong amino acid into the protein. This causes the protein to be nonfunctional or to function improperly.

Consider the human genetic disease sickle cell anemia. In the normal β-chain of hemoglobin, the sixth amino acid is glutamic acid. In the β-chain of sickle cell hemoglobin, the sixth amino acid is valine. How did this amino acid substitution arise? The answer lies in examination of the codons for glutamic acid and valine:

Glutamic acid: GAA or GAG

Valine: GUG, GUC, GUA, or GUU

A point mutation of A → U in the second nucleotide changes some codons for glutamic acid into codons for valine:

GA A → GUA

GAG → GUG

Glutamic acid codon   Valine codon

This mutation in a single codon leads to the change in amino acid sequence at position 6 in the β-chain of human hemoglobin from glutamic acid to valine. The result of this seemingly minor change is sickle cell anemia in individuals who inherit two copies of the mutant gene.

Question 20.11

The sequence of a gene on the mRNA is normally AUGCCCGACUUU. A point mutation in the gene results in the mRNA sequence AUGCGCGACUUU. What are the amino acid sequences of the normal and mutant proteins? Would you expect this to be a silent mutation?

Question 20.12

The sequence of a gene on the mRNA is normally AUGCCCGACUUU. A point mutation in the gene results in the mRNA sequence AUGCCGGACUUU. What are the amino acid sequences of the normal and mutant proteins? Would you expect this to be a silent mutation?

Mutagens and Carcinogens

Any chemical that causes a change in the DNA sequence is called a *mutagen*. Often, mutagens are also **carcinogens,** cancer-causing chemicals. Most cancers result from mutations in a single normal cell. These mutations result in the loss of normal growth control, causing the abnormal cell to proliferate. If that growth is not controlled or destroyed, it will result in the death of the individual. We are exposed to many carcinogens in the course of our lives. Sometimes we are exposed to a carcinogen by accident, but in some cases it is by choice. There are about three thousand chemical components in cigarette smoke, and several are potent mutagens. As a result, people who smoke have a much greater chance of lung cancer than those who don't.

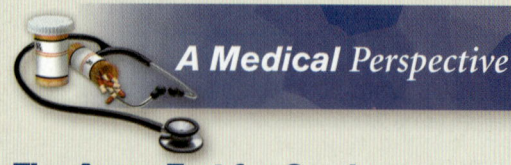

A Medical Perspective

The Ames Test for Carcinogens

Each day we come into contact with a variety of chemicals, including insecticides, food additives, hair dyes, automobile emissions, and cigarette smoke. Some of these chemicals have the potential to cause cancer. How do we determine whether these agents are harmful? More particularly, how do we determine whether they cause cancer?

If we consider the example of cigarette smoke, we see that it can be years, even centuries, before a relationship is seen between a chemical and cancer. Europeans and Americans have been smoking since Sir Walter Raleigh introduced tobacco into England in the seventeenth century. However, it was not until three centuries later that physicians and scientists demonstrated the link between smoking and lung cancer. Obviously, this epidemiological approach takes too long, and too many people die. Alternatively, we can test chemicals by treating laboratory animals, such as mice, and observing them for various kinds of cancer. However, this, too, can take years, is expensive, and requires the sacrifice of many laboratory animals. How, then, can chemicals be tested for carcinogenicity (the ability to cause cancer) quickly and inexpensively? In the 1970s it was recognized that most carcinogens are also mutagens. That is, they cause cancer by causing mutations in the DNA, and the mutations cause the cells of the body to lose growth control. Bruce Ames, a biochemist and bacterial geneticist, developed a test using mutants of the bacterium *Salmonella typhimurium* that can demonstrate in 48–72 hours whether a chemical is a mutagen and thus a suspected carcinogen.

Ames chose several mutants of *S. typhimurium* that cannot grow unless the amino acid histidine is added to the growth medium. The Ames test involves subjecting these bacteria to a chemical and determining whether the chemical causes reversion of the mutation. In other words, the researcher is looking for a mutation that reverses the original mutation. When a reversion occurs, the bacteria will be able to grow in the absence of histidine.

The details of the Ames test are shown in the accompanying figure. Both an experimental and a control test are done. The control test contains no carcinogen and will show the number of spontaneous revertants that occur in the culture. If there are many colonies on the surface of the experimental plate and only a few colonies on the negative control plate, it can be concluded that the chemical tested is a mutagen. It is therefore possible that the chemical is also a carcinogen.

The Ames test has greatly accelerated our ability to test new compounds for mutagenic and possibly carcinogenic effects. However, once the Ames test identifies a mutagenic compound, testing in animals must be done to show conclusively that the compound also causes cancer.

For Further Understanding

A researcher carried out the Ames test in which an experimental sample was exposed to a suspected mutagen and a control sample was not. A sample from each tube was grown on a medium containing no histidine. On the experimental plate, he observed forty-three colonies and on the control plate, he observed thirty-one colonies. He concluded that the substance is a mutagen. When he reported his data and conclusion to his supervisor, she told him that his conclusions were not valid. How can the researcher modify his experimental procedure to obtain better data?

Suppose that the mutation in a strain of *S. typhimurium* produces the codon UUA instead of UUC. What is the amino acid change caused by this mutation? What base substitutions could correct the mutant codon so that it once again calls for the correct amino acid? What base substitutions would not correct the mutant codon?

Ultraviolet Light Damage and DNA Repair

Ultraviolet (UV) light is another agent that causes damage to DNA. Absorption of UV light by DNA causes adjacent pyrimidine bases to become covalently linked. The product is called a **pyrimidine dimer.** As a result of pyrimidine dimer formation, there is no hydrogen bonding between these pyrimidine molecules and the complementary bases on the other DNA strand. This stretch of DNA cannot be replicated or transcribed!

Bacteria such as *Escherichia coli* have four different mechanisms to repair ultraviolet light damage. However, even a repair process can make a mistake. Mutations occur when the UV damage repair system makes an error and causes a change in the nucleotide sequence of the DNA.

In medicine, the pyrimidine dimerization reaction is used to advantage in hospitals where germicidal (UV) light is used to kill bacteria in the air and on environmental surfaces, such as in a vacant operating room. This cell death is caused by

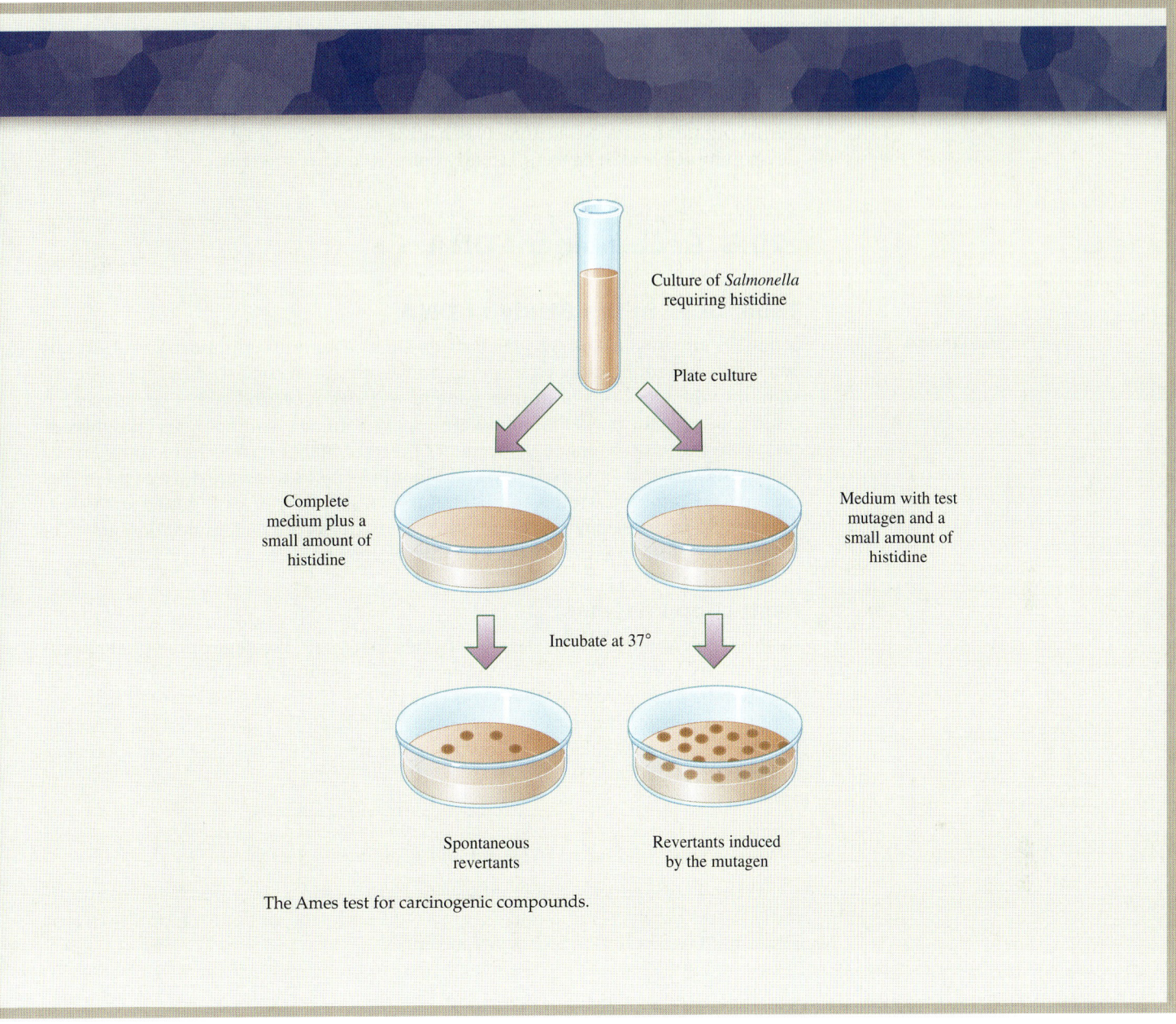

The Ames test for carcinogenic compounds.

pyrimidine dimer formation on a massive scale. The repair systems of the bacteria are overwhelmed, and the cells die.

Of course, the same type of pyrimidine dimer formation can occur in human cells as well. Lying out in the sun all day to acquire a fashionable tan exposes the skin to large amounts of UV light. This damages the skin by formation of many pyrimidine dimers. Exposure to high levels of UV from sunlight or tanning booths has been linked to a rising incidence of skin cancer in human populations.

Consequences of Defects in DNA Repair

The human repair system for pyrimidine dimers is quite complex, requiring at least five enzymes. The first step in repair of the pyrimidine dimer is the cleavage of the sugar-phosphate backbone of the DNA near the site of the damage. The enzyme that performs this is called a *repair endonuclease*. If the gene encoding this enzyme is defective, pyrimidine dimers cannot be repaired. The accumulation of mutations combined with a simultaneous decrease in the efficiency of DNA repair mechanisms leads to an increased incidence of cancer. For example, a mutation in the repair endonuclease gene, or in other genes in the repair pathway, results in the genetic skin disorder called *xeroderma pigmentosum*. People who suffer from xeroderma pigmentosum are extremely sensitive to the ultraviolet rays of sunlight and develop multiple skin cancers, usually before the age of twenty.

20.8 Recombinant DNA

Tools Used in the Study of DNA

Scientists are often asked why they study such seemingly unimportant subjects as bacterial DNA replication. One very good reason is that such studies often lend insight into the workings of human genetic systems. A second is that such research often produces the tools that allow great leaps into new technologies. Nowhere is this more true than in the development of recombinant DNA technology. Many of the techniques and tools used in recombinant DNA studies were developed or discovered during basic studies on bacterial DNA replication and gene expression. These include many enzymes that catalyze reactions of DNA molecules, gel electrophoresis, cloning vectors, and hybridization techniques.

Restriction Enzymes

Restriction enzymes are bacterial enzymes that "cut" the sugar-phosphate backbone of DNA molecules at specific nucleotide sequences. The first of these enzymes to be purified and studied was called EcoR1. The name is derived from the genus and species name of the bacteria from which it was isolated, in this case *Escherichia coli*, or *E. coli*. The following is the specific nucleotide sequence recognized by EcoR1:

5′ ----------------GAATTC----------------3′
3′ ----------------CTTAAG----------------5′

When EcoR1 cuts the DNA at this site, it does so in a staggered fashion. Specifically, it cuts between the G and the first A on both strands. Cutting produces two DNA fragments with the following structure:

5′------------------G   AATTC----------------3′
3′------------------CTTAA   G----------------5′

These staggered termini are called *sticky ends* because they can reassociate with one another by hydrogen bonding. This is a property of the DNA fragments generated by restriction enzymes that is very important to gene cloning.

Examples of other restriction enzymes and their specific recognition sequences are listed in Table 20.1. The sites on the sugar-phosphate backbone that are cut by the enzymes are indicated by slashes.

These enzymes are used to digest large DNA molecules into smaller fragments of specific size. Because a restriction enzyme always cuts at the same site, DNA from a particular individual generates a reproducible set of DNA fragments. This is convenient for the study or cloning of DNA from any source.

Agarose Gel Electrophoresis

One means of studying the DNA fragments produced by restriction enzyme digestion is agarose gel electrophoresis. The digested DNA sample is placed in a sample well in the gel, and an electrical current is applied. The negative charge of the

TABLE 20.1 Common Restriction Enzymes and Their Recognition Sequences

Restriction Enzyme	Recognition Sequence
BamHI	5'-G/GATCC-3'
	3'-CCTAG/G-5'
HindIII	5'-A/AGCTT-3'
	3'-TTCGA/A-5'
SalI	5'-G/TCGAC-3'
	3'-CAGCT/G-5'
BglII	5'-A/GATCT-3'
	3'-TCTAG/A-5'
PstI	5'-CTGCA/G-3'
	3'-G/ACGTC-5'

phosphoryl groups in the sugar-phosphate backbone causes the DNA fragment to move through the gel away from the negative electrode (cathode) and toward the positive electrode (anode). The smaller DNA fragments move more rapidly than the larger ones, and as a result the DNA fragments end up distributed throughout the gel according to their size. The sizes of each fragment can be determined by comparison with the migration pattern of DNA fragments of known size.

Hybridization

Agarose gel electrophoresis allows the determination of the size of a DNA fragment. However, in recombinant DNA research it is also important to identify what gene is carried by a particular DNA fragment.

Hybridization is a technique used to identify the presence of a gene on a particular DNA fragment. This technique is based on the fact that complementary DNA sequences will hydrogen bond, or hybridize, to one another. In fact, even RNA can be used in hybridization studies. RNA can hybridize to DNA molecules or to other RNA molecules.

One technique, called Southern blotting, involves hybridization of DNA fragments from an agarose gel (Figure 20.22a). DNA digested by a restriction enzyme is run on an agarose gel. Next the DNA fragments are transferred by blotting onto a special membrane filter. Figure 20.22b shows an apparatus that uses an electric field to transfer DNA from a gel onto a filter. In the next step, the DNA molecules on the filter are "melted" into single DNA strands so that they are ready for hybridization. The filter is then bathed in a solution containing a radioactive DNA or RNA molecule. This probe will hybridize to any DNA fragments on the filter that are complementary to it. X-ray film is used to detect any bands where the radioactive probe hybridized, thus locating the gene of interest.

DNA Cloning Vectors

DNA cloning experiments combine these technologies with a few additional tricks to isolate single copies of a gene and then produce billions of copies. To produce multiple copies of a gene, it may be joined to a **cloning vector.** A cloning vector is a piece of DNA having its own replication origin so that it can be replicated inside a host cell. Often the bacterium *E. coli* serves as the host cell in which the vector carrying the cloned DNA is replicated in abundance.

There are two major kinds of cloning vectors. The first are bacterial virus or phage vectors. These are bacterial viruses that have been genetically altered to

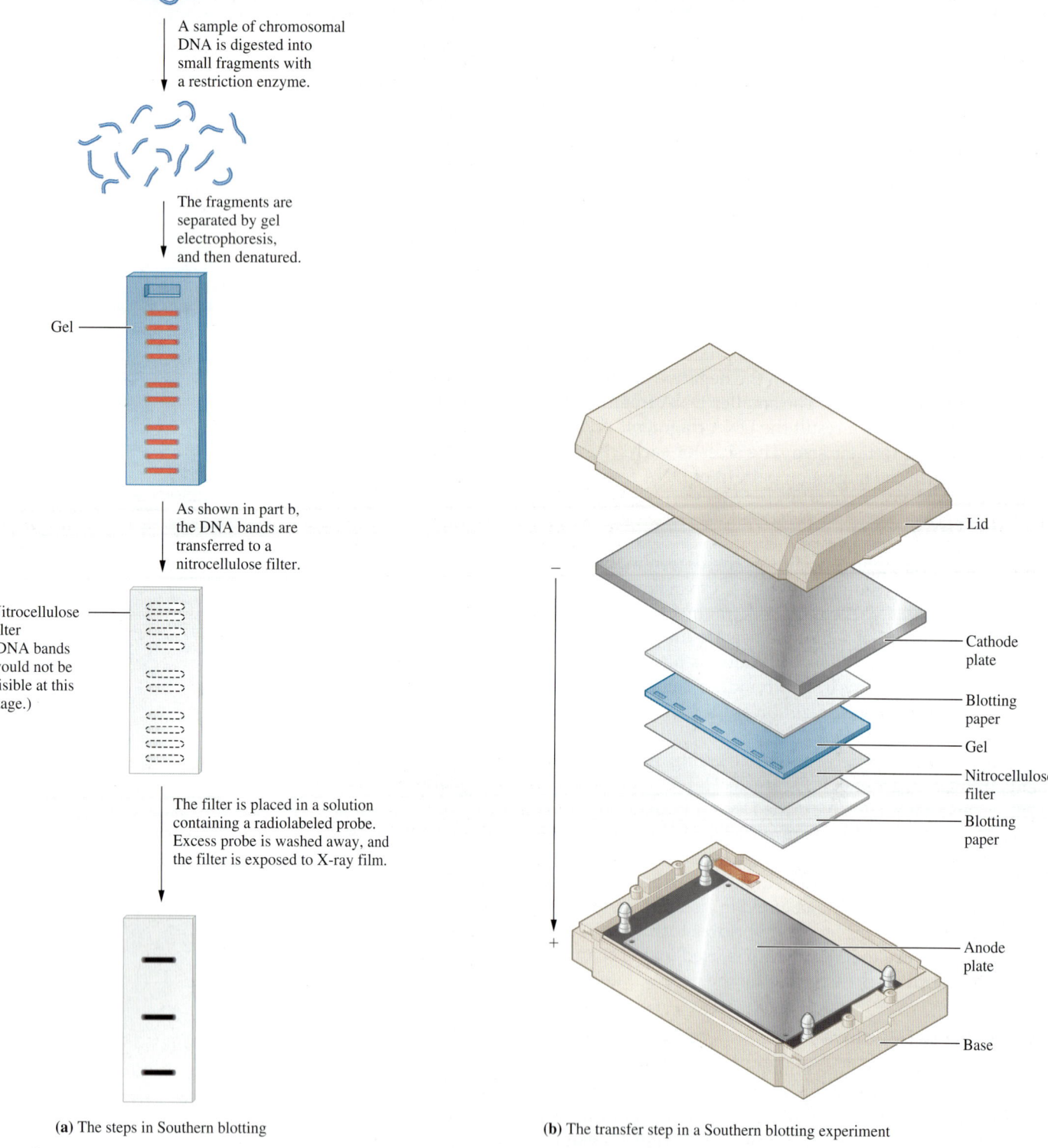

(a) The steps in Southern blotting

(b) The transfer step in a Southern blotting experiment

Figure 20.22

Southern blot hybridization. (a) DNA is digested and the fragments are separated by gel electrophoresis. The DNA is transferred from the gel onto a filter and "melted" into single strands. A radioactive probe is applied and the filter exposed to X-ray film. (b) An electroblotting apparatus for the transfer of DNA from a gel onto a filter.

allow the addition of cloned DNA fragments. These viruses have all the genes required to replicate one hundred to two hundred copies of the virus (and cloned fragment) per infected cell.

The second commonly used vector is a plasmid vector. Plasmids are extra pieces of circular DNA found in most kinds of bacteria. The plasmids that are used as cloning vectors often contain antibiotic resistance genes that are useful in the selection of cells containing a plasmid.

Each plasmid has its own replication origin to allow efficient DNA replication in the bacterial host cell. Most plasmid vectors also have a *selectable marker*, often a gene for resistance to an antibiotic. Finally, plasmid vectors have a gene that has several restriction enzyme sites useful for cloning. The valuable feature of this gene is that it is inactivated when a DNA fragment has been cloned into it. Thus, cells containing a plasmid carrying a cloned DNA fragment can be recognized by their ability to grow in the presence of antibiotic and by loss of function of the gene into which the cloned DNA has been inserted.

Antibiotic resistance causes countless problems in the treatment of bacterial infections.

Genetic Engineering

Now that we have assembled most of the tools needed for a cloning experiment, we must decide which gene to clone. The example that we will use is the cloning of the β-globin genes for normal and sickle cell hemoglobin. DNA from an individual with normal hemoglobin is digested with a restriction enzyme. This is the target DNA. The vector DNA must be digested with the same enzyme (Figure 20.23). In our example, the restriction enzyme cuts within the *lacZ* gene, which codes for the enzyme β-galactosidase.

The digested vector and target DNA are mixed together under conditions that encourage the sticky ends of the target and vector DNA to hybridize with one another. The sticky ends are then covalently linked by the enzyme DNA ligase. This enzyme catalyzes the formation of phosphoester bonds between the two pieces of DNA.

Now the recombinant DNA molecules are introduced into bacterial cells by a process called transformation. Next, the cells of the transformation mixture are plated on a solid nutrient agar medium containing the antibiotic ampicillin and the β-galactosidase substrate X-gal (5-bromo-4-chloro-3-indolyl-β-D-galactoside). Only those cells containing the antibiotic resistance gene will survive and grow into bacterial colonies. Cells with an intact *lacZ* gene will produce the enzyme β-galactosidase. The enzyme will hydrolyze X-gal to produce a blue product that will cause the colonies to appear blue. If the *lacZ* gene has been inactivated by insertion of a cloned gene, no β-galactosidase will be produced and the colonies will be white.

Now hybridization can be used to detect the clones that carry the β-globin gene. A replica of the experimental plate is made by transferring some cells from each colony onto a membrane filter. These cells are gently broken open so that the released DNA becomes attached to the membrane. When hybridization is carried out on these filters, the radioactive probe will hybridize only to the complementary sequences of the β-globin gene. When the membrane filter is exposed to X-ray film, a "spot" will appear on the developed film only at the site of a colony carrying the desired clone. By going back to the original plate, we can select cells from that colony and grow them for further study (Figure 20.24).

The same procedure can be used to clone the β-chain gene of sickle cell hemoglobin. Then the two can be studied and compared to determine the nature of the genetic defect.

This simple example makes it appear that all gene cloning is very easy and straightforward. This has proved to be far from the truth. Genetic engineers have had to overcome many obstacles to clone eukaryotic genes of particular medical interest. One of the first obstacles encountered was the presence of introns within eukaryotic genes. Bacteria that are used for cloning lack the enzymatic machinery

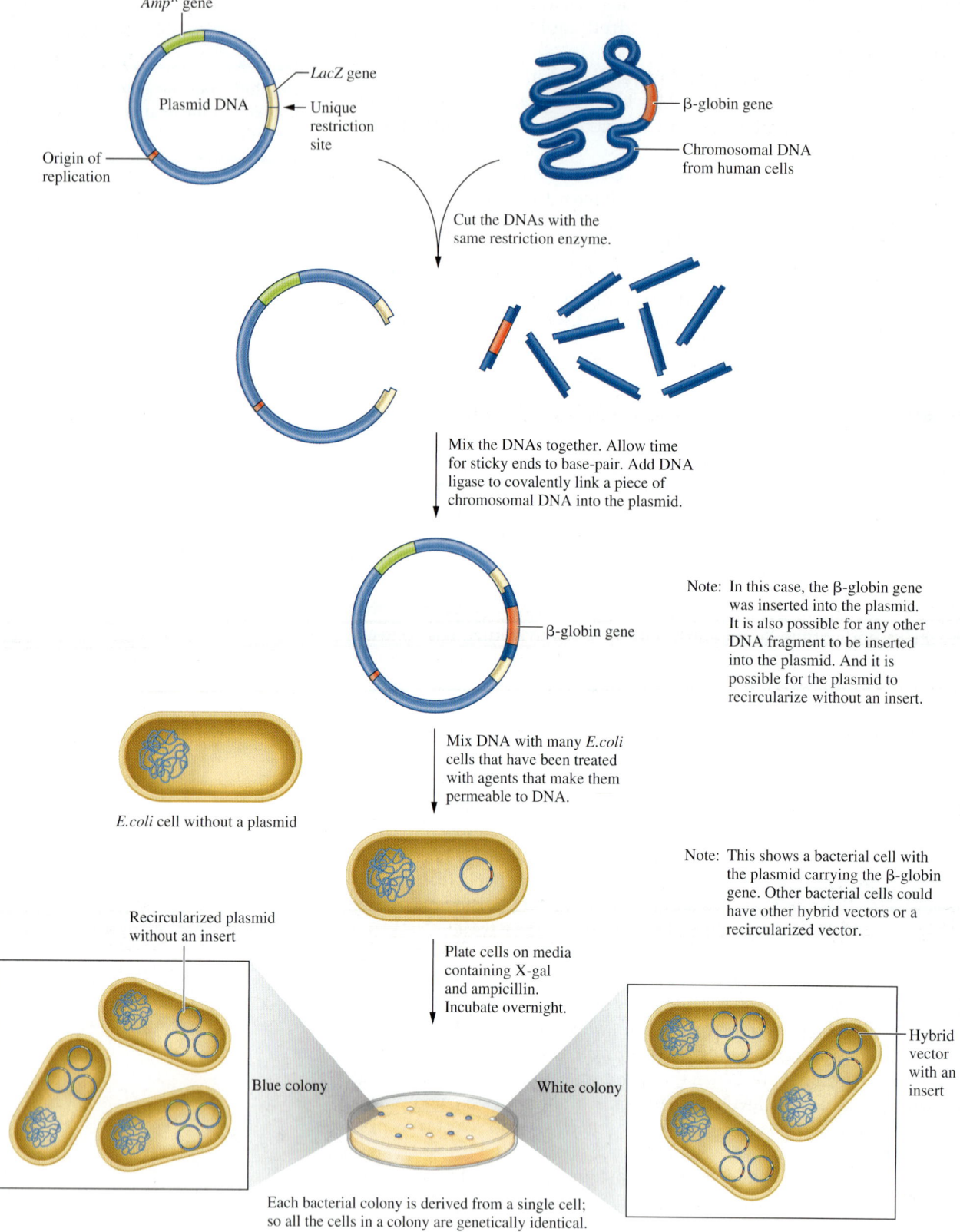

Figure 20.23
Cloning of eukaryotic DNA into a plasmid cloning vector.

TABLE 20.2 A Brief List of Medically Important Proteins Produced by Genetic Engineering

Protein	Medical Condition Treated
Insulin	Insulin-dependent diabetes
Human growth hormone	Pituitary dwarfism
Factor VIII	Type A hemophilia
Factor IX	Type B hemophilia
Tissue plasminogen factor	Stroke, myocardial infarction
Streptokinase	Myocardial infarction
Interferon	Cancer, some virus infections
Interleukin-2	Cancer
Tumor necrosis factor	Cancer
Atrial natriuretic factor	Hypertension
Erythropoietin	Anemia
Thymosin α-1	Stimulate immune system
Hepatitis B virus (HBV) vaccine	Prevent HBV viral hepatitis
Influenza vaccine	Prevent influenza infection

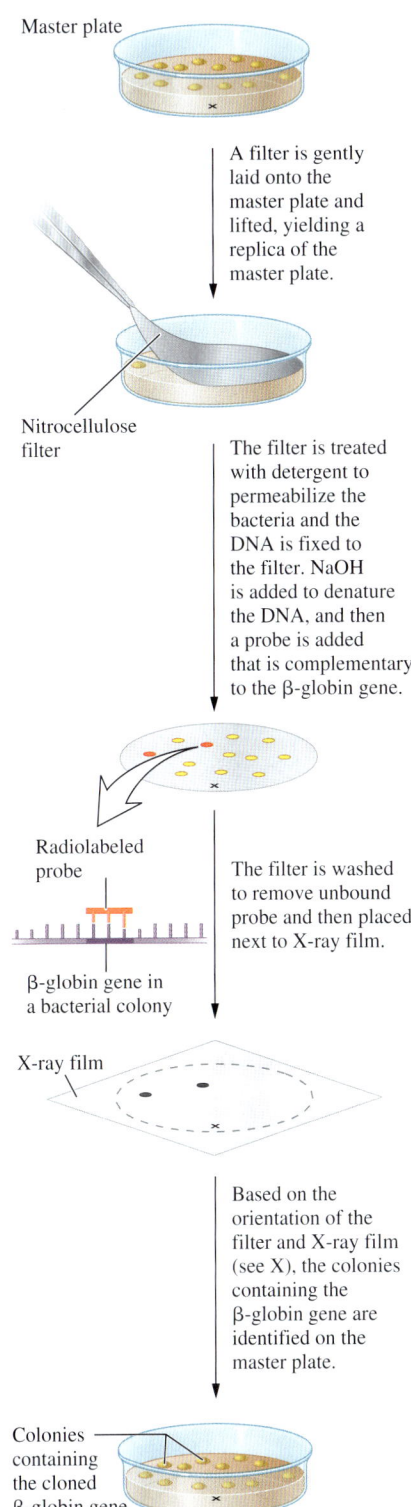

Figure 20.24
Colony blot hybridization for detection of cells carrying a plasmid clone of the β-chain gene of hemoglobin.

to splice out introns. Molecular biologists found that a DNA copy of a eukaryotic mRNA could be made by using the enzyme reverse transcriptase from a family of viruses called retroviruses (see A Medical Perspective: Fooling the AIDS Virus with "Look-Alike" Nucleotides, Section 20.2). Such a DNA copy of the mRNA carries all the protein-coding sequences of a gene but none of the intron sequences. Thus bacteria are able to transcribe and translate the cloned DNA and produce valuable products for use in medicine and other applications.

This is only one of the many technical problems that have been overcome by the amazing developments in recombinant DNA technology. A brief but impressive list of medically important products of genetic engineering is presented in Table 20.2.

20.9 Polymerase Chain Reaction

A bacterium originally isolated from a hot spring in Yellowstone National Park provides the key to a powerful molecular tool for the study of DNA. Polymerase chain reaction (PCR) allows scientists to produce unlimited amounts of any gene of interest and the bacterium *Thermus aquaticus* produces a heat-stable DNA polymerase (Taq polymerase) that allows the process to work.

The human genome consists of approximately three billion base pairs of DNA. But suppose you are interested in studying only one gene, perhaps the gene responsible for muscular dystrophy or cystic fibrosis. It's like looking for a needle in a haystack. Using PCR, a scientist can make millions of copies of the gene of interest, while ignoring the thousands of other genes on human chromosomes.

The secret to this specificity is the synthesis of a DNA primer, a short piece of single-stranded DNA that will specifically hybridize to the beginning of a particular gene. DNA polymerases require a primer for initiation of DNA synthesis because they act by adding new nucleotides to the 3'—OH of the last nucleotide of the primer.

To perform PCR, a small amount of DNA is mixed with Taq polymerase, the primers, and the four DNA nucleotide triphosphates. The mixture is then placed in an instrument called a *thermocycler*. The temperature in the thermocycler is

A Human Perspective

DNA Fingerprinting

Four U.S. Army helicopters swept over the field of illicit coca plants (*Erythroxylum* spp.) growing in a mountainous region of northern Colombia. When the soldiers were certain that the fields were unguarded, a fifth helicopter landed. From it emerged Dr. Jim Saunders, currently director of the Molecular Biology, Biochemistry, and Bioinformatics Program at Towson University, and a team of researchers from the Agricultural Research Service of the U.S. Department of Agriculture (ARS-USDA). Quickly the scientists gathered leaves from mature plants, as well as from seedlings growing in a coca nursery, and returned to the helicopter with their valuable samples. With a final sweep over the field, the Army helicopters sprayed herbicides to kill the coca plants.

From 1997 to 2001, this scene was repeated in regions of Colombia known to have the highest coca production. The reason for these collections was to study the genetic diversity of the coca plants being grown for the illegal production of cocaine. The tool selected for this study was DNA fingerprinting.

DNA fingerprinting was developed in the 1980s by Alec Jeffries of the University of Leicester in England. The idea grew out of basic molecular genetic studies of the human genome. Scientists observed that some DNA sequences varied greatly from one person to the next. Such hypervariable regions are made up of variable numbers of repeats of short DNA sequences. They are located at many sites on different chromosomes. Each person has a different number of repeats and when his or her DNA is digested with restriction enzymes, a unique set of DNA fragments is generated. Jeffries invented DNA fingerprinting by developing a set of DNA probes that detect these variable number tandem repeats (VNTRs) when used in hybridization with Southern blots.

Coca nursery next to a mature field in Colombia.

Although several variations of DNA fingerprinting exist, the basic technique is quite simple. DNA is digested with restriction enzymes, producing a set of DNA fragments. These are separated by electrophoresis through an agarose gel. The DNA fragments are then transferred to membrane filters and hybridized with the radioactive probe DNA. The bands that hybridize the radioactive probe are visualized by exposing the membrane to X-ray film and developing a "picture" of the gel. The result is what Jeffries calls a *DNA fingerprint*, a set of twenty-five to sixty DNA bands that are unique to an individual.

raised to 94–96°C for several minutes to separate the two strands of DNA. The temperature is then dropped to 50–56°C to allow the primers to hybridize to the target DNA. Finally, the temperature is raised to 72°C to allow Taq polymerase to act, reading the template DNA strand and polymerizing a daughter strand extended from the primer. At the end of this step, the amount of the gene has doubled (Figure 20.25).

Now the three steps are repeated. With each cycle the amount of the gene is doubled. Theoretically after thirty cycles, you have one billion times more DNA than you started with!

PCR can be used in genetic screening to detect the gene responsible for muscular dystrophy. It can also be used to diagnose disease. For instance, it can be used to amplify small amounts of HIV in the blood. It can also be used by forensic scientists to amplify DNA from a single hair follicle or a tiny drop of blood at a crime scene.

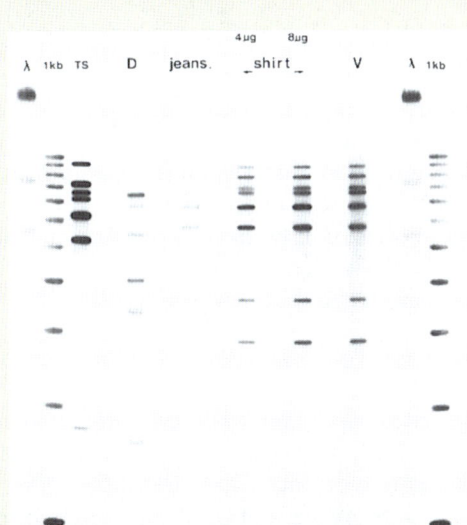

An example of a DNA fingerprint used in a criminal case. The DNA sample designated *V* is that of the victim and the sample designated *D* is that of the defendant. The samples labeled *jeans* and *shirt* were taken from the clothing of the defendant. The DNA bands from the defendant's clothing clearly match the DNA bands of the victim, providing evidence of the guilt of the defendant.

DNA fingerprinting is now routinely used for paternity testing, testing for certain genetic disorders, and identification of the dead in cases where no other identification is available. DNA fingerprints are used as evidence in criminal cases involving rape and murder. In such cases, the evidence may be little more than a hair with an intact follicle on the clothing of the victim.

Less widely known is the use of DNA fingerprinting to study genetic diversity in natural populations of plants and animals. The greater the genetic diversity, the healthier the population is likely to be. Populations with low genetic diversity face a far higher probability of extinction under adverse conditions. Customs officials have used DNA fingerprinting to determine whether confiscated elephant tusks were taken illegally from an endangered population of elephants or were obtained from a legally harvested population.

In the case of the coca plants, ARS wanted to know whether the drug cartels were developing improved strains that might be hardier or more pest resistant or that have a higher concentration of cocaine. Their conclusions, which you can read in *Phytochemistry* (*64:* 187–197, 2003), were that the drug cartels have introduced significant genetic modification into coca plants in Colombia in the last two decades. In addition, some of these new variants, those producing the highest levels of cocaine, have been transplanted to other regions of the country. All of this indicates that the cocaine agribusiness is thriving.

For Further Understanding

As this sampling of applications suggests, DNA fingerprinting has become an invaluable tool in law enforcement, medicine, and basic research. What other applications of this technology can you think of?

Do some research on the development of DNA fingerprinting as a research and forensics tool. What is the probability that two individuals will have the same DNA fingerprint? How are these probabilities determined?

20.10 The Human Genome Project

 LEARNING GOAL

In 1990 the Department of Energy and the National Institutes of Health began the Human Genome Project (HGP), a multinational project that would extend into the next millennium. The goals of the HGP were to identify all of the genes in human DNA and to sequence the entire three billion nucleotide pairs of the genome. In order to accomplish these goals, enormous computer databases had to be developed to store the information and computer software had to be designed to analyze it.

Initially, the HGP planned to complete the work by the year 2005. However, as a result of technological advances made by those in the project, a working draft of the human genome was published in February 2001 and the successful completion of the project was announced on April 14, 2003.

Genetic Strategies for Genome Analysis

The strategy for HGP was rather straightforward. In order to determine the DNA sequence of the human genome, genomic libraries had to be produced. A *genomic library* is a set of clones representing the entire genome. The DNA sequences of each of these clones could then be determined. Of course, once the sequence of each of these clones is determined, there is no way to know how they are arranged along the chromosomes.

A second technique, called *chromosome walking,* provides both DNA sequence information, as well as a method for identifying the DNA sequences next to it on the chromosome. This method requires clones that are overlapping. To accomplish this, libraries of clones are made using many different restriction enzymes. The DNA sequence of a fragment is determined. Then that information is used to develop a probe for any clones in the library that are overlapping. Each time a DNA fragment is sequenced, the information is used to identify overlapping clones. This process continues, allowing scientists to walk along the chromosome in two directions until the entire sequence is cloned, mapped, and sequenced.

DNA Sequencing

The method of DNA sequencing that is used is based on a technique developed by Frederick Sanger. A cloned piece of DNA is separated into its two strands. Each of these will serve as a template strand to carry out DNA replication in test tubes. A primer strand is also needed. This is a short piece of DNA that will hybridize to the template strand. The primer is the starting point for addition of new nucleotides during DNA synthesis.

The DNA is then placed in four test tubes with all of the enzymes and nucleotides required for DNA synthesis. In addition, each tube contains an unusual nucleotide, called a dideoxynucleotide. These nucleotides differ from the stan-

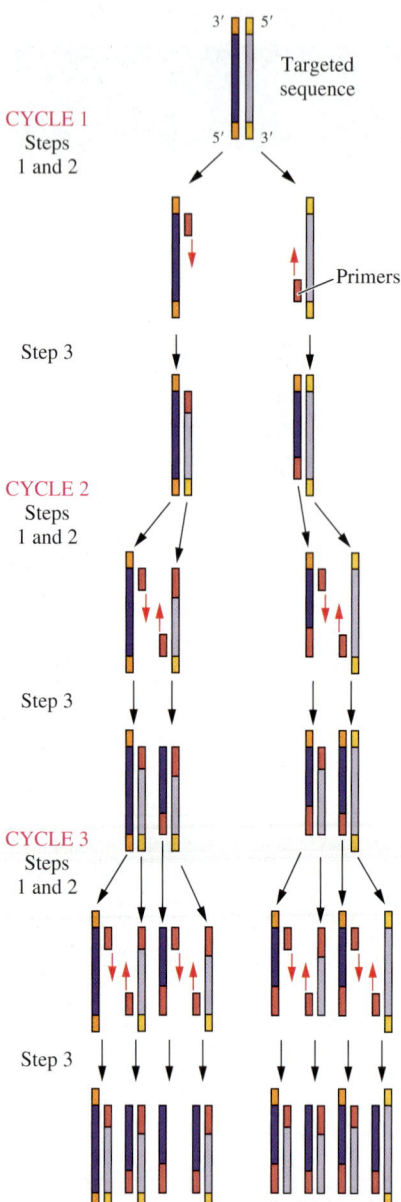

Figure 20.25
Polymerase chain reaction.

Figure 20.26
DNA sequencing by chain termination requires a template DNA strand and a radioactive primer. These are placed into each of four reaction mixtures that contain DNA polymerase, the four DNA nucleotides (dATP, dCTP, dGTP, and TTP), as well as one of the four dideoxynucleotides. Following the reaction, the products are separated on a DNA sequencing gel. The sequence is read from an autoradiograph of the gel.

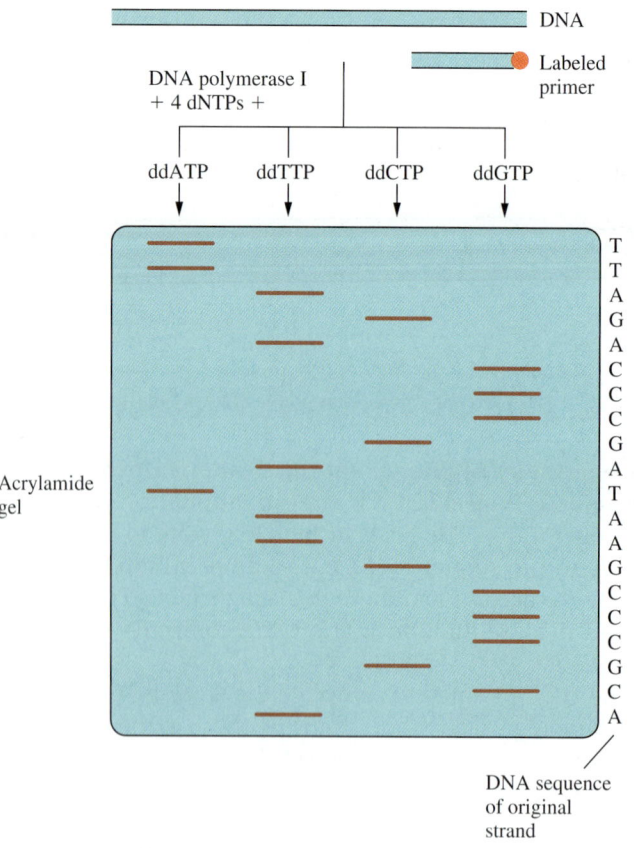

A Medical Perspective

A Genetic Approach to Familial Emphysema

Familial emphysema is a human genetic disease resulting from the inability to produce the protein α_1-antitrypsin. See also A Medical Perspective: α_1-Antitrypsin and Familial Emphysema, in Chapter 19. In individuals who have inherited one or two copies of the α_1-antitrypsin gene, this serum protein protects the lungs from the enzyme elastase. Normally, elastase fights bacteria and helps in the destruction and removal of dead lung tissue. However, the enzyme can also cause lung damage. By inhibiting elastase, α_1-antitrypsin prevents lung damage. Individuals who have inherited two defective α_1-antitrypsin genes do not produce this protein and suffer from familial, or A1AD, emphysema. In the absence of α_1-antitrypsin, the elastase and other proteases cause the severe lung damage characteristic of emphysema.

A1AD is the second most common genetic disorder in Caucasians. It is estimated that there are 100,000 sufferers in the United States and that one in five Americans carries the gene. The disorder, discovered in 1963, is often misdiagnosed as asthma or chronic obstructive pulmonary disease. In fact, it is estimated that fewer than 5% of the sufferers are diagnosed with A1AD.

The α_1-antitrypsin gene has been cloned. Early experiments with sheep showed that the protein remains stable when administered as an aerosol and remains functional after it has passed through the pulmonary epithelium. This research offers hope of an effective treatment for this disease.

The current treatment involves weekly IV injections of α_1-antitrypsin. The supply of the protein, purified from human plasma that has been demonstrated to be virus free, is rather limited. Thus, the injections are expensive. In addition, they are painful. These two factors cause some sufferers to refuse the treatment.

Recently Dr. Terry Flotte and his colleagues at the University of Florida have taken a new approach. They have cloned the gene for α_1-antitrypsin into the DNA of adeno-associated virus. This virus is an ideal vector for human gene replacement therapy because it replicates only in cells that are not dividing and it does not stimulate a strong immune or inflammatory response. The researchers injected the virus carrying the cloned α_1-antitrypsin gene into the muscle tissue of mice, then tested for the level of α_1-antitrypsin in the blood. The results were very promising. Effective levels of α_1-antitrypsin were produced in the muscle cells of the mice and secreted into the bloodstream. Furthermore, the level of α_1-antitrypsin remained at therapeutic levels for more than four months. The research team is planning tests with larger animals and eventually will confirm their results in human trials.

For Further Understanding

Of the three treatments described in this perspective, which do you think has the highest probability of success in the long term? Defend your answer.

Why is it impractical to "replace" the defective gene in an adult suffering from a genetic disease such as A1AD?

---

dard nucleotides by having a hydrogen atom at the 3′ position of the deoxyribose, rather than a hydroxyl group. When a dideoxynucleotide is incorporated into a growing DNA chain, it acts as a chain terminator. Because it does not have a 3′-hydroxyl group, no phosphoester bond can be formed with another nucleotide and no further polymerization can occur.

Each of the four tubes containing the DNA, enzymes, and an excess of the nucleotides required for replication will also have a small amount of one of the four dideoxynucleotides. In the tube that receives dideoxyadenosine triphosphate (ddA), for example, DNA synthesis will begin. As replication proceeds, either the standard nucleotide or ddA will be incorporated into the growing strand. Since the standard nucleotide is present in excess, the dideoxynucleotide will be incorporated infrequently and randomly. This produces a family of DNA fragments that terminate at the location of one of the deoxyadenosines in the molecule.

The same reaction is done with each of the dideoxynucleotides. The DNA fragments are then separated by gel electrophoresis on a DNA sequencing gel. The four reactions are placed in four wells, side by side, on the gel. Following electrophoresis, the DNA sequence can be read directly from the gel, as shown in Figure 20.26.

When chain termination DNA sequencing was first done, radioactive isotopes were used to label the DNA strands. However, new technology has resulted in automated systems that employ dideoxynucleotides that are labeled with fluorescent dyes, a different color for each dideoxynucleotide. Because each reaction (A, G, C, and T) will be a different color, all the reactions can be done in a single reaction mixture and the products separated on a single lane of a sequencing gel. A computer then "reads" the gel by distinguishing the color of each DNA band. The sequence information is directly stored into a databank for later analysis.

There is currently a vast amount of DNA information available on the Internet. The complete genomes of many bacteria have been reported, as well as the sequence information generated by the Human Genome Project. Because we know the genetic code, we can predict the amino acid sequence of proteins encoded by the genes. Researchers can also compare the sequences of normal genes with those of people suffering from genetic disorders. The enormity of the DNA information available, as well as the many types of analysis that need to be carried out, have given rise to an entirely new branch of science. The field of **bioinformatics** is a marriage of computer information sciences and DNA technology that is helping to devise methods for understanding, analyzing, and applying the DNA sequence information that we are gathering.

SUMMARY

20.1 The Structure of the Nucleotide

DNA and RNA are polymers of nucleotides, which are composed of a five-carbon sugar (ribose in RNA and 2'-deoxyribose in DNA), a nitrogenous base, and one, two, or three phosphoryl groups. There are two kinds of nitrogenous bases, the *purines* (adenine and guanine) and the *pyrimidines* (cytosine, thymine, and uracil). *Deoxyribonucleotides* are the subunits of DNA. *Ribonucleotides* are the subunits of RNA.

20.2 The Structure of DNA and RNA

Nucleotides are joined by 3'–5' phosphodiester bonds in both DNA and RNA. DNA is a *double helix,* two strands of DNA wound around one another. The sugar-phosphate backbone is on the outside of the helix, and complementary pairs of bases extend into the center of the helix. The *base pairs* are held together by hydrogen bonds. Adenine base pairs with thymine, and cytosine base pairs with guanine. The two strands of DNA in the helix are antiparallel to one another. RNA is single stranded.

20.3 DNA Replication

DNA replication involves synthesis of a faithful copy of the DNA molecule. It is semiconservative; each daughter molecule consists of one parental strand and one newly synthesized strand. *DNA polymerase III* "reads" each parental strand and synthesizes the complementary daughter strand according to the rules of base pairing.

20.4 Information Flow in Biological Systems

The *central dogma* states that the flow of biological information in cells is DNA → RNA → protein. There are three classes of RNA: *messenger RNA, transfer RNA,* and *ribosomal RNA*. *Transcription* is the process by which RNA molecules are synthesized. *RNA polymerase* catalyzes the synthesis of RNA. Transcription occurs in three stages: initiation, elongation, and termination. Eukaryotic genes contain *introns,* sequences that do not encode protein. These are removed from the primary transcript by the process of *RNA splicing.* The final mRNA contains only the protein coding sequences or *exons.* This final mRNA also has an added 5' cap structure and 3' poly(A) tail.

20.5 The Genetic Code

The genetic code is a triplet code. Each code word is called a *codon* and consists of three nucleotides. There are sixty-four codons in the genetic code. Of these, three are *termination codons* (UAA, UAG, and UGA), and the remaining sixty-one specify an amino acid. Most amino acids have several codons. As a result, the genetic code is said to be *degenerate.*

20.6 Protein Synthesis

The process of protein synthesis is called *translation.* The genetic code words on the mRNA are decoded by tRNA. Each tRNA has an *anticodon* that is complementary to a codon on the mRNA. In addition the tRNA is covalently linked to its correct amino acid. Thus hydrogen bonding between codon and anticodon brings the correct amino acid to the site of protein synthesis. Translation also occurs in three stages called initiation, chain elongation, and termination.

20.7 Mutation, Ultraviolet Light, and DNA Repair

Any change in a DNA sequence is a *mutation*. Mutations are classified according to the type of DNA alteration, including *point mutations*, *deletion mutations*, and *insertion mutations*. Ultraviolet light (UV) causes formation of *pyrimidine dimers*. Mistakes can be made during pyrimidine dimer repair, causing UV-induced mutations. Germicidal (UV) lamps are used to kill bacteria on environmental surfaces. UV damage to skin can result in skin cancer.

20.8 Recombinant DNA

Several tools are required for genetic engineering, including *restriction enzymes, agarose gel electrophoresis, hybridization,* and *cloning vectors.* Cloning a DNA fragment involves digestion of the target and vector DNA with a restriction enzyme. DNA ligase joins the target and vector DNA covalently, and the recombinant DNA molecules are introduced into bacterial cells by transformation. The desired clone is located by using antibiotic selection and hybridization. Many eukaryotic genes have been cloned for the purpose of producing medically important proteins.

20.9 Polymerase Chain Reaction

Using a heat-stable DNA polymerase produced by the bacterium *Thermus aquaticus* and specific DNA primers, polymerase chain reaction allows the amplification of DNA sequences that are present in small quantities. This technique is useful in genetic screening, diagnosis of viral or bacterial disease, and forensic science.

20.10 The Human Genome Project

The Human Genome Project has identified and mapped the genes of the human genome and determined the complete DNA sequence of each of the chromosomes. To do this, genomic libraries were generated and the DNA sequences of the clones were determined. To map the sequences along each chromosome, chromosome walking was used. DNA sequencing involves reactions in which DNA polymerase copies specific DNA sequences. Nucleotide analogues that cause chain termination (dideoxynucleotides) are incorporated randomly into the growing DNA chain. This generates a family of DNA fragments that differ in size by one nucleotide. DNA sequencing gels separate these fragments and provide DNA sequence data.

KEY TERMS

aminoacyl tRNA (20.6)
aminoacyl tRNA binding site of ribosome (A-site) (20.6)
aminoacyl tRNA synthetase (20.6)
anticodon (20.4)
antiparallel strands (20.2)
base pairs (20.2)
bioinformatics (20.10)
cap structure (20.4)
carcinogen (20.7)
central dogma (20.4)
chromosome (20.2)
cloning vector (20.8)
codon (20.4)
complementary strands (20.2)
degenerate code (20.5)
deletion mutation (20.7)
deoxyribonucleic acid (DNA) (20.1)
deoxyribonucleotide (20.1)
DNA polymerase III (20.3)
double helix (20.2)
elongation factor (20.6)
eukaryote (20.2)
exon (20.4)
genome (20.2)
hybridization (20.8)
initiation factor (20.6)
insertion mutation (20.7)
intron (20.4)
lagging strand (20.3)
leading strand (20.3)
messenger RNA (mRNA) (20.4)
mutagen (20.7)
mutation (20.7)
nucleosome (20.2)
nucleotide (20.1)
peptidyl tRNA binding site of ribosome (P-site) (20.6)
point mutation (20.7)
poly(A) tail (20.4)
polysome (20.6)
post-transcriptional modification (20.4)
primary transcript (20.4)
prokaryote (20.2)
promoter (20.4)
purine (20.1)
pyrimidine (20.1)
pyrimidine dimer (20.7)
release factor (20.6)
replication fork (20.3)
replication origin (20.3)
restriction enzyme (20.8)
ribonucleic acid (RNA) (20.1)
ribonucleotide (20.1)
ribosomal RNA (rRNA) (20.4)
ribosome (20.6)
RNA polymerase (20.4)
RNA splicing (20.4)
semiconservative replication (20.3)
silent mutation (20.7)
termination codon (20.6)
transcription (20.4)
transfer RNA (tRNA) (20.4)
translation (20.4)
translocation (20.6)

QUESTIONS AND PROBLEMS

The Structure of the Nucleotide

Foundations

20.13 What is a heterocyclic amine?
20.14 What components of nucleic acids are heterocyclic amines?

Applications

20.15 Draw the structure of the purine ring, and indicate the nitrogen that is bonded to sugars in nucleotides.
20.16 a. Draw the ring structure of the pyrimidines.
 b. In a nucleotide, which nitrogen atom of pyrimidine rings is bonded to the sugar?
20.17 ATP is the universal energy currency of the cell. What components make up the ATP nucleotide?
20.18 One of the energy-harvesting steps of the citric acid cycle results in the production of GTP. What is the structure of the GTP nucleotide?

The Structure of DNA and RNA

Foundations

20.19 The two strands of a DNA molecule are antiparallel. What is meant by this description?
20.20 List three differences between DNA and RNA.

Applications

20.21 What is the significance of the following repeat distances in the structure of the DNA molecule: 0.34 nm, 3.4 nm, and 2 nm?

20.22 Except for the functional groups attached to the rings, the nitrogenous bases are largely flat, hydrophobic molecules. Explain why the arrangement of the purines and pyrimidines found in DNA molecules is very stable.

20.23 How many hydrogen bonds link the adenine-thymine base pair?

20.24 How many hydrogen bonds link the guanine-cytosine base pair?

20.25 Write the structure that results when deoxycytosine-5'-monophosphate is linked by a 3' → 5' phosphodiester bond to thymidine-5'-monophosphate.

20.26 Write the structure that results when adenosine-5'-monophosphate is linked by a 3' → 5' phosphodiester bond to uridine-5'-monophosphate.

20.27 Describe the structure of the prokaryotic chromosome.

20.28 Describe the structure of the eukaryotic chromosome.

DNA Replication

Foundations

20.29 What is meant by semiconservative DNA replication?

20.30 Draw a diagram illustrating semiconservative DNA replication.

Applications

20.31 What are the two primary functions of DNA polymerase III?

20.32 a. Why is DNA polymerase said to be template-directed?
b. Why is DNA replication a self-correcting process?

20.33 If a DNA strand had the nucleotide sequence

5'-ATGCGGCTAGAATATTCCA-3'

what would the sequence of the complementary daughter strand be?

20.34 If the sequence of a double-stranded DNA is

5'- G A A T T C C T T A A G G A T C G A T C -3'
     | | | | | | | | | | | | | | | | | | | |
3'- C T T A A G G A A T T C C T A G C T A G -5'

what would the sequence of the two daughter DNA molecules be after DNA replication? Indicate which strands are newly synthesized and which are parental.

20.35 What is the replication origin of a DNA molecule?

20.36 What is occurring at the replication fork?

20.37 What is the function of the enzyme helicase?

20.38 What is the function of the enzyme primase?

20.39 What role does the RNA primer play in DNA replication?

20.40 Explain the differences between leading strand and lagging strand replication.

Information Flow in Biological Systems

Foundations

20.41 What is the central dogma of molecular biology?

20.42 What are the roles of DNA, RNA, and protein in information flow in biological systems?

20.43 On what molecule is the anticodon found?

20.44 On what molecule is the codon found?

Applications

20.45 If a gene had the nucleotide sequence

5'-TACCTAGCTCTGGTCATTAAGGCAGTA-3'

what would the sequence of the mRNA be?

20.46 If a mRNA had the nucleotide sequence

5'-AUGCCCUUUCAUUACCCGGUA-3'

what was the sequence of the DNA strand that was transcribed?

20.47 What is meant by the term *RNA splicing*?

20.48 The following is the unspliced transcript of a eukaryotic gene:

exon 1   intron A   exon 2   intron B   exon 3   intron C   exon 4

What would the structure of the final mature mRNA look like, and which of the above sequences would be found in the mature mRNA?

20.49 List the three classes of RNA molecules.

20.50 What is the function of each of the classes of RNA molecules?

20.51 What is the function of the spliceosome?

20.52 What are snRNPs? How do they facilitate RNA splicing?

20.53 What is a poly(A) tail?

20.54 What is the purpose of the poly(A) tail on eukaryotic mRNA?

20.55 What is the cap structure?

20.56 What is the function of the cap structure on eukaryotic mRNA?

The Genetic Code

Foundations

20.57 How many codons constitute the genetic code?

20.58 What is meant by a triplet code?

20.59 What is meant by the reading frame of a gene?

20.60 What happens to the reading frame of a gene if a nucleotide is deleted?

Applications

20.61 Which two amino acids are encoded by only one codon?

20.62 Which amino acids are encoded by six codons?

20.63 An essential gene has the codon 5'-UUU-3' in a critical position. If this codon is mutated to the sequence 5'-UUA-3', what is the expected consequence for the cell?

20.64 An essential gene has the codon 5'-UUA-3' in a critical position. If this codon is mutated to the sequence 5'-UUG-3', what is the expected consequence for the cell?

Protein Synthesis

Foundations

20.65 What is the function of ribosomes?

20.66 What are the two tRNA binding sites on the ribosome?

Applications

20.67 Explain the relationship between the sequence of nucleotides of a gene in the DNA and the sequence of amino acids in the protein encoded by that gene.

20.68 Explain how a change in the sequence of nucleotides of a gene, a mutation, may alter the sequence of amino acids in the protein encoded by that gene.

20.69 Briefly describe the three stages of translation: initiation, elongation, and termination.

20.70 What peptide sequence would be formed from the mRNA 5'-AUGUGUAGUGACCAACCGAUUUCACUGUGA-3'?

The following diagram shows the reaction that produces an aminoacyl tRNA, in this case methionyl tRNA. Use the following diagram to answer Questions 20.71 and 20.72.

20.71 By what type of bond is an amino acid linked to a tRNA molecule in an aminoacyl tRNA molecule?

20.72 Draw the structure of an alanine residue bound to the 3' position of adenine at the 3' end of alanyl tRNA.

$$ATP \longrightarrow AMP + PP_i$$

$$\underset{\text{tRNA}}{\underset{|}{\underset{3'OH}{\underset{|}{\underset{O^-}{\underset{+}{O}}}}}} \overset{N^+H_3}{\underset{|}{C}} - CH_2CH_2SCH_3 \quad \underset{\text{tRNA}}{\underset{|}{\underset{O}{\underset{|}{\underset{O}{\underset{\|}{C}}}}}} \overset{N^+H_3}{\underset{|}{C}} - CH_2CH_2SCH_3$$

The amino acyl linkage is formed between the 3'—OH of the tRNA and the carboxylate group of the amino acid, methionine

Mutation, Ultraviolet Light, and DNA Repair

Foundations
20.73 Define the term *point mutation*.
20.74 What are deletion and insertion mutations?

Applications
20.75 Why are some mutations silent?
20.76 Which is more likely to be a silent mutation, a point mutation or a deletion mutation? Explain your reasoning.
20.77 What damage does UV light cause in DNA, and how does this lead to mutations?
20.78 Explain why UV lights are effective germicides on environmental surfaces.
20.79 What is a carcinogen? Why are carcinogens also mutagens?
20.80 a. What causes the genetic disease xeroderma pigmentosum?
   b. Why are people who suffer from xeroderma pigmentosum prone to cancer?

Recombinant DNA

Foundations
20.81 What is a restriction enzyme?
20.82 Of what value are restriction enzymes in recombinant DNA research?
20.83 What is a selectable marker?
20.84 What is a cloning vector?

Applications
20.85 Name three products of recombinant DNA that are of value in the field of medicine.
20.86 a. What is the ultimate goal of genetic engineering?
   b. What ethical issues does this goal raise?

Polymerase Chain Reaction

20.87 After ten cycles of polymerase chain reaction, how many copies of target DNA would you have for each original molecule in the mixture?
20.88 List several practical applications of polymerase chain reaction.

The Human Genome Project

Foundations
20.89 What are the major goals of the Human Genome Project?
20.90 What are the potential benefits of the information gained in the Human Genome Project?
20.91 What is a genome library?
20.92 What is meant by the term *chromosome walking*?

20.93 What is a dideoxynucleotide?
20.94 How does a dideoxynucleotide cause chain termination in DNA replication?

Applications
20.95 A researcher has determined the sequence of the following pieces of DNA. Using this sequence information, map the location of these pieces relative to one another.
   a. 5' AGCTCCTGATTTCATACAGTTTCTACTACCTACTA 3'
   b. 5' AGACATTCTATCTACCTAGACTATGTTCAGAA 3'
   c. 5' TTCAGAACTCATTCAGACCTACTACTATACCTTGG GAGCTCCT 3'
   d. 5' ACCTACTAGACTATACTACTACTAAGGGGACTATT CCAGACTT 3'
20.96 Draw a DNA sequencing gel that would represent the sequence shown below. Be sure to label which lanes of the gel represent each of the four dideoxynucleotides in the chain termination reaction mixture.

5' GACTATCCTAG 3'

CRITICAL THINKING PROBLEMS

1. It has been suggested that the triplet genetic code evolved from a two-nucleotide code. Perhaps there were fewer amino acids in the ancient proteins. Examine the genetic code in Figure 20.17. What features of the code support this hypothesis?
2. The strands of DNA can be separated by heating the DNA sample. The input heat energy breaks the hydrogen bonds between base pairs, allowing the strands to separate from one another. Suppose that you are given two DNA samples. One has a G + C content of 70% and the other has a G + C content of 45%. Which of these samples will require a higher temperature to separate the strands? Explain your answer.
3. A mutation produces a tRNA with a new anticodon. Originally the anticodon was 5'-CCA-3'; the mutant anticodon is 5'-UCA-3'. What effect will this mutant tRNA have on cellular translation?
4. You have just cloned an EcoR1 fragment that is 1650 base pairs (bp) and contains the gene for the hormone leptin. Your first job is to prepare a restriction enzyme map of the recombinant plasmid. You know that you have cloned into a plasmid vector that is 805 bp and that has only one EcoR1 site (the one into which you cloned). There are no other restriction enzyme sites in the plasmid. The following table shows the restriction enzymes used and the DNA fragment sizes that result. Draw a map of the circular recombinant plasmid and a representation of the gel from which the fragment sizes were obtained.

Restriction Enzymes	DNA Fragment Sizes (bp)
EcoR1	805, 1650
EcoR1 + BamHI	450, 805, 1200
EcoR1 + SalI	200, 805, 1450
BamHI + SalI	200, 250, 805, 1200

5. A scientist is interested in cloning the gene for blood clotting factor VIII into bacteria so that large amounts of the protein can be produced to treat hemophiliacs. Knowing that bacterial cells cannot carry out RNA splicing, she clones a complementary DNA copy of the factor VIII mRNA and introduces this into bacteria. However, there is no transcription of the cloned factor VIII gene. How could the scientist engineer the gene so that the bacterial cell RNA polymerase will transcribe it?

BIOCHEMISTRY

21

Carbohydrate Metabolism

Some familiar fermentation products.

Learning Goals

1. Discuss the importance of ATP in cellular energy transfer processes.

2. Describe the three stages of catabolism of dietary proteins, carbohydrates, and lipids.

3. Discuss glycolysis in terms of its two major segments.

4. Looking at an equation representing any of the chemical reactions that occur in glycolysis, describe the kind of reaction that is occurring and the significance of that reaction to the pathway.

5. Describe the mechanism of regulation of the rate of glycolysis. Discuss particular examples of that regulation.

6. Discuss the practical and metabolic roles of fermentation reactions.

7. List several products of the pentose phosphate pathway that are required for biosynthesis.

8. Compare glycolysis and gluconeogenesis.

9. Summarize the regulation of blood glucose levels by glycogenesis and glycogenolysis.

Outline

Chemistry Connection:
The Man Who Got Tipsy from Eating Pasta

21.1 ATP: The Cellular Energy Currency

21.2 Overview of Catabolic Processes

21.3 Glycolysis

A Medical Perspective:
Genetic Disorders of Glycolysis

21.4 Fermentations

A Human Perspective:
Fermentations: The Good, the Bad, and the Ugly

21.5 The Pentose Phosphate Pathway

21.6 Gluconeogenesis: The Synthesis of Glucose

21.7 Glycogen Synthesis and Degradation

A Medical Perspective:
Diagnosing Diabetes

A Human Perspective:
Glycogen Storage Diseases

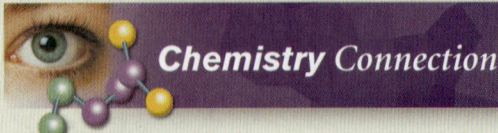

Chemistry Connection

The Man Who Got Tipsy from Eating Pasta

Imagine becoming drunk after eating a plate of spaghetti or a bag of potato chips. That is exactly what happened to Charles Swaart while he was stationed in Tokyo after World War II. Suddenly, he would be completely drunk—without having swallowed a drop of alcohol.

During the next two decades, Swaart continued to have unexplainable bouts of drunkenness and horrible hangovers. The problem was so serious that his liver was being destroyed. But in 1964, Swaart heard of a man in Japan who suffered from the same mysterious—and embarrassing—symptoms. After twenty-five years, physicians diagnosed the problem. There was a mutant strain of the yeast *Candida albicans* living in the gastrointestinal tract of the Japanese man. These yeast cells were using carbohydrates from the man's diet to make ethanol. The metabolic pathways used by these yeast cells were glycolysis and alcohol fermentation, two of the pathways that we will study in this chapter.

Swaart took advantage of the therapy used in Japan. He had to try several antibiotics over the years. But finally, in 1975, all of the mutant yeast cells in his intestine were killed, and his life returned to normal.

Why was it so difficult for physicians to solve this medical mystery? Nonfermenting *Candida albicans* is a regular inhabitant of the human gut. It took some very clever scientific detective work to find this mutant ethanol-producing strain. The scientists even have a hypothesis about where the mutant yeast came from. They think that the radiation released in one of the atomic bomb blasts at Nagasaki or Hiroshima may have caused the mutation.

In Chapter 20 we learned about the kinds of DNA damage that cause mutations. In this chapter we begin our study of the chemical reactions used by all organisms to provide energy for cellular work.

Introduction

Just as we need energy to run, jump, and think, the cell needs a ready supply of cellular energy for the many functions that support these activities. Cells need energy for active transport, to move molecules between the environment and the cell. Energy is also needed for biosynthesis of small metabolic molecules and production of macromolecules from these intermediates. Finally, energy is required for mechanical work, including muscle contraction and motility of sperm cells. Table 21.1 lists some examples of each of these energy-requiring processes.

We need a supply of energy-rich food molecules that can be degraded, or oxidized, to provide this needed cellular energy. Our diet includes three major sources of energy: carbohydrates, fats, and proteins. Each of these types of large biological molecules must be broken down into its basic subunits—simple sugars, fatty acids and glycerol, and amino acids—before they can be taken into the cell and used to produce cellular energy. Of these classes of food molecules, carbohydrates are the most readily used. The pathway for the first stages of carbohydrate breakdown is called glycolysis. We find the same pathway in organisms as different as the simple bacterium and humans.

In this chapter we are going to examine the steps of this ancient energy-harvesting pathway. We will see that it is responsible for the capture of some of the bond energy of carbohydrates and the storage of that energy in the molecular form of adenosine triphosphate (ATP). Glycolysis actually releases and stores very little (2.2%) of the potential energy of glucose, but the pathway also serves as a source of biosynthetic building blocks. It also modifies the carbohydrates in such a way that other pathways are able to release as much as 40% of the potential energy.

Recall that the potential energy of a compound is the bond energy of that compound.

TABLE 21.1	The Types of Cellular Work That Require Energy
Biosynthesis: Synthesis of Metabolic Intermediates and Macromolecules	
Synthesis of glucose from CO_2 and H_2O in the process of photosynthesis in plants	
Synthesis of amino acids	
Synthesis of nucleotides	
Synthesis of lipids	
Protein synthesis from amino acids	
Synthesis of nucleic acids	
Synthesis of organelles and membranes	
Active Transport: Movement of Ions and Molecules	
Transport of H^+ to maintain constant pH	
Transport of food molecules into the cell	
Transport of K^+ and Na^+ into and out of nerve cells for transmission of nerve impulses	
Secretion of HCl from parietal cells into the stomach	
Transport of waste from the blood into the urine in the kidneys	
Transport of amino acids and most hexose sugars into the blood from the intestine	
Accumulation of calcium ions in the mitochondria	
Motility	
Contraction and flexion of muscle cells	
Separation of chromosomes during cell division	
Ability of sperm to swim via flagella	
Movement of foreign substances out of the respiratory tract by cilia on the epithelial lining of the trachea	
Translocation of eggs into the fallopian tubes by cilia in the female reproductive tract	

21.1 ATP: The Cellular Energy Currency

The degradation of fuel molecules, called **catabolism,** provides the energy for cellular energy-requiring functions, including **anabolism,** or biosynthesis. Actually, the energy of a food source can be released in one of two ways: as heat or, more important to the cell, as chemical bond energy. We can envision two alternative modes of aerobic degradation of the simple sugar glucose. Imagine that we simply set the glucose afire. This would result in its complete oxidation to CO_2 and H_2O and would release 686 kcal/mol of glucose. Yet in terms of a cell, what would be accomplished? Nothing. All of the potential energy of the bonds of glucose is lost as heat and light.

 LEARNING GOAL

The cell uses a different strategy. With a series of enzymes, biochemical pathways in the cell carry out a step-by-step oxidation of glucose. Small amounts of energy are released at several points in the pathway and that energy is harvested and saved in the bonds of a molecule that has been called the *universal energy currency.* This molecule is **adenosine triphosphate (ATP).**

ATP serves as a "go-between" molecule that couples the *exergonic* (energy releasing) reactions of catabolism and the endergonic (energy requiring) reactions of anabolism. To understand how this molecule harvests the energy and releases it for energy-requiring reactions, we must take a look at the structure of this amazing molecule (Figure 21.1). ATP is a **nucleotide,** which means that it is a molecule composed of a nitrogenous base; a five-carbon sugar; and one, two, or three phosphoryl groups.

Nitrogenous bases are heterocyclic amines. Their structure and functions are discussed in Section 15.2. The five-carbon sugars are discussed in Section 16.4. More information on the structure of nucleotides is found in Section 20.1.

714   Chapter 21 Carbohydrate Metabolism

Figure 21.1
The structure of the universal energy currency, ATP.

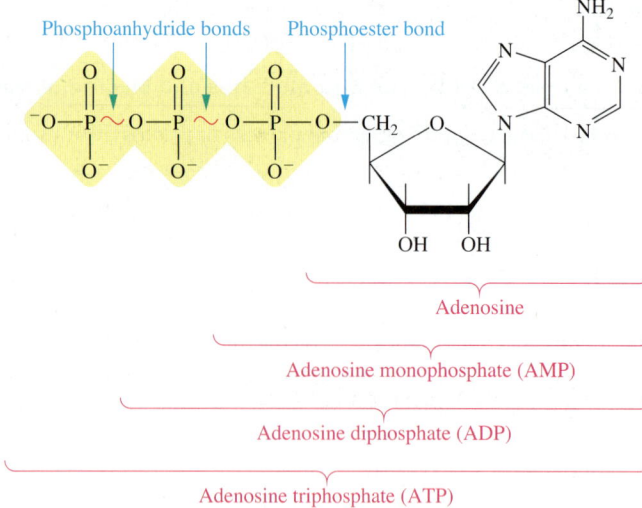

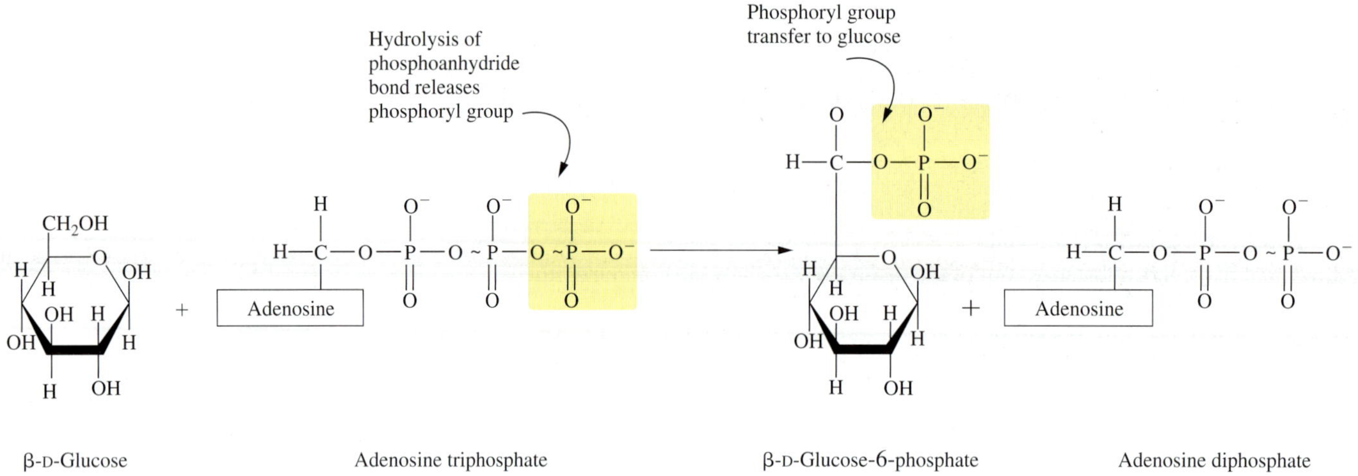

Figure 21.2
The hydrolysis of the phosphoanhydride bond of ATP releases inorganic phosphate and energy. In this coupled reaction catalyzed by an enzyme, the phosphoryl group and some of the released energy are transferred to β-D-glucose.

Nature's high energy bonds, including phosphoanhydride and phosphoester bonds, are discussed in Section 14.4.

See Sections 14.4 and 20.1 for further information on the structure of ATP and hydrolysis of the phosphoanhydride bonds.

In ATP, a phosphoester bond joins the first phosphoryl group to the five-carbon sugar ribose. The next two phosphoryl groups are joined to one another by phosphoanhydride bonds (Figure 21.1). Recall that the phosphoanhydride bond is a *high-energy bond*. When it is broken or hydrolyzed, a large amount of energy is released. When the phosphoanhydride bond of ATP is broken, the energy that is released can be used for cellular work. These high-energy bonds are indicated as squiggles (~) in Figure 21.1.

Hydrolysis of ATP yields adenosine diphosphate (ADP), an inorganic phosphate group (P_i), and energy (Figure 21.2). The energy released by this hydrolysis of ATP is then used to drive biological processes, for instance, the phosphorylation of glucose or fructose.

An example of the way in which the energy of ATP is used can be seen in the first step of glycolysis, the anaerobic degradation of glucose to harvest chemical energy. The first step involves the transfer of a phosphoryl group, $-PO_3^{2-}$, from ATP to the C-6 hydroxyl group of glucose (Figure 21.2). This reaction is catalyzed by the enzyme hexokinase.

This reaction can be dissected to reveal the role of ATP as a source of energy. Although this is a coupled reaction, we can think of it as a two-step process. The first step is the hydrolysis of ATP to ADP and phosphate, abbreviated P_i. This is an exergonic reaction that *releases* about 7 kcal/mol of energy:

$$ATP + H_2O \longrightarrow ADP + P_i + 7 \text{ kcal/mol}$$

The second step, the synthesis of glucose-6-phosphate from glucose and phosphate, is an endergonic reaction that *requires* 3.0 kcal/mol:

$$3.0 \text{ kcal/mol} + \text{glucose} + P_i \longrightarrow \text{glucose-6-phosphate} + H_2O$$

These two chemical reactions can then be added to give the equation showing the way in which ATP hydrolysis is *coupled* to the phosphorylation of glucose:

$$ATP + H_2O \longrightarrow ADP + P_i + 7 \text{ kcal/mol}$$
$$3.0 \text{ kcal/mol} + \text{glucose} + P_i \longrightarrow \text{glucose-6-phosphate} + H_2O$$
$$\text{Net: } ATP + \text{glucose} \longrightarrow \text{glucose-6-phosphate} + ADP + 4 \text{ kcal/mol}$$

Because the hydrolysis of ATP releases more energy than is required to synthesize glucose-6-phosphate from glucose and phosphate, there is an overall energy release in this process and the reaction proceeds spontaneously to the right. The product, glucose-6-phosphate, has more energy than the reactant, glucose, because it now carries some of the energy from the original phosphoanhydride bond of ATP.

The primary function of all catabolic pathways is to harvest the chemical energy of fuel molecules and to store that energy by the production of ATP. This continuous production of ATP is what provides the stored potential energy that is used to power most cellular functions.

Question 21.1

Why is ATP called the universal energy currency?

Question 21.2

List five biological activities that require ATP.

21.2 Overview of Catabolic Processes

 **LEARNING GOAL**

Although carbohydrates, fats, and proteins can all be degraded to release energy, carbohydrates are the most readily used energy source. We will begin by examining the oxidation of the hexose glucose. In Chapters 22 and 23 we will see how the pathways of glucose oxidation are also used for the degradation of fats and proteins.

Any catabolic process must begin with a supply of nutrients. When we eat a meal, we are eating quantities of carbohydrates, fats, and proteins. From this point the catabolic processes can be broken down into a series of stages. The three stages of catabolism are summarized in Figure 21.3.

Stage I: Hydrolysis of Dietary Macromolecules into Small Subunits

The purpose of the first stage of catabolism is to degrade large food molecules into their component subunits. These subunits—simple sugars, amino acids, fatty acids, and glycerol—are then taken into the cells of the body for use as an energy source. The process of digestion is summarized in Figure 21.4.

Polysaccharides are hydrolyzed to monosaccharides. This process begins in the mouth, where the enzyme amylase begins the hydrolysis of starch. Digestion continues in the small intestine, where pancreatic amylase further hydrolyzes the

Figure 21.3
The three stages of the conversion of food into cellular energy in the form of ATP.

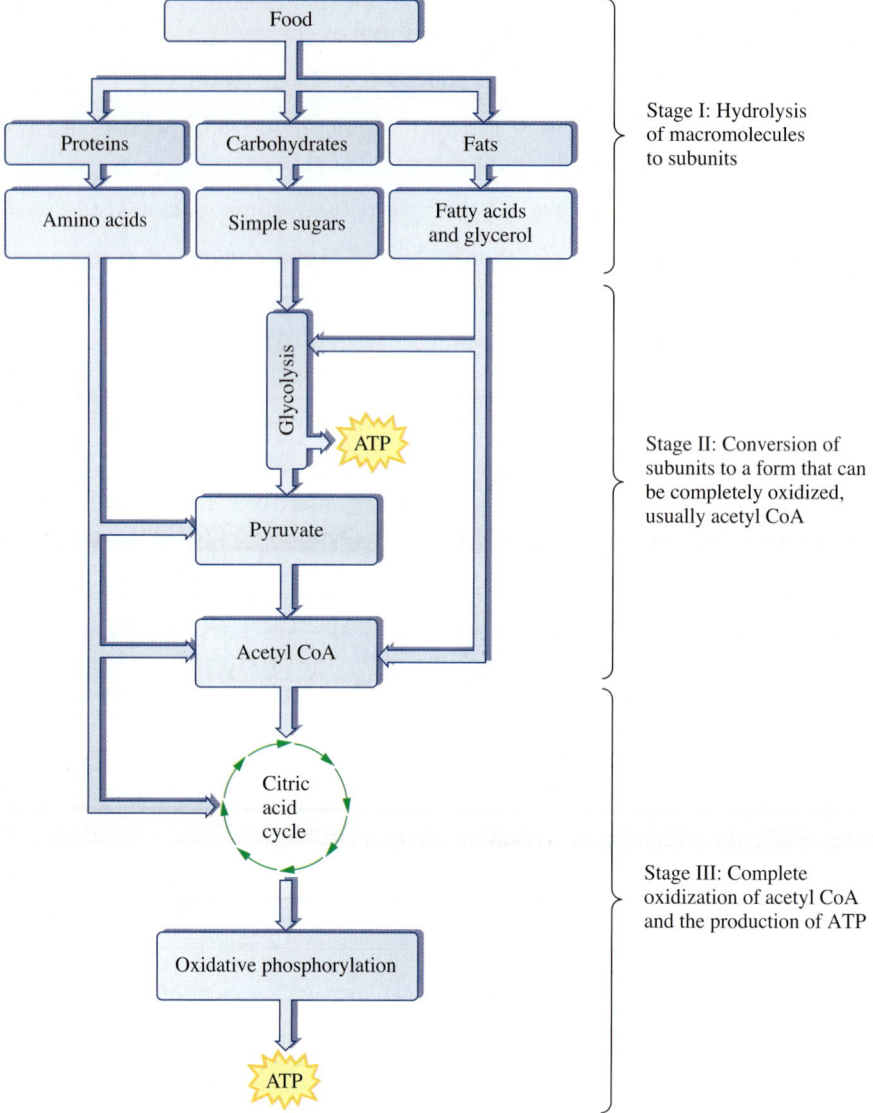

In the laboratory, a strong acid or base and high temperatures are required for hydrolysis of amide bonds (Section 15.3). However, this reaction proceeds quickly under physiological conditions when catalyzed by enzymes (Section 19.11).

The digestion and transport of fats are considered in greater detail in Chapter 23.

starch into maltose (a disaccharide of glucose). Maltase catalyzes the hydrolysis of maltose, producing two glucose molecules. Similarly, sucrose is hydrolyzed to glucose and fructose by the enzyme sucrase, and lactose (milk sugar) is degraded into the monosaccharides glucose and galactose by the enzyme lactase in the small intestine. The monosaccharides are taken up by the epithelial cells of the intestine in an energy-requiring process called *active transport.*

The digestion of proteins begins in the stomach, where the low pH denatures the proteins so that they are more easily hydrolyzed by the enzyme pepsin. They are further degraded in the small intestine by trypsin, chymotrypsin, elastase, and other proteases. The products of protein digestion—amino acids and short oligopeptides—are taken up by the cells lining the intestine. This uptake also involves an active transport mechanism.

Digestion of fats does not begin until the food reaches the small intestine, even though there are lipases in both the saliva and stomach fluid. Fats arrive in the duodenum, the first portion of the small intestine, in the form of large fat globules. Bile salts produced by the liver break these up into an emulsion of tiny fat droplets. Because the small droplets have a greater surface area, the lipids are now more accessible to the action of pancreatic lipase. This enzyme hydrolyzes the fats into fatty acids and glycerol, which are taken up by intestinal cells by a transport

21.2 Overview of Catabolic Processes

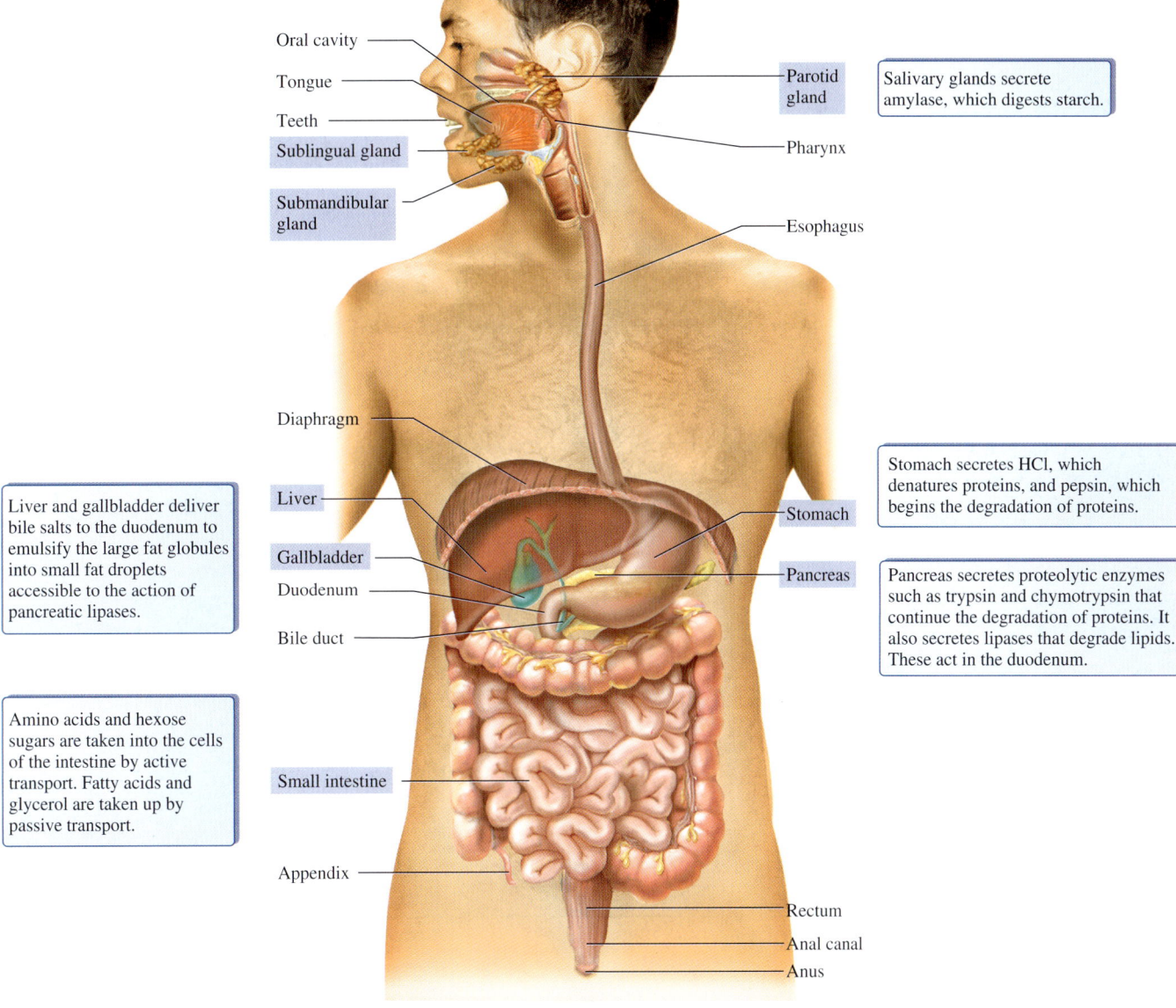

Figure 21.4
An overview of the digestive processes that hydrolyze carbohydrates, proteins, and fats.

process that does not require energy. This process is called *passive transport*. A summary of these hydrolysis reactions is shown in Figure 21.5.

Active and passive transport are discussed in Section 17.6.

Stage II: Conversion of Monomers into a Form That Can Be Completely Oxidized

The monosaccharides, amino acids, fatty acids, and glycerol must now be assimilated into the pathways of energy metabolism. The two major pathways are glycolysis and the citric acid cycle (see Figure 21.3). Sugars usually enter the glycolysis pathway in the form of glucose or fructose. They are eventually converted to acetyl CoA, which is a form that can be completely oxidized in the citric acid cycle. Amino groups are removed from amino acids, and the remaining carbon skeletons enter the catabolic processes at many steps of the citric acid cycle. Fatty acids are converted to acetyl CoA and enter the citric acid cycle in that form. Glycerol, produced by the hydrolysis of fats, is converted to glyceraldehyde-3-phosphate, one of the intermediates of glycolysis, and enters energy metabolism at that level.

The citric acid cycle is considered in detail in Section 22.4.

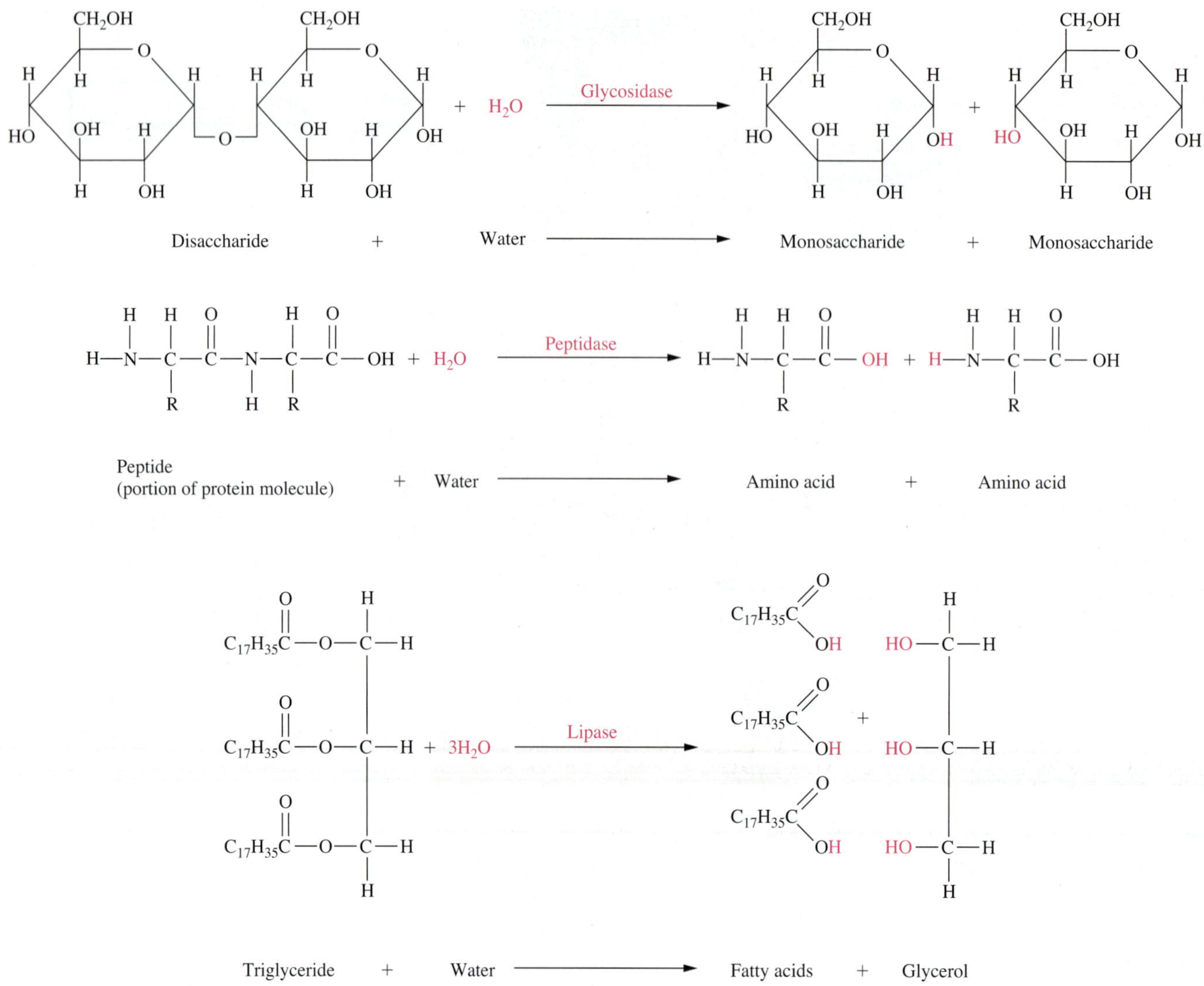

Figure 21.5
A summary of the hydrolysis reactions of carbohydrates, proteins, and fats.

Stage III: The Complete Oxidation of Nutrients and the Production of ATP

Oxidative phosphorylation is described in Section 22.6.

Acetyl CoA carries acetyl groups, two-carbon remnants of the nutrients, to the citric acid cycle. Acetyl CoA enters the cycle, and electrons and hydrogen atoms are harvested during the complete oxidation of the acetyl group to CO_2. Coenzyme A is released (recycled) to carry additional acetyl groups to the pathway. The electrons and hydrogen atoms that are harvested are used in the process of oxidative phosphorylation to produce ATP.

Question 21.3 Briefly describe the three stages of catabolism.

Question 21.4 Discuss the digestion of dietary carbohydrates, lipids, and proteins.

21.3 Glycolysis

An Overview

Glycolysis, also known as the Embden-Meyerhof Pathway, is a pathway for carbohydrate catabolism that begins with the substrate D-glucose. The very fact that all organisms can use glucose as an energy source for glycolysis suggests that glycolysis was the first successful energy-harvesting pathway that evolved on the earth. The pathway evolved at a time when the earth's atmosphere was *anaerobic;* no free oxygen was available. As a result, glycolysis requires no oxygen; it is an anaerobic process. Further, it must have evolved in very simple, single-celled organisms, much like bacteria. These organisms did not have complex organelles in the cytoplasm to carry out specific cellular functions. Thus glycolysis was a process carried out by enzymes that were free in the cytoplasm. To this day, glycolysis remains an anaerobic process carried out by cytoplasmic enzymes, even in cells as complex as our own.

The ten steps of glycolysis, catalyzed by ten enzymes, are outlined in Figure 21.6. The first reactions of glycolysis involve an energy investment. ATP molecules are hydrolyzed, energy is released, and phosphoryl groups are added to the hexose sugars. In the remaining steps of glycolysis, energy is harvested to produce a net gain of ATP.

The three major products of glycolysis are seen in Figure 21.6. These are chemical energy in the form of ATP, chemical energy in the form of NADH, and two three-carbon pyruvate molecules. Each of these products is considered below:

- **Chemical energy as ATP.** Four ATP molecules are formed by the process of **substrate-level phosphorylation.** This means that a high-energy phosphoryl group from one of the substrates in glycolysis is transferred to ADP to form ATP. The two substrates involved in these transfer reactions are 1,3-bisphosphoglycerate and phosphoenolpyruvate (see Figure 21.6, steps 7 and 10). Although four ATP molecules are produced during glycolysis, the *net* gain is only two ATP molecules because two ATP molecules are used early in glycolysis (Figure 21.6, steps 1 and 3). The two ATP molecules produced represent only 2.2% of the potential energy of the glucose molecule. Thus glycolysis is not a very efficient energy-harvesting process.
- **Chemical energy in the form of reduced NAD$^+$, NADH. Nicotinamide adenine dinucleotide (NAD$^+$)** is a coenzyme derived from the vitamin niacin. The reduced form of NAD$^+$, NADH, carries hydride anions, hydrogen atoms with two electrons (H:$^-$), removed during the oxidation of glyceraldehyde-3-phosphate (see Figure 21.6, step 6). Under aerobic conditions, the electrons and hydrogen atom are transported from the cytoplasm into the mitochondria. Here they enter an electron transport system for the generation of ATP by **oxidative phosphorylation.** Under anaerobic conditions, NADH is used as a source of electrons in fermentation reactions.
- **Two pyruvate molecules.** At the end of glycolysis the six-carbon glucose molecule has been converted into two three-carbon pyruvate molecules. The fate of the pyruvate also depends on whether the reactions are occurring in the presence or absence of oxygen. Under aerobic conditions it is used to produce acetyl CoA destined for the citric acid cycle and complete oxidation. Under anaerobic conditions it is used as an electron acceptor in fermentation reactions.

In any event these last two products must be used in some way so that glycolysis can continue to function and produce ATP. There are two reasons for this. First, if pyruvate were allowed to build up, it would cause glycolysis to stop,

The structure of NAD$^+$ and the way it functions as a hydride anion carrier are shown in Figure 19.8 and described in Section 19.7.

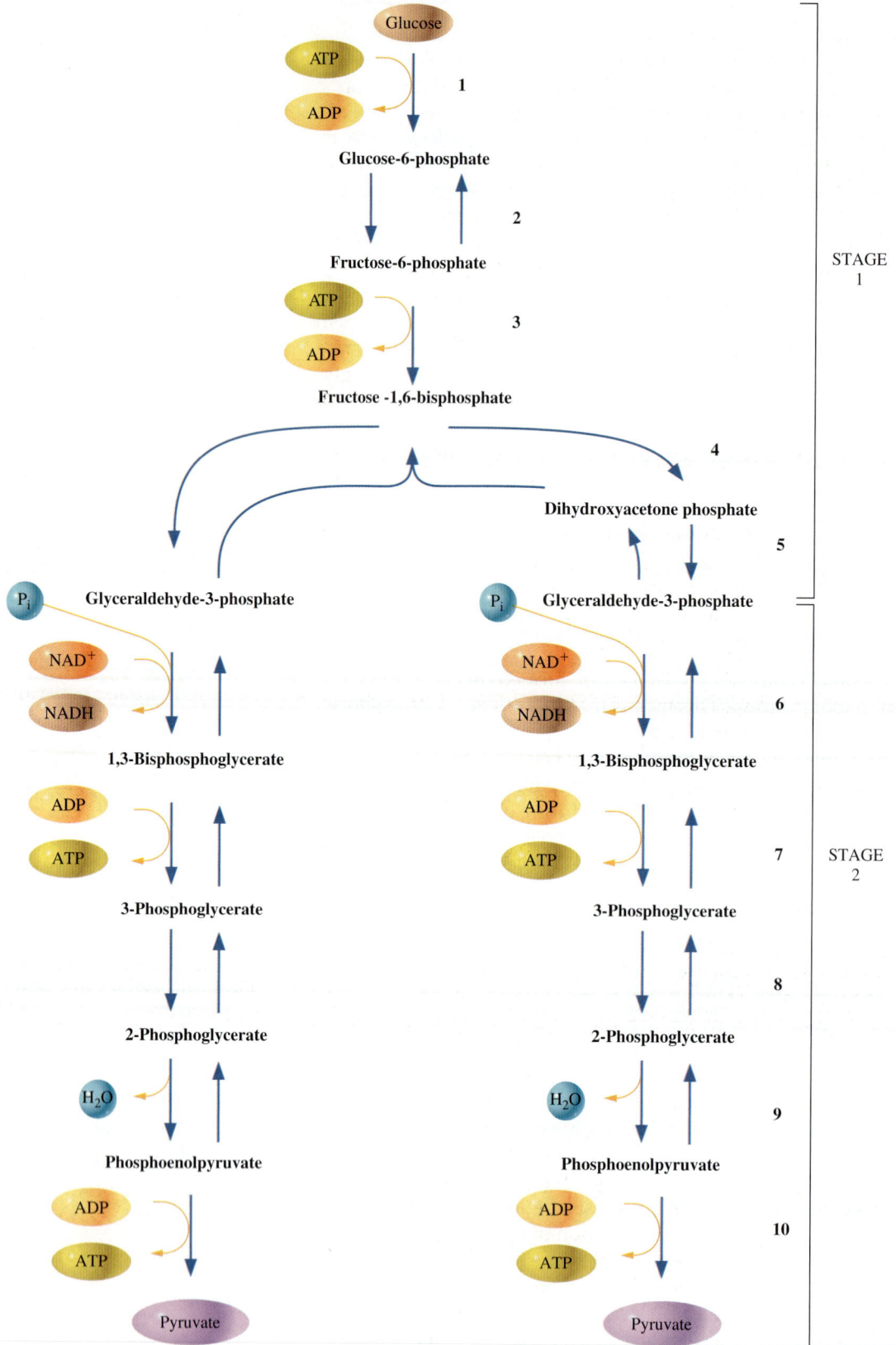

Figure 21.6
A summary of the reactions of glycolysis. These reactions occur in the cell cytoplasm.

thereby stopping the production of ATP. Thus pyruvate must be used in some kind of follow-up reaction, aerobic or anaerobic. Second, in step 6, glyceraldehyde-3-phosphate is oxidized and NAD$^+$ is reduced (accepts the hydride anion). The cell has only a small supply of NAD$^+$. If all the NAD$^+$ is reduced, none will be available for this reaction, and glycolysis will stop. Therefore NADH must be reoxidized so that glycolysis can continue to produce ATP for the cell.

Reactions of Glycolysis

Glycolysis can be divided into two major segments. The first is the investment of ATP energy. Without this investment, glucose would not have enough energy for glycolysis to continue, and there would be no ATP produced. This segment includes the first five reactions of the pathway. The second major segment involves the remaining reactions of the pathway (steps 6–10), those that result in a net energy yield.

3 LEARNING GOAL

4 LEARNING GOAL

Reaction 1

The substrate, glucose, is phosphorylated by the enzyme *hexokinase* in a coupled phosphorylation reaction. The source of the phosphoryl group is ATP. At first this reaction seems contrary to the overall purpose of catabolism, the *production* of ATP. The expenditure of ATP in these early reactions must be thought of as an "investment." The cell actually goes into energy "debt" in these early reactions, but this is absolutely necessary to get the pathway started.

Glucose + ATP →(Hexokinase) Glucose-6-phosphate + ADP + H$^+$

The enzyme name can tell us a lot about the reaction (see Section 19.1). The suffix *-kinase* tells us that the enzyme is a transferase that will transfer a phosphoryl group, in this case from an ATP molecule to the substrate. The prefix *hexo-* gives us a hint that the substrate is a six-carbon sugar. Hexokinase predominantly phosphorylates the six-carbon sugar glucose.

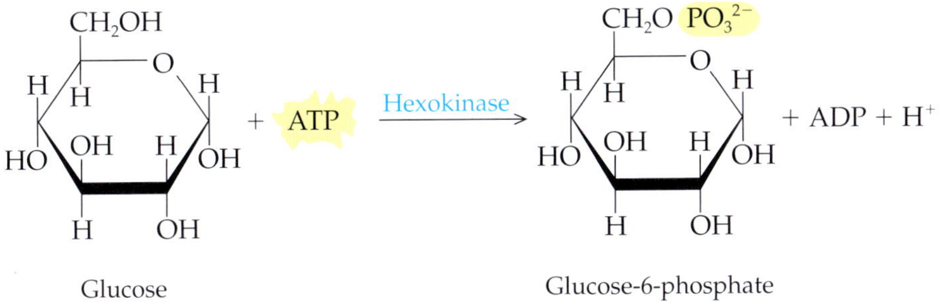

Reaction 2

The glucose-6-phosphate formed in the first reaction is rearranged to produce the structural isomer fructose-6-phosphate. The enzyme *phosphoglucose isomerase* catalyzes this isomerization. The result is that the C-1 carbon of the six-carbon sugar is exposed; it is no longer part of the ring structure. Examination of the open-chain structures reveals that this isomerization converts an aldose into a ketose.

Glucose-6-phosphate ⇌(Phosphoglucose isomerase) Fructose-6-phosphate

The enzyme name, phosphoglucose isomerase, provides clues to the reaction that is being catalyzed (Section 19.1). *Isomerase* tells us that the enzyme will catalyze the interconversion of one isomer into another. *Phosphoglucose* suggests that the substrate is a phosphorylated form of glucose.

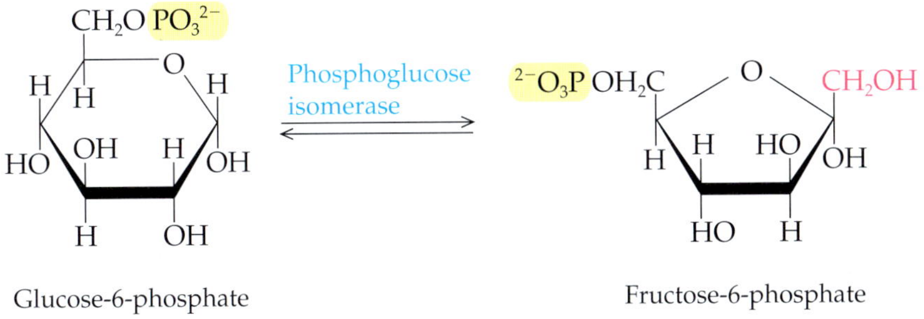

A Medical Perspective

Genetic Disorders of Glycolysis

Imagine always having difficulty with physical exercise. Imagine the coach telling you to get tough and run that lap again, since you were last! Imagine being accused of being lazy because you didn't carry your share of the camping gear. Imagine all that and not knowing why it is that you can't keep up with your friends or the others in your physical education class. This has been the fate of thousands of people who suffer from a metabolic myopathy—a muscle (*myo-*) disorder (*-pathy*) caused by an inability to extract the energy from the food that they eat.

The onset of fatigue during exercise is called *exercise intolerance*. It is one of the major symptoms of a metabolic myopathy. But simple fatigue is just the mildest of the symptoms. Overexertion may cause episodes of muscle breakdown (rhabdomyolysis) in which the muscle cells, unable to provide enough ATP energy for themselves, begin to die. The muscle breakdown causes greatly elevated blood levels of creatine kinase. Creatine kinase is an abundant enzyme in muscle that is critical in energy metabolism (see A Human Perspective: Exercise and Energy Metabolism, in Chapter 22). When muscle cells die, this enzyme is released into the bloodstream. Another symptom is myoglobinuria (myoglobin in the urine). Recall that myoglobin is the oxygen storage protein in muscle. When muscles die, myoglobin ends up in the urine, turning it the color of cola soft drinks. Myoglobinuria may even cause kidney damage. Accompanying these clinical symptoms, many people describe intense muscle pain. They describe it as a cramp, but it is not a cramp, since the muscle is not able to contract because of the lack of energy. Rather, the pain is caused by cell death and tissue damage that result from an inability to produce enough ATP.

There are three glycolytic enzyme deficiencies that lead to metabolic myopathy. The first is phosphofructokinase deficiency or Tarui's disease. Although this is not a sex-linked disorder, the great majority of sufferers are males (nine males to one female). The disorder is most frequently found in U.S. Ashkenazi Jews and Italian families. Onset of symptoms typically occurs between the ages of twenty and forty, although some severe cases have been reported in infants and young children. Patients experiencing the late-onset form of Tarui's disease typically experienced exercise intolerance when they were younger. Vigorous exercise results in myoglobinuria and severe muscle pain. Meals high in carbohydrates worsen the exercise intolerance. Early-onset disease is often associated with respiratory failure, cardiomyopathy (heart muscle disease), seizures, and cortical blindness.

Phosphoglycerate kinase deficiency is a sex-linked genetic disorder (located on the X chromosome). As a result, far more males than females suffer from this disease. There are many clinical features associated with this deficiency, although only rarely are they all found in the same patient. These symptoms range from mental retardation and seizures to a slowly progressive myopathy.

Phosphoglycerate mutase deficiency has been mapped on chromosome 7. The disorder is found predominantly in U.S. African American, Italian, and Japanese families. The clinical features include exercise intolerance, muscle pain, and myoglobinuria following more intense exercise.

Since each of these disorders is caused by the lack of an enzyme, scientists are trying to design a way to replace the lost activity. Oral medication will not work because enzymes are proteins. They would simply be digested, like any other dietary protein. Enzyme replacement therapy is one approach that is being studied. This would involve periodic injections of the enzyme into the bloodstream, a treatment just like the injection of insulin by diabetics. Enzyme replacement therapy would require a large supply of the enzyme. Following the model of insulin, the gene for the enzyme could be cloned into bacteria. The bacteria would then produce the protein, which would be purified for use by humans.

Another strategy is to introduce the gene for the missing enzyme into the patient's cells. This is called gene therapy. This method would also require that the gene for the enzyme be cloned. The DNA would then have to be introduced into the body using a safe procedure that would promote entry into target cells. There are still many obstacles to overcome before this type of treatment will be a reality for sufferers of metabolic myopathy.

A number of physicians are using a commonsense approach to the management of these disorders. Logic tells us that if a person cannot harvest energy from carbohydrates in the diet, perhaps a diet high in protein and lipids might be beneficial. As with any condition of this sort, it is important to consult a physician who understands the metabolic disorder and who will design and supervise a customized diet.

For Further Understanding

Write equations showing the reactions catalyzed by each of the enzymes discussed in this perspective and explain how the absence of each will impair ATP production.

Applying what you learned in Chapter 20, describe an experiment to clone the gene for phosphoglycerate kinase.

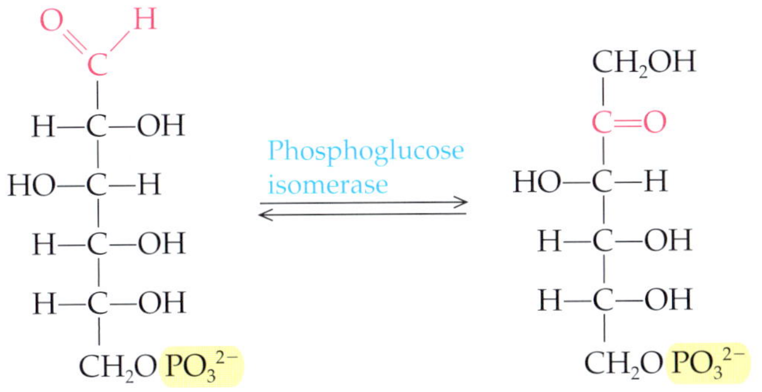

21.3 Glycolysis

[Structure: Glucose-6-phosphate (an aldose) ⇌ (Phosphoglucose isomerase) Fructose-6-phosphate (a ketose)]

This is an enediol reaction. It occurs through exactly the same steps as the conversion of fructose to glucose that we discussed in Section 16.4.

Reaction 3

A second energy "investment" is catalyzed by the enzyme *phosphofructokinase*. The phosphoanhydride bond in ATP is hydrolyzed, and a phosphoester linkage between the phosphoryl group and the C-1 hydroxyl group of fructose-6-phosphate is formed. The product is fructose-1,6-bisphosphate.

The suffix *-kinase* in the name of the enzyme tells us that this is a coupled reaction: ATP is hydrolyzed and a phosphoryl group is transferred to another molecule. The prefix *phosphofructo-* tells us the other molecule is a phosphorylated form of fructose.

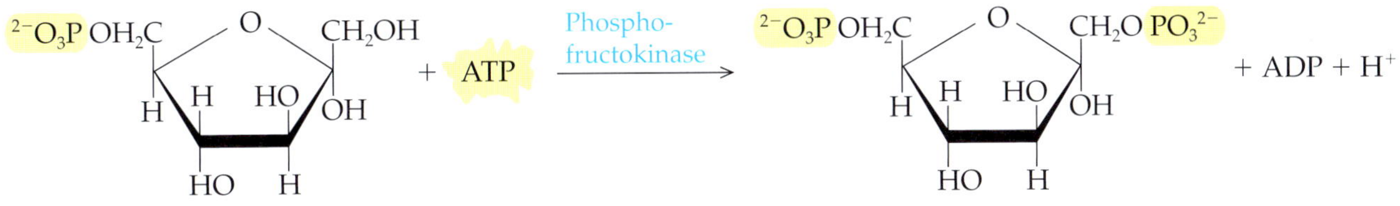

[Structure: Fructose-6-phosphate + ATP → (Phosphofructokinase) Fructose-1,6-bisphosphate + ADP + H⁺]

Reaction 4

Fructose-1,6-bisphosphate is split into two three-carbon intermediates in a reaction catalyzed by the enzyme *aldolase*. The products are glyceraldehyde-3-phosphate (G3P) and dihydroxyacetone phosphate.

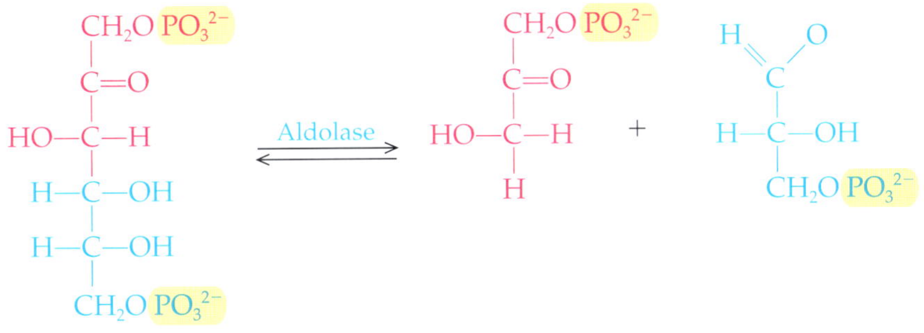

[Structure: Fructose-1,6-bisphosphate ⇌ (Aldolase) Dihydroxyacetone phosphate + Glyceraldehyde-3-phosphate]

In aldol condensation, aldehydes and ketones react to form a larger molecule (Section 13.4). This reaction is a reverse aldol condensation. The large ketone sugar fructose-1,6-bisphosphate is broken down into dihydroxyacetone phosphate (a ketone) and glyceraldehyde-3-phosphate (an aldehyde).

The double reaction arrows tell us that this is a reversible reaction. The reverse reaction is an aldol condensation (Section 13.4) that we will study in the pathway for glucose synthesis called gluconeogenesis (Section 21.6).

Reaction 5

Because G3P is the only substrate that can be used by the next enzyme in the pathway, the dihydroxyacetone phosphate is rearranged to become a second molecule of G3P. The enzyme that mediates this isomerization is *triose phosphate isomerase*.

The enzyme name hints that two isomers of a phosphorylated three-carbon sugar are going to be interconverted (Section 19.1). The ketone dihydroxyacetone phosphate and its isomeric aldehyde, phosphoglyceraldehyde-3-phosphate are interconverted through an enediol intermediate.

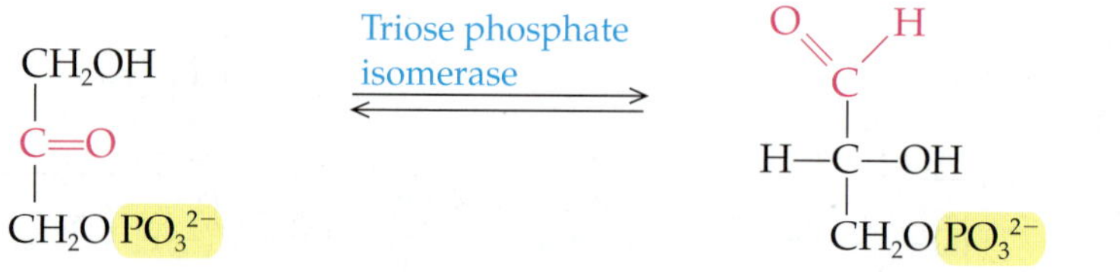

$$\text{Dihydroxyacetone phosphate} \underset{}{\overset{\text{Triose phosphate isomerase}}{\rightleftharpoons}} \text{Glyceraldehyde-3-phosphate}$$

Reaction 6

In this reaction the aldehyde glyceraldehyde-3-phosphate is oxidized to a carboxylic acid in a reaction catalyzed by *glyceraldehyde-3-phosphate dehydrogenase*. This is the first step in glycolysis that harvests energy, and it involves the reduction of the coenzyme nicotinamide adenine dinucleotide (NAD^+). This reaction occurs in two steps. First, NAD^+ is reduced to NADH as the aldehyde group of glyceraldehyde-3-phosphate is oxidized to a carboxyl group. Second, an inorganic phosphate group is transferred to the carboxyl group to give 1,3-bisphosphoglycerate. Notice that the new bond is denoted with a squiggle (~), indicating that this is a high-energy bond. This, and all remaining reactions of glycolysis, occur twice for each glucose because each glucose has been converted into two molecules of glyceraldehyde-3-phosphate.

The name glyceraldehyde-3-phosphate dehydrogenase tells us that the substrate glyceraldehyde-3-phosphate is going to be oxidized. In this reaction, we see that the aldehyde group has been oxidized to a carboxylate group (Section 13.4).

Actually the intermediate of the oxidation reaction is a high-energy thioester formed between the enzyme and the substrate (Section 14.4). When this bond is hydrolyzed, enough energy is released to allow the formation of a bond between an oxygen atom of an inorganic phosphate group and the substrate.

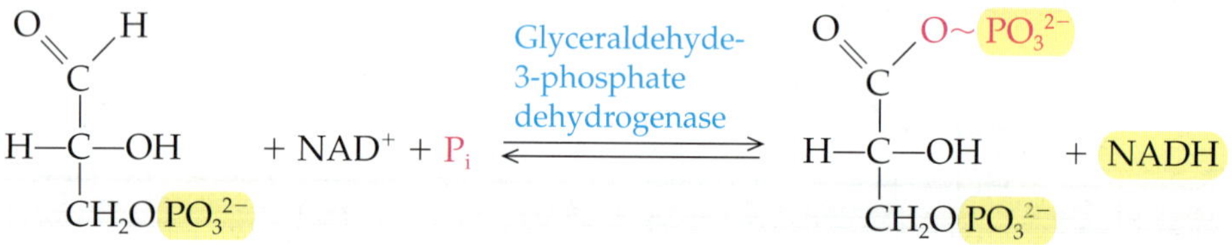

Glyceraldehyde-3-phosphate + NAD^+ + P_i ⇌ (Glyceraldehyde-3-phosphate dehydrogenase) 1,3-Bisphosphoglycerate + NADH

Reaction 7

In this reaction, energy is harvested in the form of *ATP*. The enzyme *phosphoglycerate kinase* catalyzes the transfer of the phosphoryl group of 1,3-bisphosphoglycerate to ADP. This is the first substrate-level phosphorylation of glycolysis, and it produces ATP and 3-phosphoglycerate. It is a coupled reaction in which the high-energy bond is hydrolyzed and the energy released is used to drive the synthesis of ATP.

Once again, the enzyme name reveals a great deal about the reaction. The suffix -kinase tells us that a phosphoryl group will be transferred. In this case, a phosphoester bond in the substrate 1,3-bisphosphoglycerate is hydrolyzed and ADP is phosphorylated. Note that this is a reversible reaction.

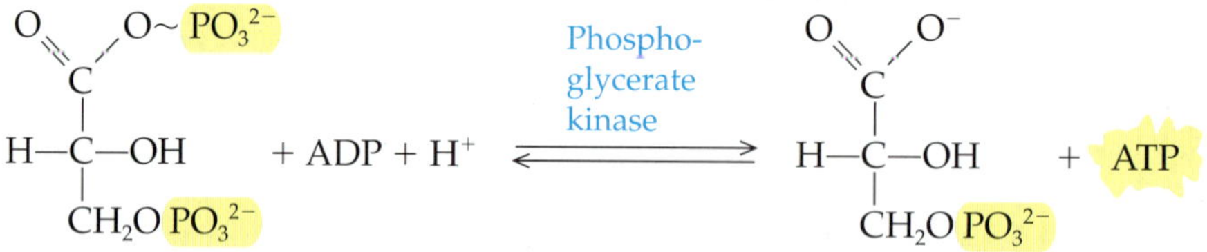

1,3-Bisphosphoglycerate + ADP + H^+ ⇌ (Phosphoglycerate kinase) 3-Phosphoglycerate + ATP

Reaction 8

3-Phosphoglycerate is isomerized to produce 2-phosphoglycerate in a reaction catalyzed by the enzyme *phosphoglycerate mutase*. The phosphoryl group attached to the third carbon of 3-phosphoglycerate is transferred to the second carbon.

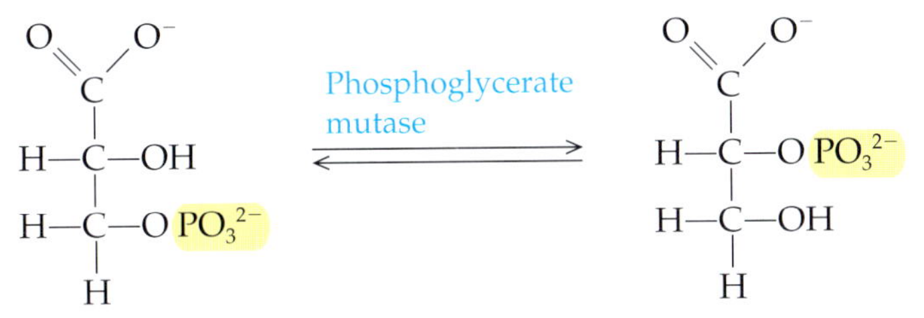

21.3 Glycolysis

[3-Phosphoglycerate → 2-Phosphoglycerate, catalyzed by Phosphoglycerate mutase]

The suffix -*mutase* indicates another type of isomerase. Notice that the chemical formulas of the substrate and reactant are the same. The only difference is in the location of the phosphoryl group.

Reaction 9

In this step the enzyme *enolase* catalyzes the dehydration (removal of a water molecule) of 2-phosphoglycerate. The energy-rich product is phosphoenolpyruvate, the highest energy phosphorylated compound in metabolism.

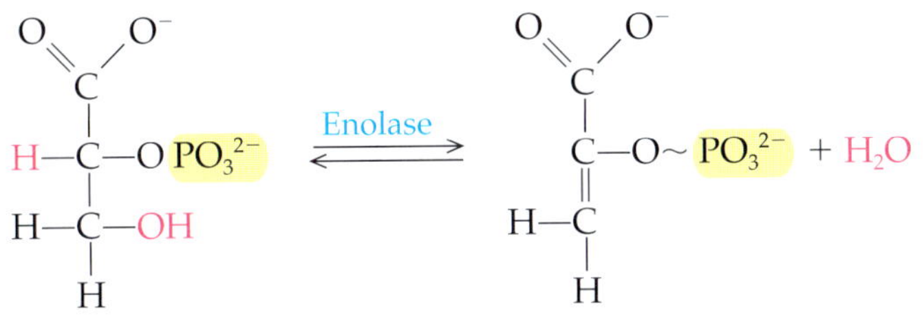

[2-Phosphoglycerate → Phosphoenolpyruvate + H_2O, catalyzed by Enolase]

In Section 13.4 we learned that aldehydes and ketones exist in an equilibrium mixture of two tautomers called the keto and enol forms. The dehydration of 2-phosphoglycerate produces the molecule phosphoenolpyruvate, which is in the enol form. In this case, the enol is extremely unstable. Because of this instability, the phosphoester bond in the product is a high-energy bond; in other words, a great deal of energy is released when this bond is broken.

Reaction 10

Here we see the final substrate-level phosphorylation in the pathway, which is catalyzed by *pyruvate kinase*. Phosphoenolpyruvate serves as a donor of the phosphoryl group that is transferred to ADP to produce ATP. This is another coupled reaction in which hydrolysis of the phosphoester bond in phosphoenolpyruvate provides energy for the formation of the phosphoanhydride bond of ATP. The final product of glycolysis is pyruvate.

The enzyme name indicates that a phosphoryl group will be transferred (kinase) and that the product will be pyruvate. Pyruvate is a keto tautomer and is much more stable than the enol (Section 13.4).

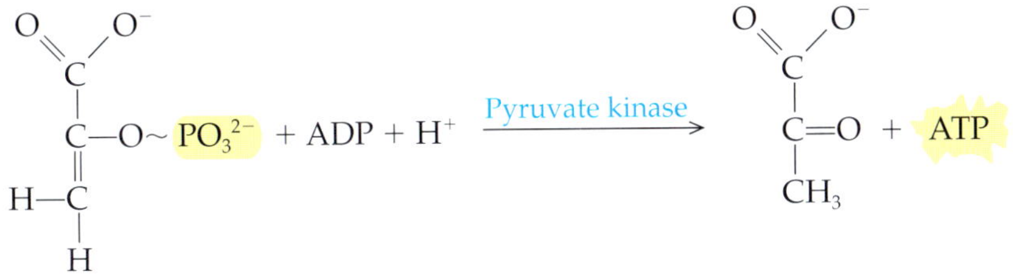

[Phosphoenolpyruvate + ADP + H^+ → Pyruvate + ATP, catalyzed by Pyruvate kinase]

It should be noted that reactions 6 through 10 occur twice per glucose molecule, because the starting six-carbon sugar is split into two three-carbon molecules. Thus in reaction 6, two NADH molecules are generated, and a total of four ATP molecules are made (steps 7 and 10). The net ATP gain from this pathway is, however, only two ATP molecules because there was an energy investment of two ATP molecules in the early steps of the pathway. This investment was paid back by the two ATP molecules produced by substrate-level phosphorylation in step 7. The actual energy yield is produced by substrate-level phosphorylation in reaction 10.

Question 21.5

What is substrate-level phosphorylation?

Question 21.6
What are the major products of glycolysis?

Question 21.7
Describe an overview of the reactions of glycolysis.

Question 21.8
How do the names of the first three enzymes of the glycolytic pathway relate to the reactions they catalyze?

LEARNING GOAL

Regulation of Glycolysis

Energy-harvesting pathways, such as glycolysis, are responsive to the energy needs of the cell. Reactions of the pathway speed up when there is a demand for ATP. They slow down when there is abundant ATP to meet the energy requirements of the cell.

One of the major mechanisms for the control of the rate of glycolysis is the use of *allosteric enzymes.* In addition to the active site, which binds the substrate, allosteric enzymes have an effector binding site, which binds a chemical signal that alters the rate at which the enzyme catalyzes the reaction. Effector binding may increase (positive allosterism) or decrease the rate of reaction (negative allosterism).

The chemical signals, or effectors, that indicate the energy needs of the cell include molecules such as ATP. When the ATP concentration is high, the cell must have sufficient energy. Similarly, ADP and AMP, which are precursors of ATP, are indicators that the cell is in need of ATP. In fact, all of these molecules are allosteric effectors that alter the rate of irreversible reactions catalyzed by enzymes in the glycolytic pathway.

The enzyme hexokinase, which catalyzes the phosphorylation of glucose, is allosterically inhibited by the product of the reaction it catalyzes, glucose-6-phosphate. A buildup of this product indicates that the reactions of glycolysis are not proceeding at a rapid rate, presumably because the cell has enough energy.

Phosphofructokinase, the enzyme that catalyzes the third reaction in glycolysis, is a key regulatory enzyme in the pathway. ATP is an allosteric inhibitor of phosphofructokinase, whereas AMP and ADP are allosteric activators. Another allosteric inhibitor of phosphofructokinase is citrate. As we will see in the next chapter, citrate is the first intermediate in the citric acid cycle, a pathway that results in the complete oxidation of the pyruvate. A high concentration of citrate signals that sufficient substrate is entering the citric acid cycle. The inhibition of phosphofructokinase by citrate is an example of *feedback inhibition:* the product, citrate, allosterically inhibits the activity of an enzyme early in the pathway.

There are additional mechanisms that regulate the rate of glycolysis, but we will focus on those that involve principles studied previously (Section 19.9).

The last enzyme in glycolysis, pyruvate kinase, is also subject to allosteric regulation. In this case, fructose-1,6-bisphosphate, the product of the reaction catalyzed by phosphofructokinase, is the allosteric activator. Thus, activation of phosphofructokinase results in the activation of pyruvate kinase. This is an example of *feedforward activation* because the product of an earlier reaction causes activation of an enzyme later in the pathway.

21.4 Fermentations

LEARNING GOAL

In the overview of glycolysis, we noted that the pyruvate produced must be used up in some way so that the pathway will continue to produce ATP. Similarly, the NADH produced by glycolysis in step 6 (see Figure 21.6) must be reoxidized at a

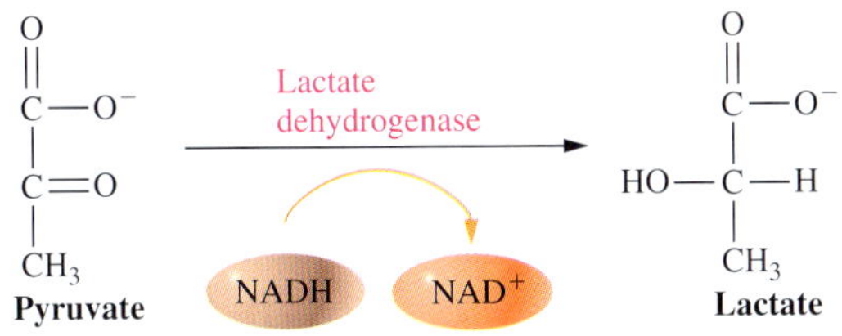

Figure 21.7
The final reaction of lactate fermentation.

later time, or glycolysis will grind to a halt as the available NAD^+ is used up. If the cell is functioning under aerobic conditions, NADH will be reoxidized, and pyruvate will be completely oxidized by aerobic respiration. Under anaerobic conditions, however, different types of fermentation reactions accomplish these purposes. **Fermentations** are catabolic reactions that occur with no net oxidation. Pyruvate or an organic compound produced from pyruvate is reduced as NADH is oxidized. We will examine two types of fermentation pathways in detail: lactate fermentation and alcohol fermentation.

Aerobic respiration is discussed in Chapter 22.

Lactate Fermentation

Lactate fermentation is familiar to anyone who has performed strenuous exercise. If you exercise so hard that your lungs and circulatory system can't deliver enough oxygen to the working muscles, your aerobic (oxygen-requiring) energy-harvesting pathways are not able to supply enough ATP to your muscles. But the muscles still demand energy. Under these anaerobic conditions, lactate fermentation begins. In this reaction, the enzyme *lactate dehydrogenase* reduces pyruvate to lactate. NADH is the reducing agent for this process (Figure 21.7). As pyruvate is reduced, NADH is oxidized, and NAD^+ is again available, permitting glycolysis to continue.

The lactate produced in the working muscle passes into the blood. Eventually, if strenuous exercise is continued, the concentration of lactate becomes so high that this fermentation can no longer continue. Glycolysis, and thus ATP production, stops. The muscle, deprived of energy, can no longer function. This point of exhaustion is called the **anaerobic threshold.**

Of course, most of us do not exercise to this point. When exercise is finished, the body begins the process of reclaiming all of the potential energy that was lost in the form of lactate. The liver takes up the lactate from the blood and converts it back to pyruvate. Now that a sufficient supply of oxygen is available, the pyruvate can be completely oxidized in the much more efficient aerobic energy-harvesting reactions to replenish the store of ATP. Alternatively, the pyruvate may be converted to glucose and used to restore the supply of liver and muscle glycogen. This exchange of metabolites between the muscles and liver is called the *Cori Cycle.*

The Cori Cycle is described in Section 21.6 and shown in Figure 21.13.

A variety of bacteria are able to carry out lactate fermentation under anaerobic conditions. This is of great importance in the dairy industry, because these organisms are used to produce yogurt and some cheeses. The tangy flavor of yogurt is contributed by the lactate produced by these bacteria. Unfortunately, similar organisms also cause milk to spoil.

As we saw in A Human Perspective: Tooth Decay and Simple Sugars (Chapter 16), the lactate produced by oral bacteria is responsible for the gradual removal of calcium from tooth enamel and the resulting dental cavities.

Alcohol Fermentation

Alcohol fermentation has been appreciated, if not understood, since the dawn of civilization. The fermentation process itself was discovered by Louis Pasteur during his studies of the chemistry of winemaking and "diseases of wines." Under anaerobic conditions, yeast are able to ferment the sugars produced by fruit and grains. The sugars are broken down to pyruvate by glycolysis. This is followed by the two reactions of alcohol fermentation. First, *pyruvate decarboxylase* removes CO_2 from the pyruvate producing acetaldehyde (Figure 21.8). Second, *alcohol dehydrogenase* catalyzes the reduction of acetaldehyde to ethanol but, more

These applications and other fermentations are described in A Human Perspective: Fermentations: The Good, the Bad, and the Ugly.

A Human Perspective

Fermentations: The Good, the Bad, and the Ugly

In this chapter we have seen that fermentation is an anaerobic, cytoplasmic process that allows continued ATP generation by glycolysis. ATP production can continue because the pyruvate produced by the pathway is utilized in the fermentation and because NAD^+ is regenerated.

The stable end products of alcohol fermentation are CO_2 and ethanol. These have been used by humankind in a variety of ways, including the production of alcoholic beverages, bread making, and alternative fuel sources.

If alcohol fermentation is carried out by using fruit juices in a vented vat, the CO_2 will escape, and the result will be a still wine (not bubbly). But conditions must remain anaerobic; otherwise, fermentation will stop, and aerobic energy-harvesting reactions will ruin the wine. Fortunately for vintners (wine makers), when a vat is fermenting actively, enough CO_2 is produced to create a layer that keeps the oxygen-containing air away from the fermenting juice, thus maintaining an anaerobic atmosphere.

Now suppose we want to make a sparkling wine, such as champagne. To do this, we simply have to trap the CO_2 produced. In this case the fermentation proceeds in a sealed bottle, a very strong bottle. Both the fermentation products, CO_2 and ethanol, accumulate. Under pressure within the sealed bottle the CO_2 remains in solution. When the top is "popped," the pressure is released, and the CO_2 comes out of solution in the form of bubbles.

In either case the fermentation continues until the alcohol concentration reaches 12–13%. At that point the yeast "stews in its own juices"! That is, 12–13% ethanol kills the yeast cells that produce it. This points out a last generalization about fermentations. The stable fermentation end product, whether it is lactate or ethanol, eventually accumulates to a concentration that is toxic to the organism. Muscle fatigue is the early effect of lactate buildup in the working muscle. In the same way, continued accumulation of the fermentation product can lead to concentrations that are fatal if there is no means of getting rid of the toxic product or of getting away from it. For single-celled organisms the result is generally death. Our bodies have evolved in such a way that lactate buildup contributes to muscle fatigue that causes the exerciser to stop the exercise. Then the lactate is removed from the blood and converted to glucose by the process of gluconeogenesis.

The production of bread, wine, and cheese depends on fermentation processes.

Figure 21.8
The final two reactions of alcohol fermentation.

Another application of alcohol fermentation is the use of yeast in bread making. When we mix the water, sugar, and dried yeast, the yeast cells begin to grow and carry out the process of fermentation. This mixture is then added to the flour, milk, shortening, and salt, and the dough is placed in a warm place to rise. The yeast continues to grow and ferment the sugar, producing CO_2 that causes the bread to rise. Of course, when we bake the bread, the yeast cells are killed, and the ethanol evaporates, but we are left with a light and airy loaf of bread.

Today, alcohol produced by fermentation is being considered as an alternative fuel to replace the use of some fossil fuels. Geneticists and bioengineers are trying to develop strains of yeast that can survive higher alcohol concentrations and thus convert more of the sugar of corn and other grains into alcohol.

Bacteria perform a variety of other fermentations. The propionibacteria produce propionic acid and CO_2. The acid gives Swiss cheese its characteristic flavor, and the CO_2 gas produces the characteristic holes in the cheese. Other bacteria, the clostridia, perform a fermentation that is responsible in part for the horrible symptoms of gas gangrene. When these bacteria are inadvertently introduced into deep tissues by a puncture wound, they find a nice anaerobic environment in which to grow. In fact, these organisms are *obligate anaerobes*; that is, they are killed by even a small amount of oxygen. As they grow, they perform a fermentation called the *butyric acid, butanol, acetone fermentation*. This results in the formation of CO_2, the gas associated with gas gangrene. The CO_2 infiltrates the local tissues and helps to maintain an anaerobic environment because oxygen from the local blood supply cannot enter the area of the wound. Now able to grow well, these bacteria produce a variety of toxins and enzymes that cause extensive tissue death and necrosis. In addition, the fermentation produces acetic acid, ethanol, acetone, isopropanol, butanol, and butyric acid (which is responsible, along with the necrosis, for the characteristic foul smell of gas gangrene). Certainly, the presence of these organic chemicals in the wound enhances tissue death.

Gas gangrene is very difficult to treat. Because the bacteria establish an anaerobic region of cell death and cut off the local circulation, systemic antibiotics do not infiltrate the wound and kill the bacteria. Even our immune response is stymied. Treatment usually involves surgical removal of the necrotic tissue accompanied by antibiotic therapy. In some cases a hyperbaric oxygen chamber is employed. The infected extremity is placed in an environment with a very high partial pressure of oxygen. The oxygen forced into the tissues is poisonous to the bacteria, and they die.

These are but a few examples of the fermentations that have an effect on humans. Regardless of the specific chemical reactions, all fermentations share the following traits:

- They use pyruvate produced in glycolysis.
- They reoxidize the NADH produced in glycolysis.
- They are self-limiting because the accumulated stable fermentation end product eventually kills the cell that produces it.

For Further Understanding

Write condensed structural formulas for each of the fermentation products made by clostridia in gas gangrene. Identify the functional groups and provide the I.U.P.A.C. name for each.

Explain the importance of utilizing pyruvate and reoxidizing NADH to the ability of a cell to continue producing ATP.

important, reoxidizes NADH in the process. The regeneration of NAD^+ allows glycolysis to continue, just as in the case of lactate fermentation.

The two products of alcohol fermentation, then, are ethanol and CO_2. We take advantage of this fermentation in the production of wines and other alcoholic beverages and in the process of breadmaking.

Question 21.9 How is the alcohol fermentation in yeast similar to lactate production in skeletal muscle?

Question 21.10 Why must pyruvate be used and NADH be reoxidized so that glycolysis can continue?

Figure 21.9
Summary of the major stages of the pentose phosphate pathway.

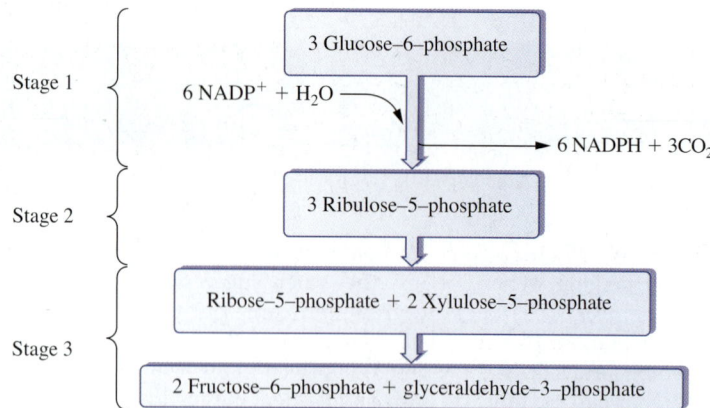

21.5 The Pentose Phosphate Pathway

The **pentose phosphate pathway** is an alternative pathway for glucose oxidation. It provides the cell with energy in the form of reducing power for biosynthesis. Specifically, NADPH is produced in the oxidative stage of this pathway. NADPH is the reducing agent required for many biosynthetic pathways.

The details of the pentose phosphate pathway will not be covered in this text. But an overview of the key reactions will allow us to understand the importance of the pathway (Figure 21.9). We can consider the pathway in three stages. The first is the oxidative stage, which can be summarized as

glucose-6-phosphate + 2NADP$^+$ + H$_2$O ⟶

ribulose-5-phosphate + 2NADPH + CO$_2$

These reactions provide the NADPH required for biosynthesis.

The second stage involves isomerization reactions that convert ribulose-5-phosphate into ribose-5-phosphate or xylulose-5-phosphate. The pathway's name reflects the production of these phosphorylated five-carbon sugars (pentose phosphates).

The third stage is a complex series of reactions involving C—C bond breakage and formation. The result of these reactions is the formation of two molecules of fructose-6-phosphate and one molecule of glyceraldehyde-3-phosphate from three molecules of pentose phosphate.

In addition to providing reducing power (NADPH), the pentose phosphate pathway provides sugar phosphates that are required for biosynthesis. For instance, ribose-5-phosphate is used for the synthesis of nucleotides such as ATP. The four-carbon sugar phosphate, erythrose-4-phosphate, produced in the third stage of the pentose phosphate pathway is a precursor of the amino acids phenylalanine, tyrosine, and tryptophan.

The pentose phosphate pathway is most active in tissues involved in cholesterol and fatty acid biosynthesis. These two processes require abundant NADPH. Thus the liver, which is the site of cholesterol synthesis and a major site for fatty acid biosynthesis, and adipose (fat) tissue, where active fatty acid synthesis also occurs, have very high levels of pentose phosphate pathway enzymes.

The pathway for fatty acid biosynthesis is discussed in Section 23.4.

21.6 Gluconeogenesis: The Synthesis of Glucose

Under normal conditions, we have enough glucose to satisfy our needs. However, under some conditions the body must make glucose. This is necessary following strenuous exercise to replenish the liver and muscle supplies of glycogen. It also

21.6 Gluconeogenesis: The Synthesis of Glucose

occurs during starvation so that the body can maintain adequate blood glucose levels to supply the brain cells and red blood cells. Under normal conditions these two tissues use only glucose for energy.

Glucose is produced by the process of **gluconeogenesis,** the production of glucose from noncarbohydrate starting materials (Figure 21.10). Gluconeogenesis, an anabolic pathway, occurs primarily in the liver. Lactate, all the amino acids except leucine and lysine, and glycerol from fats can all be used to make glucose. However, the amino acids and glycerol are generally used only under starvation conditions.

At first glance, gluconeogenesis appears to be simply the reverse of glycolysis (compare Figures 21.10 and 21.6), because the intermediates of the two pathways are identical. But this is not the case, because steps 1, 3, and 10 of glycolysis are irreversible, and therefore the reverse reactions must be carried out by other enzymes. In step 1 of glycolysis, hexokinase catalyzes the phosphorylation of glucose. In gluconeogenesis the dephosphorylation of glucose-6-phosphate is carried out by the enzyme *glucose-6-phosphatase,* which is found in the liver but not in muscle. Similarly, reaction 3, the phosphorylation of fructose-6-phosphate catalyzed by phosphofructokinase, is irreversible. That step is bypassed in gluconeogenesis by using the enzyme *fructose-1,6-bisphosphatase.* Finally, the phosphorylation of ADP catalyzed by

Under extreme conditions of starvation the brain eventually switches to the use of ketone bodies. Ketone bodies are produced, under certain circumstances, from the breakdown of lipids (Section 23.3).

Figure 21.10

Comparison of the reactions of glycolysis and gluconeogenesis.

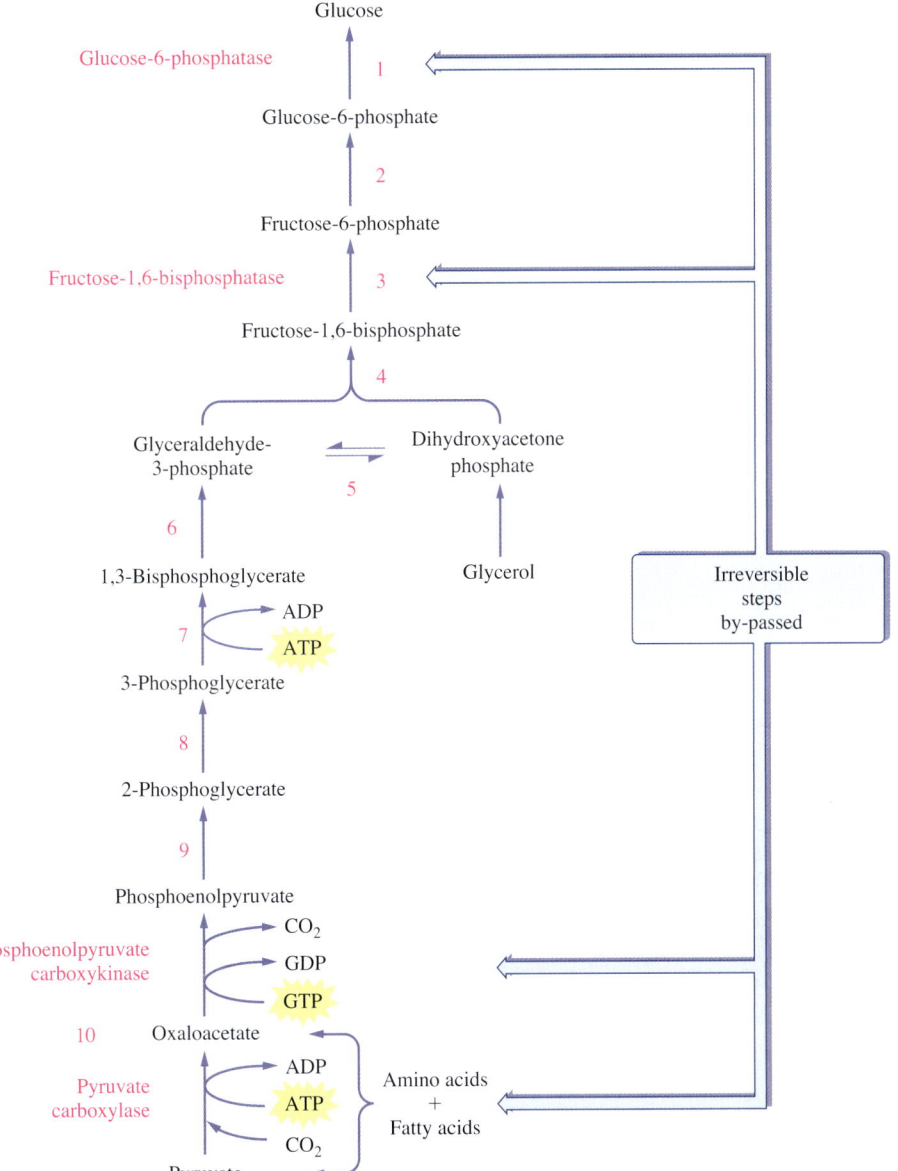

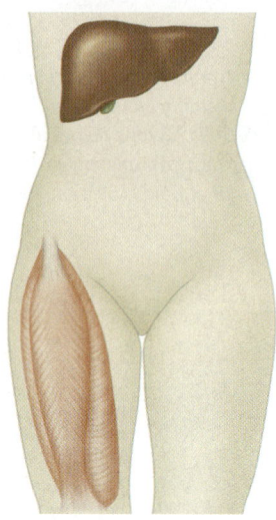
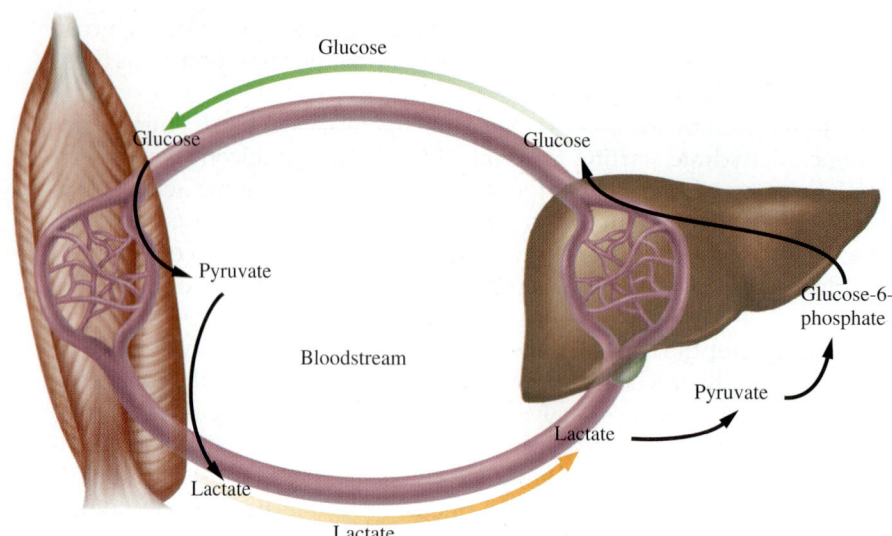

Figure 21.11
The Cori Cycle.

pyruvate kinase, step 10 of glycolysis, cannot be reversed. The conversion of pyruvate to phosphoenolpyruvate actually involves two enzymes and some unusual reactions. First, the enzyme *pyruvate carboxylase* adds CO_2 to pyruvate. The product is the four-carbon compound oxaloacetate. Then *phosphoenolpyruvate carboxykinase* removes the CO_2 and adds a phosphoryl group. The donor of the phosphoryl group in this unusual reaction is **guanosine triphosphate (GTP)**. This is a nucleotide like ATP, except that the nitrogenous base is guanine.

This last pair of reactions is complicated by the fact that pyruvate carboxylase is found in the mitochondria, whereas phosphoenolpyruvate carboxykinase is found in the cytoplasm. As we will see in Chapters 22 and 23, mitochondria are organelles in which the final oxidation of food molecules occurs and large amounts of ATP are produced. A complicated shuttle system transports the oxaloacetate produced in the mitochondria through the two mitochondrial membranes and into the cytoplasm. There, phosphoenolpyruvate carboxykinase catalyzes its conversion to phosphoenolpyruvate.

If glycolysis and gluconeogenesis were not regulated in some fashion, the two pathways would occur simultaneously, with the disastrous effect that nothing would get done. Three convenient sites for this regulation are the three bypass reactions. Step 3 of glycolysis is catalyzed by the enzyme phosphofructokinase. This enzyme is stimulated by high concentrations of AMP, ADP, and inorganic phosphate, signals that the cell needs energy. When the enzyme is active, glycolysis proceeds. On the other hand, when ATP is plentiful, phosphofructokinase is inhibited, and fructose-1,6-bisphosphatase is stimulated. The net result is that in times of energy excess (high concentrations of ATP), gluconeogenesis will occur.

As we have seen, the conversion of lactate into glucose is important in mammals. As the muscles work, they produce lactate, which is converted back to glucose in the liver. The glucose is transported into the blood and from there back to the muscle. In the muscle it can be catabolized to produce ATP, or it can be used to replenish the muscle stores of glycogen. This cyclic process between the liver and skeletal muscles is called the **Cori Cycle** and is shown in Figure 21.11. Through this cycle, gluconeogenesis produces enough glucose to restore the depleted muscle glycogen reservoir within forty-eight hours.

Question 21.11

What are the major differences between gluconeogenesis and glycolysis?

What do the three irreversible reactions of glycolysis have in common?

Question 21.12

21.7 Glycogen Synthesis and Degradation

Glucose is the sole source of energy of mammalian red blood cells and the major source of energy for the brain. Neither red blood cells nor the brain can store glucose; thus a constant supply must be available as blood glucose. This is provided by dietary glucose and by the production of glucose either by gluconeogenesis or by **glycogenolysis,** the degradation of glycogen. Glycogen is a long-branched-chain polymer of glucose. Stored in the liver and skeletal muscles, it is the principal storage form of glucose.

The total amount of glucose in the blood of a 70-kg (approximately 150-lb) adult is about 20 g, but the brain alone consumes 5–6 g of glucose per hour. Breakdown of glycogen in the liver mobilizes the glucose when hormonal signals register a need for increased levels of blood glucose. Skeletal muscle also contains substantial stores of glycogen, which provide energy for rapid muscle contraction. However, this glycogen is not able to contribute to blood glucose because muscle cells do not have the enzyme glucose-6-phosphatase. Because glucose cannot be formed from the glucose-6-phosphate, it cannot be released into the bloodstream.

9 LEARNING GOAL

The Structure of Glycogen

Glycogen is a highly branched glucose polymer in which the "main chain" is linked by α (1 → 4) glycosidic bonds. The polymer also has numerous α (1 → 6) glycosidic bonds, which provide many branch points along the chain. This structure is shown schematically in Figure 21.12. **Glycogen granules** with a diameter of 10–40 nm are found in the cytoplasm of liver and muscle cells. These granules exist in complexes with the enzymes that are responsible for glycogen synthesis and degradation. The structure of such a granule is also shown in Figure 21.12.

Glycogenolysis: Glycogen Degradation

Two hormones control glycogenolysis, the degradation of glycogen. These are **glucagon,** a peptide hormone synthesized in the pancreas, and *epinephrine,* produced in the adrenal glands. Glucagon is released from the pancreas in response to low blood glucose, and epinephrine is released from the adrenal glands in response to a threat or a stress. Both situations require an increase in blood glucose, and both hormones function by altering the activity of two enzymes, glycogen phosphorylase and glycogen synthase. *Glycogen phosphorylase* is involved in glycogen degradation and is activated; *glycogen synthase* is involved in glycogen synthesis and is inactivated. The steps in glycogen degradation are summarized as follows.

Step 1. The enzyme glycogen phosphorylase catalyzes *phosphorolysis* of a glucose at one end of a glycogen polymer (Figure 21.13). The reaction involves the displacement of a glucose unit of glycogen by a phosphate group. As a result of phosphorolysis, glucose-1-phosphate is produced without using ATP as the phosphoryl group donor.

Step 2. Glycogen contains many branches bound to the α (1 → 4) backbone by α (1 → 6) glycosidic bonds. These branches must be removed to allow the complete degradation of glycogen. The extensive action of glycogen phosphorylase produces a smaller polysaccharide with a single glucose bound by an α (1 → 6) glycosidic bond to the main chain. The enzyme α (1 → 6) *glycosidase,* also called the *debranching enzyme,* hydrolyzes the α (1 → 6) glycosidic bond at a branch point and frees one molecule of glucose (Figure 21.14). This molecule of glucose can be phosphorylated

Figure 21.12
The structure of glycogen and a glycogen granule.

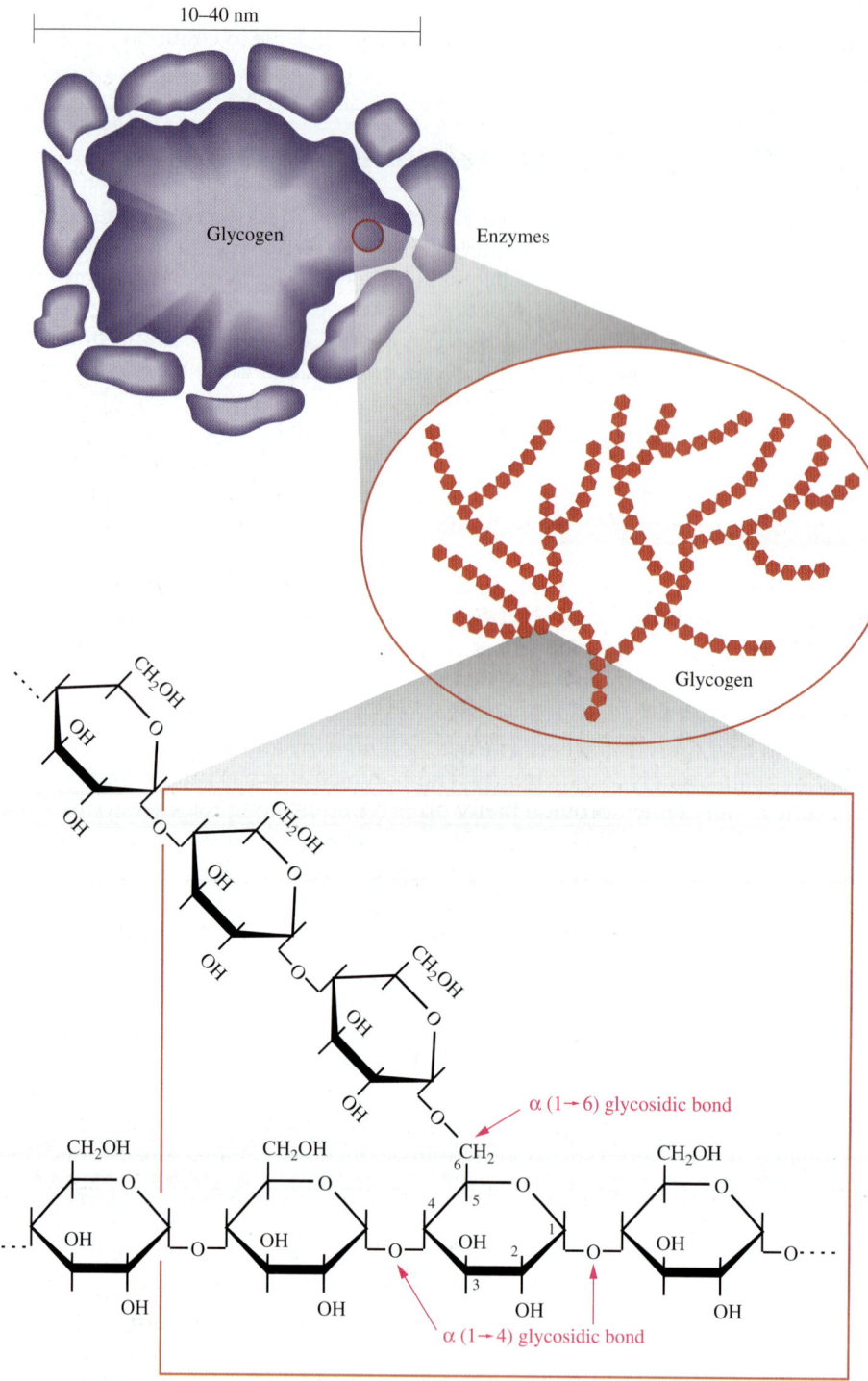

and utilized in glycolysis, or it may be released into the bloodstream for use elsewhere. Hydrolysis of the branch bond liberates another stretch of α (1 → 4)-linked glucose for the action of glycogen phosphorylase.

Step 3. Glucose-1-phosphate is converted to glucose-6-phosphate by *phosphoglucomutase* (Figure 21.15). Glucose originally stored in glycogen enters glycolysis through the action of phosphoglucomutase. Alternatively, in the liver and kidneys it may be dephosphorylated for transport into the bloodstream.

Question 21.13

Explain the role of glycogen phosphorylase in glycogenolysis.

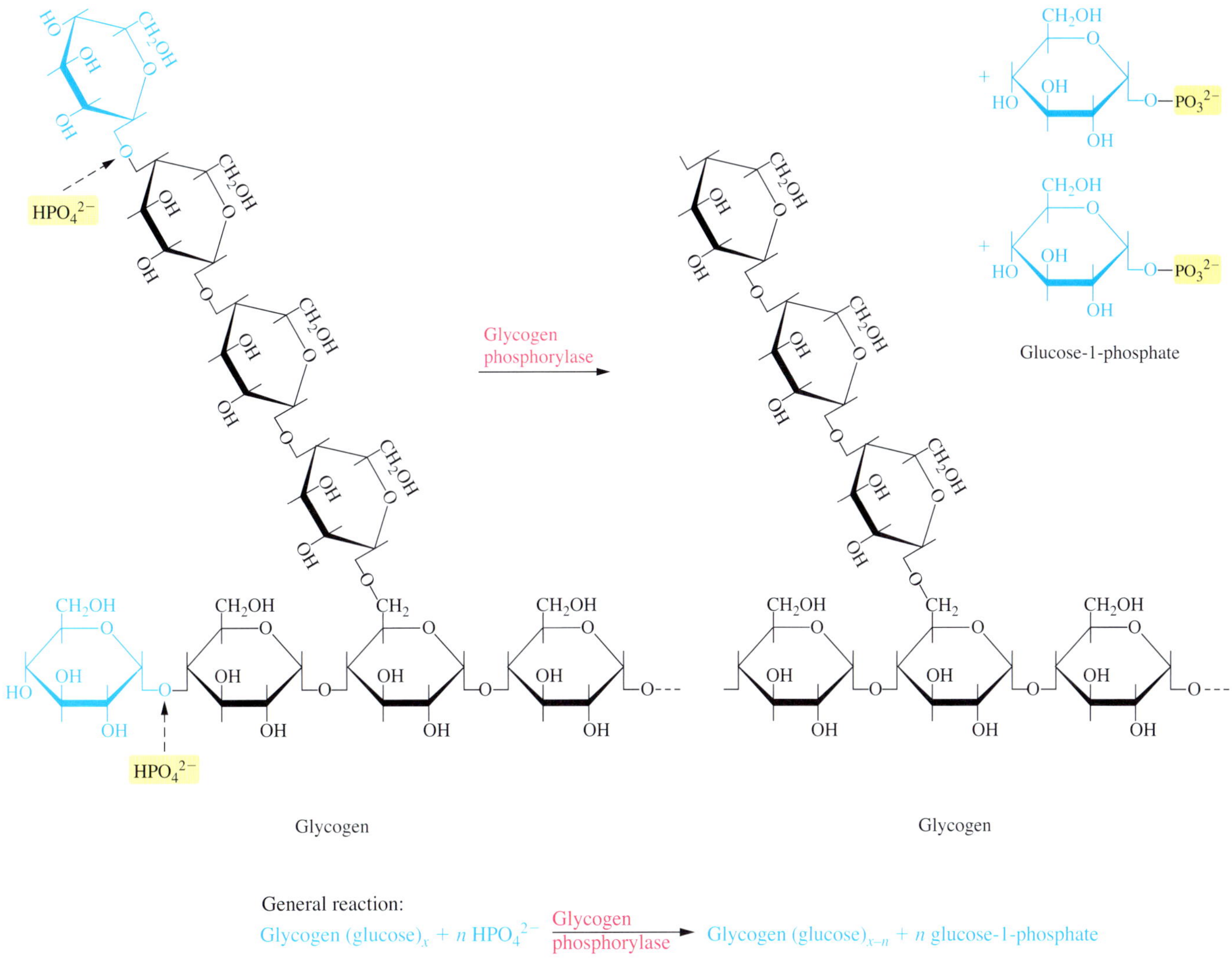

21.7 Glycogen Synthesis and Degradation

Figure 21.13
The action of glycogen phosphorylase in glycogenolysis.

How does the action of glycogen phosphorylase and phosphoglucomutase result in an energy savings for the cell if the product, glucose-6-phosphate, is used directly in glycolysis?

Question 21.14

Glycogenesis: Glycogen Synthesis

The hormone **insulin**, produced by the pancreas in response to high blood glucose levels, stimulates the synthesis of glycogen, **glycogenesis.** Insulin is perhaps one of the most influential hormones in the body because it directly alters the metabolism and uptake of glucose in all but a few cells.

When blood glucose rises, as after a meal, the beta cells of the pancreas secrete insulin. It immediately accelerates the uptake of glucose by all the cells of the body except the brain and certain blood cells. In these cells the uptake of glucose is insulin-independent. The increased uptake of glucose is especially marked in the liver, heart, skeletal muscle, and adipose tissue.

In the liver, insulin promotes glycogen synthesis and storage by inhibiting glycogen phosphorylase, thus inhibiting glycogen degradation. It also stimulates glycogen synthase and glucokinase, two enzymes that are involved in glycogen synthesis.

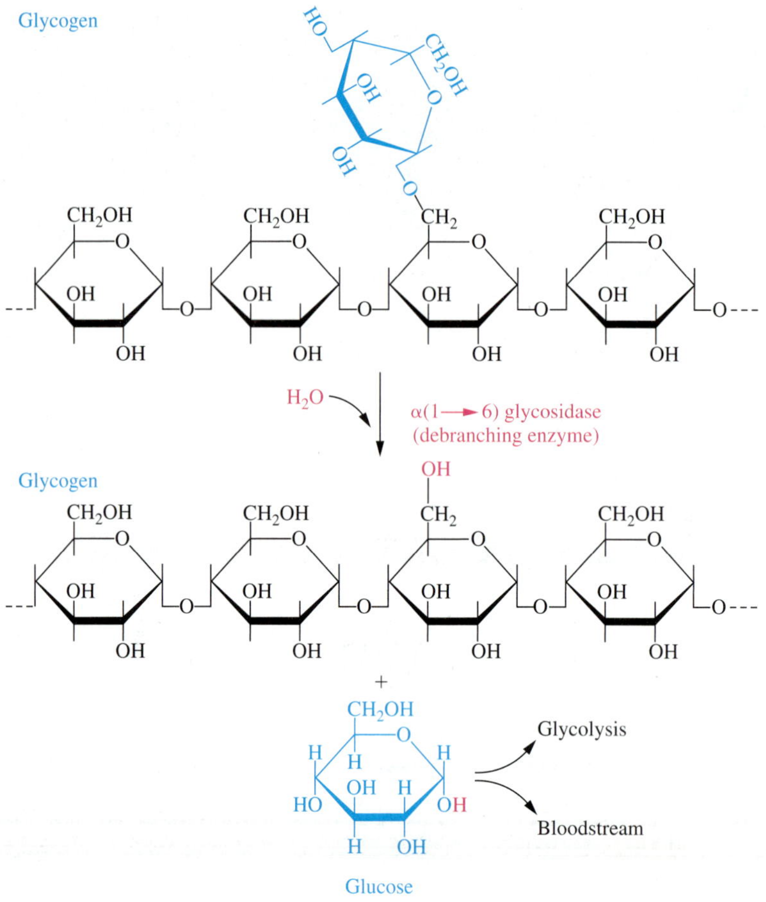

Figure 21.14
The action of α (1 → 6) glycosidase (debranching enzyme) in glycogen degradation.

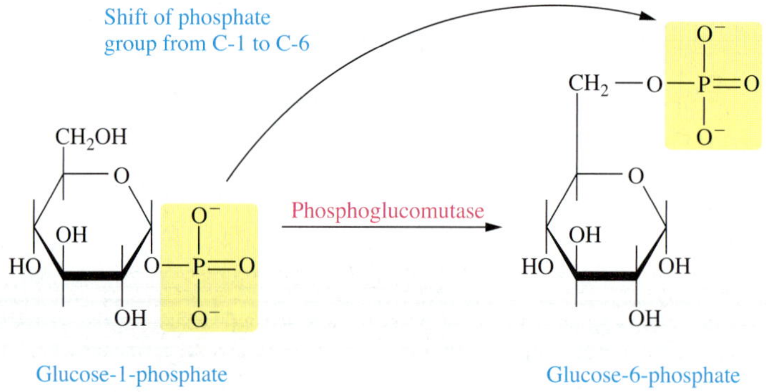

Figure 21.15
The action of phosphoglucomutase in glycogen degradation.

Although glycogenesis and glycogenolysis share some reactions in common, the two pathways are not simply the reverse of one another. Glycogenesis involves some very unusual reactions, which we will now examine in detail.

The first reaction of glycogen synthesis in the liver traps glucose within the cell by phosphorylating it. In this reaction, catalyzed by the enzyme *glucokinase*, ATP serves as a phosphoryl donor, and glucose-6-phosphate is formed:

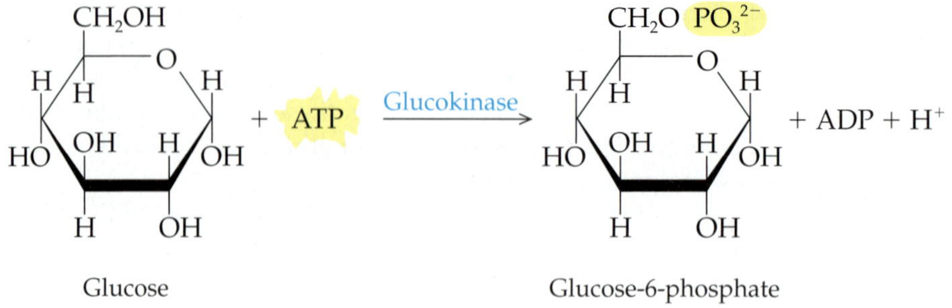

21.7 Glycogen Synthesis and Degradation

The second reaction of glycogenesis is the reverse of one of the reactions of glycogenolysis. The glucose-6-phosphate formed in the first step is isomerized to glucose-1-phosphate. The enzyme that catalyzes this step is phosphoglucomutase:

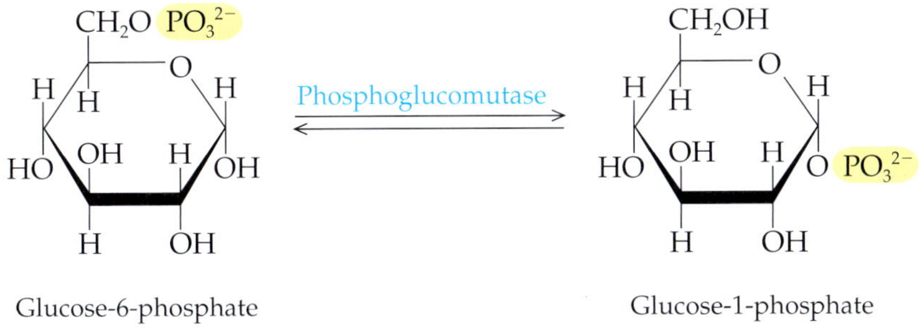

Glucose-6-phosphate ⇌ (Phosphoglucomutase) Glucose-1-phosphate

The glucose-1-phosphate must now be activated before it can be added to the growing glycogen chain. The high-energy compound that accomplishes this is the nucleotide **uridine triphosphate (UTP)**. In this reaction, mediated by the enzyme *pyrophosphorylase*, the C-1 phosphoryl group of glucose is linked to the α-phosphoryl group of UTP to produce UDP-glucose:

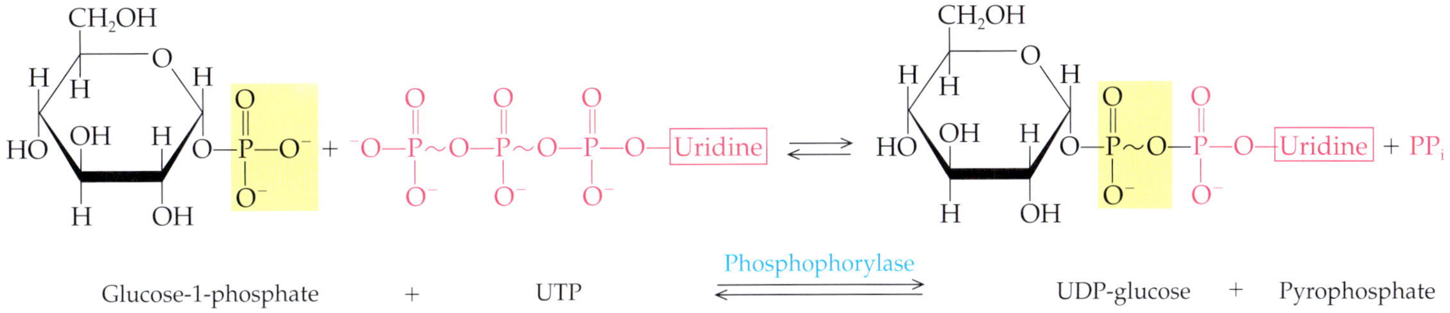

Glucose-1-phosphate + UTP ⇌ (Phosphophorylase) UDP-glucose + Pyrophosphate

This is accompanied by the release of a pyrophosphate group (PP_i). The structure of UDP-glucose is seen in Figure 21.16.

The UDP-glucose can now be used to extend glycogen chains. The enzyme glycogen synthase breaks the phosphoester linkage of UDP-glucose and forms an α (1 → 4) glycosidic bond between the glucose and the growing glycogen chain. UDP is released in the process.

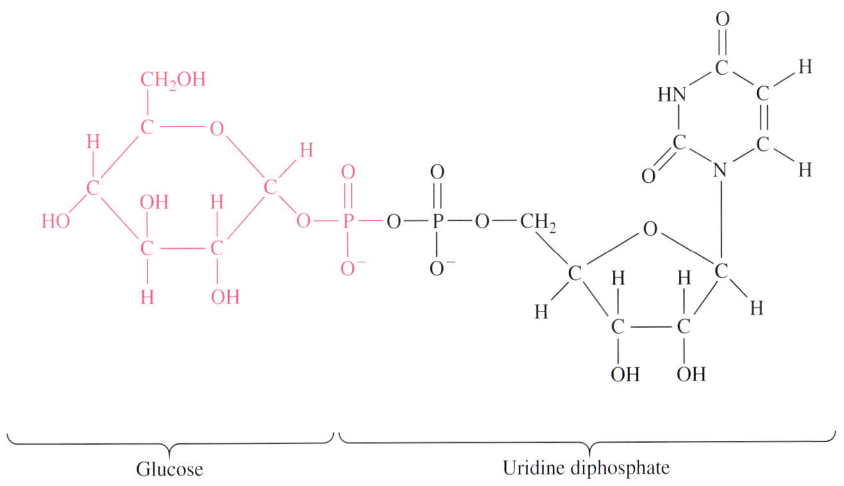

Figure 21.16
The structure of UDP-glucose.

A Medical Perspective

Diagnosing Diabetes

When diagnosing diabetes, doctors take many factors and symptoms into consideration. However, there are two primary tests that are performed to determine whether an individual is properly regulating blood glucose levels. First and foremost is the fasting blood glucose test. A person who has fasted since midnight should have a blood glucose level between 70 and 110 mg/dL in the morning. If the level is 140 mg/dL on at least two occasions, a diagnosis of diabetes is generally made.

The second commonly used test is the glucose tolerance test. For this test the subject must fast for at least ten hours. A beginning blood sample is drawn to determine the fasting blood glucose level. This will serve as the background level for the test. The subject ingests 50–100 g of glucose (40 g/m² body surface), and the blood glucose level is measured at thirty minutes, and at one, two, and three hours after ingesting the glucose.

A graph is made of the blood glucose levels over time. For a person who does not have diabetes, the curve will show a peak of blood glucose at approximately one hour. There will be a reduction in the level, and perhaps a slight hypoglycemia (low blood glucose level) over the next hour. Thereafter, the blood glucose level stabilizes at normal levels.

An individual is said to have impaired glucose tolerance if the blood glucose level remains between 140 and 200 mg/dL two hours after ingestion of the glucose solution. This suggests that there is a risk of the individual developing diabetes and is reason to prescribe periodic testing to allow early intervention.

If the blood glucose level remains at or above 200 mg/dL after two hours, a tentative diagnosis of diabetes is made. However, this result warrants further testing on subsequent days to rule out transient problems, such as the effect of medications on blood glucose levels.

It was recently suggested that the upper blood glucose level of 200 mg/dL should be lowered to 180 mg/dL as the standard to diagnose impaired glucose tolerance and diabetes. This would allow earlier detection and intervention. Considering the grave nature of long-term diabetic complications, it is thought to be very beneficial to begin treatment at an early stage to maintain constant blood glucose levels. For more information on diabetes, see A Medical Perspective: Diabetes Mellitus and Ketone Bodies, in Chapter 23.

For Further Understanding

Draw a graph representing blood glucose levels for a normal glucose tolerance test.

Draw a similar graph for an individual who would be diagnosed as diabetic.

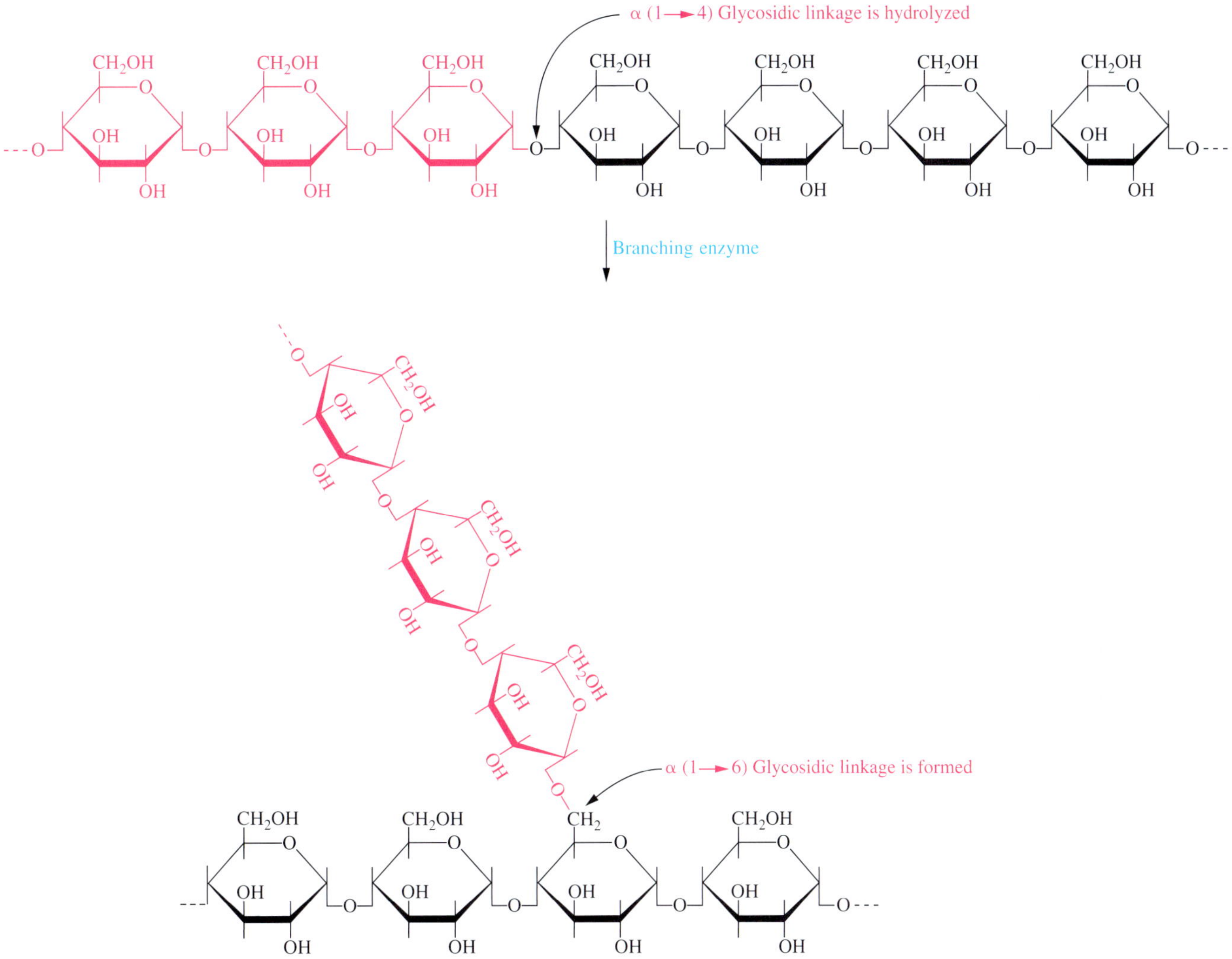

21.7 Glycogen Synthesis and Degradation

Figure 21.17
The action of the branching enzyme in glycogen synthesis.

Finally, we must introduce the α (1 → 6) glycosidic linkages to form the branches. The branches are quite important to proper glycogen utilization. As Figure 21.17 shows, the *branching enzyme* removes a section of the linear α (1 → 4) linked glycogen and reattaches it in α (1 → 6) glycosidic linkage elsewhere in the chain.

Question 21.15
Describe the way in which glucokinase traps glucose inside liver cells.

Question 21.16
Describe the reaction catalyzed by the branching enzyme.

Compatibility of Glycogenesis and Glycogenolysis

As was the case with glycolysis and gluconeogenesis, it would be futile for the cell to carry out glycogen synthesis and degradation simultaneously. The results achieved by the action of one pathway would be undone by the other. This problem is avoided by a series of hormonal controls that activate the enzymes of one pathway while inactivating the enzymes of the other pathway.

A Human Perspective

Glycogen Storage Diseases

Glycogen metabolism is important for the proper function of many aspects of cellular metabolism. Many diseases of glycogen metabolism have been discovered. Generally, these are diseases that result in the excessive accumulation of glycogen in the liver, muscle, and tubules of the kidneys. Often they are caused by defects in one of the enzymes involved in the degradation of glycogen.

One example is an inherited defect of glycogen metabolism known as *von Gierke's disease*. This disease results from a defective gene for glucose-6-phosphatase, which catalyzes the final step of gluconeogenesis and glycogenolysis. People who lack glucose-6-phosphatase cannot convert glucose-6-phosphate to glucose. As we have seen, the liver is the primary source of blood glucose, and much of this glucose is produced by gluconeogenesis. Glucose-6-phosphate, unlike glucose, cannot cross the cell membrane, and the liver of a person suffering from von Gierke's disease cannot provide him or her with glucose. The blood sugar level falls precipitously low between meals. In addition, the lack of glucose-6-phosphatase also affects glycogen metabolism. Because glucose-6-phosphatase is absent, the supply of glucose-6-phosphate in the liver is large. This glucose-6-phosphate can also be converted to glycogen. A person suffering from von Gierke's disease has a massively enlarged liver as a result of enormously increased stores of glycogen.

Defects in other enzymes of glycogen metabolism also exist. *Cori's disease* is caused by a genetic defect in the debranching enzyme. As a result, individuals who have this disease cannot completely degrade glycogen and thus use their glycogen stores very inefficiently.

On the other side of the coin, *Andersen's disease* results from a genetic defect in the branching enzyme. Individuals who have this disease produce very long, unbranched glycogen chains. This genetic disorder results in decreased efficiency of glycogen storage.

A final example of a glycogen storage disease is *McArdle's disease*. In this syndrome, the muscle cells lack the enzyme glycogen phosphorylase and cannot degrade glycogen to glucose. Individuals who have this disease have little tolerance for physical exercise because their muscles cannot provide enough glucose for the necessary energy-harvesting processes. It is interesting to note that the liver enzyme glycogen phosphorylase is perfectly normal, and these people respond appropriately with a rise of blood glucose levels under the influence of glucagon or epinephrine.

For Further Understanding

Write equations showing the reactions catalyzed by the enzymes that are defective in each of the genetic disorders described in this perspective.

There are different isoenzymes of glycogen phosphorylase, one found in the liver and the other in skeletal muscle. Discuss the differences you would expect between a defect in the muscle isoenzyme and a defect in the liver isoenzyme. (Isoenzymes were described in Chapter 19 in A Medical Perspective: Enzymes, Isoenzymes, and Myocardial Infarction.)

When the blood glucose level is too high, a condition known as **hyperglycemia**, insulin stimulates the uptake of glucose via a transport mechanism. It further stimulates the trapping of the glucose by the elevated activity of glucokinase. Finally, it activates glycogen synthase, the last enzyme in the synthesis of glycogen chains. To further accelerate storage, insulin *inhibits* the first enzyme in glycogen degradation, glycogen phosphorylase. The net effect, seen in Figure 21.18, is that glucose is removed from the bloodstream and converted into glycogen in the liver. When the glycogen stores are filled, excess glucose is converted to fat and stored in adipose tissue.

Glucagon is produced in response to low blood glucose levels, a condition known as **hypoglycemia**, and has an effect opposite to that of insulin. It stimulates glycogen phosphorylase, which catalyzes the first stage of glycogen degradation. This accelerates glycogenolysis and release of glucose into the bloodstream. The effect is further enhanced because glucagon inhibits glycogen synthase. The opposing effects of insulin and glucagon are summarized in Figure 21.18.

This elegant system of hormonal control ensures that the reactions involved in glycogen degradation and synthesis do not compete with one another. In this way they provide glucose when the blood level is too low, and they cause the storage of glucose in times of excess.

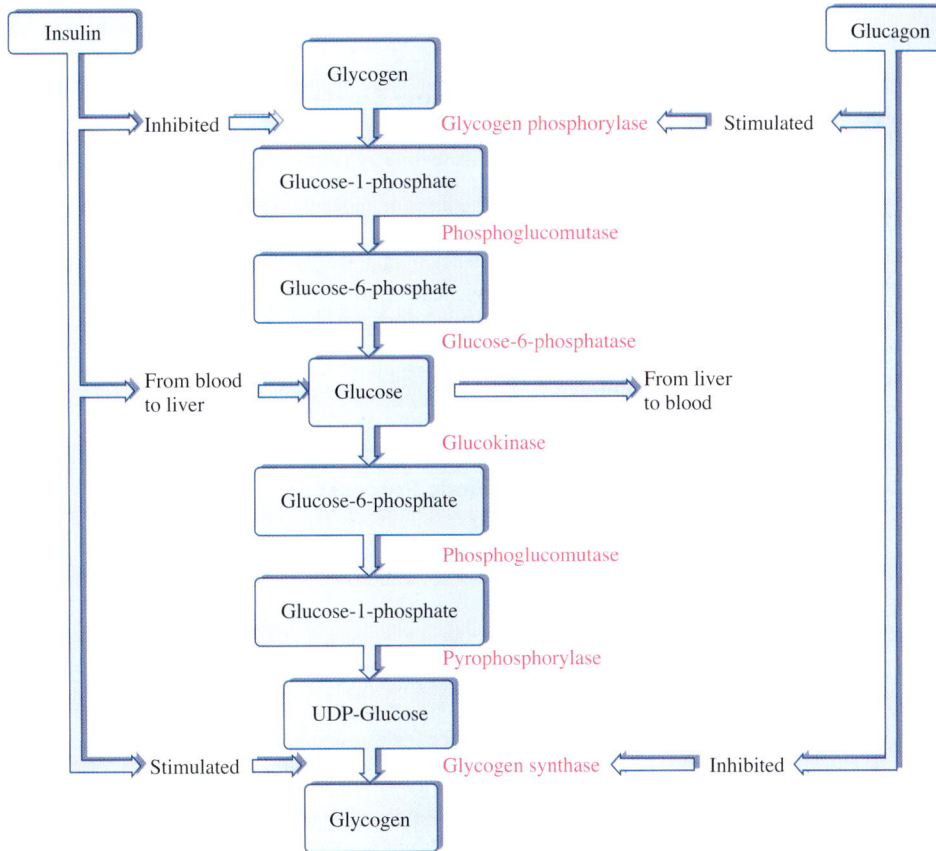

Figure 21.18
The opposing effects of the hormones insulin and glucagon on glycogen metabolism.

Question 21.17
Explain how glucagon affects the synthesis and degradation of glycogen.

Question 21.18
How does insulin affect the storage and degradation of glycogen?

SUMMARY

21.1 ATP: The Cellular Energy Currency

Adenosine triphosphate, ATP, is a *nucleotide* composed of adenine, the sugar ribose, and a triphosphate group. The energy released by the hydrolysis of the phosphoanhydride bond between the second and third phosphoryl groups provides the energy for most cellular work.

21.2 Overview of Catabolic Processes

The body needs a supply of ATP to carry out life processes. To provide this ATP, we consume a variety of energy-rich food molecules: carbohydrates, lipids, and proteins. In the digestive tract these large molecules are degraded into smaller molecules (monosaccharides, glycerol, fatty acids, and amino acids) that are absorbed by our cells. These molecules are further broken down to generate ATP.

21.3 Glycolysis

Glycolysis is the pathway for the *catabolism* of glucose that leads to pyruvate. It is an anaerobic process carried out by enzymes in the cytoplasm of the cell. The net harvest of ATP during glycolysis is two molecules of ATP per molecule of glucose. Two molecules of NADH are also produced. The rate of glycolysis responds to the energy demands of the cell. The regulation of glycolysis occurs through the allosteric enzymes hexokinase, phosphofructokinase, and pyruvate kinase.

21.4 Fermentations

Under anaerobic conditions the NADH produced by glycolysis is used to reduce pyruvate to lactate in skeletal muscle (lactate *fermentation*) or to convert acetaldehyde to ethanol in yeast (alcohol fermentation).

21.5 The Pentose Phosphate Pathway

The *pentose phosphate pathway* is an alternative pathway for glucose degradation that is particularly abundant in the liver and adipose tissue. It provides the cell with a source of NADPH to serve as a reducing agent for biosynthetic reactions. It also provides ribose-5-phosphate for nucleotide synthesis and erythrose-4-phosphate for biosynthesis of the amino acids tryptophan, tyrosine, and phenylalanine.

21.6 Gluconeogenesis: The Synthesis of Glucose

Gluconeogenesis is the pathway for glucose synthesis from noncarbohydrate starting materials. It occurs in mammalian liver. Glucose can be made from lactate, all the amino acids except lysine and leucine, and glycerol. Gluconeogenesis is not simply the reversal of glycolysis. Three steps in glycolysis in which ATP is produced or consumed are bypassed by different enzymes in gluconeogenesis. All other enzymes in gluconeogenesis are shared with glycolysis.

21.7 Glycogen Synthesis and Degradation

Glycogenesis is the pathway for the synthesis of glycogen, and *glycogenolysis* is the pathway for the degradation of glycogen. The concentration of blood glucose is controlled by the liver. A high blood glucose level causes secretion of *insulin*. This hormone stimulates glycogenesis and inhibits glycogenolysis. When blood glucose levels are too low, the hormone *glucagon* stimulates gluconeogenesis and glycogen degradation in the liver.

KEY TERMS

adenosine triphosphate (ATP) (21.1)
anabolism (21.1)
anaerobic threshold (21.4)
catabolism (21.1)
Cori Cycle (21.6)
fermentation (21.4)
glucagon (21.7)
gluconeogenesis (21.6)
glycogen (21.7)
glycogenesis (21.7)
glycogen granule (21.7)
glycogenolysis (21.7)
glycolysis (21.3)
guanosine triphosphate (GTP) (21.6)
hyperglycemia (21.7)
hypoglycemia (21.7)
insulin (21.7)
nicotinamide adenine dinucleotide (NAD^+) (21.3)
nucleotide (21.1)
oxidative phosphorylation (21.3)
pentose phosphate pathway (21.5)
substrate-level phosphorylation (21.3)
uridine triphosphate (UTP) (21.7)

QUESTIONS AND PROBLEMS

ATP: The Cellular Energy Currency

Foundations
21.19 What molecule is primarily responsible for conserving the energy released in catabolism?
21.20 Describe the structure of ATP.

Applications
21.21 Write a reaction showing the hydrolysis of the terminal phosphoanhydride bond of ATP.
21.22 What is meant by the term *high-energy bond*?
21.23 What is meant by a coupled reaction?
21.24 Compare and contrast anabolism and catabolism in terms of their roles in metabolism and their relationship to ATP.

Overview of Catabolic Processes

Foundations
21.25 What is the most readily used energy source in the diet?
21.26 What is a hydrolysis reaction?

Applications
21.27 Write an equation showing the hydrolysis of maltose.
21.28 Write an equation showing the hydrolysis of sucrose.
21.29 Write an equation showing the hydrolysis of lactose.
21.30 How are monosaccharides transported into a cell?
21.31 Write an equation showing the hydrolysis of a triglyceride consisting of glycerol, oleic acid, linoleic acid, and stearic acid.
21.32 How are fatty acids taken up into the cell?
21.33 Write an equation showing the hydrolysis of the dipeptide alanyl-leucine.
21.34 How are amino acids transported into the cell?

Glycolysis

Foundations
21.35 Define glycolysis and describe its role in cellular metabolism.
21.36 What are the end products of glycolysis?
21.37 Why does glycolysis require a supply of NAD^+ to function?
21.38 Why must the NADH produced in glycolysis be reoxidized to NAD^+?
21.39 What is the net energy yield of ATP in glycolysis?
21.40 How many molecules of ATP are produced by substrate-level phosphorylation during glycolysis?
21.41 Explain how muscle is able to carry out rapid contraction for prolonged periods even though its supply of ATP is sufficient only for a fraction of a second of rapid contraction.
21.42 Where in the muscle cell does glycolysis occur?
21.43 Write the balanced chemical equation for glycolysis.
21.44 Write a chemical equation for the transfer of a phosphoryl group from ATP to fructose-6-phosphate.
21.45 Which glycolysis reactions are catalyzed by each of the following enzymes?
   a. Hexokinase
   b. Pyruvate kinase
   c. Phosphoglycerate mutase
   d. Glyceraldehyde-3-phosphate dehydrogenase
21.46 Fill in the blanks:
   a. _____ molecules of ATP are produced per molecule of glucose that is converted to pyruvate.
   b. Two molecules of ATP are consumed in the conversion of _____ to fructose-1,6-bisphosphate.
   c. NAD^+ is _____ to NADH in the first energy-releasing step of glycolysis.
   d. The second substrate-level phosphorylation in glycolysis is phosphoryl group transfer from phosphoenolpyruvate to _____ .

Applications

21.47 Examine the following pair of reactions and use them to answer Questions 21.47–21.50. What type of enzyme would catalyze each of these reactions?

(a)
```
      O
      ‖
      C—H
      |
   H—C—OH
      |
  HO—C—H
      |
   H—C—OH      ⇌
      |
   H—C—OH
      |
   H—C—H
      |
      O
      |
   ⁻O—P—O⁻
      ‖
      O
```

(b)
```
      CH₂OH
      |
      C=O
      |
  HO—C—H
      |
   H—C—OH
      |
   H—C—OH
      |
   H—C—H
      |
      O
      |
   ⁻O—P—O⁻
      ‖
      O
```

(c)
```
      CH₂OH
      |
      C=O
      |
   H—C—H       ⇌
      |
      O
      |
   ⁻O—P—O⁻
      ‖
      O
```

(d)
```
      O
      ‖
      C—H
      |
   H—C—OH
      |
   H—C—H
      |
      O
      |
   ⁻O—P—O⁻
      ‖
      O
```

21.48 To which family of organic molecules do a and d belong? To which family of organic molecules do b and c belong?

21.49 What is the name of the type of intermediate formed in each of these reactions?

21.50 Draw the intermediate that would be formed in each of these reactions.

21.51 When an enzyme has the term *kinase* in the name, what type of reaction do you expect it to catalyze?

21.52 What features do the reactions catalyzed by hexokinase and phosphofructokinase share in common?

21.53 What is the role of NAD⁺ in a biochemical oxidation reaction?

21.54 Write the equation for the reaction catalyzed by glyceraldehyde-3-phosphate dehydrogenase. Highlight the chemical changes that show this to be an oxidation reaction.

21.55 The enzyme that catalyzes step 9 of glycolysis is called enolase. What is the significance of that name?

21.56 Draw the enol tautomer of pyruvate.

21.57 What is the importance of the regulation of glycolysis?

21.58 Explain the role of allosteric enzymes in control of glycolysis.

21.59 What molecules serve as allosteric effectors of phosphofructokinase?

21.60 What molecule serves as an allosteric inhibitor of hexokinase?

21.61 Explain the role of citrate in the feedback inhibition of glycolysis.

21.62 Explain the feedforward activation mechanism that results in the activation of pyruvate kinase.

Fermentations

Foundations

21.63 Write a balanced chemical equation for the conversion of acetaldehyde to ethanol.

21.64 Write a balanced chemical equation for the conversion of pyruvate to lactate.

Applications

21.65 After running a 100-m dash, a sprinter had a high concentration of muscle lactate. What process is responsible for production of lactate?

21.66 If the muscle of an organism had no lactate dehydrogenase, could anaerobic glycolysis occur in those muscle cells? Explain your answer.

21.67 What food products are the result of lactate fermentation?

21.68 Explain the value of alcohol fermentation in bread making.

21.69 What enzyme catalyzes the reduction of pyruvate to lactate?

21.70 What enzymes catalyze the conversion of pyruvate to ethanol and carbon dioxide?

21.71 A child was brought to the doctor's office suffering from a strange set of symptoms. When the child exercised hard, she became giddy and behaved as though drunk. What do you think is the metabolic basis of these symptoms?

21.72 A family started a batch of wine by adding yeast to grape juice and placing the mixture in a sealed bottle. Two weeks later, the bottle exploded. What metabolic reactions—and specifically, what product of those reactions—caused the bottle to explode?

The Pentose Phosphate Pathway

21.73 Describe the three stages of the pentose phosphate pathway.

21.74 Write an equation to summarize the pentose phosphate pathway.

21.75 Of what value are the ribose-5-phosphate and erythrose-4-phosphate that are produced in the pentose phosphate pathway?

21.76 Of what value is the NADPH that is produced in the pentose phosphate pathway?

Gluconeogenesis

21.77 Define gluconeogenesis and describe its role in metabolism.

21.78 What is the role of guanosine triphosphate in gluconeogenesis?

21.79 What organ is primarily responsible for gluconeogenesis?

21.80 What is the physiological function of gluconeogenesis?

21.81 Lactate can be converted to glucose by gluconeogenesis. To what metabolic intermediate must lactate be converted so that it can be a substrate for the enzymes of gluconeogenesis?

21.82 L-Alanine can be converted to pyruvate. Can L-alanine also be converted to glucose? Explain your answer.

21.83 Explain why gluconeogenesis is not simply the reversal of glycolysis.

21.84 In step 10 of glycolysis, phosphoenolpyruvate is converted to pyruvate, and ATP is produced by substrate-level phosphorylation. How is this reaction bypassed in gluconeogenesis?

21.85 Which steps in the glycolysis pathway are irreversible?

21.86 What enzymatic reactions of gluconeogenesis bypass the irreversible steps of glycolysis?

Glycogen Synthesis and Degradation

Foundations

21.87 What organs are primarily responsible for maintaining the proper blood glucose level?

21.88 Why must the blood glucose level be carefully regulated?

21.89 What does the term *hypoglycemia* mean?

21.90 What does the term *hyperglycemia* mean?

Applications

21.91 a. What enzymes involved in glycogen metabolism are stimulated by insulin?
b. What effect does this have on glycogen metabolism?
c. What effect does this have on blood glucose levels?

21.92
  a. What enzyme is stimulated by glucagon?
  b. What effect does this have on glycogen metabolism?
  c. What effect does this have on blood glucose levels?
21.93 Explain how a defect in glycogen metabolism can cause hypoglycemia.
21.94 What defects of glycogen metabolism would lead to a large increase in the concentration of liver glycogen?

CRITICAL THINKING PROBLEMS

1. An enzyme that hydrolyzes ATP (an ATPase) bound to the plasma membrane of certain tumor cells has an abnormally high activity. How will this activity affect the rate of glycolysis?
2. Explain why no net oxidation occurs during anaerobic glycolysis followed by lactate fermentation.
3. A certain person was found to have a defect in glycogen metabolism. The liver of this person could (a) make glucose-6-phosphate from lactate and (b) synthesize glucose-6-phosphate from glycogen but (c) could not synthesize glycogen from glucose-6-phosphate. What enzyme is defective?
4. A scientist added phosphate labeled with radioactive phosphorus (^{32}P) to a bacterial culture growing anaerobically (without O_2). She then purified all the compounds produced during glycolysis. Look carefully at the steps of the pathway. Predict which of the intermediates of the pathway would be the first one to contain radioactive phosphate. On which carbon of this compound would you expect to find the radioactive phosphate?
5. A two-month-old baby was brought to the hospital suffering from seizures. He deteriorated progressively over time, showing psychomotor retardation. Blood tests revealed a high concentration of lactate and pyruvate. Although blood levels of alanine were high, they did not stimulate gluconeogenesis. The doctor measured the activity of pyruvate carboxylase in the baby and found it to be only 1% of the normal level. What reaction is catalyzed by pyruvate carboxylase? How could this deficiency cause the baby's symptoms and test results?

BIOCHEMISTRY

22

Aerobic Respiration and Energy Production

Downhill skiing demands a great deal of energy.

Learning Goals

1. Name the regions of the mitochondria and the function of each region.

2. Describe the reaction that results in the conversion of pyruvate to acetyl CoA, describing the location of the reaction and the components of the pyruvate dehydrogenase complex.

3. Summarize the reactions of aerobic respiration.

4. Looking at an equation representing any of the chemical reactions that occur in the citric acid cycle, describe the kind of reaction that is occurring and the significance of that reaction to the pathway.

5. Explain the mechanisms for the control of the citric acid cycle.

6. Describe the process of oxidative phosphorylation.

7. Describe the conversion of amino acids to molecules that can enter the citric acid cycle.

8. Explain the importance of the urea cycle and describe its essential steps.

9. Discuss the cause and effect of hyperammonemia.

10. Summarize the role of the citric acid cycle in catabolism and anabolism.

Outline

Chemistry Connection:
Mitochondria from Mom

22.1 The Mitochondria

A Human Perspective:
Exercise and Energy Metabolism

22.2 Conversion of Pyruvate to Acetyl CoA

22.3 An Overview of Aerobic Respiration

22.4 The Citric Acid Cycle (The Krebs Cycle)

22.5 Control of the Citric Acid Cycle

22.6 Oxidative Phosphorylation

A Human Perspective:
Brown Fat: The Fat That Makes You Thin?

22.7 The Degradation of Amino Acids

22.8 The Urea Cycle

A Medical Perspective:
Pyruvate Carboxylase Deficiency

22.9 Overview of Anabolism: The Citric Acid Cycle as a Source of Biosynthetic Intermediates

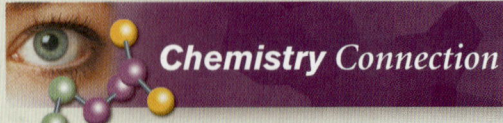

Chemistry Connection

Mitochondria from Mom

In this chapter we will be studying the amazing, intricate set of reactions that allow us to completely degrade fuel molecules such as sugars and amino acids. These oxygen-requiring reactions occur in cellular organelles called *mitochondria*.

We are used to thinking of the organelles as a collection of membrane-bound structures that are synthesized under the direction of the genetic information in the nucleus of the cell. Not so with the mitochondria. These organelles have their own genetic information and are able to make some of their own proteins. They grow and multiply in a way very similar to the simple bacteria. This, along with other information on the structure and activities of mitochondria, has led researchers to conclude that the mitochondria are actually the descendants of bacteria captured by eukaryotic cells millions of years ago.

Recent studies of the mitochondrial genetic information (DNA) have revealed fascinating new information. For instance, although each of us inherited half our genetic information from our mothers and half from our fathers, each of us inherited all of our mitochondria from our mothers. The reason for this is that when the sperm fertilizes the egg, only the sperm nucleus enters the cell.

The observation that all of our mitochondria are inherited from our mothers led Dr. A. Wilson to study the mitochondrial DNA of thousands of women around the world. He thought that by looking for similarities and differences in the mitochondrial DNA he would be able to identify a "Mitochondrial Eve"—the mother of all humanity. He didn't really think that he could identify a single woman who would have lived tens of thousands of years ago. But he hoped to determine the location of the first population of human women to help answer questions about the origin of humankind. Although the idea was a good one, the study had a number of experimental flaws. Currently, a hot debate is going on among hundreds of scientists about the Mitochondrial Eve. This controversy should encourage better experiments and analysis to help us identify our origins and to better understand the workings of the mitochondria.

Like the mitochondria themselves, some genetic diseases of energy metabolism are maternally inherited. One such disease, Leber's hereditary optic neuropathy (LHON), causes blindness and heart problems. People with LHON have a reduced ability to make ATP. As a result, sensitive tissues that demand a great deal of energy eventually die. LHON sufferers eventually lose their sight because the optic nerve dies from lack of energy.

Researchers have identified and cloned a mutant mitochondrial gene that is responsible for LHON. The defect is a mutant form of *NADH dehydrogenase*. NADH dehydrogenase is a huge, complex enzyme that accepts electrons from NADH and sends them on through an electron transport system. Passage of electrons through the electron transport system allows the synthesis of ATP. If NADH dehydrogenase is defective, passage of electrons through the electron transport system is less efficient, and less ATP is made. In LHON sufferers, the result is eventual blindness.

In this chapter and the next, we will study some of the important biochemical reactions that occur in the mitochondria. A better understanding of the function of healthy mitochondria will eventually allow us to help those suffering from LHON and other mitochondrial genetic diseases.

Introduction

An organelle is a compartment within the cytoplasm that has a specialized function.

As we have seen, the anaerobic glycolysis pathway begins the breakdown of glucose and produces a small amount of ATP and NADH. But it is aerobic catabolic pathways that complete the oxidation of glucose to CO_2 and H_2O and provide most of the ATP needed by the body. In fact, this process, called aerobic respiration, produces thirty-six ATP molecules using the energy harvested from each glucose molecule that enters glycolysis. These reactions occur in metabolic pathways located in mitochondria, the cellular "power plants." Mitochondria are a type of membrane-enclosed cell organelle.

Here, in the mitochondria, the final oxidations of carbohydrates, lipids, and proteins occur. Here, also, the electrons that are harvested in these oxidation reactions are used to make ATP. In these remarkably efficient reactions, nearly 40% of the potential energy of glucose is stored as ATP.

22.1 The Mitochondria

Mitochondria are football-shaped organelles that are roughly the size of a bacterial cell. They are surrounded by an **outer mitochondrial membrane** and an **inner mitochondrial membrane** (Figure 22.1). The space between the two membranes is the **intermembrane space**, and the space inside of the inner membrane is the **matrix space**. The enzymes of the citric acid cycle, of the β-oxidation pathway for the breakdown of fatty acids, and for the degradation of amino acids are all found in the mitochondrial matrix space.

Structure and Function

The outer mitochondrial membrane has many small pores through which small molecules (less than 10,000 g/mol) can pass. Thus, the small molecules to be oxidized for the production of ATP can easily enter the mitochondrial intermembrane space.

The inner membrane is highly folded to create a large surface area. The folded membranes are known as **cristae**. The inner mitochondrial membrane is almost completely impermeable to most substances. For this reason it has many transport proteins to bring particular fuel molecules into the matrix space. Also embedded within the inner mitochondrial membrane are the protein electron carriers of the *electron transport system* and *ATP synthase*. ATP synthase is a large complex of many proteins that catalyzes the synthesis of ATP.

Origin of the Mitochondria

Not only are mitochondria roughly the size of bacteria, they have several other features that have led researchers to suspect that they may once have been free-living bacteria that were "captured" by eukaryotic cells. They have their own genetic information (DNA). They also make their own ribosomes that are very similar to those of bacteria. These ribosomes allow the mitochondria to synthesize some of their own proteins. Finally, mitochondria are actually self-replicating; they grow in size and divide to produce new mitochondria. All of these characteristics suggest that the mitochondria that produce the majority of the ATP for our cells evolved from bacteria "captured" perhaps as long as 1.5×10^9 years ago.

As we saw in Chapter 20, ribosomes are complexes of protein and RNA that serve as small platforms for protein synthesis.

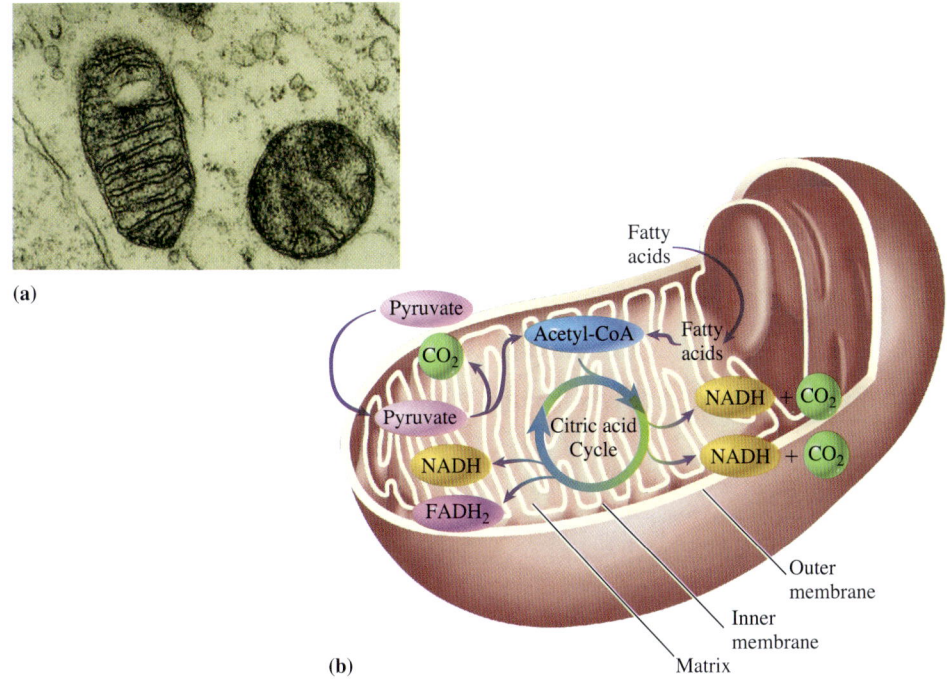

Figure 22.1
Structure of the mitochondrion.
(a) Electron micrograph of mitochondria.
(b) Schematic drawing of the mitochondrion.

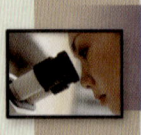

A Human Perspective

Exercise and Energy Metabolism

The Olympic sprinters get set in the blocks. The gun goes off, and roughly ten seconds later the 100-m dash is over. Elsewhere, the marathoners line up. They will run 26 miles and 385 yards in a little over two hours. Both sports involve running, but they utilize very different sources of energy.

Let's look at the sprinter first. The immediate source of energy for the sprinter is stored ATP. But the quantity of stored ATP is very small, only about three ounces. This allows the sprinter to run as fast as he or she can for about three seconds. Obviously, another source of stored energy must be tapped, and that energy store is *creatine phosphate*:

The structure of creatine phosphate.

Creatine phosphate, stored in the muscle, donates its high-energy phosphate to ADP to produce new supplies of ATP.

This will keep our runner in motion for another five or six seconds before the store of creatine phosphate is also depleted. This is almost enough energy to finish the 100-m dash, but in reality, all the runners are slowing down, owing to energy depletion, and the winner is the sprinter who is slowing down the least!

Consider a longer race, the 400-m or the 800-m. These runners run at maximum capacity for much longer. When they have depleted their ATP and creatine phosphate stores, they must synthesize more ATP. Of course, the cells have been making ATP all the time, but now the demand for energy is much greater. To supply this increased demand, the anaerobic energy-generating reactions (glycolysis and lactate fermentation, Chapter 21) and aerobic processes (citric acid cycle and oxidative phosphorylation) begin to function much more rapidly. Often, however, these athletes are running so strenuously that they cannot provide enough oxygen to the exercising muscle to allow oxidative phosphorylation to function efficiently. When this happens, the muscles must rely on glycolysis and lactate fermentation to provide *most* of the energy requirement. The chemical by-product of these anaerobic processes, lactate, builds up in the muscle and diffuses into the bloodstream. However, the concentration of lactate inevitably builds up in the working muscle and causes muscle fatigue and, eventually, muscle failure. Thus, exercise that depends primarily on anaerobic ATP production cannot continue for very long.

The marathoner presents us with a different scenario. This runner will deplete his or her stores of ATP and creatine phosphate as quickly as a short-distance runner. The anaerobic glycolytic pathway will begin to degrade glucose provided by the blood at a more rapid rate, as will the citric acid cycle and oxidative phosphorylation. The major difference in ATP production between the long-distance runner and the short- or middle-distance runner is that the muscles of the long-distance runner derive almost all the energy through aerobic pathways. These individuals continue to run long distances at a pace that allows them to supply virtually all the oxygen needed by the exercising muscle. In fact, only aerobic pathways can provide

Phosphoryl group transfer from creatine phosphate to ADP is catalyzed by the enzyme creatine kinase.

Question 22.1 What is the function of the mitochondria?

Question 22.2 How do the mitochondria differ from the other components of eukaryotic cells?

22.1 The Mitochondria

Sprinters at the starting block.

a constant supply of ATP for exercise that goes on for hours. Theoretically, under such conditions our runner could run indefinitely, utilizing first his or her stored glycogen and eventually stored lipids. Of course, in reality, other factors such as dehydration and fatigue place limits on the athlete's ability to continue.

From this we can conclude that long-distance runners must have a great capacity to produce ATP aerobically, in the mitochondria, whereas short- and middle-distance runners need a great capacity to produce energy anaerobically, in the cytoplasm of the muscle cells. It is interesting to note that the muscles of these runners reflect these diverse needs.

When one examines muscle tissue that has been surgically removed, one finds two predominant types of muscle fibers. *Fast twitch muscle fibers* are large, relatively plump, pale cells. They have only a few mitochondria but contain a large reserve of glycogen and high concentrations of the enzymes that are needed for glycolysis and lactate fermentation. These muscle fibers fatigue rather quickly because fermentation is inefficient, quickly depleting the cell's glycogen store and causing the accumulation of lactate.

Slow twitch muscle fiber cells are about half the diameter of fast twitch muscle cells and are red. The red color is a result of the high concentrations of myoglobin in these cells. Recall that myoglobin stores oxygen for the cell (Section 18.9) and facilitates rapid diffusion of oxygen throughout the cell. In addition, slow twitch muscle fiber cells are packed with mitochondria. With this abundance of oxygen and mitochondria these cells have the capacity for extended ATP production via aerobic pathways—ideal for endurance sports like marathon racing.

It is not surprising, then, that researchers have found that the muscles of sprinters have many more fast twitch muscle fibers and those of endurance athletes have many more slow twitch muscle fibers. One question that many researchers are trying to answer is whether the type of muscle fibers an individual has is a function of genetic makeup or training. Is a marathon runner born to be a long-distance runner, or are his or her abilities due to the type of training the runner undergoes? There is no doubt that the training regimen for an endurance runner does indeed increase the number of slow twitch muscle fibers and that of a sprinter increases the number of fast twitch muscle fibers. But there is intriguing new evidence to suggest that the muscles of endurance athletes have a greater proportion of slow twitch muscle fibers before they ever begin training. It appears that some of us truly were born to run.

For Further Understanding

It has been said that the winner of the 100-m race is the one who is slowing down the least. Explain this observation in terms of energy-harvesting pathways.

Design an experiment to safely test whether the type of muscle fibers a runner has are the result of training or genetic makeup.

Question 22.3 Draw a schematic diagram of a mitochondrion, and label the parts of this organelle.

Question 22.4 Describe the evidence that suggests that mitochondria evolved from free-living bacteria.

22.2 Conversion of Pyruvate to Acetyl CoA

LEARNING GOAL

As we saw in Chapter 21, under *anaerobic* conditions, glucose is broken down into two pyruvate molecules that are then converted to a stable fermentation product. This limited degradation of glucose releases very little of the potential energy of glucose. Under *aerobic* conditions the cells can use oxygen and completely oxidize glucose to CO_2 in a metabolic pathway called the *citric acid cycle*.

This pathway is often referred to as the *Krebs cycle* in honor of Sir Hans Krebs who worked out the steps of this cyclic pathway from his own experimental data and that of other researchers. It is also called the *tricarboxylic acid (TCA) cycle* because several of the early intermediates in the pathway have three carboxyl groups.

Once pyruvate enters the mitochondria, it must be converted to a two-carbon acetyl group. This acetyl group must be "activated" to enter the reactions of the citric acid cycle. Activation occurs when the acetyl group is bonded to the thiol group of coenzyme A. **Coenzyme A** is a large thiol derived from ATP and the vitamin pantothenic acid (Figure 22.2). It is an acceptor of acetyl groups (in red in Figure 22.2), which are bonded to it through a high-energy thioester bond. The acetyl coenzyme A **(acetyl CoA)** formed is the "activated" form of the acetyl group.

Figure 22.3 shows us the reaction that converts pyruvate to acetyl CoA. First, pyruvate is decarboxylated, which means that it loses a carboxyl group that is released as CO_2. Next it is oxidized, and the hydride anion that is removed is accepted by NAD^+. Finally, the remaining acetyl group, CH_3CO-, is linked to coenzyme A by a thioester bond. This very complex reaction is carried out by three enzymes and five coenzymes that are organized together in a single bundle called the **pyruvate dehydrogenase complex** (see Figure 22.3). This organization allows the substrate to be passed from one enzyme to the next as each chemical reaction occurs. A schematic representation of this "disassembly line" is shown in Figure 22.3b.

This single reaction requires four coenzymes made from four different vitamins, in addition to the coenzyme lipoamide. These are thiamine pyrophosphate, derived from thiamine (Vitamin B_1); FAD, derived from riboflavin (Vitamin B_2); NAD^+, derived from niacin; and coenzyme A, derived from pantothenic acid. Obviously, a deficiency in any of these vitamins would seriously reduce the amount of acetyl CoA that our cells could produce. This, in turn, would limit the amount of ATP that the body could make and would contribute to vitamin-deficiency diseases. Fortunately, a well-balanced diet provides an adequate supply of these and other vitamins.

Coenzyme A is described in Sections 12.9 and 14.4.

The structure and function of pantothenic acid are found online at www.mhhe.com/denniston5e in Water-Soluble Vitamins.

Thioester bonds are discussed in Section 14.4.

These vitamins are discussed online at www.mhhe.com/denniston5e in Water-Soluble Vitamins.

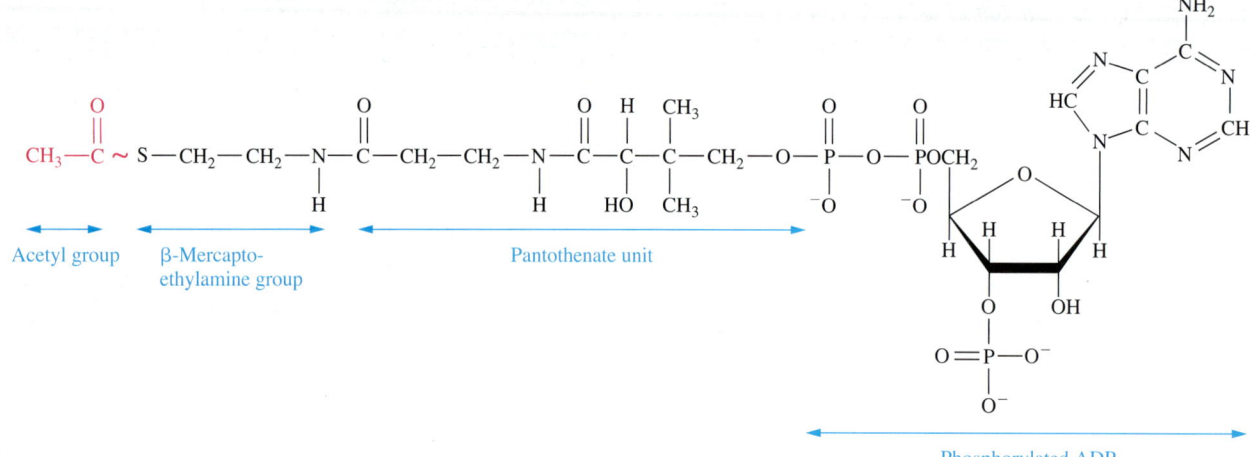

Figure 22.2
The structure of acetyl CoA. The bond between the acetyl group and coenzyme A is a high-energy thioester bond.

22.2 Conversion of Pyruvate to Acetyl CoA

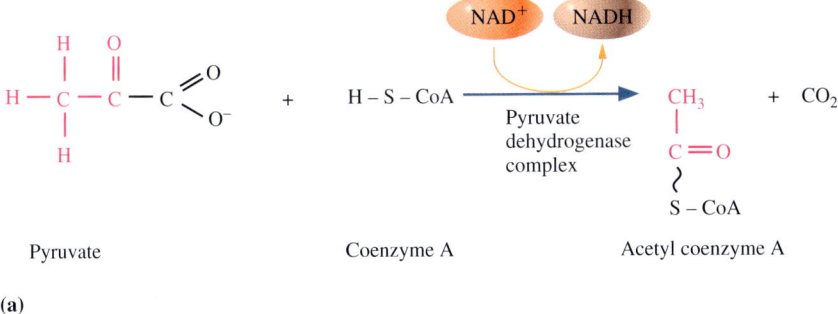

(a)

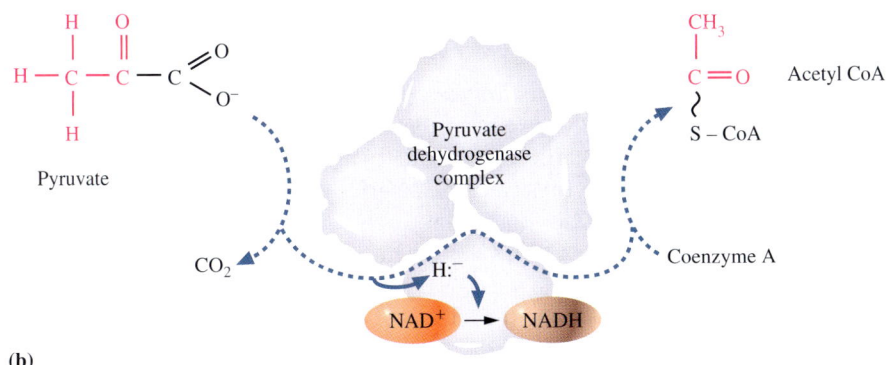

(b)

Figure 22.3
The decarboxylation and oxidation of pyruvate to produce acetyl CoA. (a) The overall reaction in which CO_2 and an $H:^-$ are removed from pyruvate and the remaining acetyl group is attached to coenzyme A. This requires the concerted action of three enzymes and five coenzymes. (b) The pyruvate dehydrogenase complex that carries out this reaction is actually a cluster of enzymes and coenzymes. The substrate is passed from one enzyme to the next as the reaction occurs.

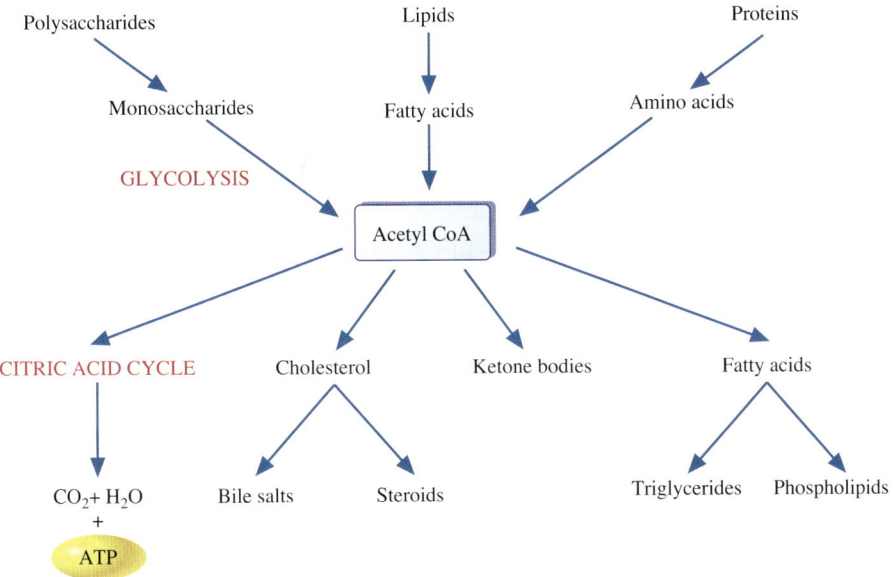

Figure 22.4
The central role of acetyl CoA in cellular metabolism.

In Figure 22.4, we see that acetyl CoA is a central character in cellular metabolism. It is produced by the degradation of glucose, fatty acids, and some amino acids. The major function of acetyl CoA in energy-harvesting pathways is to carry the acetyl group to the citric acid cycle, in which it will be used to produce large amounts of ATP. In addition to these catabolic duties, the acetyl group of acetyl CoA can also be used for *anabolic* or biosynthetic reactions to produce cholesterol and fatty acids. It is through this intermediate, acetyl CoA, that all the energy sources (fats, proteins, and carbohydrates) are interconvertible.

Question 22.5 What vitamins are required for acetyl CoA production from pyruvate?

Question 22.6 What is the major role of coenzyme A in catabolic reactions?

22.3 An Overview of Aerobic Respiration

LEARNING GOAL 3

Aerobic respiration is the oxygen-requiring breakdown of food molecules and production of ATP. The different steps of aerobic respiration occur in different compartments of the mitochondria.

The enzymes for the citric acid cycle are found in the mitochondrial matrix space. The first enzyme catalyzes a reaction that joins the acetyl group of acetyl CoA (two carbons) to a four-carbon molecule (oxaloacetate) to produce citrate (six carbons). The remaining enzymes catalyze a series of rearrangements, decarboxylations (removal of CO_2), and oxidation–reduction reactions. The eventual products of this cyclic pathway are two CO_2 molecules and oxaloacetate—the molecule we began with.

At several steps in the citric acid cycle, a substrate is oxidized. In three of these steps, a pair of electrons is transferred from the substrate to NAD^+, producing NADH (three NADH molecules per turn of the cycle). At another step a pair of electrons is transferred from a substrate to FAD, producing $FADH_2$ (one $FADH_2$ molecule per turn of the cycle).

Remember (Section 19.7) that it is really the hydride anion with its pair of electrons ($H:^-$) that is transferred to NAD^+ to produce NADH. Similarly, a pair of hydrogen atoms (and thus two electrons) are transferred to FAD to produce $FADH_2$.

The electrons are passed from NADH or $FADH_2$, through an electron transport system located in the inner mitochondrial membrane, and finally to the terminal electron acceptor, molecular oxygen (O_2). The transfer of electrons through the electron transport system causes protons (H^+) to be pumped from the mitochondrial matrix into the intermembrane compartment. The result is a high-energy H^+ reservoir.

In the final step, the energy of the H^+ reservoir is used to make ATP. This last step is carried out by the enzyme complex ATP synthase. As protons flow back into the mitochondrial matrix through a pore in the ATP synthase complex, the enzyme catalyzes the synthesis of ATP.

This long, involved process is called *oxidative phosphorylation*, because the energy of electrons from the *oxidation* of substrates in the citric acid cycle is used to *phosphorylate* ADP and produce ATP. The details of each of these steps will be examined in upcoming sections.

Question 22.7 What is meant by the term *oxidative phosphorylation*?

Question 22.8 What does the term *aerobic respiration* mean?

22.4 The Citric Acid Cycle (The Krebs Cycle)

Reactions of the Citric Acid Cycle

The **citric acid cycle** is the final stage of the breakdown of carbohydrates, fats, and amino acids released from dietary proteins (Figure 22.5). To understand this important cycle, let's follow the fate of the acetyl group of an acetyl CoA as it passes through the citric acid cycle. The numbered steps listed below correspond to the steps in the citric acid cycle that are summarized in Figure 22.5.

LEARNING GOAL

The formation of acetyl CoA was described in Section 22.2.

Reaction 1. This is a condensation reaction between the acetyl group of acetyl CoA and oxaloacetate. Actually, this is another biological example of an aldol condensation reaction. It is catalyzed by the enzyme *citrate synthase*. The product that is formed is citrate:

Aldol condensation reactions are reactions between aldehydes and ketones to form larger molecules (Section 13.4).

$$\begin{array}{c}COO^-\\|\\C=O\\|\\CH_2\\|\\COO^-\end{array} + H_3C-\overset{O}{\overset{\|}{C}}\sim S-CoA + H_2O \xrightarrow{\text{Citrate synthase}} \begin{array}{c}COO^-\\|\\CH_2\\|\\HO-C-COO^-\\|\\H-C-H\\|\\COO^-\end{array} + HS-CoA + H^+$$

Oxaloacetate   Acetyl CoA   Citrate   Coenzyme A

Reaction 2. The enzyme *aconitase* catalyzes the dehydration of citrate, producing *cis*-aconitate. The same enzyme, aconitase, then catalyzes addition of a water molecule to the *cis*-aconitate, converting it to isocitrate. The net effect of these two steps is the isomerization of citrate to isocitrate:

Notice that the conversion of citrate to cis-aconitate is a biological example of the dehydration of an alcohol to produce an alkene (Section 12.5). The conversion of cis-aconitate to isocitrate is a biochemical example of the hydration of an alkene to produce an alcohol (Sections 11.5 and 12.5).

$$\begin{array}{c}COO^-\\|\\CH_2\\|\\HO-C-COO^-\\|\\H-C-H\\|\\COO^-\end{array} \xrightarrow{\text{Aconitase}} \begin{array}{c}COO^-\\|\\CH_2\\|\\C-COO^-\\\|\\C-H\\|\\COO^-\end{array} + H_2O \xrightarrow{\text{Aconitase}} \begin{array}{c}COO^-\\|\\CH_2\\|\\H-C-COO^-\\|\\HO-C-H\\|\\COO^-\end{array}$$

Citrate   *cis*-Aconitate   Isocitrate

Reaction 3. The first oxidative step of the citric acid cycle is catalyzed by *isocitrate dehydrogenase*. It is a complex reaction in which three things happen:
  a. the hydroxyl group of isocitrate is oxidized to a ketone,
  b. carbon dioxide is released, and
  c. NAD^+ is reduced to NADH.

The product of this oxidative decarboxylation reaction is α-ketoglutarate:

The oxidation of a secondary alcohol produces a ketone (Sections 12.5 and 13.4).

The structure of NAD^+ and its reduction to NADH are shown in Figure 19.8.

$$\begin{array}{c}COO^-\\|\\CH_2\\|\\H-C-COO^-\\|\\HO-C-H\\|\\COO^-\end{array} + NAD^+ \xrightarrow{\text{Isocitrate dehydrogenase}} \begin{array}{c}COO^-\\|\\CH_2\\|\\CH_2\\|\\C=O\\|\\COO^-\end{array} + CO_2 + \text{NADH}$$

Isocitrate   α-Ketoglutarate

Remember, in organic (and thus biochemical) reactions, oxidation can be recognized as a gain of oxygen or loss of hydrogen (Section 12.6).

Figure 22.5
The reactions of the citric acid cycle.

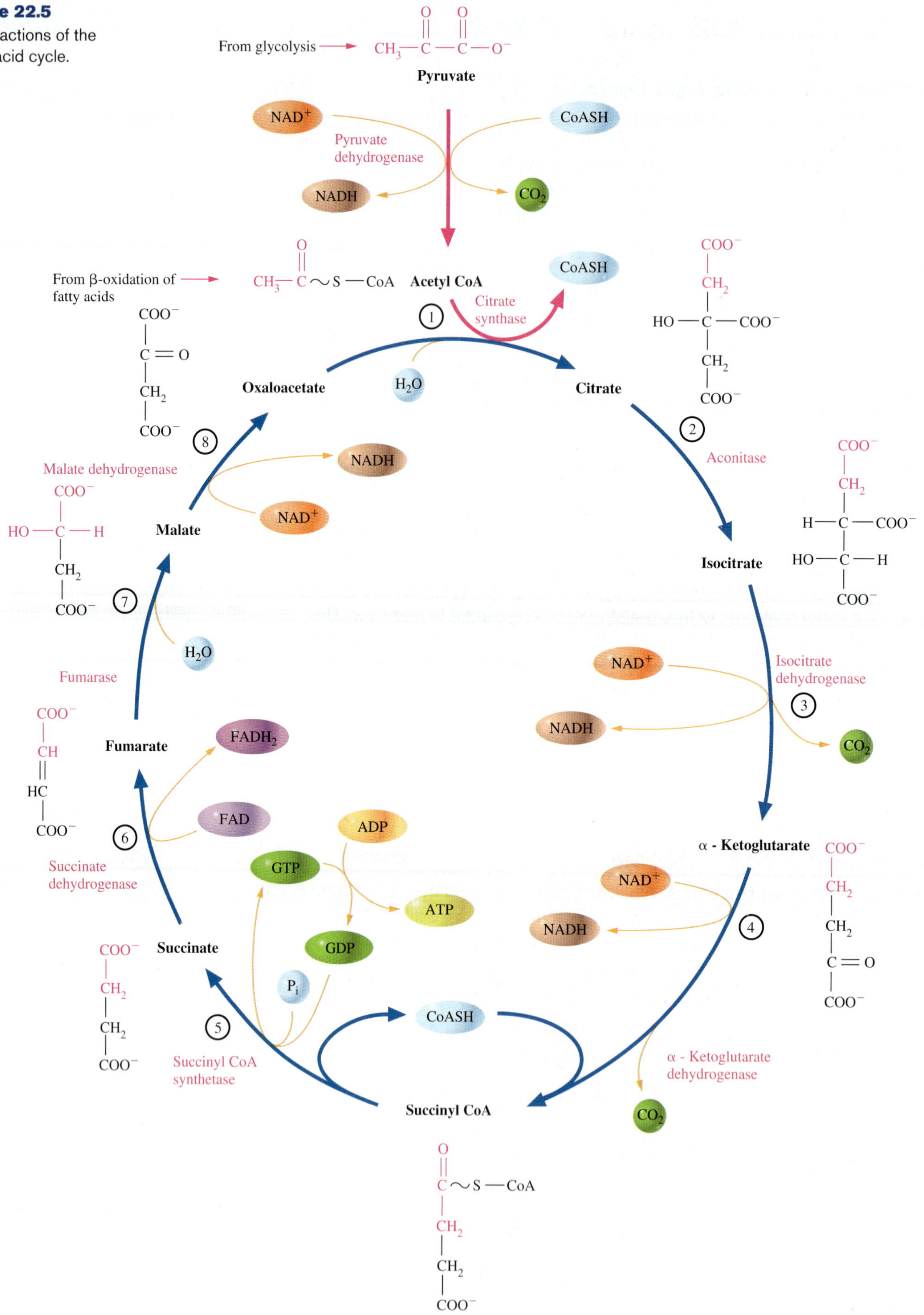

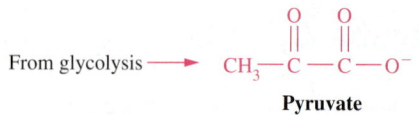

22.4 The Citric Acid Cycle (The Krebs Cycle)

Reaction 4. Coenzyme A enters the picture again as the α-*ketoglutarate dehydrogenase* complex carries out a complex series of reactions similar to those catalyzed by the pyruvate dehydrogenase complex. The same coenzymes are required and, once again, three chemical events occur:

a. α-ketoglutarate loses a carboxylate group as CO_2,

b. it is oxidized and NAD^+ is reduced to NADH, and

c. coenzyme A combines with the product, succinate, to form succinyl CoA. The bond formed between succinate and coenzyme A is a high-energy thioester bond.

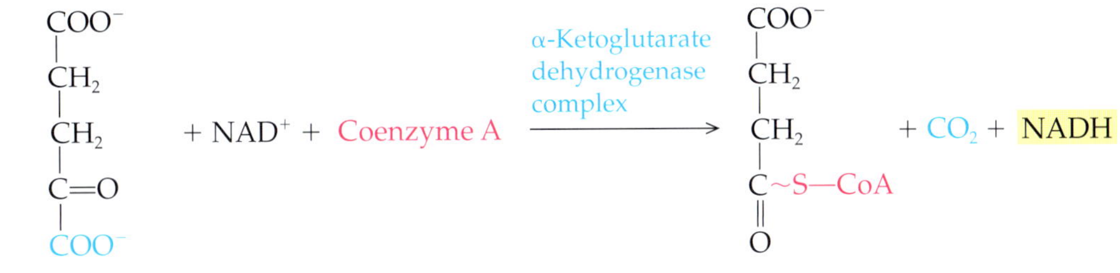

The pyruvate dehydrogenase complex was described in Section 22.2 and shown in Figure 22.3.

$$\underset{\alpha\text{-Ketoglutarate}}{\begin{array}{c}COO^-\\|\\CH_2\\|\\CH_2\\|\\C=O\\|\\COO^-\end{array}} + NAD^+ + \text{Coenzyme A} \xrightarrow{\substack{\alpha\text{-Ketoglutarate} \\ \text{dehydrogenase} \\ \text{complex}}} \underset{\text{Succinyl CoA}}{\begin{array}{c}COO^-\\|\\CH_2\\|\\CH_2\\|\\C\sim S-CoA\\\|\\O\end{array}} + CO_2 + \text{NADH}$$

Reaction 5. Succinyl CoA is converted to succinate in this step, which once more is chemically very involved. The enzyme *succinyl CoA synthase* catalyzes a coupled reaction in which the high-energy thioester bond of succinyl CoA is hydrolyzed and an inorganic phosphate group is added to GDP to make GTP:

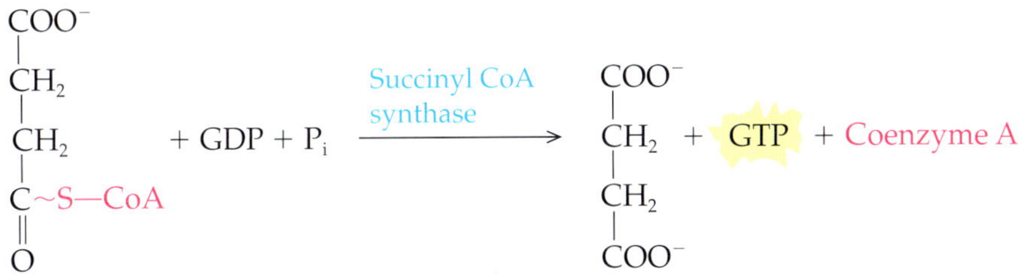

$$\underset{\text{Succinyl CoA}}{\begin{array}{c}COO^-\\|\\CH_2\\|\\CH_2\\|\\C\sim S-CoA\\\|\\O\end{array}} + GDP + P_i \xrightarrow{\substack{\text{Succinyl CoA} \\ \text{synthase}}} \underset{\text{Succinate}}{\begin{array}{c}COO^-\\|\\CH_2\\|\\CH_2\\|\\COO^-\end{array}} + \text{GTP} + \text{Coenzyme A}$$

Another enzyme, *dinucleotide diphosphokinase*, then catalyzes the transfer of a phosphoryl group from GTP to ADP to make ATP:

$$\text{GTP} + \text{ADP} \xrightarrow{\text{Dinucleotide diphosphokinase}} \text{GDP} + \text{ATP}$$

Reaction 6. *Succinate dehydrogenase* then catalyzes the oxidation of succinate to fumarate in the next step. The oxidizing agent, *flavin adenine dinucleotide (FAD)*, is reduced in this step:

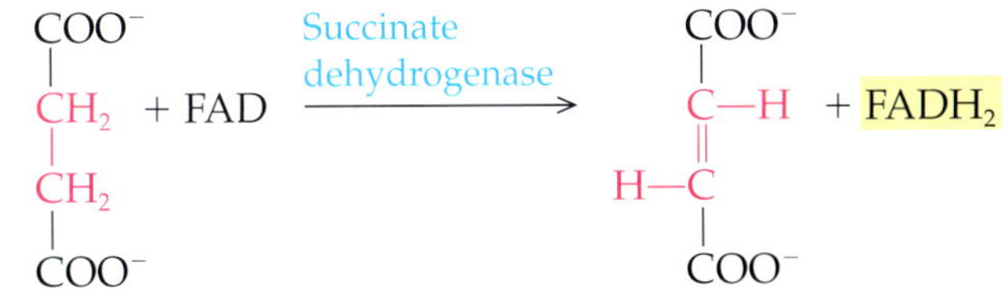

The structure of FAD was shown in Figure 19.8.

$$\underset{\text{Succinate}}{\begin{array}{c}COO^-\\|\\CH_2\\|\\CH_2\\|\\COO^-\end{array}} + FAD \xrightarrow{\text{Succinate dehydrogenase}} \underset{\text{Fumarate}}{\begin{array}{c}COO^-\\|\\C-H\\\|\\H-C\\|\\COO^-\end{array}} + \text{FADH}_2$$

We studied hydrogenation of alkenes to produce alkanes in Section 11.5. This is simply the reverse.

Reaction 7. Addition of H_2O to the double bond of fumarate gives malate. The enzyme *fumarase* catalyzes this reaction:

$$\underset{\text{Fumarate}}{\begin{array}{c} COO^- \\ | \\ C-H \\ \| \\ H-C \\ | \\ COO^- \end{array}} + H_2O \xrightarrow{\text{Fumarase}} \underset{\text{Malate}}{\begin{array}{c} COO^- \\ | \\ HO-C-H \\ | \\ H-C-H \\ | \\ COO^- \end{array}}$$

This reaction is a biological example of the hydration of an alkene to produce an alcohol (Sections 11.5 and 12.5).

Reaction 8. In the final step of the citric acid cycle, *malate dehydrogenase* catalyzes the reduction of NAD^+ to NADH and the oxidation of malate to oxaloacetate. Because the citric acid cycle "began" with the addition of an acetyl group to oxaloacetate, we have come full circle.

$$\underset{\text{Malate}}{\begin{array}{c} COO^- \\ | \\ HO-C-H \\ | \\ CH_2 \\ | \\ COO^- \end{array}} + NAD^+ \xrightarrow{\text{Malate dehydrogenase}} \underset{\text{Oxaloacetate}}{\begin{array}{c} COO^- \\ | \\ C=O \\ | \\ CH_2 \\ | \\ COO^- \end{array}} + NADH$$

This reaction is a biochemical example of the oxidation of a secondary alcohol to a ketone, which we studied in Sections 12.5 and 14.4.

22.5 Control of the Citric Acid Cycle

LEARNING GOAL 5

Allosteric enzymes bind to *effectors*, such as ATP or ADP, that alter the shape of the enzyme active site, either stimulating the rate of the reaction (positive allosterism) or inhibiting the reaction (negative allosterism). For more detail, see Section 19.9.

Just like glycolysis, the citric acid cycle is responsive to the energy needs of the cell. The pathway speeds up when there is a greater demand for ATP, and it slows down when ATP energy is in excess. In the last chapter we saw that several of the enzymes that catalyze the reactions of glycolysis are *allosteric enzymes*. Similarly, four enzymes or enzyme complexes involved in the complete oxidation of pyruvate are allosteric enzymes. Because the control of the pathway must be precise, there are several enzymatic steps that are regulated. These are summarized in Figure 22.6 and below:

1. *Conversion of pyruvate to acetyl CoA.* The pyruvate dehydrogenase complex is inhibited by high concentrations of ATP, acetyl CoA, and NADH. Of course, the presence of these compounds in abundance signals that the cell has an adequate supply of energy, and thus energy metabolism is slowed.
2. *Synthesis of citrate from oxaloacetate and acetyl CoA.* The enzyme citrate synthase is an allosteric enzyme. In this case, the negative effector is ATP. Again, this is logical because an excess of ATP indicates that the cell has an abundance of energy.
3. *Oxidation and decarboxylation of isocitrate to α-ketoglutarate.* Isocitrate dehydrogenase is also an allosteric enzyme; however, the enzyme is controlled by the positive allosteric effector, ADP. ADP is a signal that the levels of ATP must be low, and therefore the rate of the citric acid cycle should be increased. Interestingly, isocitrate dehydrogenase is also *inhibited* by high levels of NADH and ATP.
4. *Conversion of α-ketoglutarate to succinyl CoA.* The α-ketoglutarate dehydrogenase complex is inhibited by high levels of the products of the reactions that it catalyzes, namely, NADH and succinyl CoA. It is further inhibited by high concentrations of ATP.

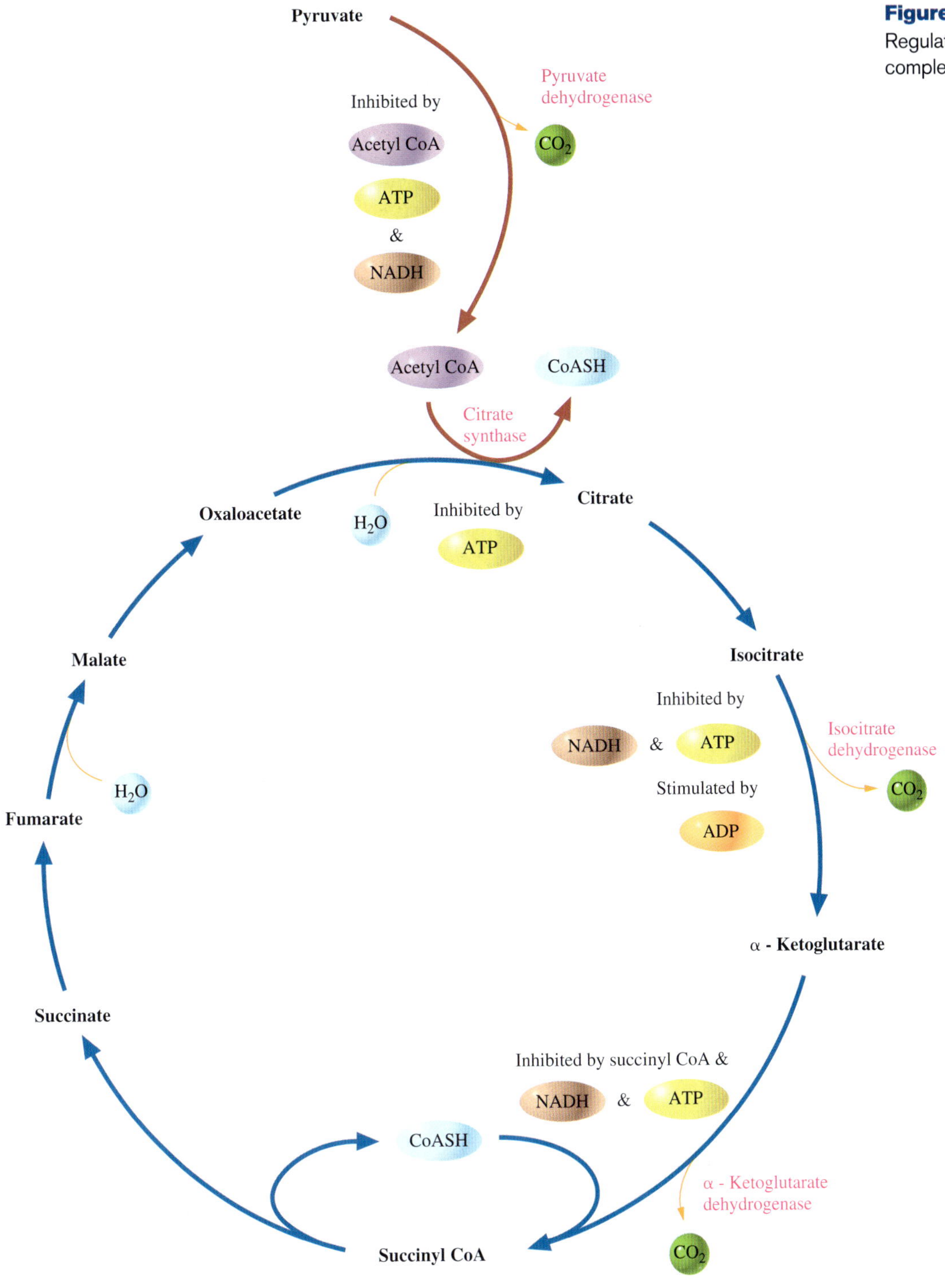

Figure 22.6
Regulation of the pyruvate dehydrogenase complex and the citric acid cycle.

22.6 Oxidative Phosphorylation

In Section 22.3 we noted that the electrons carried by NADH can be used to produce three ATP molecules, and those carried by FADH$_2$ can be used to produce two ATP molecules. We turn now to the process by which the energy of electrons carried by these coenzymes is converted to ATP energy. It is a series of reactions called **oxidative phosphorylation,** which couples the oxidation of NADH and FADH$_2$ to the phosphorylation of ADP to generate ATP.

LEARNING GOAL 6

A Human Perspective

Brown Fat: The Fat That Makes You Thin?

Humans have two types of fat, or adipose, tissue. *White fat* is distributed throughout the body and is composed of aggregations of cells having membranous vacuoles containing stored triglycerides. The size and number of these storage vacuoles determine whether a person is overweight or not. The other type of fat is *brown fat*. Brown fat is a specialized tissue for heat production, called *nonshivering thermogenesis*. As the name suggests, this is a means of generating heat in the absence of the shivering response. The cells of brown fat look nothing like those of white fat. They do contain small fat vacuoles; however, the distinguishing feature of brown fat is the huge number of mitochondria within the cytoplasm. In addition, brown fat tissue contains a great many blood vessels. These provide oxygen for the thermogenic metabolic reactions.

Brown fat is most pronounced in newborns, cold-adapted mammals, and hibernators. One major difficulty faced by a newborn is temperature regulation. The baby leaves an environment in which he or she was bathed in fluid of a constant 37°C, body temperature. Suddenly, the child is thrust into a world that is much colder and in which he or she must generate his or her own warmth internally. By having a good reserve of active brown fat to generate that heat, the newborn is protected against cold shock at the time of birth. However, this thermogenesis literally burns up most of the brown fat tissue, and adults typically have so little brown fat that it can be found only by using a special technique called *thermography*, which detects temperature differences throughout a body. However, in some individuals, brown fat is very highly developed. For instance, the Korean diving women who spend 6–7 hours every day diving for pearls in cold water have a massive amount of brown fat to warm them by nonshivering thermogenesis. Thus, development of brown fat is a mechanism of cold adaptation.

When it was noticed that such cold-adapted individuals were seldom overweight, a correlation was made between the amount of brown fat in the body and the tendency to become overweight. Studies done with rats suggest that, to some degree, fatness is genetically determined. In other words, you are as lean as your genes allow you to be. In these studies, cold-adapted and non-cold-adapted rats were fed cafeteria food—as much as they wanted—and their weight gain was monitored. In every case the cold-adapted rats, with their greater quantity of brown fat, gained significantly less weight than their non-cold-adapted counterparts, despite the fact that they ate as much as the non-cold-adapted rats. This and other studies led researchers to conclude that brown fat burns excess fat in a highly caloric diet.

How does brown fat generate heat and burn excess calories? For the answer we must turn to the mitochondrion. In addition to the ATP synthase and the electron transport system proteins that are found in all mitochondria, there is a protein in the inner mitochondrial membrane of brown fat tissue called *thermogenin*. This protein has a channel in the center through which the protons (H^+) of the intermembrane space could pass back into the mitochondrial matrix. Under normal conditions, this channel is plugged by a GDP molecule so that it remains closed and the proton gradient can continue to drive ATP synthesis by oxidative phosphorylation.

When brown fat is "turned on," by cold exposure or in response to certain hormones, there is an immediate increase in the rate of glycolysis and β-oxidation of the stored fat (Chapter 23). These reactions produce acetyl CoA, which then fuels the citric acid cycle. The citric acid cycle, of course, produces NADH and $FADH_2$, which carry electrons to the electron transport system. Finally, the electron transport system pumps protons into the intermembrane space. Under usual conditions, the energy of the proton gradient would be used to synthesize ATP. However, when brown fat is stimulated, the GDP that had plugged the pore in thermogenin is lost. Now protons pass freely back into the matrix space, and the proton gradient is dissipated. The energy of the gradient, no longer useful for generating ATP, is released as *heat*, the heat that warms and protects newborns and cold-adapted individuals.

Brown fat is just one of the body's many systems for maintaining a constant internal environment regardless of the conditions in the external environment. Such mechanisms, called *homeostatic mechanisms*, are absolutely essential to allow the body to adapt to and survive in an ever-changing environment.

For Further Understanding

Hibernators eat a great deal in preparation for their long winter nap. Explain their lifestyle in terms of energy requirements and heat production.

Explain why cold-adapted mammals can eat a high caloric diet and remain thin.

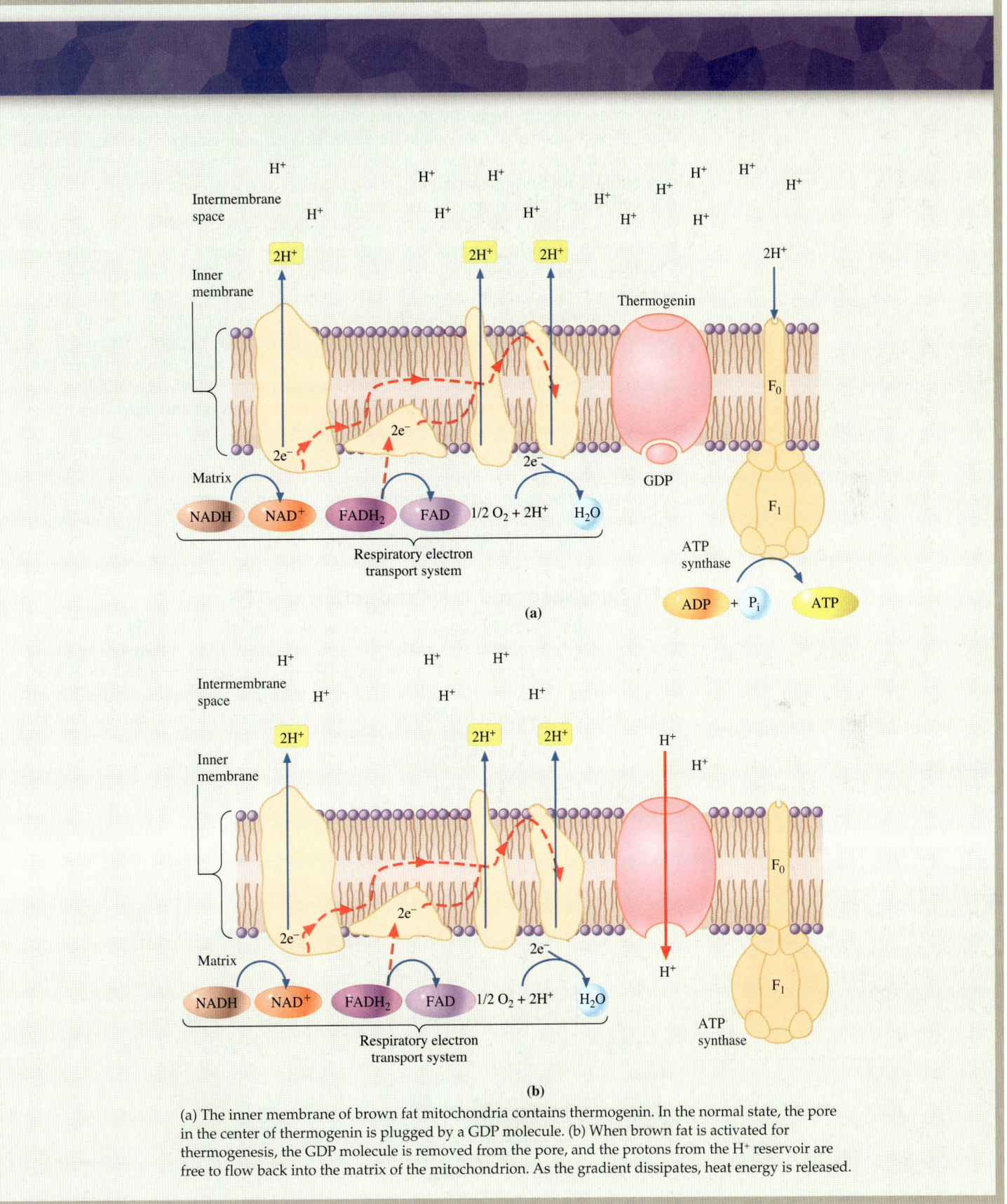

(a) The inner membrane of brown fat mitochondria contains thermogenin. In the normal state, the pore in the center of thermogenin is plugged by a GDP molecule. (b) When brown fat is activated for thermogenesis, the GDP molecule is removed from the pore, and the protons from the H+ reservoir are free to flow back into the matrix of the mitochondrion. As the gradient dissipates, heat energy is released.

Electron Transport Systems and the Hydrogen Ion Gradient

Before we try to understand the mechanism of oxidative phosphorylation, let's first look at the molecules that carry out this complex process. Embedded within the mitochondrial inner membrane are **electron transport systems.** These are made up of a series of electron carriers, including coenzymes and cytochromes. All these molecules are located within the membrane in an arrangement that allows them to pass electrons from one to the next. This array of electron carriers is called the *respiratory electron transport system* (Figure 22.7). As you would expect in such sequential oxidation-reduction reactions, the electrons lose some energy with each transfer. Some of this energy is used to make ATP.

At three sites in the electron transport system, protons (H^+) can be pumped from the mitochondrial matrix to the intermembrane space. These H^+ contribute to a high-energy H^+ reservoir. At each of the three sites, enough H^+ are pumped into the H^+ reservoir to produce one ATP molecule. The first site is NADH dehydrogenase. Because electrons from NADH enter the electron transport system by being transferred to NADH dehydrogenase, all three sites actively pump H^+, and three ATP molecules are made (see Figure 22.7). $FADH_2$ is a less "powerful" electron donor. It transfers its electrons to an electron carrier that follows NADH dehydrogenase. As a result, when $FADH_2$ is oxidized, only the second and third sites pump H^+, and only two ATP molecules are made.

The last component needed for oxidative phosphorylation is a multiprotein complex called **ATP synthase,** also called the F_0F_1 **complex** (see Figure 22.7). The F_0 portion of the molecule is a channel through which H^+ pass. It spans the inner mitochondrial membrane, as shown in Figure 22.7. The F_1 part of the molecule is an enzyme that catalyzes the phosphorylation of ADP to produce ATP.

ATP Synthase and the Production of ATP

How does all this complicated machinery actually function? NADH carries electrons, originally from glucose, to the first carrier of the electron transport system, NADH dehydrogenase (see Figure 22.7). There, NADH is oxidized to NAD^+, which returns to the site of the citric acid cycle to be reduced again. As the dashed red line shows, the pair of electrons is passed to the next electron carrier, and H^+

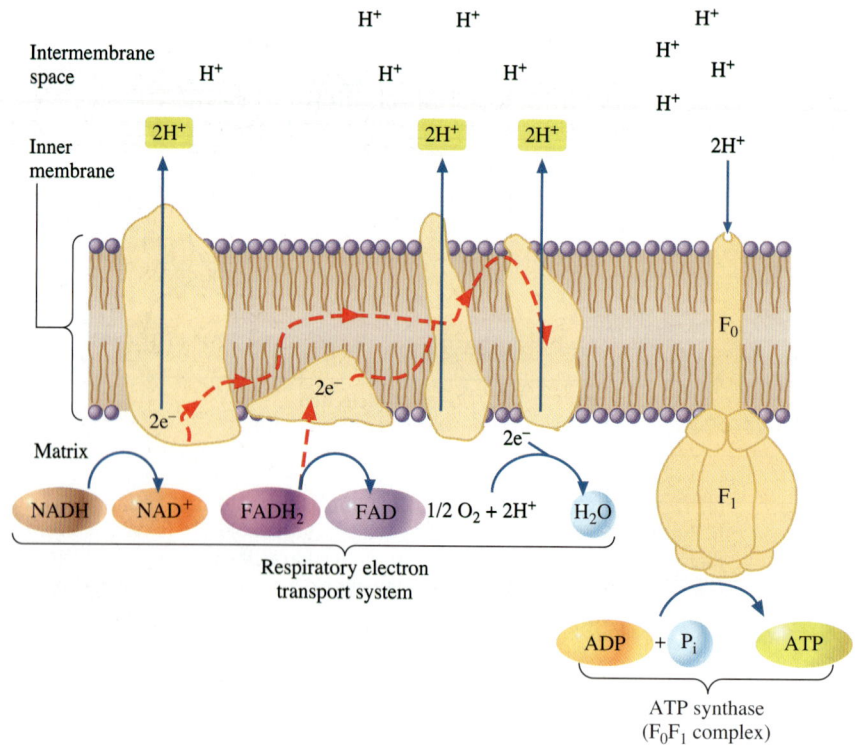

Figure 22.7
Electrons flow from NADH to molecular oxygen through a series of electron carriers embedded in the inner mitochondrial membrane. Protons are pumped from the mitochondrial matrix space into the intermembrane space. This results in a hydrogen ion reservoir in the intermembrane space. As protons pass through the channel in ATP synthase, their energy is used to phosphorylate ADP and produce ATP.

22.6 Oxidative Phosphorylation

are pumped to the intermembrane compartment. The electrons are passed sequentially through the electron transport system, and at two additional sites, H^+ from the matrix are pumped into the intermembrane compartment. With each transfer the electrons lose some of their potential energy. It is this energy that is used to transport H^+ across the inner mitochondrial membrane and into the H^+ reservoir. As mentioned above, $FADH_2$ donates its electrons to a carrier of lower energy and fewer H^+ are pumped into the reservoir.

Finally, the electrons arrive at the last carrier. They now have too little energy to accomplish any more work, but they *must* be donated to some final electron acceptor so that the electron transport system can continue to function. In aerobic organisms the **terminal electron acceptor** is molecular oxygen, O_2, and the product is water.

As the electron transport system continues to function, a high concentration of protons builds up in the intermembrane space. This creates an H^+ gradient across the inner mitochondrial membrane. Such a gradient is an enormous energy source, like water stored behind a dam. The mitochondria make use of the potential energy of the gradient to synthesize ATP energy.

ATP synthase harvests the energy of this gradient by making ATP. H^+ pass through the F_0 channel back into the matrix. This causes F_1 to become an active enzyme that catalyzes the phosphorylation of ADP to produce ATP. In this way the energy of the H^+ reservoir is harvested to make ATP.

> The importance of keeping the electron transport system functioning becomes obvious when we consider what occurs in cyanide poisoning. Cyanide binds to the heme group iron of cytochrome oxidase, instantly stopping electron transfers and causing death within minutes!

Question 22.9

Write a balanced chemical equation for the reduction of NAD^+.

Question 22.10

Write a balanced chemical equation for the reduction of FAD.

Summary of the Energy Yield

One turn of the citric acid cycle results in the production of two CO_2 molecules, three NADH molecules, one $FADH_2$ molecule, and one ATP molecule. Oxidative phosphorylation yields three ATP molecules per NADH molecule and two ATP molecules per $FADH_2$ molecule. The only exception to these energy yields is the NADH produced in the cytoplasm during glycolysis. Oxidative phosphorylation yields only two ATP molecules per cytoplasmic NADH molecule. The reason for this is that energy must be expended to shuttle electrons from NADH in the cytoplasm to $FADH_2$ in the mitochondrion.

Knowing this information and keeping in mind that two turns of the citric acid cycle are required, we can sum up the total energy yield from the complete oxidation of one glucose molecule.

In some tissues of the body there is a more efficient shuttle system that results in the production of three ATP per cytoplasmic NADH. This system is described online at www.mhhe.com/denniston5e in Energy Yields from Aerobic Respiration: Some Alternatives.

EXAMPLE 22.1

Determining the Yield of ATP from Aerobic Respiration

Calculate the number of ATP produced by the complete oxidation of one molecule of glucose.

Solution

Glycolysis:
Substrate-level phosphorylation	2 ATP
2 NADH × 2 ATP/cytoplasmic NADH	4 ATP

Continued—

> ### EXAMPLE 22.1 —Continued
>
> Conversion of 2 pyruvate molecules to 2 acetyl CoA molecules:
>
> | | |
> |---|---:|
> | 2 NADH × 3 ATP/NADH | 6 ATP |
> | Citric acid cycle (two turns): | |
> | 2 GTP × 1 ATP/GTP | 2 ATP |
> | 6 NADH × 3 ATP/NADH | 18 ATP |
> | 2 FADH$_2$ × 2 ATP/FADH$_2$ | 4 ATP |
> | | **36 ATP** |
>
> This represents an energy harvest of about 40% of the potential energy of glucose.

Aerobic metabolism is very much more efficient than anaerobic metabolism. The abundant energy harvested by aerobic metabolism has had enormous consequences for the biological world. Much of the energy released by the oxidation of fuels is not lost as heat but conserved in the form of ATP. Organisms that possess abundant energy have evolved into multicellular organisms and developed specialized functions. As a consequence of their energy requirements, all multicellular organisms are aerobic.

22.7 The Degradation of Amino Acids

LEARNING GOAL Carbohydrates are not our only source of energy. As we saw in Chapter 21, dietary protein is digested to amino acids that can also be used as an energy source, although this is not their major metabolic function. Most of the amino acids used for energy come from the diet. In fact, it is only under starvation conditions, when stored glycogen has been depleted, that the body begins to burn its own protein, for instance from muscle, as a fuel.

The fate of the mixture of amino acids provided by digestion of protein depends upon a balance between the need for amino acids for biosynthesis and the need for cellular energy. Only those amino acids that are not needed for protein synthesis are eventually converted into citric acid cycle intermediates and used as fuel.

The degradation of amino acids occurs primarily in the liver and takes place in two stages. The first stage is the removal of the α-amino group, and the second is the degradation of the carbon skeleton. In land mammals the amino group generally ends up in urea, which is excreted in the urine. The carbon skeletons can be converted into a variety of compounds, including citric acid cycle intermediates, pyruvate, acetyl CoA, or acetoacetyl CoA. The degradation of the carbon skeletons is summarized in Figure 22.8. Deamination reactions and the fate of the carbon skeletons of amino acids are the focus of this section.

Removal of α-Amino Groups: Transamination

The first stage of amino acid degradation, the removal of the α-amino group, is usually accomplished by a **transamination** reaction. **Transaminases** catalyze the transfer of the α-amino group from an α-amino acid to an α-keto acid:

22.7 The Degradation of Amino Acids

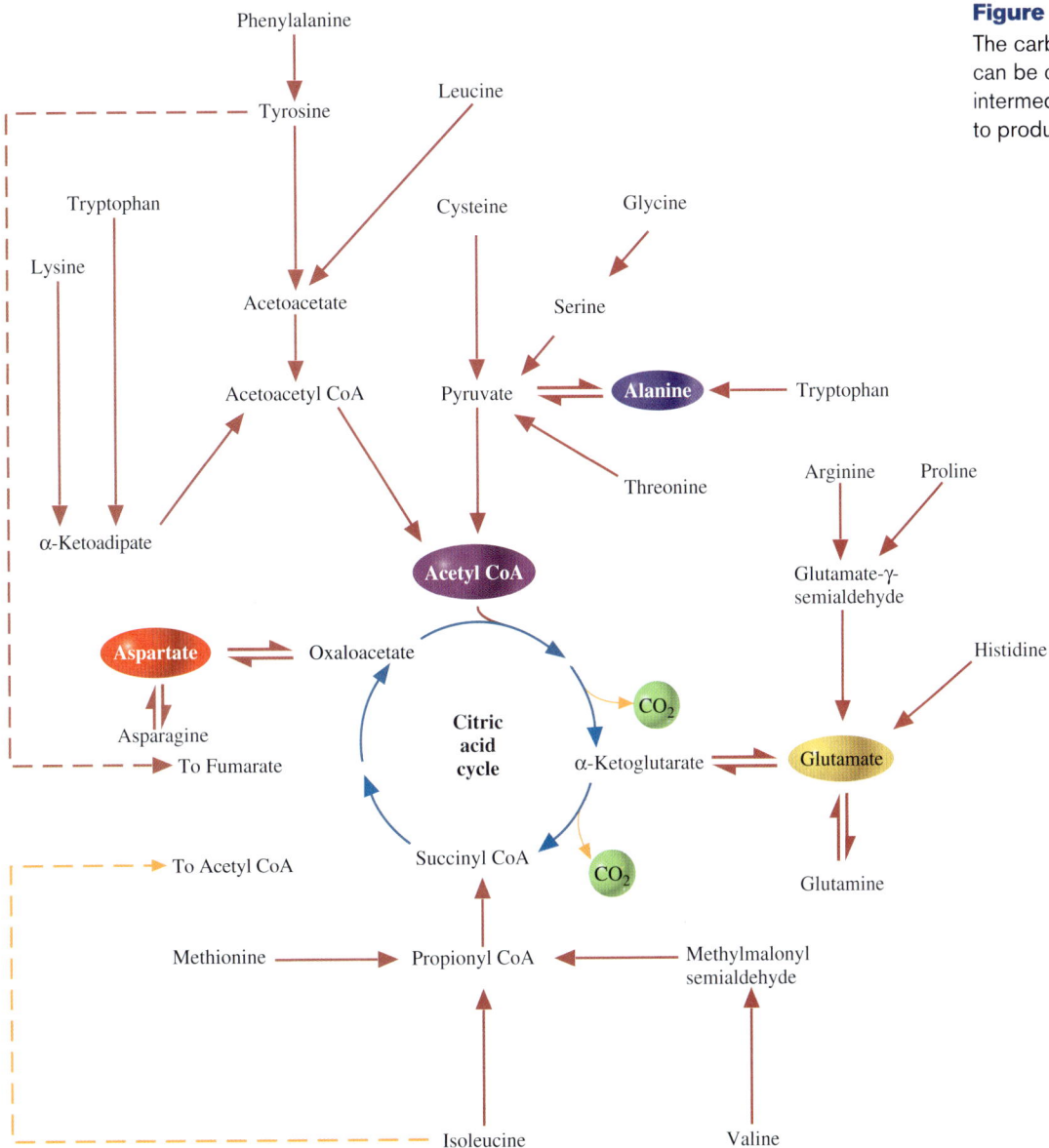

The α-amino group of a great many amino acids is transferred to α-ketoglutarate to produce the amino acid glutamate and a new keto acid. This glutamate family of transaminases is especially important because the α-keto acid corresponding to glutamate is α-ketoglutarate, a citric acid cycle intermediate. The glutamate transaminases thus provide a direct link between amino acid degradation and the citric acid cycle.

Figure 22.8
The carbon skeletons of amino acids can be converted to citric acid cycle intermediates and completely oxidized to produce ATP energy.

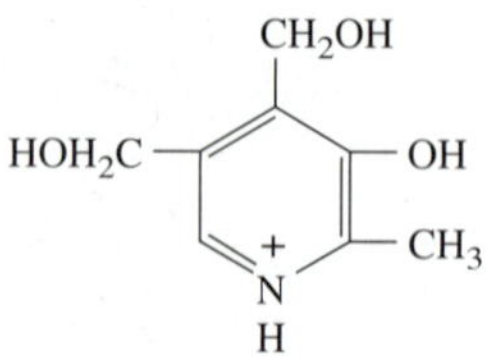

Pyridoxine (vitamin B₆)

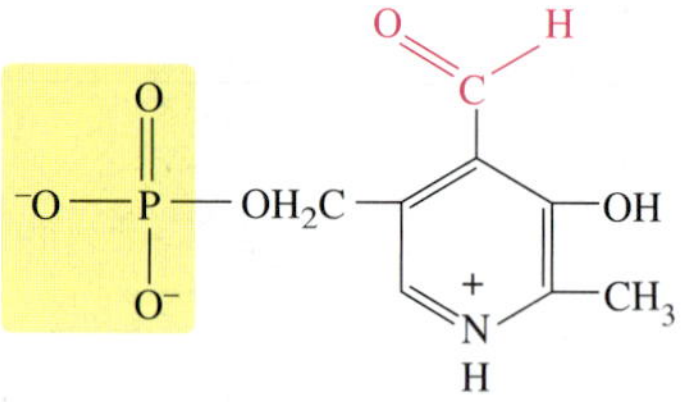

Pyridoxal phosphate

Figure 22.9
The structure of pyridoxal phosphate, the coenzyme required for all transamination reactions, and pyridoxine, vitamin B$_6$, the vitamin from which it is derived.

For more information on these vitamins and the coenzymes that are made from them, look online at www.mhhe.com/denniston5e in Water-Soluble Vitamins.

Aspartate transaminase catalyzes the transfer of the α-amino group of aspartate to α-ketoglutarate, producing oxaloacetate and glutamate:

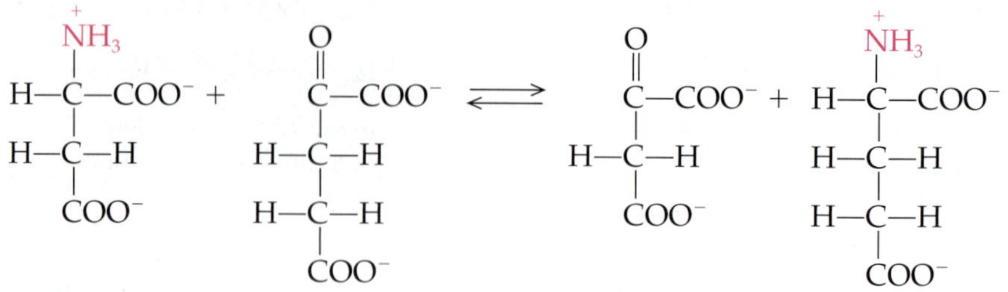

Aspartate + α-Ketoglutarate ⇌ Oxaloacetate + Glutamate

Another important transaminase in mammalian tissues is *alanine transaminase*, which catalyzes the transfer of the α-amino group of alanine to α-ketoglutarate and produces pyruvate and glutamate:

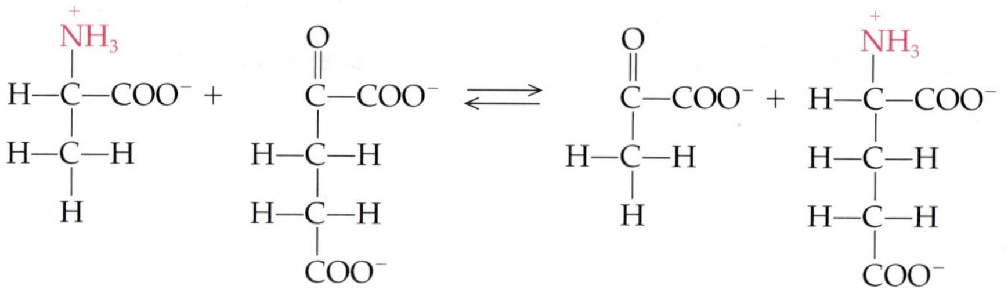

Alanine + α-Ketoglutarate ⇌ Pyruvate + Glutamate

All of the more than fifty transaminases that have been discovered require the coenzyme **pyridoxal phosphate.** This coenzyme is derived from vitamin B$_6$ (pyridoxine, Figure 22.9).

The transamination reactions shown above appear to be a simple transfer, but in reality, the reaction is much more complex. The transaminase binds the amino acid (aspartate in Figure 22.10a) in its active site. Then the α-amino group of aspartate is transferred to pyridoxal phosphate, producing pyridoxamine phosphate and oxaloacetate (Figure 22.10b). The amino group is then transferred to an α-keto acid, in this case, α-ketoglutarate (Figure 22.10c), to produce the amino acid glutamate (Figure 22.10d). Next we will examine the fate of the amino group that has been transferred to α-ketoglutarate to produce glutamate.

Question 22.11 What is the role of pyridoxal phosphate in transamination reactions?

Question 22.12 What is the function of a transaminase?

Removal of α-Amino Groups: Oxidative Deamination

In the next stage of amino acid degradation, ammonium ion is liberated from the glutamate formed by the transaminase. This breakdown of glutamate, catalyzed by the enzyme *glutamate dehydrogenase*, occurs as follows:

22.7 The Degradation of Amino Acids

Glutamate + NAD$^+$ + H$_2$O ⇌ NH$_4^+$ + α-Ketoglutarate + NADH

This is an example of an **oxidative deamination,** an oxidation-reduction process in which NAD$^+$ is reduced to NADH and the amino acid is deaminated (the amino group is removed). A summary of the deamination reactions described is shown in Figure 22.11.

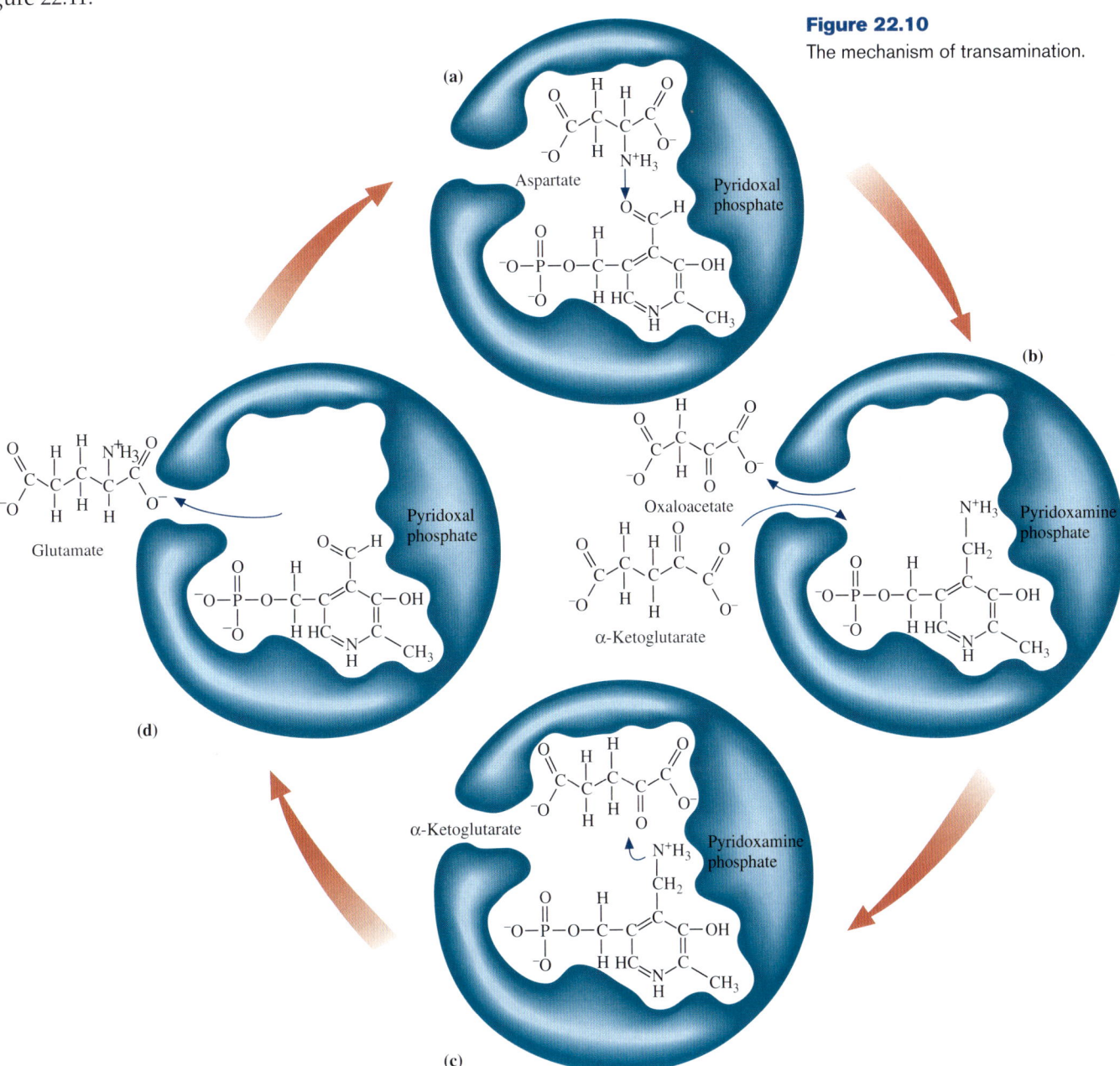

Figure 22.10
The mechanism of transamination.

Figure 22.11
Summary of the deamination of an α-amino acid and the fate of the ammonium ion (NH_4^+).

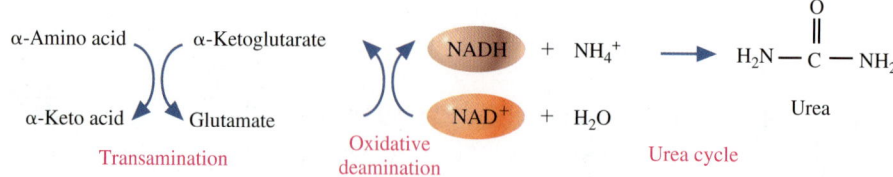

The Fate of Amino Acid Carbon Skeletons

The carbon skeletons produced by these and other deamination reactions enter glycolysis or the citric acid cycle at many steps. For instance, we have seen that transamination converts aspartate to oxaloacetate and alanine to pyruvate. The positions at which the carbon skeletons of various amino acids enter the energy-harvesting pathways are summarized in Figure 22.8.

22.8 The Urea Cycle

  LEARNING GOAL 8

Oxidative deamination produces large amounts of ammonium ion. Because ammonium ions are extremely toxic, they must be removed from the body, regardless of the energy expenditure required. In humans, they are detoxified in the liver by converting the ammonium ions into urea. This pathway, called the **urea cycle,** is the method by which toxic ammonium ions are kept out of the blood. The excess ammonium ions incorporated in urea are excreted in the urine (Figure 22.12).

Reactions of the Urea Cycle

The five reactions of the urea cycle are shown in Figure 22.12, and details of the reactions are summarized as follows.

Step 1. The first step of the cycle is a reaction in which CO_2 and NH_4^+ form carbamoyl phosphate. This reaction, which also requires ATP and H_2O, occurs in the mitochondria and is catalyzed by the enzyme *carbamoyl phosphate synthase*.

$$CO_2 + NH_4^+ + 2ATP + H_2O \longrightarrow H_2N-\overset{\overset{O}{\|}}{C}-O-\overset{\overset{O}{\|}}{\underset{\underset{O^-}{|}}{P}}-O^- + 2ADP + P_i + 3H^+$$

Carbamoyl phosphate

The urea cycle involves several unusual amino acids that are not found in polypeptides.

Step 2. The carbamoyl phosphate now condenses with the amino acid ornithine to produce the amino acid citrulline. This reaction also occurs in the mitochondria and is catalyzed by the enzyme *ornithine transcarbamoylase*.

$$\begin{array}{c}\overset{+}{NH_3}\\|\\H-C-H\\|\\H-C-H\\|\\H-C-H\\|\\H-\overset{+}{C}-NH_3\\|\\COO^-\end{array} + H_2N-\overset{\overset{O}{\|}}{C}-O-\overset{\overset{O}{\|}}{\underset{\underset{O^-}{|}}{P}}-O^- \longrightarrow \begin{array}{c}H-N-\overset{\overset{O}{\|}}{C}-NH_2\\|\\H-C-H\\|\\H-C-H\\|\\H-C-H\\|\\H-\overset{+}{C}-NH_3\\|\\COO^-\end{array} + P_i$$

Ornithine   Carbamoyl phosphate   Citrulline

22.8 The Urea Cycle

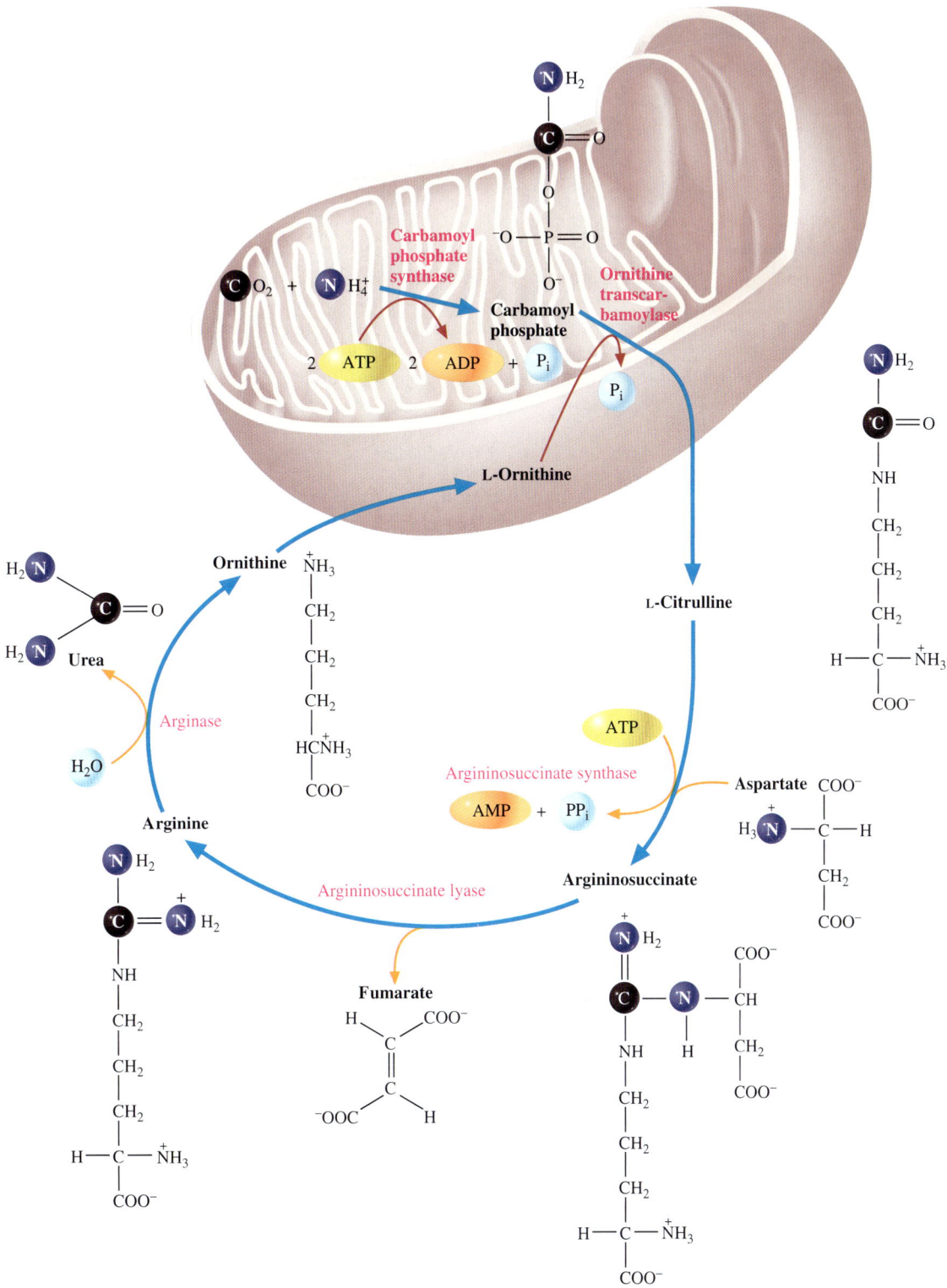

Figure 22.12
The urea cycle converts ammonium ions into urea, which is less toxic. The intracellular locations of the reactions are indicated. Citrulline, formed in the reaction between ornithine and carbamoyl phosphate, is transported out of the mitochondrion and into the cytoplasm. Ornithine, a substrate for the formation of citrulline, is transported from the cytoplasm into the mitochondrion.

The abbreviation PP$_i$ represents the pyrophosphate group, which consists of two phosphate groups joined by a phosphoanhydride bond:

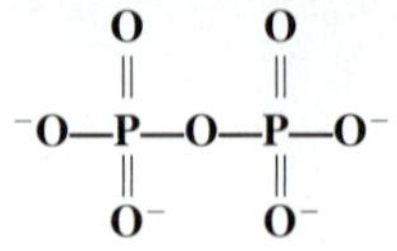

$$^-O-\overset{\overset{O}{\|}}{P}-O-\overset{\overset{O}{\|}}{P}-O^-$$
$$\;\;\;\;\;\;O^-\;\;\;\;\;\;O^-$$

Step 3. Citrulline is transported into the cytoplasm and now condenses with aspartate to produce argininosuccinate. This reaction, which requires energy released by the hydrolysis of ATP, is catalyzed by the enzyme *argininosuccinate synthase*.

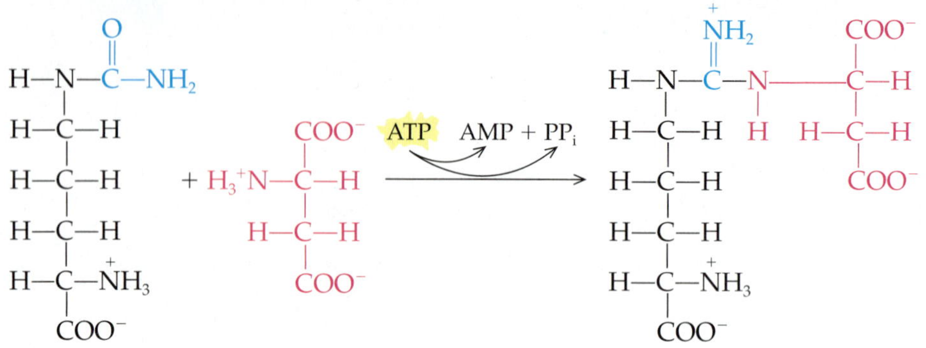

Citrulline + Aspartate $\xrightarrow{\text{ATP} \rightarrow \text{AMP} + \text{PP}_i}$ Argininosuccinate

Step 4. Now the argininosuccinate is cleaved to produce the amino acid arginine and the citric acid cycle intermediate fumarate. This reaction is catalyzed by the enzyme *argininosuccinate lyase*.

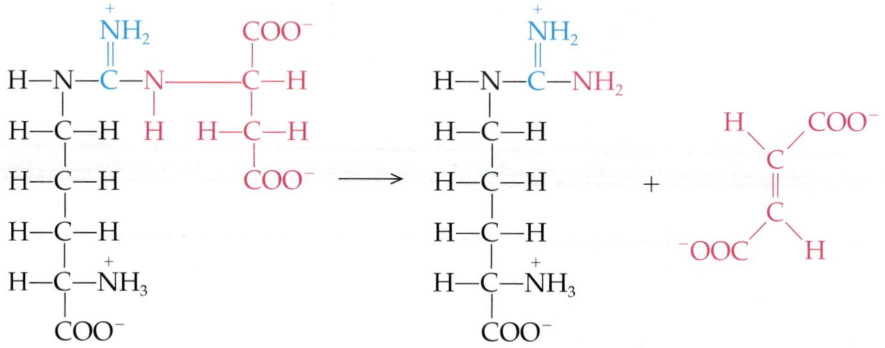

Argininosuccinate $\longrightarrow$ Arginine + Fumarate

Step 5. Finally, arginine is hydrolyzed to generate urea, the product of the reaction to be excreted, and ornithine, the original reactant in the cycle. *Arginase* is the enzyme that catalyzes this reaction.

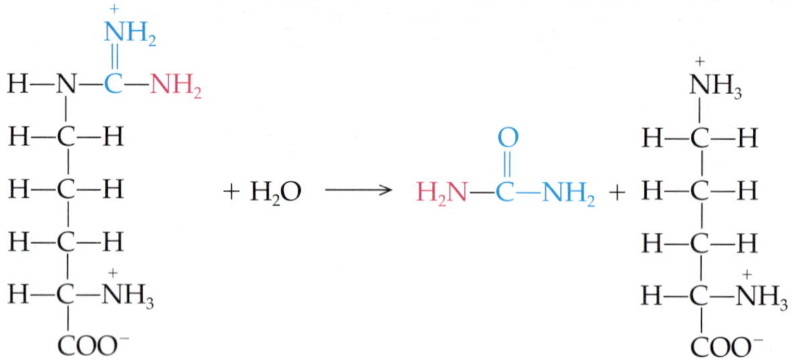

Arginine + H$_2$O $\longrightarrow$ Urea + Ornithine

Note that one of the amino groups in urea is derived from the ammonium ion and the second is derived from the amino acid aspartate.

There are genetically transmitted diseases that result from a deficiency of one of the enzymes of the urea cycle. The importance of the urea cycle is apparent

LEARNING GOAL 9

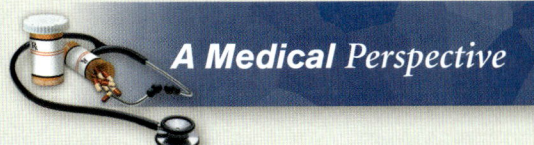

A Medical Perspective

Pyruvate Carboxylase Deficiency

Pyruvate carboxylase is the enzyme that converts pyruvate to oxaloacetate.

$$\text{Pyruvate} + CO_2 + ATP + H_2O \longrightarrow \text{Oxaloacetate} + ADP + 2H^+$$

This reaction is important because it provides oxaloacetate for the citric acid cycle when the supplies have run low because of the demands of biosynthesis. It is also the enzyme that catalyzes the first step in gluconeogenesis, the pathway that provides the body with needed glucose in times of starvation or periods of exercise that deplete glycogen stores. But somehow these descriptions don't fill us with a sense of the importance of this enzyme and its jobs. It is not until we investigate a case study of a child born with pyruvate carboxylase deficiency that we see the full impact of this enzyme.

Pyruvate carboxylase deficiency is found in about 1 in 250,000 births; however, there is an increased incidence in native North American Indians who speak the Algonquin dialect and in the French. There are two types of genetic disorders that have been described. In the neonatal form of the disease, there is a complete absence of the enzyme. Symptoms are apparent at birth and the child is born with brain abnormalities. In the infantile form, the patient develops symptoms early in infancy. Again, it is neurological symptoms that draw attention to the condition. The infants do not develop mental or psychomotor skills. They may develop seizures and/or respiratory depression. In both cases, it is the brain that suffers the greatest damage. In fact, this is the case in most of the disorders that reduce energy metabolism because the brain has such high energy requirements.

Biochemically, patients exhibit quite a variety of symptoms. They show acidosis (low blood pH) due to accumulations of lactate and extremely high pyruvate concentrations in the blood. Blood levels of alanine are also high and large doses of alanine do not stimulate gluconeogenesis. Furthermore, a patient's cells accumulate lipid.

We can understand each of these symptoms by considering the pathways affected by the absence of this single enzyme. Lactic acidosis results from the fact that the body must rely on glycolysis and lactate fermentation for most of its energy needs. Alanine levels are high because it isn't being transaminated to pyruvate efficiently, because pyruvate levels are so high. In addition, alanine can't be converted to glucose by gluconeogenesis. Although the excess alanine is taken up by the liver and converted to pyruvate, the pyruvate can't be converted to glucose. Lipids accumulate because a great deal of pyruvate is converted to acetyl CoA. However, the acetyl CoA is not used to produce citrate as a result of the absence of oxaloacetate. So, the acetyl CoA is thus used to synthesize fatty acids, which are stored as triglycerides.

Dietary intervention has been tried. One such regimen is to supplement with aspartic acid and glutamic acid. The theory behind this treatment is as complex as the many symptoms of the disorder. Both amino acids can be aminated (amino groups added) in non-nervous tissue. This produces asparagine and glutamine, both of which are able to cross the blood-brain barrier. Glutamine is deaminated to glutamate, which is then transaminated to α-ketoglutarate, indirectly replenishing oxaloacetate. Asparagine can be deaminated to aspartate, which can be converted to oxaloacetate. This serves as a second supply of oxaloacetate. To date, these attempts at dietary intervention have not proved successful. Perhaps in time research will provide the tools for enzyme replacement therapy or gene therapy that could alleviate the symptoms.

For Further Understanding

Write an equation showing the reaction catalyzed by pyruvate carboxylase using structural formulas for pyruvate and oxaloacetate.

Supplementing the diet with asparagine and glutamine was tried as a treatment for pyruvate carboxylase deficiency. Write equations showing the reactions that convert these amino acids into citric acid cycle intermediates.

---

when we consider the terrible symptoms suffered by afflicted individuals. A deficiency of urea cycle enzymes causes an elevation of the concentration of NH_4^+, a condition known as **hyperammonemia**. If there is a complete deficiency of one of the enzymes of the urea cycle, the result is death in early infancy. If there is a partial deficiency of one of the enzymes of the urea cycle, the result may be retardation, convulsions, and vomiting. In these milder forms of hyperammonemia, a low-protein diet leads to a lower concentration of NH_4^+ in blood and less severe clinical symptoms.

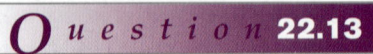

Question 22.13 What is the purpose of the urea cycle?

Question 22.14 Where do the reactions of the urea cycle occur?

22.9 Overview of Anabolism: The Citric Acid Cycle as a Source of Biosynthetic Intermediates

LEARNING GOAL

So far, we have talked about the citric acid cycle only as an energy-harvesting mechanism. We have seen that dietary carbohydrates and amino acids enter the pathway at various stages and are oxidized to generate NADH and FADH$_2$, which, by means of oxidative phosphorylation, are used to make ATP.

However, the role of the citric acid cycle in cellular metabolism involves more than just **catabolism.** It plays a key role in **anabolism,** or biosynthesis, as well. Figure 22.13 shows the central role of glycolysis and the citric acid cycle as energy-harvesting reactions, as well as their role as a source of biosynthetic precursors.

As you may already suspect from the fact that amino acids can be converted into citric acid cycle intermediates, these same citric acid cycle intermediates can also be used as starting materials for the synthesis of amino acids. Oxaloacetate provides the carbon skeleton for the one-step synthesis of the amino acid aspartate by the transamination reaction:

$$\text{oxaloacetate} + \text{glutamate} \rightleftharpoons \text{aspartate} + \alpha\text{-ketoglutarate}$$

Aside from providing aspartate for protein synthesis, this reaction provides aspartate for the urea cycle.

Asparagine is made from aspartate by the amination reaction

$$\text{aspartate} + \text{NH}_4^+ + \text{ATP} \longrightarrow \text{asparagine} + \text{AMP} + \text{PP}_i + \text{H}^+$$

The nine amino acids not shown in Figure 22.13 (histidine, isoleucine, leucine, lysine, methionine, phenylalanine, threonine, tryptophan, and valine) are called the essential amino acids (Section 18.11) because they cannot be synthesized by humans. Arginine is an essential amino acid for infants and adults under physical stress.

α-Ketoglutarate serves as the starting carbon chain for the family of amino acids including glutamate, glutamine, proline, and arginine. Glutamate is especially important because it serves as the donor of the α-amino group of almost all other amino acids. It is synthesized from NH$_4^+$ and α-ketoglutarate in a reaction mediated by glutamate dehydrogenase. This is the reverse of the reaction shown in Figure 22.11 and previously described. In this case the coenzyme that serves as the reducing agent is NADPH.

$$\text{NH}_4^+ + \alpha\text{-ketoglutarate} + \text{NADPH} \rightleftharpoons \text{L-glutamate} + \text{NADP}^+ + \text{H}_2\text{O}$$

Glutamine, proline, and arginine are synthesized from glutamate.

Examination of Figure 22.13 reveals that serine, glycine, and cysteine are synthesized from 3-phosphoglycerate; alanine is synthesized from pyruvate; and tyrosine is produced from phosphoenolpyruvate and the four-carbon sugar erythrose-4-phosphate, which, in turn, is synthesized from glucose-6-phosphate in the pentose phosphate pathway. In addition to the amino acid precursors, glycolysis and the citric acid cycle also provide precursors for lipids and the nitrogenous bases that are required to make DNA, the molecule that carries the genetic information. They also generate precursors for heme, the prosthetic group that is required for hemoglobin, myoglobin, and the cytochromes.

Actually, in humans tyrosine is made from the essential amino acid phenylalanine.

Clearly, the reactions of glycolysis and the citric acid cycle are central to both anabolic and catabolic cellular activities. Metabolic pathways that function in both anabolism and catabolism are called **amphibolic pathways.** Consider for a moment the difficulties that the dual nature of these pathways could present to the

22.9 Overview of Anabolism

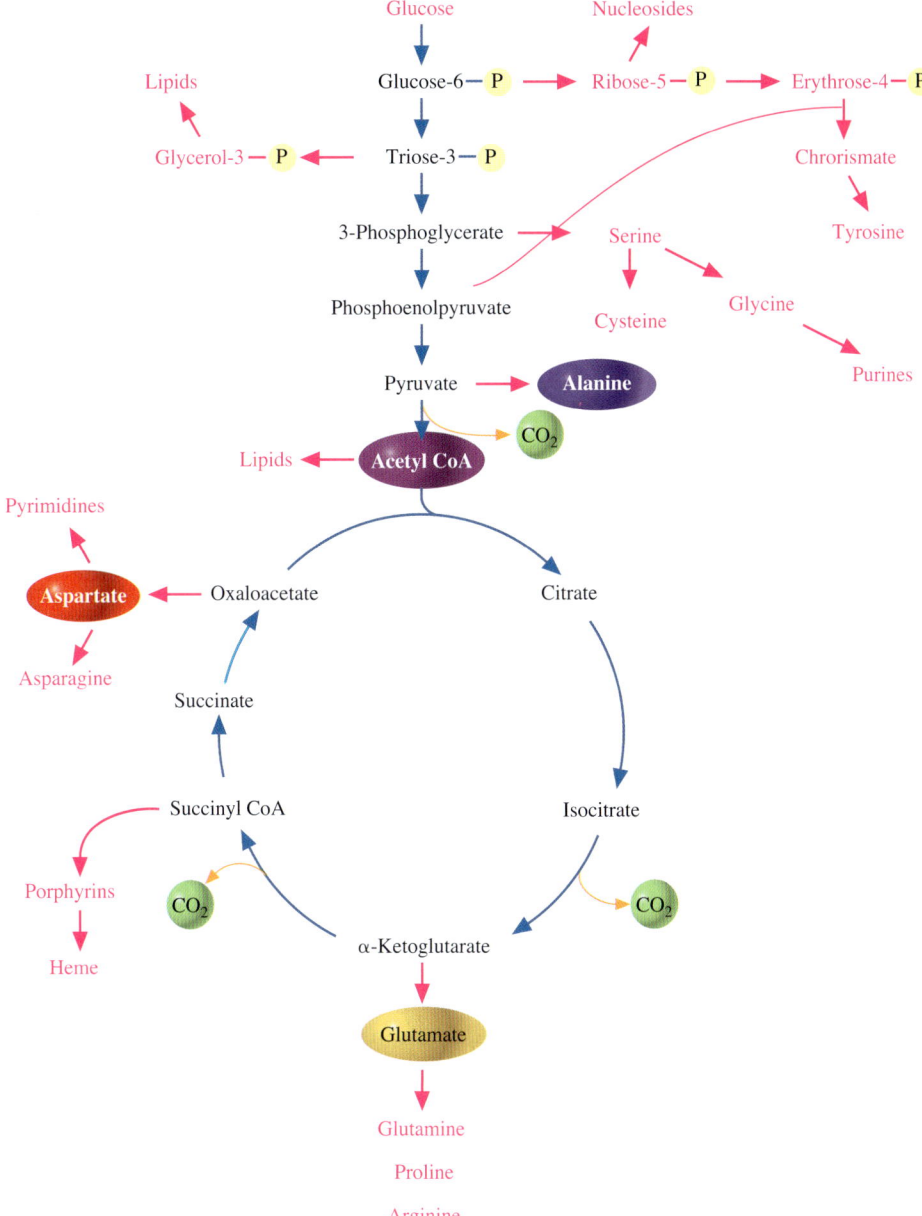

Figure 22.13
Glycolysis, the pentose phosphate pathway, and the citric acid cycle also provide a variety of precursors for the biosynthesis of amino acids, nitrogenous bases, and porphyrins.

cell. When the cell is actively growing, there is a great demand for biosynthetic precursors to build new cell structures. A close look at Figure 22.13 shows us that periods of active cell growth and biosynthesis may deplete the supply of citric acid cycle intermediates. The problem is, the processes of growth and biosynthesis also require a great deal of ATP!

The solution to this problem is to have an alternative pathway for oxaloacetate synthesis that can produce enough oxaloacetate to supply the anabolic and catabolic requirements of the cell. Although bacteria and plants have several mechanisms, the only way that mammalian cells can produce more oxaloacetate is by the carboxylation of pyruvate, a reaction that is also important in gluconeogenesis. This reaction is

$$\text{pyruvate} + CO_2 + ATP \longrightarrow \text{oxaloacetate} + ADP + P_i$$

The enzyme that catalyzes this reaction is *pyruvate carboxylase*. It is a conjugated protein having as its covalently linked prosthetic group the vitamin *biotin*. This enzyme is "turned on" by high levels of acetyl CoA, a signal that the cell requires

Carboxylation of pyruvate during gluconeogenesis is discussed in Section 21.6.

See also A Medical Perspective: Pyruvate Carboxylase Deficiency earlier in this chapter.

high levels of the citric acid cycle intermediates, particularly oxaloacetate, the beginning substrate.

The reaction catalyzed by pyruvate carboxylase is called an **anaplerotic reaction**. The term *anaplerotic* means to fill up. Indeed, this critical enzyme must constantly replenish the oxaloacetate and thus indirectly all the citric acid cycle intermediates that are withdrawn as biosynthetic precursors for the reactions summarized in Figure 22.13.

Question 22.15 Explain how the citric acid cycle serves as an amphibolic pathway.

Question 22.16 What is the function of an anaplerotic reaction?

SUMMARY

22.1 The Mitochondria

The *mitochondria* are aerobic cell organelles that are responsible for most of the ATP production in eukaryotic cells. They are enclosed by a double membrane. The outer membrane permits low-molecular-weight molecules to pass through. The *inner mitochondrial membrane,* by contrast, is almost completely impermeable to most molecules. The inner mitochondrial membrane is the site where *oxidative phosphorylation* occurs. The enzymes of the *citric acid cycle,* of amino acid catabolism, and of fatty acid oxidation are located in the *matrix space* of the mitochondrion.

22.2 Conversion of Pyruvate to Acetyl CoA

Under aerobic conditions, pyruvate is oxidized by the pyruvate dehydrogenase complex. Acetyl CoA, formed in this reaction, is a central molecule in both catabolism and anabolism.

22.3 An Overview of Aerobic Respiration

Aerobic respiration is the oxygen-requiring degradation of food molecules and production of ATP. *Oxidative phosphorylation* is the process that uses the high-energy electrons harvested by oxidation of substrates of the citric acid cycle to produce ATP.

22.4 The Citric Acid Cycle (The Krebs Cycle)

The *citric acid cycle* is the final pathway for the degradation of carbohydrates, amino acids, and fatty acids. The citric acid cycle occurs in the matrix of the mitochondria. It is a cyclic series of biochemical reactions that accomplishes the complete oxidation of the carbon skeletons of food molecules.

22.5 Control of the Citric Acid Cycle

Because the rate of ATP production by the cell must vary with the amount of available oxygen and the energy requirements of the body at any particular time, the citric acid cycle is regulated at several steps. This allows the cell to generate more energy when needed, as for exercise, and less energy when the body is at rest.

22.6 Oxidative Phosphorylation

Oxidative phosphorylation is the process by which NADH and $FADH_2$ are oxidized and ATP is produced. Two molecules of ATP are produced when $FADH_2$ is oxidized, and three molecules of ATP are produced when NADH is oxidized. The complete oxidation of one glucose molecule by glycolysis, the citric acid cycle, and oxidative phosphorylation yields thirty-six molecules of ATP versus two molecules of ATP for anaerobic degradation of glucose by glycolysis and fermentation.

22.7 The Degradation of Amino Acids

Amino acids are oxidized in the mitochondria. The first step of amino acid catabolism is deamination, the removal of the amino group. The carbon skeletons of amino acids are converted into molecules that can enter the citric acid cycle.

22.8 The Urea Cycle

In the *urea cycle* the toxic ammonium ions released by deamination of amino acids are incorporated in urea, which is excreted in the urine.

22.9 Overview of Anabolism: The Citric Acid Cycle as a Source of Biosynthetic Intermediates

In addition to its role in *catabolism*, the citric acid cycle also plays an important role in cellular *anabolism*, or biosynthetic reactions. Many of the citric acid cycle intermediates are precursors for the synthesis of amino acids and macromolecules required by the cell. A pathway that functions in both catabolic and anabolic reactions is called an *amphibolic pathway*.

KEY TERMS

acetyl CoA (22.2)
aerobic respiration (22.3)
amphibolic pathway (22.9)
anabolism (22.9)
anaplerotic reaction (22.9)
ATP synthase (22.6)
catabolism (22.9)
citric acid cycle (22.4)
coenzyme A (22.2)
cristae (22.1)
electron transport system (22.6)
F_0F_1 complex (22.6)
hyperammonemia (22.8)
inner mitochondrial membrane (22.1)
intermembrane space (22.1)
matrix space (22.1)
mitochondria (22.1)
outer mitochondrial membrane (22.1)
oxidative deamination (22.7)
oxidative phosphorylation (22.6)
pyridoxal phosphate (22.7)
pyruvate dehydrogenase complex (22.2)
terminal electron acceptor (22.6)
transaminase (22.7)
transamination (22.7)
urea cycle (22.8)

QUESTIONS AND PROBLEMS

The Mitochondria

Foundations

22.17 Define the term *mitochondrion*.
22.18 Define the term *cristae*.

Applications

22.19 What is the function of the intermembrane compartment of the mitochondria?
22.20 What biochemical processes occur in the matrix space of the mitochondria?
22.21 In what important way do the inner and outer mitochondrial membranes differ?
22.22 What kinds of proteins are found in the inner mitochondrial membrane?

Conversion of Pyruvate to Acetyl CoA

Foundations

22.23 What is coenzyme A?
22.24 What is the role of coenzyme A in the reaction catalyzed by pyruvate dehydrogenase?
22.25 In the reaction catalyzed by pyruvate dehydrogenase, pyruvate is decarboxylated. What is meant by the term *decarboxylation*?
22.26 In the reaction catalyzed by pyruvate dehydrogenase, pyruvate is also oxidized. What substance is reduced when pyruvate is oxidized? What is the product of that reduction reaction?

Applications

22.27 Under what metabolic conditions is pyruvate converted to acetyl CoA?
22.28 Write a chemical equation for the production of acetyl CoA from pyruvate. Under what conditions does this reaction occur?
22.29 How could a deficiency of riboflavin, thiamine, niacin, or pantothenic acid reduce the amount of ATP the body can produce?
22.30 In what form are the vitamins riboflavin, thiamine, niacin, and pantothenic acid needed by the pyruvate dehydrogenase complex?

The Citric Acid Cycle

Foundations

22.31 The reaction catalyzed by citrate synthase is an aldol condensation. Define the term *aldol condensation*.
22.32 The pair of reactions catalyzed by aconitase results in the conversion of isocitrate to its isomer citrate. What are isomers?
22.33 How is oxidation of an organic molecule often recognized?
22.34 The hydroxyl group of isocitrate is oxidized to a ketone, α-ketoglutarate, in the third reaction of the citric acid cycle. Write this reaction and circle the chemical change that reveals that this is an oxidation reaction.
22.35 The reaction catalyzed by succinate dehydrogenase is a dehydrogenation reaction. What is meant by the term *dehydrogenation reaction*?
22.36 The reaction catalyzed by fumarase is an example of the hydration of an alkene to produce an alcohol. Write the equation for this reaction. What is meant by the term *hydration reaction*?
22.37 Label each of the following statements as true or false:
   a. Both glycolysis and the citric acid cycle are aerobic processes.
   b. Both glycolysis and the citric acid cycle are anaerobic processes.
   c. Glycolysis occurs in the cytoplasm, and the citric acid cycle occurs in the mitochondria.
   d. The inner membrane of the mitochondrion is virtually impermeable to most substances.
22.38 Fill in the blanks:
   a. The proteins of the electron transport system are found in the _____, the enzymes of the citric acid cycle are found in the _____, and the hydrogen ion reservoir is found in the _____ of the mitochondria.
   b. The infoldings of the inner mitochondrial membrane are called _____.
   c. Energy released by oxidation in the citric acid cycle is conserved in the form of phosphoanhydride bonds in _____.
   d. The purpose of the citric acid cycle is the _____ of the acetyl group.
22.39 a. To what metabolic intermediate is the acetyl group of acetyl CoA transferred in the citric acid cycle?
   b. What is the product of this reaction?
22.40 To what final products is the acetyl group of acetyl CoA converted during oxidation in the citric acid cycle?
22.41 How many ions of NAD^+ are reduced to molecules of NADH during one turn of the citric acid cycle?
22.42 How many molecules of FAD are converted to $FADH_2$ during one turn of the citric acid cycle?
22.43 What is the net yield of ATP for anaerobic glycolysis?
22.44 How many molecules of ATP are produced by the complete degradation of glucose via glycolysis, the citric acid cycle, and oxidative phosphorylation?

22.45 What is the function of acetyl CoA in the citric acid cycle?
22.46 What is the function of oxaloacetate in the citric acid cycle?
22.47 GTP is formed in one step of the citric acid cycle. How is this GTP converted into ATP?
22.48 What is the chemical meaning of the term *decarboxylation?*

Applications
22.49 The first reaction in the citric acid cycle is an aldol condensation. Write the equation for this reaction and explain its significance.
22.50 A bacterial culture is given ^{14}C-labeled pyruvate as its sole source of carbon and energy. The following is the structure of the radiolabeled pyruvate.

$$*CH_3-\overset{O}{\underset{\|}{C}}-\overset{O}{\underset{\|}{C}}-O^-$$

Follow the fate of the radioactive carbon through the reactions of the citric acid cycle.
22.51 The enzyme aconitase catalyzes the isomerization of citrate into isocitrate. Discuss the two reactions catalyzed by aconitase in terms of the chemistry of alcohols and alkenes.
22.52 A bacterial culture is given ^{14}C-labeled pyruvate as its sole source of carbon and energy. The following is the structure of the radiolabeled pyruvate.

$$CH_3-\overset{O}{\underset{\|}{C^*}}-\overset{O}{\underset{\|}{C}}-O^-$$

Follow the fate of the radioactive carbon through the reactions of the citric acid cycle.
22.53 In the oxidation of malate to oxaloacetate, what is the structural evidence that an oxidation reaction has occurred? What functional groups are involved?
22.54 In the oxidation of succinate to fumarate, what is the structural evidence that an oxidation reaction has occurred? What functional groups are involved?
22.55 To what class of enzymes does dinucleotide diphosphokinase belong? Explain your answer.
22.56 To what class of enzymes does succinate dehydrogenase belong? Explain your answer.

Control of the Citric Acid Cycle
Foundations
22.57 Define the term *allosteric enzyme.*
22.58 Define the term *effector.*
22.59 Define *negative allosterism.*
22.60 Define *positive allosterism.*

Applications
22.61 What four allosteric enzymes or enzyme complexes are responsible for the regulation of the citric acid cycle?
22.62 Which of the four allosteric enzymes or enzyme complexes in the citric acid cycle are under negative allosteric control? Which are under positive allosteric control?
22.63 What is the importance of the regulation of the citric acid cycle?
22.64 Explain the role of allosteric enzymes in control of the citric acid cycle.
22.65 What molecule serves as a signal to increase the rate of the reactions of the citric acid cycle?
22.66 What molecules serve as signals to decrease the rate of the reactions of the citric acid cycle?

Oxidative Phosphorylation
Foundations
22.67 Define the term *electron transport system.*
22.68 What is the terminal electron acceptor in aerobic respiration?

Applications
22.69 How many molecules of ATP are produced when one molecule of NADH is oxidized by oxidative phosphorylation?
22.70 How many molecules of ATP are produced when one molecule of FADH$_2$ is oxidized by oxidative phosphorylation?
22.71 What is the source of energy for the synthesis of ATP in mitochondria?
22.72 What is the name of the enzyme that catalyzes ATP synthesis in mitochondria?
22.73 What is the function of the electron transport systems of the mitochondria?
22.74 What is the cellular location of the electron transport systems?
22.75 a. Compare the number of molecules of ATP produced by glycolysis to the number of ATP molecules produced by oxidation of glucose by aerobic respiration.
  b. Which pathway produces more ATP? Explain.
22.76 At which steps in the citric acid cycle do oxidation–reduction reactions occur?

The Degradation of Amino Acids
Foundations
22.77 What chemical transformation is carried out by transaminases?
22.78 Write a chemical equation for the transfer of an amino group from alanine to α-ketoglutarate, catalyzed by a transaminase.
22.79 Why is the glutamate family of transaminases so important?
22.80 What biochemical reaction is catalyzed by glutamate dehydrogenase?

Applications
22.81 Into which citric acid cycle intermediate is each of the following amino acids converted?
  a. Alanine
  b. Glutamate
  c. Aspartate
  d. Phenylalanine
  e. Threonine
  f. Arginine
22.82 What is the net ATP yield for degradation of each of the amino acids listed in Problem 22.81?
22.83 Write a balanced equation for the synthesis of glutamate that is mediated by the enzyme glutamate dehydrogenase.
22.84 Write a balanced equation for the transamination of aspartate.

The Urea Cycle
22.85 What metabolic condition is produced if the urea cycle does not function properly?
22.86 What is hyperammonemia? How are mild forms of this disease treated?
22.87 The structure of urea is

$$NH_2-\overset{O}{\underset{\|}{C}}-NH_2$$

  a. What substances are the sources of each of the amino groups in the urea molecule?
  b. What substance is the source of the carbonyl group?
22.88 What is the energy source used for the urea cycle?

Overview of Anabolism: The Citric Acid Cycle as a Source of Biosynthetic Intermediates

Foundations
22.89 Define the term *anabolism*.
22.90 Define the term *catabolism*.

Applications
22.91 From which citric acid cycle intermediate is the amino acid glutamate synthesized?
22.92 What amino acids are synthesized from α-ketoglutarate?
22.93 What is the role of the citric acid cycle in biosynthesis?
22.94 How are citric acid cycle intermediates replenished when they are in demand for biosynthesis?
22.95 What is meant by the term *essential amino acid*?
22.96 What are the nine essential amino acids?
22.97 Write a balanced equation for the reaction catalyzed by pyruvate carboxylase.
22.98 How does the reaction described in Problem 22.97 allow the citric acid cycle to fulfill its roles in both catabolism and anabolism?

CRITICAL THINKING PROBLEMS

1. A one-month-old baby boy was brought to the hospital showing severely delayed development and cerebral atrophy. Blood tests showed high levels of lactate and pyruvate. By three months of age, very high levels of succinate and fumarate were found in the urine. Fumarase activity was absent in the liver and muscle tissue. The baby died at five months of age. This was the first reported case of fumarase deficiency and the defect was recognized too late for effective therapy to be administered. What reaction is catalyzed by fumarase? How would a deficiency of this mitochondrial enzyme account for the baby's symptoms and test results?

2. A certain bacterium can grow with ethanol as its only source of energy and carbon. Propose a pathway to describe how ethanol can enter a pathway that would allow ATP production and synthesis of precursors for biosynthesis.

3. Fluoroacetate has been used as a rat poison and can be fatal when eaten by humans. Patients with fluoroacetate poisoning accumulate citrate and fluorocitrate within the cells. What enzyme is inhibited by fluoroacetate? Explain your reasoning.

4. The pyruvate dehydrogenase complex is activated by removal of a phosphoryl group from pyruvate dehydrogenase. This reaction is catalyzed by the enzyme pyruvate dehydrogenase phosphate phosphatase. A baby is born with a defect in this enzyme. What effects would this defect have on the rate of each of the following pathways: aerobic respiration, glycolysis, lactate fermentation? Explain your reasoning.

5. Pyruvate dehydrogenase phosphate phosphatase is stimulated by Ca^{2+}. In muscles, the Ca^{2+} concentration increases dramatically during muscle contraction. How would the elevated Ca^{2+} concentration affect the rate of glycolysis and the citric acid cycle?

6. Liver contains high levels of nucleic acids. When excess nucleic acids are degraded, ribose-5-phosphate is one of the degradation products that accumulate in the cell. Can this substance be used as a source of energy? What pathway would be used?

7. In birds, arginine is an essential amino acid. Can birds produce urea as a means of removing ammonium ions from the blood? Explain your reasoning.

BIOCHEMISTRY

Fatty Acid Metabolism

23

A tasty source of energy.

Learning Goals

1. Summarize the digestion and storage of lipids.
2. Describe the degradation of fatty acids by β-oxidation.
3. Explain the role of acetyl CoA in fatty acid metabolism.
4. Understand the role of ketone body production in β-oxidation.
5. Compare β-oxidation of fatty acids and fatty acid biosynthesis.
6. Describe the regulation of lipid and carbohydrate metabolism in relation to the liver, adipose tissue, muscle tissue, and the brain.
7. Summarize the antagonistic effects of glucagon and insulin.

Outline

Chemistry Connection:
Obesity: A Genetic Disorder?

23.1 Lipid Metabolism in Animals
23.2 Fatty Acid Degradation

A Human Perspective:
Losing Those Unwanted Pounds of Adipose Tissue

23.3 Ketone Bodies

A Medical Perspective:
Diabetes Mellitus and Ketone Bodies

23.4 Fatty Acid Synthesis
23.5 The Regulation of Lipid and Carbohydrate Metabolism
23.6 The Effects of Insulin and Glucagon on Cellular Metabolism

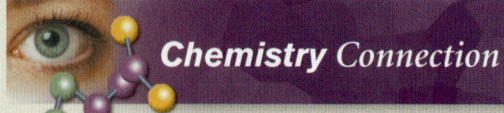

Chemistry Connection

Obesity: A Genetic Disorder?

Approximately a third of all Americans are obese; that is, they are more than 20% overweight. One million are morbidly obese; they carry so much extra weight that it threatens their health. Many obese people simply eat too much and exercise too little, but others actually gain weight even though they eat fewer calories than people of normal weight. This observation led many researchers to the hypothesis that obesity in some people is a genetic disorder.

This hypothesis was supported by the 1950 discovery of an obesity mutation in mice. Selective breeding produced a strain of genetically obese mice from the original mutant mouse. The hypothesis was further strengthened by the results of experiments performed in the 1970s by Douglas Coleman. Coleman connected the circulatory systems of a genetically obese mouse and a normal mouse. The obese mouse started eating less and lost weight. Coleman concluded that there was a substance in the blood of normal mice that signals the brain to decrease the appetite. Obese mice, he hypothesized, can't produce this "satiety factor," and thus they continue to eat and gain weight.

In 1987, Jeffrey Friedman assembled a team of researchers to map and then clone the obesity gene that was responsible for appetite control. In 1994, after seven years of intense effort, the scientists achieved their goal, but they still had to demonstrate that the protein encoded by the cloned obesity gene did, indeed, have a metabolic effect. The gene was modified to be compatible with the genetic system of bacteria so that they could be used to manufacture the protein. When the engineered gene was then introduced into bacteria, they produced an abundance of the protein product. The protein was then purified in preparation for animal testing.

The researchers calculated that a normal mouse has about 12.5 mg of the protein in its blood. They injected that amount into each of ten mice that were so fat they couldn't squeeze into the feeding tunnels used for normal mice. The day after the first injection, graduate student Jeff Halaas observed that the mice had eaten less food. Injections were given daily, and each day the obese mice ate less. After two weeks of treatment, each of the ten mice had lost about 30% of its weight. In addition, the mice had become more active and their metabolisms had speeded up.

When normal mice underwent similar treatment, their body fat fell from 12.2% to 0.67%, which meant that these mice had no extra fat tissue. The 0.67% of their body weight represented by fat was accounted for by the membranes that surround each of the cells of their bodies! Because of the dramatic results, Friedman and his colleagues called the protein leptin, from the Greek word *leptos,* meaning slender.

The leptin protein is a hormone, and ongoing research is aimed at understanding how leptin works to control metabolism and food intake. Friedman has hypothesized that it is a signal in a metabolic thermostat. Fat cells produce leptin and secrete it into the bloodstream. As a result, the leptin concentration in a normal person is proportional to the amount of body fat. The blood concentration of the hormone is monitored by a center in the brain, probably the hypothalamus, a region known to control appetite and set metabolic rates. When the concentration reaches a certain level, it triggers the hypothalamus to suppress the appetite. If a genetically obese person, or mouse, produces no leptin or only small amounts of it, the hypothalamus "thinks" that the individual has too little body fat or is starving. Under these circumstances it does not send a signal to suppress hunger and the individual continues to eat.

The human leptin gene also has been cloned and shown to correct genetic obesity in mice. But what about obesity in humans? Leptin has been tested in a small number of obese individuals. Unfortunately, the dramatic results achieved with mice were *not* observed with humans. Why? It seems that nearly all of the obese volunteers already produced an abundance of leptin. In fact, fewer than ten people have been found, to date, who do not produce leptin. For these individuals, leptin injections do, indeed, reduce their appetites and lead to significant weight loss.

For the majority of obese humans, who produce leptin, perhaps a genetic defect exists in the leptin receptor. Or perhaps the genetics of obesity in humans is more complex than in mice. Clearly lipid metabolism in animals is a complex process and is not yet fully understood. The discovery of the leptin gene, and the hormone it produces, is just one part of the story. In this chapter we will study other aspects of lipid metabolism: the pathways for fatty acid degradation and biosynthesis and the processes by which dietary lipids are digested and excess lipids are stored.

Introduction

The metabolism of fatty acids and lipids revolves around the fate of acetyl CoA. We saw in Chapter 22 that, under aerobic conditions, pyruvate is converted to acetyl CoA, which feeds into the citric acid cycle. Fatty acids are also degraded to acetyl CoA and oxidized by the citric acid cycle, as are certain amino acids. Moreover, acetyl CoA is itself the starting material for the biosynthesis of fatty acids, cholesterol, and steroid hormones. Acetyl CoA is thus a key intermediary in lipid metabolism.

23.1 Lipid Metabolism in Animals

Digestion and Absorption of Dietary Triglycerides

Triglycerides are highly hydrophobic ("water fearing"). Because of this they must be processed before they can be digested, absorbed, and metabolized. Because processing of dietary lipids occurs in the small intestine, the water soluble **lipases**, enzymes that hydrolyze triglycerides, that are found in the stomach and in the saliva are not very effective. In fact, most dietary fat arrives in the duodenum, the first part of the small intestine, in the form of fat globules. These fat globules stimulate the secretion of bile from the gallbladder. **Bile** is composed of micelles of lecithin, cholesterol, protein, bile salts, inorganic ions, and bile pigments. **Micelles** (Figure 23.1) are aggregations of molecules having a polar region and a nonpolar region. The nonpolar ends of bile salts tend to bunch together when placed in water. The hydrophilic ("water loving") regions of these molecules interact with water. Bile salts are made in the liver and stored in the gallbladder, awaiting the stimulus to be secreted into the duodenum. The major bile salts in humans are cholate and chenodeoxycholate (Figure 23.2).

LEARNING GOAL

See Sections 14.1 and 17.2 for a discussion of micelles.

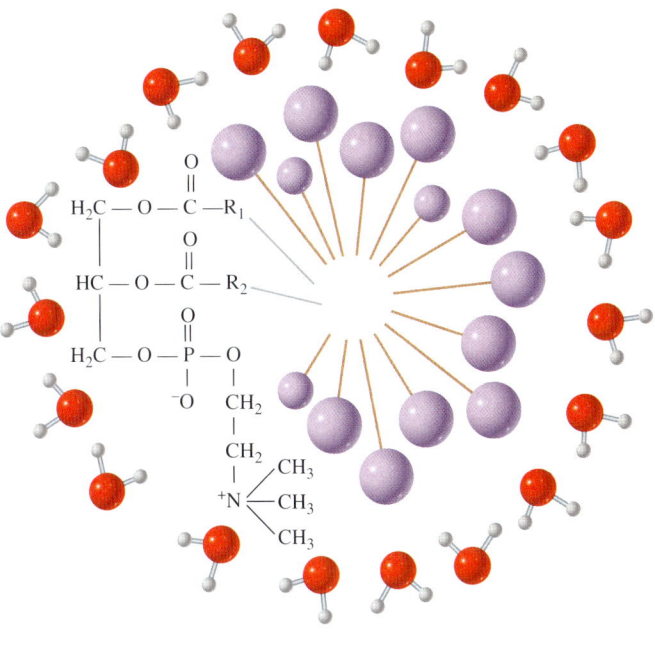

Figure 23.1
The structure of a micelle formed from the phospholipid lecithin. The straight lines represent the long hydrophobic fatty acid tails, and the spheres represent the hydrophilic heads of the phospholipid.

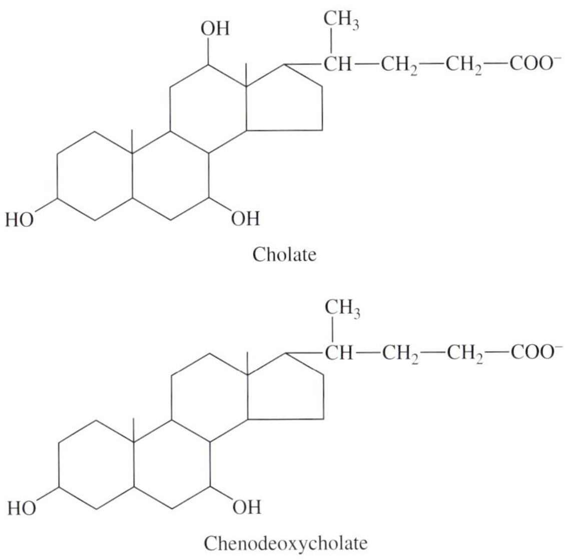

Figure 23.2
Structures of the most common bile acids in human bile: cholate and chenodeoxycholate.

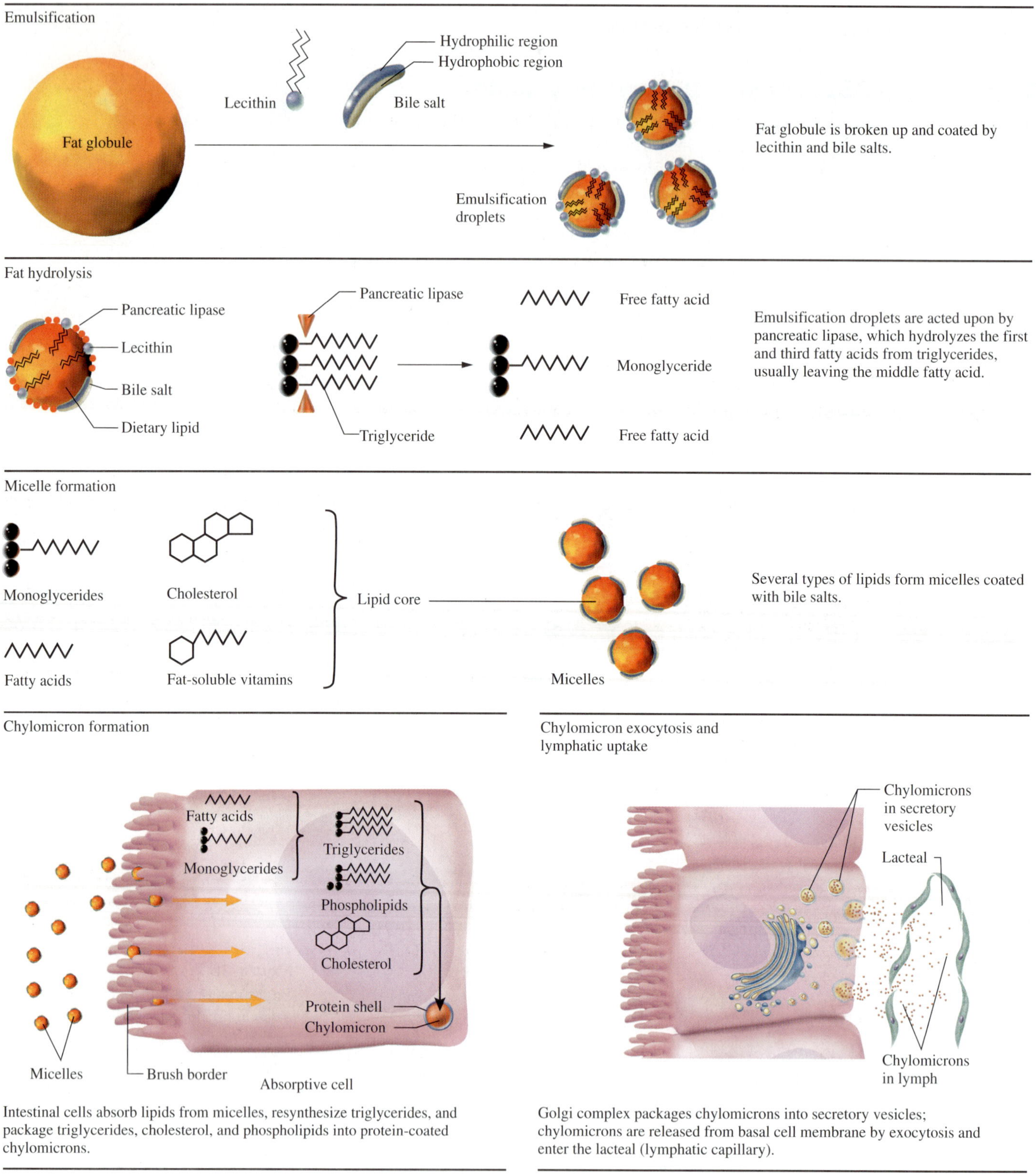

Emulsification
Hydrophilic region
Hydrophobic region
Lecithin
Bile salt
Fat globule
Emulsification droplets
Fat globule is broken up and coated by lecithin and bile salts.
Fat hydrolysis
Pancreatic lipase
Lecithin
Bile salt
Dietary lipid
Pancreatic lipase
Triglyceride
Free fatty acid
Monoglyceride
Free fatty acid
Emulsification droplets are acted upon by pancreatic lipase, which hydrolyzes the first and third fatty acids from triglycerides, usually leaving the middle fatty acid.
Micelle formation
Monoglycerides
Cholesterol
Fatty acids
Fat-soluble vitamins
Lipid core
Micelles
Several types of lipids form micelles coated with bile salts.
Chylomicron formation
Fatty acids
Monoglycerides
Triglycerides
Phospholipids
Cholesterol
Protein shell
Chylomicron
Micelles
Brush border
Absorptive cell
Intestinal cells absorb lipids from micelles, resynthesize triglycerides, and package triglycerides, cholesterol, and phospholipids into protein-coated chylomicrons.
Chylomicron exocytosis and lymphatic uptake
Chylomicrons in secretory vesicles
Lacteal
Chylomicrons in lymph
Golgi complex packages chylomicrons into secretory vesicles; chylomicrons are released from basal cell membrane by exocytosis and enter the lacteal (lymphatic capillary).

Figure 23.3
Stages of lipid digestion in the intestinal tract.

Cholesterol is almost completely insoluble in water, but the conversion of cholesterol to bile salts creates *detergents* whose polar heads make them soluble in the aqueous phase and whose hydrophobic tails bind triglycerides. After a meal is eaten, bile flows through the common bile duct into the duodenum, where bile salts emulsify the fat globules into tiny droplets. This increases the surface area of the lipid molecules, allowing them to be more easily hydrolyzed by lipases (Figure 23.3).

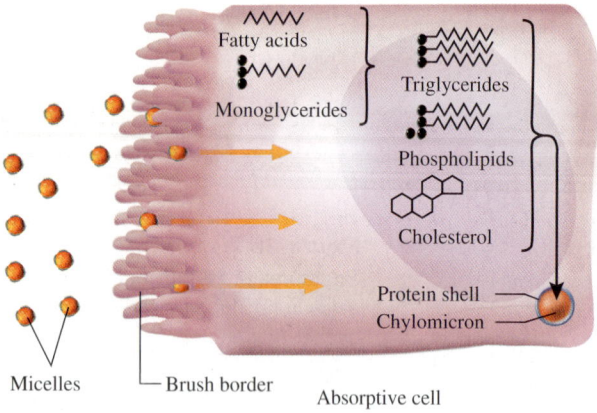

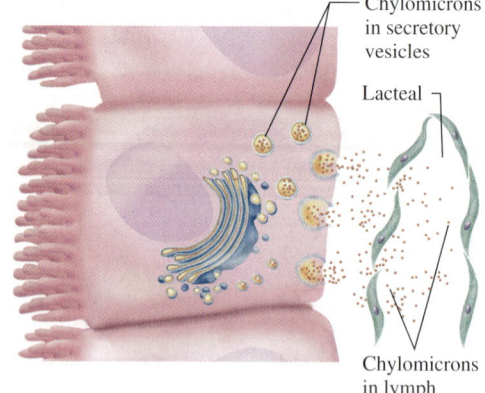

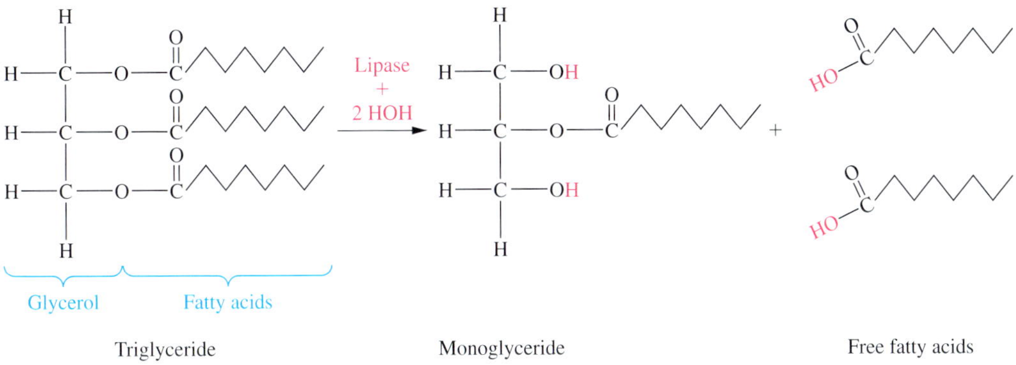

23.1 Lipid Metabolism in Animals

Figure 23.4
The action of pancreatic lipase in the hydrolysis of dietary lipids.

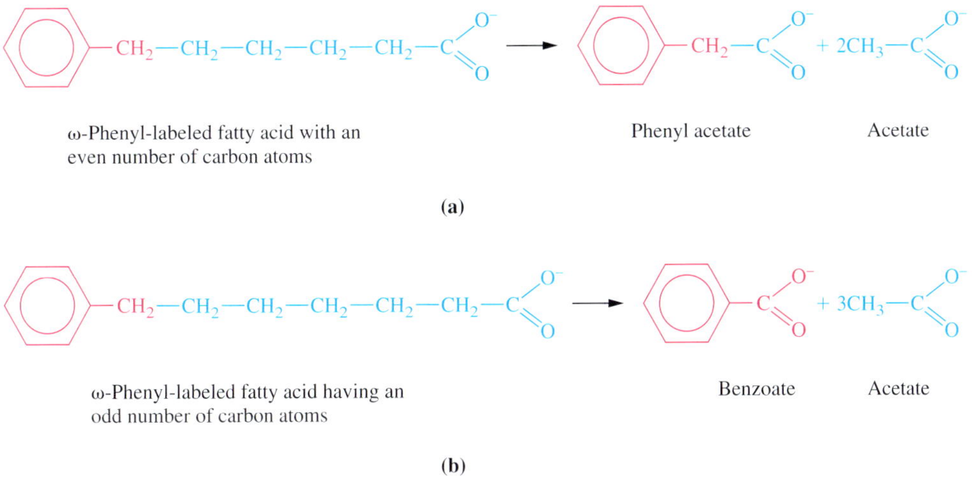

Figure 23.5
The last carbon of the chain is called the ω-carbon (omega-carbon), so the attached phenyl group is an ω-phenyl group. (a) Oxidation of ω-phenyl-labeled fatty acids occurs two carbons at a time. Fatty acids having an even number of carbon atoms are degraded to phenyl acetate and "acetate." (b) Oxidation of ω-phenyl-labeled fatty acids that contain an odd number of carbon atoms yields benzoate and "acetate."

Much of the lipid in these droplets is in the form of **triglycerides,** or triacylglycerols, which are fatty acid esters of glycerol. A protein called **colipase** binds to the surface of the lipid droplets and helps pancreatic lipases to stick to the surface and hydrolyze the ester bonds between the glycerol and fatty acids of the triglycerides (Figure 23.4). In this process, two of the three fatty acids are liberated, and the monoglycerides and free fatty acids produced mix freely with the micelles of bile. These micelles are readily absorbed through the membranes of the intestinal epithelial cells (Figure 23.5).

Surprisingly, the monoglycerides and fatty acids are then reassembled into triglycerides that are combined with protein to produce the class of plasma lipoproteins called **chylomicrons** (Figure 23.3). These collections of lipid and protein are secreted into small lymphatic vessels and eventually arrive in the bloodstream. In the bloodstream the triglycerides are once again hydrolyzed to produce glycerol and free fatty acids that are then absorbed by the cells. If the body needs energy, these molecules are degraded to produce ATP. If the body does not need energy, these energy-rich molecules are stored.

Triglycerides are described in Section 17.3.

Plasma lipoproteins are described in Section 17.5.

Lipid Storage

Fatty acids are stored in the form of triglycerides. Most of the body's triglyceride molecules are stored as fat droplets in the cytoplasm of **adipocytes** (fat cells) that make up **adipose tissue.** Each adipocyte contains a large fat droplet that accounts for nearly the entire volume of the cell. Other cells, such as those of cardiac muscle, contain a few small fat droplets. In these cells the fat droplets are surrounded by mitochondria. When the cells need energy, triglycerides are hydrolyzed to release fatty acids that are transported into the matrix space of the mitochondria. There the fatty acids are completely oxidized, and ATP is produced.

The fatty acids provided by the hydrolysis of triglycerides are a very rich energy source for the body. The complete oxidation of fatty acids releases much more energy than the oxidation of a comparable amount of glycogen.

Question 23.1

How do bile salts aid in the digestion of dietary lipids?

Question 23.2

Why must dietary lipids be processed before enzymatic digestion can be effective?

23.2 Fatty Acid Degradation

An Overview of Fatty Acid Degradation

LEARNING GOAL 2

LEARNING GOAL 3

Early in the twentieth century, a very clever experiment was done to determine how fatty acids are degraded. Recall from Chapter 9 that radioactive elements can be attached to biological molecules and followed through the body. A German biochemist, Franz Knoop, devised a similar kind of labeling experiment long before radioactive tracers were available. Knoop fed dogs fatty acids in which the usual terminal methyl group had a phenyl group attached to it. Such molecules are called ω-labeled (omega-labeled) fatty acids (Figure 23.5). When he isolated the metabolized fatty acids from the urine of the dogs, he found that phenyl acetate was formed when the fatty acid had an even number of carbon atoms in the chain. But benzoate was formed when the fatty acid had an odd number of carbon atoms. Knoop interpreted these data to mean that the degradation of fatty acids occurs by the removal of two-carbon acetate groups from the carboxyl end of the fatty acid. We now know that the two-carbon fragments produced by the degradation of fatty acids are not acetate, but acetyl CoA. The pathway for the breakdown of fatty acids into acetyl CoA is called **β-oxidation.**

This pathway is called β-oxidation because it involves the stepwise oxidation of the β-carbon of the fatty acid.

The β-oxidation cycle (steps 2–5, Figure 23.6) consists of a set of four reactions whose overall form is similar to the last four reactions of the citric acid cycle. Each trip through the sequence of reactions releases acetyl CoA and returns a fatty acyl CoA molecule that has two fewer carbons. One molecule of $FADH_2$, equivalent to two ATP molecules, and one molecule of NADH, equivalent to three ATP molecules, are produced for each cycle of β-oxidation.

Review Section 22.6 for the ATP yields that result from oxidation of $FADH_2$ and NADH.

EXAMPLE 23.1 Predicting the Products of β-Oxidation of a Fatty Acid

What products would be produced by the β-oxidation of 10-phenyldecanoic acid?

Solution

This ten-carbon fatty acid would be broken down into four acetyl CoA molecules and one phenyl acetate molecule. Because four cycles through β-oxidation are required to break down a ten-carbon fatty acid, four NADH molecules and four $FADH_2$ molecules would also be produced.

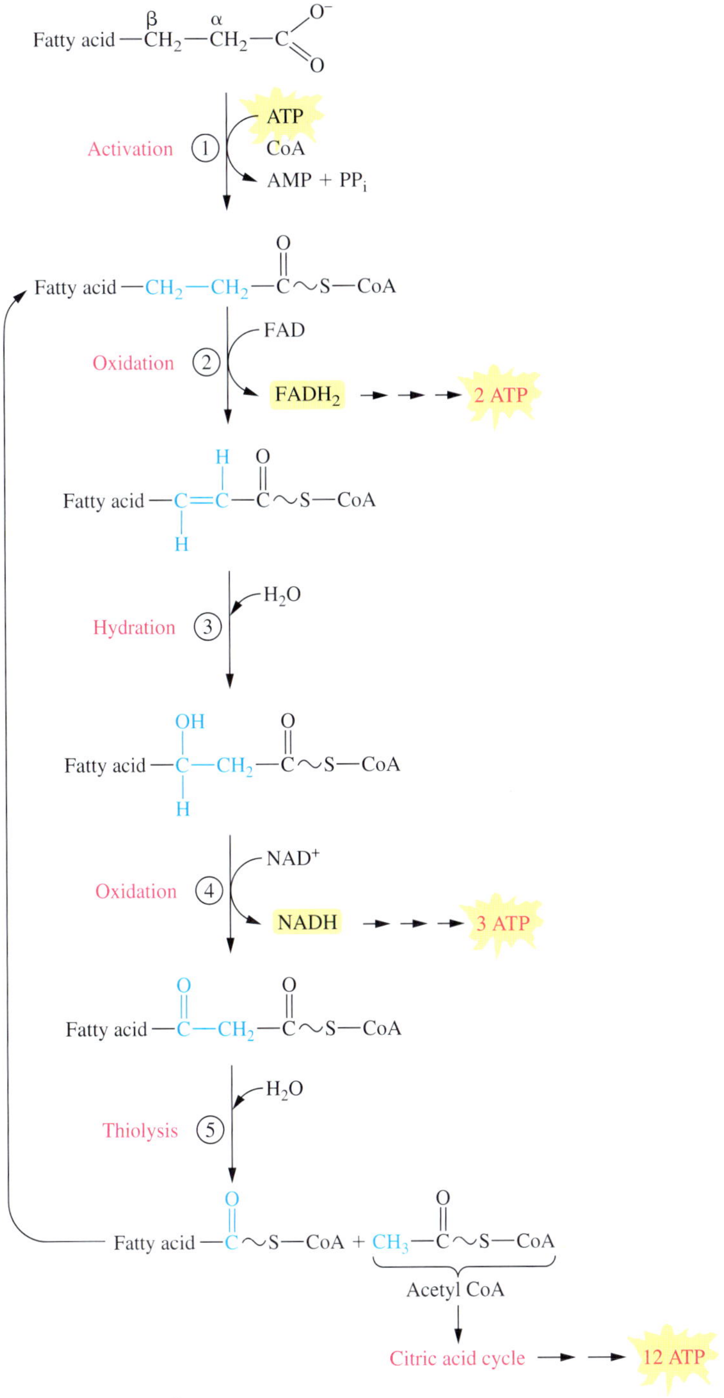

23.2 Fatty Acid Degradation

Figure 23.6
The reactions in β-oxidation of fatty acids.

Fatty acid—$\overset{\beta}{CH_2}$—$\overset{\alpha}{CH_2}$—$C\overset{O^-}{\underset{O}{\diagup}}$

Activation ① ATP, CoA → AMP + PP$_i$

Fatty acid—CH_2—CH_2—$\overset{O}{\underset{\|}{C}}$~S—CoA

Oxidation ② FAD → FADH$_2$ → → → 2 ATP

Fatty acid—$\overset{H}{\underset{H}{C}}$=$C$—$\overset{O}{\underset{\|}{C}}$~S—CoA

Hydration ③ H$_2$O

Fatty acid—$\overset{OH}{\underset{H}{C}}$—$CH_2$—$\overset{O}{\underset{\|}{C}}$~S—CoA

Oxidation ④ NAD$^+$ → NADH → → → 3 ATP

Fatty acid—$\overset{O}{\underset{\|}{C}}$—$CH_2$—$\overset{O}{\underset{\|}{C}}$~S—CoA

Thiolysis ⑤ H$_2$O

Fatty acid—$\overset{O}{\underset{\|}{C}}$~S—CoA + CH_3—$\overset{O}{\underset{\|}{C}}$~S—CoA

Acetyl CoA

Citric acid cycle → → 12 ATP

Question 23.3

What products would be formed by β-oxidation of each of the following fatty acids? (*Hint:* Refer to Example 23.1.)

a. 9-Phenylnonanoic acid
b. 8-Phenyloctanoic acid
c. 7-Phenylheptanoic acid
d. 12-Phenyldodecanoic acid

A Human Perspective

Losing Those Unwanted Pounds of Adipose Tissue

Weight, or overweight, is a topic of great concern to the American populace. A glance through almost any popular magazine quickly informs us that by today's standards, "beautiful" is synonymous with "thin." The models in all these magazines are extremely thin, and there are literally dozens of ads for weight-loss programs. Americans spend millions of dollars each year trying to attain this slim ideal of the fashion models.

Studies have revealed that this slim ideal is often below a desirable, healthy body weight. In fact, the suggested weight for a 6-foot tall male between 18 and 39 years of age is 179 pounds. For a 5'6" female in the same age range, the desired weight is 142 pounds. For a 5'1" female, 126 pounds is recommended. Just as being too thin can cause health problems, so too can obesity.

What is obesity, and does it have disadvantages beyond aesthetics? An individual is considered to be obese if his or her body weight is more than 20% above the ideal weight for his or her height. The accompanying table lists desirable body weights, according to sex, age, height, and body frame.

Overweight carries with it a wide range of physical problems, including elevated blood cholesterol levels; high blood pressure; increased incidence of diabetes, cancer, and heart disease; and increased probability of early death. It often causes psychological problems as well, such as guilt and low self-esteem.

Many factors may contribute to obesity. These include genetic factors, a sedentary lifestyle, and a preference for high-calorie, high-fat foods. However, the real concern is how to lose weight. How can we lose weight wisely and safely and

Men*					Women**				
Height					Height				
Feet	Inches	Small Frame	Medium Frame	Large Frame	Feet	Inches	Small Frame	Medium Frame	Large Frame
5	2	128–134	131–141	138–150	4	10	102–111	109–121	118–131
5	3	130–136	133–143	140–153	4	11	103–113	111–123	120–134
5	4	132–138	135–145	142–156	5	0	104–115	113–126	122–137
5	5	134–140	137–148	144–160	5	1	106–118	115–129	125–140
5	6	136–142	139–151	146–164	5	2	108–121	118–132	128–143
5	7	138–145	142–154	149–168	5	3	111–124	121–135	131–147
5	8	140–148	145–157	152–172	5	4	114–127	124–138	134–151
5	9	142–151	148–160	155–176	5	5	117–130	127–141	137–155
5	10	144–154	151–163	158–180	5	6	120–133	130–144	140–159
5	11	146–157	154–166	161–184	5	7	123–136	133–147	143–163
6	0	149–160	157–170	164–188	5	8	126–139	136–150	146–167
6	1	152–164	160–174	168–192	5	9	129–142	139–153	149–170
6	2	155–168	164–178	172–197	5	10	132–145	142–156	152–173
6	3	158–172	167–182	176–202	5	11	135–148	145–159	155–176
6	4	162–176	171–187	181–207	6	0	138–151	148–162	158–179

*Weights at ages 25–59 based on lowest mortality. Weight in pounds according to frame (in indoor clothing weighing 5 lb, shoes with 1" heels).
**Weights at ages 25–59 based on lowest mortality. Weight in pounds according to frame (in indoor clothing weighing 3 lb, shoes with 1" heels).
Reprinted with permission of the Metropolitan Life Insurance Companies *Statistical Bulletin*.

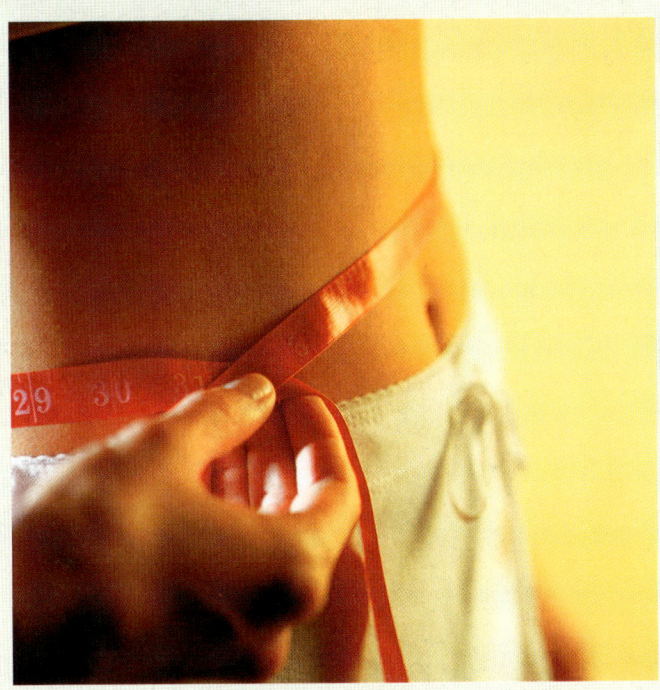

Reduced caloric intake and exercise are the keys to permanent weight loss.

keep the weight off for the rest of our lives? Unfortunately, the answer is *not* the answer that most people want to hear. The prevalence and financial success of the quick-weight-loss programs suggest that the majority of people want a program that is rapid and effortless. Unfortunately, most programs that promise dramatic weight reduction with little effort are usually ineffective or, worse, unsafe. The truth is that weight loss and management are best obtained by a program involving three elements.

1. *Reduced caloric intake.* A pound of body fat is equivalent to 3500 Calories (kilocalories). So if you want to lose 2 pounds each week, a reasonable goal, you must reduce your caloric intake by 1000 Calories per day. Remember that diets recommending fewer than 1200 Calories per day are difficult to maintain because they are not very satisfying and may be unsafe because they don't provide all the required vitamins and minerals. The best way to decrease Calories is to reduce fat and increase complex carbohydrates in the diet.

2. *Exercise.* Increase energy expenditures by 200–400 Calories each day. You may choose walking, running, or mowing the lawn; the type of activity doesn't matter, as long as you get moving. Exercise has additional benefits. It increases cardiovascular fitness, provides a psychological lift, and may increase the base rate at which you burn calories after exercise is finished.

3. *Behavior modification.* For some people, overweight is as much a psychological problem as it is a physical problem, and half the battle is learning to recognize the triggers that cause overeating. Several principles of behavior modification have been found to be very helpful.
   a. Keep a diary. Record the amount of foods eaten and the circumstances—for instance, a meal at the kitchen table or a bag of chips in the car on the way home.
   b. Identify your eating triggers. Do you eat when you feel stress, boredom, fatigue, joy?
   c. Develop a plan for avoiding or coping with your trigger situations or emotions. You might exercise when you feel that stress-at-the-end-of-the-day trigger or carry a bag of carrot sticks for the midmorning-boredom trigger.
   d. Set realistic goals, and reward yourself when you reach them. The reward should not be food related.

As you can see, there is no "quick fix" for safe, effective weight control. A commitment must be made to modify existing diet and exercise habits. Most important, those habits must be avoided forever and replaced by new, healthier behaviors and attitudes.

For Further Understanding

In terms of the energy-harvesting reactions we have studied in Chapters 22 and 23, explain how reduced caloric intake and increase in activity level contribute to weight loss.

If you increased your energy expenditure by 200 Calories per day and did not change your eating habits, how long would it take you to lose 10 pounds?

Question 23.4

What does ω refer to in the naming of ω-phenyl-labeled fatty acids?

The Reactions of β-Oxidation

LEARNING GOAL 2

The enzymes that catalyze the β-oxidation of fatty acids are located in the matrix space of the mitochondria. Special transport mechanisms are required to bring fatty acid molecules into the mitochondrial matrix. Once inside, the fatty acids are degraded by the reactions of β-oxidation. As we will see, these reactions interact with oxidative phosphorylation and the citric acid cycle to produce ATP.

Reaction 1. The first step is an *activation* reaction that results in the production of a fatty acyl CoA molecule. A thioester bond is formed between coenzyme A and the fatty acid:

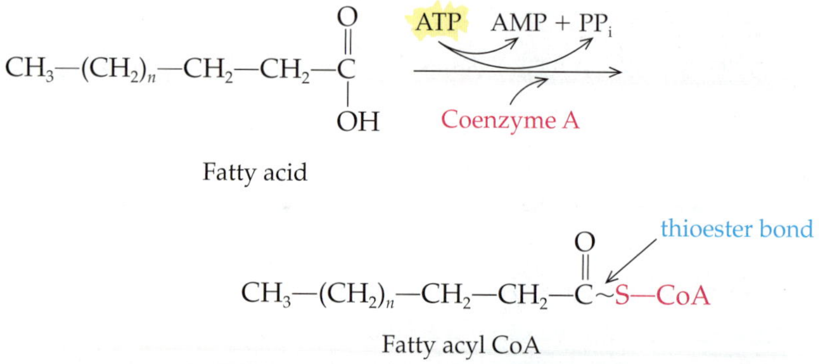

$$CH_3-(CH_2)_n-CH_2-CH_2-\overset{O}{\underset{OH}{C}} \xrightarrow[\text{Coenzyme A}]{ATP \quad AMP + PP_i}$$

Fatty acid

$$CH_3-(CH_2)_n-CH_2-CH_2-\overset{O}{C}\sim S-CoA \quad \text{thioester bond}$$

Fatty acyl CoA

This reaction requires energy in the form of ATP, which is cleaved to AMP and pyrophosphate. This involves hydrolysis of two phosphoanhydride bonds. Here again we see the need to invest a small amount of energy so that a much greater amount of energy can be harvested later in the pathway. Coenzyme A is also required for this step. The product, a fatty acyl CoA, has a *high-energy* thioester bond between the fatty acid and coenzyme A. *Acyl-CoA ligase*, which catalyzes this reaction, is located in the outer membrane of the mitochondria. The mechanism that brings the fatty acyl CoA into the mitochondrial matrix involves a carrier molecule called *carnitine*. The first step, catalyzed by the enzyme *carnitine acyltransferase I*, is the transfer of the fatty acyl group to carnitine, producing acylcarnitine and coenzyme A. Next a carrier protein located in the mitochondrial inner membrane transfers the acylcarnitine into the mitochondrial matrix. There *carnitine acyltransferase II* catalyzes the regeneration of fatty acyl CoA, which now becomes involved in the remaining reactions of β-oxidation.

Acyl group transfer reactions are described in Section 14.4.

Reaction 2. The next reaction is an *oxidation* reaction that removes a pair of hydrogen atoms from the fatty acid. These are used to reduce FAD to produce FADH$_2$. This *dehydrogenation* reaction is catalyzed by the enzyme *acyl-CoA dehydrogenase* and results in the formation of a carbon-carbon double bond:

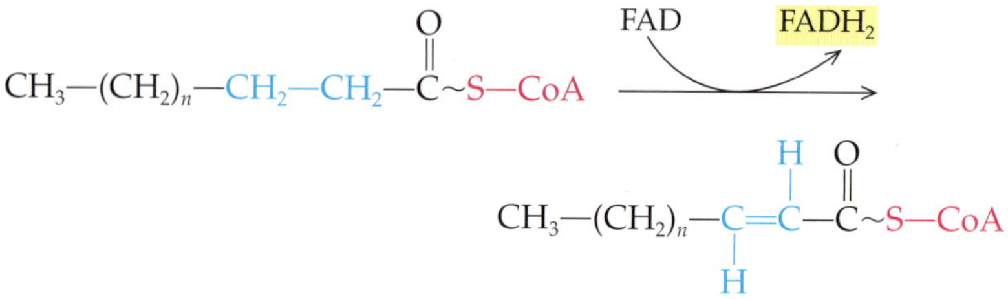

$$CH_3-(CH_2)_n-CH_2-CH_2-\overset{O}{C}\sim S-CoA \xrightarrow{FAD \quad FADH_2}$$

$$CH_3-(CH_2)_n-\underset{H}{\overset{H}{C}}=C-\overset{O}{C}\sim S-CoA$$

23.2 Fatty Acid Degradation

Oxidative phosphorylation yields two ATP molecules for each molecule of FADH$_2$ produced by this oxidation–reduction reaction.

Reaction 3. The third reaction involves the *hydration* of the double bond produced in reaction 2. As a result the β-carbon is hydroxylated. This reaction is catalyzed by the enzyme *enoyl-CoA hydrase*.

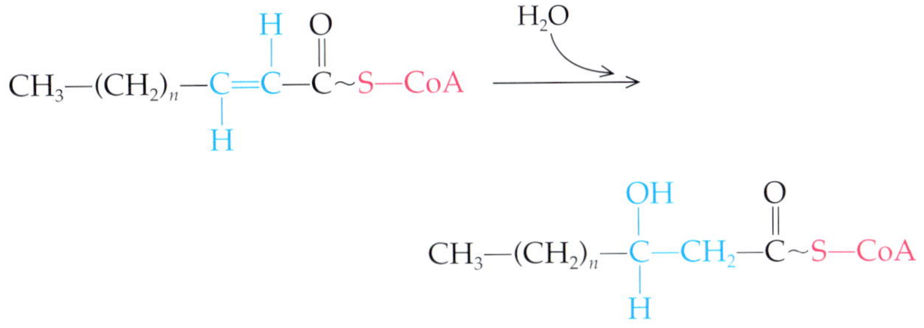

$$CH_3-(CH_2)_n-\underset{H}{\overset{H}{C}}=\underset{}{\overset{}{C}}-\overset{O}{\underset{}{\overset{\|}{C}}}\sim S-CoA \xrightarrow{H_2O}$$

$$CH_3-(CH_2)_n-\underset{H}{\overset{OH}{\underset{|}{C}}}-CH_2-\overset{O}{\underset{}{\overset{\|}{C}}}\sim S-CoA$$

Reaction 4. In this *oxidation* reaction the hydroxyl group of the β-carbon is now dehydrogenated. NAD$^+$ is reduced to form NADH that is subsequently used to produce three ATP molecules by oxidative phosphorylation. L-β-*Hydroxyacyl-CoA dehydrogenase* catalyzes this reaction.

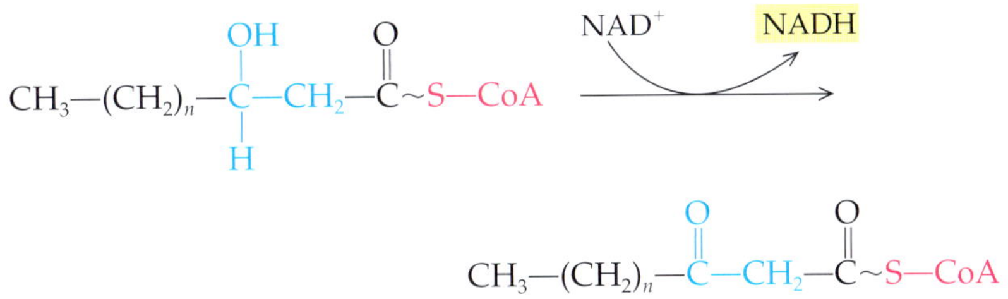

$$CH_3-(CH_2)_n-\underset{H}{\overset{OH}{\underset{|}{C}}}-CH_2-\overset{O}{\overset{\|}{C}}\sim S-CoA \xrightarrow{NAD^+ \quad NADH}$$

$$CH_3-(CH_2)_n-\overset{O}{\overset{\|}{C}}-CH_2-\overset{O}{\overset{\|}{C}}\sim S-CoA$$

Reaction 5. The final reaction, catalyzed by the enzyme *thiolase*, is the cleavage that releases acetyl CoA. This is accomplished by *thiolysis*, attack of a molecule of coenzyme A on the β-carbon. The result is the release of acetyl CoA and a fatty acyl CoA that is two carbons shorter than the beginning fatty acid:

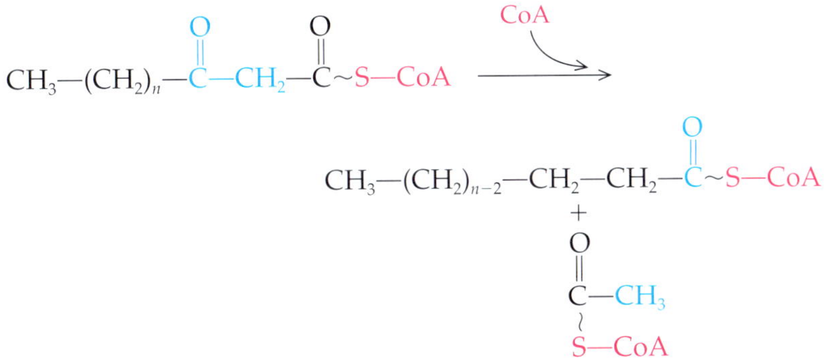

$$CH_3-(CH_2)_n-\overset{O}{\overset{\|}{C}}-CH_2-\overset{O}{\overset{\|}{C}}\sim S-CoA \xrightarrow{CoA}$$

$$CH_3-(CH_2)_{n-2}-CH_2-CH_2-\overset{O}{\overset{\|}{C}}\sim S-CoA$$
$$+$$
$$\overset{O}{\overset{\|}{C}}-CH_3$$
$$\underset{S-CoA}{|}$$

The shortened fatty acyl CoA is further oxidized by cycling through reactions 2–5 until the fatty acid carbon chain is completely degraded to acetyl CoA. The acetyl CoA produced by β-oxidation of fatty acids then enters the reactions of the citric acid cycle. Of course, this eventually results in the production of 12 ATP molecules per molecule of acetyl CoA released during β-oxidation.

As an example of the energy yield from β-oxidation, the balance sheet for ATP production when the sixteen-carbon-fatty acid palmitic acid is degraded by β-oxidation is summarized in Figure 23.7. Complete oxidation of palmitate results in production of 129 molecules of ATP, *three and one half times more energy than results from the complete oxidation of an equivalent amount of glucose.*

Figure 23.7
Complete oxidation of palmitic acid yields 129 molecules of ATP. Note that the activation step is considered to be an expenditure of two high-energy phosphoanhydride bonds because ATP is hydrolyzed to AMP + PP$_i$.

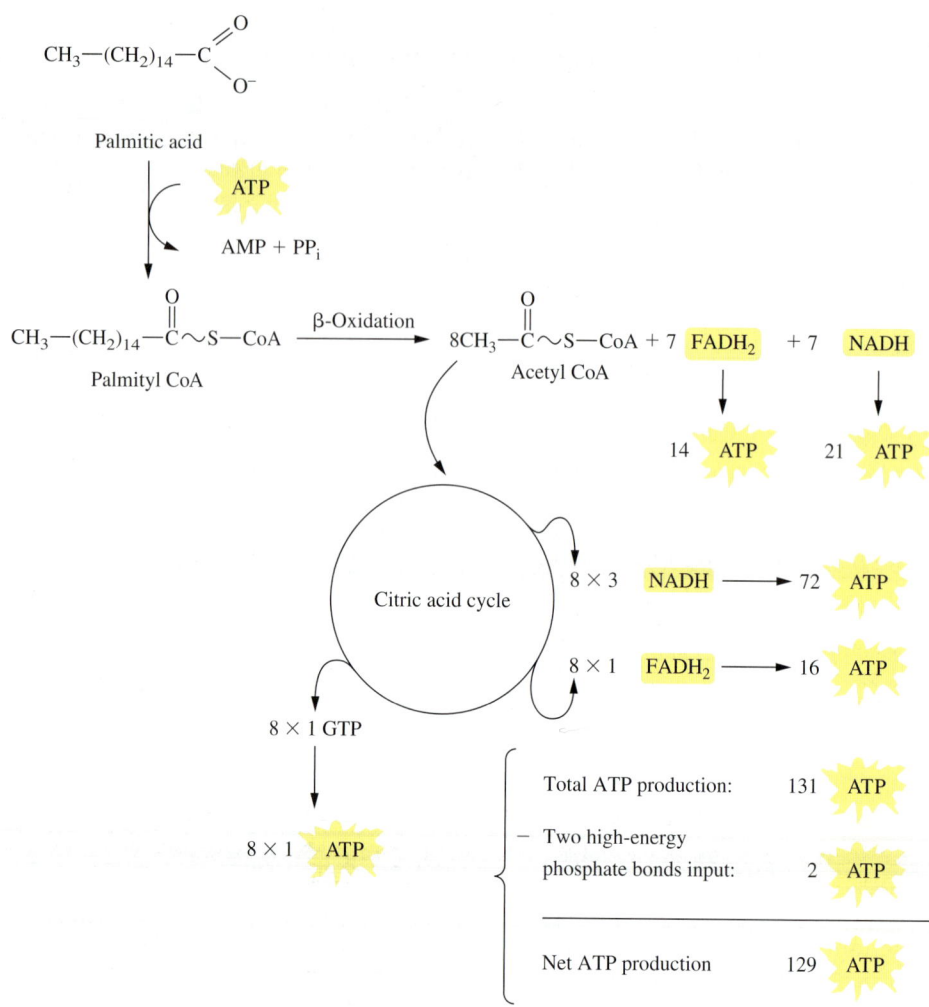

EXAMPLE 23.2 — **Calculating the Amount of ATP Produced in Complete Oxidation of a Fatty Acid**

How many molecules of ATP are produced in the complete oxidation of stearic acid, an eighteen-carbon saturated fatty acid?

Solution

Step 1 (activation)	−2 ATP
Steps 2–5:	
8 FADH$_2$ × 2 ATP/FADH$_2$	16 ATP
8 NADH × 3 ATP/NADH	24 ATP
9 acetyl CoA (to citric acid cycle):	
9 × 1 GTP × 1 ATP/GTP	9 ATP
9 × 3 NADH × 3 ATP/NADH	81 ATP
9 × 1 FADH$_2$ × 2 ATP/FADH$_2$	18 ATP
	146 ATP

Question 23.5
Write out the sequence of steps for β-oxidation of butyryl CoA.

Question 23.6
What is the energy yield from the complete degradation of butyryl CoA via β-oxidation, the citric acid cycle, and oxidative phosphorylation?

23.3 Ketone Bodies

For the acetyl CoA produced by the β-oxidation of fatty acids to efficiently enter the citric acid cycle, there must be an adequate supply of oxaloacetate. If glycolysis and β-oxidation are occurring at the same rate, there will be a steady supply of pyruvate (from glycolysis) that can be converted to oxaloacetate. But what happens if the supply of oxaloacetate is too low to allow all of the acetyl CoA to enter the citric acid cycle? Under these conditions, acetyl CoA is converted to the so-called **ketone bodies**: β-hydroxybutyrate, acetone, and acetoacetate (Figure 23.8).

4 LEARNING GOAL

See Section 22.9 for a review of the reactions that provide oxaloacetate.

Ketosis

Ketosis, abnormally high levels of blood ketone bodies, is a situation that arises under some pathological conditions, such as starvation, a diet that is extremely low in carbohydrates (as with the high-protein diets), or uncontrolled **diabetes mellitus**. The carbohydrate intake of a diabetic is normal, but the carbohydrates cannot get into the cell to be used as fuel. Thus diabetes amounts to starvation in the midst of plenty. In diabetes the very high concentration of ketone acids in the blood leads to **ketoacidosis**. The ketone acids are relatively strong acids and therefore readily dissociate to release H^+. Under these conditions the blood pH becomes acidic, which can lead to death.

Diabetes mellitus is a disease characterized by the appearance of glucose in the urine as a result of high blood glucose levels. The disease is usually caused by the inability to produce the hormone insulin.

Ketogenesis

The pathway for the production of ketone bodies (Figure 23.9) begins with a "reversal" of the last step of β-oxidation. When oxaloacetate levels are low, the enzyme that normally carries out the last reaction of β-oxidation now catalyzes the fusion of two acetyl CoA molecules to produce acetoacetyl CoA:

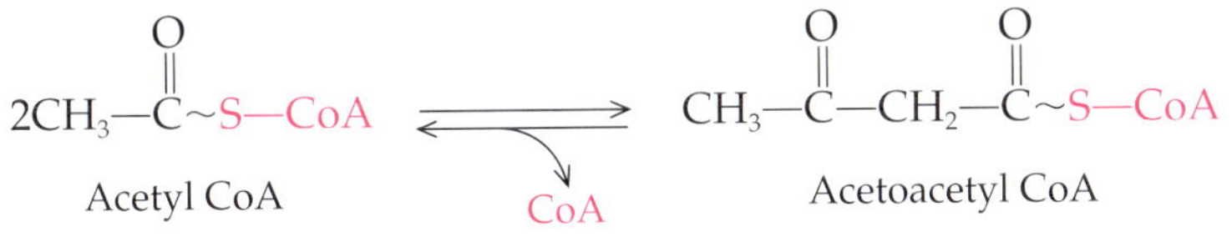

$$2CH_3-\overset{O}{\underset{\|}{C}}\sim S-CoA \rightleftharpoons CH_3-\overset{O}{\underset{\|}{C}}-CH_2-\overset{O}{\underset{\|}{C}}\sim S-CoA$$

Acetyl CoA → CoA → Acetoacetyl CoA

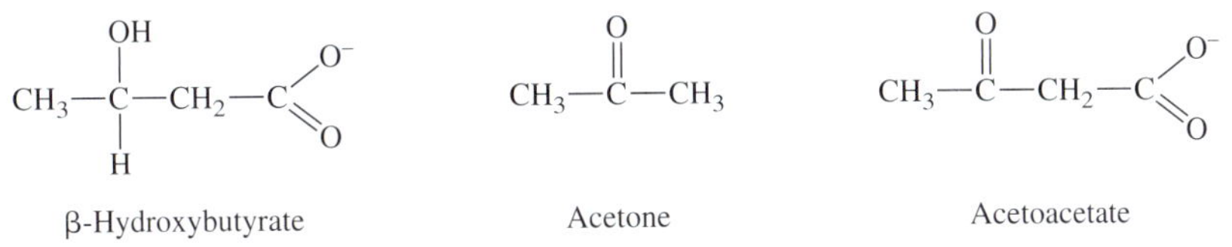

β-Hydroxybutyrate: $CH_3-\overset{OH}{\underset{H}{C}}-CH_2-C\overset{O^-}{\underset{O}{\diagdown}}$

Acetone: $CH_3-\overset{O}{\underset{\|}{C}}-CH_3$

Acetoacetate: $CH_3-\overset{O}{\underset{\|}{C}}-CH_2-C\overset{O^-}{\underset{O}{\diagdown}}$

Figure 23.8
Structures of ketone bodies.

Figure 23.9
Summary of the reactions involved in ketogenesis.

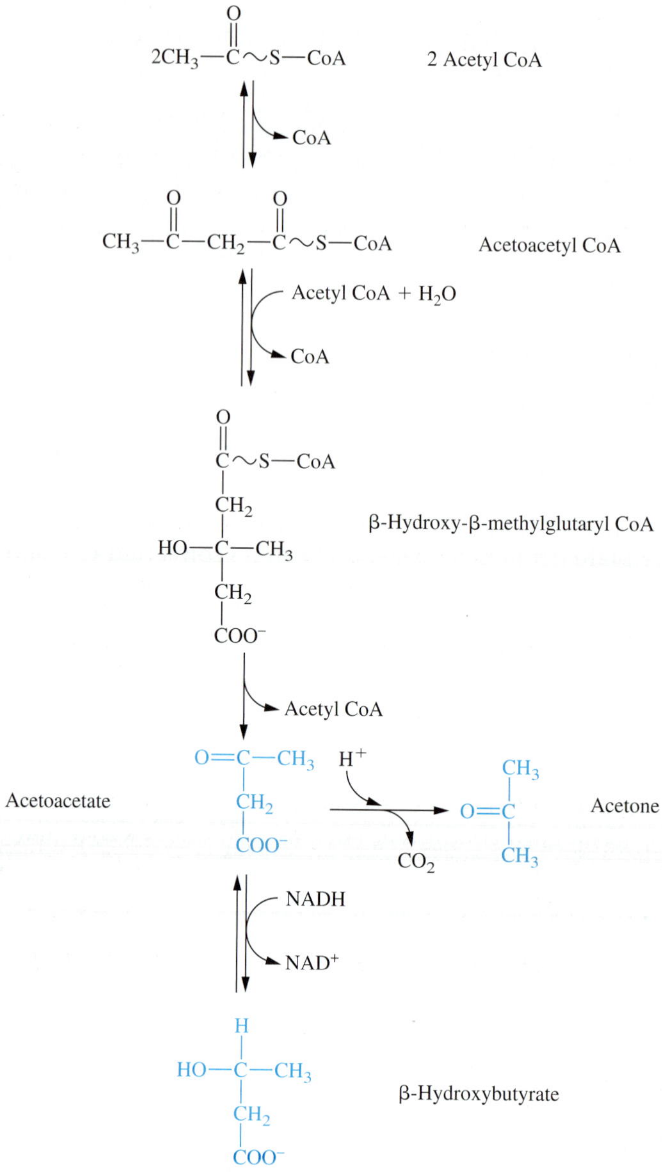

Acetoacetyl CoA can react with a third acetyl CoA molecule to yield β-hydroxy-β-methylglutaryl CoA (HMG-CoA):

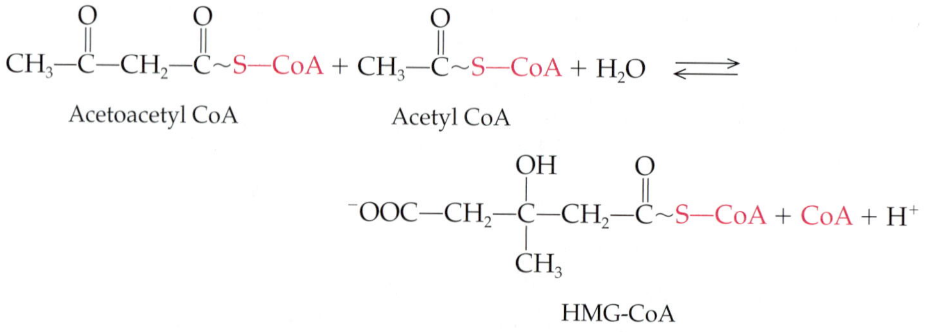

$$CH_3-\overset{O}{\underset{\|}{C}}-CH_2-\overset{O}{\underset{\|}{C}}\sim S-CoA + CH_3-\overset{O}{\underset{\|}{C}}\sim S-CoA + H_2O \rightleftharpoons$$

Acetoacetyl CoA   Acetyl CoA

$$^-OOC-CH_2-\underset{\underset{CH_3}{|}}{\overset{\overset{OH}{|}}{C}}-CH_2-\overset{O}{\underset{\|}{C}}\sim S-CoA + CoA + H^+$$

HMG-CoA

If HMG-CoA were formed in the cytoplasm, it would serve as a precursor for cholesterol biosynthesis. But ketogenesis, like β-oxidation, occurs in the mitochondrial matrix, and here HMG-CoA is cleaved to yield acetoacetate and acetyl CoA:

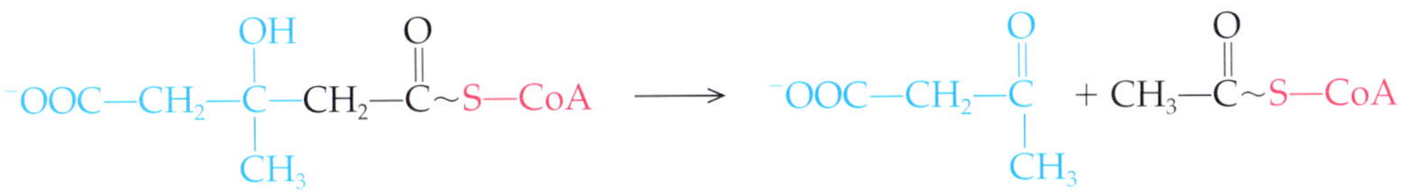

$$^-OOC-CH_2-\underset{\underset{CH_3}{|}}{\overset{\overset{OH}{|}}{C}}-CH_2-\overset{\overset{O}{\|}}{C}\sim S-CoA \longrightarrow {}^-OOC-CH_2-\underset{\underset{CH_3}{|}}{\overset{\overset{O}{\|}}{C}} + CH_3-\overset{\overset{O}{\|}}{C}\sim S-CoA$$

HMG-CoA   Acetoacetate   Acetyl CoA

In very small amounts, acetoacetate spontaneously loses carbon dioxide to give acetone. This is the reaction that causes the "acetone breath" that is often associated with uncontrolled diabetes mellitus.

$$^-OOC-CH_2-\underset{\underset{CH_3}{|}}{\overset{\overset{O}{\|}}{C}} + H^+ \xrightarrow{\searrow CO_2} CH_3-\overset{\overset{O}{\|}}{C}-CH_3$$

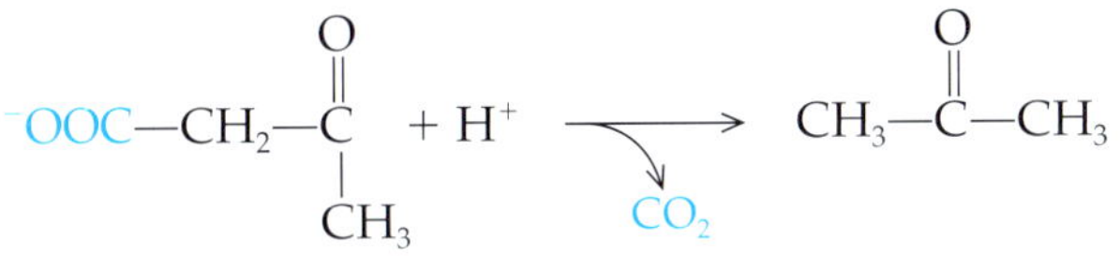

Acetoacetate   Acetone

More frequently, it undergoes NADH-dependent reduction to produce β-hydroxybutyrate:

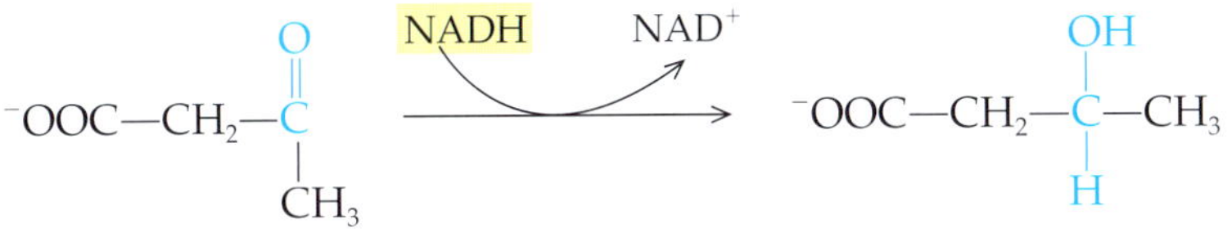

$$^-OOC-CH_2-\underset{\underset{CH_3}{|}}{\overset{\overset{O}{\|}}{C}} \xrightarrow{\text{NADH} \quad \text{NAD}^+} {}^-OOC-CH_2-\underset{\underset{H}{|}}{\overset{\overset{OH}{|}}{C}}-CH_3$$

Acetoacetate   β-Hydroxybutyrate

Acetoacetate and β-hydroxybutyrate are produced primarily in the liver. These metabolites diffuse into the blood and are circulated to other tissues, where they may be reconverted to acetyl CoA and used to produce ATP. In fact, the heart muscle derives most of its metabolic energy from the oxidation of ketone bodies, not from the oxidation of glucose. Other tissues that are best adapted to the use of glucose will increasingly rely on ketone bodies for energy when glucose becomes unavailable or limited. This is particularly true of the brain.

What conditions lead to excess production of ketone bodies?

Question **23.7**

What is the cause of the characteristic "acetone breath" that is associated with uncontrolled diabetes mellitus?

Question **23.8**

23.4 Fatty Acid Synthesis

All organisms possess the ability to synthesize fatty acids. In humans the excess acetyl CoA produced by carbohydrate degradation is used to make fatty acids that are then stored as triglycerides.

 **LEARNING GOAL**

A Comparison of Fatty Acid Synthesis and Degradation

On first examination, fatty acid synthesis appears to be simply the reverse of β-oxidation. Specifically, the fatty acid chain is constructed by the sequential addition of two-carbon acetyl groups (Figure 23.10). Although the chemistry of fatty acid synthesis and breakdown are similar, there are several major differences between β-oxidation and fatty acid biosynthesis. These are summarized as follows.

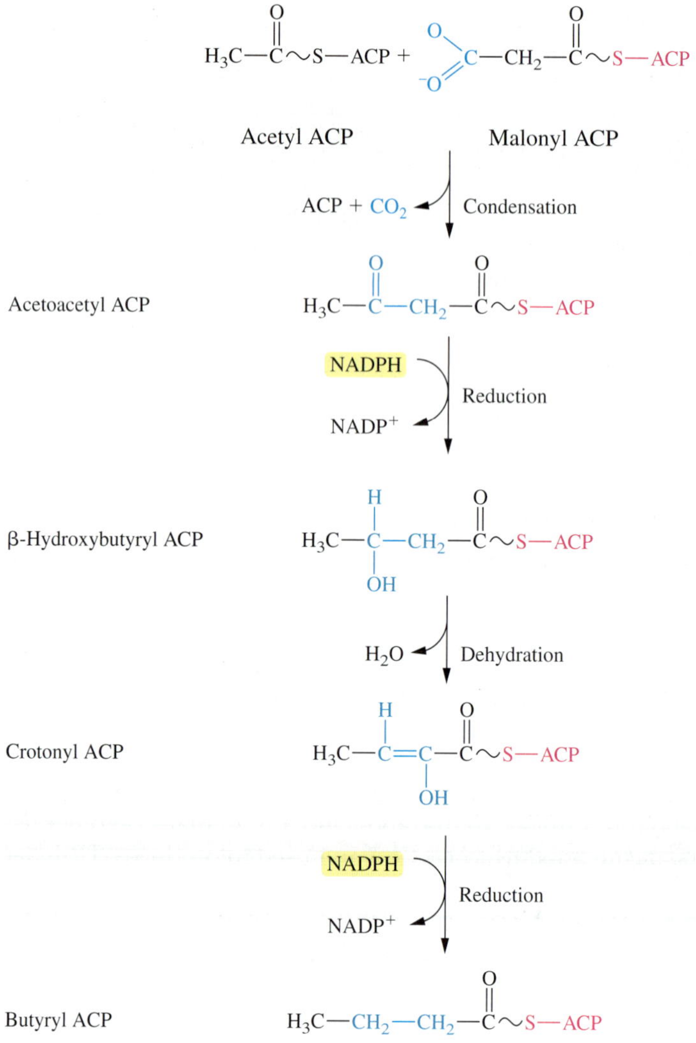

Figure 23.10
Summary of fatty acid synthesis. Malonyl ACP is produced in two reactions: carboxylation of acetyl CoA to produce malonyl CoA and transfer of the malonyl acyl group from malonyl CoA to ACP.

$$H_3C-\overset{O}{\underset{\|}{C}}\sim S-ACP + {}^-O-\overset{O}{\underset{\|}{C}}-CH_2-\overset{O}{\underset{\|}{C}}\sim S-ACP$$

Acetyl ACP   Malonyl ACP

ACP + CO_2 ← | Condensation

Acetoacetyl ACP $H_3C-\overset{O}{\underset{\|}{C}}-CH_2-\overset{O}{\underset{\|}{C}}\sim S-ACP$

NADPH ↘ | Reduction
NADP⁺ ↙

β-Hydroxybutyryl ACP $H_3C-\underset{OH}{\overset{H}{\underset{|}{\overset{|}{C}}}}-CH_2-\overset{O}{\underset{\|}{C}}\sim S-ACP$

H_2O ← | Dehydration

Crotonyl ACP $H_3C-\overset{H}{\underset{\|}{C}}=\underset{OH}{\overset{}{\underset{|}{C}}}-\overset{O}{\underset{\|}{C}}\sim S-ACP$

NADPH ↘ | Reduction
NADP⁺ ↙

Butyryl ACP $H_3C-CH_2-CH_2-\overset{O}{\underset{\|}{C}}\sim S-ACP$

- **Intracellular location.** The enzymes responsible for fatty acid biosynthesis are located in the cytoplasm of the cell, whereas those responsible for the degradation of fatty acids are in the mitochondria.
- **Acyl group carriers.** The activated intermediates of fatty acid biosynthesis are bound to a carrier molecule called the **acyl carrier protein (ACP)** (Figure 23.11). In β-oxidation the acyl group carrier was coenzyme A. However, there are important similarities between these two carriers. Both contain the **phosphopantetheine** group, which is made from the vitamin pantothenic acid. In both cases the fatty acyl group is bound by a thioester bond to the phosphopantetheine group.
- **Enzymes involved.** Fatty acid biosynthesis is carried out by a multienzyme complex known as *fatty acid synthase*. The enzymes responsible for fatty acid degradation are not physically associated in such complexes.
- **Electron carriers.** NADH and $FADH_2$ are produced by fatty acid oxidation, whereas NADPH is the reducing agent for fatty acid biosynthesis. As a general rule, *NADH is produced by catabolic reactions, and NADPH is the reducing agent of biosynthetic reactions*. These two coenzymes differ only by the presence of a phosphate group bound to the ribose ring of NADPH (Figure 23.12). The enzymes that use these coenzymes, however, are easily able to distinguish them on this basis.

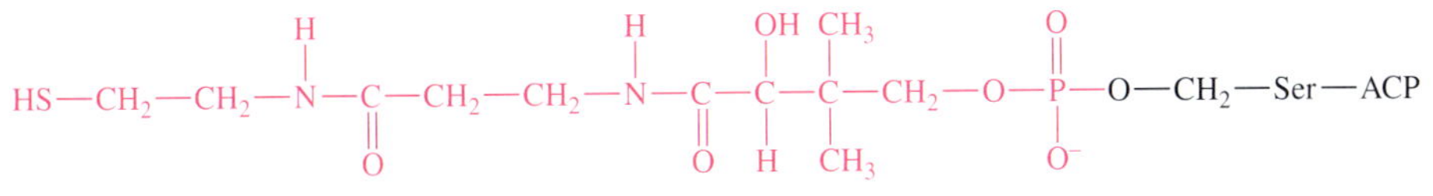

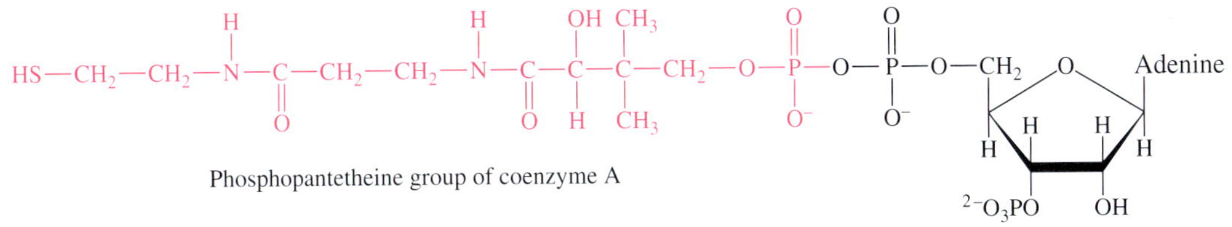

23.5 The Regulation of Lipid and Carbohydrate Metabolism

Figure 23.11
The structure of the phosphopantetheine group, the reactive group common to coenzyme A and acyl carrier protein, is highlighted in red.

Phosphopantetheine prosthetic group of ACP

Phosphopantetheine group of coenzyme A

Question 23.9

List the four major differences between β-oxidation and fatty acid biosynthesis that reveal that the two processes are not just the reverse of one another.

Question 23.10

What chemical group is part of coenzyme A and acyl carrier protein and allows both molecules to form thioester bonds to fatty acids?

23.5 The Regulation of Lipid and Carbohydrate Metabolism

The metabolism of fatty acids and carbohydrates occurs to a different extent in different organs. As we will see in this section, the regulation of these two related aspects of metabolism is of great physiological importance.

The Liver

The liver provides a steady supply of glucose for muscle and brain and plays a major role in the regulation of blood glucose concentration. This regulation is under hormonal control. Recall that the hormone insulin causes blood glucose to be taken up by the liver and stored as glycogen *(glycogenesis)*. In this way the liver reduces the blood glucose levels when they are too high.

The hormone glucagon, on the other hand, stimulates the breakdown of glycogen and the release of glucose into the bloodstream. Lactate produced by muscles under anaerobic conditions is also taken up by liver cells and is converted to glucose by gluconeogenesis. Both glycogen degradation *(glycogenolysis)* and *gluconeogenesis* are pathways that produce glucose for export to other organs when energy is needed (Figure 23.13).

The liver also plays a central role in lipid metabolism. When excess fuel is available, the liver synthesizes fatty acids. These are used to produce triglycerides that are transported from the liver to adipose tissues by very low density lipoprotein (VLDL) complexes. In fact, VLDL complexes provide adipose tissue with its major source of fatty acids. This transport is particularly active when more calories are eaten than are burned! During fasting or starvation conditions, however, the liver converts fatty acids to acetoacetate and other ketone bodies. The liver cannot use these ketone bodies because it lacks an enzyme for the conversion of acetoacetate to acetyl CoA. Therefore the ketone bodies produced by the liver are exported to other organs where they are oxidized to make ATP.

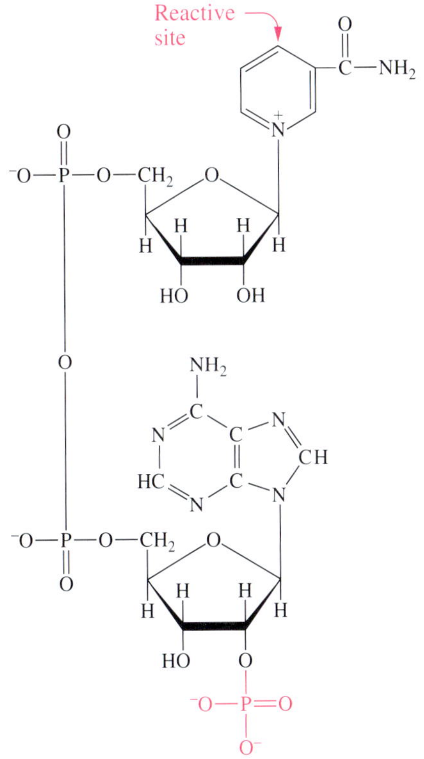

Figure 23.12
Structure of NADPH. The phosphate group shown in red is the structural feature that distinguishes NADPH from NADH.

A Medical Perspective

Diabetes Mellitus and Ketone Bodies

More than one person, found unconscious on the streets of some metropolis, has been carted to jail only to die of complications arising from uncontrolled diabetes mellitus. Others are fortunate enough to arrive in hospital emergency rooms. A quick test for diabetes mellitus–induced coma is the odor of acetone on the breath of the afflicted person. Acetone is one of several metabolites produced by diabetics that are known collectively as *ketone bodies*.

The term *diabetes* was used by the ancient Greeks to designate diseases in which excess urine is produced. Two thousand years later, in the eighteenth century, the urine of certain individuals was found to contain sugar, and the name *diabetes mellitus* (Latin: *mellitus,* sweetened with honey) was given to this disease. People suffering from diabetes mellitus waste away as they excrete large amounts of sugar-containing urine.

The cause of insulin-dependent diabetes mellitus is an inadequate production of insulin by the body. Insulin is secreted in response to high blood glucose levels. It binds to the membrane receptor protein on its target cells. Binding increases the rate of transport of glucose across the membrane and stimulates glycogen synthesis, lipid biosynthesis, and protein synthesis. As a result, the blood glucose level is reduced. Clearly, the inability to produce sufficient insulin seriously impairs the body's ability to regulate metabolism.

Individuals suffering from diabetes mellitus do not produce enough insulin to properly regulate blood glucose levels. This generally results from the destruction of the β-cells of the islets of Langerhans. One theory to explain the mysterious disappearance of these cells is that a virus infection stimulates the immune system to produce antibodies that cause the destruction of the β-cells.

In the absence of insulin the uptake of glucose into the tissues is not stimulated, and a great deal of glucose is eliminated in the urine. Without insulin, then, adipose cells are unable to take up the glucose required to synthesize triglycerides. As a result, the rate of fat hydrolysis is much greater than the rate of fat resynthesis, and large quantities of free fatty acids are liberated into the bloodstream. Because glucose is not being efficiently taken into cells, carbohydrate metabolism slows, and there is an increase in the rate of lipid catabolism. In the liver this lipid catabolism results in the production of ketone bodies: acetone, acetoacetate, and β-hydroxybutyrate.

A similar situation can develop from improper eating, fasting, or dieting—any situation in which the body is not provided with sufficient energy in the form of carbohydrates. These ketone bodies cannot all be oxidized by the citric acid cycle, which is limited by the supply of oxaloacetate. The acetone concentration in blood rises to levels so high that acetone can be detected in the breath of untreated diabetics. The elevated concentration of ketones in the blood can overwhelm the buffering capacity of the blood, resulting in ketoacidosis. Ketones, too, will be excreted through the kidney. In fact, the presence of excess ketones in the urine can raise the osmotic concentration of the urine so that it behaves as an "osmotic diuretic," causing the excretion of enormous amounts of water. As a result, the patient may become severely dehydrated. In extreme cases the combination of dehydration and ketoacidosis may lead to coma and death.

It has been observed that diabetics also have a higher than normal level of glucagon in the blood. As we have seen, glucagon stimulates lipid catabolism and ketogenesis. It may be that the symptoms previously described result from both the deficiency of insulin and the elevated glucagon levels. The absence of insulin may cause the elevated blood glucose and fatty acid levels, whereas the glucagon, by stimulating ketogenesis, may be responsible for the ketoacidosis and dehydration.

There is no cure for diabetes. However, when the problem is the result of the inability to produce active insulin, blood glucose levels can be controlled moderately well by the injection of human insulin produced from the cloned insulin gene. Unfortunately, one or even a few injections of insulin each day cannot mimic the precise control of blood glucose accomplished by the pancreas.

As a result, diabetics suffer progressive tissue degeneration that leads to early death. One primary cause of this degeneration is atherosclerosis, the deposition of plaque on the walls of blood vessels. This causes a high frequency of strokes, heart attack, and gangrene of the feet and lower extremities, often necessitating amputation. Kidney failure causes the death of about 20% of diabetics under forty years of age, and diabetic retinopathy (various kinds of damage to the retina of the eye) ranks fourth among the leading causes of blindness in the United States. Nerves are also damaged, resulting in neuropathies that can cause pain or numbness, particularly of the feet.

23.5 The Regulation of Lipid and Carbohydrate Metabolism

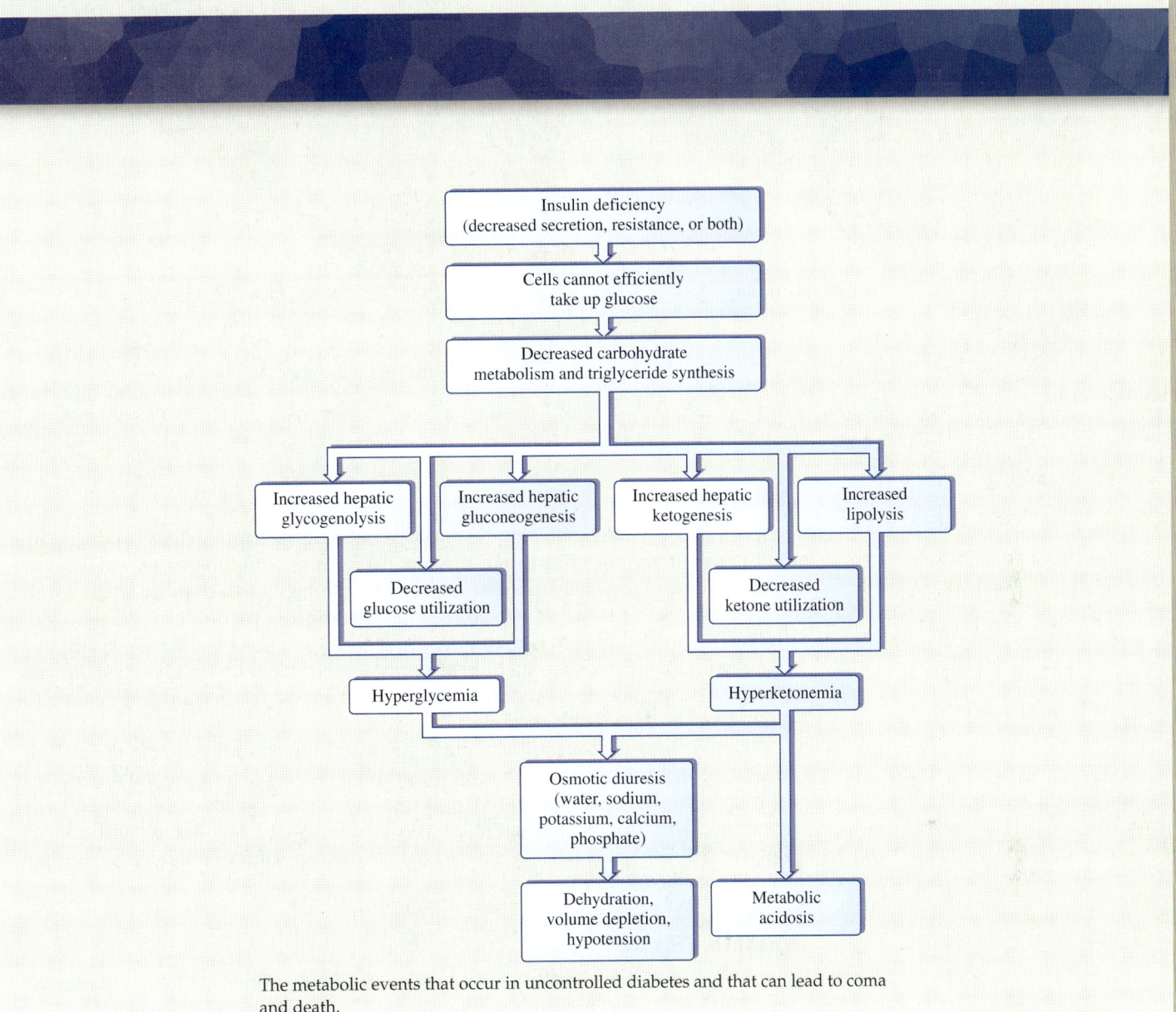

The metabolic events that occur in uncontrolled diabetes and that can lead to coma and death.

There is no doubt that insulin injections prolong the life of diabetics, but only the presence of a fully functioning pancreas can allow a diabetic to live a life free of the complications noted here. At present, pancreas transplants do not have a good track record. Only about 50% of the transplants are functioning after one year. It is hoped that improved transplantation techniques will be developed so that diabetics can live a normal life span, free of debilitating disease.

For Further Understanding

The Atkins' low carbohydrate diet recommends that dieters test their urine for the presence of ketone bodies as an indicator that the diet is working. In terms of lipid and carbohydrate metabolism, explain why ketone bodies are being produced and why this is an indication that the diet is working.

An excess of ketone bodies in the blood causes ketoacidosis. Consider the chemical structure of the ketone bodies and explain why they are acids.

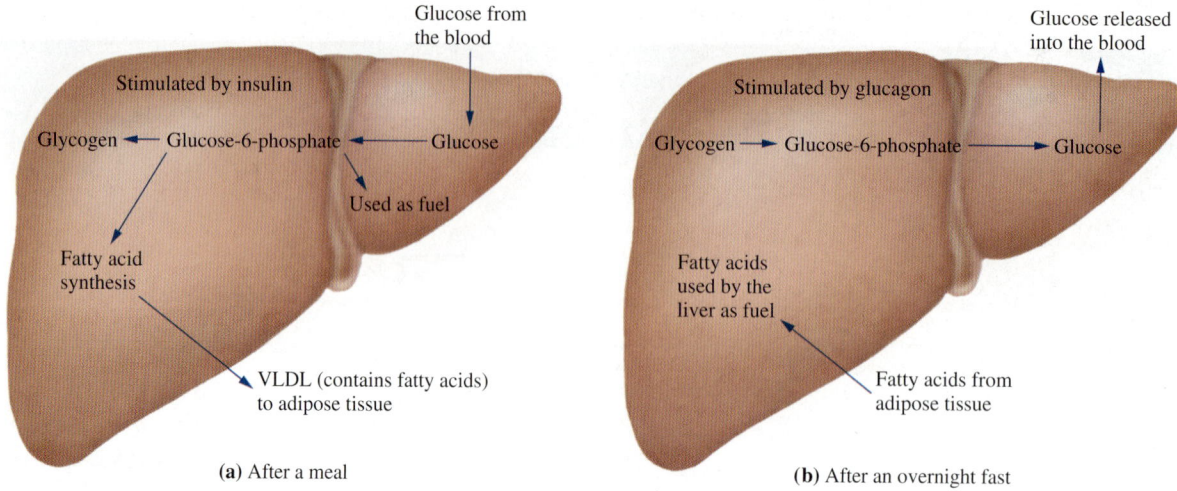

Figure 23.13
The liver controls the concentration of blood glucose.

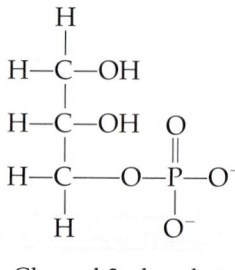

Glycerol-3-phosphate

Gluconeogenesis is described in Section 21.6.

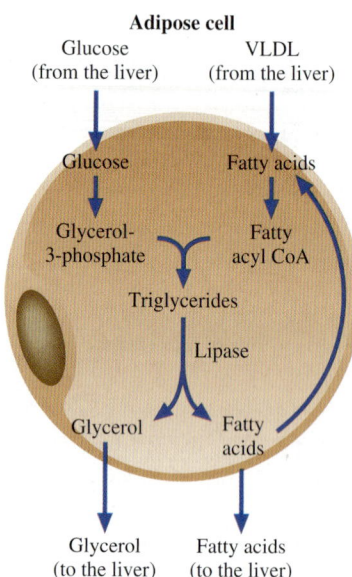

Figure 23.14
Synthesis and degradation of triglycerides in adipose tissue.

Adipose Tissue

Adipose tissue is the major storage depot of fatty acids. Triglycerides produced by the liver are transported through the bloodstream as components of VLDL complexes. The triglycerides are hydrolyzed by the same lipases that act on chylomicrons, and the fatty acids are absorbed by adipose tissue. The synthesis of triglycerides in adipose tissue requires glycerol-3-phosphate. However, adipose tissue is unable to make glycerol-3-phosphate and depends on glycolysis for its supply of this molecule. Thus adipose cells must have a ready source of glucose to synthesize and store triglycerides.

Triglycerides are constantly being hydrolyzed and resynthesized in the cells of adipose tissue. Lipases that are under hormonal control determine the rate of hydrolysis of triglycerides into fatty acids and glycerol. If glucose is in limited supply, there will not be sufficient glycerol-3-phosphate for the resynthesis of triglycerides, and the fatty acids and glycerol are exported to the liver for further processing (Figure 23.14).

Muscle Tissue

The energy demand of resting muscle is generally supplied by the β-oxidation of fatty acids. The heart muscle actually prefers ketone bodies over glucose.

Working muscle, however, obtains energy by degradation of its own supply of glycogen. Glycogen degradation produces glucose-6-phosphate, which is directly funneled into glycolysis. If the muscle is working so hard that it doesn't get enough oxygen, it produces large amounts of lactate. This fermentation end product, as well as alanine (from catabolism of proteins and transamination of pyruvate), is exported to the liver. Here they are converted to glucose by gluconeogenesis (Figure 23.15).

The Brain

Under normal conditions the brain uses glucose as its sole source of metabolic energy. When the body is in the resting state, about 60% of the free glucose of the body is used by the brain. Starvation depletes glycogen stores, and the amount of glucose available to the brain drops sharply. The ketone bodies acetoacetate and β-hydroxybutyrate are then used by the brain as an alternative energy source. Fatty acids are transported in the blood in complexes with proteins and cannot cross the blood-brain barrier to be used by brain cells as an energy source. But ketone bodies, which have a free carboxyl group, are soluble in blood and can enter the brain.

23.6 The Effects of Insulin and Glucagon on Cellular Metabolism

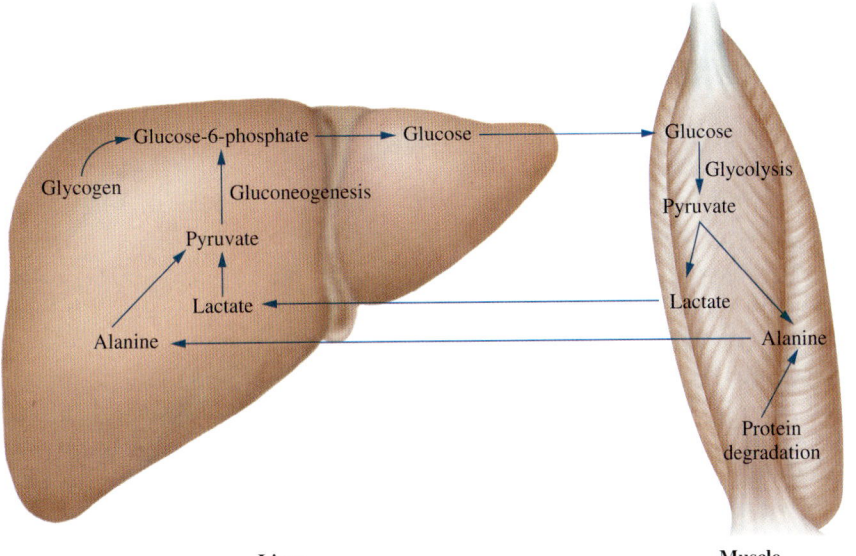

Figure 23.15
Metabolic relationships between liver and muscle.

Question 23.11
How does the liver regulate blood glucose levels?

Question 23.12
Why is regulation of blood glucose levels important to the efficient function of the brain?

23.6 The Effects of Insulin and Glucagon on Cellular Metabolism

The hormone **insulin** is produced by the β-cells of the islets of Langerhans in the pancreas. It is secreted from these cells in response to an increase in the blood glucose level. Insulin lowers the concentration of blood glucose by causing a number of changes in metabolism (Table 23.1).

The simplest way to lower blood glucose levels is to stimulate storage of glucose, both as glycogen and as triglycerides. *Insulin therefore activates biosynthetic processes and inhibits catabolic processes.*

Insulin acts only on those cells, known as *target cells,* that possess a specific insulin receptor protein in their plasma membranes. The major target cells for insulin are liver, adipose, and muscle cells.

The blood glucose level is normally about 10 mM. However, a substantial meal increases the concentration of blood glucose considerably and stimulates insulin secretion. Subsequent binding of insulin to the plasma membrane insulin receptor increases the rate of transport of glucose across the membrane and into cells.

Insulin exerts a variety of effects on all aspects of cellular metabolism:

- **Carbohydrate metabolism.** Insulin stimulates glycogen synthesis. At the same time it inhibits glycogenolysis and gluconeogenesis. The overall result of these activities is the storage of excess glucose.
- **Protein metabolism.** Insulin stimulates transport and uptake of amino acids, as well as the incorporation of amino acids into proteins.

 **LEARNING GOAL**

The effect of insulin on glycogen metabolism is described in Section 21.7.

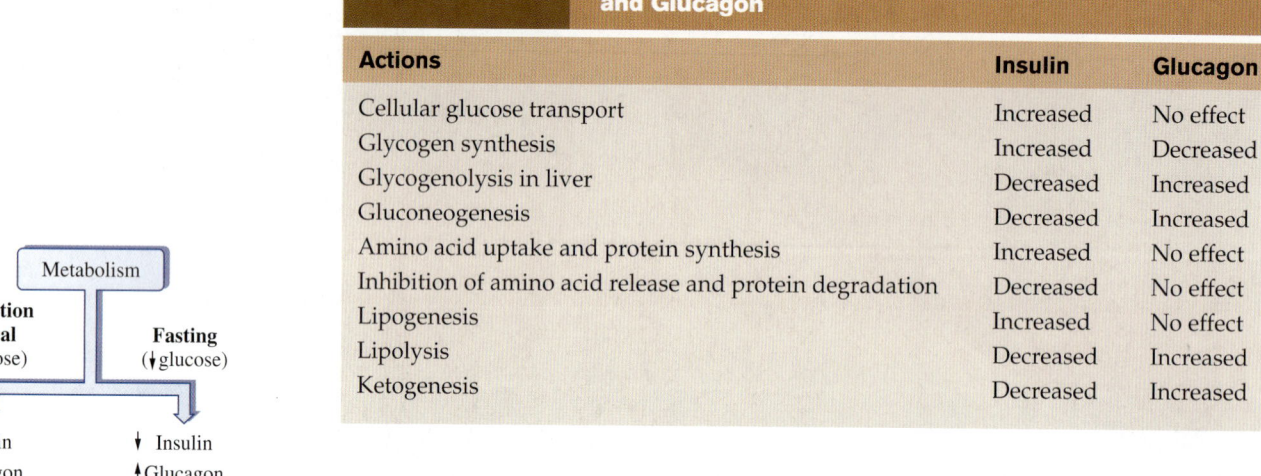

TABLE 23.1 Comparison of the Metabolic Effects of Insulin and Glucagon

Actions	Insulin	Glucagon
Cellular glucose transport	Increased	No effect
Glycogen synthesis	Increased	Decreased
Glycogenolysis in liver	Decreased	Increased
Gluconeogenesis	Decreased	Increased
Amino acid uptake and protein synthesis	Increased	No effect
Inhibition of amino acid release and protein degradation	Decreased	No effect
Lipogenesis	Increased	No effect
Lipolysis	Decreased	Increased
Ketogenesis	Decreased	Increased

- **Lipid metabolism.** Insulin stimulates uptake of glucose by adipose cells, as well as the synthesis and storage of triglycerides. As we have seen, storage of lipids requires a source of glucose, and insulin helps the process by increasing the available glucose. At the same time, insulin inhibits the breakdown of stored triglycerides.

As you may have already guessed, insulin is only part of the overall regulation of cellular metabolism in the body. A second hormone, **glucagon,** is secreted by the α-cells of the islets of Langerhans in response to decreased blood glucose levels. The effects of glucagon, generally the opposite of the effects of insulin, are summarized in Table 23.1. Although it has no direct effect on glucose uptake, glucagon inhibits glycogen synthesis and stimulates glycogenolysis and gluconeogenesis. It also stimulates the breakdown of fats and ketogenesis.

The antagonistic effects of these two hormones, seen in Figure 23.16, are critical for the maintenance of adequate blood glucose levels. During fasting, low blood glucose levels stimulate production of glucagon, which increases blood glucose by stimulating the breakdown of glycogen and the production of glucose by gluconeogenesis. This ensures a ready supply of glucose for the tissues, especially the brain. On the other hand, when blood glucose levels are too high, insulin is secreted. It stimulates the removal of the excess glucose by enhancing uptake and inducing pathways for storage.

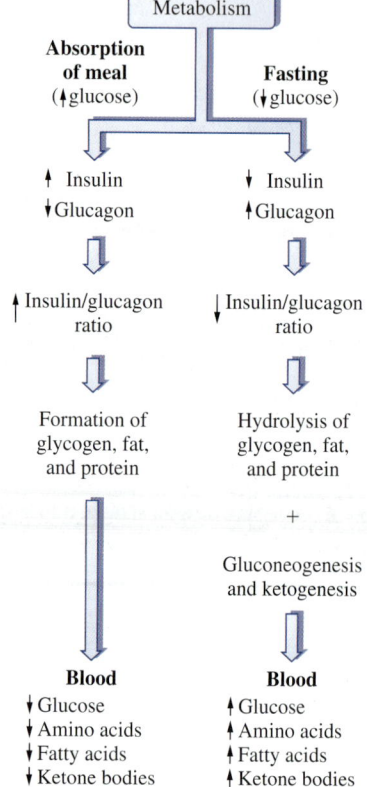

Figure 23.16
A summary of the antagonistic effects of insulin and glucagon.

Question 23.13 Summarize the effects of the hormone insulin on carbohydrate, lipid, and amino acid metabolism.

Question 23.14 Summarize the effects of the hormone glucagon on carbohydrate and lipid metabolism.

SUMMARY

23.1 Lipid Metabolism in Animals

Dietary lipids (*triglycerides*) are emulsified into tiny fat droplets in the intestine by the action of *bile* salts. Pancreatic *lipase* catalyzes the hydrolysis of triglycerides into monoglycerides and fatty acids. These are absorbed by intestinal epithelial cells, reassembled into triglycerides, and combined with protein to form *chylomicrons*. Chylomicrons are transported to the cells of the body through the bloodstream. Fatty acids are stored as triglycerides (triacylglycerols) in fat droplets in the cytoplasm of *adipocytes*.

23.2 Fatty Acid Degradation

Fatty acids are degraded to acetyl CoA in the mitochondria by the β-*oxidation* pathway, which involves five steps: (1) the production of a fatty acyl CoA molecule, (2) oxidation of the fatty acid by an FAD-dependent dehydrogenase, (3) hydration, (4) oxidation by an NAD^+-dependent dehydrogenase, and (5) cleavage of the chain with release of acetyl CoA and a fatty acyl CoA that is two carbons shorter than the beginning fatty acid. The last four reactions are repeated until the fatty acid is completely degraded to acetyl CoA.

23.3 Ketone Bodies

Under some conditions, fatty acid degradation occurs more rapidly than glycolysis. As a result, a large amount of acetyl CoA is produced from fatty acids, but little oxaloacetate is generated from pyruvate. When oxaloacetate levels are too low, the excess acetyl CoA is converted to the *ketone bodies* acetone, acetoacetate, and β-hydroxybutyrate.

23.4 Fatty Acid Synthesis

Fatty acid biosynthesis occurs by the sequential addition of acetyl groups and, on first inspection, appears to be a simple reversal of the β-oxidation pathway. Although the biochemical reactions are similar, fatty acid synthesis differs from β-oxidation in the following ways: It occurs in the cytoplasm, utilizes *acyl carrier protein* and NADPH, and is carried out by a multienzyme complex, fatty acid synthase.

23.5 The Regulation of Lipid and Carbohydrate Metabolism

Lipid and carbohydrate metabolism occur to different extents in different organs. The liver regulates the flow of metabolites to brain, muscle, and adipose tissue and ultimately controls the concentration of blood glucose. *Adipose tissue* is the major storage depot for fatty acids. Triglycerides are constantly hydrolyzed and resynthesized in adipose tissue. Muscle oxidizes glucose, fatty acids, and ketone bodies. The brain uses glucose as a fuel except in prolonged fasting or starvation, when it will use ketone bodies as an energy source.

23.6 The Effects of Insulin and Glucagon on Cellular Metabolism

Insulin stimulates biosynthetic processes and inhibits catabolism in liver, muscle, and adipose tissue. Insulin is synthesized in the β-cells of the pancreas and is secreted when the blood glucose levels become too high. The insulin receptor protein binds to the insulin. This binding mediates a variety of responses in target tissues, including the storage of glucose and lipids. *Glucagon* is secreted when blood glucose levels are too low. It has the opposite effects on metabolism, including the breakdown of lipids and glycogen.

KEY TERMS

acyl carrier protein (ACP) (23.4)
adipocyte (23.1)
adipose tissue (23.1)
bile (23.1)
chylomicron (23.1)
colipase (23.1)
diabetes mellitus (23.3)
glucagon (23.6)
insulin (23.6)
ketoacidosis (23.3)
ketone bodies (23.3)
ketosis (23.3)
lipase (23.1)
micelle (23.1)
β-oxidation (23.2)
phosphopantetheine (23.4)
triglyceride (23.1)

QUESTIONS AND PROBLEMS

Lipid Metabolism in Animals

Foundations

23.15 Describe the composition of bile.
23.16 Why are bile salts referred to as detergents?
23.17 Define the term *micelle*.
23.18 In Figure 23.1, a micelle composed of the phospholipid lecithin is shown. Why is lecithin a good molecule for the formation of micelles?
23.19 Define the term *triglyceride*.
23.20 Draw the structure of a triglyceride composed of glycerol, palmitoleic acid, linolenic acid, and oleic acid.
23.21 Review the information on chylomicrons in Chapter 17. Describe the composition of chylomicrons?
23.22 What is an adipocyte?

Applications

23.23 What is the major storage form of fatty acids?
23.24 What tissue is the major storage depot for lipids?
23.25 What is the outstanding structural feature of an adipocyte?
23.26 What is the major metabolic function of adipose tissue?
23.27 What is the general reaction catalyzed by lipases?
23.28 Why are the lipases that are found in saliva and in the stomach not very effective at digesting triglycerides?
23.29 List three major biological molecules for which acetyl CoA is a precursor.
23.30 Why are triglycerides more efficient energy-storage molecules than glycogen?
23.31 What are chylomicrons, and what is their function?
23.32 a. What are very low density lipoproteins?
b. Compare the function of VLDLs with that of chylomicrons.
23.33 What is the function of the bile salts in the digestion of dietary lipids?

23.34 What is the function of colipase in the digestion of dietary lipids?
23.35 Describe the stages of lipid digestion.
23.36 Describe the transport of lipids digested in the lumen of the intestines to the cells of the body.

Fatty Acid Degradation

Foundations

23.37 What is the energy source for the activation of a fatty acid in preparation for β-oxidation?
23.38 Which bond in fatty acyl CoA is a high energy bond?
23.39 What is carnitine?
23.40 Explain the mechanism by which a fatty acyl group is brought into the mitochondrial matrix.
23.41 Explain why the reaction catalyzed by acyl-CoA dehydrogenase is an example of an oxidation reaction.
23.42 What is the reactant that is oxidized in the reaction catalyzed by acyl-CoA dehydrogenase? What is the reactant that is reduced in this reaction?
23.43 What is the product of the hydration of an alkene?
23.44 Which reaction in β-oxidation is a hydration reaction? What is the name of the enzyme that catalyzes this reaction? Write an equation representing this reaction.

Applications

23.45 What products are formed when the ω-phenyl-labeled carboxylic acid 14-phenyltetradecanoic acid is degraded by β-oxidation?
23.46 What products are formed when the ω-phenyl-labeled carboxylic acid 5-phenylpentanoic acid is degraded by β-oxidation?
23.47 Calculate the number of ATP molecules produced by complete β-oxidation of the fourteen-carbon saturated fatty acid tetradecanoic acid (common name: myristic acid).
23.48 a. Write the sequence of steps that would be followed for one round of β-oxidation of hexanoic acid.
   b. Calculate the number of ATP molecules produced by complete β-oxidation of hexanoic acid.
23.49 How many molecules of ATP are produced for each molecule of $FADH_2$ that is generated by β-oxidation?
23.50 How many molecules of ATP are produced for each molecule of NADH generated by β-oxidation?
23.51 What is the fate of the acetyl CoA produced by β-oxidation?
23.52 How many ATP molecules are produced from each acetyl CoA molecule generated in β-oxidation that enters the citric acid cycle?

Ketone Bodies

Foundations

23.53 What are ketone bodies?
23.54 What are the chemical properties of ketone bodies?
23.55 What is ketosis?
23.56 Define *ketoacidosis*.
23.57 In what part of the cell does ketogenesis occur? Be specific.
23.58 What would be the fate of HMG-CoA produced in ketogenesis if it were produced in the cell cytoplasm?

Applications

23.59 Draw the structures of acetoacetate and β-hydroxybutyrate.
23.60 Describe the relationship between the formation of ketone bodies and β-oxidation.
23.61 Why do uncontrolled diabetics produce large amounts of ketone bodies?
23.62 How does the presence of ketone bodies in the blood lead to ketoacidosis?
23.63 When does the heart use ketone bodies?
23.64 When does the brain use ketone bodies?

Fatty Acid Synthesis

Foundations

23.65 Where in the cell does fatty acid biosynthesis occur?
23.66 What is the acyl group carrier in fatty acid biosynthesis?
23.67 What enzyme is involved in fatty acid biosynthesis?
23.68 What is the reducing agent for fatty acid biosynthesis?

Applications

23.69 a. What is the role of the phosphopantetheine group in fatty acid biosynthesis?
   b. From what molecule is phosphopantetheine made?
23.70 What molecules involved in fatty acid degradation and fatty acid biosynthesis contain the phosphopantetheine group?
23.71 How does the structure of fatty acid synthase differ from that of the enzymes that carry out β-oxidation?
23.72 In what cellular compartments do fatty acid biosynthesis and β-oxidation occur?

The Regulation of Lipid and Carbohydrate Metabolism

Foundations

22.73 Define the term *glycogenesis*.
22.74 Define the term *glycogenolysis*.
22.75 Define the term *gluconeogenesis*.
22.76 How are the fatty acids synthesized in the liver transported to adipose tissue?

Applications

22.77 Which pathway provides the majority of the ATP for *resting* muscle?
22.78 Why is the liver unable to utilize ketone bodies as an energy source?
23.79 What is the major metabolic function of the liver?
23.80 What is the fate of lactate produced in skeletal muscle during rapid contraction?
23.81 What are the major fuels of the heart, brain, and liver?
23.82 Why can't the brain use fatty acids as fuel?
23.83 Briefly describe triglyceride metabolism in an adipocyte.
23.84 What is the source of the glycerol molecule that is used in the synthesis of triglycerides?

The Effects of Insulin and Glucagon on Cellular Metabolism

Foundations

23.85 In general, what is the effect of insulin on catabolic and anabolic or biosynthetic processes?
23.86 What is the trigger that causes insulin to be secreted into the bloodstream?
23.87 What is meant by the term *target cell*?
23.88 What are the primary target cells of insulin?
23.89 What is the trigger that causes glucagon to be secreted into the bloodstream?
23.90 What are the primary target cells of glucagon?

Applications

23.91 Where is insulin produced?
23.92 Where is glucagon produced?
23.93 How does insulin affect carbohydrate metabolism?
23.94 How does glucagon affect carbohydrate metabolism?

23.95 How does insulin affect lipid metabolism?
23.96 How does glucagon affect lipid metabolism?
23.97 Why is it said that diabetes mellitus amounts to starvation in the midst of plenty?
23.98 What is the role of the insulin receptor in controlling blood glucose levels?

CRITICAL THINKING PROBLEMS

1. Suppose that fatty acids were degraded by sequential oxidation of the α-carbon. What product(s) would Knoop have obtained with fatty acids with even numbers of carbon atoms? What product(s) would he have obtained with fatty acids with odd numbers of carbon atoms?
2. Oil-eating bacteria can oxidize long-chain alkanes. In the first step of the pathway, the enzyme monooxygenase catalyzes a reaction that converts the long-chain alkane into a primary alcohol. Data from research studies indicate that three more reactions are required to allow the primary alcohol to enter the β-oxidation pathway. Propose a pathway that would convert the long-chain alcohol into a product that could enter the β-oxidation pathway.
3. A young woman sought the advice of her physician because she was 30 pounds overweight. The excess weight was in the form of triglycerides carried in adipose tissue. Yet when the woman described her diet, it became obvious that she actually ate very moderate amounts of fatty foods. Most of her caloric intake was in the form of carbohydrates. This included candy, cake, beer, and soft drinks. Explain how the excess calories consumed in the form of carbohydrates ended up being stored as triglycerides in adipose tissue.
4. Olestra is a fat substitute that provides no calories, yet has a creamy, tongue-pleasing consistency. Because it can withstand heating, it can be used to prepare foods such as potato chips and crackers. Recently the Food and Drug Administration approved olestra for use in prepared foods. Olestra is a sucrose polyester produced by esterification of six, seven, or eight fatty acids to molecules of sucrose. Develop a hypothesis to explain why olestra is not a source of dietary calories.
5. Carnitine is a tertiary amine found in mitochondria that is involved in transporting the acyl groups of fatty acids from the cytoplasm into the mitochondria. The fatty acyl group is transferred from a fatty acyl CoA molecule and esterified to carnitine. Inside the mitochondria, the reaction is reversed and the fatty acid enters the β-oxidation pathway.

    A seventeen-year-old male went to a university medical center complaining of fatigue and poor exercise tolerance. Muscle biopsies revealed droplets of triglycerides in his muscle cells. Biochemical analysis showed that he had only one-fifth of the normal amount of carnitine in his muscle cells.

    What effect will carnitine deficiency have on β-oxidation? What effect will carnitine deficiency have on glucose metabolism?
6. Acetyl CoA carboxylase catalyzes the formation of malonyl CoA from acetyl CoA and the bicarbonate anion, a reaction that requires the hydrolysis of ATP. Write a balanced equation showing this reaction.

    The reaction catalyzed by acetyl CoA carboxylase is the rate-limiting step in fatty acid biosynthesis. The malonyl group is transferred from coenzyme A to acyl carrier protein; similarly, the acetyl group is transferred from coenzyme A to acyl carrier protein. This provides the two beginning substrates of fatty acid biosynthesis shown in Figure 23.10.

    Consider the following case study. A baby boy was brought to the emergency room with severe respiratory distress. Examination revealed muscle pathology, poor growth, and severe brain damage. A liver biopsy revealed that the child didn't make acetyl CoA carboxylase. What metabolic pathway is defective in this child? How is this defect related to the respiratory distress suffered by the baby?

Glossary

A

absolute specificity (19.5) the property of an enzyme that allows it to bind and catalyze the reaction of only one substrate

accuracy (1.4) the nearness of an experimental value to the true value

acetal (13.4) the family of organic compounds formed via the reaction of two molecules of alcohol with an aldehyde in the presence of an acid catalyst; acetals have the following general structure:

$$R^1-\underset{\underset{H}{|}}{\overset{\overset{OR^2}{|}}{C}}-OR^3$$

acetyl coenzyme A (acetyl CoA) (14.4, 22.2) a molecule composed of coenzyme A and an acetyl group; the intermediate that provides acetyl groups for complete oxidation by aerobic respiration

acid (8.1) a substance that behaves as a proton donor

acid anhydride (14.3) the product formed by the combination of an acid chloride and a carboxylate ion; structurally they are two carboxylic acids with a water molecule removed:

$$(Ar)\ R-\overset{\overset{O}{\|}}{C}-O-\overset{\overset{O}{\|}}{C}-R\ (Ar)$$

acid-base reaction (4.3) reaction that involves the transfer of a hydrogen ion (H^+) from one reactant to another

acid chloride (14.3) member of the family of organic compounds with the general formula

$$(Ar)\ R-\overset{\overset{O}{\|}}{C}-Cl$$

activated complex (7.3) the arrangement of atoms at the top of the potential energy barrier as a reaction proceeds

activation energy (7.3) the threshold energy that must be overcome to produce a chemical reaction

active site (19.4) the cleft in the surface of an enzyme that is the site of substrate binding

active transport (17.6) the movement of molecules across a membrane against a concentration gradient

acyl carrier protein (ACP) (23.4) the protein that forms a thioester linkage with fatty acids during fatty acid synthesis

acyl group (14: Intro, 15.3) the functional group found in carboxylic acid derivatives that contains the carbonyl group attached to one alkyl or aryl group:

$$(Ar)\ R-\overset{\overset{O}{\|}}{C}-$$

addition polymer (11.5) polymers prepared by the sequential addition of a monomer

addition reaction (11.5, 13.4) a reaction in which two molecules add together to form a new molecule; often involves the addition of one molecule to a double or triple bond in an unsaturated molecule; e.g., the addition of alcohol to an aldehyde or ketone to form a hemiacetal or hemiketal

adenosine triphosphate (ATP) (15.4, 21.1) a nucleotide composed of the purine adenine, the sugar ribose, and three phosphoryl groups; the primary energy storage and transport molecule used by the cells in cellular metabolism

adipocyte (23.1) a fat cell

adipose tissue (23.1) fatty tissue that stores most of the body lipids

aerobic respiration (22.3) the oxygen-requiring degradation of food molecules and production of ATP

alcohol (12.1) an organic compound that contains a hydroxyl group (—OH) attached to an alkyl group

aldehyde (13.1) a class of organic molecules characterized by a carbonyl group; the carbonyl carbon is bonded to a hydrogen atom and to another hydrogen or an alkyl or aryl group. Aldehydes have the following general structure:

$$(Ar)-\overset{\overset{O}{\|}}{C}-H \qquad R-\overset{\overset{O}{\|}}{C}-H$$

aldol condensation (13.4) a reaction in which aldehydes or ketones react to form a larger molecule

aldose (16.2) a sugar that contains an aldehyde (carbonyl) group

aliphatic hydrocarbon (10.1) any member of the alkanes, alkenes, and alkynes or the substituted alkanes, alkenes, and alkynes

alkali metal (2.4) an element within Group IA (1) of the periodic table

alkaline earth metal (2.4) an element within Group IIA (2) of the periodic table

alkaloid (15.2) a class of naturally occurring compounds that contain one or more nitrogen heterocyclic rings; many of the alkaloids have medicinal and other physiological effects

alkane (10.2) a hydrocarbon that contains only carbon and hydrogen and is bonded together through carbon-hydrogen and carbon-carbon single bonds; a saturated hydrocarbon with the general molecular formula C_nH_{2n+2}

alkene (11.1) a hydrocarbon that contains one or more carbon-carbon double bonds; an unsaturated hydrocarbon with the general formula C_nH_{2n}

alkyl group (10.2) a hydrocarbon group that results from the removal of one hydrogen from the original hydrocarbon (e.g., methyl, —CH_3; ethyl, —CH_2CH_3)

alkyl halide (10.5) a substituted hydrocarbon with the general structure R—X, in which R— represents any alkyl group and X = a halogen (F—, Cl—, Br—, or I—)

alkylammonium ion (15.1) the ion formed when the lone pair of electrons of the nitrogen atom of an amine is shared with a proton (H^+) from a water molecule

alkyne (11.1) a hydrocarbon that contains one or more carbon-carbon triple bonds; an unsaturated hydrocarbon with the general formula C_nH_{2n-2}

allosteric enzyme (19.9) an enzyme that has an effector binding site and an active site; effector binding changes the shape of the active site, rendering it either active or inactive

alpha particle (9.1) a particle consisting of two protons and two neutrons; the alpha particle is identical to a helium nucleus

amide bond (15.3) the bond between the carbonyl carbon of a carboxylic acid and the amino nitrogen of an amine

G-1

amides (15.3) the family of organic compounds formed by the reaction between a carboxylic acid derivative and an amine and characterized by the amide group

amines (15.1) the family of organic molecules with the general formula RNH_2, R_2NH, or R_3N (R— can represent either an alkyl or aryl group); they may be viewed as substituted ammonia molecules in which one or more of the ammonia hydrogens has been substituted by a more complex organic group

α-amino acid (18.2) the subunits of proteins composed of an α-carbon bonded to a carboxylate group, a protonated amino group, a hydrogen atom, and a variable R group

aminoacyl group (15.4) the functional group that is characteristic of an amino acid; the aminoacyl group has the following general structure:

$$\overset{+}{H_3N}-\overset{\overset{\displaystyle H}{|}}{\underset{\underset{\displaystyle R}{|}}{C}}-\overset{\overset{\displaystyle O}{\parallel}}{C}-$$

aminoacyl tRNA (20.6) the transfer RNA covalently linked to the correct amino acid

aminoacyl tRNA binding site of ribosome (A-site) (20.6) a pocket on the surface of a ribosome that holds the aminoacyl tRNA during translation

aminoacyl tRNA synthetase (20.6) an enzyme that recognizes one tRNA and covalently links the appropriate amino acid to it

amorphous solid (5.3) a solid with no organized, regular structure

amphibolic pathway (22.9) a metabolic pathway that functions in both anabolism and catabolism

amphiprotic (8.1) a substance that can behave either as a Brønsted acid or a Brønsted base

amylopectin (16.6) a highly branched form of amylose; the branches are attached to the C-6 hydroxyl by $\alpha(1 \rightarrow 6)$ glycosidic linkage; a component of starch

amylose (16.6) a linear polymer of α-D-glucose molecules bonded in $\alpha(1 \rightarrow 4)$ glycosidic linkage that is a major component of starch; a polysaccharide storage form

anabolism (21.1, 22.9) all of the cellular energy-requiring biosynthetic pathways

anaerobic threshold (21.4) the point at which the level of lactate in the exercising muscle inhibits glycolysis and the muscle, deprived of energy, ceases to function

analgesic (15.2) any drug that acts as a painkiller, e.g., aspirin, acetaminophen

anaplerotic reaction (22.9) a reaction that replenishes a substrate needed for a biochemical pathway

anesthetic (15.2) a drug that causes a lack of sensation in part of the body (local anesthetic) or causes unconsciousness (general anesthetic)

angular structure (3.4) a planar molecule with bond angles other than 180°

anion (2.1) a negatively charged atom or group of atoms

anode (8.5) the positively charged electrode in an electrical cell

anomers (16.4) isomers of cyclic monosaccharides that differ from one another in the arrangement of bonds around the hemiacetal carbon

antibodies (18.1) immunoglobulins; specific glycoproteins produced by cells of the immune system in response to invasion by infectious agents

anticodon (20.4) a sequence of three ribonucleotides on a tRNA that are complementary to a codon on the mRNA; codon-anticodon binding results in delivery of the correct amino acid to the site of protein synthesis

antigen (18.1) any substance that is able to stimulate the immune system; generally a protein or large carbohydrate

antiparallel strands (20.2) a term describing the polarities of the two strands of the DNA double helix; on one strand the sugar-phosphate backbone advances in the $5' \rightarrow 3'$ direction; on the opposite, complementary strand the sugar-phosphate backbone advances in the $3' \rightarrow 5'$ direction

apoenzyme (19.7) the protein portion of an enzyme that requires a cofactor to function in catalysis

aqueous solution (6.1) any solution in which the solvent is water

arachidonic acid (17.2) a fatty acid derived from linoleic acid; the precursor of the prostaglandins

aromatic hydrocarbon (10.1, 11.6) an organic compound that contains the benzene ring or a derivative of the benzene ring

Arrhenius theory (8.1) a theory that describes an acid as a substance that dissociates to produce H^+ and a base as a substance that dissociates to produce OH^-

artificial radioactivity (9.6) radiation that results from the conversion of a stable nucleus to another, unstable nucleus

atherosclerosis (17.4) deposition of excess plasma cholesterol and other lipids and proteins on the walls of arteries, resulting in decreased artery diameter and increased blood pressure

atom (2.1) the smallest unit of an element that retains the properties of that element

atomic mass (2.1) the mass of an atom expressed in atomic mass units

atomic mass unit (4.1) 1/12 of the mass of a ^{12}C atom, equivalent to 1.661×10^{-24} g

atomic number (2.1) the number of protons in the nucleus of an atom; it is a characteristic identifier of an element

atomic orbital (2.3, 2.5) a specific region of space where an electron may be found

ATP synthase (22.6) a multiprotein complex within the inner mitochondrial membrane that uses the energy of the proton (H^+) gradient to produce ATP

autoionization (8.1) also known as *self-ionization*, the reaction of a substance, such as water, with itself to produce a positive and a negative ion

Avogadro's law (5.1) a law that states that the volume is directly proportional to the number of moles of gas particles, assuming that the pressure and temperature are constant

Avogadro's number (4.1) 6.022×10^{23} particles of matter contained in 1 mol of a substance

axial atom (10.4) an atom that lies above or below a cycloalkane ring

B

background radiation (9.7) the radiation that emanates from natural sources

barometer (5.1) a device for measuring pressure

base (8.1) a substance that behaves as a proton acceptor

base pair (20.2) a hydrogen-bonded pair of bases within the DNA double helix; the standard base pairs always involve a purine and a pyrimidine; in particular, adenine always base pairs with thymine and cytosine with guanine

Benedict's reagent (16.4) a buffered solution of Cu^{2+} ions that can be used to test for reducing sugars or to distinguish between aldehydes and ketones

Benedict's test (13.4) a test used to determine the presence of reducing sugars or to distinguish between aldehydes and ketones; it requires a buffered solution of Cu^{2+} ions that are reduced to Cu^+, which precipitates as brick-red Cu_2O

beta particle (9.1) an electron formed in the nucleus by the conversion of a neutron into a proton

bile (23.1) micelles of lecithin, cholesterol, bile salts, protein, inorganic ions, and bile pigments that aid in lipid digestion by emulsifying fat droplets

binding energy (9.3) the energy required to break down the nucleus into its component parts

bioinformatics (20.10) an interdisciplinary field that uses computer information sciences and DNA technology to devise methods for understanding, analyzing, and applying DNA sequence information

boat conformation (10.4) a form of a six-member cycloalkane that resembles a rowboat. It is less stable than the chair conformation because the hydrogen atoms are not perfectly staggered

boiling point (3.3) the temperature at which the vapor pressure of a liquid is equal to the atmospheric pressure

bond energy (3.4) the amount of energy necessary to break a chemical bond

Boyle's law (5.1) a law stating that the volume of a gas varies inversely with the pressure exerted if the temperature and number of moles of gas are constant

breeder reactor (9.4) a nuclear reactor that produces its own fuel in the process of providing electrical energy

Brønsted-Lowry theory (8.1) a theory that describes an acid as a proton donor and a base as a proton acceptor

buffer capacity (8.4) a measure of the ability of a solution to resist large changes in pH when a strong acid or strong base is added

buffer solution (8.4) a solution containing a weak acid or base and its salt (the conjugate base or acid) that is resistant to large changes in pH upon addition of strong acids or bases

buret (8.3) a device calibrated to deliver accurately known volumes of liquid, as in a titration

C

C-terminal amino acid (18.3) the amino acid in a peptide that has a free α-CO_2^- group; the last amino acid in a peptide

calorimetry (7.2) the measurement of heat energy changes during a chemical reaction

cap structure (20.4) a 7-methylguanosine unit covalently bonded to the 5′ end of a mRNA by a 5′–5′ triphosphate bridge

carbinol carbon (12.4) that carbon in an alcohol to which the hydroxyl group is attached

carbohydrate (16.1) generally sugars and polymers of sugars; the primary source of energy for the cell

carbonyl group (13: Intro) the functional group that contains a carbon-oxygen double bond: —C=O; the functional group found in aldehydes and ketones

carboxyl group (14.1) the —COOH functional group; the functional group found in carboxylic acids

carboxylic acid (14.1) a member of the family of organic compounds that contain the —COOH functional group

carboxylic acid derivative (14.2) any of several families of organic compounds, including the esters and amides, that are derived from carboxylic acids and have the general formula

Z = —OR or OAr for the esters, and Z = —NH₂ for the amides

carcinogen (20.7) any chemical or physical agent that causes mutations in the DNA that lead to uncontrolled cell growth or cancer

catabolism (21.1, 22.9) the degradation of fuel molecules and production of ATP for cellular functions

catalyst (7.3) any substance that increases the rate of a chemical reaction (by lowering the activation energy of the reaction) and that is not destroyed in the course of the reaction

cathode (8.5) the negatively charged electrode in an electrical cell

cathode rays (2.2) a stream of electrons that is given off by the cathode (negative electrode) in a cathode ray tube

cation (2.1) a positively charged atom or group of atoms

cellulose (16.6) a polymer of β-D-glucose linked by β(1 → 4) glycosidic bonds

central dogma (20.4) a statement of the directional transfer of the genetic information in cells: DNA → RNA → Protein

chain reaction (9.4) the process in a fission reactor that involves neutron production and causes subsequent reactions accompanied by the production of more neutrons in a continuing process

chair conformation (10.4) the most energetically favorable conformation for a six-member cycloalkane; so-called for its resemblance to a lawn chair

Charles's law (5.1) a law stating that the volume of a gas is directly proportional to the temperature of the gas, assuming that the pressure and number of moles of the gas are constant

chemical bond (3.1) the attractive force holding two atomic nuclei together in a chemical compound

chemical equation (4.3) a record of chemical change, showing the conversion of reactants to products

chemical formula (4.2) the representation of a compound or ion in which elemental symbols represent types of atoms and subscripts show the relative numbers of atoms

chemical property (1.2) characteristics of a substance that relate to the substance's participation in a chemical reaction

chemical reaction (1.2) a process in which atoms are rearranged to produce new combinations

chemistry (1.1) the study of matter and the changes that matter undergoes

chiral carbon (16.3) a carbon atom bonded to four different atoms or groups of atoms

chiral molecule (16.3) molecule capable of existing in mirror-image forms

cholesterol (17.4) a twenty-seven-carbon steroid ring structure that serves as the precursor of the steroid hormones

chromosome (20.2) a piece of DNA that carries all the genetic instructions, or genes, of an organism

chylomicron (17.5, 23.1) a plasma lipoprotein (aggregate of protein and triglycerides) that carries triglycerides from the intestine to all body tissues via the bloodstream

cis-trans isomers (10.3) isomers that differ from one another in the placement of substituents on a double bond or ring

citric acid cycle (22.4) a cyclic biochemical pathway that is the final stage of degradation of carbohydrates, fats, and amino acids. It results in the complete oxidation of acetyl groups derived from these dietary fuels

cloning vector (20.8) a DNA molecule that can carry a cloned DNA fragment into a cell and that has a replication origin that allows the DNA to be replicated abundantly within the host cell

coagulation (18.10) the process by which proteins in solution are denatured and aggregate with one another to produce a solid

codon (20.4) a group of three ribonucleotides on the mRNA that specifies the addition of a specific amino acid onto the growing peptide chain

coenzyme (19.7) an organic group required by some enzymes; it generally serves as a donor or acceptor of electrons or a functional group in a reaction

coenzyme A (22.2) a molecule derived from ATP and the vitamin pantothenic acid; coenzyme A functions in the transfer of acetyl groups in lipid and carbohydrate metabolism

cofactor (19.7) an inorganic group, usually a metal ion, that must be bound to an apoenzyme to maintain the correct configuration of the active site

colipase (23.1) a protein that aids in lipid digestion by binding to the surface of lipid droplets and facilitating binding of pancreatic lipase

colligative property (6.4) property of a solution that is dependent only on the concentration of solute particles

colloidal suspension (6.1) a heterogeneous mixture of solute particles in a solvent; distribution of solute particles is not uniform because of the size of the particles

combination reaction (4.3) a reaction in which two substances join to form another substance

combined gas law (5.1) an equation that describes the behavior of a gas when volume, pressure, and temperature may change simultaneously

combustion (10.5) the oxidation of hydrocarbons by burning in the presence of air to produce carbon dioxide and water

competitive inhibitor (19.10) a structural analog; a molecule that has a structure very similar to the natural substrate of an enzyme, competes with the natural substrate for binding to the enzyme active site, and inhibits the reaction

complementary strands (20.2) the opposite strands of the double helix are hydrogen-bonded to one another such that adenine and thymine or guanine and cytosine are always paired

complete protein (18.11) a protein source that contains all the essential and nonessential amino acids

complex lipid (17.5) a lipid bonded to other types of molecules

compound (1.2) a substance that is characterized by constant composition and that can be chemically broken down into elements

concentration (1.5, 6.2) a measure of the quantity of a substance contained in a specified volume of solution

condensation (5.2) the conversion of a gas to a liquid

condensation polymer (14.2) a polymer, which is a large molecule formed by combination of many small molecules (monomers) that results from joining of monomers in a reaction that forms a small molecule, such as water or an alcohol

condensed formula (10.2) a structural formula showing all of the atoms in a molecule and placing them in a sequential arrangement that details which atoms are bonded to each other; the bonds themselves are not shown

conformations, conformers (10.4) discrete, distinct isomeric structures that may be converted, one to the other, by rotation about the bonds in the molecule

conjugate acid (8.1) substance that has one more proton than the base from which it is derived

conjugate acid-base pair (8.1) two species related to each other through the gain or loss of a proton

conjugate base (8.1) substance that has one less proton than the acid from which it is derived

constitutional isomers (10.2) two molecules having the same molecular formulas, but different chemical structures

Cori Cycle (21.6) a metabolic pathway in which the lactate produced by working muscle is taken up by cells in the liver and converted back to glucose by gluconeogenesis

corrosion (8.5) the unwanted oxidation of a metal

covalent bond (3.1) a pair of electrons shared between two atoms

covalent solid (5.3) a collection of atoms held together by covalent bonds

crenation (6.4) the shrinkage of red blood cells caused by water loss to the surrounding medium

cristae (22.1) the folds of the inner membrane of the mitochondria

crystal lattice (3.2) a unit of a solid characterized by a regular arrangement of components

crystalline solid (5.3) a solid having a regular repeating atomic structure

curie (9.8) the quantity of radioactive material that produces 3.7×10^{10} nuclear disintegrations per second

cycloalkane (10.3) a cyclic alkane; a saturated hydrocarbon that has the general formula C_nH_{2n}

D

Dalton's law (5.1) also called the law of partial pressures; states that the total pressure exerted by a gas mixture is the sum of the partial pressures of the component gases

data (1.3) a group of facts resulting from an experiment

decomposition reaction (4.3) the breakdown of a substance into two or more substances

defense proteins (18.1) proteins that defend the body against infectious diseases. Antibodies are defense proteins

degenerate code (20.5) a term used to describe the fact that several triplet codons may be used to specify a single amino acid in the genetic code

dehydration (of alcohols) (12.5) a reaction that involves the loss of a water molecule, in this case the loss of water from an alcohol and the simultaneous formation of an alkene

deletion mutation (20.7) a mutation that results in the loss of one or more nucleotides from a DNA sequence

denaturation (18.10) the process by which the organized structure of a protein is disrupted, resulting in a completely disorganized, nonfunctional form of the protein

density (1.5) mass per unit volume of a substance

deoxyribonucleic acid (DNA) (20.1) the nucleic acid molecule that carries all of the genetic information of an organism; the DNA molecule is a double helix composed of two strands, each of which is composed of phosphate groups, deoxyribose, and the nitrogenous bases thymine, cytosine, adenine, and guanine

deoxyribonucleotide (20.1) a nucleoside phosphate or nucleotide composed of a nitrogenous base in β-N-glycosidic linkage to the 1′ carbon of the sugar 2′-deoxyribose and with one, two, or three phosphoryl groups esterified at the hydroxyl of the 5′ carbon

diabetes mellitus (23.3) a disease caused by the production of insufficient levels of insulin and characterized by the appearance of very high levels of glucose in the blood and urine

dialysis (6.6) the removal of waste material via transport across a membrane

diglyceride (17.3) the product of esterification of glycerol at two positions

dipole-dipole interactions (5.2) attractive forces between polar molecules

disaccharide (16.1) a sugar composed of two monosaccharides joined through an oxygen atom bridge

dissociation (3.3) production of positive and negative ions when an ionic compound dissolves in water

disulfide (12.9) an organic compound that contains a disulfide group (—S—S—)

DNA polymerase III (20.3) the enzyme that catalyzes the polymerization of daughter DNA strands using the parental strand as a template

double bond (3.4) a bond in which two pairs of electrons are shared by two atoms

double helix (20.2) the spiral staircase-like structure of the DNA molecule characterized by two sugar-phosphate backbones wound around the outside and nitrogenous bases extending into the center

double-replacement reaction (4.3) a chemical change in which cations and anions "exchange partners"

dynamic equilibrium (7.4) the state that exists when the rate of change in the concentration of products and reactants is equal, resulting in no net concentration change

E

eicosanoid (17.2) any of the derivatives of twenty-carbon fatty acids, including the prostaglandins, leukotrienes, and thromboxanes

electrolysis (8.5) an electrochemical process that uses electrical energy to cause nonspontaneous oxidation-reduction reactions to occur

electrolyte (3.3, 6.1) a material that dissolves in water to produce a solution that conducts an electrical current

electrolytic solution (3.3) a solution composed of an electrolytic solute dissolved in water

electromagnetic radiation (2.3) energy that is propagated as waves at the speed of light

electromagnetic spectrum (2.3) the complete range of electromagnetic waves

electron (2.1) a negatively charged particle outside of the nucleus of an atom

electron affinity (2.7) the energy released when an electron is added to an isolated atom

electron configuration (2.5) the arrangement of electrons around a nucleus of an atom, ion, or a collection of nuclei of a molecule

electron density (2.3) the probability of finding the electron in a particular location

electron transport system (22.6) the series of electron transport proteins embedded in the inner mitochondrial membrane that accept high-energy electrons from NADH and $FADH_2$ and transfer them in stepwise fashion to molecular oxygen (O_2)

electronegativity (3.1) a measure of the tendency of an atom in a molecule to attract shared electrons

element (1.2) a substance that cannot be decomposed into simpler substances by chemical or physical means

elimination reaction (12.5) a reaction in which a molecule loses atoms or ions from its structure

elongation factor (20.6) proteins that facilitate the elongation phase of translation

emulsifying agent (17.3) a bipolar molecule that aids in the suspension of fats in water

enantiomers (16.3) stereoisomers that are nonsuperimposable mirror images of one another

endothermic reaction (7.1) a chemical or physical change in which energy is absorbed

energy (1.1) the capacity to do work

energy level (2.3) one of numerous atomic regions where electrons may be found

enthalpy (7.1) a term that represents heat energy

entropy (7.1) a measure of randomness or disorder

enzyme (18.1, 19: Intro)   a protein that serves as a biological catalyst

enzyme specificity (19.5)   the ability of an enzyme to bind to only one, or a very few, substrates and thus catalyze only a single reaction

enzyme-substrate complex (19.4)   a molecular aggregate formed when the substrate binds to the active site of the enzyme

equatorial atom (10.4)   an atom that lies in the plane of a cycloalkane ring

equilibrium reaction (7.4)   a reaction that is reversible and the rates of the forward and reverse reactions are equal

equivalence point (8.3)   the situation in which reactants have been mixed in the molar ratio corresponding to the balanced equation

equivalent (6.3)   the number of grams of an ion corresponding to Avogadro's number of electrical charges

error (1.4)   the difference between the true value and the experimental value for data or results

essential amino acid (18.11)   an amino acid that cannot be synthesized by the body and must therefore be supplied by the diet

essential fatty acids (17.2)   the fatty acids linolenic and linoleic acids that must be supplied in the diet because they cannot be synthesized by the body

ester (14.2)   a carboxylic acid derivative formed by the reaction of a carboxylic acid and an alcohol. Esters have the following general formula:

$$\underset{R-C-OR}{\overset{O}{\parallel}} \quad \underset{R-C-O(Ar)}{\overset{O}{\parallel}} \quad \underset{(Ar)-C-O(Ar)}{\overset{O}{\parallel}}$$

esterification (17.2)   the formation of an ester in the reaction of a carboxylic acid and an alcohol

ether (12.8)   an organic compound that contains two alkyl and/or aryl groups attached to an oxygen atom; R—O—R, Ar—O—R, and Ar—O—Ar

eukaryote (20.2)   an organism having cells containing a true nucleus enclosed by a nuclear membrane and having a variety of membrane-bound organelles that segregate different cellular functions into different compartments

evaporation (5.2)   the conversion of a liquid to a gas below the boiling point of the liquid

exon (20.4)   protein-coding sequences of a gene found on the final mature mRNA

exothermic reaction (7.1)   a chemical or physical change that releases energy

extensive property (1.2)   a property of a substance that depends on the quantity of the substance

F

F_0F_1 complex (22.6)   an alternative term for the ATP synthase, the multiprotein complex in the inner mitochondrial membrane that uses the energy of the proton gradient to produce ATP

facilitated diffusion (17.6)   movement of a solute across a membrane from an area of high concentration to an area of low concentration through a transmembrane protein, or permease

fatty acid (14.1, 17.2)   any member of the family of continuous-chain carboxylic acids that generally contain four to twenty carbon atoms; the most concentrated source of energy used by the cell

feedback inhibition (19.9)   the process whereby excess product of a biosynthetic pathway turns off the entire pathway for its own synthesis

fermentation (12.3, 21.4)   anaerobic (in the absence of oxygen) catabolic reactions that occur with no net oxidation. Pyruvate or an organic compound produced from pyruvate is reduced as NADH is oxidized

fibrous protein (18.5)   a protein composed of peptides arranged in long sheets or fibers

Fischer Projection (16.3)   a two-dimensional drawing of a molecule, which shows a chiral carbon at the intersection of two lines and horizontal lines representing bonds projecting out of the page and vertical lines representing bonds that project into the page

fission (9.4)   the splitting of heavy nuclei into lighter nuclei accompanied by the release of large quantities of energy

fluid mosaic model (17.6)   the model of membrane structure that describes the fluid nature of the lipid bilayer and the presence of numerous proteins embedded within the membrane

formula (3.2)   the representation of the fundamental compound unit using chemical symbols and numerical subscripts

formula unit (4.2)   the smallest collection of atoms from which the formula of a compound can be established

formula weight (4.2)   the mass of a formula unit of a compound relative to a standard (carbon-12)

free energy (7.1)   the combined contribution of entropy and enthalpy for a chemical reaction

fructose (16.4)   a ketohexose that is also called levulose and fruit sugar; the sweetest of all sugars, abundant in honey and fruits

fuel value (7.2)   the amount of energy derived from a given mass of material

functional group (10.1)   an atom (or group of atoms and their bonds) that imparts specific chemical and physical properties to a molecule

fusion (9.4)   the joining of light nuclei to form heavier nuclei, accompanied by the release of large amounts of energy

G

galactose (16.4)   an aldohexose that is a component of lactose (milk sugar)

galactosemia (16.5)   a human genetic disease caused by the inability to convert galactose to a phosphorylated form of glucose (glucose-1-phosphate) that can be used in cellular metabolic reactions

gamma ray (9.1)   a high-energy emission from nuclear processes, traveling at the speed of light; the high-energy region of the electromagnetic spectrum

gaseous state (1.2)   a physical state of matter characterized by a lack of fixed shape or volume and ease of compressibility

genome (20.2)   the complete set of genetic information in all the chromosomes of an organism

geometric isomer (10.3, 11.3)   an isomer that differs from another isomer in the placement of substituents on a double bond or a ring

globular protein (18.6)   a protein composed of polypeptide chains that are tightly folded into a compact spherical shape

glucagon (21.7, 23.6)   a peptide hormone synthesized by the α-cells of the islets of Langerhans in the pancreas and secreted in response to low blood glucose levels; glucagon promotes glycogenolysis and gluconeogenesis and thereby increases the concentration of blood glucose

gluconeogenesis (21.6)   the synthesis of glucose from noncarbohydrate precursors

glucose (16.4)   an aldohexose, the most abundant monosaccharide; it is a component of many disaccharides, such as lactose and sucrose, and of polysaccharides, such as cellulose, starch, and glycogen

glyceraldehyde (16.3)   an aldotriose that is the simplest carbohydrate; phosphorylated forms of glyceraldehyde are important intermediates in cellular metabolic reactions

glyceride (17.3)   a lipid that contains glycerol

glycogen (16.6, 21.7)   a long, branched polymer of glucose stored in liver and muscles of animals; it consists of a linear backbone of α-D-glucose in α(1 → 4) linkage, with numerous short branches attached to the C-6 hydroxyl group by α(1 → 6) linkage

glycogenesis (21.7)   the metabolic pathway that results in the addition of glucose to growing glycogen polymers when blood glucose levels are high

glycogen granule (21.7)   a core of glycogen surrounded by enzymes responsible for glycogen synthesis and degradation

glycogenolysis (21.7)   the biochemical pathway that results in the removal of glucose molecules from glycogen polymers when blood glucose levels are low

glycolysis (21.3)   the enzymatic pathway that converts a glucose molecule into two molecules of pyruvate; this anaerobic process generates a net energy yield of two molecules of ATP and two molecules of NADH

glycoprotein (18.7)   a protein bonded to sugar groups

glycosidic bond (16.1)   the bond between the hydroxyl group of the C-1 carbon of one sugar and a hydroxyl group of another sugar

group (2.4)   any one of eighteen vertical columns of elements; often referred to as a family

group specificity (19.5) an enzyme that catalyzes reactions involving similar substrate molecules having the same functional groups

guanosine triphosphate (GTP) (21.6) a nucleotide composed of the purine guanosine, the sugar ribose, and three phosphoryl groups

H

half-life ($t_{1/2}$) (9.3) the length of time required for one-half of the initial mass of an isotope to decay to products

halogen (2.4) an element found in Group VIIA (17) of the periodic table

halogenation (10.5, 11.5) a reaction in which one of the C—H bonds of a hydrocarbon is replaced with a C—X bond (X = Br or Cl generally)

Haworth projection (16.4) a means of representing the orientation of substituent groups around a cyclic sugar molecule

α-helix (18.5) a right-handed coiled secondary structure maintained by hydrogen bonds between the amide hydrogen of one amino acid and the carbonyl oxygen of an amino acid four residues away

heme group (18.9) the chemical group found in hemoglobin and myoglobin that is responsible for the ability to carry oxygen

hemiacetal (13.4, 16.4) the family of organic compounds formed via the reaction of one molecule of alcohol with an aldehyde in the presence of an acid catalyst; hemiacetals have the following general structure:

$$\begin{array}{c} \text{OH} \\ | \\ R^1\text{—C—OR}^2 \\ | \\ \text{H} \end{array}$$

hemiketal (13.4, 16.4) the family of organic compounds formed via the reaction of one molecule of alcohol with a ketone in the presence of an acid catalyst; hemiketals have the following general structure:

$$\begin{array}{c} \text{OH} \\ | \\ R^1\text{—C—OR}^3 \\ | \\ R^2 \end{array}$$

hemoglobin (18.9) the major protein component of red blood cells; the function of this red, iron-containing protein is transport of oxygen

hemolysis (6.4) the rupture of red blood cells resulting from movement of water from the surrounding medium into the cell

Henderson-Hasselbalch equation (8.4) an equation for calculating the pH of a buffer system:

$$pH = pKa + \log \frac{[\text{conjugate base}]}{[\text{weak acid}]}$$

Henry's law (6.1) a law stating that the number of moles of a gas dissolved in a liquid at a given temperature is proportional to the partial pressure of the gas

heterocyclic amine (15.2) a heterocyclic compound that contains nitrogen in at least one position in the ring skeleton

heterocyclic aromatic compound (11.7) cyclic aromatic compound having at least one atom other than carbon in the structure of the aromatic ring

heterogeneous mixture (1.2) a mixture of two or more substances characterized by nonuniform composition

hexose (16.2) a six-carbon monosaccharide

high-density lipoprotein (HDL) (17.5) a plasma lipoprotein that transports cholesterol from peripheral tissue to the liver

holoenzyme (19.7) an active enzyme consisting of an apoenzyme bound to a cofactor

homogeneous mixture (1.2) a mixture of two or more substances characterized by uniform composition

hybridization (20.8) a technique for identifying DNA or RNA sequences that is based on specific hydrogen bonding between a radioactive probe and complementary DNA or RNA sequences

hydrate (4.2) any substance that has water molecules incorporated in its structure

hydration (11.5, 12.5) a reaction in which water is added to a molecule, e.g., the addition of water to an alkene to form an alcohol

hydrocarbon (10.1) a compound composed solely of the elements carbon and hydrogen

hydrogen bonding (5.2) the attractive force between a hydrogen atom covalently bonded to a small, highly electronegative atom and another atom containing an unshared pair of electrons

hydrogenation (11.5, 13.4, 17.2) a reaction in which hydrogen (H_2) is added to a double or a triple bond

hydrohalogenation (11.5) the addition of a hydrohalogen (HCl, HBr, or HI) to an unsaturated bond

hydrolase (19.1) an enzyme that catalyzes hydrolysis reactions

hydrolysis (14.2) a chemical change that involves the reaction of a molecule with water; the process by which molecules are broken into their constituents by addition of water

hydronium ion (8.1) a protonated water molecule, H_3O^+

hydrophilic amino acid (18.1) "water loving"; a polar or ionic amino acid that has a high affinity for water

hydrophobic amino acid (18.2) "water fearing"; a nonpolar amino acid that prefers contact with other nonpolar amino acids over contact with water

hydroxyl group (12.1) the —OH functional group that is characteristic of alcohols

hyperammonemia (22.8) a genetic defect in one of the enzymes of the urea cycle that results in toxic or even fatal elevation of the concentration of ammonium ions in the body

hyperglycemia (21.7) blood glucose levels that are higher than normal

hypertonic solution (6.4, 17.6) the more concentrated solution of two separated by a semipermeable membrane

hypoglycemia (21.7) blood glucose levels that are lower than normal

hypothesis (1.1) an attempt to explain observations in a commonsense way

hypotonic solution (6.4, 17.6) the more dilute solution of two separated by a semipermeable membrane

I

ideal gas (5.1) a gas in which the particles do not interact and the volume of the individual gas particles is assumed to be negligible

ideal gas law (5.1) a law stating that for an ideal gas the product of pressure and volume is proportional to the product of the number of moles of the gas and its temperature; the proportionality constant for an ideal gas is symbolized R

incomplete protein (18.11) a protein source that does not contain all the essential and nonessential amino acids

indicator (8.3) a solute that shows some condition of a solution (such as acidity or basicity) by its color

induced fit model (19.4) the theory of enzyme-substrate binding that assumes that the enzyme is a flexible molecule and that both the substrate and the enzyme change their shapes to accommodate one another as the enzyme-substrate complex forms

initiation factors (20.6) proteins that are required for formation of the translation initiation complex, which is composed of the large and small ribosomal subunits, the mRNA, and the initiator tRNA, methionyl tRNA

inner mitochondrial membrane (22.1) the highly folded, impermeable membrane within the mitochondrion that is the location of the electron transport system and ATP synthase

insertion mutation (20.7) a mutation that results in the addition of one or more nucleotides to a DNA sequence

insulin (21.7, 23.6) a hormone released from the pancreas in response to high blood glucose levels; insulin stimulates glycogenesis, fat storage, and cellular uptake and storage of glucose from the blood

intensive property (1.2) a property of a substance that is independent of the quantity of the substance

intermembrane space (22.1) the region between the outer and inner mitochondrial membranes, which is the location of the proton (H^+) reservoir that drives ATP synthesis

intermolecular force (3.5) any attractive force that occurs between molecules

intramolecular force (3.5) any attractive force that occurs within molecules

intron (20.4) a noncoding sequence within a eukaryotic gene that must be removed from the primary transcript to produce a functional mRNA

ion (2.1)   an electrically charged particle formed by the gain or loss of electrons

ionic bonding (3.1)   an electrostatic attractive force between ions resulting from electron transfer

ionic solid (5.3)   a solid composed of positive and negative ions in a regular three-dimensional crystalline arrangement

ionization energy (2.7)   the energy needed to remove an electron from an atom in the gas phase

ionizing radiation (9.1)   radiation that is sufficiently high in energy to cause ion formation upon impact with an atom

ion pair (3.1)   the simplest formula unit for an ionic compound

ion product for water (8.1)   the product of the hydronium and hydroxide ion concentrations in pure water at a specified temperature; at 25°C, it has a value of 1.0×10^{-14}

irreversible enzyme inhibitor (19.10)   a chemical that binds strongly to the R groups of an amino acid in the active site and eliminates enzyme activity

isoelectric point (18.10)   a situation in which a protein has an equal number of positive and negative charges and therefore has an overall net charge of zero

isoelectronic (2.6)   atoms, ions, and molecules containing the same number of electrons

isomerase (19.1)   an enzyme that catalyzes the conversion of one isomer to another

isotonic solution (6.4, 17.6)   a solution that has the same solute concentration as another solution with which it is being compared; a solution that has the same osmotic pressure as a solution existing within a cell

isotope (2.1)   atom of the same element that differs in mass because it contains different numbers of neutrons

I.U.P.A.C. Nomenclature System (10.2)   the International Union of Pure and Applied Chemistry (I.U.P.A.C.) standard, universal system for the nomenclature of organic compounds

K

α-keratin (18.5)   a member of the family of fibrous proteins that form the covering of most land animals; major components of fur, skin, beaks, and nails

ketal (13.4)   the family of organic compounds formed via the reaction of two molecules of alcohol with a ketone in the presence of an acid catalyst; ketals have the following general structure:

$$\begin{array}{c} OR^3 \\ | \\ R^1{-}C{-}OR^4 \\ | \\ R^2 \end{array}$$

ketoacidosis (23.3)   a drop in the pH of the blood caused by elevated levels of ketone bodies

ketone (13.1)   a family of organic molecules characterized by a carbonyl group; the carbonyl carbon is bonded to two alkyl groups, two aryl groups, or one alkyl and one aryl group; ketones have the following general structures:

$$\underset{R-C-R}{\overset{O}{\|}} \quad \underset{R-C-(Ar)}{\overset{O}{\|}} \quad \underset{(Ar)-C-(Ar)}{\overset{O}{\|}}$$

ketone bodies (23.3)   acetone, acetoacetone, and β-hydroxybutyrate produced from fatty acids in the liver via acetyl CoA

ketose (16.2)   a sugar that contains a ketone (carbonyl) group

ketosis (23.3)   an abnormal rise in the level of ketone bodies in the blood

kinetic energy (1.5)   the energy resulting from motion of an object [kinetic energy = 1/2 (mass)(velocity)^2]

kinetic-molecular theory (5.1)   the fundamental model of particle behavior in the gas phase

kinetics (7.3)   the study of rates of chemical reactions

L

lactose (16.5)   a disaccharide composed of β-D-galactose and either α- or β-D-glucose in β(1 → 4) glycosidic linkage; milk sugar

lactose intolerance (16.5)   the inability to produce the digestive enzyme lactase, which degrades lactose to galactose and glucose

lagging strand (20.3)   in DNA replication, the strand that is synthesized discontinuously from numerous RNA primers

law (1.1)   a summary of a large quantity of information

law of conservation of mass (4.3)   a law stating that, in chemical change, matter cannot be created or destroyed

leading strand (20.3)   in DNA replication, the strand that is synthesized continuously from a single RNA primer

LeChatelier's principle (7.4)   a law stating that when a system at equilibrium is disturbed, the equilibrium shifts in the direction that minimizes the disturbance

lethal dose (LD$_{50}$) (9.8)   the quantity of toxic material (such as radiation) that causes the death of 50% of a population of an organism

Lewis symbol (3.1)   representation of an atom or ion using the atomic symbol (for the nucleus and core electrons) and dots to represent valence electrons

ligase (19.1)   an enzyme that catalyzes the joining of two molecules

linear structure (3.4)   the structure of a molecule in which the bond angles about the central atom(s) is (are) 180°

line formula (10.2)   the simplest representation of a molecule in which it is assumed that there is a carbon atom at any location where two or more lines intersect, there is a carbon at the end of any line, and each carbon is bonded to the correct number of hydrogen atoms

linkage specificity (19.5)   the property of an enzyme that allows it to catalyze reactions involving only one kind of bond in the substrate molecule

lipase (23.1)   an enzyme that hydrolyzes the ester linkage between glycerol and the fatty acids of triglycerides

lipid (17.1)   a member of the group of biological molecules of varying composition that are classified together on the basis of their solubility in nonpolar solvents

liquid state (1.2)   a physical state of matter characterized by a fixed volume and the absence of a fixed shape

lock-and-key model (19.4)   the theory of enzyme-substrate binding that depicts enzymes as inflexible molecules; the substrate fits into the rigid active site in the same way a key fits into a lock

London forces (5.2)   weak attractive forces between molecules that result from short-lived dipoles that occur because of the continuous movement of electrons in the molecules

lone pair (3.4)   an electron pair that is not involved in bonding

low-density lipoprotein (LDL) (17.5)   a plasma lipoprotein that carries cholesterol to peripheral tissues and helps to regulate cholesterol levels in those tissues

lyase (19.1)   an enzyme that catalyzes a reaction involving double bonds

M

maltose (16.5)   a disaccharide composed of α-D-glucose and a second glucose molecule in α(1 → 4) glycosidic linkage

Markovnikov's rule (11.5)   the rule stating that a hydrogen atom, adding to a carbon-carbon double bond, will add to the carbon having the larger number of hydrogens attached to it

mass (1.5)   a quantity of matter

mass number (2.1)   the sum of the number of protons and neutrons in an atom

matrix space (22.1)   the region of the mitochondrion within the inner membrane; the location of the enzymes that carry out the reactions of the citric acid cycle and β-oxidation of fatty acids

matter (1.1)   the material component of the universe

melting point (3.3, 5.3)   the temperature at which a solid converts to a liquid

messenger RNA (20.4)   an RNA species produced by transcription and that specifies the amino acid sequence for a protein

metal (2.4)   an element located on the left side of the periodic table (left of the "staircase" boundary)

metallic bond (5.3)   a bond that results from the orbital overlap of metal atoms

metallic solid (5.3)   a solid composed of metal atoms held together by metallic bonds

metalloid (2.4) an element along the "staircase" boundary between metals and nonmetals; metalloids exhibit both metallic and nonmetallic properties

metastable isotope (9.2) an isotope that will give up some energy to produce a more stable form of the same isotope

micelle (23.1) an aggregation of molecules having nonpolar and polar regions; the nonpolar regions of the molecules aggregate, leaving the polar regions facing the surrounding water

mitochondria (22.1) the cellular "power plants" in which the reactions of the citric acid cycle, the electron transport system, and ATP synthase function to produce ATP

mixture (1.2) a material composed of two or more substances

molality (6.4) the number of moles of solute per kilogram of solvent

molar mass (4.1) the mass in grams of 1 mol of a substance

molar volume (5.1) the volume occupied by 1 mol of a substance

molarity (6.3) the number of moles of solute per liter of solution

mole (4.1) the amount of substance containing Avogadro's number of particles

molecular formula (10.2) a formula that provides the atoms and number of each type of atom in a molecule but gives no information regarding the bonding pattern involved in the structure of the molecule

molecular solid (5.3) a solid in which the molecules are held together by dipole-dipole and London forces (van der Waals forces)

molecule (3.2) a unit in which the atoms of two or more elements are held together by chemical bonds

monatomic ion (3.2) an ion formed by electron gain or loss from a single atom

monoglyceride (17.3) the product of the esterification of glycerol at one position

monomer (11.5) the individual molecules from which a polymer is formed

monosaccharide (16.1) the simplest type of carbohydrate consisting of a single saccharide unit

movement protein (18.1) a protein involved in any aspect of movement in an organism, for instance actin and myosin in muscle tissue and flagellin that composes bacterial flagella

mutagen (20.7) any chemical or physical agent that causes changes in the nucleotide sequence of a gene

mutation (20.7) any change in the nucleotide sequence of a gene

myoglobin (18.9) the oxygen storage protein found in muscle

N

N-terminal amino acid (18.3) the amino acid in a peptide that has a free α-N^+H_3 group; the first amino acid of a peptide

natural radioactivity (9.6) the spontaneous decay of a nucleus to produce high-energy particles or rays

negative allosterism (19.9) effector binding inactivates the active site of an allosteric enzyme

neurotransmitter (15.5) a chemical that carries a message, or signal, from a nerve cell to a target cell

neutral glyceride (17.3) the product of the esterification of glycerol at one, two, or three positions

neutralization (8.3) the reaction between an acid and a base

neutron (2.1) an uncharged particle, with the same mass as the proton, in the nucleus of an atom

nicotinamide adenine dinucleotide (NAD^+) (21.3) a molecule synthesized from the vitamin niacin and the nucleotide ATP and that serves as a carrier of hydride anions; a coenzyme that is an oxidizing agent used in a variety of metabolic processes

noble gas (2.4) elements in Group VIIIA (18) of the periodic table

nomenclature (3.2) a system for naming chemical compounds

nonelectrolyte (3.3, 6.1) a substance that, when dissolved in water, produces a solution that does not conduct an electrical current

nonessential amino acid (18.11) any amino acid that can be synthesized by the body

nonmetal (2.4) an element located on the right side of the periodic table (right of the "staircase" boundary)

nonreducing sugar (16.5) a sugar that cannot be oxidized by Benedict's or Tollens' reagent

normal boiling point (5.2) the temperature at which a substance will boil at 1 atm of pressure

nuclear equation (9.2) a balanced equation accounting for the products and reactants in a nuclear reaction

nuclear imaging (9.6) the generation of images of components of the body (organs, tissues) using techniques based on the measurement of radiation

nuclear medicine (9.6) a field of medicine that uses radioisotopes for diagnostic and therapeutic purposes

nuclear reactor (9.6) a device for conversion of nuclear energy into electrical energy

nucleosome (20.2) the first level of chromosome structure consisting of a strand of DNA wrapped around a small disk of histone proteins

nucleotide (20.1, 21.1) a molecule composed of a nitrogenous base, a five-carbon sugar, and one, two, or three phosphoryl groups

nucleus (2.1) the small, dense center of positive charge in the atom

nuclide (9.1) any atom characterized by an atomic number and a mass number

nutrient protein (18.1) a protein that serves as a source of amino acids for embryos or infants

nutritional Calorie (7.2) equivalent to one kilocalorie (1000 calories); also known as a large Calorie

O

octet rule (2.6) a rule predicting that atoms form the most stable molecules or ions when they are surrounded by eight electrons in their highest occupied energy level

oligosaccharide (16.1) an intermediate-sized carbohydrate composed of from three to ten monosaccharides

order of the reaction (7.3) the exponent of each concentration term in the rate equation

osmolarity (6.4) molarity of particles in solution; this value is used for osmotic pressure calculations

osmosis (6.4, 17.6) net flow of a solvent across a semipermeable membrane in response to a concentration gradient

osmotic pressure (6.4, 17.6) the net force with which water enters a solution through a semipermeable membrane; alternatively, the pressure required to stop net transfer of solvent across a semipermeable membrane

outer mitochondrial membrane (22.1) the membrane that surrounds the mitochondrion and separates it from the contents of the cytoplasm; it is highly permeable to small "food" molecules

β-oxidation (23.2) the biochemical pathway that results in the oxidation of fatty acids and the production of acetyl CoA

oxidation (8.5, 12.6, 13.4, 14.1) a loss of electrons; in organic compounds it may be recognized as a loss of hydrogen atoms or the gain of oxygen

oxidation-reduction reaction (4.3, 9.5) also called redox reaction, a reaction involving the transfer of one or more electrons from one reactant to another

oxidative deamination (22.7) an oxidation-reduction reaction in which NAD^+ is reduced and the amino acid is deaminated

oxidative phosphorylation (21.3, 22.6) production of ATP using the energy of electrons harvested during biological oxidation-reduction reactions

oxidizing agent (8.5) a substance that oxidizes, or removes electrons from, another substance; the oxidizing agent is reduced in the process

oxidoreductase (19.1) an enzyme that catalyzes an oxidation-reduction reaction

P

pancreatic serine proteases (19.11) a family of proteolytic enzymes, including trypsin, chymotrypsin, and elastase, that arose by divergent evolution

parent compound or parent chain (10.2) in the I.U.P.A.C. Nomenclature System the parent compound is the longest carbon-carbon chain containing the principal functional group in the molecule that is being named

partial pressure (5.1) the pressure exerted by one component of a gas mixture

particle accelerator (9.6) a device for production of high-energy nuclear particles based on the interaction of charged particles with magnetic and electrical fields

passive transport (17.6) the net movement of a solute from an area of high concentration to an area of low concentration

pentose (16.2) a five-carbon monosaccharide

pentose phosphate pathway (21.5) an alternative pathway for glucose degradation that provides the cell with reducing power in the form of NADPH

peptide bond (15.4, 18.3) the amide bond between two amino acids in a peptide chain

peptidyl tRNA binding site of ribosome (P-site) (20.6) a pocket on the surface of the ribosome that holds the tRNA bound to the growing peptide chain

percent yield (4.5) the ratio of the actual and theoretical yields of a chemical reaction multiplied by 100%

period (2.4) any one of seven horizontal rows of elements in the periodic table

periodic law (2.4) a law stating that properties of elements are periodic functions of their atomic numbers (Note that Mendeleev's original statement was based on atomic masses.)

peripheral membrane protein (17.6) a protein bound to either the inner or the outer surface of a membrane

phenol (12.7) an organic compound that contains a hydroxyl group (—OH) attached to a benzene ring

phenyl group (11.6) a benzene ring that has had a hydrogen atom removed, C_6H_5—

pH optimum (19.8) the pH at which an enzyme catalyzes the reaction at maximum efficiency

phosphatidate (17.3) a molecule of glycerol with fatty acids esterified to C-1 and C-2 of glycerol and a free phosphoryl group esterified at C-3

phosphoester (14.4) the product of the reaction between phosphoric acid and an alcohol

phosphoglyceride (17.3) a molecule with fatty acids esterified at the C-1 and C-2 positions of glycerol and a phosphoryl group esterified at the C-3 position

phospholipid (17.3) a lipid containing a phosphoryl group

phosphopantetheine (23.4) the portion of coenzyme A and the acyl carrier protein that is derived from the vitamin pantothenic acid

phosphoric anhydride (14.4) the bond formed when two phosphate groups react with one another and a water molecule is lost

pH scale (8.2) a numerical representation of acidity or basicity of a solution; pH = $-\log [H_3O^+]$

physical change (1.2) a change in the form of a substance but not in its chemical composition; no chemical bonds are broken in a physical change

physical property (1.2) a characteristic of a substance that can be observed without the substance undergoing change (examples include color, density, melting and boiling points)

plasma lipoprotein (17.5) a complex composed of lipid and protein that is responsible for the transport of lipids throughout the body

β-pleated sheet (18.5) a common secondary structure of a peptide chain that resembles the pleats of an Oriental fan

point mutation (20.7) the substitution of one nucleotide pair for another within a gene

polar covalent bonding (3.4) a covalent bond in which the electrons are not equally shared

polar covalent molecule (3.4) a molecule that has a permanent electric dipole moment resulting from an unsymmetrical electron distribution; a dipolar molecule

poly(A) tail (20.4) a tract of 100–200 adenosine monophosphate units covalently attached to the 3′ end of eukaryotic messenger RNA molecules

polyatomic ion (3.2) an ion containing a number of atoms

polymer (11.5) a very large molecule formed by the combination of many small molecules (called monomers) (e.g., polyamides, nylons)

polyprotic substance (8.3) a substance that can accept or donate more than one proton per molecule

polysaccharide (16.1) a large, complex carbohydrate composed of long chains of monosaccharides

polysome (20.6) complexes of many ribosomes all simultaneously translating a single mRNA

positive allosterism (19.9) effector binding activates the active site of an allosteric enzyme

positron (9.2) particle that has the same mass as an electron but opposite (+) charge

post-transcriptional modification (20.4) alterations of the primary transcripts produced in eukaryotic cells; these include addition of a poly(A) tail to the 3′ end of the mRNA, addition of the cap structure to the 5′ end of the mRNA, and RNA splicing

potential energy (1.5) stored energy or energy caused by position or composition

precipitate (6.1) an insoluble substance formed and separated from a solution

precision (1.4) the degree of agreement among replicate measurements of the same quantity

pressure (5.1) a force per unit area

primary (1°) alcohol (12.4) an alcohol with the general formula RCH_2OH

primary (1°) amine (15.1) an amine with the general formula RNH_2

primary (1°) carbon (10.2) a carbon atom that is bonded to only one other carbon atom

primary structure (of a protein) (18.4) the linear sequence of amino acids in a protein chain determined by the genetic information of the gene for each protein

primary transcript (20.4) the RNA product of transcription in eukaryotic cells, before post-transcriptional modifications are carried out

product (4.3, 19.2) the chemical species that results from a chemical reaction and that appears on the right side of a chemical equation

proenzyme (19.9) the inactive form of a proteolytic enzyme

prokaryote (20.2) an organism with simple cellular structure in which there is no true nucleus enclosed by a nuclear membrane and there are no true membrane-bound organelles in the cytoplasm

promoter (20.4) the sequence of nucleotides immediately before a gene that is recognized by the RNA polymerase and signals the start point and direction of transcription

properties (1.2) characteristics of matter

prostaglandins (17.2) a family of hormonelike substances derived from the twenty-carbon fatty acid, arachidonic acid; produced by many cells of the body, they regulate many body functions

prosthetic group (18.7) the nonprotein portion of a protein that is essential to the biological activity of the protein; often a complex organic compound

protein (18: Intro) a macromolecule whose primary structure is a linear sequence of α-amino acids and whose final structure results from folding of the chain into a specific three-dimensional structure; proteins serve as catalysts, structural components, and nutritional elements for the cell

protein modification (19.9) a means of enzyme regulation in which a chemical group is covalently added to or removed from a protein. The chemical modification either turns the enzyme on or turns it off

proteolytic enzyme (19.11) an enzyme that hydrolyzes the peptide bonds between amino acids in a protein chain

proton (2.1) a positively charged particle in the nucleus of an atom

pure substance (1.2) a substance with constant composition

purine (20.1) a family of nitrogenous bases (heterocyclic amines) that are components of DNA and RNA and consist of a six-sided ring fused to a five-sided ring; the common purines in nucleic acids are adenine and guanine

pyridoxal phosphate (22.7) a coenzyme derived from vitamin B_6 that is required for all transamination reactions

pyrimidine (20.1) a family of nitrogenous bases (heterocyclic amines) that are components of nucleic acids and consist of a single six-sided ring; the common pyrimidines of DNA are cytosine and thymine; the common pyrimidines of RNA are cytosine and uracil

pyrimidine dimer (20.7) UV-light induced covalent bonding of two adjacent pyrimidine bases in a strand of DNA

pyruvate dehydrogenase complex (22.2) a complex of all the enzymes and coenzymes required for the synthesis of CO_2 and acetyl CoA from pyruvate

Q

quantization (2.3) a characteristic that energy can occur only in discrete units called quanta

quaternary ammonium salt (15.1) an amine salt with the general formula $R_4N^+A^-$ (in which R— can be an alkyl or aryl group or a hydrogen atom and A^- can be any anion)

quaternary (4°) carbon (10.2) a carbon atom that is bonded to four other carbon atoms

quaternary structure (of a protein) (18.7) aggregation of more than one folded peptide chain to yield a functional protein

R

rad (9.8) abbreviation for *radiation absorbed dose*, the absorption of 2.4×10^{-3} calories of energy per kilogram of absorbing tissue

radioactivity (9.1) the process by which atoms emit high-energy particles or rays; the spontaneous decomposition of a nucleus to produce a different nucleus

radiocarbon dating (9.5) the estimation of the age of objects through measurement of isotopic ratios of carbon

Raoult's law (6.4) a law stating that the vapor pressure of a component is equal to its mole fraction times the vapor pressure of the pure component

rate constant (7.3) the proportionality constant that relates the rate of a reaction and the concentration of reactants

rate equation (7.3) expresses the rate of a reaction in terms of reactant concentration and a rate constant

rate of chemical reaction (7.3) the change in concentration of a reactant or product per unit time

reactant (1.2, 4.3) starting material for a chemical reaction, appearing on the left side of a chemical equation

reducing agent (8.5) a substance that reduces, or donates electrons to, another substance; the reducing agent is itself oxidized in the process

reducing sugar (16.4) a sugar that can be oxidized by Benedict's or Tollens' reagents; includes all monosaccharides and most disaccharides

reduction (8.5, 12.6) the gain of electrons; in organic compounds it may be recognized by a gain of hydrogen or loss of oxygen

regulatory proteins (18.1) proteins that control cell functions such as metabolism and reproduction

release factor (20.6) a protein that binds to the termination codon in the empty A-site of the ribosome and causes the peptidyl transferase to hydrolyze the bond between the peptide and the peptidyl tRNA

rem (9.8) abbreviation for *roentgen equivalent for man*, the product of rad and RBE

replication fork (20.3) the point at which new nucleotides are added to the growing daughter DNA strand

replication origin (20.3) the region of a DNA molecule where DNA replication always begins

representative element (2.4) member of the groups of the periodic table designated as A

resonance (3.4) a condition that occurs when more than one valid Lewis structure can be written for a particular molecule

resonance form (3.4) one of a number of valid Lewis structures for a particular molecule

resonance hybrid (3.4) a description of the bonding in a molecule resulting from a superimposition of all valid Lewis structures (resonance forms)

restriction enzyme (20.8) a bacterial enzyme that recognizes specific nucleotide sequences on a DNA molecule and cuts the sugar-phosphate backbone of the DNA at or near that site

result (1.3) the outcome of a designed experiment, often determined from individual bits of data

reversible, competitive enzyme inhibitor (19.10) a chemical that resembles the structure and charge distribution of the natural substrate and competes with it for the active site of an enzyme

reversible, noncompetitive enzyme inhibitor (19.10) a chemical that binds weakly to an amino acid R group of an enzyme and inhibits activity; when the inhibitor dissociates, the enzyme is restored to its active form

reversible reaction (7.4) a reaction that will proceed in either direction, reactants to products or products to reactants

ribonucleic acid (RNA) (20.1) single-stranded nucleic acid molecules that are composed of phosphoryl groups, ribose, and the nitrogenous bases uracil, cytosine, adenine, and guanine

ribonucleotide (20.1) a ribonucleoside phosphate or nucleotide composed of a nitrogenous base in β-*N*-glycosidic linkage to the 1′ carbon of the sugar ribose and with one, two, or three phosphoryl groups esterified at the hydroxyl of the 5′ carbon of the ribose

ribose (16.4) a five-carbon monosaccharide that is a component of RNA and many coenzymes

ribosomal RNA (rRNA) (20.4) the RNA species that are structural and functional components of the small and large ribosomal subunits

ribosome (20.6) an organelle composed of a large and a small subunit, each of which is made up of ribosomal RNA and proteins; the platform on which translation occurs and that carries the enzymatic activity that forms peptide bonds

RNA polymerase (20.4) the enzyme that catalyzes the synthesis of RNA molecules using DNA as the template

RNA splicing (20.4) removal of portions of the primary transcript that do not encode protein sequences

röentgen (9.8) the dose of radiation producing 2.1×10^9 ions in 1 cm^3 of air at 0°C and 1 atm of pressure

S

saccharide (16.1) a sugar molecule

saponification (14.2, 17.2) a reaction in which a soap is produced; more generally, the hydrolysis of an ester by an aqueous base

saturated fatty acid (17.2) a long-chain monocarboxylic acid in which each carbon of the chain is bonded to the maximum number of hydrogen atoms

saturated hydrocarbon (10.1) an alkane; a hydrocarbon that contains only carbon and hydrogen bonded together through carbon-hydrogen and carbon-carbon single bonds

saturated solution (6.1) one in which undissolved solute is in equilibrium with the solution

scientific method (1.1) the process of studying our surroundings that is based on experimentation

scientific notation (1.4) a system used to represent numbers as powers of ten

secondary (2°) alcohol (12.4) an alcohol with the general formula R_2CHOH

secondary (2°) amine (15.1) an amine with the general formula R_2NH

secondary (2°) carbon (10.2) a carbon atom that is bonded to two other carbon atoms

secondary structure (of a protein) (18.5) folding of the primary structure of a protein into an α-helix or a β-pleated sheet; folding is maintained by hydrogen bonds between the amide hydrogen and the carbonyl oxygen of the peptide bond

semiconservative DNA replication (20.3) DNA polymerase "reads" each parental strand of DNA and produces a complementary daughter strand; thus, all newly synthesized DNA molecules consist of one parental and one daughter strand

semipermeable membrane (6.4, 17.6) a membrane permeable to the solvent but not the solute; a material that allows the transport of certain substances from one side of the membrane to the other

shielding (9.7) material used to provide protection from radiation

sickle cell anemia (18.9) a human genetic disease resulting from inheriting mutant hemoglobin genes from both parents

significant figures (1.4) all digits in a number known with certainty and the first uncertain digit

silent mutation (20.7) a mutation that changes the sequence of the DNA but does not alter the amino acid sequence of the protein encoded by the DNA

single bond (3.4) a bond in which one pair of electrons is shared by two atoms

single-replacement reaction (4.3) also called substitution reaction, one in which one atom in a molecule is displaced by another

soap (14.2) any of a variety of the alkali metal salts of fatty acids

solid state (1.2) a physical state of matter characterized by its rigidity and fixed volume and shape

solubility (3.5, 6.1) the amount of a substance that will dissolve in a given volume of solvent at a specified temperature

solute (6.1) a component of a solution that is present in lesser quantity than the solvent

solution (6.1) a homogeneous (uniform) mixture of two or more substances

solvent (6.1) the solution component that is present in the largest quantity

specific gravity (1.5) the ratio of the density of a substance to the density of water at 4°C or any specified temperature

specific heat (7.2) the quantity of heat (calories) required to raise the temperature of 1 g of a substance one degree Celsius

spectroscopy (2.3) the measurement of intensity and energy of electromagnetic radiation

speed of light (2.3) 2.99×10^8 m/s in a vacuum

sphingolipid (17.4) a phospholipid that is derived from the amino alcohol sphingosine rather than from glycerol

sphingomyelin (17.4) a sphingolipid found in abundance in the myelin sheath that surrounds and insulates cells of the central nervous system

standard solution (8.3) a solution whose concentration is accurately known

standard temperature and pressure (STP) (5.1) defined as 273 K and 1 atm

stereochemical specificity (19.5) the property of an enzyme that allows it to catalyze reactions involving only one enantiomer of the substrate

stereochemistry (16.3) the study of the spatial arrangement of atoms in a molecule

stereoisomers (10.3, 16.3) a pair of molecules having the same structural formula and bonding pattern but differing in the arrangement of the atoms in space

steroid (17.4) a lipid derived from cholesterol and composed of one five-sided ring and three six-sided rings; the steroids include sex hormones and anti-inflammatory compounds

structural analog (19.10) a chemical having a structure and charge distribution very similar to those of a natural enzyme substrate

structural formula (10.2) a formula showing all of the atoms in a molecule and exhibiting all bonds as lines

structural isomers (10.2) molecules having the same molecular formula but different chemical structures

structural protein (18.1) a protein that provides mechanical support for large plants and animals

sublevel (2.5) a set of equal-energy orbitals within a principal energy level

substituted hydrocarbon (10.1) a hydrocarbon in which one or more hydrogen atoms is replaced by another atom or group of atoms

substitution reaction (10.5, 11.6) a reaction that results in the replacement of one group for another

substrate (19.1) the reactant in a chemical reaction that binds to an enzyme active site and is converted to product

substrate-level phosphorylation (21.3) the production of ATP by the transfer of a phosphoryl group from the substrate of a reaction to ADP

sucrose (16.5) a disaccharide composed of α-D-glucose and β-D-fructose in (α1 → β2) glycosidic linkage; table sugar

supersaturated solution (6.1) a solution that is more concentrated than a saturated solution (Note that such a solution is not at equilibrium.)

surface tension (5.2) a measure of the strength of the attractive forces at the surface of a liquid

surfactant (5.2) a substance that decreases the surface tension of a liquid

surroundings (7.1) the universe outside of the system

suspension (6.1) a heterogeneous mixture of particles; the suspended particles are larger than those found in a colloidal suspension

system (7.1) the process under study

T

temperature (1.5) a measure of the relative "hotness" or "coldness" of an object

temperature optimum (19.8) the temperature at which an enzyme functions optimally and the rate of reaction is maximal

terminal electron acceptor (22.6) the final electron acceptor in an electron transport system that removes the low-energy electrons from the system; in aerobic organisms the terminal electron acceptor is molecular oxygen

termination codon (20.6) a triplet of ribonucleotides with no corresponding anticodon on a tRNA; as a result, translation will end, because there is no amino acid to transfer to the peptide chain

terpene (17.4) the general term for lipids that are synthesized from isoprene units; the terpenes include steroids, bile salts, lipid-soluble vitamins, and chlorophyll

tertiary (3°) alcohol (12.4) an alcohol with the general formula R_3COH

tertiary (3°) amine (15.1) an amine with the general formula R_3N

tertiary (3°) carbon (10.2) a carbon atom that is bonded to three other carbon atoms

tertiary structure (of a protein) (18.6) the globular, three-dimensional structure of a protein that results from folding the regions of secondary structure; this folding occurs spontaneously as a result of interactions of the side chains or R groups of the amino acids

tetrahedral structure (3.4) a molecule consisting of four groups attached to a central atom that occupy the four corners of an imagined regular tetrahedron

tetrose (16.2) a four-carbon monosaccharide

theoretical yield (4.5) the maximum amount of product that can be produced from a given amount of reactant

theory (1.1) a hypothesis supported by extensive testing that explains and predicts facts

thermodynamics (7.1) the branch of science that deals with the relationship between energies of systems, work, and heat

thioester (14.4) the product of a reaction between a thiol and a carboxylic acid

thiol (12.9) an organic compound that contains a thiol group (—SH)

titration (8.3) the process of adding a solution from a buret to a sample until a reaction is complete, at which time the volume is accurately measured and the concentration of the sample is calculated

Tollens' test (13.4) a test reagent (silver nitrate in ammonium hydroxide) used to distinguish aldehydes and ketones; also called the Tollens' silver mirror test

tracer (9.6) a radioisotope that is rapidly and selectively transmitted to the part of the body for which diagnosis is desired

transaminase (22.7) an enzyme that catalyzes the transfer of an amino group from one molecule to another

transamination (22.7) a reaction in which an amino group is transferred from one molecule to another

transcription (20.4) the synthesis of RNA from a DNA template

transferase (19.1)   an enzyme that catalyzes the transfer of a functional group from one molecule to another

transfer RNA (tRNA) (15.4, 20.4)   small RNAs that bind to a specific amino acid at the 3' end and mediate its addition at the appropriate site in a growing peptide chain; accomplished by recognition of the correct codon on the mRNA by the complementary anticodon on the tRNA

transition element (2.4)   any element located between Groups IIA (2) and IIIA (13) in the long periods of the periodic table

transition state (19.6)   the unstable intermediate in catalysis in which the enzyme has altered the form of the substrate so that it now shares properties of both the substrate and the product

translation (20.4)   the synthesis of a protein from the genetic code carried on the mRNA

translocation (20.6)   movement of the ribosome along the mRNA during translation

transmembrane protein (17.6)   a protein that is embedded within a membrane and crosses the lipid bilayer, protruding from the membrane both inside and outside the cell

transport protein (18.1)   a protein that transports materials across the cell membrane or throughout the body

triglyceride (17.3, 23.1)   triacylglycerol; a molecule composed of glycerol esterified to three fatty acids

trigonal pyramidal molecule (3.4)   a nonplanar structure involving three groups bonded to a central atom in which each group is equidistant from the central atom

triose (16.2)   a three-carbon monosaccharide

triple bond (3.4)   a bond in which three pairs of electrons are shared by two atoms

U

uncertainty (1.4)   the degree of doubt in a single measurement

unit (1.3)   a determinate quantity (of length, time, etc.) that has been adopted as a standard of measurement

unsaturated fatty acid (17.2)   a long-chain monocarboxylic acid having at least one carbon-to-carbon double bond

unsaturated hydrocarbon (10.1, 11: Intro)   a hydrocarbon containing at least one multiple (double or triple) bond

urea cycle (22.8)   a cyclic series of reactions that detoxifies ammonium ions by incorporating them into urea, which is excreted from the body

uridine triphosphate (UTP) (21.7)   a nucleotide composed of the pyrimidine uracil, the sugar ribose, and three phosphoryl groups and that serves as a carrier of glucose-1-phosphate in glycogenesis

V

valence electron (2.5)   electron in the outermost shell (principal quantum level) of an atom

valence shell electron pair repulsion theory (VSEPR) (3.4)   a model that predicts molecular geometry using the premise that electron pairs will arrange themselves as far apart as possible, to minimize electron repulsion

van der Waals forces (5.2)   a general term for intermolecular forces that include dipole-dipole and London forces

vapor pressure of a liquid (5.2)   the pressure exerted by the vapor at the surface of a liquid at equilibrium

very low density lipoprotein (VLDL) (17.5)   a plasma lipoprotein that binds triglycerides synthesized by the liver and carries them to adipose tissue for storage

viscosity (5.2)   a measure of the resistance to flow of a substance at constant temperature

vitamin (19.7)   an organic substance that is required in the diet in small amounts; water-soluble vitamins are used in the synthesis of coenzymes required for the function of cellular enzymes; lipid-soluble vitamins are involved in calcium metabolism, vision, and blood clotting

voltaic cell (8.5)   an electrochemical cell that converts chemical energy into electrical energy

W

wax (17.4)   a collection of lipids that are generally considered to be esters of long-chain alcohols

weight (1.5)   the force exerted on an object by gravity

weight/volume percent [% (W/V)] (6.2)   the concentration of a solution expressed as a ratio of grams of solute to milliliters of solution multiplied by 100%

weight/weight percent [% (W/W)] (6.2)   the concentration of a solution expressed as a ratio of mass of solute to mass of solution multiplied by 100%

Z

Zaitsev's rule (12.5)   states that in an elimination reaction, the alkene with the greatest number of alkyl groups on the double-bonded carbon (the more highly substituted alkene) is the major product of the reaction

Answers to Odd-Numbered Problems

Chapter 1

1.1 a. Physical property
  b. Chemical property
  c. Physical property
  d. Physical property
  e. Physical property
1.3 a. Pure substance
  b. Heterogeneous mixture
  c. Homogeneous mixture
  d. Pure substance
1.5 a. 1.0×10^3 mL
  b. 1.0×10^6 µL
  c. 1.0×10^{-3} kL
  d. 1.0×10^2 cL
  e. 1.0×10^{-1} daL
1.7 a. 1.3×10^{-2} m
  b. 0.71 L
  c. 2.00 oz
  d. 1.5×10^{-4} m^2
1.9 a. Three
  b. Three
  c. Four
  d. Two
  e. Three
1.11 a. 2.4×10^{-3}
  b. 1.80×10^{-2}
  c. 2.24×10^2
1.13 a. 8.09
  b. 5.9
  c. 20.19
1.15 a. 51
  b. 8.0×10^1
  c. 1.6×10^2
1.17 a. 61.4
  b. 6.17
  c. 6.65×10^{-2}
1.19 a. 0°C
  b. 273 K
1.21 23.7 g
1.23 a. Chemistry is the study of matter and the changes that matter undergoes.
  b. Matter is the material component of the universe.
  c. Energy is the ability to do work.
1.25 a. Potential energy is stored energy, or energy due to position or composition.
  b. Kinetic energy is the energy resulting from motion of an object.
  c. Data are a group of facts resulting from an experiment.
1.27 a. Gram (or kilogram)
  b. Liter
  c. Meter
1.29 Mass is an independent quantity while weight is dependent on gravity.
1.31 Density is mass per volume. Specific gravity is the ratio of the density of a substance to the density of water at 4°C.
1.33 The scientific method is an organized way of doing science.
1.35 A theory.
1.37 A physical property is a characteristic of a substance that can be observed without the substance undergoing a change in chemical composition.
1.39 Chemical properties of matter include flammability and toxicity.
1.41 A pure substance has constant composition with only a single substance whereas a mixture is composed of two or more substances.
1.43 Mixtures are composed of two or more substances. A homogeneous mixture has uniform composition while a heterogeneous mixture has non-uniform composition.
1.45 a. Chemical reaction
  b. Physical change
  c. Physical change
1.47 a. Physical property
  b. Chemical property
1.49 a. Pure substance
  b. Pure substance
  c. Mixture
1.51 a. Homogeneous
  b. Homogeneous
  c. Homogeneous
1.53 A gas is made up of particles that are widely separated. A gas will expand to fill any container and it has no definite shape or volume.
1.55 a. Extensive property
  b. Extensive property
  c. Intensive property
1.57 An element is a pure substance that cannot be changed into a simpler form of matter by any chemical reaction. An atom is the smallest unit of an element that retains the properties of that element.
1.59 a. Iron, oxygen, carbon are just a few of the more than 100 possible elements
  b. Sodium chloride, water, sucrose, ethyl alcohol
1.61 a. 32 oz
  b. 1.0×10^{-3} t
  c. 9.1×10^2 g
  d. 9.1×10^5 mg
  e. 9.1×10^1 da

AP-1

1.63
  a. 6.6×10^{-3} lb
  b. 1.1×10^{-1} oz
  c. 3.0×10^{-3} kg
  d. 3.0×10^{2} cg
  e. 3.0×10^{3} mg

1.65
  a. 10.0°C
  b. 283.2 K

1.67
  a. 293.2 K
  b. 68.0°F

1.69 4 L

1.71 101°F

1.73 5 cm is shorter than 5 in.

1.75 5.0 μg is smaller that 5.0 mg.

1.77
  a. 3
  b. 3
  c. 3
  d. 4
  e. 4
  f. 3

1.79
  a. 3.87×10^{-3}
  b. 5.20×10^{-2}
  c. 2.62×10^{-3}
  d. 2.43×10^{1}
  e. 2.40×10^{2}
  f. 2.41×10^{0}

1.81
  a. Precision is a measure of the agreement of replicate results.
  b. Accuracy is the degree of agreement between the true value and measured value.

1.83
  a. 1.5×10^{4}
  b. 2.41×10^{-1}
  c. 5.99
  d. 1139.42
  e. 7.21×10^{3}

1.85
  a. 1.23×10^{1}
  b. 5.69×10^{-2}
  c. -1.527×10^{3}
  d. 7.89×10^{-7}
  e. 9.2×10^{7}
  f. 5.280×10^{-3}
  g. 1.279×10^{0}
  h. -5.3177×10^{2}

1.87
  a. 3,240
  b. 0.000150
  c. 0.4579
  d. −683,000
  e. −0.0821
  f. 299,790,000
  g. 1.50
  h. 602,200,000,000,000,000,000,000

1.89 6.00 g/mL

1.91 1.08×10^{3} g

1.93 teak

1.95 0.789

1.97 Lead has the lowest density and platinum has the greatest density.

1.99 12.6 mL

Chapter 2

2.1
  a. 16 protons, 16 electrons, 16 neutrons
  b. 11 protons, 11 electrons, 12 neutrons

2.3 20.18 amu

2.5 Electron density is the probability that an electron will be found in a particular region of an atomic orbital.

2.7
  a. Zr (zirconium)
  b. 22.99
  c. Cr (chromium)
  d. Bi (bismuth)

2.9
  a. Helium, atomic number = 2, mass = 4.00 amu
  b. Fluorine, atomic number = 9, mass = 19.00 amu
  c. Manganese, atomic number = 25, mass = 54.94 amu

2.11
  a. Total electrons = 11, valence electrons = 1
  b. Total electrons = 12, valence electrons = 2
  c. Total electrons = 16, valence electrons = 6
  d. Total electrons = 17, valence electrons = 7
  e. Total electrons = 18, valence electrons = 8

2.13
  a. Sulfur: $1s^2, 2s^2, 2p^6, 3s^2, 3p^4$
  b. Calcium: $1s^2, 2s^2, 2p^6, 3s^2, 3p^6, 4s^2$

2.15
  a. [Ne] $3s^2, 3p^4$
  b. [Ar] $4s^2$

2.17
  a. Ca^{2+} and Ar are isoelectronic
  b. Sr^{2+} and Kr are isoelectronic
  c. S^{2-} and Ar are isoelectronic
  d. Mg^{2+} and Ne are isoelectronic
  e. P^{3-} and Ar are isoelectronic

2.19
  a. (Smallest) F, N, Be (largest)
  b. (Lowest) Be, N, F, (highest)
  c. (Lowest) Be, N, F, (highest)

2.21
  a. 8 protons, 8 electrons, 8 neutrons.
  b. 16 neutrons.

2.23

Particle	Mass	Charge
a. electron	5.4×10^{-4} amu	−1
b. proton	1.00 amu	+1
c. neutron	1.00 amu	0

2.25
  a. An ion is a charged atom or group of atoms formed by the loss or gain of electrons.
  b. A loss of electrons by a neutral species results in a cation.
  c. A gain of electrons by a neutral species results in an anion.

2.27 From the periodic table, all isotopes of Rn have 86 protons. Isotopes differ in the number of neutrons.

2.29
  a. 34
  b. 46

2.31
  a. $^{1}_{1}H$
  b. $^{14}_{6}C$

2.33

	Atomic Symbol	# Protons	# Neutrons	# Electrons	Charge
a.	$^{23}_{11}Na$	11	12	11	0
b.	$^{32}_{16}S^{2-}$	16	16	18	2−
c.	$^{16}_{8}O$	8	8	8	0
d.	$^{24}_{12}Mg^{2+}$	12	12	10	2+
e.	$^{39}_{19}K^{+}$	19	20	18	1+

2.35
  a. Neutrons
  b. Protons
  c. Protons, neutrons
  d. Ion
  e. Nucleus, negative

2.37
  • All matter consists of tiny particles called atoms.
  • Atoms cannot be created, divided, destroyed, or converted to any other type of atom.
  • All atoms of a particular element have identical properties.
  • Atoms of different elements have different properties.
  • Atoms combine in simple whole-number ratios.
  • Chemical change involves joining, separating, or rearranging atoms.

2.39
  a. Chadwick—demonstrated the existence of the neutron in 1932.
  b. Goldstein—identified positive charge in the atom.

2.41 **a.** Dalton—developed the Law of Multiple Proportions; determined the relative atomic weights of the elements known at that time; developed the first scientific atomic theory.
b. Crookes—developed the cathode ray tube and discovered "cathode rays;" characterized electron properties.

2.43 Our understanding of the nucleus is based on the gold foil experiment performed by Geiger and interpreted by Rutherford. In this experiment, Geiger bombarded a piece of gold foil with alpha particles, and observed that some alpha particles passed straight through the foil, others were deflected and some simply bounced back. This led Rutherford to propose that the atom consisted of a small, dense nucleus (alpha particles bounced back), surrounded by a cloud of electrons (some alpha particles were deflected). The size of the nucleus is small when compared to the volume of the atom (alpha particles were able to pass through the foil).

2.45 Crookes used the cathode ray tube. He observed particles emitted by the cathode and traveling toward the anode. This ray was deflected by an electric field. Thomson measured the curvature of the ray influenced by the electric and magnetic fields. This measurement provided the mass to charge ratio of the negative particle. Thomson also gave the particle the name, electron.

2.47 A cathode ray is the negatively charged particle formed in a cathode ray tube.

2.49 Radiowave ↑
Microwave
Infrared   Increasing
Visible   Wavelength
Ultraviolet
X-ray
Gamma ray

2.51 Infrared radiation has greater energy than microwave radiation.

2.53 Spectroscopy is the measurement of intensity and energy of electromagnetic radiation.

2.55 According to Bohr, Planck, and others, electrons exist only in certain allowed regions, quantum levels, outside of the nucleus.

2.57 • Electrons are found in orbits at discrete distances from the nucleus.
• The orbits are quantized—they are of discrete energies.
• Electrons can only be found in these orbits, never in between (they are able to jump instantaneously from orbit to orbit).
• Electrons can undergo transitions—if an electron absorbs energy, it will jump to a higher orbit; when the electron falls back to a lower orbit, it will release energy.

2.59 Bohr's atomic model was the first to successfully account for electronic properties of atoms, specifically, the interaction of atoms and light (spectroscopy).

2.61 **a.** Sodium
b. Potassium
c. Magnesium

2.63 Group IA (or 1) is known collectively as the alkali metals and consists of lithium, sodium, potassium, rubidium, cesium, and francium.

2.65 Group VIIA (or 17) is known collectively as the halogens and consists of fluorine, chlorine, bromine, iodine, and astatine.

2.67 **a.** True
b. True

2.69 **a.** Na, Ni, Al
b. Na, Al
c. Na, Ni, Al
d. Ar

2.71 **a.** One
b. One
c. Three
d. Seven
e. Zero (or eight)
f. Zero (or two)

2.73 A principal energy level is designated $n = 1, 2, 3$, and so forth. It is similar to Bohr's orbits in concept. A sublevel is a part of a principal energy level and is designated s, p, d, and f.

2.75 The s orbital represents the probability of finding an electron in a region of space surrounding the nucleus.

2.77 Three p orbitals (p_x, p_y, p_z) can exist in a given principal energy level.

2.79 A $3p$ orbital is a higher energy orbital than a $2p$ orbital because it is a part of a higher energy principal energy level.

2.81 $2\ e^-$ for $n = 1$
$8\ e^-$ for $n = 2$
$18\ e^-$ for $n = 3$

2.83 **a.** $3p$ orbital
b. $3s$ orbital
c. $3d$ orbital
d. $4s$ orbital
e. $3d$ orbital
f. $3p$ orbital

2.85 **a.** Not possible
b. Possible
c. Not possible
d. Not possible

2.87 **a.** Li^+
b. O^{2-}
c. Ca^{2+}
d. Br^-
e. S^{2-}
f. Al^{3+}

2.89 **a.** Isoelectronic
b. Isoelectronic

2.91 **a.** Na^+
b. S^{2-}
c. Cl^-

2.93 **a.** $1s^2, 2s^2, 2p^6, 3s^2, 3p^6$
b. $1s^2, 2s^2, 2p^6$

2.95 **a.** (Smallest) F, O, N (Largest)
b. (Smallest) Li, K, Cs (Largest)
c. (Smallest) Cl, Br, I (Largest)

2.97 **a.** (Smallest) O, N, F (Largest)
b. (Smallest) Cs, K, Li (Largest)
c. (Smallest) I, Br, Cl (Largest)

2.99 A positive ion is always smaller than its parent atom because the positive charge of the nucleus is shared among fewer electrons in the ion. As a result, each electron is pulled closer to the nucleus and the volume of the ion decreases.

2.101 The fluoride ion has a completed octet of electrons and an electron configuration resembling its nearest noble gas.

Chapter 3

3.1 **a.** LiBr
b. $CaBr_2$
c. Ca_3N_2

3.3 **a.** Potassium cyanide
b. Magnesium sulfide
c. Magnesium acetate

3.5 **a.** $CaCO_3$
b. $NaHCO_3$
c. Cu_2SO_4

3.7
  a. Diboron trioxide
  b. Nitrogen oxide
  c. Iodine chloride
  d. Phosphorus trichloride

3.9
  a. P_2O_5
  b. SiO_2

3.11
  a. 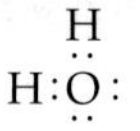 H:Ö:
  b. 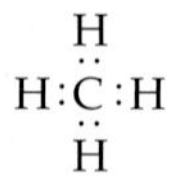 H:C:H with H above and below

3.13
  a. 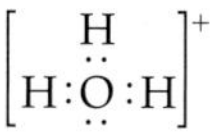 $[H:\ddot{O}:H]^+$ with H above
  b.  $[:\ddot{O}:H]^-$

3.15
  a. 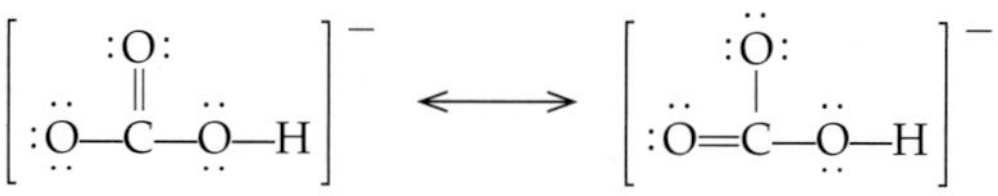 $[:\ddot{O}:C(=O):\ddot{O}-H]^- \longleftrightarrow [:\ddot{O}=C:\ddot{O}-H]^-$ (resonance structures of HCO_3^-)
  b. 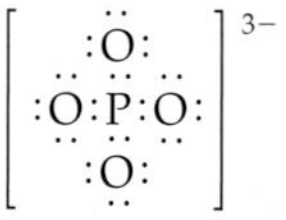 $[:\ddot{O}:P(\ddot{O}:)(\ddot{O}:):\ddot{O}:]^{3-}$

3.17
  a. The bonded nuclei are closer together when a double bond exists, in comparison to a single bond.
  b. The bond strength increases as the bond order increases. Therefore, a double bond is stronger than a single bond.

3.19
  a.  $[:\ddot{O}:Se::O: \longleftrightarrow :\ddot{O}::Se:\ddot{O}:]$

3.21
  a. 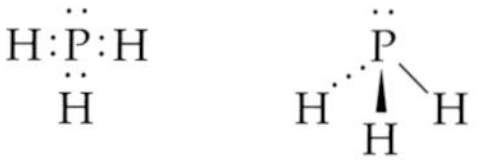 H:P:H with H below; P with three H's (pyramidal)
  b. 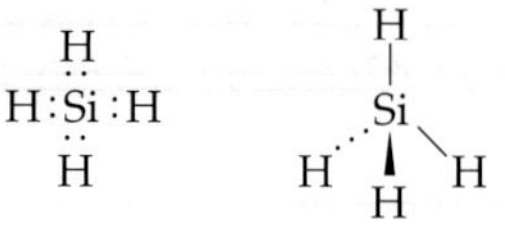 H:Si:H with H below; Si with three H's

3.23
  a. Oxygen is more electronegative than sulfur; the bond is polar. The electrons are pulled toward the oxygen atom.
  b. Nitrogen is more electronegative than carbon; the bond is polar. The electrons are pulled toward the nitrogen atom.
  c. There is no electronegativity difference between two identical atoms; the bond is nonpolar.
  d. Chlorine is more electronegative than iodine; the bond is polar. The electrons are pulled toward the chlorine atom.

3.25
  a. Nonpolar
  b. Polar
  c. Polar
  d. Nonpolar

3.27
  a. H_2O
  b. CO
  c. NH_3
  d. ICl

3.29
  a. Ionic
  b. Covalent
  c. Covalent
  d. Covalent

3.31
  a. Covalent
  b. Covalent
  c. Covalent
  d. Ionic

3.33
  a. 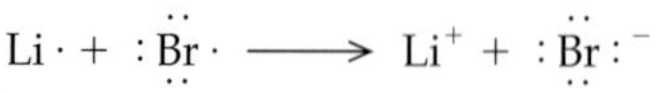 $Li\cdot + :\ddot{Br}\cdot \longrightarrow Li^+ + :\ddot{Br}:^-$
  b.  $\cdot Mg \cdot + 2 :\ddot{Cl}\cdot \longrightarrow Mg^{2+} + 2 :\ddot{Cl}:^-$

3.35
  a. 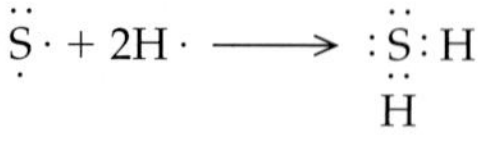 $:\ddot{S}\cdot + 2H\cdot \longrightarrow :\ddot{S}:H$ with H below
  b. 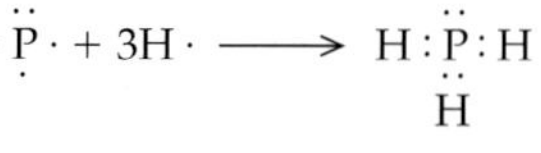 $\cdot \ddot{P} \cdot + 3H \cdot \longrightarrow H:\ddot{P}:H$ with H below

3.37 He has two valence electrons (electron configuration $1s^2$) and a complete N=1 level. It has a stable electron configuration, with no tendency to gain or lose electrons, and satisfies the octet rule (2 e$^-$ for period 1). Hence, it is nonreactive.
  He:

3.39
  a. Sodium ion
  b. Copper(I) ion (or cuprous ion)
  c. Magnesium ion
  d. Iron(II) ion (or ferrous ion)
  e. Iron(III) ion (or ferric ion)

3.41
  a. Sulfide ion
  b. Chloride ion
  c. Carbonate ion

3.43
  a. K^+
  b. Br^-

3.45
  a. SO_4^{2-}
  b. NO_3^-

3.47
  a. NaCl
  b. $MgBr_2$

3.49
  a. AgCN
  b. NH_4Cl

3.51
  a. Magnesium chloride
  b. Aluminum chloride

3.53
  a. Nitrogen dioxide
  b. Sulfur trioxide

3.55
  a. Al_2O_3
  b. Li_2S

3.57
  a. CO or CO_2
  b. SO_2 or SO_3

3.59
  a. $NaNO_3$
  b. $Mg(NO_3)_2$

3.61
  a. NH_4I
  b. $(NH_4)_2SO_4$

3.63
  a. Copper(II) sulfide
  b. Copper(II) sulfate

3.65
  a. Sodium hypochlorite
  b. Sodium chlorite

3.67 Ionic solid state compounds exist in regular, repeating, three-dimensional structures; the crystal lattice. The crystal lattice is made up of positive and negative ions. Solid state covalent compounds are made up of molecules which may be arranged in a regular crystalline pattern or in an irregular (amorphous) structure.

3.69 The boiling points of ionic solids are generally much higher than those of covalent solids.

3.71 KCl would be expected to exist as a solid at room temperature; it is an ionic compound, and ionic compounds are characterized by high melting points.

3.73 Water will have a higher boiling point. Water is a polar molecule with strong intermolecular attractive forces, whereas carbon tetrachloride is a nonpolar molecule with weak intermolecular attractive forces. More energy, hence, a higher temperature is required to overcome the attractive forces among the water molecules.

3.75
  a. H·
  b. He:
  c. 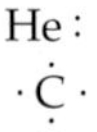 ·Ċ·
  d.  ·N̈·

3.77
  a. Li^+
  b. Mg^{2+}
  c. 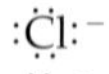 $:\ddot{Cl}:^-$
  d. 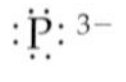 $:\ddot{P}:^{3-}$

3.79 a. 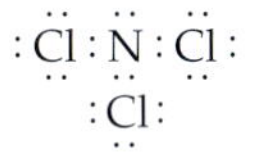 :Cl:N:Cl:
:Cl:

b. 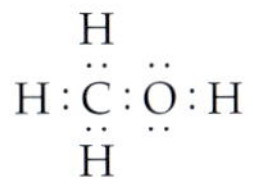 H:C:O:H
H

c. :S::C::S:

3.81 a. 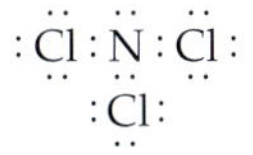 :Cl:N:Cl:
:Cl:
Pyramidal,
polar
water soluble

b. 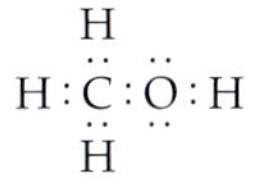 H:C:O:H
H
Tetrahedral around C,
angular around O
polar
water soluble

c.  :S::C::S:
Linear
nonpolar
not water soluble

3.83 Resonance can occur when more than one valid Lewis structure can be written for a molecule. Each individual structure which can be drawn is a resonance form. The true nature of the structure for the molecule is the resonance hybrid, which consists of the "average" of the resonance forms.

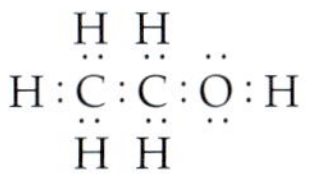

3.85 [H:O: / H:C:C:O: / H]⁻ ⟷ [H:O: / H:C:C::O / H]⁻

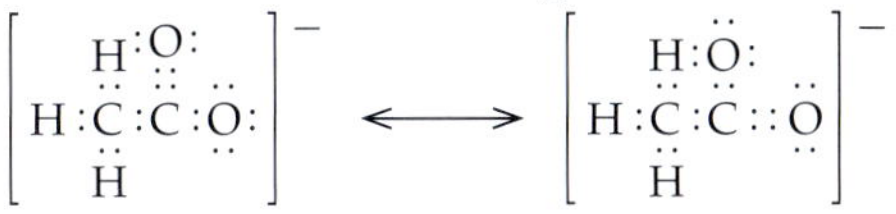

3.87 H:C:C:O:H
H H

3.89 H:C:C:C:H (with O above middle C)
H H

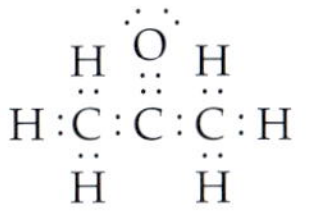

3.91 a. Polar covalent
b. Polar covalent
c. Ionic
d. Ionic
e. Ionic

3.93 a. [:C≡N:]⁻
b. [:Si≡P:]⁻
c), d), and e) are ionic compounds.

3.95 A molecule containing no polar bonds *must* be nonpolar. A molecule containing polar bonds may or may not itself be polar. It depends upon the number and arrangement of the bonds.

3.97 Polar compounds have strong intermolecular attractive forces. Higher temperatures are needed to overcome these forces and convert the solid to a liquid; hence, we predict higher melting points for polar compounds when compared to non-polar compounds.

3.99 Yes

Chapter 4

4.1 26.98 g Al/mol Al

4.3 a. 1.51×10^{24} oxygen atoms
b. 3.01×10^{24} oxygen atoms

4.5 14.0 g He

4.7 a. 17.04 g/mol.
b. 180.18 g/mol.
c. 237.95 g/mol.

4.9 a. DR
b. SR
c. DR
d. D

4.11 a. KCl(aq) + AgNO₃(aq) → KNO₃(aq) + AgCl(s)
A precipitation reaction occurs.
b. CH₃COOK(aq) + AgNO₃(aq) → no reaction
No precipitation reaction occurs.

4.13 a. 4Fe(s) + 3O₂(g) → 2Fe₂O₃(s)
b. 2C₆H₆(l) + 15O₂(g) → 12CO₂(g) + 6H₂O(g)

4.15 a. 90.1 g H₂O
b. 0.590 mol LiCl

4.17 a. 3 mol O₂
b. 96.00 g O₂

4.19 a. 4Fe(s) + 3O₂(g) → 2Fe₂O₃(s)
b. 3.50 g Fe

4.21 a. 132.0 g SnF₂
b. 3.79 % yield

4.23 Examples of other packaging units include a *ream* of paper (500 sheets of paper), a six-pack of soft drinks, a case of canned goods (24 cans), to name a few.

4.25 a. 28.09 g.
b. 107.9 g.

4.27 39.95 g.

4.29 40.36 g Ne

4.31 4.00 g He/mol He

4.33 a. 5.00 mol He
b. 1.7 mol Na
c. 4.2×10^{-2} mol Cl₂

4.35 1.62×10^3 g Ag

4.37 A molecule is a single unit comprised of atoms joined by covalent bonds. An ion-pair is composed of positive and negatively charged ions joined by electrostatic attraction, the ionic bond. The ion pairs, unlike the molecule, do not form single units; the electrostatic charge is directed to other ions in a crystal lattice, as well.

4.39 a. 58.44 g/mol
b. 142.04 g/mol
c. 357.49 g/mol

4.41 32.00 g/mol O₂

4.43 249.70 grams

4.45 a. 0.257 mol NaCl
b. 0.106 mol Na₂SO₄

4.47 a. 18.02 g H₂O
b. 116.9 g NaCl

4.49 a. 40.0 g He
b. 2.02×10^2 g H

4.51 a. 2.43 g Mg
b. 10.0 g CaCO₃

4.53 a. 4.00 g NaOH
b. 9.81 g H₂SO₄

4.55 a. 0.420 mol KBr
b. 0.415 mol MgSO₄

4.57 a. 6.57×10^{-1} mol CS_2
b. 2.14×10^{-1} mol $Al_2(CO_3)_3$
4.59 The ultimate basis for a correct chemical equation is the law of conservation of mass. No mass may be gained or lost in a chemical reaction, and the chemical equation must reflect this fact.
4.61 The subscript tells us the number of atoms or ions contained in one unit of the compound.
4.63 a. $MgCO_3 (s) \xrightarrow{\Delta} MgO (s) + CO_2 (g)$
b. $Zn(s) + CuSO_4(aq) \rightarrow ZnSO_4(aq) + Cu(s)$
4.65 $2NaOH(aq) + FeCl_2(aq) \rightarrow Fe(OH)_2(s) + 2NaCl(aq)$
4.67 Heat is necessary for the reaction to occur.
4.69 If we change the subscript we change the identity of the compound.
4.71 A reactant is the starting material for a chemical reaction.
4.73 A product is the chemical species that results from a chemical reaction.
4.75 a. $2C_2H_6(g) + 7O_2(g) \rightarrow 4CO_2(g) + 6H_2O(g)$
b. $6K_2O(s) + P_4O_{10}(s) \rightarrow 4K_3PO_4(s)$
c. $MgBr_2(aq) + H_2SO_4(aq) \rightarrow 2HBr(g) + MgSO_4(aq)$
4.77 a. $Ca(s) + F_2(g) \rightarrow CaF_2(s)$
b. $2Mg(s) + O_2(g) \rightarrow 2MgO(s)$
c. $3H_2(g) + N_2(g) \rightarrow 2NH_3(g)$
4.79 a. $2C_4H_{10}(g) + 13O_2(g) \rightarrow 10H_2O(g) + 8CO_2(g)$
b. $Au_2S_3(s) + 3H_2(g) \rightarrow 2Au(s) + 3H_2S(g)$
c. $Al(OH)_3(s) + 3HCl(aq) \rightarrow AlCl_3(aq) + 3H_2O(l)$
d. $(NH_4)_2Cr_2O_7(s) \rightarrow Cr_2O_3(s) + N_2(g) + 4H_2O(g)$
e. $C_2H_5OH(l) + 3O_2(g) \rightarrow 2CO_2(g) + 3H_2O(g)$
4.81 a. $N_2(g) + 3H_2(g) \rightarrow 2NH_3(g)$
b. $HCl(aq) + NaOH(aq) \rightarrow NaCl(aq) + H_2O(l)$
4.83 a. $C_6H_{12}O_6(s) + 6O_2(g) \rightarrow 6H_2O(l) + 6CO_2(g)$
b. $Na_2CO_3(s) \xrightarrow{\Delta} Na_2O(s) + CO_2(g)$
4.85 50.3 g B_2O_3
4.87 104 g $CrCl_3$
4.89 a. $N_2(g) + 3H_2(g) \rightarrow 2NH_3(g)$
b. Three moles of H_2 will react with one mole of N_2
c. One mole of N_2 will produce two moles of the product NH_3
d. 1.50 mol H_2
e. 17.0 g NH_3
4.91 a. 149.21 g/mol.
b. 1.20×10^{24} O atoms
c. 32.00 g O
d. 10.7 g O
4.93 7.39 g O_2
4.95 6.14×10^4 g O_2
4.97 70.6 g $C_{10}H_{22}$
4.99 9.13×10^2 g N_2
4.101 92.6%
4.103 6.85×10^2 g N_2

Chapter 5

5.1 a. 0.954 atm
b. 0.382 atm
c. 0.730 atm
5.3 a. 38 atm
b. 25 atm
5.5 a. 3.76 L
b. 3.41 L
c. 2.75 L
5.7 0.200 atm
5.9 4.46 mol H_2
5.11 9.00 L
5.13 0.223 mol N_2
5.15 In all cases, gas particles are much further apart than similar particles in the liquid or solid state. In most cases, particles in the liquid state are, on average, farther apart than those in the solid state. Water is the exception; liquid water's molecules are closer together than they are in the solid state.
5.17 Pressure is a force/unit area. Gas particles are in continuous, random motion. Collisions with the walls of the container results in a force (mass × acceleration) on the walls of the container. The sum of these collisional forces constitutes the pressure exerted by the gas.
5.19 Gases are easily compressed simply because there is a great deal of space between particles; they can be pushed closer together (compressed) because the space is available.
5.21 Gas particles are in continuous, random motion. They are free (minimal attractive forces between particles) to roam, up to the boundary of their container.
5.23 Gases exhibit more ideal behavior at low pressures. At low pressures, gas particles are more widely separated and therefore the attractive forces between particles are less. The ideal gas model assumes negligible attractive forces between gas particles.
5.25 The kinetic molecular theory states that the average kinetic energy of the gas particles increases as the temperature increases. Kinetic energy is proportional to (velocity)^2. Therefore, as the temperature increases the gas particle velocity increases and the rate of mixing increases as well.
5.27 The volume of the balloon is directly proportional to the pressure the gas exerts on the inside surface of the balloon. As the balloon cools, its pressure drops, and the balloon contracts. (Recall that the speed, hence the force exerted by the molecules decreases as the temperature decreases.)
5.29 Boyle's law states that the volume of a gas varies inversely with the gas pressure if the temperature and the number of moles of gas are held constant.
5.31 Volume will decrease according to Boyle's law. Volume is inversely proportional to the pressure exerted on the gas.
5.33 1 atm
5.35 5 L-atm
5.37 5.23 atm
5.39 Charles's law states that the volume of a gas varies directly with the absolute temperature if pressure and number of moles of gas are constant.
5.41 The Kelvin scale is the only scale that is directly proportional to molecular motion, and it is the motion that determines the physical properties of gases.
5.43 No. The volume is proportional to the temperature in K, not Celsius.
5.45 0.96 L
5.47 1.51 L
5.49 • Volume and temperature are *directly* proportional; increasing T *increases* V.
• Volume and pressure are *inversely* proportional; decreasing P *increases* V.
Therefore, both variables work together to *increase* the volume.
5.51 $V_f = \dfrac{P_i V_i T_f}{P_f T_i}$
5.53 1.82×10^{-2} L
5.55 Avogadro's law states that equal volumes of any ideal gas contain the same number of moles if measured at constant temperature and pressure.
5.57 6.00 L
5.59 No. One mole of an ideal gas will occupy exactly 22.4 L; however, there is no completely ideal gas and careful measurement will show a different volume.

5.61 Standard temperature is 273K.
5.63 0.80 mol
5.65 22.4 L
5.67 0.276 mol
5.69 5.94×10^{-2} L
5.71 22.4 L
5.73 9.08×10^3 L
5.75 172°C
5.77 Dalton's law states that the total pressure of a mixture of gases is the sum of the partial pressures of the component gases.
5.79 0.74 atm
5.81 Intermolecular forces in liquids are considerably stronger than intermolecular forces in gases. Particles are, on average, much closer together in liquids and the strength of attraction is inversely proportional to the distance of separation.
5.83 The vapor pressure of a liquid increases as the temperature of the liquid increases.
5.85 Evaporation is the conversion of a liquid to a gas at a temperature lower than the boiling point of the liquid. Condensation is the conversion of a gas to a liquid at a temperature lower than the boiling point of the liquid.
5.87 Viscosity is the resistance to flow caused by intermolecular attractive forces. Complex molecules may become entangled and not slide smoothly across one another.
5.89 All molecules exhibit London forces.
5.91 Only methanol exhibits hydrogen bonding. Methanol has an oxygen atom bonded to a hydrogen atom, a necessary condition for hydrogen bonding.
5.93 Solids are essentially incompressible because the average distance of separation among particles in the solid state is small. There is literally no space for the particles to crowd closer together.
5.95 a. High melting temperature, brittle
      b. High melting temperature, hard
5.97 Beryllium.
5.99 Mercury.

Chapter 6

6.1 A chemical analysis must be performed in order to determine the identity of all components, a qualitative analysis. If only one component is found, it is a pure substance; two or more components indicates a true solution.
6.3 After the container of soft drink is opened, CO_2 diffuses into the surrounding atmosphere; consequently the partial pressure of CO_2 over the soft drink decreases and the equilibrium
$$CO_2 (g) \leftrightarrow CO_2 (aq)$$
shifts to the left, lowering the concentration of CO_2 in the soft drink.
6.5 16.7% NaCl
6.7 7.50% KCl
6.9 2.56×10^{-2} % oxygen
6.11 20.0 % oxygen
6.13 2.00×10^5 ppm
6.15 0.125 mol HCl
6.17 To prepare the solution, dilute 1.7×10^{-2} L of 12 M HCl with sufficient water to produce 1.0×10^2 mL of total solution.
6.19 Pure water
6.21 1.0×10^{-2} osmol
6.23 0.24 atm
6.25 Polar carbon monoxide is more soluble in water.
6.27 1.54×10^{-2} mol Na^+/L
6.29 a. 2.00% NaCl
     b. 6.60% $C_6H_{12}O_6$
6.31 a. 5.00% ethanol
     b. 10.0% ethanol
6.33 a. 21.0% NaCl
     b. 3.75% NaCl
6.35 a. 2.25 g NaCl
     b. 3.13 g $NaC_2H_3O_2$
6.37 19.5% KNO_3
6.39 1.00 g sugar
6.41 0.04% (w/w) solution is more concentrated.
6.43 2.0×10^{-3} ppt
6.45 a. 0.342 M NaCl
     b. 0.367 M $C_6H_{12}O_6$
6.47 a. 1.46 g NaCl
     b. 9.00 g $C_6H_{12}O_6$
6.49 0.146 M $C_{12}H_{22}O_{11}$
6.51 5.00×10^{-2} L
6.53 20.0 M
6.55 0.900 M
6.57 158 g glucose
6.59 0.0500 M
6.61 A colligative property is a solution property that depends on the concentration of solute particles rather than the identity of the particles.
6.63 Salt is an ionic substance that dissociates in water to produce positive and negative ions. These ions (or particles) lower the freezing point of water. If the concentration of salt particles is large, the freezing point may be depressed below the surrounding temperature, and the ice would melt.
6.65 Chemical properties depend on the identity of the substance, whereas colligative properties depend on concentration, not identity.
6.67 Raoult's law states that when a solute is added to a solvent, the vapor pressure of the solvent decreases in proportion to the concentration of the solute.
6.69 One mole of $CaCl_2$ produces three moles of particles in solution whereas one mole of NaCl produces two moles of particles in solution. Therefore, a one molar $CaCl_2$ solution contains a greater number of particles than a one molar NaCl solution and will produce a greater freezing-point depression.
6.71 Sucrose
6.73 Sucrose
6.75 24 atm
6.77 A → B
6.79 No net flow
6.81 Hypertonic
6.83 Hypotonic
6.85 Water is often termed the "universal solvent" because it is a polar molecule and will dissolve, at least to some extent, most ionic and polar covalent compounds. The majority of our body mass is water and this water is an important part of the nutrient transport system due to its solvent properties. This is true in other animals and plants as well. Because of its ability to hydrogen bond, water has a high boiling point and a low vapor pressure. Also, water is abundant and easily purified.
6.87 The shelf life is a function of the stability of the ammonia-water solution. The ammonia can react with the water to convert to the extremely soluble and stable ammonium ion. Also, ammonia and water are polar molecules. Polar interactions, particularly hydrogen bonding, are strong and contribute to the long-term solution stability.

6.89 [Lewis structure: Na⁺ with two H₂O molecules showing lone pairs on O]

6.91 Polar; like dissolves like (H₂O is polar)

6.93 [Lewis structure: H—N(H)(H) with δ⁻ on N, δ⁺ on H, interacting with H—O—H showing δ⁻ on O]

6.95 In dialysis, sodium ions move from a region of high concentration to a region of low concentration. If we wish to remove (transport) sodium ions from the blood, they can move to a region of lower concentration, the dialysis solution.

6.97 Elevated concentrations of sodium ion in the blood may cause confusion, stupor, or coma.

6.99 Elevated concentrations of sodium ion in the blood may occur whenever large amounts of water are lost. Diarrhea, diabetes, and certain high-protein diets are particularly problematic.

Chapter 7

7.1 a. Exothermic
b. Exothermic
c. Exothermic

7.3 He(g)

7.5 $\Delta G = (+) - T(-)$
ΔG must always be positive.

7.7 13°C

7.9 2.7×10^3 J

7.11 $\dfrac{2.1 \times 10^2 \text{ nutritional Cal}}{\text{candy bar}}$

7.13 Heat energy produced by the friction of striking the match provides the activation energy necessary for this combustion process.

7.15 If the enzyme catalyzed a process needed to sustain life, the substance interfering with that enzyme would be classified as a poison.

7.17 a. rate = $k[N_2]^n[O_2]^{n'}$
b. rate = $k[C_4H_6]^n$

7.19 At rush hour, approximately the same number of passengers enter and exit the train at any given stop. Throughout the trip, the number of passengers on the train may be essentially unchanged, but the identity of the individual passengers is continually changing.

7.21 Measure the concentrations of products and reactants at a series of times until no further concentration change is observed.

7.23 a. $K_{eq} = \dfrac{[N_2][O_2]^2}{[NO_2]^2}$
b. $K_{eq} = [H_2]^2[O_2]$

7.25 Product formation

7.27 8.2×10^{-2}

7.29 a. A would decrease
b. A would increase
c. A would decrease
d. A would remain the same

7.31 joule

7.33 An exothermic reaction is one in which energy is released during chemical change.

7.35 A fuel must release heat in the combustion (oxidation) process.

7.37 The temperature of the water (or solution) is measured in a calorimeter. If the reaction being studied is exothermic, release energy heats the water and the temperature increases. In an endothermic reaction, heat flows from the water to the reaction and the water temperature decreases.

7.39 Double-walled containers, used in calorimeters, provide a small airspace between the part of the calorimeter (inside wall) containing the sample solution and the outside wall, contacting the surroundings. This makes heat transfer more difficult.

7.41 Free energy is the combined contribution of entropy and enthalpy for a chemical reaction.

7.43 The first law of thermodynamics, the law of conservation of energy, states that the energy of the universe is constant.

7.45 Enthalpy is a measure of heat energy.

7.47 1.20×10^3 cal

7.49 5.02×10^3 J

7.51 a. Entropy increases.
b. Entropy increases.

7.53 $\Delta G = (-) - T(+)$
ΔG must always be negative and the process is always spontaneous.

7.55 Isopropyl alcohol quickly evaporates (liquid → gas) after being applied to the skin. Conversion of a liquid to a gas requires heat energy. The heat energy is supplied by the skin. When this heat is lost, the skin temperature drops.

7.57 Decomposition of leaves and twigs to produce soil.

7.59 The activated complex is the arrangement of reactants in an unstable transition state as a chemical reaction proceeds. The activated complex must form in order to convert reactants to products.

7.61 The rate of a reaction is the change in concentration of a reactant or product per unit time. The rate constant is the proportionality constant that relates rate and concentration. The order is the exponent of each concentration term in the rate equation.

7.63 Increase

7.65 A catalyst increases the rate of a reaction without itself undergoing change.

7.67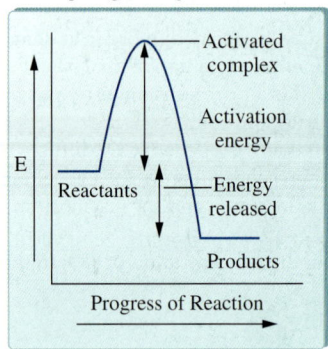
Non-catalyzed reaction
Higher activation energy

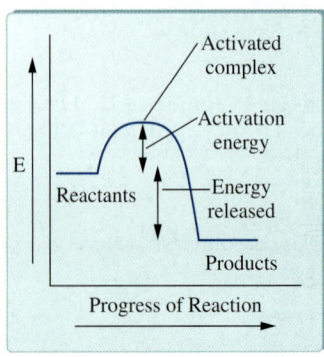
Catalyzed reaction
Lower activation energy

7.69 Enzymes are biological catalysts. The enzyme lysozyme catalyzes a process that results in the destruction of the cell walls of many harmful bacteria. This helps to prevent disease in organisms. The breakdown of foods to produce material for construction and repair of body tissue, as well as energy, is catalyzed by a variety of enzymes. For example, amylase begins the hydrolysis of starch in the mouth.

7.71 An increase in concentration of reactants means that there are more molecules in a certain volume. The probability of collision is enhanced because they travel a shorter distance before meeting another molecule. The rate is proportional to the number of collisions per unit time.

7.73 Rate = $k[N_2O_4]$

7.75 A catalyst speeds up a chemical reaction by facilitating the formation of the activated complex, thus lowering the activation energy, the energy barrier for the reaction.

7.77 Products

7.79 Ice and water at 0°C

7.81 Le Chaltelier's principle states that when a system at equilibrium is disturbed, the equilibrium shifts in the direction that minimizes the disturbance.

7.83 A dynamic equilibrium has fixed concentrations of all reactants and products - these concentrations do not change with time. However, the process is dynamic because products and reactants are continuously being formed and consumed. The concentrations do not change because the rates of production and consumption are equal.

7.85 $K_{eq} = \dfrac{[NO_2]^2}{[N_2O_4]}$

7.87 A physical equilibrium describes physical change; examples include the equilibrium between ice and water, or the equilibrium vapor pressure of a liquid. A chemical equilibrium describes chemical change; examples include the reactions shown in questions 7.93 and 7.94.

7.89 a. Equilibrium shifts to the left.
     b. No change
     c. No change

7.91 a. False
     b. False

7.93 a. PCl_3 increases
     b. PCl_3 decreases
     c. PCl_3 decreases
     d. PCl_3 decreases
     e. PCl_3 remains the same

7.95 Decrease

7.97 $K_{eq} = \dfrac{[CO][H_2]}{[H_2O]}$

7.99 False

7.101 Removing the cap allows CO_2 to escape into the atmosphere. This corresponds to the removal of product (CO_2):
$$CO_2(l) \rightleftharpoons CO_2(g)$$
The equilibrium shifts to the right, dissolved CO_2 is lost, and the beverage goes "flat".

7.103 $K_{eq} = \dfrac{[NH_2]^2}{[N_2][H_2]^3}$

Chapter 8

8.1 a. $HF(aq) + H_2O(l) \rightleftharpoons H_3O^+(aq) + F^-(aq)$
    b. $NH_3(aq) + H_2O(l) \rightleftharpoons NH_4^+(aq) + OH^-(aq)$

8.3 a. HF and F^-; H_2O and H_3O^+
    b. NH_3 and NH_4^+; H_2O and OH^-

8.5 a. NH_4^+
    b. H_2SO_4

8.7 1.0×10^{-11} M

8.9 12.00

8.11 3.2×10^{-9} M

8.13 0.1000 M NaOH

8.15 $CO_2 + H_2O \rightleftharpoons H_2CO_3 \rightleftharpoons H_3O^+ + HCO_3^-$
An increase in the partial pressure of CO_2 is a stress on the left side of the equilibrium. The equilibrium will shift to the right in an effort to decrease the concentration of CO_2. This will cause the molar concentration of H_2CO_3 to increase.

8.17 In Question 8.15, the equilibrium shifts to the right. Therefore the molar concentration of H_3O^+ should increase.
    In Question 8.16, the equilibrium shifts to the left. Therefore the molar concentration of H_3O^+ should decrease.

8.19 4.87

8.21 4.74

8.23 4.87

8.25 $Ca \rightarrow Ca^{2+} + 2\,e^-$ (oxidation ½ reaction)
    $S + 2\,e^- \rightarrow S^{2-}$ (reduction ½ reaction)
    $Ca + S \rightarrow CaS$ (complete reaction)

8.27 a. An Arrhenius acid is a substance that dissociates, producing hydrogen ions.
    b. A Brønsted-Lowry acid is a substance that behaves as a proton donor.

8.29 The Brønsted-Lowry theory provides a broader view of acid-base theory than does the Arrhenius theory. Brønsted-Lowry emphasizes the role of the solvent in the dissociation process.

8.31 a. $HNO_2(aq) + H_2O(l) \rightleftharpoons H_3O^+(aq) + NO_2^-(aq)$
    b. $HCN(aq) + H_2O(l) \rightleftharpoons H_3O^+(aq) + CN^-(aq)$

8.33 a. HNO_2 and NO_2^-; H_2O and H_3O^+
    b. HCN and CN^-; H_2O and H_3O^+

8.35 a. Weak
    b. Weak
    c. Weak

8.37 a. CN^- and HCN; NH_3 and NH_4^+
    b. CO_3^{2-} and HCO_3^-; Cl^- and HCl

8.39 Concentration refers to the quantity of acid or base contained in a specified volume of solvent. Strength refers to the degree of dissociation of the acid or base.

8.41 a. Bronsted acid
    b. Bronsted base
    c. Both

8.43 a. Bronsted acid
    b. Both
    c. Bronsted base

8.45 HCN

8.47 I^-

8.49 a. 1.0×10^{-7} M
    b. 1.0×10^{-11} M

8.51 a. Neutral
    b. Basic

8.53 a. 7.00
    b. 5.00

8.55 a. $[H_3O^+] = 1.0 \times 10^{-1}$ M
       $[OH^-] = 1.0 \times 10^{-13}$ M
    b. $[H_3O^+] = 1.0 \times 10^{-9}$ M
       $[OH^-] = 1.0 \times 10^{-5}$ M

8.57 a. $[H_3O^+] = 5.0 \times 10^{-2}$ M
       $[OH^-] = 2.0 \times 10^{-13}$ M
    b. $[H_3O^+] = 2.0 \times 10^{-10}$ M
       $[OH^-] = 5.0 \times 10^{-5}$ M

8.59 A neutralization reaction is one in which an acid and a base react to produce water and a salt (a "neutral" solution).

8.61 a. $[H_3O^+] = 1.0 \times 10^{-6}$ M
$[OH^-] = 1.0 \times 10^{-8}$ M
b. $[H_3O^+] = 6.3 \times 10^{-6}$ M
$[OH^-] = 1.6 \times 10^{-9}$ M
c. $[H_3O^+] = 1.6 \times 10^{-8}$ M
$[OH^-] = 6.3 \times 10^{-7}$ M

8.63 a. 1×10^2
b. 1×10^4
c. 1×10^{10}

8.65 a. $[H_3O^+] = 1 \times 10^{-5}$
b. $[H_3O^+] = 1 \times 10^{-12}$
c. $[H_3O^+] = 3.2 \times 10^{-6}$

8.67 a. pH = 6.00
b. pH = 8.00
c. pH = 3.25

8.69 pH = 3.12

8.71 pH = 10.74

8.73 a. NH_3 and NH_4Cl can form a buffer solution.
b. HNO_3 and KNO_3 cannot form a buffer solution.

8.75 a. A buffer solution contains components (a weak acid and its salt or a weak base and its salt) that enable the solution to resist large changes in pH when acids or bases are added.
b. Acidosis is a medical condition characterized by higher-than-normal levels of CO_2 in the blood and lower-than-normal blood pH.

8.77 a. Addition of strong acid is equivalent to adding H_3O^+. This is a stress on the right side of the equilibrium and the equilibrium will shift to the left. Consequently the $[CH_3COOH]$ increases.
b. Water, in this case, is a solvent and does not appear in the equilibrium expression. Hence, it does not alter the position of the equilibrium.

8.79 $[H_3O^+] = 2.32 \times 10^{-7}$ M

8.81 The very fact that an acid has a K_a value means that it is a weak acid. The smaller the value of K_a, the weaker the acid.

8.83 pH = 4.74

8.85 $11.2 = \dfrac{[HCO_3^-]}{[H_2CO_3]}$

8.87 a. Oxidation is the loss of electrons, loss of hydrogen atoms, or gain of oxygen atoms.
b. An oxidizing agent removes electrons from another substance. In doing so the oxidizing agent becomes reduced.

8.89 The species oxidized *loses* electrons.

8.91 During an oxidation-reduction reaction the species *oxidized* is the reducing agent.

8.93 Cl_2 + $2KI$ → $2KCl + I_2$
substance reduced — substance oxidized
oxidizing agent — reducing agent

8.95 $2KI \rightarrow 2K^+ + 2e^- + I_2$ (oxidation ½ reaction)
$Cl_2 + 2e^- \rightarrow 2Cl^-$ (reduction ½ reaction)

8.97 An oxidation-reduction reaction must take place to produce electron flow in a voltaic cell.

8.99 Storage battery.

Chapter 9

9.1 X-ray, ultraviolet, visible, infra-red, microwave, and radiowave.

9.3 a. $^{85}_{36}Kr \rightarrow ^{85}_{37}Rb + ^{0}_{-1}e$
b. $^{226}_{88}Ra \rightarrow ^{4}_{2}He + ^{222}_{86}Rn$

9.5 6.3 ng of sodium-24 remain after 2.5 days

9.7 1/4 of the radioisotope remains after 2 half-lives

9.9 Isotopes with short half-lives release their radiation rapidly. There is much more radiation per unit time observed with short half-life substances; hence, the signal is stronger and the sensitivity of the procedure is enhanced.

9.11 The rem takes into account the relative biological effect of the radiation in addition to the quantity of radiation. This provides a more meaningful estimate of potential radiation damage to human tissue.

9.13 Natural radioactivity is the spontaneous decay of a nucleus to produce high-energy particles or rays.

9.15 Two protons and two neutrons

9.17 An electron with a –1 charge.

9.19 A positron has a positive charge and a beta particle has a negative charge.

9.21 • charge, α = +2, β = –1
• mass, α = 4 amu, β = 0.000549 amu
• velocity, α = 10% of C, β = 90% of C

9.23 Chemical reactions involve joining, separating and rearranging atoms; valence electrons are critically involved. Nuclear reactions involve only changes in nuclear composition.

9.25 $^{4}_{2}He$

9.27 $^{235}_{92}U$

9.29 $^{1}_{1}H$ 1 – 1 = 0 neutrons
$^{2}_{1}H$ 2 – 1 = 1 neutron
$^{3}_{1}H$ 3 – 1 = 2 neutrons

9.31 $^{15}_{7}N$

9.33 Alpha and beta particles are matter; gamma radiation is pure energy. Alpha particles are large and relatively slow moving. They are the least energetic and least penetrating. Gamma radiation moves at the speed of light, highly energetic, and most penetrating.

9.35 A helium atom has two electrons; an α particle has no electrons.

9.37 $^{60}_{27}Co \rightarrow ^{60}_{28}Ni + ^{0}_{-1}\beta + \gamma$

9.39 $^{23}_{11}Na + ^{2}_{1}H \rightarrow ^{24}_{11}Na + ^{1}_{1}H$

9.41 $^{24}_{10}Ne \rightarrow \beta + ^{24}_{11}Na$

9.43 $^{140}_{55}Cs \rightarrow ^{140}_{56}Ba + ^{0}_{-1}e$

9.45 $^{209}_{83}Bi + ^{54}_{24}Cr \rightarrow ^{262}_{107}Bh + ^{1}_{0}n$

9.47 $^{27}_{12}Mg \rightarrow ^{0}_{-1}e + ^{27}_{13}Al$

9.49 $^{12}_{7}N \rightarrow ^{12}_{6}C + ^{0}_{1}e$

9.51 Natural radioactivity is a spontaneous process; artificial radioactivity is nonspontaneous and results from a nuclear reaction that produces an unstable nucleus.

9.53 • Nuclei for light atoms tend to be most stable if their neutron/proton ratio is close to 1.
• Nuclei with more than 84 protons tend to be unstable.
• Isotopes with a "magic number" of protons or neutrons (2, 8, 20, 50, 82, or 126 protons or neutrons) tend to be stable.
• Isotopes with even numbers of protons or neutrons tend to be more stable.

9.55 $^{20}_{8}O$; Oxygen-20 has 20 – 8 = 12 neutrons, an n/p of 12/8, or 1.5. The n/p is probably too high for stability even though it does have a "magic number" of protons and an even number of protons and neutrons.

9.57 $^{48}_{24}Cr$; Chromium-48 has 48 – 24 = 24 neutrons, an n/p of 24/24, or 1.0. It also has an even number of protons and neutrons. It would probably be stable.

9.59 0.40 mg of iodine-131 remains

9.61 13 mg of iron-59 remains

9.63 201 hours

9.65 Fission

9.67 a. The fission process involves the breaking down of large, unstable nuclei into smaller, more stable nuclei. This process releases some of the binding energy in the form of heat and/or light.
  b. The heat generated during the fission process could be used to generate steam, which is then used to drive a turbine to create electricity.
9.69 $^{3}_{1}H + ^{1}_{1}H \rightarrow ^{4}_{2}He$ + energy
9.71 A "breeder" reactor creates the fuel which can be used by a conventional fission reactor during its fission process.
9.73 The reaction in a fission reactor that involves neutron production and causes subsequent reactions accompanied by the production of more neutrons in a continuing process.
9.75 High operating temperatures
9.77 Radiocarbon dating is a process used to determine the age of objects. The ratio of the masses of the stable isotope, carbon-12, and unstable isotope, carbon-14, is measured. Using this value and the half-life of carbon-14, the age of the coffin may be calculated.
9.79 $^{108}_{47}Ag + ^{4}_{2}He \rightarrow ^{112}_{49}In$
9.81 a. Technetium-99 m is used to study the heart (cardiac output, size, and shape), kidney (follow-up procedure for kidney transplant), and liver and spleen (size, shape, presence of tumors).
  b. Xenon-133 is used to locate regions of reduced ventilation and presence of tumors in the lung.
9.83 Radiation therapy provides sufficient energy to destroy molecules critical to the reproduction of cancer cells.
9.85 Background radiation, radiation from natural sources, is emitted by the sun as cosmic radiation, and from naturally radioactive isotopes found throughout our environment.
9.87 Level decreases
9.89 Positive effect
9.91 Positive effect
9.93 Yes
9.95 A film badge detects gamma radiation by darkening photographic film in proportion to the amount of radiation exposure over time. Badges are periodically collected and evaluated for their level of exposure. This mirrors the level of exposure of the personnel wearing the badges.
9.97 Relative biological effect is a measure of the damage to biological tissue caused by different forms of radiation.
9.99 a. The curie is the amount of radioactive material needed to produce 3.7×10^{10} atomic disintegrations per second.
  b. The roentgen is the amount of radioactive material needed to produce 2×10^{9} ion-pairs when passing through 1 cc of air at 0°C.

Chapter 10

10.1 The student could test the solubility of the substance in water and in an organic solvent, such as hexane. Solubility in hexane would suggest an organic substance; whereas solubility in water would indicate an inorganic compound. The student could also determine the melting and boiling points of the substance. If the melting and boiling points are very high, an inorganic substance would be suspected.

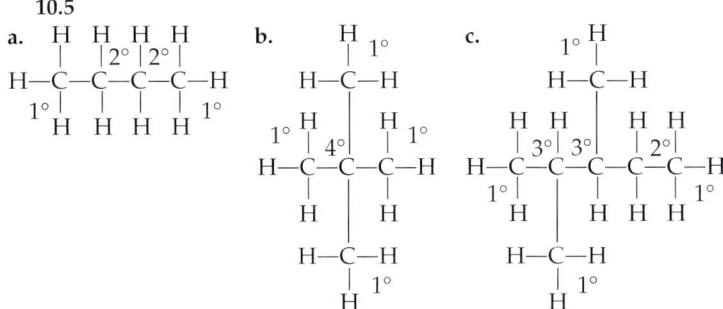

10.7 a. 2,3-Dimethylbutane
  b. 2,2-Dimethylpentane
  c. 2,2-Dimethylpropane
  d. 1,2,3-Tribromopropane
10.9 a. The straight chain isomers of molecular formula C_4H_9Br:

  b. The straight chain isomers of molecular formula $C_4H_8Br_2$:

Answers

10.11 a. *trans*-1-Bromo-2-ethylcyclobutane
b. *trans*-1,2-Dimethylcyclopropane
c. Propylcyclohexane

10.13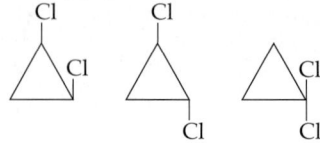

10.15 Three of the six axial hydrogen atoms of cyclohexane lie above the ring. The remaining three hydrogen atoms lie below the ring.

10.17 a. The combustion of cyclobutane:

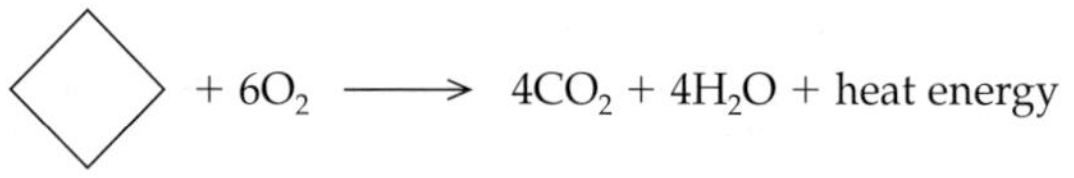

cyclobutane + 6O$_2$ ⟶ 4CO$_2$ + 4H$_2$O + heat energy

b. The monobromination of propane will produce two products as shown in the following two equations:

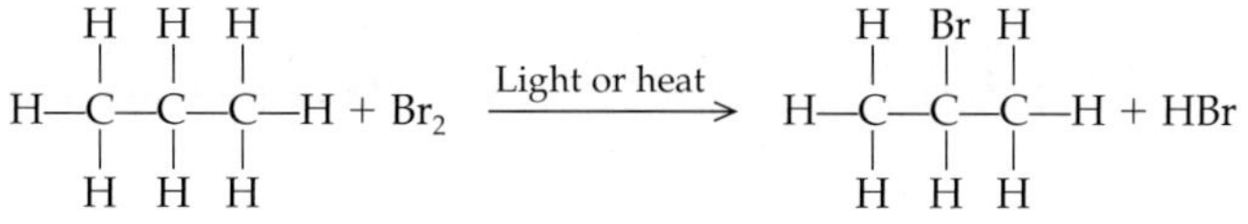

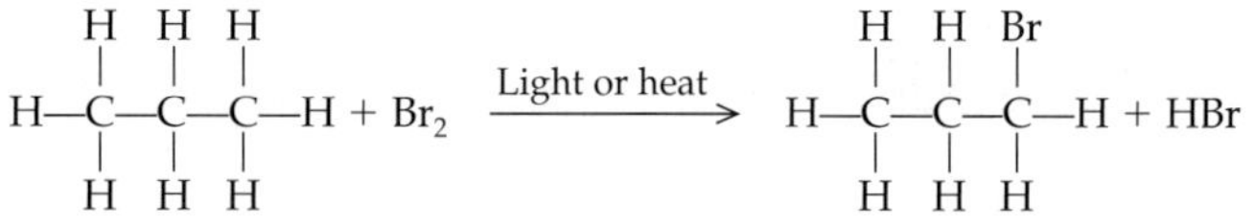

CH$_3$CH$_2$CH$_3$ + Br$_2$ —Light or heat→ CH$_3$CH$_2$CH$_2$Br + HBr

CH$_3$CH$_2$CH$_3$ + Br$_2$ —Light or heat→ CH$_3$CHBrCH$_3$ + HBr

c. 2CH$_3$CH$_3$ + 7O$_2$ → 4CO$_2$ + 6H$_2$O + energy

d. CH$_3$CH$_2$CH$_2$CH$_3$ + Cl$_2$ —Heat or light→ CH$_3$CH$_2$CH$_2$CH$_2$Cl + HCl

CH$_3$CH$_2$CH$_2$CH$_3$ + Cl$_2$ —Heat or light→ CH$_3$CH$_2$CHClCH$_3$ + HCl

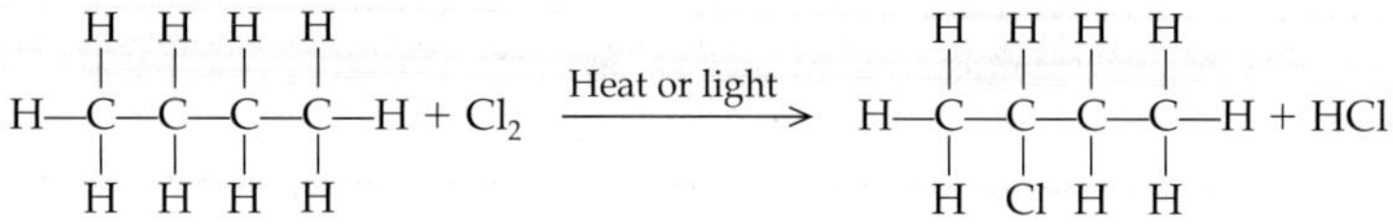

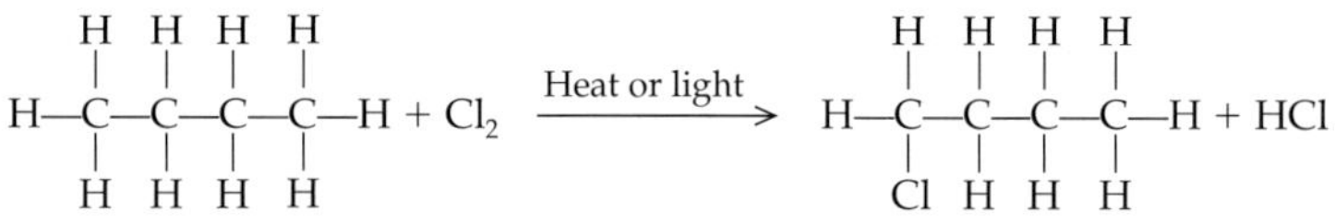

10.19 The products in the reactions in Problem 10.17b are 1-bromopropane and 2-bromopropane. The products of the reactions in Problem 10.17d are 1-chlorobutane and 2-chlorobutane.

10.21 The number of organic compounds is nearly limitless because carbon forms stable covalent bonds with other carbon atoms in a variety of different patterns. In addition, carbon can form stable bonds with other elements and functional groups, producing many families of organic compounds, including alcohols, aldehydes, ketones, esters, ethers, amines, and amides. Finally, carbon can form double or triple bonds with other carbon atoms to produce organic molecules with different properties.

10.23 The allotropes of carbon include graphite, diamond, and buckminsterfullerene.

10.25 Because ionic substances often form three-dimensional crystals made up of many positive and negative ions, they generally have much higher melting and boiling points than covalent compounds.

10.27 a. LiCl > H$_2$O > CH$_4$
b. NaCl > C$_3$H$_8$ > C$_2$H$_6$

10.29 a. LiCl would be a solid; H$_2$O would be a liquid; and CH$_4$ would be a gas.
b. NaCl would be a solid; both C$_3$H$_8$ and C$_2$H$_6$ would be gases.

10.31 a. Water-soluble inorganic compounds
b. Inorganic compounds
c. Organic compounds
d. Inorganic compounds
e. Organic compounds

10.33 a. 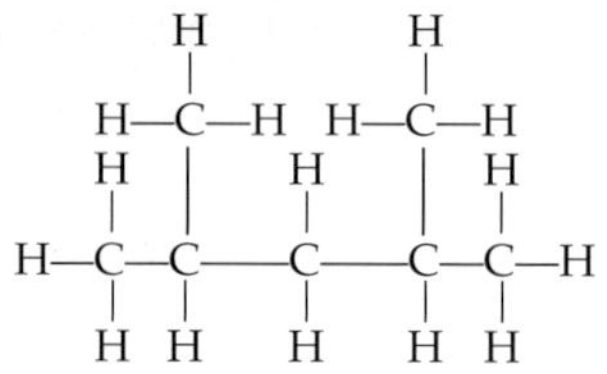 (structure shown) b. 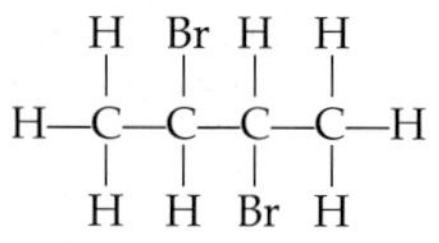 (structure shown)

10.35 a.  CH$_3$CH$_2$CHCH$_3$ with CH$_3$ branch
b. 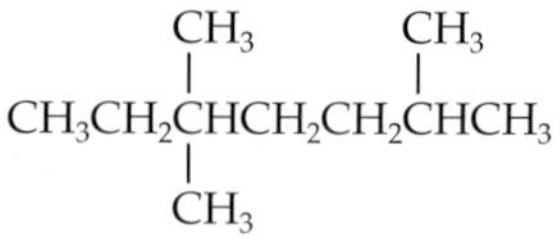 CH$_3$CH$_2$CHCH$_2$CH$_2$CHCH$_3$ with CH$_3$ branches

10.37 a. 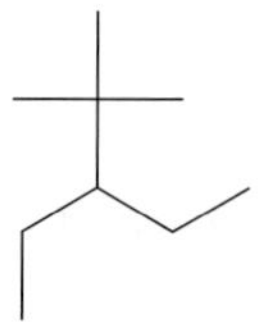 b. 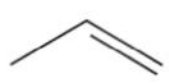 c. 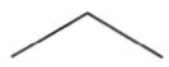 (structures shown)

10.39 a. 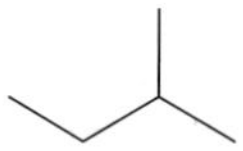 b.  c. 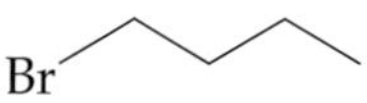 (structures shown, c contains Br)

10.41 Structure b is not possible because there are five bonds to carbon-2. Structure d is not possible because there are five bonds to carbon-3. Structure e is not possible because there are five bonds to carbon-3. Structure f is not possible because there are five bonds to carbon-3.

10.43 a. 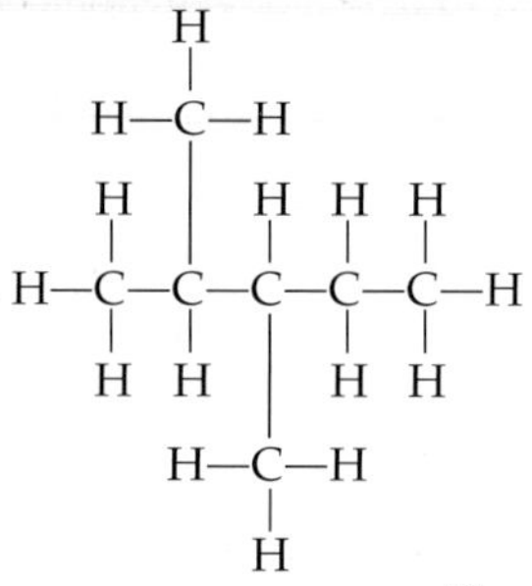 (structure shown) b. 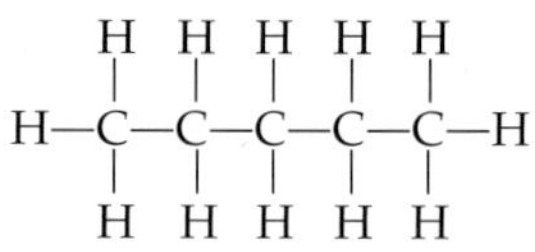 (structure shown)

10.45 a. 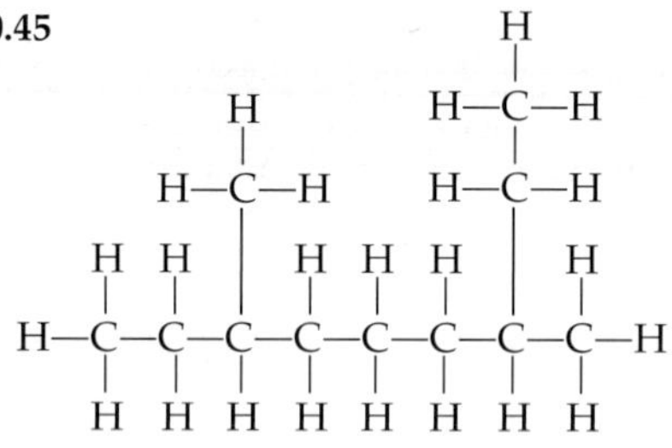 (structure shown) b. 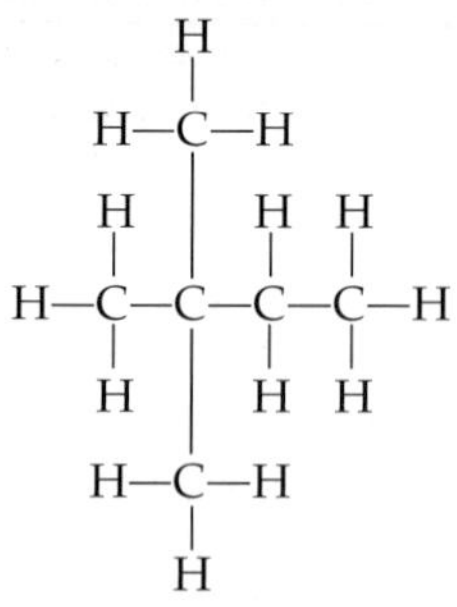 (structure shown)

10.47 An alcohol

H—C—C—OH structure

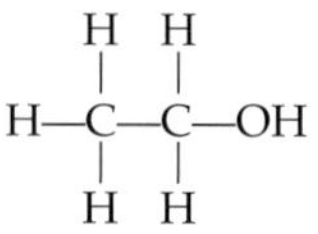

An aldehyde

H—C—C(=O)—H structure

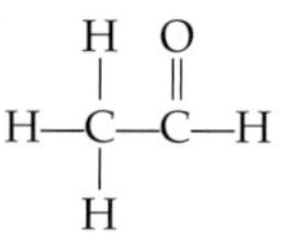

A ketone

```
    H O H
    | ‖ |
H—C—C—C—H
    |   |
    H   H
```

A carboxylic acid

```
    H O
    | ‖
H—C—C—OH
    |
    H
```

An amine

```
    H H      H
    | |     /
H—C—C—N
    | |     \
    H H      H
```

10.49
  a. C_nH_{2n+2}
  b. C_nH_{2n-2}
  c. C_nH_{2n}
  d. C_nH_{2n}
  e. C_nH_{2n-2}

10.51 Alkanes have only carbon-to-carbon and carbon-to-hydrogen single bonds, as in the molecule ethane:

```
    H H
    | |
H—C—C—H
    | |
    H H
```

Alkenes have at least one carbon-to-carbon double bond, as in the molecule ethene:

```
    H       H
     \     /
      C=C
     /     \
    H       H
```

Alkynes have at least one carbon-to-carbon triple bond, as in the molecule ethyne:

$$H—C≡C—H$$

10.53 a. A carboxylic acid:

```
         O
         ‖
CH_3CH_2—C—OH
```

  b. An amine: $CH_3CH_2CH_2—NH_2$
  c. An alcohol: $CH_3CH_2CH_2—OH$
  d. An ether: $CH_3CH_2—O—CH_2CH_3$

10.55

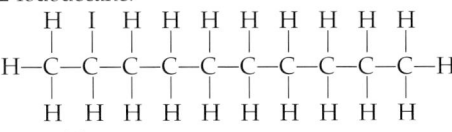

Aspirin
Acetylsalicylic acid

10.57 Hydrocarbons are nonpolar molecules, and hence are not soluble in water.

10.59 a. heptane > hexane > butane > ethane
  b. $CH_3CH_2CH_2CH_2CH_2CH_2CH_2CH_2CH_3$ > $CH_3CH_2CH_2CH_2CH_3$ > $CH_3CH_2CH_3$

10.61 a. Heptane and hexane would be liquid at room temperature; butane and ethane would be gases.
  b. $CH_3CH_2CH_2CH_2CH_2CH_2CH_2CH_2CH_3$ and $CH_3CH_2CH_2CH_2CH_3$ would be liquids at room temperature; $CH_3CH_2CH_3$ would be a gas.

10.63 Nonane: $CH_3CH_2CH_2CH_2CH_2CH_2CH_2CH_2CH_3$
  Pentane: $CH_3CH_2CH_2CH_2CH_3$
  Propane: $CH_3CH_2CH_3$

10.65 a.
```
      Br
      |
CH_3CHCH_2CH_3
```

  b.
```
        Cl
        |
CH_3—C—CH_3
        |
        CH_3
```

  c.
```
        CH_3
        |
CH_3—C—CH_2CH_2CH_2CH_3
        |
        CH_3
```

10.67 a. 2,2-Dibromobutane:
```
    H Br H H
    | |  | |
H—C—C—C—C—H
    | |  | |
    H Br H H
```

  b. 2-Iododecane:
```
    H I  H H H H H H H H
    | |  | | | | | | | |
H—C—C—C—C—C—C—C—C—C—C—H
    | |  | | | | | | | |
    H H  H H H H H H H H
```

  c. 1,2-Dichloropentane:
```
    H Cl H H H
    | |  | | |
Cl—C—C—C—C—C—H
    | |  | | |
    H H  H H H
```

  d. 1-Bromo-2-methylpentane:
```
         H
         |
      H—C—H
         |
    H    H H H
    |    | | |
H—C—C—C—C—C—H
    |    | | |
    Br   H H H
```

10.69 a. 3-Methylpentane   c. 1-Bromoheptane
  b. 2,5-Dimethylhexane   d. 1-Chloro-3-methylbutane

10.71 a. 2-Chloropropane   d. 1-Chloro-2-methylpropane
  b. 2-Iodobutane   e. 2-Iodo-2-methylpropane
  c. 2,2-Dibromopropane

10.73 a. 2-Chlorohexane   c. 3-Chloropentane
  b. 1,4-Dibromobutane   d. 2-Methylheptane

10.75 a. The first pair of molecules are constitutional isomers: hexane and 2-methylpentane.
  b. The second pair of molecules are identical. Both are heptane.

10.77 a. Incorrect: 3-Methylhexane
  b. Incorrect: 2-Methylbutane
  c. Incorrect: 3-Methylheptane
  d. Correct

10.79 a. 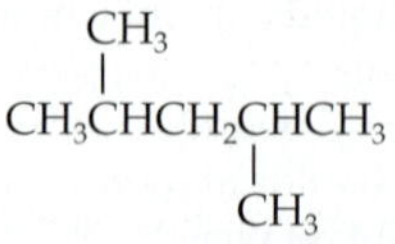 CH₃CHCH₂CHCH₃ with CH₃ on C2 and CH₃ on C4 — The name given in the problem is correct.
   b. 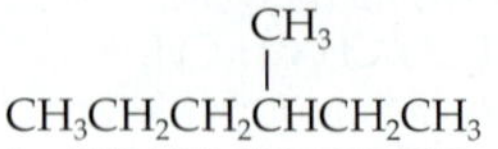 CH₃CH₂CH₂CHCH₂CH₃ with CH₃ branch — The correct name is 3-methylhexane.
   c. I—CH₂CH₂CH₂CH₂CH₂—I — The name given in the problem is correct.
   d. 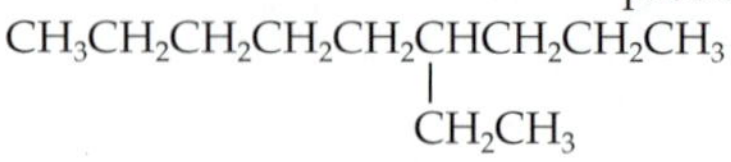 CH₃CH₂CH₂CH₂CH₂CHCH₂CH₂CH₃ with CH₂CH₃ branch — The correct name is 4-ethylnonane.
   e. 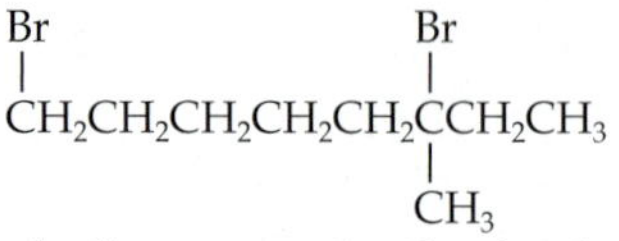 CH₂CH₂CH₂CH₂CH₂CCH₂CH₃ with Br, Br and CH₃ substituents — The name given in the problem is correct.

10.81 Cycloalkanes are a family of molecules having carbon-to-carbon bonds in a ring structure.

10.83 The general formula for a cycloalkane is C_nH_{2n}.

10.85
   a. Chlorocyclopropane
   b. *cis*-1, 2-Dichlorocyclopropane
   c. *trans*-1, 2-Dichlorocyclopropane
   d. Bromocyclobutane

10.87 a. 1-Bromo-2-methylcyclobutane: b. Iodocyclopropane:

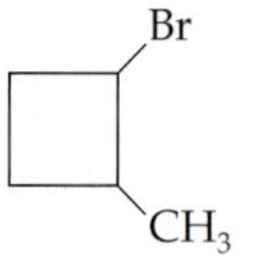

10.89
   a. Incorrect—1, 2-Dibromocyclobutane
   b. Incorrect—1, 2-Diethylcyclobutane
   c. Correct
   d. Incorrect—1, 2, 3-Trichlorocyclohexane

10.91 a. 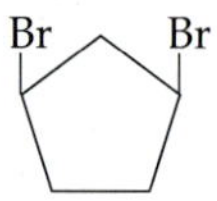 Br, Br on cyclopentane   b. 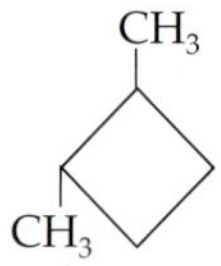 CH₃, CH₃ on cyclobutane
   c. 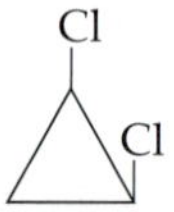 Cl, Cl on cyclopropane   d. 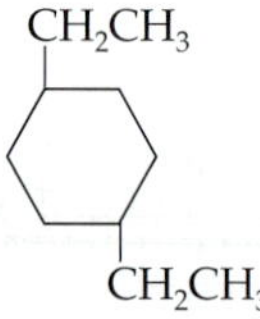 CH₂CH₃, CH₂CH₃ on cyclohexane

10.93
   a. *cis*-1, 2-Dibromocyclopentane
   b. *trans*-1, 3-Dibromocyclopentane
   c. *cis*-1, 2-Dimethylcyclohexane
   d. *cis*-1, 2-Dimethylcyclopropane

10.95 Conformational isomers are distinct isomeric structures that may be converted into one another by rotation about the bonds in the molecule.

10.97 In the chair conformation the hydrogen atoms, and thus the electron pairs of the C—H bonds, are farther from one another. As a result, there is less electron repulsion and the structure is more stable (more energetically favored). In the boat conformation, the electron pairs are more crowded. This causes greater electron repulsion, producing a less stable, less energetically favored conformation.

10.99 Because conformations are freely and rapidly interconverted, they cannot be separated from one another.

10.101 One conformation is more stable than the other because the electron pairs of the carbon-hydrogen bonds are farther from one another.

10.103 a. $C_3H_8 + 5O_2 \rightarrow 4H_2O + 3CO_2$
   b. $C_7H_{16} + 11O_2 \rightarrow 8H_2O + 7CO_2$
   c. $C_9H_{20} + 14O_2 \rightarrow 10H_2O + 9CO_2$
   d. $2C_{10}H_{22} + 31O_2 \rightarrow 22H_2O + 20CO_2$

10.105 a. $8CO_2 + 10H_2O$
   b. 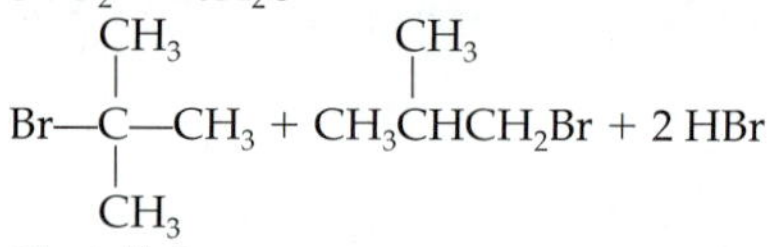 Br—C(CH₃)(CH₃)—CH₃ + CH₃CHCH₂Br + 2 HBr
   c. Cl_2 + light

10.107 The following molecules are all isomers of C_6H_{14}.

   CH₃CH₂CH₂CH₂CH₂CH₃   CH₃CHCH₂CH₂CH₃ with CH₃
   Hexane   2-Methylpentane

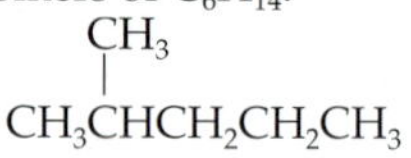

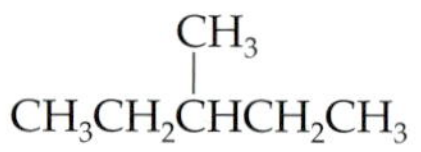

   CH₃CH₂CHCH₂CH₃ with CH₃   CH₃CHCHCH₃ with CH₃, CH₃
   3-Methylpentane 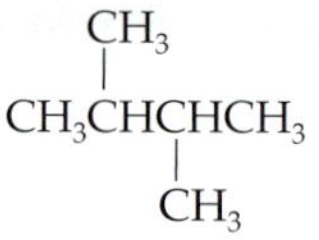   2,3-Dimethylbutane

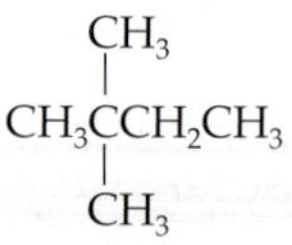

   CH₃CCH₂CH₃ with CH₃, CH₃
   2,2-Dimethylbutane

   a. 2, 3-Dimethylbutane produces only two monobrominated derivatives: 1-bromo-2, 3-dimethylbutane and 2-bromo-2, 3-dimethylbutane.
   b. Hexane produces three monobrominated products: 1-bromohexane, 2-bromohexane, and 3-bromohexane. 2, 2-Dimethylbutane also produces three monobrominated products: 1-bromo-2, 2-dimethylbutane, 2-bromo-3, 3-dimethylbutane, and 1-bromo-3, 3-dimethylbutane.
   c. 3-Methylpentane produces four monobrominated products: 1-bromo-3-methylpentane, 2-bromo-3-methylpentane, 3-bromo-3-methylpentane, and 1-bromo-2-ethylbutane.

10.109 The hydrocarbon is cyclooctane, having a molecular formula of C_8H_{16}.

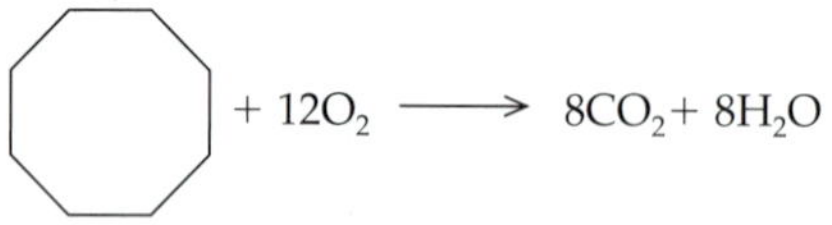

   cyclooctane + $12O_2 \longrightarrow 8CO_2 + 8H_2O$

Chapter 11

11.1 a. 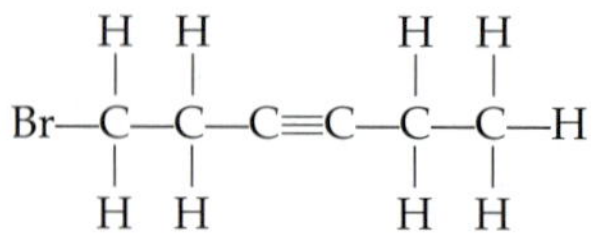 Br—C(H)(H)—C≡C—C(H)(H)—C(H)(H)—H
   b. 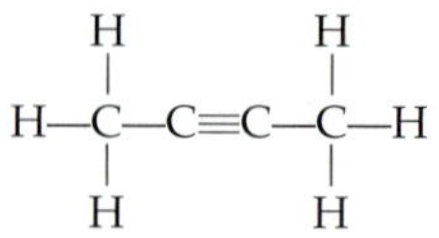 H—C(H)(H)—C≡C—C(H)(H)—H

11.3 a. Cl—C≡C—Cl

b. H—C≡C—CH₂—CH₂—CH₂—CH₂—CH₂—CH₂—CH₂—I (chain with all H's shown)

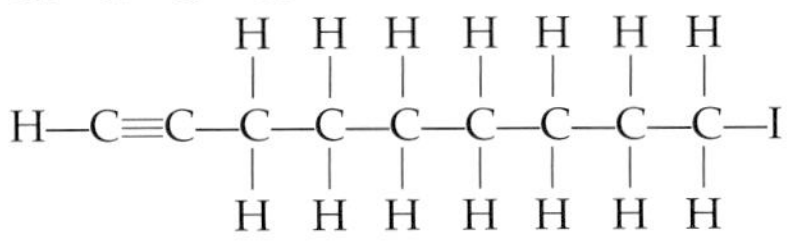

11.5 a.

cis-3-Hexene (CH₃CH₂ and CH₂CH₃ on same side)

trans-3-Hexene (CH₃CH₂ and CH₂CH₃ on opposite sides)

b.

trans-2,3-Dibromo-2-butene

cis-2,3-Dibromo-2-butene

11.7 Molecule c can exist as *cis-* and *trans-*isomers because there are two different groups on each of the carbon atoms attached by the double bond.

11.9 a. CH₃CH₂ and CH₂CH₂CH₂CH₃ on C=C with H's

b. CH₃ and H on one carbon; Cl and CH₂CHCH₃ on other

c. CH₃ and Cl on one carbon; Cl and CH₃ on other

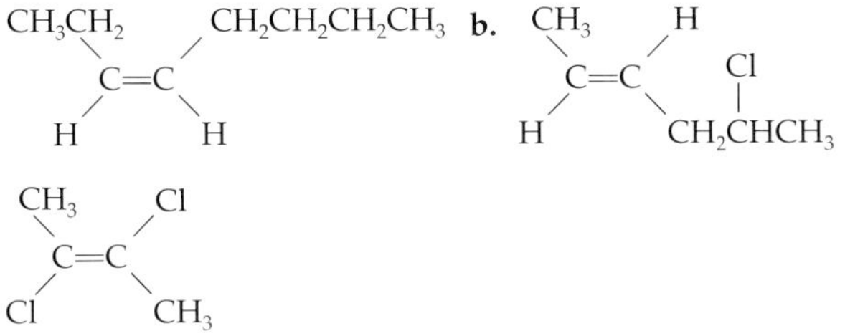

11.11 The hydrogenation of the *cis* and *trans* isomers of 2-pentene would produce the same product, pentane.

11.13 a. H₃C—C≡C—CH₃ + 2 H₂ $\xrightarrow{Ni}$ Butane

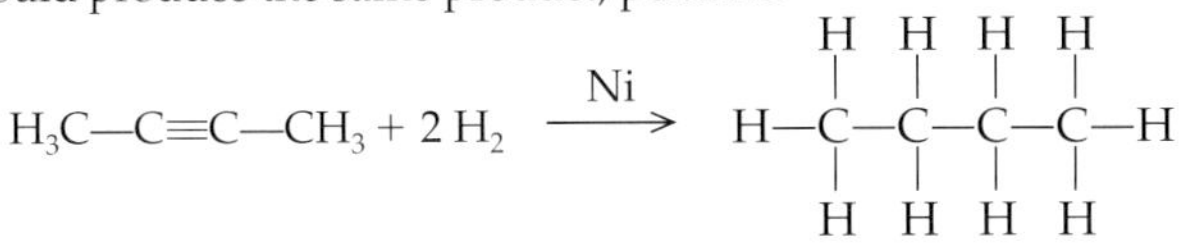

2-Butyne

b. H₃C—C≡C—CH₂CH₃ + 2 H₂ $\xrightarrow{Ni}$ Pentane

2-Pentyne

11.15 a. CH₃CH=CH₂ + Br₂ → CH₃CHBrCH₂Br

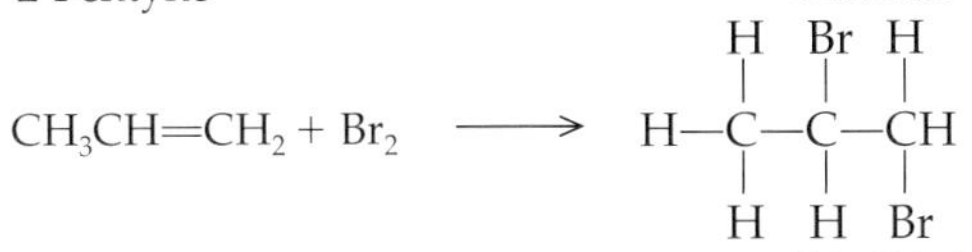

b. CH₃CH=CHCH₃ + Br₂ → CH₃CHBrCHBrCH₃

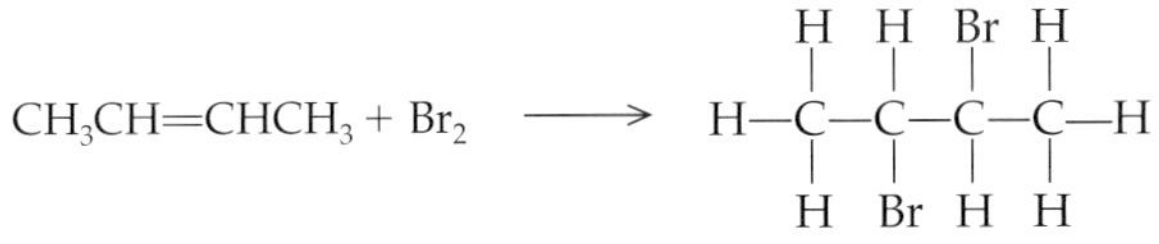

11.17 a. CH₃C≡CCH₃ + 2Cl₂ → CH₃CCl₂CCl₂CH₃

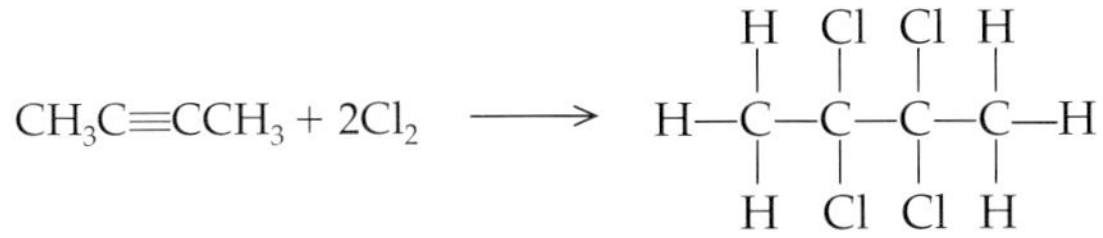

b. CH₃C≡CCH₂CH₃ + 2Cl₂ → CH₃CCl₂CCl₂CH₂CH₃

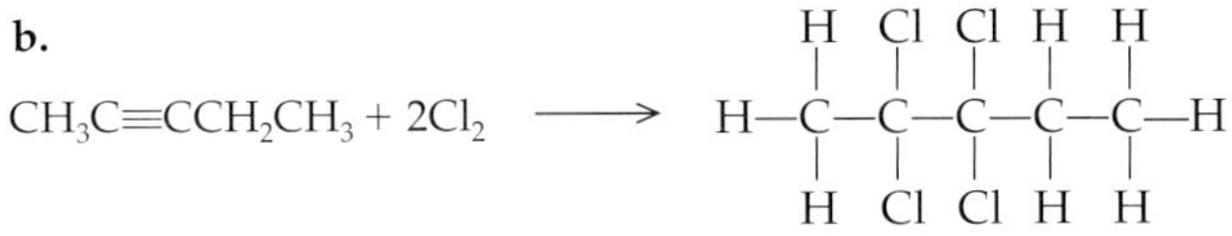

11.19 a. CH₃CH=CHCH₃ + H₂O $\xrightarrow{H^+}$ CH₃CHCH₂CH₃ (only product)
                                                    |
                                                    OH

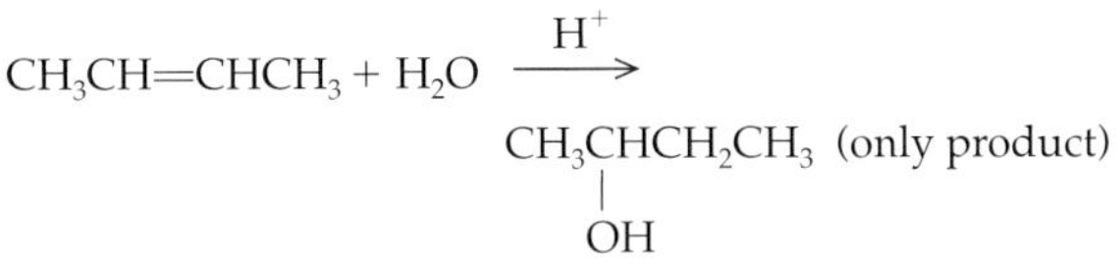

b. CH₂=CHCH₂CH₂CHCH₃ + H₂O $\xrightarrow{H^+}$
              |
              CH₃

CH₃CHCH₂CH₂CHCH₃ (major product)
  | |
  OH   CH₃

CH₂=CHCH₂CH₂CHCH₃ + H₂O $\xrightarrow{H^+}$
              |
              CH₃

CH₂CH₂CH₂CH₂CHCH₃ (minor product)
  | |
  OH   CH₃

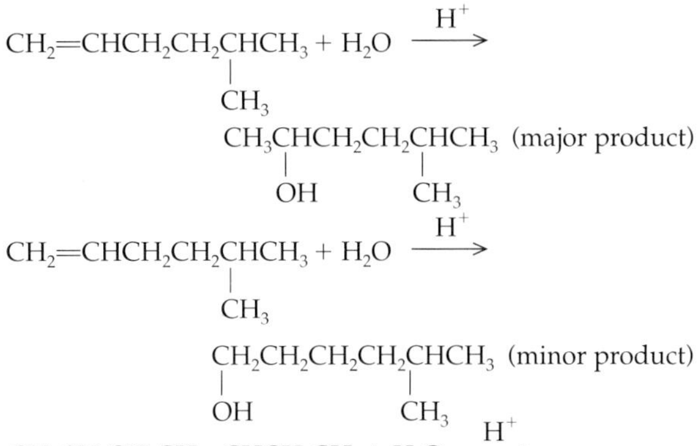

c. CH₃CH₂CH₂CH=CHCH₂CH₃ + H₂O $\xrightarrow{H^+}$

CH₃CH₂CH₂CHCH₂CH₂CH₃
         |
         OH

CH₃CH₂CH₂CH=CHCH₂CH₃ + H₂O $\xrightarrow{H^+}$

CH₃CH₂CH₂CH₂CHCH₂CH₃
             |
             OH

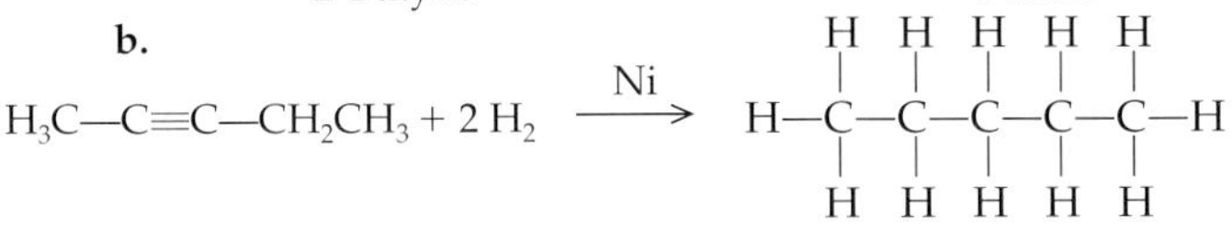

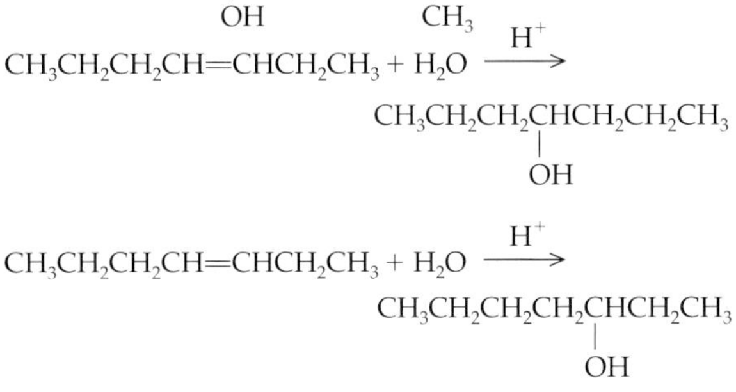

These products will be formed in approximately equal amounts.

d. CH₃CHClCH=CHCHClCH₃ + H₂O $\xrightarrow{H^+}$

CH₃CHClCHCH₂CHClCH₃ (only product)
       |
       OH

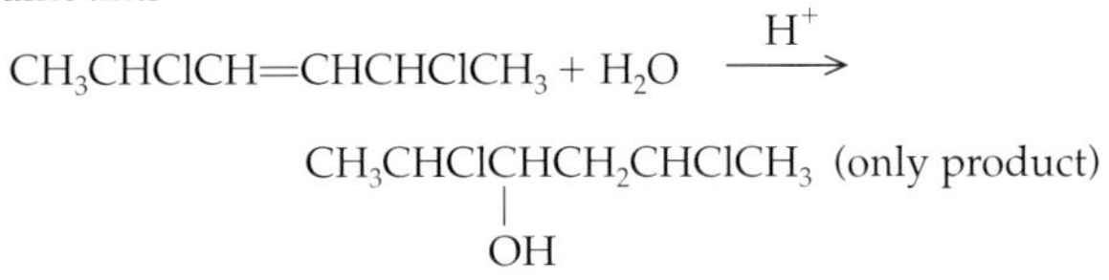

11.21 a. H₃C—C≡CH + H₂O $\xrightarrow{H^+}$ H₂C=C(OH)—CH₃ (enol) → CH₃—C(=O)—CH₃

Or

H₃C—C≡CH + H₂O $\xrightarrow{H^+}$ CH₃—C(H)=CH(OH) → CH₃CH₂CHO

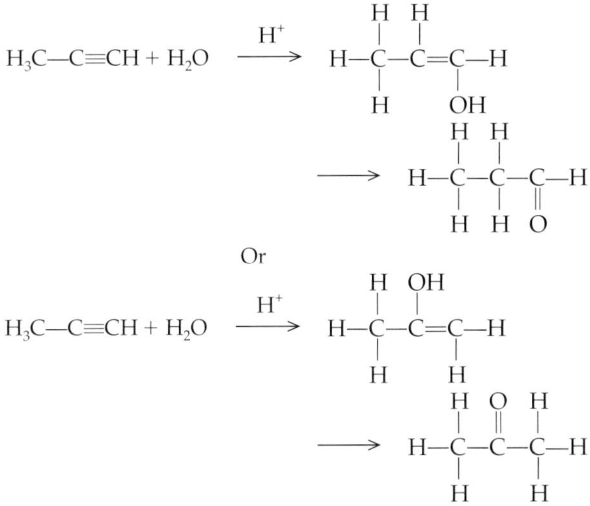

b. H₃C—C≡CCH₂CH₃ + H₂O $\xrightarrow{H^+}$ CH₃—C(OH)=CH—CH₂CH₃ → CH₃—C(=O)—CH₂—CH₂CH₃

Or

H₃C—C≡CCH₂CH₃ + H₂O $\xrightarrow{H^+}$ CH₃—CH=C(OH)—CH₂CH₃ → CH₃—CH₂—C(=O)—CH₂CH₃

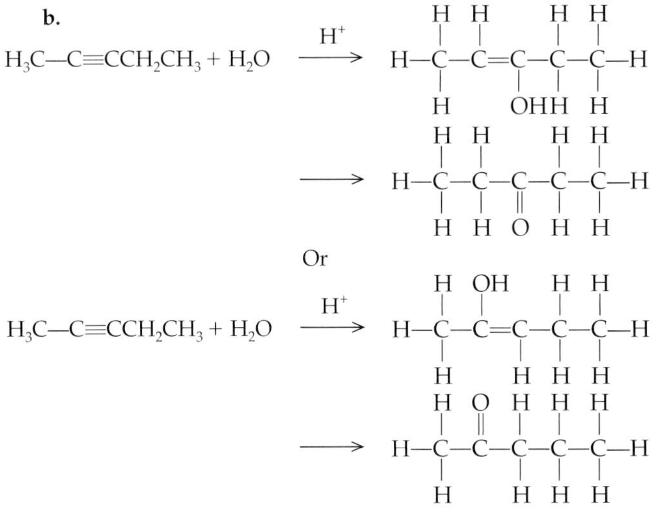

11.23
a. Reactant—cis-2-butene; Only product—butane
b. Reactant—1-butene; Major product—2-butanol
c. Reactant—2-butene; Only product—2,3-dichlorobutane
d. Reactant—1-pentene; Major product—2-bromopentane

11.25
a. 1,3,5-trichlorobenzene structure
b. 2-methylphenol (o-cresol) structure
c. 2-bromo-4-...phenol structure (OH, Br, Br)
d. 1,4-dinitrobenzene structure
e. 2-nitroaniline structure (NH₂, NO₂)
f. 3-nitrotoluene structure (CH₃, NO₂)

11.27 The longer the carbon chain of an alkene, the higher the boiling point.

11.29 The general formula for an alkane is C_nH_{2n+2}.
The general formula for an alkene is C_nH_{2n}.
The general formula for an alkyne is C_nH_{2n-2}.

11.31 Ethene is a planar molecule. All of the bond angles are 120°.

11.33 In alkanes, such as ethane, the four bonds around each carbon atom have tetrahedral geometry. The bond angles are 109.5°. In alkenes, such as ethene, each carbon is bonded by two single bonds and one double bond. The molecule is planar and each bond angle is approximately 120°.

11.35 Ethyne is a linear molecule. All of the bond angles are 180°.

11.37 In alkanes, such as ethane, the four bonds around each carbon atom have tetrahedral geometry. The bond angles are 109.5°. In alkenes, such as ethene, each carbon is bonded by two single bonds and one double bond. The molecule is planar and each bond angle is approximately 120°. In alkynes, such as ethyne, each carbon is bonded by one single bond and one triple bond. The molecule is linear and the bond angles are 180°.

11.39
a. 2-Pentyne > Propyne > Ethyne
b. 3-Decene > 2-Butene > Ethene

11.41 Identify the longest carbon chain containing the carbon-to-carbon double or triple bond. Replace the –ane suffix of the alkane name with –ene for an alkene or –yne for an alkyne. Number the chain to give the lowest number to the first of the two carbons involved in the double or triple bond. Determine the name and carbon number of each substituent group and place that information as a prefix in front of the name of the parent compound.

11.43 Geometric isomers of alkenes differ from one another in the placement of substituents attached to each of the carbon atoms of the double bond. Of the pair of geometric isomers, the cis-isomer, is the one in which identical groups are on the same side of the double bond.

11.45
a. (CH₃)(CH₂CH₂CH₃)C=C(CH₃)(H)
b. (CH₃CH₂)(H)C=C(H)(CH₂CH₂CH₃)
c. (Cl)(CH₂)C=C(CH₂CH₃)(H)... structure
d. (CH₃)(CH₃CCl)C=C(CH₂CH₂CH₃)(H)
e. (CH₃CH(CH₃))(H)C=C(H)(CH(Br)(CHCH₃CH₂CH₃))

11.47
a. 3-Methyl-1-pentene
b. 7-Bromo-1-heptene
c. 5-Bromo-3-heptene
d. 1-t-Butyl-4-methylcyclohexene

11.49
a. 1,3,5-Trifluoropentane: CH₂F—CH₂—CHF—CH₂—CH₂F
b. cis-2-Octene: (H₃C)(H)C=C(H)(CH₂CH₂CH₂CH₂CH₃)
c. Dipropylacetylene: CH₃CH₂CH₂—C≡C—CH₂CH₂CH₃

11.51
a. 2,3-Dibromobutane could not exist as cis and trans isomers.
b. cis-2-Heptene and trans-2-Heptene structures
c. cis-2,3-Dibromo-2-butene and trans-2,3-Dibromo-2-butene structures
d. Propene cannot exist as cis and trans isomers.

11.53 Alkenes b and c would not exhibit cis-trans isomerism.

11.55 Alkenes b and d can exist as both cis- and trans- isomers.

11.57
a. 1,5-Nonadiene
b. 1,4,7-Nonatriene
c. 2,5-Octadiene
d. 4-Methyl-2,5-heptadiene

11.59 R₂C=CR₂ + H₂ →(Pt, Pd, or Ni, heat or pressure)→ R₂CH—CHR₂

11.61 R₂C=CR₂ + X₂ → R₂CX—CXR₂

11.63 R₂C=CR₂ + H₂O →(H⁺)→ R₂CH—C(OH)R₂

11.65 The primary difference between complete hydrogenation of an alkene and an alkyne is that 2 moles of H_2 are required for the complete hydrogenation of an alkyne.

11.67 Addition of bromine (Br_2) to an alkene results in a color change from red to colorless. If equimolar quantities of Br_2 are added to hexene, the reaction mixture will change from red to colorless. This color change will not occur if cyclohexane is used.

11.69
a. H_2
b. H_2O
c. HBr
d. $19O_2 \rightarrow 12CO_2 + 14H_2O$
e. Cl_2
f. (cyclopentene structure)

11.71
a. $H_3C-C\equiv C-CH_3 + 2H_2 \xrightarrow[\text{heat or pressure}]{\text{Pt, Pd, or Ni}} H_3C-\underset{H}{\underset{|}{C}}-\underset{H}{\underset{|}{C}}-CH_3$ (with H H on top)

2-Butyne

b. $CH_3CH_2-C\equiv C-CH_3 + 2X_2 \longrightarrow CH_3CH_2-\underset{X}{\underset{|}{C}}-\underset{X}{\underset{|}{C}}-CH_3$ (with X X on top)

2-Pentyne

11.73 $CH_2=CHCH_2CH_2CH_3$, $CH_3CH=CHCH_2CH_3$,

$\underset{CH_3}{\underset{|}{CH_3C}}=CHCH_3$, $\underset{CH_3}{\underset{|}{CH_2=CCH_2CH_3}}$, $\underset{CH_3}{\underset{|}{CH_2=CHCHCH_3}}$

11.75
a. $\underset{}{\overset{Br}{\underset{|}{CH_3CHCH_2CH_3}}}$

b. $CH_3CH_2-\underset{CH_3}{\underset{|}{\overset{I}{\overset{|}{C}}}}-CH_2CH_2CH_3 + CH_3\underset{I}{\underset{|}{CH}}\underset{CH_3}{\underset{|}{CH}}CH_2CH_2CH_3$

(major product) (minor product)

c. (cyclopentane with Cl)

11.77 A polymer is a macromolecule composed of repeating structural units called *monomers*.

11.79
$n\ \underset{F}{\underset{|}{\overset{F}{\overset{|}{C}}}}=\underset{F}{\underset{|}{\overset{F}{\overset{|}{C}}}} \longrightarrow \left[\underset{F}{\underset{|}{\overset{F}{\overset{|}{C}}}}-\underset{F}{\underset{|}{\overset{F}{\overset{|}{C}}}}\right]_n$

Tetrafluoroethene   Teflon

11.81
a. $CH_3\underset{H}{\overset{H}{C}}=CCH_2CH_3 + H_2O \xrightarrow{H^+} CH_3\underset{OH}{\underset{|}{CH}}CH_2CH_2CH_3$

2-Pentene

$CH_3\underset{H}{\overset{H}{C}}=CCH_2CH_3 + H_2O \xrightarrow{H^+} CH_3CH_2\underset{OH}{\underset{|}{CH}}CH_2CH_3$

These products will be formed in approximately equal amounts.

b. $CH_2\overset{H}{\underset{Br}{\overset{|}{C}}}=\overset{H}{\underset{|}{C}}-H + H_2O \xrightarrow{H^+} CH_2\underset{Br\ OH}{\underset{|\ \ |}{CH}}CH_3$ (major product)

3-Bromo-1-propene

$CH_2\overset{H}{\underset{Br}{\overset{|}{C}}}H=\overset{H}{\underset{|}{C}}-H + H_2O \xrightarrow{H^+} CH_2\underset{Br}{\underset{|}{CH_2CHOH}}$ (minor product)

c. (3,4-dimethylcyclohexene) $+ H_2O \xrightarrow{H^+}$ (2,3-dimethylcyclohexanol with OH)

3,4-Dimethylcyclohexene

(3,4-dimethylcyclohexene) $+ H_2O \xrightarrow{H^+}$ (alternate dimethylcyclohexanol)

These products will be formed in approximately equal amounts.

11.83
a. $CH_2=CHCH_2\underset{CH_3}{\underset{|}{CH}}CH_3 + H_2O \xrightarrow{H^+} CH_2CH_2CH_2\underset{CH_3}{\underset{|}{CH}}CH_3$ with OH on first CH_2

(This is the minor product of this reaction.)

b. $CH_3\underset{H}{\overset{H}{C}}=CCH_2CH_2CH_3 + HBr \longrightarrow CH_3CH_2\underset{Br}{\underset{|}{CH}}CH_2CH_3$

OR

$CH_3CH_2\underset{H}{\overset{H}{C}}=CCH_2CH_3 + HBr \longrightarrow CH_3CH_2\underset{Br}{\underset{|}{CH}}CH_2CH_3$

c. (methylcyclohexene) + HBr → (bromo-methylcyclohexane)

d. (cyclopentene-CH_2CH_3) + $H_2O \xrightarrow{H^+}$ (cyclopentanol-CH_2CH_3 with OH)

11.85
a. $CH_2=CHCH_2CH=CHCH_3 + 2H_2 \xrightarrow[\text{heat}]{Pt} CH_3(CH_2)_4CH_3$

1,4-Hexadiene   Hexane

b. $CH_3CH=CHCH=CHCH=CHCH_3 + 3H_2 \xrightarrow[\text{heat}]{Ni} CH_3(CH_2)_6CH_3$

2,4,6-Octatriene   Octane

c. (1,3-cyclohexadiene) $+ 2H_2 \xrightarrow[\text{Pressure}]{Pd}$ (cyclohexane)

1,3-Cyclohexadiene   Cyclohexane

d. (1,3,5-cyclooctatriene) $+ 3H_2 \xrightarrow[\text{heat}]{Ni}$ (cyclooctane)

1,3,5-Cyclooctatriene   Cyclooctane

11.87 The term aromatic hydrocarbon was first used as a term to describe the pleasant-smelling resins of tropical trees.

11.89 Resonance hybrids are molecules for which more than one valid Lewis structure can be written.

11.91
a. (benzene ring with CH_3, Br, Br substituents)
b. (benzene ring with CH_2CH_3, CH_2CH_3, CH_2CH_3 substituents)

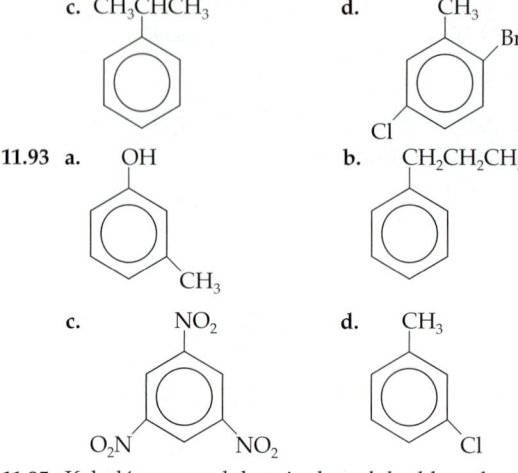

11.95 Kekulé proposed that single and double carbon-carbon bonds alternate around the benzene ring. To explain why benzene does not react like other unsaturated compounds, he proposed that the double and single bonds shift positions rapidly.

11.97 An addition reaction involves addition of a molecule to a double or triple bond in an unsaturated molecule. In a substitution reaction, one chemical group replaces another.

11.99 C₆H₆ + Cl₂ →(FeCl₃) C₆H₅Cl + HCl

11.101 Pyrimidine

11.103 Purine

Chapter 12

12.1
 a. 4-Methyl-1-pentanol
 b. 4-Methyl-2-hexanol
 c. 1, 2, 3-Propanetriol
 d. 4-Chloro-3-methyl-1-hexanol

12.3
 a. Primary
 b. Secondary
 c. Tertiary
 d. Aromatic (phenol)
 e. Secondary

12.5
 a. $CH_3CH=CH_2 + H_2O \xrightarrow{H^+} CH_3CH(OH)CH_3 + CH_3CH_2CH_2OH$
   major product   minor product
 b. $CH_2=CH_2 + H_2O \xrightarrow{H^+} CH_3CH_2OH$
 c. $CH_3CH_2CH=CHCH_2CH_3 + H_2O \xrightarrow{H^+} CH_3CH_2CH_2CH(OH)CH_2CH_3$

12.7
 a. The major product is a secondary alcohol (2-propanol) and the minor product is a primary alcohol (1-propanol).
 b. The product, ethanol, is a primary alcohol.
 c. The product, 3-hexanol, is a secondary alcohol.

12.9 The following equation represents the reduction of the ketone, butanone. This reaction requires a catalyst.

$$CH_3CH_2\overset{O}{\overset{\|}{C}}CH_3 + H_2 \rightarrow CH_3CH_2CH(OH)CH_3$$

12.11 The following equation represents the reduction of the aldehyde, butanal. This reaction requires a catalyst. The product is 1-butanol.

$$CH_3CH_2CH_2\overset{O}{\overset{\|}{C}}-H + H_2 \rightarrow CH_3CH_2CH_2CH_2OH$$

12.13
 a. Ethanol
 b. 2-Propanol (major product), 1-Propanol (minor product)
 c. 2-Butanol

12.15
 a. 2-Butanol is the major product. 1-Butanol is the minor product.
 b. 2-Methyl-2-propanol is the major product. 2-Methyl-1-propanol is the minor product.

12.17
 a. $CH_3CH=CH_2$
 b. $CH_3CH=CHCH_3 + CH_3CH_2CH=CH_2$

12.19
 a. $\underset{\underset{CH_3}{|}}{\overset{\overset{CH_3}{|}}{CH_3C}}CH_2CH_2OH \rightarrow \underset{\underset{CH_3}{|}}{\overset{\overset{CH_3}{|}}{CH_3C}}CH_2\overset{O}{\overset{\|}{C}}-H$
 b. $CH_3CH_2OH \rightarrow CH_3\overset{O}{\overset{\|}{C}}-H$

12.21
 a. $CH_3\underset{\underset{}{\overset{\overset{OH}{|}}{}}}{C}HCH_2CH_3 \rightarrow CH_3\overset{O}{\overset{\|}{C}}CH_2CH_3$
 b. $CH_3\underset{\underset{}{\overset{\overset{OH}{|}}{}}}{C}HCH_2CH_2CH_3 \rightarrow CH_3\overset{O}{\overset{\|}{C}}CH_2CH_2CH_3$

12.23
 a. 1-Ethoxypropane
 b. 1-Methoxypropane

12.25
 a. Ethyl propyl ether
 b. Methyl propyl ether

12.27
$CH_3CH_2OH + CH_3CH_2OH \xrightarrow{H^+} CH_3CH_2-O-CH_2CH_3 + H_2O$
Ethanol   Diethyl ether   Water

12.29 The longer the hydrocarbon tail of an alcohol becomes, the less water soluble it will be.

12.31 a < d < c < b

12.33
 a. CH_3CH_2OH
 b. $CH_3CH_2CH_2CH_2OH$
 c. $CH_3\underset{\underset{OH}{|}}{C}HCH_3$

12.35 The I.U.P.A.C. rules for the nomenclature of alcohols require you to name the parent compound, that is the longest continuous carbon chain bonded to the –OH group. Replace the –e ending of the parent alkane with –ol of the alcohol. Number the parent chain so that the carbon bearing the hydroxyl group has the lowest possible number. Name and number all other substituents. If there is more than one hydroxyl group, the –ol ending will be modified to reflect the number. If there are two –OH groups, the suffix –diol is used; if it has three –OH groups, the suffix –triol is used, etc.

12.37 a. 1-Heptanol
b. 2-Propanol
c. 2,2-Dimethylpropanol

12.39 a. 3-Hexanol:

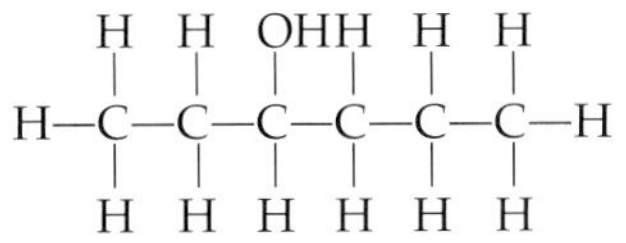

```
   H  H  OH H  H  H
   |  |  |  |  |  |
H—C—C—C—C—C—C—H
   |  |  |  |  |  |
   H  H  H  H  H  H
```

b. 1,2,3-Pentanetriol:

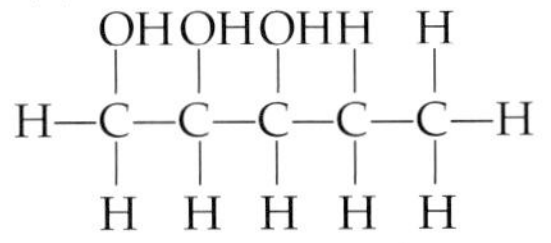

```
   OH OH OH H  H
   |  |  |  |  |
H—C—C—C—C—C—H
   |  |  |  |  |
   H  H  H  H  H
```

c. 2-Methyl-2-pentanol:

```
   H  OH H  H  H
   |  |  |  |  |
H—C—C—C—C—C—H
   |  |  |  |  |
   H     H  H  H
      |
    H—C—H
      |
      H
```

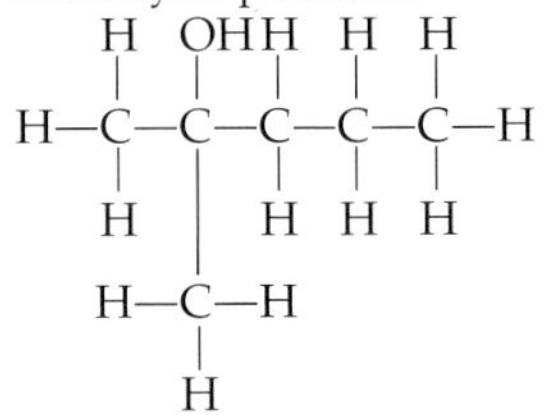

12.41 a. Cyclopentanol
b. Cyclooctanol
c. 3-Methylcyclohexanol

12.43 a. Methyl alcohol
b. Ethyl alcohol
c. Ethylene glycol
d. Propyl alcohol

12.45 a. 4-Methyl-2-hexanol
$$CH_3CHCH_2CHCH_2CH_3$$
with CH_3 and OH substituents

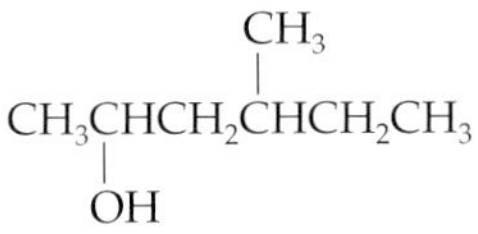

b. Isobutyl alcohol
$$CH_3CHCH_2OH$$ with CH_3

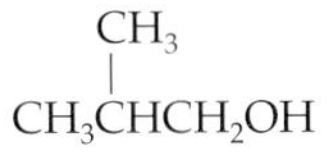

c. 1,5-Pentanediol $CH_2CH_2CH_2CH_2CH_2$ with OH on each end

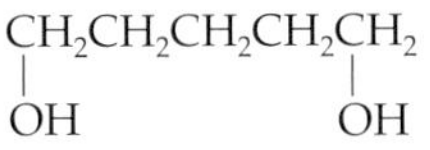

d. 2-Nonanol $CH_3CHCH_2CH_2CH_2CH_2CH_2CH_2CH_3$ with OH

e. 1,3,5-Cyclohexanetriol (cyclohexane ring with three OH groups at 1,3,5 positions)

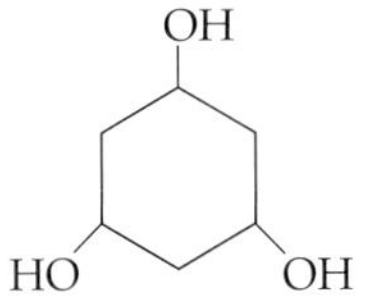

12.47 Denatured alcohol is 100% ethanol to which benzene or methanol is added. The additive makes the ethanol unfit to drink and prevents illegal use of pure ethanol.

12.49 Fermentation is the anaerobic degradation of sugar that involves no net oxidation. The alcohol fermentation, carried out by yeast, produces ethanol and carbon dioxide.

12.51 When the ethanol concentration in a fermentation reaches 12–13%, the yeast producing the ethanol are killed by it. To produce a liquor of higher alcohol concentration, the product of the original fermentation must be distilled.

12.53 The carbinol carbon is the one to which the hydroxyl group is bonded.

12.55 a. Primary   b. Secondary
c. Tertiary   d. Tertiary
e. Tertiary

12.57 a. Tertiary
b. Secondary
c. Primary
d. Tertiary

12.59 Alkene + H_2O $\xrightarrow{H^+}$ Alcohol (general hydration reaction of an alkene to form an alcohol)

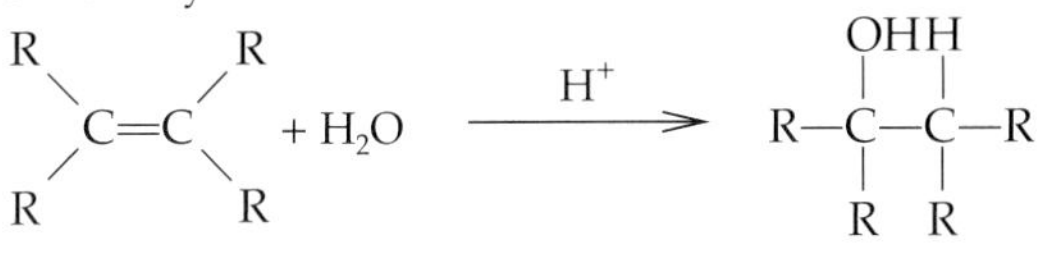

12.61 Alcohol $\xrightarrow{H^+, \text{heat}}$ Alkene + H_2O (dehydration of alcohol)

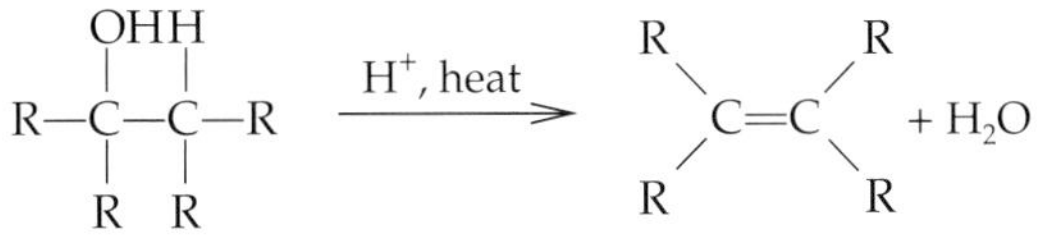

12.63 Secondary alcohol $\xrightarrow{[O]}$ Ketone

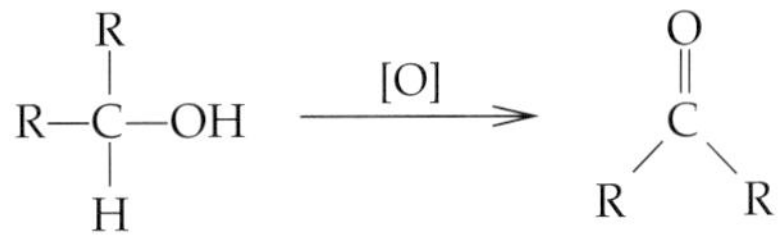

12.65 a. 2-Pentanol (major product), 1-pentanol (minor product)
b. 2-Pentanol and 3-pentanol
c. 3-Methyl-2-butanol (major product), 3-methyl-1-butanol (minor product)
d. 3,3-Dimethyl-2-butanol (major product), 3,3-dimethyl-1-butanol (minor product)

12.67 a. $CH_3CH=CCH_2CH_2CH_3 + H_2O \xrightarrow{H^+}$
2-Hexene → $CH_3CHCH_2CH_2CH_2CH_3$ with OH (2-Hexanol) or $CH_3CH_2CHCH_2CH_2CH_3$ with OH (3-Hexanol)

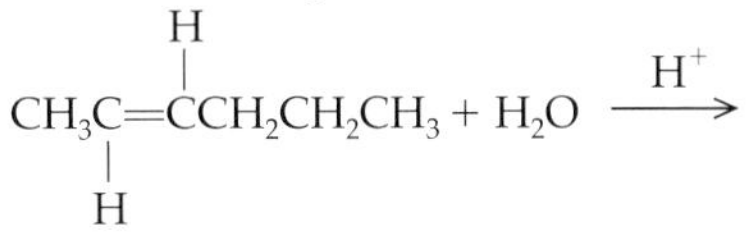

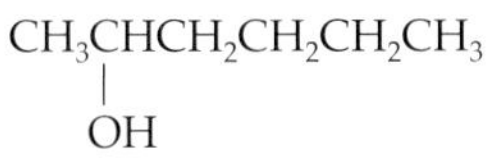

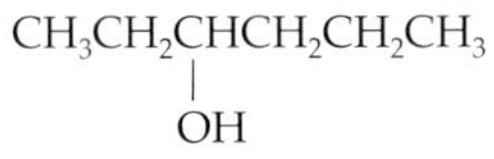

These products will be formed in approximately equal amounts.

b. Cyclopentene + $H_2O \xrightarrow{H^+}$ Cyclopentanol

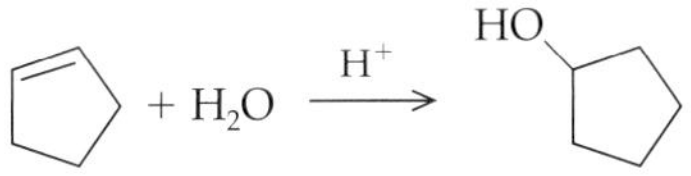

c. $CH_2=CHCH_2CH_2CH_2CH_2CH_2CH_3 + H_2O \xrightarrow{H^+}$
1-Octene → $CH_3CHCH_2CH_2CH_2CH_2CH_2CH_3$ with OH, 2-Octanol (major product)
or $CH_2CH_2CH_2CH_2CH_2CH_2CH_2CH_3$ with OH, 1-Octanol (minor product)

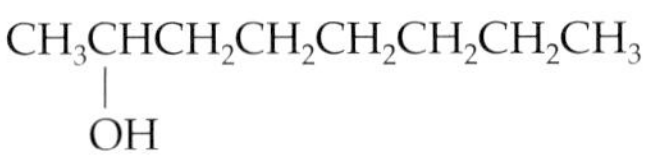

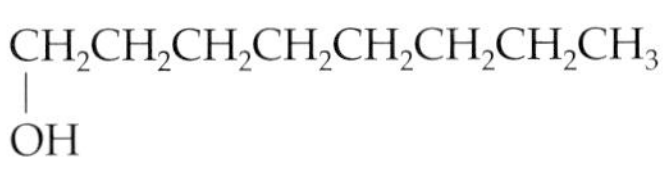

d. 1-Methylcyclohexene + H$_2$O $\xrightarrow{H^+}$ 1-Methylcyclohexanol (major product) or 2-Methylcyclohexanol (minor product)

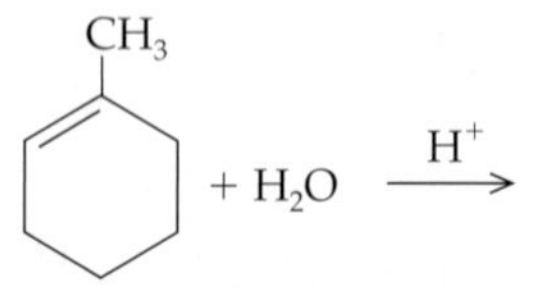

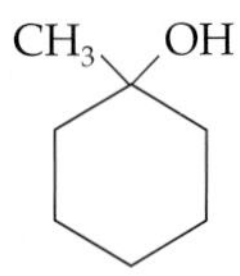

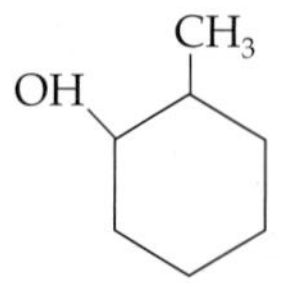

12.69 a. 2-Butanone
b. N.R.
c. Cyclohexanone
d. N.R.

12.71 a. 3-Pentanone
b. Propanal (Upon further oxidation, propanoic acid would be formed.)
c. 4-Methyl-2-pentanone
d. N.R.
e. 3-Phenylpropanal (Upon further oxidation, 3-phenylpropanoic acid will be formed.)

12.73 CH$_3$CH$_2$OH $\xrightarrow{\text{liver enzymes}}$ CH$_3$—C(=O)—H

Ethanol → Ethanal

The product, ethanal, is responsible for the symptoms of a hangover.

12.75 The reaction in which a water molecule is added to 1-butene is a hydration reaction.

CH$_3$CH$_2$CH=CH$_2$ + H$_2$O $\xrightarrow{H^+}$ CH$_3$CH$_2$CH(OH)CH$_3$

1-Butene → 2-Butanol

12.77 CH$_3$CH=CH$_2$ $\xrightarrow{H_2O, H^+}$ CH$_3$—CH(OH)—CH$_3$ $\xrightarrow{[O]}$ CH$_3$—C(=O)—CH$_3$

Propene (propylene) → 2-Propanol (isopropanol) → Propanone (acetone)

12.79 [Steroid structure shown]

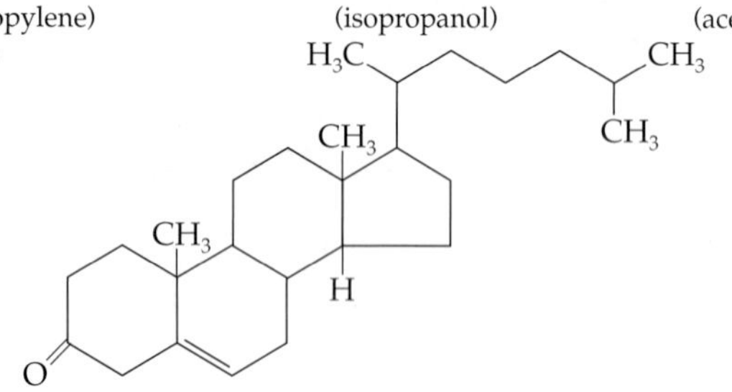

12.81 Oxidation is a loss of electrons, whereas reduction is a gain of electrons.

12.83 CH$_3$CH$_2$CH$_3$ < CH$_3$CH$_2$CH$_2$OH < CH$_3$CH$_2$C(=O)H < CH$_3$CH$_2$C(=O)OH

12.85 Phenols are compounds with an —OH attached to a benzene ring.

12.87 Picric acid: [structure with OH, three NO$_2$ groups]   2,4,6-Trinitrotoluene: [structure with CH$_3$, three NO$_2$ groups]

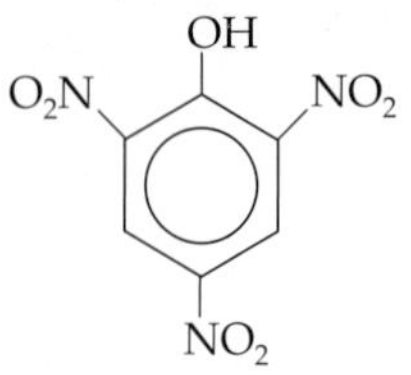

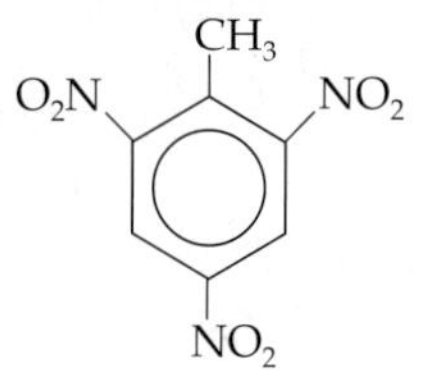

Picric acid is water-soluble because of the polar hydroxyl group that can form hydrogen bonds with water.

12.89 Hexachlorophene, hexylresorcinol, and o-phenylphenol are phenol compounds used as antiseptics or disinfectants.

12.91 Alcohols of molecular formula C$_4$H$_{10}$O

CH$_3$CH$_2$CH$_2$CH$_2$OH,   CH$_3$CH(OH)CH$_2$CH$_3$,

CH$_3$CH(CH$_3$)CH$_2$OH,   CH$_3$—C(OH)(CH$_3$)—CH$_3$

Ethers of molecular formula C$_4$H$_{10}$O
CH$_3$—O—CH$_2$CH$_2$CH$_3$ CH$_3$CH$_2$—O—CH$_2$CH$_3$
CH$_3$—O—CH(CH$_3$)CH$_3$

12.93 Penthrane: 2,2-Dichloro-1,1-difluoro-1-methoxyethane
Enthrane: 2-Chloro-1-(difluoromethoxy)-1,1,2-trifluoroethane

12.95 a. CH$_3$CH$_2$—O—CH$_2$CH$_3$ + H$_2$O
b. CH$_3$CH$_2$—O—CH$_2$CH$_3$ + CH$_3$—O—CH$_3$ + CH$_3$—O—CH$_2$CH$_3$ + H$_2$O
c. CH$_3$—O—CH$_3$ + CH$_3$—O—CH(CH$_3$) + CH$_3$CH(CH$_3$)—O—CH(CH$_3$)$_2$ + H$_2$O
d. [cyclopentyl]—CH$_2$—O—CH$_2$—[cyclopentyl]

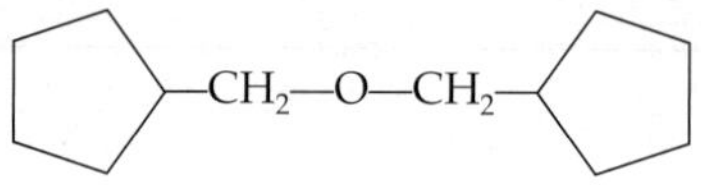

12.97 a. 2-Ethoxypentane
b. 2-Methoxybutane
c. 1-Ethoxybutane
d. Methoxycyclopentane

12.99 Cystine:
H$_3$N$^+$—CH(COO$^-$)—CH$_2$—S—S—CH$_2$—CH(COO$^-$)—NH$_3^+$

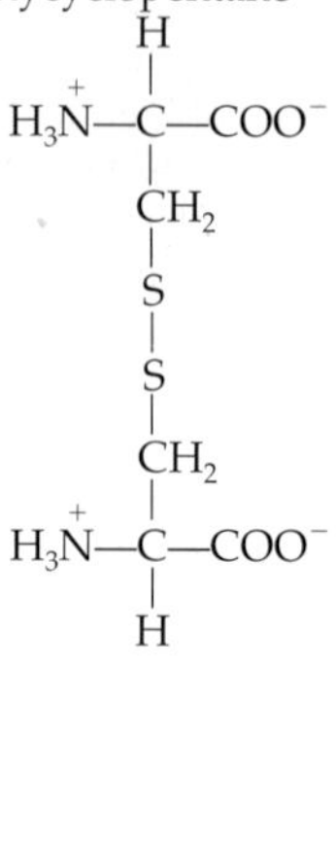

12.101
a. 1-Propanethiol
b. 2-Butanethiol
c. 2-Methyl-2-butanethiol
d. 1, 4-Cyclohexanedithiol

Chapter 13

13.1 a. $CH_3-\overset{\overset{O}{\|}}{C}-CH_3$ b. $CH_3\overset{\overset{OH}{|}}{CH}CH_2CH_2CH_3$

13.3 a. $CH_3CH_2\overset{\overset{O}{\|}}{C}-OH$ b. $CH_3\overset{\overset{O}{\|}}{C}-OH$

13.5
a. Butanal
b. 2,4-Dimethylpentanal

13.7
a. I.U.P.A.C.: 3, 4-Dimethylpentanal
   Common: β, γ-Dimethylvaleraldehyde
b. I.U.P.A.C.: 2-Ethylpentanal
   Common: α-Ethylvaleraldehyde

13.9 a. 3-Methylnonanal:

$$CH_3CH_2CH_2CH_2CH_2CH_2\overset{\overset{}{|}}{CH}CH_2\overset{\overset{O}{\|}}{C}-H$$
$$\quad\quad\quad\quad\quad\quad\quad\quad\quad CH_3$$

b. β-Bromovaleraldehyde:

$$CH_3CH_2\overset{\overset{}{|}}{CH}CH_2\overset{\overset{O}{\|}}{C}-H$$
$$\quad\quad\quad Br$$

13.11
a. 3-Iodobutanone
b. 4-Methyl-2-octanone

13.13 a. Methyl isopropyl ketone (I.U.P.A.C. name: 3-Methyl-2-butanone):

$$CH_3-\overset{\overset{O}{\|}}{C}-\overset{\overset{}{|}}{CH}CH_3$$
$$\quad\quad\quad\quad CH_3$$

b. 4-Heptanone:

$$CH_3CH_2CH_2-\overset{\overset{O}{\|}}{C}-CH_2CH_2CH_3$$

13.15
$CH_3-\overset{\overset{O}{\|}}{C}-H$

13.17
$CH_3CH_2CH_2OH \xrightarrow{H_2Cr_2O_7} CH_3CH_2\overset{\overset{O}{\|}}{C}H$
1-Propanol   Propanal

13.19
$CH_3\overset{\overset{O}{\|}}{C}-H + Ag(NH_3)_2^+ \longrightarrow CH_3\overset{\overset{O}{\|}}{C}-O^- + Ag^0$
Ethanal   Silver ammonia complex   Ethanoate anion   Silver metal

13.21
$CH_3-\overset{\overset{O}{\|}}{C}-CH_3 + H_2 \xrightarrow{Ni} CH_3\overset{\overset{OH}{|}}{CH}CH_3$
Propanone   2-Propanol

13.23
a. Reduction
b. Reduction
c. Reduction
d. Oxidation
e. Reduction

13.25
a. Hemiacetal
b. Ketal
c. Acetal
d. Hemiketal

13.27
$H-\overset{\overset{H}{|}}{\underset{\underset{H}{|}}{C}}-\overset{\overset{H}{|}}{\underset{\underset{H}{|}}{C}}-\overset{\overset{O}{\|}}{C}-H \rightleftharpoons H-\overset{\overset{H}{|}}{\underset{\underset{H}{|}}{C}}-\overset{\overset{OH}{|}}{C}=\overset{}{\underset{\underset{H}{|}}{C}}-H$

Propanal Keto form   Propanal Enol form

13.29
$2CH_3CH_2\overset{\overset{O}{\|}}{C}-H \xrightarrow{OH^-} CH_3CH_2\overset{\overset{OH}{|}}{CH}\overset{}{\underset{\underset{CH_3}{|}}{CH}}\overset{\overset{O}{\|}}{C}-H$

Propanal   3-Hydroxy-2-methylpentanal

13.31 As the carbon chain length increases, the compounds become less polar and more hydrocarbonlike. As a result, their solubility in water decreases.

13.33 A good solvent should dissolve a wide range of compounds. Simple ketones are considered to be universal solvents because they have both a polar carbonyl group and nonpolar side chains. As a result, they dissolve organic compounds and are also miscible in water.

13.35

$CH_3-\overset{\overset{O}{\|}}{C}-H \cdots H-\overset{\overset{O}{\|}}{C}-CH_3$ (with H-bonding to central H₂O)

13.37 Alcohols have higher boiling points than aldehydes or ketones of comparable molecular weights because alcohol molecules can form intermolecular hydrogen bonds with one another. Aldehydes and ketones cannot form intermolecular hydrogen bonds.

13.39 To name an aldehyde using the I.U.P.A.C. nomenclature system, identify and name the longest carbon chain containing the carbonyl group. Replace the final -*e* of the alkane name with -*al*. Number and name all substituents as usual. Remember that the carbonyl carbon is always carbon-1 and does not need to be numbered in the name of the compound.

13.41 The common names of aldehydes are derived from the same Latin roots as the corresponding carboxylic acids. For instance, methanal is formaldehyde; ethanal is acetaldehyde; propanal is propionaldehyde, *etc*.

Substituted aldehydes are named as derivatives of the straight-chain parent compound. Greek letters are used to indicate the position of substituents. The carbon nearest the carbonyl group is the α-carbon, the next is the β-carbon, and so on.

13.43 a. $H-\overset{\overset{O}{\|}}{C}-H$ b. (7-carbon chain with Br substituents on C1 and C2, aldehyde at end)

13.45
a. 3-Chloro-2-pentanone
(structure with Cl on C3, ketone at C2)

b. Benzaldehyde (phenyl-CHO structure)

13.47 a. Butanone
b. 2-Ethylhexanal
13.49 a. 3-Nitrobenzaldehyde
b. 3,4-Dihydroxycyclopentanone
13.51 a. 3-Bromobutanal
b. 2-Chloro-2-methyl-4-heptanone
13.53 a. 4,6-Dimethyl-3-heptanone
b. 3,3-Dimethylcyclopentanone
13.55 a. Acetone
b. Ethyl methyl ketone
c. Acetaldehyde
d. Propionaldehyde
e. Methyl isopropyl ketone

13.57 a. CH₃CHCH₂C(=O)—H with OH on CH
b. CH₃CH₂CH₂CHC(=O)—H with CH₃
c. CH₃CH₂CHCH₂CH₂C(=O)—H with Br
d. CH₃CH₂CHCH₂C(=O)—H with I
e. CH₃CH₂CH₂CH₂CHCHC(=O)—H with CH₃ and OH

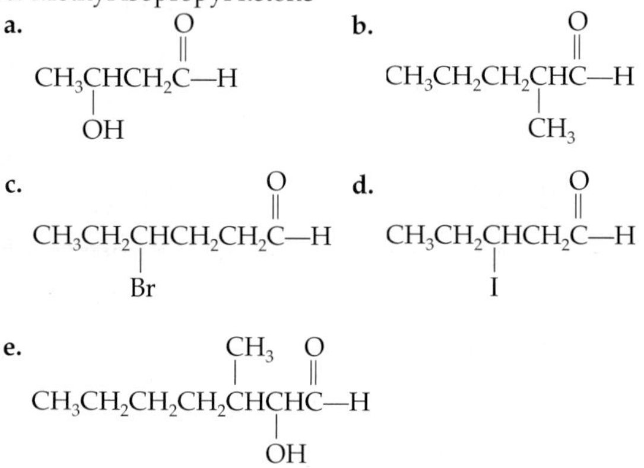

13.59 Acetone is a good solvent because it can dissolve a wide range of compounds. It has both a polar carbonyl group and nonpolar side chains. As a result, it dissolves organic compounds and is also miscible in water.

13.61 The liver

13.63 In organic molecules, oxidation may be recognized as a gain of oxygen or a loss of hydrogen. An aldehyde may be oxidized to form a carboxylic acid as in the following example in which ethanal is oxidized to produce ethanoic acid.

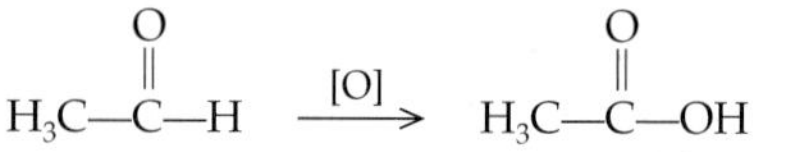

H₃C—C(=O)—H —[O]→ H₃C—C(=O)—OH
Ethanal   Ethanoic acid

13.65 Addition reactions of aldehydes or ketones are those in which a second molecule is added to the double bond of the carbonyl group. An example is the addition of the alcohol ethanol to the aldehyde ethanal.

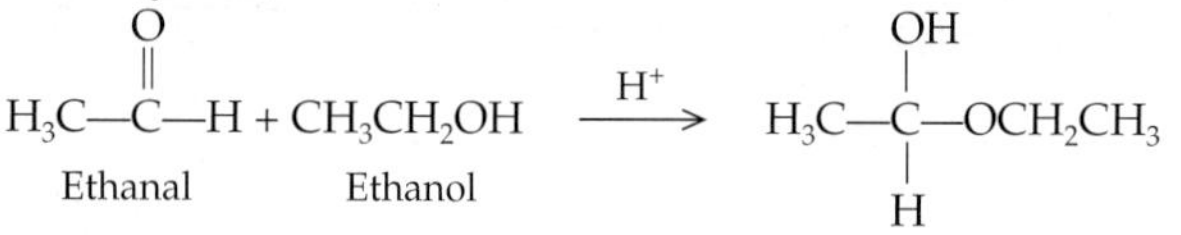

H₃C—C(=O)—H + CH₃CH₂OH —H⁺→ H₃C—C(OH)(OCH₂CH₃)—H
Ethanal   Ethanol   Hemiacetal

13.67 a. CH₃CH₂CH(OH)CH₃ —[O]→ CH₃CH₂C(=O)CH₃
2-Butanol   Butanone

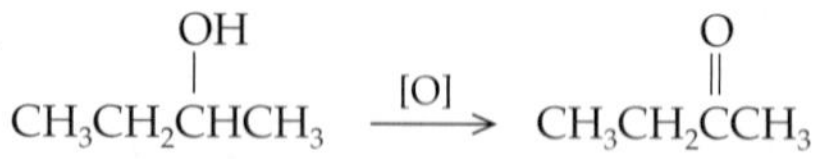

b. CH₃CH(CH₃)CH₂OH —[O]→ CH₃CH(CH₃)—C(=O)—H
2-Methyl-1-propanol   Methylpropanal

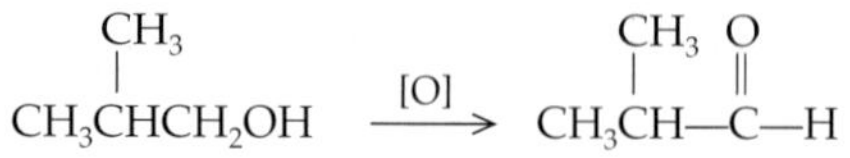

Note that methylpropanal can be further oxidized to methylpropanoic acid.

c. Cyclopentanol —[O]→ Cyclopentanone

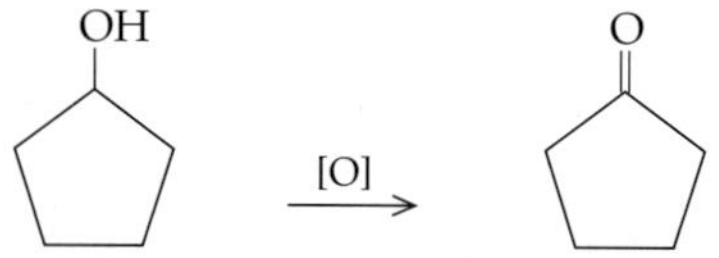

13.69 R—CH₂OH —[O]→ R—C(=O)—H —[O]→ R—C(=O)—OH
Primary alcohol   Aldehyde   Carboxylic acid

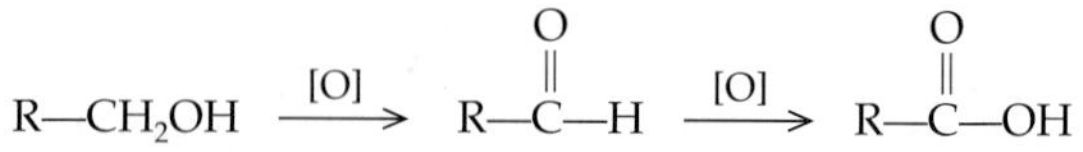

13.71 a. Reduction reaction 
CH₃—C(=O)—H → CH₃CH₂OH
Ethanal   Ethanol

b. Reduction reaction 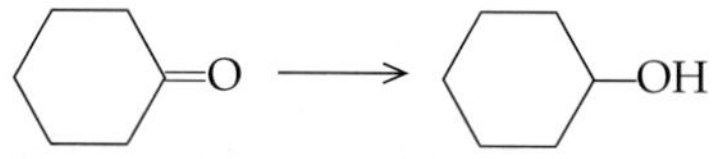
Cyclohexanone → Cyclohexanol

c. Oxidation reaction 
CH₃CH(OH)CH₃ → CH₃—C(=O)—CH₃
2-Propanol   Propanone

13.73 Only (c) 3-methylbutanal and (f) acetaldehyde would give a positive Tollens' test.

13.75 a. CH₃—C(=O)—CH₃ + CH₃CH₂OH —H⁺→ CH₃—C(OH)(OCH₂CH₃)—CH₃

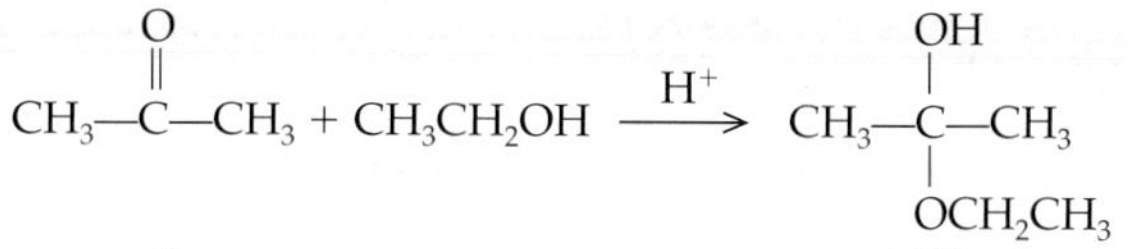

b. CH₃—C(=O)—H + CH₃CH₂OH —H⁺→ CH₃—C(OH)(OCH₂CH₃)—H

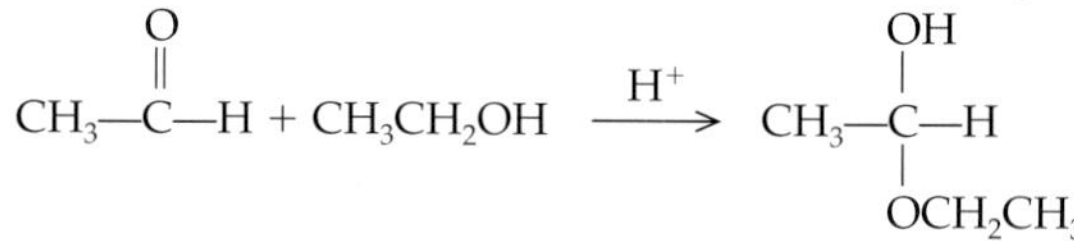

13.77 Hemiacetal
13.79 Acetal
13.81 a. CH₃—C(=O)—CH₃ + 2 CH₃OH —H⁺→ CH₃—C(OCH₃)₂—CH₃ + H₂O

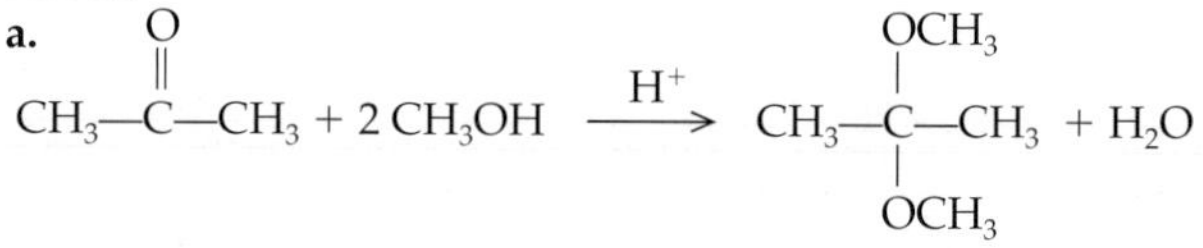

b. CH₃—C(=O)—H + 2 CH₃OH —H⁺→ CH₃—C(OCH₃)₂—H + H₂O

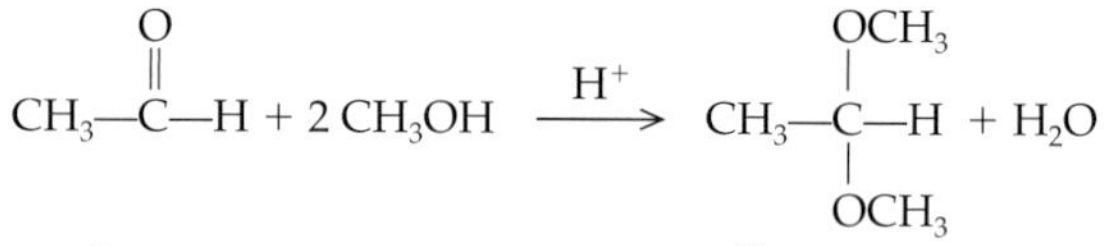

13.83 a. H—C(=O)—OH   **b.** CH₃—C(=O)—OH

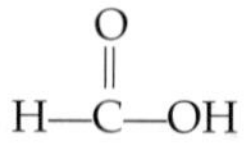

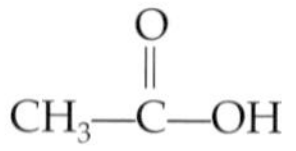

13.85 a. Methanal
b. Propanal
13.87 a. False **c.** False
b. True **d.** False

13.89 2 CH₃—C(=O)—H —OH⁻→ CH₃CH(OH)CH₂C(=O)—H
Ethanal   3-Hydroxybutanal

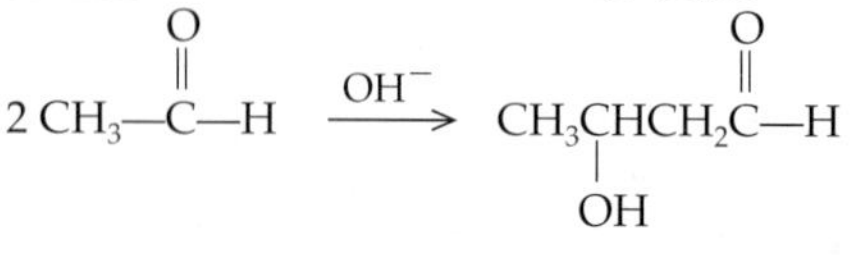

13.91

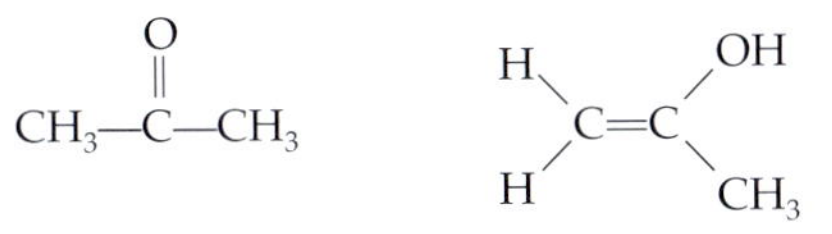

$CH_3-\overset{\overset{O}{\|}}{C}-CH_3$

Keto form of Propanone

$\underset{H}{\overset{H}{}}C=C\underset{CH_3}{\overset{OH}{}}$

Enol form of Propanone

13.93 a. $CH_3CH_2CH_2-\underset{OCH_2CH_3}{\overset{OH}{\underset{|}{C}}}-CH_3$

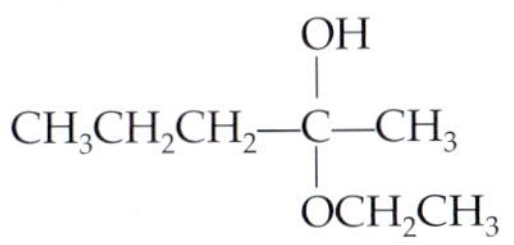

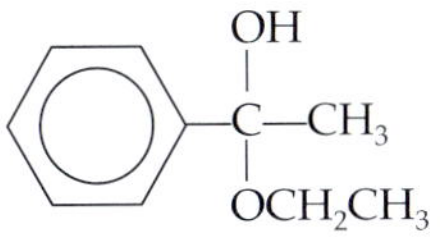

b. Phenyl-$\underset{OCH_2CH_3}{\overset{OH}{\underset{|}{C}}}$-$CH_3$

c. Cyclopentyl with $-OH$ and $-OCH_2CH_3$

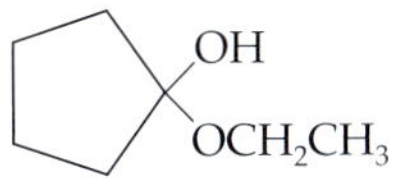

13.95 (1) $2CH_3CH_2OH$ (2) $KMnO_4/OH^-$ (3) $CH_3CH=CH_2$

Chapter 14

14.1 a. Ketone
   b. Ketone
   c. Alkane

14.3 The carboxyl group consists of two very polar groups, the carbonyl group and the hydroxyl group. Thus, carboxylic acids are very polar, in addition to which, they can hydrogen bond to one another. Aldehydes are polar, as a result of the carbonyl group, but cannot hydrogen bond to one another. As a result, carboxylic acids have higher boiling points than aldehydes of the same carbon chain length.

14.5 a. 2, 4-Dimethylpentanoic acid
   b. 2, 4-Dichlorobutanoic acid

14.7 a. 2,3-Dihydroxybutanoic acid:

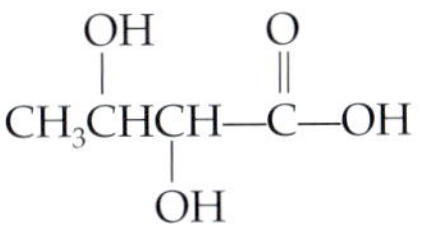

$CH_3\underset{OH}{\overset{OH}{\underset{|}{CH}}}\overset{}{\underset{|}{CH}}-\overset{\overset{O}{\|}}{C}-OH$

   b. 2-Bromo-3-chloro-4-methylhexanoic acid:

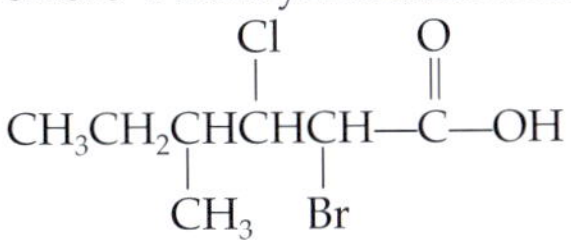

$CH_3CH_2\underset{CH_3}{\overset{Cl}{\underset{|}{CH}}}CH\underset{Br}{\overset{}{\underset{|}{CH}}}-\overset{\overset{O}{\|}}{C}-OH$

14.9 a. α,γ-Dimethylvaleric acid
   b. α,γ-Dichlorobutyric acid

14.11 a. benzene ring with −COOH and −CH_3

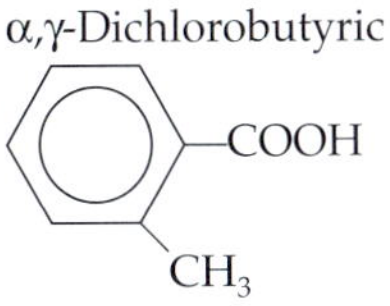

   b. benzene ring with Br, Br, Br and −COOH

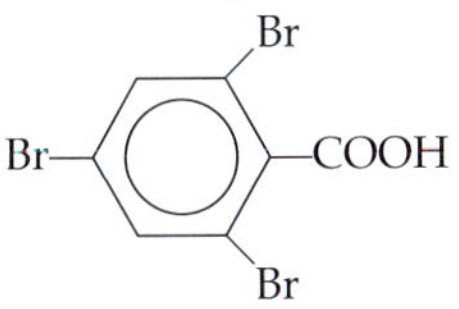

   c. triphenyl−C−COOH

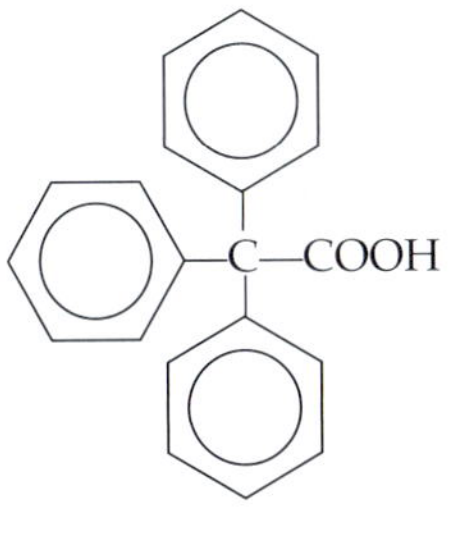

14.13 a. $CH_3CH_2-\overset{\overset{O}{\|}}{C}-H \longrightarrow CH_3CH_2-\overset{\overset{O}{\|}}{C}-OH$
   Propanal would be the first oxidation product. However, it would quickly be oxidized further to propanoic acid.

   b. $HO-\overset{\overset{O}{\|}}{C}-CH_2CH_2CH_2CH_3$

14.15 a. $CH_3-\overset{\overset{O}{\|}}{C}-H \longrightarrow CH_3-\overset{\overset{O}{\|}}{C}-OH$

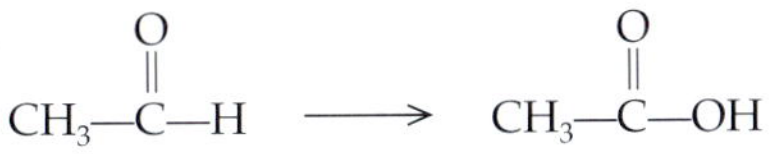

   b. $HO-\overset{\overset{O}{\|}}{C}-CH_2-CH_2-\overset{\overset{O}{\|}}{C}-OH$

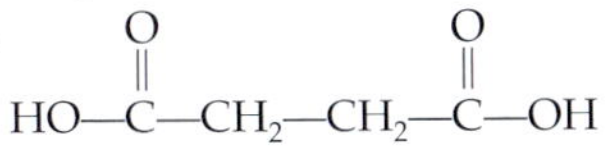

14.17 a. Potassium propanoate
   b. Barium butanoate

14.19 a. Propyl butanoate (propyl butyrate)
   b. Ethyl butanoate (ethyl butyrate)

14.21 a. The following reaction between 1-butanol and ethanoic acid produces butyl ethanoate. It requires a trace of acid and heat. It is also reversible.

$CH_3CH_2CH_2CH_2OH + CH_3COOH \leftrightarrow CH_3\overset{\overset{O}{\|}}{C}-OCH_2CH_2CH_2CH_3$

   b. The following reaction between ethanol and propanoic acid produces ethyl propanoate. It requires a trace of acid and heat. It is also reversible.

$CH_3CH_2OH + CH_3CH_2COOH \leftrightarrow CH_3CH_2\overset{\overset{O}{\|}}{C}-OCH_2CH_3$

14.23 a. $CH_3COOH + CH_3CH_2CH_2OH$
   Ethanoic acid   1-Propanol

   b. $CH_3CH_2CH_2CH_2CH_2COO^-K^+ + CH_3CH_2CH_2OH$
   Potassium hexanoate   1-Propanol

   c. $CH_3CH_2CH_2CH_2COO^-Na^+ + CH_3OH$
   Sodium pentanoate   Methanol

   d. $CH_3CH_2CH_2CH_2COOH + CH_3CHCH_2CH_2CH_3$
                                                    $|$
                                                   OH
   Hexanoic acid   2-Pentanol

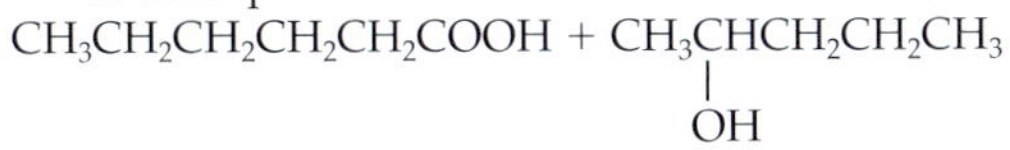

14.25 a.

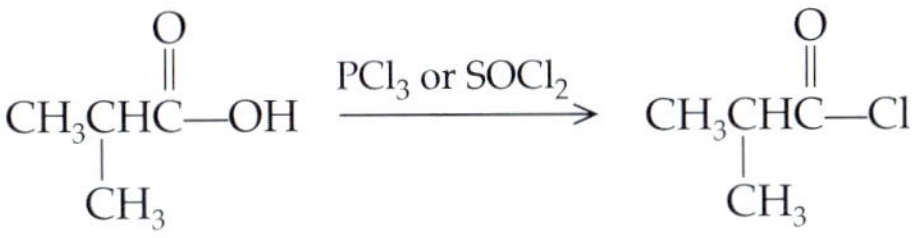

$CH_3\underset{CH_3}{\overset{}{\underset{|}{CH}}}\overset{\overset{O}{\|}}{C}-OH \xrightarrow{PCl_3 \text{ or } SOCl_2} CH_3\underset{CH_3}{\overset{}{\underset{|}{CH}}}\overset{\overset{O}{\|}}{C}-Cl$

2-Methylpropanoic acid → 2-Methylpropanoyl chloride

   b.

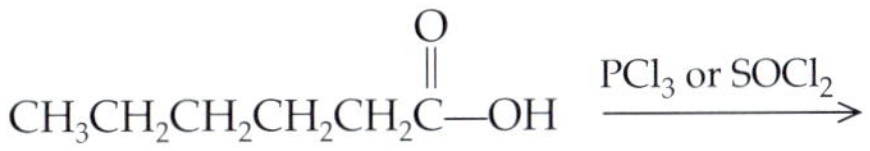

$CH_3CH_2CH_2CH_2CH_2\overset{\overset{O}{\|}}{C}-OH \xrightarrow{PCl_3 \text{ or } SOCl_2}$

Hexanoic acid

$CH_3CH_2CH_2CH_2CH_2\overset{\overset{O}{\|}}{C}-Cl$

Hexanoyl chloride

14.27 a.

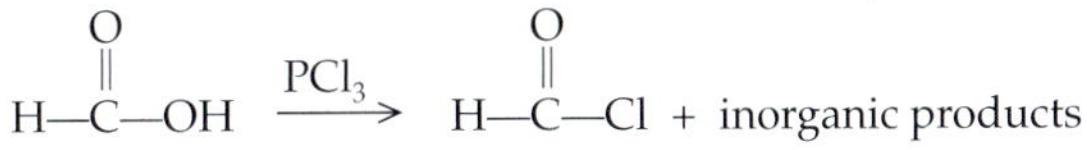

$H-\overset{\overset{O}{\|}}{C}-OH \xrightarrow{PCl_3} H-\overset{\overset{O}{\|}}{C}-Cl$ + inorganic products

Formic acid → Formyl chloride

   b.

$CH_3CH_2-\overset{\overset{O}{\|}}{C}-OH \xrightarrow{PCl_3} CH_3CH_2-\overset{\overset{O}{\|}}{C}-Cl$ + inorganic products

Propionic acid → Propionyl chloride

14.29 a.

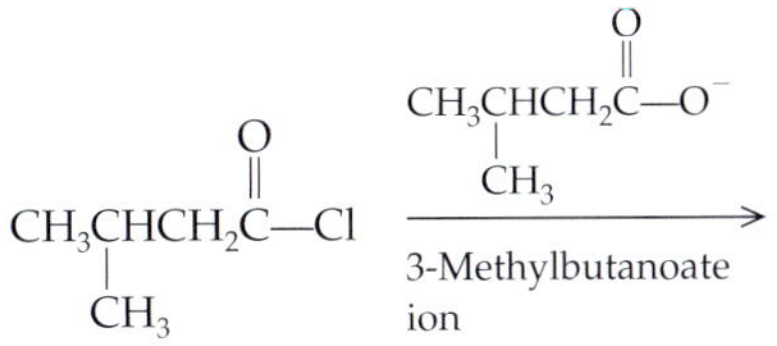

$CH_3\underset{CH_3}{\overset{}{\underset{|}{CH}}}CH_2\overset{\overset{O}{\|}}{C}-Cl \xrightarrow{CH_3\underset{CH_3}{\overset{}{\underset{|}{CH}}}CH_2\overset{\overset{O}{\|}}{C}-O^-}$

3-Methylbutanoyl chloride, 3-Methylbutanoate ion

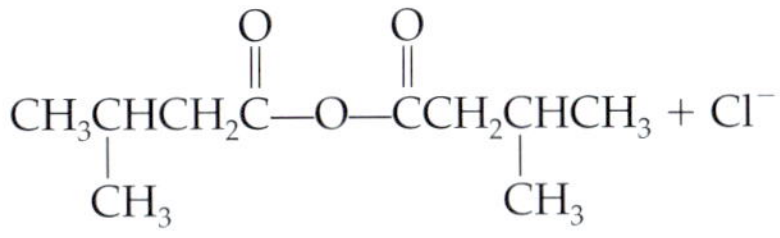

$CH_3\underset{CH_3}{\overset{}{\underset{|}{CH}}}CH_2\overset{\overset{O}{\|}}{C}-O-\overset{\overset{O}{\|}}{C}CH_2\underset{CH_3}{\overset{}{\underset{|}{CH}}}CH_3 + Cl^-$

3-Methylbutanoic anhydride

b.

$$H-\overset{\overset{O}{\|}}{C}-Cl \xrightarrow{CH_3\overset{\overset{O}{\|}}{C}-O^-} H-\overset{\overset{O}{\|}}{C}-O-\overset{\overset{O}{\|}}{C}-CH_3 + Cl^-$$

Methanoyl chloride — Ethanoate ion — Ethanoic methanoic anhydride

14.31 Aldehydes are polar, as a result of the carbonyl group, but cannot hydrogen bond to one another. Alcohols are polar and can hydrogen bond as a result of the polar hydroxyl group. The carboxyl group of the carboxylic acids consists of both of these groups: the carbonyl group and the hydroxyl group. Thus, carboxylic acids are more polar than either aldehydes or alcohols, in addition to which, they can hydrogen bond to one another. As a result, carboxylic acids have higher boiling points than aldehydes or alcohols of the same carbon chain length.

14.33
  a. 3-Hexanone
  b. 3-Hexanone
  c. Hexane

14.35 Propanoic acid > 2-Butanol > Butanal > 2-Methylbutane

14.37
  a. Heptanoic acid
  b. 1-Propanol
  c. Pentanoic acid
  d. Butanoic acid

14.39 The smaller carboxylic acids are water-soluble. They have sharp, sour tastes and unpleasant aromas.

14.41 Citric acid is found naturally in citrus fruits. It is added to foods to give them a tart flavor or to act as a food preservative and anti-oxidant. Adipic acid imparts a tart flavor to soft drinks and is a preservative.

14.43 Determine the name of the parent compound, that is the longest carbon chain containing the carboxyl group. Change the -e ending of the alkane name to -oic acid. Number the chain so that the carboxyl carbon is carbon-1. Name and number substituents in the usual way.

14.45 a.

$$H-\underset{H}{\overset{H}{C}}-\underset{H}{\overset{H}{C}}-\underset{Br}{\overset{H}{C}}-\underset{H}{\overset{H}{C}}-\overset{\overset{O}{\|}}{C}-OH$$

b.

$$H-\underset{H}{\overset{H}{C}}-\underset{H}{\overset{H-C-H}{C}}-\underset{Br}{\overset{H}{C}}-\overset{\overset{O}{\|}}{C}-OH$$

c.

Cyclohexane with —COOH and Br substituents

14.47
  a. I.U.P.A.C. name: Methanoic acid
     Common name: Formic acid
  b. I.U.P.A.C. name: 3-Methylbutanoic acid
     Common name: β-Methylbutyric acid
  c. I.U.P.A.C. name: Cyclopentanecarboxylic acid
     Common name: Cyclovalericcarboxylic acid

14.49

Butanoic acid and Methylpropanoic acid structures

14.51 a. $CH_3CH_2-\underset{CH_3}{\overset{CH_3}{C}}-CH_2CH_2COOH$

b. $CH_3CHCH\underset{Br}{\overset{CH_3}{|}}CH_2COOH$

c. Benzene ring with —COOH, O_2N, NO_2 substituents

d. Cyclohexane with —COOH and H_3C substituents

14.53
  a. I.U.P.A.C. name: 2-Hydroxypropanoic acid
     Common name: α-Hydroxypropionic acid
  b. I.U.P.A.C. name: 3-Hydroxybutanoic acid
     Common name: β-Hydroxybutyric acid
  c. I.U.P.A.C. name: 4,4-Dimethylpentanoic acid
     Common name: γ,γ-Dimethylvaleric acid
  d. I.U.P.A.C. name: 3,3-Dichloropentanoic acid
     Common name: β,β-Dichlorovaleric acid

14.55 In organic molecules, oxidation may be recognized as a gain of oxygen or a loss of hydrogen. An aldehyde may be oxidized to form a carboxylic acid as in the following example in which ethanal is oxidized to produce ethanoic acid.

$$H_3C-\overset{\overset{O}{\|}}{C}-H \xrightarrow{[O]} H_3C-\overset{\overset{O}{\|}}{C}-OH$$

Ethanal → Ethanoic acid

14.57 The following general equation represents the dissociation of a carboxylic acid.

$$R-\overset{\overset{O}{\|}}{C}-OH \rightleftharpoons R-\overset{\overset{O}{\|}}{C}-O^- + H^+$$

14.59 When a strong base is added to a carboxylic acid, neutralization occurs.

14.61 Soaps are made from water, a strong base, and natural fats or oils.

14.63
  a. CH_3COOH
  b. $CH_3CH_2CH_2-\overset{\overset{O}{\|}}{C}-O-CH_3 + H_2O$
  c. CH_3OH

14.65
  a. The oxidation of 1-pentanol yields pentanal.
  b. Continued oxidation of pentanal yields pentanoic acid.

14.67 Esters are mildly polar as a result of the polar carbonyl group within the structure.

14.69 Esters are formed in the reaction of a carboxylic acid with an alcohol. The name is derived by using the alkyl or aryl portion of the alcohol I.U.P.A.C. name as the first name. The -ic acid ending of the I.U.P.A.C. name of the carboxylic acid is replaced with -ate and follows the name of the aryl or alkyl group.

14.71
a. Methyl benzoate structure: benzene-C(=O)-OCH₃

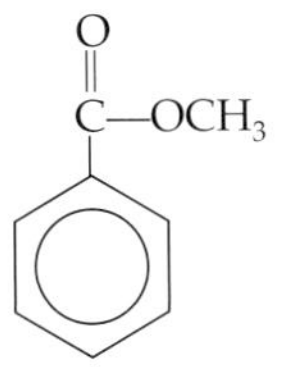

b. $CH_3CH_2CH_2CH_2CH_2CH_2CH_2CH_2CH_2-\overset{O}{\underset{\|}{C}}-O-CH_2CH_2CH_2CH_3$

c. $CH_3CH_2-\overset{O}{\underset{\|}{C}}-O-CH_3$

d. $CH_3CH_2-\overset{O}{\underset{\|}{C}}-O-CH_2CH_3$

14.73
a. Ethyl ethanoate
b. Methyl propanoate
c. Methyl-3-methylbutanoate
d. Cyclopentyl benzoate

14.75 The following equation shows the general reaction for the preparation of an ester:

$$R-\overset{O}{\underset{\|}{C}}-OH + R-OH \underset{}{\overset{H^+, \text{heat}}{\rightleftharpoons}} R-\overset{O}{\underset{\|}{C}}-OR + H_2O$$

Carboxylic acid + Alcohol ⇌ Ester + Water

14.77 The following equation shows the general reaction for the acid-catalyzed hydrolysis of an ester:

$$R-\overset{O}{\underset{\|}{C}}-OR + H_2O \underset{}{\overset{H^+, \text{heat}}{\rightleftharpoons}} R-\overset{O}{\underset{\|}{C}}-OH + R-OH$$

Ester + Water ⇌ Carboxylic acid + Alcohol

14.79 A hydrolysis reaction is the cleavage of any bond by the addition of a water molecule.

14.81
a. $CH_3CH_2CH_2-\overset{O}{\underset{\|}{C}}-O-CH_2CH_3$

b. $CH_3CH_2-\overset{O}{\underset{\|}{C}}-OH + CH_3CH_2OH$

c. $CH_3CH_2CH_2OH$

d. $CH_3CH_2\overset{Br}{\underset{|}{CH}}CH_2-\overset{O}{\underset{\|}{C}}-O^- + CH_3CH_2OH$

14.83 Saponification is a reaction in which a soap is produced. More generally, it is the hydrolysis of an ester in the presence of a base. The following reaction shows the base-catalyzed hydrolysis of an ester:

$CH_3(CH_2)_{14}-\overset{O}{\underset{\|}{C}}-O-CH_3 + NaOH \longrightarrow$

$CH_3(CH_2)_{14}-\overset{O}{\underset{\|}{C}}-O^-Na^+ + CH_3OH$

14.85

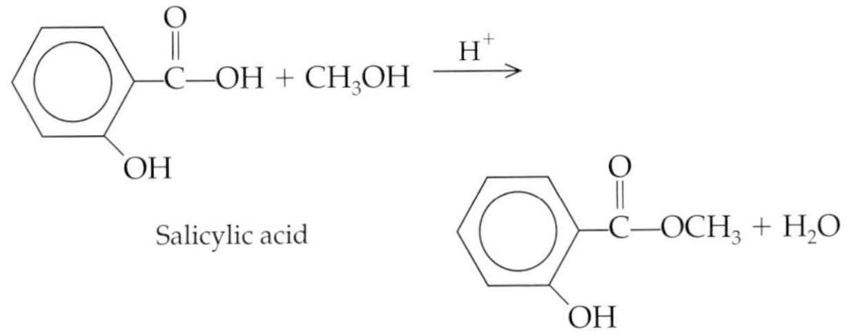

Salicylic acid + $CH_3OH \overset{H^+}{\longrightarrow}$ Methyl salicylate + H_2O

14.87 Compound A is

$CH_3CH_2CH_2CH_2-\overset{O}{\underset{\|}{C}}-O-CH_3$

Compound B is

$CH_3CH_2CH_2CH_2-\overset{O}{\underset{\|}{C}}-OH$

Compound C is CH_3OH

14.89
a. $CH_3CH_2-\overset{O}{\underset{\|}{C}}-OCH_2CH_2CH_3 \overset{H^+, \text{heat}}{\rightleftharpoons}$
Propyl propanoate

$CH_3CH_2-\overset{O}{\underset{\|}{C}}-OH + CH_3CH_2CH_2OH$
Propanoic acid + 1-Propanol

b. $H-\overset{O}{\underset{\|}{C}}-OCH_2CH_2CH_2CH_3 \overset{H^+, \text{heat}}{\rightleftharpoons}$
Butyl methanoate

$H-\overset{O}{\underset{\|}{C}}-OH + CH_3CH_2CH_2CH_2OH$
Methanoic acid + 1-Butanol

c. $H-\overset{O}{\underset{\|}{C}}-OCH_2CH_3 \overset{H^+, \text{heat}}{\rightleftharpoons}$
Ethyl methanoate

$H-\overset{O}{\underset{\|}{C}}-OH + CH_3CH_2OH$
Methanoic acid + Ethanol

d. $CH_3CH_2CH_2CH_2-\overset{O}{\underset{\|}{C}}-OCH_3 \overset{H^+, \text{heat}}{\rightleftharpoons}$
Methyl pentanoate

$CH_3CH_2CH_2CH_2-\overset{O}{\underset{\|}{C}}-OH + CH_3OH$
Pentanoic acid + Ethanol

14.91
a. PCl_3, PCl_5, or $SOCl_2$
b. $CH_3-\overset{O}{\underset{\|}{C}}-O^-$
c. cyclohexyl-$\overset{O}{\underset{\|}{C}}-O^-$

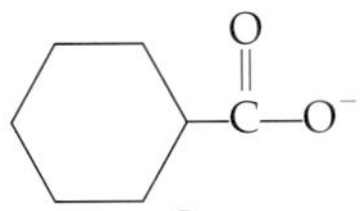

14.93
a. benzoic acid-$\overset{O}{\underset{\|}{C}}-OH + HCl$
b. $2\ CH_3-\overset{O}{\underset{\|}{C}}-OH$

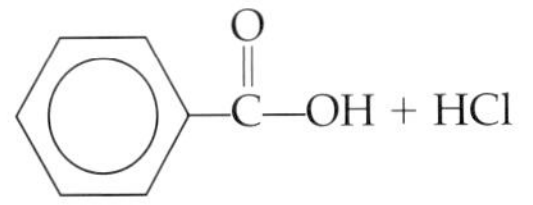

14.95
a. $CH_3(CH_2)_8-\overset{O}{\underset{\|}{C}}-O-\overset{O}{\underset{\|}{C}}-(CH_2)_8CH_3$

b. $CH_3-\overset{O}{\underset{\|}{C}}-O-\overset{O}{\underset{\|}{C}}-CH_3$

c. $CH_3(CH_2)_3-\overset{O}{\underset{\|}{C}}-O-\overset{O}{\underset{\|}{C}}-(CH_2)_3CH_3$

d. benzene-$\overset{O}{\underset{\|}{C}}-Cl$

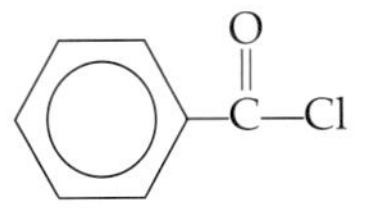

14.97 Acid chlorides are noxious, irritating chemicals. They are slightly polar and have boiling points similar to comparable aldehydes or ketones. They cannot be dissolved in water because they react violently with it.

14.99 a.

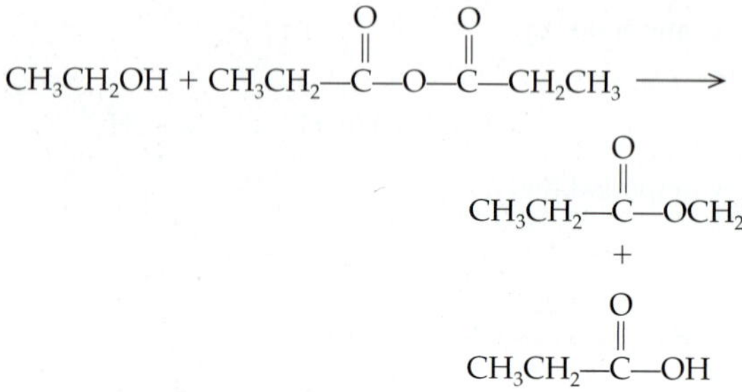

$$CH_3CH_2OH + CH_3CH_2-\overset{O}{\underset{\|}{C}}-O-\overset{O}{\underset{\|}{C}}-CH_2CH_3 \longrightarrow$$

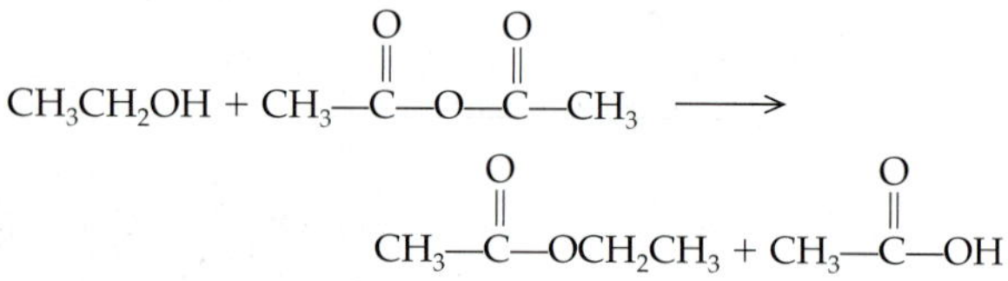

$$CH_3CH_2-\overset{O}{\underset{\|}{C}}-OCH_2CH_3$$
$$+$$
$$CH_3CH_2-\overset{O}{\underset{\|}{C}}-OH$$

b.

$$CH_3CH_2OH + CH_3-\overset{O}{\underset{\|}{C}}-O-\overset{O}{\underset{\|}{C}}-CH_3 \longrightarrow$$

$$CH_3-\overset{O}{\underset{\|}{C}}-OCH_2CH_3 + CH_3-\overset{O}{\underset{\|}{C}}-OH$$

c.

$$CH_3CH_2OH + H-\overset{O}{\underset{\|}{C}}-O-\overset{O}{\underset{\|}{C}}-H \longrightarrow$$

$$H-\overset{O}{\underset{\|}{C}}-OCH_2CH_3 + H-\overset{O}{\underset{\|}{C}}-OH$$

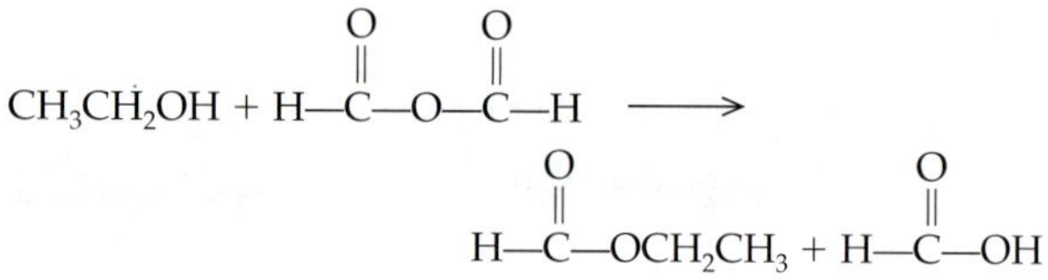

14.101 a. Monoester:

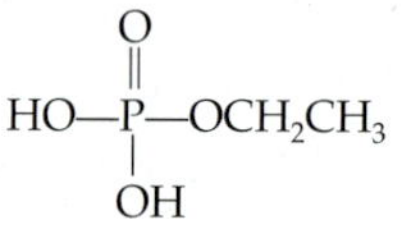

$$HO-\overset{O}{\underset{\underset{OH}{|}}{\overset{\|}{P}}}-OCH_2CH_3$$

b. Diester:

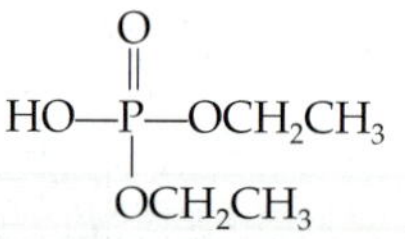

$$HO-\overset{O}{\underset{\underset{OCH_2CH_3}{|}}{\overset{\|}{P}}}-OCH_2CH_3$$

c. Triester:

$$CH_3CH_2-O-\overset{O}{\underset{\underset{OCH_2CH_3}{|}}{\overset{\|}{P}}}-OCH_2CH_3$$

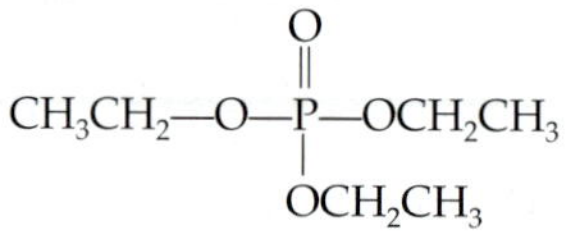

14.103 ATP is the molecule used to store the energy released in metabolic reactions. The energy is stored in the phosphoanhydride bonds between two phosphoryl groups. The energy is released when the bond is hydrolyzed. A portion of the energy can be transferred to another molecule if the phosphoryl group is transferred from ATP to the other molecule.

14.105

$$CH_3-\overset{O}{\underset{\|}{C}}{\sim}S-COENZYME\ A$$

The squiggle denotes a high energy bond.

14.107

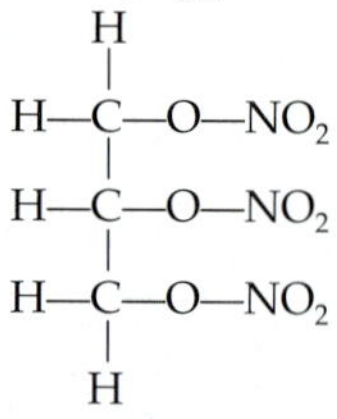

H—C—O—NO$_2$
 |
H—C—O—NO$_2$
 |
H—C—O—NO$_2$
 |
 H

Chapter 15

15.1 a. Tertiary
 b. Primary
 c. Secondary

15.3

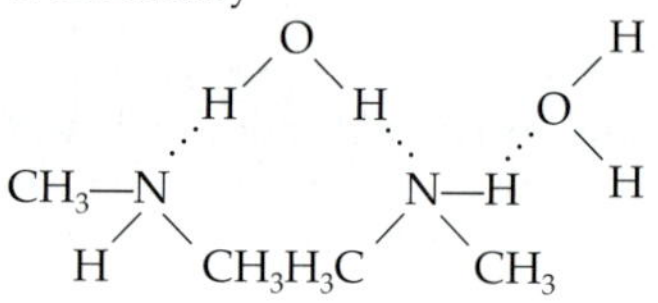

CH$_3$—N(H)(H)···O(H)(H)···O(H)(H)
         CH$_3$H$_2$C—N—H   CH$_3$

15.5 a. Methanol because the intermolecular hydrogen bonds between alcohol molecules will be stronger.
 b. Water because the intermolecular hydrogen bonds between water molecules will be stronger.
 c. Ethylamine because it has a higher molecular weight.
 d. Propylamine because propylamine molecules can form intermolecular hydrogen bonds while the nonpolar butane cannot do so.

15.7 a. C$_6$H$_5$—N(H)—CH$_3$
 b. C$_6$H$_5$—N(CH$_3$)—CH$_3$
 c. C$_6$H$_5$—N(H)—CH$_2$CH$_3$
 d. C$_6$H$_5$—N(H)—CH(CH$_3$)CH$_3$

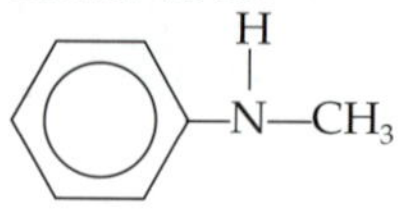

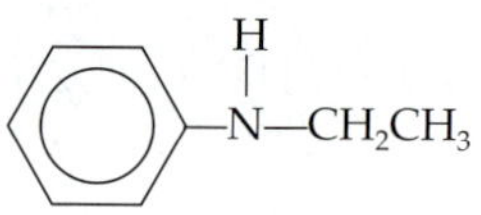

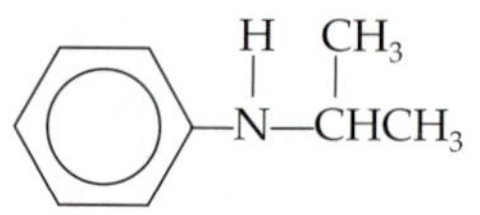

15.9 a.

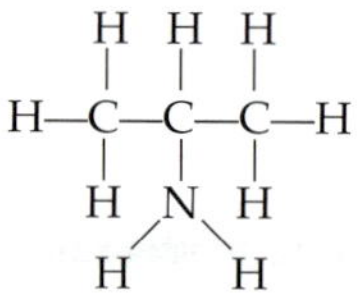

H—C(H)—C(H)—C(H)—H
       |
       N
      / \
     H   H
     with H's on carbons

b. H—C—C—C—C—C—C—C—H (7 C chain) with N—H substituent

c. H—C—C—C—C—C—C—C—H with N—H, and branch H—C—H, H—C—H

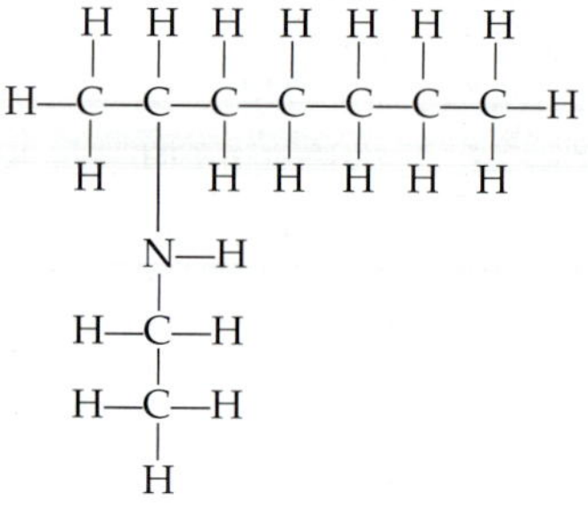

d. H—C—C—C—C—C—H with N(H)(H) branch

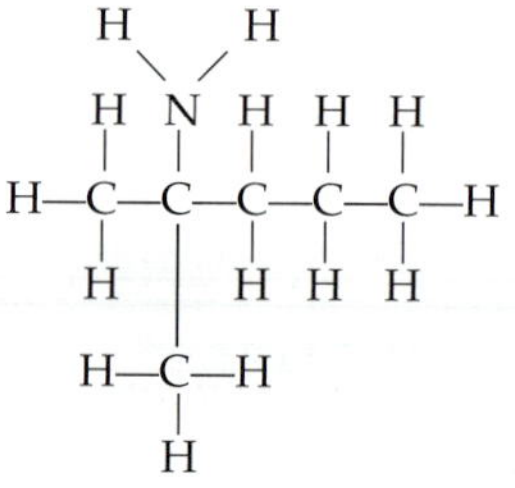

e. H—C—C—C—C—C—C—C—C—C—H with N(H)(H), Cl, I substituents

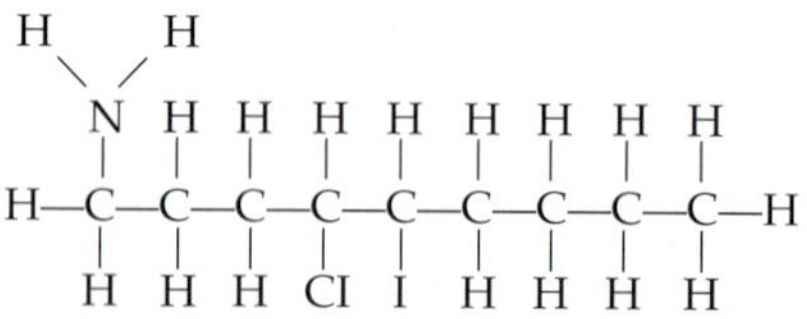

f. H—C—C—N—C—C—H with branches H—C—C—C—H and H—C—C—C—H

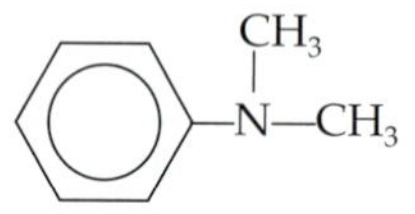

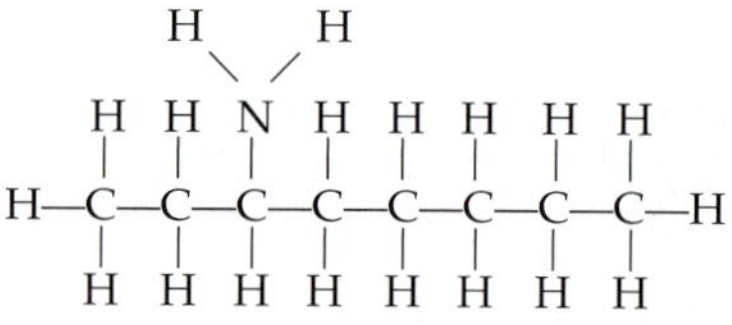

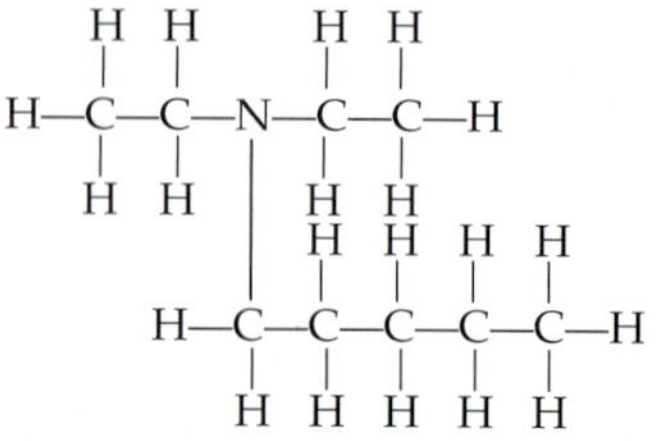

15.11 a. 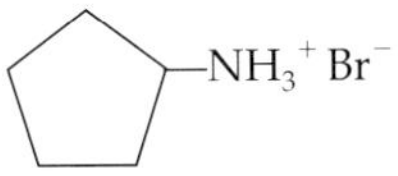 cyclopentyl-NH$_3^+$ Br$^-$

b. CH$_3$CH$_2$—N$^+$H(CH$_3$)—H + OH$^-$ (with H and H on N)

c. CH$_3$—N$^+$H$_3$ + OH$^-$

15.13 a. CH$_3$—NH$_2$

b. CH$_3$—NH—CH$_3$

15.15 The nitrogen atom is more polar than the hydrogen atom in amines; thus, the N-H bond is polar and hydrogen bonding can occur between primary or secondary amine molecules. Thus, amines have a higher boiling point than alkanes, which are nonpolar. Because nitrogen is not as electronegative as oxygen, the N-H bond is not as polar as the O-H. As a result, intermolecular hydrogen bonds between primary and secondary amine molecules are not as strong as the hydrogen bonds between alcohol molecules. Thus, alcohols have a higher boiling point.

15.17 In systematic nomenclature, primary amines are named by determining the name of the parent compound, the longest continuous carbon chain containing the amine group. The -e ending of the alkane chain is replaced with -amine. Thus, an alkane becomes an alkanamine. The parent chain is then numbered to give the carbon bearing the amine group the lowest possible number. Finally, all substituents are named and numbered and added as prefixes to the "alkanamine" name.

15.19 Amphetamines elevate blood pressure and pulse rate. They also decrease the appetite.

15.21 a. 1-Butanamine would be more soluble in water because it has a polar amine group that can form hydrogen bonds with water molecules.

b. 2-Pentanamine would be more soluble in water because it has a polar amine group that can form hydrogen bonds with water molecules.

15.23 Triethylamine molecules cannot form hydrogen bonds with one another, but 1-hexanamine molecules are able to do so.

15.25 a. 2-Butanamine
b. 3-Hexanamine
c. Cyclopentanamine
d. 2-Methyl-2-propanamine

15.27 a. CH$_3$CH$_2$—NH—CH$_2$CH$_3$
b. CH$_3$CH$_2$CH$_2$CH$_2$NH$_2$
c. CH$_3$CH$_2$CHCH$_2$CH$_2$CH$_2$CH$_2$CH$_2$CH$_3$ with NH$_2$ on CH
d. CH$_3$CHCHCH$_2$CH$_3$ with Br on one carbon and NH$_2$ on adjacent carbon
e. 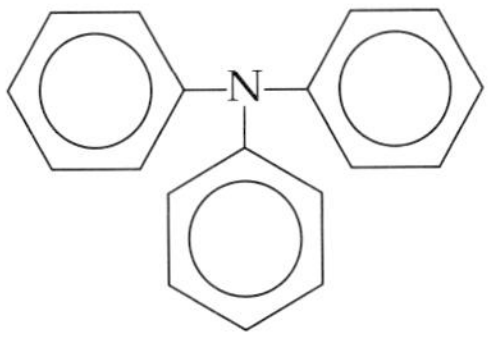 triphenylamine (N with three phenyl groups)

15.29 a. CH$_3$CHCH$_2$CH$_2$CH$_3$ with NH$_2$ on second carbon
b. CH$_3$CH$_2$CHCH$_2$NH$_2$ with Br on CH
c. CH$_3$CH$_2$—NH—CHCH$_3$ with CH$_3$ branch
d.  cyclopentyl-NH$_2$

15.31

CH$_3$CH$_2$CH$_2$CH$_2$NH$_2$ — 1-Butanamine (Primary amine)

CH$_3$CH$_2$CHCH$_3$ with NH$_2$ — 2-Butanamine (Primary amine)

CH$_3$CHCH$_2$NH$_2$ with CH$_3$ — 2-Methyl-1-propanamine (Primary amine)

CH$_3$—C(CH$_3$)(CH$_3$)—NH$_2$ — 2-Methyl-2-propanamine (Primary amine)

CH$_3$CH$_2$—N(CH$_3$)—CH$_3$ — N,N-Dimethylethanamine (Tertiary amine)

CH$_3$CH$_2$—NH—CH$_2$CH$_3$ — N-Ethylethanamine (Secondary amine)

CH$_3$CHCH$_3$ with NH—CH$_3$ — N-Methyl-2-propanamine (Secondary amine)

CH$_3$CH$_2$CH$_2$—NH—CH$_3$ — N-Methyl-1-propanamine (Secondary amine)

15.33 a. Primary
b. Secondary
c. Primary
d. Tertiary

15.35 a. 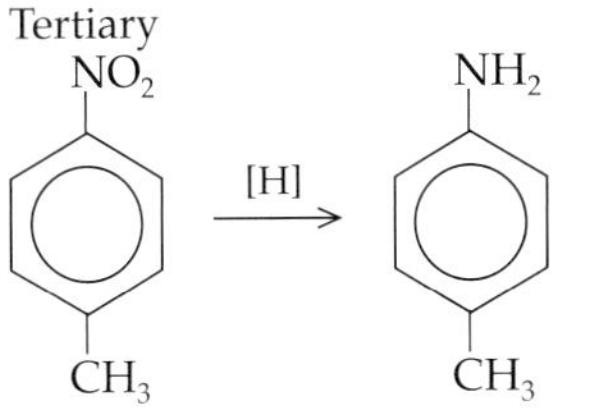 4-methyl-nitrobenzene + [H] → 4-methyl-aniline

b. 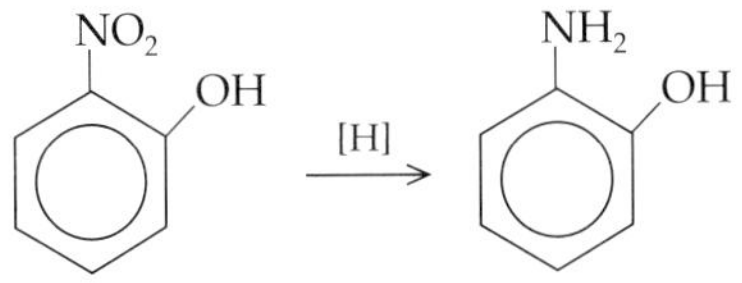 2-nitrophenol + [H] → 2-aminophenol

c. 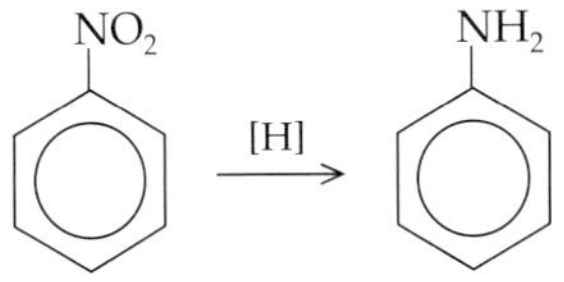 nitrobenzene + [H] → aniline

d. 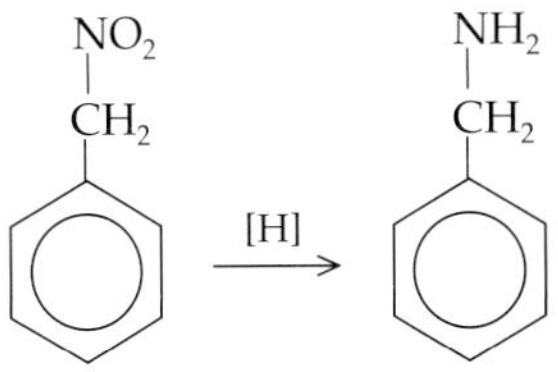 (nitromethyl)benzene + [H] → (aminomethyl)benzene

15.37
a. H_2O
b. HBr
c. $CH_3CH_2CH_2-N^+H_3$
d. $CH_3CH_2-\overset{\underset{|}{CH_2CH_3}}{N^+H_2}Cl^-$

15.39 Lower molecular weight amines are soluble in water because the N—H bond is polar and can form hydrogen bonds with water molecules.

15.41 Drugs containing amine groups are generally administered as ammonium salts because the salt is more soluble in water and, hence, in body fluids.

15.43 Putrescine (1,4-Butanediamine):
$$\underset{NH_2}{CH_2}CH_2CH_2\underset{NH_2}{CH_2}$$

Cadaverine (1,5-Pentanediamine):
$$\underset{NH_2}{CH_2}CH_2CH_2CH_2\underset{NH_2}{CH_2}$$

15.45
a. Pyridine   Indole

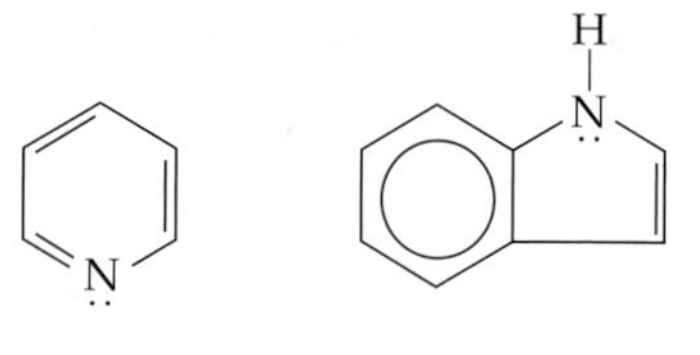

b. The indole ring is found in lysergic acid diethylamide, which is a hallucinogenic drug. The pyridine ring is found in vitamin B_6, an essential water-soluble vitamin.

15.47 Morphine, codeine, quinine, and vitamin B_6

15.49 Amides have very high boiling points because the amide group consists of two very polar functional groups, the carbonyl group and the amino group. Strong intermolecular hydrogen bonding between the N-H bond of one amide and the C=O group of a second amide results in very high boiling points.

15.51 The I.U.P.A.C. names of amides are derived from the I.U.P.A.C. names of the carboxylic acids from which they are derived. The *-oic acid* ending of the carboxylic acid is replaced with the *-amide* ending.

15.53 Barbiturates are often called "downers" because they act as sedatives. They are sometimes used as anticonvulsants for epileptics and people suffering from other disorders that manifest as neurosis, anxiety, or tension.

15.55
a. I.U.P.A.C. name: Propanamide
   Common name: Propionamide
b. I.U.P.A.C. name: Pentanamide
   Common name: Valeramide
c. I.U.P.A.C. name: N,N-Dimethylethanamide
   Common name: N,N-Dimethylacetamide

15.57
a. $CH_3-\overset{O}{\underset{\|}{C}}-NH_2$

b. $CH_3CH_2-\overset{O}{\underset{\|}{C}}-NH-CH_3$

c. $C_6H_5-\overset{O}{\underset{\|}{C}}-N(CH_2CH_3)_2$

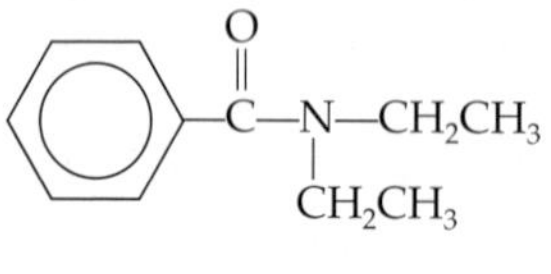

d. $CH_3CH_2\underset{Br}{\overset{CH_3}{\underset{|}{\overset{|}{C}H}}}CHCH_2-\overset{O}{\underset{\|}{C}}-NH_2$

e. $CH_3-\overset{O}{\underset{\|}{C}}-\underset{CH_3}{\overset{|}{N}}-CH_3$

15.59 N,N-Diethyl-*m*-toluamide:

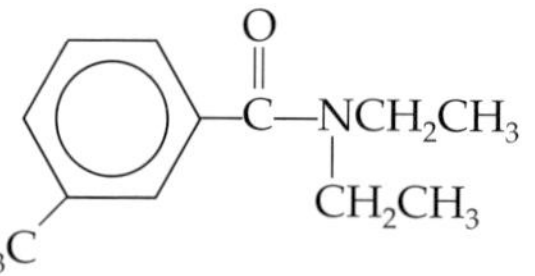

Hydrolysis of this compound would release the carboxylic acid *m*-toluic acid and the amine N-ethylethanamine (diethylamine).

15.61 Amides are not proton acceptors (bases) because the highly electronegative carbonyl oxygen has a strong attraction for the nitrogen lone pair of electrons. As a result they cannot "hold" a proton.

15.63 Amide group

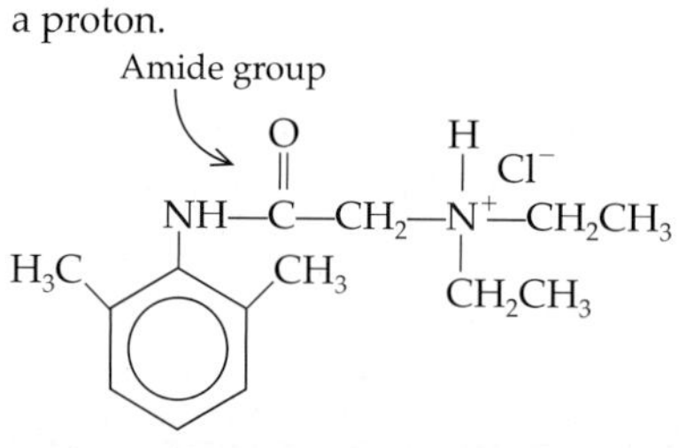

Lidocaine hydrochloride

15.65 Amide group — Carboxyl group

$CH_3(CH_2)_3SCH_2CONH$

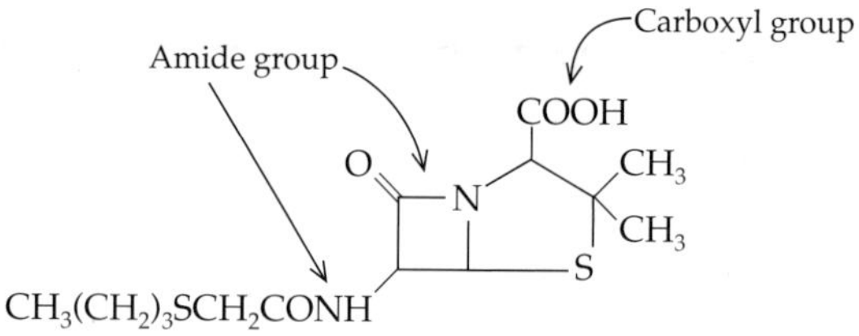

Penicillin BT

15.67
a. $CH_3-\overset{O}{\underset{\|}{C}}-NHCH_3 + H_3O^+ \longrightarrow$

N-Methylethanamide

$CH_3COOH + CH_3NH_3^+$

Ethanoic acid   Methanamine

b. $CH_3CH_2CH_2-\overset{O}{\underset{\|}{C}}-NH-CH_3 + H_3O^+ \longrightarrow$

N-Methylbutanamide

$CH_3CH_2CH_2COOH + CH_3NH_3^+$

Butanoic acid   Methanamine

c. $CH_3\underset{CH_3}{\overset{|}{C}H}CH_2-\overset{O}{\underset{\|}{C}}-NH-CH_2CH_3 + H_3O^+ \longrightarrow$

N-Ethyl-3-methylbutanamide

$CH_3\underset{CH_3}{\overset{|}{C}H}CH_2COOH + CH_3CH_2NH_3^+$

3-Methylbutanoic acid   Ethanamine

15.69 a. CH₃CH₂—C(=O)—O—C(=O)—CH₂CH₃
b. CH₃CH₂—C(=O)—NH₂ + NH₄⁺Cl⁻
c. CH₃CH₂CH₂—C(=O)—Cl + 2CH₃CH₂NH₂

15.71 H₂N—CHR—C(=O)—OH (with H on N shown)

15.73 Glycine: H₂N—CH₂—C(=O)—OH   Alanine: H₂N—CH(CH₃)—C(=O)—OH

15.75 H₂N—*CH(CH₃)—C(=O)—OH

15.77 In an acyl group transfer reaction, the acyl group of an acid chloride is transferred from the Cl of the acid chloride to the N of an amine or ammonia. The product is an amide.

15.79 A chemical that carries messages or signals from a nerve to a target cell

15.81 a. Tremors, monotonous speech, loss of memory and problem-solving ability, and loss of motor function
b. Parkinson's disease
c. Schizophrenia, intense satiety sensations

15.83 In proper amounts, dopamine causes a pleasant, satisfied feeling. This feeling becomes intense as the amount of dopamine increases. Several drugs, including cocaine, heroin, amphetamines, alcohol, and nicotine increase the levels of dopamine. It is thought that the intense satiety response this brings about may contribute to addiction to these substances.

15.85 Epinephrine is a component of the flight or fight response. It stimulates glycogen breakdown to provide the body with glucose to supply the needed energy for this stress response.

15.87 The amino acid tryptophan

15.89 Perception of pain, thermoregulation, and sleep

15.91 Promotes the itchy skin rash associated with poison ivy and insect bites; the respiratory symptoms characteristic of hay fever; secretion of stomach acid

15.93 Inhibitory neurotransmitters

15.95 When acetylcholine is released from a nerve cell, it binds to receptors on the surface of muscle cells. This binding stimulates the muscle cell to contract. To stop the contraction, the acetylcholine is then broken down to choline and acetate ion. This is catalyzed by the enzyme acetylcholinesterase.

15.97 Organophosphates inactivate acetylcholinesterase by binding covalently to it. Since acetylcholine is not broken down, nerve transmission continues, resulting in muscle spasm. Pyridine aldoxime methiodide (PAM) is an antidote to organophosphate poisoning because it displaces the organophosphate, thereby allowing acetycholinesterase to function.

Chapter 16

16.1 It is currently recommended that 45–55% of the calories in the diet should be carbohydrates. Of that amount, no more than 10% should be simple sugars.

16.3 An aldose is a sugar with an aldehyde functional group. A ketose is a sugar with a ketone functional group.

16.5 a. Ketose   **d.** Aldose
b. Aldose   **e.** Ketose
c. Ketose   **f.** Aldose

16.7 a. through f. Fischer projection structures

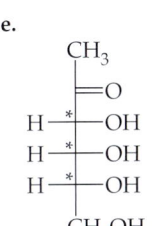

16.9 a. D- **b.** L- **c.** D- **d.** D- **e.** D- **f.** L-

16.11 D-Ribose (Fischer projection, all OH on right)

16.13 L-Ribose (Fischer projection, all OH on left)

16.15

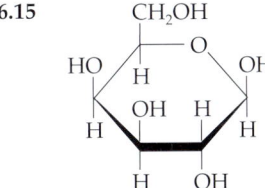   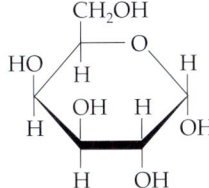

β-D-Galactose   α-D-Galactose

16.17 α-Amylase and β-amylase are digestive enzymes that break down the starch amylose. α-Amylase cleaves glycosidic bonds of the amylose chain at random, producing shorter polysaccharide chains. β-Amylase sequentially cleaves maltose (a disaccharide of glucose) from the reducing end of the polysaccharide chain.

16.19 A monosaccharide is the simplest sugar and consists of a single saccharide unit. A disaccharide is made up of two monosaccharides joined covalently by a glycosidic bond.

16.21 The molecular formula for a simple sugar is $(CH_2O)_n$. Typically n is an integer from 3 to 7.

16.23 Mashed potato flakes, rice, and corn starch contain amylose and amylopectin, both of which are polysaccharides. A candy bar contains sucrose, a disaccharide. Orange juice contains fructose, a monosaccharide. It may also contain sucrose if the label indicates that sugar has been added.

16.25 Four

16.27

$$\begin{array}{c} \text{O} \\ \parallel \\ \text{C—H} \\ \text{H—C—OH} \\ \text{HO—C—H} \\ \text{HO—C—H} \\ \text{H—C—OH} \\ \text{CH}_2\text{OH} \end{array} \qquad \begin{array}{c} \text{CH}_2\text{OH} \\ \text{C=O} \\ \text{HO—C—H} \\ \text{H—C—OH} \\ \text{H—C—OH} \\ \text{CH}_2\text{OH} \end{array}$$

D-Galactose　　　　　D-Fructose
(An aldohexose)　　　(A ketohexose)

16.29 An *aldose* is a sugar that contains an aldehyde (carbonyl) group.
16.31 A tetrose is a sugar with a four-carbon backbone.
16.33 A ketopentose is a sugar with a five-carbon backbone and containing a ketone (carbonyl) group.
16.35
　a. β-D-Glucose is a hemiacetal.
　b. β-D-Fructose is a hemiketal.
　c. α-D-Galactose is a hemiacetal.

16.37

D-Glyceraldehyde　　L-Glyceraldehyde

16.39 Stereoisomers are a pair of molecules that have the same structural formula and bonding pattern but that differ in the arrangement of the atoms in space.
16.41 A chiral carbon is one that is bonded to four different chemical groups.
16.43 A polarimeter converts monochromatic light into monochromatic plane-polarized light. This plane-polarized light is passed through a sample and into an analyzer. If the sample is optically active, it will rotate the plane of the light. The degree and angle of rotation are measured by the analyzer.
16.45 A Fischer Projection is a two-dimensional drawing of a molecule that shows a chiral carbon at the intersection of two lines. Horizontal lines at the intersection represent bonds projecting out of the page and vertical lines represent bonds that project into the page.
16.47 Dextrose is a common name used for D-glucose.
16.49 D- and L-Glyceraldehyde are a pair of enantiomers, that is, they are nonsuperimposable mirror images of one another.
16.51
　a.
$$\begin{array}{c} \text{O} \\ \parallel \\ \text{C—H} \\ \text{HO—*—H} \\ \text{H—*—OH} \\ \text{HO—*—H} \\ \text{HO—*—H} \\ \text{CH}_2\text{OH} \end{array}$$
　b.
$$\begin{array}{c} \text{O} \\ \parallel \\ \text{C—H} \\ \text{H—*—OH} \\ \text{H—*—OH} \\ \text{CH}_2\text{OH} \end{array}$$
　c.
$$\begin{array}{c} \text{O} \\ \parallel \\ \text{C—H} \\ \text{HO—*—H} \\ \text{H—*—OH} \\ \text{HO—*—H} \\ \text{H—*—OH} \\ \text{HO—*—H} \\ \text{CH}_2\text{OH} \end{array}$$

16.53 Anomers are isomers that differ in the arrangement of bonds around the hemiacetal carbon.
16.55 A hemiacetal is a member of the family of organic compounds formed in the reaction of one molecule of alcohol with an aldehyde. They have the following general structure:

$$\begin{array}{c} \text{OH} \\ | \\ \text{R—C—OR} \\ | \\ \text{H} \end{array}$$

16.57 The reaction between an aldehyde and an alcohol yields a hemiacetal. Thus, when the aldehyde portion of a glucose molecule reacts with the C-5 hydroxyl group, the product is an intramolecular hemiacetal.
16.59 When the carbonyl group at C-1 of D-glucose reacts with the C-5 hydroxyl group, a new chiral carbon is created (C-1). In the α-isomer of the cyclic sugar, the C-1 hydroxyl group is below the ring; and in the β-isomer, the C-1 hydroxyl group is above the ring.
16.61 β-Maltose and α-lactose would give positive Benedict's tests. Glycogen would give only a weak reaction because there are fewer reducing ends for a given mass of the carbohydrate.
16.63 Enantiomers are stereoisomers that are nonsuperimposable mirror images of one another. For instance:

$$\begin{array}{cc} \text{O} & \text{O} \\ \parallel & \parallel \\ \text{C—H} & \text{C—H} \\ \text{H—C—OH} & \text{HO—C—H} \\ \text{CH}_2\text{OH} & \text{CH}_2\text{OH} \end{array}$$

D-Glyceraldehyde　　L-Glyceraldehyde

16.65 An aldehyde sugar forms an intramolecular hemiacetal when the carbonyl group of the monosaccharide reacts with a hydroxyl group on one of the other carbon atoms.
16.67 A ketal is the product formed in the reaction of two molecules of alcohol with a ketone. They have the following general structure:

$$\begin{array}{c} \text{OR} \\ | \\ \text{R—C—OR} \\ | \\ \text{R} \end{array}$$

16.69 A glycosidic bond is the bond formed between the hydroxyl group of the C-1 carbon of one sugar and a hydroxyl group of another sugar.

16.71

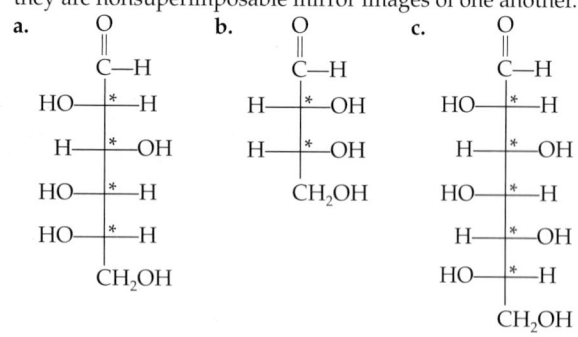

β-Maltose

16.73 Milk
16.75 Eliminating milk and milk products from the diet
16.77 Lactose intolerance is the inability to produce the enzyme lactase that hydrolyzes the milk sugar lactose into its component monosaccharides, glucose and galactose.

16.79 A polymer is a very large molecule formed by the combination of many small molecules, called monomers.

16.81 Starch

16.83 The glucose units of amylose are joined by α (1 → 4) glycosidic bonds and those of cellulose are bonded together by β (1 → 4) glycosidic bonds.

16.85 Glycogen serves as a storage molecule for glucose.

16.87 The salivary glands and the pancreas

Chapter 17

17.1
  a. $CH_3(CH_2)_7CH=CH(CH_2)_7COOH$
  b. $CH_3(CH_2)_{10}COOH$
  c. $CH_3(CH_2)_4CH=CH-CH_2-CH=CH(CH_2)_7COOH$
  d. $CH_3(CH_2)_{16}COOH$

17.3 $CH_3(CH_2)_{10}COOH$ + CH_3CH_2OH $\xrightarrow{H^+, heat}$
  Lauric acid   Ethanol
  Dodecanoic acid

$CH_3(CH_2)_{10}-\overset{O}{\underset{\|}{C}}-OCH_2CH_3$ + H_2O
  Ethyl dodecanoate

17.5
$CH_3CH_2-\overset{O}{\underset{\|}{C}}-OCH_2CH_2CH_2CH_3$ + H_2O $\xrightarrow{H^+, heat}$
  Butyl propanoate

  $CH_3CH_2CH_2CH_2OH$ + CH_3CH_2COOH
  1-Butanol   Propanoic acid

17.7
$CH_3CH_2-\overset{O}{\underset{\|}{C}}-OCH_2CH_2CH_2CH_3$ + KOH $\longrightarrow$
  Butyl propanoate

  $CH_3CH_2COO^-K^+$ + $CH_3CH_2CH_2CH_2OH$
  Potassium propanoate   1-Butanol

17.9 $CH_3(CH_2)_5CH=CH(CH_2)_7COOH$ + H_2 $\xrightarrow{Ni}$
  cis-9-hexadecenoic acid

  $CH_3(CH_2)_{14}COOH$
  Hexadecanoic acid

17.11
H—C—OH
|
H—C—OH + 2 $CH_3(CH_2)_{16}COOH$ $\longrightarrow$
|
H—C—OH
|
H

H—C—O—C—(CH_2)_{16}CH_3
|
H—C—O—C—(CH_2)_{16}CH_3
|
H—C—OH
|
H

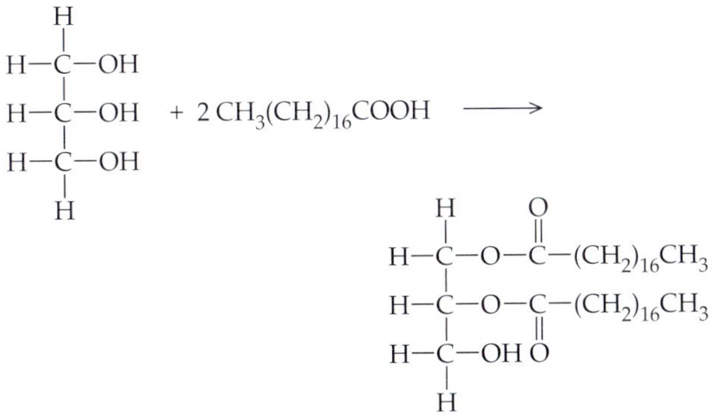

17.13 a. $CH_3(CH_2)_7CH=CH(CH_2)_7-\overset{O}{\underset{\|}{C}}-O-CH_2$
  |
  CH—OH
  |
  CH_2-OH

$CH_3(CH_2)_7CH=CH(CH_2)_7-\overset{O}{\underset{\|}{C}}-O-CH_2$
$CH_3(CH_2)_7CH=CH(CH_2)_7-\overset{O}{\underset{\|}{C}}-O-CH$
  |
  CH_2-OH

$CH_3(CH_2)_7CH=CH(CH_2)_7-\overset{O}{\underset{\|}{C}}-O-CH_2$
$CH_3(CH_2)_7CH=CH(CH_2)_7-\overset{O}{\underset{\|}{C}}-O-CH$
$CH_3(CH_2)_7CH=CH(CH_2)_7-\overset{O}{\underset{\|}{C}}-O-CH_2$

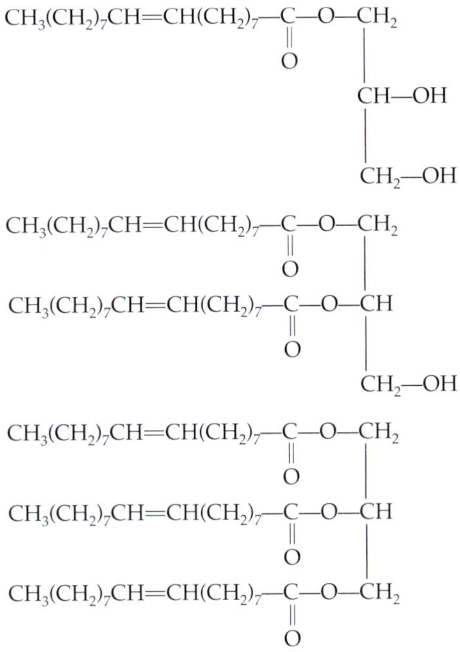

b. $CH_3(CH_2)_8-\overset{O}{\underset{\|}{C}}-O-CH_2$
  |
  CH—OH
  |
  CH_2-OH

$CH_3(CH_2)_8-\overset{O}{\underset{\|}{C}}-O-CH_2$
$CH_3(CH_2)_8-\overset{O}{\underset{\|}{C}}-O-CH$
  |
  CH_2-OH

$CH_3(CH_2)_8-\overset{O}{\underset{\|}{C}}-O-CH_2$
$CH_3(CH_2)_8-\overset{O}{\underset{\|}{C}}-O-CH$
$CH_3(CH_2)_8-\overset{O}{\underset{\|}{C}}-O-CH_2$

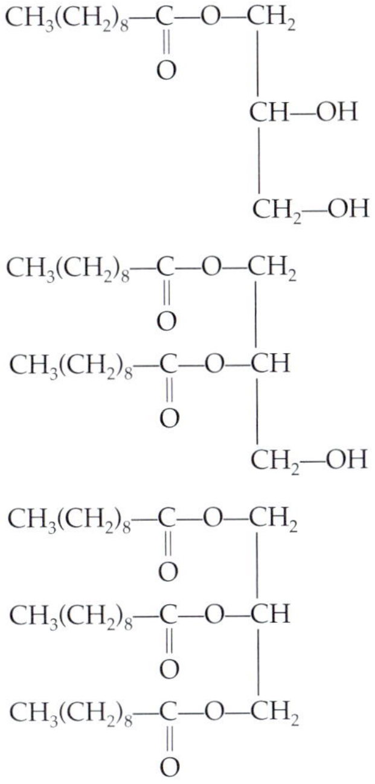

17.15

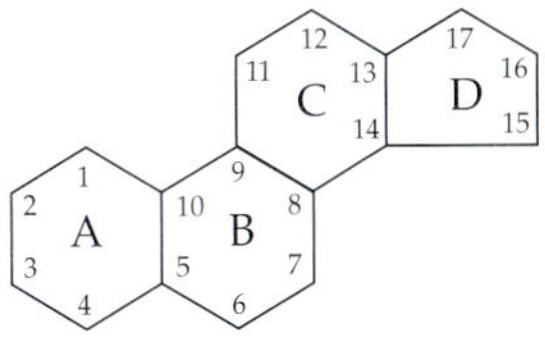

Steroid nucleus (rings A, B, C, D with carbons numbered 1–17)

17.17 Receptor-mediated endocytosis

17.19 Membrane transport resembles enzyme catalysis because both processes exhibit a high degree of specificity.

17.21 Fatty acids, glycerides, nonglyceride lipids, and complex lipids

17.23 Lipid-soluble vitamins are transported into cells of the small intestine in association with dietary fat molecules. Thus, a diet low in fat reduces the amount of vitamins A, D, E, and K that enters the body.

17.25 A saturated fatty acid is one in which the hydrocarbon tail has only carbon-to-carbon single bonds. An unsaturated fatty acid has at least one carbon-to-carbon double bond.

17.27 The melting points increase.

17.29 The melting points of fatty acids increase as the length of the hydrocarbon chains increase. This is because the intermolecular attractive forces, including van der Waals forces, increase as the length of the hydrocarbon chain increases.

17.31
  a. Decanoic acid
  $CH_3(CH_2)_8COOH$
  b. Stearic acid
  $CH_3(CH_2)_{16}COOH$

17.33
  a. I.U.P.A.C. name: Hexadecanoic acid
  Common name: Palmitic acid
  b. I.U.P.A.C. name: Dodecanoic acid
  Common name: Lauric acid

17.35
  a.
  $$\begin{array}{l} CH_2OH \\ | \\ CHOH + 3\,CH_3(CH_2)_{12}\text{C(=O)}\text{—OH} \\ | \\ CH_2OH \end{array}$$

  $\downarrow$

  $$\begin{array}{l} CH_3(CH_2)_{12}\text{—CO—O—}CH_2 \\ CH_3(CH_2)_{12}\text{—CO—O—}CH + 3H_2O \\ CH_3(CH_2)_{12}\text{—CO—O—}CH_2 \end{array}$$

  b.
  $$\begin{array}{l} CH_3(CH_2)_{16}\text{—CO—O—}CH_2 \\ CH_3(CH_2)_{16}\text{—CO—O—}CH + 3H_2O \\ CH_3(CH_2)_{16}\text{—CO—O—}CH_2 \end{array}$$

  $\downarrow$

  $$3\,CH_3(CH_2)_{16}\text{—CO—OH} + \begin{array}{l} CH_2OH \\ CHOH \\ CH_2OH \end{array}$$

  c. $CH_3CH_2CH_2CH_2CH_2CH_2CH_2CH_2CH_2\text{—CO—OH}$

  $\downarrow$ KOH

  $CH_3CH_2CH_2CH_2CH_2CH_2CH_2CH_2CH_2\text{—CO—O}^- K^+ + H_2O$

  d. $CH_3(CH_2)_4CH\!=\!CHCH_2CH\!=\!CH(CH_2)_7\text{—CO—OH} + 2H_2$

  $\downarrow$ Ni

  $CH_3(CH_2)_{16}\text{—CO—OH}$

17.37 The essential fatty acid linoleic acid is required for the synthesis of arachidonic acid, a precursor for the synthesis of the prostaglandins, a group of hormonelike molecules.

17.39 Aspirin effectively decreases the inflammatory response by inhibiting the synthesis of all prostaglandins. Aspirin works by inhibiting cyclooxygenase, the first enzyme in prostaglandin biosynthesis. This inhibition results from the transfer of an acetyl group from aspirin to the enzyme. Because cyclooxygenase is found in all cells, synthesis of all prostaglandins is inhibited.

17.41 Smooth muscle contraction, enhancement of fever and swelling associated with the inflammatory response, bronchial dilation, inhibition of secretion of acid into the stomach

17.43 A glyceride is a lipid ester that contains the glycerol molecule and from 1 to 3 fatty acids.

17.45 An emulsifying agent is a molecule that aids in the suspension of triglycerides in water. They are amphipathic molecules, such as lecithin, that serve as bridges holding together the highly polar water molecules and the nonpolar triglycerides.

17.47 A triglyceride with three saturated fatty acid tails would be a solid at room temperature. The long, straight fatty acid tails would stack with one another because of strong intermolecular and intramolecular attractions.

17.49

$$\begin{array}{l} CH_3(CH_2)_{14}\text{—CO—O—}CH_2\ \ (1) \\ \\ \underset{CH_3(CH_2)_4CH_2}{\overset{H}{\diagdown}}\!C\!=\!C\!\underset{H}{\overset{CH_2(CH_2)_6\text{—CO—O—}CH\ (2)}{\diagup}} \\ \\ \underset{CH_3(CH_2)_4CH_2}{\overset{}{\diagdown}}\!C\!=\!C\!\underset{H}{\overset{CH_2(CH_2)_6\text{—CO—O—}CH_2\ (3)}{\diagup}} \end{array}$$

17.51

$CH_3CH_2CH_2CH_2CH_2CH_2CH_2CH_2CH_2\text{—CO—O—}CH_2$ (1)
$CH_3CH_2CH_2CH_2CH_2CH_2CH_2CH_2CH_2CH_2CH_2\text{—CO—O—}CH$ (2)
$CH_2\text{—O—P(=O)(O}^-\text{)—O}^-$ (3)

17.53 Triglycerides consist of three fatty acids esterified to the three hydroxyl groups of glycerol. In phospholipids there are only two fatty acids esterified to glycerol. A phosphoryl group is esterified (phosphoester linkage) to the third hydroxyl group.

17.55 A sphingolipid is a lipid that is not derived from glycerol, but rather from sphingosine, a long-chain, nitrogen-containing (amino) alcohol. Like phospholipids, sphingolipids are amphipathic.

17.57 A glycosphingolipid or glycolipid is a lipid that is built on a ceramide backbone structure. Ceramide is a fatty acid derivative of sphingosine.

17.59 Sphingomyelins are important structural lipid components of nerve cell membranes. They are found in the myelin sheath that surrounds and insulates cells of the central nervous system.

17.61 Cholesterol is readily soluble in the hydrophobic region of biological membranes. It is involved in regulating the fluidity of the membrane.

17.63 Progesterone is the most important hormone associated with pregnancy. Testosterone is needed for development of male secondary sexual characteristics. Estrone is required for proper development of female secondary sexual characteristics.

17.65 Cortisone is used to treat rheumatoid arthritis, asthma, gastrointestinal disorders, and many skin conditions.

17.67 Myricyl palmitate (beeswax) is made up of the fatty acid palmitic acid and the alcohol myricyl alcohol—$CH_3(CH_2)_{28}CH_2OH$.

17.69 Isoprenoids are a large, diverse collection of lipids that are synthesized from the isoprene unit:

$$CH_2=C(CH_3)-CH=CH_2$$

17.71 Steroids and bile salts, lipid-soluble vitamins, certain plant hormones, and chlorophyll

17.73 Chylomicrons, high-density lipoproteins, low-density lipoproteins, and very low density lipoproteins

17.75 The terms "good" and "bad" cholesterol refer to two classes of lipoprotein complexes. The high density liproproteins, or HDL, are considered to be "good" cholesterol because a correlation has been made between elevated levels of HDL and a reduced incidence of atherosclerosis. Low density lipoproteins, or LDL, are considered to be "bad" cholesterol because evidence suggests that high levels of LDL is associated with increased risk of atherosclerosis.

17.77 Atherosclerosis results when cholesterol and other substances coat the arteries causing a narrowing of the passageways. As the passageways become narrower, greater pressure is required to provide adequate blood flow. This results in higher blood pressure (hypertension).

17.79 If the LDL receptor is defective, it cannot function to remove cholesterol-bearing LDL particles from the blood. The excess cholesterol, along with other substances, will accumulate along the walls of the arteries, causing atherosclerosis.

17.81 The basic structure of a biological membrane is a bilayer of phospholipid molecules arranged so that the hydrophobic hydrocarbon tails are packed in the center and the hydrophilic head groups are exposed on the inner and outer surfaces.

17.83 A peripheral membrane protein is bound to only one surface of the membrane, either inside or outside the cell.

17.85 Cholesterol is freely soluble in the hydrophobic layer of a biological membrane. It moderates the fluidity of the membrane by disrupting the stacking of the fatty acid tails of membrane phospholipids.

17.87 L. Frye and M. Edidin carried out studies in which specific membrane proteins on human and mouse cells were labeled with red and green fluorescent dyes, respectively. The human and mouse cells were fused into single-celled hybrids and were observed using a microscope with an ultraviolet light source. The ultraviolet light caused the dyes to fluoresce. Initially the dyes were localized in regions of the membrane representing the original human or mouse cell. Within an hour, the proteins were evenly distributed throughout the membrane of the fused cell.

17.89 If the fatty acyl tails of membrane phospholipids are converted from saturated to unsaturated, the fluidity of the membrane will increase.

17.91 In simple diffusion the molecule moves directly across the membrane, whereas in facilitated diffusion a protein channel through the membrane is required.

17.93 Active transport requires an energy input to transport molecules or ions against the gradient (from an area of lower concentration to an area of higher concentration). Facilitated diffusion is a means of passive transport in which molecules or ions pass from regions of higher concentration to regions of lower concentration through a permease protein. No energy is expended by the cell in facilitated diffusion.

17.95 An antiport transport mechanism is one in which one molecule or ion is transported into the cell while a different molecule or ion is transported out of the cell.

17.97 Each permease or channel protein has a binding site that has a shape and charge distribution that is complementary to the molecule or ion that it can bind and transport across the cell membrane.

17.99 One ATP molecule is hydrolyzed to transport 3 Na^+ out of the cell and 2 K^+ into the cell.

Chapter 18

18.1 a. Glycine (gly):

$$H_3^+N-\underset{H}{\overset{COO^-}{\underset{|}{C}}}-H$$

b. Proline (pro):

$$H_2^+N-CH-COO^-$$ with ring $H_2C-CH_2-CH_2$

c. Threonine (thr):

$$H_3^+N-\overset{COO^-}{\underset{|}{C}}-H$$
$$H-C-OH$$
$$CH_3$$

d. Aspartate (asp):

$$H_3^+N-\overset{COO^-}{\underset{|}{C}}-H$$
$$H-C-H$$
$$COO^-$$

e. Lysine (lys):

$$\begin{array}{c} COO^- \\ | \\ H_3{}^+N-C-H \\ | \\ H-C-H \\ | \\ H-C-H \\ | \\ H-C-H \\ | \\ H-C-H \\ | \\ N^+H_3 \end{array}$$

18.3 a. Alanyl-phenylalanine:

$$H_3{}^+N-\underset{CH_3}{\underset{|}{C}}H-\underset{}{\overset{O}{\overset{\|}{C}}}-\underset{H}{\overset{H}{N}}-\underset{CH_2-C_6H_5}{\underset{|}{C}}H-COO^-$$

b. Lysyl-alanine:

$$H_3{}^+N-\underset{(CH_2)_4-N^+H_3}{\underset{|}{C}}H-\underset{}{\overset{O}{\overset{\|}{C}}}-\underset{}{\overset{H}{N}}-\underset{CH_3}{\underset{|}{C}}H-COO^-$$

c. Phenylalanyl-tyrosyl-leucine:

$$H_3{}^+N-\underset{CH_2C_6H_5}{\underset{|}{C}}H-\overset{O}{\overset{\|}{C}}-N-\underset{CH_2C_6H_4OH}{\underset{|}{C}}H-\overset{O}{\overset{\|}{C}}-N-\underset{CH_2CH(CH_3)_2}{\underset{|}{C}}H-COO^-$$

18.5 The primary structure of a protein is the amino acid sequence of the protein chain. Regular, repeating folding of the peptide chain caused by hydrogen bonding between the amide nitrogens and carbonyl oxygens of the peptide bond is the secondary structure of a protein. The two most common types of secondary structure are the α-helix and the β-pleated sheet. Tertiary structure is the further folding of the regions of α-helix and β-pleated sheet into a compact, spherical structure. Formation and maintenance of the tertiary structure results from weak attractions between amino acid R groups. The binding of two or more peptides to produce a functional protein defines the quaternary structure.

18.7 Oxygen is efficiently transferred from hemoglobin to myoglobin in the muscle because myoglobin has a greater affinity for oxygen.

18.9 High temperature disrupts the hydrogen bonds and other weak interactions that maintain protein structure.

18.11 Vegetables vary in amino acid composition. No single vegetable can provide all of the amino acid requirements of the body. By eating a variety of different vegetables, all the amino acid requirements of the human body can be met.

18.13 An enzyme is a protein that serves as a biological catalyst, speeding up biological reactions.

18.15 A transport protein is a protein that transports materials across the cell membrane or throughout the body.

18.17 Enzymes speed up reactions that might take days or weeks to occur on their own. They also catalyze reactions that might require very high temperatures or harsh conditions if carried out in the laboratory. In the body, these reactions occur quickly under physiological conditions.

18.19 Transferrin is a transport protein that carries iron from the liver to the bone marrow, where it is used to produce the heme group for hemoglobin and myoglobin. Hemoglobin transports oxygen in the blood.

18.21 Egg albumin is a nutrient protein that serves as a source of protein for the developing chick. Casein is the nutrient storage protein in milk, providing protein, a source of amino acids, for mammals.

18.23 The general structure of an L-α-amino acid:

$$\begin{array}{c} COO^- \\ | \\ H_3{}^+N-C-H \\ | \\ R \end{array}$$

18.25 A zwitterion is a neutral molecule with equal numbers of positive and negative charges. Under physiological conditions, amino acids are zwitterions.

18.27 A chiral carbon is one that has four different atoms or groups of atoms attached to it.

18.29 Interactions between the R groups of the amino acids in a polypeptide chain are important for the formation and maintenance of the tertiary and quaternary structures of proteins.

18.31

Glycine, Alanine, Valine, Leucine, Isoleucine, Phenylalanine, Proline, Tryptophan, Methionine

18.33 A peptide bond is an amide bond between two amino acids in a peptide chain.

18.35 Linus Pauling and his colleagues carried out X-ray diffraction studies of protein. Interpretation of the the pattern formed when X-rays were diffracted by a crystal of pure protein led Pauling to conclude that peptide bonds are both planar (flat) and rigid and that the N-C bonds are shorter that expected. In other words, they deduced that the peptide bond has a partially double bond character because it exhibits resonance. There is no free rotation about the amide bond because the carbonyl group of the amide bond has a strong attraction for the amide nitrogen lone pair of electrons. This can best be described using a resonance model:

$$\left[\begin{array}{c} \text{R}-\underset{\underset{\text{H}}{|}}{\overset{\overset{\ddot{\text{O}}:}{\|}}{\text{C}}}-\underset{\underset{}{}}{\ddot{\text{N}}}-\text{R}' \end{array} \longleftrightarrow \begin{array}{c} \text{R}-\underset{\underset{\text{H}}{|}}{\overset{\overset{:\ddot{\text{O}}:^-}{|}}{\text{C}}}=\overset{+}{\text{N}}-\text{R}' \end{array} \right]$$

The partially double bonded character of the resonance structure restricts free rotation.

18.37 a. His-trp-cys:

[structure diagram]

b. Gly-leu-ser:

[structure diagram]

c. Arg-ile-val:

[structure diagram]

18.39 The primary structure of a protein is the sequence of amino acids bonded to one another by peptide bonds.

18.41 The primary structure of a protein determines its three dimensional shape because the location of R groups along the protein chain is determined by the primary structure. The interactions among the R groups, based on their location in the chain, will govern how the protein folds. This, in turn, dictates its three-dimensional structure and biological function.

18.43 The genetic information in the DNA dictates the order in which amino acids will be added to the protein chain. The order of the amino acids is the primary structure of the protein.

18.45 The secondary structure of a protein is the folding of the primary structure into an α-helix or β-pleated sheet.

18.47 a. α-Helix
  b. β-Pleated sheet

18.49 A fibrous protein is one that is composed of peptides arranged in long sheets or fibers.

18.51 A parallel β-pleated sheet is one in which the hydrogen bonded peptide chains have their amino-termini aligned head-to-head.

18.53 The tertiary structure of a protein is the globular, three-dimensional structure of a protein that results from folding the regions of secondary structure.

18.55

[structure of cystine with disulfide bridge]

18.57 The tertiary structure is a level of folding of a protein chain that has already undergone secondary folding. The regions of α-helix and β-pleated sheet are folded into a globular structure.

18.59 Quaternary protein structure is the aggregation of two or more folded peptide chains to produce a functional protein.

18.61 A glycoprotein is a protein with covalently attached sugars.

18.63 Hydrogen bonding maintains the secondary structure of a protein and contributes to the stability of the tertiary and quaternary levels of structure.

18.65 The peptide bond exhibits resonance, which results in a partially double bonded character. This causes the rigidity of the peptide bond.

[resonance structure diagram]

18.67 The code for the primary structure of a protein is carried in the genetic information (DNA).

18.69 The function of hemoglobin is to carry oxygen from the lungs to oxygen-demanding tissues throughout the body. Hemoglobin is found in red blood cells.

18.71 Hemoglobin is a protein composed of four subunits—two α-globin and two β-globin subunits. Each subunit holds a heme group, which in turn carries an Fe^{2+} ion.

18.73 The function of the heme group in hemoglobin and myoglobin is to bind to molecular oxygen.

18.75 Because carbon monoxide binds tightly to the heme groups of hemoglobin, it is not easily removed or replaced by oxygen. As a result, the effects of oxygen deprivation (suffocation) occur.

18.77 When sickle cell hemoglobin (HbS) is deoxygenated, the amino acid valine fits into a hydrophobic pocket on the surface of another HbS molecule. Many such sickle cell hemoglobin molecules polymerize into long rods that cause the red blood cell to sickle. In normal hemoglobin, glutamic acid is found in the place of the valine. This negatively charged amino acid will not "fit" into the hydrophobic pocket.

18.79 When individuals have one copy of the sickle cell gene and one copy of the normal gene, they are said to carry the *sickle cell trait*. These individuals will not suffer serious side effects, but may pass the trait to their offspring. Individuals with two copies of the sickle cell globin gene exhibit all the symptoms of the disease and are said to have *sickle cell anemia*.

18.81 *Denaturation* is the process by which the organized structure of a protein is disrupted, resulting in a completely disorganized, nonfunctional form of the protein.

18.83 Heat is an effective means of sterilization because it destroys the proteins of microbial life-forms, including fungi, bacteria, and viruses.

18.85 Even relatively small fluctuations in blood pH can be life threatening. It is likely that these small changes would alter the normal charges on the proteins and modify their interactions. These changes can render a protein incapable of carrying out its functions.

18.87 Proteins become polycations at low pH because the additional protons will protonate the carboxylate groups. As these negative charges are neutralized, the charge on the proteins will be contributed only by the protonated amino groups ($-N^+H_3$).

18.89 The low pH of the yogurt denatures the proteins of microbial contaminants, inhibiting their growth.

18.91 An essential amino acid is one that must be provided in the diet because it cannot be synthesized in the body.

18.93 A complete protein is one that contains all of the essential and nonessential amino acids.

18.95 Chymotrypsin catalyzes the hydrolysis of peptide bonds on the carbonyl side of aromatic amino acids.

[Structure: Phenylalanyl-alanine + H₂O → Phenylalanine + Alanine, with site of chymotrypsin-catalyzed hydrolysis indicated]

18.97 In a vegetarian diet, vegetables are the only source of dietary protein. Because individual vegetable sources do not provide all the needed amino acids, vegetables must be mixed to provide all the essential and nonessential amino acids in the amounts required for biosynthesis.

18.99 Synthesis of digestive enzymes must be carefully controlled because the active enzyme would digest, and thus destroy, the cell that produces it.

Chapter 19

19.1
   a. Transferase
   b. Transferase
   c. Isomerase
   d. Oxidoreductase
   e. Hydrolase

19.3 a. Pyruvate kinase catalyzes the transfer of a phosphoryl group from phosphoenolpyruvate to adenosine diphosphate.

[Reaction: Phosphoenolpyruvate + ADP → Pyruvate + ATP, catalyzed by Pyruvate kinase]

b. Alanine transaminase catalyzes the transfer of an amino group from alanine to α-ketoglutarate, producing pyruvate and glutamate.

[Reaction: Alanine + α-Ketoglutarate → Pyruvate + Glutamate, catalyzed by Alanine transaminase]

c. Triose phosphate isomerase catalyzes the isomerization of the ketone dihydroxyacetone phosphate to the aldehyde glyceraldehyde-3-phosphate.

[Reaction: Dihydroxyacetone phosphate → Glyceraldehyde-3-phosphate, catalyzed by Triose phosphate isomerase]

d. Pyruvate dehydrogenase catalyzes the oxidation and decarboxylation of pyruvate, producing acetyl coenzyme A and CO_2.

[Reaction: Pyruvate + H-S-CoA → Acetyl coenzyme A + CO_2, catalyzed by Pyruvate dehydrogenase]

19.5 a. Sucrose
   b. Pyruvate
   c. Succinate

19.7 The induced fit model assumes that the enzyme is flexible. Both the enzyme and the substrate are able to change shape to form the enzyme-substrate complex. The lock-and-key model assumes that the enzyme is inflexible (the lock) and the substrate (the key) fits into a specific rigid site (the active site) on the enzyme to form the enzyme-substrate complex.

19.9 An enzyme might distort a bond, thereby catalyzing bond breakage. An enzyme could bring two reactants into close proximity and in the proper orientation for the reaction to occur. Finally, an enzyme could alter the pH of the microenvironment of the active site, thereby serving as a transient donor or acceptor of H^+.

19.11 Water-soluble vitamins are required by the body for the synthesis of coenzymes that are required for the function of a variety of enzymes.

19.13 A decrease in pH will change the degree of ionization of the R groups within a peptide chain. This disturbs the weak interactions that maintain the structure of an enzyme, which may denature the enzyme. Less drastic alterations in the charge of

R groups in the active site of the enzyme can inhibit enzyme-substrate binding or destroy the catalytic ability of the active site.

19.15 Irreversible inhibitors bind very tightly, sometimes even covalently, to an R group in enzyme active sites. They generally inhibit many different enzymes. The loss of enzyme activity impairs normal cellular metabolism, resulting in death of the cell or the individual.

19.17 A structural analog is a molecule that has a structure and charge distribution very similar to that of the natural substrate of an enzyme. Generally they are able to bind to the enzyme active site. This inhibits enzyme activity because the normal substrate must compete with the structural analog to form an enzyme-substrate complex.

19.19 a.

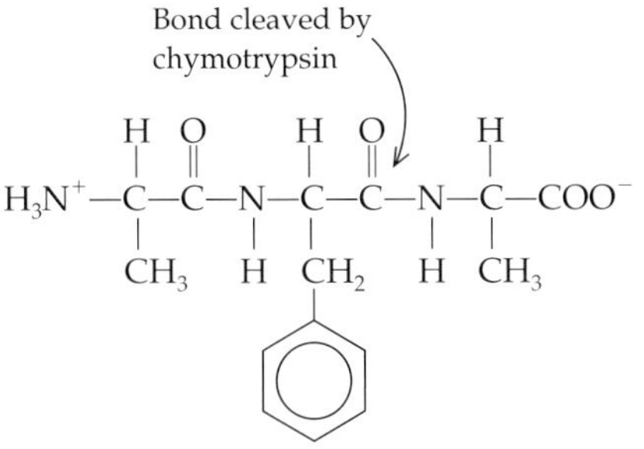

Bond cleaved by chymotrypsin

$H_3N^+-\underset{\underset{CH_3}{|}}{\overset{\overset{H}{|}}{C}}-\overset{\overset{O}{\|}}{C}-N-\underset{\underset{CH_2-C_6H_5}{|}}{\overset{\overset{H}{|}}{C}}-\overset{\overset{O}{\|}}{C}-N-\underset{\underset{CH_3}{|}}{\overset{\overset{H}{|}}{C}}-COO^-$

ala-phe-ala

b.

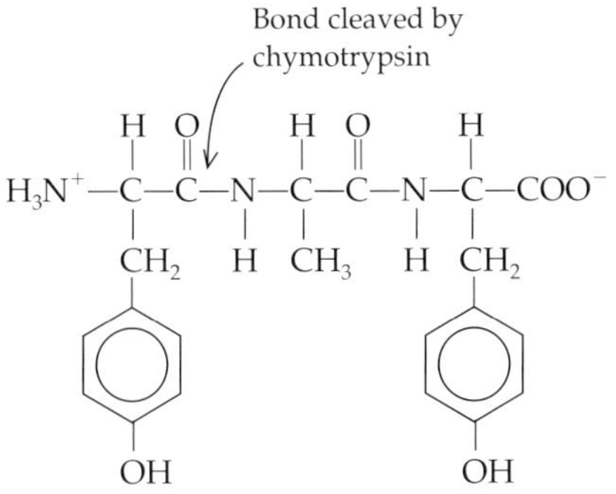

Bond cleaved by chymotrypsin

tyr-ala-tyr

19.21 Chymotrypsin   Elastase   Elastase

(peptide structure with cleavage arrows)

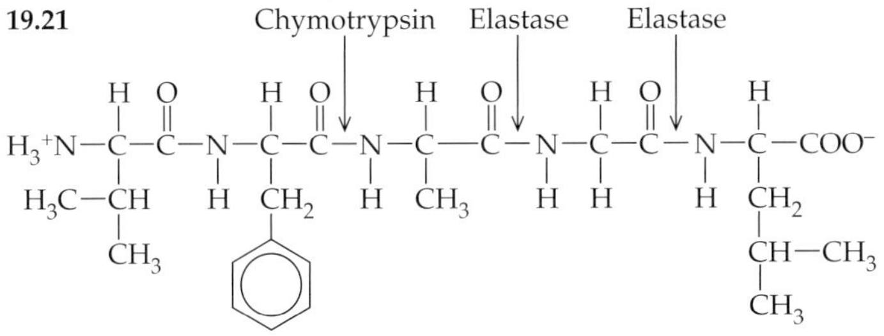

19.23 The common name of an enzyme is often derived from the name of the substrate and/or the type of reaction that it catalyzes.

19.25
1. Urease
2. Peroxidase
3. Lipase
4. Aspartase
5. Glucose-6-phosphatase
6. Sucrase

19.27 a. Citrate decarboxylase catalyzes the cleavage of a carboxyl group from citrate.
b. Adenosine diphosphate phosphorylase catalyzes the addition of a phosphate group to ADP.
c. Oxalate reductase catalyzes the reduction of oxalate.
d. Nitrite oxidase catalyzes the oxidation of nitrite.
e. *cis-trans* Isomerase catalyzes interconversion of *cis* and *trans* isomers.

19.29 A substrate is the reactant in an enzyme-catalyzed reaction that binds to the active site of the enzyme and is converted into product.

19.31 The activation energy of a reaction is the energy required for the reaction to occur.

19.33 The equilibrium constant for a chemical reaction is a reflection of the difference in energy of the reactants and products. Consider the following reaction:

$$aA + bB \rightarrow cC + dD$$

The equilibrium constant for this reaction is:

$$K_{eq} = [D]^d[C]^c/[A]^a[B]^b = [\text{products}]/[\text{reactants}]$$

Because the difference in energy between reactants and products is the same regardless of what path the reaction takes, an enzyme does not alter the equilibrium constant of a reaction.

19.35 The rate of an uncatalyzed chemical reaction typically doubles every time the substrate concentration is doubled.

19.37 The rate-limiting step is that step in an enzyme-catalyzed reaction that is the slowest, and hence limits the speed with which the substrate can be converted into product.

19.39

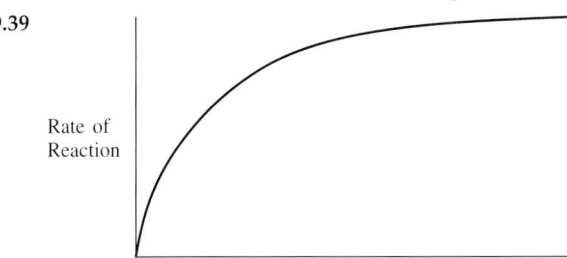

Rate of Reaction vs. Concentration of Substrate

19.41 The enzyme-substrate complex is the molecular aggregate formed when the substrate binds to the active site of an enzyme.

19.43 The catalytic groups of an enzyme active site are those functional groups that are involved in carrying out catalysis.

19.45 Enzyme active sites are pockets in the surface of an enzyme that include R groups involved in binding and R groups involved in catalysis. The shape of the active site is complementary to the shape of the substrate. Thus, the conformation of the active site determines the specificity of the enzyme. Enzyme-substrate binding involves weak, noncovalent interactions.

19.47 The lock-and-key model of enzyme-substrate binding was proposed by Emil Fischer in 1894. He thought that the active site was a rigid region of the enzyme into which the substrate fit perfectly. Thus, the model purports that the substrate simply snaps into place within the active site, like two pieces of a jigsaw puzzle fitting together.

19.49 Enzyme specificity is the ability of an enzyme to bind to only one, or a very few, substrates and thus catalyze only a single reaction.

19.51 Group specificity means that an enzyme catalyzes reactions involving similar molecules having the same functional group.

19.53 Absolute specificity means that an enzyme catalyzes the reaction of only one substrate.

19.55 Hexokinase has group specificity. The advantage is that the cell does not need to encode many enzymes to carry out the phosphorylation of six-carbon sugars. Hexokinase can carry out many of these reactions.

19.57 Methionyl tRNA synthetase has absolute specificity. This is the enzyme that attaches the amino acid methionine to the transfer RNA (tRNA) that will carry the amino acid to the site of protein synthesis. If the wrong amino acid were attached to the tRNA, it could be incorporated into the protein, destroying its correct three-dimensional structure and biological function.

19.59 The first step of an enzyme-catalyzed reaction is the formation of the enzyme-substrate complex. In the second step, the transition state is formed. This is the state in which the substrate assumes a form intermediate between the original substrate and the product. In step 3 the substrate is converted to product and the enzyme-product complex is formed. Step 4 involves the release of the product and regeneration of the enzyme in its original form.

19.61 In a reaction involving bond breaking, the enzyme might distort a bond, producing a transition state in which the bond is stressed. An enzyme could bring two reactants into close proximity and in the proper orientation for the reaction to occur, producing a transition state in which the proximity of the reactants facilitates bond formation. Finally, an enzyme could alter the pH of the microenvironment of the active site, thereby serving as a transient donor or acceptor of H^+.

19.63 A cofactor helps maintain the shape of the active site of an enzyme.

19.65 NAD^+/NADH serves an acceptor/donor of hydride anions in biochemical reactions. NAD^+/NADH serves as a coenzyme for oxidoreductases.

19.67 Changes in pH or temperature affect the activity of enzymes, as can changes in the concentration of substrate and the concentrations of certain ions.

19.69 Each of the following answers assumes that the enzyme was purified from an organism with optimal conditions for life near 37°C, pH 7.
  a. Decreasing the temperature from 37°C to 10°C will cause the rate of an enzyme-catalyzed reaction to decrease because the frequency of collisions between enzyme and substrate will decrease as the rate of molecular movement decreases.
  b. Increasing the pH from 7 to 11 will generally cause a decrease in the rate of an enzyme-catalyzed reaction. In fact, most enzymes would be denatured by a pH of 11 and enzyme activity would cease.
  c. Heating an enzyme from 37°C to 100°C will destroy enzyme activity because the enzyme would be denatured by the extreme heat.

19.71 High temperature denatures bacterial enzymes and structural proteins. Because the life of the cell is dependent on the function of these proteins, the cell dies.

19.73 A lysosome is a membrane-bound vesicle in the cytoplasm of cells that contains approximately fifty types of hydrolytic enzymes.

19.75 Enzymes used for clinical assays in hospitals are typically stored at refrigerator temperatures to ensure that they are not denatured by heat. In this way they retain their activity for long periods.

19.77 a. Cells regulate the level of enzyme activity to conserve energy. It is a waste of cellular energy to produce an enzyme if its substrate is not present or if its product is in excess.
  b. Production of proteolytic digestive enzymes must be carefully controlled because the active enzyme could destroy the cell that produces it. Thus, they are produced in an inactive form in the cell and are only activated at the site where they carry out digestion.

19.79 In positive allosterism, binding of the effector molecule turns the enzyme on. In negative allosterism, binding of the effector molecule turns the enzyme off.

19.81 A proenzyme is the inactive form of an enzyme that is converted to the active form at the site of its activity.

19.83 Blood clotting is a critical protective mechanism in the body, preventing excessive loss of blood following an injury. However, it can be a dangerous mechanism if it is triggered inappropriately. The resulting clot could cause a heart attack or stroke. By having a cascade of proteolytic reactions leading to the final formation of the clot, there are many steps at which the process can be regulated. This ensures that it will only be activated under the appropriate conditions.

19.85 *Competitive enzyme inhibition* occurs when a structural analog of the normal substrate occupies the enzyme active site so that the reaction cannot occur. The structural analog and the normal substrate compete for the active site. Thus, the rate of the reaction will depend on the relative concentrations of the two molecules.

19.87 A structural analog has a shape and charge distribution that are very similar to those of the normal substrate for an enzyme.

19.89 Irreversible inhibitors bind tightly to and block the active site of an enzyme and eliminate catalysis at the site.

19.91 The compound would be a competitive inhibitor of the enzyme.

19.93 The structural similarities among chymotrypsin, trypsin, and elastase suggest that these enzymes evolved from a single ancestral gene that was duplicated. Each copy then evolved independently.

19.95

Bond cleaved by chymotrypsin

H_3N^+—C(H)(CH$_2$-C$_6$H$_4$-OH)—C(O)—N(H)—C(H)(CH$_2$CH$_2$CH$_2$CH$_2$N$^+$H$_3$)—C(O)—N(H)—C(H)(CH$_3$)—C(O)—N(H)—C(H)(CH$_2$-C$_6$H$_5$)—COO$^-$

tyr-lys-ala-phe

19.97 Elastase will cleave the peptide bonds on the carbonyl side of alanine and glycine. Trypsin will cleave the peptide bonds on the carbonyl side of lysine and arginine. Chymotrypsin will cleave the peptide bonds on the carbonyl side of tryptophan and phenylalanine.

19.99 Creatine phosphokinase (CPK), lactate dehydrogenase (LDH), and aspartate aminotransferase (AST/SGOT)

Chapter 20

20.1 a. Adenosine diphosphate:

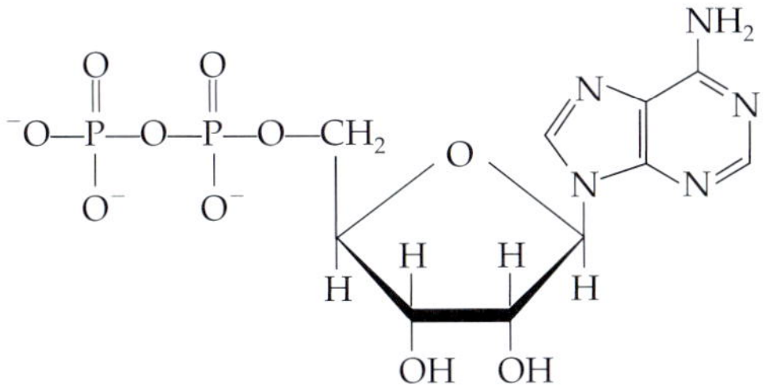

b. Deoxyguanosine triphosphate:

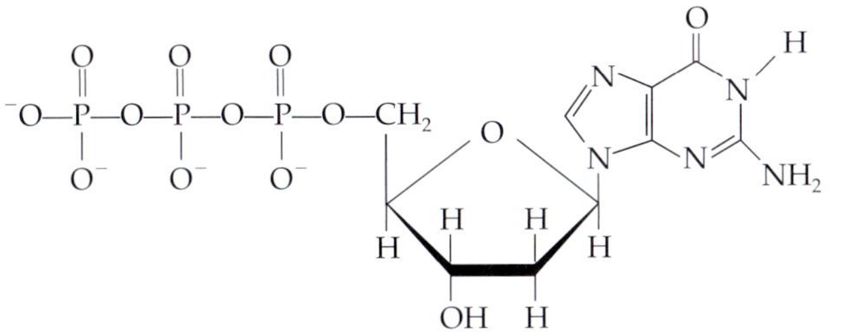

20.3 The RNA polymerase recognizes the promoter site for a gene, separates the strands of DNA, and catalyzes the polymerization of an RNA strand complementary to the DNA strand that carries the genetic code for a protein. It recognizes a termination site at the end of the gene and releases the RNA molecule.

20.5 The genetic code is said to be degenerate because several different triplet codons may serve as code words for a single amino acid.

20.7 The nitrogenous bases of the codons are complementary to those of the anticodons. As a result they are able to hydrogen bond to one another according to the base pairing rules.

20.9 The ribosomal P-site holds the peptidyl tRNA during protein synthesis. The peptidyl tRNA is the tRNA carrying the growing peptide chain. The only exception to this is during initiation of translation when the P-site holds the initiator tRNA.

20.11 The normal mRNA sequence, AUG-CCC-GAC-UUU, would encode the peptide sequence methionine-proline-aspartate-phenylalanine. The mutant mRNA sequence, AUG-CGC-GAC-UUU, would encode the mutant peptide sequence methionine-arginine-aspartate-phenylalanine. This would not be a silent mutation because a hydrophobic amino acid (proline) has been replaced by a positively charged amino acid (arginine).

20.13 A heterocyclic amine is a compound that contains nitrogen in at least one position of the ring skeleton.

20.15 It is the N-9 of the purine that forms the *N*-glycosidic bond with C-1 of the five-carbon sugar. The general structure of the purine ring is shown below:

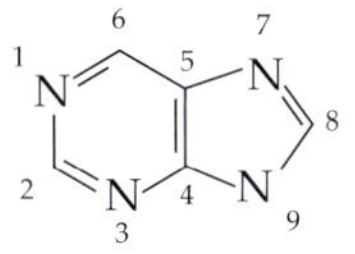

20.17 The ATP nucleotide is composed of the five-carbon sugar ribose, the purine adenine, and a triphosphate group.

20.19 The two strands of DNA in the double helix are said to be *antiparallel* because they run in opposite directions. One strand progresses in the 5' → 3' direction, and the opposite strand progresses in the 3' → 5' direction.

20.21 The DNA double helix is 2 nm in width. The nitrogenous bases are stacked at a distance of 0.34 nm from one another. One complete turn of the helix is 3.4 nm, or 10 base pairs.

20.23 Two

20.25

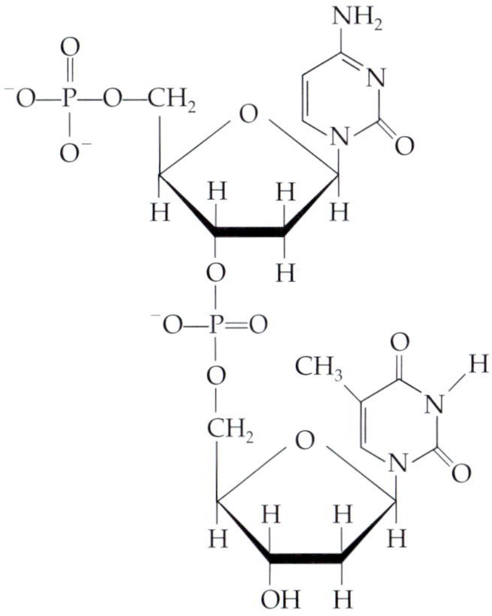

20.27 The prokaryotic chromosome is a circular DNA molecule that is supercoiled, that is, the helix is coiled on itself.

20.29 The term *semiconservative DNA replication* refers to the fact that each parental DNA strand serves as the template for the synthesis of a daughter strand. As a result, each of the daughter DNA molecules is made up of one strand of the original parental DNA and one strand of newly synthesized DNA.

20.31 The two primary functions of DNA polymerase III are to read a template DNA strand and catalyze the polymerization of a new daughter strand, and to proofread the newly synthesized strand and correct any errors by removing the incorrectly inserted nucleotide and adding the proper one.

20.33 3'-TACGCCGATCTTATAAGGT-5'

20.35 The *replication origin* of a DNA molecule is the unique sequence on the DNA molecule where DNA replication begins.

20.37 The enzyme helicase separates the strands of DNA at the origin of DNA replication so that the proteins involved in replication can interact with the nitrogenous base pairs.

20.39 The RNA primer "primes" DNA replication by providing a 3'-OH which can be used by DNA polymerase III for the addition of the next nucleotide in the growing DNA chain.

20.41 DNA → RNA → Protein

20.43 Anticodons are found on transfer RNA molecules.

20.45 3'-AUGGAUCGAGACCAGUAAUUCCGUCAU-5'.

20.47 *RNA splicing* is the process by which the noncoding sequences (introns) of the primary transcript of a eukaryotic mRNA are removed and the protein coding sequences (exons) are spliced together.

20.49 Messenger RNA, transfer RNA, and ribosomal RNA

20.51 Spliceosomes are small ribonucleoprotein complexes that carry out RNA splicing.

20.53 The *poly(A) tail* is a stretch of 100–200 adenosine nucleotides polymerized onto the 3' end of a mRNA by the enzyme poly(A) polymerase.

20.55 The *cap structure* is made up of the nucleotide 7-methylguanosine attached to the 5' end of a mRNA by a 5'-5' triphosphate bridge. Generally the first two nucleotides of the mRNA are also methylated.

20.57 Sixty-four

20.59 The reading frame of a gene is the sequential set of triplet codons that carries the genetic code for the primary structure of a protein.

20.61 Methionine and tryptophan

20.63 The codon 5'-UUU-3' encodes the amino acid phenylalanine. The mutant codon 5'-UUA-3' encodes the amino acid leucine. Both leucine and phenylalanine are hydrophobic amino acids, however, leucine has a smaller R group. It is possible that the smaller R group would disrupt the structure of the protein.

20.65 The ribosomes serve as a platform on which protein synthesis can occur. They also carry the enzymatic activity that forms peptide bonds.

20.67 The sequence of DNA nucleotides in a gene is transcribed to produce a complementary sequence of RNA nucleotides in a messenger RNA (mRNA). In the process of translation the sequence of the mRNA is read sequentially in words of three nucleotides (codons) to produce a protein. Each codon calls for the addition of a particular amino acid to the growing peptide chain. Through these processes, the sequence of nucleotides in a gene determines the sequence of amino acids in the primary structure of a protein.

20.69 In the initiation of translation, initiation factors, methionyl tRNA (the initiator tRNA), the mRNA, and the small and large ribosomal subunits form the initiation complex. During the elongation stage of translation, an aminoacyl tRNA binds to the A-site of the ribosome. Peptidyl transferase catalyzes the formation of a peptide bond and the peptide chain is transferred to the tRNA in the A-site. Translocation shifts the peptidyl tRNA from the A-site into the P-site, leaving the A-site available for the next aminoacyl tRNA. In the termination stage of translation, a termination codon is encountered. A release factor binds to the empty A-site and peptidyl transferase catalyzes the hydrolysis of the bond between the peptidyl tRNA and the completed peptide chain.

20.71 An ester bond

20.73 A point mutation is the substitution of one nucleotide pair for another in a gene.

20.75 Some mutations are silent because the change in the nucleotide sequence does not alter the amino acid sequence of the protein. This can happen because there are many amino acids encoded by multiple codons.

20.77 UV light causes the formation of pyrimidine dimers, the covalent bonding of two adjacent pyrimidine bases. Mutations occur when the UV damage repair system makes an error during the repair process. This causes a change in the nucleotide sequence of the DNA.

20.79 a. A *carcinogen* is a compound that causes cancer. Cancers are caused by mutations in the genes responsible for controlling cell division.
b. Carcinogens cause DNA damage that results in changes in the nucleotide sequence of the gene. Thus, carcinogens are also mutagens.

20.81 A *restriction enzyme* is a bacterial enzyme that "cuts" the sugar–phosphate backbone of DNA molecules at a specific nucleotide sequence.

20.83 A selectable marker is a genetic trait that can be used to detect the presence of a plasmid in a bacterium. Many plasmids have antibiotic resistance genes as selectable markers. Bacteria containing the plasmid will be able to grow in the presence of the antibiotic; those without the plasmid will be killed.

20.85 Human insulin, interferon, human growth hormone, and human blood clotting factor VIII

20.87 1024 copies

20.89 The goals of the Human Genome Project are to identify and map all of the genes of the human genome and to determine the DNA sequences of the complete three billion nucleotide pairs.

20.91 A genome library is a set of clones that represents all of the DNA sequences in the genome of an organism.

20.93 A dideoxynucleotide is one that has hydrogen atoms rather than hydroxyl groups bonded to both the 2' and 3' carbons of the five-carbon sugar.

20.95 Sequences that these DNA sequences have in common are highlighted in bold.
a. 5' **AGCTCCT**GATTTCATACAGTTTCTACT**ACCTACTA** 3'
b. 5' AGACATTCTATCTACCTAGACTATG**TTCAGAA** 3'
c. 5' **TTCAGAA**CTCATTCAGACCTACTACTATACCTTGGG **AGCTCCT** 3'
d. 5' **ACCTACTA**GACTATACTACTACTAAGGGGACTATTC CAGACTT 3'

The 5' end of sequence (a) is identical to the 3' end of sequence (c). The 3' end of sequence (a) is identical to the 5' end of sequence (d). The 3' end of sequence (b) is identical to the 5' end of sequence (c). From 5' to 3', the sequences would form the following map:

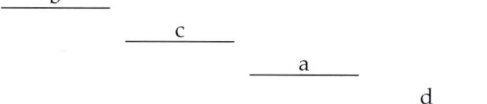

Chapter 21

21.1 ATP is called the universal energy currency because it is the major molecule used by all organisms to store energy.

21.3 The first stage of catabolism is the digestion (hydrolysis) of dietary macromolecules in the stomach and intestine.
In the second stage of catabolism, monosaccharides, amino acids, fatty acids, and glycerol are converted by metabolic reactions into molecules that can be completely oxidized.
In the third stage of catabolism, the two-carbon acetyl group of acetyl CoA is completely oxidized by the reactions of the citric acid cycle. The energy of the electrons harvested in these oxidation reactions is used to make ATP.

21.5 Substrate level phosphorylation is one way the cell can make ATP. In this reaction, a high-energy phosphoryl group of a substrate in the reaction is transferred to ADP to produce ATP.

21.7 Glycolysis is a pathway involving ten reactions. In reactions 1–3, energy is invested in the beginning substrate, glucose. This is done by transferring high-energy phosphoryl groups from ATP to the intermediates in the pathway. The product is fructose-1,6-bisphosphate. In the energy-harvesting reactions of glycolysis, fructose-1,6-bisphosphate is split into two three-carbon molecules that begin a series of rearrangement, oxidation-reduction, and substrate-level phosphorylation reactions that produce four ATP, two NADH, and two pyruvate molecules. Because of the investment of two ATP in the early steps of glycolysis, the net yield of ATP is two.

21.9 Both the alcohol and lactate fermentations are anaerobic reactions that use the pyruvate and re-oxidize the NADH produced in glycolysis.

21.11 Gluconeogenesis (synthesis of glucose from noncarbohydrate sources) appears to be the reverse of glycolysis (the first stage of carbohydrate degradation) because the intermediates in the two pathways are the same. However, reactions 1, 3, and 10 of glycolysis are not reversible reactions. Thus, the reverse reactions must be carried out by different enzymes.

21.13 The enzyme glycogen phosphorylase catalyzes the phosphorolysis of a glucose unit at one end of a glycogen molecule. The reaction involves the displacement of the glucose by a phosphate group. The products are glucose-1-phosphate and a glycogen molecule that is one glucose unit shorter.

21.15 Glucokinase traps glucose within the liver cell by phosphorylating it. Because the product, glucose-6-phosphate, is charged, it cannot be exported from the cell.

21.17 Glucagon indirectly stimulates glycogen phosphorylase, the first enzyme of glycogenolysis. This speeds up glycogen degradation. Glucagon also inhibits glycogen synthase, the first enzyme in glycogenesis. This inhibits glycogen synthesis.

21.19 ATP

21.21 Adenosine triphosphate + H_2O → Adenosine diphosphate + Inorganic phosphate group

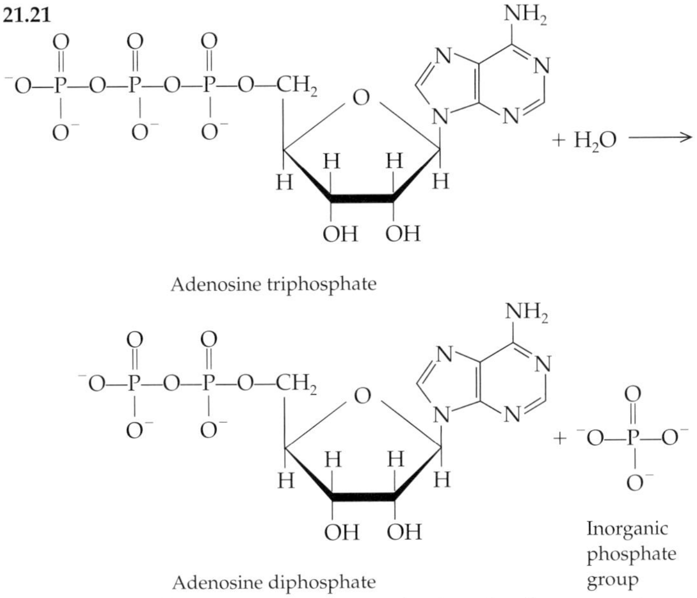

21.23 A coupled reaction is one that can be thought of as a two-step process. In a coupled reaction, two reactions occur simultaneously. Frequently one of the reactions releases the energy that drives the second, energy-requiring, reaction.

21.25 Carbohydrates

21.27 The following equation represents the hydrolysis of maltose:

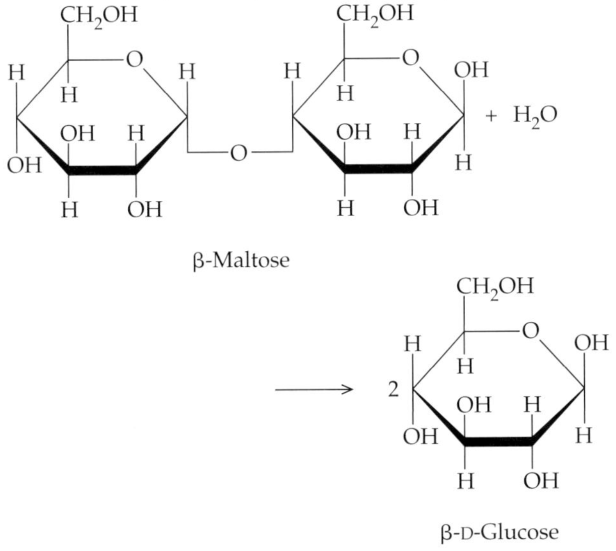

β-Maltose + H_2O → 2 β-D-Glucose

21.29 The following equation represents the hydrolysis of lactose:

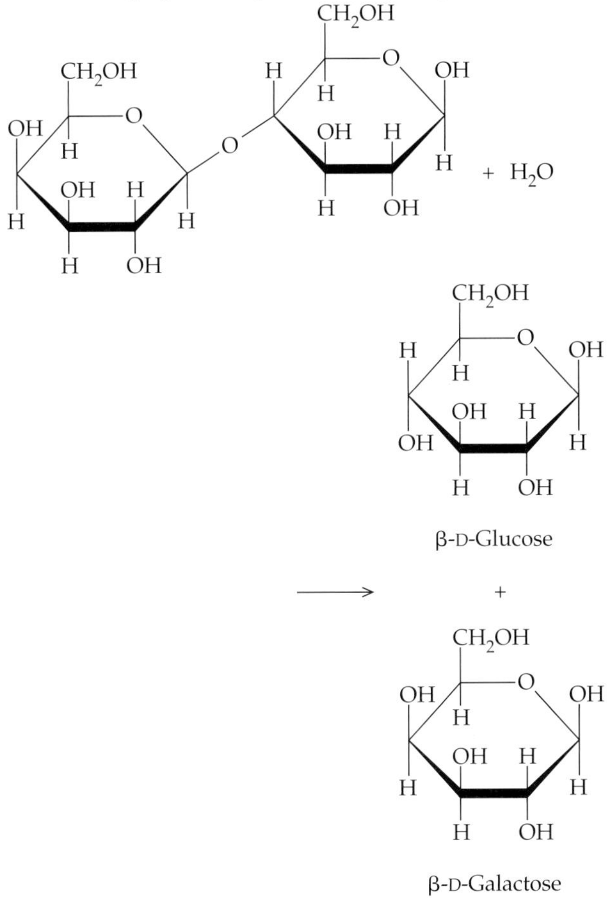

lactose + H_2O → β-D-Glucose + β-D-Galactose

21.31 The hydrolysis of a triglyceride containing oleic acid, stearic acid, and linoleic acid is represented in the following equations:

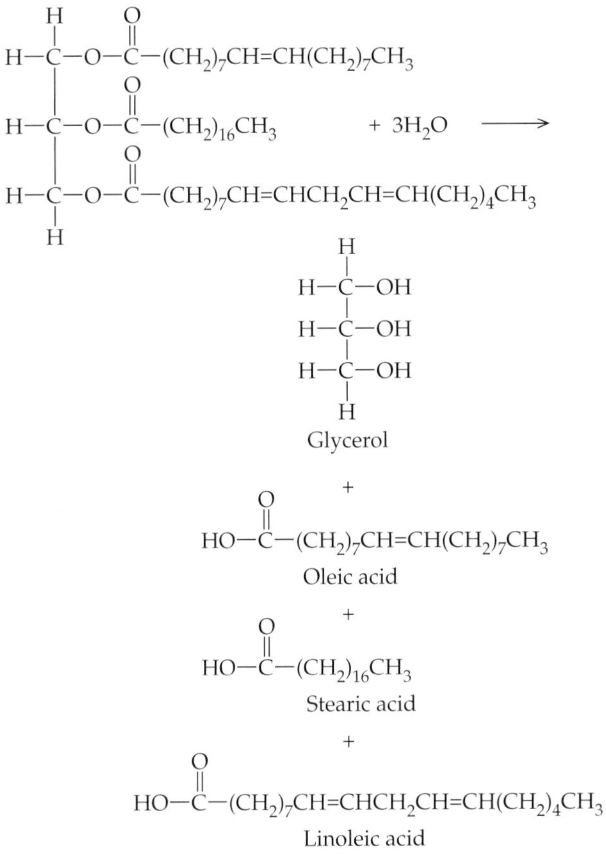

Triglyceride + $3H_2O$ → Glycerol + Oleic acid + Stearic acid + Linoleic acid

21.33 The hydrolysis of the dipeptide alanyl leucine is represented in the following equation:

$$H_3{}^+N-CH(CH_3)-C(=O)-NH-CH(CH_2CH(CH_3)CH_3)-C(=O)-O^- + H_2O \longrightarrow$$

Alanyl leucine

$$H_3{}^+N-CH(CH_3)-C(=O)-O^-$$

Alanine

$$+$$

$$H_3{}^+N-CH(CH_2CH(CH_3)CH_3)-C(=O)-O^-$$

Leucine

21.35 Glycolysis is the enzymatic pathway that converts a glucose molecule into two molecules of pyruvate. The pathway generates a net energy yield of two ATP and two NADH. Glycolysis is the first stage of carbohydrate catabolism.

21.37 Glycolysis requires NAD^+ for reaction 6 in which glyceraldehyde-3-phosphate dehydrogenase catalyzes the oxidation of glyceraldehyde-3-phosphate. NAD^+ is reduced.

21.39 Two ATP per glucose

21.41 Although muscle cells have enough ATP stored for only a few seconds of activity, glycolysis speeds up dramatically when there is a demand for more energy. If the cells have a sufficient supply of oxygen, aerobic respiration (the citric acid cycle and oxidative phosphorylation) will contribute large amounts of ATP. If oxygen is limited, the lactate fermentation will speed up. This will use up the pyruvate and re-oxidize the NADH produced by glycolysis and allow continued synthesis of ATP for muscle contraction.

21.43 $C_6H_{12}O_6 + 2ADP + 2P_i + 2NAD^+ \rightarrow$
Glucose
$\qquad 2C_3H_3O_3 + 2ATP + 2NADH + 2H_2O$
Pyruvate

21.45 a. Hexokinase catalyzes the phosphorylation of glucose.
b. Pyruvate kinase catalyzes the transfer of a phosphoryl group from phosphoenolpyruvate to ADP.
c. Phosphoglycerate mutase catalyzes the isomerization reaction that converts 3-phosphoglycerate to 2-phosphoglycerate.
d. Glyceraldehyde-3-phosphate dehydrogenase catalyzes the oxidation and phosphorylation of glyceraldehyde-3-phosphate and the reduction of NAD^+ to NADH.

21.47 Isomerase

21.49 Enediol

21.51 A kinase transfers a phosphoryl group from one molecule to another

21.53 NAD^+ is reduced, accepting a hydride anion.

21.55 Enolase catalyzes a reaction that produces an enol; in this particular reaction, it is phosphoenolpyruvate.

21.57 To optimize efficiency and minimize waste, it is important that energy-harvesting pathways, such as glycolysis, respond to the energy demands of the cell. If energy in the form of ATP is abundant, there is no need for the pathway to continue at a rapid rate. When this is the case, allosteric enzymes that catalyze the reactions of the pathway are inhibited by binding to their negative effectors. Similarly, when there is a great demand for ATP, the pathway speeds up as a result of the action of allosteric enzymes binding to positive effectors.

21.59 ATP and citrate are allosteric inhibitors of phosphofructokinase, whereas AMP and ADP are allosteric activators.

21.61 Citrate, which is the first intermediate in the citric acid cycle, is an allosteric inhibitor of phosphofructokinase. The citric acid cycle is a pathway that results in the complete oxidation of the pyruvate produced by glycolysis. A high concentration of citrate signals that sufficient substrate is entering the citric acid cycle. The inhibition of phosphofructokinase by citrate is an example of *feedback inhibition*: the product, citrate, allosterically inhibits the activity of an enzyme early in the pathway.

21.63

$$CH_2-C(=O)-H \xrightarrow{NADH \quad NAD^+} CH_3CH_2OH$$

Acetaldehyde $\qquad\qquad\qquad$ Ethanol

21.65 The lactate fermentation

21.67 Yogurt and some cheeses

21.69 Lactate dehydrogenase

21.71 This child must have the enzymes to carry out the alcohol fermentation. When the child exercised hard, there was not enough oxygen in the cells to maintain aerobic respiration. As a result, glycolysis and the alcohol fermentation were responsible for the majority of the ATP production by the child. The accumulation of alcohol (ethanol) in the child caused the symptoms of drunkenness.

21.73 The first stage of the pentose phosphate pathway is an oxidative stage in which glucose-6-phosphate is converted to ribulose-5-phosphate. Two NADPH molecules and one CO_2 molecule are also produced in these reactions. The second stage of the pentose phosphate pathway involves isomerization reactions that convert ribulose-5-phosphate into other five-carbon sugars, ribose-5-phosphate and xylulose-5-phosphate. The third stage of the pathway involves a complex series of rearrangement reactions that results in the production of two fructose-6-phosphate and one glyceraldehyde-3-phosphate molecules from three molecules of pentose phosphate.

21.75 The ribose-5-phosphate is used for the biosynthesis of nucleotides. The erythrose-4-phosphate is used for the biosynthesis of aromatic amino acids.

21.77 Gluconeogenesis is production of glucose from noncarbohydrate starting materials. This pathway can provide glucose when starvation or strenuous exercise leads to a depletion of glucose from the body.

21.79 The liver

21.81 Lactate is first converted to pyruvate.

21.83 Because steps 1, 3, and 10 of glycolysis are irreversible, gluconeogenesis is not simply the reverse of glycolysis. The reverse reactions must be carried out by different enzymes.

21.85 Steps 1, 3, and 10 of glycolysis are irreversible. Step 1 is the transfer of a phosphoryl group from ATP to carbon-6 of glucose and is catalyzed by hexokinase. Step 3 is the transfer of a phosphoryl group from ATP to carbon-1 of fructose-6-

phosphate and is catalyzed by phosphofructokinase. Step 10 is the substrate-level phosphorylation in which a phosphoryl group is transferred from phosphoenolpyruvate to ADP and is catalyzed by pyruvate kinase.

21.87 The liver and pancreas

21.89 *Hypoglycemia* is the condition in which blood glucose levels are too low.

21.91 a. Insulin stimulates glycogen synthase, the first enzyme in glycogen synthesis. It also stimulates uptake of glucose from the bloodstream into cells and phosphorylation of glucose by the enzyme glucokinase.
b. This traps glucose within liver cells and increases the storage of glucose in the form of glycogen.
c. These processes decrease blood glucose levels.

21.93 Any defect in the enzymes required to degrade glycogen or export glucose from liver cells will result in a reduced ability of the liver to provide glucose at times when blood glucose levels are low. This will cause hypoglycemia.

Chapter 22

22.1 Mitochondria are the organelles responsible for aerobic respiration.

22.3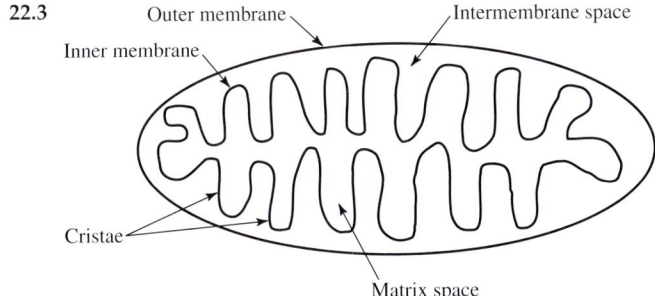

22.5 Pyruvate is converted to acetyl CoA by the pyruvate dehydrogenase complex. This huge enzyme complex requires four coenzymes, each of which is made from a different vitamin. The four coenzymes are thiamine pyrophosphate (made from thiamine), FAD (made from riboflavin), NAD^+ (made from niacin), and coenzyme A (made from the vitamin pantothenic acid). The coenzyme lipoamide is also involved in this reaction.

22.7 *Oxidative phosphorylation* is the process by which the energy of electrons harvested from oxidation of a fuel molecule is used to phosphorylate ADP to produce ATP.

22.9 $NAD^+ + H:^- \rightarrow NADH$

22.11 During transamination reactions, the α-amino group is transferred to the coenzyme pyridoxal phosphate. In the last part of the reaction, the α-amino group is transferred from pyridoxal phosphate to an α-keto acid.

22.13 The purpose of the urea cycle is to convert toxic ammonium ions to urea, which is excreted in the urine of land animals.

22.15 An amphibolic pathway is a metabolic pathway that functions both in anabolism and catabolism. The citric acid cycle is amphibolic because it has a catabolic function—it completely oxidizes the acetyl group carried by acetyl CoA to provide electrons for ATP synthesis. Because citric acid cycle intermediates are precursors for the biosynthesis of many other molecules, it also serves a function in anabolism.

22.17 The mitochondrion is an organelle that serves as the cellular power plant. The reactions of the citric acid cycle, the electron transport system, and ATP synthase function together within the mitochondrion to harvest ATP energy for the cell.

22.19 The intermembrane compartment is the location of the high-energy proton (H^+) reservoir produced by the electron transport system. The energy of this H^+ reservoir is used to make ATP.

22.21 The outer mitochondrial membrane is freely permeable to substances of molar mass less than 10,000 g/mol. The inner mitochondrial membrane is highly impermeable. Embedded within the inner mitochondrial membrane are the electron carriers of the electron transport system, and ATP synthase, the multisubunit enzyme that makes ATP.

22.23 Coenzyme A is a molecule derived from ATP and the vitamin pantothenic acid. It functions in the transfer of acetyl groups in lipid and carbohydrate metabolism.

22.25 Decarboxylation is a chemical reaction in which a carboxyl group is removed from a molecule.

22.27 Under aerobic conditions pyruvate is converted to acetyl CoA.

22.29 The coenzymes NAD^+, FAD, thiamine pyrophosphate, and coenzyme A are required by the pyruvate dehydrogenase complex for the conversion of pyruvate to acetyl CoA. These coenzymes are synthesized from the vitamins niacin, riboflavin, thiamine, and pantothenic acid, respectively. If the vitamins are not available, the coenzymes will not be available and pyruvate cannot be converted to acetyl CoA. Because the complete oxidation of the acetyl group of acetyl CoA produces the vast majority of the ATP for the body, ATP production would be severely inhibited by a deficiency of any of these vitamins.

22.31 An aldol condensation is a reaction in which aldehydes or ketones react to form larger molecules.

22.33 Oxidation in an organic molecule is often recognized as the loss of hydrogen atoms or a gain of oxygen atoms.

22.35 A dehydrogenation reaction is an oxidation reaction in which protons and electrons are removed from a molecule.

22.37 a. False
b. False
c. True
d. True

22.39 a. The acetyl group of acetyl CoA is transferred to oxaloacetate.
b. The product is citrate.

22.41 Three

22.43 Two ATP per glucose

22.45 The function of acetyl CoA in the citric acid cycle is to bring the two-carbon remnant (acetyl group) of pyruvate from glycolysis and transfer it to oxaloacetate. In this way the acetyl group enters the citric acid cycle for the final stages of oxidation.

22.47 The high-energy phosphoryl group of the GTP is transferred to ADP to produce ATP. This reaction is catalyzed by the enzyme dinucleotide diphosphokinase.

22.49
$$\begin{array}{c} COO^- \\ | \\ C=O \\ | \\ CH_2 \\ | \\ COO^- \end{array} + H_3C-\overset{O}{\underset{\|}{C}}-S-CoA + H_2O \longrightarrow$$

Oxaloacetate   Acetyl CoA

$$\begin{array}{c} COO^- \\ | \\ CH_2 \\ | \\ HO-C-COO^- \\ | \\ CH_2 \\ | \\ COO^- \end{array} + HS-CoA + H^+$$

Citrate   Coenzyme A

The importance of this reaction is that it brings the acetyl group, the two-carbon remnants of the glucose molecule, into the citric acid cycle to be completely oxidized. Through these reactions, and subsequent oxidative phosphorylation, the majority of the cellular ATP energy is provided.

22.51 The conversion of citrate to *cis*-aconitate is an example of the dehydration of an alcohol to produce an alkene (double bond). The conversion of *cis*-aconitate to isocitrate is an example of the hydration of an alkene, that is, the addition of water to the double bond, to produce an alcohol (—OH).

22.53 This reaction is an example of the oxidation of a secondary alcohol to a ketone. The two functional groups are the hydroxyl group of the alcohol and the carbonyl group of the ketone.

22.55 It is a kinase because it transfers a phosphoryl group from one molecule to another. Kinases are a specific type of transferase.

22.57 An allosteric enzyme is one that has an effector binding site and an active site. Effector binding can change the shape of the active site, causing it to be active or inactive.

22.59 Negative allosterism is a means of enzyme regulation in which effector binding inactivates the active site of the allosteric enzyme.

22.61 The citric acid cycle is regulated by the following four enzymes or enzyme complexes: pyruvate dehydrogenase complex, citrate sythase, isocitrate dehydrogenase, and the α-ketoglutarate dehydrogenase complex.

22.63 Energy-harvesting pathways, such as the citric acid cycle, must be responsive to the energy needs of the cell. If the energy requirements are high, as during exercise, the reactions must speed up. If energy demands are low and ATP is in excess, the reactions of the pathway slow down.

22.65 ADP

22.67 The electron transport system is series of electron transport proteins embedded in the inner mitochondrial membrane that accept high-energy electrons from NADH and $FADH_2$ and transfer them in stepwise fashion to molecular oxygen (O_2).

22.69 Three ATP

22.71 The oxidation of a variety of fuel molecules, including carbohydrates, the carbon skeletons of amino acids, and fatty acids provides the electrons. The energy of these electrons is used to produce an H^+ reservoir. The energy of this proton reservoir is used for ATP synthesis.

22.73 The electron transport system passes electrons harvested during oxidation of fuel molecules to molecular oxygen. At three sites protons are pumped from the mitochondrial matrix into the intermembrane compartment. Thus, the electron transport system builds the high-energy H^+ reservoir that provides energy for ATP synthesis.

22.75 a. Two ATP per glucose (net yield) are produced in glycolysis, whereas the complete oxidation of glucose in aerobic respiration (glycolysis, the citric acid cycle, and oxidative phosphorylation) results in the production of thirty-six ATP per glucose.
b. Thus, aerobic respiration harvests nearly 40% of the potential energy of glucose, and anaerobic glycolysis harvests only about 2% of the potential energy of glucose.

22.77 Transaminases transfer amino groups from amino acids to ketoacids.

22.79 The glutamate family of transaminases is very important because the ketoacid corresponding to glutamate is α-ketoglutarate, one of the citric acid cycle intermediates. This provides a link between the citric acid cycle and amino acid metabolism. These transaminases provide amino groups for amino acid synthesis and collect amino groups during catabolism of amino acids.

22.81 a. Pyruvate   d. Acetyl CoA
b. α-Ketoglutarate   e. Succinate
c. Oxaloacetate   f. α-Ketoglutarate

22.83

$$\underset{\text{α-Ketoglutarate}}{\begin{array}{c} O \\ \| \\ C-COO^- \\ | \\ CH_2 \\ | \\ CH_2 \\ | \\ COO^- \end{array}} + NADPH + N^+H_4 \longrightarrow \underset{\text{Glutamate}}{\begin{array}{c} N^+H_3 \\ | \\ H-C-COO^- \\ | \\ CH_2 \\ | \\ CH_2 \\ | \\ COO^- \end{array}} + NADP^+ + H_2O$$

Ammonia

22.85 Hyperammonemia

22.87 a. The source of one amino group of urea is the ammonium ion and the source of the other is the α-amino group of the amino acid aspartate.
b. The carbonyl group of urea is derived from CO_2.

22.89 Anabolism is a term used to describe all of the cellular energy-requiring biosynthetic pathways.

22.91 α-Ketoglutarate

22.93 Citric acid cycle intermediates are the starting materials for the biosynthesis of many biological molecules.

22.95 An essential amino acid is one that cannot be synthesized by the body and must be provided in the diet.

22.97

$$\underset{\text{Pyruvate}}{\begin{array}{c} O \\ \| \\ C-COO^- \\ | \\ CH_3 \end{array}} + CO_2 + ATP \longrightarrow \underset{\text{Oxaloacetate}}{\begin{array}{c} O \\ \| \\ C-COO^- \\ | \\ CH_2 \\ | \\ COO^- \end{array}} + ADP + P_i$$

Chapter 23

23.1 Because dietary lipids are hydrophobic, they arrive in the small intestine as large fat globules. The bile salts emulsify these fat globules into tiny fat droplets. This greatly increases the surface area of the lipids, allowing them to be more accessible to pancreatic lipases and thus more easily digested.

23.3 a. Four acetyl CoA, one benzoate, four NADH, and four $FADH_2$
b. Three acetyl CoA, one phenyl acetate, three NADH, and three $FADH_2$
c. Three acetyl CoA, one benzoate, three NADH, and three $FADH_2$
d. Five acetyl CoA, one phenyl acetate, five NADH, and five $FADH_2$

23.5

$$CH_3CH_2CH_2-\overset{O}{\underset{\|}{C}}\sim S-CoA + FAD \longrightarrow$$

$$CH_3CH=CH-\overset{O}{\underset{\|}{C}}\sim S-CoA + FADH_2$$

$$\downarrow H_2O$$

$$CH_3-\overset{OH}{\underset{H}{\overset{|}{C}}}-CH_2-\overset{O}{\underset{\|}{C}}\sim S-CoA + NAD^+ \longleftarrow$$

$$CH_3-\overset{O}{\underset{\|}{C}}-CH_2-\overset{O}{\underset{\|}{C}}\sim S-CoA + NADH$$

$$\downarrow \text{Coenzyme A}$$

$$2CH_3-\overset{O}{\underset{\|}{C}}\sim S-CoA$$

23.7 Starvation, a diet low in carbohydrates, and diabetes mellitus are conditions that lead to the production of ketone bodies.

23.9 (1) Fatty acid biosynthesis occurs in the cytoplasm whereas β-oxidation occurs in the mitochondria.
(2) The acyl group carrier in fatty acid biosynthesis is acyl carrier protein while the acyl group carrier in β-oxidation is coenzyme A.
(3) The seven enzymes of fatty acid biosynthesis are associated as a multienzyme complex called *fatty acid synthase*. The enzymes involved in β-oxidation are not physically associated with one another.
(4) NADPH is the reducing agent used in fatty acid biosynthesis. NADH and $FADH_2$ are produced by β-oxidation.

23.11 The liver regulates blood glucose levels under the control of the hormones insulin and glucagon. When blood glucose levels are too high, insulin stimulates the uptake of glucose by liver cells and the storage of the glucose in glycogen polymers. When blood glucose levels are too low, the hormone glucagon stimulates the breakdown of glycogen and release of glucose into the bloodstream. Glucagon also stimulates the liver to produce glucose for export into the bloodstream by the process of gluconeogenesis.

23.13 Insulin stimulates uptake of glucose and amino acids by cells, glycogen and protein synthesis, and storage of lipids. It inhibits glycogenolysis, gluconeogenesis, breakdown of stored triglycerides, and ketogenesis.

23.15 Bile consists of micelles of lecithin, cholesterol, bile salts, proteins, inorganic ions, and bile pigments that aid in lipid digestion by emulsifying fat droplets.

23.17 A micelle is an aggregation of molecules having nonpolar and polar regions; the nonpolar regions of the molecules aggregate, leaving the polar regions facing the surrounding water.

23.19 A triglyceride is a molecule composed of glycerol esterified to three fatty acids.

23.21 A chylomicron is a plasma lipoprotein that carries triglycerides from the intestine to all body tissues via the bloodstream. That function is reflected in the composition of the chylomicron, which is approximately 85% triglycerides, 9% phospholipids, 3% cholesterol esters, 2% protein, and 1% cholesterol.

23.23 Triglycerides

23.25 The large fat globule that takes up nearly the entire cytoplasm

23.27 Lipases catalyze the hydrolysis of the ester bonds of triglycerides.

23.29 Acetyl CoA is the precursor for fatty acids, several amino acids, cholesterol, and other steroids.

23.31 Chylomicrons are plasma lipoproteins (aggregates of protein and triglycerides) that carry dietary triglycerides from the intestine to all tissues via the bloodstream.

23.33 Bile salts serve as detergents. Fat globules stimulate their release from the gallbladder. The bile salts then emulsify the lipids, increasing their surface area and making them more accessible to digestive enzymes (pancreatic lipases).

23.35 When dietary lipids in the form of fat globules reach the duodenum, they are emulsified by bile salts. The triglycerides in the resulting tiny fat droplets are hydrolyzed into monoglycerides and fatty acids by the action of pancreatic lipases, assisted by colipase. The monoglycerides and fatty acids are absorbed by cells lining the intestine.

23.37 The energy source for the activation of a fatty acid entering β-oxidation is the hydrolysis of ATP into AMP and PP_i (pyrophosphate group), an energy expense of two high-energy phosphoester bonds.

23.39 Carnitine is a carrier molecule that brings fatty acyl groups into the mitochondrial matrix.

23.41 The following equation represents the reaction catalyzed by acyl-CoA dehydrogenase. Notice that the reaction involves the loss of two hydrogen atoms. Thus, this is an oxidation reaction.

$$H_3C-\overset{H}{\underset{(H)}{\overset{|}{C}}}-\overset{(H)}{\underset{H}{\overset{|}{C}}}-\overset{O}{\underset{\|}{C}}\sim S-CoA$$

$$FAD \searrow \text{ Acyl-CoA}$$
$$FADH_2 \nwarrow \text{ dehydrogenase}$$

$$H_3C-\overset{H}{\underset{}{\overset{|}{C}}}=\overset{}{\underset{H}{\overset{|}{C}}}-\overset{O}{\underset{\|}{C}}\sim S-CoA$$

23.43 An alcohol is the production of the hydration of an alkene.

23.45 Six acetyl CoA, one phenyl acetate, six NADH, and six $FADH_2$

23.47 112 ATP

23.49 Two ATP

23.51 The acetyl CoA produced by β-oxidation will enter the citric acid cycle.

23.53 Ketone bodies include the compounds acetone, acetoacetone, and β-hydroxybutyrate, which are produced from fatty acids in the liver via acetyl CoA.

23.55 Ketosis is an abnormal rise in the level of ketone bodies in the blood.

23.57 Matrix of the mitochondrion.

23.59

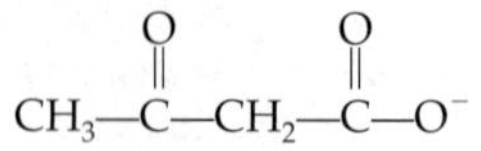

$$CH_3-\overset{O}{\underset{\|}{C}}-CH_2-\overset{O}{\underset{\|}{C}}-O^-$$

Acetoacetate

$$CH_3-\underset{\underset{OH}{|}}{CH}CH_2-\overset{O}{\underset{\|}{C}}-O^-$$

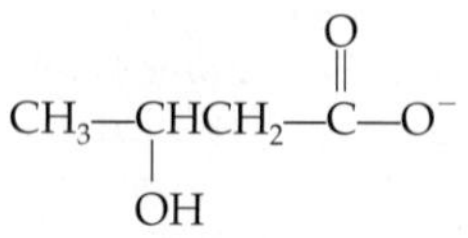

β-Hydroxybutyrate

23.61 In those suffering from uncontrolled diabetes, the glucose in the blood cannot get into the cells of the body. The excess glucose is excreted in the urine. Body cells degrade fatty acids because glucose is not available. β-oxidation of fatty acids yields enormous quantities of acetyl CoA, so much acetyl CoA, in fact, that it cannot all enter the citric acid cycle because there is not enough oxaloacetate available. Excess acetyl CoA is used for ketogenesis.

23.63 Ketone bodies are the preferred energy source of the heart.

23.65 Cytoplasm

23.67 Fatty acid synthase

23.69 The phosphopantetheine group allows formation of a high-energy thioester bond with a fatty acid. It is derived from the vitamin pantothenic acid.

23.71 Fatty acid synthase is a huge multienzyme complex consisting of the seven enzymes involved in fatty acid synthesis. It is found in the cell cytoplasm. The enzymes involved in β-oxidation are not physically associated with one another. They are free in the mitochondrial matrix space.

23.73 Glycogenesis is the synthesis of the polymer glycogen from glucose monomers.

23.75 Gluconeogenesis is the synthesis of glucose from non-carbohydrate precursors.

23.77 β-oxidation of fatty acids

23.79 The major metabolic function of the liver is to regulate blood glucose levels.

23.81 Ketone bodies are the major fuel for the heart. Glucose is the major energy source of the brain, and the liver obtains most of its energy from the oxidation of amino acid carbon skeletons.

23.83 Fatty acids are absorbed from the bloodstream by adipocytes. Using glycerol-3-phosphate, produced as a by-product of glycolysis, triglycerides are synthesized. Triglycerides are constantly being hydrolyzed and resynthesized in adipocytes. The rates of hydrolysis and synthesis are determined by lipases that are under hormonal control.

23.85 In general, insulin stimulates anabolic processes, including glycogen synthesis, uptake of amino acids and protein synthesis, and triglyceride synthesis. At the same time, catabolic processes such as glycogenolysis are inhibited.

23.87 A target cell is one that has a receptor for a particular hormone.

23.89 Decreased blood glucose levels trigger the secretion of glucagon into the bloodstream.

23.91 Insulin is produced in the β-cells of the islets of Langerhans in the pancreas.

23.93 Insulin stimulates the uptake of glucose from the blood into cells. It enhances glucose storage by stimulating glycogenesis and inhibiting glycogen degradation and gluconeogenesis.

23.95 Insulin stimulates synthesis and storage of triglycerides.

23.97 Untreated diabetes mellitus is starvation in the midst of plenty because blood glucose levels are very high. However, in the absence of insulin, blood glucose can't be taken up into cells. The excess glucose is excreted into the urine while the cells of the body are starved for energy.

Credits

Design Elements
Microscope: © Vol. 85/PhotoDisc; **Stethoscope:** © Vol. OS48/PhotoDisc; **Eye:** © Vol. OS2/PhotoDisc; **Pill Bottle:** © Vol.168/Corbis; **Pills:** © Vol. 168/Corbis.

Chapter 1
Opener: © Greg Fiume/NewSport/Corbis; **1.2a:** © Geoff Tompkinson/SPL/Photo Researchers, Inc.; **1.2b:** © T.J. Florian/Rainbow; **1.2c:** © David Parker/Segate Microelectronics Ltd./Photo Researchers, Inc.; **1.2d:** APHIS, PPQ, Otis Methods Development Center, Otis, MA/USDA; **1.3a:** © McGraw-Hill Higher Education/Louis Rosenstock, photographer; **1.3b:** © The McGraw-Hill Companies, Inc./Jeff Topping, photographer; **1.3c:** © The McGraw-Hill Companies, Inc./Louis Rosenstock, photographer; **1.4(both):** © The McGraw-Hill Companies, Inc./Ken Karp, photographer; **1.6:** © Doug Martin/Photo Researchers, Inc; **1.8a-c, 1.10a-d, 1.12:** © The McGraw-Hill Companies, Inc./Louis Rosenstock, photographer.

Chapter 2
Opener: © The McGraw-Hill Companies, Inc./Louis Rosenstock, photographer; **2.1:** © IBM Corporation–Almaden Research Center; **2.7:** © Yoav Levy/Phototake; **p. 48:** © Earth Satellite Corp./SPL/Photo Researchers, Inc.; **p. 49(right):** © The McGraw-Hill Companies, Inc./Louis Rosenstock, photographer; **p. 49(left):** © Dan McCoy/Rainbow; **p. 52:** © Vol. 2/PhotoDisc.

Chapter 3
Opener: © Chase Swift/Corbis; **3.2:** Courtesy of Trent Stephens; **3.14:** © Charles D. Winters/Photo Researchers, Inc.

Chapter 4
Opener: © Richard Megna/Fundamental Photographs; **4.1, 4.3:** © The McGraw-Hill Companies, Inc./Louis Rosenstock, photographer; **p. 144:** © Vol. 72/PhotoDisc.

Chapter 5
Opener: Courtesy of Robert Shoemaker; **p. 152:** © Hulton-Deutsch/Corbis; **5.2a,b:** © The McGraw-Hill Companies, Inc./Louis Rosenstock, photographer; **5.5:** © Peter Stef Lamberti/Stone/Getty.

Chapter 6
Opener: © Richard Megna/Fundamental Photographs; **6.1:** © Kip and Pat Peticolas/Fundamental Photographs; **p. 183:** © J.W. Mowbray/Photo Researchers, Inc.; **6.2:** © The McGraw-Hill Companies, Inc./Louis Rosenstock, photographer; **p. 197(left):** © RMF/Visuals Unlimited; **p. 197(right):** Courtesy of Rita Colwell, University of Maryland; **6.5 a-b:** © The McGraw-Hill Companies, Inc./Louis Rosenstock, photographer; **p. 199:** © The McGraw-Hill Companies, Inc./Louis Rosenstock, photographer; **p. 201:** © AJPhoto/Photo Researchers, Inc.

Chapter 7
Opener: © The McGraw-Hill Companies, Inc./Louis Rosenstock, photographer; **7.9:** © The McGraw-Hill Companies, Inc./Ken Karp, photographer; **p. 221(both):** © The McGraw-Hill Companies, Inc./Louis Rosenstock, photographer; **7.15, 7.16:** © The McGraw-Hill Companies, Inc./Ken Karp, photographer.

Chapter 8
Opener: © Richard Megna/Fundamental Photographs; **8.1:** © Dr. E.R. Degginger/Color Pic Inc.; **8.3a:** © The McGraw-Hill Companies, Inc./Louis Rosenstock, photographer; **8.3b:** © Richard Megna/Fundamental Photographs; **8.5a-b:** © The McGraw-Hill Companies, Inc./Louis Rosenstock, photographer; **8.7a-b:** © The McGraw-Hill Companies, Inc./Stephen Frisch, photographer; **p. 257, 8.8, 8.9(left):** © The McGraw-Hill Companies, Inc./Louis Rosenstock, photographer; **8.9(right):** © Tony Freeman/PhotoEdit; **8.9(middle):** © Bonnie Kamin/PhotoEdit; **p. 266:** © AAA Photo/Phototake; **8.10:** © The McGraw-Hill Companies, Inc./Stephen Frisch, photographer.

Chapter 9
Opener: © Kathy McLaughlin/The Image Works; **9.5:** © US Dept. of Energy/Photo Researchers, Inc.; **9.6:** © Gianni Tortoli/Photo Researchers, Inc.; **p. 285:** NASA; **9.7:** © Blair Seitz/Photo Researchers, Inc.; **9.8b:** Bristol-Myers Squibb Medical Imaging; **p. 292(bottom):** © SIU/Biomed/Custom Medical Stock Photo; **p. 292(top):** © The McGraw-Hill Companies, Inc./Louis Rosenstock, photographer; **9.9:** © US Dept. of Energy/Mark Marten/Photo Researchers, Inc.; **9.10:** © The McGraw-Hill Companies, Inc./Louis Rosenstock, photographer; **9.11:** © Scott Cazmine/Photo Researchers, Inc.

Chapter 10
Opener: © Corbis Royalty Free; **p. 317:** © Vol. 31/PhotoDisc; **p. 326:** © Corbis Royalty Free; **p. 331:** © Vol. 27/PhotoDisc.

Chapter 11
Opener: © Vol. 44/PhotoDisc; **p. 347:** © Buddy Mays/Corbis; **11.3(top to bottom):** © Larry Lefever/Grant Heilman Photography, Inc., © Michelle Garrett/Corbis, © Walter H. Hodge/Peter Arnold, Inc., © Hal Horwitz/Corbis; **11.5:** © The McGraw-Hill Companies, Inc./Ken Karp,photographer; **p. 362:** © Vol. 67/PhotoDisc; **p. 366:** © Alan Detrick/Photo Researchers, Inc.

Chapter 12
Opener: © David Young-Wolff/PhotoEdit; **12.3:** © Vol. 12/PhotoDisc; **p. 388:** © Vol. 18/PhotoDisc; **12 4:** © Corbis Royalty Free; **p. 401:** © Vol. 88/PhotoDisc; **12 8:** © Corbis Royalty Free; **12 9:** © Vol. OS22/PhotoDisc.

Chapter 13
Opener: © Gary Moss/Getty Images; **13.3(almonds):** © B. Borrell Cassals, Frank Lane Picture Agency/Corbis; **13.3(cinnamon):** © Rita Maas/Getty Images; **13.3(berries):** © Charles Krebs/Corbis; **13.3(vanilla):** © Eisenhut & Mayer/Getty Images; **13.3(mushroom):** © James Noble/Corbis;

13.3(lemon grass): © Corbis Royalty Free; **13.4(all):** © Richard Megna/Fundamental Photographs; **13.5:** © Rob and Ann Simpson/Visuals Unlimited; **p. 426:** © Vol. 18/PhotoDisc; **p. 430:** © Corbis Royalty Free; **p. 433:** © Hanson Carroll/Peter Arnold, Inc.

Chapter 14
Opener: © Christel Rosenfeld/Tony Stone/Corbis; **p. 456:** © Corbis Royalty Free; **p. 457:** © Paul W. Johnson/BPS; **p. 464(pineapple):** © Vol. 19/PhotoDisc; **p. 464(berries):** © Vol. OS49/PhotoDisc; **p. 464(bananas):** © Vol. 30/PhotoDisc; **p. 465(oranges, apples):** © Corbis Royalty Free; **p. 465(apricots):** © Vol. 121/Corbis; **p. 465(strawberries):** © Vol. 83/Corbis; **p. 478:** © Vol. 81/Corbis; **p. 479(left):** © William Weber/Visuals Unlimited; **p. 479(bottom right):** © Norm Thomas/Photo Researchers; **p. 479(top right):** © Vol. 101/Corbis.

Chapter 15
Opener: © Vol. 6/PhotoDisc; **15 3a-b:** © Vol. 94/Corbis; **15.3c:** © Gregory G. Dimijian/Photo Researchers; **p. 500:** © The McGraw-Hill Companies, Inc./Gary He, photographer; **p. 512a:** © Vol. 9/PhotoDisc; **p. 512b:** © Phil Larkin, CSIRO Plant Industry.

Chapter 16
Opener: © The McGraw-Hill Companies, Inc./Louis Rosenstock, photographer; **16.1:** USDA; **16.2:** © Vol. 67/PhotoDisc; **p. 528b:** © Stanley Flegler/Visuals Unlimited; **p. 530:** © Corbis Royalty Free; **p. 546:** © Jean Claude Revy-ISM / Phototake.

Chapter 17
Opener: © Lester Lefkowitz/Corbis; **p. 563:** © Roadsideamerica.com, Kirby, Smith & Wilkins; **p. 574:** © Hans Pfletschinger/Peter Arnold, Inc.; **17.10a:** © James Dennis/PhotoTake; **17.18(all):** © David M. Phillips/Visuals Unlimited.

Chapter 18
Opener: © University of Oxford/Getty Images; **p.599:** © Vol. 18/PhotoDisc; **p. 604:** © Vol. 270/Corbis; **18.14:** © Meckes/Ottawa/Photo Researchers, Inc.

Chapter 19
Opener: © J. Schmidt/National Park Service; **p. 641:** © Phil Degginger/Color-Pic, Inc.

Chapter 20
Opener: © David Young-Wolff/PhotoEdit; **p. 702:** © Dr. Charles S. Helling/US Dept. of Agriculture; **p. 703:** Courtesy of Orchid Cellmark, Germantown, Maryland.

Chapter 21
Opener: © The McGraw-Hill Companies, Inc./Louis Rosenstock, photographer; **p. 728:** © The McGraw-Hill Higher Companies, Inc./Louis Rosenstock, photographer.

Chapter 22
Opener: © PhotoDisc; **22.1a:** © CNRI/Phototake; **p. 749:** © Vol. 10/PhotoDisc.

Chapter 23
Opener: Courtesy of Katherine Denniston; **p. 785:** © Vol. 110/PhotoDisc.

Index

Note: Page numbers followed by B indicate boxed material; those followed by F indicate figures; those followed by T indicate tables.

A

ABO blood types, 546–47B
absolute specificity, of enzyme, 639
acceptable daily intake (ADI), 382B
accuracy, and measurement, 19–20
acetal, 434, 435, 437, 442, 543
acetaldehyde, 267, 430B, 728F
acetamide, 505
acetaminophen, 478B, 506
acetate ion, 99–100
acetic acid
    buffers, 257, 259, 260
    neutralization, 458
    oxidation–reduction reactions, 267
    source and nomenclature of, 453T
    vinegar, 449
    weak acids, 242, 244
acetoacetate, 791
acetoacetyl ACP, 792
acetoacetyl CoA, 789–90
acetone, 791, 794B
acetylcholine, 515–16
acetylcholinesterase, 516, 654B
acetyl CoA
    biosynthesis of biological molecules, 408
    citric acid cycle, 479–80, 718, 750–51, 754F, 756
N-acetylgalactosamine, 546B, 547B
α-D-N-acetylglucosamine, 550B
acetyl group, 480B
acetylsalicylic acid. *See* aspirin
acid(s). *See also* carboxylic acids; citric acid cycle; fatty acids
    Arrhenius theory, 241
    Brønsted–Lowry theory, 241
    buffers, 256–63
    chemical reactions, 252–55
    conjugate, 243–46
    dissociation of water, 246
    pH scale, 246–52
    properties of water, 242
    strength, 242–43

acid anhydrides, 473
acid-base reactions, 131, 458–60
acid chlorides, 470–73
acid hydrolysis, and fatty acids, 561–62
acidosis, 262B, 769B
acid rain, 252, 256–57B
aconitase, 753
acrylonitrile, 365T
actin, 597
actinide series, 55
activated complex, 220
activation energy, 220, 636
active site, and enzyme–substrate complex, 637–38
active transport, 200, 590, 713T, 716. *See also* electron transport system; transport proteins
acyl carrier protein (ACP), 792, 793F
acyl-CoA dehydrogenase, 786
acyl-CoA ligase, 786
acyl group, 476, 480B, 508
acyl group carriers, 792
Adams, Mike, 630B
addition, of significant figures, 20
addition polymers, 364–65, 374
addition reactions, 354, 374, 391
    aldehydes and ketones, 434–36, 441–42
    alkenes, 374, 391
    double bond, 354
    fatty acids, 564
adenine, 670F, 672F
adenosine diphosphate. *See* ADP
adenosine triphosphate. *See* ATP
adipic acid, 455, 456
adipocere, 563B
adipocytes, 781
adipose tissue, 568, 781, 784–85B, 796
ADP, 438, 633B, 634, 714–15, 732
aerobic respiration, 752, 753–56, 757F
African clawed frog, 5B
agarose gel electrophoresis, 696–98
agglutination, 546B
Agricultural Research Service (ARS), 702B, 703B
agriculture, and pH scale, 252. *See also* plants
AIDS (acquired immune deficiency syndrome), 641B, 676–77B

air. *See also* air pollution
    density, 30
    as solution, 178
air bags, automobile, 118B
air pollution, 295B. *See also* acid rain; greenhouse effect
alanine, 598, 601F, 602, 623T, 764
alanine aminotransferase/serum glutamate-pyruvate transaminase (ALT/SGPT), 658B, 660
alanyl-glycine-valine, 602, 603
albumin, 597, 599B, 621
Alcaligenes eutrophus, 456B
alchemists, 266B
alcohol(s). *See also* ethanol
    addition reactions, 374
    boiling points, 493T
    chemical reactions, 391–400, 409
    classification, 389–90
    functional groups, 309
    hydration, 358
    medical uses, 387–89
    nomenclature, 385–87
    oxidation, 427
    physical properties, 384–85
alcohol abuse and alcoholism, 430B
alcohol dehydrogenase, 727, 729
alcohol fermentation, 727–29
alcoholic beverages, 387, 400, 401B, 430B
aldehydes
    chemical reactions, 424–42
    functional groups, 309T
    hydration, 361
    hydrogenation, 392–93
    medical and industrial uses, 424, 425F
    nomenclature, 419–22
    oxidation reactions, 397
    physical properties, 417–19
aldohexose, 535
aldolase, 439, 723
aldol condensation, 438–39, 442, 723B
aldose, 527
aldosterone, 577
aliphatic hydrocarbons, 308
alkadienes, 344
alkali metals, 55
alkaline metals, 55

Index

alkalosis, 262B
alkanamide, 506T
alkanamine, 495T
alkanes
    alkyl groups, 313–14
    chemical reactions, 327–32
    conformations of, 325–27
    nomenclature, 315–19
    saturated hydrocarbons, 308, 310
    structure and physical properties, 310–13
alkenes
    chemical reactions, 354–65, 374, 391
    functional groups, 309T
    nomenclature, 343–45
    plants as sources of, 352–53
    structure and physical properties, 342–43
    unsaturated hydrocarbons, 308
alkoxy group, 403B
alkylammonium ion, 498
alkylammonium salts, 498, 499
alkyl dihalide, 374
alkyl groups, 313–14, 315T
alkyl halide, 329, 361, 374
alkynes
    chemical reactions, 354–65
    functional groups, 309T
    nomenclature, 343–45
    structure and physical properties, 342–43
    unsaturated hydrocarbons, 308
allosteric enzymes, 649–50, 726, 756
allotropic forms, of elemental carbon, 305
alloys, and solutions, 178
alpha decay, 280
alpha particles, 278, 279
aluminum, 85, 106
aluminum bromide, 86
aluminum nitrate, 129
alveoli, and respiration, 182
Alzheimer's disease, 599B
American Civil War, 604B
American Diabetes Association, 382B
American Medical Association, 388B
Ames, Bruce, 694B
Ames test, 694–95B
amide(s), 309T
    chemical reactions, 506–9, 517
    common functional groups, 309T
    hydrolysis, 509–10, 517
    medically important, 505–6
    nomenclature, 505
    physical properties, 504–5
amide bond, 602
amines
    chemical reactions, 497–502, 517
    common functional groups, 309T
    heterocyclic amines, 502–4
    medically important, 495–97
    nomenclature, 493–95
    physical properties, 489–93
amino acids. *See also* proteins
    amides and amide bonds, 508B, 510–11
    classes of, 600
    definition of, 407B
    degradation of, 762–66
    genetic code, 686, 687F
    stereoisomers, 598–99
    structure of, 525, 597–98
α-amino acids, 597–98
aminoacyl group, 510–11
aminoacyl tRNA, 510–11, 689–90
aminoacyl tRNA binding site (A-site), 690
γ-aminobutyric acid (GABA), 515
6–aminopenicillanic acid, 507B
amino sugars, 550B
ammonia
    amines, 489
    covalent bonding, 82
    decomposition, 210, 211
    hydrogen bonding, 171
    Lewis structures of covalent compounds, 96–97
    molecular structure, 106, 490
    nomenclature, 91
    solubility, 111–12
    synthesis, 223F, 233F
ammonium chloride, 87, 216, 507
ammonium ion, and Lewis structure, 98
ammonium nitrate, 135
ammonium sulfate, 124
ammonium sulfide, 89
amniocentesis, 668B
amorphous ionic compounds, 93
amorphous solid, 171
amphetamines, 495–96
amphibolic pathways, 770–71
amphipathic molecule, 568
amphiprotic property, of water, 242
amphotericin B, 588B, 589B
ampicillin, 507B
amylopectin, 548, 549F
amylose, 548
anabolic reactions, 751
anabolic steroids, 572
anabolism, 713, 770–72
anaerobic metabolism, 762
anaerobic threshold, 727
analgesics, 478B, 496, 504
analytical balance, 24F
analytical chemistry, 3
anaplerotic reaction, 772
Andersen's disease, 740B
anesthetics, 328B, 405, 504
angiogenesis and angiogenesis inhibitors, 596B
angiostatin, 596B
angular molecular structure, 107
aniline, 242, 369, 494
animals. *See* birds; fish; reindeer; skunk
anions, 43, 66
anisole, 369
anode, 45, 268
anomers, 537
Antabuse, 430B
anthracene, 371, 372
antibiotics, 5B, 507B, 588–89B, 699
antibodies
    defense proteins, 597
    drug-carrying liposomes, 586B
    immune response, 616B
α₁-antichymotrypsin, 599B
anticodon, 682, 690
antifreeze, 389
antigens, 597, 618B
antihistamines, 489, 514, 524B
antiknock quality, of gasoline, 326B
antiparallel β-pleated sheet, 609, 610F
antiparallel strands, 672
antiport, 587
antipyretics, 478B
antiseptics, 264B, 402. *See also* disinfectants; sterilization
α-antitrypsin, 705B
apoenzyme, 643
aqueous solution, 178
arachidic acid, 559T
arachidonic acid, 559T, 564, 566
archaeology, and radiocarbon dating, 288
arginase, 768
arginine, 601F, 623T, 768, 770
argininosuccinate, 768
argon, and electron configuration, 63
Aristotle, 669
aromatic amines, 494
aromatic compounds, 366, 373
aromatic hydrocarbons, 308, 366–72
Arrhenius theory, of acids and bases, 241
artificial kidney, 201B
artificial radioactivity, 291
artificial sweeteners, 509
aryl group, 383
ascorbic acid, 644T
asparagine, 601F, 623T, 770
aspartame, 508, 509
aspartate, 601F, 623T, 764, 768
aspartate aminotransferase/serum glutamate-oxaloacetate transminase (AST/SGOT), 658B, 659B, 660
aspartate transaminase, 764
aspirin, 315, 478B, 565, 566–67
asthma, and air pollution, 257B
atherosclerosis, 575, 580, 794
atmosphere, and methane, 307B
atmosphere (atm), as unit of measurement, 154
atom. *See also* atomic theory
    definition of, 39
    mole concept, 119–23
    periodic table and size of, 68
    structure of, 39–44, 46–49
atomic mass, 41–42, 55–56
atomic mass unit (amu), 24, 41, 119
atomic number, 39, 55–56, 70F, 277
atomic orbital, 51, 61, 79
atomic theory. *See also* atom
    Bohr atom, 49–51
    development of, 44–46
    light and atomic structure, 46–49
    modern, 51–52
ATP
    aerobic respiration, 752
    carbohydrate metabolism, 713–15, 718
    citric acid cycle, 480
    energy metabolism, 748–49B
    energy transport, 590

enzymes, 633B, 634, 646, 650
  glycolysis, 438, 477–78, 719, 732
  oxidative phosphorylation, 757, 760–62
  ribonucleotide, 671F
ATP synthase, 747, 760–61
Aufbau principle, 62–64
autoimmune reaction, 618B
autoionization, 246
automobiles, and air bags, 118B
Avogadro, Amadeo, 153
Avogadro's law, 161–62
Avogadro's number, 119–20, 191
axial atoms, 327
AZT (zidovudine), 676–77B

B

Bacillus polymyxa, 588B
background radiation, 293–94
bacteria
  biodegradable plastics, 456B
  chromosomes, 673
  DNA replication, 677–81
  extreme temperatures, 630B, 647–48
  fermentation, 729B
  heat and sterilization, 647
  magnetotactic, 78B
  membrane lipids, 582
  saliva and tooth decay, 528B
  ultraviolet light and DNA damage, 694
balance(s), 24F
balanced ionic equation, 252
balanced nuclear equation, 279–82
balancing, of chemical equations, 132–36
Bangham, Alec, 586B
barbital, 506
barbiturates, 505
barium-131, 290
barium sulfate, 129
barometer, 153–54
base(s)
  amines, 498, 517
  Arrhenius theory, 241
  Brønsted-Lowry theory, 241
  buffers, 256–63
  chemical reactions, 252–55
  conjugate, 243–46
  dissociation of water, 246
  pH scale, 246–52
  properties of water, 242
  strength, 242–43
base pairs, 672
batteries
  human body as, 270B
  voltaic cells, 269
Becquerel, Henri, 276B
beeswax, 577
behavior modification, and weight loss, 785B
Benadryl, 514
Benckiser, Reckitt, 513B
bends, and scuba diving, 183B
Benedict's reagent, 429, 431, 541–43
benzaldehyde, 369, 425F
benzalkonium chloride, 502
Benzedrine, 496

benzenamine, 494
benzene, 308, 372, 374
benzene ring, 368
benzenesulfonic acid, 372, 374
benzodiazepines, 515
benzoic acid, 369, 428, 454, 458
benzoic anhydride, 473
benzopyrene, 371, 372
benzoyl chloride, 471
benzoyl peroxide, 264B
benzyl alcohol, 371
benzyl chloride, 371
beryllium, 58, 59, 63
beryllium hydride, 103–4
Berzelius, Jöns Jakob, 304
beta decay, 280
beta-oxidation pathway, 360
beta particles, 278, 279
big bang theory, 92B
bile, 779
bile salts, 575
bilirubin, 2B
binding energy, of nucleus, 282
binding site, of enzyme, 638
biochemistry, 3. *See also* organic chemistry
biodegradable plastics, 456–57B, 470
bioinformatics, 706
biological systems. *See also* birds; fish; human body; life; plants
  DDT and biological magnification, 340B
  disaccharides, 543–47
  effects of radiation on, 293–95
  genetics and information flow in, 681–85
  monosaccharides, 535–43
  most importance elements, 54T
  oxidation-reduction reactions, 267, 400–2
biopol, 457B
biosynthesis, 512–13B, 713T
biosynthetic intermediates, and citric acid cycle, 770–72
biotin, 644T
birds, and migration, 78B
birth control, 346B, 576
1,3-bisphosphoglycerate, 724
bleach and bleaching, 264B, 266–67
blimps, 152B
blood. *See also* hemoglobin; red blood cells
  alcohol levels, 401B
  buffer solutions, 256
  control of pH, 262B
  electrolytes, 200
  glucose levels, 543, 738B, 796, 797
  proteins, 599B
  specific gravity, 31
  urea levels, 660
blood clotting, and protaglandins, 565
blood group antigens, 546–47B
blood pressure
  calcium and regulation of, 67B
  sodium ion/potassium ion ratio, 94B
blood sugar, 429
blood transfusions, 546–47B
blood types. *See* blood group antigens
blood urea nitrogen (BUN), 660
B lymphocytes, 618B, 619B

boat conformation, 326–27
body battery, 270B
body fluids, electrolytes in, 200–202
body temperature. *See also* temperature
  brown fat and thermogenesis, 758–59B
  enzymes, 620
  of mammals, 582
Bohr, Niels, 46, 49
Bohr theory, 49–51, 60
boiling point
  alcohols, 384, 493T
  aldehydes and ketones, 417
  alkanes, 311T, 313T, 343T
  alkenes and alkynes, 343T
  amines, 491, 492T, 493T
  carboxylic acids, 449
  concentration of solutions, 192–93
  ethers, 403
  ionic and covalent compounds, 93
  liquids and solids, 112–13
  physical properties, 7, 9
  vapor pressure of liquids, 169–70
  of water, 198
bomb calorimeter, 217
bond angle, 105, 311, 342
bond energies, 100–101
bond order, 101B
bone
  calcium in diet, 67B
  nuclear medicine, 290
Borgia, Lucretia, 652
Borneo, and DDT, 340B
boron, 58, 59
boron-11, 280
boron trifluoride, 105–6
Boyle, Robert, 155, 387
Boyle's law, 155–57, 160, 233
brain
  blood glucose levels, 796
  opium poppy and peptides in, 604–5B
branching enzyme, 739
bread making, and yeast, 729B
breathalyzer test, 401B
breeder reactors, 287
bridging conversion unit, 15
Bright's disease, 31B
British Anti–Lewisite (BAL), 407
bromination, 358, 359F
bromobenzene, 369, 372
o-bromobenzoic acid, 454
2-bromobenzoic acid, 472
2-bromobenzoyl chloride, 472
β-bromocaproic acid, 453
bromochlorofluoromethane, 532
2-bromo-2-chloro-1,1,1-trifluoroethane, 328B
2-bromocyclohexanol, 386
2-bromo-3,3-dimethylpentane, 317
bromoethane, 361
2-bromohexane, 316
2-bromo-3-hexyne, 344
bromomethane, 330
1-bromo-4-methylhexane, 319
2-bromo-4-methylpentanoic acid, 451
2-bromopropane, 316, 361
3-bromopropanoyl chloride, 470

γ-bromovaleraldehyde, 421
Brønsted-Lowry theory, 241
Brookhaven National Laboratory (New York), 291
brown fat, 758–59B
bubonic plague, 340B
buckminsterfullerene, 305
bucky ball, 304
buffer(s), acid-base, 256–63
buffer capacity, 258
buffer solution, 259–61
Bunsen burner, 209
buprenorphine, 512B, 513B
buret, 253
butanal, 421T, 427
butanamine, 492
butane
   boiling point, 384, 403, 417, 449
   major properties of, 306T
   molecular formula, 310, 312F
   staggered conformation, 325F
butane gas, 134–35
butanoic acid, 464B
butanoic anhydride, 475
1-butanol, 467
butanone, 422, 424
butanoyl chloride, 470, 472
1-butene and 2-butene, 348
butylated hydroxytoluene (BHT), 402
butyl group, 314T, 315T
1-butyne and 2-butyne, 343T
butyric acid, 448B, 449, 453T, 464B
butyric acid, butanol, acetone fermentation, 729B
butyryl ACP, 792
butyryl chloride, 472

C

calcium, 67B, 281
calcium carbonate, 127, 128
calcium hydroxide, 87, 124, 140–41
calcium hypochlorite, 264B
calcium oxide, 86
calcium phosphate, 126
calories, 27, 28B, 217, 785B. *See also* kilocalorie
calorimeter, 215
calorimetry, 209
cancer. *See also* carcinogens
   angiogenesis inhibitors, 596B
   calcium in diet, 67B
   mutagens and carcinogens, 693
   Pap smear test, 2B
   radiation therapy, 49B, 289
   sun exposure and skin cancer, 432–33B
Candida albicans, 712B
capillin, 346B, 347B
capric acid, 453T, 559T, 561
caproic acid, 453T
caprylic acid, 453T
cap structure, 684
carbamoyl phosphate, 766
carbinol carbon, 389

carbohydrate(s). *See also* carbohydrate metabolism
   disaccharides, 543–47
   monosaccharides, 527–29, 535–43
   polysaccharides, 548–51
   stereoisomers and stereochemistry, 529–34
   types of, 526–27
carbohydrate metabolism
   ATP, 713–15
   catabolic processes, 715–17
   fermentations, 726–29
   gluconeogenesis, 730–32
   glycogen synthesis and degradation, 733–41
   insulin and glucagon, 797
   pentose phosphate pathway, 730
   regulation, 793–97
carbolic acid, 402
carbon
   aldehydes, 420, 421
   alkyl groups, 313
   atomic mass, 42–43
   chain length and nomenclature, 316T
   chiral carbon, 530
   isotopes, 277, 278F, 280
   organic chemistry, 305–9
   radiocarbon dating, 288
   valence electrons, 58, 59
carbon-14
carbonated beverages, 177F, 181
carbonate ion, and Lewis structure, 98–99
carbon dioxide
   alcohol fermentation, 728B
   Avogadro's law, 162
   blood gases, 168B, 262B
   carbonated beverages, 177F, 181
   covalent compounds, 90
   greenhouse effect and global warming, 166B, 167, 329
   Lewis structure, 95–96
carbonic acid, 242
carbon monoxide, 90, 136B
carbon skeletons, of amino acids, 766
carbon tetrachloride, 110
carboxylases, 635
carboxyl group, 449
carboxylic acid(s). *See also* carboxylic acid derivatives
   aldehydes, 420
   chemical reactions, 457–60
   common functional groups, 309T
   esterification, 460
   foods and food industry, 455–56
   fragrance, 464B
   nomenclature, 420, 451–55
   oxidation and reduction reactions, 400
   physical properties, 449–50
   weak electrolytes, 306B
carboxylic acid derivatives, 461, 470, 478–79B
carboxypeptidase A, 651T
carcinogens, 693, 694–95B. *See also* cancer
cardiotonic steroids, 574B
cardiovascular disease, and nuclear medicine, 290

carnitine, 786
carnitine acyltransferase, 786
Carroll, Lewis, 524B
Carson, Rachel, 340B
carvone, 524B
casein, 597
catabolism, 713, 715–18, 770
catalase, 631
catalyst
   equilibrium composition, 234
   reaction rate, 222–23
catalytic cracking, 326B
catalytic groups, 638
catalytic reforming, 326B
catecholamines, 511
cathode, 45, 268
cathode rays, 45
cations, 43–44, 65, 200
cause-and-effect relationship, and mental illness, 178B
cell membrane
   antibiotics, 588–89B
   lipids, 557, 581–90
   osmotic pressure, 196
cellular metabolism
   acetyl CoA, 751
   insulin and glucagon, 797–98
cellulase, 549
cellulose, 549, 550F, 551
Celsius scale, 25, 157B
Centers for Disease Control, 264B, 430B, 641B
central dogma, of molecular biology, 681
cephalin, 568, 569F
ceramide, 571
cerebrosides, 571
Cerezyme, 661
cetyl palmitate, 577
cetylpyridinium chloride, 502
Chadwick, James, 45
chain elongation, 683, 690
chain reaction, 286
chair conformation, 325, 326F
Challenger (space shuttle), 100–101
champagne, and wine making, 728B
Chargaff, Irwin, 671
charge, of subatomic particles, 39T
Charles, Jacques, 157
Charles's law, 157–59, 160
cheese, and fermentation, 729B
chemical bonding. *See also* covalent bonding; hydrogen bonding; ionic bonding
   definition of, 78
   electronegativity, 83, 84T
   Lewis symbols, 79
   principal types of, 79–83
chemical change, 8, 126
chemical compounds. *See also* aromatic compounds; covalent compounds; inorganic compounds; ionic compounds; organic compounds
   definition of, 9
   naming and writing formulas for, 84–92
chemical control, of microbes, 264B

chemical equations. *See also* nuclear equations; rate equations
 balancing, 132–36
 calculations using, 136–45
 chemical change, 126
 chemical reactions, 128–31
 experimental basis of, 127–28
 features of, 127
chemical equilibrium, 227
chemical formulas. *See also* condensed formula; molecular formula; structural formula
 chemical compounds, 84–92
 formula weight and molar mass, 123–26
chemical properties. *See also* physical properties
 matter, 8
 periodic table, 53
chemical reactions, 8. *See also* addition reactions; reaction rates; reversible reaction; substitution reactions
 acids and bases, 252–55
 alcohols, 391–400
 aldehydes and ketones, 424–42
 alkanes and cycloalkanes, 327–32
 alkenes and alkynes, 354–65, 374
 amides, 506–9, 517
 amines, 497–502, 517
 benzene, 372
 carboxylic acids, 457–60
 citric acid cycle, 753–56
 energy, 209, 215–18
 enzymes, 636–43
 esters, 462–66
 fatty acids, 561–64
 glycolysis, 721, 723–25
 kinetics, 218–25
 β-oxidation, 786–88
 types of, 129–131
 urea cycle, 766–69
 writing, 128–29
chemical warfare, 407
chemistry. *See also* organic chemistry, pharmaceutical chemistry
 definition of, 3
 measurement in, 11–16
 models in, 5–6
 scientific method, 3–5
chenodeoxycholate, 575, 779
chiral carbon, 530
chiral molecules, 529
chlorine, 41, 263, 264B
chlorine gas, 264B
chlorobenzene, 372
4-chlorobenzoic acid, 455
4-chlorobenzoyl chloride, 470
2-chloro-2-butene, 344
chloroethane, 328B
chloroform, 328B, 331B
chloromethane, 328B
3-chloro-4-methyl-3-hexene, 344
2-chloropentane, 363
γ-chlorovaleric acid, 453
cholate, 575, 779

cholera, 197B
cholesterol, 575
choline, 502
chondroitin sulfate, 550B, 551B
chorionic villus sampling, 668B
chromic acid, 397
chromosome(s), 673–74, 675F
chromosome walking, 704
chylomicrons, 578–79, 780, 781
chymotrypsin, 622, 651T, 656–57
Cicuta maculata, 346B, 347F
cicutoxin, 346–47B
cigarette smoking. *See also* nicotine
 carcinogens and cancer, 693, 694B
 emphysema, 648B
 nicotine patch, 488B
cimetidine, 373
cinnamaldehyde, 425F
cis-trans isomers, 322–25, 347, 348. *See also* geometric isomers
citral, 425F
citrate, 754F, 756, 757F
citrate lyase, 632
citric acid, 455, 456
citric acid cycle
 aerobic respiration, 752, 753–56
 anabolism and biosynthetic intermediates, 770–72
 conversion of pyruvate to acetyl CoA, 750
 energy metabolism, 761
 hydration, 359
 oxidation and reduction reactions, 401–2, 717
 regulation of, 756, 757F
citric synthase, 753
citrulline, 766, 767F, 768
climate, and effect of water in environment, 27B
cloning and cloning vectors, 697, 699, 700F
Clostridium perfringens, 464B
coagulation, of proteins, 619
coal, and acid rain, 257B
cocaine
 amines, 499, 500F, 501–2, 503, 504
 DNA fingerprinting of coca plants, 702–3B
 drug abuse, 499, 500F, 501–2, 703B
 medical uses, 504
codeine, 503, 504, 512B
codons, 682, 686, 687F
coefficients
 chemical equation, 132
 equilibrium-constant expressions, 228
coenzyme(s). *See also* acetyl CoA
 citric acid cycle, 480
 functions of, 643–46
 nomenclature, 635
 oxidation and reduction reactions, 400
 thioesters and thioester bond, 407, 479
coenzyme A, 750
cofactors, of enzymes, 643–46
cold packs, 221B
Coleman, Douglas, 778B
colipase, 781
collagen, 551B, 612–13B

colligative properties, 191
colloid(s), 179–80
colloidal suspension, 179–80
colon cancer, 67B
colony blot hybridization, 701F
color(s)
 electromagnetic spectrum, 47
 pH indicators, 253F
colorless solutions, 179
combination reactions, 128
combined gas law, 160–61
combustion, 208B, 265–66, 327–29
comfrey, 574B
common names. *See also* nomenclature
 alcohols, 386
 aldehydes and ketones, 419–22
 nomenclature, 86, 87T, 91
competitive inhibition, 426B, 652–53
complementary strands, 672
complete protein, 623
complex carbohydrates, 526
complex lipids, 577–80
compounds. *See* chemical compounds
compressibility, of liquids, 167
computer(s), and information management, 38B
computer-activated tomography, 49B, 296
computer imaging, and diagnostic medicine, 296
concentration
 acid and base strength, 242
 properties of solutions, 191–96
 rate of chemical reaction, 222, 232
 of solutions based on mass, 182–87
 of solutions in moles and equivalent, 187–91
 systems of measurement, 27–28
condensation, 169
condensation polymers, 469–70
condensation reaction, 640, 642F
condensed formula
 alkanes, 310, 311T, 312, 313T, 342
 alkenes and alkynes, 342
conductivity, of metallic solids, 173
conformations, of alkanes and cycloalkanes, 325–27
conjugate acid, 243–46
conjugate acid-base pair, 243, 245F
conjugate base, 243–46
conservation of mass, law of, 126, 132, 141F
constitutional isomers, 319–21
conversion and conversion factors
 atoms and moles, 122
 chemical equations, 137–41
 mass, 24
 systems of measurement, 13, 15
 temperature, 26
copper
 fireworks, 52B
 as metallic solid, 173
 molar mass, 120F
 oxidation-reaction reactions, 131
 voltaic cells, 267–69
 Wilson's disease, 57B

copper sulfate, 129
copper(II) sulfate pentahydrate, 124
Cori Cycle, 727, 732
Cori's disease, 740B
corrosion, as chemical reaction, 131, 265
cortisone, 576
cosmetics industry, and liposomes, 586B
cost, of energy, 208B
covalent bonding. *See also* polar
    covalent bond
  amino acids, 610
  definition of, 79
  organic compounds, 305, 306
  process and examples of, 81–82
covalent compounds
  definition of, 82
  naming of, 89–92
  properties of, 92–93
  resonance hybrids, 102–3
covalent solids, 172
crack cocaine, 499, 500F, 501
C-reactive protein (CRP), 580
creatine kinase, 748B
creatine phosphate, 748B
creatine phosphokinase (CPK), 658B, 659B
crenation, 196
Crick, Francis, 669, 671, 686
crime, and DNA fingerprinting, 703B
cristae, 747
Crookes, William, 45
crotonyl ACP, 792
crystal lattice, 81
crystalline solids, 93, 171, 172–73
C-terminal amino acid, 602
CT scanner, 49B, 296
curie, as unit of measurement, 297
Curie, Marie & Pierre, 276B, 295B
curiosity, in science and medicine, 5B
curium–245, 281
cyanide, 761B
cyanocobalamin, 644T
cyclic AMP (cAMP), 604–5B
cycloalkanes, 308, 321–32, 347
cyclobutane, 321F
cyclohexane, 321F
cyclohexanecarboxylic acid, 452
cyclohexanol, 391
cyclohexene, 391
cyclooxygenase, 567
cyclopropane, 321F
cysteine, 407, 601F, 610, 623T
cystine, 610
cytosine, 670F, 672F

D

Dalton, John, 44, 209
Dalton's law of partial pressures, 166–67
Dalton's theory, 44
data, 11
DDT, and biological magnification, 340B
debranching enzyme, 733
decanoic acid, 562

decomposition reactions, 128
defense proteins, 597
degenerate code, 686
degradation
  of amino acids, 762–66
  of fatty acids, 782–89, 791–93
  of glycogen, 733–41
dehydration
  alcohols, 394–96, 409
  esters, 463
  ethers, 405, 409
dehydrogenases, 635
dehydrogenation reaction, 786
deletion mutations, 692
α-demascone, 425F
Demerol, 496
denaturation
  of enzymes, 646
  of proteins, 617–22
denatured alcohol, 387
density
  definition of, 28
  of gas, 163
  measurement, 29–30
dental fillings, and electrochemical
    reactions, 266B
dental plaque, 528B
deoxyribonucleotide, 670
deoxyribose, 541, 670F
detergents, 621, 780
deuterium, 92B, 277, 287
dextran, 528B
dextrorotatory, 531
diabetes mellitus
  Benedict's reagent, 542
  blood glucose levels, 738B
  electrolytes, 200
  hemodialysis, 201B
  ketosis, 789
  sucrose in diet, 382B
  urine tests, 31B
diamond, 172, 305
diarrhea, and electrolytes in body fluids, 200.
    See also oral rehydration therapy
diatomic compound, 82B
2,3-dibromobutane, 357
cis-1,2-dibromocyclopentane, 324
2,5-dibromohexane, 316
dibromomethane, 330
1,2-dibromopentane, 358
1,2-dichlorocyclohexane, 323
1,2-dichloroethene, 348
dideoxyadenosine triphosphate (ddA), 705
diet. *See also* estimated safe and adequate
    daily dietary intake; foods and
    food industry
  atherosclerosis, 580
  carbohydrates, 526
  calcium, 67B
  cholesterol, 575
  copper deficiency, 57B
  fiber, 526
  lipids, 556

  proteins, 622–23
  sodium and potassium, 94B
  sucrose, 545
  weight loss, 28B, 784–85B
diethyl ether, 404, 405
differentially permeable membranes, 194F
diffusion
  of gases, 155
  membrane transport, 584–85
digestion and digestive tract
  cellulose, 549, 551
  dietary triglycerides, 779–81
  hydrolysis, 715–17
  proenzymes, 651T
  proteins, 622–23
digitalis, 574B
Digitalis purpurea, 574B
digitoxin, 574B
diglycerides, 567
dihydroxyacetone (DHA), 433B, 439B
dihydroxyacetone phosphate, 439B, 634, 723
diisopropyl fluorophosphate (DIFP), 516
dilution, of concentrated solutions, 188–90
dimensional analysis, 13
dimethylamine, 494
2,3-dimethylbutane, 313T
dimethyl ether, 108, 405
1,1-dimethylethyl, 315T
2,2-dimethyl-3-hexyne, 344
N,N-dimethylmethanamine, 490, 491, 492,
    493, 495T
2,6-dimethyl-3-octene, 344
cis-3,4-dimethyl-3-octene, 349
2,4-dimethylpentanal, 420
2,2-dimethylpropanal, 397
dinitrogen monoxide, 90
dinitrogen tetroxide, 90, 91
dinucleotide diphosphokinase, 755
dipeptide, 602
dipole, 109
dipole-dipole attraction, 417
dipole-dipole interactions, 170
directional covalent bond, 105
disaccharides, 436, 527, 543–47
disease, as chemical system, 128. *See also*
    AIDS; cancer; cardiovascular
    disease; cholera; diabetes mellitus;
    genetic disorders; heart disease;
    medicine
disinfectants, 264B, 402. *See also* antiseptics;
    sterilization
disorder, in chemical systems, 211
dissociation
  polyprotic substances, 255
  solutions of ionic and covalent
    compounds, 93
  strong acids, 242, 244
  water, 246
distillation, 387B
disulfide bond, 407, 408F
disulfiram, 430B
divergent evolution, and enzymes, 656
division, of significant figures, 21–22

DNA (deoxyribonucleic acid)
  cloning, 700F
  damage repair, 695–96
  nucleotides, 669
  purines and pyrimidines, 373, 525
  replication, 674–81
  structure of, 541, 671–74
  ultraviolet light, 694–95
DNA cloning vectors, 697, 699
DNA fingerprinting, 702–3B
DNA helicase, 679F
DNA ligase, 633, 679F
DNA polymerase I, 679F, 680
DNA polymerase III, 678–79, 680, 681F
DNA sequencing, 704–6
cis-7-dodecenyl acetate, 479B
Domagk, Gerhard, 653
dopamine, 500, 501, 511
d orbitals, 62
double bond, 96, 101
double helix, of DNA, 671–72, 673F
double-replacement reaction, 129
drug(s). *See also* antibiotics; pharmaceutical chemistry
  HIV protease inhibitors, 641B
  mechanisms of delivery, 240B, 586B
  pharmaceutical chemistry, 144B
drug abuse
  cocaine, 499, 500F, 501–2
  dopamine levels, 511
  methamphetamine, 496, 500–501B
  opiates, 513B, 605B
dry heat, and sterilization, 647
DuPont Corporation, 364B
Duve, Christian de, 647
dynamic equilibrium, 181, 226–27

E

EcoR1, 696
effective collision, 219
eicosanoids, 564–67
Einstein, Albert, 284–85
elaidic acid, 351
elastase, 622, 651T, 656–57
electrochemical reactions, 266B
electrolysis, 269–70, 271F
electrolytes, 93, 179, 200–202
electrolytic solution, 93
electromagnetic radiation, 47–49
electromagnetic spectrum, 47
electron(s)
  atomic theory, 45
  composition of atom, 39–40
  properties based on structure of, 111–13
electron acceptor, 80
electron affinity, 70
electron arrangement, and periodic table, 56, 58–65
electron carriers, and fatty acid synthesis, 792
electron configuration, 56, 58–65
electron density, and atomic theory, 51

electron donor, 80
electronegativity, 83, 84T
electronic balance, 24F
electronic transitions, and Bohr atom, 50
electron pairs, 95
electron spin, 62
electron transport system, 747, 760
electrostatic force, 80
elements. *See also* periodic table
  biological systems, 54T
  definition of, 9
elimination reaction, 394
elongation factors, 690
Embden-Meyerhof Pathway, 719
emission spectrum, 48, 50F, 51
emphysema, 648B, 705B
emulsification, of lipids, 780
emulsifying agent, 568
emulsion, 468
enantiomers, 524B, 529–30
endosomes, 579
endostatin, 596B
endothermic reactions, 209–10, 214
enediol reaction, 542
energy. *See also* energy metabolism; fossil fuels; petroleum industry
  chemical reactions, 209, 215–18
  cost of, 208B
  definition of, 3
  equilibrium, 226–35
  food calories, 28
  kinetics, 218–25
  lipids, 557, 568
  membrane transport, 590
  systems of measurement, 27
  thermodynamics, 208–14
energy levels, and Bohr atom, 49
energy metabolism
  anabolism and citric acid cycle, 770–72
  degradation of amino acids, 762–66
  exercise, 748–49B
  oxidative metabolism, 757, 760–62
  urea cycle, 766–69
English system, of measurement, 11–12, 13, 14T, 15, 154
enkephalins, 604B, 605B
enolase, 725
enol form, 361, 436, 437, 442
enoyl-CoA hydrase, 787
enthalpy, 211
enthrane, 405
entropy, 211–13
environment. *See also* air pollution; climate, and effect of water in environment; global warming; greenhouse effect
  acid rain, 256B
  biodegradable plastics and waste disposal, 456–57B
  DDT and biological magnification, 340B
  enzymes, 646–48
  frozen methane, 307B
  nuclear waste disposal, 285B
  oil-eating microbes, 317B

petroleum industry and gasoline production, 326B
plastic recycling, 366–67B
enzyme(s). *See also* allosteric enzymes; coenzymes
  activation reaction of, 636
  body temperature, 620
  cellular functions, 597
  classification, 631–35
  cofactors, 643–46
  definition of, 630
  environmental effects, 646–48
  extreme temperatures, 630B, 647–48
  fatty acid biosynthesis, 792
  inhibition of activity, 652–56
  medical uses, 660–61
  myocardial infarction, 658–59B
  nerve transmission and nerve agents, 654–55B
  nomenclature, 635–36
  oxidation-reduction reactions, 400
  proteolytic enzymes, 656–57
  regulation of activity, 649–52
  specificity, 630–31, 639
  stereospecific, 524B
  substrate concentration and catalyzed reactions, 637
  transition state and product formation, 639–43
enzyme assays, 660
enzyme inhibitors, 652–56
enzyme replacement therapy, 660–61, 722B
enzyme-substrate complex, 637–38, 639
EPA (Environmental Protection Agency), 317B
ephedra, 496–97
ephedrine, 489, 496, 501B
epinephrine, 511, 631
equatorial atoms, 327
equilibrium
  energy, 226–35
  solubility, 181
equilibrium constant, 227–31
equilibrium reactions, 143, 226
equivalence point, 254
equivalent, and concentration of solutions, 191
error, 19–20
erythrulose, 433B
Escherichia coli, 676, 694, 696
essential amino acids, 622, 623, 770B
essential fatty acids, 564
ester(s), 309T
  chemical reactions, 462–66
  common functional groups, 309T
  hydrolysis, 466–68
  nomenclature, 461–62
  physical properties, 461
esterification, 460, 463, 561
estimated safe and adequate daily dietary intake (ESADDI), 57B, 94B
estrone, 576
ethanal, 400, 419, 421T, 424

ethanamide, 505, 506T
ethanamine, 491, 492T, 493T, 495T
ethane
  bond angle, 342
  molecular formula, 310, 312F
  nomenclature, 343
  staggered conformation, 325F
1,2-ethanediol, 386, 389
ethanoate anion, 428
ethanoic acid
  boiling point, 449
  chemical reactions, 457
  hydrolysis, 467, 472
  nomenclature, 451
  oxidation, 400
ethanoic anhydride, 473, 474
ethanoic propanoic anhydride, 474
ethanol
  alcohol fermentation, 728
  boiling point, 491, 493T
  breathalyzer test, 401B
  chemical equations, 140
  functional groups, 309
  hydration, 358
  medical use, 387
  molecular structure, 389
  nomenclature, 386
  oxidation-reduction reactions, 267
  rate equations, 225
ethanoyl chloride, 470, 471
ethene
  bond angle, 342
  dehydration, 394
  nomenclature, 343
  physical properties, 343T
  plants, 352
ethers, 309T, 403–5, 409
ethyl acetate, 461
ethylamine, 494
ethylbenzene, 369
ethyl butanoate, 463, 464B
2-ethyl-1-butanol, 390
ethyl dodecanoate, 562
ethylene, 193, 362
ethyl group, 314T
5-ethylheptanal, 420
3-ethyl-3-hexene, 350
ethylmethylamine, 494
4-ethyl-3-methyloctane, 317
4-ethyloctane, 316
6-ethyl-2-octanone, 422
ethyl pentyl ketone, 423
ethyl propanoate, 466
ethyne, 342, 343
17-ethynylestradiol, 346B
eukaryotes, and genetics, 673–74, 675F, 681, 684F, 685, 700F
evaporation, 169
evolution, and enzymes, 656
exact numbers, 22
excited state, of atom, 49, 50
exercise
  anaerobic threshold, 727
  energy metabolism, 748–49B
  weight loss, 28B, 568, 785B

exercise intolerance, 722B
exons, 684–85
EXO-SURF Neonatal, 556B
exothermic reactions, 209–10
expanded octet, 103
experimental basis, of chemical equation, 127–28
experimental quantities, 23–31
experimentation, and scientific method, 4
exponential notation, 18
extensive property, 9
Exxon Valdez (ship), 317B
eye, and chemistry of vision, 440–41B

F

Fabry's disease, 573B
facilitated diffusion, 585–87
factor-label method, 13
Fahrenheit scale, 25–26, 157B
falcarinol, 346B, 347B
familial emphysema, 648B, 705B
farnesol, 352, 353F
fast twitch muscle fibers, 749B
fat(s)
  brown fat, 758–59B
  digestion of, 716–17
fatty acid(s)
  carboxylic acids, 341F, 449, 455
  chemical reactions, 561–64
  degradation, 782–89, 791–93
  eicosanoids, 564–67
  geometric isomers, 351
  membrane phospholipids, 582
  structure and properties, 558–60
  synthesis, 791–93
fatty acid metabolism, 789–91. *See also* lipid metabolism
fatty acid synthase, 792
feedback inhibition
  of enzymes, 650–51
  of glycolysis, 726
feedforward activation, 726
fermentation, 383, 387, 456, 726–29
fetal alcohol syndrome (FAS), 388B, 401B
fetal hemoglobin, 616
F_0F_1 complex, 760
fibrils, 549
fibrinogen, 599B, 621
fibrous proteins, 608
filling order, for electrons in atoms, 63F, 64
film badges, 296
fireworks, 52B
Fischer, Emil, 524–25, 532, 534, 638
Fischer, Hermann Otto Laurenz, 525
Fischer projection formulas, 532–33
fish
  acid rain, 256B
  migration, 78B
fission, nuclear, 286
flavin adenine dinucleotide (FAD), 645, 646, 755, 761
flavors, and esters, 464–65B
Fleming, Alexander, 4B, 507B
Flotte, Terry, 705B

fluid mosaic structure, of cell membrane, 582–83
fluorine, 58T, 59, 66, 171
3-fluoro-2,4-dimethylexane, 317
folic acid, 644T, 652–53
folklore, and ethylene, 352, 362B
Food and Drug Administration (FDA), 496, 506, 509
foods and food industry. *See also* diet
  calories, 28B
  carboxylic acids, 455–56
  copper, 57T
  esters, 461
  ethylene and fruit ripening, 352, 362B
  fuel value, 217, 218
  hydrogenation, 356
  potassium, 94B
  sodium, 94B
  sugars and sugar substitutes, 382B
f orbitals, 62
forests, and acid rain, 256B
formaldehyde, 421T, 426B. *See also* methanal
formalin, 424
formic acid, 449, 453T
formula unit, 124
formula weight, 124–26
fossil fuels. *See also* oil spills; petroleum industry
  acid rain, 256B, 257B
  combustion, 265–66
  global warming, 329
  nonrenewable resources, 304B
foxglove plant, 574B
fragrances, and esters, 464–65B
Franklin, Rosalind, 671
free energy, 214
free rotation, of carbon–carbon single bond, 325
freeze–fracture, 582
freezing point, and concentration of solutions, 192–93
Friedman, Jeffrey, 778B
fructose, 539–40, 640, 642F
β-fructose, 545
D-fructose, 529
fructose-1,6-biphosphatase, 731
fructose-1,6-bisphosphate, 439, 650, 723
fructose-6-phosphate, 650, 721, 723
fruit flavors, 464B
fucose, 546B, 547B
fuel value, of foods, 217, 218
Fuller, Buckminster, 305
fumarase, 632, 756
fumarate, 754F, 757F, 768
functional group, 308–9
furan, 373
fused rings, 574, 575
fusion, nuclear, 287
fusion reactions, 92B

G

GABA, 515
galactocerebroside, 571
galactose, 540–41, 546B, 547B

galactosemia, 544–45
gallium-67, 291
gallon, units of measurement, 11
gamma production, 280–81
gamma radiation, 48B, 278–79, 280, 289
gamma rays, 278
Gane, R., 362B
gangliosides, 572
gas(es)
    Avogadro's law, 161–62
    blood and respiration, 168B
    Boyle's law, 155–57
    Charles's law, 157–59
    combined gas law, 160–61
    Dalton's law of partial pressures, 166–67
    densities, 163
    ideal gas concept, 153
    ideal gases versus real gases, 167
    ideal gas law, 163–65
    kinetic molecular theory, 154–55
    measurement of, 153–54
    molar volume, 163
    solubility of, 181–82
    states of matter, 7
gas gangrene, 464B, 729B
gasoline, petroleum industry and production of, 326B
gastrointestinal tract, and prostaglandins, 565. *See also* digestion
Gaucher's disease, 573B, 660–61
GDP molecule, 758B, 759B
Geiger, Hans, 45–46
Geiger counter, 296
gene therapy, 668B, 722B
genetic(s). *See also* DNA; genome; molecular genetics; RNA
    DNA repair, 695–96
    DNA replication, 674–81
    genetic code, 685–86, 687F
    information flow in biological systems, 681–85
    mitochondria, 746B
    mutations, 692–93
    polymerase chain reaction, 701–702
    primary amino acid sequences, 606
    protein synthesis, 687–92
    recombinant DNA, 696–701
    structure of DNA and RNA, 671–74
    structure of nucleotide, 669–71
    ultraviolet light damage, 694–95
genetic code, 685–86, 687F
genetic counseling, 668B
genetic disorders
    energy metabolism, 746B
    familial emphysema, 648B, 705B
    glycolysis, 722B
    molecular genetics and detection of, 668B
    obesity, 778B
    pyruvate carboxylase deficiency, 769B
    sickle cell anemia, 448B, 617, 693
    sphingolipid metabolism, 573B
genetic engineering, 699, 701. *See also* recombinant DNA technology
genome, 673, 701, 703–6. *See also* Human Genome Project

genome analysis, 704
genomic library, 704
Genzyme Corporation, 661
geometric isomers, 322, 346–52. *See also cis-trans* isomers
geraniol, 352, 353F, 405, 406F
Glaxo-Wellcome Company, 556B
global warming, 131, 166B, 329. *See also* greenhouse effect
globular proteins, 609–11
α-globulins, β-globulins, and γ-globulins, 599B
glucagon, 794B, 797–98
glucocerebrosidase, 573B, 660, 661
glucocerebroside, 571, 660, 661
glucokinase, 736
gluconeogenesis, 439, 730–32, 793
D-glucosamine, 550B, 551B
glucose
    Benedict's reagent, 429, 542–43
    biological systems, 535–39
    blood levels, 543, 738B, 796, 797
    calculating molarity from mass, 187–88
    condensation reaction, 640, 642F
    conformation of cycloalkanes, 325
    dehydration, 396
    facilitated diffusion, 586–87
    gluconeogenesis, 730–32
    glycolysis, 721
    Haworth projection, 538–39
    osmosis, 194
    urine levels, 542–43
    weight/volume percent, 183
D-glucose, 529, 586
L-glucose, 586
glucose-6-phosphatase, 731
glucose-6-phosphate, 721, 740B
glucose tolerance test, 738B
glucosuria, 542
glucosyl transferase, 528B
α-D-glucuronate, 550B
glutamate, 516–17, 601F, 623T, 764, 770
glutamate dehydrogenase, 764
glutamic acid, 693
glutamine, 601F, 623T
glyceraldehyde, 530, 533
D-glyceraldehyde, 529, 531F, 533, 536, 598
L-glyceraldehyde, 531F, 533, 598
glyceraldehyde-3-phosphate, 634, 723
glyceraldehyde-3-phosphate dehydrogenase, 724
glycerides, 567–70
glycerol, 167
glycerol-3-phosphate, 796
glycine, 515, 601F, 602, 612B, 623T
glycogen, 548–49, 733–41
glycogen granules, 733
glycogenolysis, 733–41, 793
glycogen phosphorylase, 652, 733, 735F
glycogen storage diseases, 740B
glycolysis
    aldol condensation, 439
    ATP, 714
    chemical reactions, 721, 723–25
    enzymes, 634

genetic disorders, 722B
    hydroxyl group, 383
    overview of, 719–21
    phosphoenolpyruvate, 438
    phosphoesters, 476–77
    regulation of, 726
glycoproteins, 583, 612
glycosaminoglycans, 550B
glycosidase, 733
glycosides, 543
glycosidic bonds, 436, 543, 545F
glycosphingolipids, 571
glycyl-alanine, 602
gold, 120F, 291
Goldstein, Eugene, 45
Gore-Tex, 364B
gram, 12, 137–41
gravity, 23
Greek language, and nomenclature, 91
greenhouse effect, 131, 166B, 167, 266, 307B. *See also* global warming
ground state, of Bohr atom, 50–51
group(s), and periodic table, 55
Group A elements, 55
Group B elements, 55
Group IA elements, 55, 59
Group IIA elements, 55, 59
Group IIIA elements, 59
Group VIIA elements, 55, 59
Group VIIIA elements, 55
group specificity, of enzyme, 639
guanine, 670F, 672F
guanosine triphosphate (GTP), 732

H

Haber process, 223F, 233F
Halaas, Jeff, 778B
half cells, 268
half-life, of radioactive isotopes, 282–84, 285B, 293–94
halide, 309T, 330
haloalkane, 329
halobenzene, 374
halogen(s), 55
halogenated ethers, 405
halogenation, 329–30, 356–58, 374
halothane, 328B
hangovers, and alcoholic beverages, 400
hard water, 563
Harrison Act (1914), 604B
Haworth projection, 538–39
heartburn, 514–15
heart disease, 94B, 270B, 574B, 658–59B, 660
heavy metals, 622
heavy water, 277
helicase, 678
helium
    combined gas law, 161
    density, 163
    electron configuration, 63
    molar volume, 164
    nuclear fusion, 287
    origin of elements, 92B
    valence electrons, 58T, 59

α-helix, of amino acid, 607–9
heme group, 615, 616F
hemiacetal
　addition reactions, 434, 435, 436, 442
　glucose, 537
　keto-enol tautomers, 437
hemiketal, 435, 436, 441, 540
hemodialysis, 201B
hemoglobin
　blood pH, 262B
　carbon monoxide poisoning, 136B
　human genetic diseases, 448B
　oxygen transport, 597, 616
　sickle cell anemia, 617
hemolysis, 196
Henderson-Hasselbalch equation, 261–63
Henry's law, 181–82
heparin, 551B
2,4-heptadiene, 344
heroin, 503, 504, 604–5B
heterocyclic amines, 488B, 502–4
heterocyclic aromatic compounds, 373
heterogeneous mixture, 10
heteropolymers, 457B, 469
heteropolysaccharides, 550–51B
hexachlorophene, 403
2,4-hexadiene, 344
hexane, 312, 313T
2-hexanone, 427
hexokinase, 634, 721, 726
hexose, 527
hexylresorcinal, 403
high-density lipoproteins (HDL), 578, 579F, 580
high-density polyethylene (HDPE), 366B, 367B
high-energy bond, 407B, 714
high entropy, 211
Hindenburg (airship), 152B
histamine, 489, 514–15
histidine, 601F, 623T
HIV (human immunodeficiency virus), 264B, 641B, 676–77B
HMG-CoA (β-hydroxy-β-methylglutaryl CoA), 790–91
holoenzyme, 643
homogeneous mixture, 10
homopolymer, 469
homostatic mechanisms, 758
hormones. *See also* insulin; steroids
　cellular metabolism, 798
　definition of, 564B
　lipids, 557
hot-air balloon, 159
hot packs, 221B
hot springs, in Yellowstone National Park, 582, 629F, 648, 701
Hughes, John, 604B
human body. *See also* biological systems; bone; brain; digestion and digestive tract; infants; medicine; reproductive system; respiratory tract
　chromosomes, 673
　genome, 701
　lipids, 557
　magnetite in brain of, 78B

Human Genome Project (HGP), 703–6
hyaluronic acid, 551B
hybridization, and recombinant DNA, 697, 698F, 699, 701F
hydrates, 124, 128
hydration, 358–61, 374, 391
hydrocarbon(s), 266, 308, 311
hydrocarbon tails, of membrane phospholipids, 581
hydrochloric acid
　Arrhenius and Brønsted–Lowry theories, 241
　calculation of pH, 248
　concentration of solution, 253–54
　energy involved in calorimeter reactions, 215–16
　hydrohalogenation, 363
　molarity of solutions, 189, 190
　neutralization, 252
　replacement reactions, 129
　as strong acid, 242, 244
hydrogen
　acid-base reactions, 131
　airships and blimps, 152B
　balancing of chemical equations, 134
　chemical equilibrium, 227
　chemical formula, 124
　covalent bonding, 81–82
　electron configuration, 63
　emission spectrum, 50F, 51
　Lewis structure and stability, 100–101
　molar mass, 120
　nuclear fusion, 287
　origin of elements, 92B
　polarity, 109
　valence electrons, 58, 59
hydrogenases, 635
hydrogenation
　alcohols, 392–93
　aldehydes and ketones, 431
　alkenes and alkynes, 354–56, 374
　catalyst, 222
　fatty acids, 564
hydrogen bonding
　alcohols, 384
　aldehydes and ketones, 418F
　amides, 505F
　amines, 491F
　amino acids, 600B, 610
　carboxylic acids, 450F
　liquids, 170–71, 173F
hydrogen chloride, 128, 132, 133
hydrogen fluoride, 82, 110, 171
hydrogen iodide, 231, 234
hydrogen ion gradient, 760
hydrogen peroxide, 132B, 264B
hydrohalogenation, 361–63, 374
hydrolases, 632, 633
hydrolysis. *See also* acid hydrolysis
　amides, 509–10, 517
　digestion, 715–17
　esters, 466–68
　lipids, 780
hydrometer, 31B
hydronium ion, 241, 249

hydrophilic molecules, 385, 468, 600
hydrophobic molecules, 384–85, 468, 600, 601F
hydrophobic pocket, 656
hydroxide ion, 249
L-β-hydroxyacyl-CoA dehydrogenase, 787
hydroxyapatite, 612B
3-hydroxybutanal, 438
β-hydroxybutyrate, 791
β-hydroxybutyric acid, 453, 457B
β-hydroxybutyryl ACP, 792
hydroxyl group, 383
4-hydroxylysine, 613B
4-hydroxy-4-methyl-2-pentanone, 439
4-hydroxyproline, 613B
α-hydroxypropionic acid, 453
β-hydroxyvaleric acid, 457B
hyperammonemia, 769
hypercholesterolemia, 579
hyperglycemia, 740
hypertension, 94B, 575
hypertonic solution, 196
hyperventilation, 262B
hypoglycemia, 740
hypothalamus, 778B
hypothesis, 4
hypotonic solution, 196, 589

I

ibuprofen, 478B, 524B
ice
　characteristics of water, 199B
　endothermic reactions, 214
　molecular solids, 172, 173F
ichthyothereol, 346B, 347B
ideal gas, 153, 163, 167
ideal gas law, 163–65
imidazole, 373, 502
immune system, 618B
immunoglobulin(s), 618–19B, 621
immunoglobulin G (IgG), 619B
incomplete octet, 103
incomplete protein, 623
indicator, of pH, 253
indole, 503
induced fit model, of enzyme activity, 638
industry, and uses of aldehydes and ketones, 424, 425F. *See also* foods and food industry; pharmaceutical chemistry; petroleum industry
inert gases, and octet rule, 65
inexact numbers, 22
infants
　brown fat and thermogenesis, 758B
　essential amino acids, 623T
　premature, 556B
inflammatory response, and prostaglandins, 565, 566
information management, and computers, 38B
infrared lamps, 48B
initiation, of transcription, 682–83, 690
initiation factors, 690, 691F

inner mitochondrial membrane, 747
inorganic chemistry, 3
inorganic compounds, compared to organic compounds, 306–7
insect(s), and pheromones, 479B
insecticides, 328B, 340B
insertion mutations, 692
instantaneous dipole, 170
insulin
    cellular metabolism, 797–98
    diabetes mellitus, 794B
    glucose and glucose transport, 535, 587
    glycogenesis, 735
    thiols and synthesis of, 408F
intensive properties, 8–9
intermembrane space, 747
intermolecular forces, 111
intermolecular hydrogen bonds, 385B
intermolecular reactions, 436
International Union of Pure and Applied Chemistry (IUPAC), 55, 315
intracellular location, of fatty acid biosynthesis, 792
intramolecular forces, 111
intramolecular hemiacetal, 537
intramolecular hemiketal, 540
intramolecular hydrogen bonds, 385B
intramolecular reactions, 436
intravenous (IV) solutions, 589–90
introns, 684–85
iodine, and nuclear medicine, 281, 284, 289, 290T
m-iodobenzoic acid, 454
ion(s)
    composition of atom, 43–44
    concentration of solutions, 190–91
    octet rule and formation, 65–68
    periodic table and size of, 68–69
ionic bonding
    amino acids, 610
    definition of, 79
    inorganic compounds, 306
    process and examples of, 79–81
ionic compounds
    formulas and nomenclature, 84–89
    properties of, 92–93
    solubilities of common, 130T
ionic solids, 172
ionization energy, 69, 70F
ionizing radiation, 279, 289
ion pairs, 80, 81
ion product, for water, 246
iron, 121, 144–45
irreversible enzyme inhibitors, 652
ischemia, 658B
isobutyl methanoate, 465B
isocitrate, 753, 754F, 756, 757F
isocitrate dehydrogenase (ICD), 658B, 753
isoelectric point, 620
isoelectronic ions, 66
isoenzymes, 658–59B
isoleucine, 601F, 623T
D-isomer, 536
isomerases, 633, 634, 721B
isoprene, 352, 364B, 572

isoprenoids, 352, 353F, 572
isopropyl alcohol, 359
isopropyl benzoate, 461
isotonic solutions, 196, 589
isotope
    nuclear structure and stability, 282
    nuclide, 277
    radioactive tracers and nuclear medicine, 289, 290T, 291–93
    structure of atom, 40–43
I.U.P.A.C. Nomenclature System, 315–21. *See also* nomenclature

J

jaundice, 2B
Jeffries, Alec, 702B
joules, 27

K

Kekulé, Friedrich, 368
Kekulé structures, 368
Kelvin scale, 25–26, 157B
keratin, 597
α-keratins, 608, 609F, 668
ketal, 435, 436, 441, 543
α–keto acid, 762, 763
ketoacidosis, 789
keto-enol tautomers, 436–38, 442
ketogenesis, 789–91
α–ketoglutarate
    amino acid degradation, 763, 764
    citric acid cycle, 753, 754F, 755, 756, 757F, 770
ketone(s)
    aldehydes and ketones, 424–42
    common functional groups, 309T
    hydration of alkenes, 361
    hydrogenation, 392–93
    medical and industrial uses, 424, 425F
    nomenclature, 422–24
    physical properties, 417–19
ketone bodies, 789–91, 794–95B
ketoses, 527, 542
ketosis, 789
9-keto-*trans*-2-decenoic acid, 479B
kidneys
    diabetes, 794B
    hemodialysis, 201B
    nuclear medicine, 290T
    prostaglandins, 566
kidney stones, and precipitation reactions, 130
kilocalorie, 27B, 526B
kilopascal, 154
kinase, 631
kinetic(s), and energy, 218–25
kinetic energy (K.E.), 27, 155B
kinetic molecular theory, 154–55, 209
Knoop, Franz, 782
Koshland, Daniel E., 638
Krebs, Sir Hans, 750
Krebs cycle, 750, 753–56

L

lactase, 545
β–lactam ring, 507B
lactate dehydrogenase (LDH)
    heart attack and enzyme assays, 660
    lactate fermentation, 727
    myocardial infarction, 658B, 659B
    oxidoreductases, 631
    reduction of ketones, 434
lactate fermentation, 727
lactic acid, 382B, 456, 528B
lactic acidosis, 769B
lactose, 544–45
lactose intolerance, 543
*lac*Z gene, 699
lagging strand, 679
lakes, and acid rain, 256B
Landsteiner, Karl, 546B
lanthanide series, 55
Latin, and nomenclature, 86
lauric acid, 559T
law(s)
    Avogardo's law, 161–62
    Boyle's law, 155–57, 160, 233
    Charles's law, 157–59, 160
    combined gas law, 160–61
    conservation of mass, 126, 132, 141F
    Dalton's law of partial pressures, 166–67
    Henry's law, 181–82
    ideal gas law, 163–65
    multiple proportions, 136B
    Raoult's law, 192
    scientific method, 5
law enforcement, and breathalyzer test, 401B. *See also* DNA fingerprinting; drug abuse
lead, 120, 186
leading strand, 679
LeBel, Joseph Achille, 532
Leber's hereditary optic neuropathy (LHON), 746B
LeChatelier's principle, 232, 258, 262B, 458
lecithin, 568, 569F
length, units of measurement, 11, 24
leptin, 778B
lethal dose (LD), of radiation, 297
leucine, 601F, 623T
leucine enkephalin, 604B, 605B
leukotrienes, 566
Levene, Phoebus, 669
levorotatory, 531
Lewis, G. N., 79
Lewisite, 407
Lewis structures
    alkanes, 311
    definition of and symbols, 79, 80F
    molecules, 93, 95–97
    octet rule, 103–4
    polarity, 109–11
    polyatomic ions, 97–100
    resonance, 101–3
    stability, multiple bonds, and bond energies, 100–101
    VSEPR theory, 104–9

life. *See also* biological systems
    importance of water, 198B
    origins of, 630B
ligases, 633
light
    atomic structure, 46–49
    fractured solids, 213B
    speed of, 47
limonene, 352, 353F, 524B
linear molecular structure, 105, 107T
line formula, 310–11
linkage specificity, of enzyme, 639
linoleic acid, 559T, 564
linolenic acid, 559T
lipases, 632, 779
lipid(s). *See also* lipid metabolism
    biological functions, 557
    cell membranes, 557, 581–90
    complex lipids, 577–80
    controversies concerning, 556
    diet, 556
    fatty acids, 558–67
    glycerides, 567–70
    nonglyceride lipids, 570–77
lipid bilayers, 581
lipid metabolism, 779–82, 793–97, 798
Lipkin, Dr. Martin, 67
liposomes, 586B
liquid(s)
    boiling and melting points, 112–13
    compared to gases and solids, 153T
    compressibility, 167
    hydrogen bonding, 170–71
    solutions, 179
    states of matter, 7
    surface tension, 168
    van der Waals forces, 170
    vapor pressure, 169–70
    viscosity, 167–68
Lister, Joseph, 264B, 402
liter, 12
lithium, 58T, 59, 63
lithium sulfide, 86
liver
    enzymes and diseases of, 660
    lipid and carbohydrate metabolism, 793
    lipoprotein receptors, 579–80
    metabolism and oxidation-reaction reactions, 267
    nuclear medicine, 290T
lock-and-key model, of enzyme activity, 638
London, Fritz, 170
London forces, 170
lone pair, of electrons, 97
low-density lipoproteins (LDL), 578, 579, 580
low-density polyethylene, 367B
low entropy, 211
lung cancer, 694B
lung diseases, and acid rain, 257B
lyases, 632
lysergic acid diethylamide (LSD), 503, 504
lysine, 601F, 623T
lysosomes, 579, 647

M

McArdle's disease, 740B
McCarron, Dr. David, 67B
McGrayne, Sharon Bertsch, 276B
magainins, 5B
magnesium carbonate, 128
magnetic resonance imaging (MRI), 178B, 292B
magnetism, and migration, 78B
magnetite, 78B
magnetosomes, 78B
magnetotactic bacteria, 78B
malaria, 617
malate
    citric acid cycle, 401–2, 754F, 756, 757F
    glycolysis, 632
    lyases, 632
malate dehydrogenase, 756
maltose, 543–44
mammals, and body temperature, 582. *See also* reindeer; skunk
mannitol, 382B
margarine, 564
marijuana, 511
Markovnikov, Vladimir, 359
Markovnikov's rule, 359, 361, 363
mass. *See also* mass number; molar mass
    concentration of solutions based on, 182–87
    experimental quantities, 23–24
    subatomic particles, 39T
Massachusetts General Hospital (Boston), 178B
mass number (A), 39, 277
mathematical representation, of reaction rate, 224–25
matrix space, 747
matter. *See also* gas; liquid(s); solid(s)
    chemical properties, 8
    classification of, 9–10
    definition of, 3
    physical properties, 7, 153T
measurement
    in chemistry, 11–16
    of gases, 153–54
    pH scale, 246–52
    of radiation, 295–98
mechanical stress, and proteins, 622
medicine. *See also* antibiotics; disease; drugs; genetic disorders; human body; pharmaceutical chemistry
    AIDS virus and nucleotides, 676–77B
    alcohols, 387–89
    aldehydes and ketones, 424
    alkynes, 346–47B
    Ames test for carcinogens, 694–95B
    amides, 505–6
    amines, 495–97
    blood gases and respiration, 168B
    blood pH, 262B
    blood pressure and sodium ion/potassium ion ratio, 94B
    carbon monoxide poisoning, 136B
    chloroform in swimming pools, 331B
    copper deficiency and Wilson's disease, 57B
    drug delivery, 240B, 586B
    enzymes, 654–55B, 658–59B, 660–61
    familial emphysema, 648B, 705B
    formaldehyde and methanol poisoning, 426B
    genetic engineering, 701T
    Gore-Tex, 364B
    hemodialysis, 201B
    HIV protease inhibitors, 641B
    hot and cold packs, 221B
    immunoglobulins, 618–19B
    liposomes, 586B
    monosaccharide derivatives and heteropolysaccharides, 550–51B
    nuclear medicine and radioactivity, 288–93
    oral rehydration therapy, 197B
    oxidizing agents for control of microbes, 264B
    pharmaceutical chemistry, 144B
    polyhalogenated hydrocarbons as anesthetics, 328B
    proteins in blood, 599B
    radiation therapy, 49B, 289
    role of curiosity in, 5B
    role of observation in, 2B
    semisynthetic penicillins, 507B
    sphingolipid metabolism, 573B
    steroids and heart disease, 574B
    urine tests and diagnosis, 31B
melanin, 668
melting point
    alkanes, 311T, 313T, 343T
    alkenes and alkynes, 343T
    fatty acids, 559F, 560
    ionic and covalent compounds, 93
    liquids and solids, 112–13, 171
    physical properties, 7
membranes. *See* cell membrane
membrane transport, 583–90
Mendel, Gregor, 669
Mendeleev, Dmitri, 52–54
Menkes' kinky hair syndrome, 57B
mental illness, 178B
mercury, 25, 30, 154
Meselson, Matthew, 676
messenger RNA (mRNA), 681
metabolic myopathy, 722B
metabolism, and oxidation-reduction reactions, 267. *See also* carbohydrate metabolism; cellular metabolism; energy metabolism; fatty acid metabolism
meta–Cresol, 369
metal(s)
    catalysts, 222
    corrosion, 265
    ionic compounds, 84–85
    periodic table, 55
metal hydroxides, 242

metallic bonds, 172
metallic solids, 172
metalloids, 55
metastable isotope, 280
metastasis, 596B
meta–Xylene, 369
meter, 12
methadone, 513B
methamphetamine, 496, 500–501B
methanal, 397, 419, 424. *See also* formaldehyde
methanamide, 506T
methanamine, 490, 492T, 493, 495T
methane
    bromination, 330
    carbon monoxide, 136B
    chemical reaction and energy, 209
    chemical reaction with oxygen, 131
    covalent bonding, 82
    crystal structure, 172F
    formula units, 125F
    frozen on ocean floor, 307B
    melting point, 93
    model of, 5–6
    molecular formula, 310
    molecular geometry, 106
    oxidation, 266
methane hydrate, 307B
methanogens, 307B
methanol, 387, 389, 426B, 493T
methedrine, 496
methicillin, 507B
methionine, 601F, 623T
methionine enkephalin, 604B, 605B
Meth lab, 501B
methoxyethane, 403, 417, 449
methyl alcohol, 389
methylamine, 242, 494
methylammonium chloride, 498
N-methylbutanamide, 508
3-methyl-1-butanethiol, 405, 406F
methyl butanoate, 465B
3-methylbutanoic acid, 451
3-methyl-1-butanol, 433
2-methyl-2-butene, 395
3-methyl-1-butene, 395
methylbutyl ethanoate, 465B
methyl butyl ketone, 423
β-methylbutyraldehyde, 421, 423
3-methyl-1,4-cyclohexadiene, 344
3-methylcyclopentene, 345
methyl decanoate, 562
N-methylethanamide, 506T
N-methylethanamine, 493, 495T
methyl ethanoate, 461
1-methylethyl, 315T
methyl group, 108, 314T
6-methyl-2-heptanol, 385
methyl methacrylate, 365T
N-methylmethanamide, 506T
N-methylmethanamine, 490, 492T, 495T
2-methylpentanal, 419, 420
3-methylpentanal, 420
2-methylpentane, 313T

3-methylpentane, 316
2-methyl-2-pentanol, 427
3-methyl-2-pentene, 350
N-methyl-1-phenyl-2-propanamin, 496
N-methylpropanamide, 505, 510
N-methylpropanamine, 494
2-methylpropane, 310
methyl propanoate, 463
2-methyl-2-propanol, 390, 399
2-methylpropene, 343T
1-methylpropyl, 315T
methyl propyl ether, 404
methyl propyl ketone, 423
methylurea, 639
metric system, of measurement, 12–13, 15
Meyer, Lothar, 52–53
micelles, 468, 779, 780
microbes. *See also* bacteria
    oil spills and oil-eating microbes, 317B
    oxidizing agents for chemical control, 264B
microfibril, 608, 609F
microwave radiation, 48B
Miescher, Friedrich, 669
migration, of birds, 78B
milliequivalents, 191
milliequivalents/liter (meq/L), 191
mitochondria, 746B, 747
mixture, definition of, 9–10
models, use of in chemistry, 5–6
molality, 192–93
molarity, 187–88
molar mass, 119–26
molar quantity, and chemical equations, 132
molar volume, of gas, 163
mole(s)
    Avogardo's number, 119–20
    calculating atoms, mass, and, 121–23
    chemical equations, 137–38
    concentration of solutions, 187–91
    conversion of reactants to products, 138–41
molecular formula. *See also* nuclear equations; structural formula
    alkanes, 310, 311T, 342, 343T
    alkenes and alkynes, 342, 343T
molecular genetics. *See also* genetic(s); genetic engineering; recombinant DNA technology
    central dogma of, 681
    detection of genetic disease, 668B
molecular geometry
    Lewis structures, 104–9
    properties based on, 111–13
molecular solids, 172
Molecular Targets Drug Discovery Program, 346B
molecule(s). *See also* molecular formula; molecular geometry
    covalent compounds, 89
    Lewis structures, 93, 95–97
    optical activity and structure of, 532
molybdate ion, 292–93

molybdenum–99, 292–93
monatomic ions, 86, 87T
monochromatic light, 530
monoglycerides, 567
monomers, 364, 469, 717
monoprotic acid, 255
monosaccharide(s), 527–29, 535–43
monosaccharide derivatives, 550–51B
morphine, 503, 504, 512B, 604–5B
Morton, Dr. William, 405
motility, and cellular energy, 713T
movement proteins, 597
Mulder, Johannes, 596
Müller, Paul, 340B
multilayer plastics, 367B
multiple bonds, 100–101
multiple proportions, law of, 136B
multiplication, of significant figures, 21–22
mummies, and soap, 563B
muscle tissue, 796
mutagens, 692, 693
mutations, genetic, 692–93
Mylar, 470
myocardial infarction, 658–59B
myoglobin, 597, 615–17
myoglobinuria, 722B
myosin, 597
myrcene, 352, 353F
myricyl palmitate (beeswax), 577
myristic acid, 559T

N

NAD$^+$ (nicotinamide adenine dinucleotide ion), 401–402, 645–46, 660, 719
NADH (nicotinamide adenine dinucleotide)
    citric acid cycle, 402, 761
    coenzymes, 645–46
    fatty acid synthesis, 792
    glycolysis, 724
    heart attack and enzyme assays, 660
    oxidative phosphorylation, 757
    reduction reactions, 434
NADH dehydrogenase, 746, 760–61
NADP$^+$ (nicotinamide adenine dinucleotide phosphate), 645, 646
Na$^+$-K$^+$ ATPase, 590
nalbuphine, 512B
naloxone, 512B
naming, of chemical compounds, 84, 86–88. *See also* common names; nomenclature
nanometers (nm), 24
naphthalene, 371, 372
naproxen, 478B
NASA (National Aeronautic and Space Administration), 307B
National Academy of Science, 94B
National Institutes of Health, 703
National Research Council, 94B
natural radioactivity, 45–46, 277–79, 291–93
negative allosterism, 649
negatively charged amino acids, 600, 601F

neon, 58T, 59, 63
neotame, 509
nerve agents, and enzymes, 654–55B
nerve synapse, 654B
net energy, 220
neurotransmitters, 511, 514–17, 654–55B
neutral glycerides, 567–68
neutralization, of acids and bases, 252–55, 458, 459, 498–99, 517
neutrons, 39–40, 45
niacin, 644T
nicotine, 373, 488B, 503, 504, 515–16. See also cigarette smoking
Niemann-Pick disease, 573B
nitrate ion, and resonance hybrids, 102–103
nitration, 374
nitric acid, 242, 256B, 257B
nitric oxide, 103, 257B, 516–17
nitrobenzene, 369, 372, 374
nitrogen
    automobile air bags, 118B
    gas density, 163
    Lewis structure and stability, 100–101
    scuba diving and the bends, 183B
    valence electrons, 58T, 59
nitrogen-16, 280
nitrogen dioxide, 90, 160, 227, 257B
nitrogen monoxide, 91
nitrogen oxides, and acid rain, 256B
Nobel Prize Women in Science (McGrayne), 286B
noble gases, 55
nomenclature. *See also* common names; I.U.P.A.C. Nomenclature System
    acid anhydrides, 475
    acid chlorides, 472
    alcohols, 385–87
    aldehydes, 419–22
    alkanes, 315–19
    alkenes and alkynes, 343–45
    amides, 505
    amino acids, 600T
    aromatic compounds, 368–71
    carboxylic acids, 451–55
    of chemical compounds, 84, 86–88
    cycloalkanes, 322, 324
    enzymes, 635–36
    esters, 461–62
    ethers, 403–4
    ketones, 422–24
    monosaccharides, 527, 529
    stereoisomers, 534
    thiols, 406
2,5-nonadiene, 344
nonane, 316
nondirectional covalent bond, 105
nonelectrolytes, 93, 179
nonessential amino acids, 622–23
nonglyceride lipids, 570–77
nonmetals, 55, 84–85
nonpolar molecules, 109–10
nonshivering thermogenesis, 758B
nonspontaneous reactions, 211

nonsteroidal anti-inflammatory drugs (NSAIDs), 478B, 551B
norepinephrine, 511, 631
norlutin, 576
normal boiling point, 169
19-norprogesterone, 576
novocaine, 496, 499B
N-terminal amino acid, 602
nuclear decay, 281–82
nuclear equations, 279–82. *See also* molecular formula
nuclear fission, 286–87
nuclear fusion, 287
nuclear imaging, 289, 296
nuclear medicine, 288–93
nuclear power, 284–87
nuclear power plant, 286
nuclear reactions, 92B
nuclear reactor, 291
nuclear stability, 282
nuclear symbols, 277
nuclear waste disposal, 285B, 295
nucleoside analogs, 676B
nucleosome, 674
nucleotide, 669–71, 713
nucleus, of atom, 39, 45–46, 282
nuclide, 277, 279
nutrient proteins, 597
nutritional calorie, 217, 218
nystatin, 588B, 589B

O

Oak Ridge National Laboratories (Tennessee), 287F
obesity, 382B, 778B, 784–85B
obligate anaerobes, 729B
observation, in medicine and science, 2B, 4
Occupational Safety and Health Act (OSHA), 279
ocean, and frozen methane, 307B
octadecanoic acid, 564
octane ratings, of gasoline, 326B
3-octanol, 431
2-octanone, 425F
4-octanone, 422
octet rule, 65–68, 79, 95, 103–4
octyl ethanoate, 465B
odd electron molecules, 103
oil-eating microbes (OEMs), 317B
oil spills, 317B
oleic acid, 351, 559T, 564
oligosaccharides, 546B, 547B
opiates, 512–13B
opium poppy, 512–13B, 604–5B
oral contraceptives, 346B, 576
oral rehydration therapy, 196, 197B
order of reaction, 224
organic chemistry
    alkanes, 310–21, 325–32
    carbon, 305–9
    chemical reactions, 327–32
    cycloalkanes, 321–32

organic compounds
    families of, 308–9
    inorganic compounds, 306–7
    nomenclature, 315–19
    origin of, 304B
organic solvents, 621
organophosphates, 654B
ornithine, 766, 767F, 768
ornithine transcarbamoylase, 766
ortho-Cresol, 369
ortho-Xylene, 369
osmolarity, 195, 588
osmosis, 193–94, 587–90
osmotic concentration, 588
osmotic membranes, 193
osmotic pressure, 193–96, 587–88
osteoporosis, 576
outer mitochondrial membrane, 747
oxacillin, 507B
oxaloacetate, 401–2, 634
    citric acid cycle, 401–2, 754F, 756, 757F, 764, 770, 771
    glycolysis, 634
oxidation, of nutrients, 718, 782–83, 786–88
oxidation half-reaction, 263
oxidation reactions
    alcohols, 397, 409, 427
    aldehyde and ketones, 427–28, 440–42
oxidation-reduction reactions, 131, 263–70, 400–402
oxidative deamination, 764–65
oxidative phosphorylation, 719, 752, 757, 760–62
oxidizing agents, 263, 264B, 265
oxidoreductases, 400, 631
oxycodone, 512B
oxygen
    blood gases, 168B
    chemical equations, 134, 137
    chemical reactions, 131
    formulas of ionic compounds, 85
    hemoglobin and transport of, 616
    Lewis structure, 100–101
    valence electrons, 58T, 59
oxygen-16, 280
oxygen gas, 164, 165
oxyhemoglobin, 616
oxymorphone, 512B
ozone, 264B

P

pacemakers, heart, 270B
paired electrons, 62
palmitic acid, 453T, 559T, 788F
palmitoleic acid, 341F, 558, 559T
pancreas, and diabetes, 794B, 795B. *See also* pancreatitis
pancreatic serine proteases, 656
pancreatitis, 660
pantothenic acid, 644T
Papanicolaou, Dr. George, 2B
Papaver somniferum, 512B

Pap smear test, 2B
para–aminobenzoic acid (PABA), 653
para-Cresol, 369
paraffin wax, 577
paragyline, 346B
parallel β-pleated sheet, 609
parasalamide, 346B
para-Xylene, 369
parent compound, 315–16, 385
Parkes, Alexander, 366B
Parkinson's disease, 501B, 511
partial electron transfer, 82
partial hydrogenation, 564
partial pressures, 166
particle accelerators, 291
parts per million (ppm), 185–86
parts per thousand (ppt), 185–86
Pascal, Blaise, 154
pascal, as measurement unit, 154
passive transport, 583–85
Pasteur, Louis, 532
Pauling, Linus, 83
penicillamine, 57B
penicillin, 4B, 315, 507B
Penicillium notatum, 315, 507B
1,4-pentadiene, 344
pentanal, 421T
2-pentanamine, 493
pentane, 316, 343
1,4-pentanedithiol, 406
2-pentanol, 360
2-pentanone, 428
3-pentanone, 422, 431
1-pentene, 343, 350
2-pentene, 355
penthrane, 405
pentose, 527
pentose phosphate pathway, 730
pentyl butanoate, 465B
pentyl group, 314T
1-pentyne, 343
pepsin, 646, 651T
pepsinogen, 622, 651
peptidase, 633
peptide bond, 510, 602–6
peptidyl transferase, 690
peptidyl tRNA binding site (P site), 690
percent yield, 143–45
periodicity, concept of, 53
periodic table
    development of, 52–56
    electron configuration, 56, 58–65
    octet rule, 65–68
    structural relationships, 107–8
    trends in, 68–70
periods, and periodic table, 55
peripheral membrane proteins, 582
permeases, 585
Perrine, Susan, 448B
Persian Gulf War, 317B
petroleum industry, and gasoline production, 326B. *See also* fossil fuels; oil spills

pH
    acid rain, 256B
    blood and control of, 262B
    buffer solution, 260–61
    calculating, 247–51
    color indicators, 253
    definition of, 246–47
    enzymes, 641, 646–47
    importance of, 252
    measuring, 247
    proteins, 620–21
pharmaceutical chemistry, 144B, 524B, 641B. *See also* antibiotics; drugs
pharmacology, 240B
phenacetin, 506
phenanthrene, 371, 372
phenols, 369, 402–3
phenylalanine, 601F, 623T, 770B
2-phenylbutane, 371
3-phenyl-1-butane, 371
phenylephrine, 496
2-phenylethanoic acid, 454
2-phenylethanol, 405, 406F
phenylethanolamine-N-methyltransferase (PNMT), 631
phenyl group, 371
phenylketonuria (PKU), 509
o-phenylphenol, 403
1-phenyl-2-propanamine, 496
3-phenylpropanoic acid, 454
pheromones, 479B
phosphate ion, 191
phosphatidate, 568, 569F
phosphatidylcholine, 568, 569F
phosphatidylethanolamine, 568, 569F
phosphoanhydride, 477
phosphoenolpyruvate, 396, 438, 725
phosphoenolpyruvate carboxykinase, 732
phosphoesters, 476–78
phosphofructokinase, 650, 723, 726
phosphoglucomutase, 734, 736F, 737
phosphoglucose isomerase, 721
2-phosphoglycerate, 633, 725
3-phosphoglycerate, 633, 724, 725
phosphoglycerate kinase, 724
phosphoglycerate kinase deficiency, 722B
phosphoglycerate mutase, 633, 724
phosphoglycerate mutase deficiency, 722B
phosphoglycerides, 568–70
phospholipids, 581
phosphopantetheine group, 792, 793F
phosphorolysis, 733
phosphorus pentafluoride, 104
phosphorus trichloride, 471
phosphoryl, 477B
photosynthesis, 8, 526F
phthalic acid, 454
physical change, 7
physical chemistry, 3
physical equilibrium, 226–27
physical properties. *See also* gas(es); liquid(s); matter; properties; solid(s)
    alcohols, 384–85

    aldehydes and ketones, 417–19
    alkanes, 310–13
    alkenes and alkynes, 342–43
    amides, 504–5
    amines, 489–93
    carboxylic acids, 449–50
    comparison of states, 153T
    esters, 461
    ionic and covalent compounds, 92–93
    periodic table, 53
physical state, of reactants, 222
physiological saline, 196
physiology, pH and biochemical reactions, 252
Phytochemistry (journal), 703B
planar molecule, 342
Planck, Max, 49
plane-polarized light, 530–31
plants. *See also* agriculture; biological systems; forests
    alkenes, 352–53
    cellulose in cell wall, 549
    DNA fingerprinting of coca plants, 702–3B
    opium poppy and opiates, 512–13B, 604–5B
    as sources of medications, 574B
plasma lipoproteins, 577–78
plastics. *See also* polymers
    biodegradable, 456–57B, 470
    recycling, 366–367B
platinum, 185
β-pleated sheet, 609
plutonium, 281, 285B, 287, 294
point mutation, 692, 693
poisoning, and carbon monoxide, 136B
polar covalent bonding, 82–83, 110
polar covalent molecule, 110
polarimeter, 531
polarity, and Lewis structures, 109–11
polar neutral amino acids, 600, 601F
polio virus, 426B
pollution. *See* acid rain; air pollution; global warming; greenhouse effect; oil spills; waste disposal
polonium isotope, 295B
polyacrylonitrile, 365T
polyanions, 621
poly(A) tail, 684
polyatomic ions, 86–87, 88T, 97–100
polycations, 621
polyesters, 469–70
polyethylene, 364B, 365
polyethylene naphthalate (PEN), 470
polyethylene terephthalate (PETE), 366B, 367B, 470
polyhalogenated hydrocarbons, 328B
polyhydroxyaldehydes, 527
polyhydroxybutyrate acid (PHB), 456B, 457B
polyhydroxyketones, 527
polylactic acid (PLA), 456B
polymer(s), 469–70. *See also* addition polymers; plastics

polymerase chain reaction (PCR), 701–2
polymethyl methacrylate, 365T
polymyxins, 588–89B
polynuclear aromatic hydrocarbons (PAH), 371–72
polyols, 382B
polypropylene, 365, 366–67B
polyprotic substances, 255
polysaccharides, 527, 548–51
polysomes, 688
polystyrene, 365T, 367B
polytetrafluoroethylene (Teflon), 365T
polyvinyl chloride (PVC), 365T, 366B, 367B
poppy. *See* opium poppy
p orbitals, 61
porphyrin, 373, 503
position emission, 280
positive allosterism, 649
positively charged amino acids, 600, 601F
postsynaptic membrane, 654B
post-transcriptional processing, of RNA, 684–85
potassium, 94B, 200, 281
potassium hydroxide, 242
potassium perchlorate, 52B
potassium permanganate, 397
potassium propanoate, 460
potential energy, 27
pound (lb), 11
precipitate, 181
precipitation reactions, 129–30
precision, in measurement, 20
prefixes
    metric system, 12
    naming of covalent compounds, 90
pregnancy. *See also* infants
    alcohol consumption during, 388B
    hemoglobin and oxygen transport, 616
    premature infants, 556B
preimplantation diagnosis, 668B
premature infants, 556B
premenstrual syndrome (PMS), 67B
pressure. *See also* vapor pressure
    chemical reactions, 233
    gases, 153, 166–67
    solubility, 181
primary alcohol, 389, 427
primary amine, 490
primary carbon, 313
primary structure, of proteins, 606, 613, 614F
primary transcript, 684
primase, 678, 679F
principal energy levels, 60, 61
products
    of chemical equation, 127
    of chemical reaction, 8, 636
proenzymes, 651
progesterone, 575, 576
prokaryotes, 673
proline, 600, 601F, 623T, 770
promoter, and transcription, 682
promotion, and Bohr atom, 49–50, 51
propanal, 419, 421T

propanamide, 505
propanamine, 491, 492T, 493T
1-propanamine, 493
propane
    boiling point, 491
    functional groups, 309
    molecular formula, 310
    nomenclature, 316, 343
propane gas, 134
1,2,3-propanetriol, 389
propanoic acid, 261, 451, 458, 466, 474
propanoic anhydride, 474
propanol, and boiling point, 384, 417, 449, 493T
1-propanol, 393, 403, 417, 449, 467
2-propanol, 359
    classification, 390
    hydration, 359
    hydrogenation, 393
    medical uses, 388–89
    molecular structure, 389
    nomenclature, 386
    oxidation, 398
propanone
    boiling point, 417, 449
    hydrogenation, 393
    industrial applications, 424
    naming of ketones, 422
    oxidation, 398
propene, 343, 394
properties. *See also* chemical properties; matter; physical properties
    definition of, 7
    electronic structure and molecular geometry, 111–13
    intensive and extensive properties, 8–9
    ionic and covalent compounds, 92–93
    periodic table, 53
propionamide, 505
propionibacteria, 729B
propionic acid, 453T
propyl acetate, 462
propyl decanoate, 561
propyl ethanoate, 463
propyl group, 314T
N-propylhexanamide, 505
propyne, 343
prostaglandins, 478B, 565–67
prostate specific antigen (PSA), 599B
prosthetic group, 612
protease inhibitors, 641B
protein(s). *See also* amino acids
    angiogenesis inhibitors, 596B
    blood, 599B
    cellular functions, 597
    collagen, 612–13B
    denaturation, 617–22
    dietary and digestion, 622–23, 716
    DNA replication, 679F
    Fischer's research on, 525
    myoglobin and hemoglobin, 615–17
    overview of, 510–11
    peptide bond, 602–6

    primary structure, 606, 613, 614F
    quaternary structure, 611–12, 614
    secondary structure, 606–9, 613–14
    shape of molecule, 638B
    tertiary structure, 609–11, 614
protein metabolism, 797
protein modification, and enzymes, 651–52
protein synthesis, 510–11, 687–92
proteolytic enzymes, 656–67
protium, 277
protofibril, 608
protons, 39–40, 45
Prozac, 514
pseudoephedrine, 496, 501B
pulmonary disease, and nuclear medicine, 290
pulmonary surfactant, 556B
pure substance, 9
purines
    heterocyclic amines, 502, 503
    heterocyclic aromatic compounds, 373
    structure of DNA, 525, 669, 670F
pyridine, 242, 373, 502, 503
pyridine aldoxime methiodide (PAM), 516, 655B
pyridoxal phosphate, 764
pyridoxine, 644T
pyrimidine, 373, 502, 503, 669, 670F
pyrimidine dimer, 694–95
pyrophosphorylase, 737
pyrrole, 373, 502
pyruvate
    alcohol fermentation, 728F
    citric acid cycle, 750–51, 754F, 756, 757F
    degradation of amino acids, 764
    glycolysis, 447B, 719, 721
pyruvate carboxylase, 732, 771–72
pyruvate carboxylase deficiency, 769B
pyruvate decarboxylase, 727
pyruvate dehydrogenase complex, 750
pyruvate kinase, 725, 726

Q

quantization, of energy, 49
quantum levels, and Bohr atom, 49
quantum mechanical atom, 60–61
quantum mechanics, 61
quantum number, 51
quaternary ammonium salts, 501–2
quaternary carbon, 313
quaternary structure, of proteins, 611–12, 614
quats, 502
quinine, 503, 504

R

Rabi, Isidor, 292B
rad (radiation absorbed dosage), 297
radiation. *See* gamma radiation; radioactivity
radiation therapy, 49B, 289
radioactive decay, 39

radioactive isotopes, 40–41
radioactivity
    balanced nuclear equation, 279–82
    biological effects of radiation, 293–95
    measurement of, 295–98
    medical applications and nuclear medicine, 288–93
    natural, 45–46, 277–79, 291–93
    nuclear power, 284–87
    radiocarbon dating, 288
    radioisotopes, 282–84
radiocarbon dating, 288
radioisotopes, 282–84
radio waves, 48B
radium, 276B, 295B
radon, 295B
Raleigh, Sir Walter, 694B
random error, 19
Raoult's law, 192
rate constant, 224, 225
rate equations, 224, 225
rate-limiting step, and enzyme-substrate complex, 637
reactants
    of chemical equation, 126, 127
    of chemical reaction, 8
reaction rate. See also chemical reactions
    factors affecting, 221–23
    mathematical representation, 224–25
    substrate concentration and enzyme-catalyzed reactions, 637
real gases, versus ideal gases, 167
receptor-mediated endocytosis, 579, 580F
recombinant DNA technology, 659B, 661, 696–701. See also genetic engineering
recycling, of plastics, 366–67B
red blood cells (RBCs), 262B, 546B, 581, 587
redox reactions. See oxidation-reduction reactions
reducing agent, 263, 265
reducing sugars, 541–43
reduction. See oxidation-reduction reactions
reduction half-reaction, 263
reduction reactions, of aldehydes and ketones, 431–34, 440–42. See also oxidation-reduction reactions
regulation
    of citric acid cycle, 756, 757F
    of enzyme activity, 649–52
    of glycolysis, 726
    of lipid and carbohydrate metabolism, 793–97
regulatory proteins, 597
reindeer, and body temperature, 582
relaxation, and Bohr atom, 49–50, 51
release factors, 691
rem (roentgen equivalent for man), 297
repair endonuclease, 695
replacement reactions, 129
replication, of DNA, 674–81
replication fork, 677
replication origin, 677

representative elements, 55, 58
reproductive system, and prostaglandins, 565. See also birth control; pregnancy
resonance, and Lewis structures, 101–3
resonance forms, 102
resonance hybrid, 102–3, 505
respiration. See also aerobic respiration; respiratory tract
    blood gases, 168B
    Henry's law, 181–82
respiratory distress syndrome (RDS), 556B
respiratory tract. See also lung disease; pulmonary disease; respiration
    acidosis and alkalosis, 262B
    electron transport system, 760
    hemoglobin and oxygen transport, 616
    prostaglandins, 566
restriction enzymes, 696
results, and measurement, 11
retina, of eye, 440–41B
retinol, 352
retroviruses, 676B. See also HIV
reverse aldol condensation, 723B
reverse transcriptase, 676B
reversible competitive enzyme inhibitors, 652–53
reversible noncompetitive enzyme inhibitors, 653–54
reversible reaction, 226, 229–30
rhodopsin, 440B
riboflavin, 644T, 646
ribonucleotides, 670, 674
ribose, 541, 670F, 674
ribosomal RNA (rRNA), 682
ribosomes, 687, 688F
ribozymes, 630B
RNA (ribonucleic acid). See also mitrochondria; transfer RNA (tRNA)
    classes of, 681–82
    heterocyclic aromatic compounds, 373
    nucleotides, 669
    post-transcriptional processing, 684–85
    structure of, 674
RNA polymerase, 682–83
RNA primer, 678
RNA splicing, 684–85
rock, and equilibrium composition, 234
roentgen, 297
Roman mythology, and soap, 468
rounding off numbers, 22–23
rubbing alcohol, 388
rust, as chemical reaction, 131, 265F
Rutherford, Ernest, 45, 46

S

safety
    acids and bases in laboratory, 246
    radioactivity and radiation, 279, 293–95
saliva, and bacteria, 528B
Salmonella typhimurium, 694B

salt(s), and carboxylic acid, 460. *See also* sodium; sodium chloride
salt bridge, 268
Sanger, Frederick, 704
saponification, 467, 468F, 562–63
Sarin (isopropylmethylfluorophosphate), 654–55B
saturated fatty acids, 559T, 560
saturated hydrocarbons, 308
saturated solution, 181
Saunders, Jim, 702B
schizophrenia, 511
Schröedinger, Erwin, 60–61
science
    role of curiosity in. 5B
    role of observation in, 2B
    scientific method, 3–5, 52
    scientific notation, 18–23
    serendipity in research, 586B
scientific method, 3–5, 52
scientific notation, 18–23
scuba diving, 183B
scurvy, 613B
seasonal affective disorder (SAD), 511
secondary alcohol, 389, 398, 427
secondary amine, 490
secondary carbon, 313
secondary structure, of proteins, 606–9, 613–14
selectable marker, 699
selectively permeable membranes, 194F
selective serotonin reuptake inhibitors (SSRIs), 514
self-tanning lotions, 432–33B
semiconservative replication, 676
semipermeable membranes, 193, 587
semisynthetic penicillins, 507B
Semmelweis, Ignatz, 264B
sensors, carbon monoxide, 136B
sequential synthesis, 144B
serendipity, in scientific research, 586B
serine, 601F, 623T, 651
serine proteases, 657B
serotonin, 511, 514
shared electron pair, 82
shielding, and radiation, 294
shorthand electron configuration, 64–65
Shroud of Turin, 288F
sialic acid, 546B, 547B
sickle cell anemia, 448B, 617, 693
side effects, of drugs, 240B
significant figures, 16–23
silent mutations, 692–93
Silent Spring (Carson 1962), 340B
silicon, and molecular geometry, 106
silicon dioxide, 90
silk fibroin, 609, 610F
silver, 120F, 173, 428, 429F
single bond, 96, 101
single-replacement reaction, 129
single-strand binding protein, 678, 679F
skeletal structure, of compound, 93
skin cancer, 432–33B

skunk, and scent molecules, 405, 406F
slow twitch muscle fibers, 749B
small nuclear ribonucleoproteins (snRNPs), 685
Smithsonian Museum of Natural History, 563B
smoking. *See* cigarette smoking
soaps, 460, 468, 469F, 562–63. *See also* detergents
sodium. *See also* salt(s); sodium chloride; sodium hydroxide
    blood pressure, 94B
    cations in body fluids, 200
    electron configuration, 63
    fireworks, 52B
    formulas of ionic compounds, 85
    ion formation, 66
    molar mass, 119–20
    oxidation and reduction, 263
sodium acetate, 257, 260
sodium azide, 118B
sodium benzoate, 459
sodium bicarbonate, 87
sodium chloride. *See also* salt(s); sodium
    chemical formula, 84, 124
    crystal structure, 172F
    formula units, 125F
    ionic bonding, 79–80, 81F
    major properties, 306T
    melting point, 93
    nomenclature, 86
    osmotic pressure, 195
    vapor pressure and boiling point, 193
    weight/volume percent, 184
sodium hypochlorite, 264B, 266
sodium hydroxide
    neutralization, 252, 253
    polyprotic substances, 255
    reactant quantities, 141–42
    replacement reactions, 129
    strong bases, 242
sodium oxide, 86
sodium propanoate, 261
sodium sulfate, 87, 89, 125
soil, and equilibrium composition, 234
solar energy, 48B
solid(s)
    boiling and melting points, 112–13
    compared to gases and liquids, 153T
    crystalline solids, 171, 172–73
    properties of, 171
    states of matter, 7
    triboluminescence and fracturing of, 213B
solubility. *See also* solutions
    common ionic compounds, 130T
    degree of, 180–81
    electronic structure, 111–12
solutes, 178
solutions. *See also* solubility
    concentration based on mass, 182–87
    concentration-dependent properties, 191–96
    concentration in moles and equivalents, 187–91

    electrolytes in body fluids, 200–202
    homogeneous mixture, 10
    ionic and covalent compounds, 93
    properties of, 178–82
    water as solvent, 198
solvent, 178, 198
s orbital, 61
sorbitol, 382B
Southern blotting hybridization, 697, 698F
specific gravity, 31
specific heat, 215
specificity, of enzymes, 630–31, 639
spectral lines, and Bohr theory, 51
spectrophotometer, 48B
spectroscopy, 47, 48–49B
speed of light, 47
sphingolipids, 570–72, 573B
sphingomyelin, 570
sphingomyelinase, 573B
sphingosine, 570
Spirea ulmaria, 315
spliceosomes, 685
spontaneous reactions, 211
stability, and Lewis structure, 100–101
staggered conformation, of alkanes, 325
Stahl, Franklin, 676
stalagmites and stalactites, 129–30
standard atmosphere, 154
standard mass, 23
standard solution, 252
standard temperature and pressure (STP), 163
starch, 548
states of matter. *See* gas; liquid(s); matter; solid(s)
Statue of Liberty, 266B
stearic acid, 453T, 559T, 564
stereochemical specificity, of enzyme, 639
stereochemistry, 524B, 529
stereoisomers, 322–23, 524B, 529–34, 598–99
stereospecific enzymes, 524B
sterilization, of medical instruments, 647. *See also* antiseptics; disinfectants
Stern, Otto, 292B
steroids, 572–77
stock system, of nomenclature, 86, 87T
strength, of acids and bases, 242–43
Streptococcus mutans, 528B
Streptococcus pyogenes, 659B
streptokinase, 659B
strong acids, 242–43, 244, 246
strong bases, 242–43, 458, 459
strontium, 52B
structural analogs, of enzymes, 652
structural formula. *See also* molecular formula
    alkanes, 310, 312, 342, 343T
    alkenes and alkynes, 342, 343T
structural isomers, 319–21
structural proteins, 597
strychnine, 503, 504
styrene, 365T
sublevels, and quantum mechanics, 61
Suboxone, 513B

subscripts, in chemical equations, 132B
substituted cycloalkanes, 323
substituted hydrocarbon, 308
substitution reactions, 329, 372
substrate, of enzyme, 635–36, 637. *See also* enzyme-substrate complex
substrate-level phosphorylation, 719
subtraction, of significant figures, 20–21
succinate, 754F, 755, 757F
succinate dehydrogenase, 755
succinylcholine, 516
succinyl CoA, 754F, 755, 756, 757F
succinyl CoA synthase, 755
sucrose, 198, 382B, 545, 640, 642F
suffixes
    covalent compounds, 90
    ionic compounds, 86
sugar. *see also* fructose; glucose; sucrose
    equilibrium and dissolution in water, 226–27
    Fischer's research on, 525
    reducing sugars, 541–43
    tooth decay, 382B, 528B
sugar alcohols, 382B
sulfa drugs, 497, 653
sulfanilamide, 497, 653
sulfatides, 571–72
sulfhydryl group, 383, 405
sulfur, 121, 122
sulfur dioxide, 101–2, 257B
sulfuric acid, 188, 242, 255, 256B, 257B
sulfur oxides, and acid rain, 256B, 257B
sulfur trioxide, 257B
sun exposure, and skin cancer, 432–33B
sun protection factor (SPF), 433B
super hot enzymes, 630B
superoxide radical, 264B
supersaturated solution, 181
surface tension, 168
surfactants, 168
surroundings, and energy stored in chemical system, 209
suspension, 180
sweet tastes, 382B
Sweeting, Linda M., 213B
swimming pools, chloroform in, 331B
symmetrical acid anhydrides, 473
synthesis
    acid anhydrides, 474
    acid chlorides, 471
    esters, 466
synthetic opioids, 512–13B
system(s), definition of, 209, 211B
systematic error, 19
systematic names, 86, 87T
Système International (S.I.), 12

T

Tagamet (cimetidine), 515
target cells, and insulin, 797
Tasmania, and opium poppies, 512B
Tauri's disease, 722B
tautomers, 436

Tay-Sachs disease, 572, 573B
technetium-99m, 280–81, 283F, 290, 292
Teflon, 364B, 365T
temperature. *See also* body temperature; boiling point; freezing point; melting point
    denaturation of proteins, 618–20
    enzymes, 630B, 647–48
    rate of chemical reaction, 222
    solubility, 181
    systems of measurement, 25–27, 157B
    vapor pressure of liquids, 169
    water and moderation of, 199B
terminal electron acceptor, 761
termination, of transcription, 683, 691–92
termination codons, 691
terpenes, 352, 572
tertiary alcohol, 389, 399, 427
tertiary amine, 490, 491
tertiary carbon, 313
tertiary structure, of proteins, 609–11, 614
testosterone, 576
tetrabromomethane, 330
tetrafluoroethene, 364B, 365T
tetrahedral molecular structure, 106, 107T, 311
tetrose, 527
β-thalassemia, 448B
thalidomide, 524B
thallium–201, 290
thebaine, 512B
theoretical yield, 143–45
theory, and scientific method, 4, 52
thermochemical equation, 210
thermocycler, 701–2
thermodynamics, 208–14
thermogenesis, 758–59B
thermogenin, 758B, 759B
thermography, 758B
thermometer, 25–26
Thermus aquaticus, 701
thiamine, 644T
thioester, 407, 478–80
thioester bond, 786
thiolase, 787
thiols, 384, 405–8
thiolysis, 787
thionyl chloride, 471
Third World countries, and oral rehydration therapy, 197B
Thomson, J. J., 45
thorium, 280
Three Mile Island nuclear accident (1979), 294
threonine, 601F, 623T, 651
thromboxanes, 565, 566F
thrombus, 659B
thymine, 670F, 672F
thyroid, and nuclear medicine, 290T
Thys-Jacobs, Dr. Susan, 67B
time
    radiation exposure, 294
    units of measurement, 25
tin, 64
tissue-type plasminogen activator (TPA), 659B

titration, 252, 255F
tobacco. *See* cigarette smoking; nicotine
Tollens' test, 428, 429F
toluene, 369
m-toluic acid, 454
o-toluidine and *p*-toluidine, 494
tooth decay, 382B, 528B
topoisomerase, 678, 679F
Torricelli, Evangelista, 153, 154F
Towson University, 702B
trace minerals, 57B
tracers, and nuclear medicine, 289
transaminases, 631, 762–64
transamination, 762–64, 765F
trans-2-butene, 356–57
trans-2-butene-1-thiol, 405, 406F
transcription, and genetics, 681, 682–83
trans-1,3-demethylcyclohexane, 324
trans-3,4-dichloro-3-heptene, 349
trans-fatty acids, 352
transferases, 631, 634
transferrin, 597
transfer RNA (tRNA), 510–11, 682, 688–90
trans isomer, 348
transition elements, 55, 58
transition state, of enzyme-substrate complex, 640
translation, and genetics, 681, 687–92
translocation, of ribosome, 680
transmembrane proteins, 582
transmethylase, 631
trans-3-methyl-3-pentene, 350
trans-9-octadecenoic acid, 351
transport proteins, 597, 616. *See also* active transport; electron transport system
trans-unsaturated fatty acids, 564B
triacylglycerol lipase, 651–52
triboluminescence, 213B
tribromomethane, 330
3,5,7-tribromooctanoic acid, 451
tricarboxylic acid (TCA) cycle, 750
trichloromethane, 328B
triesters, 468
triglycerides, 567–68, 632, 779–81, 796
trigonal planar molecule, 106, 107T
trigonal pyramidal molecule, 106, 107T
trihalomethanes (THMs), 331B
trimethylamine, 108, 491, 494
3,5,7-trimethyldecane, 316
2,2,3-trimethylpentane, 316
2,2,4-trimethylpentane, 326B
triose phosphate isomerase, 634, 723
tripeptide, 603
triple bond, 101
triple helix, 612
tritium, 92B, 277, 287
true solution, 179
trypsin, 622, 646, 651T, 656–57
tryptophan, 514F, 601F, 623T
tumors, and angiogenesis inhibitors, 596B
Tyndall effect, 180
type I insulin-dependent diabetes mellitus, 542
tyrosine, 512B, 601F, 623T, 651

U

UDP-glucose, 737, 738
ultraviolet lamps, 48B
ultraviolet (UV) light, 2B, 694–95
uncertainty, and measurement, 19–20
unequal electron density, 82
unit, and measurement, 11
U.S. Department of Agriculture (USDA), 702B, 703B
U.S. Department of Energy, 307B, 703
U.S. Department of Health and Human Services, 331B
U.S. Dietary Guidelines, 556
U.S. Geological Survey, 307B
U.S. Surgeon General, 388B
universal energy currency, 713
universal solvent, 198
universe, and big bang theory, 92B
University of Florida, 705B
University of Georgia, 630B
unsaturated fatty acids, 558, 559T, 560, 582
unsaturated hydrocarbons, 308, 341. *See also* alkenes; alkynes; aromatic hydrocarbons; geometric isomers; heterocyclic aromatic compounds
unshared pair, of electrons, 97
uracil, 670F, 674
uranium, 280, 286F, 295B
urea, 304, 660
urea cycle, 766–69
urease, 635, 639
uridine triphosphate (UTP), 737
urine tests, 31

V

vaccines and vaccination, 426B, 618–19B
valence electrons, 58–59, 79, 95
valence shell electron pair repulsion theory. *See* VSEPR theory
valeric acid, 453T
valine, 601F, 623T, 693
van der Waals forces, 170, 610
vanillin, 424, 425F
van't Hoff, Jacobus Hendricus, 532
vaporization, of water, 199B
vapor pressure
    of liquids, 169–70
    of solutions, 192
variable number tandem repeats (VNTRs), 702B
vegetarian diets, 623
very low density lipoproteins (VLDL), 578, 579F, 793
Vibrio cholera, 197B
vinegar, 244
vinyl chloride, 365T
viruses. *See* retroviruses
viscosity, of liquids, 167–68
vision, chemistry of, 440–41B
vital force, 304
vitamin(s)
    coenzymes, 643, 644T
    lipids, 557

vitamin A, 341F, 441B
vitamin B, 503, 644T
vitamin C, 613B, 643
vitamin D, 67B
vitamin K, 341
volatile esters, 464B
voltaic cells, 267–69, 270B, 271F
volume, units of measurement, 11, 24
von Gierke's disease, 740B
VSEPR theory, 104–9, 311

W

waste disposal
    biodegradable garbage bags, 456–57B
    nuclear, 285B, 295
    plastic recycling, 366–67B
water. *See also* acid rain; ice; lakes; ocean;
    water treatment
    acid-base properties, 242
    boiling point, 198
    chemical equations, 132B
    chemical formula, 124
    climate, 27B
    combination reactions, 128
    covalent bonding, 82, 83
    dissociation, 246
    equilibrium and sugar in, 226–27
    formula weight and molar mass, 125
    hard water, 563
    hydrogen bonding, 171F, 173F
    Lewis structure, 107
    life on Earth, 199B
    nomenclature, 91
    osmosis, 194
    pH scale, 250
    solubility, 111–12
    solvents, 198
    specific gravity, 31B
water treatment, and ozone, 264B. *See also*
    oral rehydration therapy
Watson, James, 669, 671
waves, and electromagnetic radiation, 47
waxes, 577
weak acids, 242–43, 244, 246, 458
weak bases, 242–43, 498
weight, and mass, 23
weighted average, and atomic mass, 41
weight loss, and diet, 28B, 784–85B
weight/volume percent, 182–84
weight/weight percent, 185
white fat, 758B
Wilkens, Maurice, 671
Wilson, A., 746B
Wilson's disease, 57B
wine and winemaking, 728B
withdrawal syndrome, and morphine, 605B
Withering, William, 574B
Wittgenstein, Eva, 432B
Wöhler, Friedrich, 304, 488
wonder drug, 144B
wood alcohol, 387
World Health Organization, 340B

X

xenon, 281, 290
xeroderma pigmentosum, 696
xerophthalmia, 441B
X-linked genetic disorders, 573B, 722B
X-rays, 48B
xylitol, 382

Y

yard, units of measurement, 11
yeast, and fermentation, 729B
Yellowstone National Park, and hot springs, 582, 629F, 648, 701

Z

Zaitsev, Alexander, 395
Zaitsev's rule, 395
Zasloff, Dr. Michael, 5B
zidovudine (AZT), 676–77B
zinc, 131, 268, 269
zwitterion, 598